MATERIAL SCIENCE AND ENVIRONMENTAL ENGINEERING

PROCEEDINGS OF THE 3RD ANNUAL 2015 INTERNATIONAL CONFERENCE ON MATERIAL SCIENCE AND ENVIRONMENTAL ENGINEERING (ICMSEE2015), WUHAN, HUBEI, CHINA, 5–6 JUNE 2015

Material Science and Environmental Engineering

Editor

Ping Chen

College of Engineering Electrical and Computer Engineering, Boise State University, Boise, ID, USA

CRC Press is an imprint of the
Taylor & Francis Group, an **informa** business

A BALKEMA BOOK

CRC Press/Balkema is an imprint of the Taylor & Francis Group, an informa business

Typeset by V Publishing Solutions Pvt Ltd., Chennai, India

Published by: CRC Press/Balkema
P.O. Box 11320, 2301 EH Leiden, The Netherlands
e-mail: Pub.NL@taylorandfrancis.com
www.crcpress.com – www.taylorandfrancis.com

ISBN: 978-1-138-02938-5 (Hbk)
ISBN: 978-1-315-63853-9 (eBook PDF)

Material Science and Environmental Engineering – Chen (Ed.)
© 2016 Taylor & Francis Group, London, ISBN 978-1-138-02938-5

Table of contents

Construction material and technology

Materials in mechanical process

Water resource and wastewater treatment

Environmental engineering and biotechnology

Material Science and Environmental Engineering – Chen (Ed.)
© 2016 Taylor & Francis Group, London, ISBN 978-1-138-02938-5

Preface

It is our great pleasure to welcome everyone to Wuhan, China to attend The 2015 International Conference on Material Science and Environmental Engineering (MSEE2015) held during June 5–6, 2015.

The last two conferences have achieved great success: MSEE2013 had been held successfully during August 17–18, 2013 in Wuhan. The proceedings of the MSEE2013 was published by DESTech Publications, Inc. and was indexed by Conference Proceedings Citation Index(CPCI). The MSEE2014 had been held successfully during March 21–23, 2014 in Changsha, China. The proceedings of the MSEE2014 was published by Trans Tech Publications, Inc. and was included in Ei Compendex. The MSEE2015 will be published by CRC/Balkema (Taylor & Francis Group). The MSEE series conference provides an excellent international academic forum for sharing knowledge and results in theory, methodology and applications on material science and environmental engineering.

In the proceedings, you can learn much more knowledge about the newest research results on metallic materials, chemical materials, composite materials, construction materials and technology, nanomaterials and nanotechnology, material behavior, electronics materials and electrical engineering, materials in mechanical process, environmental chemistry and biology, pollution control, water resource and wastewater treatment, city development and sustainable city, ecological environment and sustainable development, power engineering and green energy, environmental engineering and technology.

The MSEE2015 conference is co-sponsored by Advance Science, Technology and Industry Research Center (ASTIRC), Anhui Polytechnic University and Jiangsu University. The organizing committee has done lots of work to provide a successful conference. All papers accepted by the MSEE2015 have been reviewed by at least 2–3 expert referees from the technical program committee and external reviewers depending on the subject matter of the paper. After the rigorous peer review process, the submitted papers were selected on the basis of creativity, originality, organization and related to conference topics.

We would like to extend our appreciation to all participants in the conference for their great contribution to the success of the MSEE2015. We would like to thank the keynote speakers and individual speakers and all participating authors for their hard work and time. We also sincerely appreciate technical program committee and all reviewers, whose contributions make this conference possible. Finally, I would like to thank the great support from Mr. Leon Bijnsdorp, senior publisher at CRC/Balkema.

We hope everyone have a unique, rewarding and enjoyable weekend at the MSEE2015 in Wuhan, China.

Prof. Zhigang Zou
General Chair of MSEE2015
Nanjing University,
China

Metallic material

Material Science and Environmental Engineering – Chen (Ed.)

Microstructure and mechanical properties of AZ61 alloy with Sm

H.X. Zhu, Q.A. Li, Q. Zhang & J. Chen
School of Materials Science and Engineering, Henan University of Science and Technology, Luoyang, China
Collaborative Innovation Center of Nonferrous Metals, Henan Province, Luoyang, China

ABSTRACT: The effect of Sm on the microstructure and mechanical properties of magnesium alloy AZ61 has been studied by optical microscopy, SEM, EDS, XRD and other technologies. The results show that the addition of 1wt.% Sm can refine the grain size, cause the formation of Al2Sm phase, and enhance the mechanical properties of magnesium alloy AZ61 at room temperature and 150°C.

Keywords: Sm, Magnesium alloy, Microstructure, Mechanical properties

1 INTRODUCTION

As "the 21st green engineering materials", magnesium alloy has been widely used in many fields due to its low density, high specific strength and rigidity, high electric and thermal conductivity, excellent damping effect and good recycling potential, etc [1–4]. With advances in technology, the magnesium alloy has been greatly developed; however, its lower strength and creep resistance at elevated temperature lead to the poor heat resistance of magnesium alloy, which has been the main factor of limiting its application at elevated temperature [4–8].

Alloyed magnesium alloy has been a good option for enhancing its properties; it is believed that rare earth elements are the most effective to improve the properties of magnesium alloy at elevated temperature [9]. Many investigations have been made on the application of rare earth elements in magnesium alloy [9–12]. Sm is one of rare earth elements, but its effects in magnesium alloy are seldom researched. In this work, Sm element is introduced, and the effect of Sm on the mechanical properties of magnesium alloy AZ61 was studied.

2 EXPERIMENTAL

Metallic magnesium, aluminum, zinc, and Mg-25 wt.%Sm master alloy were used as raw materials. The base alloy AZ61 was designed as Mg-6.0 wt.%Al-1.0 wt.%Zn, and the content of Sm was 0 and 1 wt.%.

All the raw materials should be baked at 200°C before the melting started. The alloys were melted in an induction furnace with a Al_2O_3 crucible under the protection of SF6 (1vol.%) and CO_2 (bal.). The melt was held at 720°C for 3 min, and was poured into a metallic mold and the specimen was obtained. They were covered with MgO powders and were heat treated for solid solution at 420°C for 20 h and aging at 220°C for 10h.

The tensile tests were carried out at a strain rate of 1 mm/min in AG-I 250 kN precision universal material test machine at room temperature (written as RT) and 150°C (held for 5 min when tested).

The microstructure was observed by Olympus Optical Microscopy (OM) and JSM-5610 LV scanning electron microscopy (SEM). The phase analysis was performed by D8 Advance X-Ray Diffract meter (XRD) and Energy Dispersive Spectroscopy (EDS).

3 RESULTS

The microstructure of the aged alloy is shown in Figure 1. It can be seen that, the microstructure of alloy consists of white matrix and grey second phases and the second phases are on the grain boundaries or inside the grains in particle, but the amount of second phases is small. It can be known from the related phase diagrams of binary system that the microstructure of magnesium alloy AZ61 consists of α-Mg solid solution, and β-$Mg_{17}Al_{12}$ [8]. Therefore, the white matrix in AZ61 alloy should be α-Mg, and the network phase should be β-$Mg_{17}Al_{12}$. The existence of some particle phases in the grains may be explained as the formation of new phases in the alloy after 1%Sm addition [11–13]. It can be also seen that the grain boundaries of the two kinds of alloys are very obvious, and the grain size is refined after 1%Sm addition. It will be helpful to the mechanical properties.

The XRD patterns of the aged alloys are shown in Figure 2. Obviously, AZ61+1%Sm alloy mainly consists of α-Mg matrix, β-$Mg_{17}Al_{12}$ and Al_2Sm

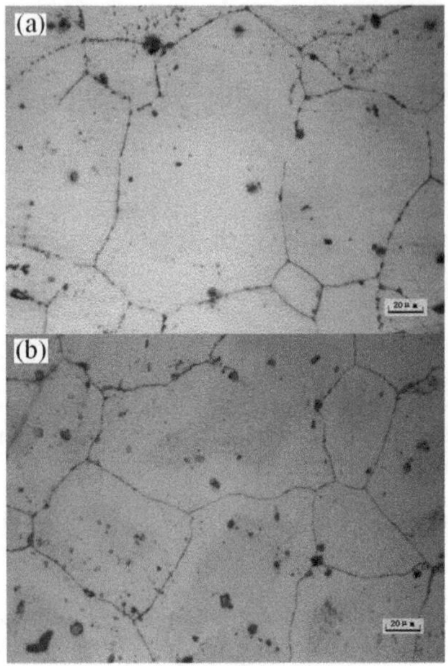

Figure 1. Microstructure of the aged alloys (a) AZ61 (b) AZ61+1%Sm.

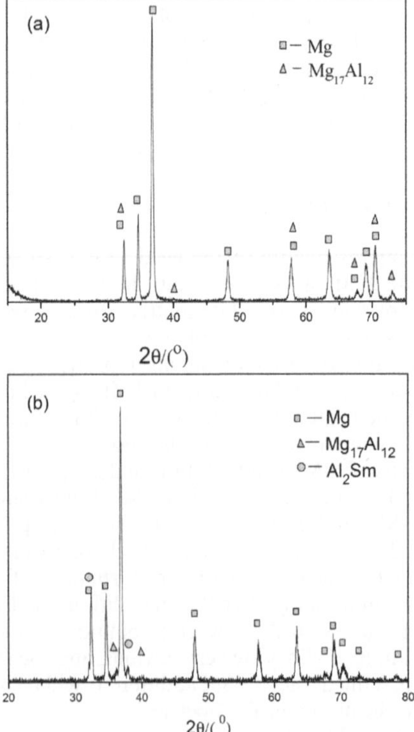

Figure 2. XRD patterns of the aged alloys (a) AZ61 (b) AZ61+1%Sm.

phase. Compared with the phases in AZ61 alloy, a new phase Al_2Sm is formed in the alloy with 1%Sm.

The SEM image and EDS analysis results of second phases in AZ61+1%Sm alloy after solid solution and aging treatment are shown in Figure 3. It can be seen that after the addition of 1%Sm, some new fine particle phases appear which distribute uniformly dispersedly in the matrix. According to the EDS analysis results, we can know that the particle phase should be $Mg_{17}Al_{12}$ or Al_2Sm which have been labelled. It also can be seen that the morphology of $Mg_{17}Al_{12}$ is improved obviously, which is very helpful to the mechanical properties of the alloy [14–17].

The mechanical properties of the alloys after solid solution and aging treatment are shown in Table 1. It is can be seen that, without the addition of Sm, the tensile strengths of AZ61 alloy at room temperature and 150°C are not very high, 229 MPa and 165 MPa respectively. After the addition of 1%Sm, the tensile strengths at room temperature and 150°C are enhanced obviously, 248 MPa and 200 MPa respectively. Compared with AZ61 without Sm addition, the strengths of the alloys are both increased by 20% or so. It also can be seen that the change of the elongation is similar to that of the tensile strength. The elongations of the alloy with Sm also increase, from 4.6% to 5.1% at room temperature, and from 5.4% to 5.8% at 150°C. It can be known from the above that the addition of 1% Sm can not only promote the strength of magnesium alloy AZ61, but also improve the plasticity at room temperature and elevated temperature.

In magnesium alloy AZ61, β-$Mg_{17}Al_{12}$ phase is the main strengthening phase at room temperature, but β-$Mg_{17}Al_{12}$ phase has poor thermal stability and its melting point is only 437°C, which is easily softened and has poor strength at elevated tem-

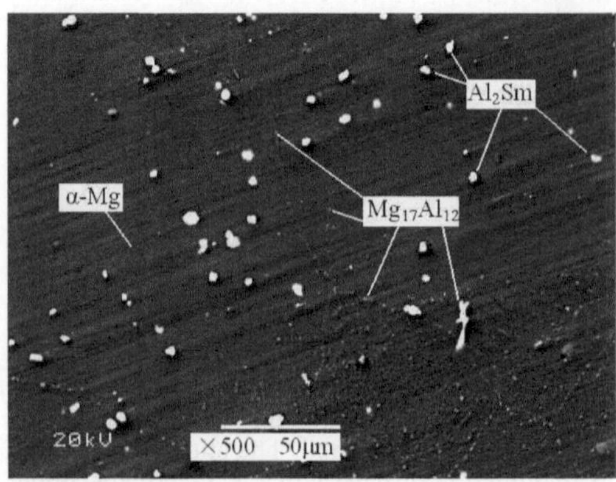

Figure 3. SEM image and EDS analysis results of aged AZ61+1%Sm alloy.

Table 1. Mechanical properties of the aged alloys.

	Tensile strength (MPa)		Elongation (%)	
Alloy	RT	150°C	RT	150°C
AZ61	229	165	4.6	5.4
AZ61+1%Sm	248	200	5.1	5.8

peratures, thus worsening the mechanical properties of magnesium alloy at elevated temperatures.

In this work, after the addition of Sm, the mechanical properties of magnesium alloy AZ61 are improved obviously, especially at elevated temperature. The main reasons for Sm to enhance the properties of magnesium alloy AZ61 at elevated temperature are the function of the dispersion strengthening of Al_2Sm and the improved morphology of β-$Mg_{17}Al_{12}$ phase [13]. Al_2Sm phase has a higher melting point (about 1500°C) and thermal stability than β-$Mg_{17}Al_{12}$ phase, which distributes dispersedly at grain boundaries, can prevent the near grain from moving at elevated temperature and hinder effectively the sliding of grain boundaries and the movement of dislocations, and therefore increase the properties of magnesium alloy at elevated temperature [12–14]. In addition, the morphology of β-$Mg_{17}Al_{12}$ phase is improved, which also plays an important role in strengthening magnesium alloy.

4 CONCLUSIONS

1. With 1wt.%Sm addition, the microstructure of magnesium alloy AZ61 is obviously improved. The morphology of β-$Mg_{17}Al_{12}$ phase is changed, and Al_2Sm particle phase with a high melting point is formed.
2. The improved microstructure causes the enhanced mechanical properties of magnesium alloy AZ61. After 1wt.%Sm addition, the mechanical properties of the alloy are obviously improved at room temperature and 150°C.

REFERENCES

[1] Jain, J.W.J. & Poole, C.W.M.A., Reducing the tension-compression yield asymmetry in a Mg-8. Al-0.5Zn alloy via precipitation, Scripta Materialia, 62(2), pp. 20–25, 2010.

[2] Lin, J.B., Peng, L.M. & Wang, Q.D., Anisotropic plastic deformation behavior of as-extruded ZK60 magnesium alloy at room temperature, Science in China E, 52(3), pp. 161–165, 2009.

[3] Li, B.M., Shailendra, P., Joshi, O. & Almagri, Q., Rate-dependent hardening due to twinning in an ultrafine-grained magnesium alloy, Acta Materialia, 60(3), pp. 1818–1826, 2012.

[4] Lapovok, R., Thomson, P.F., Cottam, R., Akihiko, W. & Yoshiki, N., The effect of grain refinement by warm equal channel angular extrusion on room temperature twinning in magnesium alloy ZK60, J. Mater. Sci. 40(3), pp. 1699–1708, 2005.

[5] Kaiser, F., Letzig, D., Bohlen, J. & Styczynski, A., Anisotropic properties of magnesium sheet AZ31, Materials Science Forum, 419–422(2), pp. 315–320, 2003.

[6] Ding, W.J., Li, J. & Wu, W.X., Texture and texture optimization of wrought Mg alloy, The Chinese Journal of Nonferrous Metals, 21(5), pp. 2371–2381, 2011. (in Chinese).

[7] Lu, L.L., Yang, P. & Wang, F.Q., Effects of thermomechanical treatments on microstructure and mechanical properties of AZ80 magnesium alloy, The Chinese Journal of Nonferrous Metals, 16(3), pp. 1034–1037, 2006. (in Chinese).

[8] Qu, J.H., Li, S.J. & Zhang, Z.G., Effect of extrusion and annealing technology on microstructure and texture of AZ31 magnesium alloy, The Chinese Journal of Nonferrous Metals, 17(2), pp. 434–438, 2007. (in Chinese).

[9] Zhang, Q., Li, Q.A., Chen, J. & Zhang, X.Y., Microstructure and mechanical properties of Mg-Y-Sm-Ca alloys, Applied Mechanics and Materials, 488(1), pp. 197–200, 2014.

[10] Moreno, I.P., Nandy, T.K. & Jonse, W.J., Microstructural stability and creep of rare-earth containing magnesium alloys. Scripta Materialia, 48(3), pp. 1029–1034, 2003.

[11] Jun, C., Li, Q.A., Zhang, Q., Li, K.J., Li, X.F. & Zhang, X.Y., Effects of Sn on Microstrutures and Mechanical Properties of Mg-6 Al-1.2Y-09 Nd Alloys, Rare Metal Materials and Engineering, 39(4), pp. 1180–1183, 2010. (in Chinese).

[12] Zhang, Q., Li, Q.A., Jing, X.T. & Zhang, X.Y., Microstructure and mechanical properties of Mg-10Y-2.5Sm alloy, Journal of Rare Earths, 28(2), pp. 375–377, 2010.

[13] Lu, Y.Z., Wang, Q.D. & Zeng. X.Q., Effects of rare earths on the microstructure, properties and fracture behavior of Mg2 Al alloys[J]. Materials Science and Engineering A, 288(1), pp. 66–76, 2000.

[14] Li, X., Yang, P. & Li, J.Z., Effect of Equal Channel Angular Extrusion on Microstructure and Texture of AZ80 Magnesium Alloy, Hot Working Technology, 39(1), pp. 85–88, 2010. (in Chinese).

[15] Zhang, Q., Li, Q.A., Jing, X.T. & Zhang, X.Y., Microstructure and mechanical properties of magnesium alloy AZ81 with yttrium, Advanced Materials Research, 152(1), pp. 197–201, 2011.

[16] Guo, L.L., Chen, Z.C. & Gao, L., Effects of grain size, texture and twinning on mechanical properties and work-hardening behavior of AZ31 magnesium alloy, Materials Science and Engineering A, 528, pp. 8537–8545, 2011.

[17] Robert, G., Matthias, M.F., & Gunter G., Texture effects on plastic deformation of magnesium, Materials Science and Engineering A, 395, pp. 338–349, 2005.

Material Science and Environmental Engineering – Chen (Ed.)
ISBN 978-1-138-02938-5

Effect of TiC addition on microstructure and mechanical properties of Ti(C, N)-based cermets

G.X. Zhong
College of Mechanical and Power Engineering, China Three Gorges University, Yichang City, China

H.Z. Yu, L. Dai, Y.H. Sun & P. Feng
College of Materials and Chemical Engineering, China Three Gorges University, Yichang City, China

ABSTRACT: Four series of Ti(C, N)-based cermets of Ti(C0.7,N0.3)-WC-Mo_2C-Ni-Co with TiC additive content of 0 wt.%, 5 wt.%, 10 wt.%, and 15 wt.% were prepared by vacuum sintering process. The microstructure and mechanical properties were studied using Scanning Electron Microscopy (SEM) and Image-Pro plus software. The results revealed that the volume fractions of both black core phase and outer rim phase increased, and the inner rim phase and the binder phase decreased with an increase in the TiC content. In addition, the binder phase is distributed evenly and wrapped around the hard phase integrally. The Vickers hardness (HV) of the cermets increased with an increase in the TiC content and showed a maximum value of 1516 HV at 15 wt.% TiC content. The Transverse Rupture Strength (TRS) and fracture toughness (KIC) of the cermets increased with an increase in the TiC content and decreased at a higher TiC content. The best comprehensive mechanical properties were found for the cermets with 10 wt.% TiC addition.

Keywords: Ti(C, N)-based cermets; titanium carbide; microstructure; mechanical properties

1 INTRODUCTION

Various materials have been investigated to promote the tool's life and mechanical properties of cermets and cutting tools that are extensively utilized. About two billion-dollar allotment in semi-finishing and finishing works has been allotted for polymers, ferrous alloys, nonferrous alloys, and advanced materials such as intermetallics and composites [1–4]. Compared to the cutting performance of the conventional tool materials made of WC–Co, cermets presented beneficial resistance to oxidation throughout metal machining and build up edge formation towing to their good chemical stability and high-temperature hardness [5, 6]. However, the TRS and the K_{IC} of cermets were still insufficient. In order to solve the problems, many means had been investigated, such as adding second phase, refining grain, changing compositions, optimizing preparation technologies and introducing advanced equipment.

Researches [7, 8, 9] showed that TiC was nitrided into Ti(C, N) by TiN, and resulted in a high hardness in Ti(C, N)-based cermets. Compared with WC, TiC had more high red hardness, lower thermal conductivity and lower friction coefficient. In ultra-fine Ti(C, N)-based cermets, the moderate additive content of TiC could improve TRS and K_{IC}. When grain size of TiC was refined to nanoscale, it improved the transverse rupture strength of the cermets.

The present work focused on the effect of TiC additive content on Ti(C, N)-based cermets microstructure and mechanical properties.

2 EXPERIMENTAL PROCEDURES

2.1 *Materials preparation*

Ti($C_{0.7}$,$N_{0.3}$)-xTiC-15 WC-8Mo_2C-12 Ni-4Co (in wt.%, x = 0, 5, 10, 15) cermets were prepared by the conventional powder metallurgy technique in this work. Table 1 shows the employed power sizes.

After weighing four groups of the raw material powers according to the mass percentage of compositions, they were milled with hard alloy

Table 1. Power size.

Component	Ti ($C_{0.7}$, $N_{0.3}$)	WC	Mo_2C	Ni	Co	TiC
Size/μm	1.5	1.5	1.95	1.0	1.0	2.0

bolls and absolute ethyl alcohol using planetary ball mill for 36 hours in nylon jar, at a rotation speed of 220 r/min. The power slurry was put into thermostat for 2 hours at 80°C, and then sieved. The dried power mixtures were compacted into a rectangular bar at 100 MPa for 60 s, and the green compacts were polished to the dimension of 21 mm × 6.5 mm × 4.8 mm. The sintering specimen was dew axed in hydrogen flow, and then was sintered at 1500°C for an hour in a vacuum furnace.

2.2 *Test method*

The TRS at room temperature was determined using the three-point bend test of ISO3327 standard with a span length of 14.5 mm and a crosshead speed of 0.5 mm/min.

Vickers hardness was measured with ISO 6507-1:2005 standard with an indentation load of 297 N. The fracture toughness, K_{IC}, was calculated from the length of the radial cracks originating in the corners of the Vickers indentations according to the formula proposed by Shetty [10]:

$$K_{IC} = 0.0889\left(\frac{HP}{4l}\right)^{\frac{1}{2}}$$

where H is the Vickers hardness, P is the indentation load, and l is the crack length.

The specimen was examined by scanning electron microscopy (SEM, FEI, and Holland) for microstructure analysis. The volume fractions of microstructures, including black core phase, white core phase, outer rim phase, inner rim phase, hard phase and binder phase, were measured using the Image-Pro plus software.

3 RESULTS AND DISCUSSION

3.1 *Effect of TiC on microstructure of Ti(C, N)-based cermets*

The microstructure of 0 wt.% TiC and 10 wt.% TiC addition was shown in Figure 1 (a) and Figure 1 (b), respectively. Both Figure 1 (a) and Figure 1 (b) showed a typical core/rim structure. The black core phase was residual Ti(C, N) or TiC and was wrapped around by the white inner rim phase. The outer rim phase was gray, and the binder phase distributed evenly and wrapped around hard phase integrally. Compared with Figure 1 (a), the grain size in Figure 1 (b) was smaller.

In order to study the effect of TiC addition on the cermets in detail, Image-Pro plus was carried out for the cermets. Each phase volume fraction of microstructure of Figure 1 (a) and Figure 1 (b) was shown independently in Table 2. Table 2 showed that the volume fraction of residual black core increased due to TiC additive, and the outer rim phase increased from 52.23% to 57.03 with an increase in the TiC content. On the contrary, the inner rim phase and the binder phase decreased at 10 wt.% TiC addition.

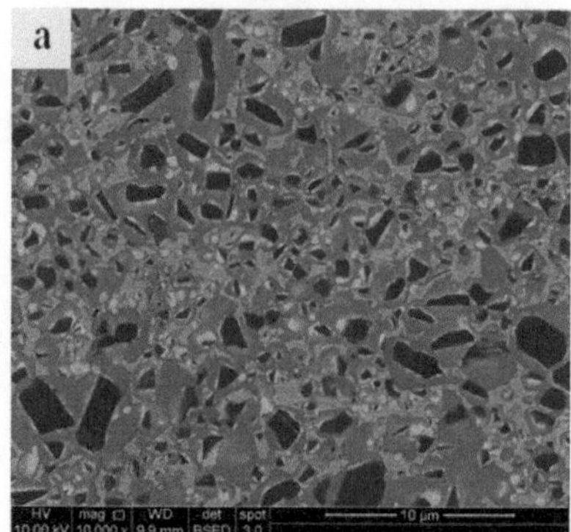

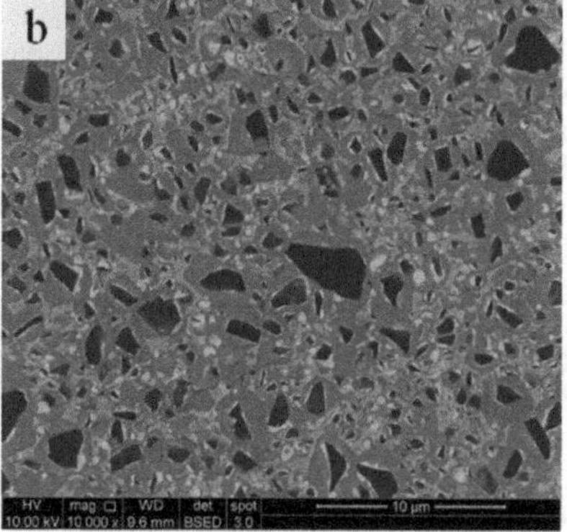

Figure 1. SEM micrographs of Ti ($C_{0.7}$, N0.3)-xTiC-WC-Mo_2C-Ni-Co. (a) x = 0, (b) x = 10.

Table 2. Each phase volume fraction.

Phase/TiC (%)	0	10
Black core	22.37	23.07
White core	1.8	2.72
Inner rim	9.34	6.75
Outer rim	53.26	57.03
Binder	13.22	10.42

3.2 *Effect of TiC on mechanical properties of Ti(C, N)-based cermets*

Mechanical properties, including the TRS, the HV, and the K_{IC}, of the cermets are shown in Figure 2.

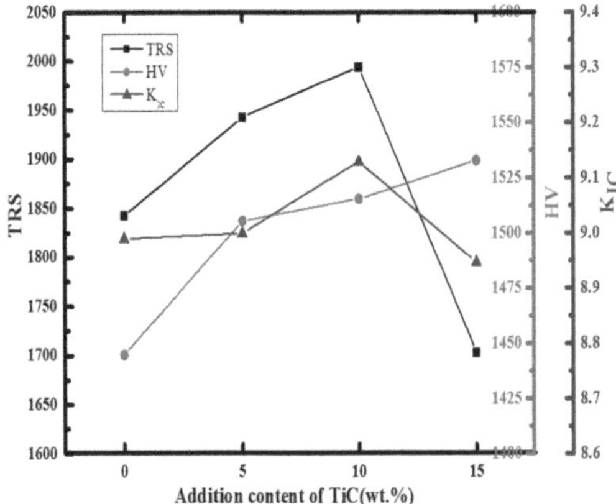

Figure 2. Effect of TiC content on Ti(C, N)-based cermets mechanical properties.

The HV of the cermets continuously increased with the increasing TiC addition and showed a maximum value of 1516 HV at 15 wt.% TiC content. This should be attributed to higher TiC hardness and rudimental TiC in microstructure as listed in Table 2. The TRS and the K_{IC} had the same variation tendency, and they increased below 10 wt.% TiC, then decreased later with excess TiC addition.

The decrease in brittle inner rim phase and the increase in tough outer rim phase were associated with the variation in the TRS and the K_{IC}, even the distributed binder phase also improved the K_{IC}. Figure 2 indicated that the additive content of TiC had a great influence on the TRS, but little on toughness. And the TRS and the K_{IC} reached the peak value at the same time with 10 wt.% TiC addition, when the comprehensive mechanical properties were best.

4 CONCLUSIONS

1. The volume fractions of both black core phase and outer rim phase increased, and the outer, inner rim phase and the binder phase decreased with 10 wt.% TiC addition.
2. TiC had a great influence on TRS, second on HV, little on K_{IC}.
3. The TRS, the HV and the K_{IC} of the cermets increased with an increase in the TiC content, and the TRS and K_{IC} decreased at a higher TiC content. The best comprehensive mechanical properties were found for the cermets with 10 wt.% TiC addition.

ACKNOWLEDGMENTS

This work was financially supported by the National Natural Science Foundation of China (51304124).

REFERENCES

[1] A. Rajabi, M.J. Ghazali, A.R. Daud. Chemical composition, microstructure and sintering temperature modifications on mechanical properties of TiC-based cermet-A review. Materials and Design. 2015, 67:95–106.

[2] Mathias Woydt, Hardy Mohrbacher. The use of niobium carbide (TiC) as cutting tools and for wear resistant tribosystems. Int. J. Refract. Met. Hard Mater. 2014.

[3] Mura G., Musu E., Delogu F. Early stages of the mechanical alloying of TiC-TiN powder ixtures. Mater Chem Phys. 2013, 137:1039–45.

[4] Gotor F.J., Bermejo R., Cordoba J., Chicardi E., Medri V., et al. Processing and characterisation of cermet/hardmetal laminates with strong interfaces. Mater Des 2014, 58:226–33.

[5] Bin Zou, Huijun Zhou, Kaitao Xu, Chuanzhen Huang, Jun Wang, Shasha Li. Study of a hot pressed sintering preparation of $Ti(C_7,N_3)$-based composite cermets materials and their performance as cutting tools. Journal of Alloys and Compounds. 2014, 611: 363–371.

[6] Wanchang Sun, Xiaolin She, LeiZ Zhang, Pan Li, Guanqun Hou, Pei Zhang. Research Progress of Strengthening and Toughening of Ti (C, N)-based Cermets. Hot Working Technology 2014, 18(43):1–4.

[7] Y. Liu, Y.Z. Jin, et al. Ultrafine (Ti, M) (C, N)-based cermet with optimal mechanical properties. Int. J. Refract. Met. Hard Mater. 2011, 29:104–107.

[8] P. Feng, Y.H. He, Y.F. Xiao, W.H. Xiong. Effect of VC addition on sinterability and microstructure of ultrafine Ti (C, N)-based cermets in spark plasma sintering. J Alloys Compd. 2008, 460 (1–2):453–459.

[9] Seongwon Kim, Jian-Min Zuo, Shinhoo Kang. Effect of WC or TiC addition on lattice parameter of surrounding structure in Ti (C0.7 N0.3)–Ni cermets investigated by TEM/CBED. Journal of the European Ceramic Society. 2010, 30: 2131–2138.

[10] S.G. Huang, L. Li, O. Van der Biest, J. Vleugels. Influence of WC addition on the microstructure and mechanical properties of NbC–Co cermets. Journal of Alloys and Compounds. 2007,430:158–164.

Material Science and Environmental Engineering – Chen (Ed.)

Thermal simulation of a net-like microstructure in weld joint of high-strength steel thick plate by Double-Sided Arc Welding

D.Y. Li, X.D. Yi, Y.Z. Li, D.Q. Yang & G.J. Zhang
State Key Laboratory of Advanced Welding and Joining, Harbin Institute of Technology, Harbin, China

ABSTRACT: In Double-Sided Arc Welding (DSAW) of high-strengthsteel thick plate, a net-like microstructure is observed along the grain boundaries of heat affected zone. In this paper, the effects of the net-like microstructure on the mechanical properties of the weld joint were studied through thermal simulation. The results indicated that there were two necessities for the net-like microstructure's coming into being and the microstructure was predominately consisted of blocky martensite-austenite constituents with high carbon and alloy elements content. A great amount of low carbon lath martensite and a few ferrites were reserved from the initial Coarse Grained Heat Affected Zone (CGHAZ). Meanwhile, Intercritically Reheated Coarse Grained Heat Affected Zone (ICCGHAZ) contained more and larger net-like microstructure, when compared with the CGHAZ and the base metal, resulting a higher tensile strength but 19.9% and 11.6% toughness loss respectively. The impact toughness of ICCGHAZ can be increased by 10.26% through high-temperature additional tempering.

Keywords: Double-Sided Arc Welding; high-strength steel; net-like microstructure; thermal simulation

1 INTRODUCTION

Double-Sided Arc Welding (DSAW) has become an effective option for thick-section low-alloy high-strength steel welding since many intermediate operations can be avoided such as back chipping, grinding and reheating, resulting in a reduced welding distortion and improved mechanical properties of weld joint (Zhang et al. 2009, Zhang & Zhang 1998, Zhang et al. 2007).

As the low-alloy high-strength steel which contains numerous alloy elements is treated by cyclic heating in multi-pass multi-layer welding, various microstructures are generated in the weld joint, which have a great influence on the mechanical performance of the weld joint (Cao et al. 2011, Murti et al. 1993). In our previous study on the DSAW experiments of thick-plate low-alloy high-strength steel, the thermal cycle curve of an ordinary weld in an intermediate layer has multi-peak temperature, which can further lead to a significant influence on the microstructure and mechanical properties of the weld joint (Zhang et al. 2008). In addition, a typical microstructure of the Heat Affected Zone (HAZ) is net-like microstructure that can be found along the grain boundaries and distributed around the fusion line. The hardness of the net-like microstructure is inhomogeneously distributed, which makes the microstructure considered as a harmful constituent that can reduce the quality of the weld joint (Lee et al. 1989). Therefore, it is necessary to have further research on the generating mechanism and effecting profiles of this constituent to the weld joint of DSAW and to verify the thereby necessity of implementing this method into practice.

However, the heating and cooling rate is fast during welding process, resulting a narrow HAZ. It is difficult to have an analysis on the attributes of the net-like microstructure from the sample of the entire weld joint directly since multiple narrowed HAZs overlap together blurring the boundary of each of them. Currently, thermal simulation is a solution to analysis the attributes of the net-like microstructure.

M-A constituent was commonly considered as the crack initiator and the main reason for the loss of toughness (You et al. 2013, Mohseni et al. 2012, Moeinifar et al. 2010). So far, many attentions have been paid on the M-A constituent in the HAZs with varied shapes. Thermal simulation was used by Davis et al. (1994) to investigate the functions of M-A constituent generating as a necklace structure along the prior austenite boundary in ICCGHAZ. It was found that toughness impairment of ICCGHAZ is not naturally related with the maximum amount of MA constituent. A near-connected grain boundary network of blocky M-A constituent is further required as well. It was also approved by Hu et al. (2014) that a necklace type of M-A constituent along the prior austenite grain boundaries was left after the subsequent cooling in multi-pass welding. Similarly, Matsuda et al. (1995) and Aihara et al. (1992) reported that complex and inhomogeneously distributed M-A constituent can be found in ICCGHAZ. The stringer M-A constituent along the lath boundary has more effect on the fracture, when comparing to the impact of blocky M-A constituent. However, not enough effort has been devoted into the research of the net-like microstructure which was generated in the DSAW of thick-section steel welding procedure.

In this paper, the attributes of the net-like microstructure which was generated in the thick-section low-alloy high-strength steel multi-pass multi-layer welding process was studied by thermal simulation experiments. The effects of the net-like microstructure on the properties of the weld joint were analyzed, which can be the theoretical foundations to the DSAW of thick-section low-alloy high-strength steel.

2 EXPERIMENTAL PROCEDURE

2.1 *Materials*

Low-alloy high-strength steel subjected to quenched and tempered condition was used as a base metal (chemical composition, wt%, is given in Table 1). The microstructure observed of the base metal presented in Figure 1 is a mixture of the isometric shaped ferrite and granular carbide, which is also called tempered sorbite. The phase transformation temperature A_{c1} is 660 °C and A_{c3} is 790 °C. Cuboids with the size of 11 mm × 11 mm × 110 mm and cylinders with the size of Φ8 mm were cut from the base metal along the direction of rolling and cleaned by acetone to remove the oil on the surface as experimental specimens.

2.2 *Thermal simulation experiment preparation*

According to the previous study on the microstructure of the high-strength steel DSAW joint, a hypothesis was made to deduce the generating mechanism of the net-like microstructure: (1) Coarse grains are the prerequisite; (2) One or two incomplete quenching thermal cycles with the peak temperature between A_{c1} to A_{c3} are needed.

Table 1. Chemical composition of base metal.

C	0.80~1.2
Si	0.2~0.5
Mn	0.5~0.7
P	0.01~0.03
S	0.01~0.03
Ni	3.0~5.5
Cr	0.5~0.9
Mo	0.3~0.7
Fe	The rest

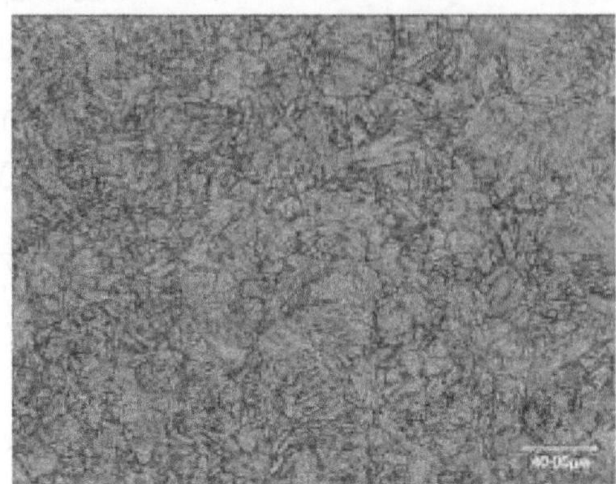

Figure 1. Microstructure of base metal.

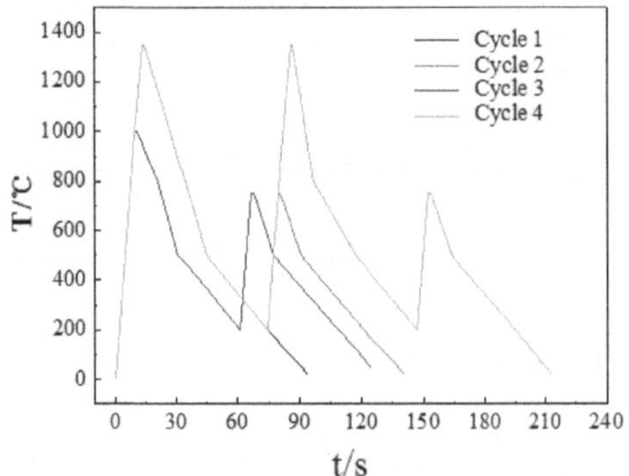

Figure 2. Pre-supposed diagram of the thermal cycles.

In order to verify the hypothesis, four thermal simulation cycles were applied to the experimental specimens. The pre-supposed curves of thermal cycles are shown schematically in Figure 2 and the detailed parameters of the cycles are shown in Table 2.

1. A coarse-grained thermal cycle, called as Cycle 1;
2. A coarse-grained thermal cycle and an incomplete quenching thermal cycle, Cycle 2;
3. A fine-grained thermal cycle and an incomplete quenching thermal cycle, Cycle 3;
4. Two coarse-grained thermal cycles and an incomplete quenching thermal cycle, Cycle 4.

There are four primary parameters in the thermal cycles: peak temperature T_m, holding time t_H, heating and cooling rate. A_{C1} and A_{C3} transformation temperatures of low-alloy high-strength steel are around 660 °C and 790 °C respectively.

Table 2. The detailed parameters of the thermal cycle. (Heating rate 100 °C/s, cooling time $t_{8/5}$ = 10 s).

	First heating		Second heating		Third heating	
No.	Tm_1/°C	t_{H1}/s	Tm_2/°C	t_{H2}/s	Tm_3/°C	t_{H3}/s
Group 1	1350	1				
Group 2	1350	1	750	1		
Group 3	1000	1	750	1		
Group 4	1350	1	1350	1	750	1

Therefore, the peak temperature of coarse-grained heat-affected zone, fine-grained heat-affected zone and incomplete quenching zone were designated as 1350 °C, 1000 °C and 750 °C respectively. It is known that the holding time in high temperature have a significant influence on the grain size, uniformity of austenite and the precipitation and dissolution of the constituents. In this case, 1 s was decided as the holding time in order to simulate the particular heat treatment process of DSAW. 100 °C/*s* was set as the average heat rate. Cooling rate is the crucial factor to the constituent formation and property performance of HAZ. In this paper, the cooling rate of $t_{8/5}$ through the range 800 °C to 500 °C was selected to evaluate the influence of the cooling rate on the HAZ. According to the actual thermal cycle curve of DSAW, $t_{8/5}$ was set as 10 s.

2.3 *Thermal simulation*

The thermal simulation was performed by a thermal simulated test machine (Gleeble 1500D). Platinum-rhodium thermocouples were attached onto the middle of the experimental specimens in the same cross section within 1 mm. Then, specimen was attached in the thermal simulated test machine by copper clips. Simulation program of heat treatment pre-designed in Figure 2 was finally executed. Three specimens were used for each situation. The simulated test curves basically agreed well with the supposed ones, which reveal the parameters of the thermal cycles were chosen reasonably.

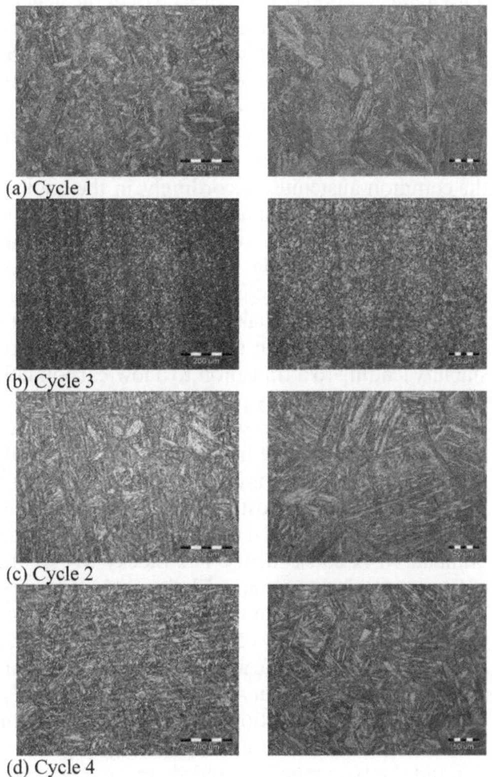

Figure 3. Microstructures of the thermal simulation specimens by optical microscopy.

3 RESULTS AND DISCUSSION

3.1 *Microstructures observation*

After the thermal cycles, microstructures of the specimens were observed using optical microscopy after etching with 3% nital and the results of the four cycles are shown in Figure 3. The net-like microstructure was not generated in the specimens who were treated by a coarse-grained thermal cycle (Fig. 3a) or a fine-grained thermal cycle and an incomplete quenching thermal cycle (Fig. 3b). On the contrary, the net-like microstructure, similar to the one in practical weld joint, was found in the specimens which were both treated by a coarse-grained thermal cycle and an incomplete quenching thermal cycle (Fig. 3c), and two coarse-grained thermal cycles and an incomplete quenching thermal cycle (Fig. 3d). Combined with the distributed positions of the net-like microstructure in practical welding joint and the results of the thermal simulation, the former hypothesis can be deduced to be true. That is, there were two necessities for the net-like microstructure's coming into being: (1) Coarse grains were the prerequisite; (2) One or more incomplete quenching thermal cycles with the peak temperature between A_{c1} to A_{c3} were needed.

A coarsening lath-type martensite in nonequilibrium state was produced in the specimens after being treated by one or two coarse-grained thermal cycles. Lath martensite with different orientations made a nonuniform distribution of carbide particles. When an incomplete quenching thermal cycle was applied, carbide particles precipitated to the α-γ interface in the formation process of α-Fe, resulting a higher carbon content in γ-Fe than the general monophase γ-Fe in high temperature. As the reason that the heating rate of welding was fast and the holding time in high temperature was short, austenite can hardly grow up within the short heating period making carbon-rich austenite. The carbon-rich austenite was more stable when comparing to the common austenite. Accordingly, in the cooling process, the carbon-rich austenite converted to the carbon-rich martensite-austenite (M-A) granular constituents distributing along the grain boundaries. Furthermore, the concentration of carbon in the grain boundaries makes a low carbon content in α-Fe. So, austenite in the grain revealed a poor stability leading to a transition to a low-carbon lath martensite in the cooling process. Scanning electron microscope observation was used to have further research on the net-like microstructure. The result is shown in Figure 4. It is clear to see from Figure 4 that the net-like microstructure along the grain boundaries was consisted of island-like, blocky and granular M-A constituents. The blocky M-A constituents interiorly contained lath martensite with different orientations. In the grain, a great amount of densely arranged low carbon lath martensite and a small amount of ferrite were reserved from the initial Coarse Grained Heat Affected Zone (CGHAZ). The chemical composition and microhardness of island-like microstructure along the grain boundaries and the base metal around them were detected by using electronic probe. The results are shown in Table 3. It can be analyzed from the results that the carbon content of the island-like microstructure

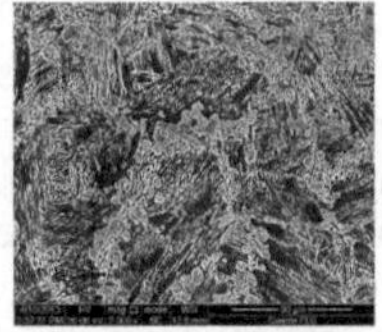

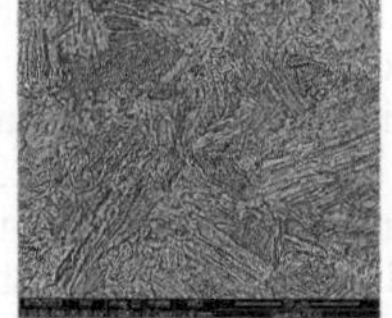

(a) Secondary electron image; (b) Shell electron scattering image

Figure 4. Fine structure of net-like microstructure and measure position of chemical composition.

Table 3. Chemical composition (w%) and microhardness of M-A constituents and nearby base.

Position	Boundary	Intragranular
C	3.57	1.11
Fe	89.05	92.965
Si	0.39	0.4
Mn	0.855	0.625
Cr	0.83	0.6
Ni	4.485	3.26
Mo	0.52	0.665
V	0.175	0.185
Ti	0.13	0.18
Microhardness	445	387

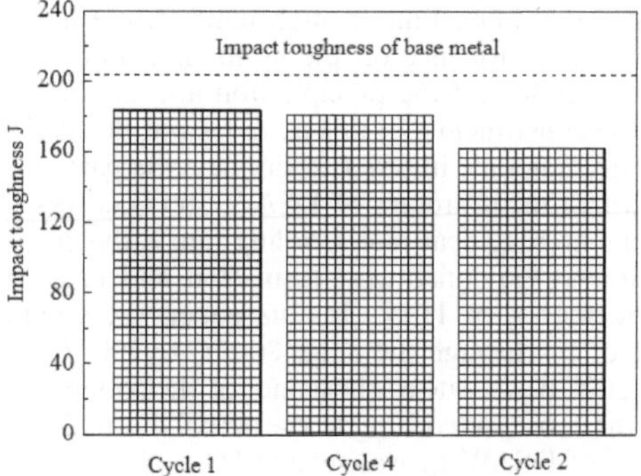

Figure 5. A comparison of the impact toughness between simulation specimens and base metal.

along the grain boundaries (4.14w%) was higher than that in the base metal (1.14w%). The content of other alloy elements and the microhardness of island-like microstructure were also higher than the base metal, revealing a further certification that the island-like microstructure is rich-carbon M-A constituents.

3.2 *Mechanical properties*

Among the HAZ subzones, the Coarse-Grained Heat-Affected Zone (CGHAZ) and Intercritically reheated CGHAZ (ICCGHAZ) are the local brittle zones, which are usually given the most attention. In this paper, tensile properties and impact toughness are used to evaluate the local brittleness.

3.2.1 *Impact toughness and fracture analysis*

A comparison between the impact toughness of simulation specimens and that of base metal is shown in Figure 5. The impact toughness of simulation specimens which experienced one or two coarse-grained thermal cycles and an incomplete quenching thermal cycle is much lower than the impact toughness of the base metal. Impact toughness of the specimen which simulated the thermal cycle of ICCGHAZ is the lowest revealing a significant brittleness. So, we focused on this specimen. The impact toughness value of base metal is 202.35 J. Compared to the impact toughness of base metal and CGHAZ, the impact toughness of ICCGHAZ reduced by 19.9% and 11.6% respectively. Obvious relationship has been found to associate the local brittleness of ICCGHAZ with the shape, size and volume fraction of M-A constituents who distributed along the grain boundaries. It can be deduced from Figure 6 that M-A constituents were not generated in the grain boundaries of the specimen which suffered a coarse-grained thermal cycle. No particular brittleness was found here as coarse grain is the main reason of brittleness. As mentioned above, the net-like microstructure can be found in the grain boundaries of the simulated specimens which experienced a coarse-grained thermal cycle and an incomplete quenching thermal cycle, or two coarse-grained thermal cycles and an incomplete quenching thermal cycle. The local brittleness was produced because of the structural hereditary phenomena. The net-like microstructure in the specimens suffered former thermal cycle with a clear sketch can be obviously seen, whereas, that in the specimens suffered latter thermal cycle had a blur and intermittent sketch. That is, more and bigger M-A constituents were produced when a specimen suffered the former thermal cycle, resulting a worse impact toughness. It can be concluded that

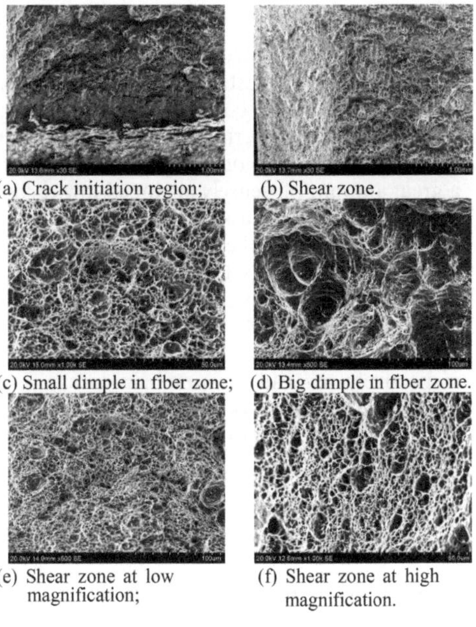

(a) Crack initiation region; (b) Shear zone.

(c) Small dimple in fiber zone; (d) Big dimple in fiber zone.

(e) Shear zone at low magnification; (f) Shear zone at high magnification.

Figure 6. The fractured surfaces by SEM.

the more and bigger M-A constituents produced in the grain boundaries are; the more easily the local brittleness phenomena can be seen.

The micro observation results of the fractured surfaces are shown in Figure 6. A local shrinkage distortion occurred along the width direction around the fracture of the specimen. Then, the fractured surfaces were observed by Scanning Electron Microscope (SEM). Dimples with varied sizes and irregular depths were found on the impact fracture revealing a typical microvoid coalescence fracture which belongs to the ductile fracture. This fact suggests that the fracture toughness of the ICCGHAZ is fine.

3.2.2 *Tensile property*

Figure 7 shows the macroscopic appearance of a cylinder specimen which was used to simulate the ICCGHAZ after being drawn in normal temperature. The fracture occurred in the non-heat-treated base metal, which reveals a conclusion that the tensile strength of ICCGHAZ is higher than that of the base metal. It can also be found that the tensile fracture is a typical cup-cone fracture, showing a obvious necking phenomena.

3.3 *Quality improvement of local brittleness*

It can be conclude that the impact toughness of ICCGHAZ is the worst in the HAZ as the net-like microstructure is generated in the grain boundaries, reducing the mechanical properties of the weld joint. In order to improve the local brittleness, additional tempering was used. The thermal cycle curve of additional tempering was designed as showing in Figure 8.

Impact test was used onto the thermal simulation specimens and the results are shown in Figure 9. After using the additional tempering, the toughness and the local brittleness of ICCGHAZ were improved. It turned better if the temperature

Figure 7. Tensile fracture.

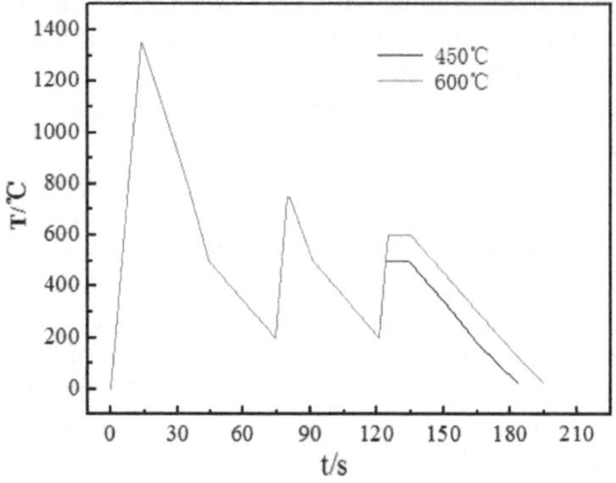

Figure 8. Thermal cycle curve of additional tempering.

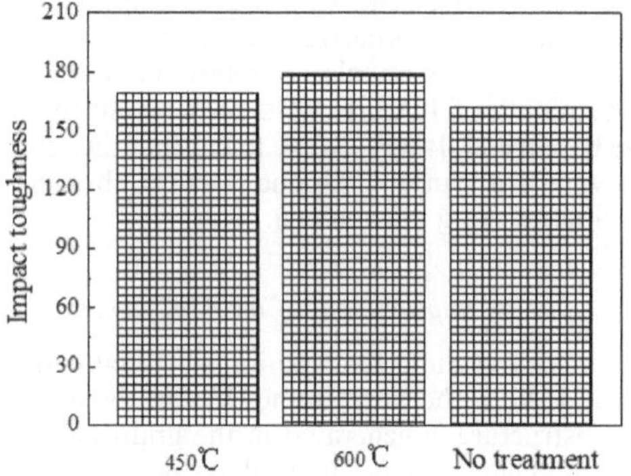

Figure 9. Impact toughness after additional tempering.

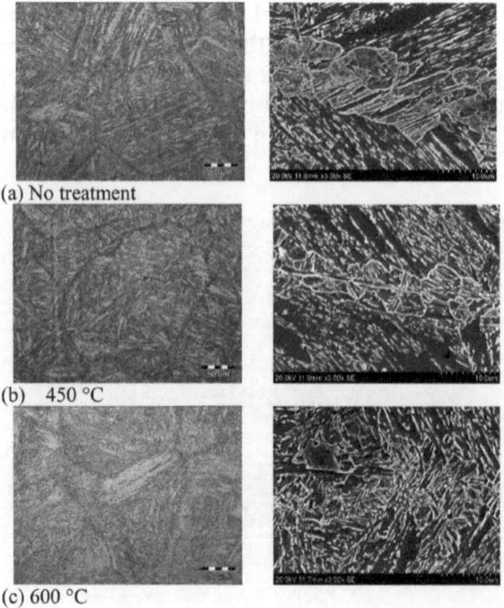

Figure 10. M-A constituents after additional tempering.

of additional tempering was chosen as 600 °C when comparing to the 450 °C. In this situation, the impact toughness of ICCGHAZ is 1.1 times as much as that of CGHAZ.

Figure 10 illustrates the comparison results of microstructure among the M-A constituents of ICCGHAZ without tempering, with 450 °C additional tempering and with 600 °C additional tempering. This fact suggests that additional tempering can reduce the local brittleness and gain the toughness of ICCGHAZ since tempering enables a transformation of M-A constituents in ICCGHAZ on the volume fraction and the size. After the additional tempering treatment, M-A constituents were refined resulting a less volume fraction. When the temperature of additional tempering was set as 600 °C, there was a significant reduction of blocky M-A constituents. The net-like microstructure in the grain boundaries can be hardly seen revealing an improvement of the toughness of ICCGHAZ. Carbon and alloy elements were diffused and precipitated, resulting a more uniformed distribution. Therefore, toughness was increased and the strength stayed the same.

4 CONCLUSIONS

In this paper, thermal simulation was used to reproduce the net-like microstructure which generally occurs in HAZ of the thick-section low-alloy high-strength steel Double-Sided Arc Welding (DSAW) joint. The generating mechanism of the net-like microstructure, fracture toughness and tensile properties of the simulated specimens were focused. The conclusions are presented as follows.

1. The net-like microstructure along the grain boundaries was consisted of island-like, blocky and granular M-A constituents. The blocky M-A constituents interiorly contained lath martensite with different orientations. In the grain, a great amount of densely arranged low carbon lath martensite and a small amount of ferrite were reserved from the initial Coarse Grained Heat Affected Zone (CGHAZ).
2. Compared to the impact toughness of base metal and CGHAZ, the impact toughness of ICCGHAZ reduced by 19.9% and 11.6% respectively, revealing a worst brittleness.
3. Impact toughness of ICCGHAZ can be increased by 10.26% through high-temperature additional tempering.

ACKNOWLEDGMENTS

The authors gratefully acknowledge the financial support provided by the National Natural Science Foundation of China under grant No. 51175119.

REFERENCES

[1] Aihara, S., Okamoto, K. 1992. Influence of local brittle zone on HAZ toughness of TMCP steels. Bulletin 373, Welding Research Council, New York: 33–43.

[2] Cao, R., Zhu, S.S., Feng, W., et al. 2011. Effects of weld metal property and fraction on the toughness of welding joints of an 8%Ni 980 MPa high strength steel. J Mater Process Technol. 211: 759–772.

[3] Davis, C.L., King, J.E. 1994. Cleavage initiation in the intercritically reheated coarse grained heat affected zone: Part I. Fractographic evidence. Metall Mater Trans A. 25 A: 563–73.

[4] Hu, J., Du, L.X., Wang, J.J., Xie, H., Gao, C.R. Misra, R.D.K., 2014. High toughness in the Intercritically Reheated Coarse-Grained (ICRCG) Heat-Affected Zone (HAZ) of low carbon microalloyed steel. Mater Sci Eng A. 590: 323–328.

[5] Lee, S., Kim, C., Lee, D.Y. 1989. Fracture mechanism in coarse grained HAZ of HSLA steel welds. Scrip Metall. 23: 995–1000.

[6] Liao, J.S., Ikeuchi, K., Matsuda, F. 1998. Toughness investigation on simulated weld HAZs of SQV-2 A pressure vessel steel. Nucl Eng Des. 183: 9–20.

[7] Matsuda, F., Ikeuchi, K., Liao, J. 1995. Effect of weld thermal cycles on the HAZ toughness of SQV-2 A pressure vessel steel. In: Trends in welding research 4th international conference. Gatlinburg, Tennessee.

[8] Moeinifar, S., Kokabi, A.H., Madaah, Hosseini. HR. 2010. Influence of peak temperature during simulation and real thermal cycles on microstructure and fracture properties of the reheated zones. Mater Des. 31: 2948–2955.

[9] Mohseni, P., Solberg, J.K., Karlsen, M., Akselsen, O.M., Østby, E. 2012. Cleavage fracture initiation at M-A constituents in intercritically coarse grained heat affected zone of a HSLA steel. Mater Sci Technol. 28: 1261–1268.

[10] Murti, V.S.R., Srinivas, P.D., Banadeki, G.H.D., Raju, K.S. 1993. Effect of heat input on the metallurgical properties of HSLA steel in multi-pass MIG welding. J Mater Process Technol. 37: 723–729.

[11] Shi, Y. & Han, Z. 2008. Effect of weld thermal cycle on microstructure and fracture toughness of simulated heat-affected zone for a 800 MPa grade high strength low alloy steel. J Mater Process Technol. 207: 30–39.

[12] You, Y., Shang, C.J., Chen, L., Subramanian, S. 2013. Investigation on the crystallography of the transformation products of reverted austenite in intercritically reheated coarse grained heat affected zone. Mater Des. 43: 485–491.

[13] Zhang, H.J., Zhang, G.J., Cai, C.B., Gao, H.M., Wu, L. 2009. Numerical simulation of three dimension stress field in double-sided double arc multipass welding process. Mater. Sci. Eng., A. 499: 309–314.

[14] Zhang, H.J., Zhang, G.J., Cai, C.B., Gao, H.M., Wu, L. 2008. Fundamental studies on in-process controlling angular distortion in asymmetrical double-sided double arc welding. J Mater Process Technol. 205: 214–223.

[15] Zhang, H.J., Zhang, G.J., Wu, L. 2007. Effects of arc distance on angular distortion by asymmetrical double sided arc welding. Sci. Technol. Weld. Join. 12: 564–571.

[16] Zhang, Y.M., Zhang, S.B. 1998. Double-sided arc welding increases weld joint penetration. Weld J. 77: 57–62.

Material Science and Environmental Engineering – Chen (Ed.)

Weld formation on overhead position for thick plate of high-strength low-alloy steel by Double-Sided Arc Welding

D.Y. Li, X.D. Yi, D.Q. Yang & G.J. Zhang
State Key Laboratory of Advanced Welding and Joining, Harbin Institute of Technology, Harbin, China

ABSTRACT: Double-Sided Arc Welding (DSAW) is a newly developed efficient method for welding thick plate of high-strength low-alloy steel. In this welding process, overhead position is inevitable. In the present work, the weld formation with the overhead position was studied. To improve the deposition rate, the necessity of weaving welding was addressed. Effects of swing width and staying time were analyzed. Optimum process parameters for excellent bead formation were acquired. DSAW on the overhead position for thick plate of high strength steel is successfully performed with high efficiency.

Keywords: double-sided arc welding; high-strength steel; overhead welding; weld formation; weaving welding

1 INTRODUCTION

Nowadays, weld structure of thick plate of high-strength low-alloy has been widely applied in high pressure vessels, nuclear, shipbuilding, heavy machinery, especially ocean engineering for its high strength and excellent toughness. The conventional welding process on site is described as follows: pre-heating, Gas Metal Arc Welding (GMAW) on one side, back chipping by means of carbon arc air gouging, then polishing, inspection with magnetic particle, preheating again, GMAW on the other side, and post-heating. It can be seen that the welding productivity is very slow. Thus, the high efficiency welding methods for thick plate on site are urgent.

In recent years, many improved GMAW welding methods are developed, such as tandem GMAW (Ueyama et al. 2005), T.I.M.E. process (Chruch, 2001), variable polarity GMAW (Harwig, 2006), and double electrode GMAW (Li & Zhang 2008). However these methods are not suitable for welding thick plates of high-strength low-alloy steel. Double Sided Arc Welding (DSAW) is a new high efficiency method without procedures of back chipping, preheating, and magnetic particle examination. The new double-power DSAW is different from the single-power DSAW designed by Zhang et al. (2001). For this new DSAW, double sided double pulse GTAW are used in backing run. Double-sided double GTAW torches, which are connected to two same welding power supplies, keep a certain arc distance and are asymmetrically located on both sides of the thick plates. Cover passes adopt the double-sided double GMAW. Two GMAW torches, supplied by two independent weld power, are symmetrically located on the upside and downside of the plate.

In our previous studies, the feasibility of DSAW technology (Zhang et al. 2007a), residual stress and no angular distortion distribution in DSAW (Zhang et al. 2009, Zhang et al. 2007b) have been performed in vertical position welding. However, overhead welding for large thick plates, the forming characteristics of which is different from that of other position welding, is inevitable in the welding process. Due to the effect of gravity, droplet transfer is difficult, and weld pool is easy to fall, resulting in a bad weld formation (Lothongkum et al. 1999, Lothongkum et al. 2001). It is well known that small process parameters are suitable for overhead welding. The problem is how to give consideration to welding efficiency and quality. To the best knowledge of the authors, nothing has been published on bead formation on overhead position for welding thick plates of high-strength low steel alloy.

This paper focuses on bead formation on overhead position in double-sided double GMAW in DSAW. Section 2 presents the description of experimental setup, followed by section 3 which depicts the necessity of weaving welding. Effects of process parameters on weaving welding are addressed in section 4. Section 5 gives the result on final weld formation of thick plate on flat-overhead position by DSAW. Section 6 ends the paper with some main conclusions.

2 EXPERIMENTAL DETAILS

The schematic diagram of double-sided double GMAW in DSAW is shown in Figure 1. One GMAW torch is on the flat welding position; the other is on the overhead position. Fronius TPS5000 was used as the welding power supply. The wire electrode diameter was 1.0 mm. The welding torches were moved by a traveller device. The thickness of low-alloy high-tensile steel plates was 50 mm. Keep double-V symmetry groove (angle = 60°), and a root gap of 3 mm. The base metal is high strength low-alloy steel, chemical composition of which is given in Table 1.

In this study, preliminary welding experiments were carried out on the flat plate with a groove, for providing the necessary information to determine limitations of process parameters. As one parameter was kept constant, the other was changed till a bad weld formation was found based on visual evaluation.

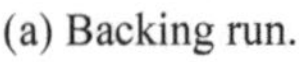

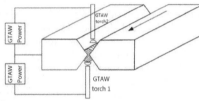

(a) Backing run.

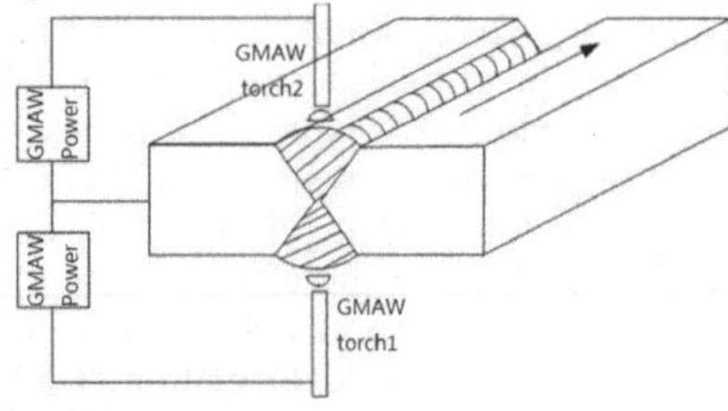

(b) Filling and cover pass.

Figure 1. Schematic diagram of double-sided double arc welding on overhead position.

Table 1. Chemical composition of base metal.

C	0.80~1.2
Si	0.2~0.5
Mn	0.5~0.7
P	0.01~0.03
S	0.01~0.03
Ni	3.0~5.5
Cr	0.5~0.9
Mo	0.3~0.7
Fe	The rest

3 NECESSITY OF WEAVING WELDING

3.1 *Force analysis of weld pool*

The stress state of weld pool is the key factor to influence the weld formation. As presented in Figure 2, forces on the weld pool mainly consist of gravity G, surface tension σ, arc force F_a, droplet impact F_d. Gravity on the weld pool tail points down, making molten metal flow. Surface tension, pointing above, is a beneficial factor for weld formation. Arc force and droplet impact are main holding powers, which act on the tip of the pool. It can be seen that the pool shape has a bearing on the combination of these forces. As G is larger than the sum of F_a, F_d, and σ, the weld pool will collapse, as shown in Figure 3. Therefore, the stability of the

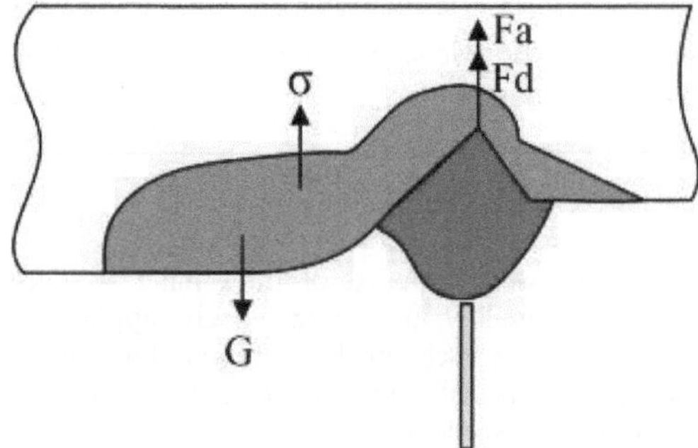

Figure 2. Force analysis of weld pool on overhead position.

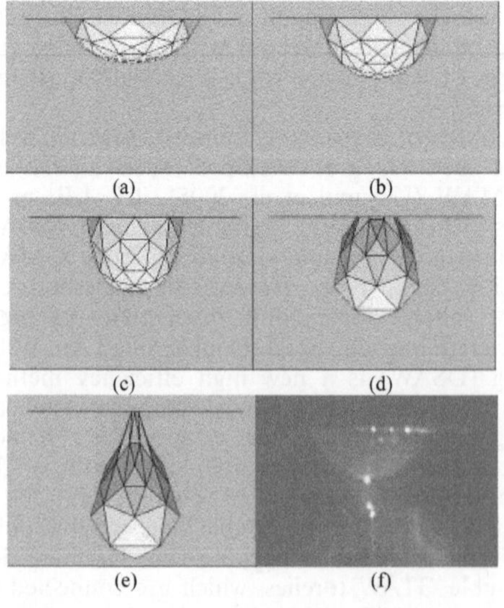

Figure 3. Simulation of drooping of weld pool on overhead position.

pool on overhead position is realized by the effect of surface tension, arc force, and droplet impact to overcome the gravity.

3.2 *Cross-section area of weld bead on overhead position with no waving*

Good weld formation can be obtained if the stress of the weld pool is well controlled. The gravity of the weld pool goes against the weld formation. Cross-section area of the weld bead has a great effect on the gravity of the weld pool. The increase of the cross-section area of the weld bead is associated with an increase in the gravity, leading to a serious droop of the weld bead. As the deposition metal reaches to a certain degree, the molten pool will detach from the base metal. The aim is how to determine the maximum cross-section area while keeping an excellent weld formation and high efficiency.

The cross-section area of the weld bead is calculated as:

$$A_m = \pi V_w D_w^2 / (4V_s) \quad (1)$$

where V_w is the wire feed rate, D_w is the wire electrode diameter, and V_s is the welding speed.

The ratio of bead height to bead width is expressed as:

$$\varphi = h/w \quad (2)$$

It can be seen from Equation 1 that the cross-sectional area of the deposited metal decreases with the decrease of welding current and the increase of the welding speed.

Figure 4 shows the effects of welding current and welding speed on weld shaping. As the arc current increases, the cross-section area and ratio of bead height to bead width increase, which results an obvious droop of the weld bead. When welding current is greater than 90 A, the cross-sectional area of the deposited metal is larger than 30 mm^2. Weld droops clearly, and ratio of bead height to bead width reaches to 0.6~0.7. So, the condition to get good weld shape is that the cross-section area should be less than 30 mm^2, and ratio of bead height to bead width should be less than 0.45. Thus, in no-waving welding, process parameters for the overhead welding should be small, which will lower the welding efficiency.

3.3 *Analysis of waving welding*

To improve the weld quality and efficiency on overhead position, a waving welding method is presented. Staying time of middle, and left and right sides are adopted in flat, horizontal, vertical

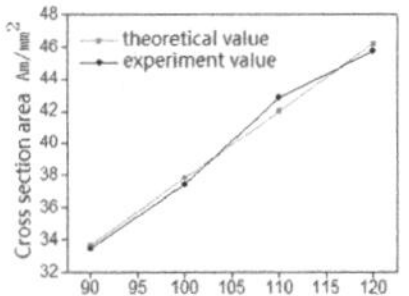

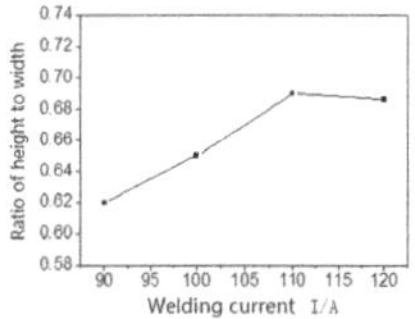

(a)Cross section area vs current; (b) Ratio of height to width vs current.

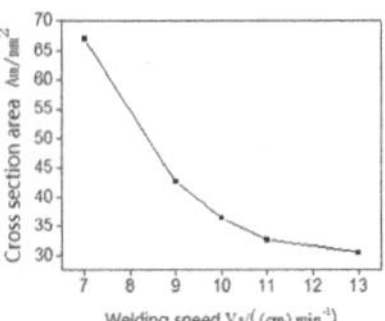

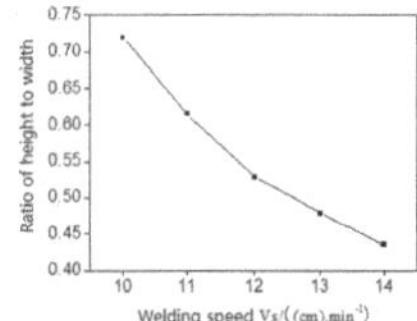

(c) Cross section area vs welding speed; (d) Ratio of height to width vs welding speed.

(e) The weld cross section in different welding current.

Figure 4. Effects of welding current and welding speed on weld shaping.

waving welding. Considering the effect of gravity, staying time of middle is not adopted in overhead welding, preventing the flow of the molten pool to the middle. The waving process for overhead welding is given in Figure 5. Staying time of both sides are t_1 and t_3. The time for once waving is t_2 and t_4. Waving width is D.

Waving on overhead position welding is composed of four procedures, namely, left staying, left waving, right staying, and right waving. During the staying procedure, the welding speed is v_1. The quantity of molten base metal on both sides and weld penetration can be controlled. During the waving procedure, the welding speed is $(v_1^2 + v_2^2)^{1/2}$, which is larger than v_1. The pool widths are pulled efficiently by waving the weld gun. The waving width and waving speed can control the flow and cooling velocity.

Weld pool images of waving welding are obtained by a passive vision sensor, as presented in Figure 6. As the arc stays on one side, the electrode wire and base metal are melted to form the pool. Then the arc waves quickly on the other side, the molten pool solidifies rapidly, preventing the pool flow to the weld center. The weld formation is improved. The weld gun waves left and right, the pool spreads sufficiently, leading to a decreased ratio of bead height to width. With the same seam length, waving welding can reduce the heat input and shorten the cooling time of the pool.

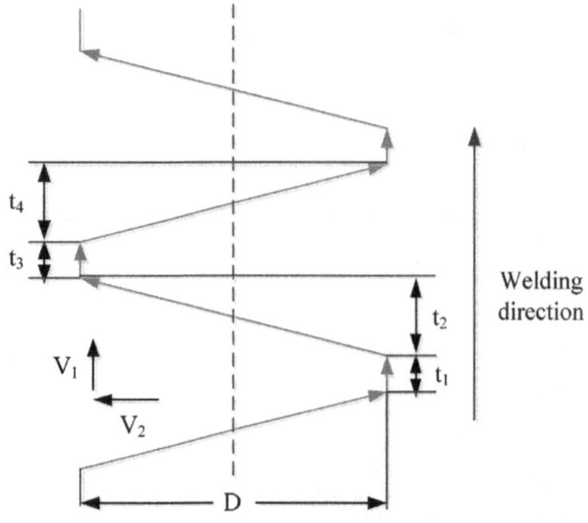

Figure 5. Schematic diagram of waving welding path.

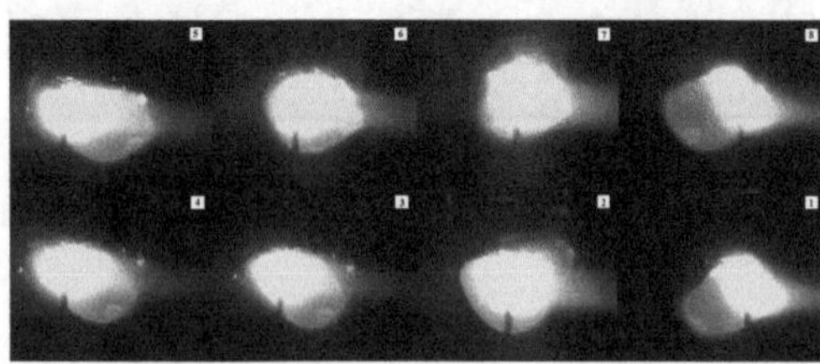

Figure 6. Weld pool image of waving welding on overhead position.

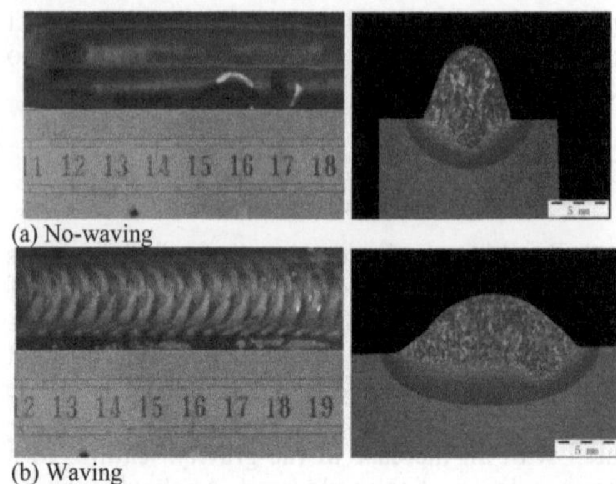

Figure 7. Comparison of weld shaping between no-waving and waving welding.

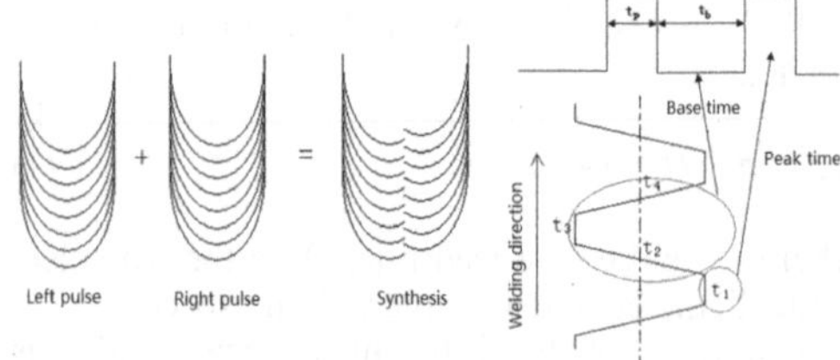

Figure 8. The corresponding relation between the waving welding parameters and pulse welding parameters.

Thus, the waving welding can adopt larger process parameters.

3.4 *Comparison analysis of no-waving welding and waving welding*

Figure 7 shows the weld shaping of overhead waving welding and overhead no-waving welding. Welding current is 90 A; welding speed is 10 cm/min. In no-waving welding, weld shape is narrow and high due to the gravity. In waving welding, weld shape is wide and short, which is better. Weld width of waving welding is double times of no-waving welding, reinforcement decreases 50%. At the same time, the cross-sectional area of the deposited metal increases 50%. So, overhead welding should employ waving welding, which can obtain good weld shape even under larger volume of deposited metal, thus greatly improves the welding production efficiency.

From weld shaping, waving welding can be equivalent to synthesis of two pulse-welding with side by side dislocation. Figure 8 shows the corresponding relationship between waving welding parameters and pulse welding parameters. One side staying duration is equivalent to peak time of pulse, welding current is the peak current, and staying time t_1 is peak time t_p. The base current time t_b of pulse is equal to the sum of the other side staying time, two waving time, which is $t_3 + t_2 + t_4$. At the peak time, metal is heated and melted, and the pool is forming. At the base time, pool cools and solidifies. The average current is

$$I_a = I \cdot \frac{t_1}{t_1 + t_2 + t_3 + t_4} < I \tag{3}$$

This can explain from the other view why waving welding can employ bigger welding current than no-waving welding.

4 EFFECT OF WAVING PARAMETERS ON WELD FORMATION

4.1 *Influence of waving width*

Waving width is one of the important process parameters on weld formation. Figure 9 displays weld formation and corresponding macro section

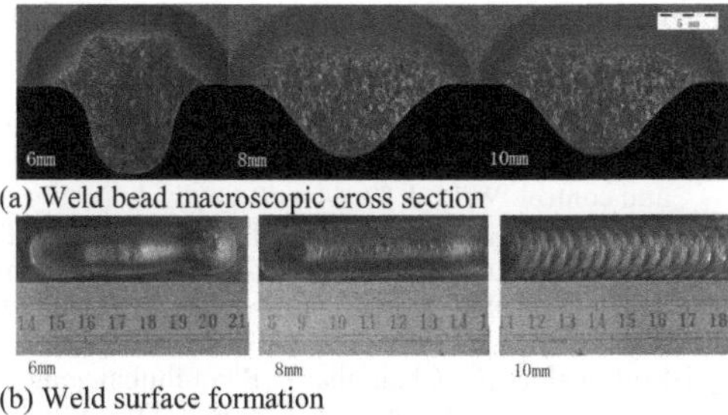

(a) Weld bead macroscopic cross section

(b) Weld surface formation

Figure 9. Weld formation with different waving widths.

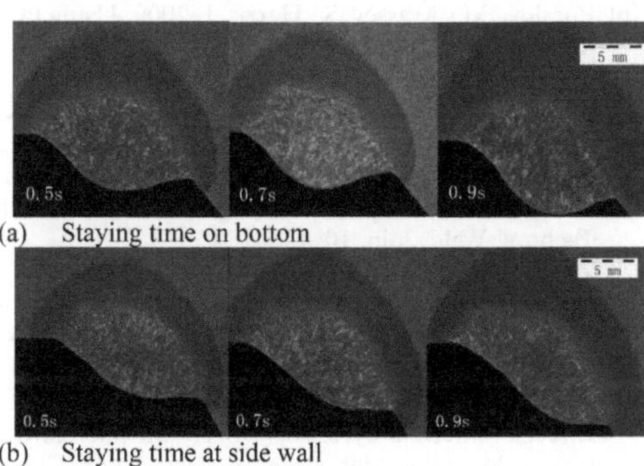

(a) Staying time on bottom

(b) Staying time at side wall

Figure 10. Weld shaping with different staying time.

of different waving widths in overhead welding. As shown in Figure 9, the weld formation is bad, and the molten pool flows to the weld center, leading to an obvious droop of the weld bead. It can be deduced that the weld formation will become better with an increase in waving width. When the waving width is 8 mm, the weld formation is smooth and attractive, as presented in Figure 9. While the waving width increases to 10 mm, the weld width increases further. However, the waving path is clearly seen on the weld surface. With the increase in waving width, the interaction between left and right pool decreases, even forming two separate weld bead. In this study, waving width with 8 mm is regarded as the optimal waving width.

4.2 *Influence of staying time*

The influences of staying time on bottom margin and side wall are shown in Figure 10. With the increase in staying time, the melted quantity and toe length increase. As the staying time increases to 0.9 s, excessive molten metal accumulates on the side, and an obvious droop happens. Thus, the staying time had better be from 0.5 s to 0.7 s.

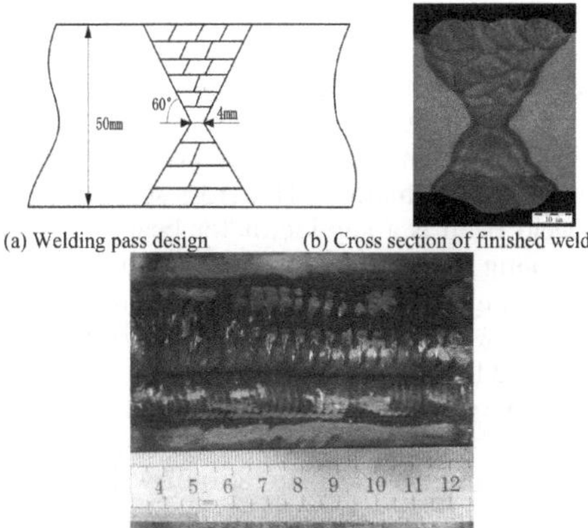

(a) Welding pass design (b) Cross section of finished weld

(c) Surface formation of overhead weld

Figure 11. Weld formation of thick plate by flat-overhead DSAW.

Table 2. Filling and cover welding parameters.

Pass	Flat welding	Overhead welding
Welding current (A)	220~280	90~120
Welding voltage (V)	25~26	20~24
Welding speed (cm/min)	18~20	4~8
Waving speed (°/s)	0	5
Staying time (s)	0	0.4~0.8

5 FINAL WELD FORMATION OF THICK PLATE ON FLAT-OVERHEAD POSITION BY DSAW

5.1 *Influence of staying time*

On the basis of the above researches, the high strength steel thick plate was successful welded by flat-overhead DSAW welding.

The groove shape and welding pass arrangement are shown in Figure 11 a. GMAW double-sided filling welding parameters are shown in Table 2. After welding, the weld surface topography is shown in Figure 11 c. The complete cross section morphology of weld by flat-overhead DSAW is shown in Figure 11 b. It can be seen that surface and inside weld forming are good, there is no lack of fusion, blowhole, slag and other macro defect. The welding efficiency of the new process is 2 times that of the old technology.

6 CONCLUSIONS

The weld formation with the overhead position for double GMAW in DSAW was studied. To improve the deposition rate, the necessity of weaving welding was addressed. Effects of welding current, welding speed, swing width, and staying time were analyzed. The major conclusions are generalized as follows.

Well controlling the pool stress is the premise for a good weld formation. The cross section area of the weld bead is a key factor for bead formation. No-waving welding on overhead position has very low welding efficiency, because only small current and high welding speed are good in order to get good weld formation.

The pool flow can be controlled by waving welding, which can shorten the cooling time of the weld pool and decrease the ratio of weld height to width. Waving welding on overhead position can simultaneously guarantee the weld shaping quality and welding efficiency.

The optimal waving width is 8 mm, and the optimum staying time ranges between 0.5 s and 0.7 s.

Double-side arc welding on the flat-overhead position for thick plate of high strength steel is well implemented, and a good weld formation is successfully achieved.

ACKNOWLEDGMENTS

The authors gratefully acknowledge the financial support provided by the National Natural Science Foundation of China under grant No. 51175119.

REFERENCES

[1] Church, J. 2001. T.I.M.E. process produces fracture-proof welds. Weld. Des. Fabr. 74: 32–35.

[2] Harwig, D.D., Dierksheide, J.E., Yapp, D., Blackman, S. 2006. Arc behavior and melting rate in the VP-GMAW process, Weld. J. 85: 52 s–62 s.

[3] Li, K.H. & Zhang, Y.M. 2008. Consumable double-electrode GMAW. Part II: monitoring, modeling and control. Weld. J. 87: 44 s–50 s.

[4] Lothongkum, G., Chaumbai, P., Bhandhubanyong, P. 1999. TIG pulse welding of 304 L austenitic stainless steel in flat, vertical and overhead position. J. Mater. Process. Technol. 89: 410–414.

[5] Lothongkum, G., Chaumbai, P., Bhandhubanyong, P. 2001. Study on the effects of pulsed TIG welding parameters on delta-ferrite content, shape factor and bead quality in orbital welding of AISI 316 L stainless steel plate. J. Mater. Process. Technol. 110: 233–238.

[6] Purslow, M., Massey, S., Harris, I. 2009. Using tandem gas metal Arc welding to create heavy weldments. Weld J. 11: 34–35.

[7] Ueyama, T., Ohnawa, T., Tanaka, M., Nakata, K. 2005. Effects of ntorch configuration and welding current on weld bead formation in high speed tandem pulsed gas metal arc welding of steel sheets. Sci. Technol. Weld. Join. 10: 750–759.

[8] Zhang, H.J., Zhang, G.J., Cai, C.B., Gao, H.M., Wu, L. 2009. Numerical simulation of three dimension stress field in double-sided double arc multipass welding process. Mater. Sci. Eng., A. 499: 309–314.

[9] Zhang, H.J., Zhang, G.J., Wang, J.H, Wu, L. 2007a. Effect of thermal cycles of DSAW on microstructure in low alloy high strength steel. Trans. China Weld. Inst. 10: 81–84.

[10] Zhang, H.J., Zhang, G.J., Wu, L. 2007b. Effects of arc distance on angular distortion by asymmetrical double sided arc welding. Sci. Technol. Weld. Join. 12: 564–571.

[11] Zhang, Y.M., Pan, C., Male A.T. 2001. Welding of austenitic stainlesssteel using double sided arc welding process. Mater Sci Technol. 17: 1280–1284.

Material Science and Environmental Engineering – Chen (Ed.)
© 2016 Taylor & Francis Group, London, ISBN 978-1-138-02938-5

Approaches for predicting the temper embrittlement tendency

S.H. Zhang, Y.L. Lv & Y. Tan
Environmental Science and Engineering, North China Electric Power University, Baoding, China

ABSTRACT: Rotor, a key component of the steam turbine, has a temper embrittlement tendency after long-running in a complex environment such as high temperature and complex alternating stress. The fracture appearance transition temperature, (referred as $FATT_{50}$) is usually used to describe temper embrittlement degree. This $FATT_{50}$ is a main parameter and must be monitored timely to revise other process parameters (e.g. warm-up time and so on) to avoid further damage. Different form conventional destructive approaches, non-destructive approaches (e.g. chemical corrosion and electrochemical method) could be employed to predict the material degradation without impairing the integrity of the components. By comparing the various methods, the paper pointed out that the electrochemical etching method has more advantages in the establishment of predicting model for the temper embrittlement.

Keywords: temper embrittlement; turbine rotor; $FATT_{50}$; non-destructive prediction

1 INTRODUCTION

Thermal power stations provide most of the electrical energy used in China due to their low running costs, reliability, safety and abundance of fuel. Compared with the rapid expansion of newly installed high productivity thermal power stations, there are considerably large numbers of aged thermal power plants operating in harsh environments of high temperature and complex alternating stress. Steam turbine rotor is the main component of the station, any failure of the rotor will cause serious consequences such as significant costs, long downtime, and even the loss of life. So demonstrated approaches to evaluate the aging process and predict the remaining service life of turbine rotor are highly desirable for safe operation and life extension necessary. [1]

There are many ways for rotor failure, the most common one is the material degradation (e.g. creep, fatigue, embrittlement, and corrosion), and the temper embrittlement is one of the most important causes for material degradation. Having operated at a changed high temperature (above 350 degrees Celsius) for a long period of time, the turbine rotor steel material would lose its flexibility and ductility. That is temper embrittlement damage. There are different views as to its cause, but the current mainstream view is that as a result of segregation of metalloid impurities (e.g. phosphorus (P), tin (Sn), and antimony (Sb)) to the grain boundaries, the segregations reduce the cohesion at grain boundaries [2, 3]. Among those impurity elements, the phosphorus is considered as the main element causing temper embrittlement [4].

Temper embrittlement in low alloy steels (e.g. Cr-Mo-V) can be described by the fracture appearance transition temperature, referred as $FATT_{50}$. At this temperature, the fracture surface of the material is 50% brittle or cleavage and 50% ductile. It is an important indicator commonly used to describe the temper embrittlement process in steamed turbine rotors [5, 6, and 7]. At the same time, it is also an important process parameter. In order to prevent the material brittle fracture, the rotor needs to be heated to above $FATT_{50}$ before the work. Once the material has a tendency to temper embrittlement, the corresponding $FATT_{50}$ will increase, and then some measured should be taken at the starting time to prolong the turbine warming time to increase the rotor temperature above its $FATT_{50}$ when the turbine is cold started and at [5]. If the increase in $FATT_{50}$ value is not detected promptly, the rotor steel could be easily damaged or even broken. Therefore, it is essential to make the accurate predictions and assessments on the temper embrittlement parameters of the turbine rotor material.

To date, different destructive and non-destructive methods have been developed to detect or predict the $FATT_{50}$ value of turbine rotors, such as small punch test [1, 8, 9, 10, and 11], electromagnetic method [12], ultrasonic [13, 14, and 15], Auger electron spectroscopy [16, 17], electrochemical method [5, 18, and 19] and chemical corrosion [20]. This paper would summarize the recent progresses in the area of predicting the temper embrittlement of turbine rotor steel with a focus on non-destructive approaches, including chemical corrosion and electrochemical methods.

2 DESTRUCTIVE METHODS

The small punch test (a miniaturized mechanical testing technique) is a typical destructive approach. It can provide accurate results of mechanical properties (e.g. yield stress, tensile strength, $FATT_{50}$, fracture toughness and creep properties). As there is a need to cut a lot of material on the practical part, the approach is destructive and generally be used only for laboratory experiments [21, 22].

Another destructive approach is to study the phosphorus grain boundary segregation on the miniature specimen by auger electron spectroscopy, and then the phosphorus concentration at the boundary can be converted into $FATT_{50}$ value [23]. It is also considered as a destructive testing method and is highly limited. Correspondingly, chemical corrosion and electrochemical methods are widely used as non-destructive predicting methods.

3 CHEMICAL CORROSION METHOD

More and more experts consider that temper embrittlement happens as a result of segregation of metalloid impurities (e.g. phosphorus (P), tin (Sn) and antimony (Sb)) to the grain boundaries. Among those impurity elements, the phosphorus is considered as the main element causing temper embrittlement [4]. The principle of measuring $FATT_{50}$ with chemical corrosion method is to etch the material surface with some kinds of acid solution. As the phosphorus would be dissolved preferentially in such an acid solution, so etch grooves would be made, then such solution is called as etch liquid. The thermal brittlement degree (indicated by $FATT_{50}$) could be predicted by measuring the groove parameters such as depth or/and width.

There are a variety of solutions that could etch the liquid. Long yuan selected a mixture of metric acid and acetic acid. He carried out a series of chemical corrosion experiments on steel turbine rotor with such mixture, recognized the relationship between width of etch groove on the grain boundary and $FATT_{50}$ which could be conducted as a multiple regression as:

$$FATT_{50} = -227 + 375W - 0.258Hv + 141Ccr + 5334Cs - 256R + 0.146J - 8.73N \quad (1)$$

where the W is the width of etch groove, Hv is the Vickers hardness, Ccr is the percentage composition of Cr, Cs is the percentage composition of S, R is the correlation coefficient, J is the brittleness index, and N is the grain size. The result shows that the errors between prediction value and measure value are in the range of ±15 °C [24].

Chen et al. employed picric acid and sodium dodecyl benzene sulfonate solutions as etch liquids to corrode the turbine rotor steel [24, 25, 26]. Phosphorus on grain boundary dissolves firstly, and width of etch groove on the grain boundary were measured and analyzed by Photoshop. The predicted model of $FATT_{50}$ was established by making multiple linear regression [24, 25] and expresses as:

$$FATT_{50} = 263.725 - 0.981T + 322.141W + 10621.77Cs + 0.119J - 16.664N \quad (2)$$

where T is temperature, w is the width of grain boundary, Cs is the concentration of impurity element S, J is Chemical content parameter (J = CSi + CMn) · (Cp + CSn), and N is the grain size. The verifying experiment shows that the model's error range is within ±20 °C [24, 25].

Zhang et al. etched steel turbine rotor under different conditions and eventually developed an etching process for the turbine steel without corrosion to the grain boundary [27]. Chen et al. selected picric with Sodium Dodecyl Benzene Sulfonate (SDBS) as etch liquid for aging turbine rotor steel. The regression equations which have been obtained from these experiments could predict the $FATT_{50}$ with the scatter of ±15 °C [28].

Chemical etching method can also be used to study other forms of material damage, such as the Intergranular corrosion. Intergranular Corrosion (IGC) is a form of corrosion where the boundaries of crystallites of the material are more susceptible to corrosion than their insides [29]. The composition of different elements (such as carbon, chromium, nickel, phosphorus, tin and antimony) determines the tendency of Intergranular corrosion in austenitic stainless steel materials. The phosphorus segregation on the grain boundary which will lead to temper embrittlement is also correlated to Intergranular corrosion [30, 31]. Bian na studied the feasibility of a method which measures the Intergranular corrosion susceptibility through the width of etch groove on the grain boundary which was generated by sulfuric acid as an etch liquid. The result showed that a good linear relationship exists between the degree of material sensitization and the width of corrosion groove generated by sulfuric acid etching method. The width of etch groove could be used to quantify the Intergranular corrosion degree of stainless steel [32].

4 ELECTROCHEMICAL METHOD

As a relatively high-precision nondestructive detecting method, electrochemical approach has drawn significant attention in predicting the material degradation in turbine rotor steels without impairing the integrity of the components. The method evaluates the temper embrittlement tendency by establishing dependency between the electrochemical signal and the extent of segregation of impurities [5].

Foreign and domestic scholars have done a substantial amount of basic research work in this area. Having investigated the electrochemical behaviors of type 321 austenitic stainless steel in sulfuric acid solution, Mao et al. found that the electrochemical polarization curves can be used to estimate the aging embrittlement degradation [33]. Kwon et al. studied the effect of thermal aging on mechanical and electrochemical behavior of pristine and degraded Cr-Mo-V alloys. In his paper, the electrochemical corrosion characteristics were investigated by the potentio-dynamic anodic polarization and the reactivation methods in a 50 wt% Ca(NO3)2 electrolyte, and the result showed that the electrochemical characteristic values have a good correlation with the rate of material degradation obtained from destructive testing [34]. Komazaki et al. developed a methodology for thermal aging embrittlement and creep damage evaluation of W alloyed 9% Cr ferritic steel. Results showed that the more serious the thermal aging and creep damage, the greater the peak current density 'Ip'. The parameter simultaneously reflected the thermal and stress effect on creep damage [32]. The increase in the Ip indicates the increase in the amount of chromium precipitated as Laves phase [35]. In the later study, the 'Ip' value was found to be increasing linearly with the degree of embrittlement as evaluated by impact absorbed energy at 0°C [36, 37]. Difference in current density (IP2-E-Ipass-E) between active second peak and passive in an anodic polarization curve of temper-embrittled specimens increases linearly with increase in $FATT_{50}$ in the range of mode transfer over transgranular cleavage to intergranular of Charpy impact fracture [38].

Zhang et al. believed that the pre-determined prediction model is too simple and the prespecified size and shape of the model are not adequate for the complex relationships between $FATT_{50}$ and its parameters, so the prediction accuracy for $FATT_{50}$ by this multiple linear regression technique is not sufficient for the life assessing of the turbine rotor. [5, 39] Genetic Programming (GP) was developed for $FATT_{50}$ prediction where the structures subject to adaptation are the hierarchically organized computer programs whose sizes and forms dynamically change during simulated evolution [5, 37–39]. Single loop Electrochemical Polarization Reactivation (EPR) test was carried out at different temperatures in 0.1 M sodium molybdate electrolyte [40]. The multiple correlation coefficient between predicted and measured $FATT_{50}$ was 0.990, indicating that the model obtained by GP can be used in predicting temper embrittlement of new rotor materials with a precision of about ±20°C [5]. Then the similar genetic programming approach was used by Zhang et al. and peak current density of reactivation measured by the potentio-dynamic anode polarization method, temperature of electrolyte, the chemical composition of steel (J-factor), Cr content and the grain size of steel were used as independent variables, while $FATT_{50}$ was used as a dependent variable. The accuracy of this model was found better than that of the model obtained using multiple linear regression method [41]. At the same time, Cao et al. improved the existing genetic programming approach with hardness as the new independent variables and the prediction error of such model is within the scatter of ±20°C, which indicated that the prediction model obtained by genetic programming is feasible and effective [41]. In addition to this, Zhang et al. proposed to model the temper embrittlement of steam turbine rotor in service with Bayesian neural network, by this method, the $FATT_{50}$ was predicted as a function of ratio of the two peak current densities (Ip/Ipr) which could be tested by electrochemical potentio-dynamic reaction method [42]. The neural network showed a more precise prediction of temper embrittlement of rotor steels than the prediction using multiple linear regressions. The training error and verifying error is with the scatter of ±20 °C [42].

Lv yaling investigated the interface performance of turbine rotor alloy (30Cr2MoV) in 0.01M Na2MoO4 by means of Electrochemical Impedance Spectroscopy (EIS) under step potential change conditions, and employed a general equivalent circuit to analyze the polarization behavior of that turbine rotor. The equivalent circuit was shown in Figure 1.

AC potential with amplitude of 5 Mv was applied to a working electrode at the corrosion potential as input signal, sixty frequencies were logarithmic sweeply selected from 100 kHz to 100 MHz. The result showed that there is a linear relationship between the interfacial impedance at a higher frequency and $FATT_{50}$ as shown as in Figure 2; the interfacial impedance could be adopted to indicate the temper brittlement degree.

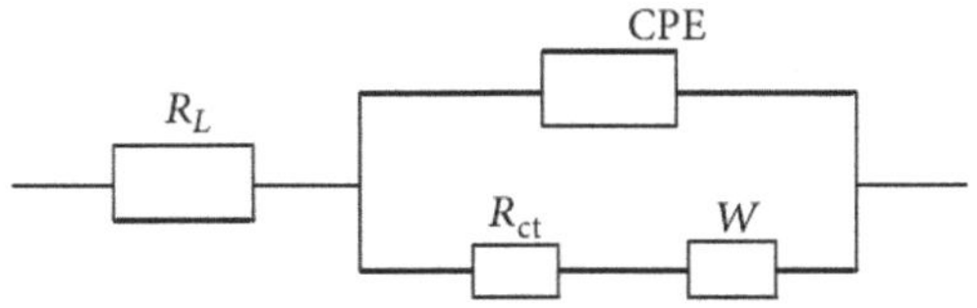

Figure 1. Constructed equivalent circuit in accordance with Nyquist diagrams of the impedance.

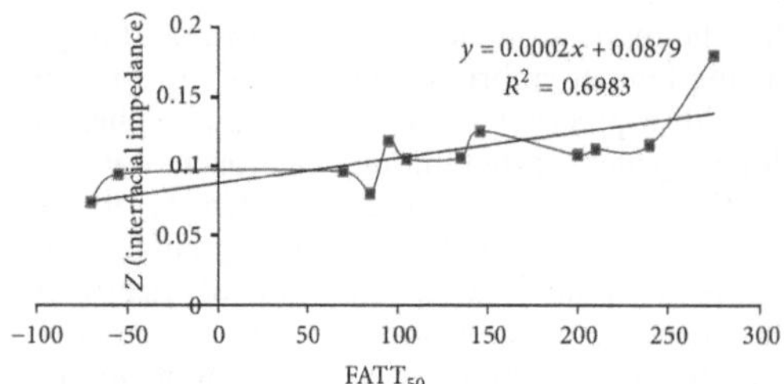

Figure 2. Relationship of interface impedance and $FATT_{50}$ values [43].

5 CONCLUSIONS

For the considerably large numbers of aged thermal power plants operating in harsh environments of high temperature and complex alternating stress, assessment of residual life is an important issue. The rotor is a key component of steam turbine in thermal power plants whose temper embrittlement tendency must be monitored timely to correct process parameters at the beginning of its running. The fracture appearance transition temperature (referred as $FATT_{50}$) is considered to be an appropriate parameter for this tendency.

Different from conventional destructive approaches, non-destructive approaches (e.g. chemical corrosion and electrochemical method) could predict the $FATT_{50}$ without impairing the integrity of the components. In recent years, many predicting models were established for $FATT_{50}$ with non-destructive approaches, especially the electrochemical methods are more widely concerned due to its simplicity and quickness. However, due to many factors, the prediction accuracy of the model established by the electrochemical methods also needs to be improved further.

ACKNOWLEDGMENTS

This paper was supported by 'the Fundamental Research Funds for the Central Universities' (NO.13MS85).

REFERENCES

[1] Ha J.S., E. Fleury, International Journal of Pressure Vessels and Piping, vol.75, no.9, pp.707–713, 1998.

[2] Chen Y.M., Y. Long, Z.G. Han, Y.H. Wang, in: Asia-Pacific Power and Energy Engineering Conference, Wuhan, 2009.

[3] Zhang S.H., Y.Z. Fan, Y.M. Chen, S.L. Zhou, in: International Conference on Industrial Technology, Hong Kong, 2005, pp.768–771.

[4] Fan Y.Z. Research on Nondestructive Test Technique for Detecting Temper Embrittlement of Steam Tuibine Rotor. [D]. North China Electric Power University, 2006. 4.

[5] Kadoya Y., T. Goto, M. Takei, N. Karuki, T. Ikuno, Y. Nishimura, Transactions of the Japan Society of Mechanical Engineers Series A, vol.57, no.537, pp.1085–1090, 1991.

[6] Uemura H., M. Kohno, Tetsu-To-Hagane/Journal of the Iron and Steel Institute of Japan, vol.93, no.4, pp.324–329, 2007.

[7] Fan Y.Z., Y.M. Chen, S.H. Zhang, Turbine Technology, vol. no.2, pp.100125884, 2004.

[8] Shekhter A., S. Kim, D.G. Carr, A.B.L. Croker, S.P. Ringer, International Journal of Pressure Vessels and Piping, vol.79, no.8–10, pp.611–615, 2002.

[9] Komazaki S.I., T. Shoji, K. Takamura, Journal of Engineering Materials and Technology, Transactions of the ASME, vol.127, no.4, pp.476–482, 2005.

[10] Turba K., R.C. Hurst, P. Hähner, Journal of Nuclear Materials, vol.428, no.1–3, pp.76–81, 2012.

[11] Mao X., T. Shoji, H. Takahashi, Journal of Testing and Evaluation, vol.15, no.1, pp.30–37, 1987.

[12] Dobmann G., International Journal of Materials and Product Technology, vol.26, no.1–2, pp.122–139, 2006.

[13] Jeong H., S.H. Nahm, K.Y. Jhang, Y.H. Nam, KSME International Journal, vol.16, no.2, pp.147–154, 2002.

[14] Jeong H., S.H. Nahm, K.Y. Jhang, Y.H. Nam, Ultrasonics, vol.41, no.7, pp.543–549, 2003.

[15] Hsu C.H., H.Y. Teng, S.C. Chiu, Materials Transactions, vol.44, no.11, pp.2363–2368, 2003.

[16] Ding R., A. Islam, S. Wu, J. Knott, Materials Science and Technology, vol.21, no.4, pp.467–475, 2005.

[17] Song S.H., J. Wu, L.Q. Weng, T.H. Xi, Materials Science and Engineering A, vol.520, no.1–2, pp.97–100, 2009.

[18] Fan Y.Z., Y.M. Chen, S.H. Zhang, Turbine Technology, vol. no.2, pp.100125884, 2004.

[19] Kadoya Y., T. Goto, M. Takei, N. Haruki, K. Ikuno, Y. Nishimura, American Society of Mechanical Engineers, Power Division (Publication) PWR, vol.13, no. pp.229–234, 1991.

[20] Lv Yaling, Zhang Shenghan, Tan Yu, Nondestructive approaches for predicting the temper embrittlement of turbine rotor steel, Journal of Chemical and Pharmaceutical Research, 2014, 6(7):1435–1439.

[21] Turba K., R.C. Hurst, P. Hähner, Journal of Nuclear Materials, vol.428, no.1–3, pp.76–81, 2012.

[22] Mao X., T. Shoji, H. Takahashi, Journal of Testing and Evaluation, vol.15, no.1, pp.30–37, 1987.

[23] R. Ding, A. Islam, S. Wu, J. Knott. Effect of phosphorus segregation on fracture properties of two quenched and tempered structural steels. Materials Science and Technology, vol.21, no.4, pp.467–475, 2005.

[24] Longyuan. The research of predicting embrittlement degree of 30Cr2MoV turbine rotor steel based on chemical etching method [D]. North China Electric Power University, 2009.12.

[25] Chen Y.M., Y.H. Wang, X.Y. Li, S.H. Zhang, Turbine Technology, vol.50, no.1, pp.74–77, 2008.

[26] Chen Y.M., Y.H. Wang, S.H. Zhang, Thermal Power Generation, vol.10, no. pp.18–21, 2007.

[27] Zhang Sheng Han, Huang xian long, Chen Chao. Research of etching behavior of turbine rotor steel in nitric acid alcohol solution.[J], Turbine Technology 2007.5.

[28] Chen Yingmin, Wangya Hui, Zhang Sheng cold. The research of chemical etching method for aging 30Cr2MoV Turbine Rotor Steels.[J]. Thermal power generation, 2007.10.

[29] Kasparova O.V., Protection of Metals, vol.40, no.5, pp.425–431, 2004.

[30] Bruemmer S.M., Corrosion, vol.42, no.3, pp.180–185, 1986.

[31] Fan Y., X. Zhai, R. Ma, A. Du, Turbine Technology, vol.50, no.4, pp.318–320, 2008.

[32] Bian ma. The research of intercrystalline corrosion quantitative expression method [D]. North China Electric Power University, 2008.12.

[33] MaoX.,W.Zhao,Corrosion,vol.49,no.4,pp.335–342, 1993.

[34] Kwon I.H., S.S. Baek, H.S. Yu, International Journal of Pressure Vessels and Piping, vol.80, no.3, pp.157–165, 2003.

[35] Komazaki S., S. Kishi, T. Shoji, K. Higuchi, K. Suzuki, Zairyo/Journal of the Society of Materials Science, Japan, vol.50, no.5, pp.503–509, 2001.

[36] Komazaki S., S. Kishi, T. Shoji, K. Higuchi, K. Suzuki, Zairyo/Journal of the Society of Materials Science, Japan, vol.50, no.5, pp.503–509, 2001.

[37] Komazaki S.I., S. Kishi, T. Shoji, H. Chiba, K. Suzuki, Materials Science Research International, vol.9, no.1, pp.42–49, 2003.

[38] Shibuya T., T. Misawa, Zairyo to Kankyo/Corrosion Engineering, vol.48, no.3, pp.146–154, 1999.

[39] Zhang S., Y. Fan, Y. Chen, Jixie Gongcheng Xuebao/ Chinese Journal of Mechanical Engineering, vol.43, no.10, pp.211–214, 2007.

[40] Zhang Sheng Han, Chen Xiaoqin, Chen Yingmin, Li Yuhong. EPR conduct research of 30Cr2MoV Turbine Rotor Steel [J]. Turbine Technology, 2005.4.

[41] Zhang S.H., Y.Z. Fan, Y.M. Chen, X.Y. Li, Kang T'ieh/Iron and Steel (Peking), vol.41, no.5, pp.69–72+77, 2006.

[42] Zhang Sheng Han, Fanyong Zhe, Chen Yingmin. Predict thermal embrittlement performance of Turbine Rotor Steel based on the BP neural network [J]. China CSEE, 2005.12.

[43] Zhang Shenghan, Lv Yaling, and Tan Yu, Evaluation of Temper Embrittlement of 30Cr2MoV Rotor Steels Using Electrochemical Impedance Spectroscopy Technique, Journal of Spectroscopy Article ID 957868.

Material Science and Environmental Engineering – Chen (Ed.)

Numerical simulating of plate cold rolling with higher mismatch of work roller velocities

J.B. Zhang
College of Mechanical Engineering, Tongling University, Tongling, China

ABSTRACT: A FEM simulation study was conducted to investigate the influence of the speed asynchronous factor, frictional coefficient and thickness reduction on the asynchronous rolling with higher values of them. A plane strain rigid-plastic model was used to predict the value of the rolling load and the direction and severity of plate curvature. Using a commercial FEM code, DEFORM, a number of cases had been studied. The results show that three parameters of asynchronous rolling are significant to the rolling force and the plate curvature, and their effects are mutually dependent. This study indicates that, for a rigid-plastic hardening plate with the same initial thickness and velocity, an asynchronous rolling process can be carried out at a slower load and a relatively straight rolled plate can be obtained, by providing a suitable course of action.

Keywords: asynchronous rolling; rolling load; plate curvature; speed asynchronous rate; thickness reduction; frictional coefficient

1 INTRODUCTION

Theoretical and experiment studies on asymmetrical rolling plate and strip have been carried out, because asymmetrical rolling offers benefits such as less rolling pressure, less rolling force, less rolling torque, and more accurate dimension in thickness of the product than those obtained by symmetrical rolling. Most investigations concerning the deformation mechanical and frictional aspects have been executed.[1–6] However, during plate and strip production some asymmetrical undesirable factors occurring during the rolling process have a tendency to lead to curvature. Asymmetrical effects lead to reducing productivity and deterioration of the shape accuracy of the rolled stock. In addition, considerable damage to mill equipment may result, especially when the curvature is severe. Works have demonstrated that an improvement in the quality of plate geometry can be achieved by utilizing the asymmetrical rolling process.[7] The idea behind the asymmetrical rolling technique is to take advantage of the effects of an asymmetric deformation zone, and primarily the reduction of the total roll separating force, to avoid or at least minimize the damage brought by the curvature via right choices of the asymmetric factors with the other variable process parameters.

Practical observations suggest that the main factors causing the plate curvature include differences in roller diameter, work roller speed mismatch, pass height, work roller surface condition.[8] However, the speed mismatch of work roller is considered as the most convenient asymmetric factor on a twin-roller-driven rolling mill with a same diameter of upper and lower roll. The asymmetrical rolling, again called as asynchronous rolling, is well-suited to control the asymmetric condition. Thus, the problem, which we are looking at, differs from the above in that the applied variables of the speed asynchronous factor, frictional coefficient and thickness reduction with higher values of their influence the asynchronous rolling process on plate curvature and deformation force. In the present paper, finite element techniques are used to predict the value of the rolling load and the direction and severity of plate curvature.

2 PROBLEM DESCRIPTION

The purpose of the investigation undertaken is to develop a model for the prediction of the force of deformation and the bend of the plate flowing out of the roller gap after the asynchronous rolling process. A three-dimensional plane strain rigid-plastic finite element model of the steady state processes is used to achieve that. Predictions may be obtained using rigid-plastic facilities of the DEFORM™ suite of programs. In the model, the workpiece is a slab with an original rectangular cross section of 2×50 mm^2 and a length of 80 mm.

Due to the symmetric nature, only half of the slab is modeled with 6,318 nodes and 4,000 solid brick elements to reduce largely calculating time without losing accuracy and improve computation efficiency. The material used for the test is steel C45 and considered to be rigid-plastic and work hardening and obeying the Von-Mises yield criterion and Levy-Mises flow criterion. Isotropic plasticity is assumed with the flow stress curves obtained from compression of C45 at varying strain rates. The upper roller, as well as the lower roller, is assumed to be rigid and has a radii of 65 mm.

Three asynchronous rolling variables assigned higher values as distinct from that in other existent documents are considered for these investigations. Firstly, a speed asynchronous factor, I, is used to characterize the velocity mismatch in the circumferential velocities of upper and lower rollers by the following relationship: $I = v_d/v_c$, where, v_d and v_c severally represent the velocities of upper and lower rollers. In this paper, the speed of the lower roller is kept constant in all cases, $v_c = 0.136\ \mathrm{m \cdot s^{-1}}$, while the upper roller is given different speeds, which are all more than v_c.

Secondly, the deformation degree is denoted as a variable by the reduction in thickness, i.e., $\varepsilon = (h_0 - h)/h_0$, where h_0 and h represent the specimen thickness before and after rolling, respectively. Single-pass cold asynchronous rolling processes are severally conducted at various reductions from 30 to 50% by the designed scheme of numerical simulation.

Finally, the frictional stress between the plate and the rollers is seen as an important influential factor on the asynchronous rolling of plate. The friction in the roller gap is described by a friction model of the form following as: $\tau_f = \mu p$, where, μ is the friction coefficient and p is the pressure at the interface. In this paper, friction factor m and friction coefficient μ are chosen in such a way that represent a unique frictional behavior.[9] When no front tension, friction plays an important role in rolling process, as it is the only mechanism by which the plate is pulled through the roll stand. Due to a larger speed mismatch of upper and lower rollers, there may be a relative "slipping" between the contact of the rollers and the plate. As a result, the process loses some of the capacity to draw the plate into the roller gap, as happens in the case of lower friction coefficient.[10] Therefore, the larger values of friction coefficient are set as $0.2 \leq \mu \leq 0.4$, as well as higher speed factors of $1.2 \leq I \leq 2.2$.

Except for these variables above, an additional rolling parameter, the initial velocity of the plate is assumed constant in all cases, i.e. $v_0 = 80\ \mathrm{mm \cdot s^{-1}}$, always lower than the horizontal component of the roller surface speed. The choice of initial velocity results in a net acceleration, which minimizes the initial impact between the workpiece and the rollers.

3 SIMULATION RESULTS

3.1 *Denotation of curvature*

The radius of curvature ρ is calculated according to the definition shown in Figure 1, and is achieved by approximating a radius of curvature to the deformed mesh. So, the plate curvature δ is given by equations following as:

$$\rho = \frac{1}{r} = \frac{l_d - l_c}{h},\ l_d = \sum_{1}^{n} v_{d,j} \cdot \Delta t_j,\ l_c = \sum_{1}^{n} v_{c,j} \cdot \Delta t_j.$$

where, l_d, l_c are the arc lengths of upper and lower surfaces of rolled plate, respectively; n is the total of FEM running steps, $v_{d,j}$, $v_{d,j}$ are the velocities at upper and lower contacting points on the outlet cross-section of plate at jth step, respectively; Δt_j is the time increment at jth step.

Finite element predictions of rolling force for friction coefficients, speed asynchronous factors and relative reductions are shown in Figures 2 and 3, as well as plate curvature in Figures 4 and 5. Positive plate curvature indicates turndown, whereas negative plate curvature shows turn up. Figures 3 and 5 show the results as a function of the speed asynchronous rate. The rate is used to compare different speed asynchronous factors and elongation ratios, and is a combined measure of roller speed, plate thickness and friction. The plastic deformation region of asynchronous rolling is divided into three distinct regions according to the directions of the frictional force

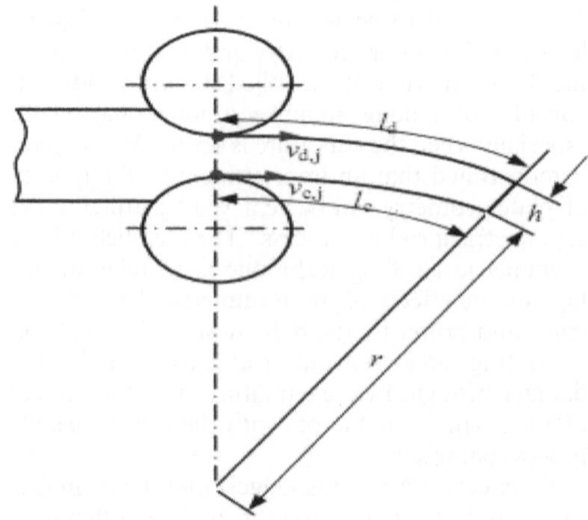

Figure 1. Definition for radius of curvature.

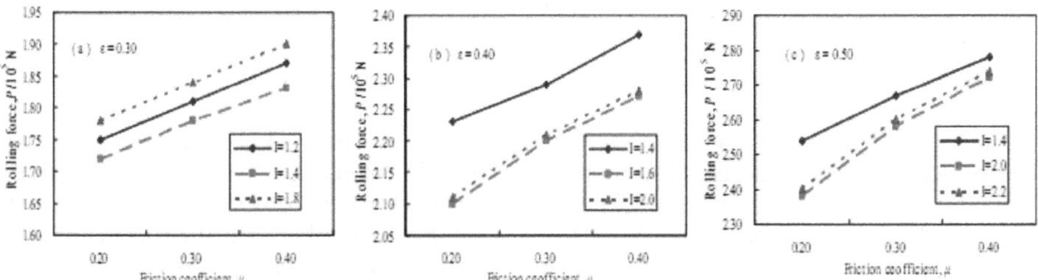

Figure 2. Effect of friction coefficient on force.

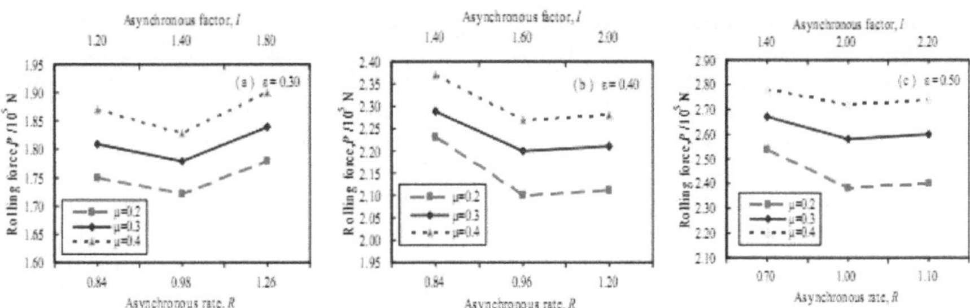

Figure 3. Effect of speed mismatch on force.

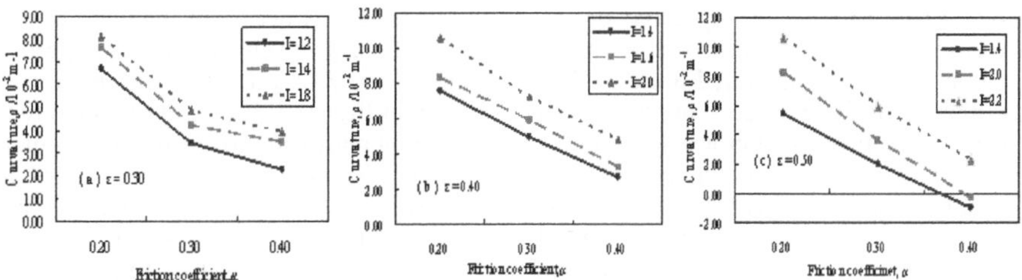

Figure 4. Effect of friction coefficient on curvature.

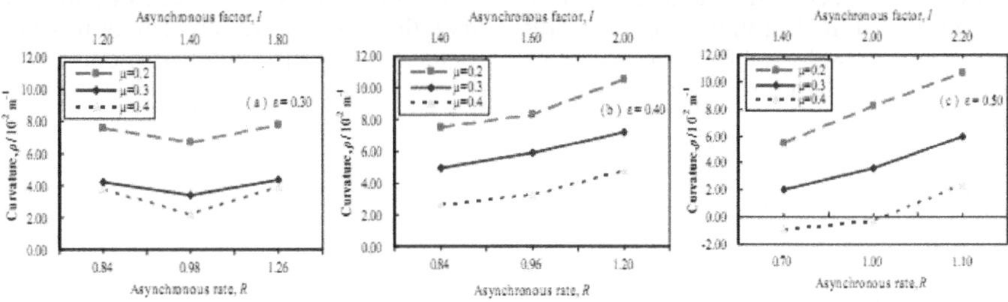

Figure 5. Effect of speed mismatch on curvature.

on the upper and lower rollers exerted on the workpiece. These are denoted forward slip zone for entry region, rubbed rolling zone for the cross shear region and backward slip zone for the exit region. A large body of evidence indicates that the rubbed rolling zone will enlarged to the full deformation region, contrarily the forward and backward zones will all decrease and ultimately vanish from plastic region, along with the increase of speed mismatch to the value of plate elongation rate. For the purpose of further investigations, a variable, R, is introduced to represent the speed mismatch ratio of deformation region, which is given by following equations: $R = I/\lambda$, where, I and λ are the speed asynchronous factor and plate elongation rate, respectively. Neglected the wide increase of plate rolled, $\lambda = H/h$, H and h are the plate thicknesses of before and after rolling, respectively.

3.2 *Effects on rolling force*

Figures 2 and 4 illustrate the effect of friction coefficient and speed asynchronous factor upon the rolling force for different thickness reductions. As expected, rolling force rapidly increases by increasing the frictional coefficient at fixed reductions, shown as Figure 2. Likely, the larger the reduction ε, the higher the force P. For example, as can be seen in Figure 6, by increasing ε from 0.30 up to 0.50, P increases considerably from 172, 223 and 254 to 183, 237 and 278 kN, respectively, for the frictional coefficient $\mu = 0.20$, 0.30 and 0.40, at fixed velocity mismatch factor $I = 1.40$. As is known to all, the higher deformation, as well as the higher friction, will arouse higher force and energy to deform the workpiece.

However, there is just an interesting change in these investigations that rolling force doesn't always decrease as the velocity mismatch rises. The change is more clearly revealed in Figure 3. With its increase, the rolling force reduces markedly when $R < 1.0$, then slightly increases (shown as Figure 3b and c), even greatly increases (shown as Figure 3a) when $R > 1.00$. Reasons for this behavior probably includes two aspects as following: on the one hand, as the speed mismatch increases, the frictional resistance brought by the work rollers on the plate in backward zone will become weaker, and the promotion effect of frictional force in rubbed rolling zone will become stronger, both resulting in reduce of rolling load. Therefore the rolling force decreases until it becomes relatively stable when $I = \lambda$ or $R = 1.00$. On the other hand, it is well known that inhomogeneous deformation will certainly enhance the flow resistance in the rolling process, resulting in increasing rolling load. The higher velocity mismatch can induce higher inhomogeneous deformation across the thickness, leading to higher rolling load. In addition, when $R > 1.00$, exorbitant mismatch of speed between upper and slower rollers will arouse slip between the rollers and workpiece. This would weaken the effectiveness of the frictional force in rubbed rolling zone.

3.3 *Effect on plate curvature*

Figures 4 and 5 shows the changes of plate curvature. It's quite clear that the curvature is positive, or the rolled plate tends down to lower (slower) roller in most of these cases. It is worth noting that under specified conditions, i.e. $I = 1.40$ and 2.00, $\mu = 0.40$ and $\varepsilon = 0.50$, the curvature becomes negative, that is to say, the rolled plate tends up to upper (faster) roller. The effect of friction coefficient μ on the plate shape is shown in Figure 4. It can be seen that the shape of rolled plate roughly becomes better owing to a reduction of curvature as μ increases. Therefore, a larger frictional coefficient can availably improve the plate shape despite inevitable increase of rolling load.

When the reduction in thickness ε is 0.30, 0.40 and 0.50, respectively, the effect of speed asynchronous factor I or rate R on the plate curvature is shown in Figure 5. It can be observed that the curvature increases significantly with an increase of I or R at $\varepsilon = 0.40$ and 0.50 (see Figure 5b and c). However, the change of plate curvature with larger values is not very significantly when I or R increases at $\varepsilon = 0.30$ (see Figure 5a). This variation of curvature above may have something to do with the deformation of the workpiece. The variety of curvature for different thickness reductions at $I = 1.40$ is illustrated in Figure 7. It appears that there are maximum curvatures at a certain value of ε for three different frictional coefficients.

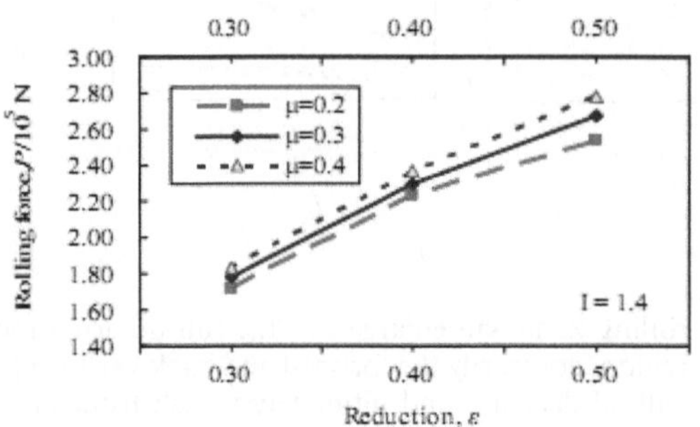

Figure 6. Effect of reduction on force.

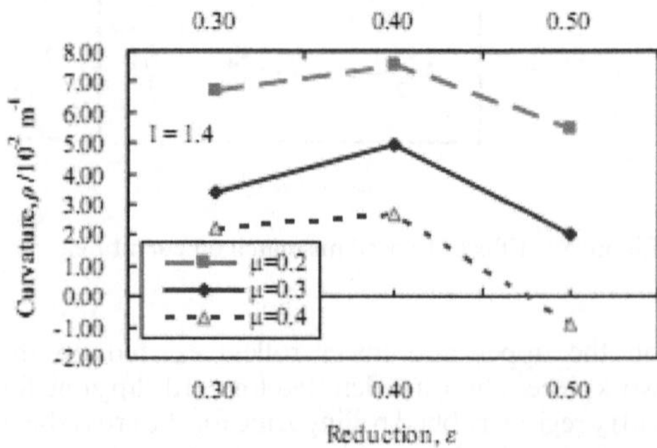

Figure 7. Effect of reduction on curvature.

When the friction coefficient μ is smaller, i.e. $\mu = 0.20$ and 0.30, the curvature is always positive and the plate bends towards the lower (slower) roller. Whereas wider curvature index depending on the thickness reduction is obtained so that in such cases, the rolled plate may be bent towards either of the work rollers. Reasons for this behavior are not obvious. An intuitive assumption would be that the rolled plate would always curve toward the slower (lower) roller, since the plate material near faster roller will tend to move faster than that near slower roller. Meanwhile, it is generally thought that the higher thickness reduction of material with a fixed inlet thickness can induce lower strain gradient across the thickness on account of a decrease of medium thickness in the deformation region, as well as the higher frictional sheer force in the rubbed rolling zone. And also, under higher mismatch of work rollers, the formability of material nearer the faster roller can't be enhanced as same as that of nearer the slower roller, on account of slipping easily arisen between the faster roller and plate. These would decrease the influence of the upper roller to deform the plate, increasing the relative effect of the lower roller. As a result, the rolled material will tend to curve toward the faster (upper) roller when it emerges from the roller gap. This result is useful to determine the values of rolling parameters and reduce the curvature of rolled plate.

4 CONCLUSIONS

A study based on pure numerical simulations was carried out to investigate the influences of speed mismatch, frictional coefficient and thickness reduction on the cold rolling asynchronous process with higher values of them. According to a series of analytical results, the following conclusions have been drawn:

1. A higher frictional coefficient will contribute to reduce the curvature of rolled plate and improve its shape, but must increase the rolling load.
2. With increasing of speed asynchronous rate, the plate curvature will basically increase, whereas the rolling force will firstly decrease, then increase, and reach its minimum when R is about 1.0.
3. As expected, the larger the thickness reduction is, the higher the rolling force is. However, the magnitude of curvature will firstly increase, then decrease with increasing the reduction in thickness of workpiece, and reach its maximum when ε is about 0.40.
4. Depending on the asynchronous rolling parameters applied, it is possible to control the curvature of rolled plate and to reduce the total rolling load by differentiating the peripheral velocities of work roller.
5. By using the appropriate relative deformation, frictional condition and speed rate of upper to lower rollers for a given feed stock thickness, it is possible to completely eliminate the unfavorable phenomenon of plate bending, or to obtain the permissible curvature which doesn't obstruct the free feed of plate to the next pass or transferring it to other facilities.

ACKNOWLEDGMENTS

This work is sponsored by the Scientific Research Starting Foundation for Talent of Tongling University, No. 2012tlxyrc06. We gratefully acknowledge the foundation's financial support and helpful advice.

REFERENCES

[1] D.Q. Ma, L.H. Wu and L.H. Zhao: Central Iron Steel Res INS Tech Bull, 1983, 3, p.43–51.
[2] J. Sha and L.C. Zhao: Central Iron Steel Res INS Tech Bull, 1985, 5, p.37–42.
[3] Z.C. Zhao, M.J. Song and M.Q. Wang: Journal of Applied Sciences, 1990, 8, 223–231.
[4] Y.L. Zhu, Y.P. Gong and Y.L. Kang: Journal of Iron and Steel Research, 1996, 8, 22–25.
[5] Y.M. Hwang and G.Y. Tzou: J. Material Engineering and performance, 1993, 2, 597–606.
[6] B. Wang, W. Hu, L.X. Kong and P. Hodgson: Metals and Materials, 1998, 4, 915–919.
[7] T. Kiefer, and A. Kugi: Mathematical and Computer Modeling of Dynamical Systems, Vol. 14 (2008), p.249–267.
[8] C.W. Knight, S.J. Hardy, A.W. Lees and K.J. Brown: Ironmaking and Steelmaking, Vol. 29 (2002), p.70–76.
[9] F. Farhat-Nia, M. Salimi, and M.R. Movahhedy: J. Mate. Pro. Techno, Vol. 177 (2006), p.525–529.
[10] A.M. Galkin, A.K. Ek, O.A. Kjselev, and K.V. Ozhmegov: Hutnik-Wiadomości Hutnicze, 2010, 5, p.210–212.

Material Science and Environmental Engineering – Chen (Ed.)
© 2016 Taylor & Francis Group, London, ISBN 978-1-138-02938-5

Structure transition of Metal-Organic Framework-5 induced by copper doping

F.M. Tian, Y.L. Chen, X.H. Zhang & M.Y. Wang
School of Science, Institute of Applied Micro-Nano Materials, Beijing Jiaotong University, Beijing, P.R. China

ABSTRACT: Copper-doped Metal-Organic Framework-5 (MOF-5) was successfully synthesized by sovothermal crystallization process for the first time. The X-ray diffraction, scanning electronic microscopy and energy diffraction spectroscopy showed that cubic MOF-5 structure can be maintained for a relative doping fraction of Cu, $\varphi_c \approx 0.125$. It was found that as the amount of Cu^{2+} ions exceeds φ_c the cubic MOF-5 structure becomes unstable, and the MIL-53-like structure emerges. The transition from the MOF-5-like structure to the MIL-53-like structure is a first order phase transition; since for the doping fraction in the interval between 0.125 and 0.75, two kinds of structures are coexist. Crystal with MOF-5-like structure is a perfect cube in micrometer scale which is a Zn-rich phase, while the crystal with MIL-53-like structure is a rod to a micrometer scale which is a Cu-rich phase.

Keywords: porous materials; metal-organic framework; doping; crystallization

1 INTRODUCTION

Metal-Organic Frameworks (MOFs) represent a new class of porous materials constructed from transition metal ions and bridging ligands. MOFs have potential applications in gas storage/separation, catalysis, nonlinear optics, sensing, and electronic devices due to their diverse properties and highly tunable construction. [1, 2] Within this wide variety of possible applications, it is of particular interest to tune the physical and chemical properties of MOFs in a certain way while fully maintaining structural features. This can basically be achieved by either manipulating the organic linker [3] or by modifying the inorganic building unit, e.g., substituting the metal species to a certain fraction. [4] Among these two strategies, the partial isomorphic substitution of the metal is the most commonly used strategy to introduce active centers into inorganic materials.

One of the most studied metal-organic frameworks is MOF-5 that originally synthesized in 1999. It is constructed by Zn_4O $(CO_2)_6$ octahedral Structure-Building Units (SBU). That is four Zn_4O tetrahedral linked by six carboxylate groups forming a simple cubic network. [5] Due to its large surface area, exceptional pore volume, and relatively high thermal stability, it therefore, may be favorable for the selective capture of gas. [6–8] However, MOF-5 is moisture-sensitive even under atmospheric conditions, because the metal-oxygen coordination bond is relative weak. This results in the structure collapse due to attack of water molecules. [9, 10] Developing new techniques to improve the structure properties of MOF-5 is a challenge for both the chemists and physicists.

Several strategies have been suggested to increase the hydrostability and gas adsorption capacity of MOF-5. [11, 12] For instance, the series materials of IRMOF-n which have larger organic linker retain the original cubic MOF-5 structure. [13, 14] Alternatively, doping by substituting metal ions partially is other way to modify MOF-5 to enhance its exceptional porosity, framework rigidity or thermal stability. To date, there are a few works reported on substituting of Zn in MOF-5 clusters with different same chemical valence. Chongli Zhong has done the work that incorporate metal species producing a spillover effect on H_2 adsorption, and post synthesis anchoring of Li or Cr metals into the aromatic rings have pointed out the possibility of carrying out an isomorphic substitution of Zn in MOF-5 clusters. [15, 16] Juan A. Botas has reported that including the cobalt and nickel succinates and zinc-cobalt and zinc-nickel 4, 4′-biphenyldicarboxylate. These results showed that the modified ligand or doping metal ions in metal clusters are both feasible ways to enhance the property of MOF-5 with maintaining its structure.

It is well known that Cu^{2+} is an alternative divalent ion, and it has similar volume size and coordination geometry with Zn^{2+}. It can be used to achieve isomorphic substitute Zn^{2+} ions in MOFs, and vice versa. One notable exception is the investigation of Zn-based analogue of HKUST-1 that was carried out recently with substitution of Zn^{2+} into HKUST-1. [17] This study indicated the existence of low level substitution of Zn^{2+} into the HKUST-1 framework. This presents a good example to show that Zn^{2+} can substitute Cu^{2+} in HKUST-1. So far, the synthesis of copper-doped MOF-5 materials has not been reported. In this work, copper-doped MOF-5 was successfully synthesized by solvothermal crystallization process for the first time. Cu doping is investigated comprehensively for both the low and high doping concentration. Moreover, it was demonstrated that the doping effect on the stability of the cubic MOF-5 structure and the phase transition of the unstable MOF-5 in high doping concentration.

2 EXPERIMENTAL SECTION

2.1 *Materials and methods*

Zinc nitrate ($Zn(NO_3)_2 \cdot 6H_2O$), Copper nitrate ($Cu(NO_3)_2 \cdot 3H_2O$), 1,4-benzenedicarboxylic acid (H_2BDC), and N, N′-Dimethylformamide (DMF) were of analytical reagent grade from commercial sources and used without further purification and modification. The images of the synthesized product were taken in an optical microscope with Powder X-Ray Diffraction (PXRD) was carried out on a Rigaku D/Max-2500 X-ray diffractometer at 40 kV, 120 mA with a scan speed of 10°/min and a step size of 0.02° in 2θ range 5 to 70°, using Cu Kα radiation (wavelength λ = 0.1543 nm). Jade from Materials Data was also used to analyze powder diffraction data and match that to previously reported structures. Scanning Electron Microscopy (SEM) studies were conducted with a HITACHI S-4800 SEM (Japan) equipped with an energy-dispersive X-ray spectroscopy (EDS) instrument operating at an accelerating voltage of 20 kV after gold deposition.

2.2 *Synthetic procedures*

Zn-based MOF-5 crystals of uniform size were synthesized following the reported procedures with a few modifications. In synthesis, the organic solvent DMF was first degassed using 4 A molecular sieve. 2.4960 g of zinc nitrate hexahydrate and 0.5280 g of 1, 4-benzenedicarboxylic acid were dissolved in 60 ml of DMF solvent. They were then transferred into Teflon-lined stainless steel autoclave immediately and heated at 398.15 K for crystallization for about 4 h. The reaction vessel was cooled to room temperature normally. At the end of the crystallization step clear white crystals of MOF-5 emerged from the wall and base of the reaction vessel. The MOF-5 crystals were separated from the reaction solution, washed with DMF to remove the unreacted zinc nitrate, followed with purification in chloroform. The chloroform purification was performed by adding chloroform into 20 mL vials containing the raw MOF-5 crystals. The vials were capped tightly and put back to the oven at 343 K for another 3 days. Solvent in vials was replenished with fresh chloroform every day. The MOF-5 was heated at 423 K under vacuum overnight in order to remove present solvents, moisture, and other volatile components. Then, the MOF-5 was cooled to room temperature and labelled for Sample I. Special care was taken to avoid exposure of the materials to air.

The synthesis of Cu-doped MOF-5s was performed following a similar synthesis procedure of pure MOF-5 as described above, in which mixed different amount of $Cu(NO_3)_2 \cdot 3H_2O$ and $Zn(NO_3)_2 \cdot 6H_2O$ (0.015625 mmol $Cu(NO_3)_2 \cdot 3H_2O$ and 0.984375 mmol $Zn(NO_3)_2 \cdot 6H_2O$ for Sample II; 0.03125 mmol $Cu(NO_3)_2 \cdot 3H_2O$ and 0.96875 mmol $Zn(NO_3)_2 \cdot 6H_2O$ for Sample III; 0.0625 mmol $Cu(NO_3)_2 \cdot 3H_2O$ and 0.9375 mmol $Zn(NO_3)_2 \cdot 6H_2O$ for Sample IV; 0.125 mmol $Cu(NO_3)_2 \cdot 3H_2O$ and 0.875 mmol $Zn(NO_3)_2 \cdot 6H_2O$ for Sample V; 0.25 mmol $Cu(NO_3)_2 \cdot 3H_2O$ and 0.75 mmol $Zn(NO_3)_2 \cdot 6H_2O$ for Sample VI; 0.5 mmol $Cu(NO_3)_2 \cdot 3H_2O$ and 0.5 mmol $Zn(NO_3)_2 \cdot 6H_2O$ for Sample VII; 1 mmol $Cu(NO_3)_2 \cdot 3H_2O$ and 0 mmol $Zn(NO_3)_2 \cdot 6H_2O$ for Sample VIII).

3 RESULTS AND DISCUSSION

The structure and the coordination environment of Zn and Cu are shown in Figure 1. The major geometries that Cu and Zn take up are octahedral and square pyramidal. In the octahedral arrangement, six ligands surround the metal atom. In the square pyramidal, four ligands surround the metal atom. Given the similarity of cation radius, valence, and coordination geometry, it is possible that Zn^{2+} and Cu^{2+} interchangeable occupy the same coordination site in the framework. Such a coordination environment is actually expected for Cu^{2+} ion and Zn^{2+} ion forming stable crystal in the as synthesized material where four oxygen atoms from the carboxylate groups forming the paddle wheels and a ligand is coordinating to the metal ion. The two different species of metal ions are often located in different coordination environments forming different crystal. Our interest in bimetallic systems came from a desire to examine the effect of two metal ions competing for the same coordination site of a comparatively simple ligand. The variation of the amount of Cu^{2+} ions in the Copper-doped MOF-5 materials of this work offers a good

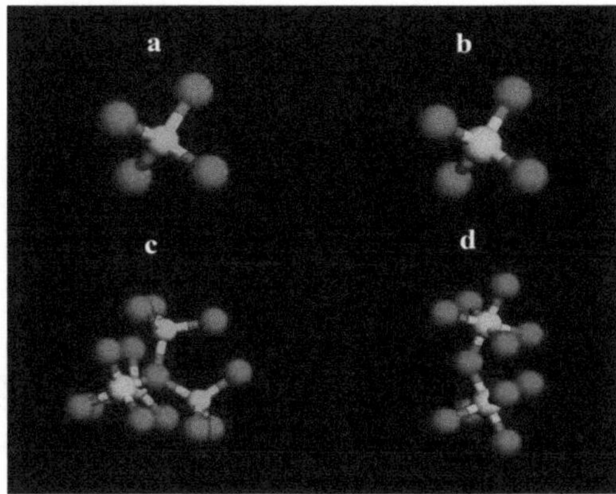

Figure 1. Three dimensional representation of coordination environment of Zn and Cu in metal-organic framework crystal structure. (a) Coordination environment of Zn atom in MOF-5; (b) Coordination environment of Cu atom; (c) Coordination environment of Zn atom was replaced by Cu atom; (d) Coordination environment of Cu atom in MIL-53(Cu). Blue, red, and green denote Cu, O, and Zn atom respectively. For clarity, all hydrogen atoms are omitted.

opportunity to study the influence of metal cluster composition on the structure and their potential applications.

A small amount of copper nitrate ($Cu(NO_3)_2 \cdot 3H_2O$) added to a DMF solution of zinc nitrate ($Zn(NO_3)_2 \cdot 6H_2O$) and 1,4-benzenedicarboxylic acid (H_2BDC) produced crystals with optical morphologies shown in Figure 2. All the images are taken at 100× magnification using normal mode. In Figure 2a, Sample I, the pure MOF-5 is well-shaped cubic crystals and exhibited clear white. In Figure 2b, Sample II, Sample III and Sample IV with a lower Cu^{2+} doping concentration $\varphi_c < 0.125$, an interesting phenomenon showed that there appears to be a coherent grain boundary between the blue copper rich and the clear zinc rich crystalline regions denoted by the gradual transition from blue to clear inside the solid. This coherent boundary is an indication of a similarity in crystal structure and the coexistence of both phases. The fact that the concentration of copper is higher than the concentration of zinc in the center indicates that the rate of nucleation of the Cu(BDC) is likely higher than that of the surrounding Zn(BDC) crystals that emanate from it. This indicates a dilution of the Cu species by the surrounding Zn species, despite the predominance of the former in the core region due to the likely faster nucleation of Cu(BDC). In Figure 2c–e, Sample V, Sample VI and Sample VII appear to be two kinds of crystals, clear white rectangle and blue cubic when the doping concentration of Cu^{2+} is higher. The results showed that when the concentration of Cu^{2+} and Zn^{2+} is comparable, the hetero-bimetallic system would form copper type crystal and zinc type crystal. The separation of these two crystal structures indicates that the coexistence of Zn-based phase and pure Cu-based phase and the transition from Zn-based phase to the Cu-based phase is a first order transition. The crystalline system containing only Cu exhibited a more intense blue color (showed in Fig. 2f) was of similar clarity. In addition, the spherical shaped red colored particles are by-products, Cu_2O obtained at a higher Cu^{2+} concentration (showed in Sample VIII).

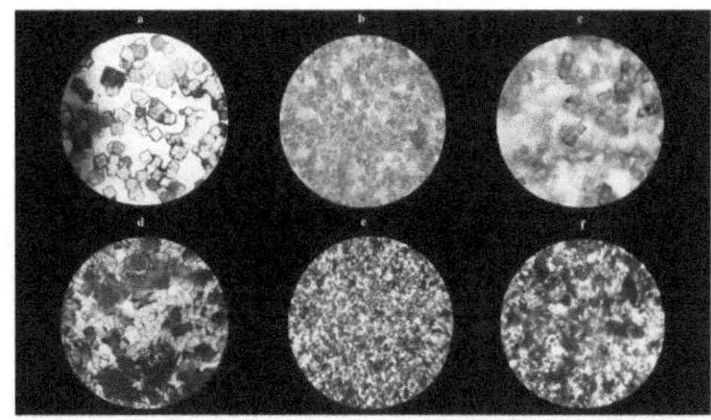

Figure 2. Optical Microscopy of Cu-doped MOF-5s (a) Sample I; (b) Sample II, Sample III and Sample IV (c) Sample V; (d) Sample VI; (e) Sample VII; (f) Sample VIII.

X-Ray Powder Diffraction (XRD) measurements of the products with different Cu compositions were performed to determine their crystal phases. XRD pattern for Sample I, pure MOF-5 sample synthesized in this work is shown in Figure 3a compared with the simulated XRD pattern for the MOF-5. The well resolved peaks imply the high crystallinity of the MOF-5 samples and the main peaks position match well with the simulated XRD pattern and the reported XRD pattern of MOF-5. [18] The intensities of the three peaks of 6.8°, 9.6 Å and 13.6° were observed different with simulated XRD pattern, probably due to some alterations of atomic orientation in crystal planes by solvent and other adsorbate molecules that fill the microspores of MOF-5. In this structure, the 13.6° peak is due to the reflections of (2, 0, –4), and the 6.8° peak is due to (1, 0, –2) planes. The position of the peaks and the location of the strongest peak were found to be shifted to a certain degree. This shift may have been caused by the partial interpenetration of crystals into MOF-5. [19, 20] The simulated XRD pattern of the non-penetrated MOF-5 based on its crystal structure is similar to its PXRD pattern, while the simulated XRD pattern of the interpenetrated MOF-5 based on the crystal structure (solvents were excluded) has different peak intensities from that of its PXRD pattern, as shown

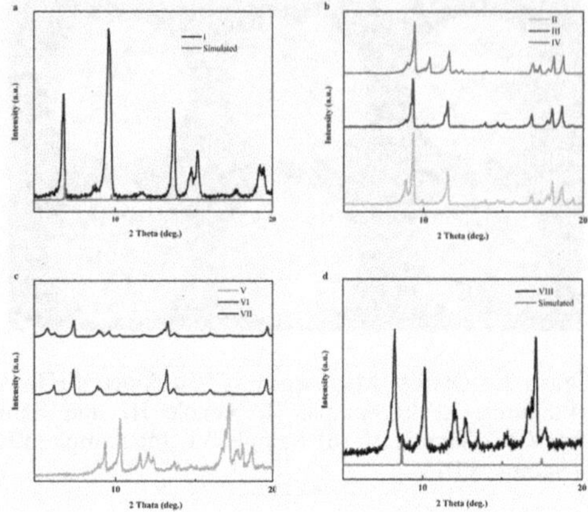

Figure 3. (a) XRD patterns of as-synthesized Sample I compared with the simulated MOF-5; (b) XRD patterns of as-synthesized Sample II, Sample III and Sample IV; (c) XRD patterns of as-synthesized Sample V, Sample VI, and Sample VII; (d) XRD patterns of as-synthesized Sample VIII compared with the simulated MIL-53(Cu).

in Figure 3a. A sharp peak below at of 9.6° was observed. This is probably caused by the residual guest molecules (chloroform solvent) that alter the crystal phase plane orientation.

The crystallinity of Cu-doped MOF-5 was probed by XRD immediately after crystal growth and without evacuation. It should be noted that the positions of the XRD peaks of MOF-5 doped with low quantities of Cu were different from pure MOF-5, as shown in Figure 3b. It can be inferred that both pure MOF-5 and MOF-5 doped with low quantities of Cu have the similar cubic structures and similar chemical compositions. Sample II, Sample III and Sample IV with a lower Cu^{2+} doping fraction $\varphi_c < 0.125$ appear isostructural to MOF-5, with a slightly larger unit cell. A sharp peak below the 2θ of 9.6° was observed in all Cu-doped MOF-5 samples. This is probably caused by the residual guest molecules (chloroform solvent) that alter the crystal phase plane orientation. The position of the peaks and the location of the strongest peak were found to be shifted to a certain degree. This shift may have been caused by the partial interpenetration of crystals into MOF-5. The main peaks of the pattern appeared at 10°, which were identical for most of the samples. Because Zn and Cu have the similarity of cation radius, valence, coordination geometry, and crystal structure, the copper and zinc salts could be synthesized together with BDC in DMF to form crystals where Zn^{2+} and Cu^{2+} interchangeable occupy the same coordination site in the framework. The results suggests that the doped Cu^{2+} ions should be well incorporated into the framework and substitute partial Zn^{2+} ions of the $[Zn_4O]^{6+}$ clusters, as well as observed in the Ni-doped MOF-5 and Co-doped MOF-5. [16, 21]

The crystal phase of Sample V, Sample VI and Sample VII with a higher Cu^{2+} doping concentration $\varphi_c > 0.125$ were confirmed to be different with pure MOF-5 and Cu-doped MOF-5 with Cu^{2+} doping concentration $\varphi_c < 0.125$ by XRD pattern, as shown in Figure 3c. It can be seen that the positions of the XRD peaks shifted when MOF-5 was doped with excessive quantities of Cu. We concluded that doping MOF-5 with excess Cu would ultimately alter the cubic structure of MOF-5 and the Cu-doped MOF-5 samples were founded to lose their cubic structure and were no longer considered as MOF-5 but show a similar XRD pattern with that of MIL-53 (Cu). [22, 23] In MIL-53(Cu) the Cu ions take up the six coordination geometry. It is simulated XRD pattern is shown is Figure 3d. Furthermore, two main peaks position of Sample V, Sample VI and Sample VII appear at 8.7° and 10.3°. The crystal phase of Cu-doped MOF-5s also suggests that the doped Cu^{2+} ions should be well incorporated into the framework and substitute partial Zn^{2+} ions of the $[Zn_4O]^{6+}$ clusters. The data present a unique crystalline phase, which is in good agreement with the simulated patterns from the MIL-53 single crystal data. Which is in good agreement with the simulated patterns from the MIL-53 single crystal data.

Powder X-ray diffraction pattern (XRD) of Sample VIII, MIL-53(Cu) is recorded in Figure 3d and shows a novel crystalline phase with reflections in the range of 5°–35°. Difference in peak intensities between simulated and experimental XRD pattern can be attributed to coordinate guest molecules. Such differences in peak intensity have previously been attributed to pore occlusion by molecular guests. The absence of peaks due to H_2BDC in the XRD pattern of MIL-53(Cu) suggests its removal from the cavities of MIL-53(Cu). [24, 25]

Scanning Electron Microscopy (SEM) images of synthesized products are shown in Figure 4. According to Figure 4a, Sample I, the pure MOF-5 morphology is well-shaped, high-quality cubic crystals ranging from approximately 50–100 μm in width. As shown in Figure 4b–d, in Sample II, Sample III and Sample IV the cubic shape which characterizes the MOF-5 structure is kept. This suggests that Cu-doping does not change the morphology of MOF-5 below a certain fraction $\varphi_c \approx 0.125$. Variations of contents of Cu in the crystal structure of the Cu-doped MOF-5 samples are the main reasons for the differences in their textural properties. It is immediately apparent that doping of a small amount of Cu below a certain fraction $\varphi_c \approx 0.125$ in the synthesis of MOF-5 changes the morphology of the resulting solid. In addition, as shown in

Figure 4. SEM images of Cu-doped MOF-5s (a) Sample I; (b) Sample II; (c) Sample III; (d) Sample IV; (e) Sample V; (f) Sample VI; (g) Sample VII; (h) Sample VIII.

Figure 4e–g, when doping concentration of Cu is increased (more than 12.5%, i.e. Sample V, Sample VI and Sample VII), there have been two kinds of crystals, which were matched with the optical images and XRD analysis, one for MOF-5 and the other for MIL-53(Cu). As shown in Figure 4h, the SEM image of MIL-53(Cu) is well-crystallized rod-like particles with size smaller than 50 μm. The shape of the crystal is from the preferential growth mode along the c-axis of the MIL-53-like structure and was accordance with the SEM images of MIL-53(Cu) in previous literature. [22, 23]

The XRD and SEM results of the samples are strong evidences of the existence of MOF-5 crystals simultaneously containing Zn and Cu, but none of them guarantee the coexistence of both metal ions within the same metal cluster. However, the certification of the presence of bimetallic MOF-5 samples rather than samples formed by segregated Zn-rich MOF-5 and all Cu-rich MOF-5 samples, has been confirmed by optical microscopy (showed in Fig. 2), which allowed us to check that the samples are composed by crystals having the same color, and by SEM-EDS elemental microanalysis (showed in Table 1), which showed practically the same Zn/Cu ratio in different crystals and in different spots

Table 1. Element analysis of as-synthesized Cu-doped MOF-5s crystals in Sample I, Sample V, Sample VI and Sample VIII (Atomic%).

Element	Sample I	Sample V	Sample VI	Sample VIII
C	40.32	41.44	40.80	36.22
O	41.55	44.19	43.02	36.65
Zn	18.13	3.47	7.86	0
Cu	0	10.90	8.32	27.14

of a particular crystal. The EDS elemental microanalysis of the Cu-doping MOF-5s was measured to study the dispersity of Cu^{2+} ions in different regions, especially in two kinds of crystals mentioned before. The SEM images completely correspond to the images of the EDS analysis of Zn content and Cu content. Both Zn content and Cu content follows the structure of Cu-doped MOF-5 crystals and the Cu content is consistent with Zn content, suggesting that Cu^{2+} ions are well dispersed in the Cu-doped MOF-5 crystals. Compared to Sample V, the Cu mapping of Sample VI shows stronger intensity, indicating the higher content of Cu elements. The EDS results also give that the average atomic ratios of Cu to Zn in all samples respectively, consistent with the ratio of them in reactants.

4 CONCLUSIONS

In conclusion, the partial doping of the MOF-5 framework with Cu^{2+} ions has been successfully carried out, showing that some of the Zn forming the metal clusters of MOF-5 can be replaced by Cu during crystallization in the synthesis process. The isomorphic substitutions were achieved in low doping concentration condition. The systematic variation of Zn/Cu ratio in the Copper-doped MOF-5 materials of this work offers a good example to show the influence of metal cluster composition on the structure. The Copper-doped MOF-5 materials show similar structures than the pure MOF-5, implying Cu-doping can prevent including zinc species into the framework and the interpenetrated structure. Moreover, an interesting phenomenon has appeared that the MIL-53 (Cu) material has been synthesized with a high concentration of Cu^{2+}. This work set up a new approach to study the doping effect of MOF and benefit to the structure design and property optimization of MOF material.

ACKNOWLEDGEMENTS

This work was supported by the National Natural Science Foundation of China No. 21376026.

REFERENCES

[1] H. Furukawa, K.E. Cordova, M. O'Keeffe, O.M. Yaghi, The Chemistry and Applications of Metal-Organic Frameworks, Science, 341 (2013).

[2] S.T. Meek, J.A. Greathouse, M.D. Allendorf, Metal-Organic Frameworks: A Rapidly Growing Class of Versatile Nanoporous Materials, Adv. Mater., 23 (2011) 249–267.

[3] Z. Wang, S.M. Cohen, Postsynthetic modification of metal-organic frameworks, Chemical Society Reviews, 38 (2009) 1315–1329.

[4] M.-H. Zeng, B. Wang, X.-Y. Wang, W.-X. Zhang, X.-M. Chen, S. Gao, Chiral Magnetic Metal-Organic Frameworks of Dimetal Subunits: Magnetism Tuning by Mixed-Metal Compositions of the Solid Solutions, InCh, 45 (2006) 7069–7076.

[5] H. Li, M. Eddaoudi, M. O'Keeffe, O.M. Yaghi, Design and synthesis of an exceptionally stable and highly porous metal-organic framework, Nature, 402 (1999) 276–279.

[6] H. Furukawa, N. Ko, Y.B. Go, N. Aratani, S.B. Choi, E. Choi, A.Ö. Yazaydin, R.Q. Snurr, M. O'Keeffe, J. Kim, O.M. Yaghi, Ultrahigh Porosity in Metal-Organic Frameworks, Science, 329 (2010) 424–428.

[7] M. Eddaoudi, J. Kim, N. Rosi, D. Vodak, J. Wachter, M. O'Keeffe, O.M. Yaghi, Systematic Design of Pore Size and Functionality in Isoreticular MOFs and Their Application in Methane Storage, Science, 295 (2002) 469–472.

[8] J.-R. Li, R.J. Kuppler, H.-C. Zhou, Selective gas adsorption and separation in metal-organic frameworks, Chemical Society Reviews, 38 (2009) 1477–1504.

[9] K.-S. Lin, A.K. Adhikari, C.-N. Ku, C.-L. Chiang, H. Kuo, Synthesis and characterization of porous HKUST-1 metal organic frameworks for hydrogen storage, International Journal of Hydrogen Energy, 37 (2012) 13865–13871.

[10] S.S. Kaye, A. Dailly, O.M. Yaghi, J.R. Long, Impact of Preparation and Handling on the Hydrogen Storage Properties of Zn4O(1,4-benzenedicarboxylate)3 (MOF-5), J. Am. Chem. Soc., 129 (2007) 14176–14177.

[11] S. Turner, O.I. Lebedev, F. Schröder, D. Esken, R.A. Fischer, G.V. Tendeloo, Direct Imaging of Loaded Metal–Organic Framework Materials (Metal@MOF-5), Chem. Mater., 20 (2008) 5622–5627.

[12] J. Park, H. Kim, S.S. Han, Y. Jung, Tuning Metal–Organic Frameworks with Open-Metal Sites and Its Origin for Enhancing CO2 Affinity by Metal Substitution, The Journal of Physical Chemistry Letters, 3 (2012) 826–829.

[13] X. Kong, H. Deng, F. Yan, J. Kim, J.A. Swisher, B. Smit, O.M. Yaghi, J.A. Reimer, Mapping of Functional Groups in Metal-Organic Frameworks, Science, 341 (2013) 882–885.

[14] H. Deng, C.J. Doonan, H. Furukawa, R.B. Ferreira, J. Towne, C.B. Knobler, B. Wang, O.M. Yaghi, Multiple Functional Groups of Varying Ratios in Metal-Organic Frameworks, Science, 327 (2010) 846–850.

[15] Q. Xu, D. Liu, Q. Yang, C. Zhong, J. Mi, Li-modified metal-organic frameworks for CO2/CH4 separation: a route to achieving high adsorption selectivity, JMCh, 20 (2010) 706–714.

[16] J.A. Botas, G. Calleja, M. Sánchez-Sánchez, M.G. Orcajo, Cobalt Doping of the MOF-5 Framework and Its Effect on Gas-Adsorption Properties, Langmuir, 26 (2010) 5300–5303.

[17] J.I. Feldblyum, M. Liu, D.W. Gidley, A.J. Matzger, Reconciling the Discrepancies between Crystallographic Porosity and Guest Access As Exemplified by Zn-HKUST-1, J. Am. Chem. Soc., 133 (2011) 18257–18263.

[18] Z. Zhao, Z. Li, Y.S. Lin, Adsorption and Diffusion of Carbon Dioxide on Metal–Organic Framework (MOF-5), Ind. Eng. Chem. Res., 48 (2009) 10015–10020.

[19] S.S. Han, D.-H. Jung, J. Heo, Interpenetration of Metal Organic Frameworks for Carbon Dioxide Capture and Hydrogen Purification: Good or Bad?, The Journal of Physical Chemistry C, 117 (2012) 71–77.

[20] J. Hafizovic, M. Bjørgen, U. Olsbye, P.D.C. Dietzel, S. Bordiga, C. Prestipino, C. Lamberti, K.P. Lillerud, The Inconsistency in Adsorption Properties and Powder XRD Data of MOF-5 Is Rationalized by Framework Interpenetration and the Presence of Organic and Inorganic Species in the Nanocavities, J. Am. Chem. Soc., 129 (2007) 3612–3620.

[21] H. Li, W. Shi, K. Zhao, H. Li, Y. Bing, P. Cheng, Enhanced Hydrostability in Ni-Doped MOF-5, InCh, 51 (2012) 9200–9207.

[22] M. Anbia, S. Sheykhi, Synthesis of nanoporous copper terephthalate [MIL-53(Cu)] as a novel methane-storage adsorbent, J. Nat. Gas Chem., 21 (2012) 680–684.

[23] M. Anbia, S. Sheykhi, Preparation of multi-walled carbon nanotube incorporated MIL-53-Cu composite metal–organic framework with enhanced methane sorption, Journal of Industrial and Engineering Chemistry, 19 (2013) 1583–1586.

[24] X.Y. Chen, H. Vinh-Thang, D. Rodrigue, S. Kaliaguine, Amine-Functionalized MIL-53 Metal–Organic Framework in Polyimide Mixed Matrix Membranes for CO2/CH4 Separation, Ind. Eng. Chem. Res., 51 (2012) 6895–6906.

[25] T.K. Trung, P. Trens, N. Tanchoux, S. Bourrelly, P.L. Llewellyn, S. Loera-Serna, C. Serre, T. Loiseau, F.o. Fajula, G.r. Férey, Hydrocarbon Adsorption in the Flexible Metal Organic Frameworks MIL-53(Al, Cr), J. Am. Chem. Soc., 130 (2008) 16926–16932.

Material Science and Environmental Engineering – Chen (Ed.)

Simulation about mold polishing pattern and test research

Z.R. Chen, X.C. Xu, X.C. Ma & T. Wang
Harbin University of Science and Technology, China

ABSTRACT: There are extensive needs of die surface polishing during the transferring process of the large die surfaces. As almost large die surfaces are free-form surfaces, currently, almost all of the large die surfaces are manually polished. In order to overcome this barrier, a floating polishing prototype has been made by our research team. This paper focuses on the polishing pattern by simulation and test. Firstly, the structure of the floating polishing heads are designed and analyzed, and three kinds of polishing patterns of polishing heads are designed: straight reciprocating motion, cross motion, and spiral motion; secondly, the polishing morphology about three polishing patterns was studied by simulation; and finally, the main affecting factors of roughness of polishing surface are contrasted and analyzed with the developed floating polishing experimental prototype, which included the abrasive grain of polishing disc, pressure on polishing head and motion pattern of polishing head. These research results are used as operating guides for applications of the floating die polisher.

Keywords: mold polishing; polishing pattern; simulation; surface roughness

1 INTRODUCTION

In the process of the development of modern industry, mould industry has gradually become the basic industry of national economy, with the development of industrial technology, the requirement to the mould becomes more and more high, especially the rapid development of automobile industry, the design and development of the auto mould is in an important position in the development of the mould. People not only attach importance to its practical use and the reliability of the car, but also attach importance to the pursuit of aesthetics, comfort, and economy. Auto mould can be used to make one of the important equipment of automobile body parts; it directly influences the quality, sale and use of the whole cars.

When large die surfaces are machined, the designed surfaces are generated. After the large dies are constructed, they cannot make qualified parts due to the design performance of the die surfaces regarding both the complexity of die surfaces and material flow prediction. The die surfaces have to be modified through a series of tryouts until qualified parts are shaped [1~2]. Surface polishing process is a major operation during die making and tryouts. Currently, almost all of the large die surfaces are manually polished, causing low working efficiency and stability of die surface quality, especially in the construction of automotive draw dies of body panels [3~4]. Therefore, the research of how to use convenient semi-automatic polishing machines is a feasible direction [5]. This paper studies on the feasibility of polishing of the three polishing heads from the perspective of polishing track.

2 MOLD POLISHING PATTERN ANALYSIS

2.1 *Traditional mold polishing pattern analysis*

Due to the complexity of the metal plastic forming, the mould manufacturing of automotive covering parts needs repeated test and revisions to complete, mold finishing in mould manufacturing, test, and the final production maintenance phase has a very important position, and mould surface grinding and polishing as mould finishing the last procedure, and also plays an important role, the polishing quality directly affects the mold production cycle and delivery time.

At present, hand polishing is a main polishing pattern for large mold polishing, usually there are two kinds of manual polishing mode, one is a kind of reciprocating polishing mode directly, another is a cross polishing mode.

2.2 *Design about three polishing discs and polishing patterns*

Manual polishing has many disadvantages. Such as manual polishing needs high degree of technical requirements for workers, especially in the final polishing process, the operator of the small mistake can cause surface damage, causing great damage, even causing the mold to scrap. Therefore, this paper designed a three polishing disc floating polishing mode. Based on analyzing the characteristics of the mould surface and the Contact situation between polishing head and mould surface, the polishing disc with 15 mm radius is designed. Each polishing disc is evenly distributed on the circumference of the radius of 20 mm, three polishing discs can have two kinds of sports, and the kind of mode is that polishing discs are fixed evenly on the same circumference of a circle, no rotation. This major is suitable to polish plane. Another way of floating polishing belongs to a complex movement that has three rotation polishing discs as a whole and has the same speed, and carries on the straight line of the reciprocating movement or cross. In addition, each polishing disc is controlled by a universal joint, can free movement within a certain angle, so as to adapt to the free surface polishing needs. This is floating polishing.

The choice of a different polishing path can get a different polishing effects under the same conditions of mold surface and machining conditions. In order to obtain a good polishing effect, in the process of polishing, the polishing disc should make every point on the processed workpiece to have the same or similar grinding and polishing shortest trip, and the polishing movement trajectories should try not to overlap at the same time. Therefore, the polishing stripes are interlaced with each other and the variable, which can reduce the surface roughness of the workpiece. In addition, the polishing force should be smooth and avoid producing too much curvature angle during polishing.

Three polishing disc fixed distribution on the same circumference of a circle as a spiral motion polishing, every grain of polishing disc is around the center of the circular to rotary motion, which is the synthesis movement of the polishing disc that has a rotation and a revolution at the same time.

3 SIMULATION AND TEST ANALYSIS

3.1 *Simulation analysis*

Abrasive, cutting speed, pressure, and polishing track have actual effect of polishing during process of polishing. Paper adopts the idea of orthogonal experiment to design simulation experiment in order to analyze the influence rule of particle size, polishing speed and floating force to the roughness of polishing surface.

Figure 1 shows a set of simulation results. The bigger the pressure of polishing head the flatter is the shape of workpiece surface after polishing.

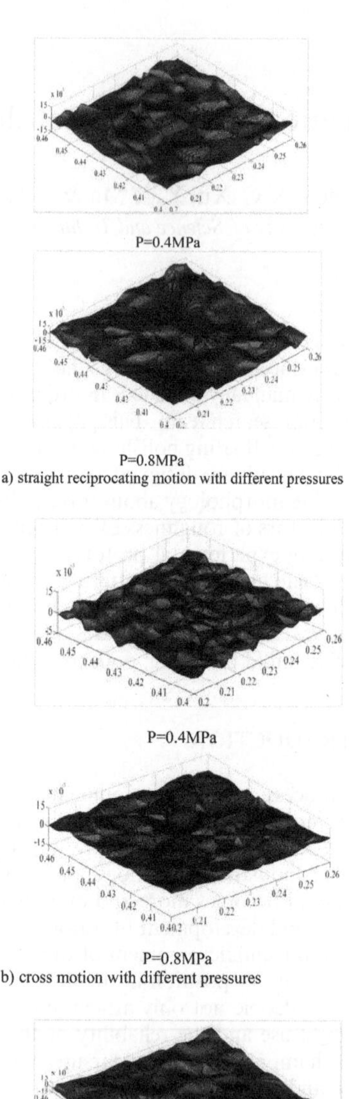

P=0.4MPa

P=0.8MPa

a) straight reciprocating motion with different pressures

P=0.4MPa

P=0.8MPa

b) cross motion with different pressures

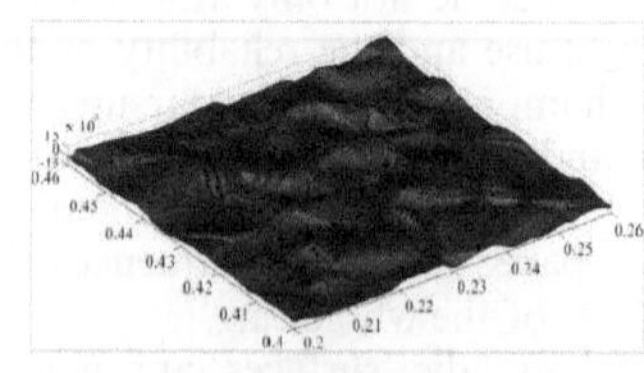

P=0.4MPa

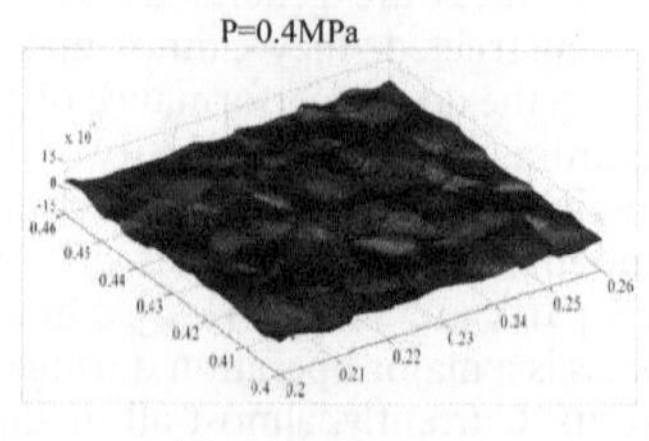

P=0.8MPa

c) spiral motion with different pressures

Figure 1. The morphology of diagram workpiece surface polished with different pressures.

When the other polishing conditions are the same, but, with different pressures, the simulation results are shown as Figure 1, with same process parameters, polishing morphology of workpiece surface polished by three discs is the best in the three polishing ways, and polishing morphology of workpiece surface polished by cross way is better than linear motion.

The morphology of diagram workpiece surface polished with different particle size and speed are researched by simulation as shown as in Figure 2 and Figure 3. The smaller the particle sizes and speeds are, the better is the morphology of diagram workpiece surface polish.

3.2 *Test analysis*

Three polishing heads move along the spiral trajectory movement, the change of the mold surface roughness after polishing is shown as Figure 4 by the test and simulation, and roughness decrease with the increase in abrasive grain size and the roughness of workpiece surface polished by the three polishing head is minimal. Comprehensive analysis of the above results, for straight line polishing, abrasive reciprocating linear motion because of reciprocating carving effect of abrasive to surface, the surface roughness is bigger. When polishing with cross reciprocating motion, the straight line reciprocating movement of unidirectional gap is destroyed, so as to obtain high surface quality. The motion of polishing with three polishing heads is a kind of composite motion of

Figure 2. Workpiece surface morphology diagram polished at different abrasive.

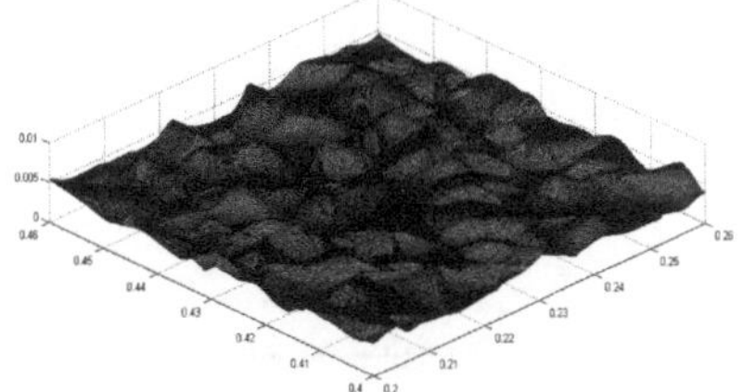

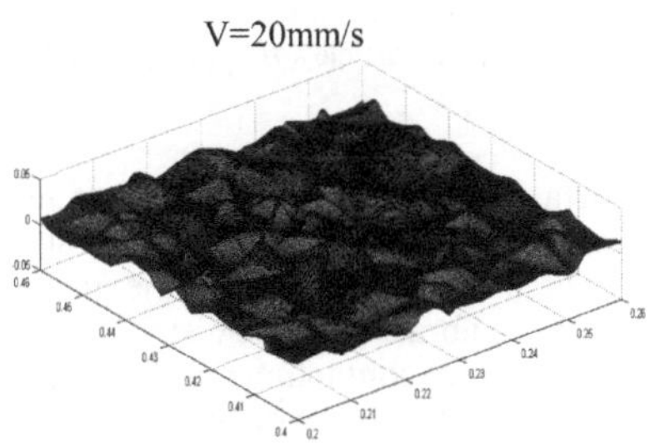

Figure 3. Workpiece machined surface morphology diagram at different speeds.

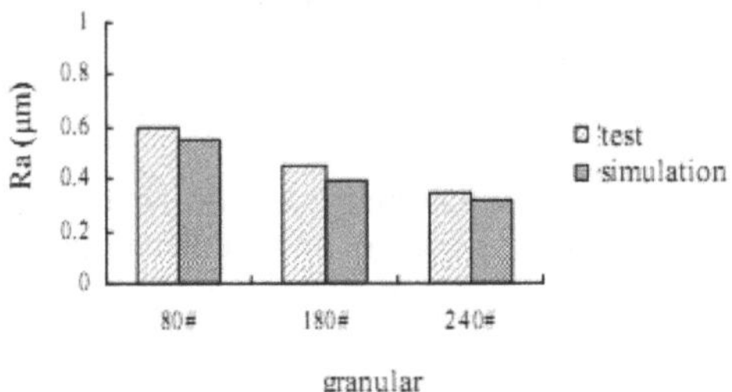

Figure 4. Effect of the abrasive grains to surface roughness with the three polishing disc.

straight line motion and spiral motion, so that the abrasive can produce more uniform, dense single point cutting effect that results in a better polishing effect, namely the polished surface roughness are smaller. At the same time, the test results and simulation results have error, this is because the simulation parameters setting has a completely ideal state, and polishing machine itself has mechanism motion errors, that have different levels of wear and tear.

The simulation and experiments also get the same results by different rotational speeds and movement speeds of polishing head: the surface quality by floating polishing is better.

The mold surface is polished by polishing head with different abrasive for 80 #, 120 #, and 270 # in order to simulate and analyze effects of particle size on the roughness. The simulation results are shown as Figure 5.

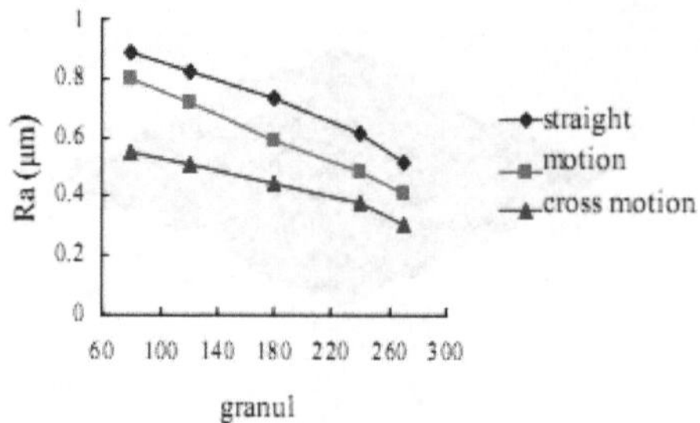

Figure 5. Effects of particle size on the roughness.

The surface roughness presents downtrends with a change in the abrasive grain size from 80 to 270 as shown as in Figure 5, because the greater that the number of abrasive particle size is, the more number of abrasive contacting mold surfaces there are, and the number of abrasive participating in polishing increases at the same time, as the surface roughness value becomes smaller so that the quality of the workpiece surface will be improved. The polishing efficiency of the three polishing discs is better when the abrasive particles are same.

4 CONCLUSION

In this paper, the polishing model of the three polishing discs was designed, under different process parameters on polishing plane that has carried on the simulation analysis. The simulation research results show that the roughness value of surface polished by three polishing disc screw is smaller than the straight line reciprocating movement and cross staggered type sports. To test and verify the simulation results of the above, the use of floating polishing experimental prototype of experiments have been carried out to verify the parameters of polishing process. The experimental results show that the simulation results agree with experimental results, and also the simulation results are dependable.

REFERENCES

[1] T. Altan and J. Mater: Processing Technology, (1996) No. 59, p. 158.
[2] T. Altan and J. Mater: Conference of Electrical Discharge Machining (Columbus, USA 1995), p. 192.
[3] C.M. Yuan, L. Zhang, Y.P. Chen and et al: High Technology Letters, (2001) No. 9, p. 76. (In Chinese).
[4] J.F. Ding, W. Li, F.J. Yu and et al: Die & Mould Industry, Vols. 32 (2006), p. 66. (In Chinese).
[5] X.C. Xu and J.W. Han: Advanced Materials Research, Vols. 97–101 (2010) p. 1883.

Material Science and Environmental Engineering – Chen (Ed.)
© 2016 Taylor & Francis Group, London, ISBN 978-1-138-02938-5

Effect of inclusions on the corrosion resistance of low carbon steel in acid solution with high chloride concentration

S. Chen & G.M. Li
College of Science, Naval University of Engineering, Wuhan, China

ABSTRACT: The pitting originates from the inclusion in the carbon steel and the corrosion resistance of the steel is influenced by the type and quantity of the inclusions. The effect of the sulfur content on the corrosion resistance of low carbon steel in the acid solution with high chloride concentration was investigated by the analysis of inclusions, electrochemical test and immersion test. The results showed that the pitting originated from the inclusion. With the increase of the sulfur content, more inclusions appeared in the steel, the corrosion potential was moved to negative, the linear polarizability was lowered and the corrosion rate of the simulated test was also improved. The sulfur content of the steel which worked in such a situation must be controlled to a low level.

Keywords: inclusion; pitting; linear polarization; EIS; corrosion resistance

1 INTRODUCTION

The corrosion problem of Cargo Oil Tanker (COT) became severe because the inner bottom plate is covered by an oil layer containing sludge, drain, and water including highly concentrated Cl^- and H_2S [1–2]. Intensive investigations have been carried out on pitting corrosion phenomena of COT steel by JFE steel company and some research institutes [3–5]. The effects of alloy elements on the corrosion resistance of COT inner plate were studied by the above organizations. The effects of impurities and microstructures on the corrosion resistance were also carried out [6–11]. Moreover, the corrosion resistance of welding joint of the COT steel was also measured by some researchers to determine the compositions of solder wires [12].

The effects of elements such as tin and antimony on the corrosion resistance of COT steel were studied by the corporation of Naval University of Engineering and Ma Anshan Steel Corporation [13]. Some researchers have revealed that the corrosion resistance of steel was affected by inclusions as sulfide inclusions [14–15]. However, there are few reports about the effects of inclusions on the corrosion properties of COT steel. The current work is devoted to investigating the effect of sulfide inclusion on the corrosion properties of the self-developed corrosion-resistant steel for COT. Its inclusion and corrosion behavior were studied by optical microscope, linear polarization, and EIS techniques. The corrosion rate was measured by couple immersion test.

2 EXPERIMENTAL PROCEDURES

2.1 *Preparation of specimens*

The compositions of common carbon steel and the self-developed steels are listed in Table 1. It was shown that the self-developed steel had a lower content of C and higher content of alloying elements, such as Cu, W, and Sn but with different sulfur contents, in comparison to common carbon steel.

All experiments were performed in 10 wt.% NaCl solution with a pH value of 0.85 at (25 ± 1) °C. The solutions were prepared with analytical pure reagents and distilled water. The pH value of the solution was adjusted using concentrated hydrochloric acid.

2.2 *Observation of inclusions*

The difference of inclusions type and quantity among the tested steels were measured according to GB/T10561-2005.

2.3 *Electrochemical experiments*

The specimens were mounted in epoxy after providing electrical contact at the black side and the exposed area was 1 cm^2. The working surfaces of samples were wet ground with SiC paper up to 1500 grit and then degreased with ethanol, cleaned with distilled water, and finally dried in air.

All electrochemical measurements were in the acid solution mentioned above. The electrochemical test cell consisted of a specimen, a Saturated

Table 1. Composition of common carbon steel and the self-developed steels (wt.%).

Number	C	Si	Mn	Cu + W + Sn	P	S	Fe
A	0.08	0.34	0.65	/	0.021	<0.002	Balance
B	0.08	0.35	0.65	0.6	0.020	0.043	Balance
C	0.08	0.34	0.65	0.6	0.022	0.07	Balance
D	0.08	0.34	0.65	0.6	0.021	<0.002	Balance

Calomel Electrode (SCE), and platinum plate as the working, reference, and counter electrodes, respectively. The linear polarization and EIS measurements were performed using an electrochemical work station (PARSTAT SS350). After the specimens had reached the stable Open-Circuit Potential (OCP), a scan rate of 0.166 mV/s was used over a range from −20 mV_{OCP} to +20 mV_{OCP} in the linear polarization test. The EIS measurements were obtained with amplitude of 10 mV at frequency ranging from 100 kHz to 10 MHz.

2.4 *Immersion test*

Weight loss measurements were carried out by the following IMO MSC 87/26/Add. 1 Annex 3 MSC.289 (87), the performance standards for alternative means of corrosion protection for cargo oil tanks of crude oil tankers, in order to simulate corrosion environment of ship plate steels equipped at flate inner bottom of COT [16]. A hole was drilled in the 60 × 25 × 5 mm plate for hanging; the specimens were polished with 600-grit SiC paper, degreased with an ultrasonic cleaner with acetone for 5 min, cleaned with distilled water and dried in hot air. The corrosion rate was calculated using the formula in the ASTM G31-71. The corrosion morphology of the plate was observed under the microscope.

3 RESULTS AND DISCUSSION

3.1 *Observation of inclusions*

Figure 1 shows the optical microscope images of the inclusion types and quantity among the tested steels. As presented in Figure 1, the inclusions in the four tested steels were composed of the main type of B fine Al_2O_3 and D type spherical inclusions. The B type inclusions in B and C steels can be divided into coarse inclusion and fine inclusion. Whereas, the existing B type fine inclusion and D type spherical inclusion in A and D steels for their lower sulfur content is presented in Figure 1(a) and Figure 1(d). By contrast, there are many fine A type fine inclusions and fine silicate inclusions in steel B and steel C for their high sulfur content, especially in steel C. Whereas the addition of copper could

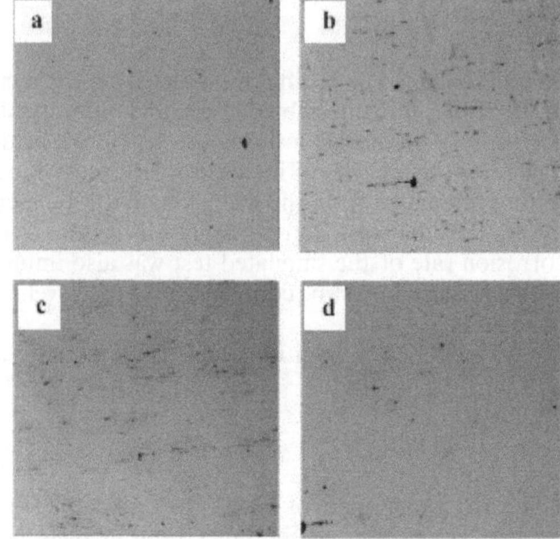

Figure 1. Optical microscope images of inclusions in tested steels (100×). ((a) The steel A, (b) the steel B, (c) the steel C, (d) the steel D).

affect the distribution of inclusions per frame area [17], the inclusions type and quantity of our self-developed steels were affected by the sulfur content for low copper addition. It's evident that the quantity of sulfide inclusions is improved with sulfur content.

3.2 *Electrochemical behavior*

Figure 2 shows the linear polarization curves of tested steel in acid solution with a high chloride concentration at (25 ± 1) °C. The corrosion potential of steel D is the most noblest. It means that the addition of alloying elements as copper and tin and tungsten can affect the corrosion potential of steel when the sulfur content is controlled to a low level. By contrast, when the sulfur content is high, the corrosion potentials of steel B and steel C become more negative even containing alloying elements as copper, tin, and tungsten. The reason is that there are more sulfide inclusions in the steels which can induce more pitting.

The polarization resistance of each linear polarization curves can be simulated by the software.

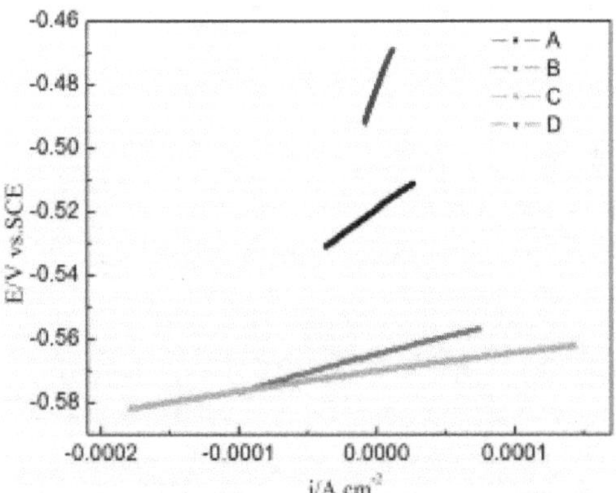

Figure 2. Linear polarization curves of tested steels. ((a) the steel A, (b) the steel B, (c) the steel C, (d) the steel D).

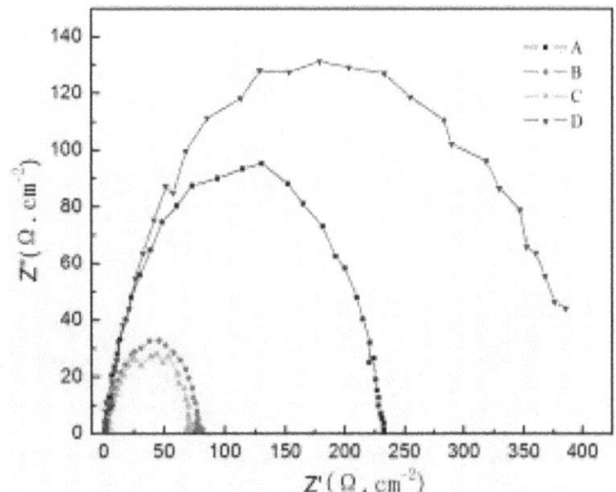

Figure 3. Ny quist plots of tested steels in acid solution with high chloride concentration. ((a) The steel A, (b) the steel B, (c) the steel C, (d) the steel D).

Table 2. The simulated Rp of tested steels.

Specimens	1#	2#	3#	4#
Rp	680.39	517.8	108.08	957.45

Table 3. Corrosion rates of tested steels (mm/y).

Specimens	A	B	C	D
Corrosion rate	2.388	2.344	6.425	0.345

The simulated results are listed in Table 2. It is evident that the polarization resistance Rp of steel D is the biggest among the tested steels. The Rp value decreases with the sulfur content because there are more sulfide inclusions in steel B and C.

Figure 3 shows EIS results of tested steel in acid solution with high chloride concentration at (25 ± 1) °C, which is presented as Nyquist plots. It shows that one semicircle exists over the measured frequency range in EIS of all tested specimens. The diameter of the arc can be regarded as the charge transform resistance (Rct), and the increase in diameter of the arc reflects an increase in the Rct value. The diameter of the arc increased with the addition of Cu, W, and Sn, whereas, the diameter of the arc decreased with the sulfur content even if the steel containing alloying elements Cu, W, and Sn.

3.3 *Corrosion test*

The average corrosion rates found from the weight loss test are listed in Table 3. The corrosion rate of steel D with addition of Cu, W, and Sn and low sulfur content is the lowest, the average corrosion rate is 0.345 mm/y. It is well known that the addition of copper (Cu) to ferritic, austenitic or duplex steels improves the resistance to uniform corrosion in sulfuric acid [17]. The reason is that the addition of copper can improve the corrosion resistance of steel matrix; it can also decrease the corrosion rate of steel for the formation of protective rust layer. The copper sulfide is formed from the dissolved copper ion and it reduces the activity of sulfur species on the inside of copper sulfide [18].

In our previous study [19], the addition of alloying element Sn can improve the overpotential of hydrogen evolution reaction and suppress the hydrogen evolution reaction. Moreover, the formation of tin oxide on the steel surface can restrain the corrosion rate of the steel in the acid solution with high chloride concentration. The corrosion resistance of plain carbon steel can be enhanced by synergistic effect of elements of Cu, W, and Sn.

The corrosion rates of steel B and C which contain Cu, W, and Sn and high sulfur content are higher than that of common carbon steel. As there are more sulfide inclusions in the steel, the inclusions can dissolve in the acid solution and accelerate the corrosion. It was reported that the MnS dissolved by following reaction [20]:

$$2MnS + 3H_2O \rightarrow 2Mn^{2+} + S_2O_3^{2-} + 6H^+ + 8e$$

This reaction decreases the pH value at the vicinity of the MnS inclusion. The pH value drop promotes the corrosion of steel and breaks the corrosion inhibition of alloying elements Cu, W, and Sn. The corrosion resistance of steel with high sulfur content is dropped evidently.

The corrosion micrographs after immersion test are shown in Figure 4. There are more corrosion

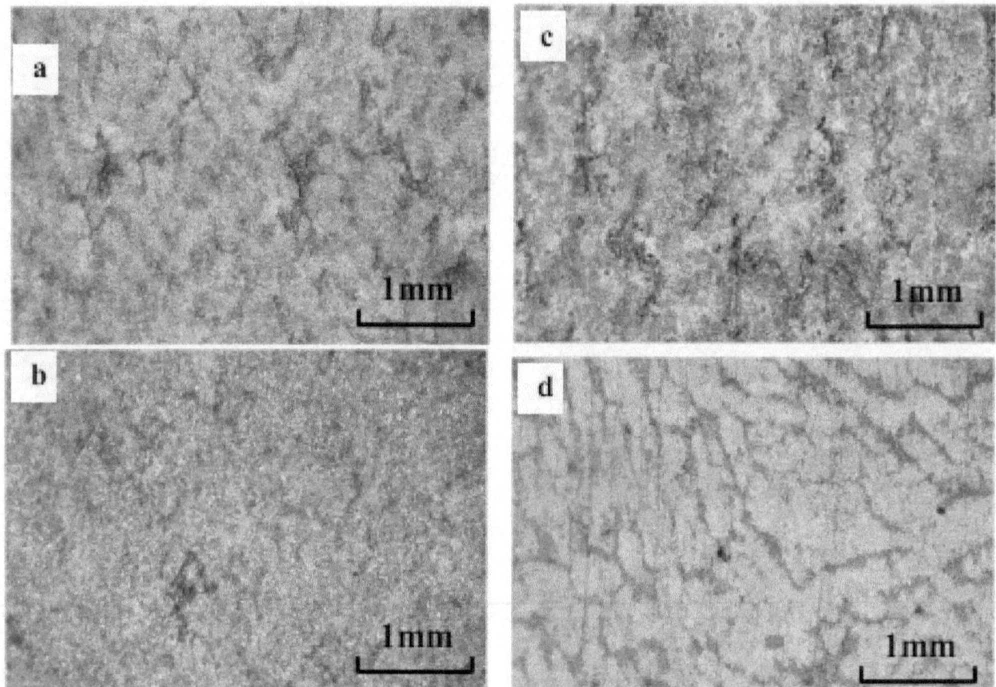

Figure 4. Stereo microscope micrograph of steel after immersion test. ((a) The steel A, (b) the steel B, (c) the steel C, (d) the steel D).

grooves that can be observed on the surface with the sulfur content increased. Furthermore, the higher the sulfur content in steel C, more inclusions exist, the corrosion rate is the biggest among the tested steels. By comparison, the corrosion rate of steel D which has low sulfur content and alloying elements Cu, W, and Sn is the lowest. There are uniform corrosions on the surface and cannot observe obvious strip corrosion except for several pits. The inclusion quantity of steel A is less than that of steel B and C, the uniform corrosion happens on the surface. It means that the effect of inclusions on the corrosion resistance of low alloy steel in acid solution with high chloride concentration is obvious. The promotion effect of inclusions is bigger than the retard effect of alloying elements.

4 CONCLUSIONS

1. The quantity and level of the inclusions were increased with sulfur content in the low carbon steel.
2. The corrosion resistance of low alloy steel in acid solution can be enhanced by addition of elements such as Cu, Sn, and W with low sulfur content, whereas, the corrosion potential became negative and polarization resistance decreased with increase in sulfur content. The corrosion rate increased remarkably with high sulfur content.
3. The corrosion originated from inclusions, and more sulfide inclusions broke the corrosion inhibition effect of alloy elements and accelerated the corrosion.

ACKNOWLEDGEMENT

This work was financially supported by the Science College, Naval University of Technology under Project No. HGDYYJJ13155.

REFERENCES

[1] Feilong Sun, Xiaogang Li, Fan Zhang, et al. Corrosion Mechanism of Corrosion-Resistant Steel Developed for Bottom Plate of Cargo Oil Tanks, Acta Metall. Sin. (Engl. Lett.) 2013, 26 (3):257–264.
[2] Ship research panel 242, Ship Research Summary Report No. 431, Tokyo, JSRA, 2002.
[3] Soares C.G., Garbatov Y., Zayed A., et al. Influence of environmental factors on corrosion of ship structures in marine atmosphere [J]. Corrosion Science, 2009, 51 (9):2014–2026.

[4] Soares C.G., Garbatov Y., Zayed A., et al. Corrosion wastage model for ship crude oil tanks [J]. Corrosion Science, 2008, 50 (11):3095–3106.

[5] Shiomi H., Kaneko M., Kashima K., et al. Development of anti-corrosion steel for cargo oil tanks[C]//TSCF 2007 Shipbuilder Meeting. Busan, Korea, 2007:1–8.

[6] B. Aberle, D.C. Agarwal, High Performance Corrosion Resistance Stainless Steels and Nickel Alloys for Oil & Gas Applications, Corrosion/2008, NACE: Houston, TX, Paper No. 08085.

[7] H. Miyuki, J. Murayama, T. Kudo, T. Moroishi, Effect of Cr, Ni and Mo on Corrosion Resistance of Highly Alloyed Materials in Sour Well Environments, Corrosion, 1984, NACE: Houston, TX, Paper No. 209.

[8] Sakashita S., Tatsumi A., Imamura H., et al. Development of anti-corrosion steel for the bottom plates of cargo oil tanks[C]/ / International Symposium on Shipbuilding Technology, Japan, Osaka, 2007:1–4.

[9] Chen Xin-hua, Dong Jun-hua, Han En-hou, et al. Effect of Ni on the ion-selectivity of rust layer on low alloy steel [J]. Materials Letters, 2007, 61:4050–4053.

[10] Nguyen Dang Nam, Min Jun Kim, oung Wook Jang, et al. Effect of tin on the corrosion behavior of low-alloy steel in an acid chloride solution [J]. Corrosion Science, 2010,52:14–20.

[11] Lea D.P., Jia W.S., Kim J.G., et al. Effect of antimony on the corrosion behavior of low-alloy steel for flue gas desulfurization system [J]. Corrosion Science, 2008, 50:1195–1204.

[12] Qi Yan-chang, Peng Yun, Tian Zhi-ling, et al. effect of Ni on corrosion resistance of weld metals for bottom plates of cargo oil tanks. Transactions of Materials and Heat Treatment (in Chinese), 2014, 34 (4):69–73.

[13] Shan Chen, Guoming Li, Xiaoyan Wang, et al. Effect of antimony on the corrosion resistance of steel in acid solution with high chloride concentration. Advanced Materials Research, 2012, 577:109–114.

[14] Huibin Wu, Jinming Liang, Di Tang, et al. Influence of Inclusion on Corrosion Behavior of E36 Grade Low-alloy Steel in Cago Oil Tank Bottom Plate Environment, Journal Of Iron And Steel Research, International, 2014, 21 (11):1016–1021.

[15] Tatyana V. Shibaeva, Veronika K. Laurinavichyute, Galina A. Tsirlina, et al. The effect of microstructure and non-metallic inclusions on corrosion behavior of low carbon steel in chloride containing solutions, Corrosion Science, 2014(80):299–308.

[16] IMO, in: Proc of the Maritime Safety Committee onits eighty-seventh session Annex 3 MSC.289 (87), London, 30 June, 2010.

[17] Soon-Hyeok Jeon, Soon-Tae Kima, In-Sung Lee, et al. Effects of copper addition on the formation of inclusions and the resistance to pitting corrosion of high performance duplex stainless steels, Corrosion Science 53 (2011) 1408–1416.

[18] Akiko Tomio, Masayuki Sagara, Takashi Doi, et al. Role of alloyed copper on corrosion resistance of austenitic stainless steel in H2S–Cl-environment, Corrosion Science, 2014, 81:144–151.

[19] Chen Shan, Li Guoming, Wang Xiaoyan, et al. Effect of Tin on corrosion behavior of low alloy steel in acid solution with high chloride concentration, Corrosion and Protection (in Chinese), 2013, 34 (6):479–481.

[20] E.G. Webb, R.C. Alkire, Electrochem. Soc. 149 (2002):B272–B279.

Material Science and Environmental Engineering – Chen (Ed.)
© 2016 Taylor & Francis Group, London, ISBN 978-1-138-02938-5

The site preference of 4d elements in ω-Ti: The computer simulation research

J.Q. Niu
Hebei Software Institute, Baoding, China

Z.R. Liu
Xuanyu Group, Baoding, China

ABSTRACT: Two un-equivalent sites exist in the martensitic ω-Ti, which has an effect on the phase stability in Ti. In this work, the site preference of 4*d* elements in *ω*-Ti was investigated by the first principles method. B site, the site that was located in the compact (0001) layer in *ω*-Ti, was found to be favorable for most of all the 4*d* elements except Y and Pd elements. This result can be understood with chemical bonding analyses, when the geometrical factor is proven to account for little.

Keywords: site preference; ω-Ti; first principles

1 INTRODUCTION

As structural and functional materials, β-Ti based alloys have been applied in various fields, due to their high strength, low elastic modulus, low density, and good corrosion resistance [1]. Their low elastic modulus is especially suitable for applications in biomedical implants [2–3]. ω phase can be formed in β-Ti based alloys upon quenching or aging, which has the higher elastic modulus than β phase and thus has a big effect on β-Ti based alloys in the biomedical field. It is well known that a small amount of the transition elements such as Nb could affect the relative phase stability between β and ω phases. This is very useful information in the design of Ti alloy and many efforts have been done to clarify it [4–9]. But, it is also a complex issue as the two un-equivalent sites exist in the martensitic ω-Ti and, unfortunately, the site preference of the alloying elements in ω-Ti has not been totally quantified so far. The previous theoretical works on the transition elements in ω-Ti are limited to group IIIB, IVB, VB, and VIB alloying elements in addition to some sp elements, no VIIB and VIII elements are included [4–6]. So is the case in other bcc transition metal alloys [10]. The purpose of this paper is to evaluate the site preference of the alloying elements in ω-Ti by the computer simulation research. In this paper, we focused on the 4d transition elements including VIIB and VIII elements.

2 METHODOLOGY

The first-principles density functional theory calculations were carried out using the ABINIT simulation package [11]. The electron-ion interaction was described using Projector Augmented Wave (PAW) method [12], the exchange correlation potential using the Generalized Gradient Approximation (GGA) is in the Perdew-Burke-Ernzerhof form [13]. The energy cutoff for the plane wave basis was set at 320 eV for all systems. The Brillouin zone integration was performed within the Monkhorst-Pack scheme. A $(3 \times 3 \times 2)$ k-mesh was used for the 48-atom cells. For partial occupancy, a Methfessel-Paxton smearing of 0.1 eV was used.

Figure 1 (a and b) showed the *β*-Ti and *ω*-Ti bulk structures. *β*-Ti has a *bcc* structure and the calculated lattice constant is 3.24 Å. A primitive cell of *ω*-Ti consists of three atoms, located at (0, 0, 0), (1/3, 1/3, 0), and (2/3, 2/3, 1/2). In a 3-atom *ω*-Ti primitive cell, one atom is in the loose-packed (0001) layer and two atoms are in the close-packed (0001) layers, called A and B sites, respectively. The calculated structural parameters of *ω*-Ti are: $a = b = 4.55$ Å, $c = 2.81$ Å, $\alpha = \beta = 90°$, and $\gamma = 120°$.

We used the 48-atom *β*-Ti and 48-atom *ω*-Ti supercells to study the relative *β* stabilization strength of the 3*d* transition elements, by substituting a 3*d* alloying element atom for a Ti atom in the supercells. The 48-atom *β*-Ti supercell is formed by stacking 12 atomic *β*-(111) layers, with

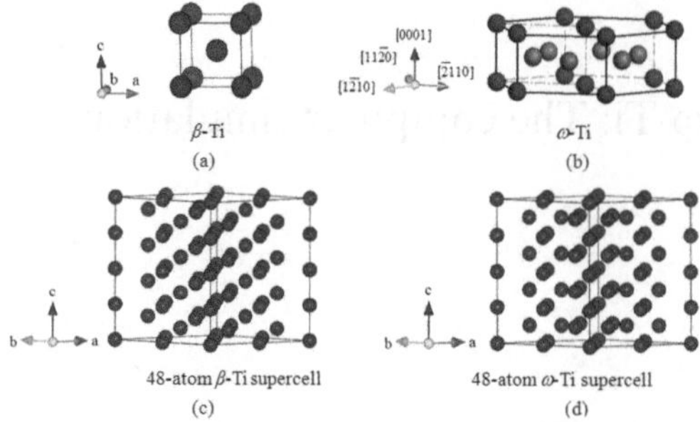

Figure 1. The atomic structures of β-Ti (a), ω-Ti (b), 48-atom β-Ti supercell (c) and 48-atom ω-Ti supercell. The structure of ω-Ti is showed by a hexagonal cell, in which a 3-atom ω-Ti primitive cell is sketched by dash dot lines. Atoms in blue and green in ω-Ti (b) represent the Ti atoms in the loose and close-packed (0001) layers, called A and B sites respectively.

4 atoms in every layer. The 48-atom ω-Ti supercell is $2 \times 2 \times 4$ supercell with respect to the 3-atom ω-Ti primitive cell in Figure 1(b), containing 8 atomic ω-(0001) layers. The structures of two 48-atom supercells are shown in Figure 1(c and d).

3 RESULTS AND DISCUSSION

The (0001) planes have different plane densities in ω lattice. Among two neighboring (0001) planes, one is loose and the next is compact. Correspondingly, two un-equivalent sites exist in ω lattice, denoted by A and B. These two sites have already been shown in Figure 1. To evaluate the site preference of 4*d* elements in ω-Ti, we calculated E_A and E_B, the total energies of 48-atom ω-Ti supercells containing a 4*d* alloying elements substituted for Ti in A and B sites respectively. Results of $\Delta E = E_A - E_B$ are listed in Figure 2, which is in a downward parabola shape with a maximum at Mo. In the figure, the positive values indicate that B site is a favorite one and the negative values indicate the other way. One can see that B site is favorable for most of all the elements except Y and Pd element.

As the most Ti-based alloys are multi-element system, we continue to investigate the site preference of 4d elements in the ternary ω-Ti system. One typical Ti-based metal alloy composition is that of Ti–36Nb–2Ta–3Zr–0.3O in mass% (Ti–23Nb–0.7Ta–2Zr–1.2O in at%) [1]. Thus, Ti_3Nb-Pd, in which the niobium composition is around 25%, was taken as the prototype of ω-Ti alloy. Pd element was taken as the ternary element, since it is an exceptional element in Figure 2. Our calculated results are listed in Table 1. It demonstrated

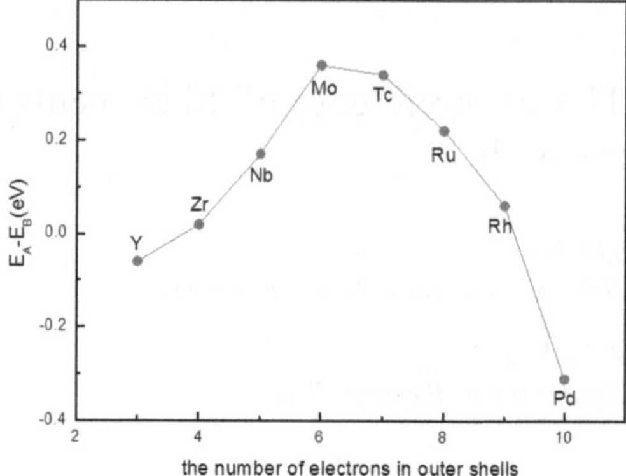

Figure 2. The relative total energies ΔE of 48-atom ω-Ti supercells containing a 4*d* alloying elements substituted for Ti in A and B sites respectively.

Table 1. Site preference of Pd element in ω-Ti_3Nb.

Substitutional sites	Ti in A-site	Ti in A-site	Nb in A-site	Nb in A-site
ΔE (eV)	−0.66	−0.71	−0.94	−0.83

that Pd prefers to substitute Nb in A-site in ω-Ti_3 Nb, in consistency with the result in Figure 2, verifying that our conclusion obtained from Figure 2 is suitable in multi-element systems too.

In the below figure, we tried to rationalize the site preference of 4d elements in ω-Ti. Generally, the electronic factor and the geometrical atom-size factor might explain the site preference. Accordingly, ΔE can be expressed as the sum of two contributions: $\Delta E = \Delta E_c + \Delta E_e$, where ΔE_c and ΔE_e respectively represent chemical energy difference, i.e, electronic factor and elastic energy difference, i.e, geometrical factor.

To give a electronic analysis, we measured and listed in Figure 3(a) and 3(b) the lowest charge density values along A-A, A-B, B-B(1), and B-B(2) directions around the alloying elements in ω-Ti supercells. To be clear, all involved directions were demonstrated in a primitive cell in the Figure 3c, too. These charge density values were used to evaluate the bonding strengths. A larger charge density value means a stronger bond. One can see a coincidence between Figure 2 and Figure 3, when both are downward curves of second order. Thus, we concluded that the electronic factor dominates the site preference of *4d* elements in ω-Ti.

To give a further confirmation, we now calculate the elastic energy in the framework of a classical elasticity theory with the *sphere and hole* model

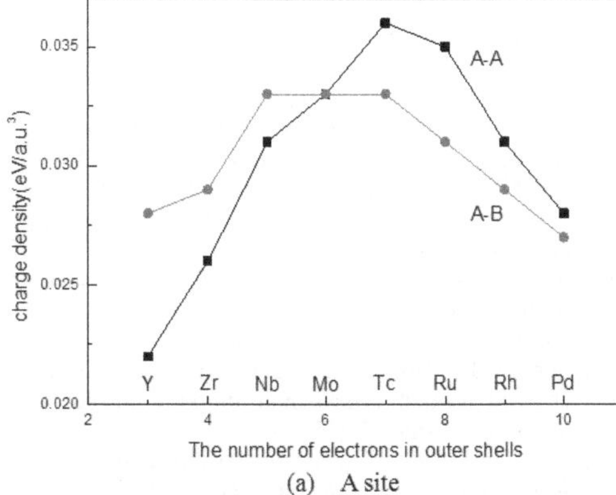

(a) A site

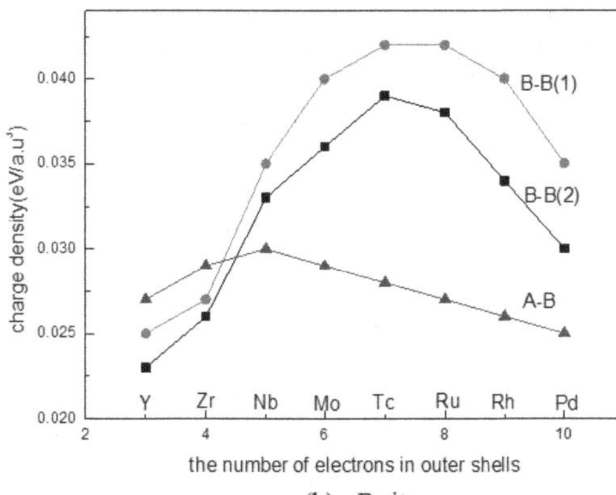

(b) B site

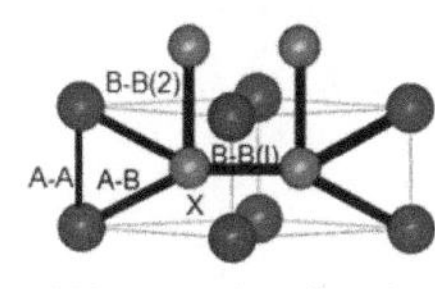

(c)The measuring directions

Figure 3. The lowest charge along the different directions from the 4d alloying elements in A (panel a) and B (panel b) sites in ω-Ti. Black thick lines in the primitive cell schematically demonstrated the different measuring directions in ω Ti (panel c). Symbol "X" showed that the location of alloying element and atoms in blue and green represent the A and B sites, respectively (panel c).

developed by Friedel and Eshelby [14–15]. In this model, the elastic energy E_e yields:

$$E_e = \frac{2K_T G_M \left(V^M - V^T\right)^2}{3K_T V^M + 4G_M V^T} \tag{1}$$

where K_T and G_M are bulk modulus of metal T and shear modulus of matrix, respectively. V_T and V_M are atomic volumes of metal T and matrix atom respectively. Clearly, 4d alloying elements correspond to metal T and Ti corresponds to matrix in

Table 2. The elastic energy difference ΔE_e(eV).

Elements	ΔE_e	Elements	ΔE_e	Elements	ΔE_e
Y	−0.02	Zr	0.006	Nb	−0.02
Mo	0.005	Tc	0.02	Ru	0.03
Rh	0.02	Pd	−0.04		

the present work. Using function 1, we gained the elastic energy ω-Ti containing 4d alloying elements in A and B sites, respectively, namely $E_{e\text{-}A}$ and $E_{e\text{-}B}$. Accordingly, we could obtain the elastic energy difference as:

$$\Delta E_e = E_{e\text{-}A} - E_{e\text{-}B} \tag{2}$$

In Table 2, we showed the calculation of ΔE_e. It turns out that ΔE_e is small enough to be negligible in the background of ΔE. It confirmed our conclusion that atom-sized factor contributes little to explain the site preference of 4d elements in ω-Ti.

4 CONCLUSION

By virtue of precise total energy calculations based on first-principles density functional theory method, we are able to evaluate with high accuracy the site preference of 4d elements in ω-Ti. We obtained the following conclusion:

1. B site is favorable for most of the 4d elements except Y and Pd elements.
2. The atom-size factor accounts for little to explain the site preference of 4d elements in ω-Ti. In contrast, the electronic factor could explain it well.

REFERENCES

[1] Saito, T., Furuta, T., Hwang, J.H., Kuramoto, S., Nishino, K., Suzuki, N., & Sakuma, T. 2003. Multifunctional alloys obtained via a dislocation-free plastic deformation mechanism. *Science*, *300*(5618), 464–467.

[2] Niinomi, M., Nakai, M., & Hieda, J. 2012. Development of new metallic alloys for biomedical applications. *Acta Biomaterialia*, *8*(11), 3888–3903.

[3] Ozan, S., Lin, J., Li, Y., Ipek, R., & Wen, C. 2015. Development of Ti–Nb–Zr alloys with high elastic admissible strain for temporary orthopedic devices. *Acta biomaterialia*. (in press).

[4] Hu, Q.M., Li, S.J., Hao, Y.L., Yang, R., Johansson, B., & Vitos, L. 2008. Phase stability and elastic modulus of Ti alloys containing Nb, Zr, and/or Sn from first-principles calculations. *Applied Physics Letters*, *93*(12), 121902.

[5] Tegner, B.E., Zhu, L., & Ackland, G.J. 2012. Relative strength of phase stabilizers in titanium alloys. *Physical Review B*, *85*(21), 214106.

[6] Lazar, P., Jahnátek, M., Hafner, J., Nagasako, N., Asahi, R., Blaas-Schenner, C., & Podloucky, R. 2011. Temperature-induced martensitic phase transitions in gum-metal approximants: First-principles investigations for Ti3 Nb. *Physical Review B*, *84*(5), 054202.

[7] Hanlumyuang, Y., Sankaran, R.P., Sherburne, M.P., Morris Jr, J.W., & Chrzan, D.C. 2012. Phonons and phase stability in Ti-V approximants to gum metal. *Physical Review B*, *85*(14), 144108.

[8] Tane, M., Nakano, T., Kuramoto, S., Niinomi, M., Takesue, N., & Nakajima, H. 2013. ω Transformation in cold-worked Ti–Nb–Ta–Zr–O alloys with low body-centered cubic phase stability and its correlation with their elastic properties. *Acta Materialia*, *61*(1), 139–150.

[9] Dai, J.H., Song, Y., Li, W., Yang, R., & Vitos, L. 2014. Influence of alloying elements Nb, Zr, Sn, and oxygen on structural stability and elastic properties of the Ti2448 alloy. *Physical Review B*, *89*(1), 014103.

[10] Landa, A., Söderlind, P., Ruban, A.V., Peil, O.E., & Vitos, L. 2009. Stability in BCC transition metals: Madelung and band-energy effects due to alloying. *Physical review letters*, *103*(23), 235501.

[11] Gonze, X., Amadon, B., Anglade, P.M., Beuken, J.M., Bottin, F., Boulanger, P. ... & Zwanziger, J.W. 2009. ABINIT: First-principles approach to material and nanosystem properties. *Computer Physics Communications*, *180*(12), 2582–2615.

[12] Kresse, G., & Joubert, D. 1999. From ultrasoft pseudopotentials to the projector augmented-wave method. *Physical Review B*, *59*(3), 1758.

[13] Perdew, J.P., Burke, K., & Ernzerhof, M. 1996. Generalized gradient approximation made simple. *Physical review letters*, *77*(18), 3865.

[14] Friedel, J. 1954. Electronic structure of primary solid solutions in metals. *Advances in Physics*, *3*(12), 446–507.

[15] Eshelby, J.D. 1956. The continuum theory of lattice defects. *Solid state physics*, *3*, 79–144.

Material Science and Environmental Engineering – Chen (Ed.)

Research progress of laser cladding on surface of titanium alloy

T.G. Zhang
School of Mechanical Engineering, Tianjin Polytechnic University, Tianjin, China

R.L. Sun
Tianjin Area Major Laboratory of Advanced Mechatronics Equipment Technology, Tianjin, China

ABSTRACT: Titanium alloy with the advantages of low density, high specific strength, and good corrosion resistance is widely used in aerospace and other fields. However, the characteristics of titanium alloy with low wear resistance, poor high temperatures, and oxidation resistance limit its application. Improving the surface properties of titanium alloy by laser cladding technology have become a hot spot in the area of surface modification on titanium alloys. Present advancement in laser cladding materials on the surface of titanium alloy is introduced and summarized. Application of the laser cladding technology is analyzed and the development trend of the future is also proposed.

Keywords: titanium alloy; laser cladding; cladding material

1 INTRODUCTION

Titanium and its alloys have low density, high strength, good corrosion resistance, good biocompatibility, and many other advantages. It acts as an important strategic metal that is being used in the aerospace, petrochemical, aviation, medicine, and biological sectors for making its indispensable materials. But the poor tribological properties of titanium alloys like high performance friction coefficient, adhesive wear, serious and poor thermal conductivity, and high temperature oxidation resistance defects, which greatly limits its potential to play [1–2]. Therefore, improving the titanium surface wear resistance, high temperature oxidation resistance, etc., has become an important issue that must be widely used to overcome titanium.

Laser cladding technique is a new surface modification technology, with heating speed, combined with the base metallurgical cladding, cladding composition, dilution controlled, etc., has become one of the effective means to improve the surface properties of titanium alloys [3–5]. This paper describes the research status of Titanium by laser cladding material system, application and development trend of laser cladding technology.

2 TITANIUM BY LASER CLADDING MATERIALS RESEARCH

Laser cladding layer formation process is a complex physical and chemical process. In this process, factors affecting laser cladding forming quality and performance of complex, in which laser cladding material is a major factor in its selection of its main consideration performance, compatibility, thermal physical properties of different substrates, etc. [6–8]. Currently cladding titanium powder material is mainly a self-fluxing alloy and metal matrix ceramic composite material.

Commonly used as self-fluxing, the three kinds of alloy materials are: nickel-based alloy material, cobalt-based alloy material, and iron-based alloy material [9–10]. Aerospace, petrochemical in laser cladding process, self-fluxing alloy powder B, Si, and other elements of the tribological properties have often got oxygen and slagging features that ambient oxygen preferentially reacts with B20 in the thermal conductivity of SiO and borosilicate. It covered the surface of the bath to prevent excessive oxidation of the liquid metal and to improve the melt can play substrate wetting ability to obtain the dilution rate, fewer impurities, gas, etc. It became small and the porosity of the substrate metallurgical combined a dense coating Geng Lin et al [9] using C0: Lasers, respectively, in TC4 alloy cladding the NiCrBSi and NiCoCr-fast, cladding layer and the substrate AlY two kinds of nickel-based alloy coating. Both coatings can produce solid solution strengthening point, which has improved to a fine-grain strengthening effect, and the coating is also NiCrBsi existing in T Mindanao, TiC, CrB, Ni.B, and Ti.Al high hardness enhanced to strengthen the role of the second phase. These technologies should strengthen due to two factors that increase the coating hardness and wear resistance of nickel-based alloys than titanium alloys.

Metal matrix ceramic composites [12–14] complex by a variety of refractory oxides, carbides, nitrides, borides, and silicides of hard ceramic materials and the like made of metal. It can be of higher laser melting metal material strength, toughness, good process performance, and ceramics with excellent wear resistance, corrosion resistance, high temperature, and chemical stability that combines the laser is often used in the preparation of high temperature oxidation coatings, abrasion resistant coatings, bio-ceramic coatings, and so on. Cui Aiyong r12 j conducted titanium cladding (Ti + A1/Ni) + (Cr203 + CeO) composite coating experiments, cladding microstructure analysis showed that there is a white light unmelted spherical liquid precipitation Cr. 0. Hardening TiAl ceramic particles and the resulting reinforcement is evenly distributed over fine dendrites and eutectic matrix. The coating microstructure is hard at 1150 HV. Ma Haibo u toward the titanium surface prepared Co-situ particle reinforced ceramic matrix composite coatings. The findings were grown in columnar grain morphology mainly to dendrites, etc., within the shaft lean Co, B, and Ti-rich TiB system ceramic reinforced particles, intergranular for y-Co, each phase uniformly distributed in the melting zone. Microhardness of the coating is about three times the substrate.

3 TITANIUM BY LASER CLADDING TECHNOLOGY

With the development of science and technology in the field of aerospace, industrial production, etc., a lot of use for titanium alloy should be made with its knowledge of mechanical temperature and corrosive atmosphere and its use in the work environment. Therefore, titanium oxide's heat resistance and corrosion resistance should be known. Research mill performance is very heavy. Materials reflecting different functions can be broadly classified into the following four kinds of titanium surfaces: high temperature oxidation coatings, corrosion resistant coatings, abrasion resistant coatings, and bio-ceramic coating.

3.1 *High temperature oxidation coating*

High temperature oxidation coating applications take place in the Pan rocket engines, gas turbines, and heat exchange devices. The use of laser cladding technology titanium alloy coating, which can significantly improve the resistance to high temperature titanium alloy substrate oxygen TiN, NiCr, and TiCrAlSi such as titanium powder surface can get into a high temperature oxidation coating and laser cladding technology in terms of a titanium surface modification technology which is very promising species. Map

Liuxiu Boatel study utilizing cross-flow C02 laser cladding NiCr. Cr. C: Composite coating, studied the performance of the original 1000 °C TiAl alloy oxidation and laser cladding coating, the results showed that the coating showed better Cr. 03 high temperature oxidation resistance, the oxide layer is of an uniform dense tissue, mainly by Kuang Al: 03, TiO and Si02 components. Huang Can Ti-6A1-4V alloy surface by laser cladding process to obtain the highest degree of cracks and holes that a few TiCrAlSi-V and TicrAlSi-Ni self TiB is found with the two kinds of multi-component alloy coating. XRD result shows the TiCrAlSi-V coating. Table shows that the surface of oxide cladding layer formed by Si02, Cr. 03, Ti02, and A1.03 and a small amount of V: 0. Transition crystal structure, oxide layer TiCrAlSi-Ni coating surface by Cr. 0., Ti02, Al: 0. And a small amount of Si0: and Ni0 composition. The experimental results that are shown in Figure 1 has the oxide 3 times the two substrates. Oxidized species of coating weight gain is much smaller than the TCA substrate that can effectively improve the alloy oxidation resistance at 800 °C.

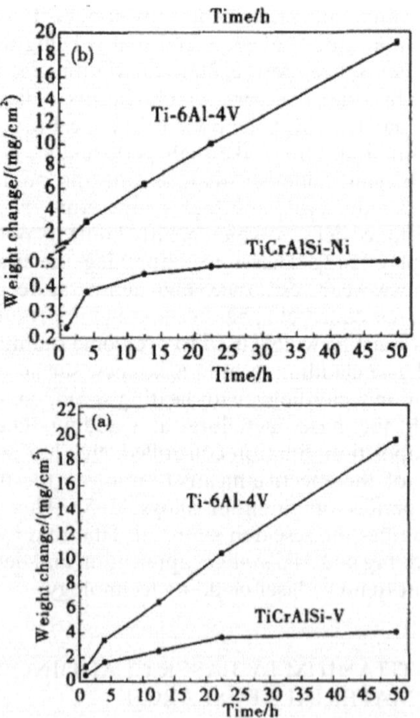

Figure 1. Oxidation weight gain curves of cladding and TC4 substrate at 800 °C.

3.2 *Corrosion resistant coating*

Laser cladding corrosion resistant coatings generally has nickel, cobalt-based self-fluxing complexes with titanium coating material performance-based body. Ni-based self-fluxing alloys containing SiC, 134C, parts long Al: 0. Other ceramic composite coating particles have a good corrosion resistance; the main component of the cobalt-based alloy material, anti-gold powder is Co, Cr, and W coating has excellent resistance to high temperature performance, anti-hot corrosion, and anti-erosion ability of laser cladding. Corrosion resistant coating of a hot research, in the titanium surface coating and laser melting mill cover coating raw trace rare earth elements or rare earth oxides can be added to significantly improve the corrosion resistance of the cladding layer.

High cedar, etc. 1 knock on Titanium by laser cladding has a non-A1.03 + (mass fraction) of 13% TiQ hot parts, components obtained homogeneous dense ceramic coating. Erosion solid alloy surface inspection system showed that the corrosion resistance of the coating can be improved compared with the base, when the erosion angle is of 30 when the effect of laser cladding corrosion resistance of ceramic coatings has significantly enhanced. Deng belated () stimulated research lines.

Studying the effect of rare earth on the corrosion resistance of the ceramic coating, respectively, titanium alloy surface excitation light cladding rare antioxidant coating composite, TCA base and by not adding rare earth composite coating made of acid, alkali and physiological salt solution corrosion resistant experimental results show that the three kinds of materials, adding rare earth coating acid, alkali, physiological salt solution of the strongest anti-corrosion ability, and adding rare earth significantly improves the protection coating to the substrate.

3.3 *Wear-resistant coating*

Wear-resistant coating is the most studied laser cladding, the most widely used as a coating lead. The wear-resistance of laser cladding coating depends on the type of reinforcement and its content and distribution in the cladding layer. Currently laser cladding layer to obtain enhancement phase is mainly done in situ method and the direct additive method.

3.3.1 *Situ method*

Formed in situ means that under certain conditions, by a chemical reaction between the elements and the element or elements and compounds, methods situ ceramic particulate phase. Prepared by in situ composite coatings, which have enhanced the thermodynamically stable phase, distribution, clean interface, combined with the advantages of a good substrate, the most studied one is the use of titanium surfaces generated in situ reaction TiC, TiB, Til32, and other enhancements phase, to improve the wear-resistance of the substrate.

Zhang Xiaodong in TCll nickel-alloy surfaces were coated graphite powder material of laser cladding experiments, uniform and continuous access to the surface of the substrate binding metallurgical cladding. Microstructure analysis showed that the laser cladding process, the substrate coating and Ti Figure 1 and C pre-coating chemical reaction in situ with petals and dendrite form of hard TiC reinforcement. As can be seen from Figure 2, along the direction of the deep layer hardness of laser cladding sample distribution presents three regions, which correspond to the Cladding Region (CZ), Binding Region (BZ) and basal Heat Affected Zone (HAZ). Hardness was tested by cladding region to reduce the heat-affected zone stepped hardness of laser cladding region (HVl000~1050) than the titanium substrate hardness (HV320~3.10) increased by 2 to 3 times, indicating that the laser cladding layer can greatly improve the surface wear TCll alloy. Mujun Shi ∞73 with NiCr-Cr. C: + 40% CaF2 (mass fraction) mixed powder as raw material, the use of laser cladding alloy surface was prepared in a composite coating microstructure consisting of primary bulk Cr, C and 7-Cr, Cs eutectic, a large amount of spherical dendritic TiC and CaF: composition, coating is 2.5 times the average microhardness average hardness of the substrate. Literature reported in the titanium substrate Table is generated in situ TiB_2/TiC composite coating, SEM found that the coating is mainly composed of black massive TiB_2, flower-like or equiaxed TiC, fine needle CrB composition, hardness tester further measured, the average value of coating hardness is HVo. 700, significantly higher than the hardness of the substrate.

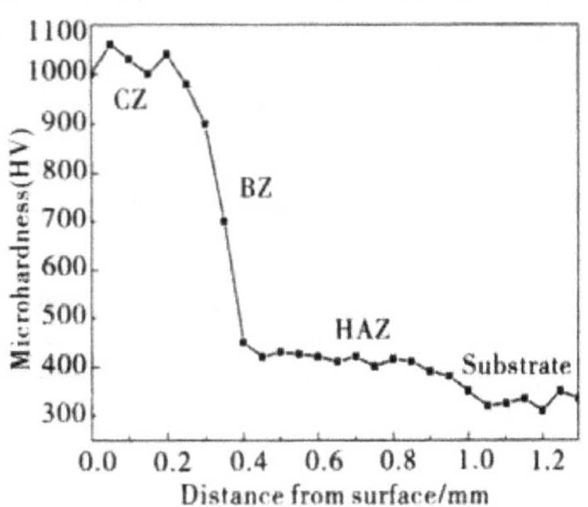

Figure 2. Distribution profile of microhardness across nickel wapped graphite coating.

Literature reported that the findings on the titanium surface in situ generated TiB and TiB2, indicating that the use of in-situ reaction of reinforcing phase with fine grain strengthening and dispersion strengthening features, improve the wear-resistance of titanium alloy; self-exothermic reaction. It is possible to prepare a cladding layer that is formed using low-power lasers.

3.3.2 *Direct addition method*

Direct addition method refers addition to cladding materials TiC, TiB, TiB2, TiN, etc. In the carbide phase, the use of laser cladding technology in the preparation of titanium surface that wear good cladding layer. In addition, the cladding material is added directly to Mos2 and WSz. During the self-lubricating phase, by adjusting the ratio of laser parameters, and cladding materials, wear-resistant cladding layer can improve self-lubrication.

P. Liu et al in titanium cladding Ni60-TiC-Mo composite coatings, SEM and TEM results show that the composite coating contains large amounts of eutectic and amorphous phases, significantly improved wear-resistance than the base. Yang plastic Creek et al using laser cladding technique TC4 alloy cladding add MoSz solid self-lubricating composite coating phase prepared better metallurgical quality of TCA/Ni/MoS2 composite. Experimental results show that the friction and wear of TC4/Ni/Mos2 composite coating significantly reduces the coefficient of friction than the TC alloy. Wear-resistance increases by about 9 times, especially when high temperature friction coefficient decreases, but with added MoS2 content, the coefficient of friction coating increases and there is a decrease in trend. Literature 343 reported under argon laser clad in titanium cladding MoS2/Ti composite coatings, the findings show that the main by ternary sulfide, molybdenum sulfide and titanium sulfide composite coating physical constitution, MoS_2/Ti composite coating friction coefficient and surface roughness were lower than Ti6A14V, composite coating showed excellent adhesion, abrasion resistance, surface hardness advantages. In addition, under the premise of not reducing the coating hardness, add dilute trace elements on the surface of titanium alloy laser cladding layer. West. one, by micro-alloying of rare earth elements.

The effect can effectively improve the density of the cladding layer by reducing the solubility of solute elements in the matrix, while enhancing the binding force of the cladding layer and the substrate, and thus significantly increasing the fracture toughness of the coating. The coating reduces cracking susceptibility, where significant improvement in the composite coating wear and corrosion resistance is seen. Thus enhancing the stability of titanium alloys in different conditions.

3.4 *Bio ceramic coating*

Titanium is a biologically inert material, and the most common element in TCA alloy of which aluminum and vanadium are harmful, so be sure to make the appropriate activation and protective treatment. This can be used for the manufacture of human bones or teeth and other alternatives. Biological ceramic titanium surface coating materials currently in preparation include hydroxyapatite (HA, HAP), 13- tricalcium phosphate (13-TCP) and fluorapatite (FA, FAP). Laser cladding biological ceramic coating both strengthen the toughness of the metal or alloy, but is also biologically active, orderly distribution of cellular tissue microcrystalline coating, structure and natural bone tissue structure of this organization are similarly led. In addition, adding the right amount of rare earth oxide cladding layer, the generation of bio-ceramic phase with a catalytic effect, help to improve its biological properties. Biological ceramic coating materials have been extensively studied and applied in the hot Titanium by laser cladding technology research in the field of development. Three samples immersed for seven days SEM image and EDS analysis in SBF.

Hu Shuhui to the HA + CaF2 and HA + Si02 as a raw material, by laser cladding technology in preparation FHA TC4 alloy and Si-HA bioceramic coating. By scanning electron microscopy found that more uniform distribution of the ceramic layer consists of fine equiaxed dendrite and composition; transition layer near the denser microstructure is conducive to combine ceramic layer and the substrate; surface tissue gradually becomes loose, helps in the growth of bone tissue structure. Simulated Body Fluid immersion test (SBF) showed (Fig. 3), the ceramic coating is deposited on the surface of Ca, P, 0 as the main element of the apatite layer, a ceramic coating formed by laser cladding with a certain biological activity. S. Yang catches 401 by laser cladding technology in the NiTi alloy substrate prepared by Hydroxyapatite/Titanium (HA/Ti) composite coating, the experimental results show that the high-power laser irradiation HA decomposition, microscopic coating organization

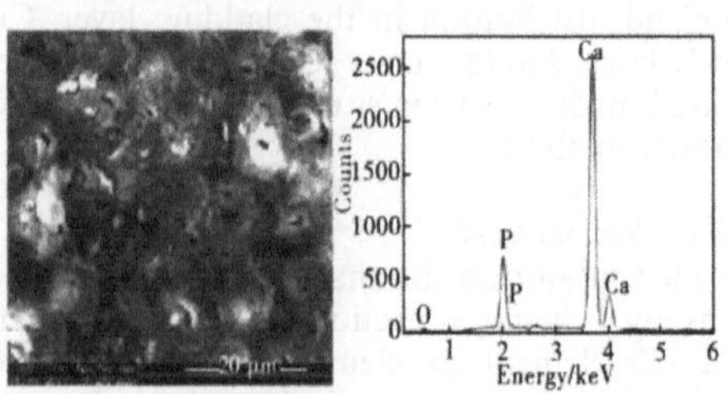

Figure 3. SEM image and energy spergy spectrum of samples sockde in SBF after 7 days.

mainly CaO, perovskite, titanium phosphide, and HA phase. In vitro experimental results show that the biological activity of NiTi alloys HA/Ti composite coating significantly improved the mechanical properties of the coating to achieve the desired biological medicine. Fan Ding et al 4 women using laser cladding technique for preparing biological ceramic composite coating on Ti6A14V alloy substrate, the results show, Ce0. Synthetic HA, 8_Ca. (P0) and other biologically active calcium phosphate-based ceramics have a significant catalyst. Wang Zhen, et al, say that using broadband laser cladding technology in the preparation of TC4 alloy containing HA + 8-TCP gradient, there is biological activity of rare earth ceramic composite coating when rare earth oxides are at Nd20. The mass fraction of 0.6%, the catalytic synthesis of HA + maximum amount of BTCP; Nd2 03 mass fraction of 0.4%~0.6%, corrosion resistant coating.

4 TITANIUM BY LASER CLADDING

By laser cladding technology for surface modification of titanium alloy research has caused people's attention. The results show that the technique is no more than a traditional surface modification proposed advantages. Some research has applications in the aerospace, medical, and other fields. But for now, the technology is still in the laboratory stage. Major yet the large-scale industrial production has been led into its future direction through the following aspects.

1. The development of new coating, the coating material to optimize the design, such as the titanium surface and more high-entropy alloys PCA cladding. The development of new laser cladding technology to reduce thermal stresses between the coating. The coating is easy to solve cracking, peeling and other issues, such as the gradient between the cladding, coating and substrate binding layer before coating, cladding.
2. In-depth study of the reaction mechanism of laser cladding process, combined with the finite element simulation software to build a laser cladding process finite element model, establish and improve its numerical simulation of temperature field and stress field, which reveals the laser and material mutual the role of machine processing.
3. To further explore the influence of rare earth elements on the surface of titanium alloy laser cladding layer performance. Existing studies show better performance composite coating containing an appropriate amount of rare earth oxides, rare earth oxides; however, excessive addition will lead to lower bath temperature, resulting in other micro-cracks.
4. Cover combines technology and energy fields. For example, in the process of laser cladding alloy surface, the electromagnetic fields or vibration field, on the one hand to promote the melting mass effect, increasing the reaction interface, reducing the contact angle, wet ability, and compatibility improvements cladding material system. On the other hand, prior to formation of dendrites breaking dendrite structure, and limiting the growth of dendrites in the fixed direction so that grain refinement performance can take place.

REFERENCES

[1] Huangzhang Hong Qu Henglei, Deng Chao, et al. FJ aviation development and application of titanium and titanium alloys. Materials Review: Summary of papers, 2011, 25 (1): 102.

[2] Wang Dongsheng, Tianzong Jun, Shen Lida, and so on. Laser surface modification of titanium alloy research on Ti-6A1-4V a study of the status quo mouth. Laser and Optoelectronics Progress, 2008 (6): 24.

[3] Zi group. Titanium New Development and Application Status EJ-i. Titanium Industry Progress, 2008 (2): 23.

[4] Song Jianli, Liyong Tang, Deng Qilin, and so on. Laser cladding forming technology research into the newspaper, 2010 (14): 29 more tool in the box.

[5] Hayden Corp. Laser cladding: One Mater Processes, 2012, 170 (11): 46 and tribological.

[6] Cattle Wei, Sun Ronglu. Research Status and Development Trend of titanium laser cladding sized on titanium interfaca IJ-.

[7] J Mater Processing Techn, 2009, 209 (5): 2237 Dong.

[8] World Games, Ma Yunzhe, Xu Bin Shi, et al. Laser Cladding Materials Research Situation mouth. Materials Review, 2006, 20 (6): 5 11 Wu C.F., Ma M.X., Liu W.J., et al. Laser cladding in-situ using laser cladding technology for titanium carbide particle rein-forced Fe-based composite coatings with.

[9] Geng Lin, Mengqing Wu, Guo Lixin. Both on the surface of titanium base alloy powder laser rare earth oxide addition EJ. J Rare Earths, 2009, 27 (6): 997.

[11] Cladding Optical Research EJ. Materials engineering, 2005(12): 4512 Cuiai Yong, Hu Fangyou back to Korea. Laser cladding titanium (Ti + A1/Ni)/(Crz03 + CeOz) and wear resistance of composite coating [J]. China.

[12] Durra Majumdar J., Manna I., Kumar A., et al. Direct laser laser, 2007 (3): 438 graded.

[13] Ma Haibo, Zhang Weiping. Cobalt-based alloy laser cladding composite coating and the substrate and the cladding process performance [J]. Metal Materials and Engineering, 2010 (12): 218914 Lei Y.W., Sun R.L., Lei J.B., et al. A new theoretical model after suitable heat treatment. For high power laser clad TiC/NiCrBSiC composite coatingson Ti6A14V alloys [J]. Opt Lasers Eng, 2010, 48 (9): 899.

Material Science and Environmental Engineering – Chen (Ed.)
© 2016 Taylor & Francis Group, London, ISBN 978-1-138-02938-5

Process control of 7075 Al alloy semi-solid slurry by thixoforming

S.D. Gan, G.S. Gan, H.S. Wang & G.Q. Meng
Chongqing Municipal Engineering Research Center of Institutions of Higher Education for Special Welding Materials and Technology (Chongqing University of Technology), Chongqing, China

ABSTRACT: The effect of temperature and holding time on microstructure of 7075 Al alloy slurry by thixoforming were investigated. The results have shown that the phenomenon of necking and remelting of dendrite grains can be found at first, then rosette grains were melted and shrinked into the globular grains. Finally the globular grains began to melt with increasing holding time. The mean size of globular grains reduced with increasing temperature, the lifetime of semi-solid slurry at 610°C, 620°C, and 630°C were at about 10 min, 5 min, and 2.5 min, respectively.

Keywords: 7075 Al alloy; semi-solid slurry; thixoforming; the grain size; lifetime

1 INTRODUCTION

It is observed that 7075 Al alloys have been widely used as structural materials in aeronautical industries due to their attractive comprehensive properties such as low density, high strength, ductility, and toughness. However, wrought aluminum alloys are difficult to cast due to hot tearing formed during the solidification process of the alloys (Eskin et al. 2004; Langlais et al. 2008). SSM, which was considered as one of the most prospective materials processing technologies in 21 century, provides the possibility of producing near-net-shape casting of wrought aluminum alloys. The key issue of semi-solid forming is how to fabricate the proper slurry containing the non-dendritic microstructure (Neag et al. 2012). Semisolid slurry of thixoforming is prepared by reheating at the solid state, which needs no stirring. Previous work has shown that SSM processing were appropriate for 7075 Al alloy, including thixoforming (Rikhtegar et al. 2010; Bolouri et al. 2011; Rokni et al. 2012; Mohammadi et al; 2013; Rogal et al. 2013; Lu et al. 2001) and rheocasting (Guo et al. 2010; Curle et al. 2010; Yang et al. 2013). But the effect of temperature and holding time on microstructure of 7075 Al alloy slurry by thixoforming has no in-depth study. The key problem to semi-solid metal forming is strict control of technical parameters, especially the liquid fraction of the alloy is sensitive with increasing temperature.

2 EXPERIMENTAL

The chemical composition of 7075 alloy used in present study was Al-5.52 wt.% Zn-2.36 wt.% Mg-1.51 wt.% Cu-0.18 wt.% Si-0.26 wt.% Fe-0.15 wt.% Mn-0.25 wt.% Cr. The liquidus and solidus temperatures of the alloy were tested by NETZSCH DSC204 Differential Scanning Calorimetry (DSC), with heating speed of 10°C/min. The molten of 7075 Al alloys at 720°C was casted in 15-mm diameter's graphite mould, then 7075 Al alloy were cut out, respectively, in height of 8 mm, finally these were put into the holding furnace at 630 ± 1°C, 620 ± 1°C, and 610 ± 1°C, respectively, and the samples can be obtained by immediately quenching with different holding times. The grain size and area of the primary solid phase were analyzed statistically by a quantitative image analysis system. A close observation of microstructure can be given by average gain diameter $D = 2(A/\pi)^{1/2}$ and shape factor $F = 4\pi A/P^2$, where A and P are average area and average perimeter of primary phase, respectively.

3 RESULTS AND DISCUSSION

The liquidus and the solidus temperatures of the 7075 alloy were 639.1°C and 477.4°C, respectively in Figure 1 (a). The liquid fraction at 630°C, 620°C, and 610°C were about 73.05%, 51.59%, and 36.58%, respectively, which suggested that the liquid fraction of the alloy was sensitive with increasing temperature in Figure 1b.

Figure 2 shows the microstructure of solid 7075 Al alloy with the different holding time at 610°C. The mean grain size can reach 113 μm and 124.7 μm for 30 min and 40 min respectively. The shape factor can reach 0.77 and 0.67 for 30 min and 40 min, respectively. α-Al grains began to melt

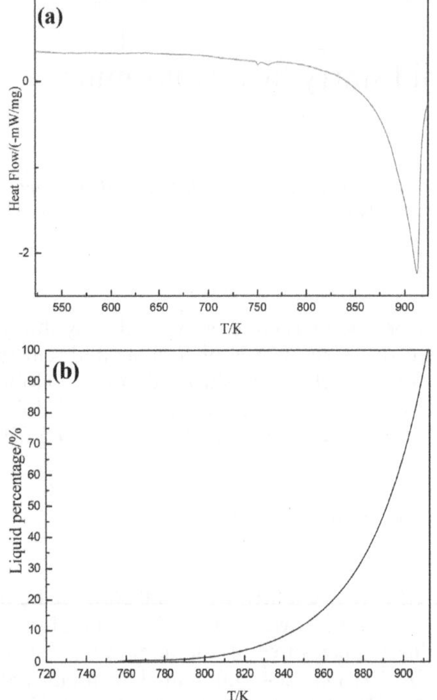

Figure 1. The DSC curve (a) and the liquid percentage as function of temperature for the 7075 Al alloy (b).

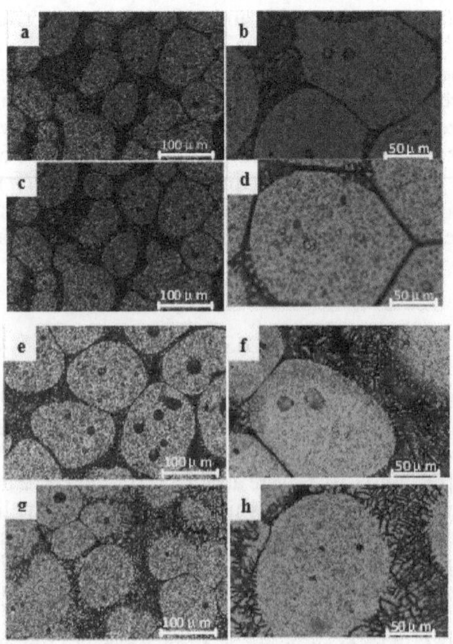

Figure 2. The microstructure of the solid 7075 Al alloy with the different holding time at 610°C. (ab) 20 min; (cd) 30 min; (ef) 40 min; (gh) 60 min.

from the interior and the outside of grains after 40 min.

It can also get good globular grains for 15~20 min at 620°C in Figure 3. The mean grain size can reach 118.7 μm and 104.1 μm, and the shape factor can reach 0.63 and 0.71 for 15 min and 20 min, respectively. Almost all of α-Al grains completely melted after 35 min.

The branch of rosette grains at 630°C began to increase in their melting gradually with time, and α-Al grains contracted to sphere because its spherical surface had the smallest surface energy

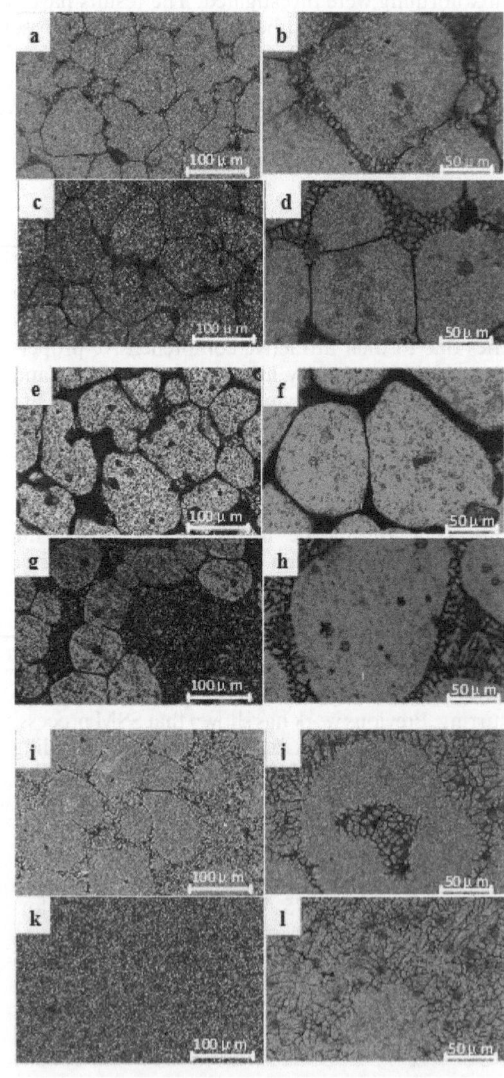

Figure 3. The microstructure of the solid 7075 Al alloy with the different holding time at 620°C. (ab) 10 min; (cd) 15 min; (ef) 20 min; (gh) 25 min; (ij) 35 min; (kl) 40 min.

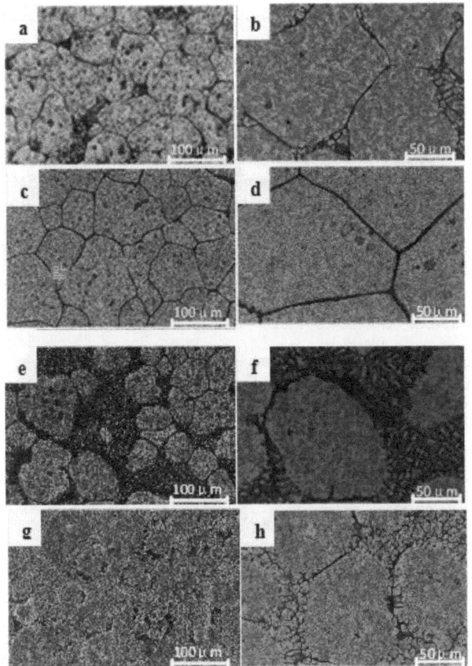

Figure 4. The microstructure of the solid 7075 Al alloy with the different holding time at 630°C. (ab) 7.5 min; (cd) 10 min; (ef) 12.5 min; (gh) 15 min.

(in Fig. 4(cd)). The mean grain size can reach 94 μm, and the shape factor can reach 0.65 and 0.78 for 7.5 min and 10 min, respectively. But α-Al grains began to melt after 10 min.

The microstructure evolution of 7075 Al alloy semi-solid slurry contained three parts: the phenomenon of necking and remelting of dendrite grains can be found at first, then the rosette grains melted and shrinked into the globular grain, and finally the globular grains began to melt with their holding time increasing because of no stable crystal nucleus. The dendrite structures of α-Al grains melted to form globular grain with holding 30 min, 20 min, and 10 min at 610°C, 620°C, and 630°C, respectively. The mean size of globular grains reduced with increasing temperature, the lifetime of semi-solid slurry at 610°C, 620°C, and 630°C were about 10 min, 5 min, and 2.5 min, respectively. This is to say that the control of holding time is the key factor in near-liquidus thixoforming.

4 CONCLUSIONS

The liquid fraction of the alloy was sensitive with increasing temperature, so it is difficult for thixoforming and rheocasting. The phenomenon of necking and remelting of dendrite grains can be found at first, then rosette grains melted and shrinked into the globular grain, and finally the globular grains began to melt with their holding time in the process of thixoforming increasing. The mean size of globular grains reduced with increasing temperature, the lifetime of semi-solid slurry at 610°C, 620°C and 630°C were about 10 min, 5 min and 2.5 min respectively.

ACKNOWLEDGMENTS

This work was supported by the National Natural Science Foundation of China (Grant No. 51505051), the Innovation Foundation of Graduate of Chongqing University of Technology (No. YCX2013101 and YCX2014215) and Chongqing Municipal Engineering Research Center of Institutions of Higher Education for Special Welding Materials and Technology (No. SWMT201502, SWMT201503 and SWMT201505), respectively.

REFERENCES

[1] Eskin D.G., Suyitno, Katgerman L. 2004. Mechanical properties in the semi-solid state and hot tearing of aluminium alloy. Progress in Materials Science 49:629–711.

[2] Langlais J., Andrade N., Lemieux A., Chen X.G., Bucher L. 2008. The semi-solid forming of an improved AA6061 wrought aluminum alloy composition. Diffusion and Defect Data. Part. B: Solid State Phenomena 141–143:511–516.

[3] Neag Adriana, Favier Véronique, Bigot Régis and Pop Mariana. 2012. Microstructure and flow behaviour during backward extrusion of semi-solid 7075 aluminium alloy. J Mater. Process. Tech. 212:1472–1480.

[4] Bolouri A., Shahmiri M., Kang C.G. 2011. Study on the effects of the compression ratio and mushy zone heating on the thixotropic microstructure of AA 7075 aluminum alloy via SIMA process. J Alloy compd. 509:402–408.

[5] Rokni M.R., Zarei-Hanzaki A., Abedi H.R., Haghdadi N. 2012. Microstructure evolution and mechanical properties of backward thixoextruded 7075 aluminum alloy. Mater. Design 36:557–563.

[6] Rikhtegar F., Ketabchi M. 2010. Investigation of mechanical properties of 7075 Al alloy formed by forward thixoextrusion process. Mater. Design 31:3943–3948.

[7] Mohammadi H. and Ketabchi M. 2013. Investigation of microstructural and mechanical properties of 7075 Al alloy prepared by sima method. Iranian Journal of Materials Science & Engineering 10:32–43.

[8] Rogal Ł., Dutkiewicz J., Atkinson H.V., Lityńska-Dobrzyńska L., Czeppe T. and Modigell M. 2013. Characterization of semi-solid processing of aluminium alloy 7075 with Sc and Zr additions. Mat. Sci. Eng. A 580:362–373.

[9] Lu G.M., Dong J., Cui J.Z., Luo S.W. 2001. As-cast microstructure and the solidifing mechanism of 7075 aluminum alloy cast by LSC. Acta metallurgica siniea 37:1045–1048. (In Chinese).

[10] Guo H.M., Yang X.J., Wang J.X., Hu B., Zhu G.L. 2010. Effects of rheoforming on microstructures and mechanical properties of 7075 wrought aluminum alloy. Transactions of Nonferrous Metals Society of China 20:355–360.

[11] Curle U.A., Govender G. 2010. Semi-solid rheocasting of grain refined aluminum alloy 7075. Transactions of Nonferrous Metals Society of China 20: 832–836.

[12] Yang B., Mao W.M., Song X.J. 2013. Microstructure evolution of semi-solid 7075 Al alloy slurry during temperature homogenization treatment. Transactions of Nonferrous Metals Society of China 23(12): 3592–3597.

Material Science and Environmental Engineering – Chen (Ed.)
© 2016 Taylor & Francis Group, London, ISBN 978-1-138-02938-5

Ni-Cu alloy anode material prepared with hard template method

H.X. You, C. Zhao, B. Qu, Y.J. Guan & J.W. Xu
Chemical Environment and Life Science Division, Dalian University of Technology, Dalian, Liaoning, China

G.Q. Guan & A. Abudula
New Graduate School of Science and Technology, Hirosaki University, Bunkyo-cho, Hirosaki, Aomori, Japan
North Japan Research Institute for Sustainable Energy, Hirosaki University, Matsubara, Aomori, Japan

ABSTRACT: Ni_xCu_{1-x} alloy powders were prepared by hard template method using Activated Carbon Fiber (ACF) as template. Phase structure of the crystalline compositions of the prepared Ni_xCu_{1-x} alloy powders were characterized based on their observed morphology and microstructure. It was found that Ni_xCu_{1-x} alloy powders are closely connected with each other and form a porous structure. The single particle had columnar shape with a cubic crystal structure. YSZ electrolyte-supported planar single cell was fabricated using LSM as cathode and Ni_xCu_{1-x}–YSZ as anode and tested using methane as fuel. The single cell with Ni_xCu_{1-x} alloy anode had better catalytic activity. Power density of the single cell increased with the increase in the molar ratio of Ni/Cu, and $Ni_{0.8}Cu_{0.2}$ cell showed the highest value of 315 $mW \cdot cm^{-2}$ at 1073 K.

Keywords: SOFC; Ni-Cu alloy; hard template method; carbon deposition; ACF

1 INTRODUCTION

Solid Oxide Fuel Cells (SOFCs) are a promising high efficient power generation tool by using hydrocarbon fuel since it can directly convert the chemical energy to electric energy. In the present researches, perovskite [1, 2] and fluorite structures [3] are considered as good anode material, most of which require to add rare earth elements due to instability of the material. Nevertheless, as China is being continuously mined for rare earth, the amount of rare earth is decreasing and the price is rising. The price of a gram of perovskite and fluorite structure anode material with rare earth elements is much higher than a nickel based anode material in China. Therefore, Ni can be used for a long-time and on a large-scale for SOFC anode material. Ni-based anode materials are widely applied due to its relative stability, high catalytic activity, and low cost. When hydrocarbon fuel is used, Nickel has a strong catalytic activity so that they can catalytically break C-H bonds [4, 5].

It is reported that adding Cu to the Ni-based ceramic powders could improve the performance of SOFCs [6, 7]. As Cu has a poor catalytic performance for hydrocarbon fuel and can hinder the formation of C-C bonds. So, preparing the superior performance of Ni-Cu alloy as SOFCs anode materials becomes a research hotspot. The performance of Ni-Cu based anode should depend on the composition and microstructure. Conventional methods for preparing Ni-Cu based anode material include mechanical mixing method [8], sol-gel method [9] and chemical precipitation method [10]. In these methods, NiO-CuO mixture will be produced at first, followed by a reduction process in order to get Ni and Cu particle. Sometimes, Ni-Cu alloy phase will not be obtained after the reduction process [11, 12].

Hard template method is an effective way to control the microstructure of the material. Using this method, it is possible to get three-dimensional microstructure with a high specific surface area in the obtained material. Dong et al. [13] prepared a Sm0.5Sr0.5CoO3 (SSC) electrolyte material with a hollow fiber network structure using egg-shell-like film as the template. Compared with the cell using the same electrolyte material prepared by sol-gel method, the maximum power density increased by 44.5%. Pinedo et al. [14] prepared $Pr_{0.6}Sr_{0.4}Fe_{0.8}Co_{0.2}O_3$ cathode nanomaterial using thin polymer film as the template, and found that its specific resistance was as low as 0.12 $\Omega \cdot cm^2$.

Activated Carbon Fiber (ACF) with pore diameters in the range of 10~30 μm is an excellent adsorption carbon material with a high specific surface area, and thus, widely used in catalysis, medicine and other fields. It is expected to be used as a template to prepare electrode material with excellent performance. Especially, ACF has reductive properties like H_2; it is also expected to obtained metal alloy material directly if it is removed in an oxygen free environment.

In the present study, Ni-Cu alloy anode material with nano-porous three-dimensional microstructure was prepared using ACF as the hard template at different conditions. X-Ray Diffraction (XRD) and Scanning Electron Microscope (SEM) were used to characterize the obtained material. YSZ electrolyte-supported planar single cell was fabricated using LSM as cathode and Ni_xCu_{1-x} alloy as anode and tested using methane as a fuel at 1073 K.

2 EXPERIMENT SECTION

2.1 *Preparing of Ni-Cu anode material*

Sol-gel method was used to prepare $Ni_{0.8}Cu_{0.2}O_x$ anode material. The preparation method was the same as our pervious study [15]. Ni-Cu alloy material was prepared using ACF (fibrous, specific surface area of 1300 $m^2 \cdot g^{-1}$) as hard template. The ACF was dried at 393 K for 1 h prior to use. Saturated mixture solution of $Ni(NO_3)_2 \cdot 6H_2O$ (>99.9%) and $Cu(NO_3)_2 \cdot 3H_2O$ (>99.9%) with a calculated molar ratio of Ni/Cu that was prepared at first, then the dried ACF was added to the saturated solution on the basis of 0.2 g ACF per milliliter solution and stirred for 30 min. The obtained slurry was dried in an oven at 373 K for 5 h, and it was calcined in Ar gas flow at 1773 K for 2 h. Finally, Ni-Cu/C alloy powders were obtained.

2.2 *Fabrication of single cell and performance test*

Electrolyte-supported single cell was fabricated as follows: Electrolyte plate with a thickness of 0.5 mm and a diameter of 20 mm was prepared by pressing dried YSZ powders (8 mol% Y_2O_3-ZrO_2, Tosoh), followed by sintered at 1623 K in the air for 4 h. Both anode and cathode were prepared using slurry coating method. The slurry for the anode was prepared by mixing $Ni_{0.8}Cu_{0.2}O_x$-YSZ prepared with sol-gel method and Ni-Cu-YSZ alloy powders with 50 wt% of binder, respectively. The binder was composed of α-pine oil and ethyl cellulose with 4:1 mass ratio, and then added 10 wt% of flour as pore-forming material. The slurry for the cathode was prepared using the similar method as the anode slurry by mixing LSM ($La_{0.85}Sr_{0.15}MnO_3$, Kojundo) with the 50 wt% of binder. After the slurry for the anode was coated on YSZ electrolyte, it was dried and sintered at 1373 K in air. Thereafter, the cathode slurry was coated on YSZ electrolyte, and then sintered at 1173 K in air. The final effective area of the anode was 0.785 cm^2 with a thickness of 0.07 mm.

Before the performance test, the anode side was reduced using H_2 at first. 15 $cm^3 \cdot min^{-1}$ of CH_4 diluted by 15 $cm^3 \cdot min^{-1}$ of N_2 was used as the fuel. 30 $cm^3 \cdot min^{-1}$ of O_2 was provided in the cathode side. The performance test method was the same as our pervious study [16] and the test temperature was 1073 K.

2.3 *Physical characterizations*

D/max–2400 X-ray power diffraction was employed to analyze crystal phase and composition. JSM-5600 LV SEM was used to observe the morphology and microstructure of the sample before and after the test. CH604D electrochemical operating instrument was employed to measure the cell voltage, and the HV–151 potentiostat/constant current instrument was used to adjust the cell current.

3 RESULTS AND DISCUSSIONS

3.1 *Phase analysis*

Figure 1 shows the XRD patterns of Ni-Cu alloy powders with different Ni/Cu molar ratios prepared by the hard template method with a calcination temperature of 1773 K.

It can be seen that Ni-Cu alloys with cubic crystalline structure were well formed in all cases, indicating that ACF not only served as the template material but also as a reducing agent for the formation of metal alloy. Herein, it should be noted that the main peaks shifted a little with the increase in Ni/Cu molar ratio, indicating minor structure changes for different Ni/Cu molar ratios in the Ni-Cu alloys.

3.2 *Microscopic structure analysis*

Figure 2 shows the SEM image of Ni-Cu alloy powders prepared with AFC as hard template. The prepared Ni-Cu alloy material copied the

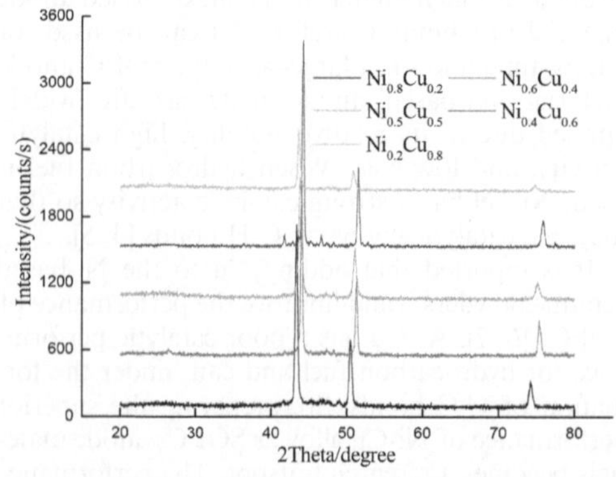

Figure 1. XRD patterns of Ni-Cu alloys with different Ni/Cu molar ratios calcined at 1773 K for 1 h.

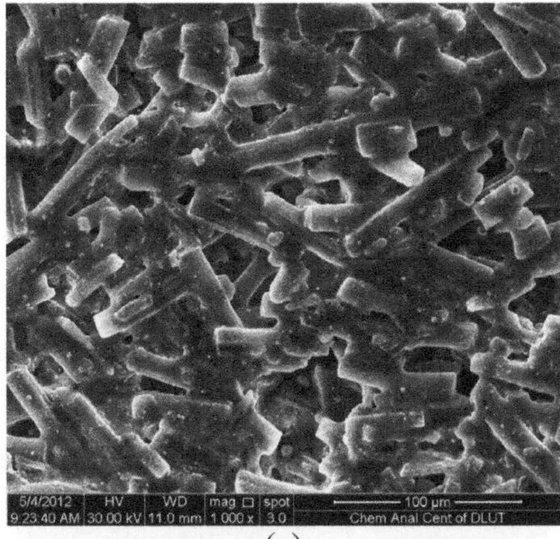

(a)

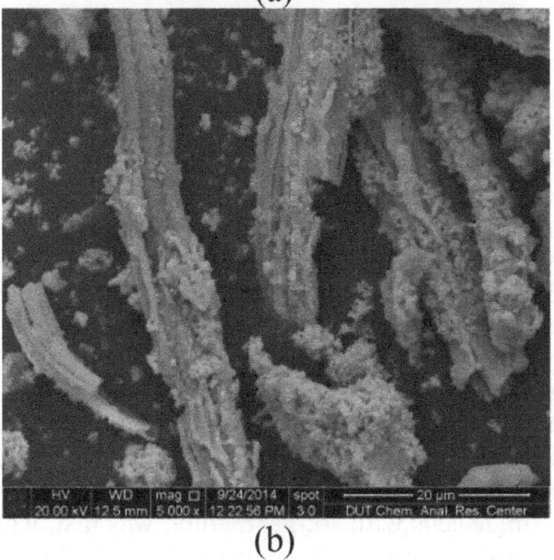

(b)

Figure 2. SEM images for Ni-Cu alloy calcined at 1773 K for 1 h.

Figure 3. SEM image for anode surface before reaction.

microstructure of AFC with fiber tubular structure. In Figure 2(b), cracked tubular structure and the fine particles existed, because Ni-Cu alloy material was calcined in high temperature and the fibrofelt was cut into small pieces. It should be noted that a porous structure was also formed in the powders, which both were a supportive skeleton, and helped to enhance the electronic conductivity.

Figure 3 shows the SEM sectional image of the cell after coating anode material. The connection was close between anode and electrolyte, which explained that the oxygen ions could move rapidly through the electrolyte and leads to a reaction at the anode. The anode sectional image shows Ni-Cu alloy of fiber tubular structure was full of the entire cell anode, which improved the electron continuous conductivity. Hence, the microstructure was able to effectively increase the contact surface of catalytic metals and electrolyte, which was also conducive to forming an independent conductive network with catalytic metals and electrolyte constituents, adding to the three phase boundary of the SOFC anode to improve the power generation performance.

Figure 4(a) shows the $Ni_{0.8}Cu_{0.2}O_x$-YSZ anode surface by SEM after the experiment, and indicates that the anode surface presents loose and porous structure after coating and sintering. There was no obvious agglomeration. Not only the contact area between particles is relatively small but also the three phase boundary is smaller. The cell with Ni_xCu_{1-x}-YSZ (x = 0.8, 0.5, 0.2) as anode has SEM on the anode surfaces after the experiment as shown in Figure 4(b). Comparing with Figure 2 before the experiment, the Figure 4(b) shows anode surface alloy materials with different column structures, which have been destroyed due to long time in high temperature operation. When the anode was sintered in the air environment, Ni_xCu_{1-x} alloy reacted with oxygen to generate nickel oxide and copper oxide, which leads to the volume of the micro tubular structure of the anode material increasing. While nickel oxide and copper oxide were reduced in the cell experiments, the volume of the tubular structure shrank and tubular structure was destroyed.

3.3 *The cell performance test*

With YSZ as electrolyte, different proportions of Ni-Cu alloys as anode material electrolyte supported cells were prepared. Figure 5 shows the current density-voltage (J-V) curve and current density-power density (J-P) curve drawn from cells with CH_4 as fuel at 1073 K. Figure 5 shows that with the increase of Ni content in alloy material, the power density of single cell increases. The open circuit voltage and maximum power density were

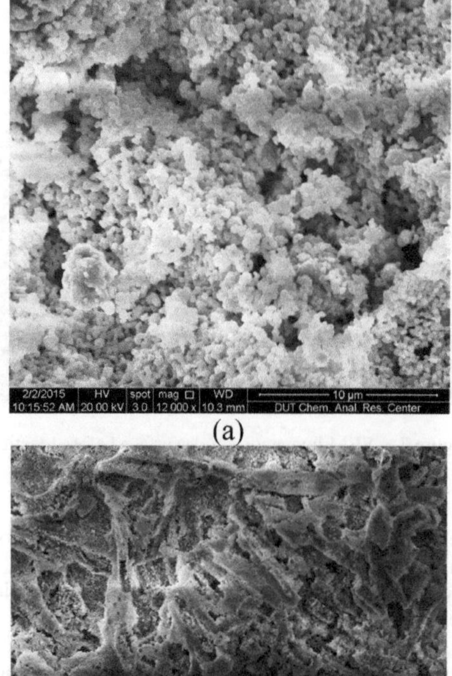

Figure 4. SEM images for anode surface after the experiment: (a) $Ni_{0.8}Cu_{0.2}O_x$-YSZ anode surface; (b) Ni_xCu_{1-x}-YSZ anode surface.

1.2005 V and 315 mW · cm^{-2}, respectively. Compared with NiO as anode material, the power density of $Ni_{0.8}Cu_{0.2}$ alloy as anode material was higher. In the Ni-Cu alloy, copper did not play a catalytic role and caused the decrease of Ni content in alloy material. However, compared to the pure Ni anode, Ni-Cu alloy anode was beneficial to increase the three phase boundary and was more conducive to the reaction for the fuel gas and the anode to improve cell performance, because alloy material had fiber tubular structure. Even if the Ni-Cu alloy anode has a certain content of Cu, the performance is still higher than that of the cell with Ni-YSZ as anode. Owing to the three-phase interface for reaction to increase, the decrease of Ni content is offset.

The fuel cell of $Ni_{0.8}Cu_{0.2}O_x$ anode by sol-gel method was tested with methane as fuel. Figure 5 shows the current density-voltage (J-V) curves and current density-power density (J-P) curves of $Ni_{0.8}Cu_{0.2}O_x$ as anode material with sol-gel method and $Ni_{0.8}Cu_{0.2}$ as anode material with the hard template method, at 1073 K. It can be seen from Figure 5 that the single cells were prepared with two methods that had a power generation performance. The open circuit voltage were 1.33 V and 1.20 V by using CH_4 as fuel, and the maximum power density were 70 mW · cm^{-2} and 315 mW · cm^{-2}, respectively. The open circuit voltage of single cells prepared by $Ni_{0.8}Cu_{0.2}O_x$ anode material with sol-gel method was greater than that with the hard template method. When the current density increased continuously, the voltage drop of single cells with sol-gel method was fast, which led to the power performance reducing. However, the power performance of single cells with the hard template method was superior. It is thus evident that nickel copper alloy in the anode formed porous structure (Fig. 2) and increased the anode porosity. So in these circumstances, the fuel gas can quickly spread to the anodic reaction region to increase reaction rate, and then power performance is improved.

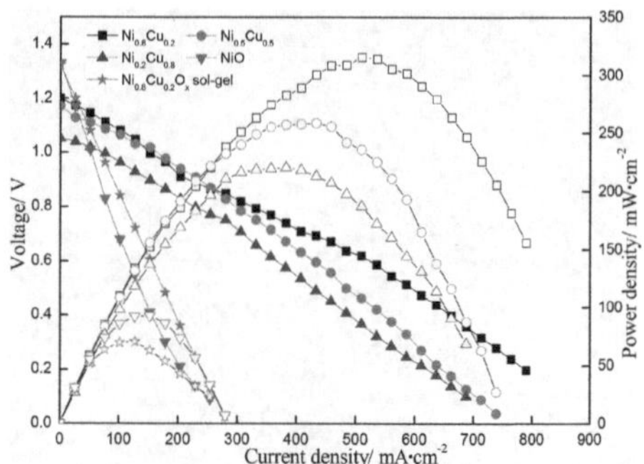

Figure 5. Current density and power density of the cells with different proportions of Ni-Cu alloy and $Ni_{0.8}Cu_{0.2}O_x$ with sol-gel method at 1073 K.

4 CONCLUSIONS

With activated carbon fiber as template, Ni_xCu_{1-x} alloy anode material had been successfully prepared in the experiment with the hard template, to make up the cell anodes. The generation performance test on the cell with Ni-Cu alloy as anode indicated that the anode alloy material forms complete cubic crystalline Ni-Cu alloy structure and the micro structure was fiber tubular. There were abundant pore structures without agglomerations. With the increase of nickel content of Ni_xCu_{1-x} alloy material, power density of single cell also increased. The maximum power density of a single cell with $Ni_{0.8}Cu_{0.2}$ alloy as anode material was 315 $mW \cdot cm^{-2}$. Under the same conditions, power performance of Ni_xCu_{1-x} anode material for the single cell was better than that of NiO anode material and $Ni_{0.8}Cu_{0.2}O_x$ anode material with sol-gel method due to the change of the anode microstructure.

The microstructure of Ni_xCu_{1-x} alloy was prepared by hard template method using activated carbon fiber, it confirmed that the porous structure was remained during the process of power generation, which helped to increase three-phase boundary and improve the performance of fuel cells. Therefore, the anode material prepared with the hard template method has high electrical performance, which is expected to be widely applied as SOFC anode material.

REFERENCES

[1] Dong D., et al. Eggshell membrane-templated synthesis of highly crystalline perovskite ceramics for solid oxide fuel cells [J]. Journal of Materials Chemistry, 2011, 21(4): 1028–32.

[2] Grgicak C.M., et al. Synergistic effects of $Ni_{1-x}Co_x$-YSZ and $Ni_{1-x}Cu_x$-YSZ alloyed cermet SOFC anodes for oxidation of hydrogen and methane fuels containing H_2S [J]. Journal of Power Sources, 2008, 183(1): 26–33.

[3] Jiang S.P., et al. GDC-Impregnated $(La_{0.75}Sr_{0.25})(Cr_{0.5}Mn_{0.5})$ O3 Anodes for Direct Utilization of Methane in Solid Oxide Fuel Cells [J]. Journal of the Electrochemical Society, 2006, 153(5): A850–A856.

[4] Jin C., et al. $La_{0.6}Sr_{1.4}MnO_4$ layered perovskite anode material for intermediate temperature solid oxide fuel cells [J]. Electrochemistry Communications, 2012, 14(1): 75–77.

[5] Jun J.H., et al. Mechanism of partial oxidation of methane over a nickel-calcium hydroxyapatite catalyst [J]. Applied Catalysis A: General, 2006, 312(0): 27–34.

[6] Liu M., et al. Performance of the nano-structured Cu–Ni (alloy) -CeO_2 anode for solid oxide fuel cells [J]. Journal of Power Sources, 2015, 274(0): 730–735.

[7] Pinedo R., et al. Synthesis of highly ordered three-dimensional nanostructures and the influence of the temperature on their application as solid oxide fuel cells cathodes [J]. Journal of Power Sources, 2011, 196(9): 4174–80.

[8] Restivo T.A.G., & de Mello-Castanho S.R.H., Nickel-Zirconia cermet processing by mechanical alloying for solid oxide fuel cell anodes [J]. Anglais, 2008, 185(2): 1262–6.

[9] Ringuedé A., et al. Electrochemical behaviour and degradation of (Ni,M)/YSZ cermet electrodes (M = Co, Cu, Fe) for high temperature applications of solid electrolytes [J]. Journal of the European Ceramic Society, 2004, 24(6): 1355–8.

[10] Triantafyllopoulos N.C., & Neophytides S.G., The nature and binding strength of carbon adspecies formed during the equilibrium dissociative adsorption of CH_4 on Ni–YSZ cermet catalysts [J]. Journal of Catalysis, 2003, 217(2): 324–33.

[11] Ueda W., et al. Nano-structuring of complex metal oxides for catalytic oxidation [J]. Catalysis Today, 2008, 132(1–4): 2–8.

[12] Wang J., et al. The basic principle of sol-gel method, development and application [J]. Chemical Industry and Engineering, 2009, 03: 273–277.

[13] Xie Z., et al. $Ni_{1-x}Cu_x$ alloy-based anodes for low-temperature solid oxide fuel cells with biomass-produced gas as fuel [J]. Journal of Power Sources, 2006, 161(2): 1056–1061.

[14] Ye X.F., et al. Improvement of Cu–CeO_2 anodes for SOFCs running on ethanol fuels [J]. Solid State Ionics, 2009, 180(2–3): 276–281.

[15] You H.X., et al.Comparison of $Ni_{0.8}Cu_{0.2}O_x$ anode material prepared with sol-gel method and $Ni_{0.8}$-$Cu_{0.2}$-coated YSZ composite anode material prepared with polyol Method [A]. 2013 International Conference on Materials Science, Machinery and Energy Engineering, MSMEE 2013, December 24, 2013–December 25, 2013[C], 2014.

[16] You H.X., et al. Reactions of Low and Middle Concentration Dry Methane over Ni/YSZ Anode of Solid Oxide Fuel Cell [J]. Journal of Power Sources, 2007, 165(2): 722–727.

Material Science and Environmental Engineering – Chen (Ed.)
 ISBN 978-1-138-02938-5

Texture and mechanical behavior of Mg-12Gd-3Y-0.5Zr alloy processed by ECAP

Y.J. Wu, X. Shen & J.W. Yang
Department of Mechanical Engineering, Nanjing Communications Institute of Technology, Nanjing, China

R. Zhu
Department of Materials Science and Engineering, Nanjing University of Science and Technology, Nanjing, China

ABSTRACT: Fine grained alloys of Mg-12Gd-3Y-0.5Zr in wt.% were obtained by an Equal Channel Angular Pressing (ECAP) in order to improve the mechanical properties of Mg alloys. Because magnesium alloy lacks sufficient ductility at low temperatures, the equal channel angular extrusion technique has been used at elevated temperature. Tensile tests were performed at room temperature at a strain rate of 5×10^{-3}/s. And the microstructures were examined by Optical Microscopy (OM), Transmission Electron Microscopy (TEM) and X-Rayed Diffraction (XRD). The warm processing temperature promotes recrystallization, which limits the capability of the ECAP technique in refining the grain size effectively. The texture has been determined experimentally for route Bc up to four passes, and it is found that the pole is at about 45° to the extrusion direction, which facilitates the activation of basal slip, and decreases the yield strength. The strength and ductility are determined by grain size and texture.

Keywords: magnesium alloy; equal channel angular pressing; thermomechanical processing; texture

1 INTRODUCTION

More attention has been devoted to the research and development of magnesium alloys in recent years as one of the effective measures to fight against energy consumption and environmental degradation. Mg alloys are especially attractive for aeronautical and automotive industry applications due to their low density and high special strength. However, the usage of Mg alloys is not as extensive as expected, due to some of their demerits (Bettles 2003), e.g. poor ductile, low strength; low creep resistance, as well as low corrosion resistance ascribed to their Hexagonal Close-Packed (HCP) crystal, which limited their practical application.

Improvement of its mechanical properties by grain refinement is one of the active research trends. Several research groups have achieved some promising results concerning improvement in the mechanical properties of Mg alloys by means of ECAP processing (Yang 2011, Ma 2009, Liu 2004), as one of the techniques of Several Plastic Deformation (SPD) (Valiev 2000), which can refine the microstructures of metals and alloys effectively. However, when the grain size of Mg alloys decreased to meso-scale, the elongation of most Mg alloys increased simultaneously, the yield strength decreased (Mehrotra 2006). Previous investigations indicated that the low value of yield strength in ECAP Mg alloys compared to extruded counterpart was due to the texture modification. Kim & Jeong (2005) showed that the most basal poles are close to 45° from the extrusion direction. As the Schmid factor on (0001) basal planes increases by the rotation of the basal poles to approximately 45° from the extrusion axis during ECAP, a lower stress is needed for yielding on the basal plane in the ECAPed Mg materials. Therefore, 45° textures will impair the strength effect by fine grains. Thus, it is essential to investigate carefully the textures in the hexagonal alloys after the ECAP processing and the relationships among the texture, microstructures and mechanical behavior in these alloys for understanding the related deformation process.

In this investigation, ECAP processing has been executed on the Mg-12Gd-3Y-0.5Zr alloy in order to improve its mechanical properties. The Mg-Gd-Y system alloy is one of the candidates for a novel Mg-based light hardenable alloy having a high creep resistance at elevated temperatures (He 2007, Peng 2007). However, this alloy lacks the strength and ductile need at room temperature for industrial applications: its yield strength in solution heat treatment about 200 MPa. The apparent effects of ECAP on microstructures and textures of the alloy are reported. And the reasons for the obvious increase in mechanical properties and the deformation mechanisms underlying the room temperature tensile tests are discussed based on the observed textures and microstructures.

2 EXPERIMENTAL PROCEDURES

As-cast Mg-12Gd-3Y-0.5Zr (wt.%) alloy is extruded into 30 mm diameter rods at a temperature of about 400°C with an extrusion ratio 11.6:1. The alloy was then solution treated at 500°C for 8 h and quenched in air immediately. The dimensions of the starting billets were 30 mm in diameter and 130 mm in length.

ECAP route Bc, which rotates the work piece 90° clockwise along its longitudinal axis between adjacent passes, was used to process the alloy billets. The die channel angle was 90°. The ECAP processing requires work piece to have certain ductility to prevent crack formation. However, this alloy lacks the strength and ductile at room temperature for industrial applications (He 2007). To ensure sufficient ductility, we carried out the ECAP processing of Mg-12Gd-3Y-0.5Zr alloy in a temperature of 370–400°C. The extrusion ram speed was approximately 15 mm/min.

Optical micrographs of the un-pressed and pressed rods were obtained after etched in a solution of 5 g picric acid +5 g acetic acid +100 ml ethyl alcohol before grinding and polishing mechanically. Transmission electron microscopy was used to observe the grain shape and microstructure. The foils were thinned by twin jet electro polishing working with a voltage 75 V and a current of 15 mA at a temperature of −30°C, using a solution of 2% HClO4 and 98% C2H5OH. A JEOL-2100 electron microscope operating at 200kV was used to examine the foils. Grain sizes were measured using the linear intercept method.

Crystallographic texture was determined by means of X-ray diffraction, using a Scintag XDS 2000 texture goniometer with CuKα radiation and a solid state high-purity germanium detector. A set of six incomplete pole figures was used to calculate the orientation distribution function, which allowed the followed recalculation of complete pole figures.

Dog-bone shaped flat samples with a gage dimension of 2 × 3 × 10 mm were cut from the longitudinal sections for tensile tests. Tensile testing of these samples was performed at room temperature at an initial strain rate of 5×10^{-3}/s. Yield strength, ultimate strength and elongation to failure was measured.

3 RESULTS AND DISCUSSION

Figure 1 shows the sample microstructures from the longitudinal sections of Mg-12Gd-3Y-0.5Zr alloy just after different ECAP passes at 400°C. In Figure 1a, the annealed sample consists of recrystallized fine and equiaxed grains with a mean grain size of approximately 17 μm, and the grain refinement effect of Zr is obvious. The dark spots in the microstructure are unsolved cuboid-shaped compound, Mg_5 (GdY), and most of the compound are distributed in the grain boundaries. As the number of ECAP passes increased, the grains are refined rapidly in the first 4 passes. And the Bc route after 4 passes yields an equiaxed grain structure. However, average grain sizes increase slightly in subsequent passes as dynamic recrystallization and grain growth allow the grains to grow. Koch (2003) has found that the grain size obtained through mechanical attrition is determined by the dynamic balance of defect creation and recovery/recrystallization during deformation. The same principle applies to the ECAP processing of Mg-12Gd-3Y-0.5Zr.

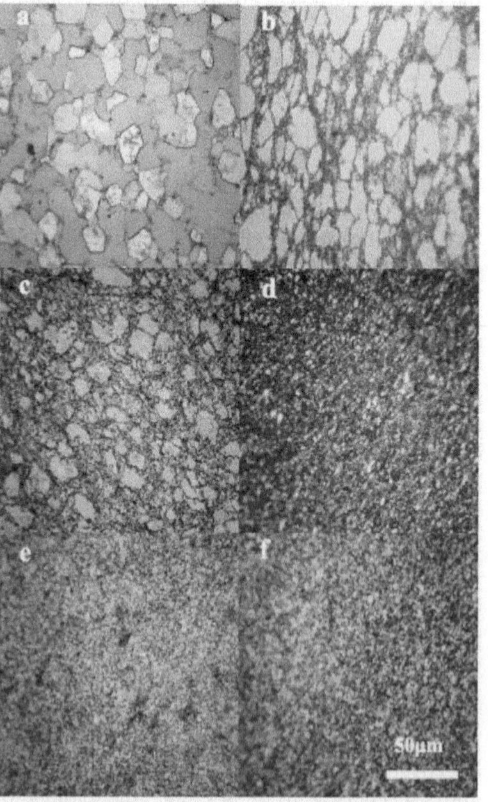

Figure 1. The optical microstructure of the ECAPed alloy for 0 p (a), 1 p (b), 2 p (c), 4 p (d), 6 p (e), 8 p (f).

That is to say, the grain refining due to the induced large strain is offset by the grain coarsening due to the dynamic recrystallization. Recrystallization is a thermally activated process, so higher processing temperature will exponentially increase the rate of dynamic recrystallization. This results in larger grains and a lower dislocation density, and consequently lower strength.

Figure 2 shows the TEM photographs of microstructures in the sample ECAPBc4. It is seen that complete recrystallization happens. Homogenous and equiaxial grains with mean size of 2.5 μm are obtained, and no obvious grain growth occurs. In Figure 2b, the arrow shows the apparent cross-slip of Mg alloys, which shows the activation of non-basal slip. Also, the twins are not observed in the samples. Koike & Kobayashi (2003) showed that the prismatic <a>, in addition to the basal <a> slip, is activated at room temperature. Chino & Kado (2008) showed that prismatic slip occurred at 83–298 K, whereas the pyramidal slip was activated at 423–559 K. Hence, the non-basal slip observed at 573 K in the Mg-12Gd-3Y-0.5Zr is probably the pyramidal slip. Also, the activation of non-basal slip and the dynamic recrystallization release the stress concentration, which suppress the activation of twinning.

Figure 3 gives the recalculated pole figures of solution-treated Mg-12Gd-3Y-0.5Zr alloys with its reflecting surface parallel to the extrusion direction. The iso-intensity contours are labeled as multiples of a random distribution (mrd). The deformation texture is distinct from other Mg alloys. The deformation texture is not very distinct from other Mg alloys. It shows the basal texture, basal plane (0001) parallel with extrusion direction, which means that the basal slip is the dominant deformation mechanism (Agnew 2001). However, the texture shows that the basal poles rotate away from the Transverse Direction (TD) towards the Rolling Direction (RD), and is more random than that of AZ31, which means that not only the basal slip and twinning, but also other deformation mechanisms such as non-basal slips are active for the Mg-12Gd-3Y-0.5Zr alloy (Agnew 2001).

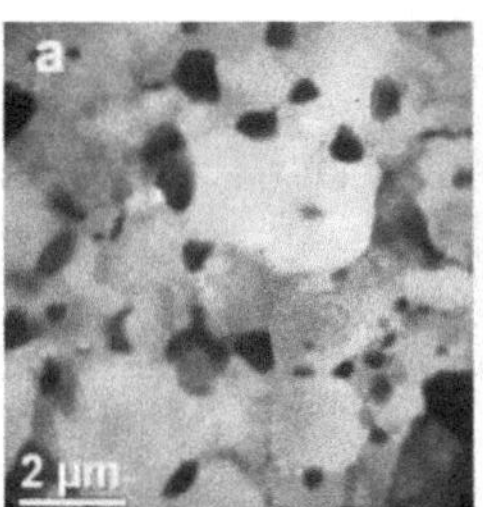

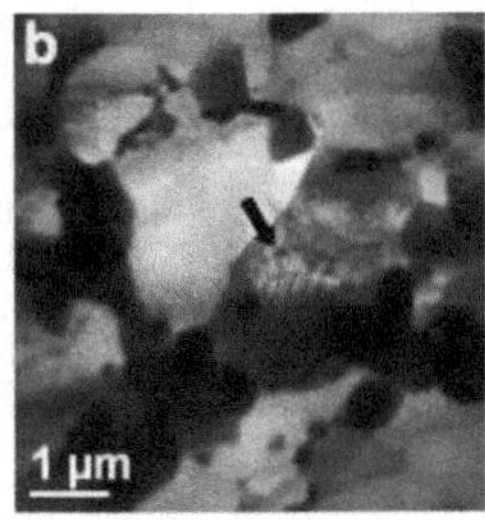

Figure 2. TEM micrographs from samples processed by ECAPBc4.

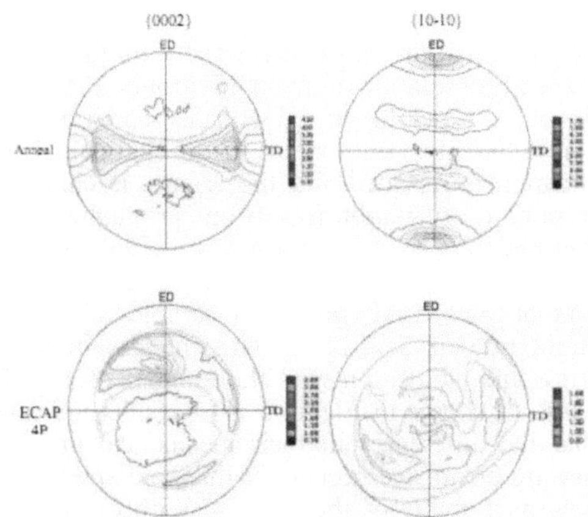

Figure 3. Basal, prismatic pole figures of Mg-12Gd-3-Y-0.5Zr alloy for extrusion, and ECAP Bc4. The reflecting surface is parallel to the extrusion axis.

Typical textures of extruded magnesium alloy, such as AZ31, have been reported in a number of related studies (Agnew 2001, Yi 2006, Barnett 2004, Jan 2007, Kleiner 2004). In all of these cases, a basal texture is presented. While the texture of Mg-12Gd-3Y-0.5Zr alloy shows a basal texture, it has a broader distribution of basal poles the RD, and is weaker than that of AZ31. The randomized texture has been associated with recrystallization (Jan 2007, Ball 1994). Typically, the recrystallization textures of Mg alloys are qualitatively similar to the deformation texture, but quantitatively weaker. And the crystallographic texture which results from recrystallization depends of the orientation of the nuclei on one hand and on any selective nature of the growth process on the other. The extrusion temperature at about 400°C in our experiment can lead to dynamic recrystallization. In Mg-12Gd-3Y-0.5Zr alloys, there are many unsolved particles, such as cuboid-shaped particles and zirconium cores, which distribute at the boundaries at high temperature. The residual particles can increase the driving force for recrystallization and act as nucleation sites by generating local in homogeneties in the strain energy and orientation (Ball 1994). This particle-stimulated nucleation can provide more randomly oriented nuclei and results in weaker recrystallization textures, which is exploited in the production of other metals, such as aluminium alloys (Bohlen 2007).

After ECAPBc4, the texture becomes weaker than that of conventional extrusion, and the main texture component is $(\bar{1}102)[1\bar{5}43]$ deduced from the measured pole figures and its corresponding OD sections. It can be seen that (0001) basal planes was rotated from parallel to the tensile axis to orientations inclined nearly 45° to the tensile axis. The deformation texture is derived from the initial texture by a rotation of about 45° around the shear direction, which lies at 45° with respect to the die axis in the intersection plane of the channels. This situation is repeated and the texture is changed from hard orientation to soft one gradually in Bc 4 passes.

Figure 4 summarizes the mechanical behaviors of the ECAP processed samples at initial strain rate of 5×10^{-3}/s. The changes in yield stress and elongation with the number of ECAP passes is shown in Figure 5. The results indicate that the differences in their mechanical properties are obvious, especially in the yield stress and deformed ability of these alloys. With decreasing of grain size of Mg-12Gd-3Y-0.5Zr alloy due to the ECAP processing, the yield stress decreases but the elongation increases. At ECAPBc6, the yield stress reached the maximum, 238 MPa. While the maximum elongation to fracture is about 14.4% at ECAPBc4. These results contradict to Hall-Petch relationship. One of reasons could be that there are different influences of grain size and texture on the mechanical properties. For example, the validities of the texture evolvement have been confirmed for numerous magnesium alloys from coarse-grained structure to about 1 μm grain size based on the ECAP technology (Su 2005).

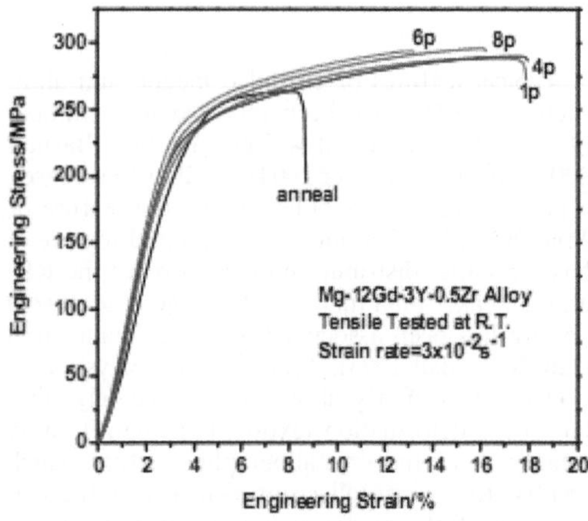

Figure 4. Tensile stress-strain curves after different ECAP passes.

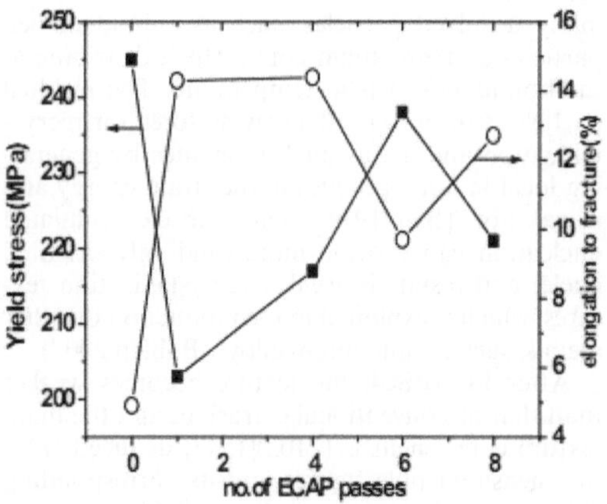

Figure 5. The variation of yield strength and.

4 CONCLUSIONS

In this paper, Equal Channel Angular Pressing (ECAP) has been used to refine the grain size of Mg-12Gd-3Y-0.5Zr billets at about 400°C because it lacks sufficient ductility at low temperatures. The warm processing temperature limits the capability of the ECAP technique in refining the grain size effectively. The texture formed at ECAP processing is about 45° to the extrusion direction, which facilitates the slip, and decreases the yield strength.

ACKNOWLEDGEMENTS

This research was supported by the senior personnel of scientific research funds of NJCI

REFERENCES

[1] Agnew, S.R. & Yoo, M.H. 2001. Application of texture simulation to understanding mechanical behavior of Mg and solid solution alloys containing Li or Y. *Acta Materialia.* 49:4277–4289.

[2] Ball, E.A. & Prangnell, P.B. 1994. Tensile-compressive yield asymmetries in high strength wrought magnesium alloys. *Scripta Metallurgical et Materialia.* 31(2):111–116

[3] Barnett, M.R. & Keshavarz, Z. 2004. Influence of grain size on the compressive deformation of wrought Mg-3 Al-1Zn. *Acta Materialia.* 52:5093–5103.

[4] Bettles, C.J. & Gibson, M.A. 2003. Microstructural design for enhanced elevated temperature properties in sand-castable magnesium alloys. *Advanced Engineering Materials.* 12:859–865.

[5] Bohlen, J. & Nurnberg, M.R. 2007. The texture and anisotropy of magnesium-zinc-rare earth sheets. *Acta Materialia.* 55:2101–2112.

[6] Chino, Y. & Kado, M. 2008. Compressive deformation behavior at room temperature –773 K in Mg-0.2 mass% (0.035at.%) Ce alloy. *Acta Materialia.* 56:387–394.

[7] He, S.M. & Zeng, X.Q. 2007. Microstructure and strengthening mechanism of high strength Mg-10Gd-2Y-0.5Zr alloy. *Journal of Alloys and Compounds.* 427:316–323.

[8] Jan, B. & Marcus, R.N. 2007. The texture and anisotropy of magnesium-zinc-rare earth sheets. *Acta Materialia.* 55:2101–2112.

[9] Kim, W.J. & Jeong, H.T. 2005. Grain-size strengthening in Equal-Channel-Angular-Pressing processed AZ31 Mg alloys with a constant texture. Materials Transactions. 46:251–258.

[10] Kleiner, S. & Uggowitzer, P.J. 2004. Mechanical anisotropy of extruded Mg-6%Al-1%Zn alloy. *Materials Science and Engineering A.* 379:258–263.

[11] Koch, C.C. 2003. Optimization of strength and ductility in nanocrystalline and ultrafine grained metals. *Scripta Materialia.* 49(7):657–662.

[12] Koike, J. & Kobayashi, T. 2003. The activity of non-basal slip systems and dynamic recovery at room temperature in fine-grained AZ31B magnesium alloys. *Acta Materialia.* 51:2055–2065.

[13] Liu, T. & Wang, Y.D. 2004. Texture and mechanical behavior of Mg-3.3%Li alloy after ECAP. *Scripta Materialia.* 51(11):1057–1061.

[14] Ma, A. & jiang, H. 2009. Improving both strength and ductility of a Mg alloy through a large number of ECAP passes. *Materials Science and Engineering A.* 513/514:122–127.

[15] Mehrotra, P. & Lillo, T.M. 2006. Ductility enhancement of a heat-treatable magnesium alloy. *Scripta Materialia.* 55(10):855–858.

[16] Peng, Q.M. & Wu, Y.M. 2007.Microtructures and properties of Mg-7Gd alloy containing Y. *Journal of Alloys and Compounds.* 430:252–256.

[17] Su, C.W. & Chua, B.W. 2005. Properties of severely plastically deformed Mg alloys. *Materials Science and Engineering A.* 402:163–169.

[18] Valiev, R.Z. & Islamgaliev, R.K. 2000. Bulk nanostructural materials from severe plastic deformation. *Progress in Materials Science.* 45(2):103–189.

[19] Yang, H.J. & An, X.H. 2011. Enhancing strength and ductility of Mg-12Gd-3Y-0.5Zr alloy by forming a bi-ultrafine microstructure. *Materials Science and Engineering A.* 528(13/14):4300–4311.

[20] Yi, S.B. & Davies, C. 2006. Deformation and texture evolution in AZ31 magnesium alloy during uniaxial loading. *Acta Materialia.* 54:549–562.

Material Science and Environmental Engineering – Chen (Ed.)
© 2016 Taylor & Francis Group, London, ISBN 978-1-138-02938-5

Effect of Sn on thermal stability and Glass Forming Ability of Fe-based amorphous alloy

Q. Fu, M.X. Fu, C. Hu, C. Wang & T. Li
College of Material Science and Engineering, Jiangsu University, Zhenjiang, China

ABSTRACT: In this work, the glass transition temperature T_g, the crystallization temperature T_x, and the peak temperature of crystallization T_p of $Fe_{68}Ni_1Al_5Ga_2P_{9.65}B_{4.6}Si_3C_{6.75}$ and $Fe_{67.5}Sn_{0.5}Ni_1Al_5Ga_2P_{9.65}B_{4.6}Si_3C_{6.75}$ were measured by DSC. According to the Kissinger formula, the crystallization activation energies were calculated by characteristic temperatures to analyze the effect of Sn on the thermal stability of the Fe-based amorphous alloy. The research shows that the thermal stability decreases in accordance with the crystallization activation energies of Fe-based amorphous alloy that reduces with the addition of Sn. Then, by using the method of the least squares fitting the linear equation, the results by relationship curves of T_g, T_x, T_p and $\ln(\beta)$ indicate that the glass forming ability enhances with the addition of Sn.

Keywords: thermal stability; glass forming ability; amorphous alloy

1 INTRODUCTION

Fe-based amorphous alloy has attracted majority of the attention of researchers with excellent properties. The technique to seek large Glass Forming Ability (GFA) and thermal stability is one of the research hot spots [1–3]. The study by Xu Min [4] showed that a small amount of Nb can effectively improve the thermal stability and the glass forming ability of the Fe-based amorphous alloy. Fu Mingxi [5–6] researched that moderate amount of Ni (less than 4%) is conducive to improve the glass forming ability and the thermal stability of the Fe-based amorphous alloy, and proper amount of Zr (almost 2%) will make Fe-based amorphous alloy having a strong glass forming ability. Sn can make the atomic arrangement more chaotic and complex because it has the effect of solid solution in the interior of amorphous alloy [7]. Therefore, it is often added to many amorphous alloy systems by researchers. In this paper, a small amount of Sn is added to the FeNiAlGaPBSiC alloy to discuss the influence of Sn on the formation and thermal stability of alloy systems under different heating rates.

2 SPECIMEN PREPARATION AND TEST METHOD

2.1 *Preparation*

$Fe_{68}Ni_1Al_5Ga_2P_{9.65}B_{4.6}Si_3C_{6.75}$($Fe_{67.5}Sn_{0.5}Ni_1Al_5Ga_2P_{9.65}B_{4.6}Si_3C_{6.75}$): This experiment puts pure Fe, pure Ni, pure Al, pure Ga, P-Fe, B-Fe, Si-Fe and high purity graphite particles (adding pure Sn), which had been prepared in high vacuum tungsten arc melting furnace to prepare master alloy. Alloy needs to be melted for 5~6 times repeatedly with the current from 400 A to 500 A in order to ensure uniform melting. The single roller melt spinning method was used to lining-up amorphous strips with the width of 3 mm and the thickness of 20 um. Aperture of Quartz Nozzle is 3 mm × 0.3 mm, spray pressure of Ar is 4×10^3 Pa and speed of copper roller is 30 m/s, respectively. The sample was made a differential thermal analysis by Germany Nietzsche instruments like Nietzsche DSC404 differential scanning calorimetry that drew its characteristic temperatures from it.

3 EXPERIMENTS AND RESULTS

Figure 1 is a DSC curve graph of amorphous alloy $Fe_{68}Ni_1A_{15}Ga_2P_{9.65}B_{4.6}Si_3C_{6.75}$(①) and $Fe_{67.5}Ni_1A_{15}Ga_2P_{9.65}B_{4.6}Si_3C_{6.75}Sn_{0.5}$ (②) at different heating rates. Table 1 shows the characteristic temperatures of Fe-based amorphous alloy, the glass transition temperature T_g; the crystallization temperature T_x and the crystallization peak temperature T_{p1} and T_{p2}, which are measured according to DSC curve. It can also be seen that T_g, T_x and T_p of alloys move to the high-temperature region with the increasing heating rate, which shows that the alloy glass transition and crystallization have a significant kinetic effect.

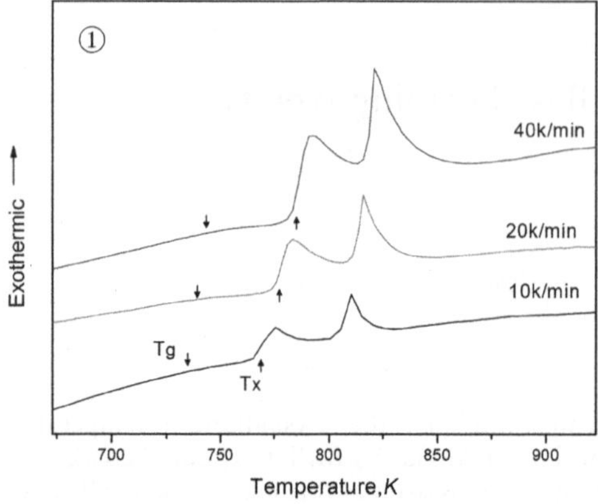

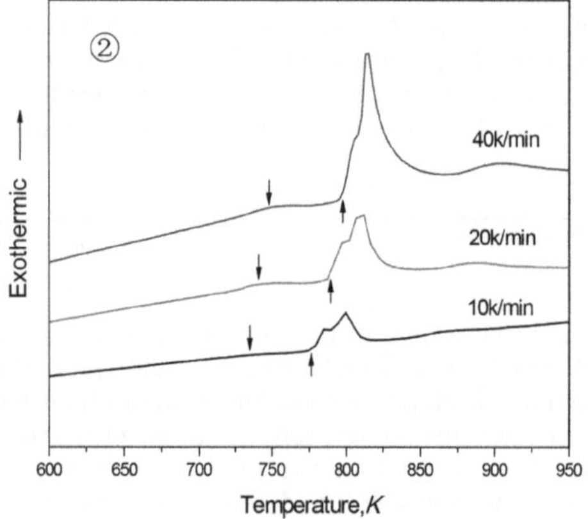

Figure 1. DSC curves of amorphous alloys continuously heated at different heating rates.

Table 1. Thermal parameters of amorphous alloys at different heating rates.

Alloys	B (K/min)	T_g (K)	T_x (K)	T_{p1} (K)	T_{p2} (K)	ΔTx
①	10	735.8	768.4	775.5	810.5	32.6
	20	737.2	777.8	783.6	816.1	40.6
	40	743.5	786.4	790.9	820.9	42.9
②	10	734.4	775.3	799.3	–	40.9
	20	742.7	787	811.9	–	44.3
	40	747.9	796.4	815.4	–	48.5

The crystallization activation energy E reflects the thermal stability of amorphous alloy more essentially. The required energy accretes with the increase of the activation energy E, and the alloy is more stable. In contrast, the smaller the activation energy E is easier to crystallize and more unstable than the amorphous alloy [8]. The activation energy for continuously heating is also known as non-isothermal crystallization activation energy that can be calculated by the Kissinger methods [9]:

$$\ln\frac{T^2}{\beta} = \frac{\Delta E}{RT} + \ln\frac{\Delta E}{R\nu_0}$$

In the formula, β (Unit K/min) is the heating rate, T is the characteristic temperature of DSC, which may be T_g, T_x or T_p (Unit K), ΔE is the activation energy, R is the gas constant, and frequency factor ν_0 is obtained by the slope and intercept. A straight line with a slope of–E/R can be obtained by drawing the curve of the $\ln(\beta/T^2)$ and 1/T at different heating rates. The activation energies of the crystallization process of amorphous alloy can be got by the multiplying slope of the lines by the gas constant R. Figure 2 shows the relation curves of the alloy $Fe_{68}Ni_1Al_5Ga_2P_{9.65}B_{4.6}Si_3C_{6.75}$ and $Fe_{67.5}Ni_1Al_5Ga_2P_{9.65}B_{4.6}Si_3C_{6.75}Sn_{0.5}$ with 1000/T.

The Kissinger curves shown in Figure 2 have a good linear relationship in the continuous heating

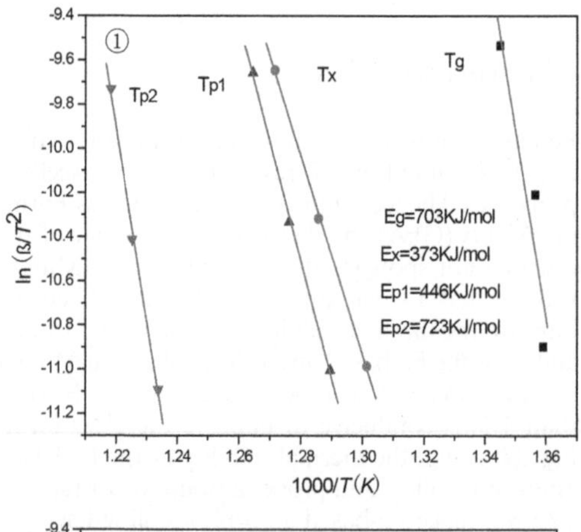

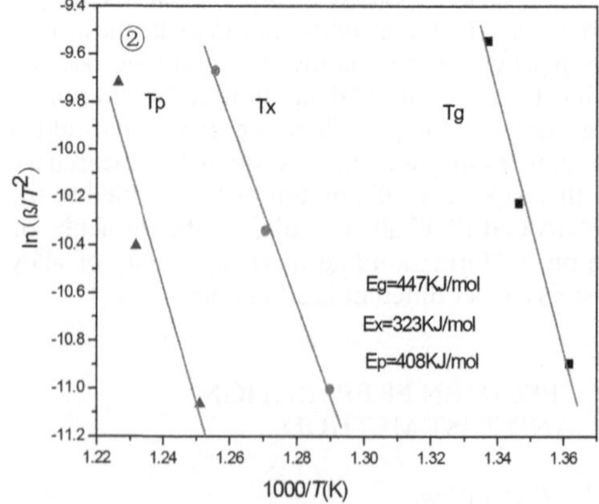

Figure 2. Kissinger plots obtained from continuously heating for amorphous alloys.

Table 2. Activation energy of amorphous alloys at different heating rates.

Alloys	E_g	E_x	E_{p1}	E_{p2}
①	703	374	446	723
②	447	323	408	–

process of amorphous alloys. The activation energies of two kinds of Fe-based amorphous alloys at different heating rates can be obtained by calculating.

The activation energy values corresponding to characteristic temperatures are shown in Table 2. The initial crystallization temperature is associated with the crystallization nucleation, and peak temperature relates to the growth of the crystals [10]. E_{p1} is greater than E_x, which shows that the first that precipitates crystal growth is more difficult than the nucleation process. However, the E_{p1} is much smaller than E_{p2} which shows that the rate of the first that precipitates crystal growth is much faster than the latter that precipitates. As can be seen from Table 2, E_{p1} of alloy ① and alloy ② are both greater than E_x, which indicates that the crystal growth is more difficult than its nucleation in the first precipitates during the crystallization process of the series of alloys. By comparing E_{p1} between the alloy ① and alloy ②, it can be seen that E_{p1} of alloy with an addition of Sn decreases from 446 KJ/mol to 408 KJ/mol before adding, so the first precipitates crystal growth adding element Sn is relatively easier than that before adding, and the growth rate is faster than added before.

The activation energies corresponding to each characteristic temperature of amorphous alloys with an addition of Sn are smaller than that before adding. It indicates that the thermal stability of alloy decreases with the addition of trace Sn.

The experimental results show that T_g, T_x, T_{p1}, and T_{p2} of alloy ① and alloy ② accrete with the increase of the heating rate β of DSC, which indicates that both the glass transition and the crystallization process of these alloys have a dynamic effect. Figure 3 shows the relationship curves of the T_g, T_x, T_{p1}, and T_{p2} versus ln (β) in the two kinds of Fe-based amorphous alloys. It is shown that T_g, T_x, T_{p1}, and T_{p2} versus ln (β) in the series of amorphous alloys have a good linear relationship. By using the method of the least squares fitting linear equation, the relationship of T_g, T_x, T_{p1}, and T_{p2} versus ln (β) can be expressed as: $T = A_T + B_T \ln(\beta)$, in which A_T and B_T are constants [11].

Table 3 lists the values of alloy A_T and B_T. It can be seen from Table 3 that B_T values of T_x, T_{p1}, and T_{p2} of alloy are greater than the B_T values of T_g, which illustrates that the dependence on the heating rate for crystallization of these alloys is greater than the glass transition, namely the dynamic effect of crystallization is more obvious than that of glass transition.

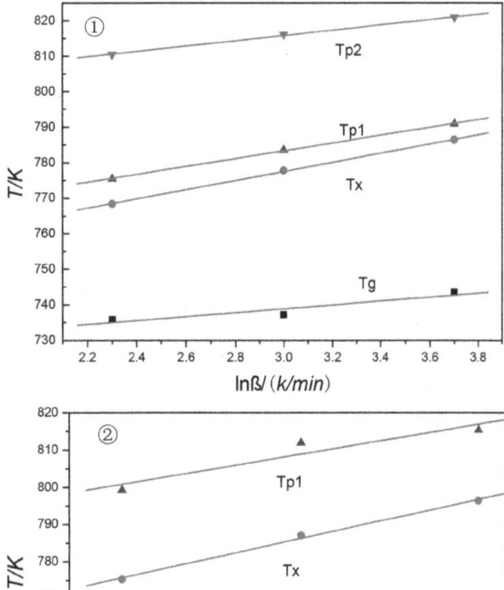

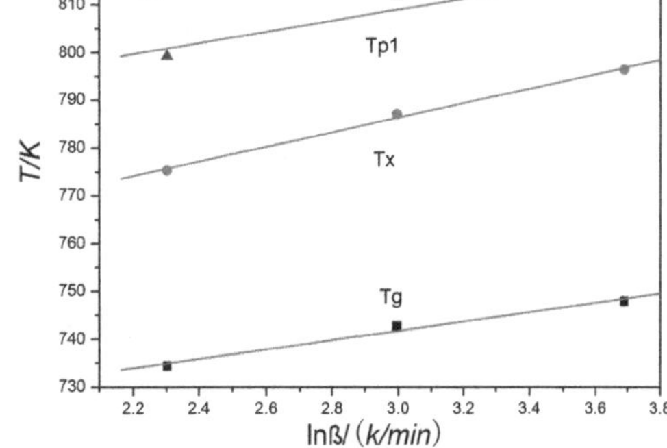

Figure 3. Plots of T_g, T_x and T_{p1} versus ln(β) for ① and ② amorphous alloy.

Table 3. Values of A_T and B_T for ① and ② amorphous alloy.

Alloys	Values	T_g	T_x	T_{p1}	T_{p2}
①	A_T	722.3	738.9	750.3	793.5
	B_T	5.5	12.9	11.1	7.4
②	A_T	712.5	740.6	774.1	–
	B_T	9.7	15.2	11.6	–

The B_T values corresponding to T_g values reflect the strength of the Glass Forming Ability (GFA). Compared with B_T values corresponding to T_g values of the two kinds of amorphous alloys in Table 3, it can be seen that the GFA of the alloy ② is stronger than that of the alloy ①. At the same time, the difference between the T_x and T_g (i.e. $\Delta Tx = T_x - T_g$), supercooled liquid region ΔTx [12], characterize the stability of supercooled liquid, usually as a criterion of metallic glass GFA.

Inoue [13–15] et al. considered that bigger ΔTx means stronger GFA of metallic glass system. In Table 1, it can be seen that all ΔTx values of Fe-based amorphous alloy with addition of trace element Sn increase under different heating rates, which also reflects that the GFA of alloy ② is stronger than that of alloy ①. It also can be seen from DSC curves of alloy in Figure 1, the alloy has two crystallization peaks before joining Sn, namely the alloy is of two-stage crystallization mechanism, but after joining Sn, the first crystallization peak of alloy disappeared and only one peak of crystallization is left, which illustrates that the alloy changes from two-stage crystallization mechanism to one-stage crystallization mechanism. It can be concluded that after adding element Sn the alloy composition becomes close to eutectic composition, which also indirectly indicates that the addition of element Sn improves the GFA of the alloy.

4 ANALYSIS

According to the experimental results, the activation energy of amorphous alloys under each characteristic temperature is less than that before the addition of a small amount of Sn, which indicates that the thermal stability of alloy decreases with adding trace elements of Sn. The reasons are as follows:

First, the mixing heat of Sn and other components are mostly positive values, which is not beneficial to the formation of multi-element bond pairs. So it may make long-range atomic diffusion occur easily and reduce the stability of the alloy.

Further, like the fact that the atomic radius of Sn atom is larger, atoms with smaller atomic radius may be attached to the gaps between the Sn atoms, making the atomic cluster with a larger size in the alloy retain in the chilled microstructure. The existence of larger atomic clusters may decrease the nucleation activation energy in the crystallization process so as to make it easier to crystallize, which means that it reduces the thermodynamic stability of alloys.

Lastly, due to the segregation and first melting function of Sn. The melting point of Sn is relatively low. Owing to concentration fluctuations in the process of cooling and exclusion by other alloy elements with high melting points to form the short program, it makes the neighboring Sn atoms cluster in the alloy. In the subsequent heating process, Sn-Sn metal bond fracture occurs first, similar to the effect of melting, to reduce viscosity and increase other atomic activities, thus to make the crystallization activation energy low.

At the same time, it can be concluded that the ΔTx values of amorphous alloys increase with a small amount addition of Sn by analyzing experimental data. B_T values corresponding to T_g values also increase. In addition, the alloy is turned from two-stage crystallization mechanism into one-stage crystallization mechanism, which shows that the GFA of the alloy has improved.

The function of solid solution of Sn makes arrangement of atoms in amorphous alloy more *chaotic* and complex, leading to atom diffusion resistance increased and some intermetallic compounds restrained to separate out. Then, it can improve the glass forming ability of amorphous alloy.

5 CONCLUSIONS

T_g, T_x, T_p, and ΔTx values in supercooled liquid region increase with the growth of heating rate, which indicates that the glass transition and crystallization both have a significant kinetic effect. Simultaneously, the kinetic effect of crystallization is more obvious than that of glass transformation.

The adjunction of Sn in alloy system of FeNiAlGaPBSiC may make T_g, T_x, and T_p move to high-temperature region. The ΔTx values increase. Activation energy reduces, and the thermal stability of the alloy decreases. Two-stage crystallization changes into one-stage crystallization, and the ability of glass transition enhances.

REFERENCES

[1] Hoque M.S, Hakim Ma, Khan Fa, et al. Ultra-soft magnetic properties of devitrified $Fe_{75.5}Cu_{0.6}Nb_{2.4}Si_{13}B_{8.5}$ alloy [J]. Materials Chemistry and Physics, 2007, 101(1):112–117.

[2] Szewieczek D, Baron A. Electrochemical corrosion properties of amorphous $Fe_{78}Si_{13}B_9$ alloy [J]. Journal of Materials Processing Technology, 2004, 157–158:442–445.

[3] Xiao Huaxing, Chen Guang, La Peiqing. Study on tribological properties of Fe-based bulk amorphous alloy [J]. Tribology, 2006, 26(2):140–144.

[4] Xu Min, Quan Minxiu, Hu Zhuangqi. Multicomponent fe-based amorphous alloys with wide supercooled liquid region [J]. Acta Metallurgica Sinica, 2001, 37(6).

[5] Fu Mingxi, Li Yan, Cha Yanqing et al. Effects of Cu and Ni on Glass Forming Ability and Thermal Stability of Fe-based Alloys [J]. Special Casting & Nonferrous Alloys, 2006, 26(4).

[6] Fu Mingxi, Zhao Jiang, Zhou Hongjun et al. Effects of Zr on Glass Forming Ability and Thermal Stability of Fe-Al-Ga-P-C-B-Si Amorphous Alloy [J]. Special Casting & Nonferrous Alloys, 2007, 27(7).

[7] Zhao Jiang, Fu Mingxi, Wang Fangfang et al. Effects of Sn on Glass Forming Ability (GFA) and Magnetic Performance of Multi-component Fe-based Amorphous Alloys [J]. Special Casting & Nonferrous Alloys, 2007, 27(11).

[8] Xing Dawei. The formation and crystallization kinetics of ZrCuNiAlTi amorphous alloy [D]. Doctoral Dissertation of Harbin Institute of Technology, 2003.
[9] Gun B., Laws K.J., Ferry M. Static and dynamic crystallization in Mg–Cu–Y bulk metallic glass [J]. Non-Crystalline Solids, 2006, 35:3887~3895.
[10] Wang Fangfang, Structure and thermodynamic parameters of Fe based amorphous alloy [D]. Master Dissertation of Jiangsu University, 2008.
[11] Zhao Deqian, C.H. Shek, Wang Weihua. Crystallization and kinetics of Zr-Ti-Cu-Ni-Be-Fe bulk metallic glass [J]. Acta Metallurgica Sinica. 2001, (37), 7:754–758.
[12] Y Li, S.C. Ng, C.K. Ong, H.H. Hng et al. Glass forming ability of bulk glass forming alloys [J]. Scripta Materialia, 1997, 37(7):783–787.
[13] Inoue A, Gook J.S. Fe-based ferromagnetic glassy alloys with wide supercooled liquid region [J]. Mater. Trans. JIM, 1995, 36:1180~1183.
[14] Inoue A. Bulk amorphous alloys with soft and hard magnetic properties [J]. Mater. Sci. Eng. A, 1997: 226~228, 357~363.
[15] Shen T.D, Schwarz R.B. Bulk ferromagnetic glasses prepared by flux melting and water quenching [J]. Appl Phys Lett, 1999:75, 49~51.

Material Science and Environmental Engineering – Chen (Ed.)

Effect of laser processing parameters on the quality of Ni60+Ni/MoS_2 laser clad layer on titanium alloy surface

M. Fan
College of Science, Tianjin Polytechnic University, Tianjin, China

R.L. Sun
School of Mechanical Engineering, Tianjin Polytechnic University, Tianjin, China

ABSTRACT: Laser cladding of Ni60/Ni/MoS_2 powders on Ti-6Al-4V alloy substrate were performed. The impact of the laser processing parameters was discussed. Then the effects of different quality of cladding on different dilution rates were evaluated. Results show that with the increase of the scanning speed, the surface of titanium alloy by laser cladding layer width, height, and the penetration depth in basal were reduced. The laser energy density had a great influence on the microstructure of the cladding layer surface, the laser energy density is different, the microstructure of the laser cladding layer has much difference.

Keywords: laser cladding; dilution rate; nickel base; MoS_2

1 INTRODUCTION

Laser cladding technology has been successfully used in wear and corrosion resistance, oxidation resistance, and thermal barrier function of cladding layer. It shows the especial process advantages and attractive application prospect. Titanium alloy has high specific strength, high specific modulus, high corrosion resistance and good oxidation resistance and high temperature performance etc. It is widely used in aviation, aerospace, petrochemical and other fields. But its poor wear resistance and friction coefficient limit its scope of application in a certain extent. The surface of titanium alloy by laser cladding is one of the effective means to improve the wear resistance of titanium alloy surface to reduce the friction coefficient [1~4]. In this paper, it changed the melting degree of substrate of titanium alloy in laser cladding process by changing the laser processing parameters (scanning speed) way.

2 EXPERIMENTAL

2.1 *Experimental material*

TC4 (Ti-6Al-4V) alloy was used as the substrate material in the test. The composite powder of TC4 matrix consisted of Al, V, Fe, S, C, O, N, and Ti by atomic ratio of 6.01Al, 3.84V, 0.30Fe, 0.15S, 0.10C, 0.15O, and 0.15N and balance Ti. The size of samples was 60 mm × 30 mm × 10 mm. The clad materials were NiCrBSi and Ni/MoS_2 (see Table 1) (The powder was Ni and MoS_2. The volume fraction of Ni was 75% the volume fraction of MoS_2 was 25%). The weight fractions of Ni/MoS_2 in the clad materials were fixed at 0, 5, 10, and 20 wt%, respectively. The size of powders were in the range of 75 μm. Clad powers were per-placed on the surface of substrates using an organic binder, and their thickness was 1.0 mm.

2.2 *Test equipment*

NCLT CW-1000 YAG laser was used, providing a beam that was directed onto the sample surface. The parameters of laser processing were selected as: the power P = 800~900 W, the beam diameter

Table 1. Chemical composition and phase transformation temperatures.

Material	Mass fraction (%)	Melting point (°C)	Density ($g \cdot cm^{-3}$)	Size (μm)
Ni60	16% Cr, 3.5% B, 4.5% Si, 0.8% C, Fe ≤ 15%, Ni: Bal	980–1050	8.4	50–100
MoS_2	Purity ≥ 99.9%	1185	4.8	<74
Ni	Purity ≥ 99.9%	1453	8.902	<74

D ≈ 3 mm and the scanning speed V = 2~4 mm/s. The laser cladding was carried out in argon.

The microstructure and phase constitutions of laser clad layers were examined using Quanta 200 scanning electron microscope and Olympus GX-51 optical microscope. The microhardness of the clad layers was measured by an HXD-1000T type electronic microhardomere with a load of 0.5 kg and a dwell time of 10.

3 RESULT AND DISCUSSION

3.1 *Influence of laser power on the cladding layer*

The laser power has a direct impact on an important indicator of the cladding quality. When the output power of the laser is larger, the more powder is melted in the cladding layer. The probability of bubble, the substrate temperature, the deformation of the cladding layer, and the cracking of the cladding layer is greater. When the power is low, the cladding layer of powder can't be completely melted. With lubrication of melt and the matrix in the molten pool decreased, surface tension was too large. It leads to melt condensation, which is like the tears.

Figure 1 showed that in the three kinds of laser power, the cladding channels were smooth and continuous. But there were corrugation. The direction of corrugated and scanning is consistent. With the increase of scanning speed the surface of the cladding layer was smoother and then the spherical splash gradually reduced. As Figure 1(b) showed, when the layer power was P = 900 W and V = 3 mm/s, there were less spatter and the corrugated gap surface melting was smaller. The cladding layer quality is the best.

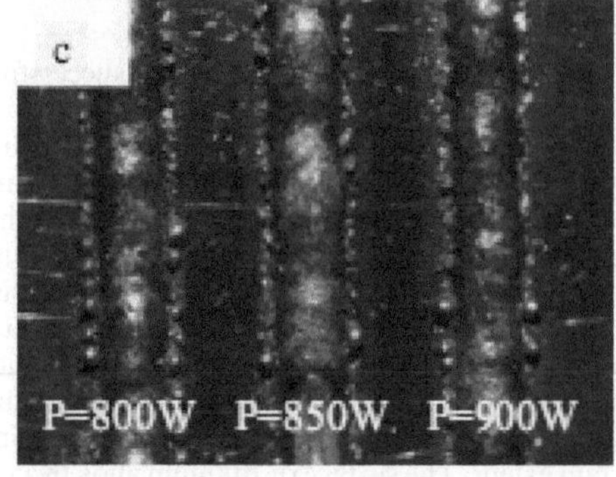

Figure 1. The morphology of cladding surface in different laser power. (a) V = 2 mm/s, (b) V = 3 mm/s, (c) V = 4 mm/s.

3.2 *Influence of scanning speed on the cladding layer*

Figure 2 showed that the scanning speed was faster and the surface of the cladding layer was smoother. Corrugated gap was small. Spherical splash formation gradually decreased. The increase of the scanning speed had great effect on the improvement of the cladding layer surface morphology. Scanning speed determined the time of laser irradiation. The slower rate of laser molten pool indicated the more time existed and the more amount of substrate melted. Cladding layer height increased with the decreasing scanning speed. The amount of melting base materials decreased with increasing scanning speed. With the decreasing scan rate, the rate of dilution of cladding layer increased.

At present, there are two methods of dilution rate calculation (see Equation 1 and Equation 2). Where A_1 = substrate melting section area, A_2 = cladding layer cross section area in Equation 1. Where η = dilution rate, ρ_p = cladding layer of alloy density, ρ_s = substrate density, X_p, X_s, X_{p+s} are the weight percentages of X elements (wt-%) in the alloy powder, the substrate alloy and the cladding layer are seen in Equation 2. Because of the laser beam spot diameter, the width of cladding layer with laser process parameters remains unchanged. Through experiments, it can be used in Equation 3. That the dilution rate of cladding layer. As shown in Figure 3, Where h = penetration depth of the substrate and H = Cladding layer height [5].

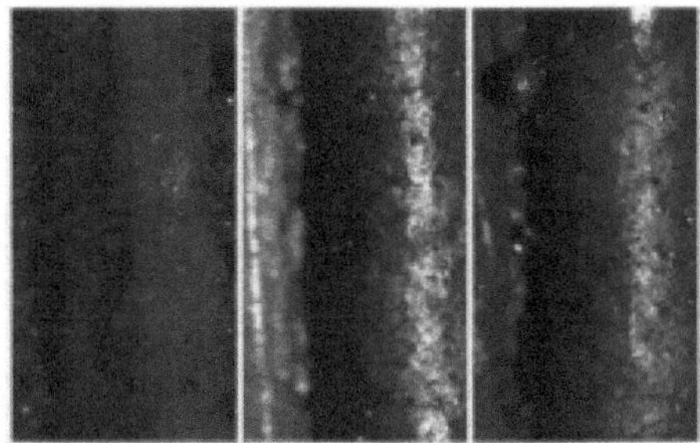

Figure 2. The morphology of cladding surface in different scanning speed (P = 900 W, D = 3.0 mm).

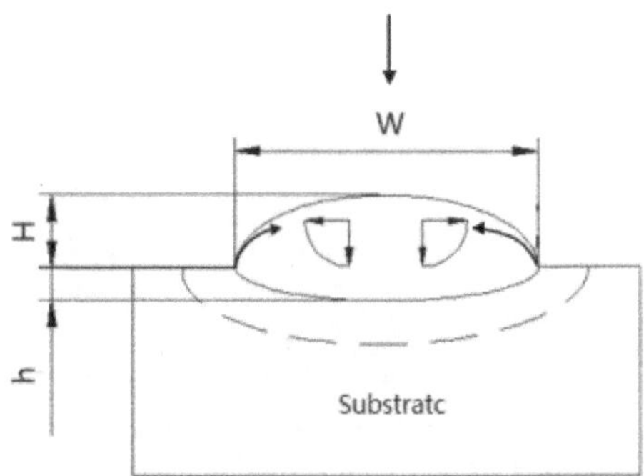

Figure 3. Geometry of the laser clad layer and the bath convection [6].

$$\eta = \frac{A_1}{A_1 + A_2} \tag{1}$$

$$\eta = \frac{\rho_P(X_{P+S} - X_P)}{\rho_s(X_S - X_{P+n}) + \rho_P(X_{P+n} - X_P)} \tag{2}$$

$$\eta = \frac{h}{H + h} \tag{3}$$

Figure 4(a) showed that the cross section morphology of laser cladding layer, the clad layer was divided into three redions: the Ni60/Ni/MoS_2 alloy Clad Zone (CZ), the Dilution Zone (DZ) and the Heat-Affected Zone (HAZ) in the titanium alloy [7]. The matrix materials have melting and the cladding layer and the substrate material formed a good metallurgical bonding. Figure 4(b), (c) and (d) showed that there are pores, cracks, and other defects are less in the laser cladding layer.

As shown in Figure 5, when other laser parameters maintain a constant, scanning speed determines the powder and the substrate by laser irradiation time. With scanning velocity increasing, the laser irradiation time was shorter and the preset powder absorbed fewer calories. The width

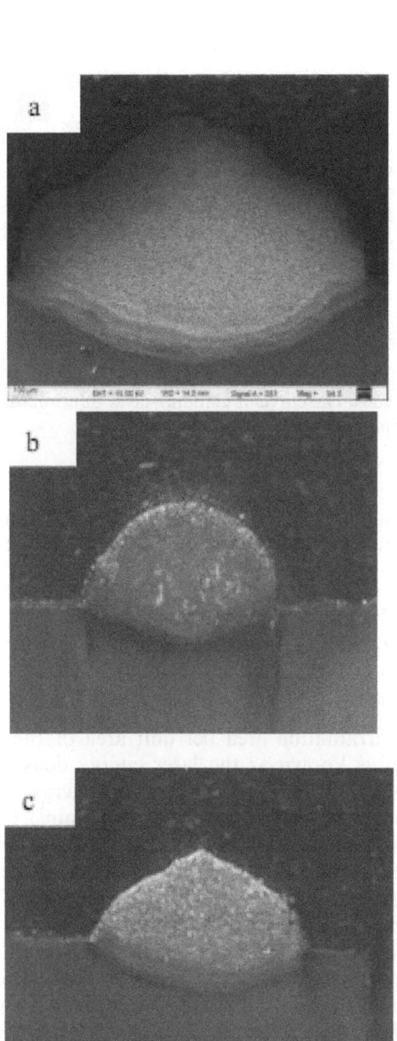

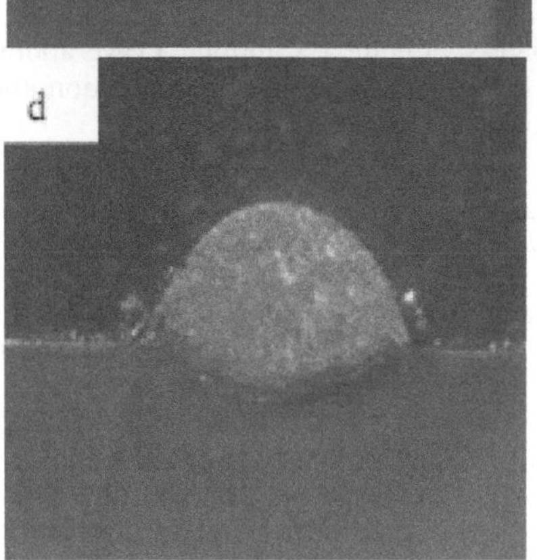

Figure 4. OM Morphology of cross-section of laser cladding coatings (a) three the clad layer redions (b) V = 2.0 mm/s; (c) V = 3.0 mm/s; (d) V = 4.0 mm/s (P = 850 W).

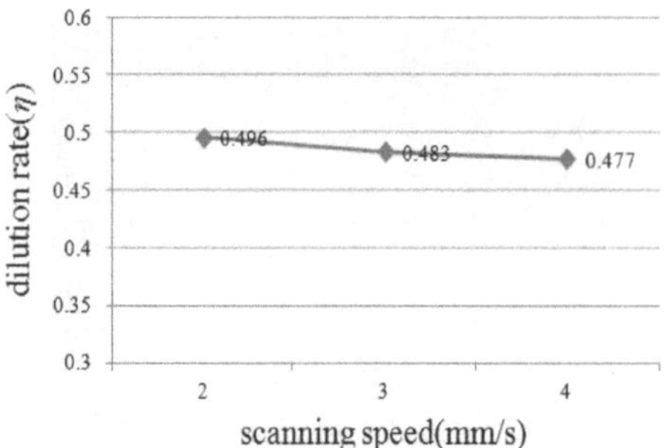

Figure 5. The curve of cladding layer dilution rate with the scanning speed.

of cladding layer, cladding layer height, and the penetration depth of basement were smaller. The dilution rate was decreed.

3.3 *Effect of laser energy density on the properties and microstructure of cladding layer*

Laser irradiation area per unit area of the input energy is known as the laser energy density (see Equation 4). Where K is the laser energy density per unit area on the irradiated area (J/mm^2), P is the output laser energy (W), D is the laser beam spot diameter (mm), V is the laser beam scanning speed (mm/s), and t is the time (s) [8]. Figure 4 (b, c, d) showed the effect of laser power and the dilution of cladding layer rate. With the increase of laser power, the width of the cross section increases, dilution of the substrate of titanium alloy is smaller. Always use the Figure caption style tag (10 points size on 11 points line space). Place the caption underneath the figure (see Section 5). Type as follows: 'Figure 1 Caption.' Leave about two lines of space between the figure caption and the text of the paper.

$$K = \frac{P}{D \cdot V} \tag{4}$$

Figure 6 (a) shows it mainly consists of lath staggered phase, acicular phase, and a small amount of spherical phase. Figure 6 (b) showed it is mainly composed of acicular phase, bulk phase, lath phase, and eutectic staggered. Figure 6 (c) showed it is mainly composed of irregular phase, the two phase, acicular dendrite, and eutectic structure. Figure 6 (d) showed that the microstructure of the coating is mainly composed of alternating phase and phase composition of ball shaped needle. Figure 7 (a) showed that the microstructure of cladding layer is uniformly distributed, the black

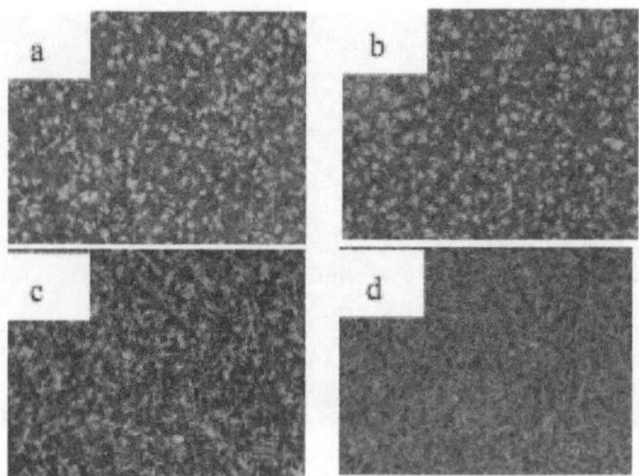

Figure 6. The central microstructure of laser cladding coating on titanium (a) $K = 0.067$ kJ·mm^{-2}; (b) $K = 0.089$ kJ·mm^{-2}; (c) $K = 0.150$ kJ·mm^{-2}; (d) $K = 0.174$ kJ·mm^{-2}.

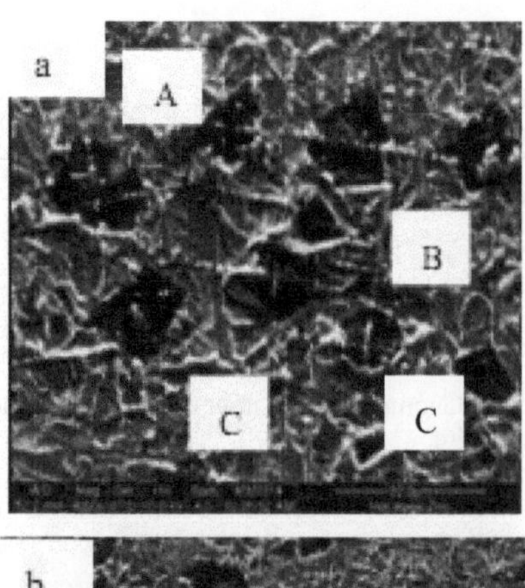

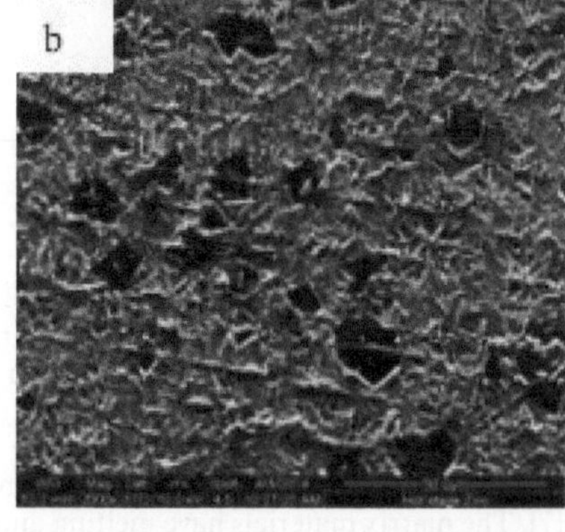

Figure 7. SEM photos of Ni60+Ni/MoS$_2$ laser cladding (P = 900 kW, D = 3.0 mm, V = 3 mm/s, K = 0.094 kJ·mm^{-2}).

bulk phase (A), the ball shaped particles (B), and dendrite phase (C). Massive black particles in the cladding layer are distributed uniformly in size between 6 to 10 μm. The white spherical particles sizes are between 1 to 2 μm, and it usually present the branches or linear distribution. On the surface of TC4 titanium alloy by laser cladding coating on, Black block (A) is the main component of Ti and B for TiB2 titanium boride. The ball shaped particles (B) is the main component of Ti and C, and TiC. The branch crystal (C) is the main component of Ni, is −Ni dendrites. By comparison with the increase of dilution rate (see in Fig. 7), the number of the cladding layer block and ball shaped particles.

4 CONCLUSIONS

When the laser power was 900 W, spot diameter was 3 mm, with the scanning speed decreased the surface of titanium alloy by laser cladding W = laser cladding width, H = laser cladding height, and h = penetration depth of substrate and dilution rate were increased (see Equation 3). The Ni60+Ni/MoS_2 laser clad layer on titanium alloy substrates was divided into clad, dilution and heat-affected zones. The microstructure of small spherical particles of titanium alloy surface by laser cladding coating (TiC) significantly increased the number of bulk phase (TiB_2), little change in the number. This is due to the formation of TiB_2 Gibbs free energy of G that is relatively low, and the melting point of TiB_2 is high, and it is easy to form and precipitation. So in the lower dilution rate, there was still a lot of bulk phase (TiB_2), while the small spherical particles (TiC) generate less.

REFERENCES

[1] Wu P., Zhou C.Z., Tang X.N. 2003. Microstructural characterization and wear behavior of laser clad nickel-based and tungsten carbide composite coatings [J]. Surf. Coat. Technol. 166:84–88.

[2] Zhu S.Y., Bi Q.L., Yang Jun, et al. 2011. Influence of Cr content on tribological properties of Ni3Al matrix high temperature self-lubricating composites [J]. Tribology International, 44(10):1182–1187.

[3] Chiu K.y., Cheng F.T., Man H.C. 2006. Corrosion behavior of AISI 316 L stainless steel surface modified with NiTi [J]. Technol. 200:6054–6061.

[4] Gao Y.X., Du L.Z., Huang C.B., et al. 2011. Wear behavior of sintered hexagonal boronnitride under atmosphere and water vapor ambiences [J]. Applied Surface Science, 257(23): 10195–10200.

[5] Zhu B.D., Zeng X.Y., Cui k. 1994. Effect of laser processing parameters on dilution the cladded coating [J]. Chinese Journal of Materials Research 8[4]:315–317.

[6] Chen Jian-min, Wang Ling-qian, Zhou Jian-song, etc. 2011 Recent development in nickel-based laser cladding China Surface Engineering, (02):13–21.

[7] Sun R.L., Yang D.Z., Guo L.X., et al. 2000. Microstructure and wear resistance of NiCrBSi laser clad layer on titanium alloy substrate [J]. Surface and Coatings Technology. 132(2000):251–255.

[8] Zhengzhong Guan. 2005. Handbook of laser processing technology [M] beijing: China metrology press, 87–89.

the branches of inner distribution. On the surface of TC4 titanium alloy by laser cladding coating, the block block (A) is the main component of Ti and B for TiB2 titanium boride. The ball shaped particles (B) is the main component of Ti and C, and TiC. The branch crystal (C) is the main component of Ni, as Ni dendrites. By comparison with the increase of dilution rate [illegible] in Fig. [illegible] the number of the cladding layer block and ball shaped particles.

4 CONCLUSIONS

When the laser power was 900 W, spot diameter was 3 mm, with the scanning speed increased the surface of titanium alloy by laser cladding W = laser cladding width, H = laser cladding height, and h = penetration depth of substrate and dilution rate were increased (see Equation 3). The Ni60+Ni/MoS layer in the cladding on titanium alloy substrate was divided into clad, dilution and heat-affected zones. The content of small spherical particles of titanium alloy surface by laser cladding coating (TiC) significantly increased the number of bulk phase (TiB), little change in the number. This is due to the formation of TiC enthalpy energy of G that is relatively low and the melting point of TiB is high and it is easy to form and precipitation. So in the lower dilution rate, there was still a lot of bulk phase (TiB), while the small spherical particles (TiC) generate less.

REFERENCES

[1] Wu P., Zhou C.Z., Tang X.N. 2003. Microstructural characterization and wear behavior of laser cladded nickel-based and tungsten carbide composite coatings [J]. Surf Coat Technol 166:84–88.

[2] Zhu S.X., [illegible] Yang J.B., et al. 2011. Influence of Cr content on tribological properties of Ni3Al matrix high temperature self-lubricating composites [J]. Tribology International 44 (0):1182–1187.

[3] Chiu K.Y., Cheng F.T., Man H.C. 2005. Corrosion behavior of AISI 316 L stainless steel surface-modified with NiTi [J]. Technol 200:1873–1879.

[4] Guo Y.X., [illegible] Huang W., et al. 2013. Wear behavior of sintered [illegible] in ambient atmosphere and water vapor ambiences [J]. Applied Surface Science 270:[illegible].

[5] Zhao P.D., Zhang X.Y., [illegible] 1993. Effects of laser processing parameters on [illegible] coating [J]. Chinese Journal of Materials Research 7(3):[illegible].

[6] Chen Jianmin, Wang [illegible], Zhou Jiansong, etc. 2011. Recent development of nickel-based laser cladding. China Surface Engineering 20(1):13–21.

[7] Sun R.L., Yang D.Z., Guo L.X., et al. 2000. Microstructure and wear resistance of NiCrBSi laser clad layer on titanium alloy substrate [J]. Surface and Coatings Technology, 132 (2000): 251–255.

[8] Zhang Zhang Shan. 2003. Handbook of laser processing technology [M]. Beijing: China metrology press, 87–89.

Material Science and Environmental Engineering – Chen (Ed.)
© 2016 Taylor & Francis Group, London, ISBN 978-1-138-02938-5

The optimization of the fake porthole die design for aluminum extrusion process

C.T. Sun
Linyi University, Linyi, Shandong Province, China

ABSTRACT: Fake porthole die is proposed to improve the ordinary die stress for aluminum extrusion process. But it is usually hard to gain uniform velocity of fake porthole die. In this manuscript, some schemes are given to optimize the fake porthole die. And after optimization, a more uniform velocity is gained.

Keywords: fake porthole die; optimization; baffle plate; bearing

1 INTRODUCTION

In recent years, with the increasing demand of lightweight products, aluminum alloy extrusion technology has developed quickly. It is widely applied in cars, ships, railway, aerospace, architecture, etc. In comparison with solid profiles, hollow profiles have wider applications due to their characteristics of large, hollow, and complex cross-section.

In practice, hollow profiles are usually produced using a conventional conical die or a porthole die. The conventional conical die is limited to extrude simple cross-section profiles, while porthole dies can produce complex cross-section profiles. Compared to the former method, the manufacturing process of porthole dies is much easier with lower cost. So 90% hollow profiles are produced by porthole dies. By this method, the billet metal is divided into several streams when entering portholes of an extrusion die. Then these streams are rejoined and welded together under high pressure in welding chamber. Finally, the profile is extruded out the die bearing and kept good shape and dimension.

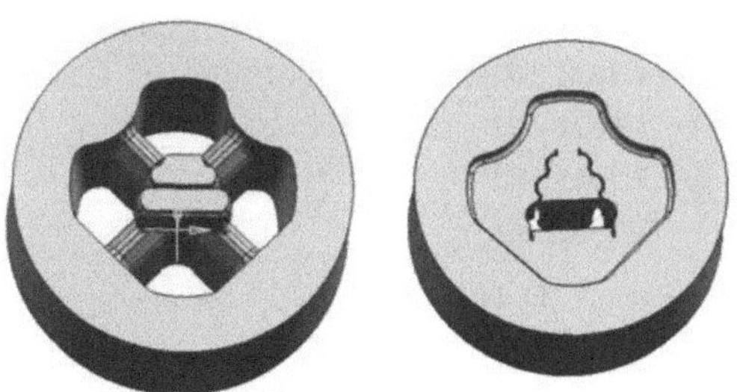

Figure 1. Fake porthole die structure.

It is known that the fake porthole die design will improve the die stress. When the ordinary porthole die design cannot satisfy the die stress need, porthole die design is proposed to meet the die stress. A lot of other problems are brought by the fake porthole die design, such as non-uniform velocity and extrudate deformation. So it is important to optimize the fake porthole die.

A fake porthole die structure is given by Figure 1. And the velocity distribution of the extrusion process is shown in Figure 2. It is clearly seen that the flow velocity in the cross-section of the bearing exit is non-uniform. The flowing velocities in Part 1 and Part 2 are larger than those in other parts and the maximum velocity is up to 68.0 mm/s, while the ribs have relatively smaller flow velocities and the minimum velocity is only 14.8 mm/s.

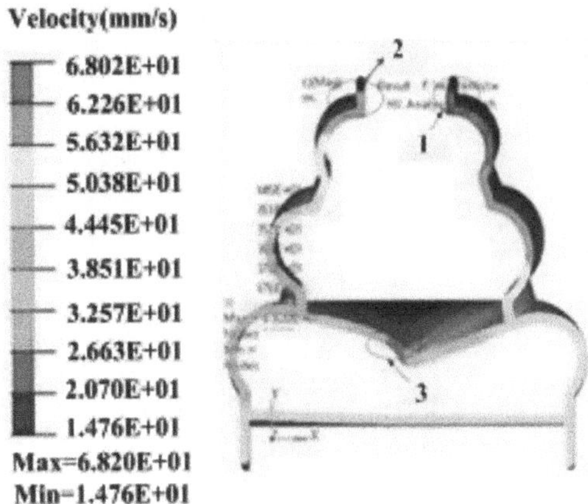

Figure 2. Velocity distribution in the cross-section of profiles.

2 DIE OPTIMIZATION

In the whole section of the extrudate, the regions with high velocities continuously squeeze those with low velocities, which cause the extrudate bent and twisted. Therefore, the initial die design is not acceptable, and some measures should be taken to optimize the die design and balance material flow during the extrusion process.

2.1 *The first optimization scheme and analysis of simulation results*

The portholes in the upper die play a very effective role in balancing the metal flow, promoting welding, reducing the extrusion pressure, and enhancing the die life. From the velocity distribution, the material supplement to the middle lips is not enough, and the lips deformation is seriously. According to the analysis, the optimization schemes are given as followings:

1. Figure 3 (a) shows the specific locations of the porthole which supply materials to the lips. In terms of the above numerical simulation results, it is clearly seen that material in part 1 and 2 of the extrudate flows slower than that in the other adjacent parts. This phenomenon illustrates that the material supply for these parts is insufficient. To make the flow velocity distribution in the cross-section of the extrudate more uniform, it is necessary to enlarge the section areas of these portholes.
2. Additionally, the dimension of the drainage channel directly influences the velocity in the rib. In the initial design the relatively slower flow velocity in the rib illustrates that the material supply is not enough. Thus the width of the drainage channel is added to increase material supply of part 3, which is shown in Figure 3 (b). By this modification, the materials will enter the porthole more fluently to accelerate the material velocity of the ribs. At meanwhile, the deformation will be decreased.

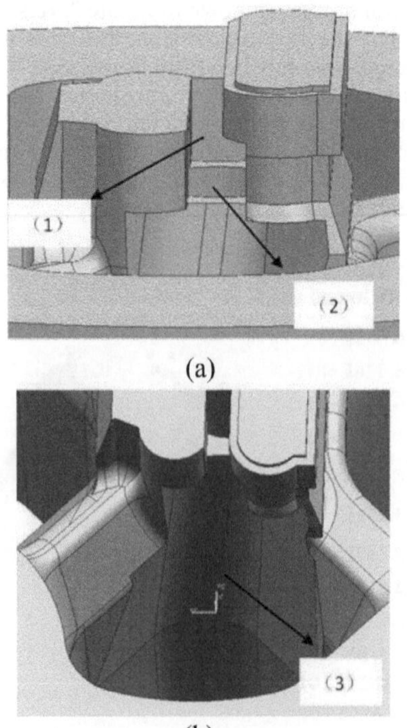

Figure 3. The first modification optimization scheme.

Figure 3 shows the 3D model for the first modification scheme. Keeping the same process parameters to the initial extrusion process, the numerical simulation is executed with the modified die design scheme, and the velocity distribution in the cross-section of the extrudate is shown in Figure 4.

In comparison with the results for the initial die design scheme, the maximum velocity is decreased from 68.0 mm/s to 69.7 mm/s, and the minimum velocity is increased from 14.8 mm/s to 27.3 mm/s in the cross-section of the extrudate. However, there is still a large difference in velocity between ribs and other regions, where serious twist and bend occur. This is because most material flows into the die orifice directly, and only a fraction of material flows into the drainage channel to supply for the rib owing to high resistance between the material and the drainage channel. Though the deflection

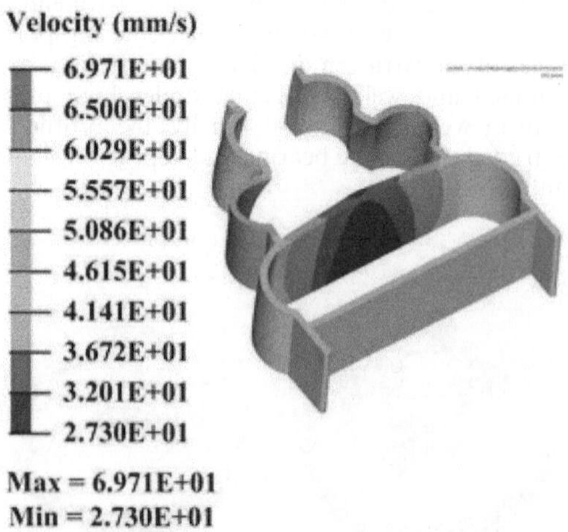

Figure 4. Velocity distribution in the cross-section of the extrudate for the first optimization scheme.

Figure 5. Deformation of the profile for first optimization scheme.

of the extrudate becomes smaller and the velocity distribution is more even after the first modification, the extrudate is still not acceptable. Thus it is necessary to take some further measures to optimize the die design and enhance the dimensional accuracy of the extrudate.

2.2 *The second optimization scheme and analysis of simulation results*

After optimization in scheme 1, Velocity distribution on the cross-section of the extrudate is more uniform. The maximum velocity is reduced, while the velocity of the ribs is greatly improved. The maximum velocity is still appeared at the extrudate end. And the end is also gains the maximum deformation. So some methods will be taken to reduce the velocity.

In this modification, the baffle plate is adopted in lower die in order to obtain more uniform metal flow. A higher and thicker baffle plate is adopted in the region with faster metal flow; otherwise, a lower and thinner one is used. The specific shape and dimension of the baffle plates are shown in Figure 6, and the velocity distribution of the extrudate for the second modification scheme is shown in Figure 7. With the baffle plate, the resistance of metal flow increases in the region with a relatively higher velocity, and thus the metal flow is slowed down noticeably. On the contrary, the metal flow is accelerated in the region with a relatively lower velocity. Therefore, it makes the velocity distribution in the cross-section more uniform. Meanwhile the baffle plate under the drainage channel makes it difficult for the metal to flow into the die orifice directly. When encountering the baffle plate, most material will change the original direction of metal flow and flow into the drainage channel in advance, thus the condition of insufficient metal supply for ribs being improved. In contrast to the first modification scheme, the maximum velocity is decreased to 59.4 mm/s, and the minimum velocity is increased to 27.3 mm/s in the cross-section of the extrudate. It indicates that the velocity distribution becomes more even in the cross-section of

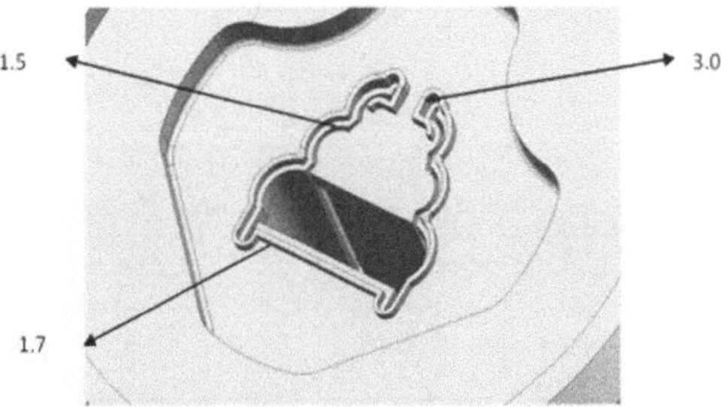

Figure 6. The locations, shapes and dimensions of the baffle plates.

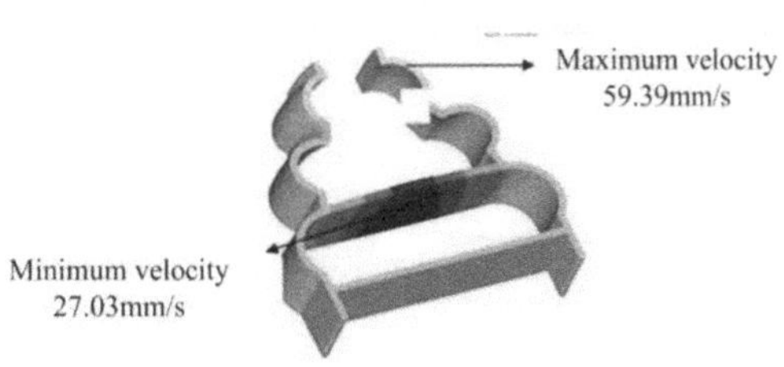

Figure 7. Velocity distribution in the cross-section of the extrudate for the second modification optimization scheme.

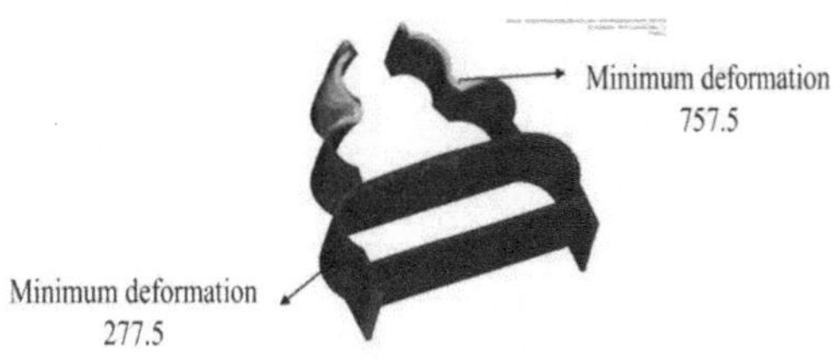

Figure 8. Veins deformation of the profile for the second modification optimization scheme.

the extrudate. However, there is still deformation in local regions of the extrudate, and the dimension accuracy of the extrudate needs to be improved further. From the Figure 8, we will see that the maximum deformation is reduced from 0.082 to 0.075 and the rib deformation is improved.

2.3 *The third optimization scheme and analysis of simulation results*

In the third modification scheme, the local bearing lengths are adjusted to balance the metal flow further in the die orifice. The die bearing is considered as a barrier when the material flows through it. Increasing the bearing length can increase the local friction resistance and slow down the metal flow. Otherwise, shorter bearing length makes a

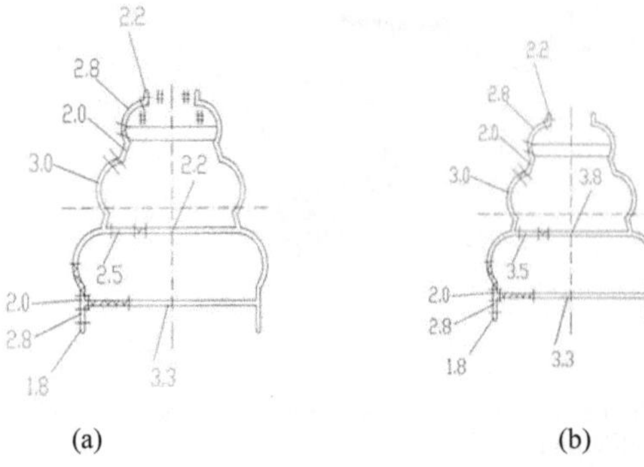

Figure 9. The bearing before and after the third modification optimization scheme (unit: mm). (a) Before modification optimization; (b) After modification optimization.

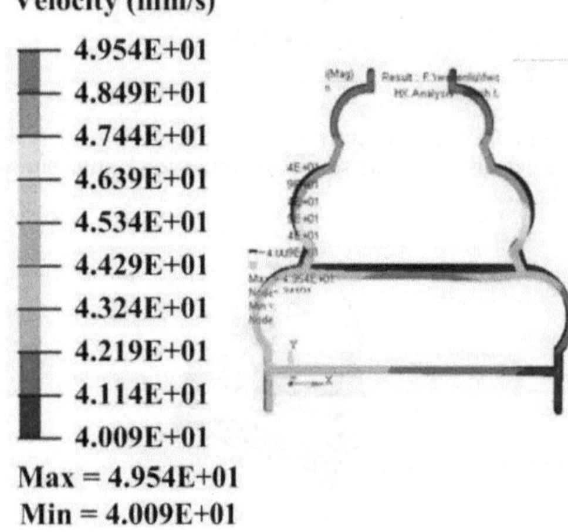

Figure 10. The velocity distribution in the cross-section of the extrudate for the third modification optimization scheme.

higher flow velocity. Therefore, by varying bearing length, more uniform velocity distribution in the cross-section of the extrudate can be obtained. The local bearing lengths are modified and the specific modification schemes are shown in Figure 9. The corresponding velocity distribution in the cross-section of the extrudate is shown in Figure 10. It is seen that the maximum velocity is decreased to 49.5 mm/s and the minimum velocity is increased to 40.1 mm/s.

REFERENCES

[1] Wu, X.H.; Zhao, G.Q.; Luan, Y.G.; Ma, X.W. Numerical simulation and die structure optimization of an aluminum rectangular hollow pipe extrusion process. Materials Science and Engineering A 2006, 435–436, 266–274.

[2] Jafarzadeh, H.; Zadshakoyan, M.; Abdi Sobbouhi, E. Numerical studies of some important design factors in radial-forward extrusion process. Materials and Manufacturing Processes 2010, 25, 857–863.

[3] Mehtaa, B.V.; Al-Zkeri, I.; Gunasekera, J.S.; Buijk, A. 3D flow analysis inside shear and streamlined extrusion dies for feeder plate design. Journal of Materials Processing Technology 2001, 113, 93–97.

Material Science and Environmental Engineering – Chen (Ed.)
 ISBN 978-1-138-02938-5

Comparisons of aluminum extrusion process with fake porthole die structure and ordinary die structure

X.M. Sun
Linyi University, Linyi, Shandong Province, China

ABSTRACT: There are solid profile, hollow profile and semi hollow profile of aluminum extrudate according to the section shape. An aluminum extrudate with hollow and semi hollow section is studied in this manuscript. Two die design schemes are proposed as fake porthole die design and ordinary porthole die design. Then the velocity, temperature, stress and die displacement of aluminum extrusion process are analyzed of the two die design schemes.

Keywords: fake porthole die; aluminum extrusion process; semi hollow profile

1 INTRODUCTION

In recent years, with the increasing demand of lightweight products, aluminum alloy extrusion technology has developed quickly. It is widely applied in cars, ships, railway, aerospace, architecture, etc. In comparison with solid profiles, hollow profiles have wider applications due to their characteristics of large, hollow, and complex cross-section.

According to the section shape, aluminum extrudate will be divided into solid profile, hollow profile and semi hollow profile. In this manuscript, an aluminum extrudate with hollow and semi hollow section is studied. For this special extrudate, the extrusion die have two design schemes, including fake porthole die design and ordinary porthole die design. And the extrusion process is analyzed with numerical simulation method. Numerical simulation can describe the extrusion process on the computer and gain the information of stress, strain, temperature, and velocity distribution of an aluminum alloy profile. This information is usually immeasurable in the production site. Thus the velocity, temperature, stress and die displacement of aluminum extrusion process are analyze of two die design schemes.

2 DIE DESIGN

Figure 1 gives the dimension and geometry of the profile cross-section. As is shown, it has two parts, the hollow part and semi hollow part. Thus two die designs are proposed. Figures 2 and 3 are the die structures of the two schemes. The semi hollow part is designed as porthole structure, which is the fake porthole die shown in Figure 3.

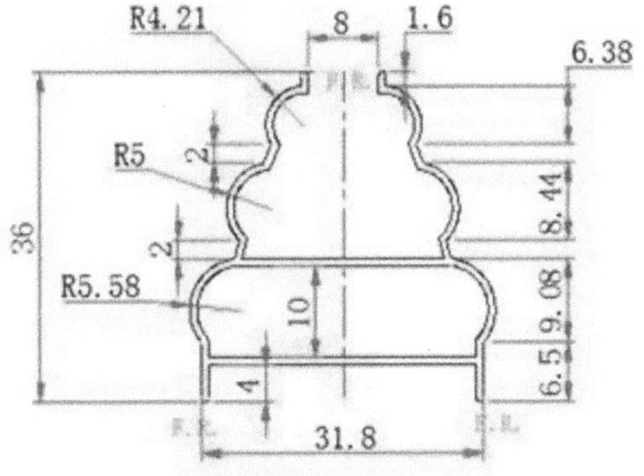

Figure 1. Dimension and geometry of the profile.

3 NUMERICAL SIMULATION RESULT

The simulation result is shown from Figure 4 to Figure 8, and a series conclusion will be gained as followings.

3.1 *Velocity*

In real extrusion process, the uniformity of flow velocity distribution in the cross-section of the bearing exit greatly influences the quality of the extrudate. With the above simulation results, we can analyze the rules of non-uniform velocity distribution in the cross-section for the porthole die. The parts 1–2 are directly toward the centers of portholes, so the material flows from the porthole to the die orifice directly without any extra resistance during the whole extrusion process.

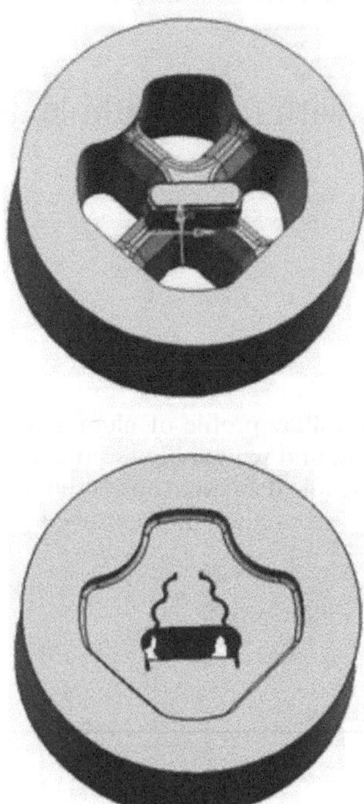

Figure 2. Ordinary porthole die structure.

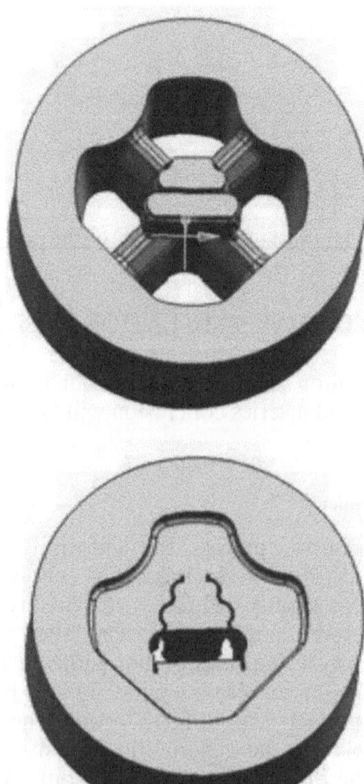

Figure 3. Fake porthole die structure.

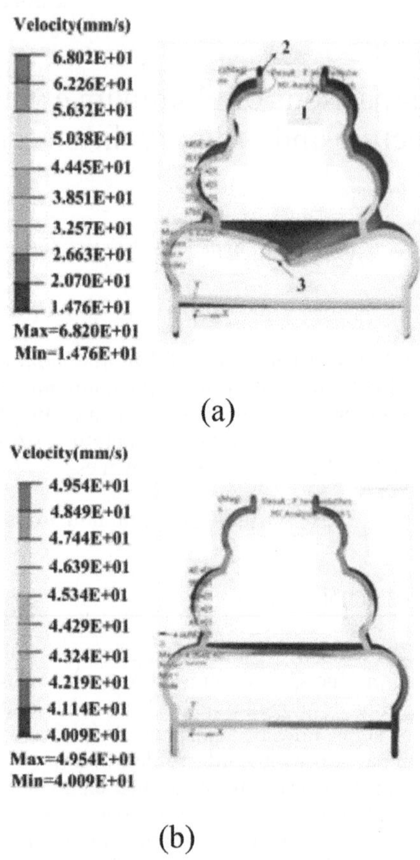

Figure 4. Comparison of velocity distribution in the cross-section of profiles (a) Fake porthole die; (b) Ordinary die.

Therefore, the flow velocities are faster than that in other parts of the extrudate. Part 3 is under the port bridge, the material has to overcome the additional friction resistance between the material and the port bridge. In addition, the metal streaming from the neighboring portholes will be rewelded under the port bridge. As a result of severe collision of two metal streams, the flow velocity will be partially counteracted. Therefore, the material in these parts flows slower than other parts in the extrudate. Finally, in order to form the rib of the extrudate, the high temperature metal needs to flow along a very tortuous and complex path. Especially the direction of the metal flow in the drainage channel is almost vertical to the extrusion direction. In the path that material flows through, the resistance of metal flow is so high that the rib is difficult to form. Thus the flow velocity in the rib of the extrudate is smallest in the whole cross-section of the extrudate. Meanwhile due to the insufficient material supply, the ribs are also undersize in comparison with the standard dimension.

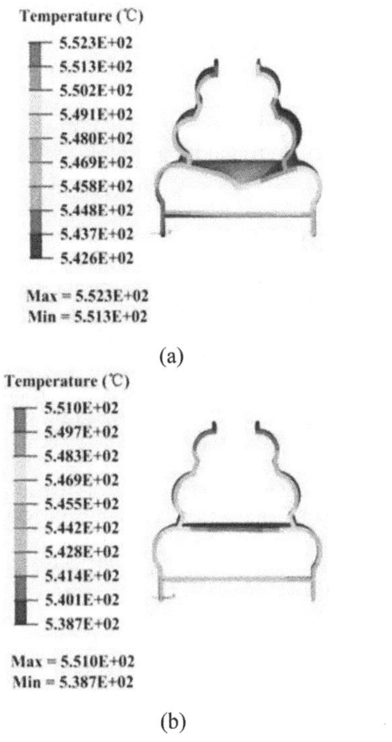

Figure 5. Comparison of temperature distribution in the cross-section of profiles. (a) Fake porthole die; (b) Ordinary die.

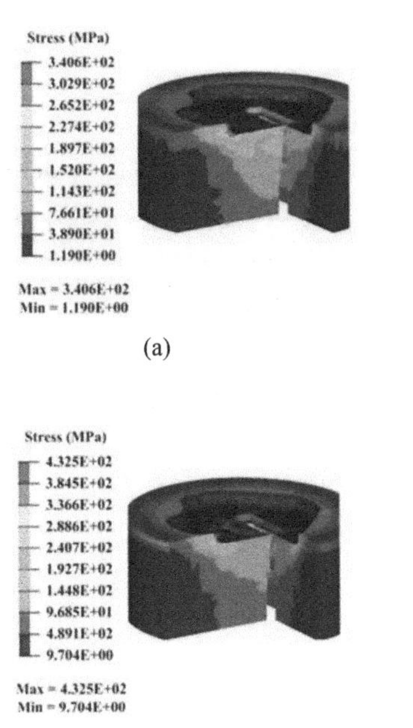

Figure 6. Comparison of stress distribution in lower die. (a) Fake porthole die; (b) Ordinary die.

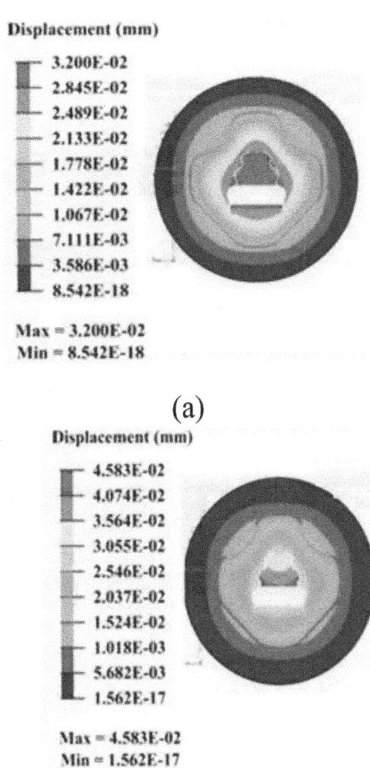

Figure 7. Comparison of displacement in lower die (a) Fake porthole die; (b) Ordinary die.

3.2 *Temperature*

The average temperature is higher of the fake porthole die, while the temperature difference is smaller. This is because a more mandrel is designed for the fake porthole die, the friction heat generated more than that of the ordinary die. Thus the average temperature is higher. The mandrel also will balance the material flowing which makes a smaller temperature difference.

3.3 *Stress and displacement*

The extrusion die is under high temperature, high pressure and high friction in the actual extrusion process. The die may scrap for plastic deformation, fatigue failure and fracture. So it is important to analyze the die stress. As is shown in Figure 6, the stress of the upper die for fake porthole die is smaller than that for the ordinary die. For the ordinary die design, there is a cantilever, which undergoes the maximum stress. And it is easy to scrap. The maximum displacement of the fake porthole and ordinary porthole are 0.032 mm and

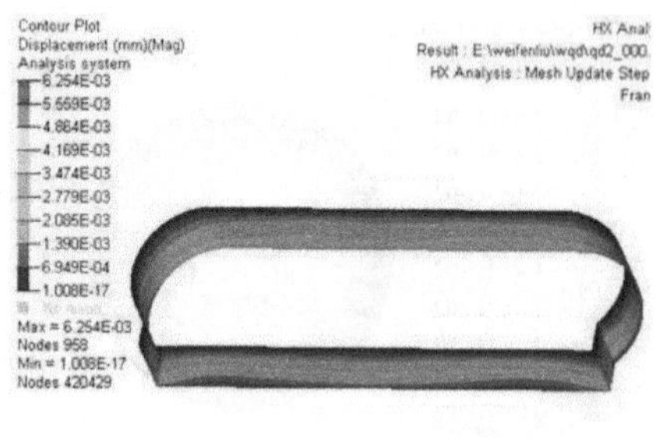

(a)

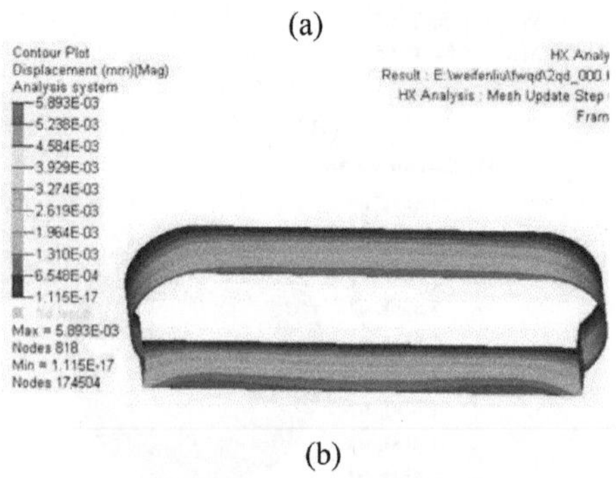

(b)

Figure 8. Comparison of displacement of the core. (a) Fake porthole die; (b) Ordinary die.

0.045 mm, respectively. So the dimension precision accuracy is fine for the fake porthole die design.

4 CONCLUSIONS

1. Fake porthole die is usually proposed for the cantilever extrusion process.
2. Fake porthole die will improved the die stress.
3. Many problems will be brought by fake porthole die, such as non-uniform velocity, high average temperature.
4. It is important to resolve the non-uniform velocity problem to get an available extrudate.

REFERENCES

[1] Bauser, M.; Sauer, G.; Siegert, K. Extrusion, 2nd Ed.; ASM International: Materials Park, OH, 2006.

[2] Liu, G.; Zhou, J.; Duszczyk, J. FE analysis of metal flow and weld seam formation in a porthole die during the extrusion of amagnesium alloy into a square tube and the effect of ram speed on weld strength. Journal of Materials Processing Technology 2008, 200, 185–198.

[3] Fang, G.; Zhou, J.; Duszczyk, J. Extrusion of 7075 aluminum alloy through double-pocket dies to manufacture a complex profile. Journal of Materials Processing Technology 2009, 201, 3050–3059.

[4] Jo, H.H.; Lee, S.K.; Lee, S.B.; Kim, B.M. Prediction of welding pressure in the non-steady state porthole die extrusion of Al7003 tubes. International Journal of Machine Tool & Manufacture 2002, 22, 753–759.

Material Science and Environmental Engineering – Chen (Ed.)
 ISBN 978-1-138-02938-5

Effect on the micro-structure and properties of AM60B co-doped with rare earth

Y.Q. Yu, Y. Li & J.J. Yang
The School of Electromechanical and Architectural Engineering, Jianghan University, Wuhan, Hubei Province, China

ABSTRACT: An experimental test based on heat treatment was done for the alloy mechanical properties of the AM60B magnesium alloy, and an analysis was made on the precipitated phase under different conditions of the mischmetal with Y, Nd and Gd. The experiment test showed that the addition of rare-earth elements could refine the structure of alloys in the AM60B alloy, and reduce the $Mg_{17}Al_{12}$ phase to disperse, which could effectively improve the tensile strength and elongation of the alloy. However, an excess of rare earth can result in coarsening the phase of $Al_{11}RE_3$ in the alloy, and lowing the mechanical properties of alloy. The experimental test explored that, with the mixed rare earth, the elongation of AM60B could increase 10%, and its tensile strength could exceed 260 Mpa.

Keywords: AM60B magnesium alloy; rare earth; mechanical properties; microstructure

1 INTRODUCTION

Magnesium and its alloys as one system of the light alloy structural materials at this stage are characterised with a low density, high specific strength and stiffness, good thermal conductivity, excellent machinability, and high recycled ability [1]. It has important application value and broad prospects in the aerospace, defense industry, automobile industry, biomedicine, electronic products and other industries [2]. However, this magnesium alloys has huge limitation on the development of commercial production due to its some characterizations such as high chemical activity, low corrosion resistance and ductility. It is difficult to obtain high tensile strength and high elongation rate at room temperature, which limit its application in our life. So how to overcome those problems and further promote the application are wildly studied by researchers, The current research has shown that adding appropriate amounts of rare earth element can effectively improve the performance of magnesium alloy [3–7], which possesses vital significance to promote the development on the application of magnesium alloy.

AM60B is one kind of cast magnesium alloy applied widely. It has been commonly used in higher ductility, toughness and corrosion resistance of the occasion. Because AM60B magnesium alloy has low aluminum content, and it decreases the compound with alloy aluminum in the amount of precipitation phase. So, the plasticity and toughness of the alloy is higher, but the intensity decreases. It is very significant to study how to make use of rare earth to increase AM60B intensity based on maintaining the high elongation, which can be very helpful to promote the application of magnesium alloy. The author and his co-workers studied how to make use of rare earth to increase AM60B intensity based on maintaining the high elongation rate, and those results provided some references for future works.

2 EXPERIMENTAL PROCEDURE

The experiment raw materials are commercial AM60B alloy ingot and the mixed rare earth with yttrium, neodymium and gadolinium. Firstly, put the magnesium alloy ingot into the furnace with a protecting gas of N_2, and then heat to 720 °C until completely melted. Then quickly add different proportions of mixed rare earth subsequently. After the rare earth is melted completely, stir them to homogenize the composition. With the heat preservation for 10 minutes, then cool it to 650 °C to cast into tensile samples via J1125B die casting machine. Samples are designed according to the national standard GB 6397-86, showed in Figure 1.

The testing of tensile strength and elongation of each group magnesium alloys co-doped different contents of mischmetal were carried out through RGM-50 electronic tensile test machine with the

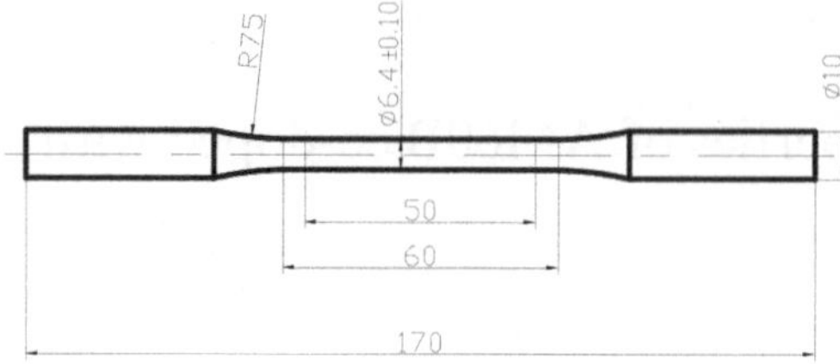

Figure 1. AM60B test specimens.

speed of 3 mm/min. In the case of microstructural characterization, the original specimens were etched by using 4% HNO_3 ethanol solution to reveal the constituents and general microstructure. The specimens were examined with MDS optical microscope and the JEM-2100F Scanning Electron Microscope (SEM). The analysis of the components was detected by X-Ray Diffractometer (XRD). Heat-treated the same composition of the sample, and finished its solid solution treatment in SX-8-10 box-type resistance furnace with KSY temperature controller. The temperature of Solid solution treatment and natural aging (T4) treatment was 420 °C. Heat preservation for 12 hours, and then repeated to test sample's mechanical properties and made microstructure analysis of those samples.

3 RESULTS AND DISCUSSIONS

3.1 *Mechanical properties*

Effects of adding amount of Rare Earth on the tensile strength and elongation were showed in Figure 2.

As can be seen from above figures, with the increase of mixed RE, the tensile strength and elongation of AM60B had been increased. When RE addition amount was 1.6%, the tensile strength reached the highest value of 263 MPa, and would be decreased with more RE. While in the case of elongation rate, it could be increased more than 10% after adding RE amount between 0.5%~1.8%wt, compared with AM60B magnesium without RE, the alloy specimen elongation increased by about 30%. However, when rare earth content was more than 2%, both the tensile strength and the elongation of the alloy would significantly decrease. And the sample was easy to produce brittle fracture during the tensile test.

3.2 *Microstructure*

Through the analysis of the metallographic sample, the microstructure of AM60B magnesium alloy is mainly composed of the matrix α-Mg phrase and

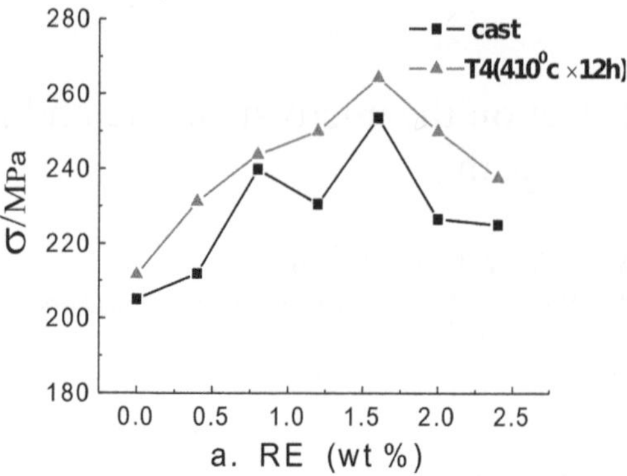

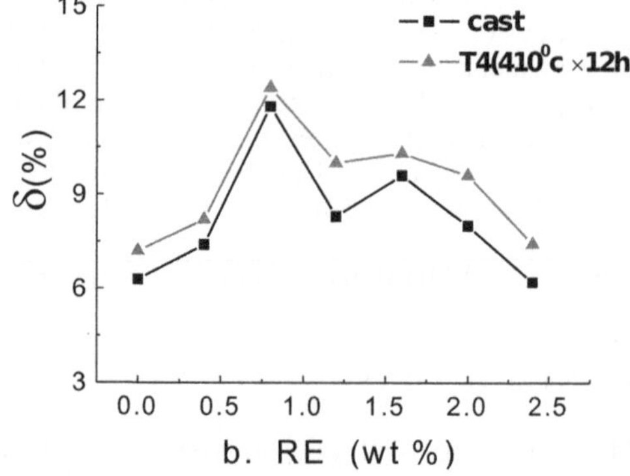

Figure 2. Mechanical properties of magnesium alloy specimens.

divorced eutectic β phrase ($Mg_{17}Al_{12}$). β phrase is distributed in the form of coarse continuous network and intermittent network along grain boundaries as gray blob, Al-Mn phase in the organization in black granule (see Fig. 3a). After adding RE elements, AM60B alloy matrix organization does not change, but there is a new Al_{11}-RE_3 phase, that is, RE and Al first formed acicular Al_{11}-RE_3 phase [8] due to their bigger electronegativity difference. There is also a part of RE concentration in the Al-Mn phase.

With the $Mg_{17}Al_{12}$ phase reducing, there was a large number of Al_{11}-RE_3 phase distribution in grain boundary, which effectively segmented alloy matrix and made matrix grains more fine (Fig. 3b and c). During the process of alloy crystallization, Al_{11}-RE_3 would restrict the grain growth by reducing the diffusion of Mg and Al, and directly cause the microstructure refinement. And the tensile strength and elongation of AM60B could be improved after these changes in the microstructure, which can be proved by the Hall-Petch equation [9]: $\sigma_s = \sigma_0 + kd^{-1/2}$. In which, σ_s is yield strength, σ_0 is a constant, k is coefficient, and d is the mean size of grain. This equation obviously explains the mechanical properties of equivalent to the relationship between particles.

However, once rare earth content is more than 2.0%, this magnesium alloy substrate would be divided into irregular area grains in larger size by Al_{11}-RE_3 phase, and also coarse rare earth compound would appear. So, it is easy to generate tiny cracks (as shown in Fig. 3d) between the Al_{11}-RE_3 phase and α-Mg matrix. In addition to this, those rare earth compounds would be enrichment to cause the crystal defect due to its low solubility with the matrix, which explains the reason why the weakening phenomenon existed in mechanical performance after adding overmuch RE.

After solid solution treatment, the mechanical property of alloy will be improved more or less, and the suitable coarse grains with more homogeneous distribution could be observed. This is because $Mg_{17}Al_{12}$ phase is dissolved into α-Mg substrate, and cause the primary phase of alloy enlarges in the period of heat treatment (refer with: Fig. 4). While RE compound is difficult to dissolve into the Al-Mn phase for its high thermostability, it will be more dispersive during the solution treatment process, and result in higher mechanical properties based on precipitate strengthen mechanism.

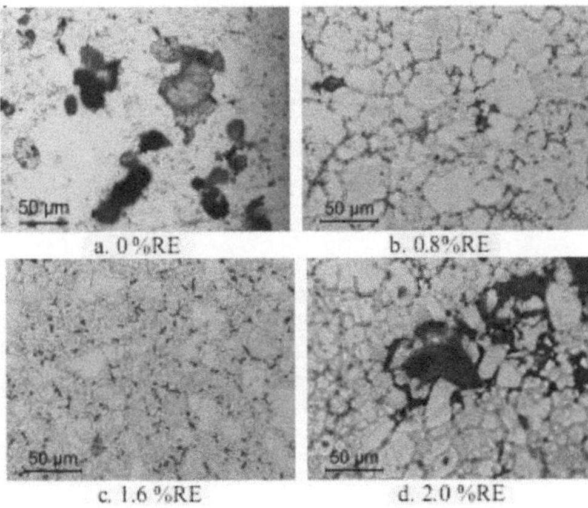

Figure 3. Microstructure of AM60B alloy specimens.

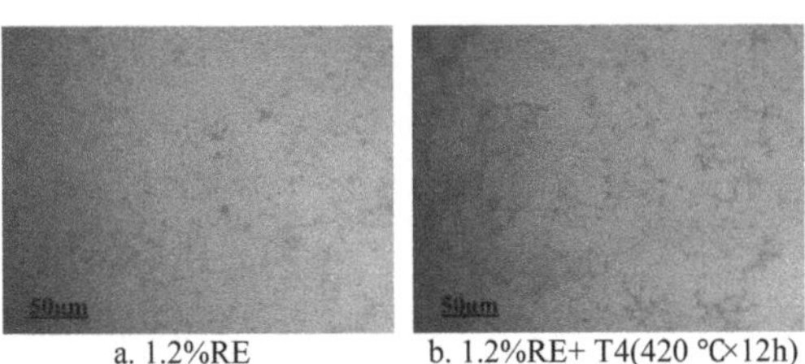

Figure 4. Influence of solid-solution on microstructures to the alloy specimens.

In addition, the shape of Al_{11}-RE_3 phase turns from long spicula into short spicula as the time of solution treatment goes on. And some spicula fuse into several pieces. As a result, the ductility and tensile property are enhanced after decreasing the possibility of dissevering the substrate.

However, the organization refining effect of solution treatment couldn't be so obvious in the case of adding excessive amounts of RE for its low solubility, and this alloy substrate was severed by vast of radial Al_{11}-RE_3 phase. Stress concentration would be aroused at the tip of spicula, which could result in flaws and reduce mechanical properties.

4 CONCLUSIONS

Based on the above experimental test and analysis, it can be concluded that, with adding proper quantity of RE, the mechanical properties of AM60B alloy can be increased significantly and the micro-organization can be more stable. The strength of anti-stress can exceed 260Mpa when the content of RE in alloy is 1.6% RE. However it should be noted that if the percentage of RE is more than 1.6%, a mass of $Al_{11}RE_3$ spicula will emerge and sever the alloy substrate, which will decline the mechanical property. Compared with wrought phase, appropriate solid solution treatment can improve the phase distribution of the alloy, and increase the strength of anti-stress in a certain extent. However, the effect can't be observed obviously if adding over much RE, which brings inspiration into the future work to further improve the mechanical property of the AM60B-RE alloy through controlling the Al_{11}-RE_3.

ACKNOWLEDGEMENT

This work was financially supported by the Wuhan Planning Project of Science and Technology (201250499145-16, 2013071004010467).

REFERENCES

[1] Chamini L. Mendisalok Singh. Magnesium Recycling: To the Grave and Beyond. J. JOM. Vol. 65 (2013), p. 1283.

[2] Yuan G.Y., Liu Z.L., Wang Q.D., et al. Microstructure refinement of Mg-Al-Zn-Sialloys. J. Materials Letters. Vol. 56 (2002), p. 53–56.

[3] Hu Yong, Yan Hong, Chen Guo-xiang, et al. Effects of Si on Microstructure and properties of in situ synthesized Mg2Si/AM60 composites [J] Rare Metal Materials and Engineering. n. Science. Vol. 38 (2009), p. 343.

[4] Dingfei Zhang, Xia Chen, Fusheng Pan, et al. Effect of RE on mechanical properties of magnesium alloys. J. Functional Materials. Vol. 5 (2014), p. 5001.
[5] Fugang Qi, Dingfei Zhang, Xihua Zhang, et al. Effect of RE Y on microstructure and mechanical properties of the Mg zinc Mn magnesium alloys [J]. Transactions of Nonferrous Metals Society of China. Vol. 24 (2014), p. 1352.
[6] Yan Jingli, Sun Yangshan, XueFeng, et al. Creep behavior of Mg–2wt% Nd binary alloy. J. Materials Science and Engineering, Vol. 524 (2009), p. 102–107.
[7] Yijin Tang, Zhenyan Zhang, Li Jin, et al. Research progress on ageing precipitation of Mg-Gd alloys. J. Transactions of Nonferrous Metals Society of China. Vol. 1 (2014), p. 8.
[8] Zhang Jinwang, Wang Shebin, Zhang Junyuan. Effects of Nd on Microstructures and Mechanical Properties of AM60 Magnesium Alloy in Vacuum Melting [J]. Rare Metal Materials and Engineering. Vol. 38 (2009). p. 1141.
[9] Mima G. Tanaka Y: Trans Jap. Inst Metals. Vol. 12 (1971), p. 317.

Chemical material

Material Science and Environmental Engineering – Chen (Ed.)

Research on efficient chlorine removing agents

T. Zhang, S.Z. Wang & C.H. Dong
Center for Engineering Design and Research under the Headquarters of General Equipment, Beijing, China

L. Han
School of Chemical Engineering and Energy, Zhengzhou University, Zhengzhou, China

ABSTRACT: Efficient solid adsorbents are provided for the removal of chlorine in air. The major materials include Ca $(OH)_2$, NaOH and a small amount of additives. The agents were characterized by Scanning Electron Microscopy (SEM) and nitrogen adsorption-desorption method, etc. The results show that the optimum chlorine removing agent with a weight ratio of 1Ca $(OH)_2$: 1 NaOH has the largest surface area and largest pore volume. The removal efficiency was determined and the results indicate that removal efficiency of the optimum agent can reach almost 100% within 10 min.

Keywords: chlorine; removing agents; adsorption

1 INTRODUCTION

As a basic chemical raw material, chlorine is widely used in the preparation of hydrochloric acid, pesticides, explosives and organic dyes, synthesizing plastic and rubber, paper bleaching, cloth and drinking water disinfection, as well as some industrial wastewater treatment [1]. Chlorine is usually loaded in metallic cylinders or other containers. Occasionally during storage and transport, chlorine leakage may happen. Chlorine is a toxic gas and the major hazardous air pollutant, which can cause not only human and animal poisoning and even sudden death, but also serious environmental problems, etc.

Recently, molecular sieves or silica gels were used as adsorbents for the removal of chlorine. Chlorine adsorption on the A, X and SBA-15 molecular sieves were studied at different temperatures and pressures [2]. The results showed that the skeleton structure and the pore volume have concerned with molecular sieve adsorption of chlorine. The larger pore volume, the greater adsorption capacity. The smaller orifice, the better chlorine removal efficiency when the pore volume is similar. But there are still some problems to be solved for molecular sieves in the loop stability, persistence and so on. Wang [3] tested chlorine adsorption properties of activated carbon, but it did not take place at room temperature and atmospheric pressure. The adsorption capacity of activated carbon for chlorine decreased with increasing temperature and decreasing pressure. Luo [4] used modified powder of Ca $(OH)_2$ as removal sorbent for chlorine, but it can't remove chlorine rapidly.

In this study, we report the fabrication of removing agents in which Ca $(OH)_2$ and NaOH are chosen as mainly materials in detail. The removal efficiency was determined and the effect of NaOH content and moisture content on removal efficiency have been discussed.

2 EXPERIMENTAL SECTION

2.1 *Materials*

Calcium hydroxide, sodium hydroxide, sodium silicate (dispersant), sodium dodecyl benzene sulfonate (surfactant), sodium stearate (moisture-proof agent) were used as received without further purification.

2.2 *Preparation of chlorine removing agents*

Typically, proper amount of sodium hydroxide, dispersant and surfactant were added into a 100 mL beaker in that order with some deionized water. The mixture was stirred vigorously until the solution became clear. Calcium hydroxide was then added to the obtained solution under stirring and the solution became turbid at this time. Then the solution was continuously stirred for 15 min at room temperature. Afterwards, the final mixtures were heated in a convection oven at 90 °C for different periods of time and the samples were collected by grinding. Finally, sodium stearate was added in the samples and the powders were well mixed.

The weight ratios of the final mixtures were 0.06 sodium silicate: 0.03 surfactant: 0.03 sodium stearate: x calcium hydroxide: y Sodium hydroxide.

2.3 *Characterization*

Scanning Electron Microscopy (SEM) images were taken with a JSM-7500F microscope (JEOL). The powder samples were coated with Au in order to reduce the beam charging. Nitrogen adsorption–desorption measurements were carried out with a model NOVA4200e Instrument (U.S. Contador) to determine the Brunauer–Emmett–Teller (BET) surface areas of the samples.

2.4 *Performance*

Chlorine used here is made in laboratory with potassium permanganate and concentrated hydrochloric acid. The generated Cl_2 was collected in 1.0 L vacuum gas collecting bag. The concentration of Cl_2 in the bag can be determined by iodometric method [5]. The quality of adsorbent is four times that of chlorine in a 5.0 L vessel and the quality of chlorine is 0.5 g. So the initial concentration of Cl_2 is about 0.1 g/L. The adsorbent powders were sprayed into the 5.0 L vessel under reduced pressure. Sampling was carried out by using air sampler at a flow rate of 0.3 L/min for 2 min when adsorption time was 3 min and 10 min, respectively. Finally, the chlorine content after adsorption in the vessel was determined by using the iodometric method.

3 RESULTS AND DISCUSSION

3.1 *The influence of Ca $(OH)_2$ and NaOH content*

Mass fraction of Ca $(OH)_2$ and NaOH in the obtained samples was 88%. When the weight ratio of Ca $(OH)_2$ and NaOH was 1:1, 2:1 and 3:1, the samples were denoted X-1, X-2 and X-3, respectively. When adsorption time was 3 min and 10 min, removal efficiency of the samples were detected respectively under similarly moisture. The results were shown in Table 1.

Table 1 shows that the three chlorine removing agents can remove Cl_2 rapidly in three minutes. It is indicated that X-1 has the highest removal efficiency, and the Cl_2 concentration is 0 g/L after 10 min. The removal efficiencies increased with increasing NaOH content. The greater the basicity, the higher adsorption efficiency of the samples. However, the adsorbent powders weren't collected if the basicity of the sample was too high. Moreover, the particle sizes of removing agents became larger than that of the samples before capturing Cl_2, which made it so easy to subside and be collected in air. Furthermore, different basicity has different specific surface area (BET), which is shown in Table 2. X-1 has the largest surface area and largest pore volume for the existence of more NaOH. Large surface area and appropriate pore size are good for the adsorption of chlorine because of the existence of physical adsorption.

3.2 *The influence of drying time*

X-1 was chosen as the optimum chlorine removing agent to study the removal efficiency of the samples heated for different periods of time. The moisture contents of the samples were different due to the different drying time. The results were shown in Table 3. It can be seen that the samples heated for shorter time had the higher removal efficiency. However, too short drying time is not propitious to the other performance such as fluidity and stability. The optimum drying time of the samples was 10 h.

3.3 *The SEM of the samples*

Figure 1 shows the SEM images of the sample X-1. The particle size of X-1 is more than 10 μm, and most of the particles are irregular globular particles. There are numerous uneven grooves on

Table 1. Adsorption performance of the chlorine removing agents.

	Adsorption time/3 min			
Removing agent	Removing agent dosage/g	Initial concentration of Cl_2/(g/L)	Final concentration of Cl_2/($\times 10^{-2}$ g/L)	Removal efficiency/%
X-1	2.008	0.1	0.52	94.8
X-2	2.012	0.1	1.05	89.5
X-3	2.009	0.1	1.76	82.4
Removing agent	Adsorption time/10 min			
X-1	2.013	0.1	0	100
X-2	2.006	0.1	0.47	95.3
X-3	2.011	0.1	0.82	91.8

Table 2. BET datas of the chlorine removing agents.

Removing agent	Surface area (m^2/g)	Pore volume (cm^3/g)	Pore radius (nm)
X-1	68.05	0.06	1.981
X-2	61.67	0.05	1.912
X-3	54.40	0.05	1.886

Table 3. Adsorption performance of X-1 heated for different periods of time.

	Adsorption time/3 min			
Drying time/h	Removing agent dosage/g	Initial concentration of Cl_2/(g/L)	Final concentration of $Cl_2/(\times 10^{-2}$ g/L)	Removal efficiency/%
10	2.010	0.1	0.55	94.5
13	2.003	0.1	0.73	92.7
16	2.013	0.1	1.08	89.2
19	2.006	0.1	1.22	87.8
Drying time/h	Adsorption time/10 min			
10	2.005	0.1	0	100
13	2.011	0.1	0.15	98.5
16	2.009	0.1	0.42	95.8
19	2.001	0.1	0.77	92.3

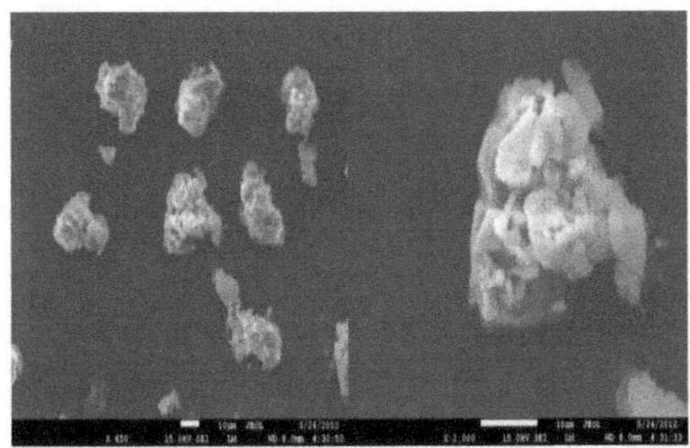

Figure 1. SEM images of the sample X-1.

the particles which can increase the surface area and then greatly promote chemical reactions and physical adsorption of Cl_2. It is suspected that the small particle size will have better removal efficiency for its bigger surface area that can increase the contact area between Cl_2 and removal sorbent.

4 CONCLUSIONS

In Conclusion, the results presented in this work show that the chlorine removing agents have a good removal efficiency for Cl_2 and they can quickly reduce the concentration of Cl_2 leaked in the air in a short time. The proper basicity and ideal moisture were very important for the adsorption performance of the agents.

ACKNOWLEDGEMENT

This work was supported by the NSFC-Henan Talent Development Joint Fund (Grant No. U1204215).

REFERENCES

[1] W.L. Wei, L.h. Cui, X. Li, J. Shen, H. Chang, Z.M. Kou. Research progress of chlorine solid absorbent, Fire Science and Technology Vol. 30 (2011), p. 371.

[2] J.W. Xue, H.K. Zhu, Z.P. Lu, F.X. Li, T. Dou. Chlorie adsorption on A, X and SBA-15 molecular sieves, Acta Petrolei Sinica (Petroleum Processing Section) Vol. 1001–8719 (2008), p. 102.

[3] H.J. Wang, W.Q. Xiao, J.W. Xue, Z.P. Lv, F.X. Li. Chlorine adsorption properties of activated carbon, Shan Xi Chemical Industry Vol. 30 (2010), p. 1.

[4] Y.C. Luo, W.J. Zhang, H.F. Li, J.L. Deng. Development of modified powder of Ca(OH)2 and its characteristics, China Safety Science Journal Vol. 8 (1998), p. 69.

[5] HJ 547-2009, Stationary source emission—determination of chlorine—iodometric method, China standard, (2010).

Material Science and Environmental Engineering – Chen (Ed.)

Synthesis, structure characterization and magnetic property of $(Mg_{1-y}Fe_y)(Al_{0.4}Cr_xFe_{1.6-x})O_4$ spinel solid solution

X.R. Wu, C. Chen, H.H. Lü & L.S. Li

Anhui Provincial Key Laboratory of Metallurgical Engineering and Resources Recycling, Anhui University of Technology, Ma'anshan, Anhui, China

ABSTRACT: A series of compounds $(Mg_{1-y}Fe_y)(Al_{0.4}Cr_xFe_{1.6-x})O_4$ (x = 0.4, 0.8, 1.2, y = 0, 0.1) were prepared by solid state reaction and characterized by means of XRD, FT-IR and SEM. Magnetic properties were measured by VSM at room temperature. Results show that spinel is formed as a single phase in all samples and solid solution is produced by incorporation of Cr^{3+} and Fe^{2+} into structure. It is also found that the lattice parameter decrease and that magnetization structure shows a transition from ferromagnetic to paramagnetic with increasing of Cr^{3+} content. However, when coupled with Fe^{2+} doping, lattice parameter and magnetization are improved.

Keywords: spinel; solid solution; magnetic property

1 INTRODUCTION

Spinel has the general structure of AB_2O_4 and contains two cation sites for metal cation occupancy. When the A-sites are occupied by M^{2+} cations and the B-sites are occupied by M^{3+} cations, it is called a normal spinel. If the A-sites are completely occupied by M^{3+} cations and the B-sites are randomly occupied by M^{2+} and M^{3+} cations, the structure is referred to as an inverse spinel. As a matter of fact, most spinels were formed as solid solutions where both sites contain a fraction of the M^{2+} (Mg^{2+}, Fe^{2+}, Ni^{2+}, Mn^{2+}, i.e.) and M^{3+} (Al^{3+}, Cr^{3+}, Fe^{3+}) and even M^{4+} cations (Ti^{4+}, Si^{4+}) (Bragg W.H. 1915, Lavina B. 2002). Spinel-type solid solutions, which differ from the terms and quantities of cations, synthesis methods, etc, have been attracted much attention due to their flexible structures and physical properties (Grimes N.W. 1975).

(Mg, Fe^{2+})(Al, Cr, Fe^{3+})O_4 spinel solid solution was mostly found in natural chromites (Osborne M.D. et al. 1981, Oka Y. et al. 1984) and in solidified stainless steel slag (Shen H. et al. 2004). In order to beneficiate chromites from natural ores and recover chromium from slag, the structural and magnetic properties of (Mg, Fe^{2+}) (Al, Cr, Fe^{3+}) O_4 are fundamental. Among binary spinels, $FeAl_2O_4$, $MgCr_2O_4$ and $FeCr_2O_4$ are normal spinels, whereas, $MgFe_2O_4$ is inverse spinel with a collinear ferromagnetic structure (Hill R.J. et al. 1979, Lenaz D. et al. 2006). Numerous papers have been reported about structural and magnetic property of chemically more complex spinel. In the system of $FeCr_2O_4$-$FeFe_2O_4$ (Robinns M. et al. 1971, Wasilewski P. 1975, Shukla S.J. et al. 2001), with Cr_2O_3 increasing, there was the transition between the inverse spinel Fe_3O_4 with an antiferromagnetic structure and the normal spinel $FeCr_2O_4$ with conical spiral spin structure, resulting the decrease in saturation magnetism. Recently, Mg^{2+}-Fe^{2+} substitution in the tetrahedral site (Lenaz D. et al. 2004, Maksimochkin V.I. et al. 2013) and Cr^{3+}-Fe^{3+} substitution in the octahedral site (Lenaz D. et al. 2006, Nesa F. et al. 2012, Lü H. et al. 2013) in the (Mg, Fe) (Fe, Cr)$_2$ O_4 system was relatively well studied and found that the lattice parameters and magnetic characteristics were much dependent on the doping cations, the contents, and the doping sites. However, when (Mg, Fe^{2+})(Al, Cr, Fe^{3+})O_4 system is considered, the effects of the substitution of Fe^{2+}-Mg^{2+} and Cr^{3+}-Fe^{3+} on the structural and magnetic properties are yet to be explored.

The present study is designed to prepare solid solution series $(Mg_{1-y}Fe_y)(Al_{0.4}Cr_xFe_{1.6-x})O_4$ (x = 0.4, 0.8, 1.2, y = 0, 0.1) by solid state reaction with an aim to investigate the evolution of structure and their magnetic property.

2 EXPERIMENTAL

2.1 *Sample preparation*

The batch compositions to prepare $(Mg_{1-y}Fe_y)(Al_{0.4}Cr_xFe_{1.6-x})O_4$ (x = 0.4, 0.8, 1.2, y = 0, 0.1) were shown in Table 1. Reagent-grade powders of MgO, Al_2O_3, Cr_2O_3, Fe_2O_3 and $FeC_2O_4 \cdot 2H_2O$ were used for synthesis of samples.

Table 1. The batch compositions to prepare $(Mg_{1-y}Fe_y)(Al_{0.4}Cr_xFe_{1.6-x})O_4$.

Sample	MgO (g)	Al_2O_3 (g)	Fe_2O_3 (g)	Cr_2O_3 (g)	$FeCO_3 \cdot 2H_2O$ (g)
x = 0.4, y = 0	2.57	1.31	6.17	1.95	0
x = 0.8, y = 0	2.59	1.32	4.15	3.94	0
x = 1.2, y = 0	2.62	1.33	2.09	5.96	0
x = 0.4, y = 0.1	2.16	1.20	5.76	1.80	1.08
x = 0.8, y = 0.1	2.16	1.20	3.84	3.72	1.08
x = 1.2, y = 0.1	2.16	1.20	1.92	5.52	1.08

These powders were accurately weighed in the required proportions and then ground well in an agate mortar, and then were transferred to electric ball mill for 2 hours. The dried powder was pressed into pellets with 1 cm in diameter and 1 cm thick under 10 MPa. The pellets in an alumina crucible were loaded in an alumina tube and placed into a furnace which was equipped with $MoSi_2$ heating elements and the even temperature zone of the furnace was 50 mm. The y = 0 series of samples were heated in air, while the y = 0.1 series of samples were heated under N_2 atmosphere to 1550 °C for 3 h. Finally they were cooled to room temperature at the rate of 5 °C/min.

2.2 *Characterization*

X-ray diffractometer (D8 Advance, Bruker AXS), which uses Cu K_α as a radiation source (λ = 1.5406Å), is performed to collect the data in the 2θ range from 10° to 80° with a step size of 0.02°. Diffraction peaks are used to identify the structure of the samples by matching their observed patterns with the standard pattern. Infrared spectrum is recorded using FT-IR instrument (Nicolet 6700, Thermo-Nicolet) for all samples in KBr medium between 400–800 cm^{-1}. After as-obtained samples are coated with carbon, the SEM and backscattering electron images are obtained using field emission scanning electron microscope (FEI Quanta 450).

Magnetization in the samples is measured with a vibrating sample magnetometer (VSM, LakeShore 7407, USA) at ambient temperature (25 °C).

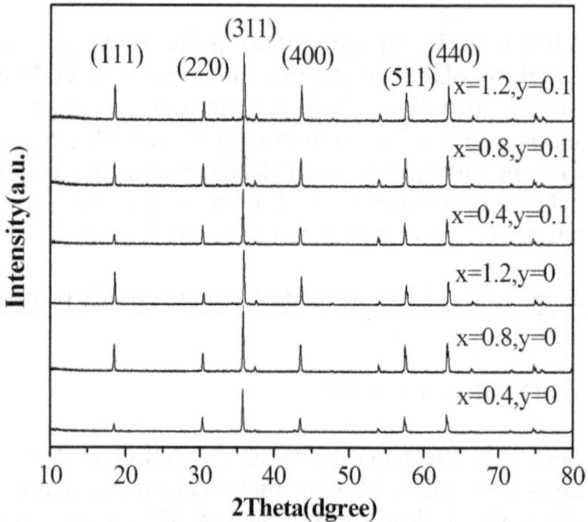

Figure 1. Typical X-ray diffraction pattern for $(Mg_{1-y}Fe_y)(Al_{0.4}Cr_xFe_{1.6-x})O_4$ system.

3 RESULTS AND DISCUSSION

3.1 *XRD analysis*

The crystal structures of the as-prepared products were examined by XRD, and the corresponding results were shown in Figure 1. The XRD patterns of the samples show a face-centered cubic structure of spinel. The peaks at round 2θ = 18.5°, 30.3°, 35.7°, 43.4°, 57.4°, and 63.0° are assigned to the (111), (220), (311), (400), (511) and (440) reflections, respectively. Reflections from the planes (222) at 37.3°, (422) at 53.8° and (533) at 74.6° have been observed with weak intensities. Well resolved peaks in XRD pattern clearly indicate that the single phase is produced in all samples and no other phase is detected.

It is also found that with increasing of x (Cr^{3+} content), the XRD spectra show that the diffraction peak becomes narrower. Similar to the crystal face (311), the position of the diffraction peak tends to shift to the high-angle region. After doping with Cr^{3+}, the principal crystalline phase remains unchanged and continues to show the characteristic diffraction peaks of the spinel structure, suggesting the substitution of Cr^{3+} for Fe^{3+} within spinel lattice.

Compared the sample series y = 0 with y = 0.1, almost the similar XRD patterns indicate that cubic structure is preserved and suggest that Fe^{2+} replaces a part of Mg^{2+} and incorporates into spinel structure.

The lattice parameters are calculated from the XRD patterns using software JADE 5.0.

Table 2. Lattice parameter and crystal size of $(Mg_{1-y}Fe_y)(Al_{0.4}Cr_xFe_{1.6-x})O_4$ samples.

	Y = 0			Y = 0.1		
Samples	X = 0.4	X = 0.8	X = 1.2	X = 0.4	X = 0.8	X = 1.2
a [Å]	8.322	8.309	8.287	8.363	8.350	8.341
D [nm]	79.63	74.76	72.73	43.48	44.65	43.94

The averaged crystal size for all samples is calculated by Scherrer's equation:

$$D = \frac{0.9\lambda}{\beta\cos\theta} \tag{1}$$

where D = averaged crystal size, λ = 0.15406 nm, β = integral breadth of a reflection (in radians 2θ) located at 2θ, and θ = Bragg angle.

As shown in Table 2, the crystal lattice constant decreases with the increase in Cr^{3+} content for samples with y = 0 and 0.1. According to crystal theory, lattice constant is much dependent on the cation size (M^{x+}) and M^{x+}-O bonding distance in the relative site. In comparison, the ionic radius of Cr^{3+} (0.063 nm) is smaller than Fe^{3+} (0.067 nm) and bonding distance of Cr^{3+}-O (0.1995 nm) is also smaller than that of Fe^{3+}-O (0.2025 nm) (Shannon R.D. 1976). Therefore, the decrease in crystal lattice constant is ascribable to the subsititution of Cr^{3+} for Fe^{3+} in spinel structure.

However, when coupled with Fe^{2+} doping, the lattice parameter slightly increases due to the replacement of smaller Mg^{2+} (0.072 nm) ions by larger Fe^{2+} (0.078 nm) ions, but the averaged crystal size decreases significantly.

3.2 *FT-IR characterization*

Figure 2 indicates the FT-IR spectra of all specimens. According to the report (Pradeep A. & Chandrasekaren G. 2006), the higher frequency band (υ_1) and lower frequency band (υ_2) are assigned to the vibration of tetrahedral (A) site and octahedral (B) site, respectively, because the band length of M-O in tetrahedral (A) site is shorter than that in octahedral (B) site. As seen from Table 3, with increasing of x (Cr^{3+} content), the lower frequency band (υ_2) shows a shift from 449 to 508 cm^{-1} (y = 0), 450 to 510 cm^{-1} (y = 0.1), due to that smaller Cr^{3+} ions replaces a part of larger Fe^{3+} ions. This further confirms that Cr^{3+} incorporates into the octahedral (B) site in spinel structure as mentioned in XRD analysis.

However, there is no obvious shift of higher frequency band (υ_1) between sample of y = 0 and y = 0.1 with the same x, as shown in Table 3.

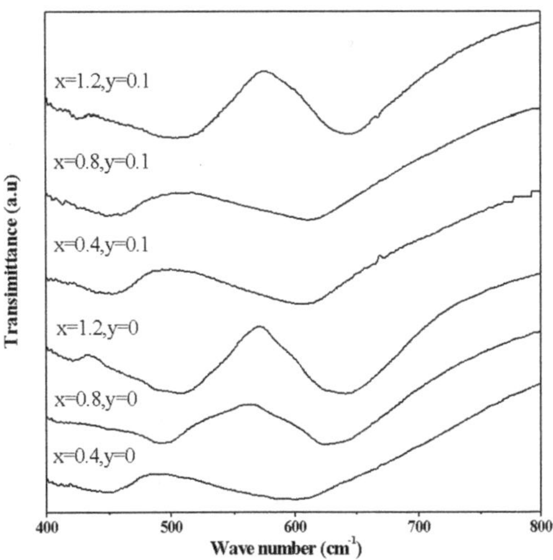

Figure 2. FT-IR spectra of $(Mg_{1-y}Fe_y)(Al_{0.4}Cr_xFe_{1.6-x})O_4$ samples.

It is may be due to that the doping content of Fe^{2+} is low.

3.3 *Magnetic property by VSM*

Figure 3 shows the hysteresis loops of $(Mg_{1-y}Fe_y)(Al_{0.4}Cr_xFe_{1.6-x})O_4$ systems, which are obtained at the ambient temperature (25 °C) under an applied field using the VSM technique. As demonstrated in Figure 3, all samples have small coercivities and remanence. For samples (x = 0.4, y = 0) and (x = 0.4, y = 0.1), magnetization (M) is easily saturated in the field of H = 1.8 T, with the saturation magnetization (M_s) of 10.59 emu/g and 15.29 emu/g, respectively. The hysteresis shows ferromagnetic behavior.

However, for samples x = 0.8 (y = 0 and 0.1) and x = 1.2 (y = 0 and 0.1), the magnetization does not saturate at H = 1.8 T. The hysteresis loops are indicative of a transition from ferromagnetic to paramagnetic characteristics, which is consistent with the investigation in $MgCr_xFe_{2-x}O_4$ spinel system (Nesa F. et al. 2012).

It was found from Figure 3 that with x increasing, the magnetization sharply reduced for sample

Table 3. IR frequency of $(Mg_{1-y}Fe_y)(Al_{0.4}Cr_xFe_{1.6-x})O_4$ samples.

Frequencies (cm^{-1})	Y = 0			Y = 0.1		
	X = 0.4	X = 0.8	X = 1.2	X = 0.4	X = 0.8	X = 1.2
(υ_2)	449	492	507	450	454	505
(υ_1)	599	624	641	606	612	643

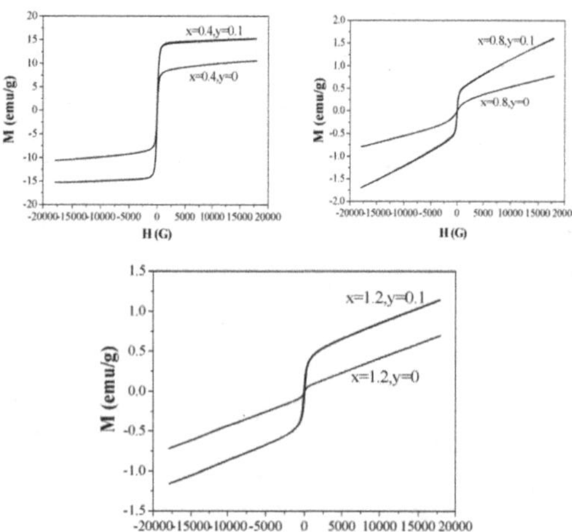

Figure 3. The specific magnetization of the $(Mg_{1-y}Fe_y)(Al_{0.4}Cr_xFe_{1.6-x})O_4$ samples.

series with y = 0 and y = 0.1, meaning that the magnetization is much dependent on Cr^{3+} doping. The result is assumed that the substitution of Cr^{3+} ($3\ \mu_B$) for Fe^{3+} ($5\ \mu_B$) has resulted in a decrease of the magnetic spins.

As compared to the samples with y = 0, the samples with y = 0.1 display a marked increase in magnetization, this is due to the subsititution of magnetic Fe^{2+} ions ($4\ \mu_B$) for non-magnetic Mg^{2+} ions ($0\ \mu_B$) into the parent lattice. All these results are correlated with the investigations by XRD and FT-IR.

3.4 *Investigation by SEM*

The backscattering images of the $(Mg_{1-y}Fe_y)(Al_{0.4}Cr_xFe_{1.6-x})O_4$ samples are shown in Figure 4. It is clear from these electron micrographs that, with increasing of Cr^{3+} content, agglomeration of particles tends to weaken and the microstructure is consisted of smaller cubic particles, which also implies smaller surface contact area and therefore reduces the orbital magnetic momentum and the intensity of paramagnetic doublet. It is an additional evident to explain the decrease of magnetization with doping of Cr^{3+}.

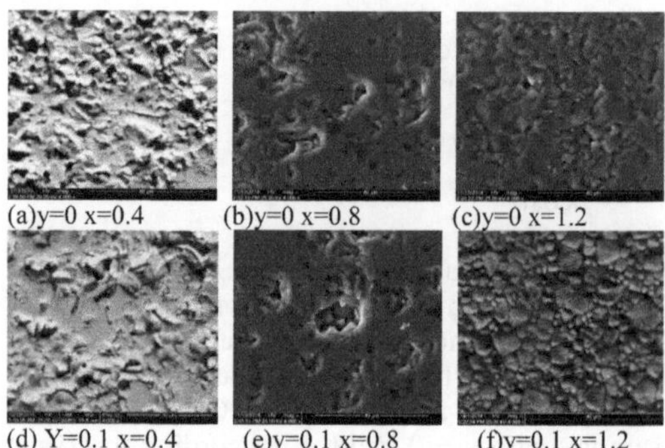

Figure 4. Scanning electron micrographs for $(Mg_{1-y}Fe_y)(Al_{0.4}Cr_xFe_{1.6-x})O_4$ samples.

Because the studied spinel solid solution series, which occur in earth's curst and in stainless steel slag, cover different content of chromium and result in the change of structure and magnetic behaviour. This study will be helpful for separating chromium-spinel from chromite and slag.

4 CONCLUSIONS

$(Mg_{1-y}Fe_y)(Al_{0.4}Cr_xFe_{1.6-x})O_4$ spinel solid solutions are synthesized by solid state reaction at high temperature 1550 °C. With Cr^{3+} doping, the lattice parameter decreases and magnetization transites from ferromagnetic into paramagnetic. When coupled with Fe^{2+} doping, however, the lattice parameter and magnetization are somewhat improved.

ACKNOWLEDGEMENTS

The authors gratefully appreciate for the financial support by the National Natural Science Foundation of China (Grant No. 51274006, No. 51204004).

REFERENCES

[1] Bragg, W.H. 1915. The structure of the spinel group of crystals. *Philosophical Magazine*, 176 (30): 305–315.

[2] Hill, R.J. Craig, J.R., & Gibbs, G.V. 1979. Systematics of the spinel structure type. *Physics and Chemistry of Minerals*, 4 (4): 317–339.

[3] Grimes, N.W. 1975. The spinels: versatile materials. *Physics in Technology*, 6 (1): 22–27.

[4] Lavina, B., Salviulo, G. & Giusta. A.D. 2002. Cation distribution and structure modelling of spinel solid solutions. *Physics and Chemistry of Minerals*, 29 (1): 10–18.

[5] Lenaz, D., Skogby, H. & Princivalle, F. 2004. Structural changes and valence states in the $MgCr_2O_4$–$FeCr_2O_4$ solid solution series. *Physics and Chemistry of Minerals*, 31 (9): 633–642.

[6] Lenaz, D., Skogby, H., Princivalle, F. & Halenius, U. 2006. The $MgCr_2O_4$–$MgFe_2O_4$ solid solution series: effects of octahedrally coordinated Fe^{3+} on T–O bond lengths. *Physics and Chemistry of Minerals*, 33 (7): 465–474.

[7] Lü, H., Zhou, G., Ma, W., Zhang, Y., Liang, P., Li, K. & Gao, L. 2013. Synthesis and characterization of $MgFe_xCr_{2-x}O_4$ by a microwave method. *Materials and Manufacturing Processes*, 28 (6): 621–625.

[8] Maksimochkin, V.I., Gubaidullin, R.R. & Gareeva, M. Ya. 2013. Magnetic Properties and Structure of $Fe_{2-x}Mg_xCrO_4$ Chromites. *Moscow University Physics Bulletin*, 68 (3): 241–248.

[9] Nesa, F., Zakaria, A.K.M., Saeed khan, M.A., Yunus, S.M., Das, A.K., Eriksson, S.-G., Khan, M.N.I. & Hakim, M.A. 2012. Structural and magnetic properties of Cr^{3+} doped Mg ferrites. *World Journal of Condensed Matter Physics*, 2 (1): 27–35.

[10] Oka, Y., Petra, S. & Chatterjee, N.D. 1984. Thermodynamic mixing properties of $Mg(Al,Cr)_2O_4$ spinel crystalline solution at high temperatures and pressures. *Contributions to Mineralogy and Petrology*, 87 (2): 196–204.

[11] Osborne, M.D., Fleet, M.E. & Bancroft G.M. 1981. Fe^{2+}-Fe^{3+} ordering in chromite and Cr-bearing spinels. *Contribution to Mineralogy and Petrology*, 77 (3): 251–255.

[12] Pradeep, A. & Chandrasekaren, G. 2006. FTIR study of Ni, Cu and Zn substituted nano-particles of $MgFe_2O_4$. *Materials Letters*, 60 (3): 371–374.

[13] Robinns, M., Wertheim, R.C., Sherwood, R.C. & Buchanan, D.N.E. 1971. Magnetic properties and site distributions in the system $FeCr_2O_4$-Fe_3O_4 ($Fe^{2+}Cr_{2-x}Fe_x^{3+}O_4$). *Journal of Physics and Chemistry of Solids*, 32 (3): 717–729.

[14] Shannon, R.D. 1976. Revised effective ionic radii and systematic studies of interatomic distances in halides and chalcogenides. *Acta Crystallographica Section A*, 32 (5): 751–767.

[15] Shen, H., Forssberg, E. & Nordström U. 2004. Physicochemical and mineralogical properties of stainless steel slags oriented to metal recovery. *Resources, Conservation and Recycling*, 40 (3): 245–272.

[16] Shukla, S.J., Jadhav, K.M. & Bichile, G.K. 2001. Study of bulk magnetic properties of the mixed spinel $MgCr_xFe_{2-x}O_4$. *Indian Journal of Pure and Applied Physics*, 39 (4): 226–230.

[17] Wasilewski, P., Virgo, D., Ulmer, G.C. & Schwerer, F.C. 1975. Magnetochemical characterization of Fe (Fe_xCr_{2-x}) O_4 spinels. *Geochimica et Cosmochimica Acta*, 39 (6–7): 889–902.

Material Science and Environmental Engineering – Chen (Ed.)
 ISBN 978-1-138-02938-5

Removal of BPA through a goethite-photocatalyzed fenton-like reaction with the enhancement of oxalic acid

T. Li & G.S. Zhang
School of Municipal and Environmental Engineering, Harbin Institute of Technology, Harbin, China

P. Wang
School of Municipal and Environmental Engineering, Harbin Institute of Technology, Harbin, China
State Key Laboratory of Urban Water Resource and Environment, Harbin, China

ABSTRACT: Goethite was synthesized successfully by the precipitation method. In order to characterize the crystal phase and micro morphology of the goethite, X Ray Diffraction (XRD) and Scanning Electron Microscope (SEM) were employed. An oxalic acid-enhanced Fenton-like (goethite/H_2O_2/oxalic acid) process in the UV irradiation was studied by the degradation of Bisphenol A (BPA) in wastewater. The results showed that the BPA removal rate and the amount of hydroxyl radicals (HO·) in goethite/H_2O_2/oxalic acid system were higher than that in goethite/H_2O_2 system. The maximum BPA removal efficiency was about 96% at pH 6 when the BPA initial concentration was 10 mg · L^{-1}, oxalic acid dosage was 0.33 mmol · L^{-1}, goethite dosage was 0.7 g · L^{-1} and H_2O_2 dosage was 5.5 mmol · L^{-1}. Moreover, the catalyst had a good stability and reusability. The BPA removal rate maintained 91% after five reaction cycles.

Keywords: goethite; photodegradation; oxalic acid; BPA

1 INTRODUCTION

Fenton process, as one of effective advanced oxidation processes, can degrade recalcitrant organic pollutants. A mass of HO· can produced with the reaction between Fe^{2+} and H_2O_2 (Lucas et al. 2007; Tunç et al. 2012; Ghiselli et al. 2004). In order to intensify the effects of treating organic pollutants, many researchers introduced photo in Fenton process. Fe^{2+} can be regenerated and additional hydroxyl radicals will be produced in the UV irradiation (Tokumura et al. 2013). Regretfully, the conventional Fenton systems can be efficient in a limited pH range (about 2.0–4.0) (Sun et al. 2009). And the the presence of dissolved iron in the solution after the treatment can not be recycled.

To overcome the drawbacks of the conventional Fenton process, a great deal of attention was focused on the development of a heterogeneous catalyst, such as α-FeOOH (goethite) (Bonnissel-Gissinger et al. 1999; Zhu et al. 2010; Krehula & Musić 2006), γ-FeOOH (lepidocrocite) (Ram et al. 1996), α-Fe_2O_3 (hematite) (Ristić & Musić 2006; Ristić et al. 2006), γ-Fe_2O_3 (maghemite) (Zhu et al. 2010; Ray et al. 2008). It is also called Fenton-like process. Those iron oxides exist extensively in the soil, aquatic sediment and waste water of mountain. And they are harmless for our environment. And they are all the high environment-friendly material. As most abundant iron oxide, goethite has stable chemical characteristics and high specific surface area (Guo et al. 2013a), and its particle structure is fine (Wang et al. 2012). The use of α-FeOOH catalyst made it possible to recycle catalyst. And it also can used efficiently in a wide pH range.

On the other hand, to reduce the cost of treatment, researchers attempted to reduce the use of H_2O_2 via using some new method. For example, some researchers found that oxalic acid which is the highest content of all the carboxylic acids can combine with Fe^{3+} and form stable complexes, and those stable complexes accelerate HO· producing in the UV irradiating (Manenti et al. 2015; Wei et al. 2013; Souza et al. 2014; Huang et al. 2010; Monteagudo et al. 2010).

In the present study, BPA was employed as a model compound of endocrine-disrupting compound. The influencing factors and the stable of goethite in the UV/Fenton-like/oxalic acid process were optimized by kinetics study.

2 MATERIALS AND METHODS

2.1 *Chemical reagents*

BPA was purchased from Tianjin Kermel Chemical Reagent Co., Ltd., China. Hydrogen peroxide (30%, w/w) and Fe $(NO_3)_3 \cdot 9H_2O$ was purchased from Sinopharm Chemical Reagent Co., Ltd., China. All other chemicals used in the experiments were purchased from Aladdin Reagent Co., Ltd. They were all of analytical grade and used without further purification. All solutions were prepared using Deionized Water (DI) at room temperature.

Photocatalytic reaction apparatus was purchased from Xi-an BL Biotechnology Co., Ltd.

2.2 *Preparation of the catalyst (α-FeOOH)*

It was prepared in a beaker and 0.5 mol·L^{-1} Fe $(NO_3)_3$ was slowly titrated with 2.5 mol·L^{-1} NaOH at a constant rate of 5 ml·min^{-1} to with a magnetically stirring up to a solution pH of 12. The suspension was then left to age in a drying oven at 60 °C for 24 h. The resulting suspensions were washed repeatedly with deionized water. After centrifugal separation, obtained solid was dried at 60 °C in vacuum dryer. The dried samples were ground in a mortar stored in a desiccator.

2.3 *Characterization of catalyst*

The crystal phase was characterized by X-Ray Diffraction (XRD), XRD was performed with a diffractometer (X'Pert PROThermo, PANalytical, Holland) using a Cu Kα radiation source under a voltage of 40 kV and current of 200 mA. XRD patterns were recorded with a scanning rate of 6°·min^{-1} from 10° to 80°. The morphology of goethite was characterized by scanning electron microscopy. The sample powder was spread on a carbon-coated sample mount and coated with gold to prevent surface charging effects.

2.4 *Photocatalytic experiments*

The photocatalytic experiments were conducted in a cylindrical reactor. The light source is a high-pressure mercury lamp (main emission lines at λ = 365 nm). A quartz tube with water recirculation was used for cooling between light and solution. Taking the desired concentration of α-FeOOH and oxalic acid to a BPA solution with an initial concentration of 10 mg·L^{-1}. The mixture was stirred by magnetic force for 30 min to achieve adsorption equilibrium respectively in the dark. Turn on the power of condensation and high-pressure mercury lamp (100 W) and wait for stabilization. Then, pour all suspension into the reactor, after adding a certain concentration of H_2O_2 to mixture. Samples were taken from the reaction solution after different periods and the total time of the irradiation for each experiment was 120 min. the supernatant was filtered through a 0.22 μm membrane filter. The sample was analyzed by UV-vis spectroscopy at 224 nm.

The removal ratio (X%) of BPA was calculated using the following equations:

$$X(\%) = \frac{C_0 - C}{C_0} \times 100\% \tag{1}$$

where C_0 and C are concentration before and after the reaction, respectively.

The effects of some parameters such as concentration of oxalic acid (from 0.11 to 0.56 mmol·L^{-1}), mass fraction of goethite in the mixture (from 0.3 to 1.1 g·L^{-1}), concentration of H_2O_2 (from 1.5 to 13.5 mmol·L^{-1}), pH (from 3 to 9), initial concentration (from 2 to 20 mg·L^{-1}) and the stability of the catalyst were studied.

2.5 *Determination of hydroxyl radicals (HO·)*

To reflect the rest of the amount of hydroxyl radicals in system, we use benzoic acid as capture agent to react with HO·. Taking 1.0 mL benzoic acid solution (1.0 × 10^{-2} mol·L^{-1}) into 10 mL colorimetric cylinder, and then adding into 1.0 mL reaction solution at different times. Finally, the solution was diluted to 5.0 mL and shaking well. The absorbency of hydroxyl radicals was determined by FP-6500 luminoscope (λex = 300 nm, λem = 407 nm, Ex = 5 nm).

2.6 *Catalyst regeneration*

For regeneration, the used catalyst was washed by deionized water and separated from the solution by centrifugation. The regenerated catalyst was placed back into the BPA solution for next photocatalytic reaction cycle. To evaluate the reusability of the catalyst, the removal rate of BPA was examined at the end of each photocatalytic reaction cycle.

3 RESULTS AND DISCUSSION

3.1 *Catalyst characterization*

Figure 1a displays the XRD patterns of the catalyst. Three major peaks were clearly observed

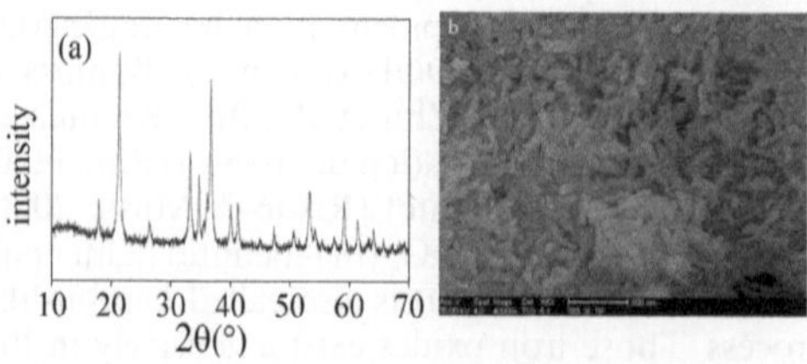

Figure 1. (a) XRD pattern and (b) SEM image of goethite catalyst.

at 2θ values of 21.28°, 33.28° and 36.74°, which were similar to the pure goethite pattern reported in the XRD standard data base library. It's demonstrated that the sample is goethite and has good crystallinity. The morphology of the α-FeOOH catalyst was characterized by SEM (Fig. 1b). The α-FeOOH particles are uniform and rod-like, and the length of particle was approximately 400~500 nm and the width was about 25~50 nm.

3.2 *Performance of different processes*

Figure 2 shows BPA removal rates at different conditions. It was certified that the combination of goethite and oxalic acid can greatly accelerate the removal of BPA. More specifically, BPA removal rates were respectively 85% and 64% in UV/goethite/H_2O_2/oxalic acid system and UV/goethite/H_2O_2 system. Oxalic acid adding could improve about 21% removal rates. The Fe^{3+} and oxalic acid formed many complexes, such as $Fe(C_2O_4)_2^-$, $Fe(C_2O_4)_3^{3-}$, which were high photochemical activity (Quici et al. 2005; Chen & Zhu 2011; Lan et al. 2010).

Figure 3 shows the amount of hydroxyl radical with benzoic acid under different conditions. It was observed that the amount of hydroxyl radical in UV/goethite/H_2O_2/oxalic acid system was more than that in UV/goethite/H_2O_2 system. And the amount of hydroxyl radical increased over time, but it decreased when the amount of hydroxyl radical reached a certain degree. Therefore, the mechanic of the photo-degradation involved the hydroxyl radical.

3.3 *Effect of oxalic acid dosage*

In order to determine the effect of oxalic acid on photo-degradation of BPA in goethite/H_2O_2/oxalic acid system, a set of experiments with different dosages of oxalic acid in the range of 0~0.56 mmol·L^{-1} without pH control (pH = 6) were treated under UV irradiation. The results were given in Figure 4. It is shown that the removal rate of BPA was increased slightly after the addition of 0.11 mmol·L^{-1} oxalic acid compared with no oxalic acid. But the removal rate of BPA was about 84% when the concentration of oxalic acid was 0.33 mmol·L^{-1}, it is raised by almost 20% compared with no oxalic acid. The removal rates of 81% and 78% were observed at an initial concentration of 0.44 mmol·L^{-1} and 0.55 mmol·L^{-1}. The removal of BPA was increased with the increase of oxalic acid concentration in the low range, but was inhibited with an excess of oxalic acid. The result was similar to some studies previously (Lee et al. 2014; Tokumura et al. 2013).

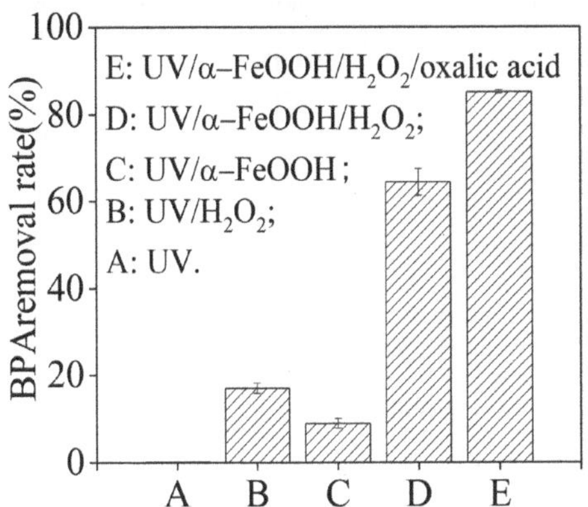

Figure 2. BPA removal rate at different conditions. (BPA initial concentration = 10 mg·L^{-1}; goethite dosage = 0.5 g·L^{-1}; H_2O_2 dosage = 1.5 mmol·L^{-1}; pH = 6.0; mercury lamp power = 100 W).

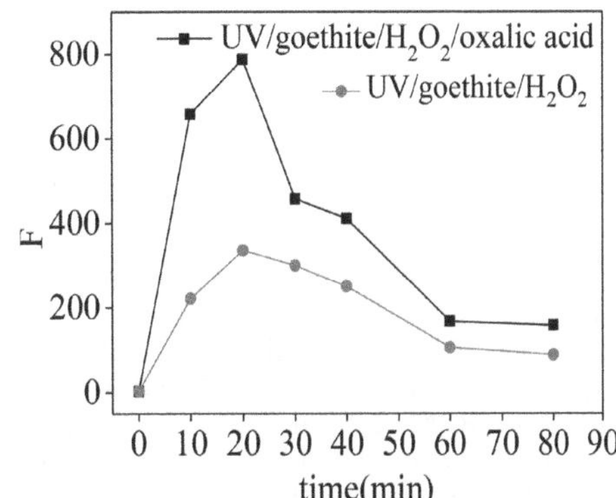

Figure 3. Fluorescent method for determination of hydroxyl radical with benzoic acid under different conditions.

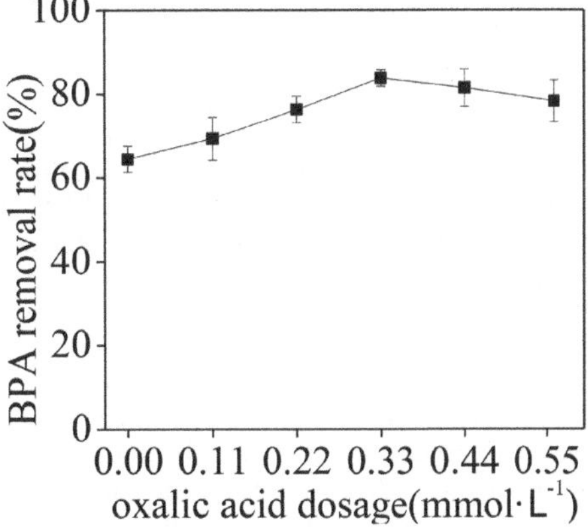

Figure 4. BPA removal rate at different oxalic acid dosage. (BPA initial concentration = 10 mg·L^{-1}; goethite dosage = 0.5 g·L^{-1}; H_2O_2 dosage = 1.5 mmol·L^{-1}; pH = 6.0; mercury lamp power = 100 W).

As a kind of organic matter, oxalic acid could be able to compete the adsorptive sites with BPA, resulting in a decrease in the amount of BPA adsorbed.

3.4 *Effect of photocatalyst dosage*

Figure 5 describes the effect of goethite dosage on the removal of BPA. According to Figure 5, in the case of reaction with goethite dosage of 0.3 g·L^{-1}, 0.5 g·L^{-1}, 0.7 g·L^{-1}, about 77%, 84% and 87% of BPA were removed after 120 min of irradiation. But compared with 0.7 g·L^{-1} goethite dosage, the removal rate of BPA decreased by 5% and 6% approximately, for 0.9 g·L^{-1} and 1.1 g·L^{-1}. Obviously, the BPA removal rate was increased significantly with the increase of goethite dosage at low level, and excess goethite could decrease the removal rate.

The reasons for the phenomenon could be that high goethite concentrations led to the arrival of less photon at the catalyst surface, since turbid solution could reduce the light penetration (Mittal et al. 2014; Xiao et al. 2015; Zhong et al. 2007). Thus, there should be an optimal dosage of goethite to achieve the maximum removal rate of BPA. The optimal goethite dosage was at 0.7 g·L^{-1} and the corresponding removal rate of BPA was 87%.

3.5 *Effect of H_2O_2 concentration*

It is important to investigate the effect of H_2O_2 concentration because it is directly connected with the amount of hydroxyl radicals and the treatment efficiency of BPA in the photocatalytic reaction. The effect of H_2O_2 dosage on the degradation of BPA by process was investigated over the range from 1.5 to 13.5 mmol·L^{-1}, and the result was given in Figure 6.

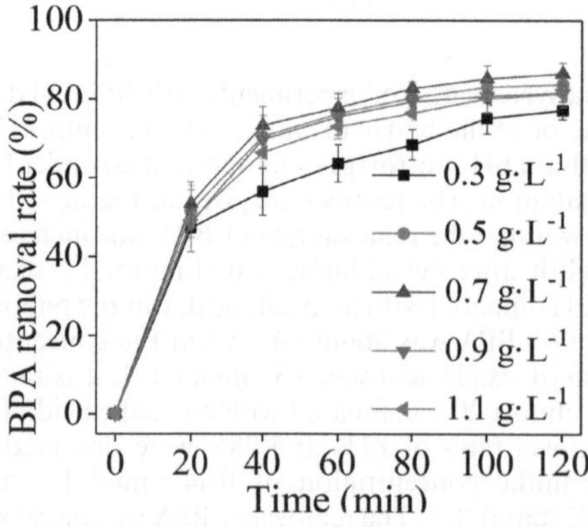

Figure 5. BPA removal rate in different goethite dosage. (BPA initial concentration = 10 mg·L^{-1}; oxalic acid dosage = 0.33 mmol·L^{-1}; H_2O_2 dosage = 1.5 mmol·L^{-1}; pH = 6.0; mercury lamp power = 100 W).

As shown in Figure 6 and Table 2, with increase of the H_2O_2 concentration from 1.5 to 5.5 mmol·L^{-1}, the BPA removal rate increased from 86% to 96%, and the second-order rate constants increased obviously from 0.0051 to 0.0225 mol^{-1}·L·min^{-1}. After that, a further increase of H_2O_2 from 5.5 to 13.5 mmol·L^{-1} could not caused the increase of BPA removal rate and k. On the contrary, it caused k decreasing from 0.0225 to 0.0119 mol^{-1}·L·min^{-1}, and the efficiency of BPA simultaneously removal

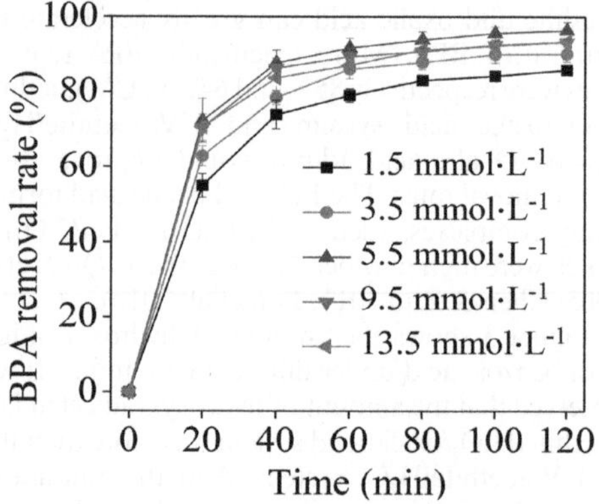

Figure 6. (a) BPA removal rate in different H_2O_2 concentration. (BPA initial concentration = 10 mg·L^{-1}; goethite dosage = 0.7 g·L^{-1}; oxalic acid dosage = 0.33 mmol·L^{-1}; pH = 6.0; mercury lamp power = 100 W).

Table 1. Second-order kinetics constants of removing BPA in different goethite dosage.

Goethite dosage (g·L^{-1})	k (mol^{-1}·L·min^{-1})	R^2
0.3	0.0028	0.9823
0.5	0.0049	0.9735
0.7	0.0055	0.9877
0.9	0.0039	0.9357
1.1	0.0037	0.9781

Table 2. Second-order kinetics constants of removing BPA in different H_2O_2 concentration.

H_2O_2 concentration (mmol·L^{-1})	k (mol^{-1}·L·min^{-1})	R^2
1.5	0.0051	0.9645
3.5	0.0080	0.9566
5.5	0.0225	0.9861
9.5	0.0116	0.9777
13.5	0.0119	0.9928

fell to 93%. According to Equation 2 and Equation 3, the increase of the H_2O_2 could lead to the increase of HO· concentration in the UV irradiation. And as a strong oxidant, HO· could remove BPA efficiently (Neamtu et al. 2003; Karci et al. 2012; Papić et al. 2009).

$$H_2O_2 + h\nu \rightarrow 2HO\cdot \quad (2)$$

$$H_2O_2 + Fe^{2+} \rightarrow HO\cdot + OH^- + Fe^{3+} \quad (3)$$

$$H_2O_2 + HO\cdot \rightarrow H_2O + HO_2\cdot \quad (4)$$

However, according to Equation 4, an excess of H_2O_2 could consume hydroxyl radical in the system, causing the reduction of BPA removal (Lunar et al. 2000; Martins et al. 2013). Thus, the optimal H_2O_2 concentration was at 5.5 mmol·L^{-1} and the corresponding removal rate of BPA was 96%.

3.6 *Effect of pH concentration*

The initial pH value of the reaction is an important factor on removing BPA. Figure 7 depicts the removal of BPA at initial pH of 3~9. According to Figure 7, the rate of BPA removal slightly decreased with the decrease of initial pH from 6 to 3. This result does not correspond with the maximum efficiency at 3 reported in the literature (Lin et al. 2014; Hermosilla et al. 2009), and can be explained by the iron species dissolved in the water. With the enhancement of acidity in the reaction, the dissolved iron species would become excessive and could cause reduction of chromaticity color. On the other hand, the hydrogen ions (H^+) would react with hydroxyl radical producted in the photocatalytic process.

The efficiency of BPA removal dropped significantly when the initial pH increased from 6 to 9. Specifically, the removal rate of BPA was only 13% at pH 9. The zeta potential of goethite decreases as the pH value of the suspension increases (Fig. 7b). Increasing pH would lead to solution carrying more negative charge, which had similar charged with BPA molecules. Thus, the precipitation of ferric ion and the electrostatic repulsion would lead to the decrease removal rate (Ruales-Lonfat et al. 2015; Guo et al. 2013b; He et al. 2005). The results indicated that removal efficiency of BPA was remarkable without the control of pH, and the initial pH was closely 6.

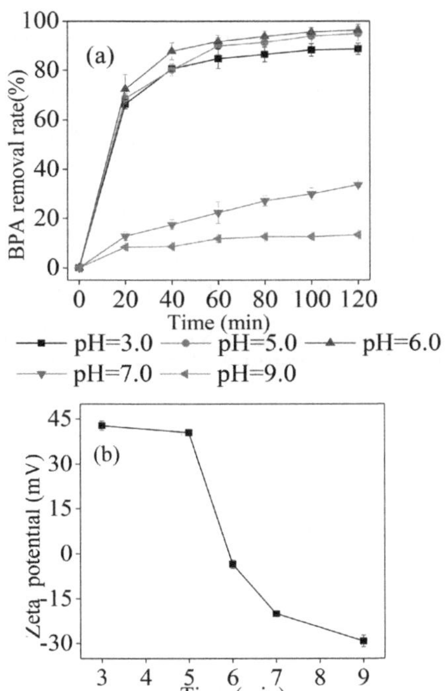

Figure 7. (a) BPA removal rate and (b) Zeta potential of goethite in different pH. (BPA initial concentration = 10 mg·L^{-1}; goethite dosage = 0.7 g·L^{-1}; H_2O_2 dosage = 5.5 mmol·L^{-1}; oxalic acid dosage = 0.33 mmol·L^{-1}; mercury lamp power = 100 W).

3.7 *Effect of initial BPA concentration*

Figure 8 shows the removal curves at different initial BPA concentrations. The BPA removal rates were 90%, 97%, 97%, 92% and 81% at BPA concentrations of 2, 8, 10, 15 and 20 mg·L^{-1}.

Table 3 lists the rate constants (k) of the reaction. It is indicated that the rate constants decreased from 0.0386 to 0.0018 mol^{-1}·L·min^{-1} with the increasing of the initial BPA concentrations from 2 to 20 mg·L^{-1}. The result is similar to many literatures (Zhou et al. 2014; Amiri & Nezamzadeh-Ejhieh

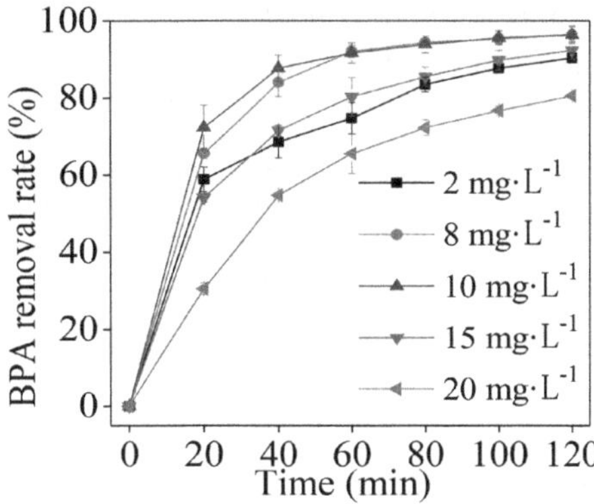

Figure 8. BPA removal rate in different initial BPA concentration. (goethite dosage = 0.7 g·L^{-1}; H_2O_2 dosage = 5.5 mmol·L^{-1}; oxalic acid dosage = 0.33 mmol·L^{-1}; pH = 6; mercury lamp power = 100 W).

Table 3. Second-order kinetics constants of removing BPA in different BPA concentration.

BPA concentration ($mg \cdot L^{-1}$)	k ($mol^{-1} \cdot L \cdot min^{-1}$)	R^2
2	0.0386	0.9533
8	0.0289	0.9824
10	0.0225	0.9861
15	0.0068	0.9537
20	0.0018	0.9943

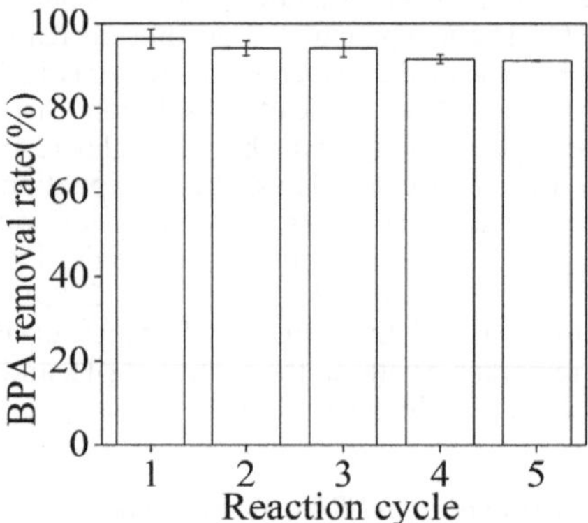

Figure 9. BPA removal rate in different reaction cycle. (BPA initial concentration = 10 $mg \cdot L^{-1}$; goethite dosage = 0.7 $g \cdot L^{-1}$; oxalic acid dosage = 0.33 $mmol \cdot L^{-1}$; H_2O_2 dosage = 5.5 $mmol \cdot L^{-1}$; pH = 6.0; mercury lamp power = 100 W).

2015; Lin & Hsu 2013). There might be many intermediates producted in reaction at high concentration of BPA. And the intermediates would consume ·OH, which was species removing BPA.

3.8 *Stability of the catalyst*

Figure 9 shows BPA removal rates in different reaction cycles. The catalyst stability during wastewater treatment over multiple reaction cycles is significant for practice applications. The catalyst was regenerated by washing with DI water and drying in oven at 60 °C after each cycle. As shown in Figure 9, the BPA removal rate decreased slightly after each regeneration. When the catalyst was reused five times, the BPA removal efficiency still remained more than 90%.

4 CONCLUSIONS

The treatment of BPA wastewater by goethite/H_2O_2/oxalic acid system in mercury lamp irradiation was studied via optimization of oxidation process and calculation of kinetics study. The presence of oxalic acid greatly increased the removal rate of BPA due to synergistic effect with the Fe3+ formed by goethite. The removal rate of BPA in goethite/H_2O_2/oxalic acid process increased by 21% compared with condition without oxalic acid. Under the optimal experiment conditions (oxalic acid dosage was 0.33 $mmol \cdot L^{-1}$, goethite dosage was 0.7 $g \cdot L^{-1}$, H_2O_2 concentration was 5.5 $mmol \cdot L^{-1}$ and pH was 6), the BPA removal rate was able to reach more than 96% and the corresponding k was equal to 0.225 $mol^{-1} \cdot L \cdot min^{-1}$. The catalyst was stable, and the BPA removal rate maintains about 91% after five reaction cycles.

REFERENCES

[1] Amiri, M., & A. Nezamzadeh-Ejhieh. 2015. Improvement of the photocatalytic activity of cupric oxide by deposition onto a natural clinoptilolite substrate. *Materials Science in Semiconductor Processing* 31:501–508.

[2] Bonnissel-Gissinger, P., M et al. Ehrhardt, & P. Behra. 1999. Modeling the Adsorption of Mercury (II) on (Hydr)oxides II: alpha-FeOOH (Goethite) and Amorphous Silica. *Journal of Colloid and Interface Science* 215 (2):313–322.

[3] Chen, J., & L. Zhu. 2011. Oxalate enhanced mechanism of hydroxyl-Fe-pillared bentonite during the degradation of Orange II by UV-Fenton process. *Journal of Hazardous Materials* 185 (2–3):1477–1481.

[4] Ghiselli, G. et al. 2004. Destruction of EDTA using Fenton and photo-Fenton-like reactions under UV-A irradiation. *Journal of Photochemistry and Photobiology A: Chemistry* 167 (1):59–67.

[5] Guo, X. et al. 2013a. Sorption thermodynamics and kinetics properties of tylosin and sulfamethazine on goethite. *Chemical Engineering Journal* 223:59–67.

[6] Guo, X. et al. 2013b. Sorption thermodynamics and kinetics properties of tylosin and sulfamethazine on goethite. *Chemical Engineering Journal* 223:59–67.

[7] He, J. et al. 2005. Photoreaction of aromatic compounds at α-FeOOH/H_2O interface in the presence of H_2O_2: evidence for organic-goethite surface complex formation. *Water Research* 39 (1):119–128.

[8] Hermosilla, D. et al. 2009. Optimizing the treatment of landfill leachate by conventional Fenton and photo-Fenton processes. *Science of the Total Environment* 407 (11):3473–3481.

[9] Huang, Y. et al. 2010. Degradation of phenol using low concentration of ferric ions by the photo-Fenton process. *Journal of the Taiwan Institute of Chemical Engineers* 41 (6):699–704.

[10] Karci, A. et al. 2012. Transformation of 2,4-dichlorophenol by H_2O_2/UV-C, Fenton and photo-Fenton processes: Oxidation products and toxicity evolution. *Journal of Photochemistry and Photobiology A: Chemistry* 230 (1):65–73.

[11] Krehula, S., & S. Musić. 2006. Influence of ruthenium ions on the precipitation of α-FeOOH, α-Fe_2O_3 and Fe_3O_4 in highly alkaline media. *Journal of Alloys And Compounds* 416 (1–2):284–290.

[12] Lan, Q. et al. 2010. Heterogeneous photodegradation of pentachlorophenol and iron cycling with goethite, hematite and oxalate under UVA illumination. *Journal Of Hazardous Materials* 174 (1–3):64–70.

[13] Lee, J. et al. 2014. Oxidation of aquatic pollutants by ferrous–oxalate complexes under dark aerobic conditions. *Journal Of Hazardous Materials* 274: 79–86.

[14] Lin, C., & L. Hsu. 2013. Removal of polyvinyl alcohol from aqueous solutions using P-25 TiO_2 and ZnO photocataysts: A comparative study. *Powder Technology* 246:351–355.

[15] Lin, Z. et al. 2014. Kinetics and products of PCB28 degradation through a goethite-catalyzed Fenton-like reaction. *Chemosphere* 101:15–20.

[16] Lucas, M.S. et al. Amaral, & J.A. Peres. 2007. Degradation of a textile reactive Azo dye by a combined chemical–biological process: Fenton's reagent-yeast. *Water Research* 41 (5):1103–1109.

[17] Lunar, L. et al. 2000. Degradation of photographic developers by Fenton's reagent: condition optimization and kinetics for metol oxidation. *Water Research* 34 (6):1791–1802.

[18] Manenti, D.R. et al. 2015. Insights into solar photo-Fenton process using iron (III)–organic ligand complexes applied to real textile wastewater treatment. *Chemical Engineering Journal* 266:203–212.

[19] Martins, R.C. et al. 2013. Catalytic activity of low cost materials for pollutants abatement by Fenton's process. *Chemical Engineering Science* 100:225–233.

[20] Mittal, M. et al. 2014. UV–Visible light induced photocatalytic studies of Cu doped ZnO nanoparticles prepared by co-precipitation method. *Solar Energy* 110:386–397.

[21] Monteagudo, J.M. et al. 2010. Catalytic degradation of Orange II in a ferrioxalate-assisted photo-Fenton process using a combined UV-A/C–solar pilot-plant system. *Applied Catalysis B: Environmental* 95 (1–2):120–129.

[22] Neamtu, M. et al. 2003. Oxidation of commercial reactive azo dye aqueous solutions by the photo-Fenton and Fenton-like processes. *Journal of Photochemistry and Photobiology A: Chemistry* 161 (1):87–93.

[23] Papić, S. et al. 2009. Decolourization and mineralization of commercial reactive dyes by using homogeneous and heterogeneous Fenton and UV/Fenton processes. *Journal of Hazardous Materials* 164 (2–3):1137–1145.

[24] Quici, N. et al. 2005. Oxalic acid destruction at high concentrations by combined heterogeneous photocatalysis and photo-Fenton processes. *Catalysis Today* 101 (3–4):253–260.

[25] Ram, S. et al. 1996. Structural features in the formation of the green rust intermediate and γ-FeOOH. *Colloids and Surfaces* 113:97–105.

[26] Ray, I. et al. 2008. Room temperature synthesis of γ-Fe_2O_3 by sonochemical route and its response towards butane. *Sensors and Actuators B: Chemical* 130 (2):882–888.

[27] Ristić, M. et al. 2006. Properties of γ-FeOOH, α-FeOOH and α-Fe_2O_3 particles precipitated by hydrolysis of $Fe3^+$ ions in perchlorate containing aqueous solutions. *Journal of Alloys And Compounds* 417 (1–2):292–299.

[28] Ristić, M., & S. Musić. 2006. Formation of porous α-Fe_2O_3 microstructure by thermal decomposition of $Fe(IO_3)_3$. *Journal of Alloys And Compounds* 425 (1–2):384–389.

[29] Ruales-Lonfat, C. et al. 2015. Iron oxides semiconductors are efficients for solar water disinfection: A comparison with photo-Fenton processes at neutral pH. *Applied Catalysis B: Environmental* 166–167:497–508.

[30] Souza, B.M. et al. 2014. Intensification of a solar photo-Fenton reaction at near neutral pH with ferrioxalate complexes: A case study on diclofenac removal from aqueous solutions. *Chemical Engineering Journal* 256:448–457.

[31] Sun, S. et al. 2009. Decolorization of an azo dye Orange G in aqueous solution by Fenton oxidation process: Effect of system parameters and kinetic study. *Journal of Hazardous Materials* 161 (2–3):1052–1057.

[32] Tokumura, M. et al. 2013. Dynamic simulation of degradation of toluene in waste gas by the photo-Fenton reaction in a bubble column. *Chemical Engineering Science* 100:212–224.

[33] Tunç, S. et al. 2012. On-line spectrophotometric method for the determination of optimum operation parameters on the decolorization of Acid Red 66 and Direct Blue 71 from aqueous solution by Fenton process. *Chemical Engineering Journal* 181–182:431–442.

[34] Wang, S. et al. 2012. Fabrication of superhydrophobic spherical-like α-FeOOH films on the wood surface by a hydrothermal method. *Colloids and Surfaces A: Physicochemical and Engineering Aspects* 403:29–34.

[35] Wei, S. et al. 2013. Decolorization of organic dyes by zero-valent iron in the presence of oxalic acid and influence of photoirradiation and hexavalent chromium. *Journal of Molecular Catalysis A: Chemical* 379:309–314.

[36] Xiao, J. et al. 2015. Organic pollutants removal in wastewater by heterogeneous photocatalytic ozonation, 1–17.

[37] Zhong, J. et al. 2007. Photocatalytic degradation of gaseous benzene over TiO_2/Sr_2CeO_4: Preparation and photocatalytic behavior of TiO_2/Sr_2CeO_4. *Journal of Hazardous Materials* 140 (1–2):200–204.

[38] Zhou, Y. et al. 2014. Visible-light-driven photocatalytic properties of layered double hydroxide supported-Bi_2O_3 modified by Pd(II) for methylene blue. *Advanced Powder Technology*.

[39] Zhu, B. et al. 2010. Enhanced third-order nonlinear optical susceptibilities of α-FeOOH nanorods. *Optical Materials* 32 (9):1244–1247.

[40] Zhu, H.Y. et al. 2010. Preparation, characterization, adsorption kinetics and thermodynamics of novel magnetic chitosan enwrapping nanosized γ-Fe_2O_3 and multi-walled carbon nanotubes with enhanced adsorption properties for methyl orange. *Bioresource Technology* 101 (14):5063–5069.

Material Science and Environmental Engineering – Chen (Ed.)
© 2016 Taylor & Francis Group, London, ISBN 978-1-138-02938-5

Investigation on the phase stability and thermal conductivity of Sc_2O_3, Y_2O_3 co-doped ZrO_2 coatings

H.F. Liu, X. Xiong, Y.L. Wang & X. Chen
State Key Laboratory of Powder Metallurgy, Central South University, Changsha, China

ABSTRACT: The phase stability and thermal-physical properties of Sc_2O_3 and Y_2O_3 co-doped ZrO_2 (ScYSZ) plasma sprayed coatings were investigated and the data were compared with those of the state-of-art Y_2O_3 stabilized ZrO_2 (YSZ). The results indicated that (5–7 mol.%) Sc_2O_3-0.5 mol.%Y_2O_3-ZrO_2 shows higher resistance to destabilization of metastable tetragonal phase in comparison with YSZ. The increase in Sc_2O_3 content can effectively promote the phase stability and reduce thermal conductivity. After heat treatment at 1500°C for 100h, all the ScYSZ coatings are composed of 100 mol.% metastable tetragonal (t′) phase, while 22.4 mol monoclinic phase appears in YSZ and that increases it to 31.6 mol.% after 300h heat treatment. No monoclinic phases are formed of 6Sc0.5YSZ and 7Sc0.5YSZ coatings even after it was heat treated for 300h, exhibiting outstanding phase stability. ScYSZ also shows significantly lower thermal conductivity than YSZ, which can be mainly attributed to the stable tetragonal structure, increased oxygen vacancies and substitution atoms.

Keywords: thermal barrier coating; Sc_2O_3; Y_2O_3-ZrO_2; phase stability; thermal conductivity

1 INTRODUCTION

Thermal Barrier Coatings (TBCs) have been found on an increasing number of applications in aircraft and gas-turbine engines to protect high-temperature metallic components by reducing the temperature of metal alloy substrates [1,2]. Up to now, the most successful TBC materials, in use, are based on 6–8wt% Y_2O_3 stabilized ZrO_2 (YSZ), which shows the best phase stability and mechanical property [3]. However, the destructive tetragonal-to-monoclinic phase transformation, which is accompanied by 3–5 vol.% volume expansion limit its long-term service is at a temperature above 1200°C [4]. Moreover, the YSZ coating shows poor sintering resistance at high temperature, resulting in an increase in thermal conductivity and Young's modulus, which can lead to the spallation and failure of TBC [5]. Co-doping rare earth oxides is considered as an effective method to develop alternative ceramic materials with high phase stability and low thermal conductivity. In recent years, a great deal of effort has been devoted to explore new TBC materials by further doping traditional YSZ with another metal oxide, such as Ta_2O_5, Sc_2O_3, La_2O_3, Ta_2O_5, Gd_2O_3, Yb_2O_3, etc, [6–15]. Among these multiple oxide systems, Sc_2O_3-Y_2O_3-ZrO_2 (ScYSZ) has been considered as one of the most promising one, owing to its satisfactory properties, including high phase stability, sintering resistance and low thermal conductivity. Jones et al. found that Sc_2O_3, Y_2O_3 co-doped ZrO_2 exhibited significantly higher tetragonal phase stability at 1400°C than YSZ, when the stabilizer's content is about 7 mol.% and the percent of Sc_2O_3 is about 90–95% [8]. X. Huang et al. found Sc_2O_3 also plays an important role in decreasing thermal conductivity that the addition of 2.71wt.%Sc_2O_3 can effectively reduce the thermal conductivity of YSZ [16].

In our former work, we reported that 8 mol.%Sc_2O_3 and 0.6 mol.%Y_2O_3 co-doped ZrO_2 exhibits improved phase stability, promoted sintering resistance as well as reduced thermal conductivity in comparison with YSZ [17]. In this paper, (5–7 mol.%) Sc_2O_3-0.5 mol.%Y_2O_3-ZrO_2 coatings were prepared by plasma spraying method. The effect of Sc_2O_3 content on the phase stability and thermo-physical properties of the coatings were investigated in detail. The relationship between the phase structure and thermal conductivity was also discussed.

2 EXPERIMENTAL PROCEDURES

2.1 *Coatings preparation*

Xmol.% Sc_2O_3-0.5 mol.%Y_2O_3-ZrO_2 (XSc0.5YS-Z, X = 5, 6, 7) ceramic powders obtained by chemical co-precipitation, then agglomerated via spray drying, were used as coating materials. An aluminum plated machine with grooves (φ10 × 1 mm) was used as substrate. The ScYSZ powders were plasma sprayed onto the substrate until the grooves were deposited entirely. Then the disk-shaped coatings were detached from the aluminum substrate using 50/50 NaOH-H_2O solution. For comparison, the traditional 8 wt.%Y_2O_3-ZrO_2 (YSZ) coating was also prepared under the same condition. The Plasma Spraying (APS) uses a Plasma-Technik A3000S system (Sulzer Metco AG F4-MB gun, Switzerland). The detailed parameters of plasma spraying are listed in Table 1.

2.2 *Characterization*

For the phase stability assessment, the disk-shaped coatings were set in alumina crucible and heat treated at 1500°C using muffle furnace. The phase structures were identified by X-ray diffraction (CuKα radiation, 40 KV, 300 mA). The mole fraction of monoclinic phase M_m, tetragonal phase $M_{t/t'}$, and cubic phase M_c were determined from the common equation (1) and (2) [6]:

$$M_m/M_{c,t'} = 0.82\times[I_m(\bar{1}11)+I_m(111)]/I_{c,t'}(111) \quad (1)$$

$$M_c/M_{t'} = 0.88\times I_c(400)/[I_{t'}(400)+I_{t'}(004)] \quad (2)$$

The disc-shaped coatings with dimensions of φ10 × 1 mm were used for thero-physical properties measurements. The density of the sample was measured using Archimedes's method. The specific heat capacity was measured by laser radiation method as reported in literature [17]. Before the measurements, the test faces of the specimen were coated with a thin film of gold to increase the absorption of laser pulses. The thermal diffusivity between 20°C to 700°C was measured using a laser flash apparatus. The thermal conductivity λ was calculated from the density ρ, heat capacity C_p and thermal diffusivity α, using the equation (3):

$$\lambda = \rho\alpha C_p \quad (3)$$

Because the samples were not fully dense, the measured thermal conductivity λ was modified for the actual data λ_0 using equation (4) [18]:

Table 1. Parameters of plasma spraying.

Parameters	ScYSZ, YSZ
Current (A)	560
Voltage (V)	72
Primary gas (Ar, L/min)	35
Secondary gas (H_2, L/min)	11
Powder feed rate (g/min)	40
Spray distance (mm)	60

$$\lambda/\lambda_0 = 1-4\varphi/3 \quad (4)$$

where φ is the fraction of porosity inside the sintered sample which was determined using the relationship $\varphi = 1-\rho/\rho_t$, ρ_t is the theoretical density.

3 RESULTS AND DISCUSSION

3.1 *Phase stability*

Figure 1 shows the XRD patterns of ScYSZ and YSZ coatings after it was heat treated at 1500°C for 100h. Table 2 listed the phase composition, cell parameters, and tetragonality. It can be seen that all the ScYSZ coatings are composed of tetragonal ZrO_2. According to literature [19], the non-transformable tetragonal phase t′ and transformable tetragonal phase t can be distinguished by tetragonality '$c/\sqrt{2}$ a'. The ratio tends to be 1.010 for t′ phase while it is superior to 1.010 for t phase. Table 2 shows that the tetragonalities of ScYSZ coatings are all below 1.010, so the tetragonal phase of ScYSZ coatings is metastable t′ phase. No monoclinic ZrO_2 (m) is detected, indicating ScYSZ coatings have good resistance to destabilization of metastable tetragonal phase. The phase compositions in YSZ are t′, c, and m. The content of m phase is about 22.4 mol% calculated by Eq. (1). The formation of m phase is consistent with the transformation from t′ to m via the process t′ → c + t → c + m. The XRD patterns of ScYSZ and YSZ coatings after it was heat treated at 1500°C for 300h are shown in Figure 2. Comparing with the diffraction patterns of the coatings after it was heat treated for 100h, the most significant difference is that m phase (10.8 mol.%) appears in 5Sc0.5YSZ. In addition, the content of m phase in YSZ coating increases from 22.4 mol.% to 31.6 mol.%, which

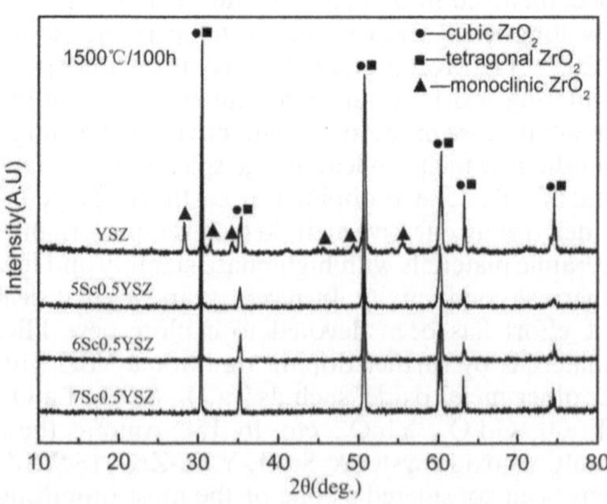

Figure 1. XRD patterns of ScYSZ and YSZ coatings after heat treatment at 1500°C for 100h.

Table 2. Phase composition and cell parameters of ScYSZ and YSZ coatings.

	Phase composition (mol.%)		
Sample ID	m	t′/c	Tetragonality
YSZ	22.4	77.6	1.0025 (100 h)
	31.6	68.4	1.0008 (300 h)
5Sc0.5YSZ	–	100	1.0037 (100 h)
	10.8	89.2	1.0042 (300 h)
6Sc0.5YSZ	–	100	1.0045 (100 h)
	–	100	1.0051 (300 h)
7Sc0.5YSZ	–	100	1.0047 (100 h)
	–	100	1.0048 (300 h)

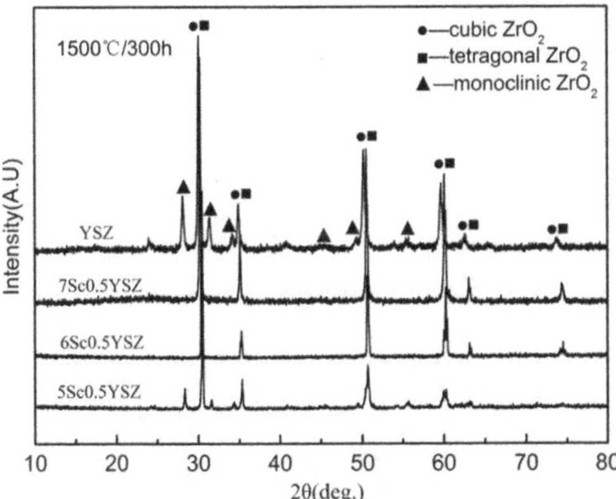

Figure 2. XRD patterns of ScYSZ and YSZ coatings after heat treatment at 1500°C for 300h.

can be attributed to the partitioning of t′ phase. At high temperatures, t′ phase undergoes a diffusion-controlled partitioning to c phase and t phase, with t phase transforming to m phase on cooling. The presence of t′ phase is beneficial to the durability of materials used as TBCs due to its high toughness. No m phase is detected in 6Sc0.5YSZ and 7Sc0.5YSZ coatings even after 300h heat treatment, which indicates that Sc_2O_3 and Y_2O_3 co-stabilized ZrO_2 exhibits better phase stability in comparison with the traditional YSZ.

3.2 *Thermal conductivity*

Figure 3 and Figure 4 show the specific heat capacities and thermal diffusivities of ScYSZ and YSZ coatings, respectively. It can be seen that the specific heat capacities of all the coatings increase as the temperature rises. This is because the higher the temperature, the faster the lattice vibration

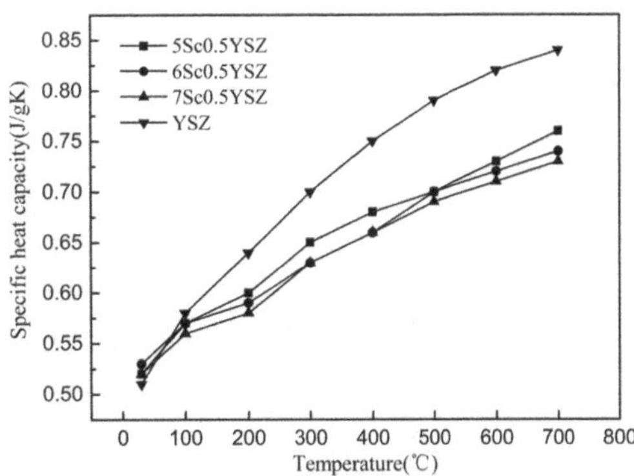

Figure 3. Specific heat capacity of ScYSZ and YSZ coatings.

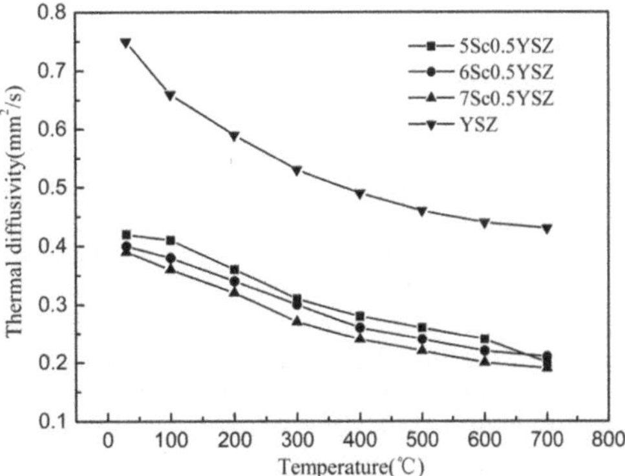

Figure 4. Thermal diffusivity of ScYSZ and YSZ coatings.

frequency. The thermal diffusivities of all the coatings decrease as the temperature rises, which reflects the typical characteristics of the phonon conduction. The specific heat capacity and thermal diffusivity of ScYSZ are both lower than that of YSZ, which directly contributes the lower thermal conductivity of ScYSZ.

Figure 5 shows the thermal conductivities of both porous and dense samples of ScYSZ and YSZ coatings. It can be seen that the thermal conductivities (both λ and λ_0) of both YSZ and ScYSZ decrease with rising temperatures. All the porous and dense samples of ScYSZ coatings have lower thermal conductivities than those of YSZ. The increase in Sc_2O_3 content can effectively reduce the thermal conductivity of ScYSZ. The porous thermal conductivity (λ) of YSZ is 1.89–1.74 W/m·K at room temperature and those of 5Sc0.5YSZ, 6Sc0.5YSZ, and 7Sc0.5YSZ are 1.06–0.74 W/m·K, 1.02–0.73 W/m·K, and 1.00–0.67 W/m·K, respectively. Taking no account of the effect of

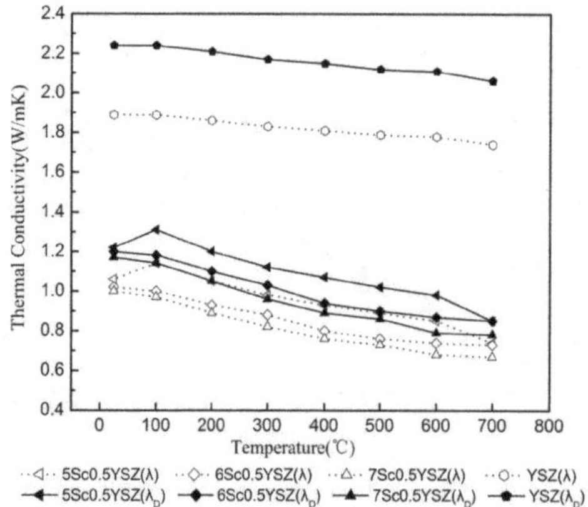

Figure 5. Thermal conductivity of ScYSZ and YSZ coatings.

pores, the thermal conductivities (λ_0) of the fully dense samples of 5Sc0.5YSZ (1.22–0.85 W/m·K), 6Sc0.5YSZ (1.20–0.85 W/m·K), and 7Sc0.5YSZ (1.17–0.78 W/m·K) are also much lower than that of YSZ (2.24–2.06 W/m·K).

The lower thermal conductivity of ScYSZ could be attributed to the high phase stability, more oxygen vacancies, and substitution atoms. For ZrO_2 crystal structure, the thermal conductivity of m-ZrO_2 is about 25 W/m·K, which is much higher than that of t-ZrO_2 (2–3 W/m·K), so the higher content of m-ZrO_2 in YSZ is an important reason for its higher thermal conductivity. For ScYSZ solid solution, Sc has a lower valence than Zr, so the additional oxygen vacancies must be introduced to maintain the electrical neutrality of the lattice in the same manner as Y_2O_3. Each vacancy can act as scattering center and enhance phonons scattering resulting in a decrease in the thermal conductivity. The mean free path of phonons which can be significantly scattered by point defects such as lattice strain, cracks, and porosity is calculated using the following equation [20]:

$$\frac{1}{l_P} = \frac{a^3}{4\pi v_m^4} w^4 c \left(\frac{\Delta M}{M} \right) \tag{5}$$

where a^3 is the volume per atom, c is the defect concentration per atom, $v_m{}^3$ is the transverse wave speed, w is the phonon frequency, M is the average mass of the host atom, and ΔM is the mass differences between the substituted and substituting atoms. So the substitution atom is another factor affecting the thermal conductivity [14]. The atomic mass of Sc (44.96) is much greater than that of Zr (91) and Y (89), which results in the higher scattering of phonon, so Sc_2O_3 has a more significant effect than Y_2O_3 on reducing the thermal conductivity by reducing phonon mean free path.

4 CONCLUSION

The phase stability and thermal conductivity of ScYSZ and YSZ plasma sprayed coatings were investigated. ScYSZ coatings show outstanding phase stability with no monoclinic phase forming after it was heat treated at 1500°C for 100h. Even after it was heat treated for 300h, 6Sc0.5YSZ and 7Sc0.5YSZ can still maintain metastable tetragonal phase, while the monoclinic phase in YSZ has been up to 31.6 mol.%, indicating ScYSZ has higher resistance to destabilization of metastable tetragonal phase than YSZ. ScYSZ also shows much lower thermal conductivity. In addition, increasing the content of Sc_2O_3 can effectively decrease the thermal conductivity of ScYSZ. The lower thermal conductivity of ScYSZ could be mainly attributed to the high phase stability, more oxygen vacancies and substitution atoms.

ACKNOWLEDGMENTS

This work is supported by the National Natural Science Foundation of China (NSFC, 51202295) and the National Basic Research Program of China (No. 2011CB605805).

REFERENCES

[1] Robert, V., Maria, O.J, Tanja, S., Daniel, E.M., Detlev. S. Overview on advanced thermal barrier coatings. *Surface & Coatings Technology* 205(2010): 938–942.

[2] Xu, Z.H., He, L.M., Mu, R.D., Zhou, X. $(Y_{0.05}La_{0.95})_2(Zr_{0.7}Ce_{0.3})_2O_7$ ceramics as a thermal barrier coating material for High-temperature applications. *Material Letters* 116(2014): 182–184.

[3] Zhang, Y.l., Yang, Y.P., Guo, H.B., Zhang, H.J., Gong, S.K. Influence of Gd_2O_3 and Yb_2O_3 Co-doping on Phase Stability, Thermo-physical Properties and Sintering of 8YSZ. *Chinese Journal of Aeronautics* 25(2012): 948–953.

[4] Liu, Z.G., Ouyang, J.H., Zhou, Y. Effect of on stability and thermal conductivity of ZrO_2-4.5 mol%Y_2O_3 ceramics, *Material Letters* 62(2008): 3524–3526.

[5] Matsumoto, M., Aoyama, K., Matsubara, H., K. Takayama. Thermal conductivity and phase stability of plasma sprayed ZrO_2-Y_2O_3-La_2O_3 coatings. *Surface & Coatings Technology* 194(2005): 31–35.

[6] Mineaki, M., Norio, Y., Hideaki, M. Low thermal conductivity and high temperature stability of ZrO_2-Y_2O_3-La_2O_3 coatings produced by electron beam PVD. *Scripta Materialia* 50(2004): 867–871.

[7] Felicia, M., Pitek, C., Levi, G. Opportunities for TBCs in the ZrO_2-$YO_{1.5}$-$TaO_{2.5}$, *Surface & Coatings Technology* 201(2007): 6044–6050.

[8] Jones, R.L., Mess, D. Improved tetragonal phase stability at 1400°C with scandia, yttria-stabilized zirconia, *Surface and Coatings Technology* 96–87(1996): 94–101.

[9] Kan, Y.M., Zhang, G.J., Wang, P.L., Omer, V.B., Jef, V. Yb_2O_3 and Y_2O_3 co-doped zirconia ceramics, *Journal of the European Ceramic Society* 26(2006): 3607–3612.

[10] Song, X.W., Xie, M., An, S.L., Hao, X.H. Mu, R.D.. Structure and thermal properties of ZrO_2-Ta_2O_5-Y_2O_3-Ln_2O_3(Ln = Nd, Sm or Gd) ceramics for thermal barrier coatings, *Scripta Materialia* 62(2010): 879–882.

[11] Sun, LL.L., Guo, H.B., Peng, H., Gong, S.K., Xu, H.B. Influence of partial substitution of Sc_2O_3 with Gd_2O_3 on the phase stability and thermal conductivity of Sc_2O_3-doped ZrO_2, *Ceramic International* 39(2013): 3447–3451.

[12] Rahaman, M.N., Gross, J.R., Dutton, R.E., Wang, H. Phase stability, sintering, and thermal conductivity of plasma-sprayed ZrO_2-Gd_2O_3 compositions for potential thermal barrier coating applications, *Acta Materialia* 54(2006): 1615–1621.

[13] Niu, X.Q., Xie, M., Zhou, F., Mu, R.D., Song, X.X., An, S.L. Substituent Influence of Yttria by Gadolinia on the Tetragonal Phase Stability for Y_2O_3-Ta_2O_5-ZrO_2 Ceramics at 1300°C, *J. Mater.Sci. Technol* 2014,30(4): 381–386.

[14] Liu, Q., Kwang, L.C. Thermophysical and thermochemical properties of new thermal barrier materials based on Dy_2O_3-Y_2O_3 co-doped zirconia, *Ceramic International* 40(2014): 11593–11599.

[15] Feing, J., Ren, X.R., Wang, X.Y., Zhou, R., Pan, W. Thermal conductivity of ytterbia-stabilized zirconia, *Scripta Materialia* 66(2012): 41–44.

[16] Huang, X., Wang, D.M., Lamontagne, M., Moreau, C. Experimental of the thermal conductivity of metal oxides co-doped yttria stabilized zirconia, *Material Science and Engineering* B 149(2008): 63–72.

[17] Liu, H.F., Li, S.L., Li, Q.L., Li, Y.M. Investigation on the phase stability, sintering and thermal conductivity of Sc_2O_3-Y_2O_3-ZrO_2 for thermal barrier coating application, *Materials and Design* 31(2010): 2972–2977.

[18] Guo, L., Guo, H.B., Gong, S.K., Xu, H.B. Improvement on the phase stability, mechanical properties and thermal insulation of Y_2O_3-stabilized ZrO_2 by Gd_2O_3 and Yb_2O_3 co-doping, *Ceramic International* 39(2013): 9009–9015.

[19] Viazzi, C., Jean-Pierre, B., Florence, A., Antoine, B. Structural study of metastable tetragonal YSZ powders produced via sol-gel route, *Jounal of Alloys and Compounds*, 2008, 452(2): 377–383.

[20] Zhang, H.S., Chen, X.G., Li, G., Wang, X.L., Dang, X.D. Influence of Gd_2O_3 addition on thermophysical properties of $La_2Ce_2O_7$ ceramics for thermal barrier coatings, *Journal of the European Ceramic Society* 32(2012): 3693–3700. volume number (issue number if necessary): page numbers.

Material Science and Environmental Engineering – Chen (Ed.)
 ISBN 978-1-138-02938-5

Investigation of the effectiveness of a chemical dispersant under different mixing conditions

J. Sun, C.C. Zhao, Z.J. Xie & G.B. Xu
College of Chemistry Engineering, China University of Petroleum, Qingdao, Shandong Province, China

ABSTRACT: Application of chemical dispersant has been recognized as a widely used oil spill response operation, especially since the *Deepwater Horizon spill*, as it can rapidly breakup the spilled oil into micro-sized droplets and thereby facilitate the removal of oil from the environment through biodegradation. A laboratory study was conducted to investigate the effectiveness of a widely used chemical dispersant in China under different mixing conditions, aiming to determine the optimum condition to apply this dispersant. In this study, TOPO crude oil and diesel oil were selected as the test oil. Filtered natural seawater, baffled flasks and a reciprocating shaker were used for the controlled experiment. The roles of oil type, and different environmental factors like mixing time, salinity and temperature in dispersant efficiency were studied systematically. The dispersant efficiency was evaluated based on the emulsification rate with a settling time of 30 s and 10 min, respectively. The dispersed oil in the aqueous phase was characterized using an ultraviolet spectrophotometer. The results showed that oil type, the mixing energy applied, and temperature were key factors influencing the effectiveness of the dispersant. A better performance of the dispersant was observed when applying to TOPO crude oil. The highest emulsification rate with a settling time of 30 s was 91%, and 65% with a settling time of 10 min. The dispersant efficiency increased with increase of the duration of the mixing energy applied, and also with the increase of temperature of the seawater from 10 to 30 °C.

Keywords: chemical dispersant; oil spill; oil properties

1 INTRODUCTION

During the last few decades, the worldwide demand for petroleum and their refined products has grown significantly. Unfortunately, large amounts of oil spill into the aquatic ecosystem as a consequence of the increasing volume of marine transportation of crude oil from the production sites (many in the Middle East countries). The oil spilled in the aquatic environment undergoes complex transformations immediately after it is released in the water. Application of chemical dispersant has been recognized as a widely used oil spill response operation, especially since the *Deepwater Horizon spill*, as it can rapidly breakup the spilled oil into micro-sized droplets and thereby facilitate the removal of oil from the environment through biodegradation. Emulsification is one of the most important processes after oil is spilled in the ocean and it can significantly affect the fate of oil spills at sea (Gong et al. 2014, Jernelov 2010, Kirby et al. 2010, Stephens et al. 2013, Sun et al. 2009, 2014).

Many studies have been conducted to understand the formation mechanisms and stability of emulsions. Different states of the emulsions are differentiated by the emulsion properties, and different methods and techniques were developed to determine the stability of the emulsion (Blondina et al. 1999, Fingas et al. 2003, 2004, Moles et al. 2002).

Dispersants may interact with the oil spills in the aquatic environment to form oil-in-water emulsion under the influence of the coastal hydrodynamic conditions (Ferguson et al. 2001, Lewis et al. 2010, Tambe and Sharma 1993). It is significant to determine the optimum conditions to apply the dispersant when oil spills happen for the oil spill remediation community (King et al. 2013, Li et al. 2010, Tamis et al. 2012).

A laboratory study was conducted to investigate the effectiveness of a widely used chemical dispersant in China under different mixing conditions. This paper presents findings from experiments conducted with GM-2 dispersant, two crude oil samples, and filtered natural seawater. The findings are expected to determine the optimum condition to apply this dispersant, and provide insights into the rate and extent of the influence of chemical dispersant on the spilled oil, and also the transfer and fate of oil spilled in the marine environment.

2 EXPERIMENTAL METHOD

TOPO crude oil and diesel oil, and dispersant of GM-2 (provided by Qingdao Guangming Environmental Technology Corporation) were used in this study. Natural seawater was collected from Tangdao Bay, China and filtered for the controlled experiment.

Density of TOPO crude oil, diesel oil, and GM-2 dispersant was measured by a density meter at 20 °C.

The emulsification rate was measured by dividing the oil quantity in the emulsion by the original oil added to the reaction chamber. Oil-in-water emulsions were prepared in 150 mL Erlenmeyer flasks containing mixtures of 50 mL seawater, 0.5 g oil, and different volumes of dispersant. The flasks were placed on a temperature controlled (20°C) reciprocating shaker at the shaking rate of 200 cycles/min for different times to provide the agitation to break the air-liquid interface and form oil droplets (Weise et al. 1999). The 150 mL Erlenmeyer flasks were sealed with silicone stoppers wrapped with aluminium foil. The oil concentration in the lower level of the emulsion at a settling time of 30 s and 10 min, respectively, was measured by an ultraviolet light spectrometer at 256 nm after extracted with petroleum ether.

3 RESULTS AND DISCUSSION

3.1 *Properties of the sample*

Density of TOPO crude oil, diesel oil were measured to be 0.9424, and 0.8543 g/cm³, respectively; density of GM-2 dispersant, and seawater measured at 20°C were 1.0103, and 1.0250 g/cm³, respectively.

The optimum dispersant to oil ratio was determined to be 1:5.

3.2 *Efficiency of the dispersant with different shaking time*

The emulsification rate of diesel oil with different shaking time of the reciprocating shaker was shown in Figure 1. As shown in Figure 1, the emulsification rate of diesel oil increased with increase of the shaking time, and reached a maximum at a shaking time of 10 min at 58.81%, and 28.21%, when the settling time was 30 s and 10 min, respectively.

As shown in Figure 2, the emulsification rate of TOPO crude oil increased with increase of the shaking time, and reached a maximum at a shaking time of 10 min at 91.10%, and 65.34%, when the settling time was 30 s and 10 min, respectively.

Therefore, the efficiency of the dispersant is higher with the TOPO crude oil than with the diesel oil.

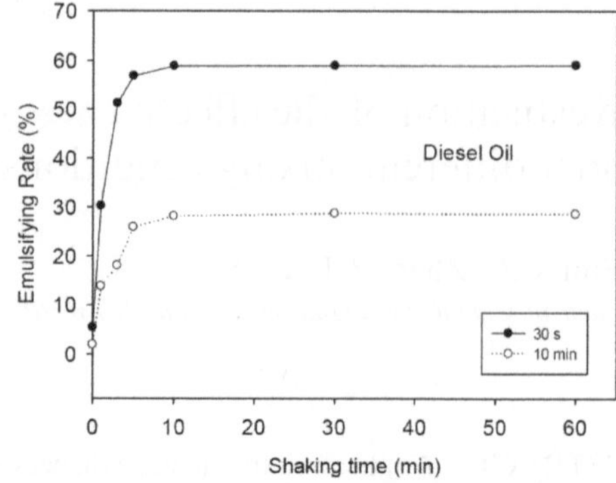

Figure 1. Variance of emulsification rate of diesel oil with shaking time.

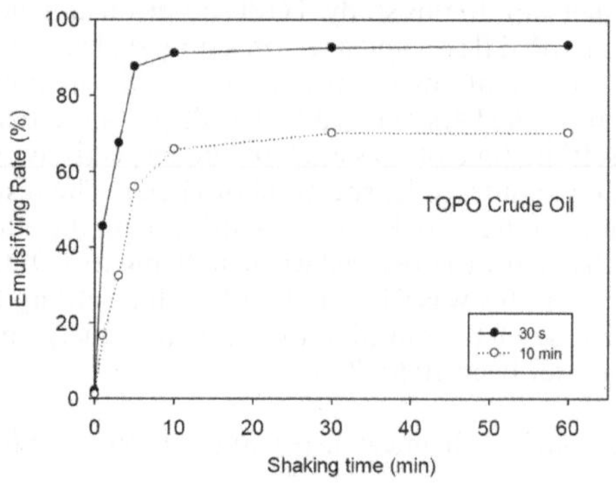

Figure 2. Variance of emulsification rate of TOPO crude oil with shaking time.

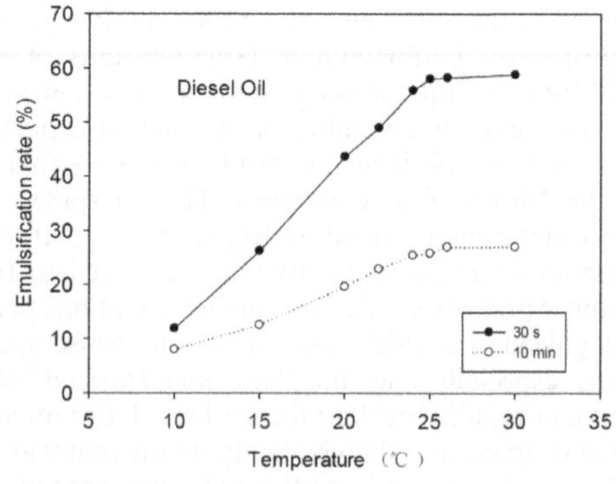

Figure 3. Variance of emulsification rate of diesel oil with temperature.

3.3 *Efficiency of the dispersant with different temperature*

The influence of temperature on the efficiency of the oil was also tested at a shaking time of 10 min. As shown in Figure 3, the emulsification rate of diesel oil increased with increase of the temperature, and reached a maximum at a temperature 25°C at 58.50%, and 24.93%, when the settling time was 30 s and 10 min, respectively.

4 CONCLUSIONS

This paper presents new data from a study on the efficiency of chemical dispersant under different mixing conditions. Findings show that efficiency of the dispersant was highly influenced by spilled oil properties and the environmental conditions. The dispersant is more efficacious when applied to TOPO crude oil than to diesel oil. The maximum emulsification rate was 91.10%, and 65.34% at a settling time of 30 s and 10 min, which indicates both the emulsification capacity and the emulsification stability of the dispersant were higher with TOPO crude oil than with diesel oil. The efficiency of the dispersant can be increased by increasing the mixing time of the oil-dispersant-seawater system to 10 min. Moreover, the efficiency of the tested dispersant can be enhanced by increase of the environment temperature especially when the settling time was 30 s, which indicates the emulsification capacity was highly increased by increase of temperature, whereas the emulsification stability increased slightly with increase of temperature. Nevertheless, further study needs to be conducted to verify the efficiency of chemical dispersant with a wider range of spilled oil and dispersant samples, and also the environmental conditions.

ACKNOWLEDGEMENTS

This work was financially supported by the Shan-dong Provincial Natural Science Foundation, China (No. ZR2012BQ012), Fundamental Research Funds for the Central Universities, Ministry of Chinese Education (No. 12CX04040 A), and Research Funds for the Scientific and Technological Development Program of Qingdao City (No. 12-1-4-7-(3)-jch), and National Natural Science Foundation of China (41201303).

REFERENCES

[1] Blondina, G.J., Singer, M.M., Lee, I., Ouano, M.T., Hodgins, M., Tjeerdema, R.S. & Sowby, M.L. 1999. Influence of salinity on petroleum accommodation by dispersants. Spill Science and Technology Bulletin 5: 127–134.

[2] Ferguson, P.L., Iden, C.R., & Brownawell, B.J. 2001. Distribution and fate of neutral alkylphenol ethoxylate metabolites in a sewage-impacted urban estuary. Environmental Science and Technology 35: 2428–2435.

[3] Fingas, M. & Fieldhouse B. 2003. Studies of the formation process of water-in-oil emulsions. Marine Pollution Bulletin 47: 369–396.

[4] Fingas, M. & Fieldhouse, B. 2004. Formation of water-in-oil emulsions and application to oil spill modeling. Journal of Hazardous m Material 107: 37–50.

[5] Gong, Y., Zhao, X., Cai, Z., O'Reilly S., Hao, X., Zhao, D. 2014. A review of oil, dispersed oil and sediment interactions in the aquatic environment: Influence on the fate, transport and remediation of oil spills. Marine Pollution Bulletin 79: 16–33.

[6] Jernelov A. 2010. The threats from oil spills: now, then, and in the future. AMBIO 39: 353–366.

[7] Kirby, M.F., Law, R.J. 2010. Accidental spills at sea-risk, impact, mitigation and the need for co-ordinated post-incident monitoring. Marine Pollution Bulletin 60: 797–803.

[8] King, T.L., Clyburne, J.C., Lee, K., Robinson, B.J. 2013. Interfacial film formation: influence on oil spreading rates in lab basin tests and dispersant effectiveness testing in a wave tank. Marine Pollution Bulletin 71: 83–91.

[9] Lewis, A., Trudel, B.K., Belore, R.C., Mullin, J.V. 2010. Large-scale dispersant leaching and effectiveness experiments with oils on calm water. Marine Pollution Bulletin 62: 244–254.

[10] Li, Z., Lee, K., King, T., Boufadel, M.C., Venosa, A.D. 2010. Effects of temperature and wave conditions on chemical dispersion efficacy of heavy fuel oil in an experimental flow-through wave tank. Marine Pollution Bulletin 60: 1550–1559.

[11] Moles, A., Holland, L. & Short, J. 2002. Effectiveness in the laboratory of Corexit 9527 and 9500 in dispersing fresh, weathered, and emulsion of Alaska North Slope crude oil under subarctic conditions. Spill Science and Technology Bulletin 7: 241–247.

[12] Stephens, E.L., Molina, V., Cole, K.M., Laws, E., Johnson, C.N. 2013. In situ and in vitro impacts of the deepwater Horizon oil spill on Vibrio parahaemolyticus. Marine Pollution Bulletin 75: 90–97.

[13] Sun, J. & Zheng, X. 2009. A review of oil-suspended particulate matter aggregation- a natural process of cleansing spilled oil in the aquatic environment. Journal of Environmental Monitoring 11: 1801–1809.

[14] Sun, J., Khelifa, A., Zhao, C., Zhao, D., Wang, Z. 2014. Laboratory investigation of oil-suspended particulate matter aggregation under different mixing conditions. Science of the Total Environment 473–474: 742–749.

[15] Tambe, D.E. & Sharma, M.M. 1993. Factors controlling the stability of colloid-stabilized emulsions. Colloid and Interface Science 157 (1): 244–253.

[16] Tamis, J.E., Jongbloed, R.H., Karman, C.C., Koops, W., Murk, A.J. 2012. Rational application of chemicals in response to oil spills may reduce environmental damage. Integrated Environmental Assessment and Management 8 (2): 231–241.

[17] Weise, A.M., Nalewajko, C., & Lee, K. 1999. Oil-mineral fine interactions facilitate oil biodegradation in seawater. Environmental Technology 20 (8): 811–824.

Material Science and Environmental Engineering – Chen (Ed.)

Absorption of NO into aqueous coal slurry with added $Fe^{II}(EDTA)$ in the presence of SO_2

W.S. Sun, J.Y. Chen, J.M. Zhou & F. Pan
College of Environmental Science and Engineering, Qingdao University, Qingdao, P.R. China

ABSTRACT: Comparative experiments were carried out using a bubbling reactor. NO, SO_2, O_2 and N_2 were used to simulate flue gas. The presence of gaseous SO_2 can result in enhanced leaching of coal pyrite producing Fe^{2+}, while Fe^{2+} is the desired ion in absorption of NO by Fe^{II} (EDTA) chelate. In this paper the effects of coal on the removal of NO were studied in a process of absorption of NO into aqueous coal slurry with added $Fe^{II}(EDTA)$ in the presence of SO_2. NO removal, nitrogen compounds produced and variations of ferrous iron concentrations were determined. Experiments showed that the absorbed NO was mainly reduced to ammonium. It was deduced that the reducing reagent was coal pyrite. The added coal in $Fe^{II}(EDTA)$ solution could inhibit the decrease of ferrous ion concentration, increase the pH of the solution and therefore increase the removal efficiency of NO. Chemical reactions accompanying Fe^{II} (EDTA) complexed NO reduction were discussed.

Keywords: nitric oxide; EDTA; reduction; leaching; coal pyrite

1 INTRODUCTION

About 90–95% of the harmful NO_x in flue gas is NO which is almost insoluble in water or alkaline solution. This is why ferrous chelates are used to absorb NO (Demmink et al. 1997, Chien et al. 2009, Jin et al. 2008, Wu et al. 2008, Santiago et al. 2010). The complexing reaction can be expressed as:

$$[Fe^{II}(EDTA)]^{2-} + NO \rightarrow [Fe^{II}(EDTA)(NO)]^{2-} \quad (1)$$

where EDTA denote ethylenediamine tetraacetate. $Fe^{II}(EDTA)$ can be regenerated from nitrosyl complex by biological reduction (Manconi et al. 2006, Lu et al. 2011).

$Fe^{II}(EDTA)$ can be oxidized to $Fe^{III}(EDTA)$ during absorption of NO. Because $Fe^{III}(EDTA)$ is not capable of binding NO, the issue of how to reduce $Fe^{III}(EDTA)$ to $Fe^{II}(EDTA)$ is one of the key issues of this process. Several $Fe^{III}(EDTA)$ reduction methods have been proposed (Suchecki et al. 2005, Zhang et al. 2012, Santiago et al. 2010, Ma et al. 2004). Many of these have concentrated on the biological reduction of $Fe^{III}(EDTA)$. Activated carbon has been used to catalyze the reduction of $Fe^{III}(EDTA)$ to $Fe^{II}(EDTA)$ to maintain solution NO removal capability, and the absorbed NO may be reduced to N_2 (Zhu et al. 2010, Long et al. 2014, Yang et al. 2011). Metal powder has also been used to reduce $Fe^{III}(EDTA)$ (Ma et al. 2005, Suchecki et al. 2014). According to the study of Ma et al. (2005), iron powder can not only regenerate $Fe^{II}(EDTA)$, but also reduce the absorbed NO to NH_3 which can be further used to produce ammonium sulfate fertilizer.

It has been known that coal slurry can be used to scrub SO_2 from flue gas (Sundaram et al. 2001, Sun et al. 2013). In this process, coal pyrite is leached and ferrous ions are produced according to reaction (2):

$$FeS_2 + H_2O + 7H_2SO_3 + 7O_2 \rightarrow Fe^{2+} + 9SO_4^{2-} + 16H^+ \quad (2)$$

where H_2SO_3 represents the aqueous SO_2. The Fe^{2+} leached through reaction (2) can catalyze the oxidation of SO_2. Fe^{2+} is the desired ion in the absorption of NO with $Fe^{II}(EDTA)$. If $Fe^{II}(EDTA)$ solution with added coal particles is used to scrub gas containing NO and SO_2, the continuously leached ferrous ions may contribute to the NO absorption process and SO_2 and NO can be simultaneously removed from flue gas. In this paper, comparative experiments were conducted to investigate the functions of coal pyrite in the process of absorption of NO into aqueous coal slurry with added $Fe^{II}(EDTA)$ in the presence of SO_2.

2 EXPERIMENTAL SECTION

The coal used was obtained from a coal mine in Shandong province. The coal was dried, ground and screened to produce 30–80 (200–600 μm)

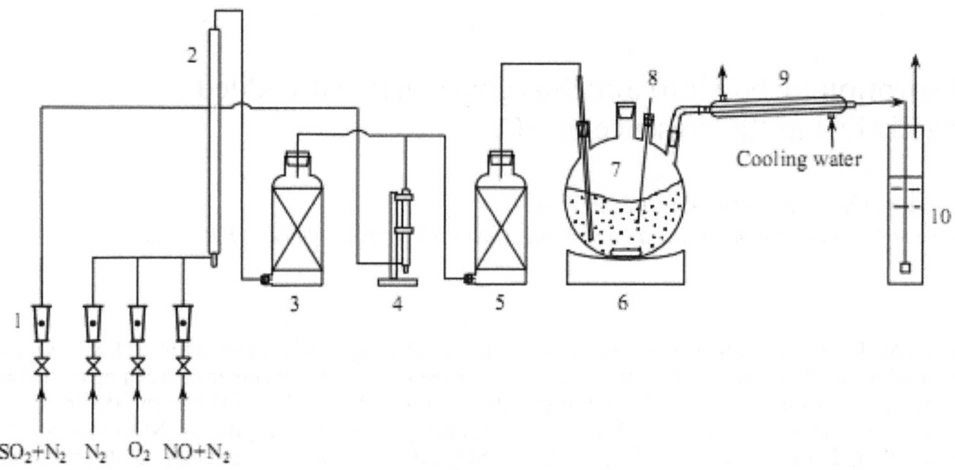

Figure 1. Schematic diagram of experimental apparatus. (1) gas rotameter, (2) soap-foam flowmeter, (3) and (5) gas mixer, (4) electronic soap-film flowmeter, (6) electromagnetic stirring heater, (7) reactor, (8) thermometer, (9) cooler, and (10) absorption bottle.

mesh fraction. The iron content in the coal sample was determined by the 1, 10-phenanthroline spectrophotometry method after nitric acid leaching. The total iron content was 2.0 wt.%.

A schematic diagram of the experimental apparatus used in this study is shown in Figure 1. Gases were delivered from four pressurized gas cylinders (N_2, O_2, 1% NO and 1% SO_2). To avoid interference from SO_2 when measuring the NO concentration in the influent gas, the 1% NO gas was firstly mixed with N_2 and O_2, and then mixed with the 1% SO_2 gas after the NO concentration in the mixed gases was measured. A soap-foam flowmeter and an electronic soap-film flowmeter were used to measure the flow rate of the mixed gases without SO_2 and the flow rate of the 1% SO_2 gas, respectively. An inclined manometer was used to measure the gas pressure in the soap-foam flowmeter. Using the ideal gas equation, the measured gas flow rate was converted to the value at 273.15 K and 1 atm. The total mixed gases were introduced into a 500-mL bubble reactor. The slurry in the reactor was heated and thoroughly stirred using an electromagnetic stirring heater. The reactor was operated batch-wise with respect to the liquid phase. There was 300 mL liquid and 30 g coal in the reactor. The ferrous chelate Fe^{II} (EDTA) was prepared by adding equimolar amounts of $FeSO_4$ and Na_2EDTA. The reaction temperature was controlled at 323 K. The gas leaving the reactor was cooled to room temperature through a cooler and was introduced into an absorption bottle containing sulfuric acid to absorb ammonia possibly remaining in the effluent gas.

In China, many coal-fired power plants use circulating Fluidized Bed (CFB) boiler. Usually, the concentration of NO in the flue gas emitted from CFB boiler is less than 400 ppm, while the concentration of O_2 in CFB boiler flue gas is about 5%. Taking this into consideration, the concentrations of O_2, NO and SO_2 in the simulated gas were controlled to be 5%, 400 ppm and 740 ppm (v/v), respectively, with the balance being N_2. The SO_2 and NO concentrations in the gases were measured using the iodine titration method and the ultraviolet spectrophotometric method, respectively. The total gas flow rate was 14.4 mL/s (at 273.15 K, 1 atm). In all the experiments, SO_2 was totally removed in the reactor according to analysis of the effluent gases. The ammonia nitrogen produced was analyzed for using Nessler's reagent spectrophotometry after the liquid samples were pretreated using micro kjeldahl distiller, while nitrite nitrogen and nitrate nitrogen were determined by N-(1-naphthyl)-1,2-diaminoethane dihydrochloride spectrophotometry and 2-isopropyl-5-methyphenol spectrophotometry, respectively. N_2O in the outlet gas was measured using a Nicolet iN10 infrared microscope. The concentration of the ferrous iron in the solution was determined by the 1,10-phenanthroline spectrophotometry method.

3 RESULTS AND DISCUSSION

3.1 *NO removal and reaction products*

For comparison, three experiments were carried out. In experiment 1 (Labeled as "Coal + Fe^{II}(EDTA) + NO + SO_2 + O_2" in Figure 2a to c), the slurry contained coal and Fe^{II}(EDTA) (10 mM). The liquid-to-solid (coal) ratio (w/w) was 10:1. In the gas there was NO, SO_2 and O_2,

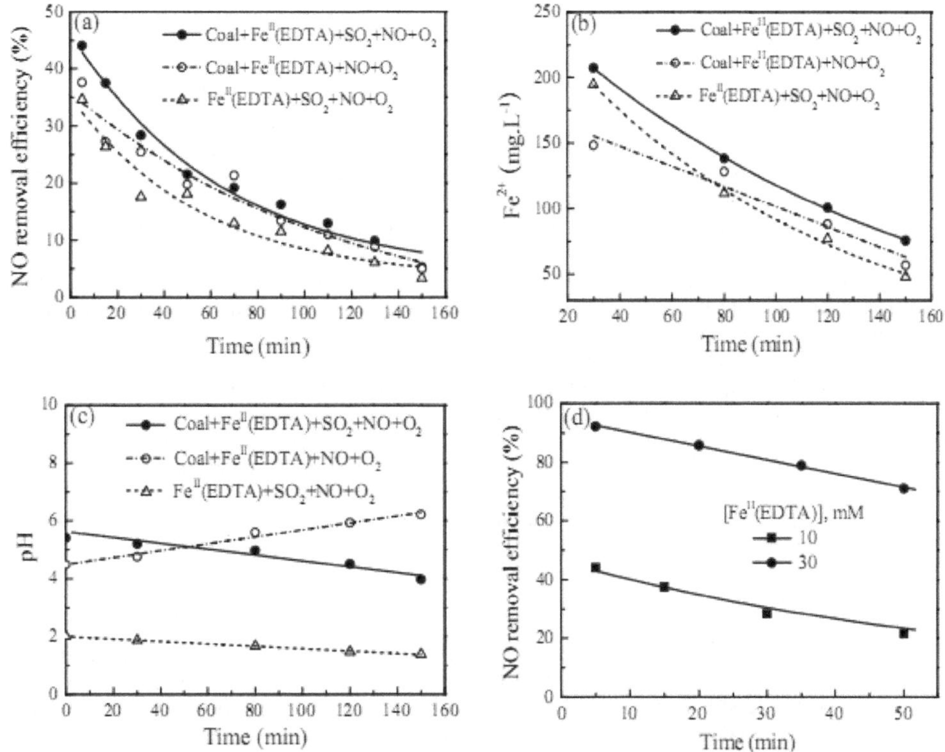

Figure 2. Results of comparative experiments and effect of Fe^{II}(EDTA) concentration. (a) Plot of NO removal efficiencies versus time. (b) Plot of ferrous ion concentrations versus time. (c) Plot of pH values versus time. (d) Effect of Fe^{II}(EDTA) concentration on NO removal efficiency. [*T*, 323 K; NO, 400 ppm (v/v); SO_2, 740 ppm (v/v); O_2, 5% (v/v); Fe^{II}(EDTA), 10 mM (in comparative experiments); liquid-to-solid ratio, 10:1 (w/w); and coal particle size fraction, 200–600 μm].

balanced by N_2. In experiment 2 (Labeled as "Coal + Fe^{II}(EDTA) + NO + O_2" in Figure 2a to c), there was no SO_2 in the gas. In experiment 3 (Labeled as "Fe^{II}(EDTA) + NO + SO_2 + O_2" in Figure 2a to c), the liquid was just the Fe^{II}(EDTA) solution, no coal was added. The concentrations of the components, if used, in the gases or slurries were the same across the three experiments.

The total NO, *N* (mol), removed during 150 min of absorption was calculated according to the following equation:

$$N = \frac{PVy}{100RT}\int_0^{150} \eta(t)\,\mathrm{d}t \tag{3}$$

where *P* = gas pressure (pa); *V* = volume flow rate of the simulated flue gas ($m^3\ min^{-1}$); *y* = inlet NO concentration ($mol\ mol^{-1}$); *R* = universal gas constant ($J\ K^{-1}\ mol^{-1}$); *T* = absolute temperature (K); η = NO removal efficiency (%); *t* = time (min). Equation (4) is the expression of $\eta(t)$ used for non-linear least squares curve fitting in which *a*, *b* and *c* are the regression coefficients. The symbols shown in Figure 2a are experimental results and lines are the fit curves. The coefficients of determination r^2 for the fit curves of experiment 1, 2 and 3 are 0.98, 0.94 and 0.95 respectively.

$$\eta(t) = a \cdot e^{(-t/b)} + \mathrm{c} \tag{4}$$

According to the calculations, the total removed NO during experiments 1, 2 and 3 were 0.432, 0.437 and 0.319 mmol respectively. Compared with experiment 3, the total NO removed with added coal (experiment 1) increased by 35%. The total NO removed in experiment 2 was almost the same as that of experiment 1. It has been found that the NO complexation capacities of Fe^{II}(EDTA) increases with increasing pH within the pH range measured (Ma et al. 2007, Jing et al. 2007). Although the ferrous iron concentration of experiment 2 was relatively low compared with that of experiment 1 (Fig. 2b), the pH value was more favorable for the absorption of NO (Fig. 2c).

Under the experimental conditions, SO_2 was totally removed due to the Fe^{3+}/Fe^{2+}-catalyzed oxidation of SO_2 and the reaction between SO_2 and coal pyrite, while the removal efficiency of NO was relatively low. But if the initial $Fe^{II}(EDTA)$ concentration was increased from 10 mM to 30 mM, the removal efficiency of NO could be increased by nearly 50 percentage points as shown in Figure 2d. So, the concentration of $Fe^{II}(EDTA)$ is critical to the removal of NO.

After 150 min reaction, ammonia nitrogen, nitrite nitrogen and nitrate nitrogen that may be produced in comparative experiments with added coal were measured in the reactor and in the absorption bottle. For experiment 1 and 2, the ratio of the total of ammonia nitrogen plus nitrite nitrogen, N_T (mol), to total NO removed, N (mol), were 0.92 and 0.49 respectively, while the ratio of ammonia nitrogen to total NO removed were 0.89 and 0.46 respectively. So under the experimental conditions the main nitrogenous compound produced was ammonia nitrogen, while the amount of nitrite nitrogen was small. Nitrate nitrogen was not detected. Compared with experiment 1, the ratio N_T/N of experiment 2, in which there was no SO_2 in the gas, is almost halved. The rest of the absorbed NO should still exist as $Fe^{II}(EDTA)(NO)$. So it may be deduced that the coexistence of SO_2 is favorable to the conversion of $Fe^{II}(EDTA)$ complexed NO to ammonia.

N_2O was not detected in the outlet gas of the simultaneous removal process under the experimental conditions examined. In the presence of SO_2, N_2 may be produced through reaction (5). But according to recent studies, even if it exists, this reaction rate should be very slow (Yan et al. 2014, Demmink et al. 1997, Wang et al. 2007).

$$2[Fe^{II}(EDTA)(NO)]^{2-} + 2SO_3^{2-} \rightarrow 2[Fe^{II}(EDTA)]^{2-} + N_2 + 2SO_4^{2-} \quad (5)$$

It has been reported that S^0 may be produced from pyrite according to equation (6) (Chandra & Gerson 2010):

$$FeS_2 \rightarrow Fe^{2+} + 2S^0 + 2e^- \quad (6)$$

$Fe^{II}(EDTA)$ complexed NO could obtain the electrons and be reduced to ammonium. The reaction could be assumed to be:

$$5FeS_2 + 2[Fe^{II}(EDTA)(NO)]^{2-} + 12H^+ \rightarrow 5Fe^{2+} + 10S^0 + 2\,NH_4^+ + 2[Fe^{II}(EDTA)]^{2-} + 2H_2O \quad (7)$$

In the presence of SO_2 and O_2, S^0 produced through reaction (7) will be ultimately oxidized to sulfate as it has been well demonstrated that the combination of SO_2 and O_2 is a strong oxidation reagent (Sundaram et al. 2001). Actually, even Fe^{3+} can oxidize S^0 to sulfate via a number of intermediate steps (Chandra & Gerson 2010). According to reaction (7), reduction of the absorbed NO promotes the leaching of ferrous ion from coal pyrite. It can also be seen from reaction (7) that acidic aqueous condition or low pH value may promote the production of ammonium.

In the gas, a small amount of NO may be oxidized to NO_2 by O_2. Then HNO_2 may be produced according to reaction (8). This reaction can explain the small amount of nitrite nitrogen detected in the liquids.

$$NO + NO_2 + H_2O = 2HNO_2 \quad (8)$$

Coal is a porous material. It can absorb various gases and organic compounds (Li et al. 2015, Tarasevich 2001). Then, it is worth studying on how the porosity of the coal affects the NO absorption and reduction process in the aqueous environment.

3.2 *Ferrous iron concentration*

To compare the effects of coal pyrite and SO_2 on the variation of ferrous iron in the solution, the concentrations of ferrous iron at different time were measured in comparative experiments as shown in Figure 2b.

It can be seen that experiment 1 had the highest concentration of Fe^{2+} during the whole 150 min reaction. The reactions that could produce the desired ferrous iron are reaction (2), (7), (9) and (10) (Sundaram et al. 2001, King & Lewis 1980, Descostes et al. 2004, Holmes & Crundwell 2000, Ou et al. 2007).

$$2FeS_2 + 7O_2 + 2H_2O \rightarrow 2Fe^{2+} + 4SO_4^{2-} + 4H^+ \quad (9)$$

$$FeS_2 + 14Fe^{3+} + 8H_2O \rightarrow 15Fe^{2+} + 2SO_4^{2-} + 16H^+ \quad (10)$$

Fe^{3+} and Fe^{2+} in reaction (10) may also be in the form of $[Fe^{III}(EDTA)]^-$ and $[Fe^{II}(EDTA)]^{2-}$ respectively as shown in reaction (11).

$$FeS_2 + 14[Fe^{III}(EDTA)]^- + 8H_2O \rightarrow 14[Fe^{II}(EDTA)]^{2-} + Fe^{2+} + 2SO_4^{2-} + 16H^+ \quad (11)$$

So, Fe^{3+} and $[Fe^{III}(EDTA)]^-$ can be reduced to Fe^{2+} and $[Fe^{II}(EDTA)]^{2-}$ respectively by coal pyrite.

The ferrous iron concentration of experiment 3 was greater than that of experiment 2 for about 70 minutes. This may be due to the effect of SO_2 which can reduce Fe^{3+} to Fe^{2+} according to reaction (12) (Cho 1986).

$$SO_2 + 2Fe^{3+} + 2H_2O \rightarrow SO_4^{2-} + 4H^+ + 2Fe^{2+} \quad (12)$$

Similarly, the ferric ion and ferrous ion in reaction (12) may also be coordinated to EDTA.

After about 70 min reaction, the ferrous iron concentration of experiment 2 became greater than that of experiment 3. Although there was SO_2 to serve as a reducing reagent in experiment 3, there was no coal pyrite to be leached to supply additional Fe^{2+}, whereas in experiment 2 coal pyrite could be leached to supply additional Fe^{2+} through reactions (7), (9), (10) and (11) despite the absence of SO_2 in the gas.

3.3 *The pH value of the liquid*

In comparative experiments, the time dependences of liquid pH are shown in Figure 2c. Except for experiment 2, the pH values decreased slowly as the absorption proceeded owing to the continuous formation of sulfuric acid. For experiment 2, there was no SO_2 in the gas, and the production of ammonium could cause the pH to increase. The absorption process without coal (experiment 3) had the lowest pH during the whole 150 min absorption. This phenomenon was caused by both the hydrolysis reaction of iron ion and the production of sulfuric acid. As coal was added to the liquids, the pH values were much greater, which was favorable to the absorption of both NO and SO_2. Therefore, besides the functions discussed above, coal can also buffer the solution pH. Some mineral substances or humus contained in coal can act as buffers and cause the pH value to be relatively high (Ma et al. 1993, Wang & Lu 1994).

4 CONCLUSION

The functions of coal pyrite in the process of absorption of NO into aqueous coal slurry with added Fe^{II}(EDTA) in the presence of SO_2 have been investigated in this work. Experiments showed that the absorbed NO was mainly reduced to ammonium under the experimental conditions examined. The reducing reagent should be coal pyrite and the reduction promoted the leaching of ferrous ion from coal pyrite. The added coal in Fe^{II}(EDTA) solution could inhibit the decrease of ferrous ion concentration in the absorption solution due to the continuously leaching of coal pyrite to supply additional Fe^{2+}. The added coal could also increase the pH of the solution. Therefore adding coal into Fe^{II}(EDTA) solution could increase the removal efficiency of NO. In addition, the removal efficiency of NO could be effectively increased by increasing the concentration of Fe^{II} (EDTA).

Problems such as the function of the porosity of the coal in the absorption and reduction process need to be studied further.

ACKNOWLEDGMENT

This work was supported financially by Qingdao Municipal Science and Technology Program, China (No. 11-2-4-2-(2)-jch).

REFERENCES

[1] Chandra, A.P. & Gerson, A.R. 2010. The mechanisms of pyrite oxidation and leaching: A fundamental perspective. *Surf. Sci. Rep*. 65: 293–315.

[2] Chien, T.W.; Hsueh, H.T.; Chu, B.Y.; Chu, H. 2009. Absorption kinetics of NO from simulated flue gas using Fe(II)EDTA solutions. *Process Saf. Environ.* 87: 300–306.

[3] Cho, E.H. 1986. Removal of SO_2 with oxygen in the presence of Fe(III). *Metall. Mater. Trans. B* 17: 745–753.

[4] Demmink, J.F.; Gils, I.C.F.G.; Beenackers, A.A.C.M. 1997. Absorption of nitric oxide into aqueous solutions of ferrous chelates accompanied by instantaneous reaction. *Ind. Eng. Chem. Res*. 36: 4914–4927.

[5] Descostes, M.; Vitorge, P.; Beaucaire, C. 2004. Pyrite dissolution in acidic media. *Geochim. Cosmochim. Acta* 68: 4559–4569.

[6] Holmes, P.R. & Crundwell, F.K. 2000. The kinetics of the oxidation of pyrite by ferric ions and dissolved oxygen: an electrochemical study. *Geochim. Cosmochim. Acta* 64: 263–274.

[7] Jin, H.F.; Santiago D.E.O.; Park, J.; Lee, K. 2008. Enhancement of nitric oxide solubility using Fe(II) EDTA and its removal by green algae Scenedesmus sp. *Biotechnol. Bioprocess Eng*. 13: 48–52.

[8] Jing, G.; Li, L.; Tang, S. 2007. Absorption capacity and absorption velocity of nitric oxide by Fe^{II}(EDTA) solution. *J. Huaqiao Univ*. (Natural Science) (Chinese). 28: 166–169.

[9] King, W.E. & Lewis, J.A. 1980. Simultaneous effects of oxygen and ferric iron on pyrite oxidation in an aqueous slurry. *Ind. Eng. Chem. Process Des. Dev.* 19: 719–722.

[10] Li, Q.; Lin, B.; Wang, K.; Zhao, M.; Ruan, M. 2015. Surface properties of pulverized coal and its effects on coal mine methane adsorption behaviors under ambient conditions. *Powder Technol*. 270: 278–286.

[11] Long, X.L.; Yang, L.; Chou, X.; Li, C.; Yuan, W. 2014. Reduction of $[Fe(III)EDTA]^-$ catalyzed by activated carbon modified with ammonia solution. *Environ. Prog. Sustain*. 33: 99–105.

[12] Lu, B.H.; Jiang, Y.; Cai, L.L.; Liu, N.; Zhang, S.H.; Li, W. 2011. Enhanced biological removal of NOx from flue gas in a biofilter by Fe(II)Cit/Fe(II)EDTA absorption. *Bioresource Technol*. 102: 7707–7712.

[13] Ma, B.; Li, W.; Jing, G.; Shi, Y. 2004. Dissimilatory reduction of Fe^{III}(EDTA) with microorganisms in the system of nitric oxide removal from the flue gas by metal chelate absorption. *J. of Environ. Sci.—China* 16: 428–430.

[14] Ma, L.; Tong, Z.; Zhang, J. 2005. Mechanism study of NOx removal from flue gas with the recovery process of absorption with acid following complex in aqueous solution and reduction with iron powder. *Acta Scientiae Circumstantiae* (Chinese) 5: 637–642.

[15] Ma, L.; Yin, Q.; Guo, Z.; Song, J. 2007. Complexing NO with Fe^{2+}EDTA aqueous solution. *Ecol. Environ.* (Chinese) 16: 26–30.

[16] Manconi, I.; Maas, P.; Lens, P. 2006. Effect of sulfur compounds on biological reduction of nitric oxide in aqueous Fe(II)$EDTA^{2-}$ solutions. *Nitric Oxide* 15: 40–49.

[17] Ma, Y.; Qi, Z.; Wang, Z.; Gao, J. 1993. Study on the coal demineralization and desulfurization by dilute alkali/acid treatment. *J. Fuel Chem. Technol.* (Chinese) 21: 61–67.

[18] Ou, L.; He, R.; Feng, Q. 2007. Influencing factors of pyrite leaching in germ-free system. *J. Cent. South Univ. Technol.* 14: 28–31.

[19] Santiago, D.E.O; Jin, H.-F.; Lee, K. 2010. The influence of ferrous-complexed EDTA as a solubilization agent and its auto-regeneration on the removal of nitric oxide gas through the culture of green alga Scenedesmus sp. *Process Biochem.* 45: 1949–1953.

[20] Suchecki, T.T.; Mathews, B.; Kumazawa, H. 2005. Kinetic study of ambient-temperature reduction of Fe^{III}edta by $Na_2S_2O_4$. *Ind. Eng. Chem. Res.* 44: 4249–4253.

[21] Suchecki, T.T.; Mathews, B.; Augustyniak, A.W.; Kumazawa, H. 2014. Applied kinetics aspects of ferric EDTA complex reduction with metal powder. *Ind. Eng. Chem. Res.* 53: 14234–14240.

[22] Sundaram, H.P; Cho, E.H.; Miller, A. 2001. SO_2 removal by leaching coal pyrite. *Energy Fuels* 15: 470–476.

[23] Sun, W.; Wang, L.; Liu, J.; Wang, L.; Zhang, Y. 2013. SO_2 removal with coal slurry in a double-stirred vessel. *Environ. Technol.* 34: 2497–2501.

[24] Tarasevich, Y.I. 2001. Porous structure and adsorption properties of natural porous coal. *Colloid. Surface. A* 176: 267–272.

[25] Wang, L.; Zhao, W.; Wu, Z. 2007. Simultaneous absorption of NO and SO_2 by Fe^{II}EDTA combined with Na_2SO_3 solution. *Chem. Eng. J.* 132: 227–232.

[26] Wang, Y. & Lu, X. 1994. Feasibility of taking coal humic acids as urease inhibitors. *Res. soil water conserv.* (Chinese) 1: 96–100.

[27] Wu, Z.B.; Wang, L.; Zhao, W.R. 2008. Kinetic study on regeneration of Fe^{II}EDTA in the wet process of NO removal. *J. Chem. Eng.* 140: 130–135.

[28] Yan, B.; Yang, J.; Guo, M.; Chen, G.; Li, Z.; Ma, S. 2014. Study on NO enhanced absorption using Fe^{II}EDTA in $(NH_4)_2SO_3$ solution. *J. Ind. Eng. Chem.* 20: 2528–2534.

[28] Yang, X.-J.; Yang, L.; Dong, L.; Long, X.-L.; Yuan, W.-K. 2011. Kinetics of the $[Fe(III)\text{-}EDTA]^-$ reduction by sulfite under the catalysis of activated carbon. *Energy Fuels* 25: 4248–4255.

[29] Zhang, S.H.; Shi, Y.; Li, W. 2012. Biological and chemical interaction of oxygen on the reduction of Fe(III)EDTA in a chemical absorption–biological reduction integrated NOx removal system. *Appl. Microbiol. Biotechnol.* 93: 2653–2659.

[30] Zhu, H.; Mao, Y.; Yang, X.; Chen, Y.; Long, X.; Yuan, W. 2010. Simultaneous absorption of NO and SO_2 into Fe^{II}-EDTA solution coupled with the Fe^{II}-EDTA regeneration catalyzed by activated carbon. *Sep Purif Technol.* 74: 1–6.

Material Science and Environmental Engineering – Chen (Ed.)

Overview and analysis of partial-nitrification control technology and its application

R.M. Liu, J.C. Chen & L.H. Zang

College of Environmental Science and Engineering, Qilu University of Technology, Ji'nan, Shandong, China

ABSTRACT: ANAMOX technology has been defined as a green and sustainable process due to without any need for carbon source. Partial-nitrification is the key step in this process to keep nitrite as a final product. Therefore, the present paper reviewed the state-of-art technologies to control partial-nitrification which could be classified into reaction condition control, chemical agent addition, and reactor configuration. Moreover, the pros and cons of these technologies in application are also analyzed and discussed.

Keywords: sewage sustainable treatment; partial-nitrification; accumulation of nitrite; real-time control; ANAMMOX

1 INTRODUCTION

Nowadays, more stringent requirements are proposed for removal of waste water treatment process efficiently due to the shortage of water resources and deterioration of water pollution. At the same time, the national level also strengthened the corresponding discharge standards of sewage treatment [1], especially the provisions of removal efficiency of nutrient elements in sewage. As we all know, nutrients in waste water including nitrogen and phosphorus is the main reason which causes eutrophication and decrease in the content of dissolved oxygen in water. Then the algae and other plants soar while in the water environment other biological entities grow, even as the whole aquatic ecosystem is destructed and it is difficult to be repaired [3]. In addition, the high concentration of nitrate nitrogen in water (NO^{2-} and NO^{3-}) also induces infant methemoglobinemia, cancer, and other diseases. In view of this, the existing technology faces enormous challenges to effectively remove nutrients discharged into the sewage water.

Currently, conventional active system processes are the mainstream of urban sewage treatment plant processes such as oxidation ditch, A^2/O and SBR reactors [4]. In the activated sludge systems, there are two complete processes to remove nitrogen including aerobic nitrification and denitrification, and in the aerobic process oxygen is required and in the anoxic reaction a carbon source is demanded as an electron donor. Therefore, the traditional denitrification process requires adequate oxygen and sufficient carbon source [2]. However, for the sewage of low C/N ratio, such a treatment of industrial waste water uses anaerobic digestion sludge supernatant and sludge dewatering leachate. Traditional denitrification way seems useless, quantitative carbon should be added to enhance denitrification process; the way of denitrification is uneconomical and unsustainable.

ANAMMOX technology provides a new path for waste water denitrification, especially for sewage of the low C/N ratio. Since Mulder [5] found ANAMMOX technology in 1985, this new technology continuously developed in the past 30 years, and there are even some practical application cases [6–9]. Scholars favor it because it does not need a carbon source. In ANAMMOX technology, about half of the NH^{4+} firstly oxidized to NO^{2-}, which is called shortcut nitrification process; then, produced NO^{2-} that can react with unoxidized residual NH^{4+} to complete autotrophic denitrification process. Compared with conventional denitrification process, ANAMMOX technology will save 100% of COD and about 60% of the oxygen demand [10].

Although this technology has been successfully applied in Germany's Hattingen sewage treatment plant [11], Strass of Austria sewage treatment plant, and the Swiss Glanerland sewage treatment plant [7], but has not been widely applied. Wherein, the key step of completed anammox is to effectively suppress the activity of nitrifying bacteria and get NO^{2-} accumulation. However, the generation of NO^{2-} is easily affected by DO and concentration of NH^{4+} and other external conditions; and once the concentration of DO, it is too high and nitrification bacteria are often at a disadvantage in the competition with denitrifying bacteria and heterotrophic bacteria [12], which greatly limits the promotion of this technology.

Based on this, the paper reviewed partial nitrification control methods reported in the international arena, and summarized based on the reaction conditions to control. Out of which three were chosen to be introduced based on chemical control and reactors, and described in three aspects, including how to control the reaction conditions, how to control chemicals, and how to choose the reactor. The advantages and disadvantages of each control, technology, and methods and application potential were analyzed and summarized.

2 CONTROL TECHNOLOGY BASED ON THE REACTION CONDITIONS

In this paper, the reaction conditions of shortcut nitrification and denitrification mainly consider the influence of the external environment condition of the reaction (temperature), the parameters of the reaction phase (SRT, NLR, pH, DO etc.) and reaction phase may affect other ionic shortcut nitrification (free ammonia concentration, FA [13]). If we can achieve the purpose of controlling nitrification smoothly through these parameters, it will be the most economical control method effectively. Although these factors are in different areas, but the principle is through the limit or inhibits the growth of NOB to help AOB gain a competitive advantage, so as to lay the foundation for the shortcut nitrification and denitrification in biological processes.

2.1 *Real-time control technique based on the change of ammonia nitrogen and dissolved oxygen*

As mentioned above, in the traditional removal of nitrogen process, in order to prevent the Dissolved Oxygen (DO) concentration becoming the limiting condition, aerobic pool DO concentration is controlled in general more than 2 mg/L. In the environment of sufficient oxygen, the growth rate of nitrobacteria is faster than nitrosobacteria. Nitrosobacteria as an intermediate state only stay for a short time, so it will soon be oxidized to nitrobacteria. From the reaction kinetics, oxygen affinity coefficient of nitrosobacteria is much smaller than denitrifying bacteria [14]; this shows that in low DO concentrations, nitrosobacteria will be easier to win in competition with denitrifying bacteria. Tokutomi found that when the concentration of DO is maintained at 1.0 mg/L, AOB growth rate is 2.6 times NOB [15]; thus, the different situation is showed compared with a high concentration of DO nitrification process. This is the reason that by controlling the concentration of DO to complete shortcut nitrification. Generally, when the concentration of DO in the reactor was maintained at 2 mg/L or less [17], there will be a higher conversion rate of ammonia and accumulation rate of nitrite (Table 1).

However, concentration of ammonia is not constant due to the dynamic changes in the actual sewage. Therefore, concentration of DO in the reactor is difficult to maintain the appropriate concentration range in order to maintain NO^{2-} accumulation, and sometimes even leading to the destruction to shortcut nitrification process. Jubany [16] maintained the concentration of DO at 3 ± 0.2 mg/L in the reactor, successfully achieved partial nitrification in activated sludge system, but after the level of the water decreased by 17 L/d to 12 L/d, the generation amount of NO^{3-} in the water increased by 1.5 g/L to 2.6 g/L. Ammonia load reduced mainly due to the reduction of the water intake, while the DO concentration in the reactor remains unchanged, resulting in a surplus of oxygen, resulting in damage to the environment shortcut nitrification. Therefore, micro-oxygen environment as an important measure to achieve partial nitrification, dissolved oxygen concentration should be adjusted in real time according to the influent ammonia load. So on the one hand energy can be saved; on the other hand neither the conversion of nitrite is affected due to the relative lack of oxygen [18] nor is the nitrification process undermined because of the relative excess oxygen. Although dissolved oxygen concentration can be well controlled by dissolved oxygen monitoring and control, but the time a slight lag, and nitrogen compounds cannot be characterized in the system, some monitoring and control technology is introduced based on real-time measurement of ammonia and dissolved oxygen.

2.2 *PH control technology*

In the short-nitrification process, it will release 2 mole H^+ with the formation of 1 mole NO^{2-}; whereas there won't be acidity when NO^{2-}

Table 1. Summarization of DO concentration to achieve NO^{2-} accumulation.

	DO (mg/L)	NO^{2-}	NH^{4+} (kg N/(m^3·d))
Antileo [21]	1.0	84%	0.1
Chuang [12]	0.42–1.2	75%	0.67
Xue [24]	0.3–0.5	95%	$NO_2^-/NH_4^+ \approx 1.37$
Feng [22]	0.3–0.8	>99%	$NO_2^-/NH_4^+ \approx 1.0$
Okabe [25]	0.0–2.0	>99%	$NO_2^-/NH_4^+ \approx 1.0$

converts into NO^{3-}. Therefore, this reaction will constantly decrease the pH in the reactor before all of the NH^{4+} gets converted when we make the short-nitrification process by controlling the DO concentration. It can be seen from the pH curve that after a uniform decline it pumps up and a trough appears, which just corresponds to the end of the NH^{4+}conversion (Fig. 1). This provides a theoretical possibility for real-time pH control technology used in short-nitrification process. After finding this point by pH, we can stop aeration so to avoid NO^{2-}, which has been accumulated converting to NO^{3-}.

In practical applications, the control of short-nitrification process by pH can be divided into two specific, one is directly based on the feature point of pH curve (the highest point or the lowest point) to control the aeration time, another is through the calculation of feature characteristics such as slope values to control. Guo et al. [19] succeeded in getting short-nitrification process by pH real-time control technology with 200 days experiment. Throughout the operation, the temperature, DO were controlled at 26 °C and 2.5 mg/L, respectively, according to the monitor, the changes of pH curves and "ammonia Valley" point to control aeration time, the effluent ammonia and ammonia conversion rates were less than or equal to 2 mg/L, greater than or equal to 90%, respectively. Yang et al. [17] conducted a trial of low-temperature short-nitrification process in step feed SBR reactor, the reaction temperature and DO concentrations were controlled at 11.9–26.5 and 2.5 mg/L. Take the second pH control method and the specific operation as follows: before applying the real-time control we should confirm the shortest and longest aeration time Tmax and Tmin. After the beginning of the experiment, one pH value can be determined at intervals of 5 s and automatically generate smooth pH images, when the reaction time T > Tmin image slope is counted; if it detects $dpH/dt \geq 0$ over 15 min, then we can consider that the ammonia nitrogen has been transformed completely, then stop aeration to avoid the accumulation of O_2.

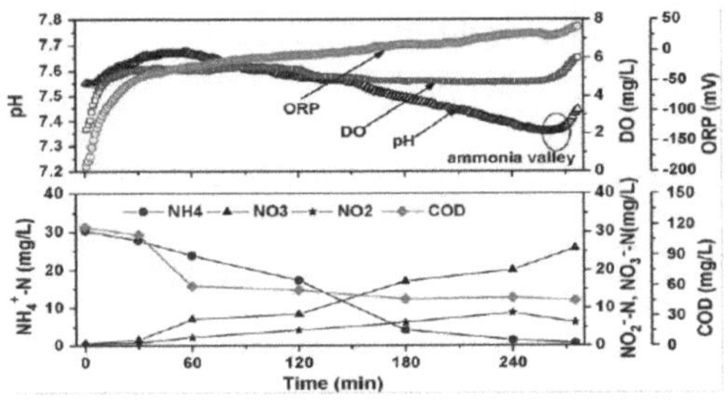

Figure 1. Typical pH curve in partial-nitrification.

Compared with the ORP and DO controls, real-time control of technology performs more stably as it has less disturbance factors and can solve the shortcomings that the controlling factors of DO measuring parameters of oxygen can't be applied to the hypoxic environment. However, there are some problems to use pH as real-time control parameters. Due to the accurate determination of the true pH value at intervals within the system, the entire control process can't add additional reagents which can affect the pH of the system, and therefore the pH of the system can not timely be adjusted to suit the optimum reaction conditions of the short-nitrification process.

3 LAYOUT OF TEXT

In order to get the nitrite nitrogen's accumulation, we can use the chemical agents to inhibit the activity of nitrifying bacteria at the beginning of the shortcut nitrification reaction or when the conditions such as temperature could not reach the requirement of eliminating nitrifying bacteria. These chemicals can be the substances contained in sewage or the specific chemical reagent, the former mainly includes the free ammonia or free nitric acid molecules which have a different effect of AOB and NOB, the latter is mainly relied on some other specific inhibitors to limit the activity of NOB. Domestic sewage contains a large amount of ammonia nitrogen. A certain concentration of free ammonia or free nitric acid molecules (FA) can be acquired by adjusting the pH, temperature, and they can be used to restrain nitrifying bacteria, as for denitrifying bacteria, the inhibitions were relatively small [20]. Many scholars have conducted related tests [21, 13], the NH^{4+} have different best load conditions because of the different reaction conditions. However, some scholars have found that in the long-term experiments, NOB can adapt to the inhibition of FA environment gradually. Li et al [22] in the trials which used FA to eliminate NOB found that the yield ratio of nitrate nitrogen was gradually reduced at the end of the test. By increasing concentrations of FA to improve shortcut nitrification process, but when the concentration of NH^{4+} up to 30 mg/L, it still cannot reach the ammonia nitrogen accumulation rate. Brockmann et al. [23] pointed out that it is not necessary, although FA inhibition method can speed up the process of eliminating NOB, especially in the long time of shortcut nitrification process.

The best additive quantity of using the free ammonia or nitrate to eliminate free nitrifying bacteria in the system is the most important. Too much or too little of it cannot reach the purpose of removing NOB. However, free ammonia or free nitric acid concentrations are affected by temperature, pH, and concentration of water, ammonia, and nitrogen comprehensive influence. How to control the ideal concentration under the condition of a multiple still needs a lot of research. Although this can be done by real-time control technology, what has been discussed above is that some scholars found adaptability of NOB limits and the application of this method in the test. In addition, the Ni^{2+} and $KClO_3$ in optimizing the nitrate accumulation of cash is good, but these two substances are toxic, and will affect subsequent disposal of sludge treatment, the risk of application remains to be studied.

4 CHOICE OF THE REACTOR

The impact of the reactor in the form of nitrite accumulation is mainly reflected in the number of microorganisms, mass transfer mechanisms, and controllability, besides different forms of the corresponding reactor startup mode selected and the control condition is not the same.

SHARON process as the world's first practical application to the sewage treatment process of partial nitrification reactor uses a form of suspended sludge reactor. From the above discussion we can see, such a reactor is easy to eliminate denitrifying bacteria by controlling the temperature and SRT. However, in the SHARON reactor, equal SRT and HRT cause the concentration of microorganisms in the reactor which leads to the desired value that cannot be accumulated and limit its processing capacity [26]. Therefore, some scholars have proposed the use of biofilm reactor to complete partial nitrification, since microorganisms attached to an immobilizing filler without being affected by HRT, relatively high concentration of microorganisms can be enriched and the processing capacity can be improved. In addition, Biofilm reactor reduces mass transfer coefficient of the water - membrane boundary layer, which is also a great contribution for the accumulation of nitrite [21]. However, biological is attached to the filler in biofilm reactor; the SRT system is not dependent on HRT as a suspended sludge but on attachment and homeostasis of shedding. Therefore, by controlling the SRT, it can not achieve the purpose of the removal of denitrifying bacteria [27], and the conversion rate of ammonia becomes lower because of the extension of SRT.

So suspended sludge reactors and biofilm reactors have advantages and disadvantages, but now the amount of water into the sewage plant is increasing and the site of sewage plant is inadequate, which determines the prevalence of short HRT process. Chuang et al. [12] uses down flow hanging sponge reactor successfully to test a stable process partial nitrification.

5 CONCLUSION

The partial nitrification control technology summarized above has advantages and disadvantages. Without having to function independently, it achieves short-nitrification process with interdependent and interacting together. The fluctuations of PH can cause changes in the FA, what's more, the size of the FA is one of the factors that influences the population structure of biological; the change of NLR affects DO concentration in the reactor, and then hinders the accumulation of nitrite. The world's first application to practical engineering nitrification process—SHARON process, uses the joint control of temperature and SRT to achieve the goal of elimination of denitrifying bacteria. So shortcut nitrification reactor is a complex multi-control process. We should first select the reactor according to the characteristics such as sewage and site and so on. And then take the appropriate control technology according to the characteristics of the reactor and its water, in order to achieve a stable accumulation of nitrite.

Shortcut nitrification-anaerobic ammonium oxidation process, have been identified sustainable wastewater treatment process in the environmental and economic aspects, which can not only reduce the amount of excess sludge, but also take the advantage in dealing with industrial wastewater with low C/N ratio. Therefore, development of an efficient shortcut nitrification control technology is still one of the urgent problems that need to solve.

REFERENCES

[1] State Environmental Protection Administration of China, emission standards for urban sewage treatment plant, 2003.

[2] Xiaodi Hao, Joseph J. Hei, Mark C.M. van. 2002. Loosdrecht. Sensitivity Analysis of a Biofilm Model Describing a One-Stage Completely Autotrophic Nitrogen Removal (CANON) Process. *Biotechnology and Bioengineering*. 77(3):266–277.

[3] K.A. Third, J. Paxman, M. Schmid, M. Strous, M.S.M. Jetten and R. Cord-Ruwisch. 2005. Enrichment of Anammox from Activated Sludge and Its Application in the CANON Process. *Microbial Ecology*. 49:236–244.

[4] HAO Xiaodi. Sustainable Sewage-Waste treatment technology. 2006. *China Building Industry Press*.

[5] Mulder, A., van de Graaf, A.A., Robertson, L.A., Kuenen, J.G. 2008. Anaerobic ammonium oxidation discovered in a denitrifying fluidized bed reactor [J]. FEMS Microbiology Ecology. 1995, 16(3):177–183.

[6] Sliekers, A.O., Third, K.A., Abma, W., Kuenen, J.G., Jetten, M.S.M. 2003. CANON and Anammox in a gas-lift reactor. *FEMS Microbiology*. 218(2):339–344.

[7] Van der star, W.R.L., Abma, W.R., Blommers, D., Mulder, J.W., Tokutomi, T., Strous, M., Picioreanu, C., Van Loosdrecht, M.C.M. 2007. Startup of reactors for anoxic ammonium oxidation: experiences from the first full-scale anammox reactor in Rotterdam. *Water Research*. 41(18):4149–4163.

[8] B. Wett. 2007. Development and implementation of a robust deammonification process. *Water Science and Technology*. 56(7):81–88.

[9] Abma, W.R., Driessen, W., Haarhuis, R., Van Loosdrecht, M.C.M. 2010. Upgrading of sewage treatment plant by sustainable and cost-effective separate treatment of industrial wastewater. *Water Science and Technology*. 61(7):1715–1722.

[10] K.A. Third, J. Paxman, M. Strous, M.S.M. Jetten and R. Cord-Ruwisch. 2005. Treatment of nitrogen-rich wastewater using partial nitrification and Anammox in the CANON process. *Water Science and Technology*. 52(4):47–54.

[11] K.H. Rosenwinkel, A. Cornelius. 2005. Deammonification in the Moving-Bed Process for the Treatment of Wastewater with High Ammonia Content. *Chemical Engineering and Technology*. 28(1):49–52.

[12] Liang Zhang, Shujun Zhang, Yiping Gan, Yongzhen Peng. 2012. Bio-augmentation to rapid realizes partial nitrification of real sewage. *Chemosphere*. 88:1097–1102.

[13] Shan Li, You-peng Chen, Chun Li, Jin-Song Guo, Fang Fang, Xu Gao. 2012. Influence of Free Ammonia on Completely Autotrophic Nitrogen Removal over Nitrite (CANON) Process. *Applied Biochemistry and Biotechnology*. 167(4):694–704.

[14] Susanne Lackner, Claus Lindenblatt, Harald Horn. 2012. 'Swing ORP' as operation strategy for stable reject water treatment by nitritation-anammox in sequencing batch reactors. *Chemical Engineering Journal*. 180:190–196.

[15] J.H. Guo, Y.Z. Peng, S.Y. Wang, Y.N. Zheng, H.J. Huang and S.J. Ge. 2009. Effective and robust partial nitrification to nitrite by real-time aetation duration control in an SBR treating domestic wastewater. *Process Biochemistry*. 44:979–985.

[16] Qing Yang, Shengbo Gu, Yongzhen Peng, Shuying Wang and Xiuhong Liu. 2010. Process in the Development of Control Strategies for the SBR Process. *CLEAN-Soil, Air, Water*. 38(8):732–749.

[17] Alghusain I, Hao Oj. 1995. Use of pH as control parameter for aerobic/anoxic sludge digestion. *Journal of Environmental Engineering*. 121(3):225–235.

[18] joanna surmacz-gorska, krist gernaey, carl demuynck, peter vanrolleghem and willy verstraete. Nitrification monitoring in acticated sludge by oxygen uptake rate (our) measurements. *Water Research*. 1996, 30(5):1228–1236.

[19] U. Sollfrank and W. Gujer. 1990. Simultaneous determination of oxygen uptake rate and oxygen transfer coefficient in activated sludge systems by an on-line method. *Water Research*. 24(6):725–732.

[20] Sebastia Puig, Lluis Corominas, Joan Colomer, Maria D. Balaguar and Jesus Colprim. On-line Oxygen Uptake as s new tool for monitoring and controlling the SBR process.

[21] Richard Blackburne, Zhiguo Yuan, Jurg Keller. 2008. Demonstration of nitrogen removal via nitrite in a sequencing batch reator treating domestic wastewater. *Water Research*. 42:2166–2176.

[22] A. C. Anthonisen, R. C. Loehr, T. B. S. Prakasam and E. G. Srinath. 1976. Inhibition of nitrification by ammonia and nitrous acid. *Water Pollution Control Federation*. 48(5):835–852.

[23] Sen Qiao, Noriko Matsumoto, Takehiko Shinohara, Takashi Nishiyama, Takao Fujii, Zafar Bhatti, Kenji Furukawa. 2010. High-rate partial nitrification performance of high ammonium containing wastewater under low temperatures. *Bioresource Technology*. 101:111–117.

[24] Sitong Liu, Fenglin Yang, Zheng Gong, Zhencheng Su. 2008. Assessment of the positive effect of salinity on the nitrogen removal performance and microbial composition during the start-up of CANON process. *Environmental Biotechnology*. 80:339–348.

[25] D. Brockmann, E. Morgenroth. 2010. Evaluating operating conditions for outcompeting nitrite oxidizers and maintaining partial nitrification in biofilm systems using biofilm modeling and Monte Carlo filtering. *Water Research*. 44(6):1995–2009.

[26] Sukru Aslan, Burhanettin Gurbuz. 2011. Influence of Operational Parameter and Low Nickel Concentrations on Partial Nitrification in a Submerged Biofilter. *Applied Biochemistry and Biotechnology*. 1543–1555.

[27] Guangjing Xu, Xiaochen Xu, Fenglin Yang, Sitong Liu. 2011. Selective inhibition of nitrite oxidation by chlorate dosing in aerobic guanules. *Journal of Hazardous Materials*. 185:249–254.

Material Science and Environmental Engineering – Chen (Ed.)
 ISBN 978-1-138-02938-5

Solvent extraction of iron (III) from acid chloride solutions by Trioctylamine and Decanol

X.H. Mao
College of Biological and Chemical Engineering, Panzhihua University, Panzhihua, Sichuan, China

ABSTRACT: The solvent extraction of iron (III) from acidic chloride solutions by Trioctylamine (TOA) and decanol in kerosene has been investigated. The effects of TOA, decanol, chloride ion and hydrogen ion concentrations on the extraction rate of iron (III) were studied. And the extraction mechanism was discussed. The solvent extraction results demonstrated that the addition of decanol eliminated the third phase in the extraction of TOA for iron (III) from acid chloride solutions and that the extraction rate of iron (III) was improved. The extracted iron is present as $HFeCl_4 \cdot TOA \cdot 2Decanol$ and $HFeCl_4 \cdot 2Decanol$. Kinetics of the extraction process was very fast, since the equilibrium was reached in 3 min. Titanium (IV), aluminum (III), calcium (II) and magnesium (II) were not extracted under the experimental condition. Iron (III) was easy to be stripped. Using water as the stripping agents iron (III) nearly could be totally stripped from the extraction complex. And the potential for the extraction and stripping of iron (III) from acid chloride liquors has been assessed through extraction and stripping isotherms.

Keywords: TOA; decanol; solvent extraction; iron (III)

1 INTRODUCTION

Titanium dioxide is widely used as pigment, as filler in paper, plastics and rubber industries and as flux in glass manufacture. There are a number of commercialized or proposed processes to produce TiO_2 (Zhang *et al.*, 2011). Among which in recent years a hydrometallurgical process has been proposed by Duyvesteyn *et al* for the production of pigment grade TiO_2 from titaniferous ores (Duyvesteyn *et al.*, 2002). The process involves a two-step solvent extraction process, first with TRPO for the extraction of titanium (IV) and iron (III), leaving iron (II) in the raffinate, with Alamine 336 in a second step for the selective removal of iron (III), leaving titanium (IV) in the raffinate.

Saji and Saji John *et al* respectively studied the extraction of titanium (IV) and iron (III) from acid chloride solutions by Cyanex 923 (Saji *et al.*, 1998; Saji John et al., 1999). Narayanan Remya and Lakshmipathy Reddy reported solvent extraction separation of titanium (IV), vanadium (V) and iron (III) from simulated waste chloride liquors of titanium minerals processing industry by Cyanex 923 (Narayanan Remya and Lakshmipathy Reddy, 2004).

The extraction of iron (III) with amines has been widely investigated. Bargeev *et al* reported the simultaneous extraction of micro and macro amounts of In, Fe, Ga, Cd, Co and Zn from HCl medium using Tri-n-octylamine (TOA) and Aliquat 336 in benzene (Bagreev *et al.*, 1978). The result showed that the extraction percentage of microelements decreased in presence of the extractable macroelements. For Fe with TOA the extracted complex was $TOAHFeCl_4$; with Aliquat 336 the extracted complex was R_4 $NFeCl_4$. Miroslav *et al* reported effect of the solvent on the extraction of iron (III) chloride by tri-n-octylamine (Miroslav *et al.*, 1978). The study of chemical analysis, viscosity and IR measurements of the organic phase concluded that the formation of cyclic polymer complexes of structures had the ratio of TOA: HCl: $FeCl_3$ as 2:2:1 in extractions carried out at low acid concentrations which changed to 1:1:1 for extractions carried out at higher acid concentrations.

However in the extraction of iron (III) with amines the important problem is the appearance of the third phase. The synergistic extraction of N235-TBP for iron (III) in sulfate solutions was studied (Liu *et al.*, 2005). The addition of TBP eliminated the third phase and that the extraction rate of iron (III) was improved. Mishra *et al* reported solvent extraction of Fe (III) from the chloride leach liquor of low grade iron ore tailings using Aliquat 336 in kerosene. p-Nonyl phenol was used as the third phase modifier (Mishra *et al.*, 2011). Zhou *et al* reported the effects of modifiers on the extraction of cobalt (II) and iron (II) with tertiary amine (Zhou *et al.*, 2001). It was found that the addition of modifiers, oxygen-containing organic compound and hexanol both could eliminated the third phase. But the mechanism didn't been covered. And the extraction of alcoholic solvents for iron (III) has not been reported. Early days the author studied and compared the extraction behavior of different extractants for titanium (IV) and iron (III) extraction from acidic chloride solutions (Mao *et al.*, 2011). The results showed that TOA was an effective extractant for iron (III) and no extraction for titanium (IV). Titanium and iron both can be extracted by TBP or D_2EHPA in concentrated HCl solution. However the third phase appeared especially for concentrated iron (III) when using TOA as extractant. To eliminate the third phase and explore the extraction process solvent extraction of iron (III) from acid chloride solutions by trioctylamine and decanol was investigated in this paper.

2 EXPERIMENTAL

2.1 *Reagents and apparatus*

Iron (III) solution was prepared from $FeCl_3$ by diluting to the required concentration with hydrochloric acid. Other metal ion solutions were prepared by dissolving their salts in hydrochloric acid and diluting to the required concentration with distilled water. TOA (98%) and decanol (98%) were procured from AODA Chemical Co., Ltd. (Luoyang, Henan, China). Kerosene was kindly supplied by Sichuan MAX-TOP petrochemical science and technology Ltd. All other chemicals used were of analytical reagent grade.

Apparatus: A 721 UV-visible spectrophotometer and a Nicolet 6700 IR spectrophotometer were employed.

2.2 *Extraction and analytical procedures*

Solvent extraction and stripping experiments were carried out by shaking required volumes of aqueous and organic phases at an O/A phase ratio of 1 for 10 minute at 298 ± 1 K. After phase separation, the concentration of the specific metal ion remaining in the aqueous phase was determined by standard procedures. Thus, iron (III), titanium (IV), aluminum (III), magnesium (II) and Calcium (II) were analyzed spectrophotometrically using respectively 1,10-phenanthroline, hydrogen peroxide, Eriochrome Cyanine R, Eriochrome black-T and ACBK. The concentration of the metal ion in the organic phase was attained by mass balance.

3 EXTRACTION EQUILIBRIUM

3.1 *Extraction equilibrium*

The third phase was found during the extraction of iron (III) from acid chloride solutions by TOA. As the addition of decanol the third phase eliminated. To explore the extraction mechanism the following investigation was studied.

The effect of TOA concentration (0.01~0.05 mol dm^{-3}) on the extraction of iron (III) has been studied by maintaining constant concentrations of decanol (0.86 mol dm^{-3}) and iron (III) (0.1 mol dm^{-3}) in 8 mol dm^{-3} HCl (Fig. 1). The result showed that the extraction of iron (III) increases with increased TOA concentration. From the slope of log D versus log [TOA], it can be inferred that one molecule of TOA is involved in the extracted complex of iron (III).

The effect of decanol concentration (0.52~2.62 mol dm^{-3}) on the extraction of iron (III) has been studied by maintaining constant concentrations of TOA (0.02 mol dm^{-3}) and iron (III) (0.1 mol dm^{-3}) in 8 mol dm^{-3} hydrochloric acid (Fig. 2). The result showed that the extraction of iron (III) increases with increased decanol concentration. From the slope of log D versus log [Decanol], it can be inferred that two molecules of decanol are involved in the extracted complex of iron (III).

The effect of chloride ion concentration (4~7 mol dm^{-3}) on the extraction of iron (III) with 0.05 mol dm^{-3} TOA and 0.86 mol dm^{-3} decanol in kerosene was investigated at a given hydrogen ion concentration (4 mol dm^{-3}) using HCl+$CaCl_2$ mixtures (Fig. 3). It is clear from the result that the extraction of iron (III) increases with increase in chloride concentration in the aqueous phase.

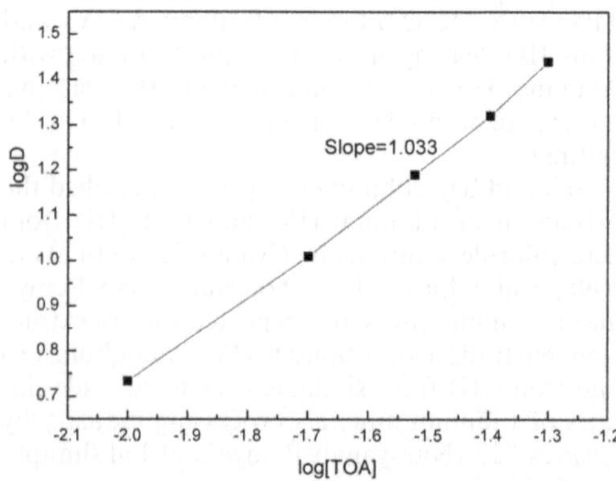

Figure 1. Effect of TOA concentration on the extraction of iron (III) (0.1 mol · dm^{-3}) from 8 mol · dm^{-3} hydrochloric acid solutions, [Decanol] = 0.86 mol · dm^{-3}.

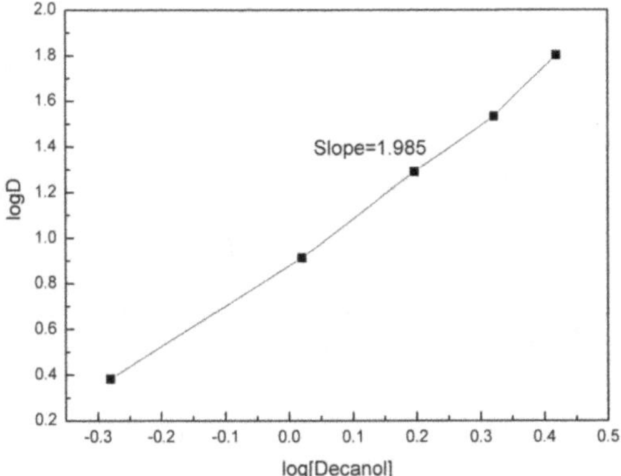

Figure 2. Effect of decanol concentration on the extraction of iron (III) (0.1 mol·dm^{-3}) from 8 mol·dm^{-3} hydrochloric acid solutions, [TOA] = 0.02 mol·dm^{-3}.

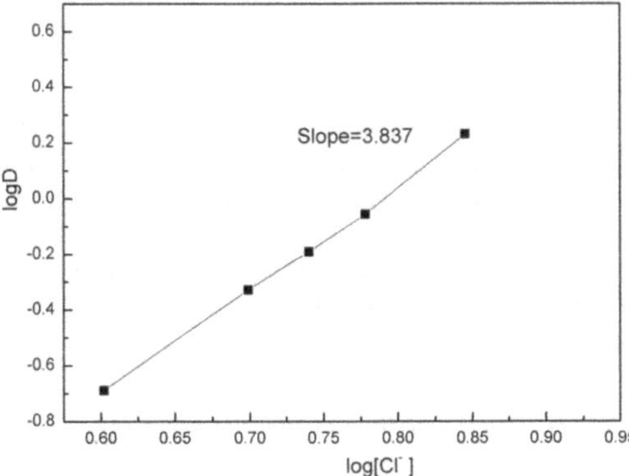

Figure. 3. Effect of chloride ion concentration on the extraction iron (III) (0.1 mol·dm^{-3}) at constant hydrogen ion concentration (4 mol·dm^{-3}), [TOA] = 0.05 mol·dm^{-3}, [decanol] = 0.86 mol·dm^{-3}.

The log D versus log [Cl$^-$] plots gave slopes of 3.837 in the case of iron (III), indicating the involvement of four chloride ions in the extracted complex of iron (III).

The extraction behavior of iron (III) with 0.05 mol dm^{-3} TOA and 0.86 mol dm^{-3} decanol in kerosene was investigated as a function of hydrogen ion concentration (4~8 mol dm^{-3}) at a given chloride ion concentration (8 mol dm^{-3}) using HCl+$CaCl_2$ mixtures (Fig. 4). The result showed that the extraction of iron (III) increases with increase in hydrogen ion concentration in the aqueous phase. The log D versus log [H$^+$] plots gave slopes of 1.035 in the case of iron (III), indicating the involvement of one hydrogen ion in the extracted complex of iron (III).

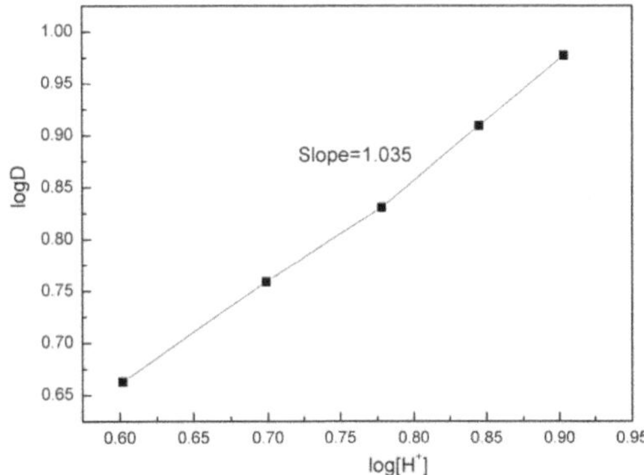

Figure 4. Effect of hydrogen ion concentration on the extraction of iron (III) (0.1 mol·dm^{-3}) at constant chloride ion concentration (8 mol·dm^{-3}), [TOA] = 0.05 mol·dm^{-3}, [decanol] = 0.86 mol·dm^{-3}.

Based on the preceding studies the extraction equilibrium of iron (III) with TOA and decanol in kerosene can be expressed as Equation 1.

$$H^+_{aq} + Fe^{3+}_{aq} + 4Cl^-_{aq} + A_{org} + 2B_{org} = HFeCl_4 \cdot A \cdot 2B_{org} \quad (1)$$

A—TOA, B—decanol.

The extraction of TOA, decanol and TOA-decanol for iron (III) in 8 mol·dm^{-3} hydrochloric acid solutions was studied further to contrast the extraction performance (Table 1).

Extraction rate of iron (III) by TOA-decanol was much greater than that by TOA or Decanol, less than the extraction addition by individual TOA and decanol. If the extraction engaged only as Equation 1, it was impossible to attain 92.14% extraction rate. As the concentration of TOA is only half of iron (III) concentration.

To explore further the extraction mechanism the extraction equilibrium of decanol for iron (III) in acid chloride solutions was investigated.

The effect of decanol concentration (0.05~1 mol dm^{-3}) on the extraction of iron (III) has been studied in 8 mol dm^{-3} hydrochloric acid (Fig. 5). The result showed that the extraction of iron (III) increases with increased decanol concentration. From the slope of log D versus log [Decanol], it can be inferred that two molecules of decanol are involved in the extracted complex of iron (III).

The effect of chloride ion concentration (4~7 mol dm^{-3}) on the extraction of iron (III) with 0.86 mol dm^{-3} decanol in kerosene was investigated at a given hydrogen ion concentration (4 mol dm^{-3}) using HCl+$CaCl_2$ mixtures (Fig. 6). It is clear from the result that the extraction of iron (III) increases

Table 1. Extraction performance of TOA, decanol and TOA-decanol for iron (III).

Extractant	TOA	Decanol	TOA-decanol
% extraction	29.02	77.44	92.14

$[Fe^{3+}]$ = 0.1 mol·dm^{-3}, [TOA] = 0.05 mol·dm^{-3}, [Decanol] = 0.86 mol·dm^{-3}.

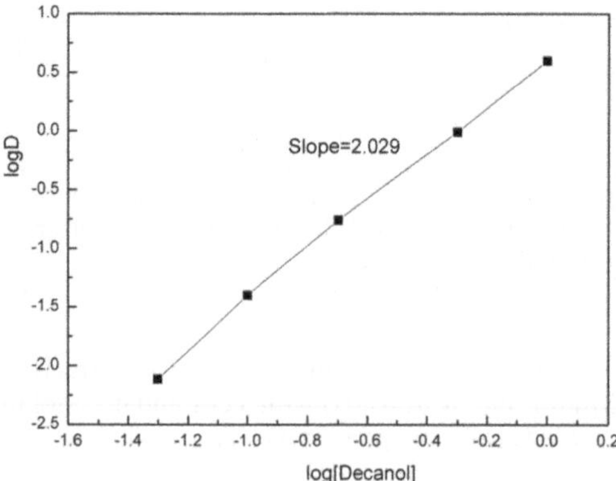

Figure 5. Effect of decanol concentration on the extraction of iron (III) (0.1 mol·dm^{-3}) in 8 mol·dm^{-3} hydrochloric acid solutions.

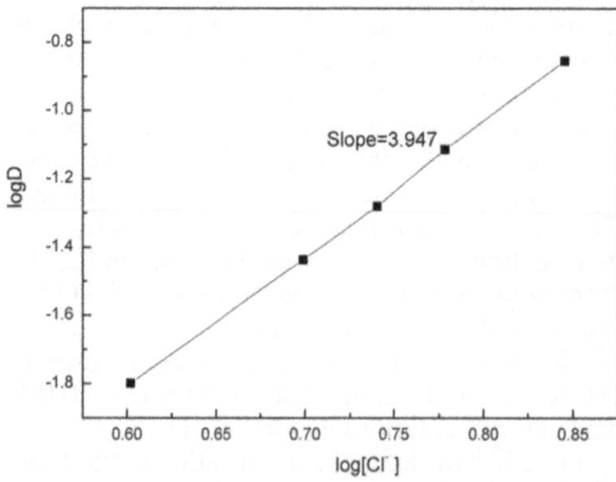

Figure 6. Effect of chloride ion concentration on the extraction iron (III) (0.1 mol·dm^{-3}) at constant hydrogen ion concentration (4 mol·dm^{-3}), [decanol] = 0.86 mol·dm^{-3}.

with increase in chloride concentration in the aqueous phase. The log D versus log $[Cl^-]$ plots gave slopes of 3.947 in the case of iron (III), indicating the involvement of four chloride ions in the extracted complex of iron (III).

The extraction behavior of iron (III) with 0.86 mol dm^{-3} decanol in kerosene was investigated as a function of hydrogen ion concentration (4~8 mol dm^{-3}) at a given chloride ion concentration (8 mol dm^{-3}) using HCl+$CaCl_2$ mixtures (Fig. 7). The result showed that the extraction of iron (III) increases with increase in hydrogen ion concentration in the aqueous phase. The log D versus log $[H^+]$ plots gave slopes of 1.034 in the case of iron (III), indicating the involvement of one hydrogen ion in the extracted complex of iron (III).

Based on the preceding studies the extraction equilibria of iron (III) with decanol in kerosene can be expressed as Equation 2.

$$H^+_{aq} + Fe^{3+}_{aq} + 4Cl^-_{aq} + 2B_{org} = HFeCl_4 \cdot 2B_{org} \quad (2)$$

All the results showed that when the concentration of decanol was more than twice of TOA concentration the extraction equilibria of iron (III) with TOA-decanol in kerosene also engage as Equation 2 except Equation 1.

3.2 *Effect of phase contact time*

Iron (III) was contacted with 0.05 mol dm^{-3} TOA and 0.86 mol dm^{-3} decanol in kerosene for a period of 1–30 min. Quantitative extraction of iron (III) by TOA and decanol was found within 3 min. There was no adverse effect on the extraction yield up to 30 min (Fig. 8). From the result it was found that the extraction of iron (III) by TOA and decanol from hydrochloric acid solutions is a quick process, applicable to industrial applications.

3.3 *IR spectra of the extracted complexes of iron (III)*

The IR spectra of the extracted complexes showed that the stretching frequency of O-H was shifted from 3334 cm^{-1} in decanol to 3506 cm^{-1} in $FeCl_4$·TOA·2Decanol and 3523 cm^{-1} in

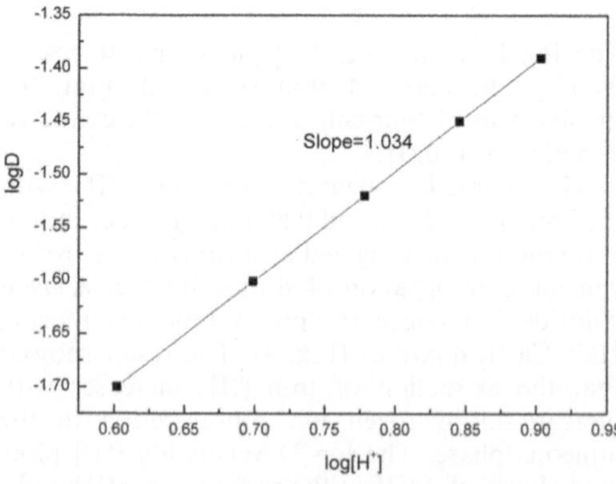

Figure 7. Effect of hydrogen ion concentration on the extraction of iron (III) (0.1 mol·dm^{-3}) at constant chloride ion concentration (8 mol·dm^{-3}), [decanol] = 0.86 mol·dm^{-3}.

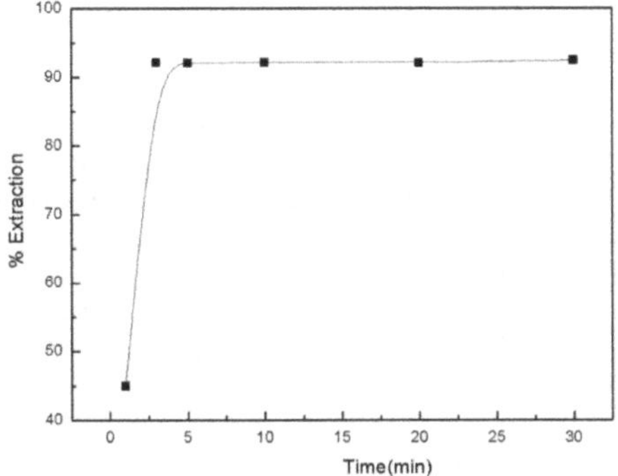

Figure 8. Dependence of extraction rate on contact time. $[Fe^{3+}] = 0.1\ mol \cdot dm^{-3}$, $[TOA] = 0.05\ mol \cdot dm^{-3}$, $[decanol] = 0.86\ mol \cdot dm^{-3}$, $[HCl] = 8\ mol \cdot dm^{-3}$.

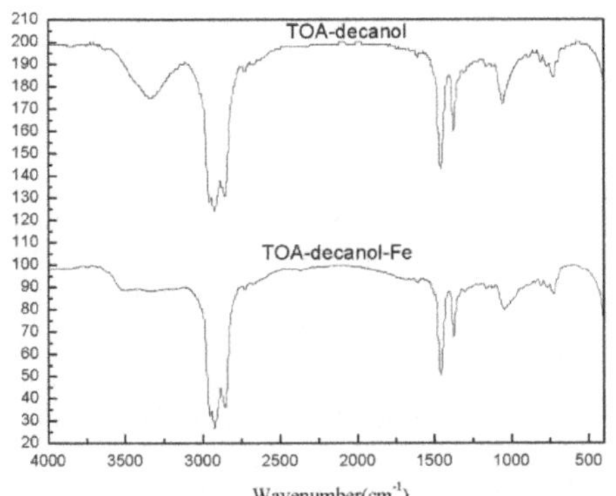

Figure 9. IR spectra of TOA-decanol and TOA-decanol-Fe.

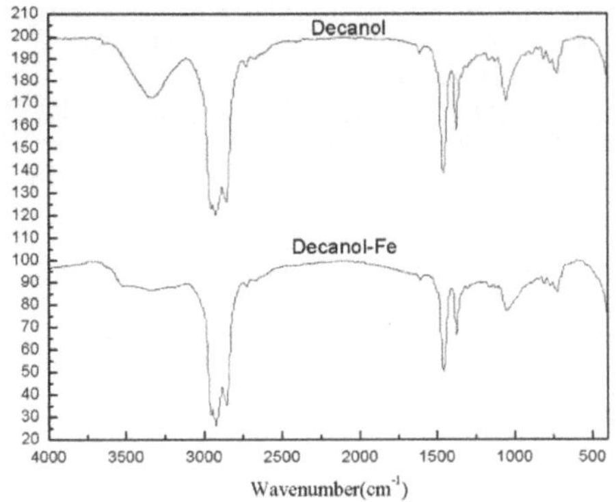

Figure 10. IR spectra of Decanol and Decanol-Fe.

$FeCl_4 \cdot 2$ Decanol. The stretching frequency of C–O was shifted from 1049 cm^{-1} in decanol to 1057 cm^{-1} in $FeCl_4 \cdot TOA \cdot 2$Decanol and 1051 cm^{-1} in $FeCl_4 \cdot 2$Decanol. The IR spectra of decanol and TOA-decanol in kerosene were agreement as TOA has no obvious infrared absorption in kerosene. These changes indicate the possibility of bonding between Fe and TOA and decanol.

3.4 *Extraction behavior of other metal ions*

The extraction behavior of other associated metal ions, namely, titanium (IV) aluminum (III), calcium (II) and magnesium (II), was studied in 8 mol dm^{-3} hydrochloric acid solutions using 0.05 mol dm^{-3} TOA and 0.86 mol dm^{-3} decanol in kerosene as extractant, which often exsist along with iron (III) in hydrochloric acid leaching solutions of titanium mineral. Through the concentration determination of the metal ions in the aqueous phase before and after the extraction it was found that titanium (IV) aluminum (III), calcium (II) and magnesium (II) were not extracted under these experimental conditions. Namely iron (III) can be separated from the associated metal ions through the extraction process.

3.5 *Extraction isotherms*

Extraction isotherms have been generated for a typical feed solution containing 0.08 mol dm^{-3} iron (III) in 8 mol dm^{-3} hydrochloric acid using 0.16 mol dm^{-3} TOA and 0.86 mol dm^{-3} decanol in kerosene (Fig. 11). The McCabe–Thiele plot for the feed solution containing 0.08 mol dm^{-3} iron (III) shows almost quantitative extraction was possible in one-stage extraction at a 1:1 aq:org phase ratio.

3.6 *Iron (III) stripping and stripping isotherms*

The stripping behavior of iron (III) has been investigated using water and low concentration of HCl as the stripping agent. Figure 12 showed that with 0~1.0 mol dm^{-3} hydrochloric acid as the stripping agents the stripping rate of iron (III) nearly attains 100%. When the concentration of hydrochloric acid is greater than 1.0 mol dm^{-3}, the stripping rate of iron (III) decreases evidently. It is clear that

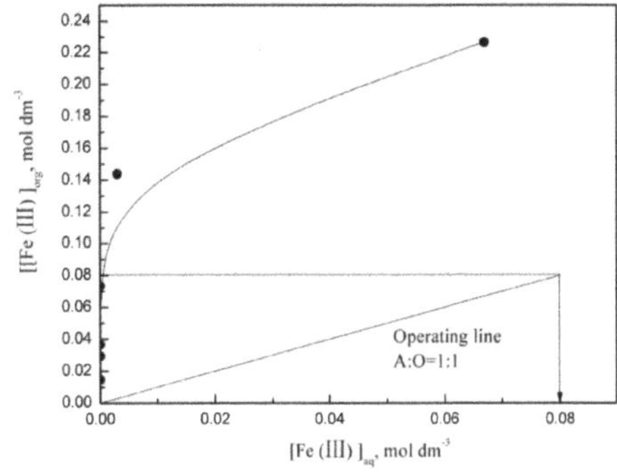

Figure 11. McCabe–Thiele plot for iron (III) extraction. $[Fe^{3+}] = 0.1\ mol \cdot dm^{-3}$, $[HCl] = 8\ mol\ dm^{-3}$, $[TOA] = 0.16\ mol \cdot dm^{-3}$, $[decanol] = 0.86\ mol \cdot dm^{-3}$.

iron (III) is easy to be stripped from the extraction complexes and water is the proper stripping agent.

Based on these results, a stripping isotherm for iron (III) was generated using water as stripping agent from a loaded organic phase consisting of 0.16 mol dm^{-3} TOA and 0.86 mol dm^{-3} decanol in kerosene containing 0.08 mol dm^{-3} iron (III). A McCabe–Thiele plot (Fig. 13) of iron (III) stripping showed that quantitative stripping is possible in three stages with an 1:1 org:aq phase ratio.

4 CONCLUSIONS

TOA and decanol extraction system is effective and selective for the extraction of iron (III) in acidic chloride solutions. The addition of decanol eliminated the third phase in the extraction of TOA for iron (III) and the extraction rate of iron (III) was increased. The extraction equilibrium equation was inferred. The extracted iron is present as $HFeCl_4 \cdot TOA \cdot 2Decanol$ and $HFeCl_4 \cdot 2Decanol$.

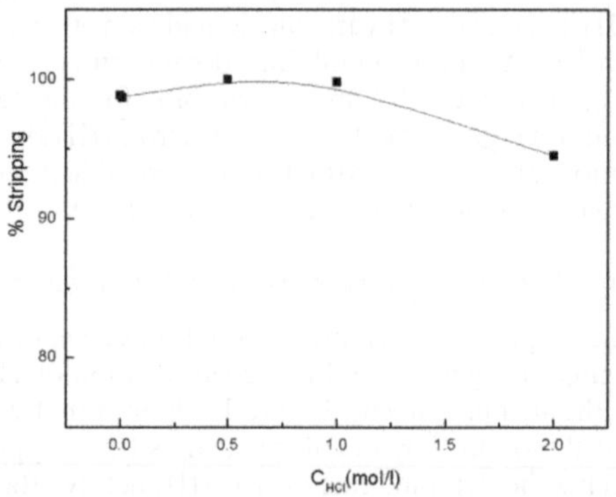

Figure 12. Stripping behavior of iron (III) from loaded TOA-decanol phase using water and low concentration of HCl.

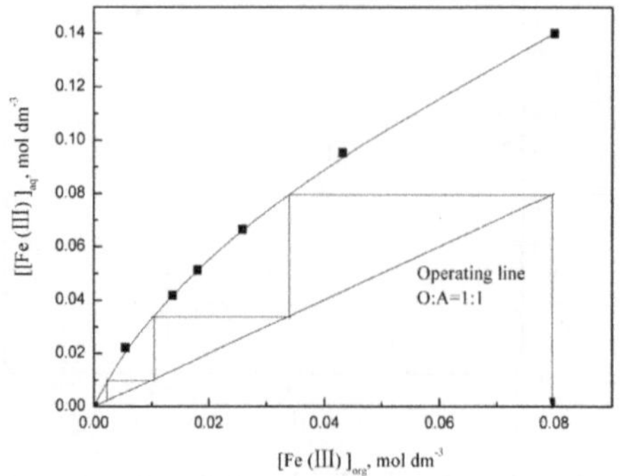

Figure 13. McCabe–Thiele plot for iron (III) stripping with water. $[Fe^{3+}]_{org} = 0.08$ mol dm^{-3}.

Extracted iron (III) is easy to be stripped and water is the proper stripping agent. Thus this study clearly highlights that TOA and decanol can be used as a potential extractant for the separation of iron (III) from titania chloride liquors containing titanium (IV), iron (III), aluminum (III), calcium (II) and magnesium (II). However, the practical utility of the current process requires further optimization of the various process parameters and also economic assessment of the process.

REFERENCES

[1] Bagreev, V.V., Fischer, C., Yudushkina, L.M., Zolotov, Y.A., 1978. Mutual influence of metals in the extraction of their chloride complexes with tri-n-octylamine and aliquat 336 in benzene. Journal of Inorganic and Nuclear Chemistry 40(3), 553–557.

[2] Duyvesteyn, W., Sabachy, B., Edmund, V., Verhulst, D., West-sells, P.G., Spitler, T.M., Vince, A., Burkholder, J.R., Paulus, B.J., and Huls, M., 2002. Processing titaniferrous ore to titanium dioxide pigment, US Patent 2002/6375923 B1.

[3] Liu, M., Zhou, Y.M., 2005. Removal of Fe (III) from sulphate solutions by synergistic extraction using N235-TBP mixed solvent systems. The Chinese Journal of Nonferrous Metals 15(10), 1648–1654.

[4] Mao, X.H., Liu, D.J., Zhang, P., 2011. Solvent extraction and separation of titanium and iron by a variety of extractants from HCl solutions. Journal of Sichuan University: Engineering Science Edition 43(Supp.1), 204–207.

[5] Miroslav, M., Jana, S., Miroslav, C., Jiri, C., 1978. Effect of the solvent on the extraction of iron(III) chloride by tri-n-octylamine. Chemicky Prumysl 28(1), 16–23.

[6] Mishra, R.K., Rout, P.C., Sarangi, K., Nathsarma, K.C., 2011. Solvent extraction of Fe(III) from the chloride leach liquor of low grade iron ore tailings using Aliquat 336. Hydrometallurgy 108(1–2), 93–99.

[7] Narayanan Remya, P., Lakshmipathy Reddy, M., 2004. Solvent extraction separation of titanium (IV), vanadium (V) and iron (III) from simulated waste chloride liquors of titanium minerals processing industry by the trialkylphosphine oxide Cyanex 923. Journal of Chemical Technology & Biotechnology 79(7), 734–741.

[8] Saji, J., Rao, T.P., Iyer, C.S.P., Reddy, M.L.P., 1998. Extraction of iron (III) from acidic chloride solutions by Cyanex 923. Hydrometallurgy 49(3), 289–296.

[9] Saji John, K., Saji, J., Reddy, M.L.P., Ramamohan, T.R., Rao, T.P., 1999. Solvent extraction of titanium (IV) from acidic chloride solutions by Cyanex 923. Hydrometallurgy 51(1), 9–18.

[10] Zhang, W., Zhu, Z., Cheng, C.Y., 2011. A literature review of titanium metallurgical processes. Hydrometallurgy 108(3–4), 177–188.

[11] Zhou X.X., Du X.N., Zhu T., 2001. Solvent Extractive Separation of Cobalt (II) and Iron (II) with Tertiary Amine. The Chinese Journal of Process Engineering 1(4), 360–364.

Material Science and Environmental Engineering – Chen (Ed.)
 ISBN 978-1-138-02938-5

Preparation of porous bismuth oxide by sol-gel method using citric acid

Z.S. Jiang & Y.J. Wang
State Key Laboratory of Explosion Science and Technology, Beijing Institute of Technology, Beijing, China

ABSTRACT: In this paper, uniform porous bismuth oxides were prepared using the sol-gel method. These porous granules were prepared with and without the surfactant PEG200. X-ray diffraction and scanning electron microscope were applied to characterize these porous bismuth oxides with and without the surfactant. The results show that the adding of surfactant increased the degree of crystallinity, purity and uniformity, and that many irregular pores were formed and distributed in bismuth oxides granules. Porous bismuth oxides prepared with 0.8 mL PEG200 contains high intensity monoclinic phase bismuth oxides.

Keywords: sol-gel; porous; bismuth oxide; citric acid

1 INTRODUCTION

Bismuth oxide Bi_2O_3 has a lot of peculiar physical and chemical properties, such as a wide energy gap change (from 2 to 3.96 eV), high oxide-ion conductivity properties, high refractive index, dielectric permittivity, and marked photoconductivity and photoluminescence (Leontie et al., 2002). A characteristic feature of bismuth oxide consists of its polymorphism: six modifications labeled as α, β, γ, δ, ε and ω-phase have been reported (Gualtieri et al., 1997: Cornei et al., 2006; Drache et al., 2007). Among them, the low-temperature α and the high-temperature δ phases are stable and the others are metastable under certain conditions (Deng et al., 2011). Due to the excellent properties and characteristics of bismuth, bismuth oxides can be used in a variety of areas such as solid oxide fuel cells (Sammes et al., 1999), gas sensors (Gou et al., 2009), fine structure glasses (Leontie et al., 2005), and catalysts (Schlesinger et al., 2013). Bismuth oxides have attracted much attention and have been studied widely.

Recently, bismuth oxide particles have been prepared through many kinds of methods, for example, co-precipitation, solvothermal method (Wang et al., 2014), chemical vapor deposition (Shen et al., 2007), and so on. But the above preparation methods are often expensive and complicated. Sol-gel methods have been proved to be good a way to prepare metal oxide particles, for example, TiO_2 nanoparticles can be prepared through sol-gel method (Myilsamy et al., 2015). There are not so many reports about preparing the porous bismuth oxide through sol-gel methods. In the paper, porous bismuth oxides were prepared by the sol-gel method using citric acid and characterize these porous bismuth oxides using different measurements.

2 EXPERIMENTAL

2.1 *Materials*

All reagents used in this study were of analytical grade and were purchased without further purification. Bismuth nitrate pentahydrate ($Bi(NO_3)_3 \cdot 5H_2O$) was used as received from Sinopharm Chemical Reagent Co., Ltd., Citric acid monohydrate ($C_6H_8O_7 \cdot H_2O$) and polyethylene glycol (PEG200) was purchased from Guangdong Xilong Chemical Co., Ltd. Ethylene glycol was used as received from Tianjin Fu Chen Chemical Reagents Factory. Nitric acid (HNO_3) (67% in purity) was purchased from Beijing Chemical Factory.

2.2 *Preparation of porous bismuth oxide*

Porous Bi_2O_3 was prepared by a sol-gel method using citric acid as complexing agent. In this synthesis procedure, 4.85 g of $Bi(NO_3)_3 \cdot 5H_2O$ was dissolved in a 24 mL nitric acid solution [V (HNO_3): V (H_2O) = 1:8] under magnetic stirring. And then 1.922 g of citric acid and 0.371 mL of ethylene glycol was added in the above solution, and 0.8 mL of PEG200 was added as surfactant. The above solution was constantly stirred for 1 h, and then the solution was heated to 80 °C and maintained at 80 °C for 3 h to form a transparent yellow gel.

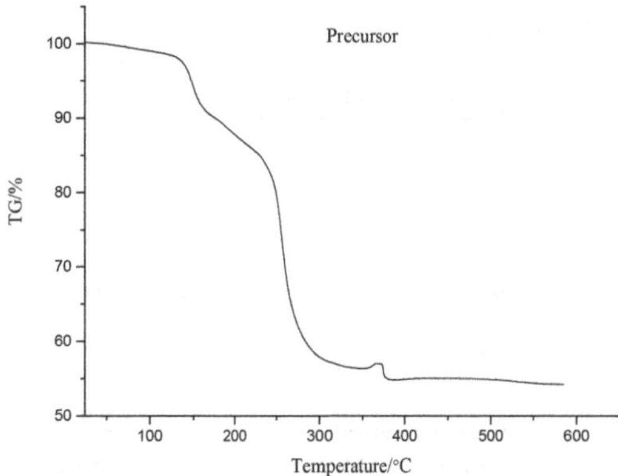

Figure 1. TG curve of the precursor prepared with 0.8 mL of PEG200 adding.

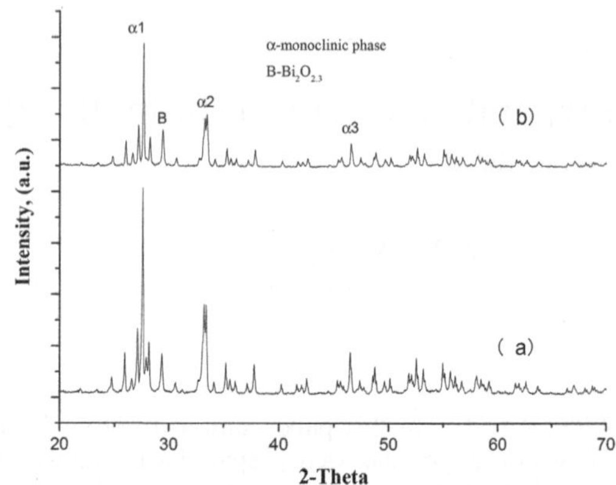

(a) Porous bismuth oxide prepared with 0.8 mL PEG200; (b) Porous bismuth oxide prepared without PEG200.

Figure 2. XRD spectra of the produced porous bismuth oxide annealed at 600 °C.

This gel was dried in an oven at 110 °C for 12 h. Subsequently, the above faint yellow fluffy precursor (foam-like) was grounded finely and subjected to thermogravimetric analysis (TG/DTA6300, SII Nano Technology Inc., Japan) in order to establish a thermal treatment schedule for the precursor. The precursor was decomposed completely when it was heated to a high temperature.

2.3 *Characterization*

The composition and crystal structure of the bismuth oxide nanoparticles was determined by X-Ray Diffraction (XRD, Bruker D8 advance) using a Cu Kα (λ = 1.54056 Å). The scans were taken at room temperatures over a wide range of 2θ = (20°–70°) at 0.05 degrees intervals. The morphologies of the as-made samples were observed on Scanning Electron Microscope (SEM, Hitachi S4800).

3 RESULTS AND DISCUSSION

3.1 *Analysis of XRD results*

Figure 2 displays the XRD spectra of the prepared porous bismuth oxide. It can be seen from the spectra that the porous bismuth oxide prepared with and without PEG200 are mainly composed of monoclinic phase of bismuth oxide (α-Bi_2O_3) (JCPDS NO. 14–0699). The small peaks 2θ = 29.3° corresponding to the diffraction plane of $Bi_2O_{2.3}$ (JCPDS NO. 76–2477). It can be also noticed that the α phase of porous bismuth oxide accounts for most of the proportion of the sample and only a small percentage of the sample is $Bi_2O_{2.3}$. The high intensity indicates that the α phase of bismuth oxides become more dominant with the adding of PEG200 while the trend is adverse for the $Bi_2O_{2.3}$, and the higher crystallinity products were acquired with the surfactant adding.

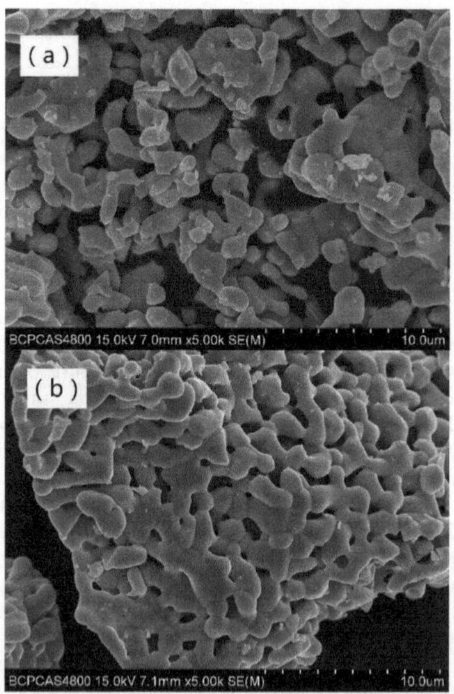

(a) Porous bismuth oxide prepared without PEG200; (b) Porous bismuth oxide prepared with 0.8 mL PEG200.

Figure 3. SEM photographs of porous bismuth oxide.

3.2 *Analysis of SEM results*

The SEM micrographs of the porous bismuth oxides prepared with surfactant adding and without surfactant are shown in Figure 3. As shown in Figure 3a, it can be noticed that the morphology

Figure 4. The process of PEG200 dissolve in water to generate micelle.

of the bismuth oxide particles are not homogeneous and like a dumbbell. The particle size is about 1 μm. While the morphology of the bismuth oxide particles prepared with adding surfactant PEG200 is more uniform, and the agglomeration phenomenon is more serious. But it is interesting to see in the Figure 3b that many irregular pores are distributed in a granule and the porous bismuth oxides form.

3.3 *Effect of surfactant on the morphology*

PEG200 is a nonionic surface active agent, and the molecular formula is HO $(CH_2CH_2O)_{200H}$. As shown in Figure 4, a special kind of micelle is formed when PEG200 is hydrolyzed in aqueous solution and plays a role of micellar solubilization to improve the solubility of bismuth salt. The surfactant PEG200 increases the uniformity of the precursor particles, and furthermore, many irregular pores form and are distributed in bismuth oxide granules when PEG200 decomposes completely and a lot of gas is produced in the heat treatment of the precursors.

4 CONCLUSION

In the present paper, porous and uniform bismuth oxides mainly composed of α phase were prepared through the sol-gel method with the surfactant PEG200 adding. Compared to the bismuth oxides prepared without surfactant, the adding of surfactant increases with the degree of crystallinity, purity, and uniformity. We can also conclude that the porous structure is formed after calcination with the surfactant adding.

ACKNOWLEDGMENTS

This paper is supported by the project of State Key Laboratory of Explosion Science and Technology (Beijing Institute of Technology, China) (No. YBKT16-06).

REFERENCES

[1] A.F. Gualtieri, S. Immovilli, M. Prudenziati, Powder X-ray diffraction data for the new polymorphic compound ω-Bi_2O_3, Powder Diffraction, 12(1997) 90–92.

[2] H. Wang, H.X. Yang, L. Lu, Topochemical synthesis of Bi_2O_3 microribbons derived from a bismuth oxalate precursor as high-performance lithium-ion batteries, RSC Advances, 4(2014) 17483–17489.

[3] H.Y. Deng, W.C. Hao, H.Z. Xu. A transition phase in the transformation from α-, β- and ε- to δ-bismuth oxide, chin. Phys. Lett, 28(2011) 056101.

[4] L. Leontie, M. Caraman, M. Alexe, C. Harnagea, Structural and optical characteristics of bismuth oxide thin films, Surf Sci, 507–510(2002) 480–485.

[5] L. Leontie, M. Caraman, A. Visinoiu, G.I. Rusu, On the optical properties of bismuth oxide thin films prepared by pulsed laser deposition, Thin Solid Films, 473(2005) 230–235

[6] M. Drache, P. Roussel, J. Wignacourt, Structures and oxide mobility in Bi-Ln-O materials: Heritage of Bi_2O_3, Chem Rev, 107(2007) 80–96.

[7] M. Myilsamy, V. Murugesan, M. Mahalakshmi, The Effect of synthesis conditions on mesoporous structure and the photocatalytic activity of TiO_2 nanoparticles, J Nanosci Nanotechno, 15(2015) 4664–4675.

[8] M. Schlesinger, M. Weber, S. Schulze, M. Hietschold, M. Mehring, Metastable β-Bi_2O_3 nanoparticles with potential for photocatalytic water purification using visible light irradiation, Chemistry Open, 2(2013) 146–155.

[9] N. Cornei, N. Tancret, F. Abraham, O. Mentré, New ε-Bi_2O_3 metastable polymorph, Inorg Chem, 45(2006) 4886–4888.

[10] N.M. Sammes, G.A. Tompsett, H. Näfe, F. Aldinger, Bismuth based oxide electrolytes—structure and ionic conductivity, J Eur Ceram Soc, 19(1999) 1801–1826.

[11] X.L. Gou, R. Li, G.X. Wang, Z.X Chen, D. Wexler. Room-temperature solution synthesis of Bi_2O_3 nanowires for gas sensing application, Nanotechnology, 20(2009) 495501.

[12] X.P. Shen, S.K. Wu, H. Zhao, Q. Liu, Synthesis of single-crystalline Bi_2O_3 nanowires by atmospheric pressure chemical vapor deposition approach, Physica E: Low-dimensional Systems and Nanostructures, 39(2007) 133–136.

Figure 4. The process of PEG200 dissolves in water to generate micelle.

of the bismuth oxide particles are not homogeneous and like a dumbbell. The particle size is about 1 μm. While the morphology of the bismuth oxide particles prepared with adding surfactant PEG200 is more uniform, and the agglomeration phenomenon is more serious. But it is interesting to see in the Figure 3b that many irregular pores are distributed in a granule and the porous bismuth oxides form.

3.3 *Effect of surfactant on the morphology*

PEG200 is a nonionic surface active agent and the molecular formula is $HO(CH_2CH_2O)_nH$. As shown in Figure 4, a special kind of micelle is formed when PEG200 is hydrolyzed in aqueous solution and plays a role of micellar solubilization to improve the solubility of bismuth salt. The surfactant PEG200 increases the uniformity of the precursor particles, and furthermore many irregular pores form and are distributed in bismuth oxide granules when PEG200 decomposes completely and a lot of gas is produced in the heat treatment of the precursors.

4 CONCLUSION

In the present paper, porous and uniform bismuth oxides mainly composed of α phase were prepared through the sol-gel method with the surfactant PEG200 adding. Compared to the bismuth oxides prepared without surfactant, the adding of surfactant increases with the degree of crystallinity, purity, and uniformity. We can also conclude that the porous structure is formed after calcination with the surfactant adding.

ACKNOWLEDGMENTS

This paper is supported by the project of State Key Laboratory of Explosion Science and Technology (Beijing Institute of Technology, China) (NO. YBKT16-06).

REFERENCES

[1] A.L. Gualtieri, S. Immovilli, M. Prudenziati, Powder X-ray diffraction data for the new polymorphic compound ω-Bi_2O_3, Powder Diffraction. 12(1997) 90–92.
[2] H. Wang, H.X. Yang, L. Lu, Topochemical synthesis of Bi_2O_3 microribbons derived from a bismuth oxalate precursor as high-performance lithium-ion batteries, RSC Advances. 4(2014) 17483–17489.
[3] H.Y. Deng, W.C. Hao, H.Z. Xu, δ-Bismuth oxide phase in the transformation from β, Planary δ-bismuth oxide, Chin. Phys. Lett. 27(2011) 056103.
[4] L. Leontie, M. Caraman, M. Alexe, C. Harnagea, Structural and optical characteristics of bismuth oxide thin films, Surf Sci. 507 (2002) 480–485.
[5] L. Leontie, M. Caraman, A. Visinoiu, G.I. Rusu, On the optical properties of bismuth oxide thin films prepared by pulsed laser deposition, Thin Solid Films. 473(2005) 230–235.
[6] M. Drache, P. Roussel, J. Wignacourt, Structures and oxide mobility in Bi-Ln-O materials: Heritage of Bi_2O_3, Chem Rev. 107(2007) 80–96.
[7] M. Mallahi, A. Shokuhfar, M. Vaezi, A. Esmaeilirad, The effect of synthesis conditions on the particle structure and visible-photocatalytic activity of Bi_2O_3 nanoparticles, J. Nanosci. Nanotechnol. 15(2015) 4604–4612.
[8] M. Schlesinger, M. Weber, S. Schulze, M. Hietschold, M. Mehring, Metastable β-Bi_2O_3 nanoparticles with potential for photocatalytic water purification using visible light irradiation, ChemistryOpen. 2(2013) 146–155.
[9] N. Cornei, N. Tancret, F. Abraham, O. Mentré, New ε-Bi_2O_3 metastable polymorph, Inorg. Chem. 45(2006) 4886.
[10] N.M. Sammes, G.A. Tompsett, H. Näfe, F. Aldinger, Bismuth based oxide electrolytes-structure and ionic conductivity, J. Eur. Ceram. Soc. 19(1999) 1801–1826.
[11] X.L. Gou, R. Li, G.X. Wang, Z.X. Chen, D. Wexler, Room-temperature solution synthesis of Bi_2O_3 nanowires for gas sensing application, Nanotechnology. 20(2009) 495501.
[12] X.P. Shen, S.K. Wu, H. Zhao, Q. Liu, Synthesis of single-crystalline Bi_2O_3 nanowires by atmospheric pressure chemical vapor deposition approach, Physica E: Low-dimensional Systems and Nanostructures. 39(2007) 133–136.

Material Science and Environmental Engineering – Chen (Ed.)
 ISBN 978-1-138-02938-5

Synthesis of spherical $SrCO_3$ powders by hydrothermal method

S.X. Wang & X.W. Zou
Qinghai Institute of Salt Lake, Chinese Academy of Science, Xining, P.R. China

ABSTRACT: A new synthetic method of strontium carbonate with hydrothermal is introduced in this paper. We got spherical $SrCO_3$ powders with $SrCl_2$ and $NaCO_3$ as raw materials through adding suitable chemical additives for crystalline control. By hydrothermal method, we got a good dispersion of spherical particles of narrow particle size distribution. The powder was characterized by SEM and XRD. We studied the effect of additives, the temperature of hydrothermal process, and pH of the solution on the morphology of strontium carbonate. The results show that due to the presence of additive-free we got rod-like structure of strontium carbonate. Added EDTA, we can got spherical particles, the size of spherical particles about 200–800 nm, but the particle dispersion is not well-gathered. Through hydrothermal treatment, morphology can be further improved.

Keywords: hydrothermal; ultrafine; spherical; strontium carbonate

1 INTRODUCTION

Recent advances have revealed that strontium carbonate powders offer many advantages as materials for use in miniaturization of high quality electronic components, high performance magnetic materials, big screen TVs, computer displays, light-emitting materials, and optical glass, etc [1, 2]. Especially as the raw materials of electronic components, the content of strontium carbonate powders needs high purity, to convert the microstructure into spherical, narrow particle size distribution, and the average particle diameter is less than 1 micron. Micro-nano strontium carbonate has a special application in catalysis and chemical sensors. The product of strontium carbonate with different morphology has a different value, so the control of strontium carbonate morphology is becoming an important part of strontium carbonate study [3].

Hydrothermal method is a commonly used method of inorganic synthesis and materials processing. The hydrothermal method has the following characteristics: relatively low energy consumption, wide applicability, and can be used for the preparation of ultrafine particles, can also get a larger crystal size, preparation of inorganic ceramic membrane. According to the research object and purpose is different, hydrothermal method can be divided into hot water crystal growth, hydrothermal synthesis, water treatment, and water thermal sintering, etc. In recent years, hot water is often used in auxiliary synthesis of some special morphology of the particles [4, 5]. In this paper, we got spherical $SrCO_3$ particles under the condition of the analysis of pure $SrCl_2$ and Na_2CO_3 as raw material, EDTA as morphology agent, and hot water auxiliary for 4 hours.

2 EXPERIMENTAL

All the reagents were of analytical grade and used without further purification. The water used in the experiment was distilled water.

The entire experimental process, including raw materials liquid preparation, reaction, and centrifugal washing and drying process is a result of: first, formulation of the 0.1 mol/L $SrCl_2$ solutions and 0.1 mol/L Na_2CO_3 solutions added to a certain amount of EDTA to 50 ml 0.1 mol/L $SrCl_2$ solutions due to magnetic stirring at room temperature for about 10 minutes; second, use of sodium hydroxide solution (1 mol/L) or hydrochloric acid solution (1 mol/L) to adjust the solution pH to a certain value, then 50 ml 0.1 mol/L Na_2CO_3 solution is to be dropped into above solution in the process of stirring; third, the total 100 ml resultant solution containing white precipitates is to be transferred into a teflon-lined stainless steel autoclave with a capacity of 100 ml, then sealed and maintained at a certain temperature for 4 h; and finally, we get $SrCO_3$ products from slurry by centrifugation, ethanol washing, drying.

The sizes and morphologies of the resulting products were studied by Scanning Electron Microscopy (SEM).

3 RESULTS AND DISCUSSION

Figure 1 shows the XRD of $SrCO_3$ sample in a typical experiment and the standard spectrum of $SrCO_3$, we know that no peak of impurities was observed, confirming the formation of pure $SrCO_3$.

3.1 *The effect of additives*

Figure 2(A) shows that the morphology of the products was composed of rod-like micro-particles of average length 5 micron in the condition of no additives. So we added EDTA in the experimental process, in order to change the morphology and size of the products. Adding a certain amount of EDTA to make $n_{EDTA}/n_{Sr^{2+}} = 1/2$, the product morphology shown in Figure 2(B) was that of spherical particles.

The results show the morphology of the particles greatly changed by addition of EDTA. EDTA is an effective spherical morphology control agent.

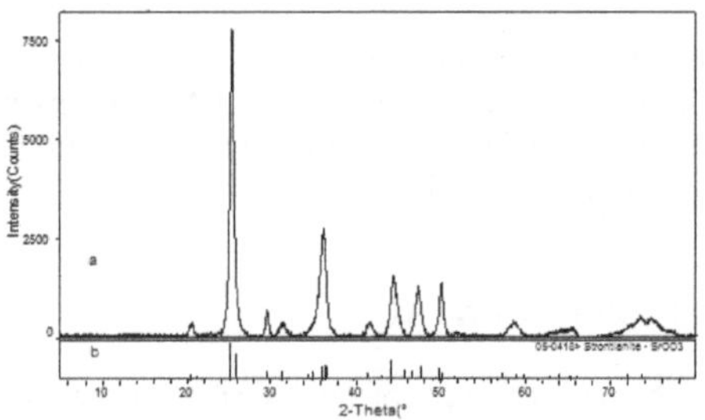

Figure 1. (a) XRD spectrum of sample (b) XRD standard spectrum of $SrCO_3$.

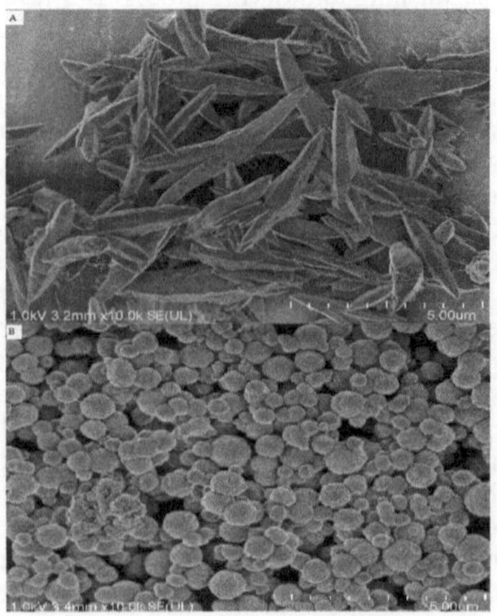

Figure 2. SEM images of sample $SrCO_3$ with different additives (A) no additive (B) in the presence of EDTA.

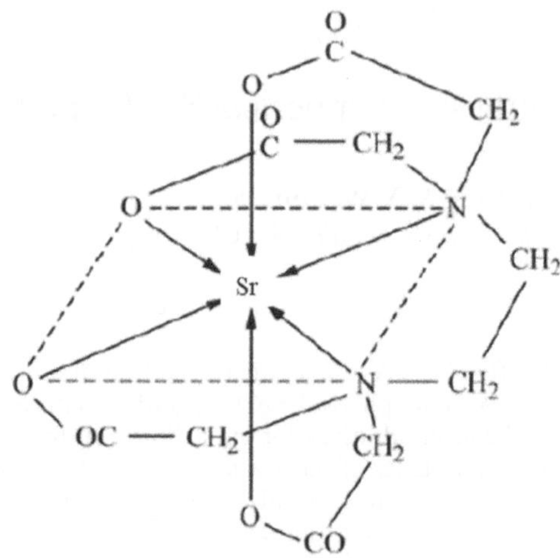

Figure 3. Molecular structural formula of Sr-EDTA complex.

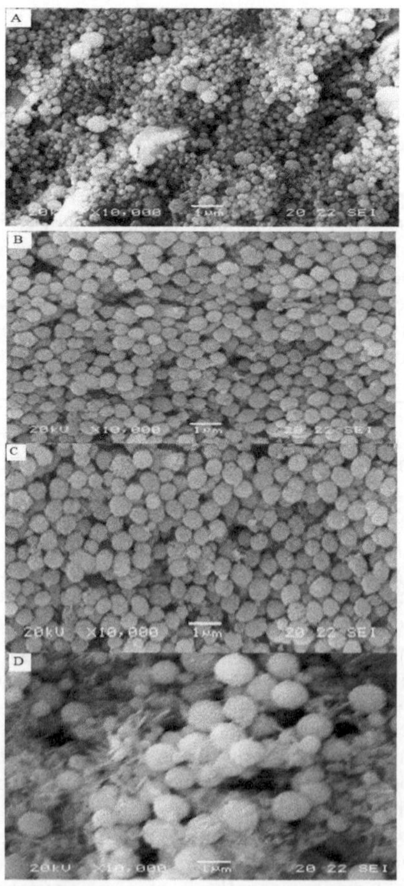

Figure 4. SEM images of sample $SrCO_3$ with different hydrothermal temperature (A) 100 °C (B) 120 °C (C) 150 °C (D) 180 °C.

EDTA can react with strontium ions from a stable Sr-EDTA complex, of which structure is shown in Figure 3. Strontium ions will release from the Sr-EDTA complex when added to Na_2CO_3 solution and form $SrCO_3$.

$$Sr^{2+} + EDTA \rightarrow Sr\text{-}EDTA \quad (1)$$

$$Sr\text{-}EDTA + CO_3^{2-} \rightarrow SrCO_3 + EDTA \quad (2)$$

3.2 *The effect of temperature*

In the condition of EDTA, by transforming the temperature of the hydrothermal process, we found that the morphology of products has not changed. We studied the effect of temperature on the morphology of strontium carbonate. Figure 4 is the SEM images of the product with different temperature. At lower temperature, strontium carbonate is given priority to globular shape, but no good dispersion. At high temperature, particle size of strontium carbonate becomes big, and the morphology changes badly. So the best hot water temperature is 120 °C.

Figure 5. SEM images of sample $SrCO_3$ with different pH of values (a) pH = 8 (b) pH = 10 (c) pH = 12 (d) pH = 14 results.

3.3 *The effect of pH*

We studied the effect of pH of the solution on the morphology of strontium carbonate. Figure 5 has the SEM images of the products with different pH values. When EDTA is added n_{EDTA}/nSr^{2+} value equal to 0.4, the effect of different pH value on morphology was investigated. The results show that when the pH value is equal to 8, 10, and 12, it results in having spherical particles and a good dispersion. When the pH value is equal to 10, the particle size of product is minimum, approximately 0.45 microns. When pH value is equal to 14, the product shape is irregular, particles are together and dispersion is bad.

4 CONCLUSIONS

By studying the preparation of strontium carbonate by hydrothermal method, we know that:

1. Due to the presence of additive-free environment, it has resulted in rod-like structures of strontium carbonate. After adding EDTA, we get spherical particles.
2. Due to the presence of EDTA, the average particle size of spherical particles increases with the temperature of hydrothermal process increasing.
3. Due to the presence of EDTA, pH change affected the spherical particle size. When the pH value is equal to 10, the particle size of $SrCO_3$ is minimum.

ACKNOWLEDGMENT

This work was supported by the Science and Technology Department of Qinghai Province (2013-G-204-2).

REFERENCES

[1] M.X. Zhang, J.C. Huo, et al. 2006. New Chemical Materials Vol. 34, p.5.

[2] S.X. Liu, H.B. Wang, J.C. Huo, et al. 2007. Inoraan Ic Chem Icals Industry Vol. 39, p.1.

[3] J.F. Chen, H.K. Zou, R.J. Liu, et al. 2001. Modern Chemical Industry Vol.21, p.9.

[4] Dirksen J A, Ring T A. 1996. Chem Eng. Sci. Vol. 51, p.1957.

[5] X. Li, J.F. Chen, G.T. Chen.1994. Acta Mechanic Sinica Vol. 26, p.266.

Material Science and Environmental Engineering – Chen (Ed.)
 ISBN 978-1-138-02938-5

The adsorption performance of sodium nitrate for vermiculite, perlite and ceramsite

R.G. Li, J.Q. Zhu & W.B. Zhou
State Key Laboratory of Silicate Materials, Wuhan University of Technology, Wuhan, P.R. China
School of Material Science and Engineering, Wuhan University of Technology, Wuhan, P.R. China

X.M. Cheng & Y.Y. Li
School of Material Science and Engineering, Wuhan University of Technology, Wuhan, P.R. China

ABSTRACT: The present work focused on the preparation and characterization of composite thermal storage materials employed in thermal storage. Composite materials were prepared by direct soaking method. The support materials were vermiculite, expanded perlite and ceramsite, and the medium was sodium nitrate. The XRD results showed that composites have good thermal stability performance. The SEM results showed that vermiculite, expanded perlite and ceramsite have porous structures. Sodium nitrate adsorbed in the inner structure, and it was stably retained in the porous structures. DSC results indicated that the mass of vermiculite, expanded perlite and ceramsite increased 589.2%, 218.5% and 54.9%, respectively. The heat enthalpies of endothermic processes were corresponding to 147.0 J/g, 118.0 J/g and 145.5 J/g. The results suggested that vermiculite, expanded perlite and ceramsite could be used as encapsulated material in the thermal storage energy fields.

Keywords: vermiculite; perlite; ceramsite; sodium nitrate; composite

1 INTRODUCTION

Thermal energy storage is widely recognized as a means to integrate renewable energies into the production on the generation side, but its applicability to the demand side is also possible. Certainly, thermal energy storage is of particular interest and significance in using this essential technique for thermal applications such as heating, hot water, cooling, air-conditioning, etc. Recently thermal energy storage is latent heat for the phase change occurs at nearly constant temperatures. However, practical difficulties usually arise in applying the latent heat material due to the density change, stability of properties under extended cycling and sometimes phase segregation [3]. Another problem of latent thermal energy storage is the low thermal conductivity and leakage [4]. Therefore it appears that molten salts have been more widely researched for heat storage applications [5]. Molten salts are likely candidates because of the low cost, stability, relatively low vapor pressure and high thermal capacity especially [6, 7].

However molten salts have very strong corrosion resistance limit the application. Recently many scholars studied this problem [8,9]. In order to solve the problem, a considerable amount of research has been carried out on the preparation of form-stable phase change materials without extra packaging, and they can still remain solid and stable without leakage even above the melting temperature [10]. Vermiculite, expanded perlite and ceramsite are usually fluffy, highly porous due to a foam-like cellular structure. They have a series of promising characteristics, such as low density, relatively low cost, good thermal stability and low moisture retention. Especially they exhibit an excellent absorbability with strong capillary force and surface tension with multiple pores so that fatty acids can be absorbed easily in pores. Sodium nitrate as one of the molten salts has been used in large scale experimental and commercial solar thermal power plants for its large heat capacity, low cost and good chemical stability. In this study, they are separated to research the absorption of sodium nitrate and the results can be used for the application.

2 EXPERIMENTAL

2.1 *Materials*

Sodium nitrate with purity ≥99.0% was supplied by Beijing Chemical Reagent Company, China.

Vermiculite was furnished by Lingshou Yixin minerals co., LTD, the grain size between 0 mm and 4 mm. Expanded graphite was purchased from XinYang Shenda thermal insulation building material Co., Ltd, and the grain size less than 1 mm. Ceramsite was supplied by Fuzhou Xincai building materials co., LTD, and the grain size from 1 mm to 4 mm.

2.2 *Methods*

Vermiculite, perlite and ceramsite were washed with ethanol and then dried at 105°C in the drying oven for 24 hours. Then materials were prepared by solution impregnation method. The method is that support materials were taken into the crucible with the medium is melting sodium nitrate at 350°C in the muffle furnace.

2.3 *Measurements*

The composites were identified by X-ray diffraction (Model D/Max-RB, Rigaku, Japan). Morphology of the fractured and the mirror-polished surfaces was observed by Scanning Electron Micrograph (JSM-5610 LV, JEOL, and Japan). Enthalpy of composite materials and phase change temperature tested through Differential Scanning Calorimeter (DSC, Pyris-1) and the testing in nitrogen atmosphere, temperature rising rate is 10°C/min.

3 RESULTS AND DISCUSSION

3.1 *XRD analysis of vermiculite, expanded perlite and ceramsite*

XRD diffraction peaks of vermiculite, expanded perlite and ceramsite were shown in Table 1. The main components of vermiculite, expanded perlite and ceramsite are silicate.

3.2 *XRD analysis of composites*

Composites are at 350°C for 10 hours and the XRD be used to analysis the crystal constitute. Figure 1 shows the XRD patterns of vermiculite, expanded perlite and ceramsite composited with sodium nitrate. All the diffraction peaks can be well indexed as the sodium nitrate (sodium nitrate, JCPDS 89–310), the d values were 3.893, 3.042, 2.811, 2.532, 2.312, 2.127 and 1.899, corresponding to sodium nitrate. The diffraction peaks indicated that no other phases were formed during the process of synthesis. The main reason is that some phase constitutes were non crystalline structure.

Table 1. Chemical composition of vermiculite, expanded perlite and ceramsite. (Wt.%).

Chemical composition	SiO_2	Al_2O_3	CaO	MgO	Fe_2O_3	Mass loss
Vermiculite	39.7	15.4	3.8	18.7	14.7	4.9
Expanded perlite	76.2	14.8	1.4	0.8	1.4	3.7
Ceramsite	65.3	16.5	1.9	2.7	4.2	6.7

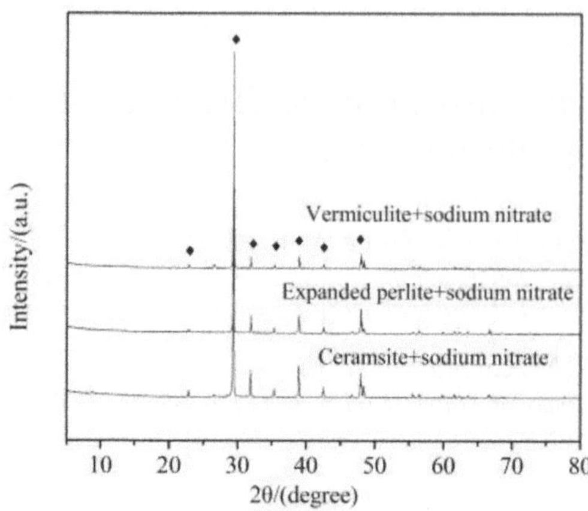

Figure 1. XRD of vermiculite, expanded perlite and ceramsite with sodium nitrate.

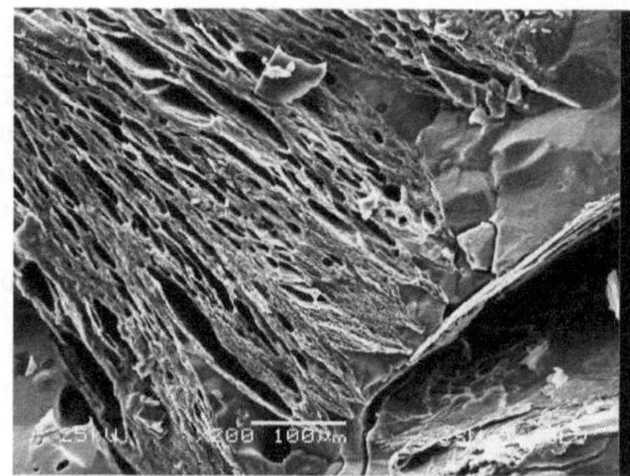

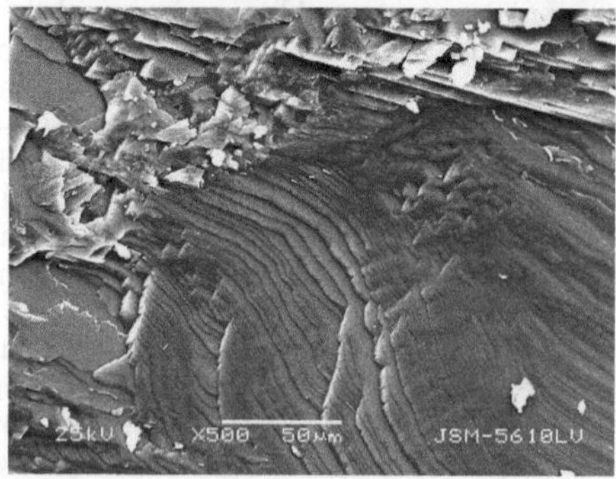

Figure 2(a). Morphology of vermiculite-sodium nitrate composite material.

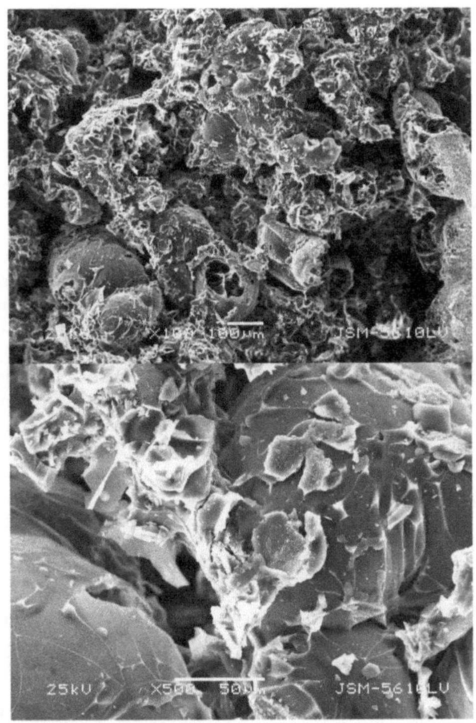

Figure 2(b). Morphology of expanded perlite-sodium nitrate composite material.

3.3 *Morphology of composite*

Morphologies of composite materials were shown in Figure 2 (a), (b) and (c). All images display the composite materials. Figure 2 showed the SEM micrographs of the fractured surfaces of vermiculite, expanded perlite and ceramsite composited with sodium nitrate. It can be seen from Figure 2(a) that the internal of vermiculite is multilayer scaly structure, and sodium nitrate was adsorbed in the layered structure crack. Figure 2(b) showed that expanded perlite was spherical in shape, usually fluffy, highly porous due to a foam-like cellular internal structure. Sodium nitrate was adsorbed in the foam-like cellular internal structure. Figure 2(c) showed that the ceramsite was a solid and massive structure, the inter space of ceramsite was less than the front of two. But the inter space be filled with sodium nitrate. The SEM morphologies of vermiculite, expanded perlite and ceramsite composites indicated that the three materials can absorb salts into their inner structure.

3.4 *DSC analysis of composites*

Thermal physical properties of composites were shown in Table 2 and the latent heat of the composite materials indicated in Figure 3. It is found that there are three endothermic peaks at about 305°C in the DSC curves. The overcooling were 19.9°C, 24.1°C and 20.5°C, respectively. The mass of vermiculite, expanded perlite and ceramsite increased 589.2%, 218.5% and 54.9%, respectively. The heat enthalpies of endothermic processes are correspond to 147.0 J/g, 118.0 J/g and 145.5 J/g. The heat enthalpy of pure sodium

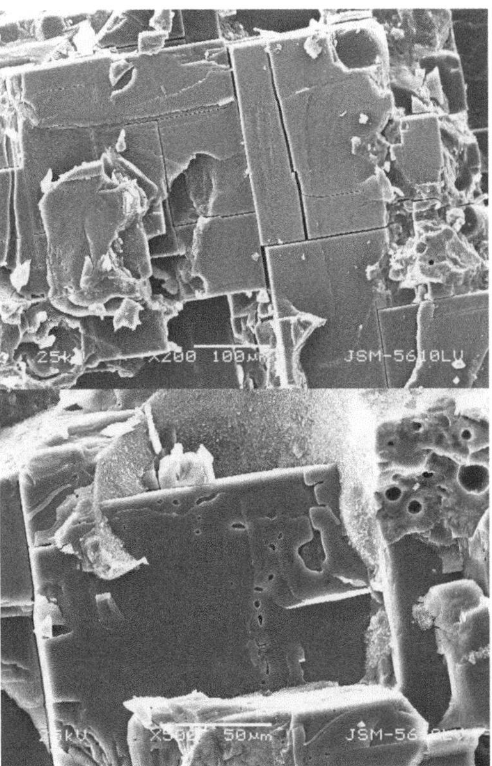

Figure 2(c). Morphology of ceramsite-sodium nitrate composite material.

Table 2. Thermal physical properties of composites.

	Melting temperature/ (°C)		Over cooling degree/(°C)	Melting enthalpy/ (J/g)
Content	Onset	End	ΔT	ΔH_m
Vermiculite-sodium nitrate	293.6	313.5	19.9	147.0
Expanded perlite-sodium nitrate	297.6	321.7	24.1	118.0
Ceramsite-sodium nitrate	296.3	316.8	20.5	145.5

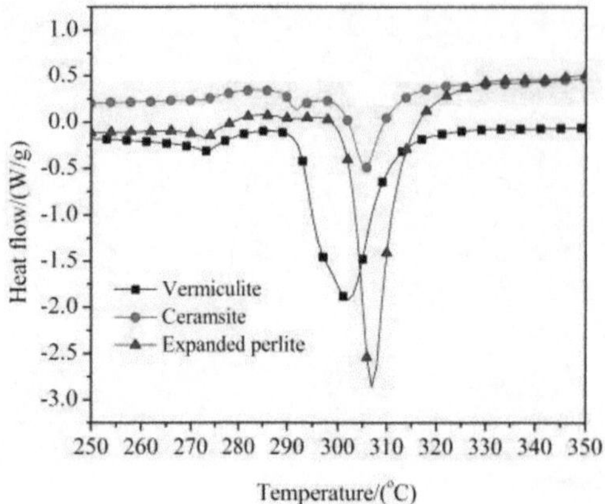

Figure 3. DSC curves of the composite heat storage materials.

nitrate is 172.0 J/g. In the composite material that three components did not undergo a phase transition process under 350°C. In the composites, only sodium nitrate can absorbs thermal energy. The mass percentage of PCM is calculated according to Eq. (1) as follows:

$$\Delta H_{composite} = (1 - wt\%)\Delta H_{PCMs} \quad (1)$$

where $\Delta H_{composite}$ represents the calculated phase change latent heat of the composite materials, ΔH_{PCMs} is the latent heat of the heat storage materials and $wt\%$ is the mass fraction of support materials. The heat enthalpy of composites commit to the mass increasing.

4 CONCLUSIONS

Composite storage heat materials were prepared by direct soaking method. Vermiculite, expanded perlite and ceramsite were as the support materials. While sodium nitrate as the phase change material, DSC results indicated that the mass of vermiculite, expanded perlite and ceramsite increased 589.2%, 218.5% and 54.9%, respectively. The heat enthalpies of endothermic processes are correspond to 147.0 J/g, 118.0 J/g and 145.5 J/g. The results indicated that vermiculite, expanded perlite and ceramsite could be used as encapsulated material in the thermal storage energy application.

ACKNOWLEDGMENTS

This work was supported by the National Science and technology support program of China (No. 2012BAA05B06).

REFERENCES

[1] Gil A., Medrano M., MartoreII I., Lazaro A., Dolapo P., Zalba B. 2010. State of the art on high temperature thermal energy storage for power generation. Part1-concepts, materials and modelisation. Renewable and Sustainable Energy Reviews, 14(1): 31–55.

[2] P. Pardo, A. Deydier, Z. Anxionnaz-Minvielle, S. Rougé, M. Cabassudb, P. Cognet. 2014. A review on high temperature thermochemical heat energy storage. Renewable and Sustainable Energy Reviews, (32): 591–610.

[3] Weihuan Zhao, David M., WenhuaYu, Taeil Kima, Dileep Singh. 2014. Phase change material with graphite foam for applications in high-temperature latent heat storage systems of concentrated solar power plants. Renewable Energy, (69): 134–146.

[4] Yu-BingTao, Ming-Jia Li, Ya-Ling He, Wen-Quan Tao. 2014. Effects of parameters on performance of high temperature molten salt latent heat storage unit. Applied Thermal Engineering, 1(38): 1–8.

[5] Bruno Cárdenas, Noel León. 2013. High temperature latent heat thermal energy storage: Phase change materials, design considerations and performance enhancement techniques. Renewable and Sustainable Energy Reviews, (27): 724–737.

[6] Zhaowen Huang, Xuenong Gao, Tao Xu, Yutang Fang, Zhengguo Zhang. 2014. Thermal property measurement and heat storage analysis of LiNO3/KCl-expanded graphite composite phase change material. Applied Energy, (115): 265–271.

[7] Zhiwei Ge, Feng Ye, Hui Cao, Guanghui Leng, Yue Qin, Yulong Ding. 2013. Carbonate-salt-based composite materials for medium-and high-temperature thermal energy storage. Particuology, (9): 1–5.

[8] Yang S., Yang N.Z., He F.Q. 2012. Investigation on Thermal Sotrage and Release Property with Organic Phase Change Heat Storage Composite Material. Wuhan Ligong Daxue Xuebao (Journal of Wuhan University of Technology), 34(2): 22–26.

[9] Farid M., Khudhair A., Razack S., Al-Hallaj S. 2004. A review on phase change energy storage: materials and applications. Energy Conversion and Management, 2004, (45): 1597–615.

[10] Chengzhou Guo, Jiaoqun Zhu, Weibing Zhou, Wen Chen. 2010. Fabrication and thermal properties of a new heat storage concrete material. Journal of Wuhan University of Technology-Mater. Sci. Ed. 2010, 25(4): 628–630.

Material Science and Environmental Engineering – Chen (Ed.)
© 2016 Taylor & Francis Group, London, ISBN 978-1-138-02938-5

Performance of Flame Retardant Hexakis (4-Aminophenoxy) Cyclotriphazene on wool fabrics lingling

L.L. Liu
Chemical Engineering and Biotechnology, College of Chemistry, Donghua University, Shanghai, China

W. Wang & D. Yu
Chemical Engineering and Biotechnology, College of Chemistry, Donghua University, Shanghai, China
Key Laboratory of Eco-Textile, Ministry of Education, Donghua University, Shanghai, China

ABSTRACT: In this paper, Hexakis (4-Aminophenoxy) Cyclotriphazene (HACP) was firstly used as Flame Retardant (FR) applied on wool fabrics. This FR wool fabric shows excellent flame resistance properties. TGA indicated that weight loss rate of FR wool fabric decreased and the maximum decomposition temperature shifted to an early point. Char residue at 600°C reaches a high value of 40.7% with weight gain of 7.28%, greatly higher than that of pristine wool fabric (28.3%). Char length is dramatically reduced. SEM shows that the char surface of FR wool fabric is full of bubbles and pores, which means that more non-flammable gases are generated of FR wool fabric. Moreover, the char surface of FR wool fabric seems to be more rigid than that of pristine wool fabric. Breaking strength of wool fabric increases after it was treated with HACP. From these results, it could be speculated that HACP plays the role of flame retardant both in gas phase and condense phase.

Keywords: hexakis (4-aminophenoxy) cyclotriphazene; flame retardant; wool fabric

1 INTRODUCTION

Wool fabric has a lot of good properties such as soft luster, good elasticity, high wear resistance, a good hygroscopic, and warmth retention [8]. Besides these, it has high nitrogen content (16%), high moisture content (10–14%) [1], high ignition temperature (570–600°C) [2] and imparts the wool fabric with a natural flame resistant property.

Although natural wool fabric has relatively high Limiting Oxygen Index (LOI) (25–28%) [9], the limiting oxygen index of processed wool fabric is usually lowered due to some agents added. So as an important textile material, flame retardant treatment is necessary for wool fabric when used in air plane, car and other areas. A lot of studies have been done on the flame retardant finish of a wool fabric. Among them, the Zirpro method developed from the 1970 s is the most common way to prepare flame retardant wool fabric [3, 4, 5, and 17]. In this method, wool fabric was treated with K_2TiF_6 or K_2ZrF_6 under acid condition and exhibits excellent flame resistance. Recently, metal oxide gel used as flame retardant on wool fabric was reported by Sun [16]. Flame resistance of wool fabric can also be obtained when wool is grafted with viny phosphate [20], Carbon Microsophere (CMSs) [13] or blended with para-aramid [6,7].

Phosphazenes contain P=N bonds derived from amination of phosphorous halides, that are rich in both phosphorus and nitrogen [18], show outstanding thermal stability. Based on these, phosphazenes has attracted a lot of interest as a new flame retardant. A series of phosphazenes derivatives were synthesized and applied in resin and textile. Two kinds of phosphazenes with –OH group were synthesized by Modestia M, and used to modify PU foams. The modified PU foams displayed improved thermal stability and char yield [14]. An aromatic–phosphate-containing cyclotriphosphazene flame retardant, (4-diphenylphosphoryloxyphenoxy)-(4-hydroxyphenoxy) cyclotriphosphazene (PPPZ) was prepared by C.Y. Yuan, and then PPPZ-PU based on PPPZ with high limiting oxygen index was obtained [19].

To our knowledge, Hexakis (4-Aminophenoxy) Cyclotriphosphazene (HACP) was firstly reported by Kober and co-workers [15]. With six –OH group in the structure, HACP is highly reactive. HACP can act as a core to produce star-shaped polymer [10, 11]. Also new flame retardant can be synthesized based on HACP [12]. As far as we know, however, HACP, acting directly as flame retardant when applied on wool fabric has not yet been reported.

In this paper, wool fabric with excellent flame resistance was prepared by using HACP as flame retardant. The Limiting Index (LOI) of prepared flame retardant wool fabric was highly improved. Char length of FR wool fabric greatly reduced. Thermal Gravimetric (TG) analysis indicates that FR wool fabric can form more non-flammable char residue than that of pristine wool fabric. SEM shows char surface of FR wool fabric is full of bubbles and pores, which means more non-flammable gas was generated. At the same time, breaking strength of FR wool fabric has increased.

2 EXPERIMENTAL

2.1 *Materials*

The wool fabric used in this experiment is plain woven, 222 g/m^2. Cyclotriphosphazene was purchased from zibo lanyin Chenical Co., Ltd., China. 4-Nitrophenol, Potassium carbonate, Tin powder, acetone, ethyl alcohol absolute, concentrated hydrochloric acid was all supplied by Sinopharm Chemical Reagent Co., Ltd., China.

2.2 *Synthesis of hexakis (4-aminophenoxy) cyclotriphosphazene*

The hexakis (4-aminophenoxy) cyclotriphosphazene was synthesized as previously reported [21].

2.3 *The preparation of flame retardant wool fabric*

In order to remove dust and other impurities, wool fabric was firstly boiled in distilled water for one hour, during which soft agitating was conducted. Then the wool fabric was rinsed with distilled water, and dried at 80°C.

A certain amount of Hexakis (4-Aminophenoxy) Cyclotriphosphazene (HACP) was dissolved in water to make flame retardant aqueous solution before it was used.

Flame retardant wool fabric was prepared with dip-pad process, then dried at 80°C for 3 min and cured at 160°C for 3 min.

2.4 *Characterization and measurement*

The morphology of char residue after combustion was observed with a TM-1000 (Hitachi, Japan) Scanning Electron Microscope (SEM). Before SEM test, the surface of char was sputter-coated with gold layer. Thermal Gravimetric Analysis (TGA) was conducted from 30°C to 600°C on a TG 209 F1 thermal instrument (NETZSCH, Germany) with a heating rate of 10°C /min under the nitrogen atmosphere and each sample was controlled to around 5 mg in primary weight. LOI was measured according to GB/T5455-1997, and the sizes of the samples were 15 cm × 5 cm. Vertical flammability test was conducted according GB/T5454-1997. Tensile properties were measured according to GB/T 3923.1-1991.

3 RESULTS AND DISCUSSION

3.1 *Sem*

Figure 2 shows the morphology observed by SEM test of char residue of pristine wool fabric (a) and FR wool fabric (b). A lot of bubbles and pores are observed in the Figure 1. (b), which can explain how HACP generated N_2 and other gases during combustion. A little pores are also observed in pristine wool fabric surface, and the diameter of the pore is lager than FR wool fabric, as seen in Figure 1 (a).

We also see that char residue is very crisp, while the surface of FR wool fabric is relatively rigid. So, it could be speculated that HACP can help wool fabric to form more char residue and non-flammable gas during combustion.

3.2 *TG analysis*

TG analysis is one of the most effective tools to evaluate the thermal stability and thermal decomposition behavior of the materials. Weight loss stages and the maximum decomposition temperature of the materials can be known from the DTG curves. Figure 1 shows the TG and DTG curves of pristine wool fabric and the FR wool fabric.

Two mass loss stages are observed in the curves of pristine wool fabric and the FR wool fabric. The first mass loss stage happened due to the loss of regained water. From Figure 2 (a), we can see that the pristine wool fabric and FR wool fabric have the same initial decomposition temperature, but the weight loss rate of FR wool fabric is lower than pristine wool fabric. In the second stage, a high slope in curves of both pristine wool fabric and FR wool fabric is observed, which is responsible for the high speed and large loss of its weight. Thermal decomposition of wool fabric took place, gas of H_2S, SO_2, CO_2, and so on generated and char was formed in this stage. The second mass loss stage of pristine wool fabric takes place at 246.3°C, while the FR wool fabric is at 232.9°C. Because of the addition

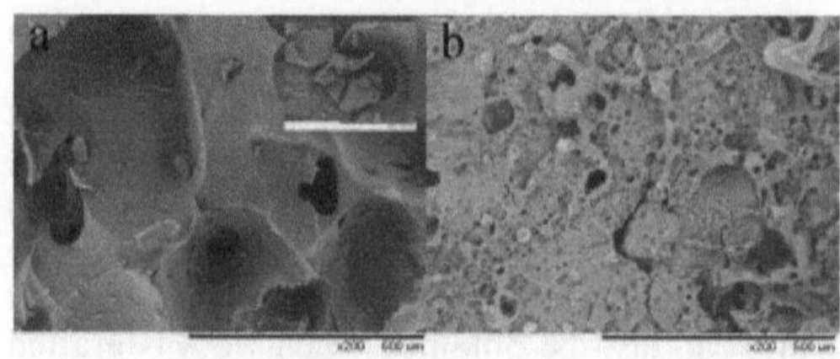

Figure 1. Char morphology of pristine wool fabric (a) and FR wool fabric (b).

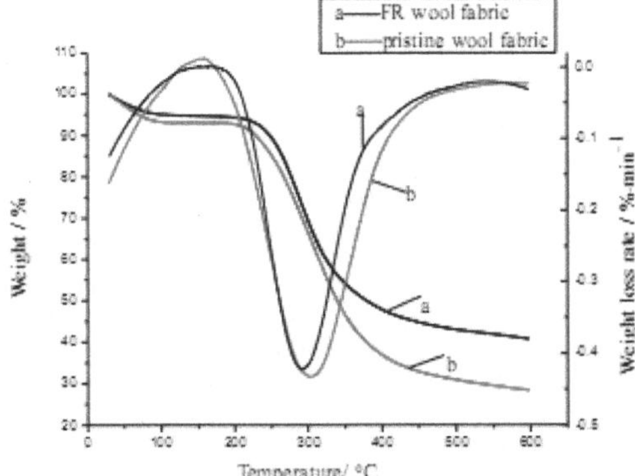

Figure 2. TG (a) and DTG (b) curves of FR wool fabric (weight gain 7.27%) and pristine wool fabric.

of THACP, the initial weight loss temperature of FR is 232.9°C, which is 12°C lower than the pristine wool fabric (246.3°C). At the same time, the maximum decomposition of FR wool fabric happened at 290°C, about 18°C lower than the pristine wool fabric (308.1°C). The early decomposition of FR wool fabric means the early forming of char, and thus more gas can be generated, which is helpful for insulating oxygen from the atmosphere as well as diluting flammable gas to slow down the decomposition rate during combustion. The weight loss rate of both pristine wool fabric and FR wool fabric can be observed in their DTG curves in Figure 1(b). FR wool fabric has a lower weight loss rate during the whole combustion, especially after the maximum decomposition temperature. Char residue at 600°C of pristine wool fabric is 28.3%, while that of FR wool fabric is 40.7%. The high amount of non-flammable char residue formed from FR wool fabric is helpful while improving flame resistance.

3.3 *LOI*

The LOI value of the FR wool fabric and pristine wool fabric is listed in the Table 1. The FR wool fabric with different weight gain of HACP has improved LOI than the pristine wool fabric. As the weight gain increased, the LOI increased accordingly. When the weight gain is above 8.6%, the LOI is higher than 35%, which means the treated wool fabric is intrinsically non-flammable.

3.4 *Vertical flammability test*

Char length and the phonomenon of both FR wool and pristine wool during the vertical flammability test are reported in Table 3. Compared with the pristine wool fabric, FR wool fabric has a shorter char length and also no after flame appeared. The FR wool fabric and pristine wool fabric samples after vertical flammability test are also showed in Figure 3.

Table 1. Thermal analysis data collected from Figure 1.

Sample	$T_{second\ onset}$ (°C)	T_{max} (°C)	Char residue at 600°C (%)
A	232.9	290.0	40.7
B	246.3	308.1	28.3

(A is FR wool fabric with weight gain of 5.09%, B is pristine wool fabric).

Table 2. Results of LOI at different HACP concentrations.

HACP concentration (g/l)	Weight gain (%)	LOI (%)
0	0	26.3
80	5.09	32.8
100	7.28	34.2
120	8.29	35.3
150	11.40	37.1

Table 3. Vertical flammability test report of samples.

Sample	Char length (cm)	After flame time (s)	After glow time (s)
A	>30	40.2	0
B	7.2	0	0

(Sample A is pristine wool fabric, sample B is FR wool fabric with weight gain of 5.09%).

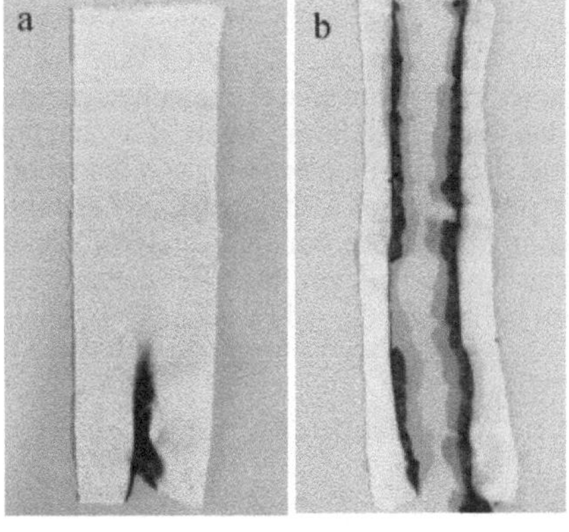

Figure 3. Photographs of FR wool fabric (a) and pristine wool fabric (b) after vertical flammability test.

3.5 *Breaking strength*

Figure 4 shows the breaking strength of pristine wool fabric and FR wool fabric, with different weight gains. The breaking strength of FR wool fabric is higher than that of pristine wool fabric.

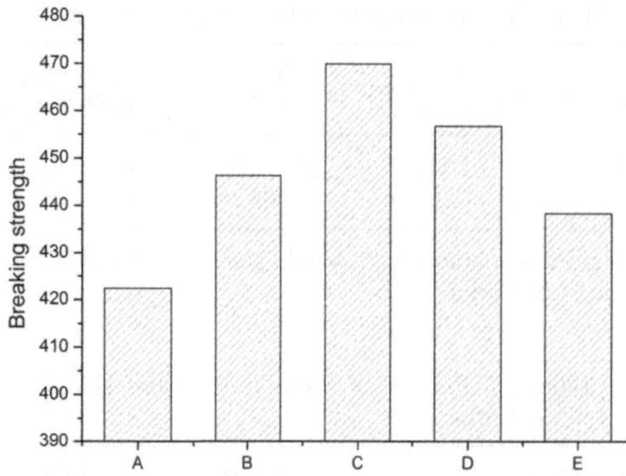

Figure 4. Breaking strength of samples (A is pristine wool fabric, B weight gain 5.09%, C weight gain 7.28%, D weight gain 8.29%, E weight gain 11.40%).

At first, as the weight gain of HACP increased, the breaking strength also increased rapidly. When the weight gain is over 7.28%, breaking tensile begins to decrease. This phenomenon can be explained as the cross-linking reaction that happened between HACP and wool fabric due to the existence of six $-NH_2$ group in the structure of HACP. But with the adding of HACP, the cross-linkage of fiber and flame retardant will decrease the breaking strength.

4 CONCLUSIONS

In this paper, flame retardant wool fabric was successfully prepared by using HACP. LOI of the produced FR wool fabric is above 35% when the weight gain is over 8.29%. HACP can help wool fabric to form more non-flammable char residue, non-flammable gas during combustion and change thermal decomposition behavior of wool fabric. FR wool fabric produced with this method has improved breaking strength. From the results of the flame resistance properties test, HACP is deduced to play its role both in condensed phase and gas phase.

ACKNOWLEDGMENTS

The research was supported by “the Fundamental Research Funds for the Central Universities”.

REFERENCES

[1] Benisek, L. 1974. Communication: improvement of the natural flame—resistance of of wool. Part I: metal-complex applications. *Journal of the Textile Institute*, *65*(2), 102–108.

[2] Benisek, L. 1974. British Patent 1372694.

[3] Benisek, L. et al. 1979. Protective Clothing—Evaluation of Wool and Other Fabrics. *Textile Research Journal*, *49*(4), 212–221.

[4] Benisek, L., & Phillips, W.A. 1983. The effect of the low smoke Zirpro treatment and wool carpet construction on flame-resistance and emission of smoke and toxic gas. *Journal of fire sciences*, *1*(6), 418–431.

[5] Benisek, L. 1984. Zirpro wool textiles. *Fire and Materials*, *8*(4), 183–195.

[6] Flambard, X. et al. 2002. Wool/para-aramid fibres blended in spun yarns as heat and fire resistant fabrics. *Polymer degradation and stability*, *77*(2), 279–284.

[7] Flambard, X. et al. 2005. Progress in safety, flame retardant textiles and flexible fire barriers for seats in transportation. *Polymer degradation and stability*, *88*(1), 98–105.

[8] Forouharshad, M. et al. 2010. Flame retardancy of wool fabric with zirconium oxychloride optimized by central composite design. *Journal of Fire Sciences*.

[9] Horrocks, A.R. 1986. Flame-retardant Finishing of Textiles. *Review of progress in coloration and related topics*, *16*(1), 62–101.

[10] Ito, Y. et al. 1997. PH-sensitive gating by conformational change of a polypeptide brush grafted onto a porous polymer membrane. *Journal of the American Chemical Society*, *119*(7), 1619–1623.

[11] Inoue, K. et al. 2003. Synthesis and Conformation of Star-Shaped Poly (γ-benzyl-L-glutamate) on a Cyclotriphosphazene Core. *Macromolecular Bioscience*, *3*(1), 26–33.

[12] Kumar, D. et al. 1983. Fire-and heat-resistant laminating resins based on maleimido-substituted aromatic cyclotriphosphazenes. *Macromolecules*, *16*(8), 1250–1257.

[13] Mei, N.I.U. et al. 2014. Study on the structure and properties of flame retardant wool fibers grafting CMSs by microwave irradiation. *Wool Textile Journal*, *42*(1).

[14] Modesti, M. et al. 2005. Thermally stable hybrid foams based on cyclophosphazenes and polyurethanes. *Polymer degradation and stability*, *87*(2), 287–292.

[15] Ottmann, G.F. et al. 1967. Aminophenoxy-and isocyanatophenoxyphosphonitriles. *Inorganic Chemistry*, *6*(2), 394–395.

[16] Sun, J.H. et al. 2009. Study on the flame retardant modification and thermal degradation of wool fiber treated with Si-Al oxide sol [J]. *Textile Auxiliaries*, *6*, 009.

[17] Wang, J. et al. 1994. The effect of heat on wool and wool treated with Zirpro by X-ray photoelectron spectroscopy. *Polymer degradation and stability*, *43*(1), 93–99.

[18] Weil, E.D. et al. 1975. Flame Retardancy of Polymeric Materials. *Vol. 3 Marcel Dekker, New York*, 185.

[19] Yuan, C.Y. et al. 2005. Thermally stable and flame-retardant aromatic phosphate and cyclotriphosphazene-containing polyurethanes: synthesis and properties. *Polymers for advanced technologies*, *16*(5), 393–399.

[20] Zhang, F. et al. 2014. Performance of Flame Retardant Wool Fabric Grafted with Vinyl Phosphate. *Journal of Engineered Fabrics & Fibers (JEFF)*, *9*(1).

[21] Zhu, J. 2011. Study of Thermal Properties of Curing of DGEBA Epoxy Resin with Hexakis-(4-Aminophenoxy)-Cyclotriphosphazene. *Advanced Materials Research*, *284*, 365–368.

Material Science and Environmental Engineering – Chen (Ed.)

ISBN 978-1-138-02938-5

The application of eco-friendly formaldehyde-free fixing agent on cotton fabric

X.P. Pei
Chemical Engineering and Biotechnology, College of Chemistry, Donghua University, Shanghai, China

J.X. He
Chemical Engineering and Biotechnology, College of Chemistry, Donghua University, Shanghai, China
Key Laboratory of Eco-Textile, Ministry of Education, Donghua University, Shanghai, China

J.L. Yao
Key Laboratory for Eco-Dyeing and Finishing of Textiles of Jiangsu Province, Jiangsu Lianfa Textile Co. Ltd., P.R. China

X. Dong
Chemical Engineering and Biotechnology, College of Chemistry, Donghua University, Shanghai, China

ABSTRACT: This work was studied to determine the effects of a new formaldehyde-free fixing agent on reactive dyed knitted cotton. The eco-friendly formaldehyde-free fixing agent using diethylenetriamine and epichlorohydrin was synthesized and applied to fixation experiment of reactive dyes. The effect of the fixative on color fastness of dyed fabric was also studied. Fastness properties including soaping and crock were all evaluated at 0.5 to 1 class. The surface structure and aggregation morphology of the fabric were also tested by Scanning Electron Microscope (SEM) and Infrared Spectroscopy (IR), from which it can be confirmed that the reaction between the fixing agent and the fabric was the reason for the improved fastness. Some relevant properties of the fixing agent were also explored in this paper.

Keywords: formaldehyde-free; fixing agent; reactive dyes

1 INTRODUCTION

Cellulose fibers can be dyed with reactive and direct dyes. Reactive dyes are widely used due to their ease of application and strong affinity with cellulose fibers. However, the main disadvantage of reactive dyes is that hydrolysis reactive dyes in dyeing and hydrolyzed dyes can be adsorbed on the fiber, which leads to poor wash fastness. The hydrolyzed dyes may be removed from the cotton fiber in a successive wash with fresh water. Darker shades can be reduced considerably in depth after only a few times of water washing.

There have been some attempts to overcome these disadvantages, for example, by composing on the fiber with metal salts, formation of the dyestuff on the fiber; treatment of the dyestuff or the fiber with formaldehyde; impregnation with resins and after-treatment with cationic auxiliaries. The use of cationic after-treatment auxiliaries has proved particularly effective.

The use of the treatment to improve the fastness properties of dyeing has a long and prolific history. Several kinds of dye fixing agents have been synthesized since the World War II, but most of them have been added to formaldehyde as a raw material or could release formaldehyde in the later process, so they are harmful to human beings and have been banned from using. In current years, more and more formaldehyde-free fixing agents are developed to replace those fixatives containing formaldehyde, and our fixative is one of them. In this paper, we use a non-formaldehyde cationic fixing agent named fixative DR (the name DR would not illustrate any additional details about the chemical structure of this fixing agent) whose raw materials are diethylenetriamine and epichlorohydrin. After-treatment of cellulosic fiber with fixing agent DR is believed that it offers an opportunity for us to increase the reactivity and substantivity of the fiber toward reactive dyes. In this paper, the different influences on fabric with different fixing agents were explored. And we also test several fastness properties such as soaping fastness and crock fastness.

2 EXPERIMENT

2.1 *Dyeing with reactive dyes*

Ten grams of the cotton fabric was dyed under liquor 15:1, and its pH value was 7. The dyes in this experiment were reactive red 3 bs, reactive yellow 3 rs, and reactive blue knr. The dye's concentration was 5% on the weight of goods (owf), with 80 g/L of Na_2SO_4 in each bath.

2.2 *Synthesis of the formaldehyde-free fixative DR*

We weighed a certain amount of diethylenetriamine and mixed it with water in a four-flask equipped with a thermometer, and a stir bar and condenser pipe. We stirred the mixture and increased its temperature to 50°C. Then we began to drop epichlorohydrin slowly. After that, we stirred it continuously for 2 hours. Then we adjusted its pH value to 6–7 after cooling it to room temperature, the fixing agent DR was then made ready for the fixation treatment.

2.3 *Fixation with fixative DR*

Three levels of concentration 0, 2.0, and 200 percent (owf), of the fixative were tested for dyed samples with a liquor ratio of 20:1 and a pH value of 7. The temperature was increased to 45°C and held for 30 minutes.

3 METHODS

3.1 *Color measurement and fastness tests*

The concerns for this work were color fastness, color change, soaping fastness, and crock fastness both dry and wet under various conditions. We used the AATCC standard Gray Scales for evaluating, staining, and change in color.

3.2 *The surfacial structure analysis of the fabric with fixative DR*

In order to explore the specific structure of fabric with fixative DR, the Scanning Electron Microscope (SEM) and Infrared Spectroscopy (IR) were used to test the surface of the fabrics.

3.3 *The solid content and viscosity of the fixative*

We can calculate the Solid Content from the equation 1 below:

$$M\% = \frac{g_1 - g_2}{g_1 - g_0} \times 100\% \tag{1}$$

where M = Solid Content; g_0 = quality of watchglass; g_1 = quality of drops fixative with watchglass before drying and g_2 = quality of drops fixative with watchglass after drying for three hours.

The viscosity of the fixative DR can be measured by SNB-2 digital rotational viscometer and its unit is mpa · s.

3.4 *The amount determination of fixative DR on fabric*

The substance of fixative DR on fabric can be divided into three parts: water soluble substance of fixative, external amount of fixative, and the fixable amount of fixative.

3.5 *The water soluble substance*

We can get the water soluble substance content of the fixative from the equation 2 below:

$$W\% = \frac{d_0 - d_h}{d_0} \times 100\% \tag{2}$$

where W = water soluble substance content; d_0 = the initial fabric weight; and d_h = the fabric weight after soaking in water.

3.6 *The external fixative*

Put the above fabric weighting d_h into the aqueous solution containing Na_2CO_3 2.5 g/L and soap flakes of 2.5 g/L at 90°C (the liquor ratio = 10:1). Next rinse the fabric and dry it till you get a constant weight, d_s.

$$S\% = \frac{d_h - d_s}{d_0} \times 100\% \tag{3}$$

where S = external fixative content; and d_s = fabric weight after soaking in alkaline liquor.

3.7 *The fixable amount of fixative*

Put the above fabric weighing d_s into aqueous solution (HCl = 2.5 g/L, 90°C, 10:1, and 30 min), then rinse and dry it to constant weight d_m whose formula is as follow:

$$F\% = \frac{d_s - d_m}{d_0} \times 100\% \tag{4}$$

where F = fixable amount of fixative; and d_m = fabric weight after soaking in acid liquor.

4 RESULTS AND DISCUSSION

4.1 *Color measurement and fastness*

To evaluate the effect of fixatives on soaping fastness and crock fastness, the dyed and finished fabric samples were assessed by gray scale values.

As shown in the Table 1, the fabric fastness was treated by fixing agent DR and increased to 0.5 class than a no-fixing fabric after it was dyed with a reactive red. Compared to fixing agent Fix, which is an excellent fixing agent abroad, DR's wet crock fastness was 0.5 class lower than Fix.

In the Table 2 and 3, the fixing agent DR has improved the same class of fastness as Table 1, including soaping fastness, both staining and fading, and dry crock fastness. Table 2 indicated that fixing agent DR has received the same effect with Fix. And in the Table 3, fabric with fixative DR also gained the same effect except that the wet crock fastness was 0.5 class lower than Fix. Among the three reactive dyes, fixing agent DR had a more obvious effect on crock fastness of fabric dyed with reactive yellow.

As the fixing agents can increase the reactivity and substantivity of the fiber towards reactive dyes and fiber to some extent, we have got better color fastness fabric than no-fixing fabric.

Table 1. The fabric fastness with reactive red.

Reactive red	Soaping fastness		Crock fastness	
	Dry	Wet	Staining	Fading
No fixing	3–4	4	4	2–3
Fixative DR	4	4–5	4–5	3
Fix*	4	4–5	4–5	3–4

*Excellent fixing agent abroad.

Table 2. The fabric fastness with reactive yellow.

Reactive yellow	Soaping fastness		Crock fastness	
	Dry	Wet	Staining	Fading
No fixing	4	4–5	3	2–3
Fixative DR	4–5	4–5	4–5	4
Fix*	4–5	4–5	4–5	4

*Excellent fixing agent abroad.

Table 3. The fabric fastness with reactive blue.

Reactive blue	Soaping fastness		Crock fastness	
	Dry	Wet	Staining	Fading
No fixing	2–3	4	3	2–3
Fixative DR	3	4–5	3–4	3
Fix*	3	4–5	3–4	3–4

*Excellent fixing agent abroad.

4.2 *SEM*

To test the surface structure of fabric with fixing agent DR, we used Scanning Electron Microscope (SEM) to do this research.

From the Figure 1, we see that fabric surface is smooth without fixing processing, and it obviously presents the flat ribbon microstructure of cotton fiber. And from the second picture, on the surface of fiber treated with fixing agent DR, there is a discontinuous layer of membrane material, thus, we can know that it's the existence form of fixative on fabric surface, and there are also some cross-linking reactions between the fixative DR and fabric. It is the cross-linking effect that improves the reactivity and substantivity between fabric and fixation, and so does the fastness after fixation treatment.

Figure 1. The surface SEM (×3000) of fabric without and with fixing agent. (The first picture is SEM with fixative and the second one is with fixative).

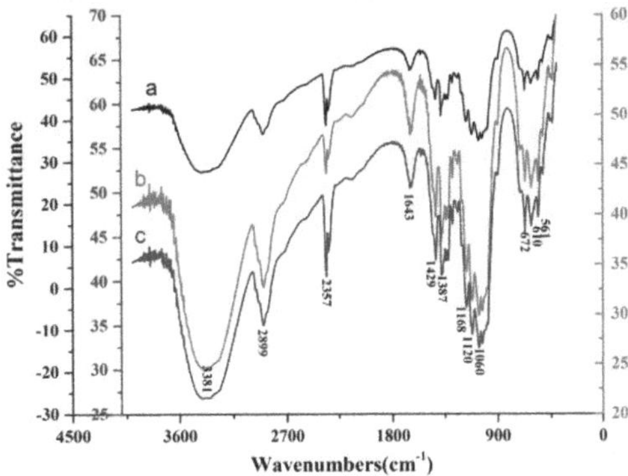

Figure 2. The IR of fabric (curve (a) is untreated fabric, curve (b) is treated fabric under drying with 2% owf fixative, and curve (c) is treated fabric under baking with 2% owf fixative).

4.3 *IR of fabric*

In order to know if there exists any reactivity between the fiber and the reactive dyes, we used Infrared Spectroscopy (IR) to test the fabrics treated by fixing agent. As shown in the Figure 2, the curve (a) shows the 3410 cm^{-1} hydroxyl absorption peak; the 2900 cm^{-1} stretching vibration absorption peak of methyl; the 1030 cm^{-1}, 1060 cm^{-1}, 1120 cm^{-1}, and 1168 cm^{-1} are the characteristic absorption peaks of the ether bond (C-O-C), which indicates that the main ingredient of fabric is celluose. After it was treated with a fixative, on the curve (b) and (c), 1345 cm^{-1} and 2150 cm^{-1} characteristic absorption peaks of the fixative emerge and contrast curve (b) and (c). The 850 cm^{-1} absorption peak (C-O) moves towards high vibration frequency, indicating that the fixative is cross-linked with fiber containing -OH. This is also the explanation for the improved color fastnesses.

4.4 *The basic parameters of fixative DR*

The Table 4 lists these basic parameters of fixative DR (the dosage of fixing agent was 2% owf). The solid content was about 45%, and the viscosity was 809 mpa·s. In the Table 2, the water soluble substance content (W%), surfacial fixative content (S%), and fixable fixative content (F%) were 0.41%, 0.09 and 0.46, respectively. Under these parameters, the fixing agent DR has excellent effect on fabrics, and from these values, we can also know the adhesion between fixative DR and fiber was the key to improve the relevant fastnesses.

Table 4. Solid content and viscosity.

Parameter	Solid content (%)	Viscosity (mpa·s)
Fixative 1*	44.35	809
Fixative 2*	44.98	813
Fixative 3*	45.64	804
Average	44.99	809

*Fixative 1, 2 and 3 were the fixing agent synthesized at the same condition.

Table 5. Parameters of fixative DR.

Parameter	W (%)	S (%)	F (%)
Fixative 1*	0.41	0.09	0.46
Fixative 2*	0.38	0.07	0.50
Fixative 3*	0.45	0.11	0.43
Average	0.41	0.09	0.46

*Fixative 1, 2 and 3 were the fixing agent synthesized at the same condition.

5 CONCLUSION

In this paper, the eco-friendly formaldehyde-free fixative DR was synthesized by diethylenetriamine and epichlorohydrin, and it was used to treat cotton fabric dyed by reactive dyes. After treatment, the soaping fastness both staining and fading, can be improved by 0.5 class, and for the crock fastness, its dry crock fastness was improved by 0.5 to 1.5 class, while the wet crock fastness was 0.5 class improved. Compared with fixing agent Fix, the fastnesses of fixing agent DR gained the same level with it except wet crock fastness, which was 0.5 class lower. In conclusion, it can meet the basic requirements of the people.

From the SEM and IR, we saw the aggregation morphology of fixative DR on the fabric surface. Among the parameters, the solid content was about 45%, and the viscosity was 809 mpa·s. The water soluble substance content, surfacial fixative content, and fixable fixative content were 0.41%, 0.09, and 0.46, respectively. Under these conditions, fixative DR can achieve the best effect on fabric.

ACKNOWLEDGMENT

The research was supported by "the Fundamental Research Funds for the Central Universities (2232014D3-12)".

REFERENCES

[1] Burness, D.M. & Bayer, H.O. 1963. Synthesis and Reactions of Quaternary Salts of Glyeidyl Amines. *The Journal of Organic Chemistry* 28 (9), 2283–2288.

[2] Cai, Y. Pailthorpe, M.T. & David, S.K. 1999. A new method for improving the dyeability of cotton with reactive dyes. *Textile research journal* 69 (6): 440–446.

[3] Chan, Z. & Jiaxi, X. 2011. Regioselective nucleophilic ring opening reactions of unsymmetric oxiranes. *Progress in Chemistry* 23 (1), 165–180.

[4] Deng, Y.M. Pi, P.H. Wen, X.F. & Cheng, 2008. Synthesis and application of formaldehyde-free cation fixing agent. *Applied Chemical Industry,* 2, 017.

[5] Haiyu, X. 1996. The Properties of Fixing Agent and Its Colour Darkening Effect. *Dyeing and Finishing*, 2, 004.

[6] Heywood, D.L. & Phillips, B. 1958. The Reaction of Epichlorohydrin with Secondary Amines. *Journal of the American Chemical Society* 80 (5), 1257–1259.

[7] Maofu, H.U. & Yuqin, Y. 2002. The Present Situation and Development Trend Of Non-formaldehyde Dye Fixative Agents. *Textile Auxiliaries*, 4, 000.
[8] Mckelvey, J.B. Webre, B.G. & Benerito, R.R. 1960. Reaction of epichlorohydrin with ammonia, aniline, and diethanolamine. *The Journal of Organic Chemistry* 25 (8), 1424–1428.
[9] Qu, J.Q. Xiang, B. & Chen, H.Q. 2012. Properties of aqueous polyurethane dispersion modified by multi-epoxy crosslinkers. *Journal of Sichuan University: Engineering Science Edition* 44 (1), 154–158.
[10] Ross, J.H. Baker, D. & Coscia, A.T. 1964. Some Reactions of epichlorohydrin with Amines. *The Journal of Organic Chemistry* 29 (4), 824–826.
[11] Sharif, S. & Ahmad, S. 2007. Role of quaternary ammonium salts in improving the fastness properties of anionic dyes on Cellulose fibres. *Coloration Technology* 123 (1): 8–17.
[12] Wei, X.U. 2001. Synthesis and Application of Non-formaldehyde Dye Fixative Agent AE. *Textile Auxiliaries*, 6, 007.
[13] Wen, S.P. & Zhang, R.S. 2011. Preparation and application of formaldehyde-free fixing agent HP. *Textile Auxiliaries*, 8, 005.
[14] Yu, Y. & Zhang, Y. 2009. Review of study on resin dye-fixatives on cotton fabrics. *Modern Applied Science* 3 (10), 9.

Material Science and Environmental Engineering – Chen (Ed.)

Different morphologies SBA-15 for extraction of uranium ion from aqueous solution

C.B. Xiong, Y. Wang, L. She, C.Y. Liu & Y. Ding

Key Subject Laboratory of National Defense for Radioactive Waste and Environmental Security, Southwest University of Science and Technology, Mianyang, P.R. China

ABSTRACT: Rod-like and plate-like SBA-15 has been synthesized by a facile method. The adsorption performances of both morphologies SBA-15 on uranium have been investigated. The effects of PH, initial uranium concentration, and contact time on adsorption behavior were also investigated. It was found that the shorter channel SBA-15 has higher adsorption capacity of uranium up to 260.8 mg/g. And the adsorption equilibrium can reach within 10 min. This implied that the short channel SBA-15 is a fascinating material for separation of uranium rapidly and efficiently from radioactive water, and the plate-like SBA-15 has a more promising performance in adsorbing uranium from solution.

Keywords: SBA-15; morphology; uranium adsorption; adsorption behavior

1 INTRODUCTION

Separation of uranium efficiently from radioactive water is always an important content of unclear energy system, as it plays a key role in the collecting of uranium and protecting of the environment [1]. The rapid development of nuclear industry brings it to more focus. In order to realize this goal, chemical precipitation, solvent extraction, electrolysis, and ion-exchange have been used to extract uranium. However, low efficiency, poor adsorption capability, and extreme operating conditions often limited their use in nuclear industry. Meanwhile, the large amount of byproducts is also difficult to handle [2–3]. Adsorption, as a high-efficiency and convenience way, has gotten widespread concern. Plenty of sorbents such as natural minerals [4–5], metal oxides [6], carbon materials [7–8], agricultural and forest wastes [9] have been employed to remove uranium from radioactive water. Among these adsorbents, SBA-15 has a more attractive ability of sorption and enrichment of uranium ions from the solution relying on its huge specific surface area, large and uniform pore size, well hydrothermal stability, environmentally benign behavior, and easy to be grafted by organic ligands [10]. Since Zhao DY synthesized the fiber-like SBA-15 in 1998 [11], different kinds of morphologies, such as SBA-15, had been reported by other researchers gradually including rod-like [12], plate-like [13], cube-like, and spherical [14] SBA-15. Researches on the adsorption properties of different morphologies of SBA-15 are significant, as the morphology itself is a key matter that impacts the adsorption ability and the morphology. It may help us to choose the framework material to graft other groups. Nevertheless, there are still lesser papers that have reported the extraction performance differences between diverse morphologies SBA-15 in radioactive water.

Therefore, in this work, plate-like SBA-15 and rod-like SBA-15 were chosen as the research object because of its more solo and regular structure compared to other typical mesoporous silica. The SBA-15 was performed at different conditions of PH, initial uranium concentration, contact time, and desorption rate.

2 EXPERIMENTAL

2.1 *Materials*

$EO_{20}PO_{70}EO_{20}$ was purchased from Sigma, Germany. Tetraethoxysilane (TEOS) come from Sinopharm Chemical Reagent limited corporation. Uranium stock solution was prepared by dissolving 1.1792 g U_3O_8 into moderate concentrated nitric acid, and then transferred to a 1000 mL volumetric flask, diluted with deionized water to particular volume. Arsenazo III (0.1%) was used to determine the concentration of uranium at wavelength 652 nm [15]. All of other chemical reagents used were of Analytical grade.

2.2 *Synthesis of SBA-15*

Rod-like SBA-15 was prepared by following the procedures made by Yi Ding [16]: 2.4 g P123 was added to 120 ml HCL (2 mol/l) solution and stirred at 40°C until the P123 was dissolve completely, and 7.5 ml TEOS was added under vigorous stirring. After 5 min stirring, the solution was kept in 40°C for 20 h. Then, the mixture was transferred into an autoclave and aged for 24 h at 100°C. The products was separated by filtration, washed with deionized water, dried, and then calcined at 550°C for 5 h to remove the surfactant. The procedures of the synthesis of plate-like SBA-15 [17] were carried out by changing the concentration of HCL to 1 mol/l and self-assembled at a temperature of 30°C.

2.3 *Adsorption experiments*

Working solution was prepared by diluted uranium stock solution with deionized water and PH was adjusted by NaOH and HNO3 solution. All batch adsorption experiments were carried out at room temperature and the effect of contact time and concentration of uranium ions was experimented at a solid-solution ratio of 0.2 mg/ml. Adsorption capacity Q (mg/g) was calculated by using the following equations in Ref [18]:

$$Q = \frac{(C_o - C_e)}{M} V \tag{1}$$

where C_o (mg/l) is the initial uranium concentration, C_e (mg/l) is the residual uranium concentration at equilibrium, V (l) is the volume of uranium solution, M is the mass of SAB-15, and Q (mg/g) is the equilibrium uranium uptake capacity.

3 RESULTS AND DISCUSSION

3.1 *Characterization*

Figure 1 shows the SEM images of rod- and plate-like SBA-15, which reveals the synthesized SBA-15 possesses a good morphology. From Figure 1A and B, sample B show a plate-like morphology, and sample A is rod-like. The plate-like SBA-15 has an average width and thickness of about 800–1200 and 200–300 nm and the length of rod-like SBA-15 is 1200 nm. With the length of SBA-15, it was implied that the channel length, which impacted the diffusion of organic groups like the plate-like SBA-15, may graft more organic groups.

Figure 1. SEM images of SBA-15: rod-like (A), plate-like (B).

3.2 *Effect of contact time*

Equilibrium time of adsorption has been detected as; the shorter the balance time, the more amount of uranium can be collected at that period. Figure 2 showed both rod-like and plate-like SBA-15 could adsorb more than 80% of uranium within 10 min. Then, the adsorption capacity changed slightly, and 10 min is thought of as equilibrium time. Furthermore, the plate-like SBA-15 has a more eminent adsorption capacity of 269.2 mg/g than rod-like SBA-15 (210.2 mg/g).

3.3 *Effect of PH*

PH is always a curial factor for adsorption experiment. Not only the PH changed the species of uranium, but also affected adsorbents' surface charge and binding sites. Figure 3 reflected the result of PH effect on uranium adsorption capacity. Over the pH range 2–9, the adsorbing capacity was increased at first and then decreased. And the maximum peak discovered at PH = 7 for SBA-15 of plate-like, and PH = 6 for rod-like. Different optimal adsorption PH for two types of SBA-15 maybe caused by the different charges of various morphologies SBA-15 in solution and the diverse species of uranium [19]. As the disappointing performance of adsorbents at low PH could be attributed to the protonation of the adsorbents, the main existence form of uranium is UO_2^{2+}, and the two factors that enhanced the electrostatic repulsion. With the PH increase, the adsorbents'

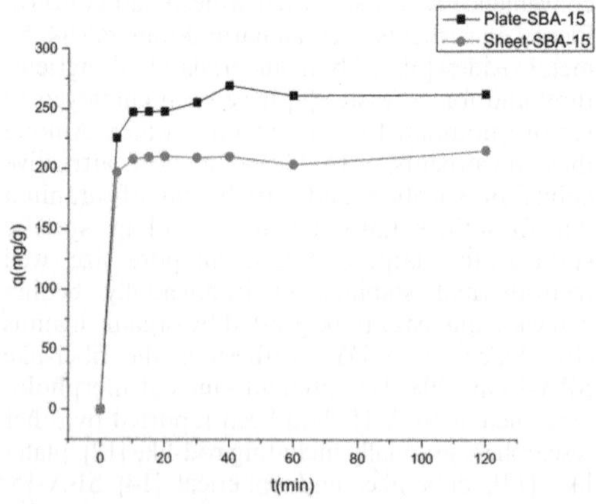

Figure 2. The effect of contact time on different morphology of SBA-15 (C_o = 100 mg/g, PH = 7).

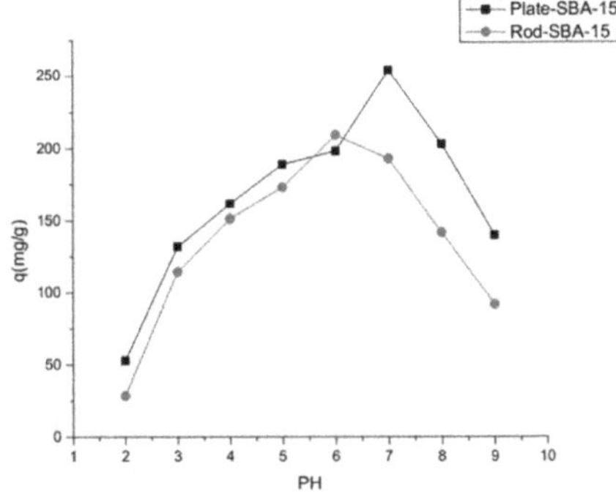

Figure 3. The adsorption capability of different morphology of SBA at varies of PH.

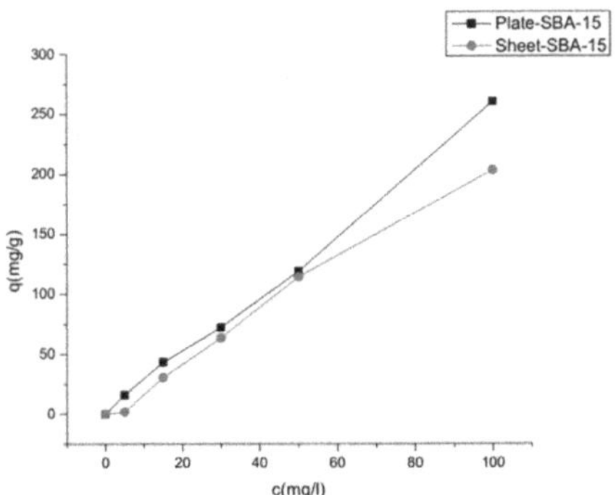

Figure 4. The influence of initial concentration on the adsorption of uranium onto the SBA-15 (PH = 7, t = 120 min).

surface charge decreases, which can be verified by Zeta potential [20, 17]. The species of uranium are diverse, and over PH = 7, and the appearance of $(UO_2)_3(OH)_7^-$ and $UO_2(OH)_3^-$ are also limited. The adsorption amount of uranium is the same with low PH [21].

3.4 *Effect of uranium concentration*

As seen in Figure 4, with the initial concentration of uranium increased, the adsorption capability also increased. And the slope of sheet-like SBA-15 increased more rapidly than rod-like SBA-15 at high concentration, and sheet-like SBA-15 is faster. This result can be ascribed to different quantities of binding sites of adsorbents.

4 CONCLUSION

The extraction performance of rod-like and plate-like SBA-15 towards the uranium have been shown. The results indicate that PH and the concentration of uranium ions highly impact the adsorptive property of SBA-15. Plate-like SBA-15 has an adsorption capacity of 260.8 mg/g (PH = 7), while the capacity of rod-like SBA-15 is 210.2 mg/g (PH = 6). Meanwhile, both the adsorptions can reach equilibrium in 10 min, and the adsorption capability be decreased with the concentration of uranium.

ACKNOWLEDGMENT

This work was supported by the Doctor research Foundation of Southwest University of Science and Technology (No. 13zx7136).

REFERENCES

[1] Ribera D., Labrot F., Tisnerat G., et al. Uranium in the environment: occurrence, transfer, and biological effects. [J]. Rev Environ Contam Toxicol, 1996, 146:53–89.

[2] Chen S., Guo B., Wang Y., et al. Study on sorption of U (VI) onto ordered mesoporous silicas [J]. Journal of Radioanalytical and Nuclear Chemistry, 2013, 295(2):1435–1442.

[3] B.L., A.G., M.L. Introduction for 20 years of research on ordered mesoporous materials. [J]. Chemical Society Reviews, 2013, 9:3661–3662.

[4] X.S., Z.C., Z.X., et al. Removal of uranium (VI) from aqueous solution by adsorption of hematite. [J]. Journal of Environmental Radioactivity, 2009, 100:162–166.

[5] Donat R. The removal of uranium (VI) from aqueous solutions onto natural sepiolite [J]. The Journal of Chemical Thermodynamics, 2009, 41(7):829–835.

[6] R.H., W.Z., Y.W., et al. Removal of uranium (VI) from aqueous solutions by manganese oxide coated zeolite: discussion of adsorption isotherms and pH effect. [J]. Journal of Environmental Radioactivity, 2007, 93(3):127–143.

[7] Xingliang L., Song. Adsorption of Uranium by Carbon Materials from Aqueous Solutions [J]. Progress in Chemistry, 2011, 23(7):1446–1453.

[8] Bin-wen Nie, Zhi-bin Zhang, Xiao-hong Cao, et al. Sorption study of uranium from aqueous solution on ordered mesoporous carbon CMK-3 [J]. Journal of Radioanalytical and Nuclear Chemistry, 2013, 295(1):663–670.

[9] Abram Farid Bishay, Environmental application of rice straw in energy production and potential adsorption of uranium and heavy metals. Journal of radioanalytical and nuclear chemistry, 286(1), 81–89.

[10] Wang X., Yuan L., Wang Y., et al. Mesoporous silica SBA-15 functionalized with phosphonate and amino groups for uranium uptake [J]. Science china-chemistry, 2012, 55(9):1705–1711.

[11] Zhao D., Huo Q., Feng J., et al. Nonionic Triblock and Star Diblock Copolymer and Oligomeric Surfactant Syntheses of Highly Ordered, Hydrothermally Stable, Mesoporous Silica Structures [J]. J. Am. Chem. Soc., 1998, 120(24):6024–6036.

[12] Ding Y., Yin G., Liao X., et al. Key role of sodium silicate modulus in synthesis of mesoporous silica SBA-15 rods with controllable lengths and diameters [J]. Materials Letters, 2012, 75(1):45–47.

[13] Chen S., Chen Y., Lee J., et al. Tuning pore diameter of platelet SBA-15 materials with short mesochannels for enzyme adsorption [J]. Journal of Materials Chemistry, 2011, 15(15):5693–5703.

[14] A.K., S.Y., P.G.S., et al. Synthesis of ordered large pore SBA-15 spherical particles for adsorption of biomolecules [J]. Journal of Chromatography A, 2006, 1122:13–20.

[15] Wang, Xinghui, Zhu, et al. Removal of uranium (VI) ion from aqueous solution by SBA-15[J]. Annals of Nuclear Energy, 2013, 56:151–157.

[16] Ding Y., Yin G., Liao X., et al. A convenient route to synthesize SBA-15 rods with tunable pore length for lysozyme adsorption [J]. Microporous and Mesoporous Materials, 2013, 170(4): 45–51.

[17] Ding Y., Yin G., Liao X., et al. A convenient route to synthesize SBA-15 rods with tunable pore length for lysozyme adsorption [J]. Microporous and Mesoporous Materials, 2013, 170(4):45–51.

[18] G.T., J.G., Y.J., et al. Sorption Of Uranium (VI) Using Oxime-Grafted Ordered Mesoporous Carbon Cmk-5 [J]. Journal of hazardous materials, 2011, 190:442–450.

[19] Liu Y., Li Q., Cao X., et al. Removal of uranium (VI) from aqueous solutions by CMK-3 and its polymer composite [J]. Applied Surface Science, 2013, 285(10):258–266.

[20] Gabriel U., Gaudet J.-, Spadini L., et al. Reactive transport of uranyl in a goethite column: an experimental and modelling study [J]. Chemical Geology, 1998, 151(1):107–128.

[21] Liu Y., Li Q., Cao X., et al. Removal of uranium (VI) from aqueous solutions by CMK-3 and its polymer composite [J]. Applied Surface Science, 2013, 285(10):258–266.

Material Science and Environmental Engineering – Chen (Ed.)
© 2016 Taylor & Francis Group, London, ISBN 978-1-138-02938-5

Synthesis and characterization of meso-microporous materials Beta-SBA-15

X.L. Wang, B. Ren, P. Du, M.H. Zhang, S.T. Song, A.J. Duan, C.M. Xu & Z. Zhao
State Key Laboratory of Heavy Oil Processing, China University of Petroleum, Beijing, P.R. China

ABSTRACT: The meso-microporous composite materials having different textural properties were synthesized by modifying the procedures mentioned in the literature. Nonionic block copolymer PEO-PPO-PEO (P123) was used as the mesostructure directing agent, and 1,3,5-trimethylbenzene (TMB) as additives. Beta-SBA-15 (TMB/P123 = 0.1, 0.2, or 0.3) meso-microporous composite materials were synthesized from zeolite Beta nanoclusters by self-assembly method using P123 as the mesostructure directing agent under the basic condition in this work. The materials were characterized using N2 adsorption-desorption isotherms (BET), X-ray diffraction, FTIR, TEM, and SEM analyses. XRD, BET, and TEM analyses had confirmed the synthesis of meso-microporous composite materials with different textural characteristics.

Keywords: meso-microporous composite materials; synthesis; characterization

1 INTRODUCTION

Compared with the traditional Al_2O_3 materials, meso- and micro- or even large porous materials have been widely applied in catalytic and absorption areas in the industry for their certain advantages in the structure composition and morphology[1]. However, there exist some unignorable shortcomings for the application of these porous materials in the petrochemical industry. For example, in the hydrodesulfrization (HDS) reaction of diesel, microporous-zeolite showed ignorable catalytic reaction performance of macromolecular containing sulfur compounds due to its extremely small aperture. Compared to the microporous-zeolite, mesoporous materials with larger pore size also have the disadvantages of poor hydrothermal stability which limits its further application [2, 3]. Therefore, it is significant to develop novel composite materials with hierarchical pore structures combining the advantages of pore structures at different scales. Fortunately, it is feasible to introduce mesoporous structures into the traditional microporous zeolite to obtain the meso-microporous composites with mesoporous and microporous structures. Combination of the advantages of these two porous structures will make the diffusion rate of the reactants and products in a catalytic reaction greatly improved, leading to an improvement of the catalyst reactivity.

Sun et al. [4] took soft organic silicon template as a guided agent to synthesize mesoporous ZSM-5 composite molecular sieve, and loaded Pt-Pd to obtain the catalyst for HDS reaction of 4,6-Dimethyldibenzothiophene (4,6-DMDBT). The experiment results indicated that the HDS catalytic activity for 4,6-DMDBT of the catalyst based on the mesoporous ZSM-5 composite material is 24 times more than that of common ZSM-5, and 2 times than more than that of Al_2O_3. Besides, the catalyst based on the mesoporous ZSM-5 composite materials changed the intermediate reaction product and the reaction route.

Meso-microporous composite materials combine the excellent performance of both porous materials and make up the shortage of single channel for the molecular sieve. Recently, meso-microporous composite materials have been applied to diesel hydrodesulfurization reaction since the hierarchical channel structure and strong acid activity center of the composite will be beneficial to eliminate the steric hindrance of DMDBT in diesel oil and make the reactant molecules easily closer to the catalyst active centers, then resulting in deep HDS to produce clean diesel with ultra low and even zero sulfur for content.

In this research, we synthesized composite materials Beta-SBA-15 with different mesoporous aperture size. Then, the materials were characterized using N_2 adsorption-desorption isotherms (BET), X-ray diffraction, and FTIR, TEM and SEM analyses.

2 EXPERIMENT

51.90 g of TEAOH (25m%) and 42.86 g of TEOS were dissolved in a beaker and stirred continuously for 30 min to make solution (1). 0.3734 g of NaOH and 1.5112 g of $NaAlO_2$,7 g of TEAOH (25m%) were dissolved in 4 g of deionized water and stirred continuously to obtain solution (2) in another beaker. The solution (2) was slowly dropped into solution (1) and stirred for 2 to 4 hours until the formed solution state was no longer separated, then transferred to an autoclave at 413 K for 24 h, cooled and stirred for 30 minutes to obtain Beta microcrystalline emulsion.

Meso-microporous composite materials Beta-SBA-15 was synthesized using nonionic block copolymer PEO-PPO-PEO (P123) as the template and 1, 3, 5-trimethylbenzene (TMB) as additives. 3.0 g of P123 and 40.5 g of 2 M HCl were dissolved in 85.5 g of deionized water, and then stirred continuously at 308 K for 4 h. To obtain different ratios of TMB/P123, 1, 3, 5-trimethylbenzene (TMB), a certain amount of TEOS, and a certain mass of Beta microcrystalline emulsion, were slowly dropped into the solution and stirred continuously for 24 h, then transferred to an autoclave at 373 K for 24 h. The obtained solid was filtrated with deionized water, dried at 353 K for 12 h, and finally calcined at 773 K for 6 h.

The composite materials were ammonium exchanged for twice: the composite materials and 1 mol/L ammonium chloride solution were mixed in accordance with the proportion of 1 g material into 10 ml ammonium chloride solution, stirred for 1 h at 85 °C in water bath and dried at a temperature of 80 °C for 12 h after filtration. Then the dried sample, ammonium, was exchanged again in the same manner, and finally calcinated to obtain H-type Beta-SBA-15(BS) materials.

3 RESULTS AND DISCUSSION

Samples of the Beta-SBA-15 composite materials synthesized under different TMB/P123 ratios were indicated as BSx, x = 1, 2, 3 represented as Beta-SBA-15 composites with different mass ratios of TMB/P123 = (0.1, 0.2, 0.3).

3.1 *XRD analysis*

Figure 1 shows the small-angle XRD spectra for the synthesized meso-microporous composite BS materials. It is obvious that the three characteristic diffraction peaks of SBA-15 belong to 100, 110, and 200 interface of the two-dimensional hexagonal system. With the increase in the ratios of TMB/P123 in the original materials, the corresponding diffraction peaks of BS material shift to small angle, which indicates that the mesoporous aperture of the synthetic composite BS increases with the increase of pore enlarging agent. Besides,

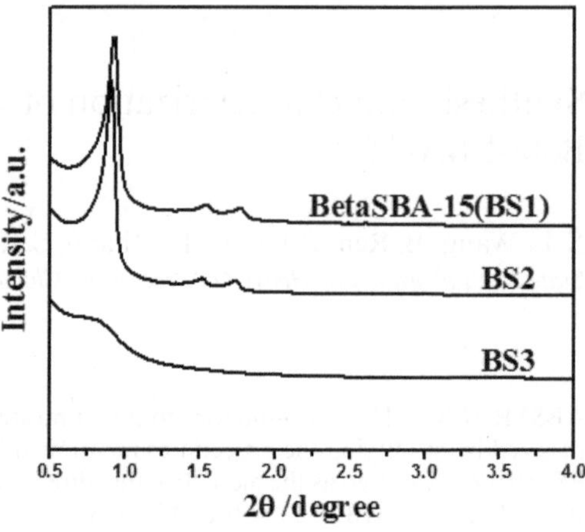

Figure 1. Small-angle XRD pattern of BS.

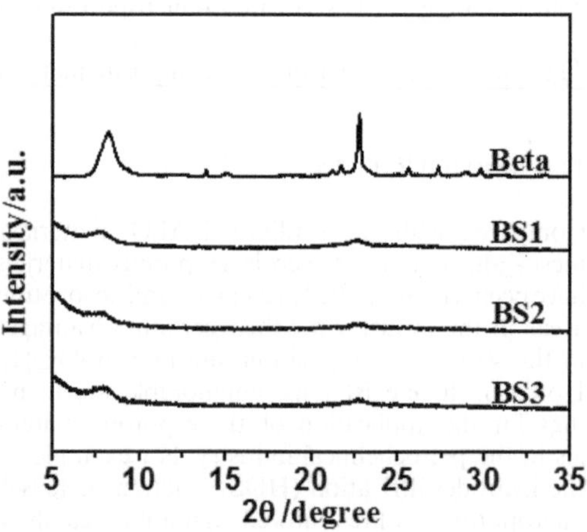

Figure 2. Wide-angle XRD pattern of BS.

the XRD diffraction peak intensity of the material decreased successively, which demonstrates that order degree of the synthesized composite declines in turn.

Figure 2 represents the XRD spectra for the synthesized BS material in a wide angle domain. The characteristic diffraction peaks of $2\theta = 7.8°$ and 22.4° are attributed to the Beta zeolite, which indicates that the synthesized composite material has the microporous structure similar to the Beta molecular sieve.

3.2 *Pore structure characterization*

The N_2 adsorption results of the specific surface area, pore volume, and aperture of meso-microporous materials BS that were synthesized under different TMB/P123 ratios were shown in Table 1. It is obvious

Table 1. Physicochemical properties of BSx.

Samples	S_{BET} ($m^2 \cdot g^{-1}$)	V_{mes} ($cm^3 \cdot g^{-1}$)	V_{mic} ($cm^3 \cdot g^{-1}$)	d_{BJH} (nm)
SBA-15	699	0.71	0.17	6.4
BS1	653	0.7	0.16	8.36
BS2	705	0.75	0.18	8.96
BS3	677	0.72	0.18	10.2
Beta	502	–	0.22	–

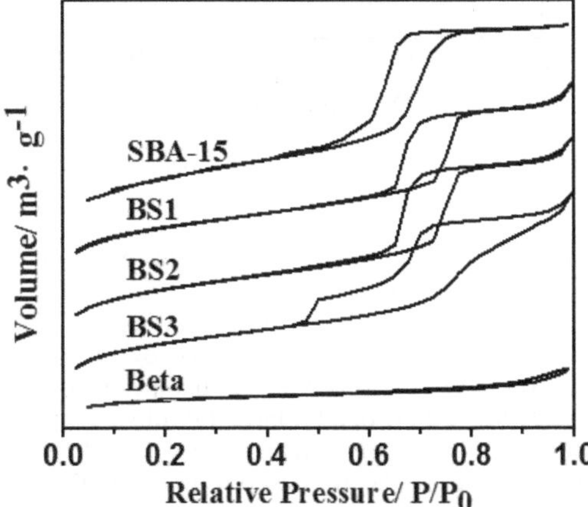

Figure 3. N_2 adsorption-desorption isotherms of BS series materials.

that the synthesized meso-microporous composite materials have large specific surface area, pore volume, and aperture, which is beneficial to improve the HDS performance of catalyst.

Figure 3 shows the N_2 adsorption–desorption isotherms for the synthesized BS composite material. The BS composite material shows a shape of N_2 adsorption-desorption isothermal curve different from Beta zeolite, but similar to the SBA-15 mesoporous materials, which shows a larger H3 hysteresis loop within the scope of 0.6 to 0.9 relative pressure, indicating the typical mesoporous structure. Furthermore, the hysteresis loops for both microporous and meso-microporous composite materials are quite steep, which suggests the narrow pore size distribution of the two materials. In addition, with the increase of TMB/P123 ratio, the hysteresis loop shifts to the direction of relative greater pressure (P/P_0), which indicates that the aperture of the meso-microporous composite increases with the increase in the mass ratio of TMB/P123. The results are in accordance with the small-angle XRD characterization results.

The pore size distribution of the synthesized BS material is shown in Figure 4. From the figure it

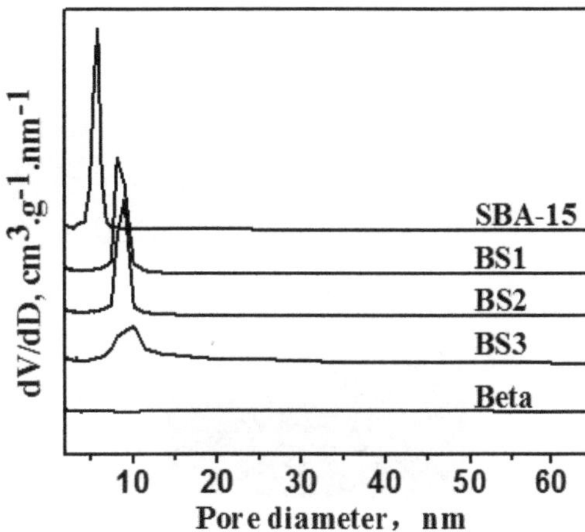

Figure 4. Pore size distribution of BS series materials.

can be seen that both of mesoporous SBA-15 and composite BS materials showed the narrow pore size distributions. The average pore size of SBA-15 mesoporous materials without expanding agent was about 6 nm. The average pore size of BS composite increases significantly with the addition of the swelling agent. No obvious peak was observed within mesoporous aperture for Beta molecular sieve, indicating that the pure Beta molecular sieve does not have the pore with mesoscopic scale. These results further demonstrated that the meso-microporous composite material with larger aperture of the mesoporous structure is successfully synthesised.

3.3 *TEM characterization*

TEM images of different samples are presented in Figure 5. BS composite has a uniform, well ordered two-dimensional hexagonal mesopore structure, indicating that Beta-SBA-15 is synthesized successfully.

3.4 *SEM characterization*

The SEM image of the BS composite is shown in Figure 6. It can be seen that the morphology Beta microcrystalline, without being fully crystallized, are nanosized flocculent particles with the diameters of a few nanometers. The crystallinity of Beta zeolite is obviously much better than that of Beta microcrystalline emulsion. The morphology of mesoporous materials SBA-15 is orderly granular particles, with particle size of about 1 μm. In contrast, Beta-SBA-15 is a worm-liked bundle of fiber with a long and rough surface and the length reach of more than 5 μm. Therefore, the emulsion of Beta microcrystalline shows a certain influence

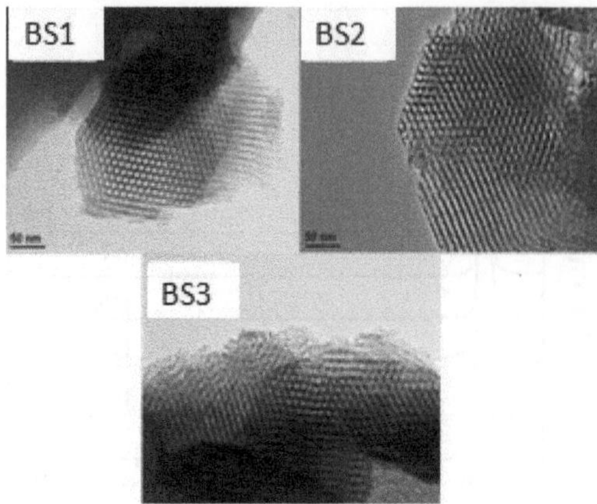

Figure 5. TEM images of BS series materials.

Figure 6. SEM images of BS series materials.

on the morphology of the synthetic meso-microporous composite.

3.5 *FTIR spectroscopy*

FTIR spectra of BS composites are shown in Figure 7. Pure silicon mesoporous material SBA-15 shows the infrared absorption peaks in the wave numbers of 460 cm^{-1}, 810 cm^{-1}, 950 cm^{-1}, and 1030 cm^{-1}, which are basically the absorption peaks of skeletal vibrations for pure silica mesoporous materials. The clear adsorption peak at 460 cm^{-1}, 810 cm^{-1} 950 cm^{-1}, and 1030 cm^{-1} are attributed to the bending, stretching, and vibration of Si-O-Si, the symmetric stretching vibration of Si-O-Si band, the existence space of Si-OH groups,

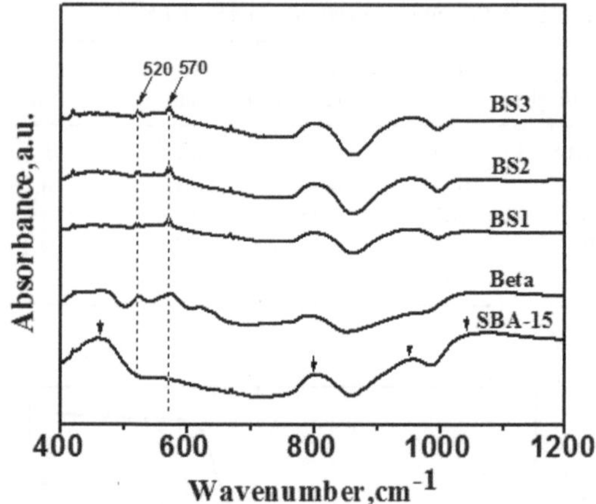

Figure 7. FTIR spectra of SBA-15, Beta, and BSx.

and the asymmetric stretching vibration peak of Si-O-Si [5], respectively.

As shown in Figure 7, compared to that of pure SBA-15, the peak positions caused by the vibration of Si-O-Si and Si-OH of the spectra of the composite Beta-SBA-15 have no obvious shift, indicating that composite materials have the skeleton of pure silicon mesoporous materials. In addition, the composite materials showed two additional infrared absorption peaks in the wave numbers of 520 cm^{-1} and 570 cm^{-1}, which are of the typical characteristic vibration absorption peaks of five-member and six-member ring in the Beta microporous zeolite skeleton structure [6]. The above results show that the synthetic composite materials contained the precursor structure of Beta molecular sieve. However, compared to the pure Beta molecular sieve, the composite materials show weak intensity of the double loop vibration peak, which was weak because the sizes of Beta microcrystalline in meso-microporous composite material are small.

4 CONCLUSION

Herein, we studied the impact of synthetic composite materials with different TMB/P123 ratios. Besides, based on the characterization and analysis of the basic physical properties of meso-microporous composite materials and corresponding catalysts, some important conclusions can be obtained as follows:

1. According to the characterization of XRD, IR, and N_2 adsorption-desorption, the composite material with both of the microporous structure of Beta zeolite and the mesoporous structure of SBA-15 has been successfully

synthesized through nano-assembly method, using Beta zeolite microcrystalline emulsion as raw material.

2. SEM and TEM images indicated that the synthesized Beta-SBA-15 showed a fiber bundle structure with a long wormlike formation and a regular ordered two-dimensional hexagonal pore structure, respectively.

ACKNOWLEDGMENTS

The authors acknowledge the financial supports from the National Natural Science Foundation of China (No. 21276277 and U1463207), CNOOC Huizhou Refinery Branch (HL00FW2012-0196), National Basic Research Program of China (Grant No. 2012CB215001 & 2012CB215002), and the CNPC-Petrochemical Research Key Project.

REFERENCES

[1] Tao Y., Kanoh H., Kaneko K. Uniform Mesopore-Donated Zeolite Y Using Carbon Aerogel Templating [J]. The Journal of Physical Chemistry: B, 2003, 107(40): 10974–10976.

[2] Jacobsen C.J.H., Madsen C., Houzvicka J., et al. Mesoporous Zeolite Single Crystals [J]. Journal of the American Chemical Society, 2000, 122(29): 7116–7117.

[3] Groen J.C., Jansen J.C., Moulijn J.A. et al. Optimal Aluminum-Assisted Mesoporosity Development in MFI Zeolites by Desilication [J]. The Journal of Physical Chemistry B, 2004, 108(35): 13062–13065.

[4] Sun Y., Prins R. Hydrodesulfurization of 4, 6-Dimethyldibenzothiophene over Noble Metals Supported on Mesoporous Zeolites [J]. Angewandte Chemie International Edition, 2008, 47(44): 8478–8481.

[5] Umamaheswari V., Palanichamy M., Murugesan V. Isopropylation of m-Cresol over Mesoporous Al-MCM-41 Molecular Sieves [J]. Journal of Catalysis, 2002, 210(2): 367–374.

[6] Joaquin P.-P., Johan A.M., Peter A.J. Crystallization mechanism of zeolite beta from $(TEA)_2O$, Na_2O and K_2O containing aluminosilicate gels [J]. Applied Catalysis, 1987, 31(1): 35–64.

Material Science and Environmental Engineering – Chen (Ed.)
© 2016 Taylor & Francis Group, London, ISBN 978-1-138-02938-5

Production of Rubberized Bitumen by oxidation of black oil

Y. Tileuberdi, Y.K. Ongarbayev, Y.I. Imanbayev & M.I. Tulepov
Al-farabi Kazakh National University, Almaty, Kazakhstan

Z.A. Mansurov & B.K. Tuleutaev
Institute of Combustion Problems, Almaty, Kazakhstan

ABSTRACT: In the paper, production process of Rubberized Bitumen (RB)-based Rubber Crumb (RC) from worn tires was investigated. During the experiment, the heavy oil vacuum residue (black oil) was oxidized with an amount of 0.3 and 0.5 wt.% of catalyst ($FeCl_3 \cdot 6H_2O$) at 240 °C. As a result it was observed, with increasing content of rubber from 10 wt.% to 20 wt.% in bitumen, the penetration and ductility of rubberized bitumen were decreased. Otherwise softening point of RB was increased in any content of rubber. Physical and mechanical characteristics of rubberized bitumen are established by standard methods. The structure of rubberized bitumen was characterized by optical microscopy and Scanning Electron Microscopy (SEM).

1 INTRODUCTION

1.1 *Scrap tire and rubber crumb*

Crumb Rubber (CR) is the recycled rubber obtained by mechanical shearing or grinding of scrap tires into small particles. Scrap tires are valuable secondary raw materials containing 65–70% rubber, 15–25% technical-grade carbon, and 10–15% high-quality metal. During the recycling process steel and fluff are removed, leaving tire rubber with a granular consistency. Eng. Vasco Pampulim, et al. offered the model of typical tire rubber mix (Fig. 1). Thus, the efficient processing of scrap tires makes it possible not only to solve environmental problems but also to perform economically rational utilization processes [1–4].

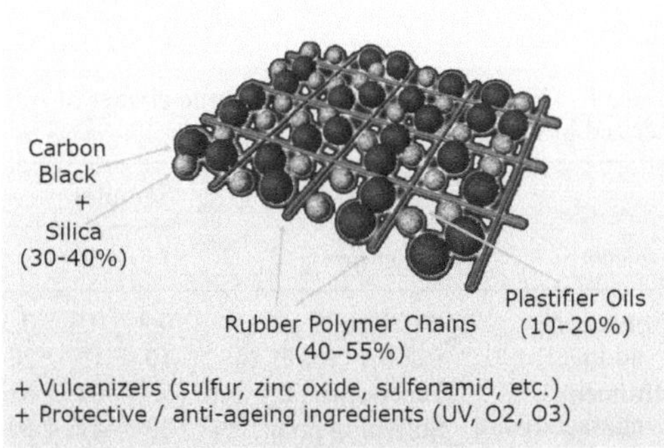

Figure 1. Model of typical tire rubber mix [4].

1.2 *Using rubber crumb and rubberized bitumen*

In the world, scientists have been offered a variety of ways to recycle and utilize the rubber crumb from worn tires. A well-known method is to burn the rubber waste to produce energy while producing cement [5]. Crumb rubber is often used in astroturf for cushioning, where it is sometimes referred to as astro-dirt. CR was used to remove ethylbenzene, toluene and xylene from aqueous solutions [6]. Rubber crumb also goes into the manufacturing of several auto parts and small percentages of crumb rubber go into manufacturing of new tires. A revolutionary nanotechnology process was developed to produce wood-replacement products paper-replacement materials from used tires [7]. Cut tires are used for the manufacture of drainage tubes, tapes for the protection of cables and pipelines, and soundproof walls. The combustion of tires generates energy and pyrolysis under conditions of relatively low temperatures. In addition, the processing of tires to obtain rubber crumbs for the manufacture of rubber bitumen compounds and asphalt-rubber compositions for insulating and roofing materials [3, 8–9].

There are many modification processes and additives that are currently used in bitumen modifications, such as Styrene Butadiene Styrene (SBS), styrene-butadiene rubber, ethylene vinyl acetate, and Crumb Rubber Modifier (CRM) [10].

Researches and applications of CRM and other modifications in the world showed that the bitumen binder has many advantages characteristics like improved resistance to rutting due to high viscosity, high softening point and better resilience, improved resistance to surface initiated, reduce fatigue/reflection cracking, reduce temperature susceptibility, reduce noise, resistance to fissure propagation, improved resistance to ultra-violet, improved resistance to oxygen or ozone, improved durability and lower pavement maintenance costs, and saving in energy and natural resource by using waste products [4, 11–12]. Asphalt concrete prepared with rubber-bitumen compounds exhibits high performance, enhanced wear and heat resistance, and resistance to aging [1, 13–14].

2 MATERIALS AND METHODS

2.1 *Production method of rubberized bitumen*

Experiments were performed to obtain rubber-bitumen compounds by mixing black oil (heavy oil residue) and rubber crumb. Mixing black oil with rubber crumb in an amount up to 20 wt.% at a temperature of 240 °C for 90 minutes and the stirring rate was 40/min. Oxidation processes of black oil in the presence of a catalyst in an amount of 0.3 and 0.5 wt% of $FeCl_3 \cdot 6H_2O$ were conducted. Air flow rate was 2.4 l/min. Then, oxidation products were stirred with a variety contents of (10; 15; 20 wt.%) rubber crumb for 60 minutes. Process was carried out at special apparatus, which was presented in Figure 2.

The apparatus were heated electrically (4), and the temperature of reactor was fixed by thermoregulator (8). Top to reactor connected stirrer (1 and 5), which allows for mechanical mixing of raw materials for process intensification.

1 - motor stirrer, 2 - pipe for gas outlet, 3 - a cylindri-

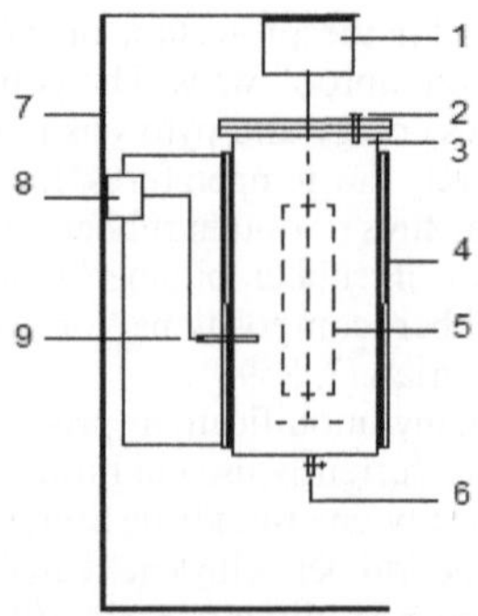

cal reactor, 4 - electric oven, 5 - agitator, 6 - a branch pipe for the withdrawal of products, 7 - Cabinet, 8 - thermo regulator, 9 - thermocouple.

Figure 2. Scheme of apparatus for preparing RB.

2.2 *Determination of physical and mechanical characteristics of bitumen materials*

Penetration involves the determination of the extent to which a standard needle penetrates a properly prepared sample of bitumen under specified conditions of temperature, load, and time. The unit of penetration is 0.1 mm, which is generally omitted in favor of reporting just the measured number. It was determined by apparatus Penetrometer PNB-03 in accordance with standard 11501-78.

Softening temperature—The temperature at which the bitumen goes from a relatively solid state into a liquid state. The softening point was determined by the method of "ring and ball" according to standard 11506-73.

The penetration index characterizes the degree of penetration of colloidal bitumen or rejection of his status from a purely viscous form. It is determined by empirical formula:

Ductility test gives a measure of adhesive property of bitumen and its ability to stretch. Tensile properties were determined by the apparatus Ductilometer CDB-974 N according to standard 11505-75.

3 RESULTS AND DISCUSSION

3.1 *Producing rubber modified bitumen*

Standard accordance of rubber modified bitumen was determined according to "Recommendation on the application of crumb rubber in road construction R RK 218-76-2008". Physical and mechanical characteristics of rubberized bitumen produced from heavy oil with rubber crumb (in same content of catalyst) are given in Table 2.

As seen from Table 1, with increasing content of rubber from 10 wt.% to 20 wt.% in bitumen, the penetration of rubberized bitumen decreased. Otherwise softening point of RB increases in any content of rubber. It was shown that the ductility

Table 1. Physical and mechanical characteristics of rubberized bitumen materials.

Objects	Indicators	Samples 1	2	3
Content of adding	Catalyst, %	0.3	0.3	0.3
	Rubber crumb, %	10	15	20
Bitumen characteristics	Penetration, 0.1 mm	75	64	62
	Softening point, °Ñ	39	47	50
	Ductility, cm	16	10	8
	Penetration index	0.9	1.0	1.3

Table 2. The influence of catalysts' content in the bitumen production.

Content of addition		Characteristics of bitumen		
Catalyst, %	Rubber crumb, %	Penetration, 0.1 mm	Softening point, °Ñ	Dutility, cm
0.3	15	64	50	10
0.5	15	56	52	7.4

Table 3. Group composition of black oil and bitumen products.

Hydrocarbon materials	Group composition of hydrocarbons		
	Asphaltene	Oil	Tar
Black oil	5.5	24.6	69.9
Oxidized bitumen	9.2	31.8	59.0

of RB is decreasing depending to the increase in content of modifier in bitumen. This is due to the action of rubber particles as stress concentrators. It means that the viscosity of bitumen compounds increases and starts to harden. These bitumen composition functions as a liquid or pseudo-thermoplastic matrix, the rubber particles provide resilient power frame in the amount of binder. The influence of catalysts' content in the bitumen production process is presented in the Table 2.

According to tabulated date, the adding content of catalyst increases (in same content of rubber crumb), and the physical and mechanical characteristics of products changed to highly viscous and harden. In the process rule, the catalyst content was impacted to increase its oxidation process. Group composition of hydrocarbon materials before and after oxidation process with catalysts were studied (Table 3) experimentally.

According to the results, asphaltene content was increased after oxidation of hydrocarbon. It is an infusible solid material with a density slightly more than 1 g/cm^3, which was contained in the bitumen in an amount of 10–25%. Asphaltene determines the structure formation processes, temperature resistance and increase in the viscosity of the bitumen.

Oil content was increased in bitumen after oxidation, too. It is liquid hydrocarbon at normal temperature with a density less than 1 g/cm^3. Usually, their content in bitumen is 40–60%. Oil determines the mobility and fluidity of the bitumen.

Tar content was decreased through oxidation. It is fusible visco-plastic material, solid or semi-solid at ordinary temperature with a density of about 1 g/cm^3. Tar contained in the bitumen is found in an amount of 20–40%, which determines the elasticity and extensibility of the binder [15].

3.2 *Microscopic study of rubber-bitumen compounds*

Surface structure of bitumen compound materials was studied using a microscopic technique. They describe interaction between bitumen and rubber crumb in rubberized bitumen. Optical microscopic images of rubber modified bitumen with 5 wt.% catalyst are presenting at Figure 3.

From the Figure 3 it is shown that the appearance of original bitumen is very smooth. The bitumen appearance becomes uneven after addition of rubber crumb and the rubber crumb becomes sticky after mixing it with base bitumen. During the preparation of rubber and bitumen, the bitumen aggregation almost covered the swelled rubber crumb while heating and stirring the mixture. The reason may be that rubber powder swelled by absorbing some light components or more liquid part of bitumen [16]. As the rubber crumb has

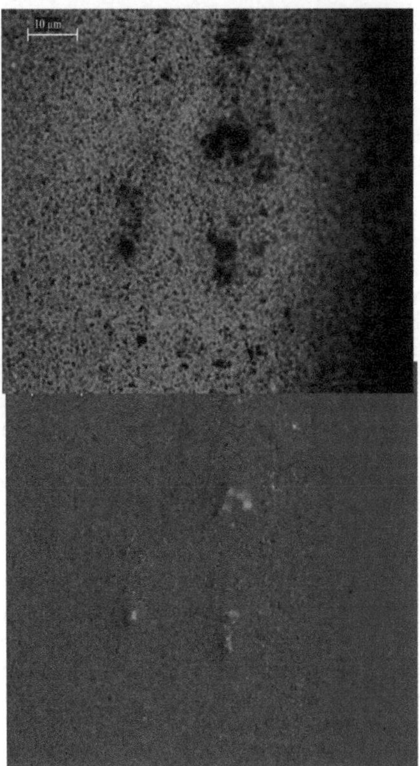

Figure 3. Optical microscopic image of rubberized bitumen.

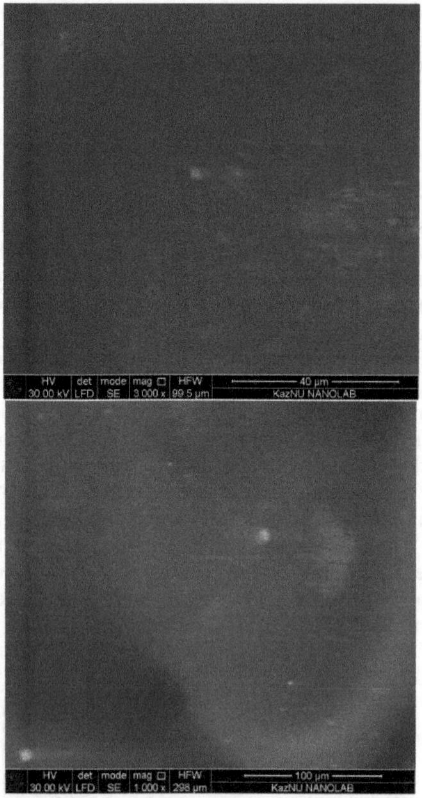

Figure 4. Scanning electron microscopic image of rubberized bitumen.

1.474 m^2/g of surface area, 0.300 nm is of medium pore size. It is important for rubber crumb to react with bitumen.

By electron microscopy images (Fig. 4) can approve that it is possible, these particles are particles of ferric chloride, which were catalysts for the oxidation of tar. In general, examples of rubber-binders are homogeneous mixtures with fully distributed components in the dispersed bitumen environment.

4 CONCLUSION

Thus, in the paper the possibility of producing rubberized bitumen from heavy oil residue with modification and catalyst. Due to the action of rubber particles as stress concentrators, the viscosity of bitumen compounds increases and starts to harden. These bitumen composition functions as a liquid or pseudo-thermoplastic matrix, the rubber particles provides resilient power frame in the amount of binder.

REFERENCES

[1] Shunin D.G., Filippova A.G., Okhotina N.A., Liakumovich A.G., and Samuilov Ya. D. 2002. Possibilities of Production and Use of Rubber-Bitumen Compounds. *Russian Journal of Applied Chemistry* 75(6): 1020–1023.

[2] Gorlova E.E., Nefedov B.K., Gorlov E.G., and Ol'gin A.A. 2008. Reprocessing of Industrial Rubber Waste in a Mixture with Shale. *Solid Fuel Chemistry* 42(2): 93–94.

[3] Gorlova E.E., Nefedov B.K., and Gorlov E.G. 2009. Manufacture of an Asphalt-Rubber Binder for Road Pavements by the Thermolysis of Tire Chips with Heavy Petroleum Residues. *Solid Fuel Chemistry*. 43(4): 224–228.

[4] Eng. Vasco Pampulim. Use of Rubber Powder cryogenically recovered from end-of-life tires in Asphalt Rubber for high performance road paving. *International Theoretical and Practical Conference on "Contemporary Approaches to Rubber Goods and Tires Recycling" (slide.) 1–2 June 2011*. Moscow.

[5] Mareike Hess, Harald Geisler, and Robert H. Schuster. 2009. Devulcanization as an opportunity to recycle rubber. *Chem. Listy 103, s1–s148 PMA 2009 & 20th SRC*: 58–60.

[6] Alamo-Nole L.A., Roman F. and Perales-Perez O. Sorption of Ethylbenzene, Toluene and Xylene onto Crumb Rubber from Aqueous Solutions. *http://www.nsti.org/BioNano2007.*

[7] Nanotech turns used tyres into building material. *http://www.dena.co.uk.*

[8] Background on Artificial Playing Fields and Crumb Rubber. *http://www.ct.gov.*

[9] Waste Tire Disposal. *http://www.state.tn.us.*

[10] Nuha S. Mashaan, Asim Hassan Ali, Mohamed Rehan Karim and Mahrez Abdelaziz. 2012. An overview of crumb rubber modified asphalt. *International Journal of the Physical Sciences* 7(2): 166–170.

[11] Ao Ying, Cao Rongji. 2010. Interaction Theory of Asphalt and Rubber. *Journal of Wuhan University of Technology-Mater. Sci.* Ed. Oct.: 853–855.

[12] Mullins O.C. 2010. The Modified Yen Model. *Energy Fuels* 24: 2179–2207.

[13] Ye Zhi-gang, Kong Xiang ming, Yu Jian-ying, Wei Lian-qi. 2003. Microstructure and Properties of Desulfurized Crumb Rubber Modified Bitumen. *Journal of Wuhan University of Technology-Mater. Sci.Ed. Mar.* 18(1): 83–85.

[14] Austruy F., Tileuberdi Ye., Ongarbaev Ye., Mansurov Z. 2012. Study of Production of Rubber-Bitumen Compounds. *Eurasian Chemico-Technological Journal*. 14(2): 133–138.

[15] Ongarbayev Ye., Golovko A., Krivtsov E., Tileuberdi E., Imanbayev Ye., Tuleutayev B., Mansurov Z. 2014. Thermocatalytic cracking of Kazakhstan's natural bitumen. *Studia UBB Chemia 59* (Lix 4): 57–64.

[16] Tileuberdi Ye., Ongarbaev Ye.K., Mansurov Z.A., Kudaybergenov K.K., Doszhanov Ye.O. 2014. Ways of Using Rubber Crumb from Worn Tire. *Applied Mechanics and Materials*. 446–447: 1512–1515.

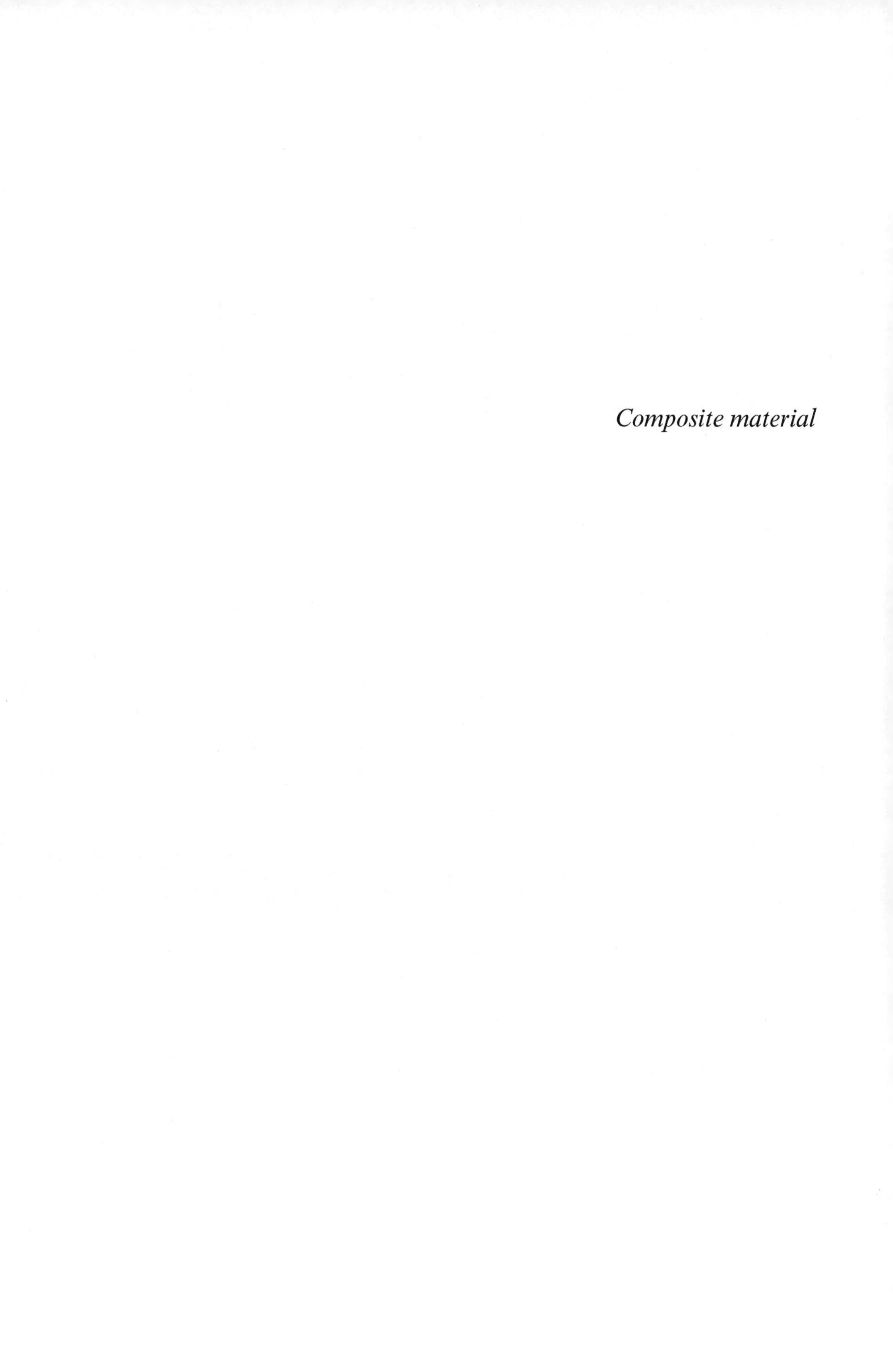

Composite material

Material Science and Environmental Engineering – Chen (Ed.)
© 2016 Taylor & Francis Group, London, ISBN 978-1-138-02938-5

Progressive failure simulation of composite snowboard bolted joints based on the energy damage evolution

R.K. Zhang
Department of Physical Education, Harbin Finance University, Heilongjiang Province, China

ABSTRACT: Composite materials have been widely used in sport equipment, e.g. composite snowboard. Bolted joints are usually used between snowboard and fixture. In this paper, based on the theory of energy damage evolution, the user material subroutine VUMAT for Abaqus has been programmed to simulate the damage accumulation and failure modes of bolted composite laminate. The effect of width to diameter ratio (W/D) ratios and end distance to diameter ratio (E/D) are analysed on the carrying capacity of bolted joint. A good agreement was obtained between numerical results and experimental results. The results show that the method used is effective.

Keywords: composite; bolted; damage accumulation; failure modes; numerical simulation

1 INTRODUCTION

Composite materials have been widely used in sport equipment, e.g. composite snowboard. Bolted joints are usually used between snowboard and fixture. Compared with general materials, composite joints have more complex failure modes, which involve numerous affecting factors. In view of the complexity, many scholars have carried out extensive and thorough research on the stress field distribution, joint strength and failure mode of composite bolted joints. Chishti M carried out experimental study on the preload, gaps and countersunk height of single lap countersunk composite joints. Sen Faruk [1] conducted experimental failure analysis on impacts of fit clearance and preload on composite bolted. Okutan Buket [2] exploited the effects of geometric parameters on the failure strength for composite pin-loaded joints through experiment and numerical simulation. Bulent Murat [3] predicted the effect of W/D, E/D on pin-loaded woven composite joints using the criteria of Hashin and Hoffman. McCarthy [4], Hühne, and Jiang Yunpeng did progressive damage analysis of composites using USDFLD module in ABAQUS. This paper adopts energy damage evolution theory and subroutine compilation to implement stiffness gradually steady degradation; conducts numerical simulation on the progressive damage of composite laminated bolted joints; solved the convergence problem in finite element method adopted in stiffness degradation, and gives the influence rule of W/D, E/D, on laminated bolted joint strength.

2 FAILURE CRITERION AND DAMAGE EVOLUTION MODEL BASED ON ENERGY THEORY

Employ Hashin failure criterion in four different failure models: matrix tension/compression failure, fiber tension/compression failure. Material goes initial failure when material stress fulfills Hashin failure criterion. For studying progressive failure, degrade corresponding material stiffness. Damage evolution defines the stiffness degradation rule after the initial failure. Constitutive equation is as follow:

$$\sigma = C(d)\varepsilon \tag{1}$$

$C(d)$ is material damage stiffness matrix. The expression is as follow:

$$C(d) = \frac{1}{D}\begin{bmatrix} (1-d_f)E_1 & (1-d_f)(1-d_m)\upsilon_{21}E_1 & 0 \\ (1-d_f)(1-d_m)\upsilon_{12}E_2 & (1-d_m)E_2 & 0 \\ 0 & 0 & D(1-d_s)G \end{bmatrix} \tag{2}$$

In formula (2), $D = 1-(1-d_f)(1-d_m)\upsilon_{12}\upsilon_{21}$, $d_s = 1-(1-d_{ft})(1-d_{fc})(1-d_{mt})(1-d_{mc})$, $d_{ft}, d_{fc}, d_{mt}, d_{mc}$ are damage variable caused by fiber, matrix tension/compression failure, shown by equivalent stress, equivalent displacement and fiber/matrix breaking energy, as been shown in Figure 1.

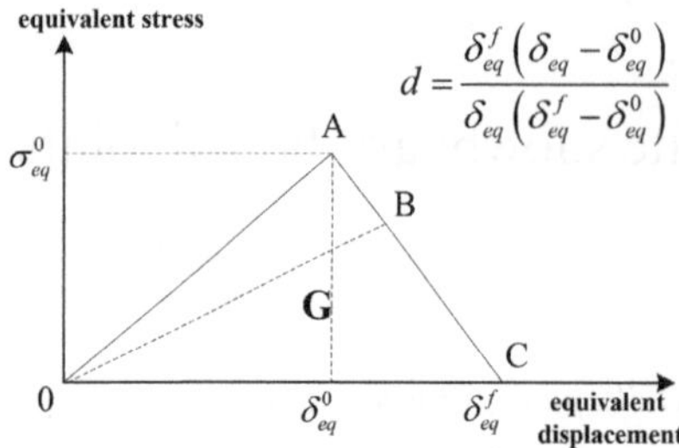

Figure 1. The calculate model of material damage variable d.

3 NUMERICAL MODEL

The geometric dimension of composite specimen is shown in Figure 2. L+E, W, D, t is respectively the length, width, bore diameter and thickness of the specimen. In the numerical model, L is 50 mm; D is 5 mm, t is 3.3 mm. The layup of the laminates is [0/90/0], and material property is shown in Table 1.

To investigate the effect of W/D and E/D on bolted joint strength and damage model, 20 operating modes is adopted—W/D ranges from 2 to 5, E/D ranges from 1 to 5. The thickness of laminates, t, is relatively small compared with length and width. Regardless of the layering along thickness direction and loading effect, the loading process fulfills plane stress condition; hence 2D finite element is adopted in the calculating example to conduct numerical analysis. There are some assumption and simplification are as follow:

1. Sufficient bolt stiffness, regardless of transformation and failure in loading process;
2. Close fit between bolt and laminate, no clearance or interference contact.

ABAQUS is adopted to build finite element model. To simulate the mutual contraction between bolt and laminate, radial boundary condition is exerted to semicircle orifice boundary. Mesh representing and boundary conditions are shown in Figure 3.

The simulation of joint progressive failure process contains mainly three aspects: failure criterion, reasonable stiffness degradation and loading analysis. Procedure is as follow: (1) Endow the model with material property and boundary condition, choosing appropriate load incremental step; (2) Conduct stress-strain analysis using laminate theory in every incremental step to acquire correspondence stress of every unit point; (3) Integrating correspondence failure criterion, determine the failure of the material unit, once found failure, the stiffness of correspondence material point is degraded in line with the stiffness, thus laminate stress is redistributed.

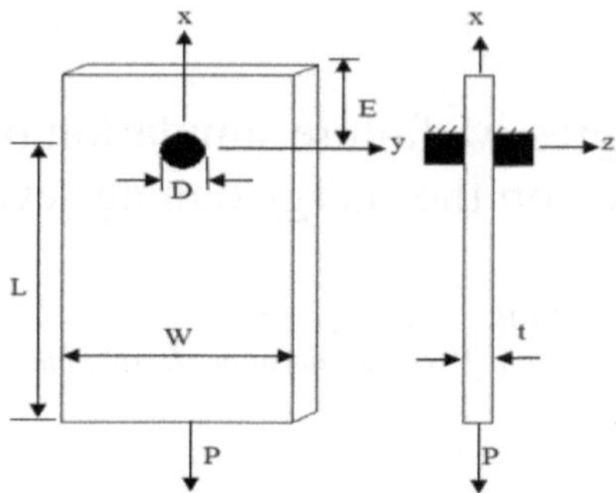

Figure 2. Geometry of specimen with single-bolt joint.

Table 1. Properties of glass-fiber/epoxy composite [2] MPa.

E1	44000
*E*2	10500
*G*12	3740
12	0.36
Xt	800
Xc	350
Yt	50
Yc	125
S	120

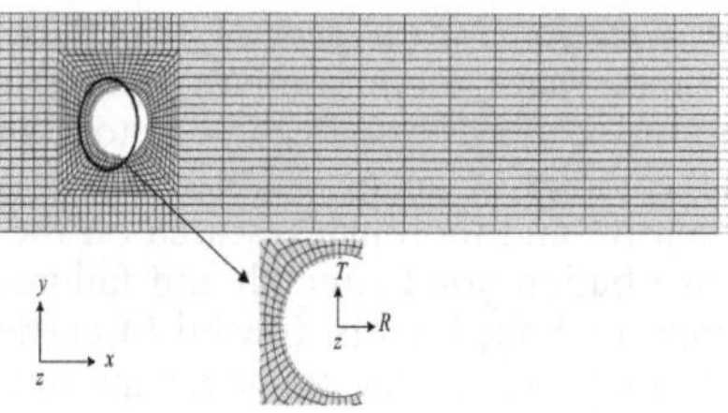

Figure 3. Mesh representing and boundary conditions.

4 NUMERICAL RESULT AND EXPERIMENT VERIFICATION

Finite element analysis is conducted on composite bolt joint using energy progressive failure method as well as the method established above. Taking the effect of geometrical parameter on laminated joint property into consideration, the first step is to define the bolt-laminate interactive bearing strength in tension, which is shown in expression (3):

$$\sigma_b = P_{\max}/Dt \tag{3}$$

Figure 4 shows influence rule of W/D and E/D on bolt-laminate bearing strength. When E/D ranges from 1 to 3, bolt-laminate interactive bears strength increases rapidly, while it slows down

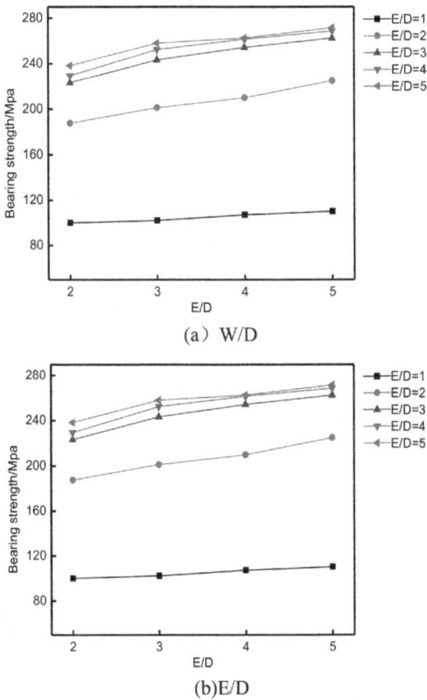

Figure 4. The effect of geometric parameters on beating strength (a) W/D (b) E/D.

when E/D is over 3. The effect of W/D is relatively stable. With the accrescent of W/D, joint bearing strength grows steadily, and the growth slows down when W/D is over 3. In general, the bearing strength of laminated bolted joint increases along with the growth of the geometrical parameter of W/D and E/D, and the effect of W/D and E/D gradually reduces when the geometrical parameter exceed a threshold value.

To verify the effectiveness of the analytical method adopted in this paper, comparison is made between the numerical simulation results with the result in literature [2], as is shown in Figure 5. It is observed that the results are corresponding. Figure 6 gives out the error analysis of numerical results and experimental result. When the numerical values of W/D and E/D are both 1, the simulation results are greater than experimental results, while in other cases, simulation results are less than the experimental results, and the relative error is within 20%. It shows that the model based on damage evolution can effectively predict the effect of geometric parameter on laminated joint.

The damage of laminated bolted joint is a process of progressive damage. It can be observed from the figure that from point A to D, there occurs matrix damage in every joint hole, and no obvi-

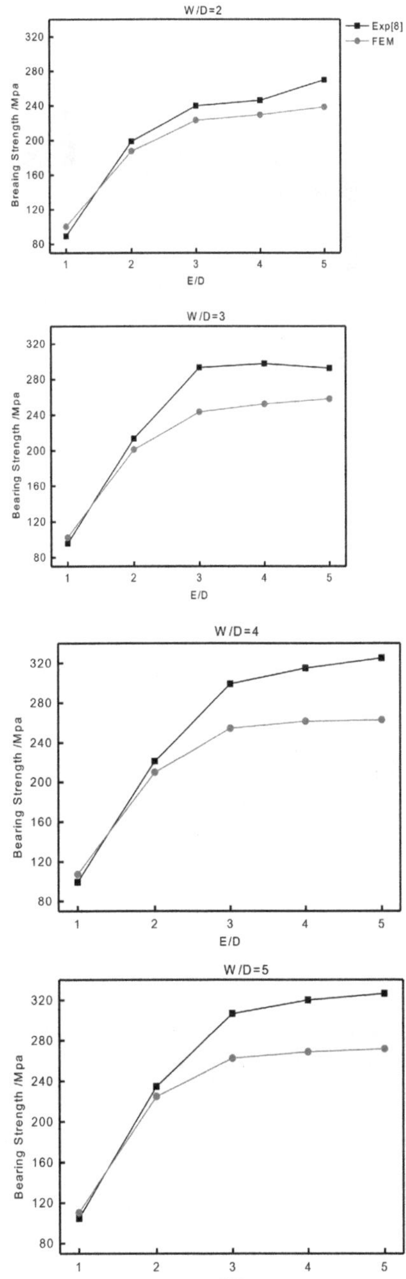

Figure 5. Comparison of [0/90/0]s laminates.

ous effect on laminated bolted joint capacity. With the increase of load, the 90°ply fiber at the point of E is the first to reach Hashin fiber compressive failure criterion, laminate load is obviously effected, then 0°ply fiber at the point of F occurs tension failure, and finally at G point, 90°ply fiber

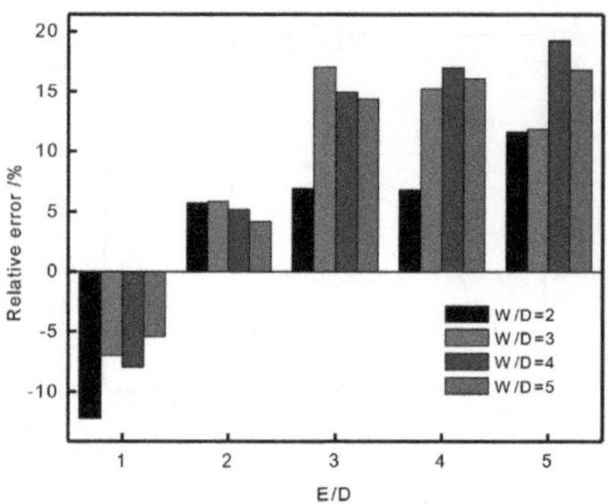

Figure 6. Error analysis of numerical results and experimental.

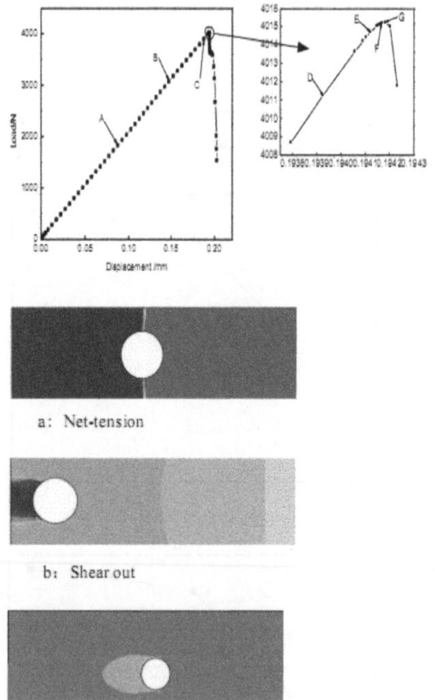

Figure 7. Typical damage modes in pined laminates.

tension failure causes the lost of laminate bearing capacity. The effect of W/D, E/D on laminated bolted joint strength is embodied in the ultimate damage mode of laminates, and three basic failure modes caused by geometric parameter are: net tension, shear-out and bearing [6]. This paper predicts three typical damage modes based on established modes, which is shown in nephogram given in Figure 7, corresponding to that was listed in literature [5]. Table 2 gives out the bolted joint failure modes and rules under the influence of W/D and E/D,

Table 2. Failure modes of specimens.

	E/D				
W/D	1	2	3	4	5
2	S	B+S	B+N	B+N	B+N
3	S	B+S	B+N	B+N	B+N
4	S	B+S	B	B	B
5	S	B	B	B	B

N: Net-tension, S: Shear-out, B: Bearing.

providing criteria for the judgment of composite bolted joint failure.

5 SUMMARY

Composite joints have numerous failure modes and complex damage process. The numerical simulation is an effective method to save the cost of experiment. With the progressive damage evolution as the basis, this paper adopts ABAQUS-UMAT subroutine to conduct numerical simulation on the progressive failure of composite bolted joints, and demonstrates some typical failure modes, the result of which is well consistent with the result of relevant experiments. The effectiveness and validity of the adopted method of numerical simulation. Calculated conclusion is as follow.

E/D, W/D have great effect on laminated bolted joints, especially E/D. When E/D and W/D are of small value, shear failure occurs. With the increase of E/D and W/D, the load capacity of laminates improve, finally tend to be stable.

REFERENCES

[1] Sen Faruk, Pakdil Murat, Sayman Onur, Benli Semih, Experimental failure analysis of mechanically fastened joints with clearance in composite laminates under preload[J]. Mater Des, 2008, 29:1159–69.

[2] Okutan Buket, The effects of geometric par ameters on the failure strength for pin-loaded multi-directional fiberglass reinforced epoxy laminate[J]. Composites Part B, 2002, 33(2):567–578.

[3] Bulent Murat Icten, Ramazan Karakuzu, Progressive failure analysis of pin-loaded carbon-epoxy woven composite plates[J]. Compos Sci Technol, 2002, 62: 1259–71.

[4] C.T. McCarthy, M.A. McCarthy, V.P. Lawlor. Progressive damage analysis of multi-bolt composite joints with variable bolt–hole clearances[J]. Composites: Part B, 2005, 36:290–305.

[5] Ramazan Karakuzu, Numan Taylak. Efects of geometric parameters on failure behavior in laminated composite plates with two parallel pin-loaded holes[J]. Composite Structures, 2008, 85:1–9.

Material Science and Environmental Engineering – Chen (Ed.)

The preparation and electrical properties of CB/UHMWPE composites

H.J. Wang & H.L. Hu
College of Materials Science and Engineering, Jilin Jianzhu University, Changchun, P.R. China

ABSTRACT: Carbon Black/Ultra High Molecular Weight Polyethylene (CB/UHMWPE) composites with a segregated structure were prepared through high-speed mechanical mixing and hot pressing at 200 °C. It was found that CB powders were well coated on the surfaces of UHMWPE particles in a very short time. The addition of 2 vol.% CB into UHMWPE matrix resulted in a dramatic increase in electrical conductivity by 14 orders of magnitude. A lower percolation threshold was achieved at the CB concentration of 1 vol.%. The electrical conductivity of CB/UHMWPE composite decreases with hot pressing time. The proposed preparation method is simple and has potential to produce low-cost polymer conductive composites with low percolation threshold.

Keywords: CB; UHMWPE; high-speed mechanical mixing; conductive network

1 INTRODUCTION

Conductive polymer composite materials have a long history of research. One major problem is that the large amount of conductive filler is needed in the composites to achieve a high conductivity. Such composites are often limited due to their high cost and poor process ability. To achieve higher conductivity at low conductive filler loadings, the conductive polymer composites with segregated network microstructures have been extensively studied in recent years. In the segregated network composites, the fillers are coated on the surface of polymer matrix particles to form an interface conductive network, which exhibits a lower electrical percolation threshold, relative to the composites that contain uniformly dispersed fillers. In general, there are three commonly used methods for fabricating the composite with a segregated structure: Liquid circumstance method [1–4], spraying method [5] and mechanical mixing method [6, 7]. Among them, mechanical mixing method is the simplest and lowest cost technique. However, some researchers consider that conductive cannot be uniformly dispersed on the surface of polymer particle by mechanical mixing, which will result in a high percolation threshold [8].

Recently, we have proposed a new method to fabricate Poly (Vinylidene Fluoride)/Carbon Annotate (PVDF/CNT) composite with a segregated structure by high-speed mechanical mixing process [9]. The results obtained showed that CNTs could be uniformly dispersed on the surface of polymer particle and a low percolation threshold could be achieved at very low CNT concentrations (0.078 vol.%). This effective new method is simple, low-cost, less time consuming, and suitable for large-scale industrial applications. Although the PVDF/CNT composite exhibits excellent electrical properties, the price of CNTs is high. It is desirable to develop conductive polymer composites with low-cost fillers. CB is a common of conductive filler with broad application and has a very low price. Therefore, in this work we prepared CB/UHMWPE composites through high-speed mechanical mixing process. The electrical properties of these composites were investigated. The aim is to understand the role of CB in the polymer matrix.

2 EXPERIMENTAL

CB was provided by Shandong Huaguang Chemical Factory, China, and its density was 1.87 g/cm^3 according to the manufacturer. UHMWPE powder (density 0.94 g/cm^3 and volume resistivity 10^{17} $\Omega\cdot$cm) was purchased from Jiangsu Shen Tai Science and Technology Development Co. Ltd, China. CB and UHMWPE powders were mechanically mixed in a QNG-9902(J) multi-purpose food mixer (from Shanghai Pumei Electrical Appliance, Co. Ltd, China) with two stainless steel grinding blades. The diameter and rotation speed of the blades were 65 mm and 23,000 rpm, respectively. The mixed powders were hot pressed into sheets of 1.0 mm in thickness under the pressure of 10 MPa at 200 °C for different time, and then cooled down to room temperature in air.

The electrical conductivity was measured by the two-terminal method using a digital multimeter for resistance $<10^7$ Ω and a high-resistance meter (ZC-36) for resistance $>10^7$ Ω. The sample surfaces were bonded with aluminum foil to reduce contact resistance. The morphology of the composites was studied by scanning electron microscopy (JEOL JSM-6700F) and optical microscopy.

3 RESULTS AND DISCUSSION

The electrical conductivity of CB/UHMWPE composites as a function of mixing time is shown in Figure 1. It can be seen that at about 40 seconds, the electrical conductivity of the composite which contains 2 vol.% CB increases 14 orders of magnitude higher than that of UHMWPE. Figure 2 shows the distribution of CB on polymer particles. It is clear that the CB powders are uniformly distributed on polymer particle surfaces. However, the CB particles appear as tiny balls and small aggregates with ca. 1 μm, indicating that the breakup and cleavage of CB aggregates cannot be sufficiently achieved by the high-speed mechanical mixing. Therefore, the dispersion of CB aggregations will be a further subject in this research. In addition, a slight decrease in electrical conductivity is observed when the mixing time exceeds 40 seconds (see Fig. 1). A similar phenomenon has also been observed in our previous work [9]. A possible explanation is that a long time mechanical mixing may cause the re-aggregation of some CB particles.

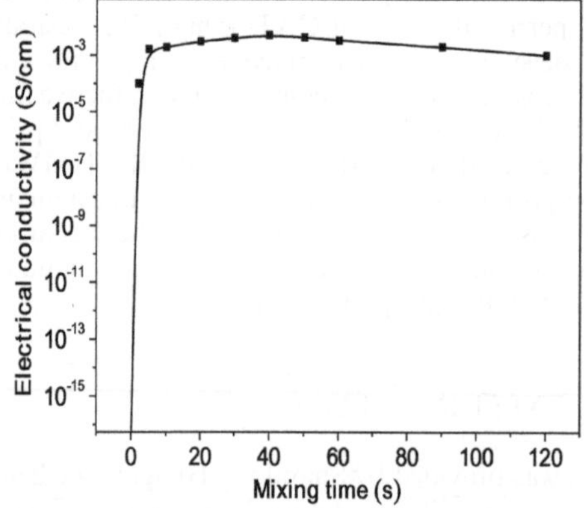

Figure 1. Electrical conductivity of CB/UHMWPE composite containing 2 vol.% CB as a function of mixing time.

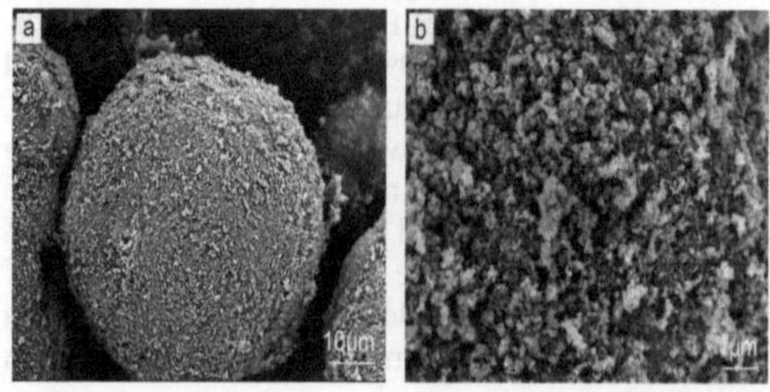

Figure 2. SEM micrographs of the mixed CB/UHMWPE composite with low (a) and high (b) magnifications. The mixing time is 40 s and the CB loading is 2 vol.%.

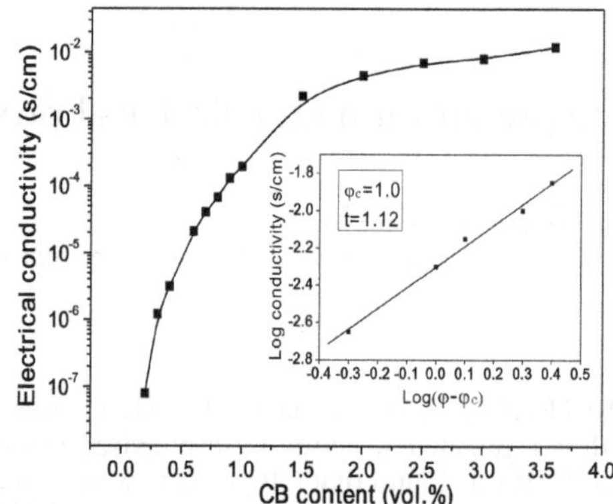

Figure 3. Electrical conductivity as a function of CB content. The insert shows the log-log plot of electrical conductivity with $\varphi - \varphi_c$ for $\varphi > \varphi_c$.

The electrical conductivity of the composites as a function of the volume fraction of the CB is shown in Figure 3. A significant change in electrical conductivity is observed between 0.25 and 1.5 vol.% which is taken as the percolation concentration range. According to the classical percolation theory, the dependence of the electrical conductivity of the composites on the conductive filler concentration can be described by a scaling law of the form [10]:

$$\sigma = \sigma_0(\varphi - \varphi_c)^t \quad (1)$$

where σ is the electrical conductivity of the composite, σ_0 is scaling factor, φ is the volume fraction of the conductive filler, φ_c is the volume fraction filler at the electrical percolation, and t is a critical exponent, which depends on the dimensionality of the system. The data in Figure 3 is fitted by log σ vs log $(\varphi - \varphi_c)$ and incrementally varying φ_c until the best linear fit is obtained. The resulting mathematical fit is inserted into Figure 3. It can be seen that φ_c and t are 1 vol.% and 1.1, respectively. Predicted values of t are about 1.1–1.3 and 1.6–2.0 for two-dimensional and three-dimensional systems [11], respectively, indicating that the composites present a two-dimensional rather than a three-dimensional conductive network, which is consistent with the interfacial localization of the CB powders.

Some investigations show that the electrical conductivity is sensitive to processing conditions [12–14]. To better understand the distribution of CB on the polymer matrix, a further investigation was made to explore the effect of hot pressing time

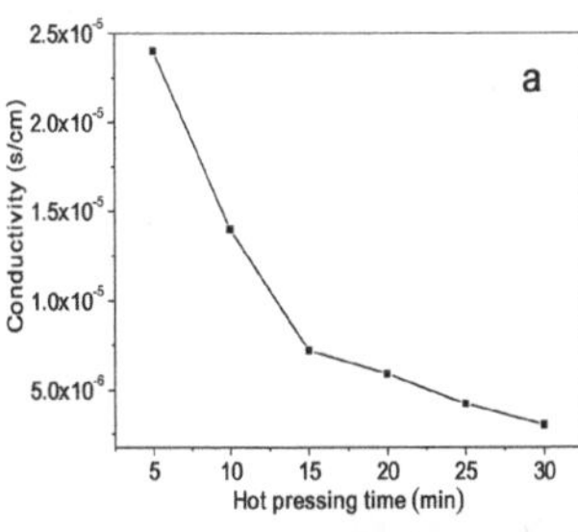

Figure 4. (a) Electrical conductivity of the hot-pressed composite as a function of hot pressing time. Optical micrographs of the composite hot-pressed at 200 °C for (b) 10 min and (c) 30 min. The CB loading is 0.6 vol.%.

on electrical conductivity. The change in electrical conductivity is shown in Figure 4a. The CB/UHMWPE composite exhibits a gradual decrease in electrical conductivity with hot pressing time. This decrease is due to the migration of CB particles from interface to the molten polymer matrix, which can be confirmed by optical micrographs in Figure 4b and c. When the hot pressing time increases from 10 to 30 min, there is an obvious broadening for the conductive paths composed of CB particles. The broadening of the conductive paths can be attributed to the migration or diffusion of CB particles in the interfacial regions owing to the lower viscosity of the polymer melt at higher temperatures, which in turn results in a decrease in the number of contact between the CB particles.

4 CONCLUSIONS

The CB/UHMWPE composites with a segregated structure were prepared by high-speed mechanical mixing method. CB can be uniformly coated on the surfaces of polymer particles in a very short time. At 2 vol.% of CB loading, the composite exhibits a dramatic enhancement in electrical conductivity by 14 orders of magnitude. A lower percolation threshold was achieved at the CB concentration of 1 vol.%. The breakup and cleavage of CB aggregates cannot be sufficiently achieved by the mechanical mixing. The CB conductive paths broaden with hot pressing time, which results in a decrease in electrical conductivity of the composite.

REFERENCES

[1] Du J.H., Zhao L., Zeng Y., Zhang L.L., Li F., Liu P.F. Comparison of electrical properties between multi-walled carbon nanotube and graphene nanosheet/high density polyethylene composites with a segregated network structure. Carbon 2011; 49: 1094–100.

[2] Hu J.W., Li M.W., Zhang M.Q., Xiao D.S., Cheng G.S., Rong M.Z. Preparation of binary conductive polymer composites with very low percolation threshold by latex blending. Macromol Rapid Commun 2003; 24: 889–93.

[3] Grunlan J.C., Mehrabi A.R., Bannon M.V., Bahr J.L. Water based single walled nanotube filled polymer composite with an exceptionally low percolation threshold. Adv Mater 2004; 16: 150–3.

[4] Yu C., Kim Y.S., Kim D., Grunlan J.C. Thermoelectric behavior of segregated-network polymer nanocomposites. Nano Lett 2008; 8: 4428–32.

[5] Zhang Q.H., Lippits D.R., Rastogi S. Dispersion and rheological aspects of SWNTs in ultrahigh molecular weight polyethylene. Macromolecules 2006; 39: 658–66.

[6] Ou R.Q., Gupta S., Parker C.A., Gerhardt R.A. Fabrication and electrical conductivity of Poly(Methyl Methacrylate) (PMMA)/Carbon Black (CB) composites: comparison between an ordered carbon black nanowire-like segregated structure and a randomly dispersed carbon black nanostructure. J Phys Chem B 2006; 110: 22365–73.

[7] Lisunova M.O., Mamunya Y.P., Lebovka N.I., Melezhyk A.V. Percolation behaviour of ultrahigh molecular weight polyethylene/multi-walled carbon nanotubes composites. Eur Polym J 2007; 43: 949–58.

[8] Gao J.F., Li Z.M., Meng Q.J., Yang Q. CNTs/UHMWPE composites with a two-dime nsional conductive network. Mater Lett 2008; 62: 3530–32.

[9] Liu Q.M., Tu J.C., Wang X., Yu W.U., Zheng W.T., Zhao Z.D. Electrical conductivity of carbon nanotube/poly (vinylidene fluoride) composites prepared by high-speed mechanical mixing. Carbon 2012; 50: 339–41.

[10] Mierczynska A., Mayne-L'Hermite M., Boiteux G., Jeszka J.K. Electrical and mechanical properties of carbon nanotube/ultrahigh-molecular-weight polyethylene composites prepared by a filler prelocalization method. J Appl Polym Sci 2007; 105: 158–68.

[11] Babinec S.J., Mussell R.D., Lundgard R.L., Cieslinski R. Electroactive thermoplastics. Adv Mater 2000; 12: 1823–33.

[12] Tchoudakov R., Narkis M., Siegmann A. Electrical conductivity of polymer blends containing liquid crystalline polymer and carbon black. Polym Eng Sci 2004; 44: 528–40.

[13] Gubbels F., Blacher S., Vanlathem E., Jerome R., Deltour R., Brouers F., Teyssie Ph. Design of electrical conductive composites: key role of the morphology on the electrical properties of carbon black filled polymer blends. Macromolecules 1995; 28: 1559–66.

[14] Mierczynska A., Friedrich J., Maneck H.-E., Boiteux G., Jeszka J.K. Segregated network polymer/carbon nanotubes composites. Cen Eur Sci J Chem 2004; 2: 363–70.

Material Science and Environmental Engineering – Chen (Ed.)
© 2016 Taylor & Francis Group, London, ISBN 978-1-138-02938-5

Preparation and performance of ceramifying EPDM rubber composite

Y. Pei, Y. Qin, X. Zhang & Z.X. Huang
School of Materials Science and Engineering, Wuhan University of Technology, Wuhan, P.R. China

ABSTRACT: Ceramifying EPDM Rubber Composites were prepared by using EPDM as matrix, Kaolin ceramic powder, and low melting point oxides as additives. The ceramics powder reacted to form ceramic residue after thermal pyrolysis under high temperature. X-Ray Diffraction (XRD) analysis of the composites at different temperatures showed the formation of new crystalline phases as a result of reactions between the degradation residue of EPDM and the ceramic powder. Mass ablating rate showed the change of the thermal stability of EPDM. Scanning Electron Microscopy (SEM) was used to explore the changes in microstructure after samples were pyrolyzed under high temperatures. The formation and growth of ceramic phase after heat-treatment enhanced the thermal stability at high temperature.

Keywords: heat-resistant EPDM; ceramic powder; active fillers; ceramifying reaction

1 INTRODUCTION

Owing to the excellent mechanical properties and stability at high temperature, ceramic material has been widely used as ideal high temperature structural materials in the aeronautic and astronautic fields. Currently, the main thermal protection methods include: (1) Endothermic method: Using a large mass of a specific heat capacity of high metal, but its large mass, and high temperature melt easily deformed, has been eliminated, (2) Radiation method: Using high rate of radiation, low absorption rate refractory metal, but limited under conditions of high heat flux applications, and (3) Ablation method: In the heat effect, material itself can occur in a variety of endothermic physical and chemical changes, such as decomposition, melting, evaporation, sublimation, etc. This is currently the most widely used anti-thermal method [1–2].

Ceramic polymer system in high-temperature aerobic state is a new type of thermal protection material. At high temperatures and aerobic conditions, the polymer matrix and ceramic powders occurs in a series of complex solid-phase reaction, forming a dense multiphase ceramic composite structure, to achieve the purpose of high-temperature-resistant ablation. Most current research of the ceramic polymer composites are in the organic silicon substrate. Mansouri et al [3] uses the silicone matrix, which by adding mica improves the strength of the ceramic products. Y.B. Cheng et al. [4–7] in Monash University of Australia added muscovite mica powders, feldspar powders and inorganic clay into the silicone rubber or silicone resin matrix to produce a ceramifiable cable. H.B. Shao et al. [8] studied the effect of glass powder amount on silicone rubber ceramics properties by adding a low melting glass powder on using the silicone matrix. Hanu et al. [9] used the silicone matrix, by adding thereto inorganic fillers, to improve the strength of porcelain objects. L.M. Su [10] using Methyl Vinyl Silicone Rubber (MVMQ) as the matrix, the modified clay mineral as powder filler, low melting glass powder/boric acid as control agent, benzoyl peroxide crosslinking agent for silicone rubber, prepared by direct blending clay/porcelain silicone rubber composites. Hanu et al. [11] studied the impact on the thermal stability of silicone system and combustion properties of mica, iron oxide, glass, or composite materials. Mansouri et al. [12] studied the influence of processing temperature on silicone composites microstructure and chemical composition. L.M. Su [13] expounded fireproof material properties of polymer composites, fireproof mechanism and research progress of the preparation. Sangita [14] with maleic anhydride grafted EPDM and EPDM as the substrate, by adding aramid fibers and nano-SiO_2 powder, prepared within a class of EPDM lightweight insulation material. H.D. Chen et al. [15] said that by adding a lower melting point it could organically modify nano-montmorillonite in EPDM prepared by a low-temperature ceramic-based nano-montmorillonite/EPDM insulation material.

As can be seen from the above literature, the study of ceramic polymer system may also focus on silicone rubber, EPDM and applications are still relatively small. By adding kaolin, try preparing ceramifiable EPDM and study the effect of temperature on the available ceramic properties.

2 EXPERIMENT

2.1 *Materials*

EPDM rubber (EPDM), the third monomer is Ethylidene Norbornene (ENB) content of 7.9%, ethylene content of 57%; Dicumyl Peroxide (DCP): the content of ≥98%, Aladdin; reinforcing filler SiO_2, industrial-grade precipitated silica; mica powder, 800 mesh, Lingshou Huajing Mica Ltd; processing aids, plasticizers, processing aids, etc. are commercially available industrial grade. The chemical compositions of the ceramic powder were given in Table 1.

2.2 *Sample preparation*

Ordering the EPDM, reinforcing filler, ceramic powder, processing aids, DCP on the open mill, mixing evenly through thin, film, placing rubber compound for a minimum of 12 h before the use of special tools to pieces, loading mode, vulcanizing. Vulcanizing pressure of 10 MPa, temperature 160°C for 30 min. Specific experimental formulations were shown in Table 2.

2.3 *Characterization*

2.3.1 *Mass ablating rate*

According to the definition of the Mass ablating rate as following:

$$M = \frac{m_1 - m_2}{t}$$

where M is the mass ablating rate (g/s), m_2 the mass of the composites after pyrolysis (g), m_1 the mass of the composites before pyrolysis (g), t the ablating time(s). Samples were heated at corresponding temperature 600°C, 800°C, 1000°C, and 1200°C for 20 min. We used analytical balance measuring to weight the samples (before pyrolysis, after pyrolysis), respectively.

Table 1. Major components of ceramic powder in terms of elemental oxide in wt.%.

Oxide	wt.%
SiO_2	52.20%
Al_2O_3	28.50%
K_2O	12.08%
Na_2O	1.45%
MgO	2.24%
Other oxides	4.53%

Table 2. Basic formula.

EPDM	100
DCP	5
S	1.5
Paraffin oil	10
White carbon	20
Ceramic powder	50

2.3.2 *Scanning Electron Microscopy (SEM)*

Samples 1, 2, 3, and 4 were heated at corresponding temperature 600°C, 800°C, 1000°C, and 1200°C. We used Japan Electronics Corporation's scanning electron microscope to observe surface and cross section morphology after the ablation.

2.3.2 *X-Ray Diffraction (XRD)*

X-ray diffraction spectra of the ceramified ash from samples after pyrolysis were obtained using the D8 Advance X-ray diffractometer. Each scan was conducted from a 2θ angle of 2 to 80°at a scan rate of 1°/min.

3 RESULTS AND DISCUSSION

3.1 *Mass ablating rate*

To explore the weight changes of the ceramifiable composites after pyrolysis, the composites were weighed by analytical balance. Figure 1 shows sample mass ablating rate at different temperatures.

Ablation rate of 1 was 0.043 g/s at 600°C, 2 was 0.049 g/s at 1000°C, 3 was 0.052 g/s at 1000°C, and 4 was 0.054 g/s. Under this range of temperature, although the mass ablating rate increases from 0.043 g/s to 0.054 g/s, the growth rate decreases from 13% to 4%. The data indicated that ceramic powder has been shown to improve the thermal stability of EPDM rubber. Under this range of temperature, ceramic powder was partly fused and

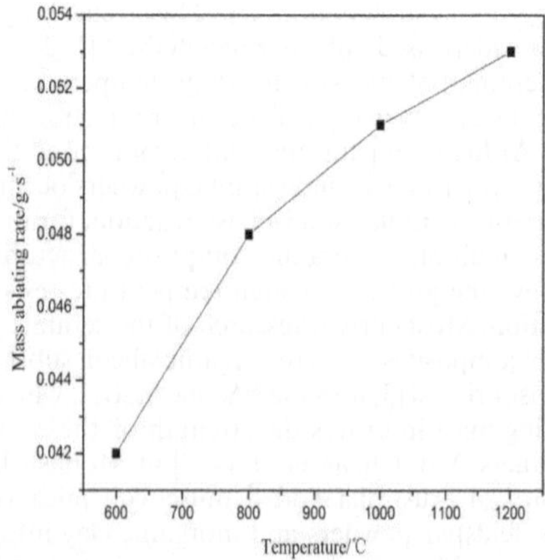

Figure 1. Mass ablating rate at different temperatures.

wrapped in the chars that were generated during the pyrolysis of EPDM rubber.

3.2 *Microstructural characteristics (SEM)*

The presence of surface of samples1,2,3, and 4 fired under 600°C, 800°C, 1000°C, and 1200°C were demonstrated using Scanning electron microscopy. (Fig. 2).

As we can see, at 600°C, on the surface of the sample it is seen that a large number of holes have been formed perpendicular to the plane of the structure. These cavities form a cracked gas diffusion transfer channel. Basic matrix and filler was separated, fluffy, when external force or internal stress that is easily broken. Therefore, the ablative property of the material is poor. As the temperature increased, the number of pores gradually decreased. At 800°C, there are still a lot of pores and cracks, and when the temperature rises to 1000°C, the surface of the melting phenomenon obviously occurred, and when the temperature rises to 1200°C, the number of holes have been very few, this shows that as the temperature continues to rise, fillers constantly melt, and the ablation layer is relatively continuous and the structure is more compact, thereby increasing the anti-ablated ability of the insulation material.

3.3 *XRD analysis*

To explore the changes in the composites after pyrolysis, the crystalline phases of the ceramified ash from the EPDM-based composites were analyzed by XRD. The pyrolysis phase transformations of samples 1, 2, 3, and 4 were illustrated in Figure 3.

From the XRD pattern, we can find that carbon layer exists in every patterns, and the difference between the four ablated products lie in the ceramic phase. The XRD pattern for sample 1 under 600°C showed SiO_2, KPO_3, $(AlPO_4)_2$. When the temperature rises to 800°C, some changes

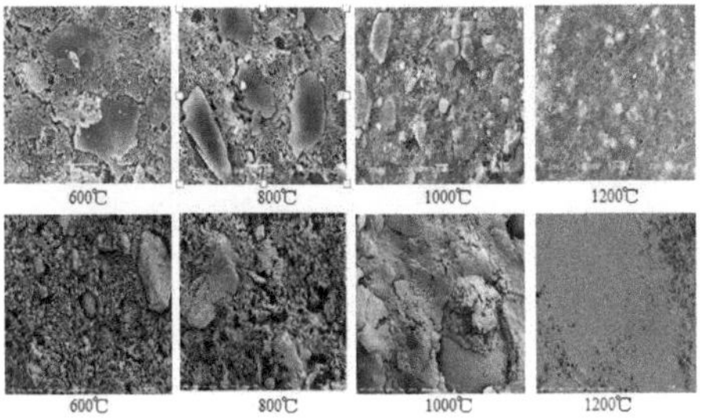

Figure 2. Ablated surface microstructure SEM spectrum of materials at different temperatures.

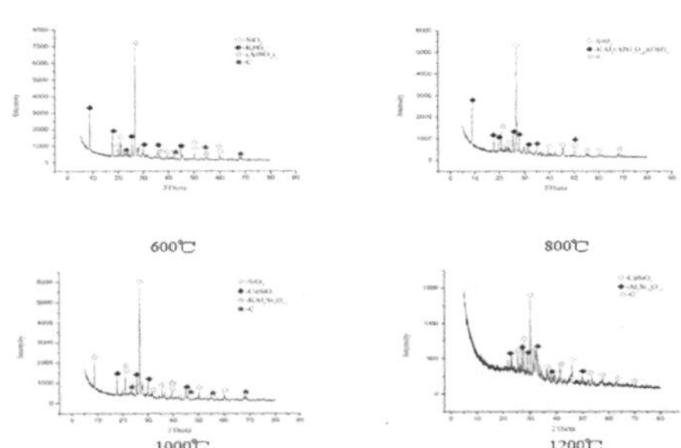

Figure 3. XRD patterns of ablation products at different temperatures.

occur in the crystalline phase, KPO_3, $(AlPO_4)_2$ disappears, and the main crystal phase stays as KAl_2 $(AlSi_3O_{10})$ $(OH)_2$. When the temperature is raised to high 1000°C, it produces a new main phase of $KAl_3Si_3O_{11}$, we can assume that at high-temperature aerobic environment crystalline phase transition occurs as follows:

$$2KAl_2(AlSi_3O_{10})(OH)_2 + O_2 \rightarrow 2KAl_3Si_3O_{11} + 2H_2O;$$
$$CaO + SiO_2 \rightarrow CaSiO_3$$

In 1200°C, we can see that there was a new crystalline phase transition occurring, the new crystal phase $Al_2Si_{16}O_{35}$. This transformation was described with further increase in temperature. The reaction between SiO_2 and Al_2O_3, resulted in a new crystalline phase. The reaction is:

$$16SiO_2 + Al_2O_3 \rightarrow Al_2Si_{16}O_{35}$$

By analyzing the XRD patterns, we can find that at low temperatures, the main reaction is that of a low-melting substance melting in the surface of the material. As the temperature rises, mica powder and oxygen react, and form a ceramic layer on the material surface. When the temperature is further raised, crystalline phase of ceramic layer also changes, forming a new crystalline phase.

4 CONCLUSION

From 800 to 1200°C in air, ceramic powders were partly fused and spread to the holes of the composites to form compact ceramics. The relatively compact ceramics stop oxygen further into the internal composites in order to restraint thermo-oxidative degradation of EPDM rubber in air atmosphere. The ceramic layer played a good seal on the micro cracks of the surface, at the same time, the structure of carbide layer was relatively continuous and so it can improve the performance of ablative resistance. The temperature had a great influence on forming the ceramic layer; it depended on the filler component.

REFERENCES

[1] Shi Zhenhai, L.I., Kezhi, L.I., Hejun, Tian Zhuo. Research Status and Application Advance of Heat Resistant Materials for Space Vehicle. Materials Review, 2007, 21(8):15–17.

[2] Fan Zhenxiang, Chen Haifeng, Zhang Changrui, Tang Gengping. Development of Thermal Protection Materials. Materials Review, 2005, 19(1):13–16.

[3] Mansouri J., Burford R.P., Cheng Y.B., et al. Formation of strong ceramified ash from silicone-based compositions [J]. Journal of Materials Science, 2005, 40(21):5741–5749.

[4] L.G., Hanu, G.P., Simon, Y.B., Cheng. Preferential orientation of muscovite in ceramifiable silicone composites. Materials Science and Engineering A, 2005, 398:180~187.

[5] Mansouri J., Wood A., Roberts K., Cheng Y-B., Burford R.P., Investigation of the ceramifying process of modified silicone–silicate compositions. J. Mater. Sci., 2007, 42:6046~6055.

[6] Alexander G., Chen Y.B., Robert S., et al. Ceramifying compositon for fire protection [P]. US 2007/0246240 A1.

[7] Alexander G., Chen Y.B., Robert S., et al. Fire-resistant silicone polymer compositions [P]. US 7652090B2.

[8] Shao Haibin, Zhang Qitu, Wu Li, Wang Tingwei. Preparation and properties of ceramifying silicone rubber [J]. Journal of Nanjing University of technology, 2011, 33(1):48–51.

[9] Hanu L.G., Simon G.P., Mansouri J., et al. Development of polymer-ceramic composites for improved fire resistance [J]. Journal of Materials Processing Technology, 2004, 153(1):401–407.

[10] L.M., Su. Preparation and Study on Thermal Stability of Porcelainized Methyl vinyl silicone rubber. Changsha: Central South University, 2011.

[11] Hanu L.G., Simon G.P., Cheng Y.B., Thermal stability and flam ability of silicone polymer composites [J]. Polymer Degradation and Stability, 2006, 91 (6):1373–1379.

[12] Mansouri J., Burford R.P., Cheng Y.B. Pyrolysis behavior of silicone-based ceramifying composites [J]. Materials Science and Engineering A, 2006, 425(1):7–14.

[13] Su Liu-mei, Cui Chang-hua, Shang Yong-jia, Zheng Feng. Research progress of ceramifiable polymer fireproof composite [J]. Materials Science and Engineering of Powder Metallurgy, 2009, 14(5):290–294.

[14] Singh S., Guchhait P.K., Bandyopadhyay G.G., et al. Development of polyimide-nano silica filled EPDM based light rocket motor insulator compound: Influence of polyimide-nano silica loading on thermal, ablation. And mechanical properties [J]. Composites Part A Applied Science and Manufacturing, 2012, 44:8–15.

[15] Chen Dehong, H.E., Yongzhu, Ling Ling, H.E., Biyan, Zhu Xuewen. Ceramifiable Thermal Insulation: A Kind of Nano-Montmorillonite/EPDM Rubber Composite [J]. Aerospace Materials & Technology, 2014 (3):25–30.

Material Science and Environmental Engineering – Chen (Ed.)

Visible light-driven floatable photocatalyst of B–N–TiO_2/EP composites

X. Wang, J.Y. Huang, J. Zhang, X.J. Wang & J.F. Zhao
State Key Laboratory of Pollution Control and Resource Reuse, College of Environmental Science and Engineering, Tongji University, Shanghai, China

ABSTRACT: Floating photocatalysts of B–N codoped TiO_2 grafted on expanded perlite (B–N–TiO_2/EP) were prepared by a facile sol-gel method. The catalysts were characterized by TG-DTA, XRD, N_2 adsorption-desorption, SEM, and UV-vis-DRS. The results showed that we could effectively obtain the photocatalysts with high BET surface area and porosity by modifying B doping contents in B–N–TiO_2/EP. Increasing the B doping contents would inhibit the transformation of anatase TiO_2 to rutile phase. Compared with that of N-TiO_2/EP, the absorption band edge of B–N–TiO_2/EP exhibits an evident red-shift and the absorption intensity of visible region increases obviously. The enhanced photodegradation rate of $B_{0.57}$-N-TiO_2/EP could reach 94% in 3 h under visible light irradiation. Moreover, the floating photocatalysts could be easily separated and reused, showing a great deal of potential for practical applications in environmental cleanup and solar energy conversion.

Keywords: B–N codoping; TiO_2; floating; visible light; photocatalyst

1 INTRODUCTION

Titanium dioxide (TiO_2) has been considered as the most promising photocatalyst for the oxidation of organic contaminants in wastewaters (Amorisco et al., 2006). However, conventional TiO_2 photocatalyst can only be activated under UV-light irradiation due to its wide band gap (e.g. $Eg \approx 3.2$ eV for anatase), resulting in a low photo-electronic transition efficiency since UV radiation accounts for only 5% of the total solar spectrum compared to the visible region (~45%) (Wen et al., 2011; Xiang et al., 2011; Yu et al., 2009). Thus, the possibility to extend the photoresponse capacity of TiO_2 to the visible region and be able to utilize a higher portion of the solar spectrum has become a topic of great interest among the scientific community. To unravel this problem, one of the most common and promising approaches is doping with impurities. In the approaches dealing with doping impurities, metals (Fe, Cr, Ni, Co or Ag) (Choi et al., 2009; Gong et al., 2012) or non-metals (N, F, S, C or B) (Gopalakrishnan et al., 2011; Pelaez et al., 2009) were utilized as dopants to narrow the band gap of TiO_2 materials and decrease the required activation energy. However, the use of metal doped TiO_2 for water treatment can lead to possible toxicity due to leakage of the metal ion into the finished water, diminishing its quality (Pelaez et al., 2010). Nonmetal doping of TiO_2 has been treated as an effective means to expand the light response range. Among non-metal-doped TiO_2 materials, co-doped TiO_2 usually shows higher photocatalytic activity in visible range because of the merits benefited from each dopant. As the co-doped elements, B–N co-doping is considered as a much effective way (Zhou et al., 2011). In et al. (2007) reported that B–N co-doped TiO_2 exhibits enhanced photocatalytic activity under UV and visible lights irradiation, which is probably attributed to the synergistic effect between boron and nitrogen, resulting in the narrowing of the band gap. Gopal et al. (Gopal et al., 2008) reported that B could fill an oxygen vacancy in the form of B- in B-TiO_2 and N could serve as a paramagnetic probe for the geometric and electronic structure of other dopants in the lattice as diamagnetic species.

Although the modified TiO_2 semiconductor is the most used for its photocatalytic activity, high stability, non-toxicity, and inexpensiveness, the interest is much more focused nowadays on the synthesis of new photocatalysts to overcome its limitations in the application. Some of the important limitations are (1) modified TiO_2 are always in the form of powder, it always sinks or suspends into the solution which decreases the using rate of light and (2) TiO_2 powder is difficult to recycle, easy to agglomerate, and causes a problem of separation from the solution (Wang et al., 2013). To overcome the limitations, much attention has been paid to the development of supported TiO_2 catalysts. In this respect, different types of supports for TiO_2 have been tested including activated carbon (Liu et al., 2007), clay (Belessi et al., 2007), silica

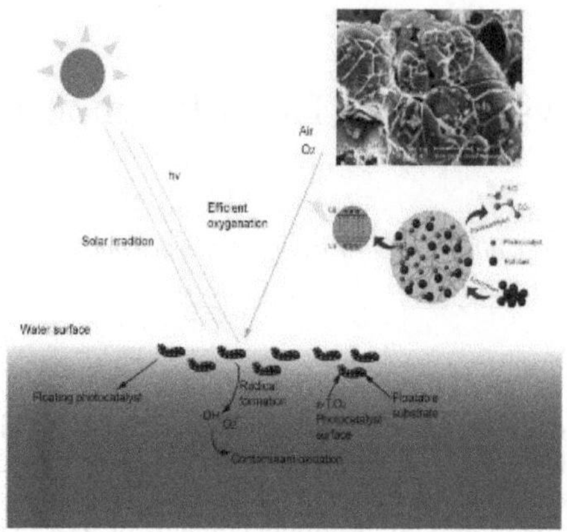

Figure 1. Schematic representation of a floating photocatalyst.

(Wang et al., 2005), and zeolite et al. (Wang et al., 2012). But these supporting substrates would still sink to the bottom of solution without mechanical agitation. Recently, some studies (Machado et al., 2006; Magalhães and Lago, 2009) developed a new concept of "floating photocatalysts", which is the TiO_2 photocatalyst synthesized on the surface of a floatable substrate (Fig. 1). The floatable photocatalysts are especially interesting for solar remediation of non-stirred and non-oxygenated reservoirs since the process maximizes the: (1) illumination/light utilization (due to their floatable ability), (2) oxygenation of the photocatalyst by the proximity with air/water interface. The optimization of illumination and oxygenation should result in higher rates of radical formation and oxidation efficiencies (Magalhães et al., 2011).

In this paper, expanded perlite particles were applied as carrier to support B–N codoped TiO_2 powder due to its floatable feature, porous structure, light mass and physical mechanics properties. Sol-gel method was employed to synthesize B–N–TiO_2/EP composite. The as-synthesized B–N–TiO_2/EP composite was characterized by SEM, TG-DTA, XRD, FT-IR, UV-vis DRS, and nitrogen adsorption analyses for Brunauer-Emmett-Teller (BET) specific surface area and mesoporous size distribution. The photocatalytic activity of this floatable photocatalysts was investigated under visible light using Rhodamine B as the pollutant model.

2 EXPERIMENTAL

2.1 *Preparation of B–N–TiO_2/EP composite*

A sol-gel preparation of B–N–TiO_2/EP was performed as follows: 18 mL tetrabutyl titanate and H_3BO_3 (0.36, 0.72, 1.8, 3.6 g corresponding to 0.11, 0.23, 0.57, 1.14 of B/Ti molar ratios) were dissolved completely into 50 mL anhydrous ethanol with stirring. Whilst still stirring, a solution containing 1.5 mL HCl (12 mol/L) and 10 mL PEG-400 (1 wt.%) were added drop by drop into the above solution to obtain solution A. Simultaneously, the solution was strongly stirred to a solution. 3 g pretreated EP was added into the solution and stirred to uniformly. Following this, 3.6 g urea was dissolved into 4.5 mL deionized water with stirring to obtain solution B. Then solution B was added to solution A while stirring to get the mixture. The resultant mixture was dispersed with ultrasonic for 1 h until a white gel was obtained. The gel was aged at room temperature for 24 h then dried at 105°C for 12 h to gain the xerogel. The resultant xerogel was crushed to obtain fine powder and further calcined at 550°C for 2 h to obtain the catalyst. In addition, N-TiO_2/EP was prepared according to the above procedure in the absence of H_3BO_3 for comparison.

2.2 *Characterization*

The TG-DTA was analyzed by Shimadzu TGA-50. For crystal structure analysis of the prepared samples, X-Ray Diffraction (XRD) analysis was carried out on a Bruker D8 ADVANCE (German) X-ray diffractometer with Cu Kα radiation (40 kV, 40 mA) with a 0.01° step and 2.5 s step time over the range $10° < 2\theta < 90°$. Nitrogen adsorption-desorption isotherms were used to determine BET surface area and pore size distribution (Micrometritics, ASAP 2020). The morphology of the synthesized materials were observed initially using scanning electron microscopy (SEM, Hitachi S4700) with the working distance of 5–12 mm and an accelerating voltage of 20 KeV. For the characterization of the light absorption features and band-gap determinations, Diffuse Reflectance Spectra (DRS) of the particles were measured in the range 200–800 nm on a UV-vis-NIR scanning spectrophotometer (Shimadzu UV-2550, Japan) equipped with an integral sphere using $BaSO_4$ as a reference. The photocatalyst powder was placed in the sample holder on an integrated sphere for the reflectance measurements.

2.3 *Photocatalytic evaluation with Rhodamine B under visible light*

The photocatalytic activities of B–N–TiO_2/EP composites were evaluated by decomposing Rhodamine B (RhB) dye under visible light irradiation at room temperature. A floating-bed photoreactor including a floating-bed of as-synthesized photocatalyst was applied in this work. Xe lamp

(XE-JY500, 500 W) with an UV cut off filter (1 M sodium nitrite solution, λ > 400 nm) was used as visible light source and irradiated from the top. Aeration and recirculation mechanisms would promote a better mass diffusion and cause the support to change their position and face the light. For the photocatalysis experiments, the concentration of RhB solution was 2.5 mg/L and the obtained photocatalyst was added at the ratio of 2 g/L. Prior to irradiation, the mixture was magnetically stirred for 30 min in the dark to reach an adsorption-desorption equilibrium. At predetermined time of 30 min intervals (30–180 min) after 30 min pre-adsorption, the samples were taken out and filtered using 0.45 μm membrane filter for analysis.

3 RESULTS AND DISCUSSION

3.1 *Characteristics of B–N–TiO$_2$/EP composite*

The analysis of TG-DTA is performed to investigate the decomposition behavior of the precursor powders due to heat treatment in N_2. The TG curve of TiO_2/EP particles show three weight loss stages. The first event occurred in the region of 50–150°C, which probably involved the desorption of physically adsorbed water and ethanol on the catalysts surface (Wang et al., 2014), leading to the major weight loss of 2%. The second and third events occurred between the temperature ranges of calcination temperature at 150–350°C and 350–500°C, recognized as fast weight loss stage and slow weight loss stage, respectively. While DTA curve showed the exothermic peak at ~210°C for TiO_2/EP samples. As discussed previously, this peak may be due to the decomposition of the organic components and then due to the formation of TiO_2 film and evolution of H_2O and CO_2 gas (Huo et al., 2010). The differential thermal of B–N–TiO_2/EP precursor in the range of 300 to 700°C was less than that of N-TiO_2/EP precursor.

The XRD patterns of the as-synthesized B–N–TiO_2/EP samples were presented in Figure 2. Comparing the XRD pattern of B–N–TiO_2/EP with different boron doping contents, there is no rutile peak in the sample of $B_{1.14}$-N-TiO_2/EP, whereas rutile appears in the samples of $B_{0.11}$-N-TiO_2/EP, $B_{0.23}$-N-TiO_2/EP, and $B_{0.57}$-N-TiO_2/EP. It is important to note that the increasing amount of boron dopant inhibits the transformation of anatase TiO_2 to rutile phase, which agrees with other literatures (Yuan et al., 2011; Zhang et al., 2014b). Even though, in this research the crystal phases of B–N–TiO_2/EP consisted of predominantly anatase. Besides the diffraction peak of anatase TiO_2, one weak peak at around 30.6° observed for $B_{1.14}$-N-TiO_2/EP, was attributed to the formation of B_2O_3 on the surface of TiO_2.

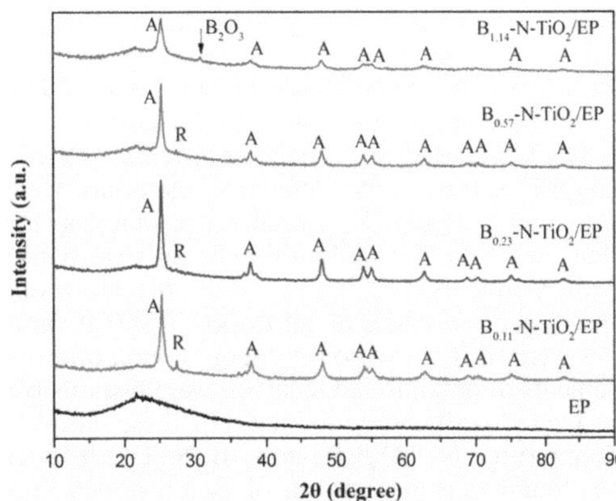

Figure 2. XRD patterns for $B_{0.11}$-N-TiO_2/EP (a) $B_{0.23}$-N-TiO_2/EP (b) $B_{0.57}$-N-TiO_2/EP (c) $B_{1.14}$-N-TiO_2/EP (d). A is stand for antase and B is stand for rutile.

Table 1. The characterization results of different samples.

Samples	S_{BET} (m^2/g)	Pore size (nm)
EP	5.174	8.528
$B_{0.11}$-N-TiO_2/EP	17.189	15.898
$B_{0.23}$-N-TiO_2/EP	28.597	27.314
$B_{0.57}$-N-TiO_2/EP	33.649	29.553
$B_{1.14}$-N-TiO_2/EP	13.049	12.503

The specific surface area and pore size of B–N–TiO_2/EP were presented in Table 1. Comparing the BET surface area of raw EP and B–N–TiO_2/EP, it is noted that B–N–TiO_2 grafted on EP was able to increase the BET surface area through a controlling boron doping amount. When B/Ti ratio increased from 0.11 to 0.57, the BET surface area increased from 17.189 to 33.649 m^2/g. However, the BET surface area decreased when boron doping amount continued to increase. Considering the BET surface of the sample, the mesopores are believed to be formed by the agglomeration and connection of adjacent nanoparticles in the sample (Tian et al., 2008). This network nanostructure offers more efficient transportation for reactant molecules to the active sites, which are expected to enhance the photocatalytic activity (Antonelli and Ying, 1995).

The morphologies of B–N–TiO_2/EP were observed by SEM (Fig. 1). It can be seen by the presence of TiO_2 particles located on the surface of EP that the particle size was in the range of nanometer scale. For B–N–TiO_2/EP composites, it can be observed TiO_2 agglomerated in higher amounts with most of the EP particle surface.

These particles were strongly attached to the surface of the composite and presented spongy-like surface, which could increase adsorption sites. The results were in agreement with BET analysis.

The UV-vis diffuse reflectance spectra of B–N–TiO_2/EP samples and reference materials were presented in Figure 3. The reference samples (EP blank and TiO_2/EP) had no significant absorbance in the visible light region ($\lambda > 420$ nm). However, the absorption edges of all doped TiO_2/EP samples were shifted to a lower energy region and the elements doped into TiO2 lattice were responsible for the red-shift absorption band of these samples. Comparing N-TiO_2/EP and B–N–TiO_2/EP, it was found that the amount of boron doped into N-TiO_2/EP affected the optical absorption of the sample in visible regions. Either too much or too less doping amount would inhibit the absorption of visible light. The samples of $B_{0.23}$-N-TiO_2/EP and $B_{0.57}$-N-TiO_2/EP showed the stronger photoabsorption in the visible light region than N-TiO_2/EP, which could imply promise for higher visible light photocatalytic activity.

3.2 *Photocatalytic activity of B–N–TiO_2/EP composite under visible light*

The photocatalytic degradation of RhB on the as-synthesized B–N–TiO_2/EP was evaluated under visible light irradiation and compared to the degradation activity of EP and N-TiO_2/EP, as displayed in Figure 4. It is clearly shown that RhB was hardly diminished under visible light irradiation with EP only. When boron content was at a lower level (B/Ti = 0.11 and B/Ti = 0.23), B–N–TiO_2/EP exhibited a poorer performance on photodegradation of RhB than N-TiO_2/EP. However, the photocatalytic activities of B–N–TiO_2/EP increased gradually with an increase in boron content. When B/N ratio was 0.57, the sample of $B_{0.57}$-N-TiO_2/EP exhibited the best performance on photodegradation of RhB (94%) after 3 h irradiation. The high photocatalytic activity of $B_{0.57}$-N-TiO_2/EP is attributed to its high specific surface (33.649 m^2/g) and the formation of Ti-N-B-O structure. The high specific area is responsible for providing strong adsorption ability toward target molecules and thus the generation of photoinduced electron-hole pairs of active sites while the Ti–N–B–O structure effectively narrowed the band gap and then easily generated electron-hole pairs. But the excess amount of B–N species would also lead to the recombination of electrons and holes.

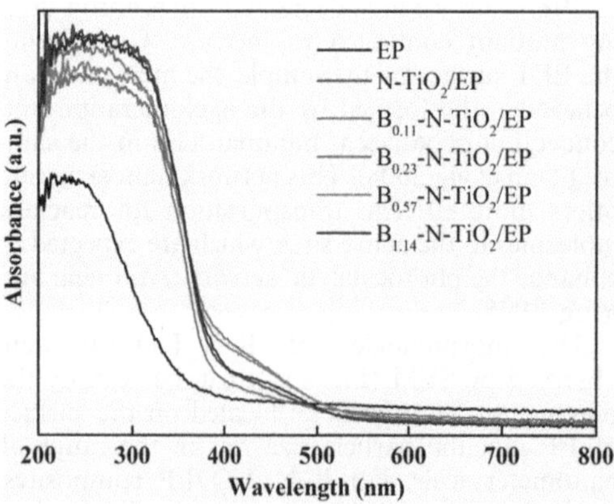

Figure 3. UV-vis absorbance of synthetic samples.

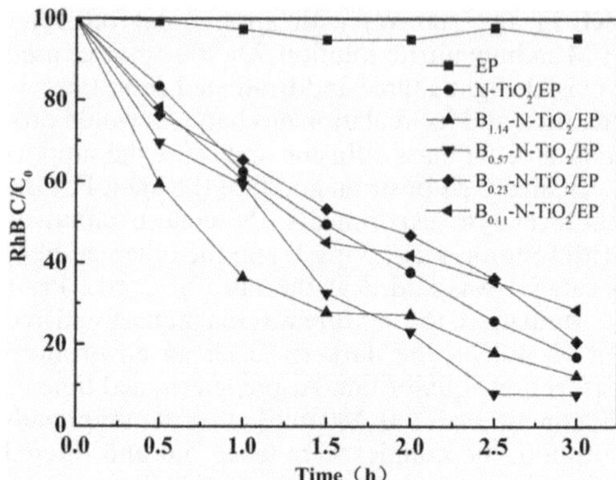

Figure 4. The photodegradation for RhB on B–N–TiO_2/EP under visible light irradiation.

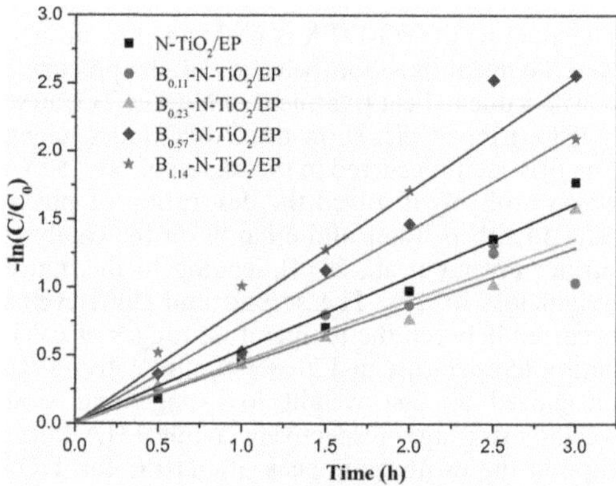

Figure 5. The pseudo-first-order kinetics of N-TiO_2/EP and B–N–TiO_2/EP under visible light irradiation.

Figure 5 showed the kinetic studies of the photodegradation of RhB over B–N–TiO_2/EP with different boron contents. It was observed that the photocatalytic reactions of RhB obeyed the pseudo-first-order kinetics according to the Langmuir-Hinselwood model and may be expressed as:

$$-\ln\left(\frac{C}{C_0}\right) = kt \quad (1)$$

where k is the observed rate constant, C_0 is the equilibrium concentration of RhB and C is the concentration at time t. Furthermore, according to the kinetics model, the k of different B–N–TiO_2/EP samples and N-TiO_2/EP were calculated and the results demonstrated that the optimum boron doping content was $B_{0.57}$-N-TiO_2/EP with a k value of 0.84960 h^{-1}, which was 1.5 times larger than that of N-TiO_2/EP.

Several factors may account for the high photocatalytic activity of the as-synthesized floating B–N–TiO_2/EP photocatalysts. First, B–N–TiO_2/EP has a favorable band structure that ensures it efficiently utilizes the visible light, as well as has strong redox ability during photocatalytic degradation. Second, it is generally accepted that the catalytic process is mainly due to the adsorption of reactant molecules to the catalytic surface (Ma et al., 2009). The as-synthesized B–N–TiO_2/EP photocatalysts' higher specific surface area with mesoporous feature allows more efficient transport for the reactant molecules to access its active sites, hence enhancing the photocatalytic efficiency. Third, high photocatalytic efficiency is related to the floating feature of EP, which provides an efficient conversion pathway for visible light (Bell, 2003; Magalhães et al., 2011).

4 CONCLUSION

In summary, floating N-B-TiO_2/EP photocatalyts with different doping contents of boron were successfully prepared by using a facile sol-gel method and confirmed by TG-DTA, XRD, BET, SEM, and UV-vis-DRS measurements. The B–N–TiO_2/EP has a specific surface area of 13~34 m^2/g, which implies a typical mesopore structure. The presence of boron and nitrogen species have been inferred to play a key role in extending the photoactivity to visible light region, effectively narrowing the band gap, and inhibiting the transformation of anatase TiO_2 to rutile phase. The investigation of photocatalytic ability showed that the B–N–TiO_2/EP activity was greatly influenced by boron doping content. At an optimal B/Ti molar ratio of 0.57, the RhB degradation reached 94% within 3 h visible light irradiation, which indicated to be very promising photocatalysts and could be employed to remediate contaminated waters.

ACKNOWLEDGMENTS

This work was supported by the National Natural Science Foundation of China (No. 21277097, 51179127).

REFERENCES

[1] Amorisco, A., Losito, I., Carbonara, T., Palmisano, F., Zambonin, P., 2006. Photocatalytic degradation of phenyl-urea herbicides chlortoluron and chloroxuron: characterization of the by-products by liquid chromatography coupled to electrospray ionization tandem mass spectrometry. *Rapid Commun Mass Sp* 20, 1569–1576.

[2] Antonelli, D.M., Ying, J.Y., 1995. Synthesis of hexagonally packed mesoporous TiO_2 by a modified sol–gel method. *Angewandte Chemie International Edition in English* 34, 2014–2017.

[3] Belessi, V., Lambropoulou, D., Konstantinou, I., Katsoulidis, A., Pomonis, P., Petridis, D., Albanis, T., 2007. Structure and photocatalytic performance of TiO_2/clay nanocomposites for the degradation of dimethachlor. *Applied Catalysis B: Environmental* 73, 292–299.

[4] Bell, A.T., 2003. The impact of nanoscience on heterogeneous catalysis. Science 299, 1688–1691.

[5] Choi, J., Park, H., Hoffmann, M.R., 2009. Effects of single metal-ion doping on the visible-light photoreactivity of TiO_2. *The Journal of Physical Chemistry C* 114, 783–792.

[6] Gong, J., Pu, W., Yang, C., Zhang, J., 2012. A simple electrochemical oxidation method to prepare highly ordered Cr-doped titania nanotube arrays with promoted photoelectrochemical property. *Electrochim Acta* 68, 178–183.

[7] Gopal, N.O., Lo, H.H., Ke, S.C., 2008. Chemical state and environment of boron dopant in B, N-codoped anatase TiO_2 nanoparticles: An avenue for probing diamagnetic dopants in TiO_2 by electron paramagnetic resonance spectroscopy. *J Am Chem Soc* 130, 2760–2761.

[8] Gopalakrishnan, K., Joshi, H.M., Kumar, P., Panchakarla, L., Rao, C., 2011. Selectivity in the photocatalytic properties of the composites of TiO_2 nanoparticles with B-and N-doped graphenes. *Chem Phys Lett* 511, 304–308.

[9] Huo, P., Yan, Y., Li, S., Li, H., Huang, W., 2010. Floating photocatalysts of fly-ash cenospheres supported AgCl/TiO_2 films with enhanced Rhodamine B photodecomposition activity. *Desalination* 256, 196–200.

[10] In, S., Orlov, A., Berg, R., García, F., Pedrosa-Jimenez, S., Tikhov, M.S., Wright, D.S., Lambert, R.M., 2007. Effective visible light-activated B-doped and B, N-codoped TiO_2 photocatalysts. *J Am Chem Soc* 129, 13790–13791.

[11] Liu, Y., Yang, S., Hong, J., Sun, C., 2007. Low-temperature preparation and microwave photocatalytic activity study of TiO_2-mounted activated carbon. *Journal of hazardous materials* 142, 208–215.

[12] Ma, D., Huang, S., Chen, W., Hu, S., Shi, F., Fan, K., 2009. Self-assembled three-dimensional hierarchical umbilicate Bi_2WO_6 microspheres from nanoplates: controlled synthesis, photocatalytic activities, and wettability. *The Journal of Physical Chemistry C* 113, 4369–4374.
[13] Machado, L.C.R., Torchia, C.B., Lago, R.M., 2006. Floating photocatalysts based on TiO_2 supported on high surface area exfoliated vermiculite for water decontamination. *Catal Commun* 7, 538–541.
[14] Magalhães, F., Lago, R.M., 2009. Floating photocatalysts based on TiO_2 grafted on expanded polystyrene beads for the solar degradation of dyes. *Solar Energy* 83, 1521–1526.
[15] Magalhães, F., Moura, F.C., Lago, R.M., 2011. TiO_2/LDPE composites: A new floating photocatalyst for solar degradation of organic contaminants. *Desalination* 276, 266–271.
[16] Pelaez, M., de la Cruz, A.A., Stathatos, E., Falaras, P., Dionysiou, D.D., 2009. Visible light-activated N-F-codoped TiO_2 nanoparticles for the photocatalytic degradation of microcystin-LR in water. *Catalysis Today* 144, 19–25.
[17] Pelaez, M., Falaras, P., Likodimos, V., Kontos, A.G., De la Cruz, A.A., O'shea, K., Dionysiou, D.D., 2010. Synthesis, structural characterization and evaluation of sol–gel-based NF-TiO_2 films with visible light-photoactivation for the removal of microcystin-LR. *Applied Catalysis B: Environmental* 99, 378–387.
[18] Sin, J.C., Lam, S.M., Satoshi, I., Lee, K.T., Mohamed, A.R., 2014. Sunlight photocatalytic activity enhancement and mechanism of novel europium-doped ZnO hierarchical micro/nanospheres for degradation of phenol. *Applied Catalysis B: Environmental* 148–149, 258–268.
[19] Tian, G., Fu, H., Jing, L., Xin, B., Pan, K., 2008. Preparation and characterization of stable biphase TiO_2 photocatalyst with high crystallinity, large surface area, and enhanced photoactivity. *The Journal of Physical Chemistry C* 112, 3083–3089.
[20] Wang, C., Shi, H., Li, Y., 2012. Synthesis and characterization of natural zeolite supported Cr-doped TiO_2 photocatalysts. *Appl Surf Sci* 258, 4328–4333.
[21] Wang, T., Yang, G., Liu, J., Yang, B., Ding, S., Yan, Z., Xiao, T., 2014. Orthogonal synthesis, structural characteristics, and enhanced visible-light photocatalysis of mesoporous Fe_2O_3/TiO_2 heterostructured microspheres. *Appl Surf Sci* 311, 314–323.
[22] Wang, X.T., Zhong, S.H., Xiao, X.F., 2005. Photocatalysis of ethane and carbon dioxide to produce hydrocarbon oxygenates over ZnO-TiO_2/SiO_2 catalyst. *Journal of Molecular Catalysis A: Chemical* 229, 87–93.
[23] Wang, X., Wu, Z., Wang, Y., Wang, W., Wang, X., Bu, Y., Zhao, J., 2013. Adsorption-photodegradation of humic acid in water by using ZnO coupled TiO_2/bamboo charcoal under visible light irradiation. *J Hazard Mater* 262, 16–24.
[24] Wen, C.Z., Jiang, H.B., Qiao, S.Z., Yang, H.G., Lu, G.Q.M., 2011. Synthesis of high-reactive facets dominated anatase TiO_2. *J Mater Chem* 21, 7052–7061.
[25] Xiang, Q., Yu, J., Wang, W., Jaroniec, M., 2011. Nitrogen self-doped nanosized TiO_2 sheets with exposed {001} facets for enhanced visible-light photocatalytic activity. *Chem. Commun.* 47, 6906–6908.
[26] Yu, J., Xiang, Q., Zhou, M., 2009. Preparation, characterization and visible-light-driven photocatalytic activity of Fe-doped titania nanorods and first-principles study for electronic structures. *Applied Catalysis B: Environmental* 90, 595–602.
[27] Yuan, J., Wang, E., Chen, Y., Yang, W., Yao, J., Cao, Y., 2011. Doping mode, band structure and photocatalytic mechanism of B–N-codoped TiO_2. *Appl Surf Sci* 257, 7335–7342.
[28] Zhang, K., Wang, X., He, T., Guo, X., Feng, Y., 2014b. Preparation and photocatalytic activity of B–N co-doped mesoporous TiO_2. *Powder Technology* 253, 608–613.
[29] Zhou, X., Peng, F., Wang, H., Yu, H., Yang, J., 2011. Effect of nitrogen-doping temperature on the structure and photocatalytic activity of the B, N-doped TiO_2. *J Solid State Chem* 184, 134–140.

Material Science and Environmental Engineering – Chen (Ed.)

Adsorption behavior of ACF/CNT composites for Cr (VI) from aqueous solution

L.P. Wang
Powder Metallurgy Research Institute, Central South University, Changsha, Hunan, China
Department of Biological and Environmental Engineering, Changsha University, Changsha, Hunan, China

M.Y. Zhang & Q.Z. Huang
Powder Metallurgy Research Institute, Central South University, Changsha, Hunan, China

Z.C. Huang
School of Minerals Processing and Bioengineering, Central South University, Changsha, Hunan, China

X.P. Long
Changsha Dingji Environmental Protection Engineering and Equipment Co. Ltd., Changsha, Hunan, China

ABSTRACT: ACF/CNT (activated carbon fiber/carbon nanotube) composites were characterized with Scanning Electron Microscope (SEM), N_2 adsorption-desorption curve and Fourier Transform Infrared Spectroscopy (FTIR) in the paper. The adsorption behavior of ACF/CNT composites for Cr (VI) from aqueous solution was investigated by the batch adsorption experiments. The results showed the equilibrium adsorption capacity decreased with increasing pH. The adsorption capacity gradually reduced with increment of adsorbent dose. The adsorption experimental data were well fitted with Freundlich adsorption isotherm equation. Furthermore, the adsorption kinetics data were well described by pseudo-second-order kinetic model. The adsorption rate was not only controlled by intra-particle diffusion. The values of ΔG were negative and the values of ΔH and E_a were caculated as 23.20 kJ/mol and 16.58 kJ/mol, respectively, which indicated the adsorption process, was a spontaneous and endothermic physisorption process. Electrostatic attraction was one of the most important adsorption mechanisms.

Keywords: ACF/CNT composites; Cr (VI); adsorption isotherms; adsorption kinetics; adsorption thermodynamics; electrostatic attraction

1 INTRODUCTION

Cr exists as Cr (III) and Cr (VI) in water environment. The toxicity of Cr is related to valence state. The toxicity of Cr (VI) is nearly 100 times bigger than Cr (III). The waste water of Cr (VI) is very poisonous to crops and human [Trezza M.A., 2003]. At present, the main treatment technologies of wastewater containing Cr (VI) include electrochemical reduction [Bhatti M.S., 2009], chemical reduction [Erdem M., 2004], membrane separation [Arslan G., 2009], biological method [Wen Yue, 2011] and adsorption [Li Jiansheng, 2008].

Through comparing and analyzing the above several treatment technologies, adsorption is thought as one of the most effective treatment technology for removing Cr (VI) from aqueous solution. ACFs have widely applied to the treatment of waste gas, but adsorption properties of ACFs for Cr (VI) are not so good because of their micropore structure. In order to improve adsorption ability of ACFs for Cr (VI) from aqueous solution, we prepared new adsorption material-ACF/CNT composites with mesoporous by modifying ACFs with chemical vapor deposition.

In this paper, the adsorption behavior of ACF/CNT composites for Cr (VI) from aqueous solution was discussed and the adsorption mechanism was obtained.

2 EXPERIMENTAL

2.1 *Materials and reagents*

New adsorbent-ACF/CNT composites were fabricated using activated carbon fibers as substrate, nickel nitrate as catalyst precursor, C_2H_2 as carbon source, H_2 as reducing gas and N_2 as carrier gas by chemical vapor deposition [Wang Liping, 2012]. All chemicals used in this study were of analytical-laboratory grade.

2.2 *Characterization methods*

The surface morphology of samples was characterized by FEI Nova Nano SEM230. NEXUS670 FTIR was used to measure surface groups. N_2 adsorption-desorption curve was gained by automated pore size analyzer (QUDRA-SORB SI). Cr (VI) concentrations were determined by spectrometry at the wavelength of maximum absorbance at 540 nm using UV-759 ultraviolet and visible spectrophotometer.

2.3 *Adsorption experimental*

Batch adsorption experiments were carried out using 250 mL glass bottles with prescribed ACF/CNT composites and 100 mL different Cr (VI) concentration solution. The glass bottles were sealed and put in a water-bathing constant temperature vibrator. Then, the glass bottles were shaken for a time at 120 r/min. The pH values of the solution were adjusted with 2 M HCl or 0.5 M NaOH and measured with pH meter. The adsorption amount of Cr (VI) at equilibrium was figured out by the following equation.

$$q_e = (C_0 - C_e)V/m \qquad (1)$$

where C_0 and C_e represent the liquid-phase concentrations of Cr (VI) at initiate and equilibrium, respectively (mg/L); V expresses the volume of the solution (L) and m is the dose of adsorbent (g).

3 RESULTS AND DISCUSSION

3.1 *Characterizations*

The apparent morphology and FTIR are shown in our previous paper [Wang Liping, 2012]. It can be seen that carbon nanotubes well distributed on the surface of activated carbon fibers and the peaks at 3447 cm^{-1} and 3741 cm^{-1} were ascribed to the existence of –OH from the paper. Figure 1 is N_2 adsorption-desorption isotherm curve and suggests that N_2 adsorption-desorption curve belonged to the type IV isotherm and the pore structure of ACF/CNT composites mainly contained micropore and mesopore. The adsorption hysteresis loop was gradational type from type B to type A, which indicated the transform of pore shape from slit shape to tube shape.

3.2 *Contrast of adsorption efficiencies of ACFs and ACF/CNT composites*

Figure 2 expresses the adsorption efficiencies of ACFs and ACF/CNT composites for Cr (VI). As can be seen from Figure 2 that the adsorption

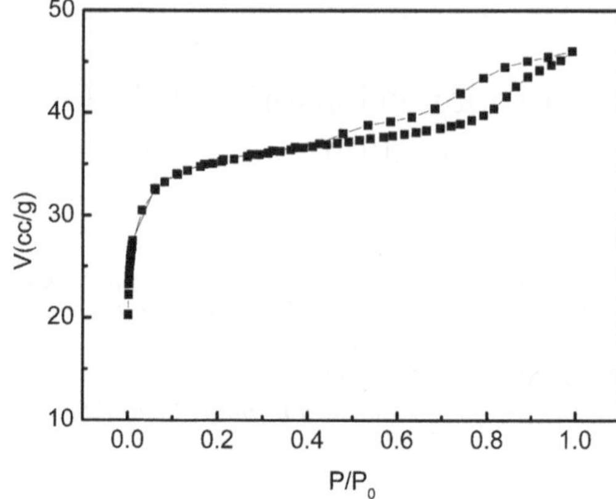

Figure 1. N_2 adsorption-desorption isotherm curve of ACF/CNT composites.

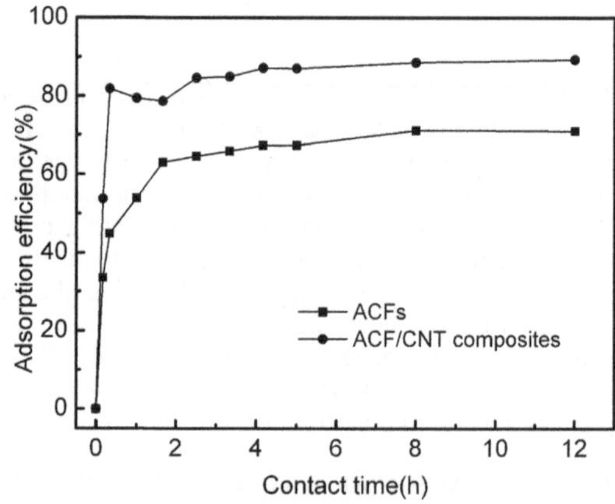

Figure 2. Adsorption efficiencies of ACFs and ACF/CNT composites (conditions: temp., 25 °C; adsorbate conc., 5 mg/L; adsorbent dose, 1 g/L; stirring rate, 120 rpm; pH, 5.5).

efficiency of ACF/CNT composites for Cr (VI) reached 84.89% and it was far greater than that of ACFs whose adsorption efficiency was 65.87% when contact time was 3.3 h.

3.3 *Effect of pH*

pH is an important factor for controlling the adsorption process. The effect of pH on the equilibrium adsorption capacity is shown in Figure 3. It was observed that adsorption capacity decreased with increase in pH. The maximum adsorption capacity occurred at pH 2, where the adsorbed quantity was 4.992 mg/g and decreased to 2.945 mg/g when pH increased to 9. It made clear that H^+ could enhance the adsorption ability of ACF/CNT composites

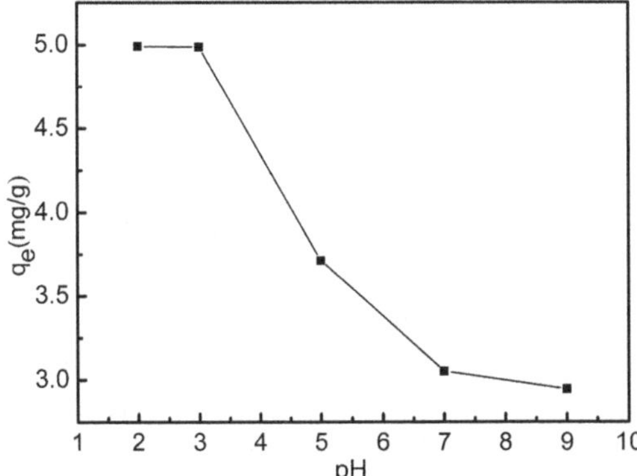

Figure 3. Effect of pH (conditions: temp., 25 °C; adsorbate conc., 5 mg/L; adsorbent dose, 1 g/L; stirring rate, 120 rpm; contact time, 300 min).

for Cr (VI). In general, Cr (VI) exists in ionic forms as CrO_4^{2-}, $HCrO_4^{2-}$ and $Cr_2O_7^{2-}$ in aqueous solution. The chemical equilibrium equation among the three negative ions is as follows:

$$2CrO_4^{2-} + 2H^+ \Leftrightarrow 2HCrO_4^- \Leftrightarrow Cr_2O_7^{2-} + H_2O \quad (2)$$

Cr (VI) mainly exists as CrO_4^{2-} in alkaline or neutral solution. However, CrO_4^{2-} becomes $HCrO_4^{2-}$ at first, then transforms into $Cr_2O_7^{2-}$ with decrease in pH. When pH is less than 2, the predominant existence form of Cr (VI) is $Cr_2O_7^{2-}$ in aqueous solution.

At lower pH (pH < 5), the surface of ACF/CNT composites was positively charged because of large number of H^+ and Cr (VI) existed as negative ion in aqueous solution, which strengthened electrostatic attraction between Cr (VI) and the surface of ACF/CNT composites and the adsorption ability was improved. In addition, redox reactions happened between Cr (VI) and carbon atoms on the surface of ACF/CNT composites and the end of carbon nanotube was opened, which greatly increased the specific surface area of ACF/CNT composites. Therefore, the adsorption ability was enhanced. However, in alkaline solution, the mass of –OH on the surface obviously increased, which resulted in negative charge characteristics of the surface of ACF/CNT composites. Therefore, the removal efficiency of ACF/CNT composites was lower for Cr (VI) due to electrostatic repulsion between adsorbate and adsorbent.

3.4 *Effect of adsorbent dose*

Experiments were carried out to find the effect of adsorbent dose on the adsorption capacity and the results are shown in Figure 4. It can be seen from Figure 4, the adsorption capacity for Cr (VI) gradually decreased from 5.139 mg/g to 2.010 mg/g with increase in adsorbent dose from 0.05 g to 0.20 g. Although more actives sites were available with increase in adsorbent dose, the adsorption saturation easily achieved with lower adsorbent dose when the mass of Cr (VI) was equal in solution.

3.5 *Adsorption isotherms*

The equilibrium adsorption isotherm is fundamental in describing the interactive behavior between adsorbate and adsorbent. Figure 5a expresses the adsorption isotherm origin data plot of ACF/CNT composites for Cr (VI). It can be seen from Figure 5a that the equilibrium adsorption capacity increased with the rise of equilibrium adsorption concentration because the increase of initial Cr (VI) concentration enhanced diffusion concentration gradient and diffusion driving force. Figure 5a also suggests that the equilibrium adsorption capacity improved when the solution temperature went up, which indicated the adsorption was an endothermic process. Langmuir and Freundlich adsorption models are commonly used to describe the adsorption isotherms.

Langmuir isotherm adsorption equation is expressed as follows:

$$\frac{C_e}{q_e} = \frac{1}{q_0 K_L} + \frac{1}{q_0} C_e \quad (3)$$

where q_e is the amount of Cr (VI) adsorbed per unit weight of the adsorbent at equilibrium *(mg/g)*, q_0 and K_L are Langmuir constants relating to adsorption capacity and adsorption rate, respectively.

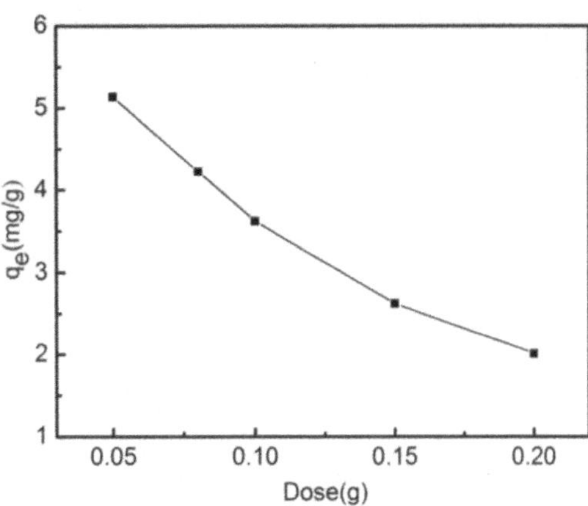

Figure 4. Effect of adsorbent dose (conditions: temp., 25 °C; adsorbate conc., 5 mg/L; pH, 5.5; stirring rate, 120 rpm; contact time, 300 min).

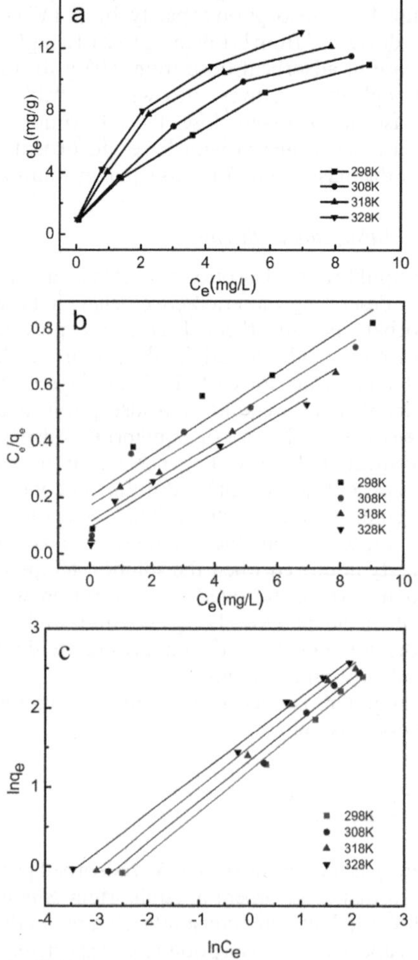

Figure 5. Adsorption isotherms: (a) adsorption isotherm initial data graph; (b) Langmuir adsorption isotherm and (c) Freundlich adsorption isotherm (conditions: pH, 5.5; stirring rate, 120 rpm; contact time, 300 min; adsorbent dose, 1 g/L).

Freundlich adsorption isotherm is represented by the following equation:

$$q_e = K_F(C_e)^{1/n} \tag{4}$$

where K_F and n are Freundlich constants.

Its linearized form was showed by below equation:

$$\ln q_e = \ln K_F + \frac{1}{n}\ln C_e \tag{5}$$

The experimental data in Figure 5a were processed according to Langmuir and Freundlich isotherm adsorption models and shown in Figure 5b and 5c, respectively. Figure 5b expresses Langmuir adsorption isotherm and the calculated q_0 and K_L are shown in Table 1 by the regression analysis. The calculated R_L is listed in Table 1 at different experimental temperatures. Table 1 shows that $R^2 > 0.86$ when adsorption data were fitted with Langmuir adsorption model. The equilibrium adsorption capacity increased from 13.5 mg/g to 15.1 mg/g when the solution temperature rose from 298 K to 328 K. It is thus clear that temperature rise was in favor of adsorption of ACF/CNT composites for Cr (VI) from aqueous solution. $0 < R_L < 1$ indicated the adsorption was favorable, in which R_L express the separation factor.

Freundlich adsorption isotherm is illustrated in Figure 5c and the calculated n and K_F are listed in Table 1. The experimental data were well described by Freundlich isotherm adsorption model due to $R^2 > 0.98$. The increment of K_F with the rise of solution temperature made clear that the higher temperature was prone to adsorption, which was consistent with the results predicted using Langmuir adsorption model.

3.6 *Adsorption kinetics*

Figure 6 shows the adsorption kinetics curves of ACF/CNT composites for Cr (VI). From Figure 6a,

Table 1. The adsorption isotherms parameters of ACF/CNT composites for Cr (VI).

		T (K)			
Isotherms	Parameters	298	308	318	328
Langmuir	q_0 (mg/g)	13.5	14.3	14.4	15.1
	K_L (L/mg)	0.364	0.409	0.612	0.713
	K_L (× 10^4 L/mol)	1.893	2.127	3.182	3.708
	R^2	0.8659	0.8664	0.9463	0.9343
	R_L	0.1208	0.1089	0.0755	0.0655
Freundlich	K_F (mg/g (L/mg)$^{1/n}$)	3.35	3.81	4.52	5.21
	n	1.89	1.92	1.93	2.04
	R^2	0.9945	0.9862	0.9902	0.9950

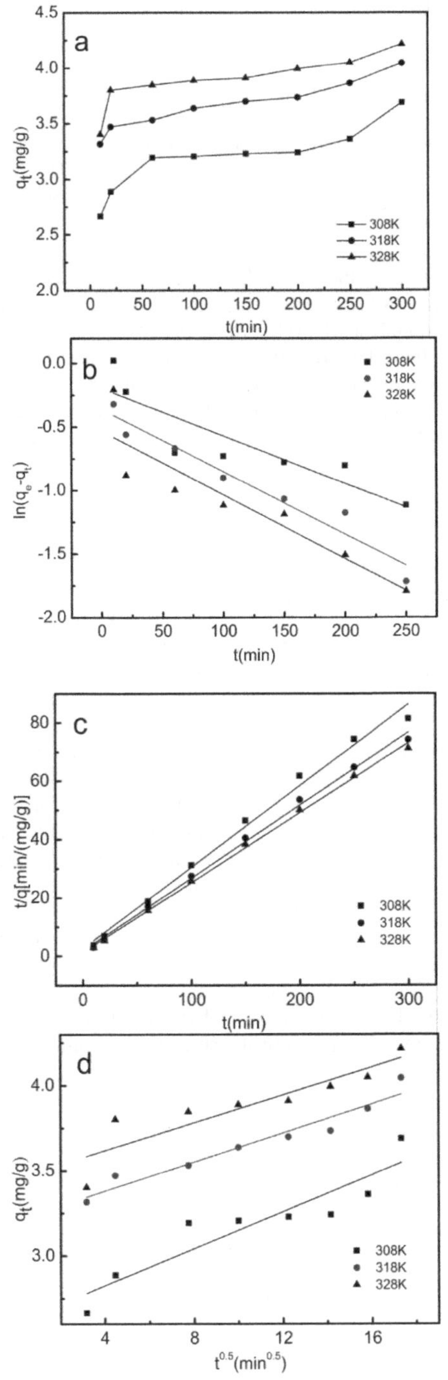

Figure 6. Adsorption kinetics: (a) adsorption kinetics original data plot; (b) pseudo first-order kinetics plot; (c) pseudo second-order kinetics plot and (d) intra-particle diffusion model plot (conditions: initial concentration: 5 mg/L; pH, 5.5; stirring rate, 120 rpm; adsorbent dose, 1 g/L).

it can be seen that the rate of adsorption was very fast during the first 10 min and became slower after that. This may be attributed to the fact that at first, there were more active sites on the surface of adsorbent and greater diffusion driving force of Cr (VI), which resulted in bigger diffusion rate and Cr (VI) ions got to the surface of adsorbent and were adsorbed in a short period. Active sites got less and less with the adsorption proceeding and the electrostatic repulsion enhanced between Cr (VI) ions on the surface of adsorbent and Cr (VI) ions in solution, so the diffusion rate declined, thus the rate of adsorption decreased. When the contact time was 300 min, active sites got saturated and the adsorption reached equilibrium. In order to clarify adsorption mechanism, pseudo-first order, pseudo-second order and intra-particle diffusion kinetics models were used to fit the experimental data.

The pseudo first-order kinetics model is given as follows:

$$\ln(q_e - q_t) = \ln q_e - k_1 t \tag{6}$$

where q_t is the amount of Cr (VI) adsorbed per unit weight of the adsorbent at t (mg/g); t is contact time; k_1 is the adsorption rate constant of pseudo-first order kinetics. By plotting a graph of ln $(q_e - q_t)$ versus t, k_1 can be estimated from the slopes.

The pseudo second-order kinetics model is described by following linear equation:

$$\frac{t}{q_t} = \frac{1}{k_2 q_e^2} + \frac{t}{q_e} \tag{7}$$

where k_2 is the adsorption rate constant of pseudo-second-order kinetics [g/(mg·min)]. The pseudo-second-order kinetics model can be applied if the plot of t/q_t against t is linear.

The intra-particle diffusion kinetics model is expressed as:

$$q_t = k_t t^{\frac{1}{2}} + C \tag{8}$$

where k_t is intra-particle diffusion rate constant [mg/(g·min$^{0.5}$)], C is a related constant to the thickness of boundary. The plots of q_t against $t^{0.5}$ are drawn based on the intra-particle diffusion kinetics model. When the regression curve is linear and the line passes through the origin, the rate-controlling step is intra-particle diffusion, while when the regression curve is linear but the line does not pass through the origin, then the intra-particle diffusion is not the only rate-controlling step. The regression curve is not linear, indicating that adsorption process may be controlled by one step or a few steps.

The pseudo first-order kinetics plot is presented in Figure 6b. The regressed parameters are listed in Table 2. Table 2 showed the liner relation between lg ($q_e - q_t$) and t was high due to $R^2 > 0.74$.

Figure 6c indicates the pseudo second-order kinetics model curve of adsorption process. The regressed parameters are listed in Table 2. A comparison between the pseudo first-order and pseudo second-order kinetic relevant coefficients suggested that adsorption of Cr (VI) by ACF/CNT composites followed closely the pseudo second-order kinetics rather than the pseudo first-order kinetics. This is obvious from Table 2, since the values of q_e obtained from pseudo second-order kinetics equation were very close to the experimental q_e values, while that from pseudo first-order kinetics equation did not agree with the experimental values.

The intra-particle diffusion model curve is shown in Figure 6d. The regressed relevant parameters are put in Table 2. From Figure 6d and Table 2, the plot of q_t to $t^{0.5}$ was linear with $R^2 > 0.77$ and the line not passed origin, indicating the intra-particle diffusion is not the only rate-controlling step. When solution temperatures changed from 308 K to 328 K, the values of C increased from 2.610 to 3.445, meaning diffusion boundary layer thickness increasing.

3.7 *Adsorption thermodynamics*

Generally, adsorption thermodynamics are used for expounding whether the adsorption process was spontaneous or not, endothermic and exothermic, physisorption or chemisorption. Thermodynamic parameters contain ΔG (kJ/mol), ΔH (kJ/mol) and ΔS [J/(mol·K)], which can be determined by the following equations:

$$\Delta G = -RT \ln K_L \tag{9}$$

$$\Delta G = \Delta H - T\Delta S \tag{10}$$

where R is the gas constant, 8.314 J/(mol·K), T is the absolute temperature (K), K_L is the Langmuir constants (L/mol). The values of ΔH and ΔS are determined from intercept and slope of the plot of ΔG versus T according to equation (10). If the value of ΔG is negative, indicating that adsorption is spontaneous in nature, or not. If the value of ΔH is negative, suggesting adsorption reaction is exothermic, or is endothermic. If/ΔH/<40 kJ/mol, adsorption process is physisorption, or is chemisorption. The positive value of ΔS demonstrates adsorption increases the disorder and chaos degree of system, conversely leads to the more order of system.

The relevant equation between the rate constants K_2 of the pseudo second-order model and temperature is obtained by changing Arrhenius equation, as follows:

$$\ln k_2 = \ln k_0 - \frac{E_a}{RT} \tag{11}$$

where k_2 is the rate constants of pseudo second-order kinetics model [g/(mg·min)], k_0 is the constant, E_a is activation energy (kJ/mol), T is the temperature (K) and R is the gas constant [8.314 J/(mol·K)]. The value of E_a is decided from slope of the plot of $\ln K_2$ versus $1/T$. Physisorption is considered with $E_a = 0 \sim 40$ kJ·mol^{-1}, while adsorption is regarded as chemisorption with $E_a > 40$ kJ·mol^{-1}.

The linear regression plot for ΔG versus T is shown in Figure 7a on the basis of equation (10). Thermodynamic parameters calculated are listed Table 3. It can be seen from Table 3 that the values of ΔG were negative and minished from −25.52 kJ/mol to −28.69 kJ/mol with increasing temperatures from 308 K to 328 K, suggesting the adsorption process was spontaneous and endothermic reaction. The value of ΔH was 23.20 kJ/mol (<40 kJ/mol), accounting for adsorption reaction was endothermic physisorption. The value of ΔS

Table 2. The adsorption kinetics parameters of ACF/CNT composites for Cr (VI).

		T (K)		
Kinetics models	Parameters	308	318	328
Pseudo first-order	q_{exp} (mg/g)	3.688	4.043	4.215
	q_e (mg/g)	0.822	0.698	0.588
	k_1 (min^{-1})	0.0037	0.0049	0.0050
	R^2	0.7404	0.9374	0.8034
Pseudo second-order	q_e (mg/g)	3.58	4.00	4.18
	k_2 [g/(mg·min)]	0.0311	0.0360	0.0462
	R^2	0.9910	0.9966	0.9981
Intra-particle diffusion model	k_t	0.0541	0.0426	0.0408
	C	2.610	3.210	3.455
	R^2	0.8281	0.9282	0.7751

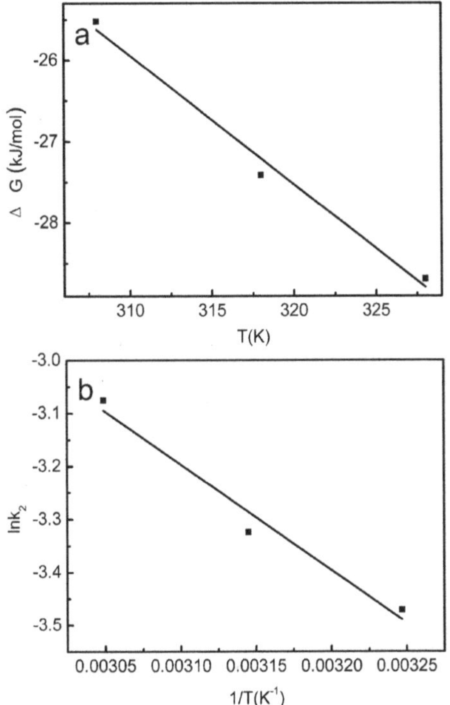

Figure 7. The regression plot for adsorption thermodynamics.

Table 3. The adsorption thermodynamics parameters of ACF/CNT composites for Cr (VI).

T (K)	ΔG (kJ/mol)	ΔH (kJ/mol)	ΔS (J/mol·K)	R^2
308	−25.52	23.20	158.5	0.9756
318	−27.41	23.20	158.5	0.9756
328	−28.69	23.20	158.5	0.9756

was 158.5 J/mol·K and the positive value of ΔS corresponds to the increased degree of freedom in the solid/liquid interface as a result of adsorption of Cr (VI) ions.

The regressed plot of in K_2 to $1/T$ is shown in Figure 7b according to equation (11). The value of E_a was calculated as 16.58 kJ·mol^{-1}, demonstrating adsorption was fast physisorption.

4 CONCLUSIONS

When pH rose from 2 to 3, the equilibrium adsorption capacity decreased from 4.992 mg/g to 4.987 mg/g, then the decrease rate increased. The adsorption capacity declined with rising adsorbent dose and reduced from 5.139 mg/g to 2.010 mg/g when adsorbent dose increased from 0.05 g to 0.2 g. The adsorption experimental data were well fitted with Freundlich equation. Furthermore, the kinetic data were well described by pseudo second-order kinetic model. The adsorption rate was not only controlled by intra-particle diffusion, but also by film diffusion.

The values of ΔG were negative, suggesting adsorption process was spontaneous. The endothermic physisorption process was considered because of the value of ΔH being 23.20 kJ/mol. The value of ΔS was 158.5 J/mol·K, indicating adsorption process was driven by entropy. Elecrtostatic attraction was one of the most important adsorption mechanisms.

ACKNOWLEDGEMENT

We acknowledge the financial supports of the Nation Natural Science Foundation of China (No. 51404041), the Natural Science Foundation of Hunan Province, China (No. 2015JJ3016), the Science and Technology Plan Project of Changsha (k1407017-11), Project Funded by China Postdoctoral Science Foundation (2015M570691) and the Postdoctoral Foundation of Central South University.

REFERENCES

[1] Arslan G. et al. 2009. Facilitated transport of Cr (VI) through a novel activated composite membrane containing Cyanex 923 as a carrier. *Journal of Membrane Science* 337(1–2): 224–231.

[2] Bhatti M.S. et al. 2009. Electrocoagulation removal of Cr (VI) from simulated wastewater using response surface methodology. *Journal of Hazardous Materials* 172(2–3): 839–846.

[3] Erdem M. et al. 2004. Cr (VI) reduction in aqueous solutions by siderite. *Journal of Hazardous Materials* 113(1–3): 217–222.

[4] Li Jiansheng et al. 2008. Synthesis, amino-functionalization of mesoporous silica and its adsorption of Cr (VI). *Journal of Colloid and Interface Science* 318(2): 309–314.

[5] Trezza M.A. et al. 2003. Hydration study of limestone blended cement in the presence of hazardous wastes containing Cr (VI). *Cement and Concrete Research* 33(7): 1039–1045.

[6] Wang Liping et al. 2012. Adsorption of methylene blue from aqueous solution on modified ACFs by chemical vapor deposition. *Chemical Engineering Journal* 189–190: 168–174.

[7] Wen Yue et al. 2011. Adsorption of Cr (VI) from aqueous solutions using chitosan-coated fly ash composite as biosorbent. *Chemical Engineering Journal* 175(0): 110–116.

Material Science and Environmental Engineering – Chen (Ed.)
© 2016 Taylor & Francis Group, London, ISBN 978-1-138-02938-5

Metal organic framework-derived ZnO/C composite for photocatalytic oxidative degradation

Y.R. Guo & W. Chang
College of Environmental and Chemical Engineering, Xi'an Polytechnic University, Shaanxi, China

W. Zhang
School of Science, Xi'an Jiaotong University, Shaanxi, China

ABSTRACT: ZnO/C composite, which was derived from a typical metal-organic framework (MOF-5), was synthesized by heat treatment in nitrogen gas. This composite material possesses remarkable photocatalytic efficiency in photodegradation of organic dyes. The crystal structure, morphology, components and spectral properties of the samples were characterized by X-Ray Diffraction (XRD), Scanning Electron Microscope (SEM), surface area analysis (BET), X-ray energy spectroscopy (EDS) and UV–Vis Diffuse Reflection Spectroscopy (UVDRS). The results showed that the carbon generated in situ retained high specific surface area and porous structure of MOF-5, which improved adsorption capacity greatly. ZnO/C composite material possessed rather remarkable photocatalytic efficiency in photodegradation of Methylene Blue (MB) compared with Degussa P25 and nano ZnO powder.

Keywords: metal organic framework; doping; adsorption; photocatalytic

1 INTRODUCTION

Metal Organic Frameworks, commonly recognized as "soft" analogues of zeolites, are new class of nanoporous materials. Due to their large specific surface areas, adjustable pore size and controllable properties, and acceptable thermal stability, MOFs are promising candidates for a wide range of application, such as, adsorption; separation; catalysis and sensing [1–2].

MOF-5 was the most representative of Metal Organic-frameworks, which has been synthesized by Yaghi's group [3] at the University of California in 1999. In 2003, the research group also reported it outstanding performance in terms of hydrogen storage [4]. After those MOFs has attracted a lot of interest in different area. Several studies of the semiconducting properties of MOFs have been reported. Seifert [5], et al. suggested in a theoretical work that MOFs are semiconductors with band gaps between 1.0 and 5.5 eV. In contrast, Bordiga [6], et al. reported that MOFs behave as metal oxide quantum microporous semiconductors with a wide band gap upon excitation with UV light, and they undergo charge separation, after which the excited delocalized electrons remain stable on the microsecond time scale [7]. So applications of MOFs to semiconductors have been explored gradually.

This paper presents a heat temperature treatment of MOF-5 under nitrogen to obtain ZnO/C composite material. It showed that amorphous carbon produced by decomposition of MOF-5 enhanced adsorption and photocatalytic activity.

2 EXPERIMENTS

2.1 *Materials preparation*

Zinc nitrate tetrahydrate (2.613 g, 9 mmol) and terephthalic acid (0.553 g, 3 mmol) were dissolved in 100 mL dimethyl formamide (DMF) in a 250 mL, three-necked round-bottomed flask. The reaction mixture was heated in 105 °C for 24 h in oil bath to yield MOF-5 white power. The reaction flask was then removed from the oil bath and cooled to room temperature. The obtained cubic crystals were repeatedly washed with DMF and anhydrous chloroform, and then soaked in anhydrous chloroform for 24 h.

The prepared MOF-5 was transferred into tube furnace and heat-treated at 700 °C under nitrogen gas with a heating rate of 1 °C/min. After reaching target temperature, keeping for 60 min and then cool to room temperature. The final product (ZnO/C) was black-colored powder [8].

2.2 *Photodegradation experiments*

To study the photocatalytic activity of the samples, the finely ground photocatalyst (10 mg) was in contact with 10 mL of methylene blue solution (2×10^{-4} mol/L) under continuous stirring in the dark with 1 h to reach adsorption equilibrium. A 300 W Xenon lamp was used to pass light to the solution through a quartz tube. Distance between the light source and the samples were 8 cm. At regular irradiation time intervals, aliquots of solution were withdrawn and centrifuged. The supernatant of these solutions was used to monitor the change of methylene blue UV–Vis absorbance spectrum [9].

2.3 *Characterization of materials*

The samples' crystal structures were confirmed using X-Ray Diffraction (XRD). A Cu-target X-ray tube was set to 40 KV and 30.0 mA Scanning Electron Microscopy (SEM) was performed on scanning electron microscope with an X-ray Energy Dispersive Spectroscopy (EDS) accessory (JEOL JSM-7000F, Japan). Scanning was performed on a sample previously sputter coated with a thin layer of platinum to avoid charging. The Brunauer-Emmett-Teller (BET) surface area of sample was measured with an automatic gas adsorption analyzer (Quantachrome Autosorb IQ, USA). The samples were pretreated at 200 °C under vacuum for 14 h. N_2 was then used to measure the adsorption and desorption on samples at different relative pressures and a temperature of −196 °C to determine the surface area and pore diameter distribution. UV–Vis diffuse reflection spectroscopy (DRUVS) was collected to get the capability of light absorption and photocatalytic activity was evaluated by photodegrading on a commercial photochemical reaction instrument (BL-GHX-V) and measuring the dye's photodegradation ratio by ultraviolet spectrophotometer (UV-2450).

3 RESULT AND DISCUSSION

The power X-ray diffraction patterns of MOF-5 and ZnO/C samples are shown in Figure 1. Figure 1a indicated that MOF-5 had high crystallinity, which was consistent with the computer simulation XRD patterns except the intensities. ZnO/C samples derived from MOF-5 under the nitrogen gas show typical peaks, attributed to ZnO with a hexagonal wurtzite structure (PDF#36-1451) (Fig. 1b). The typical peak position was (100) (002) (101). There are no peaks for carbon but it does exist in a certain amount proved by EDS analysis (Fig. 2b) which indicated that the carbonaceous specials is amorphous phase.

SEM image of MOF-5 shows the typical cubic crystal (Fig. 2a), while ZnO/C composite material derived from it remains the same structure as well (Fig. 2b). The EDS analysis based on these two

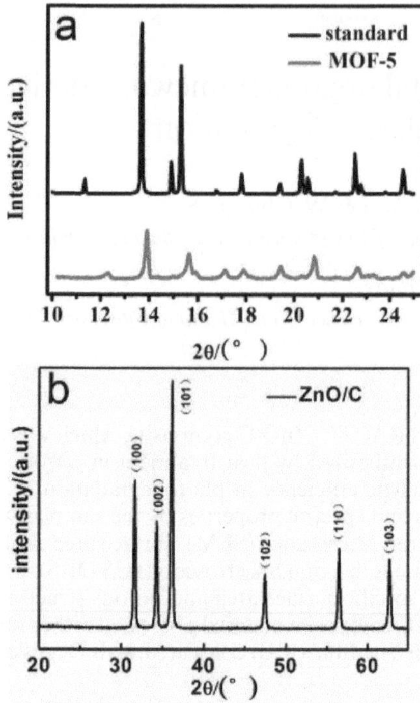

Figure 1. Power XRD patterns of MOF-5 and ZnO/C.

samples have great differents on elements ratio (Fig. 2c, 2d). MOF-5 has much higher oxygen ratio and ZnO/C has rather high amount of carbon. It indicates that heat-treatment changed part of the linker into small particles of amorphous carbon which existed on the surface mostly.

The specific surface area and pore size were analyzed by Brunauer–Emmett-Teller (BET) procedure from the nitrogen adsorption isotherms. Based on isotherms calculated specific surface areas were 766 m^2/g for MOF-5 and 390 m^2/g for the derived ZnO/C composite material. Although ZnO/C had lower specific surface area than MOF-5, it carried on the basic structure characteristics of MOF-5, which was much higher than that of P25 (commercial TiO_2, power, SSA 50 ± 15 m^2/g) and commercial nano ZnO power (SSA 50 ± 10 m^2/g). The average pore size of MOF-5 composite was approximately 0.6 nm, and the micropore volume was 0.477 cc/g. These results indicated that ZnO/C was a microporous material. It explained why ZnO/C had excellent adsorption properties than P25 and nano ZnO.

The UV–Vis diffuse reflection spectroscopy of MOF-5 and ZnO/C were shown in Figure 3. The MOF-5 was almost no absorption in the visible region from 400 to 800 nm, while ZnO/C had a rather strong adsorption, which could be attributed

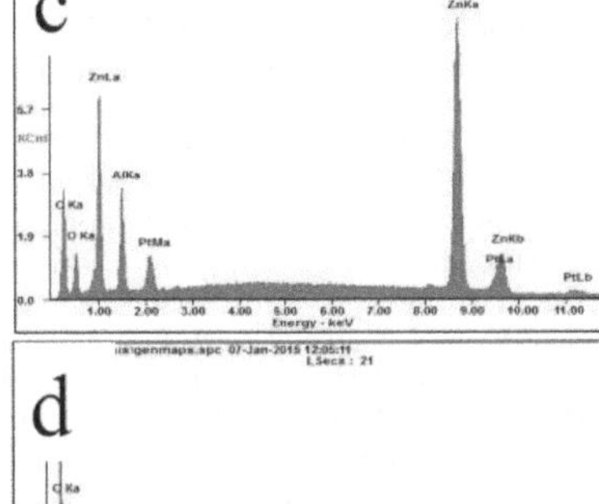

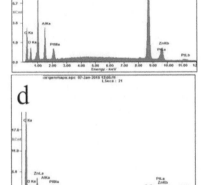

Figure 2. FE-SEM images of (a) MOF-5; (b) ZnO/C. EDS image of (c) MOF-5; (d) ZnO/C.

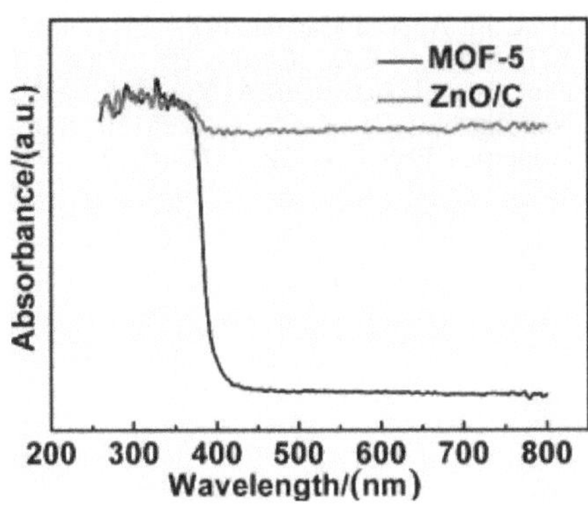

Figure 3. UV–Vis diffuse reflection spectroscopy of MOF-5 and ZnO/C.

to production of amorphous phase carbon after heat-treatment.

In order to evaluate the photocatalytic efficiency of the prepared ZnO/C, photocatalytic oxidation experiments were carried out using nano ZnO and P25 as references. ZnO/C has much higher specific surface areas than ZnO and P25, which can affect the results greatly if not eliminate the effects of adsorption. Therefore, adsorption equilibrium for Methylene Blue (MB) has been tested first. As shown in Figure 4, ZnO and P25 showed little adsorption of the MB while ZnO/C uptaked the dye molecules a lot. It reached to adsorption equilibrium about 1 h.

Photodegradation activity of ZnO, ZnO/C and P25 samples was studied by the photodegradation of methylene blue under 300 W Xenon, and the photodegradation ratio were compared with those of ZnO, ZnO/C and P25, as shown in Figure 5. The result showed that heat-treatment ZnO/C and

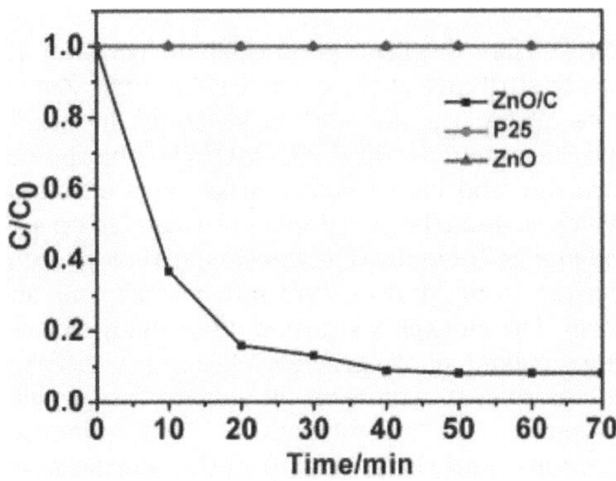

Figure 4. The MB adsorption equilibrium of ZnO, ZnO/C and P25.

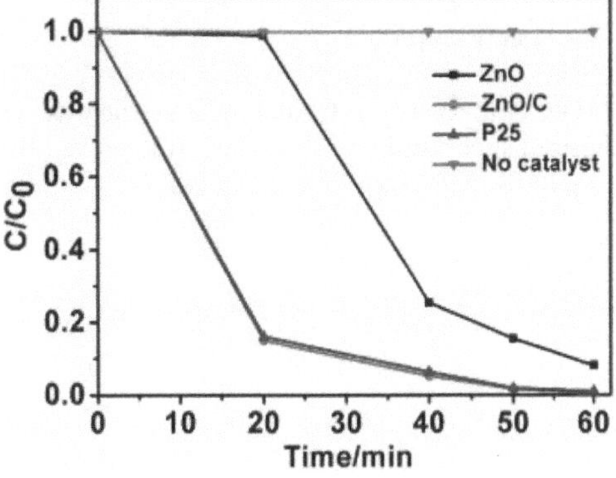

Figure 5. Photocatalytic degradation of the samples ZnO; ZnO/C; P25 for MB.

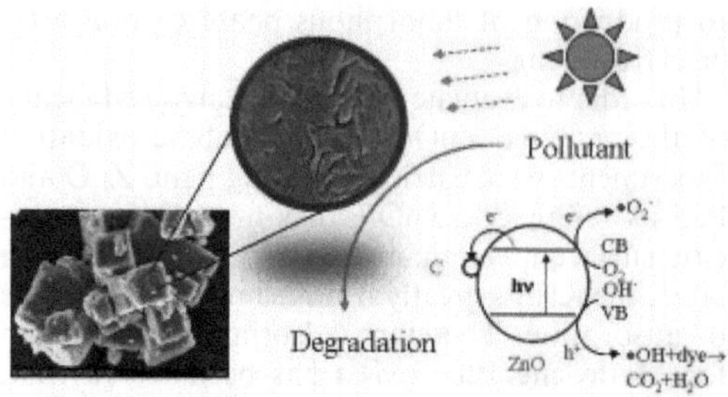

Figure 6. Scheme diagram of the photodegradation mechanism of ZnO/C.

has excellent photocatalytic activity and higher photodegradation ratio. Such as ZnO/C is 94.8%, P25 is 93.7%, but ZnO is 74.7% in 40 min.

4 THE PRINCIPLE OF PHOTOCATALYTIC DEGRADATION

ZnO/C has excellent photocatalytic performance can be attributed as these two reasons. First, ZnO/C with amorphous the carbon generated in situ by heat-treatment retained the MOF-5 cubic porous structure and high specific surface area effectively which facilitated dye molecules be catalyzed on activated sites. Secondly, The ZnO and carbon particles derived from MOF-5 were rather small and uniform. The amorphous carbon offers many electron traps, making photo-generated electrons transferred to amorphous carbon on the surface easily, which suppressed the recombination of photo-induced electrons and holes significantly. Furthermore, amorphous carbon protected ZnO to a certain extent and reduced photo-corrosion of ZnO greatly. Photocatalytic oxidized degradation mechanism [10] can be explained as follows in Figure 6.

5 CONCLUSION

ZnO/C composite photocatalytic material was synthesized successfully in this study. It is proved that the control of high temperature heat treatment of MOF-5 materials, which was not only keep original morphology and achieve higher specific surface area, but also generated ZnO in situ. The amorphous carbon suppressed recombination of photoinduced electrons and holes, and then enhanced the photocatalytic degradation rate. Its adsorption properties and catalytic activity was both superior to the same conditions of P25.

REFERENCES

[1] J.L.C. Rowsell, O.M. Yaghi. 2004. Metal-organic frame ZnO works: a new class of porous materials, Microporous and Mesoporous Materials (73): 3–14.

[2] J.R. Li, H.C. Zhou. 2009. Selective gas adsorption and separation in metal-organic frameworks, Chemical Society Reviews (38): 1477–1504.

[3] Li H., Eddaoudi M., O. Keeffe M. 1999. Design and synthesis of an exceptionally stable and highly porous metal-organic framework. Nature 402 (6759): 276–279.

[4] Rosi N.L, Eckert J. 2003. Hydrogen storage in microporous Metal-organic Framework. Science 300 (5622): 1127–1129.

[5] A. Kuc, A. Enyashin. 2007. Metal-organic framework: structural, energetic, electronic, and mechanical properties, J. Phys. Chem (111): 8179–8186.

[6] S. Bordiga, C. Lamberti, G. Ricchiardi. 2004. Electronic and vibrational properties of a MOF-5 metal-organic framework: ZnO quantum dot behavior, Chem. Commun. 2300–2301.

[7] F.X.L.I. Xamena, A. Corma. H. Garcia. 2007. Applications for Metal-Organic Frameworks (MOFs) as quantum dot semiconductors. J. Phys. Chem (111): 80–85.

[8] S.J. Yang, J.H. Im, T. Kim. 2011. MOF-derived ZnO and ZnO@C composites with high photocatalytic activity and adsorption capacity. Journal of Hazardous Materials (186): 376–382.

[9] G.L. Drisko, A. Zelcer, X.D. Wang. 2012. Synthesis and Photocatalytic Activity of Titania Monoliths Prepared with Controlled Macroand Mesopore Structure. Applied Materials (4): 4123–4130.

[10] Y.H. Zheng, C.Q. Chen, Y.Y. Zhan. 2008. Photocatalytic Activity of Ag/ZnO Heterostructure Nanocatalyst: Correlation between Structure and Property. J. Phys. C. (112): 10773–10777.

Material Science and Environmental Engineering – Chen (Ed.)
© 2016 Taylor & Francis Group, London, ISBN 978-1-138-02938-5

Study on the microstructure and properties of PTFE particles/Ni-based alloy composite coatings

Y.X. Shang
College of Materials Science and Chemistry Engineering, Harbin Engineering University, Harbin, China

X.M. Zhang
College of Materials Science and Chemistry Engineering, Harbin Engineering University, Harbin, China
College of Mechanical and Electrical Engineering, Harbin Engineering University, Harbin, China

ABSTRACT: In order to improve the friction and wear properties of nickel-based alloy coatings, PTFE particles/Ni-based alloy composite coatings with different contents of PTFE were prepared on the surface of low carbon steel by electro-brush plating method, and the coatings were heat-treated at different temperature. Surface morphologies, surface element contents and properties of the composite coatings were studied by SEM, EDS, friction-wear test and hardness test. Test and analysis results indicated that the composite coatings were obviously fined and compacted because of containing PTFE particles. In addition, properties of the composite coatings in wear resistance and hardness were improved through suitable heat treatment.

Keywords: composite coating; microstructure; heat-treated; friction-wear properties

1 INTRODUCTION

The wear resistant coatings produced by plating, such as hard chromium, electroless nickel and composite coatings of nickel or cobalt with ceramic particles, are used to improve the resistance to abrasive wear of softer metals and their alloys [1~5]. Composite electro-brush plating is a new method that makes the metal ions and the insoluble particles co-deposit to the substrate surface [6]. The function of insoluble particles in the composite coatings is consolidating. The composite structure of insoluble particles and metals makes the composite coatings possess favorable wear resistance and corrosion resistance [7].

In this paper, Ni/PTFE self-lubricating composite coatings were prepared using the method of brush plating, and their friction and wear performances were investigated.

2 EXPERIMENTAL

2.1 *Materials and pretreatment*

Low carbon steel was used as the material of coating substrates with deposition area of 15 × 15 mm. Before deposition, all substrates were mechanically polished by using waterproof abrasive paper and then pretreated with alcohol and ultrasonic treatment.

Quick nickel bath was used as substrate plating, and the compositions were listed in Table 1. White Polytetrafluoroethylene (PTFE) powers with average grain size of 5 μm were used as additives. After several attempts, the dispersion process was finally determined, that is, using sodium lauryl sulfate as surfactant, Op-10 as emulsifier, add 10 g/L surfactant, 10~16 ml/L emulsifier and a certain

Table 1. Compositions for different pretreatment baths and coatings baths.

Quick nickel electroplating bath	$NiSO_4$	180~205 g/L
	CH_3COONH	0~50 g/L
	$(NH_4)_2C_2O_4$	0~20 g/L
	$NH_3{\cdot}H_2O$	0~300 ml/L
Special nickel electroplating bath	$NiSO_4$	250~260 g/L
	$C_6H_5O_7(NH_4)_3$	0~50 g/L
	CH_3COONH	50~100 ml/L
	HCl	0~60 ml/L
Electrocleaning solution	NaOH	20~25 g/L
	Na_2CO_3	20~25 g/L
	NaCl	2~3 g/L
	Na_3PO_4	40~60 g/L
Activating solution 1	HCl	20~50 ml/L
	NaCl	50~150 g/L
Activating solution 2	$C_6H_5Na_3O_7$	50~150 g/L
	$C_6H_8O_7$	50~120 g/L
	$NiCl_2$	0~10 g/L

amount of PTFE in quick nickel bath, after ultrasonic dispersion 60 min, suspension viscous liquid with PTFE contents of 10 g/L, 20 g/L, 30 g/L and 40 g/L respectively were obtained.

2.2 *Preparation of coatings*

2.2.1 *Pretreatment*

1. DSC-30-QA type DC power supply and ZDB-1 type plating pen were used. Rectangular cold pressing graphite was used as anode, and packed the anode into flat medical absorbent cotton of 0.3 cm.
2. Electrocleaning process: The substrates were washed by acetone to remove oil and rust. Plating pen was connected to power anode, and substrate to cathode. After dipped in electrocleaning solution (the composition was shown in Table 1), the plating pen was used to wipe the substrate with the relative speed between anode and cathode as 4~6 m/min, then the substrate was rinsed with distilled water.
3. Activating treatment: The power was then reversed, after dipped in activating solution 1 (the composition was shown in Table 1), the plating pen was used to wipe the substrate with the relative speed between anode and cathode as 6~8 m/min, until the color of substrate changed into dark blond. The steps above were repeated using activating solution 2 (the composition was shown in Table 1) until the color of substrate was a little brighter.
4. Bottom coating brushing process: After dipped in special nickel bath (the composition was shown in Table 1), the plating pen was used to wipe the substrate with relative speed between anode and cathode as 8~10 m/min for 30 s under electrode positive, then the substrate was rinsed with distilled water.

2.2.2 *Preparation of PTFE/Ni composite coatings*

After stirred a few times in different plating solution respectively, the plating pen was used to brush the substrate coated bottom coating for 10 min, with relative speed between anode and cathode as 8~10 m/min under electrode positive, then the substrate was rinsed with distilled water and blow dried.

2.2.3 *Heat treatment of coatings*

The samples were vacuum sealed in quartz tubes, then heated to certain temperature and kept at that temperature for 1 hour, then cooled in air.

2.3 *Detection method*

Morphologies and microstructures of the coatings in top were carried out on FEI Quant200 Scanning Electron Microscope (SEM). The compositions of the samples were detected by Genesis Energy Dispersion Spectrometer (EDS) attached to SEM. X-Ray Diffraction (X, Pert Pro)was used to analyze the phase of composite coatings with diffraction angle of 20~80 degree. Friction-wear tester was used to test the friction and wear properties of samples at room temperature in 200 gram-force.

3 RESULTS AND DISCUSSION

3.1 *Phase and properties of the coatings before heat treatment*

3.1.1 *Surface morphologies of the coatings*

Figure 1 shows SEM surface morphologies of nickel and Ni/PTFE composite coatings prepared by brush electroplating. The observation shows that the morphologies of all the coatings made up of nodules clusters with each cluster composed of multiple small cells. Nodules of PTFE/Ni composite coatings (b, c, d and e) are relatively uniform, flat, fine and close. The difference in nodules shape leads to relatively dense microstructure of PTFE/Ni composite coating. That is related to the addition of PTFE particles, which has changed the

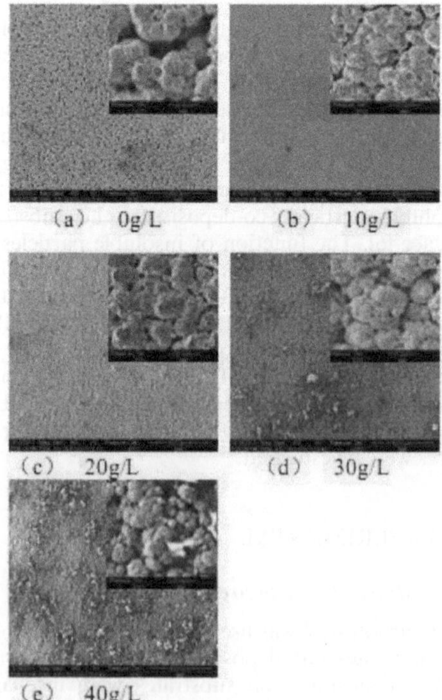

Figure 1. SEM of the surface of coatings prepared from the plating solution with different content of PTFE.

growth behavior of nickel coating. Without addition of PTFE particles, just deposited layer was thicker than previously deposited one and the middle part of each layer was thicker than the two sides, which was caused by tip effect. Therefore nodules of nickel coating are spherical. For the morphologies of PTFE/Ni composite coatings with different contents of PTFE, morphology of the composite coating prepared from the bath with 20 g/L PTFE particles is denser than that of prepared from the bath with 10 g/L PTFE particles. On the other hand, the size of agglomerate particles becomes bigger and looser with the increase of PTFE in bath when PTFE is between 20 g/L to 40 g/L. The densest and minimum of surface agglomerate particles morphology appears at the surface of the composite coating prepared from the bath with 20 g/L content of PTFE. What's more, the white agglomerate particles increase on PTFE/Ni composite coating surface with the increase of PTFE in baths. According to the results of EDS, the white agglomerate particles are PTFE. Those indicate that the more contents of PTFE particles in plating solution, the more difficult to homodisperse the PTFE in solution, which leads to the agglomerate of PTFE on the surface of coatings. Overvalued of PTFE content makes some big agglomerate particles appear, and the subsequent deposition of nickel can't completely package on them, eventually leads to the uneven dispersion of PTFE on the surface of the coatings. When PTFE content in plating solutions is less than 20 g/L, the dispersion effect of PTFE in solution is well, PTFE agglomerate particles are less and unevenly scattered on surface of the coatings.

3.1.2 *Phase analysis of composite coatings*

XRD patterns of the five kinds of coatings are shown in Figure 2. It was indicated that only the three peaks of Ni crystal indices could be identified, without the diffraction peals of F and C. At the same time, all of the three peaks of different coating prepared by plating solutions with different PTFE contents are sharp. The widths of the diffraction peaks of Ni in composite coatings are broader than that of pure Ni coatings. That is to say, by adding PTFE powders, crystal particles of Ni substrates can be refined.

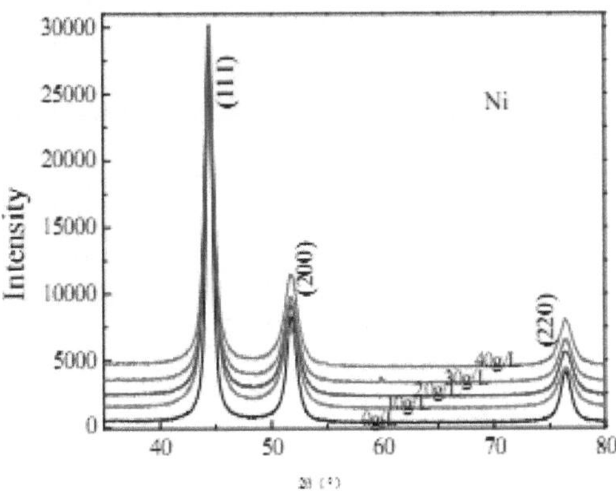

Figure 2. XRD patterns of composite coatings.

3.1.3 *Tribological characteristics of composite coatings*

The morphologies of composite coatings after wear test prepared by plating solutions with different PTFE contents are shown in Figure 3 and the coefficient of sliding friction and abrasion loss are shown in Figure 4. It is found that the wear trajectory of pure Ni coating is discontinuity, and the wear trajectory of PTFE/Ni composite coatings are on the contrary. This suggests through adding of PTFE, the coatings become softer obviously, the friction coefficient of coatings can be reduced and the wear resistance of coatings increased. For composite coatings, when the PTFE content is low

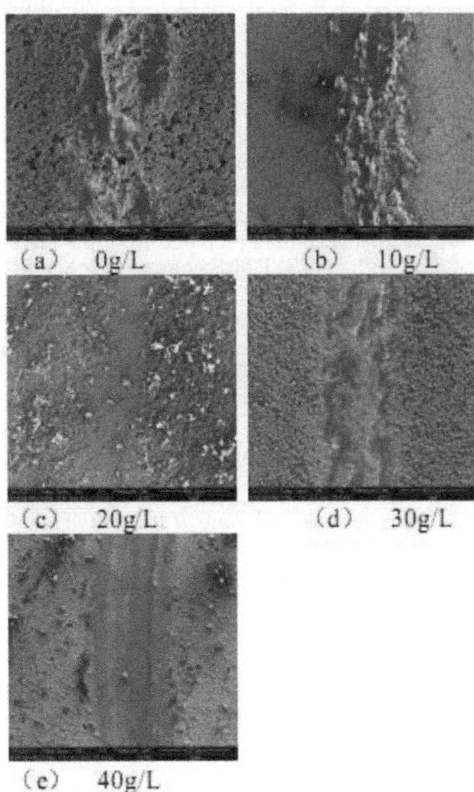

Figure 3. Surface wear morphologies of the coatings with different content of PTFE.

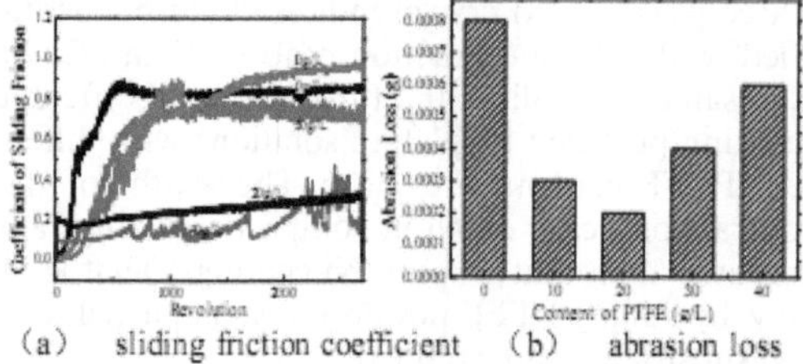

(a) sliding friction coefficient (b) abrasion loss

Figure 4. Sliding friction coefficient and abrasion loss of the coatings with different content of PTFE.

(Fig. 3b), due to the imperfect coverage of PTFE, the stability of the sliding friction coefficient can be affected. On the other side, when the content of PTFE is too high (Fig. 3d), it is difficult to disperse PTFE powders in plating solutions, which leads to the inhomogeneous distribution of PTFE in the coatings. After the friction, the abrasion is more serious in some places, so its sliding friction coefficient is not stable too. For the composite coating prepared from the plating solution with PTFE content of 20 g/L, due to the absence of aggregation and homogenous dispersion of PTFE, the sliding friction coefficient was decreased obviously and the abrasion resistance increased consequently. That is evident in Figure 4. The sliding friction coefficient of the coating prepared from the plating solution with 20 g/L PTFE content is minimum (which is about 0.2) and very stable. But for other coatings, the sliding friction coefficient is either too high or unstable. It is found by comparison that the abrasion loss of the coating prepared from the plating solution with 20 g/L PTFE content is lower than any other coating, although all of the abrasion loss of the coatings is low because that only 200 g load is used during wear experiment. From wear morphologies, friction coefficient and abrasion loss, the Ni/PTFE composite coating prepared from the plating solution with 20 g/L PTFE content has good friction and wear properties.

3.2 *Morphologies and phase of the coatings after heat treatment*

3.2.1 *Micro morphologies and composition analysis*

It is common for electro-brush plating coating to need heat treatment in order to get stable composite coatings. The coating prepared from the plating solution with 20 g/L PTFE content is selected to heat treat because of its good friction and wear properties.

In order to obtain stable coating structure, heat treatment for electro-brush plating coating was carried out. The coating prepared by the bath with 20 g/L content of PTFE was selected to heat treatment due to its good friction and wear properties. The surface morphologies of the coatings treated at different temperature were shown in Figure 5. It is illustrated that the morphologies of the coatings treated by different temperature have clear differences. The surface morphologies of the coatings were still nodules clusters; moreover, the density of the cluster has no apparent change, when the treated temperature is less than 400°C. As for the morphology of the coating treated at 600°C, the density decreased obviously and there were many micro-pits distributed on the surface of nodules clusters. That is maybe because of the volatile and cracking of PTFE under high temperature. According to the results of EDS, compared to the coating without heat treated, the content of PTFE on the surface of the coating treated at 200°C has not changed, and the contents of PTFE on the surfaces of the coatings treated at 400°C and 600°C decrease obviously, and there are white powders leaving inside of the sealed tube. This is due to the melting of PTFE at 327°C to 342°C and the boiling of PTFE at about 400°C. PTFE will volatilite in abundance when treated at above 400°C and pyrolyze at 600°C.

3.2.2 *Phase analysis*

The XRD curves of composite coatings treated at different temperature were given in Figure 6. It is illustrated that PTFEs are solubilized in the matrix. The higher heat-treatment temperature the higher height of big three peaks shows that the size of grains gets bigger when the temperature of heat-treatment becomes higher.

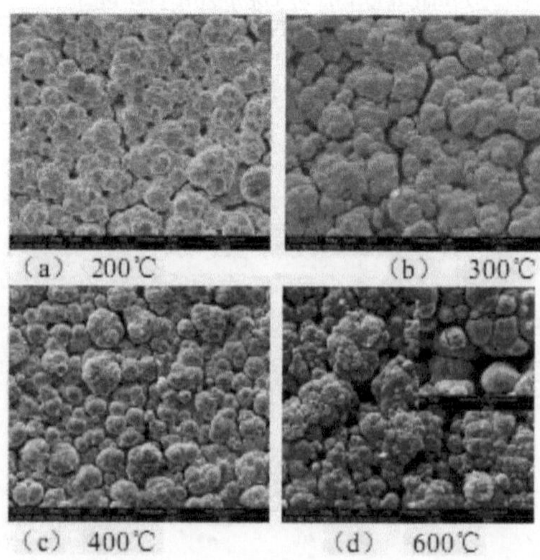

(a) 200°C (b) 300°C (c) 400°C (d) 600°C

Figure 5. SEM morphologies of coatings treated in different temperature.

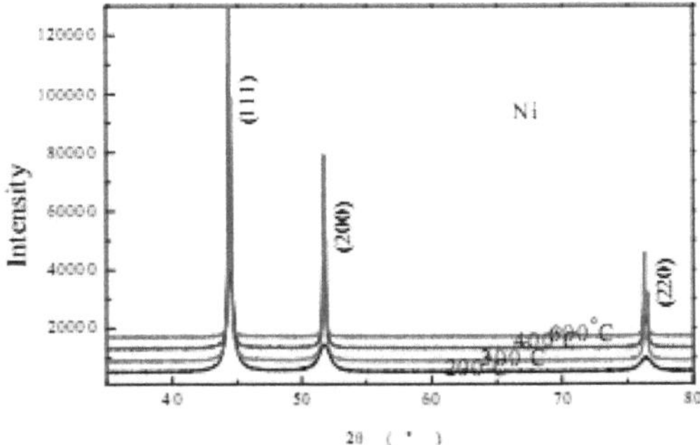

Figure 6. XRD curves of composite coating treated in different temperature.

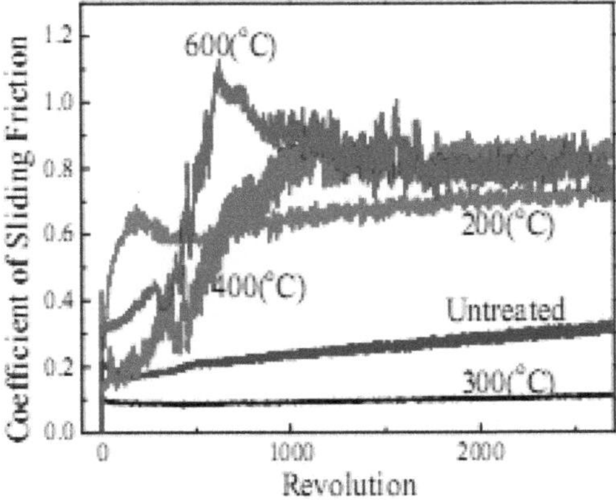

Figure 7. Sliding friction coefficients of the coatings treated at different temperature.

3.2.3 *Wear resistance of composite coatings*

Sliding friction coefficients of the coatings treated in different temperature are shown in Figure 7. Experimental results show that the friction and wear property of the coating treated at 300°C is the best. The friction and wear coefficients of the coatings get higher and the stabilities get worse when treated at 400°C and 600°C. This is because of that the volatility at 400°C and pyrolysis at 600°C of PTFE makes the content of PTFE in coatings decreased and the self-lubricating property of the coatings reduced. Meanwhile, when the coating is treated at below 200°C, there has no any physical and chemistry change in PTFE. However, when the coating is treated at 300°C, part of the PTFE presents molten state, and disperses more even after cooled. These make the friction and wear coefficient of the coating treated at 300°C is lower than others. As a result, if the coating is treated at about 300°C, its friction and wear property will be improved.

4 CONCLUSION

The results indicate that the PTFE particles/Ni-based alloy composite coatings brushed by fast nickel plating bath with 20 g/L PTFE particles (average diameter in 5 μm) had relatively uniform micro-morphologies and high wear resistance. In this condition, the sedimentary speed is fast, the grain was refined. When the coatings were heat-treated at 300°C, good friction and wear properties of the coatings were maintained.

ACKNOWLEDGEMENT

This work is supported by the China Postdoctoral Science Foundation (grant no. 2015M571392) and the National Natural Science Foundation of China (grant no. 51309067).

REFERENCES

[1] M. Pazderová, M. Bradáč and M. Valeš, Tribological behavior of composite coatings, Procedia Engineering 10 (2011) 472–477.

[2] Y.S. Huang, X.T. Zeng and X.F. Hu et al, Corrosion resistance properties of electroless nickel composite coatings, Electrochimica Acta 49 (2004) 4313–4319.

[3] A. Ramalho and J.C. Miranda, Friction and wear of electroless NiP and NIP+PTFE coatings, Wear 259 (2009) 828–834.

[4] Iman R. Mafi and Changiz Dehghanian, Comparison of the coating properties and corrosion rates in electroless Ni-P/PTFE composites prepared by different types of surfactants, Applied Surface Science 257 (2011) 8653–8658.

[5] E. Pena-Munoz, P. Berçot and A. Grosjean et al, Electrolytic and electroless coatings of Ni-PTFE composites: Study of some characteristics, Surface and Coatings Technology 107 (1998) 85–93.

[6] Yating Wu, Hezhou Liu and Bin Shen et al, The friction and wear of electroless Ni-P matrix with PTFE and/or SiC particles composite, Tribology International 39 (2006) 553–559.

[7] Hu S.B., Tu J.P. and Mei Z. et al, Adhesion strength and high temperature wear behavior of ion plating TiN composite coating with electric brush plating Ni-W interlayer, Surface and Coatings Technology 141 (2001) 174.

Material Science and Environmental Engineering – Chen (Ed.)
© 2016 Taylor & Francis Group, London, ISBN 978-1-138-02938-5

Effect of Ope-POSS on glass transition temperature of PU hybrid composites

R. Pan & L.L. Wang
Chemistry and Material Science College, Sichuan Normal University, Chengdu City, P.R. China

Y. Liu
Key Laboratory of Special Waste Water Treatment, Sichuan Province Higher Education System, Chengdu City, P.R. China

ABSTRACT: Octa (propylglycidyl ether) polyhedral oligomeric silsesquioxane (Ope-POSS) with eight glycidyl groups was incorporated in concentrations of 5, 10, 15 and 20 wt% into Toluene-2, 4-Diisocyanate (TDI) and Poly-Propylene oxide Gloycol (PPG) to construct Ope-POSS/PU hybrid composites models, respectively. The glass transition temperature of these hybrid composite models was characterized by volume-temperature behavior analysis. As the result shows: the stiff enclosed cage of Ope-POSS linked to the polymer backbones acts as a rigid node with less mobility and results in the steric hindrance of polymer chain apparently, which lead to the increasing Tg and also broaden the glass transition region of Ope-POSS/PU hybrid composites.

Keywords: POSS; polyurethane; molecular dynamics; molecular mechanics

1 INTRODUCTION

As nano-particle, polyhedral oligomeric silsesquioxane (abbreviation POSS) has received increasing attention for its inorganic-organic hybrid nature. Most of POSS consists of a totally enclosed cage with eight silica atoms linking together via oxygen atoms. [1] Polymer properties, such as permeability, friction, mechanical and thermal properties, can be significantly altered by introducing different POSS molecules at various concentrations into polymer matrices. [2] As with other polymers, POSS can improve the properties of PU in applications ranging from elastomers, coatings to adhesives. Since polyurethane has a large range of application and Ope-POSS is a good modification of filler, the mechanism about how the covalent inclusion of Ope-POSS into the PU backbone will tailor the glass transition temperature of hybrid composite is deserved to investigate.

2 SIMULATION PROCEDURES

Accelrys Amorphous Cell module and COMPASS force field in Materials Studio software were adopted in all simulation process in our research, which has been used successfully for the simulation of polymer nano-composites containing POSS [3]. As periodic boundary conditions imposed, an initial low density was used to construct the bulk cubic structures of random hybrid co-polymers. 10 initial configurations for each sample were optimized by molecular dynamics technique under the NPT (constant particle numbers, pressure and temperature) conditions at 4 Gpa with a minimization involving 30000 steps to relax and equilibrium. After this minimization procedure, the density fluctuation of each system is less than 0.05 g/cm^3 under a given condition. Since these optimized configurations might not be in a local energy minimum state, an annealing procedure from 623 K to 273 K was applied on the above optimized configurations by conducting the Velocity Verlet algorithm in NVT dynamics to reduce the possible potential energy. Finally, configurations

Table 1. Characteristics of Ope-POSS/PU hybrid composites.

Sample code	Ope-POSS: TDI: PPG (mole ratio)	Ope-POSS (wt.%)	Initial density (g/cm^3)	Final density (g/cm^3)
0	0:06:03	0	1.32	1.28
5	1:12:06	5	1.24	1.20
10	1:06:03	10	1.21	1.25
15	3:12:06	15	1.19	1.17
20	2:06:03	20	1.17	1.21

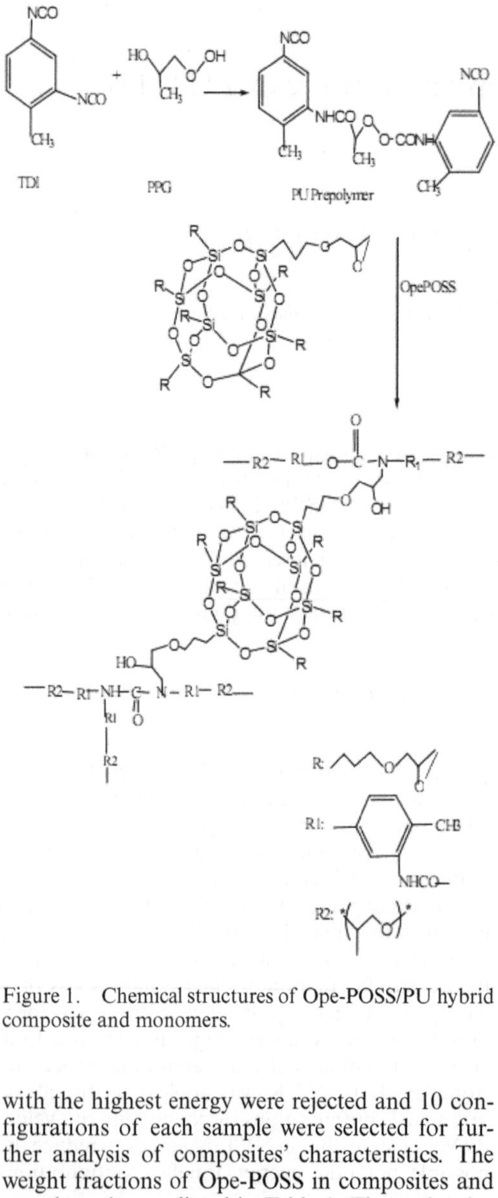

Figure 1. Chemical structures of Ope-POSS/PU hybrid composite and monomers.

with the highest energy were rejected and 10 configurations of each sample were selected for further analysis of composites' characteristics. The weight fractions of Ope-POSS in composites and sample codes are listed in Table 1. The molecular structures of Ope-POSS /PU hybrid composite are shown in Figure 1. [4]

3 3D BOUNDARY CONDITION STRUCTURE CONSTRUCTION

Recently, it was also reported that the structure and energy of the POSS/PU hybrid composites were successfully simulated by use of the COMPASS force field. [4] Model structure was generated through several cycles of molecular mechanics and

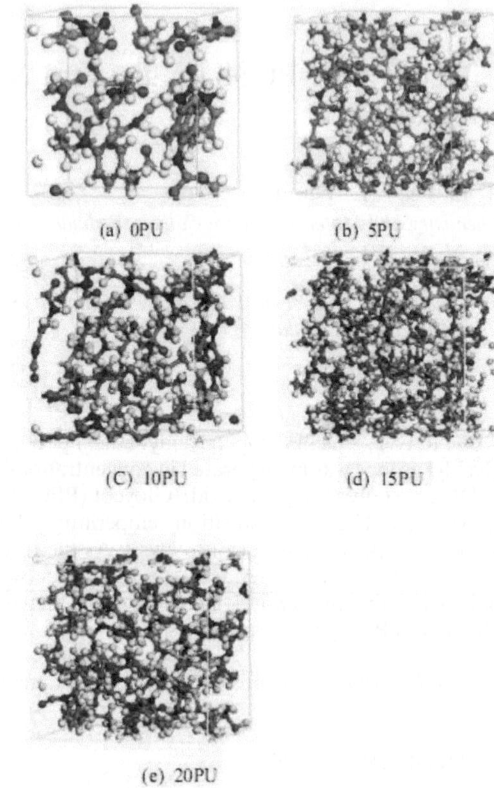

Figure 2. 3D periodic boundary conditions of Ope-POSS/PU hybrid composites: (a) 0 PU; (b) 5 PU; (c) 10 PU; (d) 15 PU; (e) 20 PU.

molecular dynamics energy minimization. After the above minimization procedure, the density fluctuation of each system is less than 0.05 g/cm^3 under a given condition, indicating that the structure generated is fully relaxed and is in the equilibrium state which can be confirmed by energy optimization.

4 VOLUME-TEMPERATURE BEHAVIOR ANALYSIS

Usually, for the estimation of Tg, the group contribution method is easily used for common polymers but useless for complex aromatic systems. For new groups, the deficient parameters will lead to the invalidity of this method. Another limitation is that the information of structural details can't be obtained from the correlation method. Molecular dynamic simulation at the atomic level overcomes these limitations and becomes a useful approach for studies of the glass transition temperature and

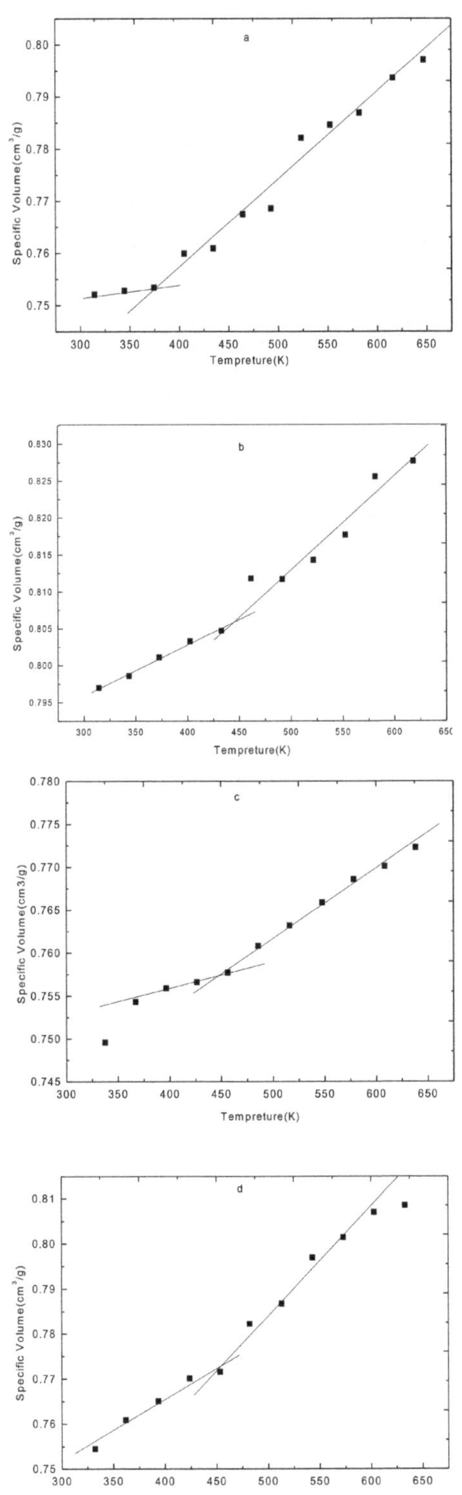

Figure 3. (*Continued*)

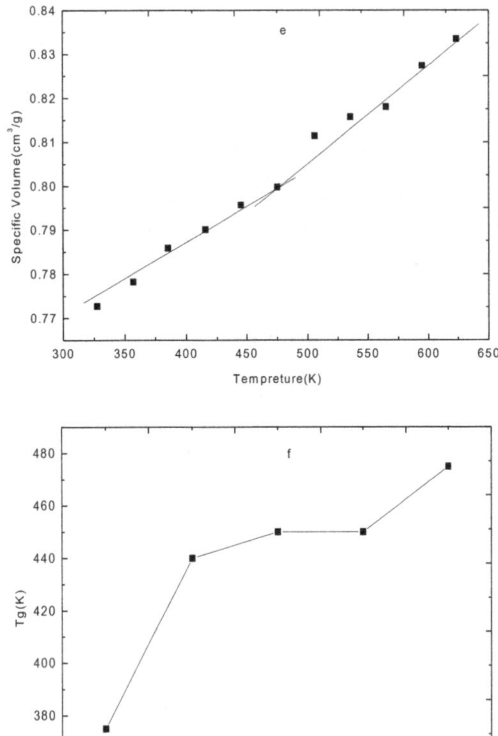

Figure 3. Specific volume vs. Temperature for (a) 0 PU; (b) 5 PU; (C) 10 PU; (d) 15 PU; (e) 20 PU; (f) Tg vs. Ope-POSS concentration in hybrid composites.

polymer structures. The basic theory is to obtain the curve of the specific volume during step-wise cooling through the trajectory file and find the turning point in the curve as Tg [5] which has been demonstrated by other reports.

In Figure 3, the simulated Tg is determined at the point where slopes intersect. The most significant observation is that the introduction of Ope-POSS into PU backbone increases the Tg of composites. 20 PU has the highest Tg of 480 K whereas 0 PU has the lowest Tg of 370 K. Combined with the chain mobility analysis in MSD, the increase of Tg can be attributed to the Ope-POSS stiff cores which rigidify polymer backbones and restrict the motions of surrounding chains.

5 CONCLUSION

In this study, molecular simulation was applied in study the effect of Ope-POSS cage structure on chain mobility and glass transition temperature of Ope-POSS/PU hybrid composites at molecular level. The fluctuation of density and optimized energy plots verified the accuracy of the models and the applied force field. As the result shows, with concentration increasing in hybrid composites, Ope-POSS as a rigid core linked to the polymer chains, apparently restricts the motion of the whole polymer chain and leads to the increase of glass transition temperature of hybrid composites.

ACKNOWLEDGMENTS

This work was supported by Sichuan Education Office Foundation (Project No. 15ZA 0040) in China and Key Laboratory of Special Waste Water Treatment in Sichuan Province Higher Education System.

REFERENCES

[1] Information on http://www.hybridplastics.com/docs/user-v2.06.pdf.

[2] Jianqing Zhao & Yi Fu et al. 2008. Polyhedral Oligomeric Silsesquioxane (POSS)-Modified thermoplastic and thermosetting nanocomposites: A review, Polym. Polym. Compos. 16 (8): 483–500.

[3] Zheng L. & Coughlin E.B. et al. 2001. X-ray Characterizations of Polyethylene Polyhedral Oligomeric Silsesquioxane Copolymers, Macromol. 35: 2375–2379.

[4] Lingling Wang & Rui Pan. 2015. Influence of trisilanolphenyl POSS on structure and thermal properties of polyurethane hybrid composites: a molecular simulation approach, Acta Polymerica Sinica. Doi: 10.11777/j.issn1000-3304.2015.14231.

[5] YinYani & Monica H. Lamm 2009. Molecular dynamics simulation of mixed matrix nanocomposites containing polyimide and Polyhedral Oligomeric Silsesquioxane (POSS), Polym. 50: 1324–1332.

Material Science and Environmental Engineering – Chen (Ed.)
© 2016 Taylor & Francis Group, London, ISBN 978-1-138-02938-5

Synthesis and structural characterization of a new Ca (II) complex with 2-Formyl-benzenesulfonate-thiosemicarbazide ligand

H. Zou
Weifang Experimental High School, Weifang, P.R. China

ABSTRACT: A new Ca (II) complex, $[Ca(H_2O)_6 \cdot L] \cdot (L) \cdot (H_2O)_4$ (L = 2-formyl-benzenesulfonate-thiosemicarbazide) has been synthesized by the reaction of $CaClO_4$ with sodium 2-(thiosemicarbazonomethyl) benzoate in the CH_3CH_2OH/H_2O solution. The structure of Ca (II) complex was determined by X-ray single crystal diffraction analysis. The results showed that the Ca (II) ion is seven-coordination with six oxygen atoms of coordinated water molecules and one oxygen atom of 2-formyl-benzenesulfonate-thiosemicarbazide ligand. The Ca (II) complex formed two dimensional layered structures by the interaction of hydrogen bonds and π-π stacking.

Keywords: Ca (II) complex; 2-Formyl-benzenesulfonate-thiosemicarbazide; preparation; structural characterization

1 INTRODUCTION

The study on metal complex materials have received much attention because of their potential application in many field, such as antitumor and antibacterial drugs, luminescent probes, gas absorption, chemical analysis, and so on [1–4]. However, the above studies are largely focused on the synthesis and properties of transition metal complex and rare earth complexes [5–7]. And the studies on Ca (II) complex material are less, so it is significance to study on the synthesis and structure of Ca (II) complex materials [8–10].

In this paper, a new Ca (II) complex was synthesized and its structure was determined by single-crystal X-ray diffraction. The Ca (II) complex formed two dimensional layered structures by the interaction of hydrogen bonds and π-π stacking.

2 EXPERIMENTIAL

2.1 *Materials and measurements*

$CaClO_4$, 2-formyl-benzenesulfonic acid sodium salt, thiosemicarbazide and all other solvents used were analytical grade. The crystal data was collected on a Bruker Smart-1000 CCD Area Detector.

2.2 *Synthesis of Ca (II) complex*

0.5 mmol 2-formyl-benzenesulfonic acid sodium salt and 0.5 mmol thiosemicarbazide were dissolved in 15 mL ethanol/water solution. The above mixture was stirred for 0.5 h at 60 °C. Then 0.5 mmol $CaClO_4$ was added to the above solution. The mixture was continually stirred for 5 h 60 °C, and a little white precipitate formed. The filtrate was standed in air for two week; the crystals of Ca (II) complex suitable for X-ray crystal determination were obtained.

2.3 *Structural determination*

A colourless block single crystal with dimensions of 0.22 mm × 0.21 mm × 0.20 mm was placed on a glass fiber and mounted on a CCD area detector. Diffraction data were collected by $\varphi \sim \omega$ scan mode using a graphite-monochromatic Mo $K\alpha$ radiation ($\lambda = 0.71073$ Å) at 296 K. A total of 4090 reflections were collected in the range of $2.17 \leq \theta \leq 25.00°$, of which 3851 were unique ($R_{int} = 0.0365$) and 3431 were observed with $I > 2\sigma(I)$. The structure was solved by direct methods using SHELXL-97 [11] and expanded using Fourier techniques. Molecular graphics were drawn with the program package SHELXTL-97 crystallographic software package [12]. The most relevant crystal data for complex are quoted in Table 1. Selected bond lengths and bond angles of the Ca (II) complex are shown in Table 2.

3 RESULTS AND DISCUSSION

3.1 *Structure characterization*

The reaction of $Ca(ClO_4)_2$ with a Schiff base ligand condensed by 2-formyl-benzenesulfonic

Table 1. Crystal data for the complex.

Complex	Ca (II) complex
Formula	$C_{16}H_{36}CaN_6O_{16}S_4$
Colour/shape	Colourless/block
Formula weight	736.83
Crystal size (mm^3)	$0.22 \times 0.21 \times 0.20$
Crystal system	Triclinic
Space group	*P*-1
a (Å)	6.894(9)
b (Å)	9.811(13)
c (Å)	24.68(3)
α (°)	99.680(16)
β (°)	94.779(16)
γ (°)	103.207(17)
Volume (Å^3)	1589(4)
Z	2
D_c/(g/cm^{-3})	1.540
μ (mm^{-1})	0.536
F(000)	772
T/K	296(2)
θ range/°	2.17/25.00
Index ranges	$-7 \le h \le 7$
	$-11 \le k \le 4$
	$-20 \le l \le 29$
Reflections collected/unique	4090/3851
Data/restraints/parameters	3851/18/401
S	1.131
$(\Delta/\sigma)_{max}$	0.001
Goodness-of-fit on F^2	1.066
R_1, wR_2 [$I > 2\sigma(I)$]	0.1359, 0.4324
R_1, wR_2 (all data)	0.1406, 0.4403
Largest diff. peak and hole (e·Å^{-3})	1.152 and −1.246
CCDC	1057697

acid sodium salt and thiosemicarbazide gave a new Ca (II) complex. The molecular structure of the Ca (II) complex is shown in Figure 1. The molecular packing arrangement is shown in Figure 2. The two dimensional layered structure is shown in Figure 3. From Figure 1, we can see that the Ca (II) complex was made up of one cation $[Ca(H_2O)_6 \cdot L]^+$, one L^- anion and four lattice water. The Ca (II) atom is seven-coordinated with six O atoms from six coordinated water molecules, and one oxygen atom from the L ligands, 2-formyl-benzenesulfonate-thiosemicarb azide. The coordination environment of Ca (II) center can be described as distorted pengonal bipyramidal CaO_7 geometry. The bond lengths of Ca-O bonds are in the range of 2.333(11) Å and 2.429(8) Å. From Figure 2, it can be seen that rich hydrogen bonds are exist in the Ca (II) complex. The bond angles around the Ca (II) atom range

Table 2. Selected bond lengths [Å] and bond angles [°] of the complex.

Ca(1)-O(9)	2.333(11)
Ca(1)-O(7)	2.340(9)
Ca(1)-O(2)	2.362(6)
Ca(1)-O(8)	2.389(12)
Ca(1)-O(10)	2.390(7)
Ca(1)-O(11)	2.427(7)
Ca(1)-O(12)	2.429(8)
S(1)-O(1)	1.435(7)
S(1)-O(2)	1.438(6)
S(1)-O(3)	1.447(7)
C(7)-N(1)	1.260(9)
C(8)-N(3)	1.331(12)
O(9)-Ca(1)-O(7)	84.3(4)
O(9)-Ca(1)-O(2)	104.3(4)
O(7)-Ca(1)-O(2)	170.6(3)
O(9)-Ca(1)-O(8)	142.4(4)
O(7)-Ca(1)-O(8)	81.1(5)
O(2)-Ca(1)-O(8)	89.6(4)
O(9)-Ca(1)-O(10)	72.1(3)
O(7)-Ca(1)-O(10)	98.9(3)
O(2)-Ca(1)-O(10)	87.6(2)
O(8)-Ca(1)-O(10)	144.4(4)
O(9)-Ca(1)-O(11)	72.4(3)
O(7)-Ca(1)-O(11)	94.8(3)
O(2)-Ca(1)-O(11)	84.3(3)
O(8)-Ca(1)-O(11)	74.5(4)
O(11)-Ca(1)-O(10)	140.3(3)
O(9)-Ca(1)-O(12)	140.1(3)
O(7)-Ca(1)-O(12)	80.8(3)
O(2)-Ca(1)-O(12)	94.5(3)
O(8)-Ca(1)-O(12)	70.9(3)
O(10)-Ca(1)-O(12)	73.9(3)
O(11)-Ca(1)-O(12)	145.4(3)
O(1)-S(1)-O(2)	109.6(4)
O(1)-S(1)-O(3)	114.2(5)
O(2)-S(1)-O(3)	112.6(4)
C(7)-N(1)-N(2)	112.8(6)
C(8)-N(2)-N(1)	115.7(6)
C(15)-N(4)-N(5)	114.5(6)
C(16)-N(5)-N(4)	120.0(6)
N(3)-C(8)-N(2)	119.9(7)
N(5)-C(16)-N(6)	118.6(6)

from 72.1(3)° to 170.6(3)°, which is in accordance with ref. [13]. The Ca (II) complex molecules form two dimensional layered structures by the interaction of hydrogen bonds and π-π stacking.

3.2 *Conclusions*

A new Ca (II) complex has been obtained by the reaction of $CaClO_4$ with sodium 2-(thiosemicarbazonomethyl) benzoate in the CH_3CH_2OH/H_2O solution. The structure of

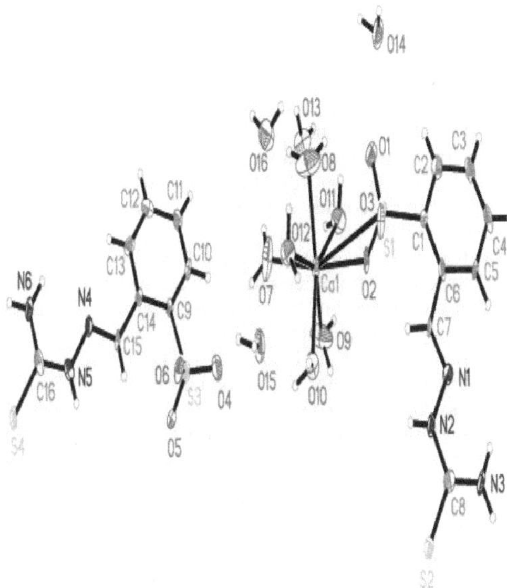

Figure 1. Molecular structure of Ca (II) complex.

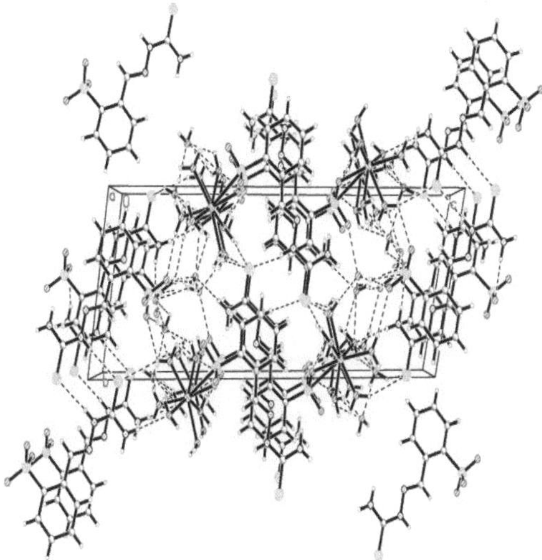

Figure 2. The molecular packing arrangement of Ca (II) complex.

Ca (II) complex was determined by X-ray single crystal diffraction analysis. The results showed that the Ca (II) ion is seven-coordination with six oxygen atoms of coordinated water molecules and one oxygen atom of 2-formylbenzenesulfonate-thiosemicarbazide ligand. The Ca (II) complex formed two dimensional layered structures by the interaction of hydrogen bonds and π-π stacking.

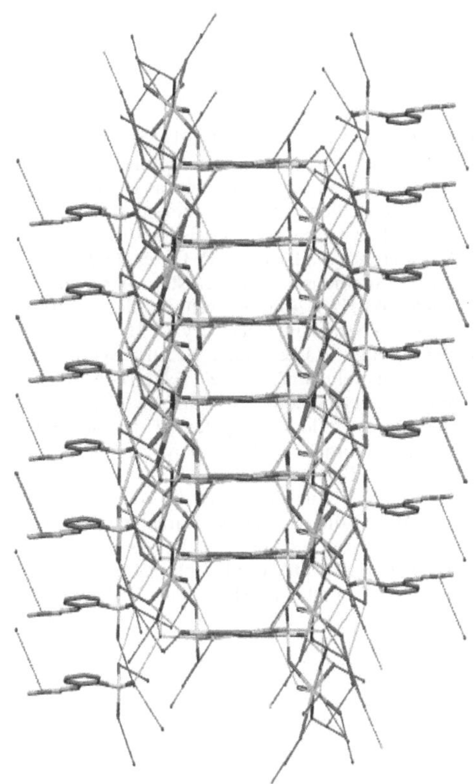

Figure 3. The two dimensional layered structure.

4 SUPPLEMENTARY MATERIAL

Crystallographic data for the structure reported in this paper has been deposited with the Cambridge Crystallographic Data Centre as supplementary publication No. CCDC1057697. Copy of the data can be obtained free of charge on application to CCDC, 12 Union Road, Cambridge CB2 1EZ, UK (Fax: +44-1223-336-033; E-Mail: deposit@ccdc.cam.ac.uk).

REFERENCES

[1] Bharti, A.; Bharati P.; Chaudhari U.K.; Singh A.; Kushawaha S.K.; Singh N.K. & Bharty M.K. 2015. Synthesis, crystal structures and photoluminescent properties of new homoleptic and heteroleptic zinc (II) dithiocarbamato complexes. *Polyhedron* 85: 712–719.

[2] Shafaatian, B.; Ozbakzaei, Z.; Notash, B. & Rezvani, S.A. 2015. Synthesis, characterization, single crystal X-ray determination, fluorescence and electrochemical studies of new dinuclear nickel (II) and oxovanadium (IV) complexes containing double schiff base ligands. *Spectrochimica Acta Part A: Molecular and Biomolecular Spectroscopy* 140: 248–255.

[3] Song, Q.B.; An, X.X.; Xia, T.; Zhou, X.C. & Shen, T.H. 2015. Catalytic asymmetric henry reaction using copper (II) chiral tridentate schiff base complexes and their polymer supported complexes. *Comptes Rendus Chimie* 18: 215–222.

[4] Tai, X.S.; Liu, L.L. & Yin, J. 2014. Synthesis, crystal structure of tetra-Nuclear macrocyclic Cu (II) complex material and its application as catalysts for A3 coupling reaction. Journal of Inorganic and Organometallic Polymers and Materials 24: 1014–1020.

[5] Alexandrescu, L.; Bourosh, P.; Oprea, O. & Jitaru, I. 2014. Synthesis and crystal structure of [La(NO3)3(H2O)2(bipy)] center dot 1.5(bipy). Journal of Structural Chemistry 55: 107–111.

[6] Mei, L.; Sun, W.J. & Chu, T.W. 2014. Synthesis and biological evaluation of novel (TcN)-Tc-99 m-labeled bisnitroimidazole complexes containing monoamine-monoamide dithiol as potential tumor hypoxia markers. Journal of Radioanalytical and Nuclear Chemistry 301: 831–838.

[7] Sharma, R.P.; Saini, A.; Kumar, S.; Venugopalan, P. & Ferretti, V. 2014. Synthesis, characterization, single crystal structure and DFT calculations of [Cu(temed)(H2O)4] (1,5-napthalenedisulphonate) center dot 2(H2O). Journal of Molecular Structure 1067: 210–215.

[8] Tai, X.S.; Zhao, W.H. & Li, F.H. 2013. Synthesis, crystal structure and antitumor activity of Ca (II) coordination polymer constructed by N-benzenesulphonyl-L-phenylalanine. Chinese Journal of Inorganic Chemistry 29: 2200–2204.

[9] Tai, X.S. & Zhao, W.H. 2013. Synthesis, structural characterization, and antitumor activity of a Ca (II) coordination polymer based on 1,6-naphthalenedisulfonate and 4,4′-bipyridyl. Materials 6: 3547–3555.

[10] Tai, X.S.; Zhao, W.H. & Li, F.H. 2013. Synthesis, structural characterization and antitumor activity of a Ca (II) schiff base complex. Chinese Journal of Inorganic Chemistry 29: 1328–1332.

[11] Scheldrick, G.M. 1997. SHELXL-97, Program for Crystal Structure Solution; University of GÖttingen: GÖttingen, Germany.

[12] Sheldrick, G.M. 1997. SHELXTL-97, Program for Crystal Structure Refinement; University of GÖttingen: GÖttingen, Germany.

[13] Tai, X.S. & Zhao, W.H. 2013. Synthesis, crystal structure and antitumor acyivity of Ca (II) coordination polymer based on 1,5-Naphthalenedisulfonate. Journal of Inorganic and Organometallic Polymers and Materials 23: 1354–1357.

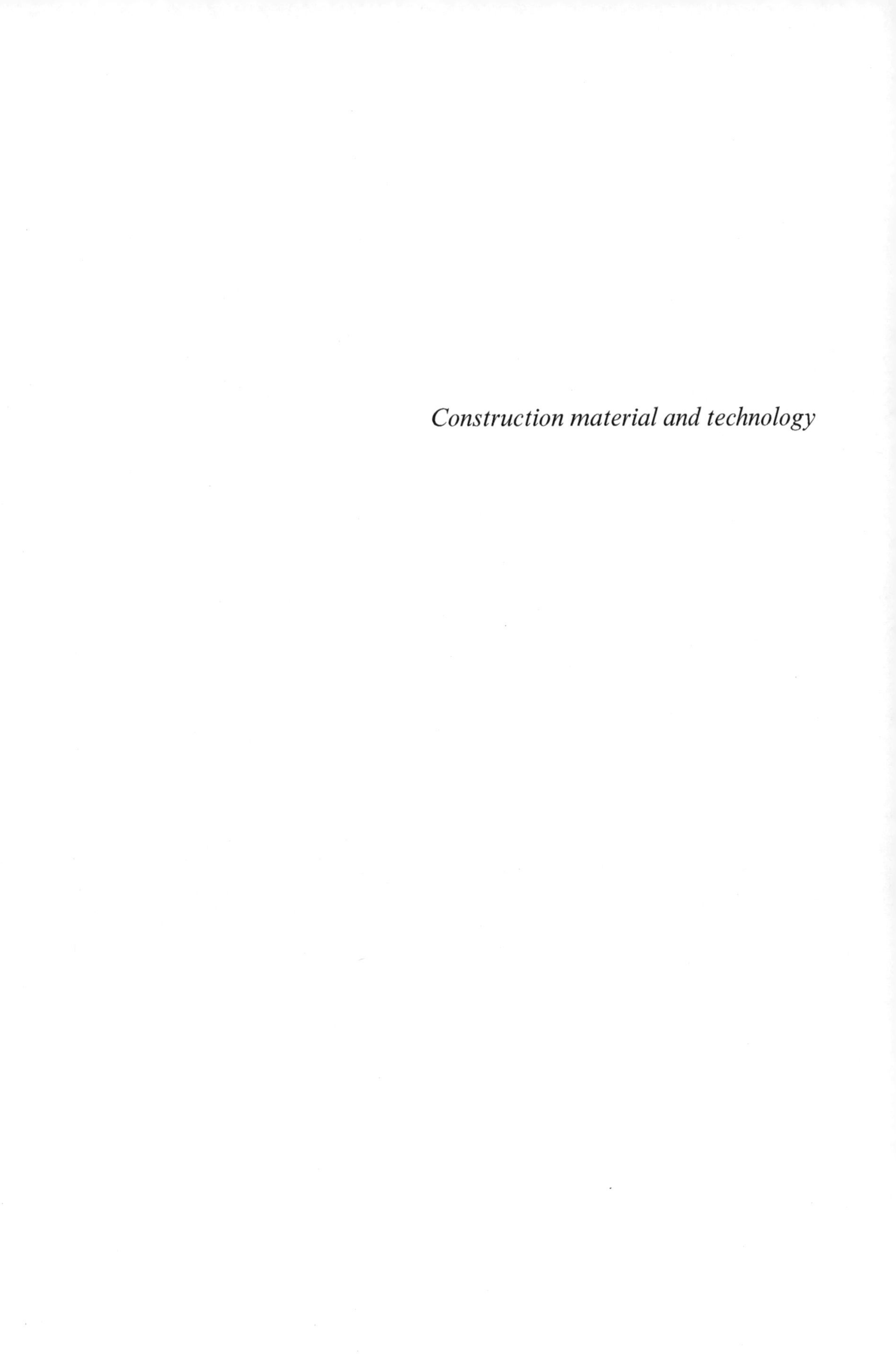

Construction material and technology

Material Science and Environmental Engineering – Chen (Ed.)

Experimental study on mechanical behavior of concretes with sea sand and sea water

L. Xing
College of Urban Construction, Zhejiang Shuren University, Hangzhou, China

ABSTRACT: With the development of marine economy, a lot of marine buildings and port structures need to be built. So amount of concrete is needed. If these buildings were built by the concretes with sea sand and sea water instead of the traditional concretes, raw materials can be obtained locally and time and cost can be saved. Based on the above considerations, the workability and strength were studied with the changing of the content of salinity and shells. The results show that with the increase of salinity, the slump is increased and the mobility is enhanced, but the cohesion is deteriorated; with the increasing content of shells, the slump is reduced, water retention is worse. The strength of concrete with sea sand and sea water is not vary with the changing of the content of the salinity and shells, but the early strength has declined comparing with conventional concrete, dropping by about 20 percent. With the increase of age, the strength of two kinds of concretes is quite.

Keywords: concrete; sea sand; sea water; strength; workability

1 INTRODUCTION

With the development of marine economy, a lot of marine buildings and port structures need to be built. So amount of concrete is needed. If all the buildings are built by the traditional concrete, a lot of sand and freshwater need to be transported from inland. Thus the construction period is affected, while the construction costs are increased. And transportation for some distance farther island construction is also very difficult, the same time, these islands or coastal areas, there are plenty of sea sand and sea resources are available. If sea sand and sea can be utilized to prepare the concrete, on the island to meet the marine economic development and national defense construction has extraordinary significance.

Sea sand is produced in the ocean (including beach sand, sea sand) and into the sea near the mouth of the sand, it compared to the natural river sand, although the quality is similar, but the density, grain size, salt, shells, etc., is difference. The type and quantity of the sea sand is varied with the different sea area [1]. The specific gravity of sea sand in the region vary, sand particle size is generally between 0.3~2.0 mm, which accounts for the ratio of coarse sand than the river sand is small. The salt content and sell content of sea sand was changes with sand mining area, the type, size, different water content and different sand. Overall, the sea sand is smooth surface, dry and well graded. But the existing problem is the high salt content and the content of shells [2].

In early 1995, the influence of different content of sand on the concrete workability, compressive strength, flexural strength and impermeability was studied [3] by contrasting the test data of two kinds of concrete with water to cement ratio of 0.5 and 0.44, one was the test concrete with fine aggregate containing shells, the other was the reference concrete which fine aggregate was river sand. The results showed that the liquidity of concrete with 4.05% shell is better than the reference concrete, but with the increase of content of shells, the fluidity is gradually worse. When the shell content reaches 18.68%, the fluidity of test concrete is very worse than the reference concrete, and the bleeding phenomenon is found when the water cement ratio of concrete is 0.5. The content of shells has little influence on the flexural strength of concrete and impermeability.

In 2005, the engineering characteristics of the sea sand were studied [4]. The results of the study indicated that, although the shell sand than ordinary sand compressibility, shear strength is slightly lower than the ordinary sand, but it's still a good engineering material.

In 2012, the influence of shell content on compressive strength and fluidity of concrete was studied [5]. The results were noted that, when the content of shells in the sea sand is no more than 3%, the fluidity of concrete is not effect, even has a positive effect. When the shell content exceeds 8%, the fluidity of concrete mixture decreased and the slump was a very small degree. When the content of shells is little, there is on influence to the compressive strength of concrete, but with the increase of content of shells, the influence on strength is gradually obvious. When the content of shells is up to 10.17%, the 28 day strength of concrete is decreased evidently.

There are lots of literatures about the effect of salt content on concrete corrosion [6–7]. The corrosion of reinforcement effect by the strength of the concrete, the protective layer thickness, rebar diameter and the content of chloride was studied mainly through the permeability test of concrete and fast durability test. Through this research, the optimal method content of chloride ion and the other to solve the corrosion of the steel can be explored. There are also many literatures [8–10] noted on sea sand concrete. The purpose of all the study is to using the sea sand in the traditional concrete by the desalination of the sea sand, removing chloride and the barrier of chloride ion. The project is supported by the Science and Technology Department of Zhejiang Province and the project research objective is not to change the composition of sea sand and seawater to use in the concrete. The content of shells and salt on the compressive strength of concrete is to be studied in the paper. And the durability and the FRP concrete with the sea sand and the seawater will be studied in other literatures.

2 EXPERIMENTAL PROGRAMS

2.1 *Material properties*

The Granite stones of well-graded gravel and the largest particle size 31.5 mm, and the minimum size of 5 mm, and clay content of 0.5%, and the apparent density of 2623 kg/m^3 were selected as the coarse aggregate. The ordinary porland cement with 28d compressive strength of 46.8 MPa, manufactured by Qianjiang cement factory was selected.

The river sand with clay content of 1.5%, and fineness modulus of 3.1, and the apparent density of 2611 kg/m^3 was used for the reference concrete. The sea sand from the Zhoushan area with a regional representative sampling points belonging to the East China Sea waters, sand mining site for nearly Qingfeng marina waters was used for the test concrete. The chloride ions, sulfate ions, shell content, clay content of sea sand is shown in Table 1.

The drinking water without affecting normal clotting, hardening substances was used for the reference concrete. The sea water from Zhoushan sea area was used for the test concrete. The salt content of sea water is shown in Table 2.

Table 1. Content of chloride ion, sulfate ion and total salt in sea sand.

Cl^- (%)	SO_4^{2-} (%)	Total salt (%)	Shells (%)	Mud (%)
0.1096	0.0091	0.1385	8.2	0.53

Table 2. Content of chloride ion, sulfate ion and total salt in sea water.

Cl^- (mg/mL)	SO_4^{2-} (mg/mL)	Total salt (mg/mL)
20.69	1.79	26.9

2.2 *Specimen preparation*

Concrete cubes of 150 mm in side length were prepared from three kinds of different strength grade concrete C30, C40, and C50. In each concrete strength grade, concrete cubes of different sets of shells and sea salt content were prepared. The influence of salt content and shell content of the test concrete were studied by measuring the mixing performance and compressive strength of the test concrete and the reference concrete.

2.3 *Effect of salt content on workability and compressive strength*

There kinds of different grade concrete were prepared from the same batch sea and sea sand. To exclude the influence of the shells, the content of shells was descent to the least by using square holes sieve to remove most of the shell. The salt content was changed by controlling the consumption of the sea the consumption of sea was 0%, 20%, 40%, 60%, 80%, 100% of 195 kg, which was shown in Table 3. The influence of concrete workability and compressive strength could be displayed by measuring the workability of the test concrete.

2.4 *Effect of shells content on workability and compressive strength*

There kinds of different grade concrete were prepared from the same batch sea which the content of shells was descent to the least by using square holes sieve to remove most of the shell. The salt content was controlled by using the drinking water. The content of shells was changed by adding the shells. The adding amount of shells was 8.2%, 10.2%, 12.2%, 14.2%, 16.2%, 18.2%, which was shown in Table 4. The influence of concrete workability and compressive strength could be displayed by measuring the workability of the test concrete.

Table 3. The mix proportion of concrete.

Sea water content	Strength grade of concrete	Shell content	Mud content	Cementitious (kg)	Fine aggregate (kg)	Coarse aggregate (kg)	Slump	Compressive strength
0%	C30	8.2%	0.53%	355	648	1203	16	28.2
	C40			398	614	1193	15	37.2
	C50			488	567	1133	16	46.6
20%	C30	8.2%	0.53%	355	648	1203	23	27.4
	C40			398	614	1193	25	29.9
	C50			488	567	1133	25	48.2
40%	C30	8.2%	0.53%	355	648	1203	50	27.4
	C40			398	614	1193	40	30.5
	C50			488	567	1133	35	48.6
60%	C30	8.2%	0.53%	355	648	1203	65	25.9
	C40			398	614	1193	60	30.1
	C50			488	567	1133	45	40
80%	C30	8.2%	0.53%	355	648	1203	90	29.9
	C40			398	614	1193	85	30
	C50			488	567	1133	75	42.3
100%	C30	8.2%	0.53%	355	648	1203	90	26.6
	C40			398	614	1193	93	28
	C50			488	567	1133	88	38

Table 4. The mix proportion of concrete.

Shell content	Strength grade of concrete	Salt content	Mud content	Cementitious (kg)	Fine aggregate (kg)	Coarse aggregate (kg)	Slump	Compressive strength
8.2%	C30	0.092%	0.53%	355	648	1203	80	26.6
	C40			398	614	1193	70	26.6
	C50			488	567	1133	65	38.9
10.2%	C30	0.092%	0.53%	355	648	1203	75	25.9
	C40			398	614	1193	70	30.9
	C50			488	567	1133	55	39.8
12.2%	C30	0.092%	0.53%	355	648	1203	80	26.6
	C40			398	614	1193	65	28.8
	C50			488	567	1133	53	37.6
14.2%	C30	0.092%	0.53%	355	648	1203	50	27.7
	C40			398	614	1193	60	30.5
	C50			488	567	1133	50	38.3
16.2%	C30	0.092%	0.53%	355	648	1203	50	26.9
	C40			398	614	1193	50	30.2
	C50			488	567	1133	48	41.1
18.2%	C30	0.092%	0.53%	355	648	1203	55	26.2
	C40			398	614	1193	55	35.1
	C50			488	567	1133	50	36.8

3 EXPERIMENTAL RESULTS

3.1 *The workability of the test concrete*

The slump of the test concrete was increased with the increasing of the salt content, but the water retention was decreased. The result was shown in Table 3. The deterioration of workability and cohesion of the test concrete was caused by the proportion of sea water greater than the conventional water. And the relative amount of mixing water was reduced with the increasing of the content of sea water. The slump and the water retention of the test concrete were decreased with the increasing of the shell content. The result was shown in Table 4.

Table 5. Slump of the reference concrete.

Strength grade of concrete	C30	C40	C50
Slump	90	112	105

Table 6. The values of the 28 days compressive of the reference concrete.

Strength grade of concrete	C30	C40	C50
Compressive strength	34	43.1	49.1

The reasons for this phenomenon may be due to the cyclotron resistance of the shells leading the mobility poor. The worsening of the workability and cohesion of the test concrete was induced by the decreasing of the effective rate of sand with the content of shells increasing.

Compared with the reference concrete, the slump and workability of the test concrete was descended because of the salt and shells. The result was shown in Table 5. The influence of the salt was bigger than the content of shells.

3.2 *The compressive strength of the test concrete*

The concrete cubes of 150 mm in side length were prepared for measuring the compressive strength of the test concrete. The cube specimens were divided into 39 groups. Six specimens of each group were tested using a 2000 kN capacity MTS machine. These values averaged from six tested specimens were listed in Table 3 and Table 4.

The values of 28 days compressive strength of the test concrete were listed in Table 3 and Table 4. And the compressive strength of the reference concrete was given in Table 6. The strength of concrete with sea sand and sea water is not vary with the changing of the content of the salinity and shells, but the early strength has declined comparing with the reference concrete, dropping by about 20 percent. With the increase of age, the strength of two kinds of concretes is quite.

4 SUMMARY AND CONCLUSIONS

This paper has presented the results of a systematic experimental program consisting by changing the content of shells and salt for the purpose of better understanding the influence of the content of shells and salt, and also to compare with the reference concrete. Based on the experimental results and the analysis, the following conclusions can be drawn:

- The slump of the test concrete approximately linearly increases, liquidity becomes larger, the cohesion deteriorates with increasing of the content of salt;
- With the increase of dosage of sea shells, the slump of the test concrete will gradually becomes smaller, water retention worse. Therefore, the amount of sand ration and mixing water should be increased and the sea sand with less shells should be choose in order to ensure the good and easy performance;
- The increase of salt content has little effect on the compressive strength of the test concrete, but compared with the reference concrete, the early strength of the test concrete declines, falling in the range of about 20%. With the age increased, the ultimate strength of the test concrete can reach the design strength;
- The compressive strength of the test concrete few decreases with the increasing of the shell content. In comparison, in the same conditions, the shell content on concrete high strength grade of concrete (such as the test of C50 sea sand concrete strength influence of seawater) should be significantly more shells, due to the existence of high grade concrete strength grade of concrete strength with more traditional shrink large prime;
- In the preparation of concrete with sea and sea sand, the curing time should be extended for ensuring the concrete that reaches the strength requirements. In addition, sand should be over 5 mm mesh sieve before use, to more than 5 mm particles. This can reduce the content of shells, the performance of the concrete with sea and sea sand can be guaranteed.

ACKNOWLEDGMENTS

The writers gratefully acknowledge the support for this work, which was funded by the Science and Technology Department of Zhejiang Province in 2013 (No. 2013C31126).

REFERENCES

[1] Ding Dehuai, Ke weixiao, Huang Xuzhi. The usage of the sea sand. Concrete.1992, (7):50–53.

[2] Hong Dinghai. The coastal area of reinforced concrete with sand problems and Countermeasures. Concrete. 2003, (2):17–18.

[3] Tan Weizhu. Present situation of development of concrete technology and the prospects for sustainable development [J]. Architecture Technology, 2006(4).

[4] Hu Jian, Hu Yi. Development and Countermeasures for sea sand resources in China [J]. Marine geological dynamic, 2005, 21(7): 4–8.

[5] Wu Zhehui. Experimental study on the influence of content of shells in the sea sand on concrete workability and compressive strength [J]. Fujian construction science and technology. 2012. No. 6:47–48.
[6] Hongnaifeng. Sand corrosion and harm of "sea sand house" [J]. Industrial Construction, 2004, 34(11): 65–68.
[7] Fuminori Tomosawa. Japan's experiences and standards on the durability problems of reinforced concrete structures. Int. J. Structural Engineering, Vol. 1, No. 1, 2009.
[8] Ismail H. Ç agatay. Experimental evaluation of buildings damaged in recent earthquakes in Turkey. Engineering Failure Analysis 12 (2005) 440–452.
[9] W.P.S. Dias, G.A.P.S.N. Seneviratne, S.M.A. Nanayakkara, offshore sand for reinforced concrete [J]. Construction and Building Materials 22 (2008) 1377–1384.
[10] J. Limeira, L. Agullo, M. Etxeberria. Dredged marine sand in concrete: An experimental section of a harbor pavement [J]. Construction and Building Materials. 24 (2010) 863–870.

Material Science and Environmental Engineering – Chen (Ed.)

Research on civil construction technology in architectural engineering

F.F. Yang
The Department of Architecture and Engineering, Qinhuangdao College, Northeast Petroleum University, Qinhuangdao, China

ABSTRACT: In the construction engineering, civil construction is the foundation of the whole project. It is characterized by multiple types of cross-link, complex construction, and a wide range. This requires the construction according to the requirements of civil engineering technology, in ensuring the quality of construction this paper discusses the preparation of civil construction technology, and a variety of construction technology.

Keywords: construction engineering; civil engineering; construction technology

1 INTRODUCTION

With the reform and opening up, China's rapid economic and social development, effectively promoted the city process in China, and increasingly large scale construction. In the whole process of construction of building engineering, all kinds of construction technology and mode and design have important influence on the whole building. Therefore, taking reasonable and effective construction technology and construction forms can greatly improve the overall quality of the building and effectively shorten the construction period.

Development and application of modern building technology, has injected new vigor and vitality to the development of China's construction industry. In this paper, construction technology of Architectural Technology in earthwork construction technology, concrete precast pile and low carbon building high-tech techniques are described and analyzed in detail.

2 ANALYSIS OF CIVIL CONSTRUCTION TECHNOLOGY

2.1 *Combining of construction technology and civil engineering*

To achieve mutual coordination of civil construction technology in the construction process of construction project, only mutual cooperation to build a high quality project, in order to ensure the quality of the project (Feng 2010). The author through the analysis that the concrete should do the following: first, mechanical equipment into the factory began to take measures to block the construction machinery and equipment from the technology. Those of construction machinery and equipment into play all sealed up, avoid using when some parts missing or damaged. Secondly, in the construction time, but also serious controlling each construction machinery equipment is what to do with. It can really bear the scope of construction and construction key. At the same time to achieve the set a professional management staff of the construction team leader, must go through professional management staff signature before use in the use of construction machinery and equipment.

Construction management personnel should strictly audit hoisting plan of machinery and equipment transportation construction engineering the declaration. To develop a detailed plan, it is necessary to take the detailed measures in implementation of mechanical equipment from the occupied space (He 2012). In addition, relevant professional and technical personnel should be reserved whole position detection, elevation and section, to ensure the pouring operation is completed in a comprehensive inspection qualified after concrete.

2.2 *The combining key in the construction engineering*

In the civil construction must carry out effective cooperation with various types of work, we must first to drainage professional construction as the key construction coordination. It should provide the leveling layer thickness and the roof insulation layer thickness. The civil construction only provide detailed and accurate data to ensure construction engineering position accurately, so as to ensure the water supply and drainage pipe the successful completion (Song 2010).

The coordination to the air-conditioning and HVAC construction aspect, in this respect also need to get the details of elevation, the specific location of the. For those tubers and louver wall to installation and decoration uniform, because only the unity of the material used, and elevation can easily control. In addition, in coordination with the electrical professional construction process should pay special attention to the laying of lighting line and line slot. Grading ring in case of high-rise building should pay attention to every three layers to setup and install lightning protection. Doing so is in order to effectively promote the building lightning protection effect.

3 MAIN CONSTRUCTION TECHNOLOGIES

3.1 *Construction technique of earthwork*

In construction engineering, mining, filling and transportation process of earthwork includes all the soil and drainage, precipitation, soil wall support and other preparatory work and aided engineering. The most common earthwork includes site formation, foundation pit, floor filling, and filling sugared excavation and foundation pit backfill. General requirements: design elevation of large engineering projects is usually to determine the site design of plane, ground leveling. Site formation is the transformation of natural ground plane adult is required.

Site elevation should meet the planning, production technology and transportation, drainage and the highest flood level requirements, and strive to make the site excavation and fill balance and the minimum earthwork. Method for determining design elevation: general methods such as site are relatively flat, no special requirements on site elevation, in accordance with the digging and filling earthwork equal principle to determine the site elevation.

In practical engineering, the design elevation calculated, also should be adjusted to consider the following factors, this is accomplished in the earthwork calculation after completion. Considering soil can eventually loose, needs to rise to reach the design elevation, the actual balance of earthwork. Considering the project more than soil or engineering with soil, the corresponding increase or decrease according to the design elevation. Economic comparison results, such as the use of OTC soil or waste soil construction plan, you should consider so earthwork volume caused by changes in need to be adjusted the design elevation. Adjust the work site design plane is heavy, such as modifying the design elevation, must re earthwork calculation. The earthwork construction scene is shown in Figure 1.

Figure 1. Earthwork construction scene.

3.2 *Construction technology of exterior wall structure*

Need to grasp the technical points of civil construction and construction of building exterior wall structure, detailed explanation on related technical problems in the construction, the only way to a greater extent to ensure the engineering quality. In the civil engineering specialty structure of air duct construction, must carry out the construction in strict accordance with the construction standard, should make full use of wall structure in construction engineering as air positive pressure ventilation. In fire protection system smoke duct and the air passage of the toilet in the construction should also pay attention to the implementation of the construction technology, attention should be paid to the steel wire mesh making a good fixation in the design of a good position, and in accordance with the corresponding proportion will glue and cement together after painting.

In the course of civil engineering construction also need to pay attention to clear mechanical equipment thrown mouth position and whirl openings way and section. Attention should be paid to the alignment check carefully building floor reserved holes and architectural design drawings on location, and the resulting data to record form related meeting minutes to do timely find out. To those new connected stand tutee detection pipe is also to see whether meet the design requirements of the general, for those who do not conform to the quality standard, must carry out rectification. Building exterior wall is shown in Figure 2.

3.3 *Construction technology of precast concrete pile*

In the modern construction, the general multistore buildings when used for natural shallow foundation are good. It has the advantages of low cost,

Figure 2. Building wall construction.

Figure 3. Construction case of precast concrete pile.

simple construction. If the natural shallow soil is weak, can use mechanical compaction, compaction, preloading, deep mixing, chemical strengthening methods such as artificial reinforcement, the formation of the artificial foundation.

Piling preparation must be done before the following preparations: remove the obstacles in the way of the construction on the ground and underground; construction site level; positioning line; set power supply, water supply system; installation piling machine etc. Positioning points and leveling points pile axis, should be set without affecting the location of piling, leveling point is not less than two. It can then check the deviation of pile position and the depth of pile in construction process. One construction case of precast concrete pile is shown in Figure 3.

3.4 *Construction technology of low carbon building*

Low carbon building refers to the use of the building materials and equipment manufacturing, construction and building of the entire life cycle, reduce the use of fossil energy, improve energy efficiency, and reduce carbon dioxide emissions. The current low carbon building has gradually become the main trend of international construction industry. "Low carbon buildings" low carbon building adopts the structure system, ground source heat pump system, intelligent wiring technology used in water use and other low-carbon power distribution system. The comprehensive utilization consists of solar energy, energy-saving doors and windows, rainwater collection. The low carbon building model is shown in Figure 4.

Low carbon building technology can not only greatly reduce energy consumption, and can make the emission level of building carbon decreases about 50%. Low carbon building mainly has the following several forms of technology, external wall energy saving technology: composite technology of wall with insulation layer, the external thermal insulation layer and a sandwich insulating layer three. China adopts the sandwich thermal insulation practice more, while in European countries, most of the use of external foamed polyphenyl board practices. In Germany, exterior insulation construction accounted for 80% of the total amount of the building, and 70% of them are used polystyrene foam board. Doors and windows energy saving technology: hollow glass, coated glass, high strength low fireproof glass using magnetron vacuum sputtering method plating metal containing silver layer of glass and the most special smart glass. Roof energy saving technology: to realize building energy saving by intelligent technology, eco technology wishes, such as solar energy roof and controlled ventilation roof.

Refrigeration and lighting is a major part of building energy consumption, such as the use of ground source heat pump system, radiant floor displacement ventilation system. The development and utilization of new energy: solar water heater,

Figure 4. A model of low carbon building.

photovoltaic roof panel, photoelectric wall board, photoelectric sunshade board, photoelectric wall between windows, skylights and photoelectric glass curtain wall etc.

4 CONCLUSION

Civil engineering is the most important part in the architectural engineering, civil construction technology is a complex and long work, is an important guarantee for the construction of the project construction quality, construction quality has a direct relationship with people's life and property security and national economic benefits. Therefore, we in the construction of civil engineering construction, summarize and research to constantly civil construction case of existing problems, and make the effective measures to improve the construction technology. Only serious grasp the civil construction quality, construction enterprises to obtain long-term development in the increasingly can fierce competition in the market.

REFERENCES

[1] Feng, K.L. 2010. Construction method of bored piles in civil construction machinery into analysis. *Heilongjiang Science and Technology Information*, 4(9): 209–210.

[2] He, R.G. 2012. The construction site management. *Modern Decoration* 3(11): 101–102.

[3] Song, H.T. 2010. The construction technology of plain concrete in civil construction. *Managers* 6(10): 371–172.

Material Science and Environmental Engineering – Chen (Ed.)
© 2016 Taylor & Francis Group, London, ISBN 978-1-138-02938-5

Coal mining on the analysis of the impact of building damage

C.L. Huang
Chongqing Vocational College of Building Engineering Management, Chongqing, China

ABSTRACT: For a long time, use of coal resources are our greatest resource, which is the economic development plays an important role. However, coal mining also caused many problems, such as groundwater damage, destruction of buildings, etc., seriously hampered the coal mining industry and other related things develop. The study carried out for coal mining damage caused by ground building issues, including coal mining caused by poor surface changes resulting in destruction of buildings, coal mining has brought a variety of changes in the surface extent of damage caused to buildings. And it achieved good results in this status quo building design, thereby reducing the extent of the damage mining buildings.

Keywords: coal mining; architectural design; architectural structure

1 INTRODUCTION

Underground mining is a common way in coal mining. After underground mining, surfaces above the mining section will deform necessarily since there is no underground support in the mining section, thus lead to some damages on buildings on surfaces, especially the fully mechanized sublevel caving mining, which has a very serious influence on buildings on surface. Aiming at this situation, appropriate mining technologies and measures with good effect shall be chosen to reduce the probability of building damages in mining cave-in area; moreover, it shall be started with architectural structures, making them more adapt to the change of surface. So the elements of surface changes and building damages caused by coal mining need to be studied and analyzed, and the resisting mining deformation architecture system with strong feasibility shall be chosen, which is of great practical significance in providing solid foundation for the long-term development of mining industry.

2 BUILDING DEFORMATION CAUSED BY SURFACE DEFORMATION

It is well known that solid foundation is the base of buildings, and deformations may occur if the foundation is not solid enough, and it will bring great negative effects on buildings. During the process of coal mining, surface changes must be occurred due to the extraction of underground mining, thus to influence the foundation of buildings, and lead to longitudinal or lateral, slant or curvature change finally. There are many kinds of surface changes caused by coal mining among them horizontal deformations are mainly tensile and compressive deformation, which produce forces on buildings in horizontal direction. Forces on buildings caused by surface deformation make the foundations of buildings that parallel to surface occur stretch or contraction in horizontal direction, and bending deformation in perpendicular direction, the specific deformations are shown in Figure 1. The foundation or wall of buildings will be damaged once the force of surface is more than a certain point.

The force on foundation is not balance when there is longitudinal deformation in surface, which cause additional bending moment and shear force, then buildings are affected. In most cases, buildings in steady sinking ranges after the stabilization of surface changes in mining sections, will occur changes due to the dynamic deformation of surface when coal mining, that is to say, buildings start to change due to the stretch (positive curvature) of

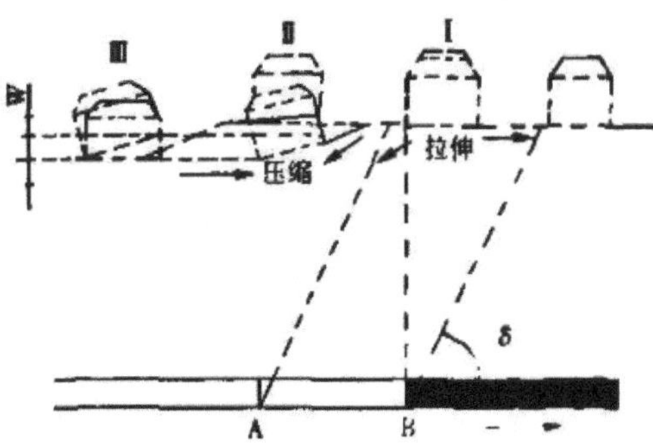

Figure 1. Building deformations.

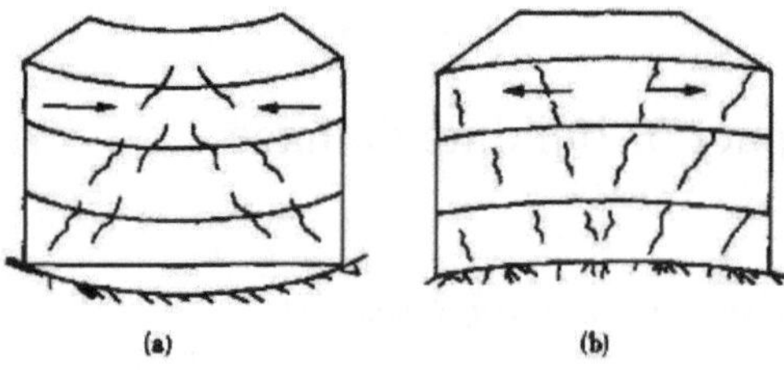

Figure 2. Building deformations caused by various kinds of surface reformations.

surface, then bear the effect of surface contraction gradually through the extension of working surface, the deformation of buildings will recover if the distance between working surface and buildings is higher than a certain specific value. So the building damage in steady sinking ranges is negligible if buildings are not damaged in the process of coal mining.

From the foregoing, there is such a relationship between building damage and coal mining, which is related to used materials of buildings, volume of buildings, depth of foundation, property of foundation, building loads, plane figure of buildings, the upper part rigidity of buildings and so on. In general, building damage is not caused by only one kind of reformation, and various kinds of reformations cause various kinds of curvatures and damage ways.

Curvature damage situation that buildings need to bear is shown in Figure 2.

3 THE MECHANISM OF BUILDING DAMAGE CAUSED BY COAL MINING

Since coal mining extracts underground mining from underground, the surface will occur deformation and translation in different degrees necessarily, and it may cause collapse, once these situations have effects on buildings, then these buildings will be damaged. Elements that cause this kind of deformation and damage mainly including the following aspects.

3.1 *Horizontal deformation*

Horizontal deformations occurred in surface usually manifest as stretch or contraction. Buildings tolerable level of surface stretch is lower than which of surface contraction, so the effect of stretch is very great, even a small degree of surface stretch may cause cracks of buildings. Holes (that is, doors and windows) are weak parts in a building, and it is the most difficult for them to resist the effect of surface deformation. Damages caused by surface contraction also manifest in the holes of buildings, which makes buildings extrusion deformation and there are also horizontal fractures in brick masonry walls. Of course, the degree of damage on buildings is also affected by other elements, especially rigidity and area of plane.

3.2 *Surface subsidence*

The degree of surface stable subsidence damage on buildings is very small, in most time there is no effect at all. It is not hard to understand this situation, and models can be used to simulate the situation of buildings in the process of surface subsidence, since the process of surface subsidence is quite gentle, buildings start to sink integrally, there is no centre-of-gravity shift caused by slant, which reduces the probability of occurrence of collapse and snap. And the amount of situations that cause surface stable subsidence is very few, it only happens in exploitation process of supercritical areas, and the probability is very low. This kind of situations can't produce high degree of damages, and buildings will not deform, there is only integral movement. If the degree of surface subsidence is very serious, and underground water level is very shallow, the rising of water level caused by coal mining would lead to the concentration of water in buildings, and result in softening of building foundation, thus lead to the appearance of safety problems of buildings.

3.3 *Surface slant*

Buildings will incline if the surface inclines, and buildings will collapse if the degree of slant exceeds what they can hold. Areas with surface slant are mostly the edge of gob surface, especially the position of inflection point. Surface slant will certainly cause building slant, which changes the gravity center of buildings, and collapses or snaps integrally when it can't afford any more. The damage of this situation on buildings with high height, such as water tower, is the most serious.

4 DESIGN AND STRENGTHENING OF BUILDINGS

4.1 *Building design*

There are some principles of building design shall be followed in coal mining section: (1) Reasonable planning, the relationship between building design and coal mining should be fully considered when designing, the range that can't be affected by deformation zones should be chosen; (2) The flat shape of buildings should avoid over complication in building design, and the main wall should be kept symmetry with the principal axis of buildings; (3) The situation that there is no flexibility in buildings should be avoided, which is used to deal with surface deformation and reduce the effect degree caused by surface deformation. Moreover, the rigidity and integrality of each constituent part should be enhanced and the affordability of surface deformation should be improved. The detailed design methods are as follows.

4.1.1 *Setting scientific deformation joint*

Deformation joint of buildings is the main mean that used to protect buildings in coal mining sections. Setting deformation joint not only has a good effect on reducing the damage degree of buildings, but also a measure with very low capital consumption. The location and size shall be determined with the structure and area of buildings and the kind of surface deformation when setting deformation joint. The size of each constituent part of buildings is limited by the horizontal deformation value and curvature deformation value of surface.

4.1.2 *Setting ring beam and constructional column*

It is necessary to set ring beam in the construction of buildings in coal gobs, which could improve the affordability of surface deformation effectively and enhance rigidity and integrality of buildings. The purpose of setting constructional column is to lower the possibility of walls being cut and improve the rigidity of buildings. Constructional column is set in the corner position of walls in buildings in most situations, and it needs to be connected with ring beam on both sides.

4.1.3 *Setting horizontal sliding layer*

Sliding layer means using building materials with strong lubricity. Setting sliding layer is for the separation of building foundation and upper structure, if there is a deform in surface, it will force on building foundations, friction appeared in sliding layer could be delivered to upper structures, among them some will be eliminated by sliding layer, reducing the force on buildings, which would lower damage degree of buildings to the limits and ensure the safety of buildings. The final purpose of setting horizontal sliding layer is to allow upper buildings to eliminate the force caused by surface deformation effectively, thus to improve the safety of structure and lower the damage degree of buildings caused by coal mining.

4.1.4 *Self-balancing high deformability buildings based on rise-fall dot foundation*

Rise-fall dot foundation is a kind of foundation architectures that rigidity is smaller than flexibility, and it can rise and fall freely; There are equipments adjusting horizontal deformation of buildings on the top of building foundations, which can not only remit horizontal deformations, but also defuse vertical deformations effectively, but the principle of choosing rigid measures in main structures does not change. The safety of buildings in coal mining sections can be ensured according to the fusion of "rigidity and flexibility", and the safety of buildings can be enhanced to the limits when fully mechanized sublevel caving mining.

4.2 *Strengthening techniques of buildings*

There are many strengthening methods to strengthen buildings in coal mining areas, and what kind of strengthening methods shall be chosen is determined by the specific situations of buildings.

Among strengthening methods, direct strengthening method is the method that improve the affordability of architectural structures by using multiple ways. After the strengthening of former structures is completed, other parts could share responsibility for partial load effect under the situation that structures are deformed, and the deformation of each strengthened component interface should be adjusted renewably; Adding building cross-sections, steel-encased, sticking steel plates, carbon fiber strengthening and many other measures can be used if direct strengthening of buildings is wanted. Indirect strengthening method, another method of strengthening methods, is the method that aiming at enhancing the safety of buildings by reducing the force of outside with different ways, among them the common measures are method of changing usage, method of changing or transferring force transmission line, shock insulation method and so on. The monolithic prestressing strengthening method is realized on the foundation of conception strengthening, it is a measure that strengthens buildings with force of steel tie rod by treating strengthening objects as a whole; this kind of strengthening method can reduce load, strengthen and coordinate the distribution of internal force and many other effects, and it especially applies to long-span structures.

5 CONCLUSION

Since underground mining is extracted in coal mining, surface collapse would occur in most situations, especially in the situation of oversize coal mining area, building damage caused by the surface depression in mining sections is more and more serious, it has become the main problem met in coal mining. The main surface deformation situations are surface depression, translation, stretch or contraction and so on when stopping, surface deformations would force on building foundations necessarily, which lead to the deformation and damage of buildings. Aiming at this situation, for reducing the damage of influenced buildings in mining cave-in area, not only suitable mining technologies, but also measures with good effects should be chosen to enhance the stationary of surface buildings, moreover, architectural structure shall be started with to make it more adapt to surface changes. So the elements of surface deformation and building damage caused by coal mining need to be studied and analyzed, and the resisting mining deformation architecture system with strong feasibility shall be chosen, which is of great practical significance in providing solid foundation for the long-term development of mining industry. In this paper, the writer starts with the elements of building deformation caused by coal mining, and finds the reasons of building deformation, hopes to provide powerful theoretical basis for the good development of mining industry.

REFERENCES

[1] Duan Jingmin, Qian Yongjiu. Anti Deformation Residential Building System in Area above Mining Goaf and Reinforced Technology [J]. Coal Engineering, 2008 (8): 44–46.

[2] Sun Wensheng. The Strip Mining of Research on Under-buildings [J]. China High-Tech Enterprises, 2012 (28): 126–127.

[3] Huang Hua. Aquifer Damage by Mining in a Coal Mine and its Prevention Measures [J]. China Science and Technology Review, 2012 (34): 10.

[4] Gao Qing, Gao Binyan, Fan Longgang, etc. Risk Evaluation and Predictive Prevention on the Ming Collapse of Suncun Coal Mine [J]. Urban Geology, 2013 (2): 28–30.

[5] Li Guijuan, Fang Fangrong, Lv Guangluo, etc. The Effect and Prevention of Ground Water in Cretaceous System of Binchang Mining Area on Coal Mining [J]. Shanxi Coal, 2013 (3): 52–53.

[6] Li Junhu. Discussions on Coal Mining Technologies in Coal Mining Under New Situations [J]. Journal of Shanxi Coal-Mining Administrators College, 2013 (2): 48–49.

[7] Hu Chun, Xu Xiaodong. Curriculum Development of Basic Practical Computer Course in Vocational Colleges Based on Vocational Ability—Taking Coal Mining Technology Specialty as an Example [J]. Science & Technology Information, 2012 (32): 10141.

[8] Ma Zhaoqiang, Ni Qiang. The Application of Continuous Miner Room-and-Pillar Mining Technology in Coal Mining [J]. Science & Technology for Development, 2012 (8): 139–140.

Material Science and Environmental Engineering – Chen (Ed.)
© 2016 Taylor & Francis Group, London, ISBN 978-1-138-02938-5

Post-earthquake dynamic stability reanalysis of Shapai arch dam abutment

H.Y. Zeng, J.H. Zhang, X.K. Liu & Z.Z. Wang
State Key Laboratory of Hydraulics and Mountain River Development and Protection, College of Water Resources and Hydropower, Sichuan University, Chengdu, China

W.G. Zhao & R.K. Wang
China Hydropower Engineering Consulting Group, Chengdu Design and Research Institute, Chengdu, China

ABSTRACT: Shapai RCC dam is a 132 m high arch dam, and it is one of the few examples withstood great earthquakes, so it is significant to reanalyze the dynamic stability of dam abutment. A new way of abutment dynamic stability analysis method combining the advantages of finite element method and limit equilibrium method is proposed, and the limit cumulative displacement formula is put forward and acted as the stability criterion of sliders. Dynamic stability analysis of Shapai arch dam's abutment is conducted under two seismic waves and the overload study of Shapai arch dam is conducted under the artificial waves reflecting the 5.12 earthquake. The results show that Shapai arch dam's abutment is stable under the impact of 5.12 earthquake waves ($a_h = 2.05$ m/s^2), which is consistent with the actual seismic effects. However, under the action of design waves with 1% exceeding probability level in 100 years ($a_h = 5.31$ m/s^2), the Shapai arch dam's left bank abutment is unstable, while on the right bank, the abutment keeps stable. When the left bank withstands the 3.2, 3.6 times artificial waves action, the instability happens on the left bank slider L1 and L2 respectively. The instability happens on the right bank slider when there is a 4.2 times overload. The slip direction and the influence of safety factor changing with time to cumulative displacement are considered. The presented new way of abutment dynamic stability analysis can not only reflect the influence of period when dynamic safety factor is less than 1.0, but also it duration time, and also the reverse sliding effect in the seismic action.

Keywords: abutment slider; dynamic stability; the cumulative displacement; overload analysis

1 INTRODUCTION

Shapai RCC arch dam is a 132 m high arch dam in Wenchuan earthquake zone, Sichuan, China. It is about 36 kilometers away from the Wenchuan 5.12 earthquake epicenter. No obvious damage is found in the Shapai arch dam body and abutment after the 5.12 earthquake, showing a good anti-seismic performance. Because the Shapai arch dam is one of a few hundred meters high dam withstood the great earthquake, the dynamic stability reanalysis of dam abutment is significant for the dam abutment serving as key support and force transmission mechanism, and being usually also weakest part of the arch dam to resist the seismic action.

The rigid body limit equilibrium method acts as a preferred method in most arch dam abutment dynamic stability analysis in existence, because this method is simple to use for adopting pseudo-static method. But the slider force changing with time and dynamic magnification effect cannot be reflected in this method. However, in the finite element method, the analysis result can reflect the time-history effect more accurately and the stress state can be computed at any time, but it is not convenient to evaluate the stability of dam abutment slider in FEM. In order to overcome the shortage of the above methods, many scholars have improved the finite element method and proposed a method by using stress and deformation value calculated by numerical calculation, the sliding force and resistance force are obtained by getting the integral value of stress on sliding surfaces, finally the safety factor is calculated. As early as 1999, Zhang et al. adopted the rigid body spring element method to get the dynamic safety factor time-history curve of dam abutment slider. Stianson and Fredlund et al.. proposed a way to solve slope slider safety factor based on interpolation technology, in this method the stress-deformation field is interpolated to slider surface grid nodes. Bao and Yang et al. put forward the three-dimensional multiple grid method, and this method is applied in the high arch dam abutment slider static safety factor calculation. Fang et al. have got the dynamic safety factor of abutment slider combining with the finite element method and limit equilibrium method. Zhang et al. put forward an abutment rock mass dynamic stability analysis method based on safety factor criterion. For the seismic dynamic analysis of Shapai arch dam, Cheng, et al., built the numerical model using ADINA software, simulated the process of seismic dynamic response, and reasonable calculation results were achieved. But the method only considered the mutation of key point displacement-time history curve, without taking into account the safety of the whole dam abutment slider. For the criterion judging the slider stability, Wieczorek and Jibson, et al., regard the critical displacement as a constant according to the engineering experience, without considering the size and mechanical parameters of slider.

2 CALCULATION THEORY

The FEM-LEM method of abutment dynamic stability is proposed based on Finite Element Method (FEM) and rigid Limit Equilibrium Method (LEM). In this method, the basic calculation steps of the method are: firstly, the grid of dam-foundation system and the sliders' grid are built, and the latter is independent from the former. Meanwhile, the static stress field and dynamic stress field for each step are computed in FEM. Secondly, the slider's surface stress fields are obtained by interpolation. Then along the slider surface, the node stress integral value is decomposed into total sliding force and resistance force using formulae (1) and (2).

$$F_H = \vec{R} \cdot \vec{s} \quad (1)$$

$$F_Z = \int_{a_c} \left(c + (\sigma_n - p) f\right) da \quad (2)$$

In which: F_H is the total sliding force, F_Z is the total resistance force, $\vec{s}$ is the sliding direction determined as Goodman, R.E, $\vec{R}$ is the resultant force, c is cohesion, σ_n is the normal stress of sliding face, p is the seepage stress acting on sliding surface nodes, f is the friction coefficient.

The dynamic safe factor F_s can be calculated by using formula (3).

$$F_S = \frac{F_Z(t)}{F_H(t)} = \frac{F_Z(t)}{\vec{R}(t)\vec{s}} \quad (3)$$

In the formula: F_s is the instant safe factor at the moment t, $F_Z(t)$ is the total resistance slide force at time t, $F_H(t)$ is the total sliding force at time t, $\vec{R}(t)$ is the resultant force at time t and $\vec{s}$ is the block sliding direction.

Thirdly, the acceleration is projected to the sliding direction when the safe factor is less than 1.0, and the projected acceleration is integrated to get the slider's cumulative displacement. There are two assumptions when calculating the cumulative displacement:

1. The slider is a rigid body with cohesion and friction on sliding surface.
2. The most dangerous direction is the same with the direction in the static state and under the action of seism, and the direction of maximum cumulative displacement is the same as the most dangerous sliding direction. Figure 1 shows the sliding direction and forces on a slider at any time.

In which:

$S(t)$ is sliding force at time t,

S is sliding force along the most dangerous sliding direction $\vec{s}$,

$R(t)$ is resistance force at time t,

R is resistance force along the most dangerous sliding direction $\vec{s}(t)$.

$\alpha\left(0 \le \alpha \le \pi\right)$ is the angle between the $\vec{s}$ and $\vec{s}$(t), the slider's total mass is m. From Figure 1, the slider's acceleration $a_m(t)$ along the most dangerous sliding direction at time t can be expressed as:

$$a_m(t) = \cos\alpha \cdot \left[S(t) - R(t)\right]/m \quad (4)$$

In the dynamic calculation, the acceleration value can be obtained at any moment, so the velocity V_m (t) at time t can be achieved using formula (5).

$$V_m(t) = V_m\left(t - \Delta t\right) + \Delta V_m \quad (5)$$

In which, ΔV_m can be calculated as follow:

$$\Delta V_m = a_m(t) \cdot \Delta t \quad (6)$$

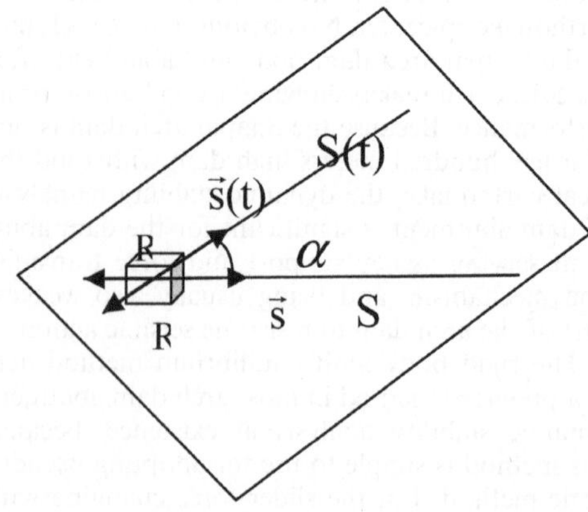

Figure 1. The slider sliding direction and force diagram.

Then the cumulative displacement $S_m(t)$ at time t can be got by formula (7).

$$S_m(t) = S_m(t-\Delta t) + \int_{t-\Delta t}^{t} V_m(t)dt \quad (7)$$

Finally, a limit cumulative displacement formula is proposed in (8) acting as the stability criteria combining the block size and mechanical property parameters, and the block's dynamic stability is judged by comparing the limit cumulative displacement and the calculated value in (7).

$$S_{\lim} = \beta R_t L / E \quad (8)$$

In formula (8), $S_{\lim}$ is the limit cumulative displacement, L is the length of slider along the sliding direction, β is an adjustment factor, usually it is 1.0, E and R_t is elastic modulus and tensile strength of weak plane of geotechnical materials. This formula shows the deformation rate $S_{\lim}/L$ of slider weak plane is proportional to the limit tensile strain.

3 THE PROJECT PROFILE OF SHAPAI ARCH DAM

The Shapai hydropower station located in Wenchuan country, Tibetan and Qiang autonomous Aba prefecture, Sichuan province, China. It is a three center single arch dam, the azimuth of dam body centre line is N55°~60°E, the vault elevation is 1867.0 meters and the maximum dam body height is 132 meters. The normal storage level is 1866.00 meters, and the dead water level is 1825 meters, the upstream silt elevation is 1796 meters. It's a reservoir with 18 million m³ total storage capacity, and the installed capacity of hydropower station is 16000 KW, the power output is 179 million kWh per year.

In the Shapai dam site, there is deep valley and both sides of the natural valley are steep slopes with some place goes up to more than 200 meters high. The downstream slope rock mass is assigned III classification, the dam shoulders of both bank and upstream slope are mainly the second rock belt classified as IV-2. The geological structure of slope rock mass is a system consisting of many joints and fissures. The dam plane cutting profile of 1830 meters elevation is shown in Figure 2.

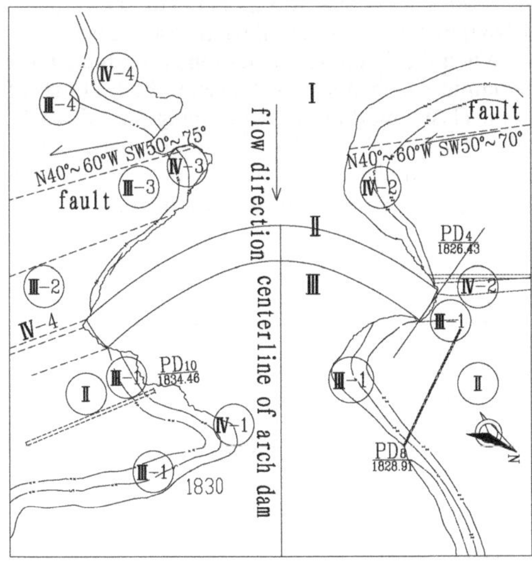

Figure 2. 1830 m elevation cutting profile of Shapai dam.

4 CALCULATION MODEL

4.1 *Numerical model of the arch dam*

The size of Shapai arch dam finite element model is 500 meters along x axis, 425 meters along y axis, the bottom elevation of the model is 1600 meters in vertical direction, and the top elevation is 2000 meters extends to the mountain surface. The dam's crest elevation is 1867.5 meters, under the foundation surface with 1740.0 meters elevation, the thickness of bedrock is 1.1 times of dam height. The whole calculation domain is decomposed into 6301 nodes and 5491 elements. The three-dimensional grid stereogram is shown as Figure 3. The rock mass mechanics parameters for calculation are shown in Table 1.

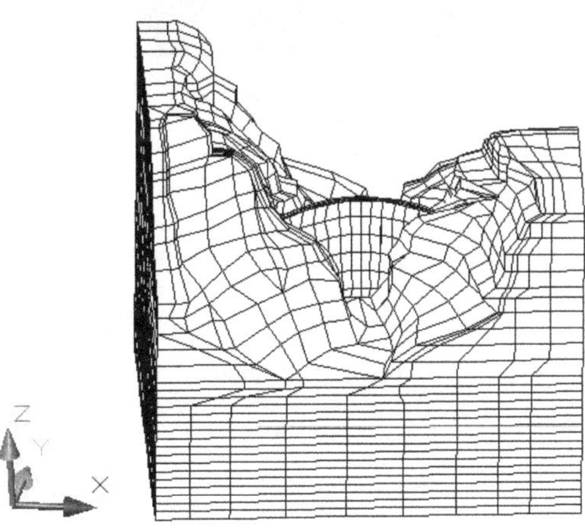

Figure 3. The finite element model of Shapai arch dam-foundation system.

4.2 *The grid and parameters of typical slider*

When studying the stability of Shapai arch dam, 5 control sliders are selected cut by geological structural surfaces, 3 of them located on the left bank, the other two double-sided sliders located on the right bank. To be concise, only sliders L1, L2 on left bank and R1 slider on right bank (Fig. 4) are discussed. In this study, the dynamic safety factor and accumulative displacement of each block were calculated under two earthquake conditions. The strength parameters and azimuth of typical sideslip surface and bottom sliding surface of each slider is given in Table 2.

Table 1. Mechanics parameters of rock and dam mass.

Rock class	Deformation modulus E (GPa)	Poisson ratio μ	Bulk density t/m^3	Shear strength f	Shear strength C (MPa)
Dam body	18.0	0.167	2.4	1.1	0.9
Class II	11.0	0.23	2.75	1.15	1.0
Class III	9.0	0.25	2.75	0.95	0.85
Class IV	4.5	0.3	2.75	0.65	0.35
F_1 fracture	3.0	0.32	2.6	0.55	0.15
Class V and overburden	0.4	0.35	2.5	0.35	0.02

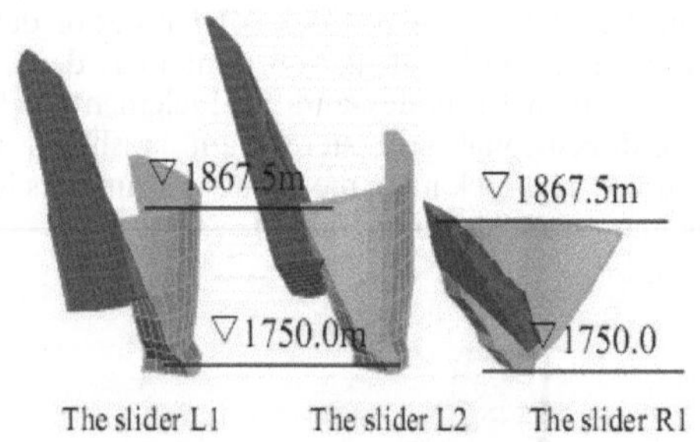

Figure 4. Typical slider grid figure.

5 CALCULATION CONDITIONS

In the static calculation, the loads considered include gravity load, water pressure, silt load and temperature load. The upstream water pressure load is 1860 meters elevation and the downstream is 1750 meters elevation. The silt elevation is 1796.0 meter; silt internal friction angle is 0°.

When the dynamic calculation is conducted, the dynamic elastic modulus is taken to be 1.3 times of static elastic modulus. The dynamic loads for two earthquake conditions is artificial seismic wave with peak acceleration $a_h = 2.05$ m/s^2, which is deemed to be peak acceleration during 5.12 earthquake in the dam site after back analysis and design wave with peak acceleration $a_h = 5.31$ m/s^2, which is 1% exceeding probability in 100 years. Artificial wave is three acceleration time-history curves based on the synthesis solution of Wenchan 5.12 earthquake epicenter mechanism and recorded acceleration of strong earthquake motion nearby. The peak acceleration value of horizontal, down to river and vertical direction are 2.05 m/s^2, –2.05 m/s^2 and 2.05 m/s^2 respectively. The design wave with 1% exceeding probability in 100 years is provided by design institute, the peak acceleration value of horizontal, down to river and vertical direction are –5.31 m/s^2, 5.31 m/s^2 and 5.31 m/s^2 respectively. The three directional acceleration time-history curves of artificial wave and design wave with 1% exceeding probability in 100 years are shown in Figure 5.

In the static calculation, the loads considered include gravity load, water pressure, silt load and temperature load. The upstream water pressure load is 1860 meters elevation and the downstream is 1750 meters elevation. The silt elevation is 1796.0 meters; silt internal friction angle is 0°.

When the dynamic calculation is conducted, the dynamic elastic modulus is taken to be 1.3 times of static elastic modulus. The damping ratio $\xi = 0.03$. The dynamic loads for two earthquake conditions is artificial seismic wave with peak acceleration $a_h = 2.05$ m/s^2, which is deemed to be peak

Table 2. The sliding orientation and strength parameters.

Number	Bottom elevation	The sideslip direction and the value of C f: Steep surface	Slow surface	Downstream surface
L1	1810	N50°E/SE75° C = 0.54 f = 0.9	N30°E/SE20° C = 0.68 f = 1	N60°W/SW∠70°
L2	1830	N50°E/SE75° C = 0.54 f = 0.9	N30°E/SE30° C = 0.68 f = 1	N40° W/SW∠70°
R1	1850	N30°E/NW60° C = 0.55 f = 0.9	N45°W/NE30° C = 0.70 f = 1	

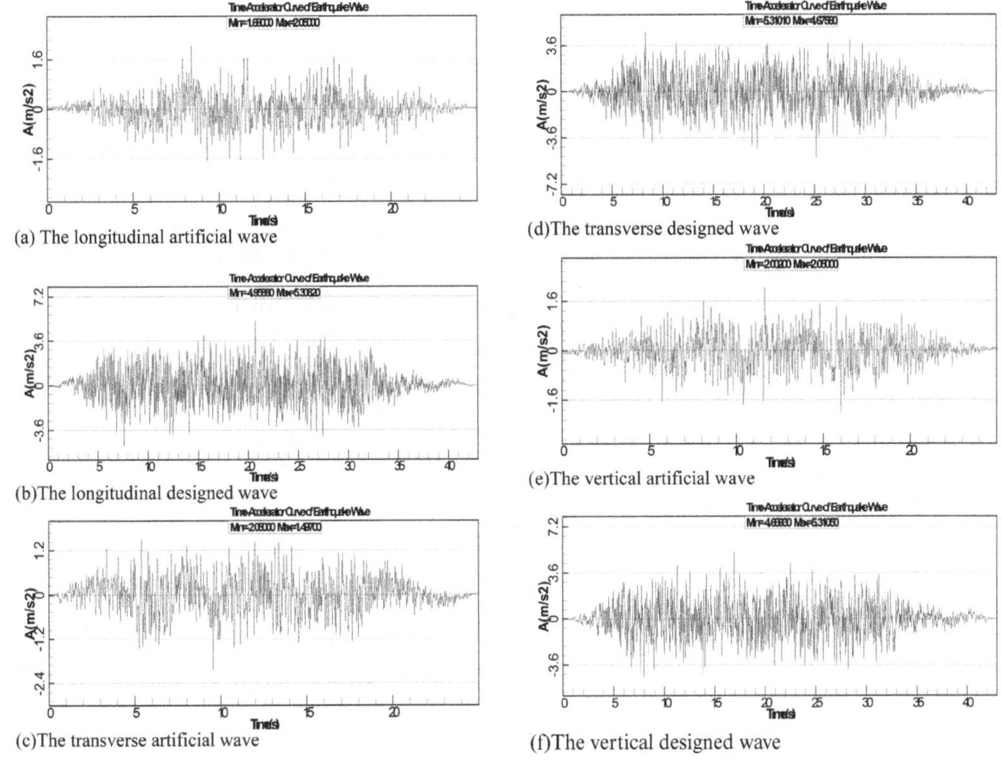

(a) The longitudinal artificial wave

(b)The longitudinal designed wave

(c)The transverse artificial wave

(d)The transverse designed wave

(e)The vertical artificial wave

(f)The vertical designed wave

Figure 5. The waves of seismic time-histories curve.

Table 3. Load combinations for calculation.

Condition	Load	Water pressure	Silt pressure	Dynamic load	Sliders
1	Dead weight Temperature decline	Normal water level	Upstream silt	Artificial wave	L1 L2 R1
2	Dead weight Temperature decline	Normal water level	Upstream silt	Designed wave	L1 L2 R1

acceleration during 5.12 earthquake in the dam site after back analysis and design wave with peak acceleration $a_h = 5.31$ m/s^2, which is 1% exceeding probability in 100 years.

Artificial wave is three acceleration time-history curves based on the synthesis solution of Wenchan 5.12 earthquake epicenter mechanism and recorded acceleration of strong earthquake motion nearby. The peak acceleration value of horizontal, down to river and vertical direction are 2.05 m/s^2, -2.05 m/s^2 and 2.05 m/s^2 respectively. The design wave with 1% exceeding probability in 100 years is provided by design institute, the peak acceleration value of horizontal, down to river and vertical direction are -5.31 m/s^2, 5.31 m/s^2 and 5.31 m/s^2 respectively. The three directional acceleration time-history curves of artificial wave and design wave with 1% exceeding probability in 100 years are shown in Figure 5. Table 3 shows the load combination of two earthquake conditions.

6 CALCULATION RESULTS

In order to determine the limit cumulative displacement for each block based on the formula of dynamic stability criterion, the formula (8) is used to get the limit cumulative displacement $S_{\lim}$

of each block, the calculation results is shown in Table 4.

The process of calculation is as follows: at first, the sliding force and resistance force of sliding surface at each moment during earthquake are computed with formula (1) and (2). Then formula (3) is used to get the dynamic safety factor for each slider at each moment. As the dynamic safety factor is always greater than 1.0 under condition 1, all the sliders are safe, while in some period it is less than 1.0 under condition 2. Here only the dynamic safety factor time-history curve under condition 2 is shown in Figure 6. For the period of safety factor less than 1.0, the cumulative displacement for each slider is obtained based on formula (7).

Table 4. The limit cumulative displacement value.

Slider	Length along sliding direction (m)	Cohesion C (MPa)	Internal friction angle φ (°)	Limit cumulative displacement $S_{\lim}$ (cm)
L1	65.60	0.68	44.71	9.31
L2	73.4	0.68	44.71	10.42
R1	140.5	0.70	44.71	20.52

Assuming Tc is the total time when dynamic safety factor F_s is less than 1.0, and T is the total earthquake time, then Tc/T represents the percentage of time when dynamic safety factor is less than 1.0. To facilitate the comparative analysis, the range of dynamic safety factor F_s and the percentage Tc/T in two conditions are summarized in Table 5.

Table 5 shows that the minimum dynamic safety factor is greater then 1.0 under artificial waves, so the limit cumulative displacement for each block

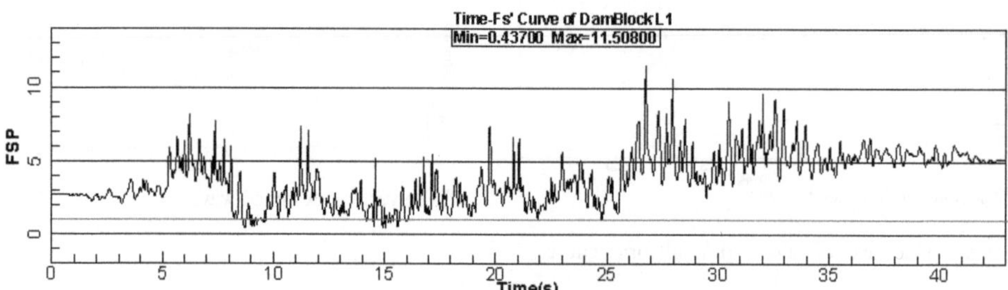

(a) The dynamic safe factor time-history curve of slider L1 in condition 2.

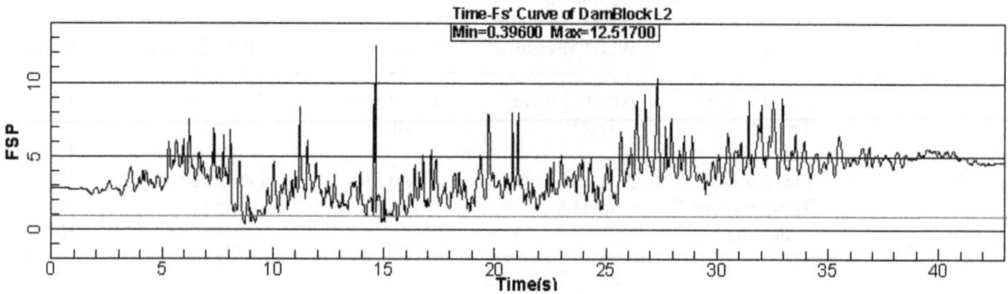

(b) The dynamic safe factor time-history curve of slider L2 in condition 2.

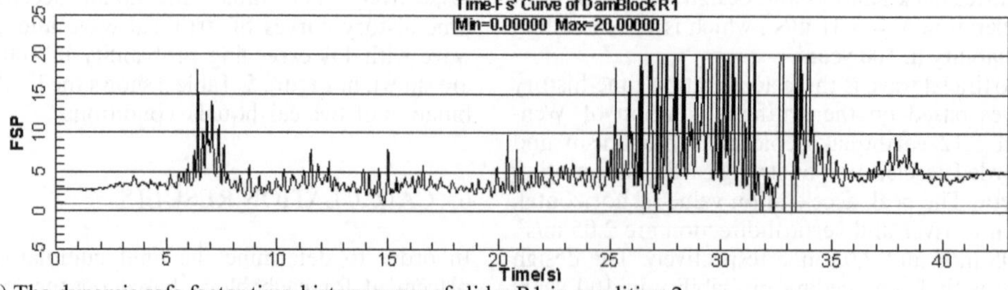

(c) The dynamic safe factor time-history curve of slider R1 in condition 2.

Figure 6. The safety factor time-history curves of dam abutment sliders.

Table 5. The safety factor result of all working condition.

Slider number	L1		L2		R1	
Calculated condition	F_s	Tc/T	F_s	Tc/T	F_s	Tc/T
1	1.20~6.63	0.00	1.35~6.36	0.00	2.41~4.38	0.00
2	0.44~11.51	2.37%	0.40~12.52	1.68%	0.00~20.00	1.49%

is 0. It is consistent with the actual situation that the Shapai arch dam abutment is safe and stable under the Wenchan 5.12 earthquake. In larger seismic waves as under condition 2, the results show that the value Tc/T of left bank abutment slider L1 is bigger than slider L2, while the Tc/T values of slider L1 and R1 are almost the same. The minimum safety factors of all sliders are less than 0.5. For the left blank sliders L1, L2, the minimum safety factor value is 0.44, 0.40 respectively, while for right bank it is 0.0. In order to judge the stability of block, the cumulative displacement of block is checked through applying twice integrals of slider acceleration. The time-history curves of limit cumulative displacement S_{lim} calculated in formula (8) and cumulative displacement in FEM-LEM method for all blocks are shown in Figure 7. The stability of block is determined by comparing if the peak cumulative displacement S_m has exceeded S_{lim}.

Figure 6 and Figure 7 show consistent increase of cumulative displacement during earthquake in condition 2, while for the right block, the cumulative displacement reach to the maximum value in 26~33 seconds, then it began to decrease gradually. For the left blocks, the reason of the result is the steep sliding surface, whose acceleration projection vector is always along the dangerous direction, so the cumulative displacement increased with time. While for the right bank, the sliding surface is relatively flat, the acceleration projection vector can be the same or opposite with the dangerous direction in different time. The presence of reversed sliding leads to decrease of displacement along sliding direction.

Under the action of designed waves in condition 2, the cumulative displacement of left bank slider L1, L2 is 68.536 cm, 55.779 cm, both of them are much bigger than the limit value in Table 4, so the two blocks are instable under this condition. For right blank block R1, the cumulative displacement is 6.623 cm, which is far less than the limit value, so the block on right bank is stable.

By using the criterion of limit cumulative displacement, it is implied that the dynamic safety factor instantaneously less then 1.0 does not mean the loss of stability of the whole slider. So the advantage of this criterion is distinct. It has not only reflected the influence of dynamic safety factor less than 1.0 but also take into account of the duration time, and also considered the slider size affection and sliding direction of slider.

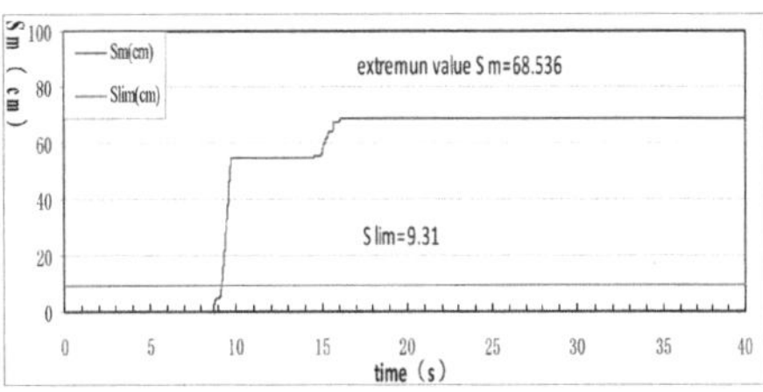

The cumulative displacement time-history curve and the S_{lim} value of slider L1

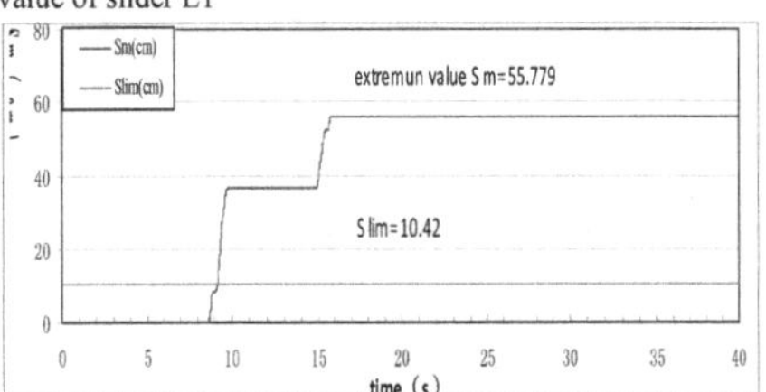

The cumulative displacement time-history curve and the S_{lim} value of slider L2

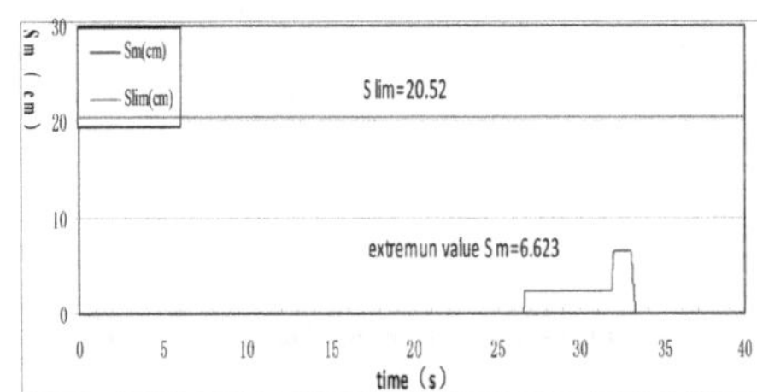

The cumulative displacement time history curve and the S_{lim} value of slider R1

Figure 7. The cumulative displacement time history curve and the S_{lim} of left and right dam abutment slider under the second working conditions (design waves).

7 DYNAMIC OVERLOAD ANALYSIS

The cumulative displacement of Shapai arch dam abutment slider is zero under the action of artificial wave. In order to further test the arch dam's anti-seismic capacity, the dynamic wave was amplified with K_d = 1.2, 1.6, 2.0, 2.4, 3.0, 3.4, 3.8, 4.2, 4.4 times respectively, the stress fields were obtained under different overload waves. The overload ratio-cumulative displacement curve of each slider were shown in Figure 8.

The instability criteria of block show that bock will be unstable when cumulative displacement is greater than the limit value. Figure 8 indicates the overload ratio K_d is 3.2 for the left bank block L1

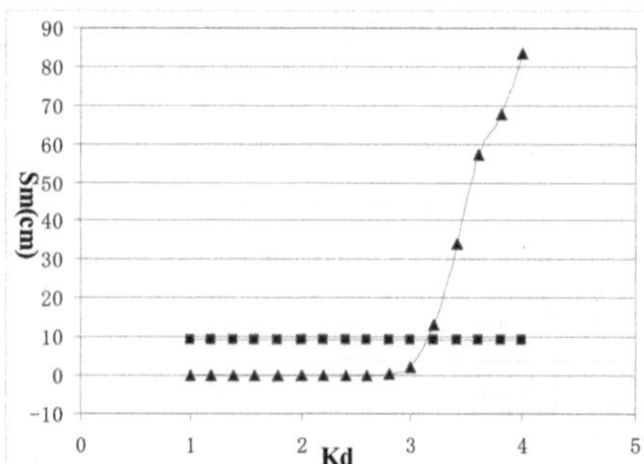

The overload ratio-cumulative displacement curve of L1

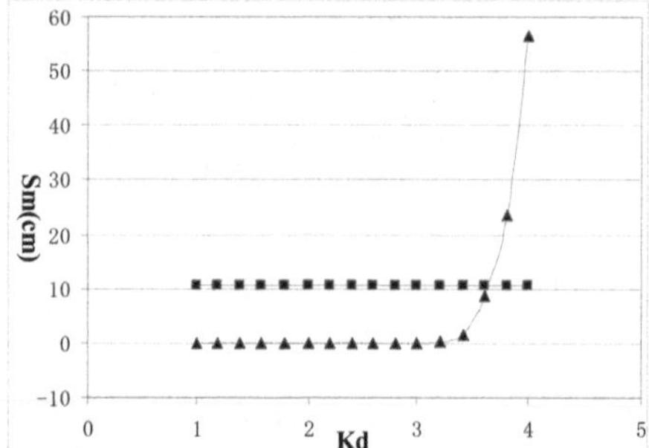

The overload ratio-cumulative displacement curve of L2

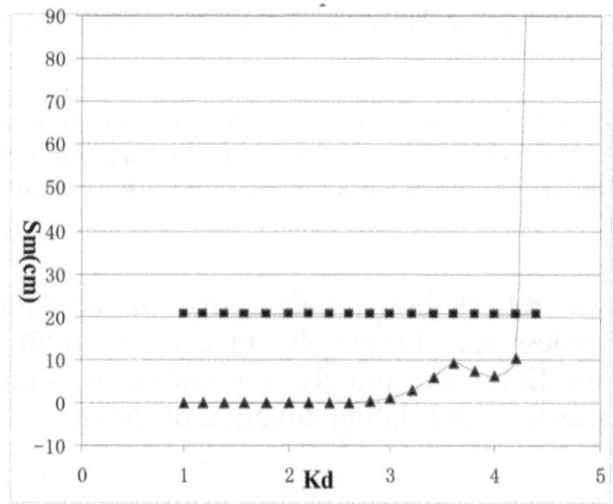

The overload ratio-cumulative displacement curve of R1

Figure 8. The overload ratio-cumulative displacement curve of dam abutment slider.

and K_d = 3.6 for the block L2. The cumulative displacement value is bigger than the limit value when the overload ratio is 4.2 for the right bank block R1. In brief, the dynamic stability of right bank is better than left bank.

Figure 8 show that L1 and L2 loss stability when overload factors are 3.2 and 3.6, or $a_h = 6.56$ m/s^2 and $a_h = 7.38$ m/s^2. And Figure 7 shows L1 and L2 loss stability under designing wave ($a_h = 5.31$ m/s^2), which is much smaller than Figure 8 indicated. The seemingly paradoxical results show that the overload factor depends not only on the value of a_h, but also the wave frequency, and the wave pattern.

8 CONCLUSION

The advantages of finite element method and limit equilibrium method are integrated, an instability criterion of slider is proposed based on the limit cumulative displacement formula. The dynamic stability analysis of Shapai arch dam abutment is carried out under the action of two seismic waves. Meanwhile, dynamic overload analysis has been carried out by inputting amplified artificial wave. Analysis of dynamic stability of Shapai arch dam shows that:

1. The cumulative displacement of each block is 0 and the minimum value of dynamic safety factor is greater then 1.0 under the action of artificial wave, it means that the abutment of Shapai arch dam is stable under the action of Wenchuan 5.12 earthquake. Under the action of design wave with 1% exceeding probability in 100 years, 2 blocks on left bank are unstable but the block on right bank is stable, showing that the stability of right bank is better than left bank.
2. By comparing the cumulative displacement S_m and limit cumulative displacement S_{lim} of each block, the new criterion of cumulative displacement method has demonstrated certain practicability and superiority. It can not only reflect the influence of period when dynamic safety factor is less then 1.0, but also its duration time, and the reversed sliding effect in the seismic action.
3. By observing intersection of cumulative displacement S_m curve with limit cumulative displacement S_{lim}, the overload factor K_d can be determined. It is found that different wave input with same peak acceleration a_h may result in different overload factors K_d, showing that the overload factor depends not only on the value of a_h, but also the wave frequency, and the wave pattern.

REFERENCES

[1] Zhang Bo-yan. 2001 Chen Hou-qun. Analysis on abutment aseismatic stability by using finite element method and rigid body limit equilibrium method [J]. Chinese Journal of Rock Mechanics and Engineering 20(5): 665–670. (in Chinese)

[2] Chai Jun-rui. 2003. Review on dynamic stability analysis of rock mass in high arch dam abutment [J]. Rock and Soil Mechanics 24:7666–673. (in Chinese)

[3] Zheng Hong. 2007. A rigorous three-dimensional limit equilibrium method [J]. Chinese Journal of Rock Mechanics and Engineering 26(8):1529–1536. (in Chinese)

[4] Su chao, Li jun-hong, Ren qing-wen. 2003. FEM-based dynamic stability analysis of abutment for high arch dams [J]. Journal of Hehai University (Natural Sciences) 31(2):144–147. (in Chinese).

[5] Li Meng, Chang Xiao-lin. 2008. FEM-based dynamic stability analysis of arch dam abutment [J]. Hubei Water Power 2:7–9. (in Chinese)

[6] Zhang Jian-hai, Fan Jin-wei, He. Jiang-da. 1999. Dynamic safety evaluation of slopes or dam foundations using rigid body-spring element method [J]. Chinese Journal of Rock Mechanics and Engineering 18(4):387–391. (in Chinese)

[7] Stianson J.R.A. 2008. Three-Dimensional Slope Stability Method Based on Finite Element Stress Analysis and Dynamic Programming [D]. University of Alberta.

[8] Bao. T.F, Xu. B.S. 2011. Hybrid Method of Limit Equilibrium and Finite Element Internal Force for Analysis of Arch Dam Stability against Sliding [J]. China Tech 54(4): 793–798.

[9] Bao Jin-qing, Yang-Qiang, Guan Fu-hai. 2011. A improved 3-D multi-grid method and its engineering application [J]. Journal of Hydroelectric Engineering. 30(2): 112–117. (in Chinese)

[10] Fang Wei, Liu Xiao-qing. 2011. Analysis of stability of Gaixia arch dam abutment [J]. J of Three Gorges Univ. (Natural Sciences) 33(1):17–20. (in Chinese)

[11] Zhang Kui-jing, Zhang Liao-jun, Zhu Yin-ru et al. 2012. A dynamic anti-sliding stability analysis method for abutment fractured rock mass based on safety index criterion [J]. Rock and Soil Mechanics. 33(7): 2160–2166. (in Chinese).

[12] Wieczorek, Wieczorek G.F., Wilson R.C., Harp E.L. et al. 1985. Map showing slope stability during earthquakes in San Mateo County, California. Miscellaneous Investigation Maps I-1257-E, U.S.G.S.

[13] Jibson R.W. and Keefer D.K. 1989. Analysis of the seismic origin of landslides examples from the New Madrid seismic zone. Geological Society of American Bulletin 105, 521–536.

[14] Goodman, R.E. 1985. Block Theory and its Application to Rock Engineering [M], N.J. Prentice-Hall.

Material Science and Environmental Engineering – Chen (Ed.)
© 2016 Taylor & Francis Group, London, ISBN 978-1-138-02938-5

Constructing healthy building's outside environment base on CPTED theories

K.Q. Wang & J.Y. Wang
Jilin Jianzhu University, Changchun, Jilin Province, China

ABSTRACT: Personal defense safety is essential premise and important foundation to construct positive and healthy living environment and it's an issue that should be taken great attentions on human habitat environmental design. Building's outside environment are carriers and conditions of crime, CPTED emphasizes on reducing the opportunities of potential offenders implementing the crime. The paper puts forward some kind of tactics on how to build a healthy living environment base on CPTED theories.

Keywords: building's outside environment; defense safety; healthy; CPTED

1 INTRODUCTION

Safety is the basic premise and important support to realize positive and healthy living environment, it is a issue that should be taken seriously on human habitat environmental design. One of building's most original appeal is to block wind and rain, to avoid cold and heat, and to prevent from damage, the original injury prevention is mainly to guard against animal attack or natural disaster, etc., now with the relationship is getting complicate, it's scope expands to prevent interpersonal injury. Today's human living environment design has responsibilities to ensure the safety of broad sense, including defense safety, behavior safety, psychology safety and disaster safety.

Building's outside environment are carriers and conditions of crime, there are close relationship between crime and building's outside environment, favorable outside environment of buildings are powerful means of crime prevention. For architecture and urban planning, personal defense safety is mainly reflected in crime prevention through environmental design, short for CPTED, it emphasizes on increasing the difficulties of potential criminals implementing crime, limiting and eliminating the conditions that facilitate access to crime, enlarging the possibility of being found and being prevented, reducing the opportunities of potential offenders implementing the crime, thereby increasing the safety of potential victims. the researches on "CPTED" went deep gradually in the developed countries in recent years, and applied to the practices of design and public safety administration. The theories and practices have proved that it could reduce the occurrence of crime through the effective environmental design.

2 THE KEY IDEA OF CPTED THEORIES

CPTED is short for crime prevention through environmental design, CPTED theory put forward a number of strategies such as natural surveillance, access control, territorial reinforcement, image building, facilities maintenance and activities support. Among them, natural surveillance strategy aim to maximize monitoring to space activities, and make all kinds of abnormal behavior can be found in time, and then let the activities within the space of residents feel more secure; Access control strategy is mainly focus on controlling the approaching ways to potential crime targets, so as to increase the risk of committing crime, to reduce crime; Territorial reinforcement is mainly to create a clear sense of space and private ownership, and do not welcome any outsiders trespassing; As for the image building, facilities maintenance and activities support strategies, the main basis points are increasing environmental vitality and popularity, and avoiding "broken Windows" environmental phenomenon. Certainly, there are not absolutely independent among various strategies, they have intersection set and common joints each other (see Fig. 1).

The key idea of CPTED theory is reducing the opportunities and increase the risks as perceived by potential offenders. It rests on the assumption that crime do not simply criminals act on impulse, they make choices through perceiving risk in a given situation and then controlling over whether they take action or not.

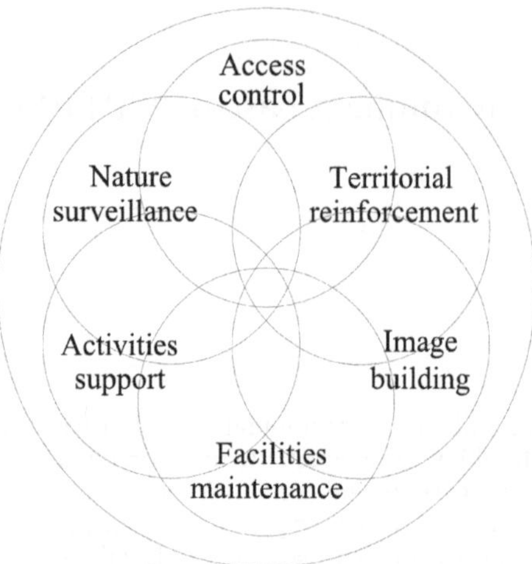

Figure 1. CPTED strategies that interweave with each other.

3 SAFETY IS THE BASIC DEMANDS OF HEALTHY LIVING ENVIRONMENT

3.1 *Healthy human habitat environment design should pay special attention to safety issues*

Safety is one of the basic demands of human beings, so defense safety design as the crucial content of safety design is a issue that should be taken seriously on human habitat environmental design (including architecture, planning, landscape, etc.). American humanistic psychologist Abraham Maslow published his book named "A Theory of Human Motivation" in 1954, in his book he divided the people's needs from low level to high level into five hierarchies which are "Physiological needs, Security needs (Safety needs), Social needs (Love and belonging), Esteem, and Self-actualization needs", security needs is one of the most original and basic requirements. It is the one of important aspect of human living environmental design. The design of architecture and planning should guarantee not only the demands of normal functions, but also strive to ensure people's safety demands. Defense safety design is the essential means of creating safety human settlements, and it is the necessary link and important content of human living environmental design, which including building's outside environment design, and as well it is the embodiment of the human-based design idea.

3.2 *Healthy building's outside environment should be defensible space*

American architect Oscar Newman put forward the concept of "defense safety design" in the book named "Defensible Space"(Newman 1972) in 1972. Healthy building's outside environment should be defensible space. With the negative influence of physical space environment protrudes increasingly, the safety characters of healthy building's outside environment getting more and more attentions gradually. People get to know the significances for crime prevention gradually through effective design of planning, architecture, landscape and other material elements. And by strengthening the natural surveillance, creating sense of territory, building positive space, achieving access control, constructing unique architectural image. As well as with the aid of the improvement and control of the physical or psychological characteristics of outside environmental factors such as location, distance, territory, visibility, accessible, attraction, cohesive force, and control force, so as to make up the space blind spots, time blind spots, society blind spots and mentality blind spots of the outside environment where easy to implementing crime, thus making the crime target do not easy be access and hurt, and the criminal process easy be discover, criminal do not convenient to escape after crime, thus to make harder on the implementation of criminal behavior, more risk, excuse decreases, and finally achieve the purpose of constructing safety environment. These above security strategies and basis points, should be as a guiding idea of healthy living environment design.

4 HEALTHY BUILDING'S OUTSIDE ENVIRONMENT STRATEGIES BASE ON CPTED THEORIES

4.1 *Experience of west developed countries*

Substantial research has shown that criminals avoid well-used residential areas where their activities might easily be observed (Jacobs 1961, Merry 1981, Rhodes & Conley 1981). So, the zones are relatively safer where have enough surveillance, or close to high-activity areas or are bound up with organized functions such as administrative office or sales booth or concession stand. Positive living environments are key factors in the development of neighborhood social ties and the discouragement of potential criminals because they encourage neighborliness and enforce informal surveillance. Contact among neighbors and informal surveillance are, in turn, linked to strength of community and lower crime rates (Taylor 1988). In addition, plant size, density, genre and form are also the important contents of constructing healthy outside environment of building. For example, Public housing residents with nearby trees and natural landscapes reported 25% fewer acts of domestic aggression and violence (Kuo & Sullivan 2001a).

Public housing buildings with greater amounts of vegetation had 56% fewer violent crimes than buildings with low amounts of vegetation (Kuo & Sullivan 2001b).

4.2 *Basic strategies of CPTED*

The potential criminal carry out a crime must have three elements which are as follow: environment conditions factors of crime be sufficient, subject conditions factors of crime be sufficient and object conditions factors of crime be sufficient. The outside environment where is vulnerable to crime is the overlap area of the three ingredients (see Fig. 2).

Healthy outside environment can reduce fear, and increase citizen's natural surveillance. Jane Jacobs suggested nearly 50 years ago that a well-used city street is safer than an empty street (Macdonald & Gifford 1989). Safety, she argued, is guaranteed by people who watch the streets every day because they use the streets every day. She also supported mixed-use buildings as a way to increase social interaction. So the simple presence of more "eyes on the street" would deter crime. Rational criminals would choose to commit a crime in a place where cues suggest weaker social organization and neighborhood involvement and avoid in the areas where with greater surveillance and greater likelihood of intervention. Surveillance related to defense safety design.

For example, in order to restrict access of potential crime subject, to stop irrelevant crowds going in, to use physical methods prohibit or delay the potential crime subject entering or access, we can control the form and the number of intersections of road, control the width of roads, set up reasonable road hierarchy and road network density,

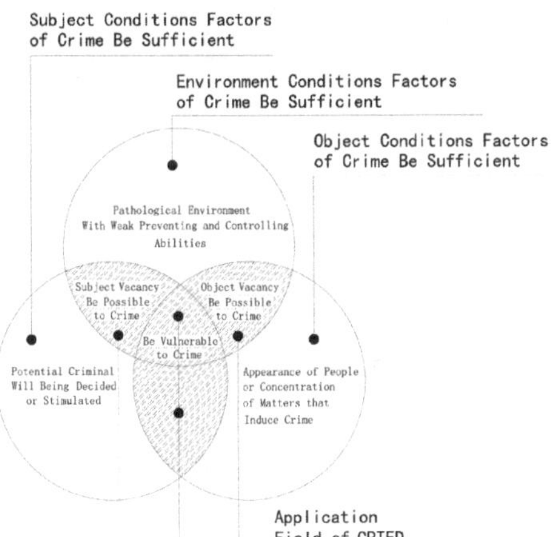

Figure 2. The outside environment that vulnerable to crime has three characteristics.

set speed obstacles, clear spatial mark character, strengthen the road illumination and neighborhood monitoring. In order to strengthen the control of domain of crime target, mainly control the boundary, entrance, scope of perceive, we can set up plants, fences or walls around particular areas to stop pedestrians crossing, using special logo to impede trespassers moochers or alcoholics; Use of public art, sculpture, flags and banners to do tag; Control the space size in order to strengthen contact among neighbors; Strengthen territoriality using landscape, decorative columns or fences to define territory or property ownership, and so on.

5 CONCLUSION

With the emergence of high-rise buildings in the city, the traditional blocks and approachable scale architectural form are gradually disappearing, because of the interaction between people is less, and lack of necessary communication among people, it is hard to form a harmonious neighborhood relationship, this lead to interpersonal indifference, city public security blind spots increases. To some extent, giving full play to crime prevention and control effects of building's outside environment design, can eliminate some induced factors of crime, reduce the occurrence of violence, criminal behavior that adverse building's outside environment bring about, to reduce the negative space in the building's outside environment, and to create a positive "humanistic space" and "defensible space", it can promote the communication and mutual cooperation between people, improve the neighborhood relationship, increase the natural supervision to strangers, build safe production and living environment, enhance people's sense of security, it can effectively improve the quality of the residents' production and living environment. So, carrying out CPTED improves the quality of human habitat environment effectively.

ACKNOWLEDGMENT

This paper is Supported by science and technology plan projects of the Ministry of Housing and Urban-Rural Development, 2013 (2013-R2-16); The Research Fund for the Doctoral Program of Jilin JIANZHU university.

REFERENCES

[1] Oscar Newman. Defensible Space: Creating Prevention through Urban Design. The Macmillan Publishing, 1972:264.

[2] Jane Jacobs. The Death and Life of Great American Cities. New York: Random House, 1961:44.
[3] Merry, S.E. 1981. Defensible Space Undefended. Urban Affairs Quarterly: 16:397–422.
[4] Rhodes, W.M. and C. Conley. 1981. Crime and Mobility: An Empirical Study. In: P.J. Brantingham, and P.L. Brantingham (eds.), Environmental Criminology. Sage, Beverly Hills, CA. 167–188.
[5] Taylor, R.B. 1988. Human Territorial Functioning: An Empirical, Evolutionary Perspective on Individual and Small Group Territorial Cognitions, Behaviors, and Consequences. Cambridge University Press, NY, 351.
[6] Kuo, F.E., and W.C. Sullivan. 2001a. Aggression and Violence in the Inner City: Effects of Environment via Mental Fatigue. Environment and Behavior 33, 4:543–571.
[7] Kuo, F.E., and W.C. Sullivan. 2001b. Environment and Crime in the Inner City: Does Vegetation Reduce Crime? Environment and Behavior 33, 3:343–367. online summary.
[8] Macdonald, J.E., and R. Gifford. 1989. Territorial Cues and Defensible Space Theory: The Burglar's Point of View. Journal of Environmental Psychology 9:193–205.
[9] Cozens Paul. Public health and the potential benefits of Crime Prevention through Environmental Design. [J]. New South Wales public health bulletin, 2007, 1811–12.

Material Science and Environmental Engineering – Chen (Ed.)
© 2016 Taylor & Francis Group, London, ISBN 978-1-138-02938-5

Technical feasibility study of unfired brick with coal gangue at the Wulanmulun site, Inner Mongolia, China

Z.B. Yu, H.T. Peng, Y.D. Zhu, J. Li, Q. Zhao, M.H. You & X.P. Zhang
College of Water Conservancy and Civil Engineering, China Agricultural University, Beijing, China

ABSTRACT: The production of unfired brick with coal gangue is an effective way for coal gangue utilization. The mineral composition of the coal gangue was ascertained by rotating anode X-ray polycrystalline diffractometry. The main mineral composition of coal gangue is quartz, plagioclase, calcite, etc. The main chemical composition is SiO_2 and Al_2O_3. Under the action of the cementing agent and admixture (alkali slag), the hydration reaction of active SiO_2 and Al_2O_3 can form C-S-H gelation, ettringite, tobermorite, etc. The strength of unfired brick increased with increasing content of alkali slag in the range of 10–16%, decreased with increasing content of coal gangue. For unfired brick with coal gangue at the Wulanmulun Site, the suggestion of suitable proportion of raw materials is: coal gangue content ranged from 60 to 75%, cement content ranged from 8 to 12%, slag content ranged from 15 to 25%.

Keywords: coal gangue; unfired brick; mechanism of strength increase; proportion

1 INTRODUCTION

Coal gangue is industrial waste from the processing of coal-mining production and coal washing. According to preliminary statistics, per tonne of coal mining can produce 0.2 tonne of coal gangue, per tonne of coal washing can produce 0.65–0.7 tonne of coal gangue (Zhang et al. 2013). At present, coal gangue have accumulated more than 5 billion tonnes in China. The amount will continually increase with 300–350 million tonnes per year. By 2020, annual emissions will increase to 729 million tonnes of coal gangue (Ge et al. 2010). The wulanmulun site has proven coal reserves of nearly 10 billion tonnes. With the increase in coal production, coal gangue emissions increases. A large amount of storage and are discharging gangue which is not only a pollution source but also a kind of available resources. As long as coal gangue is reasonably used, the waste can be turned into useful product. Coal gangue used to produce unfired brick is an effective way of coal gangue utilization.

2 MINERAL AND CHEMICAL COMPOSITIONS

There is a lot of coal gangue at the wulanmulun site, Inner Mongolia, China. In order to study the technical feasibility of the coal gangue of wulanmulun site produce unfired brick, need to analysis coal gangue mineral compositions and chemical compositions. Quantitative phase analysis of coal gangue with rotating anode X-ray polycrystalline diffractometry. Coal gangue of wulanmulun site mineral compositions and chemical compositions are shown in Table 1.

Through the analysis, the main mineral compositions of coal gangue are quartz, plagioclase, calcite, etc. Mineral containing Si elements accounted

Table 1. Mineral compositions and chemical compositions of coal gangue.

Mineral composition	Quartz	Plagioclase	Clinochlore	Calcite	Siderite	Microcline		Muscovite	
Content	33%	17%	15%	13%	11%	6%		5%	
Si and Al element	Mineral containing Si element				Mineral containing Al element				
Content	≥60%				≥30%				
Chemical composition	SiO_2	Al_2O_3	$FeCO_3$	$CaCO_3$	Fe_2O_3	CaO	Na_2O	MgO	K_2O
Content	46.0%	17.7%	10.0%	11.8%	6.3%	2.3%	1.2%	3.2%	1.5%

for more than 60%, including quartz, plagioclase, clinochlore, etc. Mineral containing Al elements accounted for more than 30% of the total, including plagioclase, clinochlore, etc. And also contains calcite which has a certain mechanical strength. The main chemical compositions of coal gangue include the SiO_2, Al_2O_3, $FeCO_3$, $CaCO_3$ and Fe_2O_3. SiO_2 and Al_2O_3 is necessary condition for forming the calcium silicate hydrate and hydration calcium aluminate. The SiO_2 accounts for 46.0% of coal gangue, the Al_2O_3 accounts for 17.7% of coal gangue, the ratio of SiO_2 to Al_2O_3 is about 3:1, this is the sufficient basis for complete hydration reaction. The $FeCO_3$ accounts for 10.0% of coal gangue, the Fe_2O_3 accounts for 6.3% of coal gangue, this can improve the efficiency of the hydration reaction. Therefore, for mineral composition and chemical compositions, it's feasible to produce unfired brick of coal gangue at wulanmulun site, Inner Mongolia, China.

3 MECHANISM OF STRENGTH OF UNFIRED BRICK

The main chemical composition of coal gangue is SiO_2 and Al_2O_3, similar to the clay, but there is no spontaneous combustion of coal gangue almost no activity. After the coal gangue mixed with water, with the introduction of alkali activator can make the solution containing a large number of hydroxyl ion. Hydroxyl ions can destroy some covalent bond, such as Si–O–Si, Al–O–Al and Al–O–Si in the high degree of polymerization glassy state network. Destroyed of covalent bonds become the unsaturated reactive bond, prompting depolymerization and solution-diffusion of silicon and aluminum (GU. 2007). In the presence of activator, dissolution of SiO_2 and Al_2O_3 active substances can participate in secondary reaction of coal gangue and cement hydration products, forming stable and does not dissolve in water of hydration products, such as calcium silicate hydrate, hydrated calcium aluminate, ettringite, tobermorite, etc. (Wu et al. 2011) All the particles in the system bonding together by the hydration products which mutual coupling by a way of contact form locked particles. And acicular ettringite fill in the pore can promote the densification of structure and compressive strength of unfired brick increased gradually. C-S-H gelation can improve the strength of unfired brick, tobermorite can improve the fire resistance of unfired brick, ettringite can improve the swelling property and stability of unfired brick. C-A-H can accelerate the hydration reaction in the formation of unfired brick. The reaction process of raw materials is shown in Figure 1.

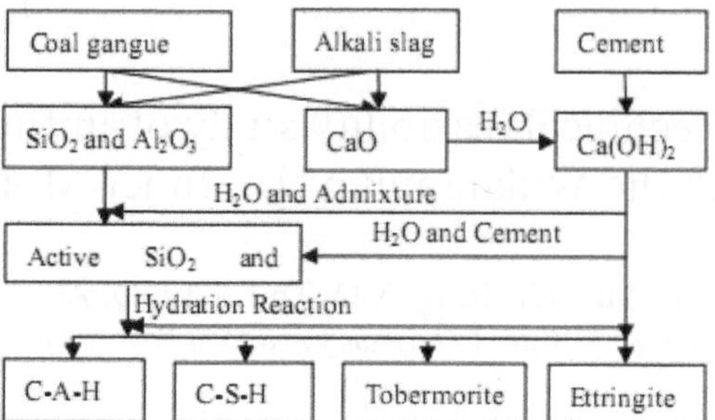

Figure 1. The process of reaction between different raw materials.

4 PROPORTION OF COAL GANGUE UNFIRED BRICK

At present, China had considerable research results of coal gangue unfired brick. Most of the coal gangue unfired brick with coal gangue as the main raw material; slag, sand and cement as admixture. Collation and analysis the research results from raw materials selection, raw material ratio, process conditions and unfired brick performance. The research results contrast of coal gangue unfired brick are shown in Table 2. Table 2 shows that the admixtures have positive effect on the strength of unfired brick, the autoclaved curing is more beneficial than normal curing to the properties of unfired brick. Table 2 shows that the compressive strength of unfired brick can fully meet the quality requirements of the highest grade (MU15-grade) of unfired bricks in the Chinese standard (JC/T422-2007), and the highest compressive strength even reached 37.25 MPa.

Unfired brick strength mainly depends on the proportion of raw materials. The relationship between compressive strength and slag content (Fig. 2) shows that the strength of unfired brick increased with increasing content of slag in the range of 10–16%. The relationship between compressive strength and cement content (Fig. 3) shows that the strength of unfired brick increased with increasing content of cement. The relationship between compressive strength and coal gangue content (Fig. 4) shows that the strength of unfired brick decreased with increasing content of coal gangue. The analysis of raw material suitable proportion is shown in Figure 5. Within a certain range, with the decrease of the proportion of coal gangue, the increase of the proportion of cement and slag, compressive strength of unfired brick generally showed a trend of increase. The quadrilateral region is the optimal proportion of raw materials as shown in Figure 5. Coal gangue content ranged from 60 to 75%, cement content

Table 2. The research results contrast of coal gangue unfired brick.

	CSL[e] (MU15)	(Wu et al. 2011)	(Y. 1993)[g]	(Zhang et al. 2014)	(Jin. 1998)	(Wu. 2005)	(Sun et al. 2014)	(Yue et al. 2014)	(Su and Gong. 2005)	(Wang et al. 2002)
Raw material										
Coalgangue (Wt.%)	/	78	78	50	60	80	40	85	80	60
Sand (Wt.%)	/	/	/	30	20	/	/	/	/	/
Cement[a] (Wt.%)	/	10		20	10	5	9	10	10	13
Alkali slag (Wt.%)	/	12	/	/	10	/	/	15		10
Admixtures (Wt.%)	/	0.05	/	/	/	2	/	/	0.01	/
Process										
Curing Condition	/	AC[f]	Nature	Nature	Nature	Nature	Nature	AC[f]	Nature	Nature
Performance of Unfired brick										
WC[b] (Wt.%)	<18	14.87	/	/	/	/	/	18.16	/	15.4
CS[c] (MPa)	>15	16.1	20.53	37.25	20	20	12	29.08	≥20	20.4
FTC[d]										
CS[b] (MPa)	>12	12	17.47	35.86	15.7	/	/	/	/	/
WC[c] (Wt.%)	<18	9.34	8	/	13.8		/	/	/	/

a. Ordinary Portland cement; b. Water content; c. Compressive Strength; d. The unfired brick performance after 15 freezing–thawing cycles; e. The highest Chinese standard for unfired rubbish gangue brick (MU15); f. Autoclaved curing at 120 °C, 20 °C/h; g. Coal gangue is 78 ± 7%, cementing agent and admixture are cement, lime and gyps.

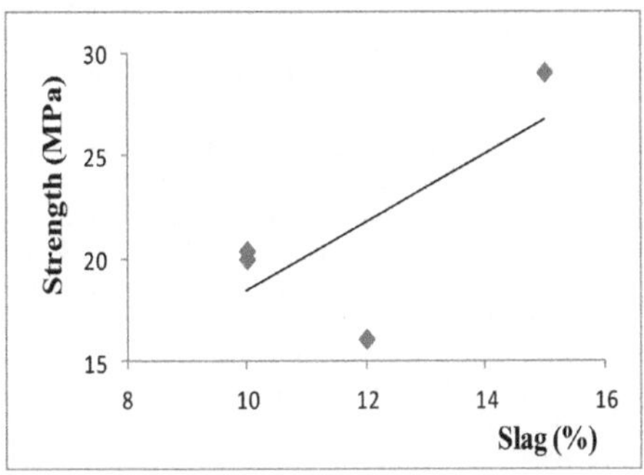

Figure 2. Relationship between strength and slag content.

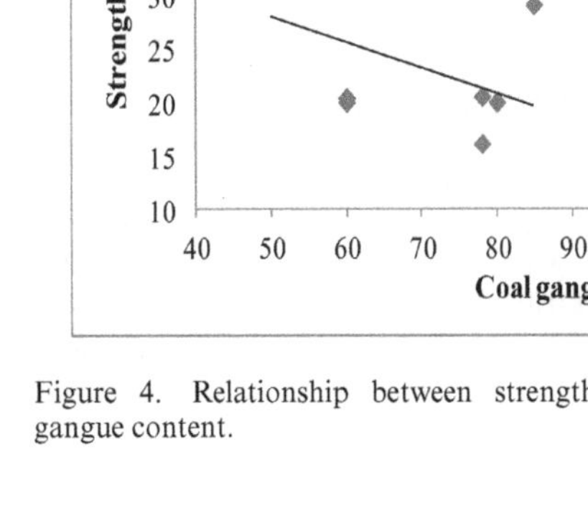

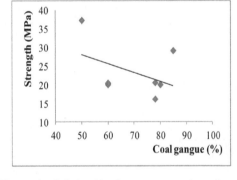

Figure 4. Relationship between strength and coal gangue content.

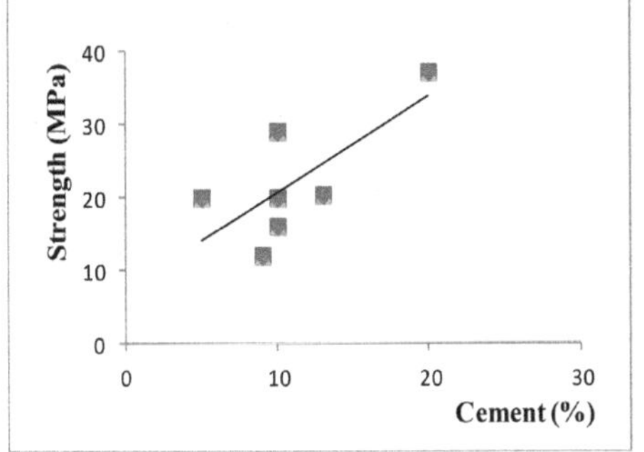

Figure 3. Relationship between strength and cement content.

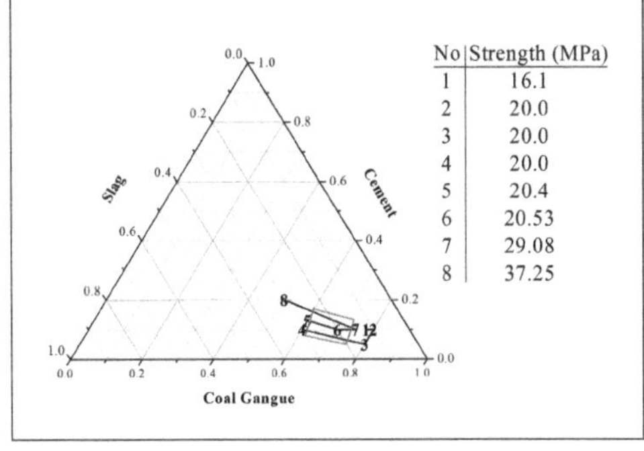

Figure 5. Analysis of raw material suitable proportion.

ranged from 8 to 12%, slag content ranged from 15 to 25%.

5 CONCLUSIONS

1. A vast amount of coal gangue of emission and accumulation not only occupies the farmland and land but also causes severe pollution to the environment due to spontaneous combustion, the rains wash, argillation, etc. The sulfide from coal gangue will pollute the atmosphere, soil and water. And coal gangue used to produce unfired brick is an effective way of coal gangue utilization.
2. Through analysis coal gangue mineral compositions and chemical compositions, it's feasible to produce unfired brick of coal gangue at wulanmulun site, Inner Mongolia, China.
3. The strength of unfired brick increased with increasing content of slag in the range of 10–16%, increased with increasing content of cement, decreased with increasing content of coal gangue.
4. For unfired brick of coal gangue at the Wulanmulun site, the suggestion of suitable proportion for raw materials is: coal gangue content ranged from 60 to 75%, cement content ranged from 8 to 12%, slag content ranged from 15 to 25%.

ACKNOWLEDGEMENTS

The authors gratefully acknowledge the fund support from Innovation and Entrepreneurship training program of Beijing Undergraduate (No. 35030010).

REFERENCES

[1] Ge, H.L., Du, H. & Zhou, C.X. 2010. Harmfulness of coal gangue and it's recycling utilization and development trend. *Coal Technology* 29 (07): 9–11.
[2] Gu, B.W. 2007. Stimulate and Forecast of China from different areas of the intrinsic characteristics of the gangue and ash activity. Materials Science and Engineering, Tong Ji University.
[3] Jin, C.G. 1998. Using coal gangue produces baking-free brick. *Coal Processing and Comprehensive Utilization* (02): 44–45.
[4] Su, J.G. & Gong, B.Y. 2005. The method of manufacturing the environmental and highpressure non-fired bricks of phosphogypsum.
[5] Sun, Z.H., Liu, K.P. & Wen, J.R. 2014. The method of manufacturing the environmental and highpressure non-fired bricks of phosphogypsum.
[6] Wang, H.S., Wang, M.W. & Huang, X.Q. 2002. Non-fired hollow brick.
[7] Wu, D.J. 2005. The method of manufacturing the environmental and highpressure non-fired bricks of phosphogypsum.
[8] Wu, H., Guo, G.Y. & Wu, S. 2011. Effect of Process Conditions on Properties of the Poor Quality Gangue Based Unfired Brick. *Non-Metallic Mines* 34 (05): 35–37.
[9] Wu, H., Kong, D.S. & Li, Z. 2011. Preparation of Unfired Brick from the Poor Quality Gangue. *Bulletin of the Chinese Ceramic Society* 30 (04): 895–898.
[10] Yi, G.X. 1993. Research non-fired coal gangue brick. *Coal Science and Technology* (11): 34–37.
[11] Yue, Q.Y., Jiang, T.T. & Yue, D.T. 2014. The method of manufacturing the environmental and high pressure non-fired bricks of phosphogypsum.
[12] Zhang, K.F., Wu, X. & Yang, W. 2013. Research progress of utilizing coal gangue as resource building materials. *Materials Review* 27 (S1): 290–293.
[13] Zhang, S.K., Zhang, X.D. & Zhang, Y. 2014. Experimental Study on No-roasting Bricks Made from Coal Gangue. *Non-Metallic Mines* 37 (05): 46–48.

Material Science and Environmental Engineering – Chen (Ed.)

Life cycle assessment of the concrete prepared by pulverized fly ash

C. Song & H. Shi
CCCC Second Harbor Engineering Co. Ltd. The 6th Branch Company, China

ABSTRACT: The cement production get widespread concern with high-energy consumption and greenhouse gas emissions. And that can reduce greenhouse gas emissions in concrete by adding Supplementary Cementitious Materials (SCM), such as slag and fly ash. The environment impact of ordinary concrete, fly ash concrete, pulverized fly ash concrete of the C30 and C50 were assessed by the method of Life Cycle Assessment (LCA). The results have indicated that the pulverized fly ash concrete had the lowest CO_2 emission among the three kinds of concrete. Compared with the ordinary concrete and fly ash concrete, the CO_2 emissions of pulverized fly ash concrete have decreased by 40%, 27% respectively in the C30 concrete. And the CO_2 emissions of pulverized fly ash concrete have decreased by 8%, 16% respectively in the C50 concrete.

Keywords: pulverized fly ash; supplementary cementitious materials; LCA; environment impact

1 INTRODUCTION

The cement and concrete industry is one of the key sources of the greenhouse gas. Its CO_2 emissions account for 15–20% CO_2 emissions associated with human activities. And the CO_2 emissions in cement production account for 80% of concrete production. That has accounted for nearly 5% CO_2 emissions associated with human activities [1–2], in China has already accounted for more than 15% [3]. The CO_2 emissions associated with concrete is mainly concentrated on the cement production, therefore it has been the focus that how to reduce the amount of cement used in concrete. Currently the main method to reduce CO_2 emissions is using mineral admixtures to replace cement such as fly ash, slag, etc., which can not only reduce the cement dosage, also have the excellent performance [4]. Shen Weiguo has invented a scattering-filling coarse aggregate technology through stratified dispersal of coarse aggregate to reduce the cement amount in the concrete, improving the coordination environment of concrete materials [5–6]. The industry solid waste is vast in our country, which could reduce the cement in the concrete through replacing a portion of cement. That is an economical and feasible way of energy saving and emission reduction to promote the comprehensive utilization of industrial solid wastes and make social benefits [7]. The fly ash concrete technology is mature, which make high quality fly ash resource utility rate high; but the abandoned original fly ash emissions is very alarming in the production process. The pulverized fly ash has been obtained by drying and grinding the waste fly ash, which can meet the grade II fly ash performance index to produce the pulverized fly ash concrete.

The LCA method is aimed to evaluate the environmental impact of a product from raw materials mining, processing, using until wasting in the whole life cycle, which has been widely used in the cement and concrete industry to analysis environmental impact. The LCA method have been used to evaluate the environmental impact deeply at home and abroad [2, 4, 7–10], but the research have been rarely through evaluating SCM adding, its energy consumption in production process by the LCA method. It has evaluating the environmental impact of PFA concrete by adding SCM to the concrete, and the concrete production.

2 RESEARCH METHOD

The LCA method is composed of four parts, including objective and boundary determination, inventory analysis, impact assessment and improvement evaluation. The objective and boundary determination is the essential part of the LCA method. The overall research system framework affects the whole evaluation process and the result directly, which must be limited to the study clearly. The inventory analysis is the fundamental part of the LCA method, which is the perfect development part currently that demonstrates and evaluates the system with the calculating objective data by summarizing the system input and output.

2.1 *Research objectives and system boundary*

Research Objectives: The study objects were the production process of C30 and C50 grade cement concrete, fly ash concrete and pulverized fly ash concrete, the production process was shown (see Fig. 1). It has been provided supporting data for the pulverized fly ash concrete production and use, which has used LCA method to evaluate resource consumption, energy consumption and waste associated with the pulverized fly ash production.

System Boundary: It has used 42.5 OPC in the concrete production, and the concrete function unit is 1 m^3. The consumption of cement production needed 1.3t limestone, 0.3t clay, 50 kg gypsum per ton, ignoring the use of other industrial waste residue. Also it was assumed that all the materials were transported by the diesel vehicles with 5t load, the cement raw material distance was 5 km, the others were calculated according to the actual transport distance, as shown in Table 1. And the environmental pollution was ignored which was caused by facilities in the production and transportation process. In addition, both water and high performance reducing agent could not be considered, which had little effect on the environment [1].

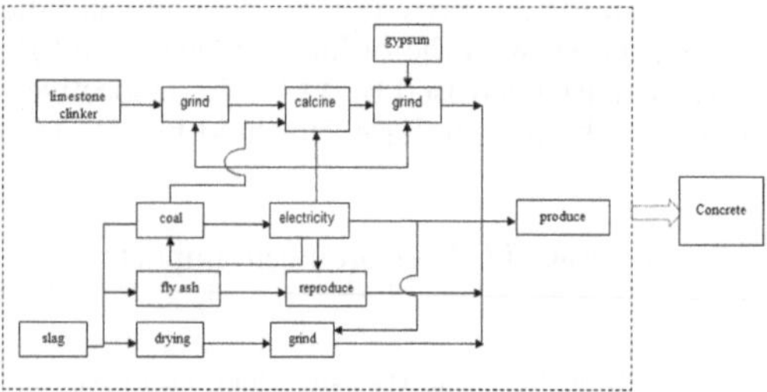

Figure 1. Concrete production process.

Table 1. Materials transport distance.

	Cem	FA	PFA	Slag	Aggregate
Distance/km	200	200	20	100	35

2.2 *Inventory analysis*

The inventory analysis quantifies the environment products load. Therefore, the product input and output throughout the life cycle of each list would be quantified objectively and process was based on system data. The inventory analysis is the key part of LCA method, including data collection and processing. By the concrete system inventory analysis, the environmental impact is directly analyzed and evaluated. The utility system environment database is the basic element of the evaluation system, public environment data seen Table 3 [5], the generated NO_x, SO_2 of electric power production and transportation system is calculated according to the corresponding average emission factor [10]. All the raw materials exploitation is in the same power consumption, without considering other pollutants. The environmental data [7] associated with cement life cycle are shown

Table 2. Concrete mix design.

		Mix design/kg/m^3						
	w/b	Cem	FA	PFA	Slag	Sand	Stone	28d Strength /MPa
C30-1	0.38	360				790	1136	50.6
C30-2	0.38	280	80			790	1136	42.3
C30-3	0.38	215		89	54	790	1136	48.2
C50-1	0.31	490				760	1048	66.2
C50-2	0.31	400	90			777	1036	64.8
C50-3	0.31	327		86	77	777	1036	68.2

The admixtures of C30 and C50 are 1%, 1.2% respectively.

Table 3. Public environment data.

	Consumption	CO_2/kg	CO/kg	C_xH_y/kg	NO_x/kg	SO_2/kg	Others/kg
Electricity/kwh	Standard coal (0.424 kg; 0.0124GJ)	0.938	4.00e-5	7.2e-5	5.1e-3	1.12-3	0. 1023
Transport/5t/km	Diesel 0.2 L(0.004GJ)	0.0673	0.002	1.1e-3	4.7e-3	1.36e-4	
Coal/t	13.89kwh						

Table 4. Energy consumption data of concrete life cycle.

Product	Electricity/kwh	Coal/kg	CO_2/kg	CO/kg	CxHy/kg	NOx/kg	SO_2/kg	Others/kg
Exploitation/t	13.89							
Cement/t	123.3	204.4	885	$1.15E^{-2}$	$2.73E^{-2}$	1.533	0.354	40
Slag/t	20	35						
PFA/t	24	25						
Concrete/m^3	2							

SCM manufacturing data is from company.

Table 5. The inventory analysis of the concrete.

	C30-1	C30-2	C30-3	C50-1	C50-2	C50-3
Material						
Limestone/kg	468	364	279.5	637	520	425.1
Clinker/kg	108	84	64.5	147	120	98.1
Gypsum/kg	18	14	10.75	24.5	20	16.35
Aggregate/kg	1926	1926	1926	1808	1813	1813
FA/kg		80			90	
PFA/kg			89			86
Slag/kg			54			77
Energy consumption						
Coal/kg	115.75	93.30	80.85	151.40	126.18	112.41
Diesel/L	5.695	5.669	4.775	6.613	6.59	5.639
Emissions						
CO_2/kg	405.58	315.87	247.44	551.66	450.73	374.16
CO/kg	6.5E-2	6.3E-2	5.3E-2	7.7E-2	7.5E-2	6.4E-2
C_xH_y/kg	5.0E-2	4.0E-2	4.0E-2	6.0E-2	5.0E-2	5.0E-2
NO_x/kg	1.15	0.92	0.747	1.54	1.29	1.09
SO_2/kg	0.23	0.18	0.15	0.32	0.26	0.22
Others/kg	23.68	18.42	14.66	32.23	26.31	22.12

in Table 4. The per ton cement material consumption for limestone, clay, gypsum are 1.3t, 0.3t and 50 kg respectively, which do not consider the other industrial waste residue. The concrete mix design is shown in Table 1. The fundamental data is seen in Table 3 and 4. The environmental load is summarized, shown in Table 5.

3 RESULTS

3.1 *Materials consumption*

The raw materials of C30-3 is reduced by 239.25 g, 107.25 kg relatively compared with the C30-1 and C30-2 concrete (see Table 5 and Fig. 2). Wherein the amount of limestone is reduced by 188.5 kg and 84.5 kg relatively, clay is reduced by 43.5 kg and 19.5 kg respectively, the gypsum is reduced by 7.25 kg and 3.25 kg respectively (see Table 5 and Fig. 2). The raw materials of C50–3 is reduced by 268.95 kg, 120.45 kg relatively compared with the C50–1 and C50–2 concrete (see Table 5 and Fig. 3). Wherein the amount of limestone is reduced by 211.9 kg and 94.9 kg relatively, clay is reduced by 48.9 kg and 21.9 kg respectively, the gypsum is reduced by 8.15 kg and 3.65 kg respectively (see Table 5 and Fig. 3).

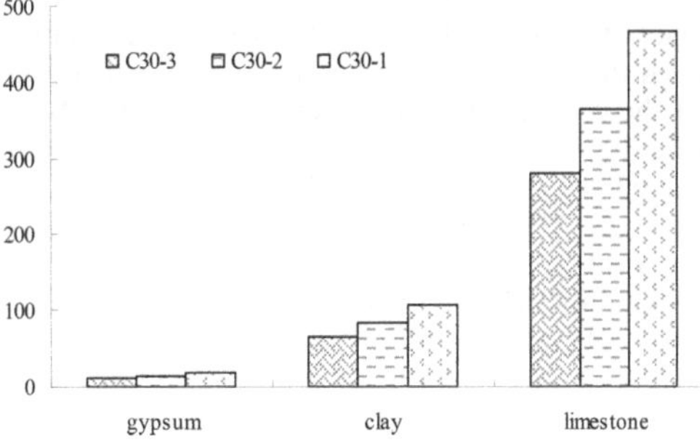

Figure 2. The raw materials of C30 concrete.

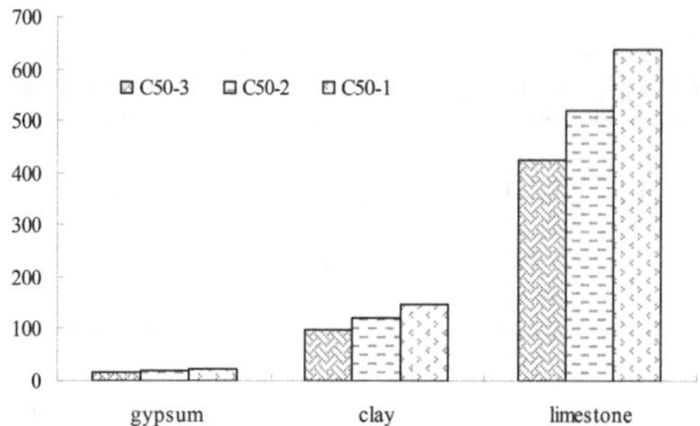

Figure 3. The raw materials of C50 concrete.

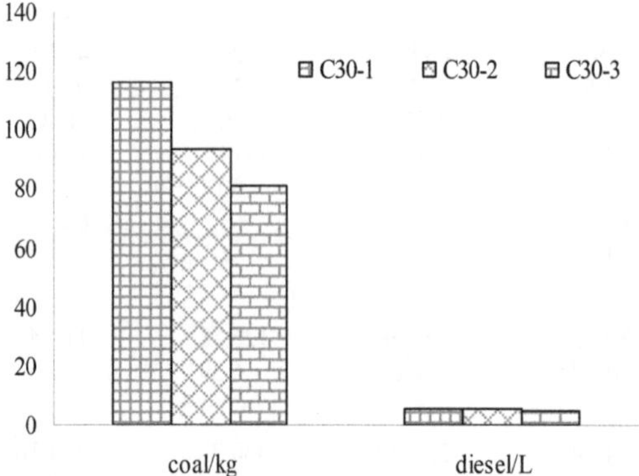

Figure 4. The energy consumption of C30 concrete.

3.2 *Energy consumption*

The coal consumption of the C30-3 is reduced by 30.2%, 13.3% c compared with C30-1 and C30-2(sees Fig. 4 and Table 5). The cement calcination need to burning coal to get much heat, which lead to the concrete high energy consumption. Therefore the C30-3 concrete with less cement has consumed less coal. And the C30-3 diesel is decreased by 16.2%, 15.8% respectively compared with C30-1 and C30-2 (sees Fig. 4 and Table 5).

And the C50 concrete have the same trend. The coal consumption of the C50–3 is reduced by 25.8%, 10.9% respectively compared with C50–1 and C50–2 (sees Fig. 5 and Table 5). And the C50–3 diesel is decreased by 14.7%, 14.4% respectively compared with C30-1 and C30-2 (sees Fig. 5 and Table 5).

3.3 *Environmental emissions*

The waste emissions of the C30-3 is reduced by 168 kg and 72 kg compared with C30-1 and C30-2; wherein the CO_2 emissions of the C30-3 is reduced by 39%, 21.7% respectively (see Fig. 6 and Table 5).

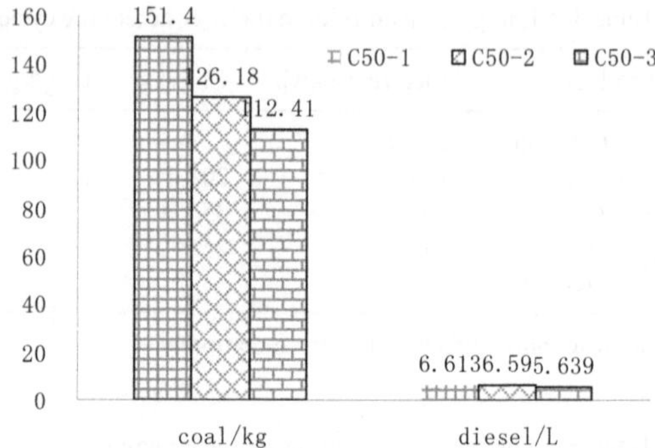

Figure 5. The energy consumption of C50 concrete.

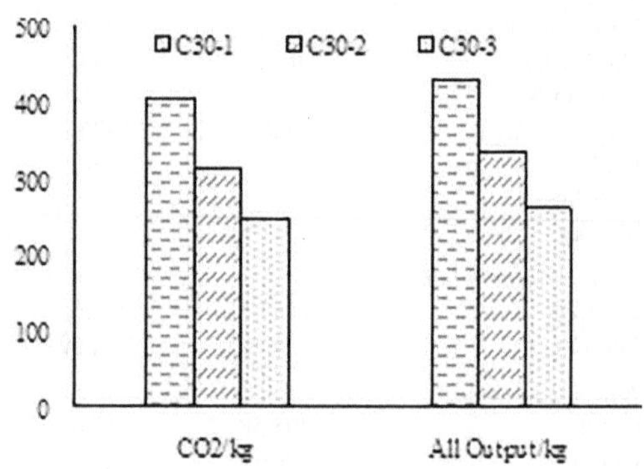

Figure 6. The emissions of C30 concrete.

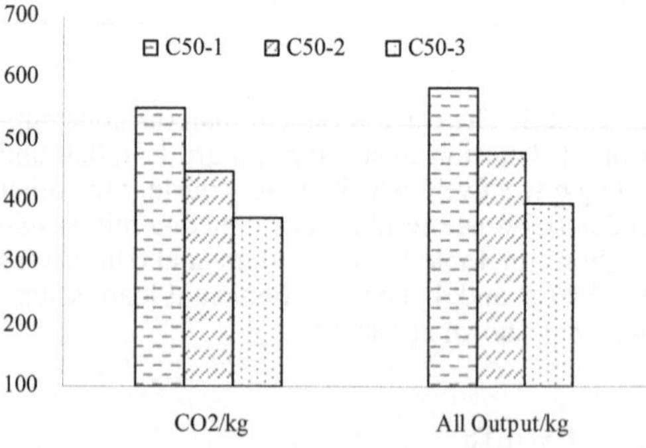

Figure 7. The emissions of C50 concrete.

The concrete emission is mainly from cement production, therefore the emission of the PFA concrete with less cement have been less than C30 cement concrete. Also that is less than C30 FA concrete. Considering the above reason, the same trend has been shown by the PFA C50 concrete. The waste emissions of C50 PFA concrete is reduced by 177.5 kg and 76.57 kg compared with C30 cement concrete and C30 FA concrete; wherein the CO_2 emissions of the PFA concrete is reduced by 32.2%, 17% respectively (see Fig. 7 and Table 5).

4 CONCLUSION

The LCA inventory analysis results have shown that the PFA concrete have directly reduced materials, energy consumption and greenhouse gas emissions compared with the traditional ordinary cement concrete. Compared with ordinary cement concrete, the materials resource consumption of C30 and C50 PFA concrete were reduced by 40.3% and 33.3% respectively; the coal consumption were decreased by 30.2% and 25.8% respectively; the CO_2 emissions were reduced by 39% and 32.2%. Compared with the fly ash concrete, the materials resource consumption of C30 and C50 PFA concrete were reduced by 23.2% and 18.3% respectively; the coal consumption were decreased by 13.3% and 10.9% respectively; the CO_2 emissions were reduced by 21.7% and 17%. The above conclusions have shown that it has been an environmental friendly way to mix concrete by the PFA.

REFERENCES

[1] David J.M. Flower, Jay G. Sanjayan. Green house gas emissions due to concrete manufacture [J]. Int J LCA, 2007, 12(5):282–288.

[2] Humphreys K., Mahasenan M. (2002): Toward a Sustainable Cement Industry. Substudy 8, Climate Change. World Business Council for Sustainable Development.

[3] Jiang Chunlai, Yang Jintian, Jin Ling etc. Tendency Analysis and Countermeasures for Total Emission Control of NO_X in Cement Industry during "The Twelfth Five-Year" Period [J]. Environment and Sustainable Development, 2013, 37(6):26–32.

[4] Liu Shunni, Lin Zongshou, Zhang Xiaowei etc. Studies on the life circle assessment of portland cement [J]. China Environmental Science, 1998, 18(4):1–5.

[5] Shen Wei-guo. Effect of Scattering-Filling Aggregate Technology on.

[6] The Mechanical Properties of Concrete [J]. Journal of Building Materials, 2007, 10(6):711–716.

[7] Shen Wei-guo, Tan Yu, Wu Lei etc. Life cycle assessment of concrete prepared by three kinds of placing process [J]. 2012, 6:21–24.

[8] Qian Jueshi. Fly Ash and Fly ash Concrete [M]. Beijing: Science Press, 2002.

[9] Kong Deyu. Preliminary study of life-cycle assessment of fly-ash concrete [J]. Journal of Zhejiang University of Technology, 2002(4):1–6.

[10] Collins F., Sanjayan J.G. The Challenge of the Cement Industry Federation Towards the Reduction of Greenhouse Emissions, Towards a Better Built Environment-Innovation, Sustainability, Information Technology. Proceedings of the International Association of Bridge and Structural Engineers Symposium, 2002.

[11] Heidrich C, Hinczak I, Ryan B. SCM's potential to lower Australia's greenhouse gas emissions profile, Iron and Steel Slag Products: A Significant Time of Scarcity. Australasian Slag Association Conference, 2005.

[12] Wang Wenxing, Wang Wei, Zhang Wanhua etc. Geographycal distribution of SO_2 and NOx emission and trends in China [J]. China environmental science, 1996, 16(3):161–165.

Material Science and Environmental Engineering – Chen (Ed.)

Influence of subway tunnel lining concrete after sodium nitrate corrosion by strength curve

C.Z. Sun
Shenyang Jianzhu University Design and Research Institute, Shenyang, China

J.K. Zhang & T.T. Sang
School of Civil Engineering, Shenyang Jianzhu University, Shenyang, China

ABSTRACT: Exploring strength curve and the corrosion mechanism of the subway tunnel lining concrete corrosion by sodium nitrate by the rebound method, the ultrasonic-rebound method and wet–dry cycle method in the laboratory. We conclude that the experiment curves of tunnel lining concrete strength after corrosion test has a good applicability, and work out the subway tunnel lining concrete strength by sodium nitrate environment corrosion after curve equation. It will provide a theoretical basis for studying strength changes of tunnel lining concrete after corrosion and the nondestructive testing of the tunnel lining concrete corrosion degree.

Keywords: ultrasonic-rebound; wet–dry cycle; rebound method; lining concrete; sodium nitrate

1 INTRODUCTION

Statistics show that the case of shallow groundwater contamination is more serious corrosion [1–2]. During the rainy season, seepage water situation of subway tunnel occurs in many of our cities. With the invasion of various corrosive contaminants, tunnel lining concrete can cause long-term corrosion damage. Eventually it cause serious risks to the subway tunnel safety and affect the normal use of the subway tunnel [3–4]. The survey found that Corrosion of concrete nitrate and sodium ions are more common in concrete corrosion. This article will Ultrasonic-rebound combined method introduce into the tunnel lining concrete strength test after corrosion. The rebound value, concrete ultrasonic speed nondestructive parameters and damage parameters compressive strength Regression relation of Shenyang subway tunnel lining concrete by sodium nitrate corrosion are analyzed [5].

2 EXPERIMENTAL DESIGN

2.1 *Test materials*

There are two types of cement in the test. C25 tunnel lining concrete block is used in Tieling iron Hsin Cement Limited production of ordinary portland cement and ordinary portland cement of C30 tunnel lining concrete block is form Shenyang Jidong shield Cement Co. Both intensity levels were 42.5. Besides, fine aggregate is produced in the Hun River fineness modulus of 2.6 in the sand and coarse aggregate is produced in Liaoyang, It is a particle size of 5~10 mm continuously graded limestone gravel. This test adopted by the water as ordinary tap water, Additives and admixtures were produced in Shenyang poly-JC-2 early strength super plasticizer and Fushun production of fly ash, Chemical reagent is from Shenyang Xinhua reagent factory $NaNO_3$ reagents in line with GB/T6009-2003 and GB672-78 request.

2.2 *Mix design*

The subway tunnel lining concrete mixture ratio is shown in Table 1, and the admixture in subway tunnel lining concrete is shown in Table 2.

2.3 *Retaining parts production and conservation*

Test Preparation tunnel lining concrete block, including the mixing, molding and curing processes.

Mixing process

First, let's rise to the surface of the water content of the sand to a predetermined value by pouring a certain amount of water (typically 15%–25%), further need to add gravel and sand with wet Stir. Then put all stirred together with cement

Table 1. Mix ratio of subway tunnel lining concrete.

	The amount of material per cubic meter of concrete			
Concrete strength	Water (kg)	Cement (kg)	Sand (kg)	Stone (kg)
C25	180	312	889	964
C30	200	408	896	896

Table 2. The admixture in subway tunnel lining concrete.

	Per cubic meter of concrete admixtures, admixtures dosage		
Concrete strength	JC-2 superplasticizer (kg)	DM-F accelerator (kg)	Fly (kg)
C25	5.90	–	55.00
C30	–	16.32	–

Table 3. Specimen grouping.

Serial number	Type of lining concrete	Lining concrete strength (MPa)	Test environment	Specimen size (mm)	Package	Number
A	C25	30.06	$NaNO_3$ (15%)	100 × 100 × 100	10 × 3	30
A	C25	30.06	Clear water	100 × 100 × 100	10 × 3	30
B	C30	37.52	$NaNO_3$ (15%)	100 × 100 × 100	10 × 3	30
B	C30	37.52	Clear water	100 × 100 × 100	10 × 3	30

and gravel. Finally, add the remaining water and additive, mixing into concrete.

Molding process

In this paper, according to GB/T50080-2002 "standard of ordinary concrete mixture performance test method" specified in the molding process, the concrete tunnel lining tryout loaded three times, and three times successively inserted pound artificial. Try using spatula around after artificial vibrating mold wall vibrators several times. Finally, the vibration test is fixed on the vibration table, and then keep the test does not produce beats in the vibration until the surface of the pulp.

Curing process

According to GB/T50081-2002 "standard of ordinary concrete mechanics performance test method" stipulated in the way of concrete block maintenance. Put test block on the temperature of (20 ± 2) °C $Ca(OH)_2$ saturated solution that it doesn't flow in the conservation 28d.

2.4 *Test grouping*

Test is divided into 40 groups of three test blocks, a total of 120 test blocks. C25, C30 two concrete tunnels lining, conducted in alternating cycles of wet and dry etching solution, soaked with water as a comparison group. Test group is shown in Table 3.

2.5 *Test instruments and methods*

Concrete cube compressive strength test equipment: Changchun pressure test limited production YB2000-A hydraulic pressure testing machine, the maximum range 2000 kN. The test equipment is shown in Figure 1.

(a)Compression testing

(b) Block testing

Figure 1. Test equipment.

Referring to GB/T50081-2002 "ordinary coagulation mechanical properties test method standards" for compressive strength tests. After curing test block 28d, we began alternating wet and dry corrosion test. First, clean the specimen surface and measure out the size (accurate to 1 mm). Then calculate the specimen force area A (mm^2) and do the test block 28d respectively standard curing. On 8 d, 16d, 24d, 32d, 40d, 48d, 56d, 64d, 72d after corrosion test block compressive strength tests are carried out.

1. The test specimen is placed on the pressure plate and the side of the test piece as a molding pressure surface keeps the specimen center and the lower platen press center alignment.
2. Continuously and evenly to pressure load of specimen, loading speed control each second 0.3 ~ 0.5 MPa. Quickly began to appear in the specimen to damage, then reduce the load to the specimen and record breaking load in the end.

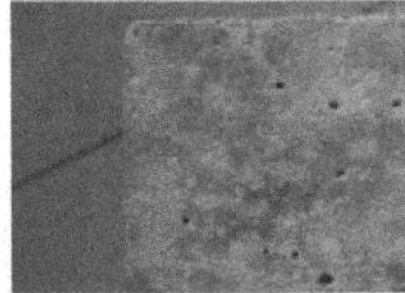

(a) The early corrosion: block surface color burn

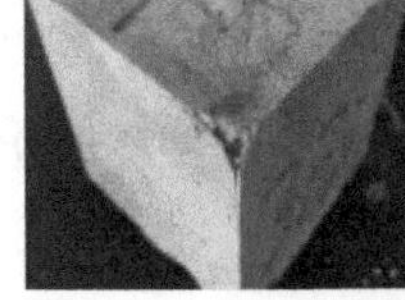

(b)Late corrosion: the corners, cracks, off angle phenomenon

Figure 2. Test phenomenon of corrosion in sodium nitrite solution.

3 WET–DRY CIRCULATION ACCELERATED CORROSION TEST

$NaNO_3$ corrosion tunnel lining concrete macroscopic description is shown in Figure 2.

In the early corrosion test, $NaNO_3$ solution specimen appearance did not change significantly and the color does not change. In the mid-term corrosion test, the appearance appeared to deepen the color, tiny holes increases, larger phenomenon, and Test block appearance of the overall changes in the obvious and serious damage such as falling edges phenomenon does not appear. In the final corrosion test, the part of the test blocks cracks in the corners and edges off, aggregate exposure.

Research has shown that the surface of test block in clean water is smooth, and its color is normal. $NaNO_3$ solution corrosion test block of color and shape change is slow. The test medium has color deepened, and the phenomenon such as tiny holes increases.

In the corrosion test, after several dry-wet alternate cycles, corrosion has accumulated at the bottom of sand, turbid solution and the solution for replacement. Under the effect of corrosion, the C25 type tunnel lining concrete block compared with C30 type tunnel lining concrete block, color change quickly and appearance of the crack damage deformation is obvious, it is more susceptible to corrosion damage.

4 ULTRASONIC-REBOUND SYNTHESIS STRENGTH CURVE ANALYSIS

In this test, corrosion age is shorter and carbonization depth is smaller, so we don't consider the impact of carbonation depth. The regression function established through the ultrasonic wave velocity and the rebound value by the use of minimum quadratic multiplication. The polynomial function, linear function, power function, exponential function was widely used in the regression function [6]. Choose those curves and analyze correlation coefficients R, average relative error δ and relative standard deviation e_r comprehensively, select the most suitable regression function.

The Table 4 shows that the polynomial functions, linear function, power function and exponential function four regression formula. Their correlation coefficient were 0.98, 0.97, 0.96, 0.97, the average relative error (δ) were 4.39%, 4.38%, 4.58%, 4.78%, relative standard deviation (e_r) were 5.98%, 5.29%, 5.81%, 6.16%, ultrasonic-rebound combined procedures average relative error was 21.49%, and relative standard deviation is 22.52%.

"Ultrasonic-rebound combined method was used to detect the strength of concrete technical regulations" provides that: Area average relative strength curve relative standard deviation (e_r) should not exceed 14%. Besides, special strength curve relative standard deviation (e_r) should not exceed 12%. Data analysis results, Ultrasound-rebound procedures after the unified strength curve of corrosion of tunnel lining concrete strength conversion, the average relative error and relative standard deviation is 21.49%, 22.52% respectively, it is far more than the precision of the specification specified value, so description specification after the unified strength curve of sodium nitrate corrosion tunnel lining concrete strength testing is not applicable.

Tunnel lining concrete comprehensive regression curve by sodium nitrate after corrosion, formula 1 regression effect than formula 2,3,4 regression effect is better Recommended formula one chosen as the subway tunnel lining concrete by nitric acid corrosion after ultrasonic rebound synthesis strength curve.

Table 4. The binary regression results of ultrasonic wave velocity and rebound value and compressive strength.

The formula	Regression formula	R	δ	e_r
1	$f = -1015.204 + 495.924v - 56.244v^2 - 3.59R - 0.0676R^2$	0.98	4.39%	5.98%
2	$f = -82.135 + 21.598v - 0.651R$	0.97	4.38%	5.29%
3	$f = 0.029v^{2.979} R^{0.761}$	0.96	4.58%	5.81%
4	$f = 0.734\, e^{(0.694v + 0.0237R)}$	0.96	4.78%	6.16%
5	$f = 0.0162v^{1.656} R^{1.41}$	–	21.49%	22.52%

Table 5. The monadic regression results of ultrasonic wave velocity and compressive strength.

The formula	Regression formula	R	δ	e_r
1	$f=-693.54+306.06v-31.86v^2$	0.96	5.06%	6.88%
2	$f=-119.77+35.468v$	0.95	5.48%	7.31%
3	$f=0.0188v^{5.113}$	0.94	6.78%	7.94%
4	$f=0.186e^{1.2v}$	0.94	6.90%	8.14%

5 ULTRASONIC STRENGTH CURVE ANALYSIS

In Table 5, regression analysis shows that: a polynomial function, linear function, power function and exponential that are four kinds of regression formulas, correlation coefficients were 0.96,0.95,0.94,0.94 average relative error (δ) were 5.06%, 5.48%, 6.78%, 6.90% and the relative standard deviation (e_r) were 6.88%, 7.31%, 7.94%, 8.14%. From the overall effect of the regression equation, formula 1, 2, regression effect is good, and formula 3, 4 regression effect is followed. In formula 1, The fitting relevance of velocity and measured compressive strength is good, the average relative error and relative standard deviation is high precision. Suggest to formula 1 as tunnel lining concrete ultrasonic measurement curves by nitrate after corrosion.

6 CONCLUSIONS

The tunnel lining concrete dry-wet alternate in sodium decline nitrate solution in the corrosion test, that it's strength significantly, and concrete severely damaged.

Use linear function, power function, exponential function and polynomial function regression formula of strength curve, when the tunnel lining concrete compressive strength conversion, the average relative error (δ) and relative standard deviation (e_r) meet the nondestructive testing specification limit value, the precision of polynomial function formula is best.

After ultrasonic-rebound synthesis regression curve of polynomial function and linear function formula, the return effect of the tunnel lining concrete after sodium nitrate solution corrosion is better. In the non-destructive testing methods, ultrasonic—rebound combined method the highest precision. In the four function regression equation, polynomial regression formula highest accuracy.

Suggested by sodium nitrate solution as the subway tunnel lining concrete corrosion after ultrasonic rebound synthesis strength curve formula is:

$$f=-1015.204+495.924v-56.244v^2-3.59R-0.0676R^2$$

ACKNOWLEDGMENTS

The Project of the National Key Technology R&D Program in the 12th Five-year Plan of China (2011BAJ10B04).

REFERENCES

[1] Kewang Tang, Yucheng Wu, Jie Hou. The current situation of the underground water quality and pollution analysis [J]. *Water resources protection, 2006, 22 (3): 1–8.*

[2] Yun Li. Both the tunnel lining damage assessment and governance decisions [D]. *Changsha: central south university, 2010.*

[3] Xiangjian Wu, Shiyou Xie. The application of efficient retarding super plasticizer in Si lin Hydropower Station when the RCC construction in high temperature season [J]. *Journal of guizhou hydropower, 2007 (6): 33–36.*

[4] Liangxi Mao, MuRu Mu, Yunbao Huang, etc. Super high-temperature retarding RCC [J]. *Jiangsu Building, 2006 (5): 53–54.*

[5] Huimin Wu. New concrete nondestructive testing technology [M]. *Beijing: China Environmental Press, 2001.*

[6] Zhean Lu, Shengchen Wang, Shuiguang Chen, Sheng Lu. Long age Gravel concrete ultrasonic—rebound synthesis strength curve study in WuHan [J]. *Concrete, 2008, 3:10–12.*

[7] Kun Wang. Qingdao subway experimental study of high performance lining concrete [D]. *Qingdao: Qingdao Technological Universi*ty, 2010.

[8] Reilly J.J. Management process for complex underground and tunneling project. Tunneling Underground [J]. *Space Technology, 2000, pp: 31–44.*

Material Science and Environmental Engineering – Chen (Ed.)
© 2016 Taylor & Francis Group, London, ISBN 978-1-138-02938-5

Degradation model and security analysis of hangers of suspension bridge

J.H. Li & J.Z. Wang
Kunming University of Science and Technology, Kunming, China

ABSTRACT: The hangers of suspension bridge are one of the most important stressed members, in order to thorough research of security analysis after the degeneration of hangers. the four stages of degradation mechanisms of sling and calculation formula of degradation time are studied, and theoretical relationship between the degree of degradation and resistance reduction coefficient is established, and then the safety analysis method of sling which broken wire considers the effect of Daniel is discussed, Finally, Taking an example of the 3rd sling of left Branch Bridge of Ma'anshan-bridge, theoretical calculations of degradation time of sling is conducted and security calculations of sling after degradation is given. The method had a certain rationality and feasibility, if combining it with the actual monitoring data to update model, it will be able to provide a reliable basis for preventive maintenance strategy better.

Keywords: hanger; degradation model; safety analysis

1 INTRODUCTION

In recent years, safety and conservation strategies of large suspension bridge are paid more and more attention, the deterioration of important components and the entire bridge has become a hotspot in engineering. James M. Stallings and Karl H. Frank analyzed the influencing factors of fatigue life of the hanger using series-parallel model [1]. According to the test results, of the New York Williamsburg main cable, J. Matteo used ductility model of hanger and the brittle model to estimate the carrying capacity of the cable [2]. In China, Lan Chengming established theoretical models calculated fatigue life of parallel wire cable, he considered that the fatigue life of the cable is controlled by a small part of the fatigue life of the shorter wire, 10% broken wires used as life termination of cable is reasonable [3]. Zhu Jinsong combined Matteo's brittle wire model and Monte Carlo method, simulated cable to have m filaments (each filament is divided into n segments) connected in parallel systems, it's also not considered strength degradation of steel due to corrosion and fatigue, the degraded condition of cable is only caused by of broken wires [4]. Based on relevant research, the paper will be further explore the theoretical degradation model of sling, and safety assessment methods of sling after degradation, and apply them to practical engineering of Ma Anshan Bridge.

2 DEGRADATION MODEL OF SLING AND THE REDUCTION FACTOR

In order to establish the precise degradation model, we must take into account as fully as possible factors leading to degradation of the sling, and establish degradation model based on different dominant factor in different time periods.

Factors leading to degradation of sling are: sheath damage; wire damage; anchor head injuries. Combined resistance reduction factor Z_1, the degradation process will degenerate into the following phases.

2.1 *Slings is installed—jacket occur cracks*

At this stage, the wire is without electrochemical corrosion and stress corrosion, therefore only considered natural aging and damage under alternating stress and fatigue damage. The natural aging damage can learn from research results of Liu Yaping to be calculated. Maanshan Bridge, located in the northern subtropical monsoon climate zone, can learn from the results of sub-tropical rainforest climate, do the least-squares fitting and simple derivation, you can get the time t_1 from installation to occur cracks of jacket that meet the required expressions:

$$\frac{T}{EA} = 6 \times 64.462e^{-0.0397t_1} \tag{1}$$

Reduction:

$$t_1 = 25.189\ln\left(\frac{386.772EA}{T}\right) \quad (2)$$

2.2 *Jacket occur cracks—galvanized coating of steel failure*

This stage jacket has not withstood cable force, the damage is mainly fatigue damage of wire, and length of time t_{II} at this stage is mainly expiration time of the galvanizing layer, can be calculated by:

$$t_{II} = \sqrt[n]{\frac{C}{C_1}} \quad (3)$$

where C_1 and n is the local parameters of rate of corrosion of galvanized layer, C can be substituted into thickness of the smallest of galvanized steel wire.

This stage the cable resistance reduction factor Z_1 is 0.90–0.95, the Z_1 can be expressed as a function of time t:

$$Z_1 = 0.95 - 0.05t/t_{II} \quad (4)$$

2.3 *Steel galvanized coating failure—fatigue cracks*

This stage mainly considers time t_{III} when corrosion fatigue occurs, according to the Faraday's formula of corrosion rate of uniform volume, the expression can be deduced:

$$t_{III} = \frac{2\pi nF\rho}{3MI_{po}}\left(a_{ci}^3 - a_0^3\right)\exp\left(\frac{\Delta H}{RT}\right) \quad (5)$$

This stage of the cable resistance reduction factor Z_1 is 0.85–0.90, the Z_1 can be expressed as a function of time t:

$$Z_1 = 0.90 - 0.05t/t_{III} \quad (6)$$

2.4 *Fatigue crack growth—the first floor of wire breakage*

This stage is mainly on account of time t_{IV} after fatigue corrosion of steel occurred. The formula of metal corrosion fatigue crack growth rate of Paris is still available the relations of time and damage can be derived:

$$s_3 = s_{ci} + \int_0^{ft_{IV}} \frac{k}{2\pi\sigma_{ff}^2}\left(\Delta K - \Delta K_{th}\right)dN \quad (7)$$

here, s_{ci} can not take 0, and need to take a stage of pitting corrosion depth. If it is determined expression of the crack tip stress σ_{FF} by the method combined experiments and experience, t_{IV} can be calculated.

If it is considered that the breakage moment of the first floor of wire is entire hanger failure time, and then this phase cable resistance reduction factor Z_1 is 0–0.85, the Z_1 can be expressed as a function of time t:

$$Z_1 = 0.85 - 0.85t/t_{IV} \quad (8)$$

Based on period of time of various stages, the wire rope jacket fatigue fracture life t from the start until age can be estimated.

$$t = t_I + t_{II} + t_{III} + t_{IV} \quad (9)$$

3 SECURITY ANALYSIS OF HANGER

3.1 *Assessment model of failure and safety of sling*

a. Sling strength model
Considering the effect of Daniel and rope strength of broken wires equal to the product of the number of wires and the average tensile strength of steel.

$$F = \tilde{n}W_n$$

where F is the tensile of sling considered effects of Daniel; ñ is the number of wire not broken; W_n is the average tensile strength of steel considered effect of Daniel.

b. Assessment Model of Safe and Failure of sling
The safety factor of actual tensile of sling under constant loads is failure criterion of cable; it can determine the probability of failure of cable. Performance function of sling failure model is:

$$f\left(A_w, \sigma_u, T_s, \eta\right) = \frac{\tilde{n}A_w\sigma_u}{(1+\eta)T_D} - \xi \quad (10)$$

where, A_ω is the cross-sectional area of individual wires; σ_u is the average tensile strength of steel wire of sling considered the effect of Daniel; T_s is cable tension generated during the operation of hanging dead load; η is ratio of cable force under live load and dead load; ñ number of no broken wire of sling; ξ is safety factor of sling for the design, in general ξ = 2.5.

The formula is performance function of safety failure of sling. Then, after determined the distribution type of each parameter, it can be used in the statistical analysis of Matlab software to generate random sample of the parameters with a certain sample size. The random sample of data used into the performance function and statistic calculations, using Monie-Carlo method to calculate the probability of failure of cable.

3.2 *Strength analysis of parallel wire of sling based the Daniel effects*

Firstly defined the ratio of the reduced average intensity wire cable and average tensile strength of the individual wires is the attenuation factor. There are:

$$\tau = \frac{R_D}{R_0} \tag{11}$$

Parallel wire sling system is composed of n filaments, when the sample size is sufficiently large (n > 150), average tensile strength obey Normal Distribution. Generated a random sample of the normal distribution in MATLAB. In order to effectively analyze the Daniel effects of hangers with different number of wires, different sample sizes were taken for analysis o, calculated correlation parameters.

Then used the following formula to calculate the average tensile strength of steel when wires have different numbers.

$$W_n = 1 - F_z(x_0) + \frac{c_n}{n} \tag{12}$$

where W_n is the tensile strength, and n is the number of wires.

Fitting the relationship between the tensile strength of steel wire and wire actual number caused of Daniel effects by regression.

4 EXAMPLES

The 3rd sling of left Branch Bridge of Maanshan Bridge is example. Maanshan sling hanged from 2 to 64 points is ordinary sling; each sling consisted by 109 ordinary steel wires. Steel wire is used φ5.0 mm high-strength galvanized wire, the standard of tensile strength is not less than 1670 MPa, elastic modulus of steel $E = 2.0 \times 10^5$MPa.

4.1 *Sling failure elapsed time*

1. Sling defense system failure experienced time can be divided into two sections
   a. HDPE jacket cracking and aging failure experienced time

   The 3rd sling of left Branch Bridge of Maanshan Bridge is example. According to their results the cable force of the 3rd sling of Ma'anshan Bridge under dead load and live loads of car, T = 1531.56 kN, $E = 2.0 \times 10^5$MPa, effective tension area $A = 2 \times 109 \times \pi (2.5 \times 10^{-3})^2 = 4.278 \times 10^{-3}$ m^2, according to:

$$t_1 = 25.189 \ln\left(\frac{386.772EA}{T}\right) \tag{13}$$

   Get $t_{11} = 218$ (weeks) = 4.2 (years).

   b. The life of anti-corrosion coating
   Based on the durable life specified in the ISO12944 standard, failure time of protective coating system is simply assumed, taking the life of durability of the coating system $t_{12} = 2$ (years).

   Failure can be obtained from the above sling protection system experienced time:

$$t_1 = t_{11} + t_{12} = 6.2 \text{ (years)}$$

2. Experienced time from corrode to fail of steel wire galvanized coating
   According to national standards [5] average quality galvanized layer of steel wire of cable and sling is not less than 300 g/m^2, you can get an average thickness of galvanization layer:

$$b = W/\rho$$

   where: W = 300 g/m^2, for the lower limits quality of galvanized layer steel wire;

   ρ = 7.2 g/cm^3, the density of pure zinc coating.

   Calculated: b = 41.67 um However, taking into account the impact of galvanized coating process itself to the minimum thickness, the coating thickness is generally smaller than the average 15 um, therefore C = 26.67 um.

   According to table of the corrosion rate of the domestic hot-dip galvanizing in all regions [6], n taken 1.088, C_1 taken 1.726.:

   According to the formula:

$$t_{II} = \sqrt[n]{\frac{C}{C_1}} \tag{14}$$

   Calculated:

$$t_{II} = \sqrt[n]{\frac{C}{C_1}} = \sqrt[1.088]{\frac{26.67}{1.726}} = 12.3 \tag{15}$$

3. Experienced time of steel wire body from corrosion pits formed to corrode fatigue occurred

According to the aforementioned degradation model:

$$t_{III} = \frac{2\pi nF\rho}{3MI_{po}}\left(a_{ci}^3 - a_0^3\right)\exp\left(\frac{\Delta H}{RT}\right) \tag{16}$$

where, M is the molar mass of material, M = 55.85;

n is the atomic valence, n = 3;

F is Fala Di constant, F = 96514 Cal/mole;

ρ is the material density, ρ = 7.8 × 106 g/m³;

R is a ventilation coefficient, R = 8.317 J/mole · K;

The average annual temperature in Ma Anshan is 15.9 °C, T is the absolute temperature, T = 237 + 15.9 = 252.9;

ΔH is steel wire redox reaction activation energy, ΔH = 59.7 J/mol;

According to the literature [7], the use of fuzzy—random method gives pits initial size a_0, and pitting flow coefficient I_{po},

$$a_0 = 1.5 \times 10^{-6} \text{ m}$$

$$I_{po} = 80.78 \text{ c/s}$$

$$a_{ci} = \pi\left(\frac{\Delta K_{th}}{2.2K_t\Delta\sigma}\right)^2 \tag{17}$$

According to the literature [7], fitting equation ΔK_{th} = −3.4324R + 5.5351, where R is a steel wire corrosion fatigue stress ratio, R here takes roughly 0.67. Whereby threshold of stress intensity factor amplitude of steel wire ΔK_{th} = 3.22, $a_{ci} = 59.3 \times 10^{-6}$ m.

The above data used into (4–3), obtained: t_{III} = 213 (days) = 0.58 (year).

4. Experienced time from corrode fatigue crack growth to wire breakage
When fatigue crack of steel wire extensions to a certain depth, the stress intensity factor of crack tip reaches the threshold K_I,

$$K_I = \sigma\sqrt{\pi a}F\left(\frac{a}{d}\right) \tag{18}$$

Assuming crack developing still by spherical approximately, according to the empirical formula:

$$K_I = \sigma\sqrt{\pi a}\left(\frac{0.865}{(a/d)+0.324} + 0.681\right) \tag{19}$$

Can be obtained from the above two equations, $K_{Imin} = 0$, $K_{Imax} = 1.77$.

Based on combining theoretical analysis and experimental data, used the Paris fitting formula to calculate $2\pi\sigma^2_{ff} = 2.158 \times 10^{10}$ MPa. According to curve of the fatigue experimental results of steel wire in artificial seawater conditions on cold drawn, A. Martin gave the environment constant k = 1.585, ΔK = 1.77 MPa. M1/2. Diameter of steel wire d = 5 mm, pitting depth $a_{ci} = 59.3 \times 10^{-6}$, crack depth of fatigue fracture of steel wire can be regarded a = 2.26 mm.

The above data substituted into the formula:

$$s_3 = s_{ci} + \int_0^{ft_{IV}} \frac{k}{2\pi\sigma_{ff}^2}(\Delta K - \Delta K_{th})dN \tag{20}$$

Regarded stress cycles when steel wire fatigue fracture

$$N = 3.36 \times 10^6 \text{ cycle}$$

Regarded stress cycle frequency of sling wire

$$f = 1/60 \text{ (cycle/s)}$$

Calculated: t_{IV} = N/f = 6.4 (years).

4.2 *Security analysis of sling wire*

1. Strength analysis of the wire sling based the Daniel effects
Steel wire Analyzed is ordinary steel wire of Ma'anshan bridge, steel wire is φ5.0 mm high-strength galvanized steel wire, standard tensile strength of steel wire is not less than 1670 MPa.
Firstly determining the distribution function of tensile strength of steel wire, y_u = 1670 MPa, according to the literature [8] whichever coefficient of variation ξ = 0.05, the standard deviation of the tensile strength of the steel wire:

$$\sigma = \xi y_u = 1670 \times 0.05 = 83.5 \tag{21}$$

Then wire tensile strength x~N(1670,6972.25)
Probability density function:

$$f_Z(x) = \frac{1}{\sqrt{2\pi} \times 83.5} e^{\frac{(x-1670)^2}{2\times 83.5^2}} \tag{22}$$

Distribution function:

$$F_Z(x) = \frac{1}{\sqrt{2\pi} \times 83.5}\int_{-\infty}^{x} e^{\frac{(t-1670)}{2\times 83.5^2}} dt \tag{23}$$

Then generate a random sample of the normal distribution in MATLAB. In order to effec-

Table 1. Daniel effect parameters.

n/ root	x_0/ MPa	F_z	f_z	a/MPa	C_n/MPa
10	1457.4	0.0054	1.8802	236.4654	548.3231
50	1466.2	0.0072	2.4423	293.2517	1043.627
100	1466.9	0.0077	2.5091	296.4251	1329.124
150	1468.7	0.0079	2.6213	301.7756	1548.899
200	1468.7	0.0079	2.6213	301.7756	1700.131

Table 2. The average tensile strength of steel wire of sling.

n/root	W_n
10	1503.325
50	1475.759
100	1468.154
150	1466.598
200	1464.237

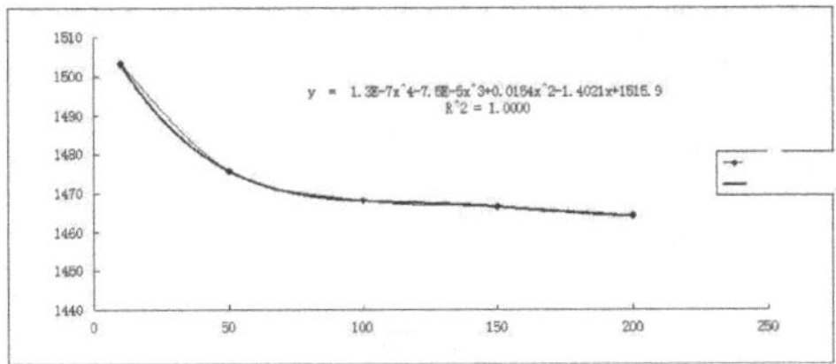

Figure 1. The average tensile strength of steel wire of sling with different numbers steel wire.

tively analyzed the effect Daniel when slings have the number of different steel wire, Samples capacity which were taken in this 10, 50, 100, 150, 200 five cases were analyzed. Sample statistic data is processed solved Daniel effect parameters as shown in Table 1.

Substituted into the formula:

$$W_n = 1 - F_Z(x_0) + \frac{c_n}{n} \tag{24}$$

Calculated the average tensile strength of steel wire of sling when sling have different number of steel wire, the results in Table 2.

Figure 1 can be obtained using polynomial regression to process data:

The relationship between the actual tensile strength and number of steel wires caused by Daniel effect can be got By regression fitting:

$$W_n = 1.3\times10^{-7}n^4 - 7.5\times10^{-5}n^3 + 0.015n^2 - 1.4021n + 1515.9 \tag{25}$$

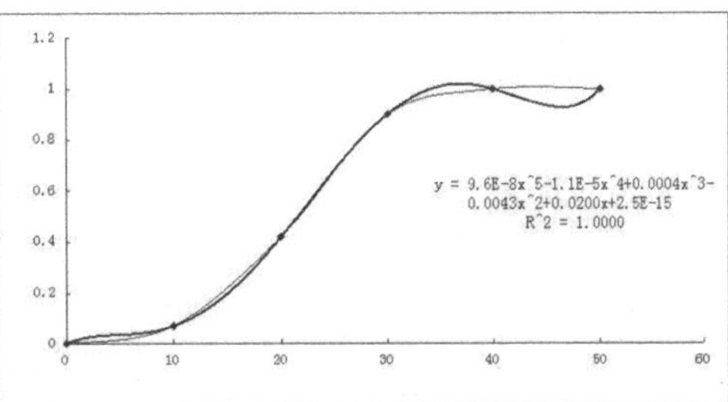

Figure 2. The plot of failure probability of broken wire of No. 3 hanger.

where, W_n is the tensile strength, n is the number of steel.

2. Safety failure probability analysis of sling wire the tensile actual safety factor of sling under dead loads is failure criterion, used it to determine the failure probability of sling. Performance function of sling failure model is:

$$f(A_w, \sigma_u, T_s, \eta) = \frac{\tilde{n}A_w\sigma_u}{(1+\eta)T_D} - \xi \tag{26}$$

Determining statistical properties of parameters of the function, that determine the distribution type, mean, standard deviation, coefficient of each parameter. According to the results, statistic properties of parameters A_w, W_n, T_D, η of the performance function will be described.

Area of the sling wire is average cross-sectional area 19.6, the tensile strength is 1670 MPa, the coefficient of variation of tensile strength and cross-sectional area is 0.05. So, the results of statistical properties of the four sample statistics of No. 3 sling of Maanshan left Branch Bridge Bridge is: A_w~N (19.6,0.98) (coefficient of variation of 0.05); σ_u~N (1670,83.5) (coefficient of variation 0.05); η~N (0.815,0.013) (coefficient of variation 0.016); TD (variation coefficient 0.02).

Distribution of the parameters is known, used Matlab to generate random samples of the parameters: the average cross-sectional area A_w = normrnd (19.6,0.98,100,1), different number of broken wires are a random sample of average tensile strength of steel, non-broken wire W_{109} = normrnd (1470.8,3.45,100,1), off 10 W_{99} = normrnd (1471.32,2.83,100,1), off 20 W_{89} = normrnd (1472.6 Wn2.56 Wn100 Wn1), off 30 W_{79} = normrnd (1474.80,2.67,100,1), off 40 W_{69} = normrnd (1478.22,3.29,100,1) and off 50 W_{59} = normrnd (1480.46,3.80,100,1) and so on.

Similarly random sample of cable tension under dead load can be produced when the number of broken wires is different.

The sample values is brought into the equation to calculate the value of the function f, due to the sample size of each parameter was 100; the number of wire broken wires followed by 0,10,20,30,40,50 six kinds situations to consider, each case strength steel and dead load cable tension was simulated 100 times. Therefore the word that random sample of data one by one brought into to the performance function to calculate the value of the performance function is very large. To simplify the calculation process, converting a random sample of data into a matrix form, using the matrix operation to complete.

$$f_{10} = (a_{10}*w_{10})/(b_1*t_{10})/1000-2.5 \tag{27}$$

where, $a_{10} = 99 * A_w$; $b_1 = \eta + 1$, generated in matlab; w_{10} is random sample of average tensile strength of steel when the number of broken wires is 10 and t_{10} is random sample of force of cable under dead load when the number of broken wires is 10.

According to the above method the value of f can be calculated when the number of broken wires are 0, 20, 30, 40, 50. To simplify the calculations, the following statement in matlab can be used to identify number which is greater than zero, and statistics.

Statistic the number of f which is greater than, equal to, less than zero. Supposed there are m $f < 0$, then the failure probability of sling can be calculated by the following equation.

$$P(f < 0) = \frac{m}{100} \tag{28}$$

The results of Statistics calculations in Table 3:

Simulating failure probability of the 3rd sling of Ma'anshan Bridge by Monte-Carlo method—broken wire curves, using the least squares to regress.

Table 3. Failure probability of broken wire of No. 3 hanger.

Broken wires	Simulation times	f > 0	f < 0	Failure probability
0	100	100	0	0
10	100	93	7	0.07
20	100	48	42	0.42
30	100	10	90	0.90
40	100	0	100	1
50	100	0	100	1

Thereby obtaining functional relationship between the probability of failure and the number of broken wires of the 3rd sling.

$$P_f = 9.6\times10^{-8}n^5 - 1.1\times10^{-5}n^4 + 0.0004n^3 - 0.0043n^2 + 0.0200n + 2.5\times10^{-15} \tag{29}$$

where, P_f is the broken wire failure probability of the 3rd sling of Ma'anshan Bridge considered the Daniel effect; n is the number of broken wires.

When the number of broken wires is between 0 to 40, you can use the above formula to estimate the probability of failure of broken wires.

5 CONCLUSION

Theoretical studies and numerical examples in this paper showed that: degradation mechanisms of the four stages and degradation time calculation formula of sling degradation and safety analysis method of the sling after broken wire considered the Daniel effect, had a certain rationality and feasibility, if combining it with the actual monitoring data to update model, it will be able to provide a reliable basis for preventive maintenance strategy better.

REFERENCES

[1] James M. Stallings, Karl H. Frank. 1991. Stay-Cable Fatigue Behavior. Journal of Structural Engineering. Vol. 117. March.

[2] John Matteo, George Deodatis, David P. Billington. 1994. Safety Analysis of Suspension-bridge Cables: Williamsburg Bridge [J]. Journal of structural Engineering.

[3] Lan Cheng Ming. 2009. Parallel wire cable theory fatigue performance. Shenyang Jianzhu University (Natural Science Edition), 25 (1).

[4] Plain Bridges. 2006. Bridge safety analysis method long-span cable. China Civil Engineering Journal, 39 (9).

[5] GB T 2973-2004. Zinc Coating quality test method [s].

[6] Cao Chunan. 2005 natural environment Cao Chunan Chinese material corrosion [M] Beijing: Chemical Industry Press.

[7] Xu Hong 2008. Bridge pull (suspended) after cord damage mechanics analysis and safety evaluation [D] Xi'an: Chang'an University.

[8] LAN Hai, Shi Jiajun. 2001. Use of Grey Incidence and Variable Weight Synthesizing in Bridge Assessment. Tongji University, 29 (1).

[9] Shang Xin Xu Yue. 2004. Bridge cable safety evaluation based on gray theory. Chang'an University (Natural Science Edition), 24 (1).

Material Science and Environmental Engineering – Chen (Ed.)

Research on the dam deformation forecast model based on material properties

F.M. Lu
Key Laboratory of Geological Hazards on Three Gorges Reservoir Area of Ministry of Education, Three Gorges University, Yichang, China

T.Y. Jiang
Computer and Information College, Three Gorges University, Yichang, China

ABSTRACT: Considering the material property of the dam, this paper conceptualizes that the deformation regularity in the different places of the dam is different. For this purpose, some forecast models are preplaced to let the computer search the forecast model whose deformation error is least. A practical example verifies the deformation forecast model close to the practical situation by means of the method. Thus, this method can be used to forecast the dam deformation.

Keywords: deformation; dam; forecast; material property

1 INTRODUCTION

So far, the statistical analysis model for the dam deformation analysis is the single model. The model considers that the dam deformation corresponds with the given model. The analysis is done on the basis of the given model, and the deformation analysis method is influenced by the human factor, and the analysis result is not ideal. In fact, the deformation of the dam is different from the different dam and the different deformation monitoring point of the dam, and the material property of the dam is considered. This paper conceptualizes that the deformation regularity in the different places of the dam is different. For this purpose, some forecast models are preplaced to let the computer search the forecast model whose deformation error is least. A practical example verifies the deformation forecast model close to the practical situation by means of the method. Thus, this method can be used to forecast the dam deformation.

2 THE MONITORING SYSTEM OF THE DAM

The height of the dam is 151 meters, and the length of the dam is 653 meters, which consists of 30 dam segments. The elevation of the dam top is 206 meters, the width of the dam top is 6 meters, the elevation of the normal water level is 200 meters, and the total reservoir capacity is 3400 million cubic meters.

The monitoring items of the dam include the external deformation monitoring, seepage and osmotic pressure monitoring, stress and strain monitoring, and temperature monitoring.

2.1 *The external deformation monitoring*

The external deformation monitoring of the dam is the important content of the safety monitoring of the dam, which can reflect the work state of the dam. It is an important foundation to erect the dam deformation forecast models and to evaluate the safety of the dam. The external deformation monitoring includes horizontal displacement monitoring and vertical displacement monitoring.

Horizontal displacement monitoring includes the horizontal displacement monitoring network, the positive perpendicular and obverse perpendicular observation, the wirework, the collimation line, and the intersection observation.

The horizontal displacement monitoring network consists of 16 points. These points are on the stable rock, which are used to monitor the deformation of the rock mass near the dam.

A total of 44 positive perpendiculars and 26 obverse perpendiculars are collocated on 6 dam segments.

A total of 3 wireworks are collocated on the adit whose elevation is 121 meters to monitor the horizontal displacement of the crown of arch and the skewback. Angles of wireworks are observed by the J_1 theodolite, and distances of wireworks are observed by the indium ruler.

A total of 5 collimation lines are collocated on the gravity dam of the left bank and the head gate of the workshop building, and the collimation line is observed by the J_1 theodolite and the active target.

A total of 8 forward intersection monitoring points are collocated at the backward position of the dam whose elevation is 120 meters to monitor the horizontal displacement of the dam. Overall, 10 distance and angle intersection monitoring points are collocated at the side slope of the right bank to monitor the horizontal displacement of the high side slope of the right bank.

The vertical displacement monitoring includes the vertical displacement monitoring network, the precise level and the statical level.

The vertical displacement monitoring network consists of the level loop whose grade is the first, and the length of the level loop is 9 kilometers.

In order to monitor the vertical displacement of the dam, leveling points for the settlement observation are collocated on some public arcades, the dam top and adits.

In order to monitor the vertical displacement of the dam foundation, some statical level instruments are collocated on some public arcades.

2.2 *The seepage and osmotic pressure monitoring*

Grouting adits are chosen as important observation positions to monitor the osmotic pressure. Main drainage curtains of the foundation public arcade of the dam are chosen as the lengthways observation section, and piezometer tubes are collocated on drainage curtains of each dam segment to test the reduction coefficient of the foundation uplift pressure. Some piezometer tubes are collocated on the main drainage curtains of grouting adits to test the antiseepage effectiveness of curtains. U-type piezometer tubes are collocated near the absorption basin of 2 dam segments to test the uplift pressure. The physical space method is used to monitor seepages, which can be used to understand the distribution of seepages and analyze the reason for seepages.

2.3 *The stress and strain and temperature monitoring*

In order to monitor the stress and strain of the dam and foundation, 363 strainmeters, 64 stress-free meters, 10 taseometers, 15 reinforcingsteel meters, 7 osmometers, 195 joint meters, 10 bedrock deformation gages, 5 multipoint displacement meters, 4 crack meters, 5 ergometers and 15 temperature meters are collocated to perform the stress and strain monitoring, deformation monitoring of the bedrock, crack monitoring, fault monitoring, pragmatic monitoring of the interlayer and temperature monitoring.

3 THE ERECTION OF THE MODEL

Because different deformation monitoring points are in different places, the deformation regularity of these deformation monitoring points is different. We can preplace some models to let the computer search the best deformation forecast model. Thus, nine models are preplaced here as follows:

$$Y_1 = a_1 + a_2\sqrt{x} + a_3 e^x + a_4 \ln(x) \tag{1}$$

$$Y_2 = a_1 + a_2\sqrt{x} + a_3\sqrt[3]{x} + a_4 \ln(x) \tag{2}$$

$$Y_3 = a_1 + a_2\sqrt[3]{x} + a_3 e^x + a_4 \ln(x) \tag{3}$$

$$Y_4 = a_1 + a_2\sqrt{x} + a_3 e^x + a_4 \sin(x) \tag{4}$$

$$Y_5 = a_1 + a_2\sqrt{x} + a_3 \ln(x) + a_4 \sin(x) \tag{5}$$

$$Y_6 = a_1 + a_2 x + a_3 x^2 + a_4 x^3 \tag{6}$$

$$Y_7 = a_1 + a_2 x + a_3 e^x + a_4 \ln(x) \tag{7}$$

$$Y_8 = a_1 + a_2\sqrt{x} + a_3 x + a_4 \sin(x) \tag{8}$$

$$Y_9 = a_1 + a_2 \ln(x) + a_3 \sin(x) + a_4 x^2 \tag{9}$$

where a_1 to a_4 are parameters of models; x is the observation time; and Y_1 to Y_9 are fitting values of nine models.

4 THE SOLUTION OF MODEL PARAMETERS

The regression equation of (1) to (9) can be formed on the basis of deformation observation values and the observation time. Model parameters of the nine models can be obtained by using the least square method.

4.1 *The choice of the weight*

We hope that the deformation forecast error is least in the deformation forecast model. We can choose large weights for the deformation observation values at the back. If multiple observations are done for the deformation monitoring values, and the deformation observation values are $y, y_2, \ldots, y_n$, respectively, and the observation time are $x_1, x_2, \ldots, x_n$, respectively, we then can obviously choose the weight of y_1 as $p_1 = \frac{x_1}{\sum_{i=1}^{n} x_i}$, …, the weight of y_n as $p_n = \frac{x_n}{\sum_{i=1}^{n} x_i}$, and the sum of y_1 to y_n as 1.

4.2 *The solution of the model*

For example, the regression equation of (1) is given by

$$
\begin{aligned}
y_1 + v_1 &= a_1 + a_2\sqrt{x_1} + a_3 e^{x_1} + a_4 \ln(x_1) \\
y_2 + v_2 &= a_1 + a_2\sqrt{x_2} + a_3 e^{x_2} + a_4 \ln(x_2) \\
y_n + v_n &= a_1 + a_2\sqrt{x_n} + a_3 e^{x_n} + a_4 \ln(x_n)
\end{aligned} \tag{10}
$$

In (10), let

$$
Y = \begin{bmatrix} y_1 \\ y_2 \\ \cdots \\ y_n \end{bmatrix}, \quad V = \begin{bmatrix} v_1 \\ v_2 \\ \cdots \\ v_n \end{bmatrix}, \quad X = \begin{bmatrix} 1 & \sqrt{x_1} & e^{x_1} & \ln(x_1) \\ 1 & \sqrt{x_2} & e^{x_2} & \ln(x_2) \\ \cdots & \cdots & \cdots & \cdots \\ 1 & \sqrt{x_n} & e^{x_n} & \ln(x_n) \end{bmatrix},
$$

$$
A = \begin{bmatrix} a_1 \\ a_2 \\ a_3 \\ a_4 \end{bmatrix}, \quad P = \begin{bmatrix} p_1 & 0 & 0 & 0 \\ 0 & p_2 & 0 & 0 \\ \cdots & \cdots & \cdots & \cdots \\ 0 & 0 & 0 & p_n \end{bmatrix},
$$

$$
Y_P^T = \begin{bmatrix} \frac{\sum_{i=1}^{n} y_i}{n} & \frac{\sum_{i=1}^{n} y_i}{n} & \cdots & \frac{\sum_{i=1}^{n} y_i}{n} \end{bmatrix}, \quad Y_t = Y - Y_P \tag{11}
$$

become

$$
Y + V = XA \tag{12}
$$

By means of the least square method, model parameters can be obtained as follows:

$$
A = (X^T P X)^{-1} X^T P Y \tag{13}
$$

The residual error is

$$
V = XA - Y \tag{14}
$$

The residual standard deviation is

$$
S = \sqrt{\frac{V^T P V}{n-4}} \tag{15}
$$

The correlation index is

$$
R = \sqrt{1 - \frac{V^T P V}{Y_t^T P Y_t}} \tag{16}
$$

4.3 *The choice of the model*

If the deformation observation values $y_1, y_2, \ldots, y_n$, y_{1+1}, y_{2+2} are observed, we can erect models to obtain parameters to use y_1 to y_n, and use y_{1+1} to decide the criterion of the best model. The model whose forecast error is least is the best model.

5 THE EXAMPLE OF THE CALCULATION

The horizontal deformation monitoring data of the monitoring point A_3 in the dam are used here, and some calculations are done, and the calculation results are summarized in Table 1.

The best model of the monitoring point A_3 is (2), the residual standard deviation is ±0.5299 mm and the correlation index is 0.97741.

Table 1. Calculation results of the monitoring point A_3.

Observation time (year-month-day)	Observation values (mm)	Water level (m)	Weight	Residual errors (mm)
2014-01-06	5.81	180.54	0.0105	−1.74
2014-02-17	5.72	174.12	0.0228	0.37
2014-03-17	6.90	182.12	0.0316	0.65
2014-05-18	7.01	194.12	0.0491	2.18
2014-06-17	9.22	193.77	0.0579	0.27
2014-07-13	7.16	193.49	0.0649	2.36
2014-08-03	9.25	199.78	0.0710	0.17
2014-08-07	11.56	202.36	0.0719	2.16
2014-08-08	11.87	203.71	0.0728	−2.49
2014-08-11	8.90	198.00	0.0737	0.45
2014-08-16	11.90	203.67	0.0746	−2.58
2014-08-17	11.63	202.72	0.0754	−2.34
2014-08-20	10.27	198.88	0.0763	−1.01
2014-08-24	8.04	196.25	0.0772	1.18
2014-09-07	7.59	197.24	0.0808	1.48
2014-10-06	5.88	198.28	0.0895	2.68

From Table 1, it can be seen that residual errors are less than 3 mm, the residual standard deviation is very small, the correlation index close to 1, and the fitting effectiveness is very good.

The deformation forecast value of the monitoring point A_3 on December 8, 2014 is 7.05 mm, the deformation monitoring value of the monitoring point A_3 on December 8, 2014 is 6.71 mm, the forecast error is 0.34 mm, and the forecast error is very small. The deformation value increases when the water level of the reservoir rises, and the deformation value decreases when the water level of the reservoir falls.

6 CONCLUSIONS

Considering the material property of the dam, for deformation monitoring points in different places, because they are in different places, the influence of the water level of the reservoir and other factors is different. The deformation regularity of these deformation monitoring points is different. Some forecast models are preplaced to let the computer search the forecast model whose deformation error is least. A practical example verifies the effectiveness of the method, and finds that the forecast error is very small.

ACKNOWLEDGMENT

This paper was funded by the State Natural Sciences Foundation (No. 41172298) and the Scientific Research Foundation for NASG Key Laboratory of Land Environment and Disaster Monitoring (No. LEDM2013B03).

REFERENCES

[1] Zongchou Yu and Lincheng Lu. 1978. *Surveying Adjustment Foundation*. Beijing: Surveying and Mapping Press.

[2] Tianxing Yuan. 1979. *The principle of the best estimation*. Beijing: Defense Industry Press.

[3] Qirui Zhang. 1988. *Practical regression analysis*. Beijing: Geology Press.

[4] Yongqi Chen. 1988. *Deformation observation data Processing*. Beijing: Surveying and Mapping Press.

[5] Xinji He and Deji Ren and Fumin Lu. 1998. Study on forecasting model of important landslides at reservoir area of Qingjiang Geheyan. *Journal of Yangtze River Scientific Research Institute* 15(5): pp 39–43.

[6] Fumin Lu and Jin Li and Shangqing Wang. 2013. Application of Kalman filter model considering Taylor series and temperature factors in the deformation forecast of Lianziya hazardous rock mountain. *Mathematics in Practice and Theory* 43 (20): pp 86–91.

Material Science and Environmental Engineering – Chen (Ed.)
 ISBN 978-1-138-02938-5

Research on the slaking deformation of earth-rock dam based on the Duncan-Chang's EB Model

X.C. Jiang
Henan Electric Power Survey and Design Institute, Zhengzhou, China

Y.Q. Li
College of Civil Engineering and Architecture, Henan University of Technology, Zhengzhou, China

ABSTRACT: Duncan-Chang's EB Model is introduced in detail in this paper. The principle of the treatment of slaking deformation is deduced. One earth-rock dam is stimulated by the finite element method during water storage. The application of Duncan-Chang's EB Model provides the basis for the discussion about the impact on stress field and displacement field by slaking deformation. The calculated results show that slaking deformation weakens the uplift displacement by buoyancy. Thus, these results gained by the numerical method may provide a reference to engineering practice.

Keywords: Duncan-Chang's EB Model; slaking deformation; initial stress increment method

1 INTRODUCTION

Earth dams, embankments and other important hydraulic foundations are generally required to contact with water directly. When inundated, soil will vary from unsaturated to saturated, which usually makes the soil structure alter, thus inducing the stress-strain relationship to change and causing slaking deformation [1–4].

Many studies have shown that in the earth-rock dam or other hydraulic foundations, whether it is filled with gravel soil or clay, even if the compacting factor is above 0.9, there exists obvious slaking deformation. As excess slaking deformation will cause problems such as stability and cracking, many scholars around the world have committed themselves to study the calculating method of slaking deformation [5–7]. Obviously, slaking deformation will occur in the earth-rock dam during water storage. However, how to compute the magnitude of slaking deformation for assessing the impact, especially by the finite element method, is critical. This paper deals with the principle of calculating the slaking deformation in detail by the infinite element method, stimulates one earth-rock dam during water storage, and assesses the impact on stress field and displacement field by slaking deformation.

2 DUNCAN-CHANG'S EB MODEL

Duncan and Chang's model is a nonlinear elastic model, which has been widely used in geotechnical engineering, especially in the numerical analyses of earth dams. It is attributed to Kondner who proposed the following hyperbolic stress-strain function to describe the deviatoric stress-axial strain curve obtained from triaxial tests.

Consider

$$\sigma_1 - \sigma_3 = \frac{\varepsilon_a}{a + b\varepsilon_a} \tag{1}$$

in which a and b are model constants. In this constitutive model, the tangential Young's modulus E_t and Poisson's ratio ν_t are used to simulate the nonlinear elastic response of soils, which are assumed to be

$$E_t = KP_a\left(\frac{\sigma_3}{P_a}\right)^n (1 - R_f S_t)^2 \tag{2}$$

$$B_t = K_b P_a\left(\frac{\sigma_3}{P_a}\right)^m \tag{3}$$

where P_a is the atmospheric pressure; K and K_b are modulus numbers; n and m are exponents determining the rate of variation of module with confining pressure; and R_f is the failure ratio with an invariable value less than 1.

The Mohr-Coulomb failure criterion is adopted in the model, and S_t is a factor defined as the shear stress level given by

$$S_t = \frac{(1 - \sin\phi)(\sigma_1 - \sigma_3)}{2c\cos\phi + 2\sigma_3 \sin\varphi} \tag{4}$$

In the unloading and reloading stage, the tangential Young's modulus is defined as

$$E_{ur} = K_{ur} P_a \left(\frac{\sigma_3}{P_a} \right)^n \tag{5}$$

So far, the model has 8 parameters, namely c, φ, K, K_{ur}, n, R_f, K_b and m. These parameters can be determined with a set of conventional triaxial tests.

In general, a curved Mohr-Coulomb failure envelop is adopted by setting $c = 0$ and letting φ vary with confining pressure according to

$$\varphi = \varphi_0 - \Delta\varphi \lg(\sigma_3/p_a) \tag{6}$$

Then, parameters c and φ are replaced by φ_0 and $\Delta\varphi$.

Duncan and Chang's EB constitutive model is quite simple; it has gained a significant success in geotechnical engineering. On the one hand, it is easy to obtain the model parameters; on the other hand, much experience has been accumulated.

3 THE TREATMENT OF SLAKING DEFORMATION

Slaking deformation can be calculated by incremental FEM [5][6][7]. Incremental FEM includes the Initial Stress Increment Method and the Initial Strain Increment Method. The Initial Stress Increment Iteration Method can turn slaking deformation into the nodal force, which is clear in concept, and therefore it is used in this paper. The basic idea is as follows.

Assume that the stress state $\{\sigma_d\}$ is accumulated by n equal stress increments before flooding, where each stress increment is as follows:

$$\delta\sigma = \{\sigma\}/n \tag{7}$$

Every strain increment can be arrived at by the rigidity matrix $[D]_d$ under the dry condition as follows:

$$\{\delta\varepsilon\} = [D]_d^{-1} \{\delta\sigma\} \tag{8}$$

The parameters in $[D]_d$ can be derived by the wind drying soil experiment, which vary with the stress state. Total strain $\{\varepsilon\}$ before flooding can be obtained by accumulating each strain increment $\{\delta\varepsilon\}$. Assume that slaking deformation is constrained and the strain after flooding is constant. Consequently, the stress $\{\sigma_\varepsilon\}$ after flooding can be derived by the stress-strain relationship under the saturation state. When $\{\varepsilon\}$ is divided into n equal increments, each increment can be derived according to Formula (8). The stress increment under the saturation state can be obtained by the rigidity matrix $[D]_w$ under the slaking condition according to Formula (9):

$$\{\delta\sigma_w\} = [D]_w \{\delta\varepsilon\} \tag{9}$$

The parameters in $[D]_w$ can be derived by the saturated soil experiment. Accumulating all the stress increments $\{\delta\sigma_w\}$, the total stress $\{\sigma_w\}$ under the saturation state can be derived when the strain is kept constant. Consequently, the stress change can be obtained according to Formula (10):

$$\{\Delta\sigma\} = \{\sigma_w\} - \{\sigma_d\} \tag{10}$$

The stress change $\{\Delta\sigma\}$ can occur when node displacement is restricted, which can be realized by applying the force of constraint on nodes. The force of constraint on nodes can be derived as follows:

$$\{F\}^e = \sum \iiint_\Omega [B]^T [D_W] \{\Delta\varepsilon\} d\Omega \tag{11}$$

For equal node load that does not exist, in order to maintain stress balance, the stress $[D_w]$ $\{\Delta\varepsilon\}$ induced by slaking deformation should be subtracted. In fact, the node displacement is not restricted, and every node serves as a hinge, not as a fixed support, so the force of constraint does not exist. In order to remove the constraint, the equal and opposite node loads must be applied. The node load, water pressure or seepage force, and other outer load such as uplift pressure should be applied on the mesh.

4 ENGINEERING APPLICATION

One earth-rock dam is 100 m high and the top of it is 10 m wide. The normal storage water level is 90 m and the slope ratio is 1:2. The top of the core wall is 6 m wide, and the slope ratio of the core wall is 1:0.2. The material parameters are listed in Table 1.

The software ABAQUS is adopted to simulate this computing model. The longitudinal thickness of the three-dimensional model is 5 m. In order to simulate the plane strain condition, the displacement of the longitudinal thickness direction is restrained. The finite element mesh is shown in Figure 1, and the upstream water pressure of the core wall applied is shown in Figure 2. By analyzing the results calculated after completing the dam construction and completion of water storage, we can obtain the following results:

1. The dam structure is bilateral, and the dam is only affected by gravity; therefore, the load, stress and displacement should also be bilateral. From the computing results, it can be seen that all the stress and displacement contours are bilateral, which completely agrees with the objective laws. In addition, based on vertical displacement contour, the utmost displacement is 61.8 cm and appears in the 1/2 position of the dam height, which is the same as the general experience.

Table 1. Parameters of soils.

Material	Dam shell (dry)	Dam shell (slaking)	Core wall
φ_0 (°)	48.2	50.1	32.8
$\Delta\varphi$ (°)	7.8	8.2	5.9
K	980	1070	695
K_{ur}	2000	2200	1460
n	0.36	0.38	0.27
R_f	0.82	0.85	0.72
K_b	530	610	416
M	0.15	0.18	0.09

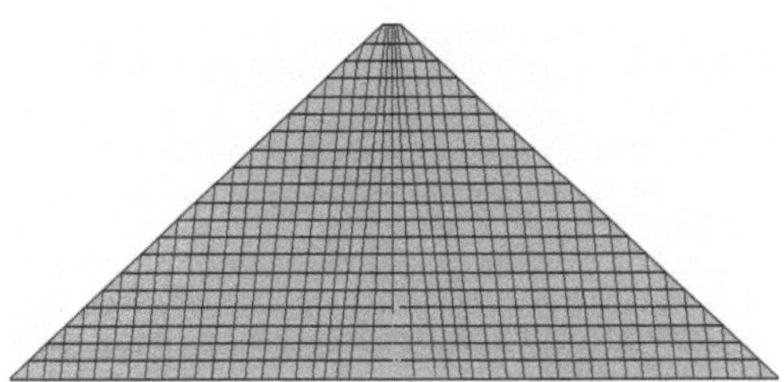

Figure 1. Finite element mesh of the dam.

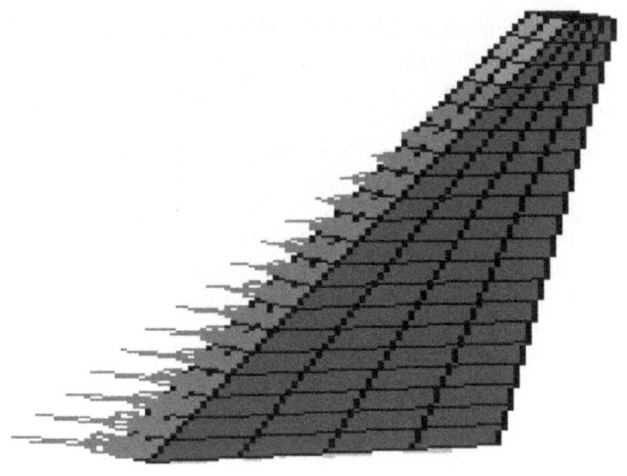

Figure 2. Water pressure applied.

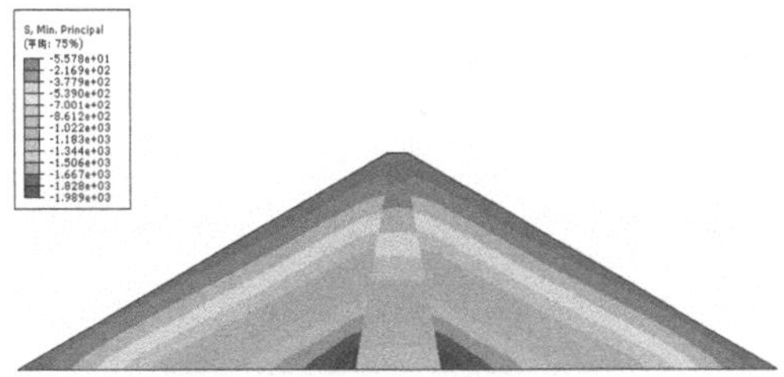

Figure 3. The minimum principal stress contour after completing the dam construction (kpa).

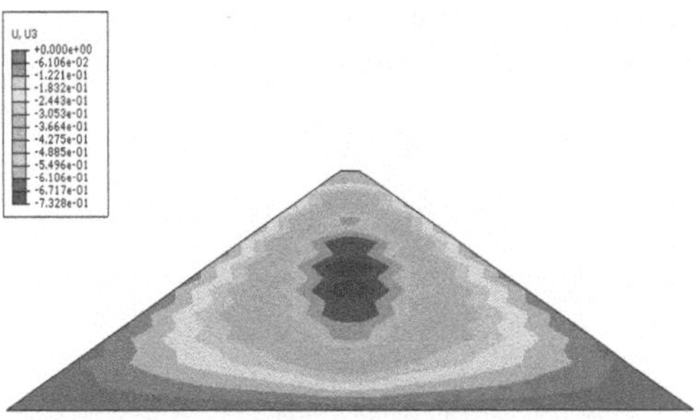

Figure 4. The vertical displacement contour after completing the dam construction (m).

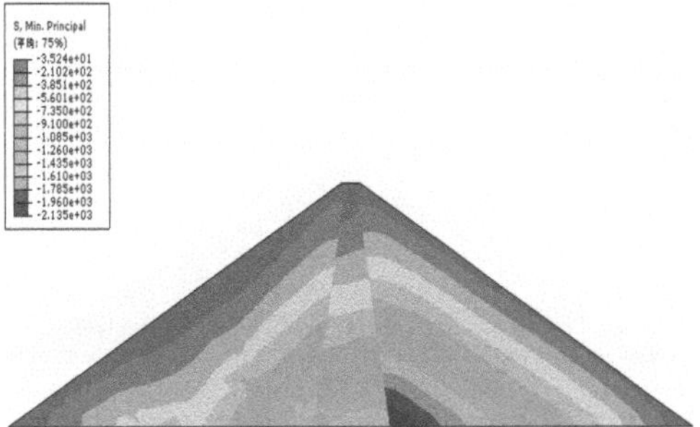

Figure 5. The minimum principal stress contour after completion of water storage (kpa).

2. From the minimum principal stress contour, it can be seen that stress distributes well. The places of the dam where the distance to the dam face is larger have greater stress. Because of the difference between the dam shell and the core wall module, the contour of minor principal stress displays the hump between the core wall and the dam shell, namely the arching effect, which is the same as the general stress computing law of the earth-rock dam.
3. According to the minimum principal stress after completion of water storage, it can be seen that the stress of the upstream dam becomes lower,

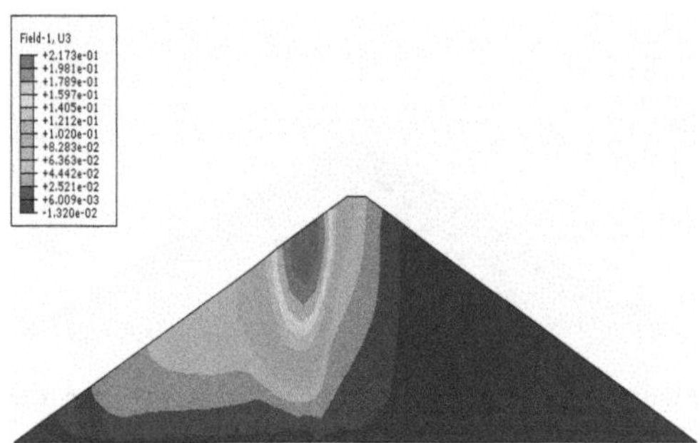

Figure 6. The vertical displacement contour after completion of water storage (m).

Figure 7. The minimum principal stress contour when considering slaking deformation (kpa).

Figure 8. The vertical displacement contour when considering slaking deformation (m).

which agrees with the common computation law of finite element software. By analyzing the vertical displacement contour after completion of water storage, we can see that the upstream dam is uplifted and the downstream dam moves downward under the horizontal water pressure, which also shows that the water pressure is loaded successfully and the computation agrees with the general law.

4. According to the minimum principal stress contour and the vertical displacement contour after completion of water storage, we can see that when the slaking deformation is not considered, the maximal uplift displacement of the upstream dam shell is 21.7 cm. When the slaking deformation is considered, the uplift move diminishes obviously, and the maximal value is only 13.4 m, which obviously contains the weakening impact caused by the slaking deformation.

5 CONCLUSIONS

All the results are in accordance with the general law of the earth-rock dam during water storage. The results show that the finite element method adopted in this paper is competent for calculating the slaking deformation and assessing the impact caused by the slaking deformation, and thus it is feasible for calculating the stress field and the displacement field. The method adopted and those results gained by the numerical method may provide a reference to engineering practice.

REFERENCES

[1] Nobari. E.S., Ducan. J.M., Movements in Dams due to Reseervior Filling[C], Performance Of Earth and Earth Supported Structures, V01.I, Partl 1973.

[2] Fu Xudong, Qiu Xiaohong. Experimental research on slaking deformation of high fill of Wushan municipal sewage treatment plant [J]. Rock and Soil Mechanics, 2004, 25(9):1385–1389.

[3] Lim Y.Y., Miller G.A. Wetting-induced compression of compacted oklahoma soils [J]. Journal of Geotechnical and Geoenvironmental Engineering, 2004, 130(10): 1014–1023.

[4] Pereira J.H.F., Fredlund D.G. Volume change behavior of collapsible compacted gneiss soil [J]. Journal of Geotechnical and Geoenvironmental Engineering, 2000, 126(10): 907–916.

[5] Li Guang-xin. Study on slaking of rock fill [J]. Chinese Journal of Geotechnical Engineering, 1990, 12(5): 58–64.

[6] Zude Liu, "Some problems about the calculation of deformation of earth-rockdams", Chinese Journal of Geotechnical Engineering, Nanjing, China, vol.5, Feb. 1983, pp1–13.

[7] Pereira J.H.F., Fredlund D.G. Volume change behavior of collapsible compacted gneiss soil [J]. Journal of Geotechnical and Geoenvironmental Engineering, 2000, 126(10): 907–916.

Material Science and Environmental Engineering – Chen (Ed.)
© 2016 Taylor & Francis Group, London, ISBN 978-1-138-02938-5

Measurement and analysis of the performance of the main ventilators of east one shaft in the Yaoqiao coal mine

G.Y. Cheng
Department of Safety Training of North China Institute of Science and Technology, Sanhe, Hebei, China

J. Cao, S. Feng & Y. Cheng
College of Safety Engineering, North China Institute of Science and Technology, Sanhe, Hebei, China

H. Wang
Shanxi China Resources Daning Energy Co. Ltd., Taiyuan, Shanxi, China

ABSTRACT: To understand the actual operating characteristic of two centrifugal fans in East One shaft of Yaoqiao Mine, the short-circuit method of ground airflow was used to measure the fan's performance, considering the need of normal mine ventilation. The velocity pressure method was used to measure the fan's air volume on the basis of the clause in the Coal Mine Safety Regulations the measurement of the fan performance measurement must be conducted once at least every 5 years. The method was used to process each measuring parameter, plot the characteristic curve of air volume, pressure and efficiency, and provide the theoretical basis for the safe and economic operation of the main ventilator and air flow adjustment. The solution design, method selection and measuring results can thus provide certain references for other coal mines with the similar situation.

Keywords: centrifugal fan; fan performance; fan characteristic curve; work condition adjustment

1 INTRODUCTION

Mine main ventilators are the main power to realize mine ventilation, and are an important part of the mine ventilation system. Their safe, highly efficient and stable operation is of great significance to the mine safe production. To achieve the above purpose, the actual operating characteristics of the main fans are needed to be grasped. The characteristic curve of large fans is usually based on test data of models or prototypes because the spot check test is used before fans come out. In fact, the reason why the actual characteristic curve has an obvious difference from the that of the models is that the quality can hardly meet the design requirements in the process of manufacturing; its performance changes constantly due to differences in external diffuser structure, the quality of the field installation, and abrasion as well as corrosion.

Coal Mine Safety Regulations states that before newly installed main ventilators are put into use, their performance must be measured once and test run must be conducted. Afterwards, the measurement of the fan performance must be conducted once at least every 5 years. Five years or so have passed since the latest measurement was made in 2009. To understand the actual operating characteristic, the determination group is organized to measure the fan's performance of two centrifugal fans in Dong Yi shaft of Yaoqiao Mine in October 2014.

Yaoqiao coal mine is located at about 82 km northwest of Xuzhou City, Jiangsu Province and 17 km to the south of Pei County. Yaoqiao coal mine has been producing for nearly 40 years, and after many reconstructions, its production has reached 4 million tons/year, while increasing to three levels: –400 *m*, –650 *m* level, and the exploiting –850 *m* level. The mixed ventilation way is four into three out, namely two pairs of the main and auxiliary shafts in the central –400 *m* level and –650 *m* level receiving the air, and the western shaft in the west boundary of the field, the East Two shaft in the east and East One shaft in the central boundary of the mine field exhausting the air.

Two G4-73-11 no. 25 *D* centrifugal fans, one working and the other backing up are installed in the East One shaft. The rated power of the matched asynchronous motor is 210 *kW*, and the rated speed is 490 *r/min*.

2 MEASURING PROCEDURE

2.1 *Measuring plan*

The short-circuit method of ground airflow was used to measure the fan's performance under the condition of no production. That is, the normal ventilation fan is responsible for the underground work, to determine the performance of the standby one, and to determine the former's performance after swap. The arrangement of determination is shown in Figure 1. The gate at 0-0 is used to block the underground airflow; boards are covered in the I-I section by increasing or decreasing the board to regulate the working conditions. Airflow comes into the centrifugal fan by the II-II section, namely the stable airflow section near the leading device; airflow passes into atmosphere through the annular diffuser.

The main performance parameters are airflow, air pressure and efficiency, of which the most important and unpredictable parameter is the work airflow. Its measurement accuracy depends mainly on the measuring section that is mainly based on the stability of air speed distribution. Considering that the air speed of air duct section measured wind is stable, but the different cross-sectional area is almost the same, the velocity pressure method should be used, but the static pressure method should be used to measure the air volume. The pitot tube is mounted in II-II different sectional locations (based on different distances from the fan housing), and then the hoses are connected to the whole pressure pipe and the static pressure tube of the pitot tube to measure the dynamic pressure.

The relational expression among airflow, pressure and area is shown:

$$Q = k \cdot s \cdot \sqrt{\frac{2h_v}{\rho}} \tag{1}$$

where h_v is the velocity pressure, *Pa*; *K* is the loss coefficient of dynamic pressure, normally 0.98; *S* is the area, m^2; and ρ is the air density, kg/m^3.

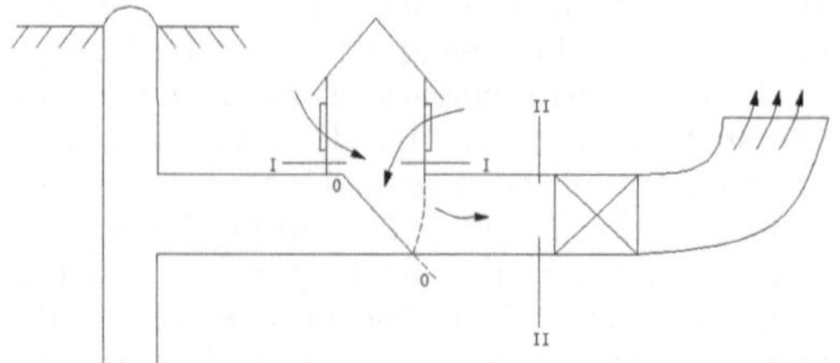

Figure 1. The schematic diagram of the fan's performance measurement.

Centrifugal fans generally require the total pressure characteristic curve. The more stable II-II section, namely, downwind of regulating conditions device and near the entrance to the fan, was selected to measure the relative static pressure H_s, and then according to Equation (2), the total pressure of the ventilator device H_t was calculated.

$$H_t = h_s - h_v + h_{vd} \tag{2}$$

where h_{vd} is the dynamic pressure of the outlet section of the diffuser, acquired by the air volume and the outlet section area.

2.2 *Measuring steps*

1. Before measurement: we organize the labor division (e.g. command group, regulating working conditions group, measuring ventilation parameters group, measuring electrical parameter group, communicating group, recording and quick count groups) based on the measurement program; prepare instruments, tools and record forms, and sign the technical parameters of the fan and motor nameplate; measure fan structure size and section size; and install measuring devices and instruments.
2. During measurement: When the aforementioned settings are ready, the fan starts. Airflow comes into the fan through the blinds, and is discharged from the diffuser. Because the shaft power of centrifugal fans increases with air volume, all the gates in the wind tunnel should be closed so as to ensure a safe start and to avoid overloading the start and burn out the motor, while centrifugal fan starts. The fan starts at a small amount of air, and then starts reading the measurement parameters after airflow stabilizes (about 3–5 minutes after the fan starts).

   While adjusting the working conditions, centrifugal fans should gradually transit from the big drag to a small air resistance by removing wood and increase the inlet section. After the measurement data of some working conditions are finished, the working conditions are regulated, and then the value of each parameter is read until the entire curve is completed (usually more than 8~9 measuring points); and then the fan stops.
3. Data collection and analysis: we use a computer or manually calculate the measuring values, plot, and then analyze to complete the measurement report.

2.3 *Precautions*

1. Safe technology is highly required in the measuring work of the main fan performance, requiring

a clear division and obeying the leader and the command.

2. Before the formal determination, the mechanical system and power supply system should be inspired comprehensively by the electrical department to ensure integrity.
3. Cleaning up sludge and debris in the wind tunnel. The quality of the gate at the regulating conditions is checked to prevent unexpected situations during the measurement.
4. While measuring performance, the temperature and current fluctuations of the motor are monitored closely. When the motor temperature or current fluctuations approach and exceed the normal range, the electromechanical driver or person in charge of the fan must stop immediately to ensure fan safety. If other dangers threaten the fan (including motor) safety, they should immediately stop and interrupt the measuring work.
5. Cut off airflow and check rigor of the gate before measurement. Leakage should be plugged in order to reduce the impact of the work on underground airflow to ensure the safety production.
6. Before measuring, a gas cylinder hose is used to inspect air tightness to prevent leakage. Then, the support rods are kept fixed with the pitot tube firm to prevent it from falling into the fan. The pitot tube in the section should be fixed firmly and straight. The hoses connected to the pitot tube should be fixed firmly to prevent jitter or loss.
7. Before measurement, the fan channel must be entered, and the wear and turbine installation quality is checked to prepare for a comparative analysis between the measuring curve and the fan factory curve.

3 MEASUREMENT RESULT AND ANALYSIS

After measuring the first fan, its atmospheric parameters, electrical parameters and wind pressure, air flow power and efficiency calculated, the fan performance curve are plotted, as shown in Tables 1 and 2 and Figures 2 and 3. The parameters and curve of the second fan are listed in Tables 3 and 4 and shown in Figures 4 and 5.

Table 1. Static parameters of the first fan.

Outlet section (m^2)	Static pressure section (m^2)	Synchronous speed (*r/min*)	Rated current (*A*)	Rated power (*kW*)	Opening angle of the leading device (°)
5.0	3.1	490	29.2	210	90

Table 2. Measuring result of the first main fan.

Number	Fan's volume (m^3/s)	Fan's pressure (*Pa*)	Axial power (*kW*)	Fan's efficiency (%)	Fan's speed (*rpm*)
1	6.7	1934.4	62.1	20.9	498.1
2	25.4	1938.0	95.5	51.5	496.5
3	28.4	1816.2	108.5	47.6	497.7
4	58.4	1534.6	125.6	71.3	494.2
5	60.9	1329.7	127.8	63.3	493.9
6	76.5	1151.2	142.6	61.7	494.3
7	90.1	977.3	139.1	63.3	494.5
8	94.0	741.6	138.2	50.4	494.4
9	93.1	618.0	136.8	42.1	495.0

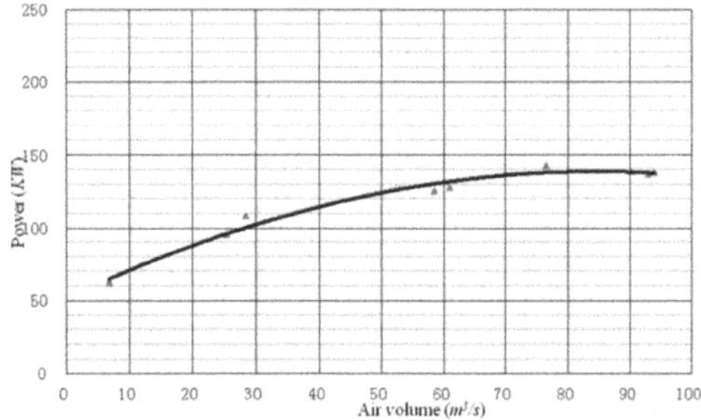

Figure 2. The characteristic curve of the output power and air volume of the first fan.

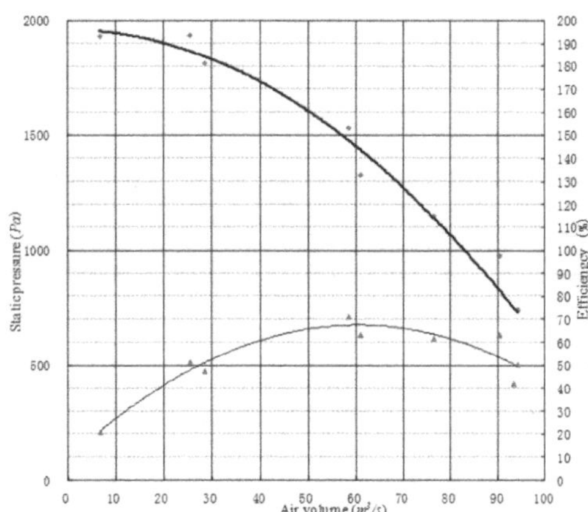

Figure 3. The characteristic curve of the air volume, air pressure and efficiency of the first fan.

1. From the measured data and the performance characteristic curve, it can be seen that the measured data are reliable and measured curve

Table 3. Static parameters of the second fan.

Outlet section (m^2)	Static pressure section (m^2)	Synchronous speed (r/min)	Rated current (A)	Rated power (kW)	Opening angle of the leading device (°)
5.0	3.1	490	29.2	210	90

Table 4. Measuring result of the second main fan.

Number	Fan's volume (m^3/s)	Fan's pressure (Pa)	Axial power (kW)	Fan's efficiency (%)	Fan's speed (rpm)
1	7.1	2214.0	67.3	23.5	496.0
2	14.9	2178.0	77.4	41.9	496.0
3	30.6	1907.0	113.2	51.6	495.0
4	61.3	1500.0	134.2	68.5	494.2
5	47.0	1698.0	112.4	70.9	493.7
6	63.9	1223.0	150.1	60.4	492.5
7	75.0	972.0	173.8	41.9	494.0
8	72.6	942.0	176.1	38.9	495.8

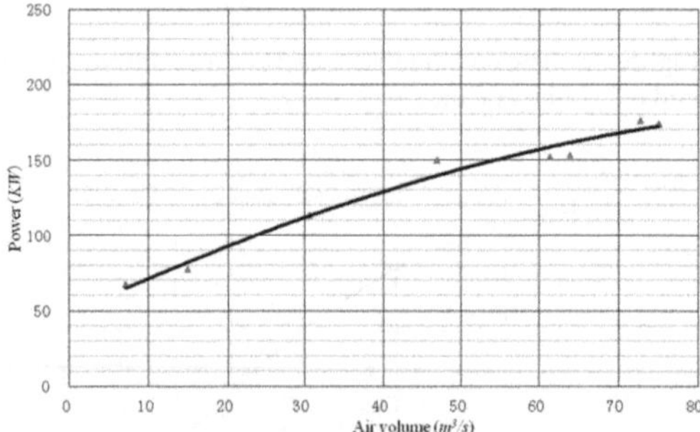

Figure 4. The characteristic curve of the output power and air volume of the second fan.

can be used for site management because of the reasonable condition regulation, appropriate section of measuring air volume and pressure, even measurement points and smooth curve.

2. From an economic point of view, the reasonable work range for the operating efficiency should not be less than 60%; considering from the safety side, centrifugal fans have no saddle section and can ensure their safe and stable operation.

   Currently, the main fan matches well with the resistance of the underground pipe network. Operating efficiency is reasonable and the lowest is up to 70%.

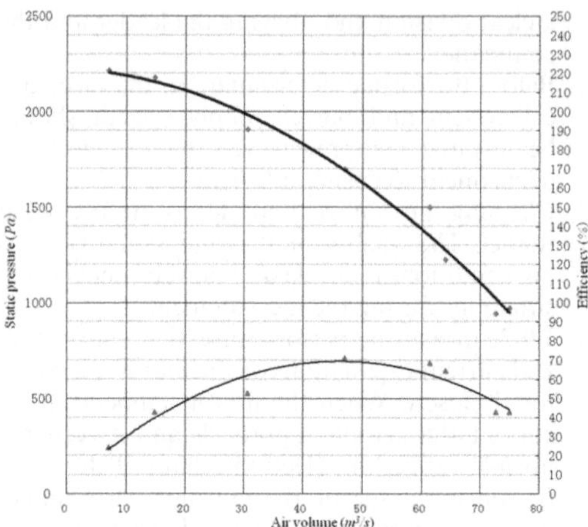

Figure 5. The characteristic curve of the air volume, air pressure and efficiency of the second fan.

3. Mounting the leading device in the suction inlet of the centrifugal fan increases a certain wind resistance on the way to the inlet, which is one of the adjustment methods called "increasing resistance". When the leading device angle ranges from 0° to 90°, pressure curves go down the fan, efficiency also reduces, and the scope of adjustable air gradually narrows.

   Figures 4 and 5 show the measured characteristic curves when the opening degree of the leading device blade is 90°. There is no room for the future regulation by using the leading device to adjust the fan operating point. Adjusting the damper and changing the fan speed is recommended for regulating the working condition.

4 CONCLUSIONS

1. Conducting the fans' performance test every 5 years and grasping their operating performance is absolutely necessary. The main fans' performance must match with the resistance of the underground ventilation network, which makes the operating point go within a reasonable scope and keep the main fan reliable and stable and to operate safely.
2. The actual performance parameters and the actual characteristic curve of the ventilator device were measured based on this determination, providing a reliable theoretical basis for air volume regulation and the fan's safe operation.
3. Accurate determination of air volume affects the reliability of the measured results of the fan's performance. Based on the operating environment of the centrifugal fan of the East

One shaft, the dynamic pressure method is chosen to measure the air volume. This method is practicable, and conclusions are practical and reliable.

4. The main centrifugal fan's power increases with its power. First, the gate or adjustable shutters should be closed, and then covered with wooden boards on the frame, so that the fan starts at the maximum wind resistance. When the speed gradually becomes normal, the operating point can be adjusted by opening the gate or windscreens and by removing the wooden boards block by block.

ACKNOWLEDGMENTS

The authors thank the support of the National Natural Science Foundation provided to this project (No. U1361130).

REFERENCES

[1] Cheng Genyin, Zhang Jinggang. Measurement and analysis of main ventilator performance in Xuzhuang Mine. China Safety Science Journal, 2010, 20(2):104–109.

[2] Cheng Genyin, Wu Huaijun, Jin Longzhe. Measurement and analysis of safety technical parameters for main ventilators of No. 3 ventilating shaft in Jinpushan Coal Mine. China Safety Science Journal, 2002, 12(4):48–54.

[3] Wang Taihui, Li Qiyuan, Bai Jianzhi, et al. Fault diagnosis of centrifugal fans. China Plant Engineering, 2011, 3:64–65.

[4] State Administration of Work Safety, State Administration of Coal Mine Safety. Coal Mine Safety Regulations. Beijing: China Coal Industry Publishing House, 2011.

[5] Xie Qidong. Application of static differential pressure method for measurement of main fan flow rate in coal mine. Journal of Safety Science and Technology, 2008, 4(4):122–125.

[6] Zhang Guoshu. Science of ventilation safety. Xuzhou: China University of Mining and Technology Press, 2011.

[7] Yao Shangwen, Lu Ping. Discussion techniques on performance test of main fan for mine. Journal of Anhui Institute of Architecture & Industry, 2014, 22(3):69–74.

[8] Gao Guoquan. Determination of main mine fan performance of a number of issues. Coal Technology, 1989, 9:17–20.

[9] Bai Shuowei, Chen Jianwei, Bai Nianyu. Analysis and research on regulating air volume of centrifugal fan by leading device. Coal Mine Machinery, 2010, 31(9):177–179.

[10] Liu Zilong. Adjusting methods and their comparison of large centrifugal fans. Coal Technology, 2009, 4(28):11–13.

Material Science and Environmental Engineering – Chen (Ed.)
© 2016 Taylor & Francis Group, London, ISBN 978-1-138-02938-5

Temperature crack control of the linear accelerator mass concrete

G.C. Yan
Infrastructure Department, Zhongnan Hospital of Wuhan University, Wuhan University, Wuhan, China

Z.X. Ma
School of Architectural and Material Engineering, Hubei University of Education, Wuhan, China

ABSTRACT: In this paper, based on the linear accelerator mass concrete in a hospital, the internal temperature stress of the mass concrete was analyzed and controlled by the method of mathematical analysis with the change in the external environment and the gradual release of hydration heat. Through the observation and summary, some measures have a guiding significance to avoid temperature cracks of mass concrete in bridges, buildings and other structures.

Keywords: mass concrete; temperature crack; thermal stress; anti-cracking measure

1 INTRODUCTION

With the development of economy, the modern architecture is increasingly toward the direction of development of large scale and complex shape. In bridge engineering, with the increasing span of bridge structure, the foundation concrete is more and more of huge volume, and the control of cracks in civil engineering becomes the key factor affecting the quality of engineering.

The construction of the medical linear accelerator in medical buildings has a positive impact on improving medical conditions and medical quality, but the construction of the project will impose a certain negative impact on the environment. Linear accelerator takes the advantage of X-ray to kill cancer cells; however, if the X-ray spills over, it will produce radiation hazards to humans. Anti-cracking measures of large volume concrete in medical linear accelerator rooms must be taken to reduce environmental pollution, to protect the environment and safeguard human health. Many factors can influence concrete cracking, and this paper focuses on the analysis of crack caused by the temperature change. The mechanism of the generation and development processes of crack in large bulk concrete will be analyzed in detail, and the effective control measures are discussed, which is of great significance for ensuring the normal use and engineering quality [1].

2 MAIN CAUSES OF TEMPERATURE CRACKS OF MASS CONCRETE

Concrete is a property of thermal expansion and cold shrinkage. When the temperature of the external environment or internal structure changes, the concrete will produce deformation. Due to the strength caused by the limitation of deformation that exceeds the anti-intensity of concrete, it leads to the phenomenon of temperature cracking. The concrete temperature deformation is the expansion and contraction of the concrete with temperature change. Temperature stress is caused by temperature deformation by the temperature difference.

The greater the temperature difference is, the greater the thermal stress is. Crack of mass concrete is mainly caused by the temperature difference. In some of the long-span bridges, temperature stress can reach and even exceed the live load stress. The dimension of the massive concrete is very huge. After the concrete is poured, a great deal of hydration heat of cement is expended, and the deleterious crack occurs [2].

The main characteristic of the temperature cracks distinguishing other cracks is opening and closing with temperature. There are many factors that can influence the temperature change: (1) year temperature difference; (2) sunshine; (3) sudden hypothermia; (4) hydration heat. Year temperature change caused by temperature change is very slow. Sunshine temperature change is mainly caused by the solar radiation effect, the temperature change and the wind speed. Strong cold air cooling temperature changes and the high–low temperature distribution outside at night after a sunset can cause sudden hypothermia. The first three temperature changes are caused by changes in natural environmental conditions, which is hard to eliminate.

Due to the heat of hydration of cement, there are three temperature stages, namely heating, cooling and stationary phases, after concrete has been poured.

Table 1. Hydration heat values of various kinds of mineral elements (J/g).

Mineral composition	Hydration ages					Full hydration
	3 days	7 days	28 days	3 months	6 months	
C_3S	463	525	553	592	644	763
C_2S	72	119	191	210	239	377
C_3A	673	754	997	1059	1169	1212
C_4AF	105	286	429	472	–	649

In the early part of the casting of concrete structures, due to the process of cement hydration heat, concrete internal temperature rises generally within 3~5 d, and a 3 d temperature is generally 55 to 59 degrees at or near the maximum temperature. Then, the temperature tends to be stable and starts to cool [3].

Concrete internal temperature gathered within the structure is not easy to lose for a long time because of poor thermal conductivity of concrete. When the internal compressive stress and the external tensile stress are greater than the allowable tensile stress, which are generated by the temperature difference between the inside and the outside in the early stage of hydration heat temperature rising rapidly due to different concrete inside and outside the cooling conditions, cracks are likely to be produced.

In the concrete cooling stage, the deformation caused by the temperature difference and the concrete volume contraction is restrained by external conditions, and cannot shrink freely, and the modulus of elasticity is relatively low at the moment. If the temperature gradient is too large, the internal tensile stress is generated in the middle of the structure section. When the tensile stress exceeds the tensile strength of concrete, the cracks in the concrete section are produced, which may produce a bad influence on concrete quality and durability. Therefore, controlling the temperature difference can reduce the temperature gradient, which is fundamental to ensuring that there are no cracks.

Cement hydration heat mainly depends on its mineral composition, and hydration heat values of various kinds of mineral elements are given in Table 1.

From Table 1, it can be seen that the calorific values of C_3S and C_3A are the biggest, so cement should be selected appropriately in mass concrete because different cement mineral elements of the same varieties and different cement grades have different hydration heats. A lower cement content of C_3S and C_3A should be considered for use in mass concrete, such as the dam concrete.

3 AN ENGINEERING PROJECT

The biggest thickness of the linear accelerator room shear wall is 2.44 m, and the maximum thickness of the roof is also up to 2.44 m. The framework support system of the linear accelerator room is complex. The concrete is poured in 3 times. The first casting ranges from 1.95 m to 1.45 m, the second casting from 1.45 m to 0.6 m, and the third casting from 0.6 m to 5.44 m. A floor plan of the linear accelerator room is shown in Figure 1.

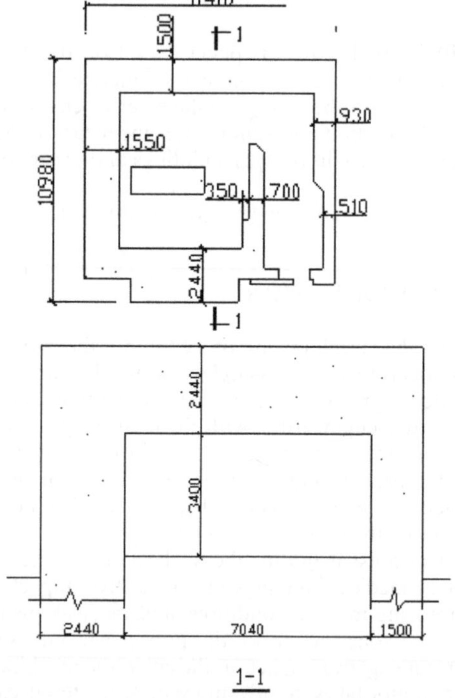

Figure 1. A floor plan of the linear accelerator room.

3.1 *Temperature measurement results*

Sensors are deployed at the representative fixed points in the temperature controlling area. The floor plan is drawn, and the thermometry circuit and system are set ready to measure the temperature. Temperature measurement data are recorded in time in the process of temperature measurement. The roof representative recorded data are provided

Table 2. The roof thermometry recorded data.

Time	1	6	11	16	21	26	31	36	41	46	51	56	61	66	71	141
Top	22.4	24.5	28.8	35.2	36.8	36.2	34.5	31.2	30.2	29.5	29.2	28.5	28.3	27.6	27.1	22.6
Middle	26.9	34.3	45.7	49.9	54	54.6	54.5	53.8	53.1	53.3	52.3	51.3	50.4	48.8	47.3	35.9
Bottom	21.9	25.6	32	35.4	37.7	37.4	37	37.1	36.3	36.6	36.8	35.8	36	35	35.1	30.5

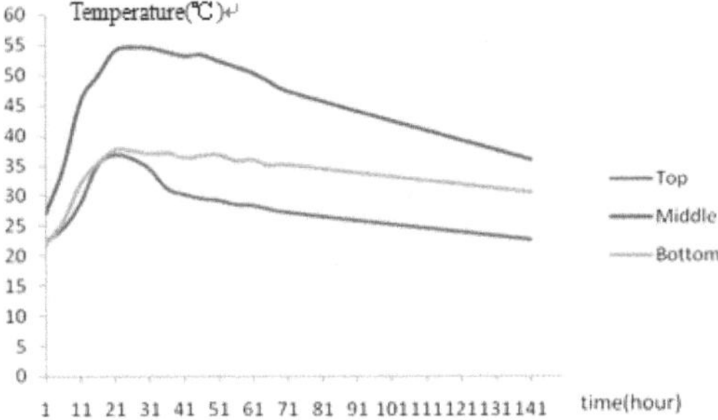

Figure 2. Thermometry data curve.

in Table 2 and in Figure 2, which are measured at a 5-hour interval.

The causes of temperature fluctuation in the mathematical graph can be divided into two categories: one is the physical and chemical changes of normal concrete; another is the cause of the error of the operator and possible instrument. If the curve can reflect the normal rule changes, it can achieve a normal forecast and control. The maximum temperature differences between the core and surface of the testing areas are given in detail in Table 3. They meet the requirements of the specifications concerned.

3.2 *Measures and methods of preventing the cracks in concrete*

1. Choosing the representative fixed points in the temperature controlling area, measuring temperature and drawing the temperature controlling curve.
2. Through mathematical calculation, choosing low hydration heat cement for dams.
3. Choosing concrete of the larger size particles and good aggregate gradation of Longquan Mountain green stone and reducing the dosage of cement.
4. Mixing retarding water reducing agent, reducing water consumption and extending the initial setting time of concrete.
5. Precooling with some ice water before the concrete mixing.
6. Through reflecting the statistical curve, when the maximum temperature differences between the core and surface approach or reach 25°C after concrete pouring, increasing the plastic film for concrete curing and controlling the concrete temperature difference between the inside and the outside by pouring hot water around a film.

Table 3. The maximum temperature difference.

The testing areas	The maximum temperature of the core (°C)	The maximum temperature differences between the core and surface (°C)	The standard requirements (°C)
1	53.2	17.3	<25
2	55.2	23.8	<25
3	56.1	24.9	<25

4 CONCLUSIONS

1. The thermometry curve can reflect the temperature change of dynamic information in the mass concrete by the mathematical analysis method, which contributes to select correctly the concrete materials and construction methods in the reinforced concrete structure design and construction according to the engineering geology, the temperature and sunshine condition to establish the effective mechanism of concrete crack control.
2. Once mass concrete cracks are formed, cracks in the structure of the important parts can be lethal, which will reduce the durability of the structures, weaken the bearing capacity of the component, at the same time may harm the safety of the structures. In this paper, anti-cracking measures combined with the actual practice are put forward by using mathematical analysis, which has a guiding significance for the crack control of mass concrete in the future.

REFERENCES

[1] Jiang Tao. 2005. Emulation analysis on large volume concrete construction. Journal of Minxi Vocational College. 2005(1):105–108.

[2] Shen Xun long, Wang Chao, Han Yan-hui. 2002. Forecast and control on temperature of massive concrete based on neural network. Journal of Hefei University of Technology (Natural Science). 25(5):758–763.

[3] Yao Junjun. 2005. The concrete construction crack control of mass pier. Railway Engineering, 2005(3): 12–14.

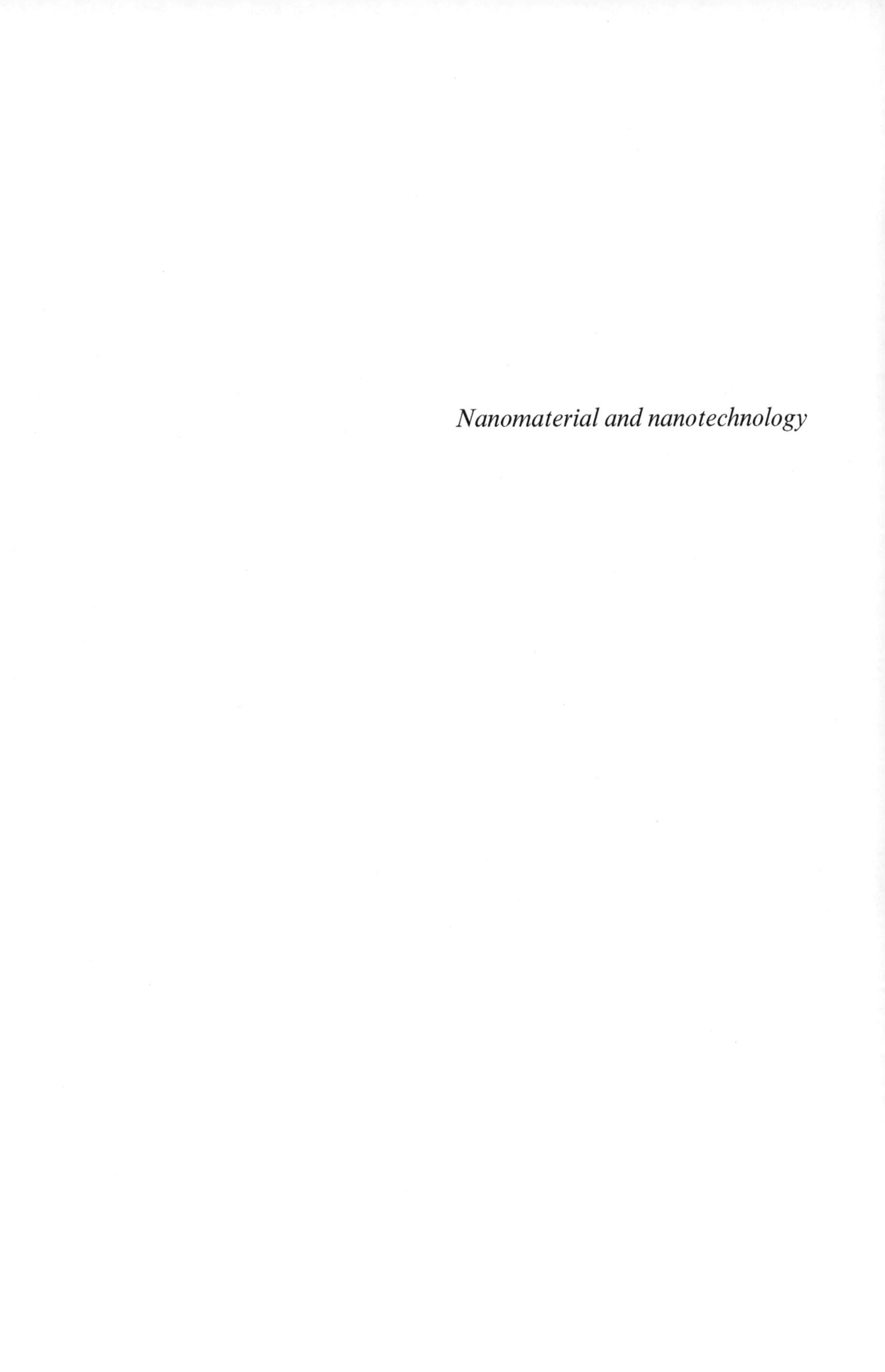

Nanomaterial and nanotechnology

Material Science and Environmental Engineering – Chen (Ed.)
© 2016 Taylor & Francis Group, London, ISBN 978-1-138-02938-5

Immobilization of GOD on fluorescent SiO_2 nanoparticles

K. Li, H.C. Liu, J. Huang & L.Y. Ding
National Engineering Laboratory for Fiber Optic Sensing Technology, Wuhan University of Technology, Wuhan, China

ABSTRACT: The fluorescent SiO_2 nanoparticles with the mean size of 420 nm were prepared and used as the carrier to immobilize Glucose Oxidase (GOD). The optimal immobilization conditions were: APTES concentration was 2% (v/v), GA concentration was 1% (v/v), pH was 6.5. The stability of GOD was improved remarkably after the immobilization. After 60 min of incubation at 60 °C, the immobilized GOD and free GOD retained 68% and 27% of its initial activity, respectively. After kept at 4°C for 18 days, immobilized GOD and free GOD maintained 81% and 65% of its initial activity, respectively. The immobilized GOD maintained 58% of its initial activity after 7 consecutive operations. Since fluorescent SiO_2 nanoparticles contain $Ru(bpy)_3Cl_2$ which is optical sensitive to oxygen, they might have the application potential in fiber optic biosensor based on enzyme catalysis and oxygen consumption.

Keywords: fluorescent SiO_2 nanoparticles; immobilized GOD; stability; fiber optic biosensor

1 INTRODUCTION

Glucose Oxidase (GOD) catalyzes the oxidation of glucose by oxygen and has been used in many fields such as wine production, canned foods storage, textile bleaching, and glucose biosensor. Compared with free GOD, the immobilized GOD has many advantages including its reuse, ease removal from the reaction medium and improved stability. SiO_2 nanoparticles are highly compatible with enzymes [1, 2] and could be a good carrier for GOD immobilization. The encapsulation of fluorescent molecules in SiO_2 nanoparticles often increases their photostability and emission quantum yield and these fluorescent SiO_2 nanoparticles can be used in many fields such as drug delivery [3, 4], bioimaging [5], cell tracking [6], and sensing [7, 8]. If the SiO_2 nanoparticles containing photosensitizer are used to immobilize enzyme, the catalysis effect of enzyme can arrive at the photosensitizer directly, which will improve the performance of fiber optic biosensor based on enzyme catalysis and oxygen consumption.

In this work, the fluorescent SiO_2 nanoparticles containing $Ru(bpy)_3Cl_2$ were prepared and used to immobilize GOD. Immobilization conditions including APTES concentration, GA concentration, pH were investigated and optimized. The properties of immobilized enzyme including thermal, storage and operational stability were also studied. The results show that when GOD is immobilized on fluorescent SiO_2 nanoparticles, it has improved catalytic performance and stability, indicating that this immobilized GOD can be used in developing fiber optic glucose sensor based on enzyme catalysis and oxygen consumption.

2 EXPERIMENTAL

2.1 *Materials and apparatus*

Glucose Oxidase (GOD, E.C. 1.1.3.4, and 100 U/mg) was purchased from Aspergillusniger. Glucose, tris(2,2-bipyridyl)dichloro-ruthenium (II) hexahydrate ($Ru(bpy)_3Cl_2$) (99.0%) were purchased from Aldrich–Sigma. Horseradish (HRP) was purchased from Sinopharm Chemical Reagent. All other reagents were of analytical grade and used without further purification. Double-distilled water was used throughout the experiments.

The morphologies of the fluorescent SiO_2 nanoparticles were observed using a field emission scanning electron microscope (JSM-5610LV, JEOLLtd., Japan) operated at 10 kV. The activity of immobilized GOD was measured by UV-2450 spectrophotometer (Shimadzu, Japan).

2.2 *Preparation of fluorescent SiO_2 nanoparticles*

At 25 °C, 19.0 mL of ethanol was mixed with 3.0 ml of ammonium hydroxide and 3.0 mL $Ru(bpy)_3Cl_2$ (0.5 mg/mL) to form solution A. 3.7 ml of ethanol was mixed with 1.3 mL of TEOS to form solution B. The two solutions were mixed and stirred for 1 h. The precipitate was collected by centrifugation and washed with ethanol three times. The obtained fluorescent SiO_2 nanoparticles were dried at 80 °C in vacuum for 6 h.

2.3 *Immobilization of GOD*

4 mL of freshly prepared APTES solution (2% (v/v)) was stirred for 30 min to complete its hydrolyzation. 40 mg of fluorescent SiO_2 nanoparticles was added into the solution and stirred for 24 h at 25 °C. The modified nanoparticles were washed with water and redispersed in 4 mL of Phosphate Buffer Solution (PBS, 0.01 M, pH 7.3). 160 μL 25% (v/v) GA was added in the mixture and stirred for 1.5 h. The mixture was washed several times with PBS buffer and water. After centrifugation, the activated fluorescent SiO_2 nanoparticles were obtained. The SiO_2 nanoparticles were dispersed in 6.5 mL PBS (0.01 M, pH = 6.5) and 2.5 mL of GOD (1 mg/mL) was added in it. The mixture was kept at 4 °C for 24 h with occasional shaking. The fluorescent SiO_2 nanoparticles were washed with PBS buffer thoroughly to remove free GOD and were stored in PBS buffer (0.2 M, pH 6.0) under 4 °C.

2.4 *Assay of GOD activity*

The activity of GOD was monitored using a colorimetric method [9]. GOD catalyzed the oxidation of glucose to form gluconic acid and H_2O_2. In the presence of peroxidase, the produced H_2O_2 reacted with phenol and 4-aminoantipyrine to form the red quinoneimine dye complex, which had the maximum adsorption at about 510 nm. By measuring this absorbance, the enzyme activity was determined.

2.5 *Stability of GOD*

To determine the thermal stability of GOD immobilized on fluorescent SiO_2 nanoparticles, the enzymatic activity of immobilized GOD was measured at different time of incubation at 60 °C. The storage stability of immobilized GOD was investigated by storing it in solution at 4 °C for 18 days and measuring its remaining activity every 3 days. For control experiments, the activity of free GOD was recorded in the same condition.

To investigate the operation stability of immobilized GOD, several consecutive operation cycles were carried out by oxidizing glucose. After finishing each oxidation cycle, the immobilized GOD was washed several times with PBS buffer (0.1 M, pH 6.5) and the procedure was repeated with a fresh aliquot of substrate.

3 RESULTS AND DISCUSSION

3.1 *Preparation of fluorescent SiO_2 nanoparticles*

We used the modified Stöber method to prepare the fluorescent SiO_2 nanoparticles. This is a green preparation method of fluorescent SiO_2 nanoparticles which is benefit to environment protection. Unlike microemulsion-based method often used to prepare fluorescent SiO_2 nanoparticles, this technique completely avoids the use of toxic organic solvents and surfactants. Since there is no need to wash the nanoparticles off surfactants molecules, these nanoparticles can be easily used in biomedical engineering field. SEM images of the fluorescent SiO_2 nanoparticles are shown in Figure 1. It can be seen that the nanoparticles are monodispersed spheres with the mean size of 420 nm. The fluorescent SiO_2 nanoparticles were photostable in water and minimal indicator leakage was observed after they were kept in water for two weeks.

3.2 *Immobilization of GOD on fluorescent SiO_2 nanoparticles*

The influence of APTES concentration on the immobilization of GOD on fluorescent SiO_2 nanoparticles is shown in Figure 2(a). It can be seen that the activity of immobilized GOD increases with increase in APTES concentration between 1~2% (v/v) and a maximal activity is observed at the APTES concentration of 2% (v/v). The immobilized enzyme activity decreases when APTES concentration is higher than 2% (v/v). The reason is probably that the higher APTES concentration causes irreversible coagulation due to the formation of complex network from APTES molecules where fluorescent SiO_2 nanoparticles were connected to each other.

The effect of GA concentration on the immobilization of GOD is shown in Figure 2(b). With the

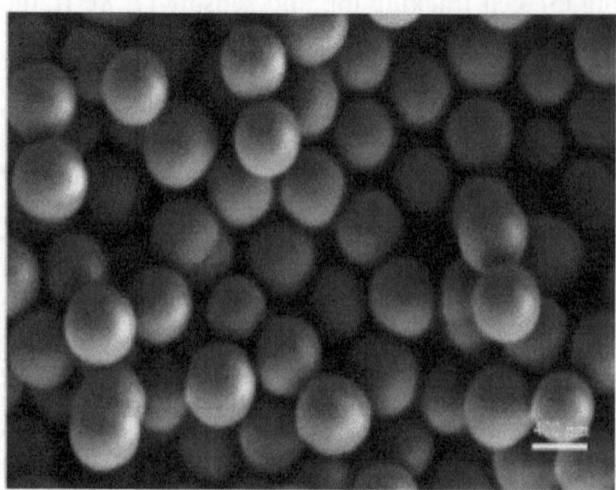

Figure 1. SEM of fluorescent SiO_2 nanoparticles.

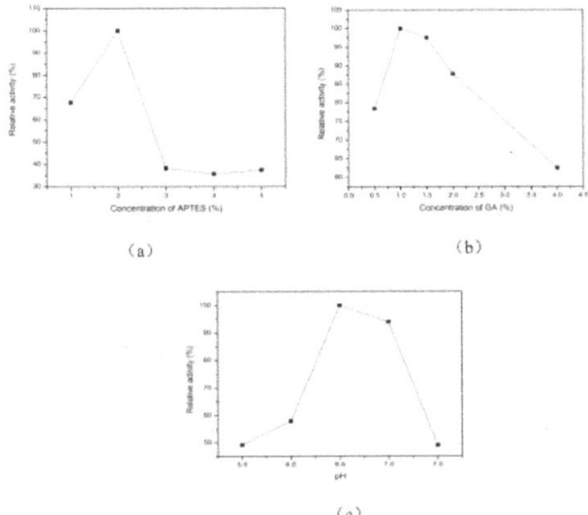

Figure 2. Effect of APTES concentration (a), GA concentration (b) and pH (c) on activity of immobilized GOD.

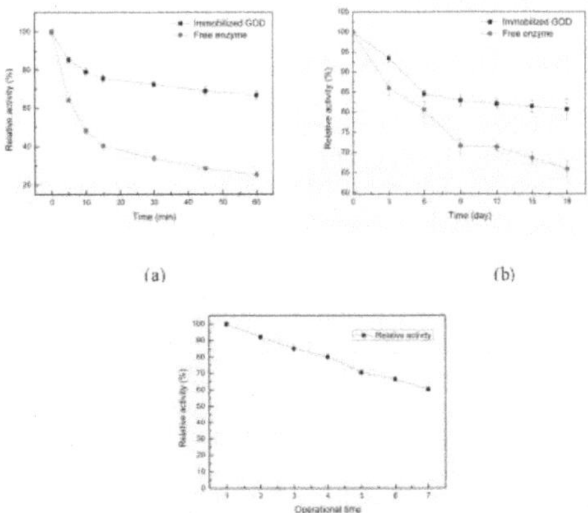

Figure 3. (a) Thermal stability of free GOD (•) and immobilized GOD (▪), (b) Storage stability of free GOD (•) and immobilized GOD (▪), (c) Option stability of immobilized GOD.

increase in GA concentration between 0.5~1% (v/v), the activity of immobilized GOD increases and reaches a maximum value at 1% (v/v). Further increase in GA concentration causes a sharp decline in the activity. Since GA is a cross-linking agent as well as a protein denaturing agent, excess GA will lead to a loss of enzyme activity.

The effect of pH on the immobilization of GOD is shown in Figure 2(c). It can be seen that relative activity of immobilized GOD has a maximal value at pH6.5, and the deviation from pH6.5 would cause the activity of immobilized GOD to decrease. GOD is a protein with a negative charge at acidic pH [10] and aminated SiO_2 nanoparticles are positively charged. In an acidic environment, free GOD molecules adsorb on silica nanoparticle surface by electrostatic interaction and therefore are easily cross-linked by GA. However, more acidic environment can hinder the formation of Schiff-base in the cross-linking process, which is unfavorable for the immobilization of GOD.

3.3 *Properties of GOD immobilized on fluorescent SiO_2 nanoparticles*

The thermal stability of the immobilized GOD was studied. As shown in Figure 3(a), with the incubation time increases at 60 °C, a sharp decline of the activity of free enzyme was observed, while the activity of immobilized GOD decreases much slowly. After 60 min of incubation, the activity of free enzyme retained 27% of its initial activity, while the immobilized GOD maintained 68% of its initial activity. The immobilization of enzyme on the fluorescent SiO_2 nanoparticles can inhibit the conformational changes and make the structure of enzyme molecules more stable at a high temperature, leading to an improved thermal stability.

The storage stabilities of free enzyme and immobilized GOD are shown in Figure 3(b). After kept at 4°C for 18 days, immobilized GOD maintained 81% of its initial activity while free GOD retained 65% of its initial activity, indicating that the immobilization of GOD on fluorescent SiO_2 nanoparticles could improve its storage stability.

The operation stability of immobilized GOD was shown in Figure 3(c). The immobilized GOD maintained 85% of its initial activity after 3 consecutive operations and 58% of its initial activity after 7 consecutive operations, showing good operation stability. The loss of immobilized GOD might be a reason for the decrease of activity because it is difficult to collect total immobilized GOD after the detection.

4 CONCLUSION

GOD was immobilized on the fluorescent SiO_2 nanoparticles. The immobilization conditions were studied and the optimal immobilization conditions were achieved. The thermal, storage and operation stability were improved remarkably after the immobilization. Since the carrier contains $Ru(bpy)_3Cl_2$ which is optical sensitive to oxygen, the immobilized GOD might be used as the sensing material for fiber optic biosensor based on enzyme catalysis and oxygen consumption.

ACKNOWLEDGEMENTS

This work was financially supported by National Natural Science Foundation of China (No: 61377092).

REFERENCES

[1] Kros, A., Gerritsen, M., Sprakel, V., Sommerdijk, N., Jansen, J., Nolte, R., Silica-based hybrid materials as biocompatible coatings for glucose sensors. Sens Actuators B, 81, pp. 68–75, 2001.

[2] Couto, C., Araujo, A., Montenegro, M., Rohwedder, J., Raimundo, I., Pasquini, C., Application of amperometric sol–gel biosensor to flow injection determination of glucose. Talanta, 56, pp. 997–1003, 2002.

[3] Li, H., Fu, Y.Q., Zhang, L.B., Liu, X.M., Qu, Y., Xu, S.T., Lü, C.L., In situ route to novel fluorescent mesoporous silica nanoparticles with 8-hydroxyquinolinate zinc complexes and their biomedical applications. Microporous Mesoporous Mater., 151, pp. 293–302, 2012.

[4] Ran, Z.P., Sun, Y., Chang, B.S., Ren, Q.G., Silica composite nanoparticles containing fluorescent solid core and mesoporous shell with different thickness as drug carrier. J Colloid Interface Sci., 410, pp. 94–101, 2013.

[5] Hooisweng, O.W., Larson, D.R., Srivastava, M., Baird, B.A., Webb, W.W. & Ulrich, W., Bright and Stable Core-Shell Fluorescent Silica Nanoparticles. Nano Lett., 5, pp. 113–117, 2005.

[6] Mumin, A.M., Barrett, J.W., Dekaban, G.A., Zhang, J., Dendritic cell internalization of foam-structured fluorescent mesoporous silica nanoparticles. J. Colloid Interface Sci., 353, pp. 156–162, 2011.

[7] Latterini L. & Amelia, M., Sensing Proteins with Luminescent Silica Nanoparticles. Langmuir, 25, pp. 4767–4773, 2009.

[8] Wang, L.L., Li, B., Zhang, L.M., Zhang, L.G., Zhao, H.F., Fabrication and characterization of a fluorescent sensor based on Rh 6G-functionlized silica nanoparticles for nitrite ion detection. Sens. Actuators B, 171–172, pp. 946–953, 2012.

[9] Huo, X.H., Liu, B.I., Deng, X.B., Zhang, B.T., Chen, H.I., Covalent immobilization of glucose oxidase onto poly(styrene-co-glycidylmethacrylate) monodisperse fluorescent microspheres synthesized by dispersion polymerization. Anal. Biochem., 368, pp. 100–110, 2007.

[10] Ahmad, A., Akhtar, M.S., Bhakuni, V., Monovalent Cation-Induced Conformational Change in Glucose Oxidase Leading to Stabilization of the Enzyme. Biochemistry, 40, pp. 1945–1955, 2001.

Material Science and Environmental Engineering – Chen (Ed.)
© 2016 Taylor & Francis Group, London, ISBN 978-1-138-02938-5

Adsorption of oil contamination on water surface with nanoscale MoS_2 powder and MoS_2 self-reclamation and reutilization

X.H. Li, P. Wang, B.S. Lv, J. He & S. Xiao
Department of Physics and Electronics, Institute of Super-Microstructure and Ultrafast Process in Advanced Materials, Central South University, Changsha, China

ABSTRACT: In this paper, the adsorption of oil contamination on water surface is studied using nanoscale MoS_2 powder. After the application of MoS_2 powder, the adsorption of oil contamination on water surface was found to be 2.1 times as its own weight at room temperature, with spheres, compound of MoS_2 powder and oil sinking to the bottom of water. MoS_2-oil spheres solidified after over 14-hour standing and can be physically recycled. After the high-heat treatment and grinding process, the recovery rate can reach up to 92%. The secondary oil absorption capacity can reach up to 90% as at the first time. The characteristics of this study are the observation of nanoscale MoS_2 adsorbing oil contamination and sinking to the bottom automatically. This method can be used as a possible application in the treatment of oil contamination in a closed pipeline. Meanwhile, this study lays the foundation for oil contamination treatment with MoS_2 in the future.

Keywords: MoS_2; oil contamination recycling; two-dimensional material; absorption

1 INTRODUCTION

In recent years, the water oil pollution problem, which is caused by the emission of the industrial and life wastewater, has attracted broad attention. The current disposal methods of oil include: adsorption, mechanical recovery method, biodegradation method, chemical dispersion method, and combustion method. The mechanical recovery method cannot effectively adsorb oil on water; the biodegradation method will destroy the ecological balance; not only the chemical method is of low cost-effectiveness, but may also cause secondary pollution; burning would be waste of resources and contributes to the pollution of the environment. In recent years, the adsorption method with so many exceptional merits, such as high cost-effectiveness, easy to conduct and no secondary pollution, has been the focal point of this work.

At present, the adsorption method of graphene, a two-dimensional (2D) material, is a research hotspot. Since 1998, Masahiro Toyoda et al. measured the maximum oil adsorption of exfoliated graphite that can reach more than 80 times its weight. In 2014, Hengchang Bi et al. measured adsorption of graphene sponge. Its maximum adsorption of organic matter reached even more than 600 times its weight. However, mass production of graphene is not yet mature, the preparation process is cumbersome, costly and of high technical content.

MoS_2 is a novel 2D material. In the MoS_2 crystal, each molybdenum atom is surrounded by six sulfur atoms located at each vertex of the triangular prism. Mo layers and S layers are held by a weak van der Waals force with a similar layered structure to graphene. And this special structure determines that the 2D material is of high relative surface area, which means it is a kind of ideal adsorbent. Compared with graphene, MoS_2 is insoluble in water and diluted acid. Due to its hydrophobicity, it will not cause secondary pollution in water. As an industrial lubricant, MoS_2 has already a mature industry chain for mass production and preparation, so molybdenum disulfide is much cheaper than graphene. The melting point of MoS_2 is 1185°C. If the temperature exceeds 400°C, slow oxidation would occur, and molybdenum trioxide will be produced. Hence, MoS_2 can be separated easily from oil contamination with heat treatment. However, research on the treatment of oil contamination with MoS_2 is still a virgin land to be reclaimed.

In addition, oil recycling by adsorption or other ways is often carried out in an open space. This is because in all these ways, manual operation requires to be above the water; however, at open pools, air pollution is inevitable. If the mixture of the adsorbent and oil can be automatically collected from water after the oil adsorption process, it is possible to conduct the whole process in a closed chamber, which can effectively minimize the pollution. Up until now, research performed in this field is rare.

For the first time, the removal of oil adsorbed on the water surface with MoS_2 powder, the novel nanoscale 2D material, is studied in this research, and we found that MoS_2 powder can adsorb oil contamination (vegetable oil) as 2.1 times as its own weight on the water surface. Thereafter, the mixture would automatically sink and be collected at the low-lying area of the bottom. This feature can effectively purify industrial and life wastewater and remove the oil contamination, and avoid the manual work above the water. This work lays the foundations for the further utilization of molybdenum disulfide for oil processing.

2 EXPERIMENTAL

2.1 *Experimental samples and instruments*

Original MoS_2 powder (content ≥ 98.5%, size 320 mesh) was obtained from Tianjin Kemiou Chemical Reagent Co., Ltd, as well as vegetable oil (0.7358 g/cm^3), illegal cooking oil (0.9772 g/cm^3), engine oil (0.6762 g/cm^3); diesel (0.7016 g/cm^3) and gasoline (0.4816 g/cm^3). Preparation and test instruments were as follows: rotary viscometer NDJ-79 (Shanghai, Scientific Instrument Equipment Co., Ltd); Electronic scale (Shimadzu Corporation, Type AW 220, Readability 0.1 mg); Resistance furnace (Beijing Zhongxing Weiye Instrument Co., Ltd, DFW-7000); Ultrasonic oscillator (Kunshan Ultrasonic Instruments Co., Ltd); 600 I double beam scanning electron microscope (FEI COMPANY, HELIOS NANOLAB 600 I); Atomic force microscope (5500; Agilent Companies, USA).

2.2 *Experimental process*

We used pipette to take 1600 ml oil (e.g. vegetable oil) and put it on the water surface. The area of the oil-covered water surface was 1194.0 mm^2 (oil weight 1.9648 g). The oil formed a barrier between the air and water, as shown in Figure 1(a). We then took an appropriate amount of original MoS_2 powder in a small beaker. After ultrasonic processing, we obtained the nanoscale MoS_2 powder. Then, we placed the nanoscale MoS_2 powder in a dried box and baked the sample for 40 min to dehydrate the nanoscale MoS_2 powder at a temperature of 80°C. We found the weight of the dry nanoscale MoS_2 powder as 1.0161 g using an electronic scale, and evenly spread the MoS_2 powder on the surface of the water pollution. Nanoscale MoS_2 powder absorbed oil quickly and significantly and gradually converged towards the center of the reservoir, and then lower recesses into a groove, as shown in Figure 1(b); the inset of the figure shows the lateral view; With the increase in the amount of nanoscale MoS_2 powder, MoS_2-oil mixture density increased and became greater than the density of water. When the gravity of MoS_2-oil mixture overcame the surface tension water-oil interface and buoyancy, MoS_2-packed oil spheres sank into the water automatically, and effectively "carried" oil contamination into the bottom of the water, which removed the barrier between the water and the air, as shown in Figure 1(c); the illustration shows the lateral view of MoS_2 sunk to the bottom. As the rest of MoS_2 powder weighed 0.0600 g, we observed that the mass of the used MoS_2 powder was 0.9561 g. After standing for 14 h, the sunk MoS_2 solidified and was extracted to a small beaker by using a spoon, tweezers or other physical methods, as shown in Figure 1(d) and 1(e).

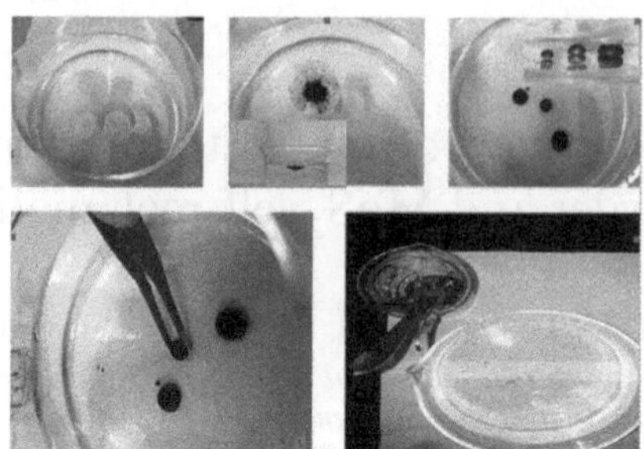

Figure 1. Nanoscale MoS_2 powder treatment for oil contamination and physical recovery. (a) The screen between the oil and the air; (b) nanoscale MoS_2 on the center and down into a groove; inset in b is the lateral view; (c) the sinking of nanoscale MoS_2 powder that carried the oil contamination; inset in c is the lateral view; (d) physically recycling oil contamination with a tweezers or spoon; (e) effective collection.

With anhydrous ethanol was used to clean the MoS_2 and the temperature of the resistance furnace was set to 385°C, after heating MoS_2 for 6 h, flake MoS_2 was obtained. Then, we ground the flake MoS_2 obtained in a mortar. The mass of the pure MoS_2 powder was found out to be 0.8796 g. We repeated the above experimental steps and tested the secondary oil absorption capacity of the nanoscale MoS_2 powder.

2.3 *Data analysis*

Figure 2(a) illustrates the Scanning Electron Microscopy (SEM) of nanoscale MoS_2. As shown in the figure, the width range of the fabricated MoS_2 flakes was 10~20 um; the inset is the optical microscope photo of the MoS_2 film sample. It is easy to note that there are obvious color interference fringes on the surface of MoS_2, which reflects the surface is of exceptional smoothness and level of interference determined by thickness. The AFM image in Figure 2(b) shows the thickness of the MoS_2 nanoflake, which is in the range of 25 nm to 40 nm.

In this experiment, 0.9561 g MoS_2 powder was used to remove the oil whose covered area on the water surface was 1194.0 mm^2 (about 1.9648 g, vegetable oil). The mass ratio of MoS_2 powder and oil was 1:2.1, which means that after the heat treatment,

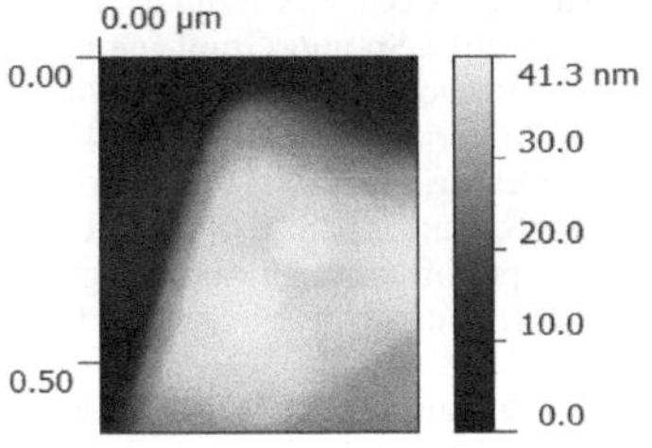

Figure 2. The characterization of nanoscale MoS_2: (a) the SEM of the treated MoS_2; (b) the AFM of the treated MoS_2.

about 0.8796 g MoS_2 powder could be recycled, with the recovery rate reaching up to 92%. And in the test of the secondary oil adsorption, about 1.040 g of MoS_2 powder was recycled, and such an amount of MoS_2 powder could continue to clean the surface area of about 1194.0 mm^2 (approximately 1.9648 g). The secondary oil absorption capacity was 90% of the first oil adsorption capacity. A possible explanation could be that a small amount of MoS_2 powder during the heat treatment was slowly oxidized to molybdenum trioxide.

When the density of the MoS_2-oil mixture was greater than the density of water, and the gravity of the MoS_2-oil mixture overcame the water surface tension and buoyancy, the oil sphere coated with MoS_2 automatically sank to the bottom. Initially, the water pollution area was larger, the volume of draining water was higher, and buoyancy was stronger. In this case, a larger amount of MoS_2 powder was needed in order to make it sink. With the increase in the mass of MoS_2 powder, MoS_2 powder carrying the oil sank, which resulted in the decrease of the area of water pollution. Figure 3 shows that with the increase in the mass of MoS_2 powder, the surface area of oil pollution decreased. The experimental results showed that when the nanoscale MoS_2 powder was spread on the surface of the water at a uniform speed, with the increase in the mass of the MoS_2 powder, the area of the surface of oil contamination was reduced, until it was clean. The process is shown in Figure 3.

Five kinds of oil were tested in the experiments, and their characteristics are summarized in Table 1. From Table 1, it can be seen that with the increase in oil density, increase percentage in the

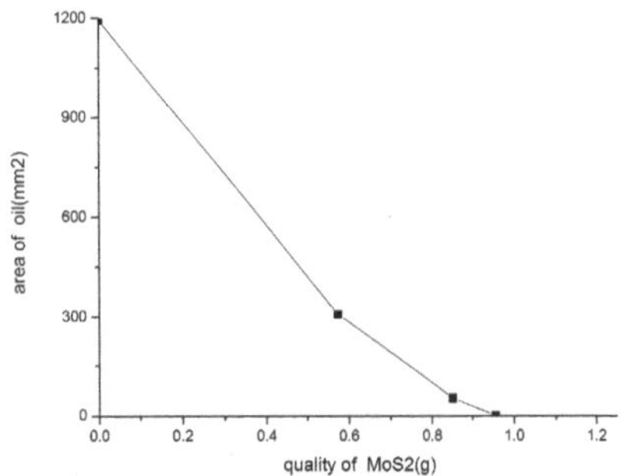

Figure 3. Relational graph of the mass of MoS_2 and area of oil contamination.

Table 1. The absorption rate of 5 kinds of oil.

Type	Viscosity at room temperature, Pa·s	Density, g/cm^3	The quality percentage increase, wt%
Vegetable oil	70	0.7358	220
Illegal cooking oil	50	0.9772	408
Engine oil	134	0.6762	158
Diesel	14	0.7016	134
Gasoline	7.5	0.4816	230

mass of MoS_2 powder increased, so the adsorption ability of the MoS_2 powder strongly depended on the oil density. Oil adsorption ability of the MoS_2 powder became stronger with the increase in the oil density. Therefore, the nanoscale MoS_2 powder was more suitable for treating the oil contamination with more impurities and higher density, such as living waste oil. From Table 1, we can see that the density of gasoline was small, but the adsorption ability of MoS_2 powder was relatively higher. Evaporation of gasoline at high speed in the adsorption process may be a reasonable explanation.

At the same time, from Table 1, we can see that the oil viscosity of oil has a little influence on the adsorption capacity of MoS_2 powder. The viscosities of diesel fuel and engine oil were 14 and 134, respectively, and the latter was 9.6 times the former, but the adsorption value of MoS_2 powder for the latter was 1.18 times the former. The internal friction of the fluid reflects its viscosity. According to the experiment results, using the nanoscale MoS_2 powder to adsorb the same mass of diesel oil and engine oil, the time needed for clear water were 30 s (second grade) and 2 min (minute grade). The removal rate of diesel was higher, respectively. The reason is that the viscosity of diesel oil is smaller, and its friction with MoS_2 micropowder is weaker, which makes MoS_2 micro powder gathering and clean-up rates quicker relatively; on the contrary, the viscos-

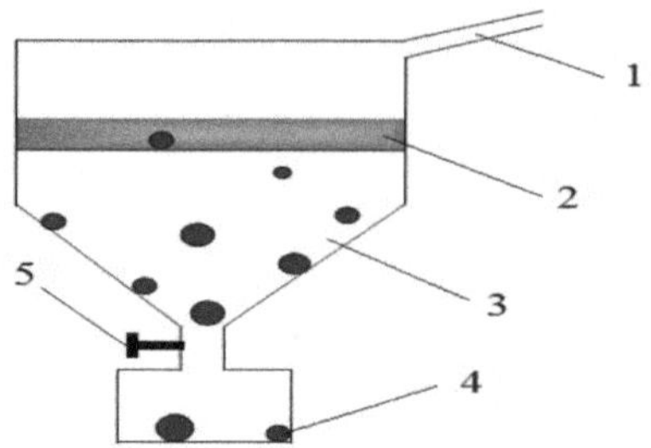

Figure 4. The diagram showing the dealing with the oil in wastewater.

ity and friction of engine oil with MoS_2 powder are larger, which means that MoS_2 powder gathering and clean-up rates would be relatively slow.

Using this method of industrial and domestic wastewater oil purification, MoS_2 carrying the oil contamination can sink automatically. And automatic recovery can be done in this way. The sunk MoS_2-oil adsorption ball can roll to collect dish and avoid working above the water surface in order to avoid further pollution to the environment, apart from its similarity. In addition, the application in a closed room can minimize the pollution to the environment. The diagram is shown in Figure 4: 1—water inlet; 2—oil; 3—water; 4—adsorption ball; 5—the valve.

3 RESULTS

For the first time, a novel material with quasi 2D MoS_2 powder whose thickness ranged from about 25 to 40 nm and width ranged from 10 to 20 microns was used to remove oil contamination on water. The experiment observed per mass of MoS_2 powder can adsorb oil (vegetable oil) on the water surface 2.1 times to its own weight, and packed oil sinks to the bottom at room temperature for the first time. The experimental mechanism for MoS_2 adsorption of surface oil is the formation of the MoS_2-oil mixed solution. With the increase in MoS_2 mass, MoS_2-oil mixture density increases and becomes greater than the density of water, and gravity of MoS_2-oil mixture becomes greater than the water surface tension. As a result, the oil-MoS_2 packed sphere sinks to the bottom. The MoS_2-oil mixture standing for 14 hours in the bottom will solidify and can be collected using a tweezers, spoon or other physical methods. The recycling process is simple and feasible to conduct. Pure MoS_2 powder can be obtained by applying the heat treatment at high temperatures followed by grinding. The recovery rate reached 92%. The secondary recovery capacity of MoS_2 powder was 90%. And it still has the potential to be improved. This study has paved the way for the further use of molybdenum disulfide for oil processing.

ACKNOWLEDGMENT

This work was financially supported by the National Nature Science Foundation of China (11104356, 11404410, 11174371 and 11204112).

REFERENCES

[1] Bi, H.C. & Xie, X. & Yin, K.B. & Zhou, Y.L. & Wan, Sh. & He, L.B. & Xu, F. & Banhart, F. & Sun, L.T. & Ruoff, R.S. 2012. Spongy Graphene as a Highly Efficient and Recyclable Sorbent for Oil and Organic Solvents. Material Views, 22:4412–4425.

[2] Bi, H.C. & Xie, X. & Yin, K.B. & Zhou, Y.L. & Wan, Sh. & Ruoff, R.S. & Sun, L.T. 2014. Highly enhanced performance of spongy graphene as an oil sorbent. Journal of Materials Chemistry A, 2: 1652–1656.

[3] Ding, Y.Y. & Wang, C.Z. & Wen, X.F. & Zhang, X.P. & Ye, L. & Zhang, A.Y. & Feng, Z.G. 2014. Preparation of Amidoxime Modified Polyacrylonitrile Nanofibers and Its Application in Metal-ion Containing Waste Water Treatment. Chemical Journal of Chinese Universitie, 34(7): 1758–1764.

[4] Fan, R.G. & Gao, Q.P. & Gao, H.J. 2014. Research progress in the removal of alkyl-phenol substances from water bodies. Industrial Water Treatment, 34(7): 10–12.

[5] Inagaki, M. & Konno, H. & Toyoda, M. Moriya, K. Kihara, T. 2000. Sorption and recovery of heavy oil by using exfoliated graphite Part II: Recovery of heavy oil and recycling of exfoliated graphite. Desalination, 128: 213–218.

[6] Gong, B.L. & Ren, L. & Yan, C. & Hu, W.Zh. 2007. Preparation of Zwitterionic Ion Exchange Packings Based on Monodisperse Hydrophilic Macroporous Resins and Their Application to Separation of Proteins. Chemical Journal of Chinese Universitie, 28(5): 831–836.

[7] Li, X.D. & Zhang, M.Q. & Wang, X.Y. & Song, M. 2002. Application of Adsorptive Biological Degradation Method in Brewery Wastewater. Jiang Su Environmental Science and Technology, 15(2): 3–4.

[8] Liu, J. 2011. The Assessment, Optimization and Quickly Decision-making of Physical Method to Clean the Oil at Sea. Dalian Maritime University.

[9] Toyoda, M. & Aizawa, J. & Inagaki, M. 1998. Sorption and recovery of heavy oil by using exfoliated graphite. Elserier Desalination, 115: 199–201.

[10] Toyoda, M. & Inchio, M. 2000. Heavy oil sorption using exfoliated graphite New application of exfoliated graphite to protect heavy oil pollution. Carbon, 38: 199–210.

[11] Xie, J. & Zhang, Y. & Wang, Q.F. & Sun, X.M. & Yu, G.R. 2013. Comparative Determination of Soil Organic Carbon with Different pH Using Combustion Method and Chemical Oxidaton Method. Chinese Journal of Soil Science, 44(2): 333–337.

[12] Wang, Y.Q. & He, J. & Xiao, S. & Yang, N.A. & Chen, H.Zh. 2014. Wavelength selective optical limiting effect on MoS_2 solution. Acta Physica Sinica, 63(14):144204.

Material Science and Environmental Engineering – Chen (Ed.)

A facile method for the preparation of hollow mesoporous organic silica nanospheres

J. Li, Z.H. Liu, L.X. Chen & X. Li
School of Chemistry, Chemical Engineer and Life Science, Wuhan University of Technology, Wuhan, P.R. China

C.C. Zhang
School of Materials Science and Engineering, Wuhan University of Technology, Wuhan, P.R. China

J. Zhang
School of Food and Pharmaceutical Engineering, Light Industry Division, Hubei University of Technology, Wuhan, P.R. China

ABSTRACT: In this study, a novel Organic functionalized Hollow Mesoporous Silica Nanospheres (OHMSNS) have been prepared by using surface hydrophobic layer protected selective etching strategy. The study shows that the protective effect of the hydrophobic layer is closely related to the type of organosilane. Meanwhile, in order to improve the protective effect of the hydrophobic layer, its derivative has also been obtained via grafting the polymer on the surface of outer layer organosilane. The properties of products are investigated in derail by TEM, FTIR and, N_2 adsorption-desorption.

Keywords: organosilica nanospheres; hollow mesostructure; surface protection etching

1 INTRODUCTION

Design and fabrication of hollow micro/nanostructures have been intensively pursued in the past decades.[1–3] Among which silica-based hollow mesoporous spheres have attracted interesting attention due to their large surface area and pore volume, excellent biocompatibility, low toxicity, high drug loading and release of guest species, and facile surface modification and functionalization.[4–7] In previous study, the most common and straightforward strategies to Hollow Mesoporous Silica Spheres (denoted HMSS) are templating methods including soft and hard template.[8–12] However, the templating methods have several drawbacks. For instance, the hard template method was typically processed in multiple steps, besides; special care must be taken to avoid shell collapse during the template removal process. On the other hand, in the soft template method, the morphology of the as-prepared products was usually difficult to control because of the deformability of the soft templates (such as micelle and vesicles.). Therefore, facile synthetic strategies to obtain high-quality HMSS are still a challenge for their practical applications.

Recently, many efforts have been devoted to fabricate HMSS. Selective etching strategy has been successfully developed to prepare HMSS, and as compared with other methods, such selective etching strategy is relative facile and economical.[13–15] However, there are rare reports about organic/inorganic hybrid HMSS prepared using this method. Very important, above-mentioned methods commonly adopted post-modification process may lead to pore blockage, resulting in decreased surface area and pore volume.

Herein, we present a facile method to prepare organic functionalized HMSS. Compared with the previous method, in this study, HMSS were obtained by a surface hydrophobic layer protected selective etching strategy using CTAB as pore-making agent. In the obtained HMSS, the hybrid silica outer shell was formed directly from organic silane via one-step method. Besides, another polymer protected HMSS have also been synthesized.

2 EXPERIMENTAL SECTION

2.1 *Chemicals and reagents*

Tetraethyl orthosilicate (TEOS), 3-thiocyanatopr opyltriethoxysilane (TCPTES), vinyltriethoxysilane (VTES), N-Vinyl-2-Pyrrolidone (NVP) were purchased from Aladdin. Pressure distillation was reduced from NVP before polymerization to remove the inhibitor. Cetyltrimethylammonium bromide

(CTAB), aqueous ammonia solution (25–28%), ethanol, hydrochloric acid, and sodium carbonate were purchased from Sinopharm Chemical Reagent Co. (Shanghai, China). All the reagents materials were used without further purification. Deionized water was used in all experiments.

2.2 *Synthesis of solid SiO_2 spheres (denoted as $sSiO_2$)*

$sSiO_2$ were synthesized by a modified StÖber method. In a typical synthesis, 3 mL TEOS were rapidly added dropwise into the mixture of 40 mL of ethanol, 2 mL of H_2O, and 2 mL of ammonia aqueous solution. And then the mixture was stirred at room temperature for 4 h, forming a white silica colloidal suspension. The silica particles were centrifugally separated from the suspension and washed with ethanol and deionized water.

2.3 *Synthesis of OHMSNS from TCPTES and VTES, respectively*

In a typical synthesis procedure, 50 mg of the above as-synthesized $sSiO_2$ was first homogeneously dispersed in 10 mL deionized water by ultrasonication for 20 min. And then the suspension was added into a mixture solution consisting of 78 mg of CTAB, 15 mL of ethanol, 15 mL of H_2O, and 0.3 mL of ammonia aqueous solution. And stirred at room temperature for 30 min. 0.15 mL of TCPTES (or VTES) was added. The resultant mixture was stirred for additional 6 h; the solid product was recovered by filtration and air-dried at room temperature overnight. Obtaining the solid core/hybrid shell structure, Herein, hybrid shell structure referred to thiocyanato functionalized silica, denoted as $sSiO_2$@CTAB/TC-SiO_2 (or $sSiO_2$@CTAB/V-SiO_2). For the preparation of HMSS, $sSiO_2$@CTAB/TC-SiO_2 (or $sSiO_2$@CTAB/V-SiO_2) nanoparticles were redispersed in 20 mL of Na_2CO_3 aqueous solution (20 mg), and then stirred at 50 °C for several hours. The obtained products were collected by filtration and washed with ethanol and water for several times. Finally, in order to remove the CTAB, the as-prepared products were extracted by refluxing in certain amount of ethanol containing concentrated aqueous HCl solution for 24 h. The extracted sample was designated as TC-HMSNS (or V-HMSNS).

2.4 *Synthesis of V-HMSNS@PVP*

For the preparation of V-HMSNS-PVP, the experiment can be divided into two parts. The first part is the preparation of $sSiO_2$@CTAB/V-SiO_2, the experimental conditions and procedures were the same as those of the corresponding $sSiO_2$@CTAB/V-SiO_2. The second part is the preparation of $sSiO_2$@CTAB/V-SiO_2@PVP. In a typical process, $sSiO_2$@CTAB/V-SiO_2 nanoparticles were dispersed in water, AIBN was added under stirring. The solution was heated at 70 °C for 30 min under N_2 protection. And then, pure NVP monomer was added into the mixture solution, and the reaction was progressed during 5~6 h until uniform composite particles were formed. For the preparation of V-HMSNS@PVP, the experimental conditions and procedures were the same as those of the corresponding V-HMSNS.

2.5 *Characterization methods*

Transmission Electron Microscopy (TEM) (JEM-2100F STEM/EDS) was conducted with a JEM 2100 F electron microscope operated at 200 kV. Fourier Transform Infrared (FTIR) spectra were collected with a Nicolet Nexus 470 IR spectrometer with KBr pellet. The nitrogen sorption experiment was performed at 77 K on Micromeritics ASAP 2020 system. The pore size distribution was calculated using desorption isotherm branch by the BJH method. Pore volume and specific surface area were calculated by using BJH and BET methods, respectively.

3 RESULTS AND DISCUSSION

3.1 *Synthesis of TC-HMSNS and V-HMSNS*

In this paper, two types of OHMSNS including TC-HMSNS and HMSNS have been successfully prepared. Figure 1A and B displays the

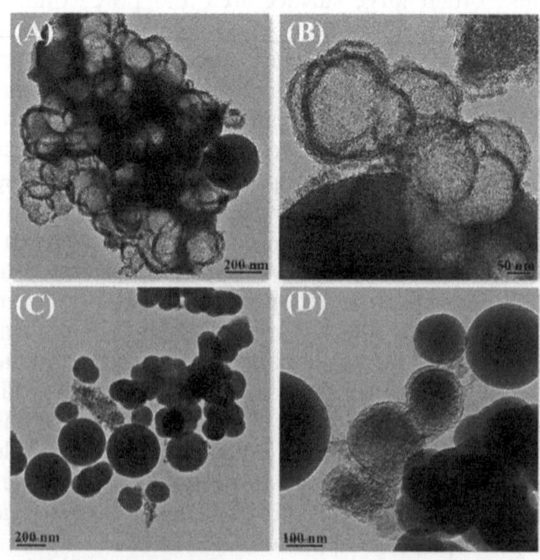

Figure 1. TEM images of (A, B) TC-HMSNS and (C, D) V-HMSNS with different magnification: (A, C) low magnification and (B, D) high magnification.

Table 1. Physicochemical parameters of the samples.

Samples	Average particles size (nm)	Average shell thickness (nm)	BET surface area ($m^2\ g^{-1}$)	Total pore volume ($cm^3\ g^{-1}$)
TC-HMSNS	150	40	59	0.12
V-HMSNS	275	11	2	0.005
V-HMSNS@PVP	316	41	60	0.28

TEM images of the typical TC-HMSNS with different magnification, respectively. As illustrated by the high-magnification TEM image (Fig. 1B), the shell of the as-prepared TC-HMSNS showed an obvious wormhole-like mesoporous structure. Meanwhile, the intact shells of TC-HMSNS suggest that a large number of thiocyanato groups from TCPTES might have been preserved after etching process. In contrast, another organosilane VTES was employed to prepare HSNSs. Figure 1C and D shows the TEM images of V-HMSNSs. It is clearly found that poor hollow structure has only been obtained, and most of spheres are still solid spheres. In addition, the porosity of TC-HMSNS and V-HMSNS are all investigated by N_2 adsorption-desorption measurements (see Table 1).

The presence of organic groups in OHMSNS can be directly proved by FTIR spectrum. As seen in Figure 2. In the s-SiO_2 spectrum, the band at 1097 cm^{-1} was due to Si–O–Si (Fig. 2a) and it is split into two peaks (1,036 and 1,134 cm^{-1}) in TC-HMSNS due to the Si-C bond destroyed the symmetry of Si–O–Si structure (Fig. 2b). On the other hand, the bands attributed to Si-C at 1345 and 1304 cm^{-1}, C_3H_6-S at 1250 cm^{-1}, and SCN at 2153 cm^{-1} were also present in the spectrum of TC-HMSNS. It is thus demonstrated that the surface of sample has been functionalized with the thiocyanato group via covalent binding. However, the FTIR spectrum of the V-HMSNS is the same as that of s-SiO_2, which the peak of vinyl groups was almost disappeared (Fig. 2c). Proved that vinyl functionalized silica crapped on the surface of s-SiO_2 has been etched first before s-SiO_2.

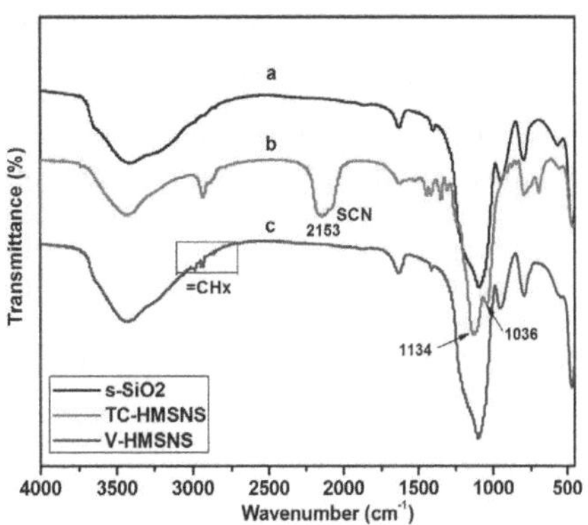

Figure 2. FTIR spectra of a) pure silica (s-SiO_2), b) TC-HMSNS and c) V-HMSNS.

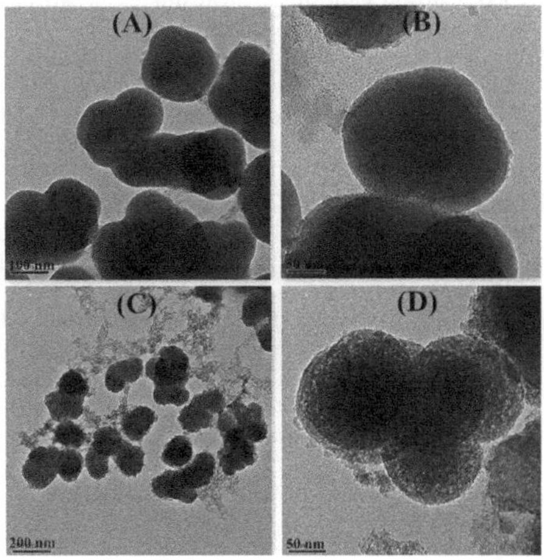

Figure 3. TEM images of (A, B) $sSiO_2$@CTAB/V-SiO_2@PVP and (C, D) V-HMSNS@PVP with different magnification: (A, C) low magnification and (B, D) high magnification.

3.2 *Grafting of PVP on the surface of $sSiO_2$@CTAB/V-SiO_2*

Based on the above discussion, it is obvious that the protective effect of the vinyl group (outer layer) is so weak that it can be easily etched out by etching agent. Therefore, we plan to graft a polymer on the surface of $sSiO_2$@CTAB/V-SiO_2 nanoparticles. Figure 3(A) and (B) gives the TEM images of $sSiO_2$@CTAB/V-SiO_2@PVP nanoparticles. It is clearly seen that core-shell structure can be observed before etching, and compared with $sSiO_2$@CTAB/V-SiO_2 nanoparticles, the $sSiO_2$@CTAB/V-SiO_2@PVP nanoparticles have obvious agglomeration due to polymerization, resulting in large spheres diameters.

More importantly, the uniform pore structure is formed after etching (Fig. 3(C)). Besides, the outer polymer shell is still retained (Fig. 3(D)). In addition, it is shown that a marked is increased in BET surface area and total pore volume (See Table 1). It is indicated that the protective effect of the outer layer can be improved by grafting polymer on the surface of $sSiO_2$@CTAB/V-SiO_2 nanoparticles.

4 CONCLUSIONS

In summary, OHMSNS with different organic groups have been synthesized via a facile surface hydrophobic layer protected selective etching strategy. The organic groups within the samples were fabricated directly from organosilane by one-step sol-gel reaction, instead of traditional co-condensation or post-grafting method. Meanwhile, our experiment also proved that the protective effect of outer layer shell can be enhanced via grafting a polymer on the outside surface.

REFERENCES

[1] Hah H.J., Kim J.S., Jeon B.J., Koo S.M., Lee Y.E. 2003. Simple preparation of monodisperse hollow silica particles without using templates. Chemical Communications 14: 1712–1713.

[2] Grzelczak M., Correa-Duarte M.A., Liz-Marzán L.A. 2006. Carbon nanotubes encapsulated in worm-like hollow silica shells. Small 2(10): 1174–1177.

[3] Yin Y.D., Erdonmez C., Aloni S., Alivisatos A.P. 2006. Faceting of nanocrystals during chemical transformation: from solid silver spheres to hollow gold octahedra. Journal of the American Chemical Society 128(39): 12671–12673.

[4] Deng T.S., Marlow F. 2012. Synthesis of monodisperse polystyrene@vinyl-SiO_2 core-shell particles and hollow SiO_2 spheres. Chemistry of Materials 24(3): 536–542.

[5] Teng M.M., Wang H.T., Li F.T., Zhang B.R. 2011. Thioether-functionalized mesoporous fiber membranes: sol-gel combined electrospun fabrication and their applications for Hg^{2+} removal. Journal of Colloid and Interface Science 355(1): 23–28.

[6] Trilla M., Cattoen X., Blanc C., Man M.W.C., Pleixats R. 2011. Silica and hybrid silica hollow spheres from imidazolium-based templating agents. Journal of Materials Chemistry 21(4): 1058–1063.

[7] Sasidharan M., Nakashima K., Gunawardhna N., Yokoi T., Ito M., Inoue M., Yusa S., Yoshio M., Tatsumi T. 2011. Periodic organosilica hollow nanospheres as anode materials for lithium ion rechargeable batteries. Nanoscale 3(11): 4768–4773.

[8] Beck J.S., Vartuli J.C., Roth W.G., Leonowicz M.E., Kresge C.T., Schmitt K.T., Chu C.T.W., Olson D.H., Sheppard E.W., McCullen S.B., Higgins J.B., Schlenker J.L. 1992. A new family of mesoporous molecular sieves prepared with liquid crystal templates. Journal of the American Chemical Society 114(27): 10834–10843.

[9] Kresge C.T., Leonowicz M.E., Roth W.J., Vartuli J.C., Beck J.C. 1992. Ordered mesoporous molecular sieves synthesized by a liquid-crystal template mechanism. Nature, 359(22): 710–712.

[10] Anil K., Yuko I., Mitsunori Y., Kenichi N. 2007. Synthesis of silica hollow nanoparticles templated by polymeric micelle with core-shell-corona structure. Journal of the American Chemical Society 129(6): 1534–1535.

[11] Liu J., Bai S.Y., Zhong H., Li C., Yang Q.H. 2010. Tunable assembly of organosilica hollow nanospheres. Journal of Physical Chemistry C 114(2): 953–961.

[12] Yang Y., Liu J., Li X.B., Liu X., Yang Q.H. 2011. Organosilane-assisted transformation from core-shell to yolk-shell nanocomposites. Chemistry of Materials 23(16): 3676–3684.

[13] Zhang Q., Zhang T.R., Ge J.P., Yin Y.D. 2008. Permeable silica shell through surface-protected etching. Nano Letteres 8(9): 2867–2871.

[14] Chen Y., Chen H.R., Guo L.M., He Q.J., Chen F., Zhou J., Feng J.W., Shi J.L. 2010. Hollow/rattle-type mesoporous nanostructures by a structural difference-based selective etching strategy. ACS nano 4(1): 529–539.

[15] Fang X.L., Chen C., Liu Z.H., Liu P.X., Zheng N.F. 2011. A cationic surfactant assisted selective etching strategy to hollow mesoporous silica spheres. Nanoscale 3(4): 1632–1639.

Material Science and Environmental Engineering – Chen (Ed.)
 ISBN 978-1-138-02938-5

Chitosan coated magnetic nanoparticles for extraction and analysis trace-level perfluorinated compounds in water solution coupled with UPLC-MS/MS

Y.H. Wang & L.F. Hu
Key Laboratory of Integrated Regulation and Resource Development on Shallow Lakes of Ministry of Education, College of Environment, Hohai University, Nanjing, P.R. China

W. Jiang
State Key Laboratory of Pollution Control and Resources Reuse, School of the Environment, Nanjing University, Nanjing, P.R. China

Y. Nie, C. Hong & G.H. Lu
Key Laboratory of Integrated Regulation and Resource Development on Shallow Lakes of Ministry of Education, College of Environment, Hohai University, Nanjing, P.R. China

ABSTRACT: In this work, Chitosan (CS) coated magnetic nanoparticles (Fe_3O_4@CS MNPs) were successfully synthesized by a one-step, one-pot method and used as adsorbents for extraction and analysis trace-level Perfluorinated Compounds (PFCs) from water solution coupled with Ultra high-Performance Liquid Chromatography-electrospray tandem Mass Spectrometry (UPLC-MS/MS). The resultant composites have nanometer size (200 nm in diameter), large surface area and superparam agnetic property. At the same time, due to the surface related groups and the positively charged of chitosan polymer coating, the Fe_3O_4@CS MNPs have high absorption efficiency to the anionic pollutants PFCs and can separate quickly from large volume water sample by external magnetic field. Affecting factors, including desorption solvent, desorption time, adsorption time and pH of the solution were carried out. Under the optimized conditions, the Limits of Detection (LOD, S/N = 3) for Perfluorohexanoic Acid (PFHxA), Pefluoroheptanoic Acid (PFHpA), Perfluorooctanoic Acid (PFOA), Perfluorononanoic Acid (PFNA), Perfluorodecanoic Acid (PFDA), Perfluoroundecanoic Acid (PFUdA) and Perfluorododecanoic Acid (PFDoA) were 0.5, 0.5, 0.5, 0.2, 0.2, 0.2 and 0.2 ng/L, respectively, and the Relative Standard Deviations (RSDs) were ranged from 5.2 to 9.8%. These results indicated that the synthesized Fe_3O_4@CS MNPs sorbents could efficiently enrich trace-level PFCs in water solution.

Keywords: Fe_3O_4/Fe_3O_4@CS/PFCs/UPLC-MS/MS

1 INTRODUCTION

Perfluorinated Compounds (PFCs) is a kind of compound that all C–H bonds are replaced by C–F groups. Among all the PFCs, perfluorinated carboxylic acid compounds have attracted much attention for their wide use. The general molecular formula of perfluorinated carboxylic acid compounds is CF_3 $(CF_2)_nCOOH$. Since fluorine element possesses strongest electronegativity, C–F bond is the most stable and largest energy of all the single bonds. Therefore, the PFCs exert most commonalities of POPs due to the introduction of fluorine atoms, for instance, good chemical stability, bioaccumulation and heat resistance, and so on [1]. Thus, the PFCs were widely used in industry, agriculture and other civilian areas, such as the industries of firefighting foam, textile, papermaking, electroplating and fluorine polymer materials, etc [2]. Therefore, the PFCs have been detected in different matrices in the global environment: river water [3], tap water [4], waste water treatment plants [5], atomosphere [6], superficial sediments [7], foodstuff packaging materials [8], and so on. With the growing demand of PFCs, it resulted in more and more serious harms, for example, accumulating easily in animals' blood, livers, kidneys, hearts, muscles and tissues, and generating toxicity to them [9]. In addition, PFCs also behaved reproductive toxicity, mutagenic toxicity, developmental toxicity, neurotoxicity and immune toxicity, etc [10]. Due to the potential hazards of

PFCs to environment and human health, it is quite urgent to develop efficient analytical methods to monitor trace level PFCs.

Chitosan (CS), a natural hydrophilic polymer, has been applied as a favorable biological material in different fields for its good biological and chemical properties. What's more, CS has a good adsorptive property to PFCs because of its abundant hydroxyl and amino groups, porous structure and positive surface charge, while it is difficult to separate it from the water solution. In recent years, nano-magnetic materials have been favored by researchers due to their low toxicity, tiny particle size, large specific surface area, easy separation property and other characteristics [11]. Therefore, the CS coated magnetic nanoparticles (Fe_3O_4@CS MNPs) have drawn much attention of researchers, which have both the advantages of CS and nano-magnetic materials. Futhermore, the CS coated magnetic nanoparticle can prevent magnetic core from being oxidized and aggregated; hence, it obtains longer storage time [12]. However, most Fe_3O_4@CS MNPs were prepared at least two steps [13], briefly, synthesis of Fe_3O_4 particles beforehand and then binding Fe_3O_4 and CS with cross-linking technique, which would be tedious and also resulted in uneven grain size of particles. So a facile method became attractive. Furthermore, it was known that Fe_3O_4@CS MNPs were mainly used in metal pollutants extraction, very few in organic compounds. Especially, there was no article that directly used Fe_3O_4@CS MNPs for the enrichment of PFCs.

Herein, we used a facile one-step, one-pot method to synthesize Fe_3O_4@CS MNPs [14], which was applied to extract seven PFCs in large volume water samples. Then, the UPLC-MS/MS was introduced to quantitatively and sensitively determine the analytes. Due to above-mentioned characteristics of polymer CS, Fe_3O_4@CS MNPs exhibited rapid separation performance, good extraction performance and high recoveries to anionic PFCs. Meanwhile, optimizing extraction conditions (desorption solvent, desorption time, adsorption time and the pH of the solution), method validations and other processes were also conducted in this study. To the best of our knowledge, this study is the first example using Fe_3O_4@CS MNPs as sorbents in combination with UPLC-MS/MS for determining trace-level PFCs in aqueous samples.

2 MATERIALS METHODS

2.1 *Materials and reagents*

Ferric chloride hexahydrate ($FeCl_3 \cdot 6H_2O$, 99%), chitosan (CS, 80–95% deacetylation, with a viscosity average molecular weight of 3.0×10^5 $gmol^{-1}$), ammonium acetate, ammonia water, ethanol, ethylene glycol, sodium acetate, hydrochloric acid, and sodium hydroxide were all purchased from Nanjing Chemical Reagent Company (Nanjing, China). All water was distilled prepared by Milli-Q water purification system (Millipore, Milford, MA, USA). Acetonitrile and methanol (HPLC grade) were obtained from Merk (Darmstadt, Germany). All glasswares and plasticwares were cleaned thoroughly by sonicating with methanol, then rinsing with distilled water and dried in vacuum at 50 °C. Seven PFCs including Perfluorohexanoic Acid (PFHxA), Pefluoroheptanoic Acid (PFHpA), Perfluorooctanoic Acid (PFOA), Perfluorononanoic Acid (PFNA), Perfluorodecanoic Acid (PFDA), Perfluoroundecanoic Acid (PFUdA) and Perfluorododecanoic Acid (PFDoA) were purchased from Alfa Aesar (Ward Hill, MA, USA).

Standard stock solutions (1 mg/ml) containing above-mentioned seven analytes were prepared with methanol as solvent and stored at 4 °C. Working solutions were diluted daily at required concentrations and also stored at 4 °C for use.

2.2 *Instrumentation*

The size and morphology of the materials were observed by Scanning Electron Microscope (SEM, type S-4800, HITACHI Co., Japan). The FT-IR spectra of the samples were recorded by Fourier Transform Infrared Spectrometer (Bruker Co., Germany). The samples and KBr pellets were dried at 50 °C in a vacuum oven for 12 h prior to grinding under the mass ratio of 1:200 in agate mortar, and the range of the scanning wave numbers was 400–4000 cm^{-1}. The separation and quantification of PFCs were performed on UPLC-MS/MS system (Agilent 1290 Series LC system coupled to an Agilent 6460 Triple Quad mass spectrometer, USA). A pH meter (Mettler Toledo group, Shanghai, China) was used to determine the pH of water solution. The glasswares and plasticwares were washed by numerical control ultrasonic cleaner (KQ-500DE, Kunshan ultrasonic instrument Co., Ltd, China). The vacuum drying oven (DZF-6050, Jinghong experimental equipment Co., Ltd, Shanghai, China) was applied to dry the samples and bottles. A gas bath thermostatic oscillator (SHZ-82, Jintan Zhongda instrument Co., Jiangsu, China) was used for enrichment and elution of analytes for the MNPs.

2.3 *Preparation of naked Fe_3O_4 MNPs and Fe_3O_4@CS MNPs*

The naked Fe_3O_4 MNPs were prepared as follows [15]. Briefly, 2.78 g of $FeSO_4 \cdot 7H_2O$ and 3.244 g of $FeCl_3 \cdot 6H_2O$ were added into 250 mL three-necked flask. And then, the mixed particles were dissolved with 100 mL distilled water that was degassed with nitrogen gas before use. After that, 10 mL strong ammonia water was added dropwise into the flask. In order to get an anaerobic reaction condition, the flask was purged with nitrogen for 30 min continuously. The mixture was stirred under the speed of 320 rpm at 80 °C in an oil bath system for 120 min. Finally, the products were settled using centrifugation (10000 rpm, 15 min) and washed with ethanol and distilled water several times, and then resuspended in 100 mL deionized water.

The chitosan coated magnetic nanoparticles (Fe_3O_4@CS MNPs) were synthesized by the one-step, one-pot method [14]. During the process, the flask was placed in an oil bath with stirrer stirring at a constant speed of 220 rpm, and maintained at 187 °C for 48 h. The nitrogen gas protection to avoid the oxidation of the particles was required for the whole process. Similarly, the nanoparticles were precipitated from the reaction mixture under centrifugation (10000 rpm, 15 min) and washed with ethanol and distilled water several times, and then resuspended in 100 mL of distilled water to form a concentration of 20 mg/mL.

The coating schematic of Fe_3O_4@CS MNPs was shown in Scheme 1. As we can see the -NH_2 of CS linked with the surface of Fe_3O_4 by coordination role, thus formed a stable core-shell structure of Fe_3O_4@CS MNPs [16].

2.4 *Extraction procedure by Fe_3O_4@CS MNPs*

The whole extraction process of Fe_3O_4@CS MNPs for PFCs was shown in Scheme 2. The detailed procedure was as follows: Firstly, 1 mL Fe_3O_4@CS MNPs solution (20 mg/mL) were added to the 100 mL water solution (containing 10 ng/L of PFCs), and then the mixture was shaken at room temperature for 20 min. After that, the Fe_3O_4@CS MNPs were separated from water solution by a magnet for about 30 s, and then the supernatant solution was discarded. Subsequently, 1.2 mL of desorption solvent was used as elute solution to desorb the analytes from the MNPs within 5 min in the gas bath thermostatic oscillator. Finally, the desorption solution volume was concentrated to 1 mL and 10 μL solution was injected into UPLC-MS/MS for analysis.

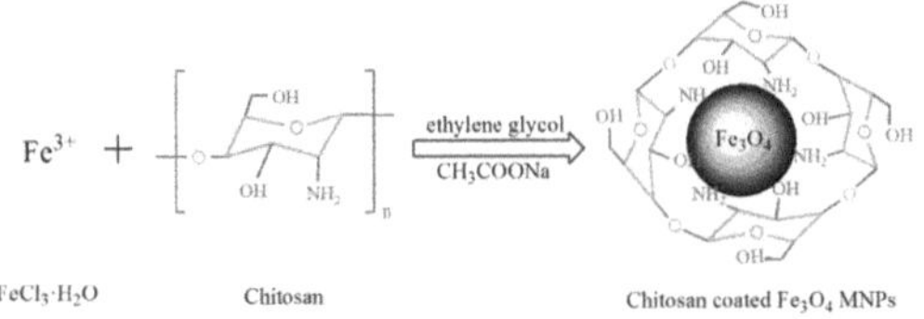

Scheme 1. Coating schematic of Fe_3O_4@CS MNPs.

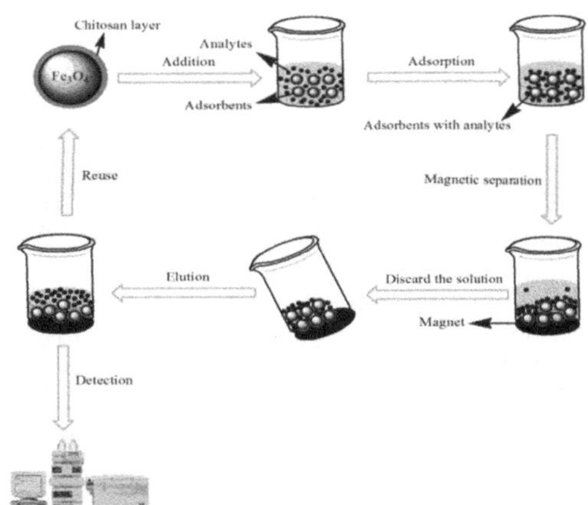

Scheme 2. Extraction process of Fe_3O_4@CS MNPs for PFCs.

2.5 *UPLC-MS/MS conditions*

An Eclipse Plus-C18 RRHD column (2.1 mm i.d. ×50 mm length, particle size of 1.8 μm, Agilent, USA) was used to separate and quantify of PFCs at the temperature of 30 °C. A 7.5 min dualistic gradient of acetonitrile and ammonium acetate (2 mmol) was performed and the flow rate was set at 0.3 mL/min constantly. The detailed gradient programs were as follows: it began with 20% acetonitrile and maintained for 0.5 min, then increased to 40% in 2 min, next rose to 60% in 3.5 min, afterwards it was linearly grew to 90% in 1 min. Finally, it was decreased to initial condition in 0.5 min.

The mass spectrometer was operated in negative electrospray ionization Multiple Reaction Monitoring (MRM) mode to identify and quantify the target analytes. The main parameters were shown below: capillary voltage, 4 KV; nozzle voltage, 500 V; pressure of nebulizing gas, 35 psi; gas temperature, 325 °C; gas flow rate, 6 L/min; sheath gas temperature, 350 °C; sheath gas flow, 11 L/min.

3 RESULTS AND DISCUSSION

3.1 *Characterization*

The SEM image of Fe_3O_4@CS MNPs is shown in Figure 1 (left). As depicted in the figure, the functionalized Fe_3O_4@CS MNPs are irregular spherical and smooth on surface [14], and the average diameter is approximately 200 nm.

Figure 1. (Left) SEM image of the synthesized Fe_3O_4@CS MNPs. (Right) the dispersion of Fe_3O_4@CS MNPs in water (a) and magnetic separation in 30 s (b).

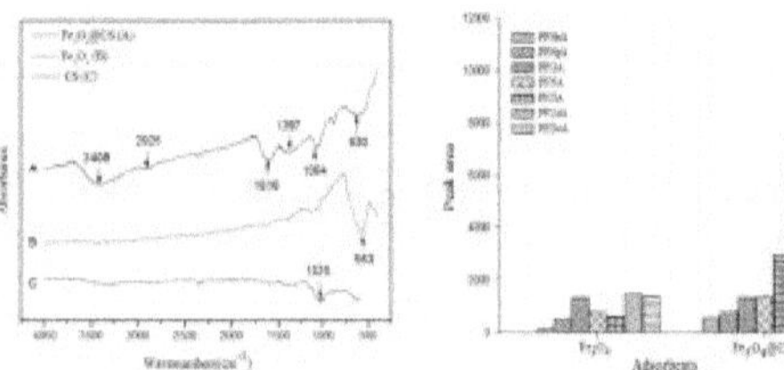

Figure 2. (Left) FT-IR spectra of pure Chitosan (CS), naked Fe_3O_4 MNPs and Fe_3O_4@CS MNPs. (Right) the adsorption performance comparison between naked Fe_3O_4 MNPs and Fe_3O_4@CS MNPs.

An excellent dispersion performance of Fe_3O_4@CS MNPs in the water solution was shown in Figure 1 (right).(a), indicating the hydrophilic chitosan polymer was homogeneously coated on the core. Then, the magnetic chitosan particles were separated by placing on external magnet for only 30 s as shown in Figure 1 (right).(b), which was far shorter than natural separation for 30 min. The results suggested that the Fe_3O_4@CS MNPs possessing fast magnetic separation property.

The FT-IR spectra of Chitosan (CS), naked Fe_3O_4 MNPs and chitosan coated Fe_3O_4 (Fe_3O_4@CS MNPs) were shown in Figure 2 (left). The Infrared (IR) spectrum adsorption peak of Fe–O group was around 563 cm^{-1} in Fe_3O_4 curve (B) [18, 19] and the Fe–O bond vibration of Fe_3O_4@CS (A) was about 630 cm^{-1}, which demonstrated that Fe_3O_4 was successfully loaded into the polymer [20]. The strong characteristic peaks around 3408 cm^{-1} was related to the N–H group bonded with O–H in CS [21, 22] What's more, the adsorption peak at 2925 cm^{-1} was attributed to the stretching vibration of C–H bond. In addition, the peaks around 1616 cm^{-1} and 1397 cm^{-1} were ascribed to the bending vibration of free N–H bond and the C–H group in CS, respectively [21]. The C–O stretching vibration peak was slightly shift from 1035 cm^{-1} in CS curve (C) to 1064 cm^{-1} in Fe_3O_4@CS curve (A). In summary, the FT-IR spectra demonstrated that the CS was coated on the Fe_3O_4 successfully.

3.2 *Possible interactions between Fe_3O_4@CS MNPs and PFCs*

The adsorption performance comparison of naked Fe_3O_4 MNPs and Fe_3O_4@CS MNPs were shown in Figure 2(right). As depicted in the bar graph, the adsorption of PFDA, PFUdA and PFDoA by Fe_3O_4@CS MNPs were all significantly better than those of naked Fe_3O_4 MNPs. Enhanced extraction efficiency of the Fe_3O_4@CS MNPs may be mainly attributed to the following reasons: Firstly, it is known that the seven PFCs all have –COOH groups and high electronegativity –F groups that can link with $-NH_2$ and –OH groups of magnetic chitosan microsphere by hydrogen bonding; Secondly, the electrostatic attraction between positive surface charged Fe_3O_4@CS MNPs and anionic PFCs [17]; Thirdly, the core-shell structure of Fe_3O_4@CS MNPs result in large surface area, whose surface adsorption ability depends primarily upon van der Waals dispersion forces [23] and hydrophobicity of PFCs, so the Fe_3O_4@CS MNPs were more attractive to longer-carbon-chain PFCs for their lower pKa [24]; Finally, the coated porous hydrophilic chitosan polymer improved the dispersibility of Fe_3O_4@CS MNPs compared with easily aggregated Fe_3O_4 MNPs so that it can touch with pollutants fully.

3.3 *Optimization of extraction conditions*

In the study, several parameters that may influence the extraction efficiency were optimized, such as the desorption solvent, desorption time, adsorption time and the pH of the solution. The optimization experiments were carried out using spiked standard PFCs aqueous solution containing 10 ng/L of each analyte.

3.3.1 *Effect of desorption solvent*

The composition of desorption solvent plays a vital role on the enrichment procedure. As we all know, the desorption solvent was used to break the fluorous affinity and release the analytes from the adsorbents [25]. Different solvents including acetonitrile [25, 26], methanol [17] and methanol containing 0.28% aqueous ammonia [27] were applied to elute PFCs from similar magnetic materials in previous studies. Thus, we chose them as preliminary desorption solvents. As shown in Figure 3 (left), acetonitrile yielded the highest recoveries for most of the PFCs (except for PFDA, PFUdA and PFDoA). It could be interpreted that the dissolving ability of acetonitrile which can break the fluorous affinity between the target molecules is higher [28].

Based on the previous report of Wang et al. [29], polarity of the desorption solvent should be taken into consideration. Due to the PFCs'

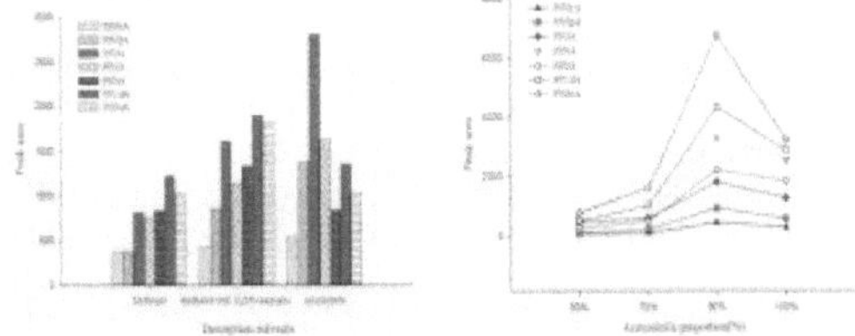

Figure 3. (Left) Effect of the desorption solvent on the extraction recoveries of PFCs (methanol, methanol with 0.28% ammonia and acetonitrile). (Right) Effect the adding of acetonitrile proportion (50, 70, 90 and 100%) in desorption solvents.

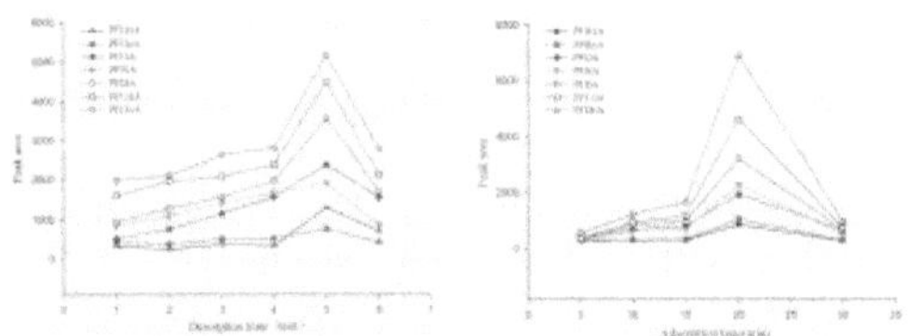

Figure 4. (Left) effect of desorption time (1, 2, 3, 4, 5 and 6 min). (Right) effect of adsorption time (5, 10, 15, 20 and 30 min).

polar properties, polar desorption solvents would get better recoveries than the less polar solvents. Hence, the proportion of acetonitrile and ultrapure water (range from 50, 70, 90 and 100%) were also investigated in this study. Finally, 90% acetonitrile was selected as desorption solvent for all the PFCs (Fig. 3 (right)).

3.3.2 *Effect of desorption time*

Desorption time is another vital factor for desorption process. In order to eliminate carryover effect, different desorption time in the range of 1–6 min were investigated in this study, and 5 min was the final selection (in Fig. 4 (left)), which is shorter than most researches [27]. After the selected desorption time, there was no carryover effect existing in our experiments.

3.3.3 *Effect of adsorption time*

The equilibrium adsorption time is also an important factor for extraction efficiency due to the short mass transfer distance and the high adsorption ability of the MNPs sorbents [27]. As shown in Figure 4 (right), the adsorption time in the range of 5–30 min was evaluated, and we could find that the adsorption performance was increased within 5–20 min to most PFCs. Further increase in the adsorption time resulted in no remarkable increase in the recoveries, which may due to the target fall off after the adsorption equilibrium. So 20 min was chosen as the best extraction time.

3.3.4 *Effect of pH of the solution*

Solution pH would change the charge property of the Fe_3O_4@CS MNPs and the ionized targets thus affect the adsorption efficiency [27] In other words, an optimum pH value can benefit the adsorption performance, and also reduces interference from complex matrix [30]. When the pH is lower than 2, the leaching Fe% of Fe_3O_4@CS MNPs is too high [14] and at pH > 8, the chitosan polymer is negatively charged, which results in the electrostatic repulsion between PFCs and Fe_3O_4@CS MNPs [17]. Thus, the effect of pH value on the extraction performance for PFCs was studied between the ranges of 3–8 (in Fig. 5). Based on the results, the extraction efficiency increased as the solution pH increased from 3 to 4. After that, it decreased rapidly when the pH ranged 4–5, and then it showed little variation in the range of 5–8. Accordingly, the optimum pH value was 4 for extraction PFCs from water solution.

3.4 *Method validation*

Under the optimized conditions, the linearity range, correlation coefficients, Limits of Detection (LODs), and precisions were investigated so as to evaluate the applicability of the aforementioned method. Calibration equations were carried out by using 100 mL ultrapure water with the seven PFCs in the range of 0.5–500 ng/L. The Correlation coefficients (r^2) were all above 0.99 except for PFHxA (0.959) and PFNA (0.989). The LODs were calculated as the concentrations of the analytes at a signal-to-noise ratio (S/N) of 3, and the results showed that the LODs of the PFCs ranged from 0.2 to 0.5 ng/L. The Relative Standard Deviations (RSDs) for the PFCs were from 5.2 to 9.8%, suggesting a good repeatability of the proposed method. These results indicated that our abovementioned methods could receive wide linear range, low Limits of Detection (LOD) and satisfactory measurement precision to determine trace-level PFCs in real water samples. The Fe_3O_4@CS MNPs synthesized by a facial one-step, one-pot method indicating is easy preparation and low cost, and it could be reused at least several times without significant loss of adsorption efficiency.

3.5 *Analysis of PFCs in the water*

According to the method, water samples from Wulong Pond, Yangtze River and domestic sewage of our school were collected to evaluate its applicability for real water samples. Among the three types of water samples, most PFCs were detected, and the concentrations in domestic sewage of our school were apparently higher than other water samples. Moreover, the concentration of PFOA was relatively higher than other PFCs, since it was the main PFC contaminant [31]. The result was consistent with results of other studies [17]

4 CONCLUSIONS

In this research, CS coated Fe_3O_4 nanoparticles (Fe_3O_4@CS MNPs) were successfully prepared and exhibited high enrichment performance of trace-level PFCs in real water samples. In combination with UPLC-MS/MS, the proposed method met the demand of environmental friendly, organic solvent saving and low cost. What's more, the method achieved high recoveries, wide linear range, low limits of detection and satisfactory measurement precision and so on. And we also believe that our core-shell magnetic beads will be one of the most promising candidates for dealing with more organic pollutants in complex environmental matrix.

This research work was financially supported by China Postdoctoral Science Special Foundation (no. 2013T60496), China Postdoctoral Science Foundation (no. 2012M511194) and the Priority Academic Program Development of Jiangsu Higher Education Institutions.

REFERENCES

[1] Wang, L., Sun, H., Yang, L., He, C., et al., Journal of Chromatography A 2010, 1217, 436–442.
[2] Martínez-Moral, M.P., Tena, M.T., Talanta 2012, 101, 104–109.
[3] Li, F., Sun, H., Hao, Z., He, N., et al., Chemosphere 2011, 84, 265–271.
[4] Llorca, M., Farré, M., Picó, Y., Müller, J., et al., Science of The Total Environment 2012, 431, 139–150.
[5] Kunacheva, C., Tanaka, S., Fujii, S., Boontanon, S.K., et al., Chemosphere 2011, 83, 737–744.
[6] Scott, B., Spncer, C., Mabury, S.A., Muir, D.C.G., Environmental Science & Technology 2006, 40, 7167–7174.
[7] Perra, G., Focardi, S.E., Guerranti, C., Marine Pollution Bulletin 2013, 76, 379–382.
[8] Zafeiraki, E., Costopoulou, D., Vassiliadou, I., Bakeas, E., Leondiadis, L., Chemosphere 2014, 94, 169–176.
[9] Houde, M., Bujas, T.A.D., Small, J., wells, r.s., Environmental Science & Technology 2006, 40, 4138–4144.
[10] Lau, C., Anitole, K., Hodes, C., Lai, D., et al., Toxicological Sciences 2007, 99, 366–394.
[11] Dung, D.T.K., Hai, T.H., Phuc, L.H., Long, B.D., et al., Journal of Physics: Conference Series 2009, 187, 012036.
[12] Donadel, K., Felisberto, M.D.V., Fávere, V.T., Rigoni, M., et al., Materials Science and Engineering: C 2008, 28, 509–514.
[13] Zhang, W., Jia, S., Wu, Q., Wu, S., et al., Materials Science and Engineering: C 2012, 32, 381–384.
[14] Jiang, W., Wang, W., Pan, B., Zhang, Q., et al., ACS Appl Mater Interfaces 2014, 6, 3421–3426.
[15] Jiang, W., Chen, X., Niu, Y., Pan, B., Journal of Hazardous Materials 2012, 243, 319–325.
[16] Liu, L., Xiao, L., Zhu, H., Shi, X., Microchimica Acta 2012, 178, 413–419.
[17] Zhang, X., Niu, Hongyun, Pan, Y., Shi, Y., Yaqi, C., Analytical Chemistry 2010, 82, 2363–2371.
[18] Shen, H.Y., Zhu, Y., Wen, X.E., Zhuang, Y.M., Analytical and bioanalytical chemistry 2007, 387, 2227–2237.
[19] Zhang, X.L., Niu, H.Y., Zhang, S.X., Cai, Y.Q., Analytical and bioanalytical chemistry 2010, 397, 791–798.
[20] Zhang, Y.L., Zhang, J., Dai, C.M., Zhou, X.F., Liu, S.G., Carbohydrate Polymers 2013, 97, 809–816.
[21] Tong, J., Chen, L., Analytical Letters 2013, 46, 1183–1197.
[22] Fan, L., Zhang, Y., Luo, C., Lu, F., et al., International Journal of Biological Macromolecules 2012, 50, 444–450.
[23] W. Potter, D., pawllszysn, j., Environmental science & technology 1994, 28, 298–305.
[24] Goss, K.U., Environment Science Technology 2008, 42, 456–458.
[25] Yang, L., Yu, W., Yan, X., Deng, C., Journal of Separation Science 2012, 35, 2629–2636.
[26] Liu, Q., Shi, J., Wang, T., Guo, F., et al., Journal of Chromatography A 2012, 1257, 1–8.
[27] Zhang, X., Niu, H., Pan, Y., Shi, Y., Cai, Y., Journal of Colloid and Interface Science 2011, 362, 107–112.
[28] Liu, X., Yu, Y., Li, Y., Zhang, H., et al., Anal Chim Acta 2014, 844, 35–43.
[29] Wang, y., Jin, s., Wang, q., Lu, g., et al., Journal of Chromatography A 2013, 1291, 27–32.
[30] Faraji, M., Yamini, Y., Rezaee, M., Talanta 2010, 81, 831–836.
[31] Zhao, X., Cai, Y., Wu, F., Pan, Y., et al., Microchemical Journal 2011, 98, 207–214.

Material Science and Environmental Engineering – Chen (Ed.)
 ISBN 978-1-138-02938-5

Crystallization and mechanical properties of Multi-Walled Carbon Nanotubes modified polyamide 6

X.T. Chen, H.J. Liu & X.Y. Zheng
College of Material Science and Chemical Engineering, Tianjin University of Science and Technology, Tianjin, China

ABSTRACT: Multi-Walled Carbon Nanotubes (MWCNTs) were functionalized with Hyperbranched Polyamides (HBPA) and their PA6 nanocomposites were prepared using a melting mixing technique. Morphology observation by scanning electron microscopy showed that grafted MWCNTs (MWCNTs-F) dispersed well in PA6 matrix. The rheological behaviors of PA6/MWCNTs-F composites were investigated by rheometer with parallel plate geometry. The storage moduli (G') of these samples increased with MWCNTs-F content at low frequency. The presence of MWCNTs-F caused nanocomposites melts to have solid-like behaviors. Differential scanning calorimetry measurement showed that the nanocomposites had higher crystallization temperature, which was attributed to the heterogeneous nucleation of MWCNTs-F. The mechanical properties were also studied and the results showed that MWCNTs-F was better than pristine MWCNTs in terms of improving mechanical properties due to its good dispersion in PA6 matrix and more strong interfacial adhesions. Adding amount of 0.5 wt% MWCNTs-F, the tensile and bending properties of the composites increased 18% and 21% than those of PA6, respectively.

Keywords: multi-walled carbon nanotubes; hyperbranched polyamides; mechanical properties; crystallization behavior

1 INTRODUCTION

Owning to the excellent electrical, thermal and mechanical properties, Carbon Nanotubes (CNTs) have received much attention since their discovery by Iijima in 1991 (Iijima, 1991) and led to the development of numerous applications. In particular, with high aspect ratio, high stiffness, and uniquely high strength, CNTs have been used as ideal reinforcing agent for structural nanocomposites (Jonathan et al. 2006). However, the high specific surface area and surface free energy lead to very strong aggregation tendency of CNTs, which result in poor dispersion of CNTs in a polymeric matrix and weak interfacial interaction, so as to affect efficient load transfer from the polymeric matrix to CNTs. In order to overcome this obstacle, small molecules or polymers are employed to modify the CNTs via physical interactions or chemical bonding (Peng et al. 2010). Among the polymer materials used for functionalization of CNTs, dendrimers and hyperbranched polymers have drawn more interest because of their unique architecture and special properties Hyperbranched polymers have a highly branched, non-entangled architecture and a large number of terminal groups (Markus et al. 2009). They exhibit much lower melt and solution viscosities and higher solubility in comparison with their linear analogues. Hence, hyperbranched polymers functionalized CNTs even with a low functionalization degree are sufficient to improve the dispersibility of CNTs in polymer matrixes (Sunit et al. 2011).

In this work, hyperbranched polyamides were grafted onto the surface of Multi-Wall Carbon Nanotube (MWCNTs), and then the modified MWCNTs (MWCNTs-F) were used to reinforce PA6. The effects of MWCNTs-F on mechanical properties and crystallization behavior of PA6 composites were investigated, and their morphological characteristics were examined by SEM.

2 EXPERIMENTAL

2.1 *Materials*

MWCNTs with diameter of 20–45 nm and length of 5–15 μm were purchased from Shenzhen Nanometer Gang Co. Ltd, with the purity higher than 97%. Hyperbranched Polyamides (HBPA) were synthesized following the method described in the literature (Huang et al. 2013). Hyperbranched polyamides were grafted onto MWCNTs by acyl chloride reaction (Lui et al. in press). The synthesis route of MWCNTs-F was shown in Figure 1.

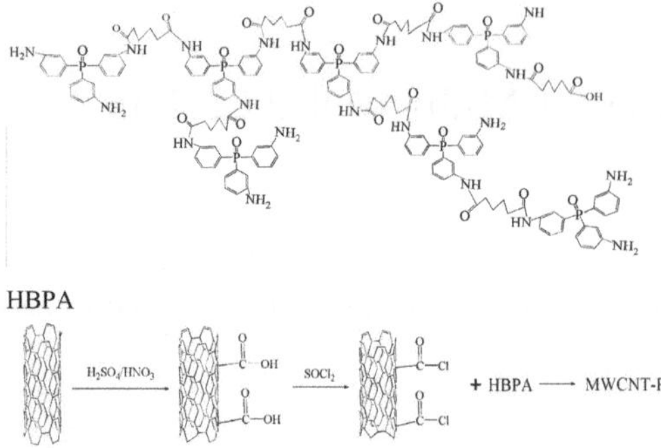

Figure 1. Preparation of MWCNTs-F.

2.2 *Fabrication of PA6/MWCNTs-F composites*

The composites with different loadings of MWCNT and MWCNTs-F were fabricated via a melt blending process using a twin-screw extruder with L/D of 25. The temperatures from hopper to die were maintained at 220°C, 240°C, 240°C and 225°C, and rotation speed of the screw was 80 rpm. The neat PA6 was also extruded at the same condition. The extrudates were pelletized, dried under vacuum at 80°C for 24 h to remove remaining water and injected into standard samples for testing.

2.3 *Measurements*

Scanning Electron Microscopy (SEM, Jeol JSM 6380) was used to explore the state of dispersion of the nanotubes in PA6 matrix. Before SEM observation, the fracture surfaces are coated with a thin layer of gold. Rheology measurements were executed on an oscillatory rheometer MARIII at 240°C using parallel plate geometry with diameter of 2.5 cm and gap of 1 mm. A frequency range of 0.1–100 Hz was used. The used strains were selected to be within the linear viscoelastic range. The crystallization behavior of PA6 and composites were studied by DSC 204F1 Phoenix (Netzsch, German) at the cooling rate of 10°C/min under nitrogen atmosphere. Mechanical properties were measured using CMT4503 microcomputer controlled electronic universal testing machine.

3 RESULTS AND DISCUSSION

3.1 *The morphology of the nanocomposites*

Figure 2 showed SEM images at magnification of 5000 to compare the morphology of composites adding MWCNTs-F or pristine MWCNTs at a same loading level at 0.5 wt%. The images revealed a homogenous dispersion of the MWCNTs-F in the PA6 matrix, while pristine MWCNTs were non-homogeneously dispersed in composite. Covalent functionalization of MWCNTs with HBPA can improve their dispersion in polymer matrix. The interface binding force between MWCNTs-F and PA6 also can be increased through the interaction between a large of terminal amide groups and PA6 chain.

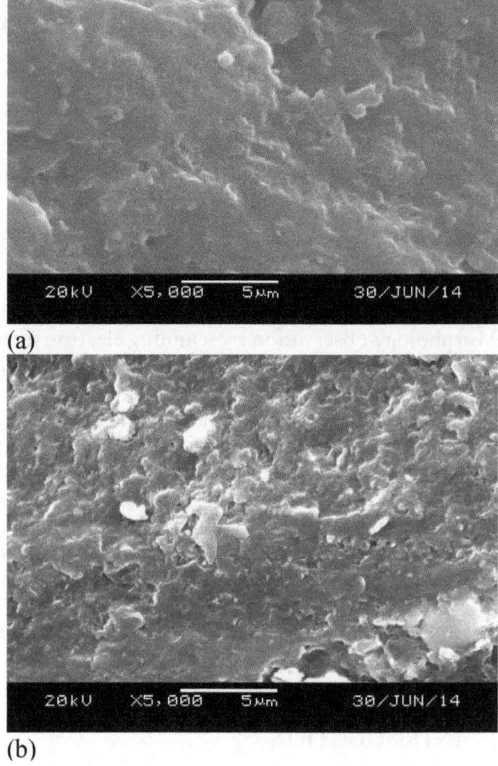

Figure 2. SEM images of PA6/MWCNTs (a) and PA6/MWCNTs-F (b) at 0.5 wt% loading.

3.2 *Rheological studies*

Figure 3 presented the storage modulus (G') of the nanocomposites with different loadings of WCNTs-F versus the frequency (ω). At low freuencies, the storage moduli of the nanocomposites were larger in comparison with that of neat PA6. The slopes of log G' versus log ω also decreased as a result of adding MWCNTs-F to the PA6. Generally, the increase of the storage modulus and the decrease of the slope of log G' versus log ω can be explained by the influence of the nanotubes on the microstructure of the polymer matrix. In this case, the nanocomposite melts present solid-like

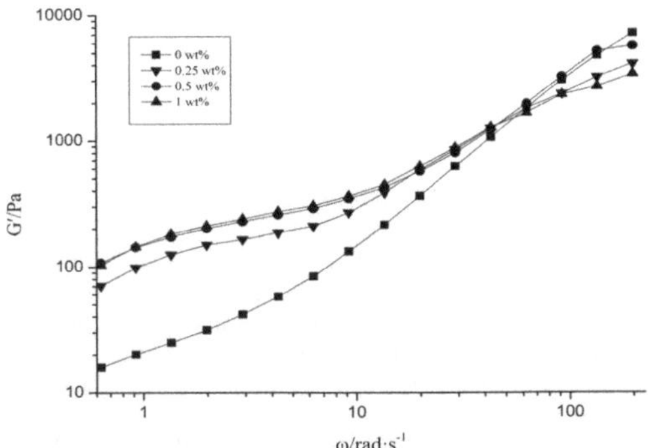

Figure 3. Storage modulus versus frequency for PA6/MWCNTs-F nanocomposites with different contents of MWCNTs-F.

Table 1. Crystallization properties of PA6/ MWCNTs-F composites with different contents of MWCNTs-F.

MWCNTs-F content (wt%)	T_c (°C)	ΔHc (J/g)	Xc (%)
0	189.6	71.3	30.9
0.25	191.7	81.7	35.4
0.5	193.5	84.7	36.8
1	193.8	74.6	32.4
1.5	195.2	54.3	23.6

behavior when the content of MWCNTs-F was 0.25%, which indicated that MWCNTs-F dispersed well in matrix and the formation of a network-like configuration impeded the motion of polymer.

3.3 *DSC analyses*

PA6 is typical semi-crystalline polymer and its mechanical properties are controlled by the crystallization process and crystallization structure. Hence the crystallization behaviors of PA6/MWCNTs-F were studied by DSC. Samples were heated to 250°C and maintained for 5 min to erase the previous therma and mechanical histories. Then the melted samples were cooled to room temperature at the rate of 10°C/min, and the cooling curves were recorded and the parameters were listed in Table 1.

A sharp and narrow crystallization peak for PA6 was observed at 189°C. For the composites, the crystallization peak gradually shifts to the higher temperatures with MWCNTs-F content increase due to the heterogeneous nucleation effect of MWCNTs-F. The crystallinity (*Xc*) of composites were calculated according to ΔHc. When MWCNTs-F adding amount was 0.5%, the crystallinity of the composite reached a maximum of 36.8%. Increase in crystallization is indirect indication of the increase in the mechanical properties due to the better dispersion of the MWCNTs-F in the PA6 matrix. Further increasing the content of MWCNTs-F, the crystallinity beganto reduce. This is probably due to the confinement of the crystals between randomly dispersed MWNTs-F, resulting retardation in the crystal growth rate.

3.4 *Mechanical properties*

The mechanical behaviors of composites are strongly decided by the properties of each component, as well as by the morphology developed during blending and the interfacial adhesion between the phases. The results of tensile and flexural tests were presented in Figure 4. All the PA6/MWCNTs-F composites showed an increase in tensile strength over neat PA6. The 0.5 wt% loading of MWCNTs-F led to 18% increase in the tensile strength compared to PA6. Pristine MWCNTs also can strengthen PA6 at low content, but the

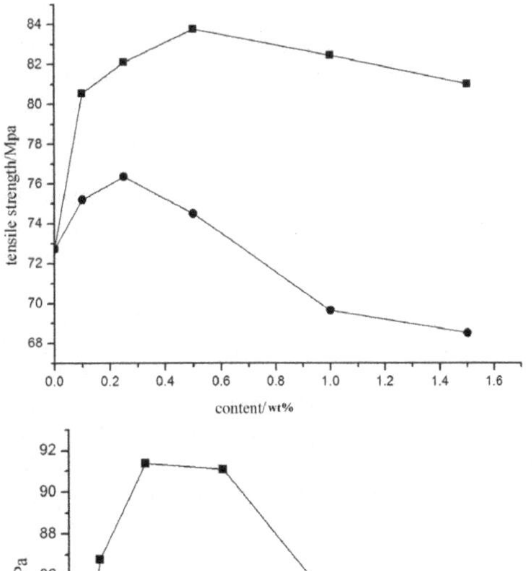

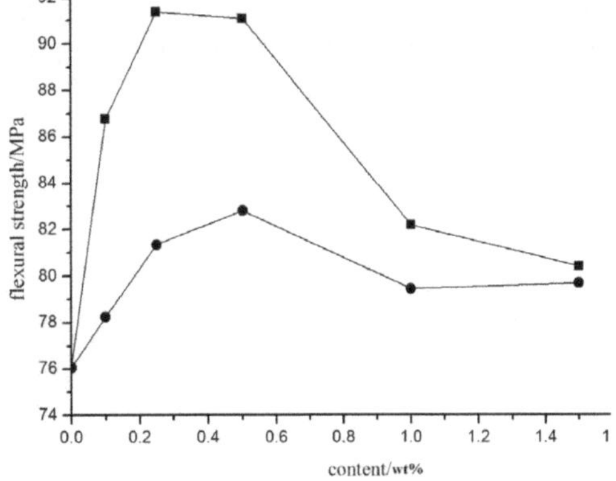

•: PA6/MWCNTs; ■:PA6/MWCNTs-F

Figure 4. Mechanical properties of PA6/MWCNTs and PA6/MWCNTs-F composites.

effect was not good. The flexural performance of PA6/MWCNTs-F was also better than pure PA6 and PA6/MWCNTs. Flexural strength of the composites loading 0.5 wt% MWCNTs-F increased by 21% than PA6. The interactions between amine groups on the MWCNTs-F surface and PA6 led to better interfacial adhesion favoring stress transfer from the matrix to CNTs. So the mechanical properties of PA6/MWCNTs-F were prior to that of PA6/MWCNTs.

4 CONCLUSIONS

PA6/MWCNTs-F nanocomposites were prepared by melt blending. SEM revealed that MWCNTs-F were more homogeneously dispersion than pristine MWCNTs in the PA6 matrix. The rheological study showed that the storage modulus of PA6/MWCNTs-F nanocomposites were higher than that of PA6, and the nanocomposites presented solid-like rheological behavior at low loading of MWCNTs-F. The incorporation of MWCNTs-F in PA6 matrix increased the crystallization temperature which was attributed to the nucleating role of nanotubes. With the loading of 0.5 wt% MWCNTs-F in PA6, the tensile stress and flexural strength improved over the neat PA6 by 18% and 21%, respectively. Further increasing MWCNTs-F content did not improve mechanical properties, which can be explained on the basis of dispersion problem of the CNTs in the PA6 matrix and also due to the crystallization behavior of the PA6.

REFERENCES

[1] Huang Y., Sun L.J., Yin H. & Chen X.T. 2013. Synthesis and characterization of phosphorus-contained hyperbranched polyamides. *Journal of Tianjin University of Science & Technology* (28):42–46.

[2] Iijima, S. 1991. Helical microtubules of graphitic carbon. *Nature* (354):56–58.

[3] Jonathan, N., Coleman, U.K., Werner J.B. & Yurii K.G. 2006. Small but strong: A review of the mechanical properties of carbon nanotube-polymer composites. *Carbon* (44):1624–1652.

[4] Markus S., Zuzana K. & Harm A.K. 2009. Dendritic and hyperbranched polyamides. *Progress in Polymer Science* (34):24–61.

[5] Peng C., Hun S.K. & Hyoung J.J. 2009. Preparation, properties and application of polyamide/carbon nanotube nanocomposites. *Macromolecular Research* (17):207–217.

[6] Sun L.J., Huang Y., Xin J.F. & Chen X.T. 2012. Preparation and properties of mwcnts modified with DOPO-HQ/EP composites. *Journal of Tianjin University of Science & Technology* (27):41–46.

[7] Sunit J.T., Hong C.Y. & Pan C.Y. 2011. Surface modification of carbon nanotubes with dendrimers or hyperbranched polymers. *Polym Chem.* (2):998–1007.

Material Science and Environmental Engineering – Chen (Ed.)
© 2016 Taylor & Francis Group, London, ISBN 978-1-138-02938-5

Fabrication and photocatalytic performance of Fe_3O_4@SiO_2@BiOCl magnetic nanocomposite

D.H. Cui, Y.Y. Li, X.D. Ruan & X.C. Song
Department of Chemistry, Fujian Normal University, Fuzhou, Fujian, China

Y. Zhang
Hangzhou Area Military Representative Office, Hangzhou, Zhejiang, China

ABSTRACT: Fe_3O_4@SiO_2@BiOCl nanocomposite photocatalyst with magnetically recyclable performance has been synthesised by a simple method. The photocatalyst showed good photocatalytic efficiency for the degradation of rhodamine B under simulated sunlight irradiation. The photodegradation of RhB over Fe_3O_4@SiO_2@BiOCl is 95% within 60 min, and the Fe_3O_4@SiO_2@BiOCl photocatalyst could be recovered due to its magnetic property. The high photocatalytic activity and magnetic recyclability of the Fe_3O_4@SiO_2@BiOCl photocatalyst can afford it promising applications in wastewater treatment and environmental remediation.

Keywords: Fe_3O_4; BiOCl; nanocomposite

1 INTRODUCTION

As a novel layered ternary oxide semiconductor, BiOCl as photocatalysis has recently attracted increasing attention as a potential environmental technology for wastewater remediation (Chen et al. 2013). For example, Li and coauthors reported that Zn^{2+} doped BiOCl microspheres exhibits better performance than other BiOCl-based photocatalyst in the photocatalytic degradation of RhB (Li et al. 2015). For the past few years, magnetic separation using magnetically recoverable nanoparticles have attracted increasing attention. Magnetically separable photocatalysts offers a promising approach for improved reusability (An et al. 2012). Here we describe the synthesis of a novel magnetic Fe_3O_4@SiO_2@BiOCl nanocomposite photocatalyst. The as-prepared photocatalyst showed highly efficient photocatalytic activities.

2 EXPERIMENTAL

2.1 *Synthesis of Fe_3O_4@SiO_2@BiOCl nanocomposite*

First, the magnetic Fe_3O_4@SiO_2 composite were synthesised by a typical sol-gel process according to the method published in the literature (Yao et al. 2014). Then, 0.12 g as-prepared magnetic Fe_3O_4@SiO_2 composite were dispersed in 30 mL of Ethylene Glycol (EG) containing 4 mmol of $Bi(NO_3)_3 \cdot 5H_2O$. After the mixture was stirred for 30 min, 10 ml of EG containing 4 mmol KBr were added dropwise to the above solution and reacted for the next 30 min under vigorous stirring. Finally, mixture was transferred into a Teflon-lined stainless steel autoclave with capacity of 50 ml and kept at 180°C for 12 h. The obtained products were separated by an external magnet, and washed with distilled water and absolute ethanol several times to remove impurities and dried at 60°C for 12 h.

2.2 *Characterization*

The morphology and size of the obtained products was characterized by transmission electron microscope (TEM, Tecnai G2 F30 S-Twin, accelerating voltage 200 kV) and scanning electron microscopy (SEM, Hitachi S-4700, accelerating voltage 15 kV). The X-Ray Diffraction (XRD) was performed on a Thermo ARL SCINTAG X'TRA X-ray diffractometer with grapite-monochromated Cu Kα irradiation (λ = 0.154056 nm). UV-vis Diffuse Reflectance Spectrum (DRS) were recorded on a Lambda 850 UV-Vis spectrometer and $BaSO_4$ was used as the reflectance standard.

2.3 *Photocatalytic experiment*

The photocatalytic activities of the magnetic Fe_3O_4@SiO_2@BiOCl nanocomposite were evaluated by degradation of rhodamine B (RhB) in an aqueous solution under simulated sunlight irradiation from a 300 W xenon lamp. In a typical procedure, 100 mg Fe_3O_4@SiO_2@BiOCl nanocomposites was poured into 200 ml 10^{-5} M RhB solution. The mixture was stirred in the dark for 30 min to reach the adsorption–desorption equilibrium of RhB on the surfaces of nanocomposite. After turning on the xenon lamp, 5 mL of the suspension was withdrawn at certain time intervals and centrifuged to remove the Fe_3O_4@SiO_2@BiOCl nanocomposite. The concentration of upper clear liquid (RhB solution) was transferred to a quartz cuvette for measuring absorbance at 553 nm with a UV759S UV-Vis spectrophotometer.

3 RESULTS AND DISCUSSION

The crystalline structure of the as-prepared magnetic Fe_3O_4@SiO_2@BiOCl nanocomposite were studied by XRD. Figure 1 shows the XRD patterns of the as-prepared photocatalyst. It clearly shows that the main diffraction peaks (002, 101, 110, 102, 112, 200, 201, 113, 211, 212, 203, 220, 214, 302) can be indexed to the tetragonal structure of BiOCl (JCPDS Card No. 06-0249). The diffraction peaks marked with "•" can be assigned to the (220) and (440) planes of face centered cubic Fe_3O_4 (JCPDS Card No. 19-0629), respectively. No relevant peaks of SiO_2 can be detected, which implies that SiO_2 kept amorphous characteristics.

Figure 2a shows a representative TEM image of Fe_3O_4@SiO_2, clearly indicating that the Fe_3O_4 nanoparticles can be coated within a silica layer. The Fe_3O_4 nanoparticles have a diameter ranged from 10 to 20 nm, and SiO_2 layer with thickness of about 50 nm coated Fe_3O_4 were observed. The morphology of Fe_3O_4@SiO_2@BiOCl nanocomposite was investigated by SEM observation as shown in Figure 2b SEM image shows that large quantity of BiOCl nanoplates were well dispersed on the surface of Fe_3O_4@SiO_2 and assembled into spheres.

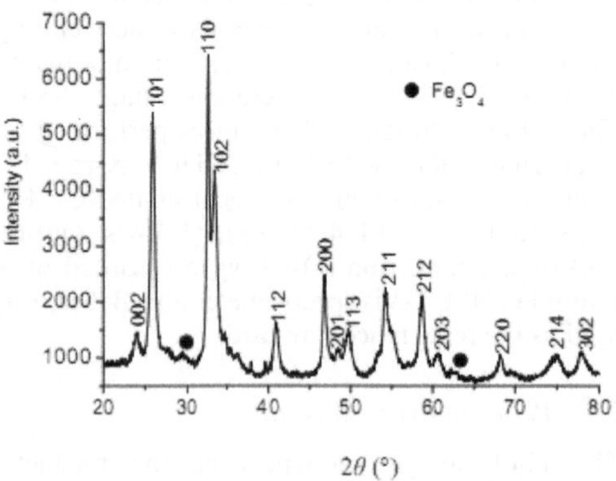

Figure 1. XRD patterns of the Fe_3O_4@SiO_2@BiOCl nanosheets.

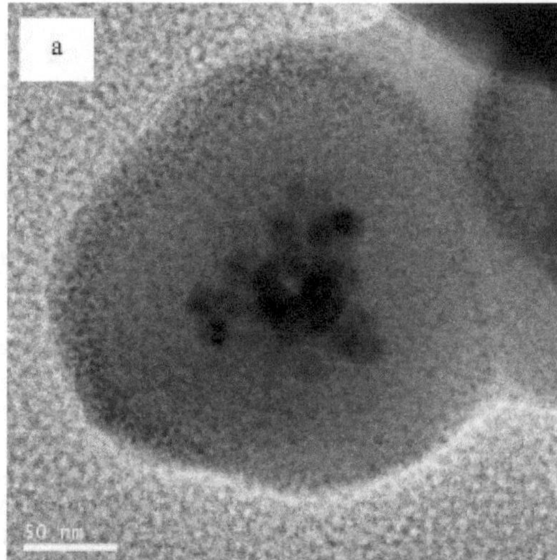

Figure 2. TEM (a) image of Fe_3O_4@SiO_2 and SEM (b) image of the Fe_3O_4@SiO_2@BiOCl nanosheets.

Since the photocatalytic performance is greatly dependent on the light absorption of the photocatalyst, the UV-Vis Diffuse Reflectance Spectra (DRS) was used to determine the ability of light absorption for the Fe_3O_4@SiO_2@BiOCl. Figure 3 shows DRS of the Fe_3O_4@SiO_2@BiOCl nanocomposite. The absorption edges of Fe_3O_4@SiO_2@BiOCl are 392 nm, and different strengths of photoabsorption in the visible light region (400–800 nm), indicating that the photocatalyst are able to absorb visible light.

The photocatalytic activity of the Fe_3O_4@SiO_2@BiOCl is evaluated under simulated solar irradiation using RhB as a probe molecule. Figure 4 shows the variation of RhB concentration (C/C_0) with irradiation time. In the absence of Fe_3O_4@SiO_2@

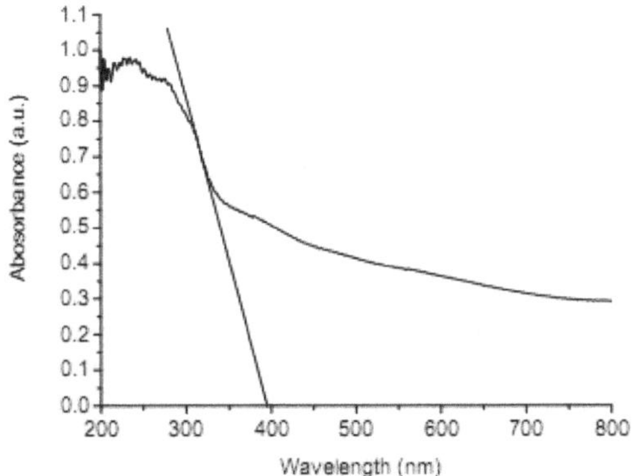

Figure 3. UV-vis diffuses reflectance spectra of Fe_3O_4@SiO_2@BiOCl.

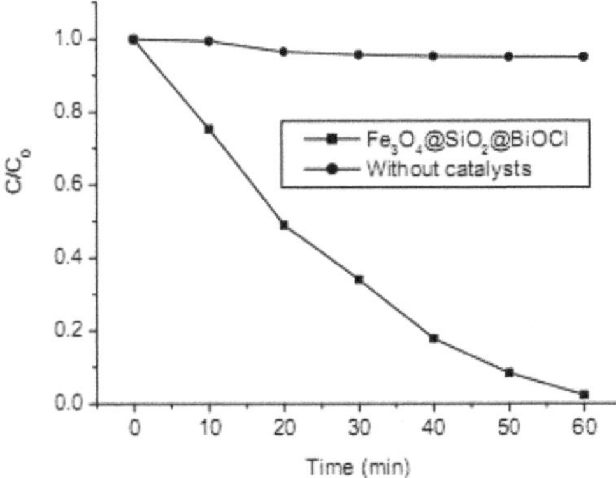

Figure 4. The photocatalytic activity of the Fe_3O_4@SiO_2@BiOCl.

BiOCl, decomposition of RhB is inappreciable, suggesting that photolysis of RhB is negligible. The photodegradation of RhB over Fe_3O_4@SiO_2@BiOCl is 95% within 60 min.

Figure 5 shows the temporal evolution of the absorption spectra of RhB degraded by Fe_3O_4@SiO_2@BiOCl under simulated solar irradiation. It was observed that the diminution of characteristic absorption of RhB is gradual. Meanwhile, the color of the RhB solution changed from initially red to transparent. It can be found that the diminishing of the main absorption peaks is accompanied by a shift of the absorption peak position to lower wavelength. The results may be attributed to a step-by-step degradation of RhB. Furthermore, the Fe_3O_4@SiO_2@BiOCl photocatalyst was attracted toward the wall of the beaker which was closer to the magnet. The results indicate that the magnetic Fe_3O_4@SiO_2@BiOCl nanocomposite show reusability.

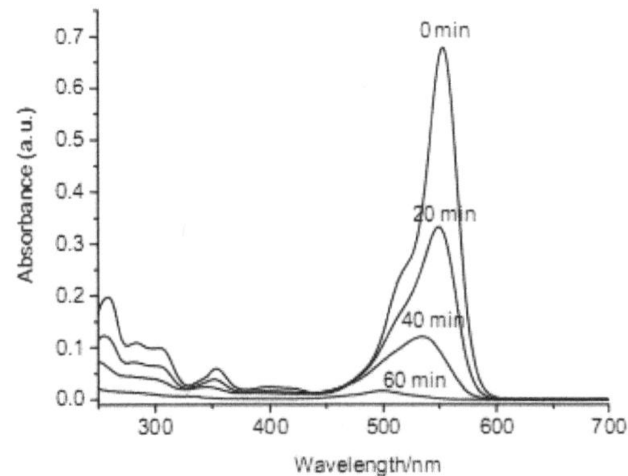

Figure 5. UV-vis spectral changes of the degradation of RhB dye by the Fe_3O_4@SiO_2@BiOCl.

4 CONCLUSIONS

A new photocatalyst Fe_3O_4@SiO_2@BiOCl has been synthesised, which not only exhibited excellent photocatalytic activity for the degradation of RhB under simulated solar irradiation, but also can be easily recovered from liquid phase photocatalytic reaction system due to its magnetic property. The high activity and magnetic recyclability of the photocatalysts make promising applications in water treatment and environmental protection.

REFERENCES

[1] An, C. Ming, X. Wang, J. Wang, S. Construction of magnetic visible-light-driven plasmonic Fe_3O_4@SiO_2@AgCl: Ag nanophotocatalyst, *J. Mater. Chem. 22 (2012) 5171–5176.*

[2] Chen, L. Huang, R. Xiong, M. Yuan, Q. He, J. Jia, J. Yao, M.Y. Luo, S.L. Au, C.T.S. Yin, F. Room-temperature synthesis of flower-like BiOX (X = Cl, Br, I) hierarchical structures and their visible-light photocatalytic activity, *Inorg. Chem. 52 (2013) 11118–11125.*

[3] Li, T.W. Huang, W.Z. Zhou, H. Yin, H.Y. Zheng, Y.F. Song, X.C. Synthesis of Zn^{2+} doped BiOCl hierarchical nanostructures and their exceptional visible light photocatalytic properties, *J. Alloy. Compd. 638 (2015) 148–154.*

[4] Yao, Y.R. Huang, W.Z. Zhou, H. Cui, X. Zheng, Y.F. Song, X.C. Synthesis of core–shell nanostructured magnetic photocatalyst Fe_3O_4@SiO_2@Ag_3PO_4 with excellent visible-light-responding photocatalytic activity, *J. Nanopart. Res. 16 (2014) 2742.*

Material Science and Environmental Engineering – Chen (Ed.)
© 2016 Taylor & Francis Group, London, ISBN 978-1-138-02938-5

Synthesis of nanosilver conductive ink for flexible circuit on cotton fabric substrate

Z.J. Rao, Z.H. Wang, W. Wang & D. Yu
College of Chemistry, Chemical Engineering and Biotechnology, Donghua University, Shanghai, P.R. China
The Key Laboratory of Science and Technology of Eco-Textile, Ministry of Education, Donghua University, Shanghai, P.R. China

ABSTRACT: Flexible circuits attract increasing attention for its advantageous applications in electronic industry. In this paper, pretreated cotton fabric acted as substrate material, guar gum selected as thickener and polyaniline was selected as additives to prepare nanosilver conductive ink. By this approach, small sized silver nanoparticles were synthesized and conductive ink with sheet resistance of 1.16 Ω/ü was acquired at temperature of 150°C. The good conductive of silver patterns indicates potential applications for flexible circuits. To demonstrate the feasibility to create flexible conductive circuits on cotton fabric, a circuit diagram with lighte emitting diodes was manufactured as well.

Keywords: nanosilver; conductive ink; flexible circuit; cotton fabric

1 INTRODUCTION

Printed electronics are applied to various high-performance electronic industry, such as Radio Frequency Identification (RFID), battery and solar cell, smart label, displays and Integrated Circuits (IC). Currently, in order to meet the lightweight and flexible properties, attentions have focused on the combination between printed circuits and flexible substrates, for instance, polymer film and paper. However, few literatures have been carried out related printed electronics on fabric, much less natural fabric, like cotton which has advantageous properties. Firstly, cotton is available, low-cost and degradable. Secondly, the lightweight and excellent flexible properties facilitate three-dimensional circuit and improve space utilization. Thirdly, the soft, comfortable, moisture, antistatic, absorption breathable and other properties bring convenience to intelligent textiles fabrication. To make the cotton fabric more suitable for fabricating conductive circuits, pretreatment was taken into consideration for the sake of elevating fabric activity.

Flexible substrate and the corresponding conductive ink are the main components of printed circuits.

Numerous researches have explored methods to acquire uniformity, stability and small sized silver nanoparticles for its high conductivity and oxidative stability compared with copper. In addition, additives, the key factors to produce high-performance conductive ink, should be compatible with silver nanoparticles and the substrate material. Here, polyaniline was selected as one of the additives for its conductivity by proton acid doping. Furthermore, it has demonstrated that silver nanoparticles can well disperse in the poly polyaniline matrix. Another significant additive acts as thickening agent to provide proper viscosity during screen printing on cotton. After trial test, guar gum, as a natural polymer thickener obtained from the seeds of the shrub cyamopsis tetragonolobus, was used to produce nanosilver conductive ink inspired by investigations in traditional printing on cotton fabric.

Based on above analysis, in this paper, nanosilver conductive ink was fabricated and silver patterns were screen printed on pretreated cotton fabric. Then the flexible circuits were sintered at maximum temperature of 150°C. The morphology, size distribution, crystalline structure and electric conductivity were investigated and the effect of pretreatment on cotton fabric was discussed. A circuit diagram with light emitting diodes was created as well.

2 EXPERIMENTAL SECTION

2.1 *Materials*

The main raw materials are silver nitrate ($AgNO_3$, Shanghai Shenbo Chemical Co., Ltd, China), guar gum (Shanghai Rgon Biological Technology Co., Ltd, China) and cotton fabric (Saintyear Holding Group Co. Ltd). Iron (II) sulfate heptahydrate ($FeSO_4 \cdot 7H_2O$), trisodium citrate dihydrate ($Na_3C_6H_5O_7 \cdot 2H_2O$), sodiumnitrate($NaNO_3$), aniline

(C_6H_7N), ammonium persulphate ($(NH_4)_2S_2O_8$), sulfuric acid (H_2SO_4), nitric acid (HNO_3) and Sodium hydroxide (NaOH) were purchased from Sinopharm Chemical Reagent Co., Ltd, China. All this reagents were of analytical grade and used as raw materials without further purification.

2.2 *Preparation of nanosilver conductive ink*

In a typical procedure, 7.5 g $FeSO_4 \cdot 7H_2O$ was completely dissolved in 17.5 g water and 14 g $Na_3C_6H_5O_7 \cdot 2H_2O$ was dissolved in another 21 g water and then mixed the two solutions. The silver inks are synthesized by uniformly adding the above mixture into 2.5 g $AgNO_3$ dissolving in 22.5 g water, followed by stirring at 25°C for 1 h.

The resulting solution was then centrifuged at 6,000 rpm for 15 min. In order to remove impurities and trisodium citrate dihydrate cladding on silver particles, the precipitation was washed by 0.4 g sodium nitrate dissolving in 40 ml water for 2 times.

Polyaniline solution was synthesized by mixing 3.4 ml H_2SO_4 acid dissolving in 250 g water with 5.8 ml aniline and 14.3 g $(NH_4)_2S_2O_8$ and stirring the mixture for 4 h at room temperature.

The viscosifying solution was prepared by dissolving 2 wt% guar gum in 100 ml water and stirring the solution at room temperature for 4 h.

The final washed product was dispersed in 1.25 ml guar gum following by adding 0.2 ml polyaniline solution. Then the mixture was homogenized (Talboys standard vortex mixer, 230 V/150 W) at 2000 rpm for 10 min.

2.3 *Preparation of nanosilver circuits*

Cotton fabric was pretreated by dipping in different concentrations of H_2SO_4 (0.05 and 0.1 mol/l), HNO_3 (0.1 and 0.2 mol/l) and NaOH (0.1 and 0.2 mol/l) for 120 min at room temperature.

The nanosilver circuits were prepared by screen printing nanosilver conductive ink on the cotton fabric. Then the samples were heated at 30°C for 2 h and then raised at rate of 2.5°C/min to 150°C which was lower than most sintering temperature. Although various literatures demonstrated that the higher sintering temperature tends to achieve lower resistivity, we defined the temperature taking substrate tolerance and operability into consideration. After keeping this sintering process for 1h, these samples were cooled to room temperature as shown in Figure 1.

2.4 *Characterization*

The shapes of the synthesized nanosilver were characterized by Transmission Electron Microscopy (TEM, JEM-2100, JEOL) and the average nanoparticle size distribution was tested by Particle Size & Zeta Potential Analyzer (Nano-ZS, Malvern Instruments Ltd.). X-Ray Diffraction (XRD) analyses of printed silver circuits on cotton were performed by using an X-ray diffractometer (D/Max-2550 PC, Rigaku). Environmental Scanning Electron Microscopy (ESEM, Quanta-250, FEI) was used to observe the surface microstructures of conductive ink. The conductibility of conductive circuits was measured by four-point-probe (RTS-9, 4 PROBES TECH) directly on the surface of conductive circuits.

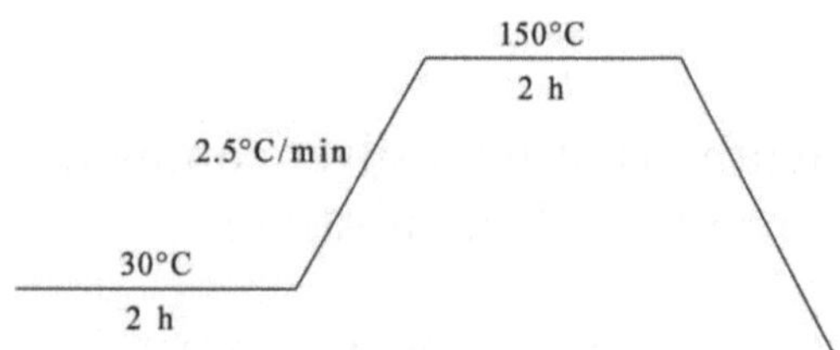

Figure 1. The scheme of sintering process.

3 RESULTS AND DISCUSSION

3.1 *Silver nanoparticles*

X-ray diffraction graph in Figure 2 displayed four characteristic peaks located at 2θ = 38.379°, 44.459°, 64.700° and 77.720° which respectively index to (111), (200), (220) and (311) plane diffraction peaks, indicating the printed circuits were composed of pure silver crystal of face centered cubic structure. Besides, the comprehensive performance of conductive circuits depends on several properties of silver nanoparticles, like silver particle size and uniformity. Thus, the shape and average nanoparticle size distribution were investigated. In this work, silver nitrate, as a metal precursor, was reduced by $FeSO_4 \cdot 7H_2O$ in the presence of $Na_3C_6H_5O_7 \cdot 2H_2O$ which functioned as dispersant. As shown in Figure 3, small sized nanoparticles were synthesized and dispersed relatively homogeneous. What should be noticed is the result of average nanoparticle size distribution by intensity depicted a bimodal distribution in Figure 4, namely the smaller average size of 7.167 nm and the larger average size of 126.4 nm. However, the result of average nanoparticle size distribution by number illustrated a unimodal distribution in Figure 5, namely the average size of 5.058 nm. This demonstrated that virtually all nanoparticles were within 10 nm except a few nanoparticles with large scale which had a much more influence on the average nanoparticle size distribution by intensity.

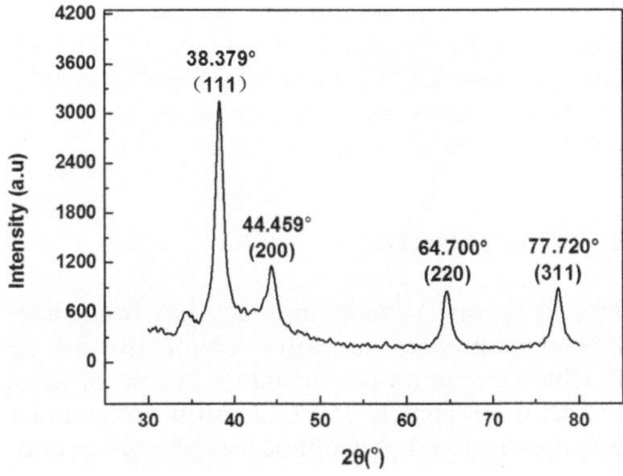

Figure 2. XRD pattern of nanosilver conductive circuits printed on cotton.

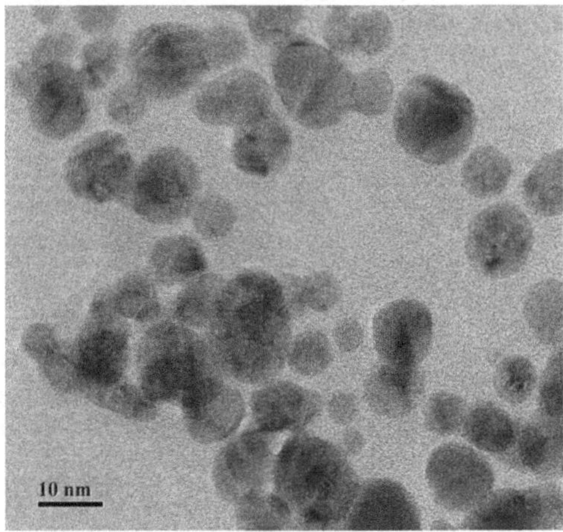

Figure 3. TEM image of synthesized silver nanoparticles.

3.2 *Pretreatment on cotton fabric*

Researches demonstrate that pretreatment on cotton by acid results in hydrolysis of cellulose, thereby bringing microscopic roughness on the fabric surface. In addition, pretreatment on cotton by base leads to fiber swelling, thus yielding increase of amorphous region in the fiber. Both impacts from acid and base will induce the elevating of hydrophilic ability and contact area. Therefore, the influence of different concentrations acid (HNO_3 and H_2SO_4) and base (NaOH) was investigated in this work. The conductibility of conductive circuits characterized by sheet resistance was represented in Table 1. It is clear that the modified cotton showed much lower sheet resistance. The sheet resistance decreased with increasing concentration of HNO_3 and NaOH and the uptrend

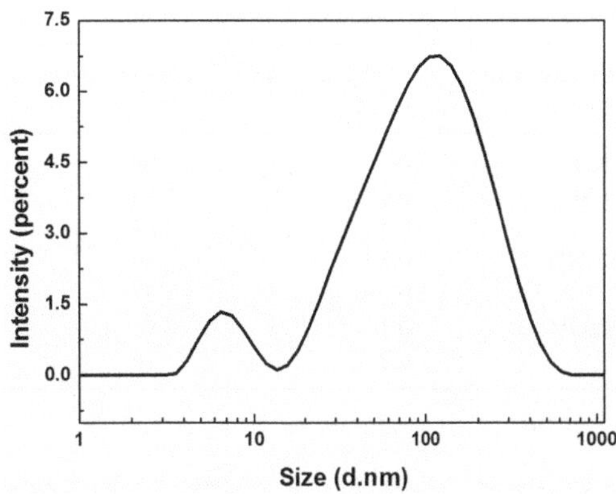

Figure 4. Size distribution of synthesized silver nanoparticles by intensity.

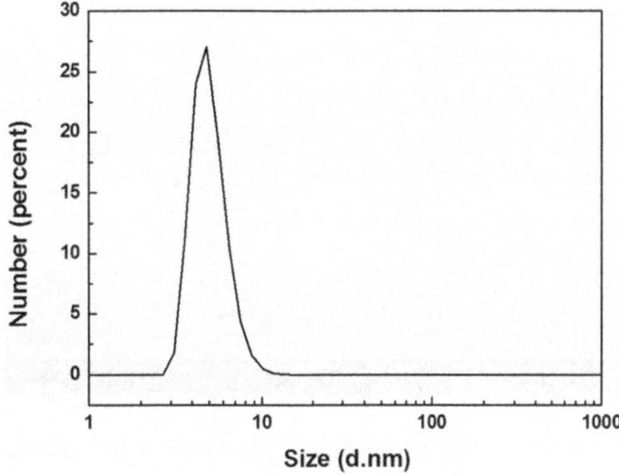

Figure 5. Size distribution of synthesized silver nanoparticles by number.

in the condition of NaOH pretreatment was more obvious than HNO_3 pretreatment. It can be explained that HNO_3 with boiling point of 83°C volatilized during 150°C in sintering process. We also noticed that higher concentration had negative effect on silver conductive circuits and cause slightly fabric carbonization.

3.3 *Nanosilver conductive circuits*

Figure 6 showed SEM image of nanosilver conductive circuits on cotton fabric after sintering process. The deposited silver nanoparticles agglomerated accompanied by volatilize and decomposition of additives in the conductive ink, formatting a compact conductive layer.

To demonstrate the feasibility to create flexible conductive circuits on cotton fabric, in this research, a circuit diagram with light emitting

Table 1. Sheet resistance with different pretreatments.

Pretreatment reagent	Concentration (mol/l)	Sheet resistance (Ω/ü)
H_2O	0	3.09
H_2SO_4	0.05	1.65
	0.1	2.69
$NaNO_3$	0.1	1.59
	0.2	1.57
NaOH	0.1	2.96
	0.2	1.16

Figure 6. SEM image of nanosilver conductive circuits after sintering process.

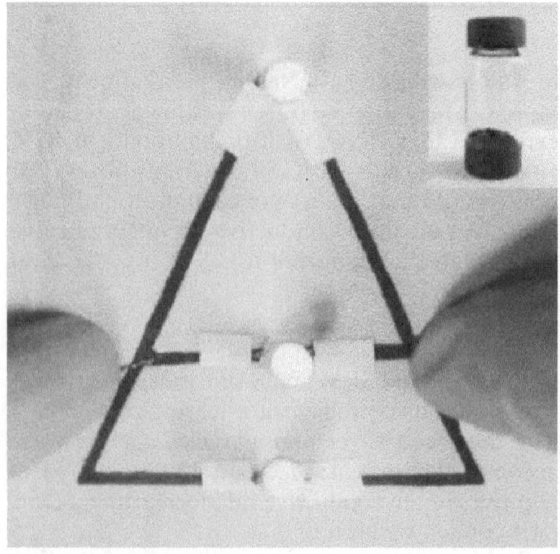

Figure 7. Optical image of conductive circuit on cotton and optical image of silver conductive ink (inset).

diodes was manufactured. As illustrated in Figure 7, light emitting diodes shined when the circuit was powered on, confirming the practicability of this conductive ink. The optical image of silver conductive ink was displayed in the inset.

4 CONCLUSIONS

In conclusion, a facile approach has been investigated to prepare nanosilver conductive ink for flexible circuits on cotton fabric at low sintering temperature, namely 150°C. Cotton pretreatment was operated and polyaniline was selected as additives to achieve lower resistivity. Guar gum, a natural polymer thickener, was environment friendly and well behaved as thickener. Therefore, it is likely that our conductive circuit fabrication method could be implemented for functional intelligent textiles fabrication.

ACKNOWLEDGEMENT

This work was supported by "the Fundamental Research Funds for the Central Universities".

REFERENCES

[1] Ahn, B.Y.; Lewis, J.A. 2014. Amphiphilic silver particles for conductive inks with controlled wetting behavior. *Materials Chemistry and Physics* 148 (3): 686–691.

[2] Auyeung, R.C.Y.; Kim, H.; Mathews, S.A.; Pique, A. 2007. Laser Direct-Write of Metallic Nanoparticle Inks. *J LASER MICRO NANOEN* 2 (1): 21–25.

[3] Baldaro, E.; Gallucci, M.; Formantici, C.; Issi, L.; Cheroni, S.; Galante, Y.M. 2012. Enzymatic improvement of guar-based thickener for better-quality silk screen printing. *Color. Technol* 128 (4): 315–322.

[4] Chang, Y.; Wang, D.Y.; Tai, Y.L.; Yang, Z.G. 2012. Preparation, characterization and reaction mechanism of a novel silver-organic conductive ink. *J. Mater. Chem* 22 (48): 25296–25301.

[5] Chou, K.S.; Lai, Y.S. 2004. Effect of polyvinyl pyrrolidone molecular weights on the formation of nanosized silver colloids. *Materials Chemistry and Physics* 83 (1): 82–88.

[6] Comiskey, B.; Albert, J.D.; Yoshizawa, H.; Jacobson, J. 1998. An electrophoretic ink for all-printed reflective electronic displays. *Nature* 394 (6690): 253–255.

[7] Denneulin, A.; Blayo, A.; Neuman, C.; Bras, J. 2011. Infra-red assisted sintering of inkjet printed silver tracks on paper substrates. *J. Nanopart. Res* 13 (9): 3815–3823.

[8] Dogome, K.; Enomae, T.; Isogai, A. 2013. Method for controlling surface energies of paper substrates to create paper-based printed electronics. *Chem. Eng. Process* 68: 21–25.

[9] Dong, T.Y.; Chen, W.T.; Wang, C.W.; Chen, C.P.; Chen, C.N.; Lin, M.C.; Song, J.M.; Chen, I.G.; Kao, T.H. 2009. One-step synthesis of uniform silver nanoparticles capped by saturated decanoate: direct spray printing ink to form metallic silver films. *PCCP* 11 (29): 6269–6275.

[10] Goo, Y.S.; Lee, Y.I.; Kim, N.; Lee, K.J.; Yoo, B.; Hong, S.J.; Kim, J.D.; Choa, Y.H. 2010. Ink-jet printing of Cu conductive ink on flexible substrate modified by oxygen plasma treatment. *Surf. Coat. Technol* 205: S369–S372.

[11] Gupta, K.; Jana, P.C.; Meikap, A.K. 2010. Optical and electrical transport properties of polyaniline-silver nanocomposite. *Synth. Met* 160 (13–14): 1566–1573.

[12] Huang, L.; Huang, Y.; Liang, J.J.; Wan, X.J.; Chen, Y.S. 2011. Graphene-based conducting inks for direct inkjet printing of flexible conductive patterns and their applications in electric circuits and chemical sensors. *Nano Res* 4 (7): 675–684.

[13] Jie, Y.; Yun, Z.; Guang-fu, Y.; Ping, Z. 2007. Preparation of nano-Ag particles and antibacterial dope loaded silver. *Key Eng. Mater* 336–338: 2115–2117.

[14] Kaler, A.; Mittal, A.K.; Katariya, M.; Harde, H.; Agrawal, A.K.; Jain, S.; Banerjee, U.C. 2014. An investigation of in vivo wound healing activity of biologically synthesized silver nanoparticles. *J. Nanopart. Res* 16 (9): 10.

[15] Khondoker, M.A.; Mun, S.C.; Kim, J. 2013. Synthesis and characterization of conductive silver ink for electrode printing on cellulose film. *APPL PHYS A-MATER* 112 (2): 411–418.

[16] Li, S.; Liu, P.; Wang, Q.S. 2012. Study on the effect of surface modifier on self-aggregation behavior of Ag nano-particle. *Appl. Surf. Sci* 263: 613–618.

[17] Muthukumar, N.; Thilagavathi, G. 2012. Development and characterization of electrically conductive polyaniline coated fabrics. *Indian J. Chem. Technol* 19 (6): 434–441.

[18] Russo, A.; Ahn, B.Y.; Adams, J.J.; Duoss, E.B.; Bernhard, J.T.; Lewis, J.A. 2011. Pen-on-Paper Flexible Electronics. *Adv. Mater* 23 (30): 3426.

[19] Siegel, A.C.; Phillips, S.T.; Dickey, M.D.; Lu, N.S.; Suo, Z.G.; Whitesides, G.M. 2010. Foldable Printed Circuit Boards on Paper Substrates. *Adv. Funct. Mater* 20 (1): 28–35.

[20] Subramanian, V.; Chang, P.C.; Lee, J.B.; Molesa, S.E.; Volkman, S.K. 2005. Printed organic transistors for ultra-low-cost RFID applications. *IEEE Trans. Compon. Packag. Technol* 28 (4): 742–747.

[21] Tokuno, T.; Nogi, M.; Karakawa, M.; Jiu, J.T.; Nge, T.T.; Aso, Y.; Suganuma, K. 2011. Fabrication of silver nanowire transparent electrodes at room temperature. *Nano Res* 4 (12): 1215–1222.

[22] Wang, Y.Q.; Li, N.; Li, D.Y.; Yu, S.Y.; Wang, C. 2015. A bio-inspired method to inkjet-printing copper pattern on polyimide substrate. *Materials Letters* 140: 127–130.

[23] Yang G.Y. 2012. Study on wet-sping of BTDA-TDI/MDI ternary copolyimide fiber. Msc Thesis, Donghua University, CN.

[24] Yan-Long, T.; Zhen-Guo, Y.; Zhi-Dong, L. 2011. A promising approach to conductive patterns with high efficiency for flexible electronics. *Appl. Surf. Sci* 257 (16): 7096–7100.

[25] Zhai, A.X.; Cai, X.H.; Jiang, X.Y.; Fan, G.Z. 2012. A novel and facile wet-chemical method for synthesis of silver microwires. *T NONFERR METAL SOC* 22 (4): 943–948.

[26] Zhu, G.; Shi, J.; Wang, W. 2010. Progress in Preparation and Applications of Silver Nano-materials. *Sci. Technol. Rev* 28 (22): 112–117.

[27] Zyung, T.; Kim, S.H.; Chu, H.Y.; Lee, J.H.; Lim, S.C.; Lee, J.I.; Oh, J. 1998. Flexible organic LED and organic thin-film transistor. *Proc. IEEE* 93 (7): 1265–1272.

Material Science and Environmental Engineering – Chen (Ed.)
 ISBN 978-1-138-02938-5

Ecological toxicity of engineered nano materials to the organisms in the environment

X.F. Cao & L.P. Liu
Weifang University of Science and Technology, Shouguang, China

ABSTRACT: The gradually increasing consumption of engineered nanomaterials in daily consumer goods will most probably lead to the release of such materials into the environment, which will increase the possible exposure of the organisms and the ecological system to a new level of pollutants and will cause a lot of unknown risks to the health of humans and wildlife. In this article, we explicit the importance to study the ecological toxicity of nano materials and present an overview of the ecotoxicity of different engineered nanoparticles to the organisms in the environment. Results from ecotoxicological studies show that certain engineered nanomaterials have effects on aquatic organisms, higher plants, and terrestrial organisms in the environment. And it's very helpful to assess health risks of man-made nanomaterials to humans. As a suggestion, the toxic mechanisms of these materials should be explored in-depth in further study.

Keywords: nanomaterials; ecological toxicity; health risk assessment; environment protection

1 INTRODUCTION

Materials in nanoscale, such as soot and organic colloid, have already existed on earth for millions of years (Nowack & Bucheli, 2007). Until recently, however, since humans have mastered the ability of synthesis this kind of materials, nanometer materials began to get extensively concerned all over the world (Liu et al., 2014). Because of its unique properties, the production and usage of man-made nanomaterials has been steadily grown (Scheringer, 2008). Presently, nanomaterials have been widely used in many industries or products such as medicine, photocatalyst, biological materials, cosmetics, dyes, environmental catalyst, fine chemical, and food. (Baan, 2007; Gratzel, 2001; Salvador et al., 2000). In March 2011, Woodrow Wilson Database gave a detailed list of 1317 kinds of consumer goods containing man-made nanomaterials on the market, the number of which rose by nearly 521% from that in March, 2006 (Liu, et al., 2014). Among these goods, the largest number of product categories is personal care products with a total of 738 kinds of products including cosmetics, sunscreen, and so on, followed by household and garden products (209 species), motor vehicle products (126 species), food and drink (105 species). With the gradually increasing consumption of man-made nanomaterials in daily consumer goods, the nano materials will be released or diffused to a variety of environmental medium, continuously (W.-M. Lee & An, 2013). This will increase the possible exposure of the organisms and the ecological system to a new level of pollutants (Wiesner et al., 2009), which will then cause a lot of unknown risks to the health of humans and wildlife (Dabrunz et al., 2011; Nel, 2006).

Man-made nanomaterial refers to the material with the scale of 100 nm at least in one dimension. Several frequently used nano materials include metal oxide nano powders, such as SiO_2, TiO_2, Al_2O_3, Fe_3O_4, Fe_2O_3, and other materials such as metal semiconductor nanoparticles, alloy, etc (Liu, et al., 2014). Materials in nanoscale tend to hold the special physical and chemical properties, such as surface effect, small size effect, and quantum effect. These properties of nanomaterials are quite different from the bulk materials with the same composition of nanomaterial, which make them show special properties such as electric conductivity, reactivity, and light sensitivity. However, these properties also bring potential bad consequences, which make them liable to produce harmful interactions with the biological system and the surroundings (Nel, 2006), causing potential ecotoxicity eventually. For example, nano materials can adsorb toxic gases (NO_2, SO_2 and etc.), toxic heavy metals (copper, lead, mercury and cadmium), refractory organics (polycyclic aromatic hydrocarbons, pesticide and etc.), and bioactive substances such as microorganisms, proteins, nucleotides in atmosphere, water, and soil. Man-made nanomaterials can be migrated and transformed in the environment, and within the long-term process, the inorganic and organic molecules adsorpted in their surface can occur complex chemical reactions, forming new contaminants. Besides, self-assembly of nanomaterials can also generate new compounds, and all these reactions will lead to large changes in the toxic effects (Nowack & Bucheli, 2007). As a result, it is necessary and urgent to take charge of the production and usage of nano material, as well as to evaluate their ecological health risks.

Before we evaluate the ecological risks brought by the use of man–made nanomaterials, their mobility, bioavailability, and biological toxicity in the environment should be studied comprehensively (Nowack & Bucheli, 2007). Besides, the detection of the toxic effects of man-made nanomaterials in different links of the food chain can bring more comprehensive understandings to the environmental and health effects of nanomaterials. Researchers in the world have conducted extensive relevant studies about the ecological toxicology effects of man-made nanomaterials, which have made remarkable achievements (Baun et al., 2008).

2 ECOTOXICITY TO AQUATIC ORGANISMS

Man-made nanomaterials can cause ecotoxicity to aquatic organisms, including single-celled organisms (bacteria, protozoa, etc.) and aquatic animals (daphnia, fish, etc.). Studies found that carbon nanotubes had the effects of growth inhibition on protozoa, and its inhibition rate showed significant dose response relationships between the concentrations of the carbon nanotubes (Zhu et al., 2006). Besides, the carbon nanotubes also showed toxicity of breathing on rainbow trout (Smith et al., 2007). However, other studies have shown that carbon nanotubes can promote the growth of protozoa in the growth medium containing yeast extract. Causes of this phenomenon is that parts peptone in medium were coupled with carbon nanotubes, which makes carbon nanotubes lose activity, but the rest of the peptone take a role in the growth increasing of protozoa (Ying et al., 2006). Purified carbon nanotubes did not have any impact to copepods. However, the unpurified carbon nanotubes increased the mortality of copepods as a result of the existence of its by-products (Templeton et al., 2006). There are relatively more researches on the toxic effects of nanomaterials C_{60} in aquatic organisms, including bacteria (Lyon et al., 2005), water flea (Lovern & Klaper, 2006) and fish (Oberdorster et al., 2006), etc. As for Ag nanoparticles, they can destroy the cell walls of bacteria, and eventually lead to the death of the organism. The interaction between cells and Ag nanoparticles are closely related to the size of the nano Ag (Jose Ruben et al., 2005), and seems to be related to the shape of the Ag nanoparticles also. Besides, Ag nanoparticles was much toxic than Ag^+ to *Escherichia coli* (Lok et al., 2006). Studies on the genotoxicity and ecological toxicity of nanomaterials CeO_2 (15 nm, 30 nm), SiO_2 (7 nm to 10 nm), TiO_2 (7 nm, 20 nm) to the big fleas and mosquito larvae showed that, nanometer CeO_2 can damage the DNA of the organisms, and CeO_2 and SiO_2 can increase the mortality of organisms. In addition, under the nanometer CeO_2 exposure, the DNA damage of the big fleas had significant correlation with its decline of reproductive capacity. However, the TiO_2 has no influence on the aquatic organisms (Lee, et al., 2009). Strigul et al. measured the acute toxicity of nano Al_2O_3, B and TiO_2 to big fleas and *vibrio fischeri*. The results showed that the TiO_2 was of low toxicity to big fleas, while nano B can kill the large flea in the content higher than 80 mg L^{-1}.

3 ECOTOXICITY TO HIGHER PLANTS

Studies on the ecological toxicology of man-made nanomaterials on higher plants are much less. Lee et al. proposed to use agar culture medium to conduct experiments of nanotoxicity on higher plants. They also use this medium to carry out the acute toxicity experiment of Cu nanoparticles on mung bean and wheat. Results showed that nanoparticles were combined with plant cells, causing the restraining of plants. The EC_{50} values of nanometer Cu to the growth of mung bean and wheat seedlings in 2-d were 335 and 570 mg L^{-1}, respectively. Therefore, the tolerance of wheat to nanometer Cu was stronger than that of mung bean (Lee et al., 2008). Studies of the perennial ryegrass by hydroponic experiment found that ZnO nanoparticles can reduce the biomass of perennial ryegrass and decrease the root elongation. Another study found that the mobility of zinc from root to the overground part is low (Lin & Xing, 2008), which is mainly due to the low content of zinc ion in aqueous solutions. Hence, the toxic effect is not produced by zinc ions, but mainly produced by Zn nanoparticles. However, results derived from the soil experiments were contradictory. Impacts of TiO_2 nanomaterials on plant oxidative stress were also researched by some scholars. Results showed that TiO_2 nanomaterials can reduce the content of H_2O_2, superoxide radicals and malondialdehyde, while increase the activities of superoxide dismutase, peroxidase, and catalase.

4 ECOTOXICITY TO TERRESTRIAL ORGANISMS

Studies on the ecological toxicity of man-made nanomaterials to terrestrial organisms were limited, mainly concentrating in the indexes of acute toxicity, growth ability of reproduction in nematodes, etc. Ecological toxicity Of ZnO, Al_2O_3, TiO_2 nanometer materials and their corresponding block material to *caenorhabditis elegans* indicated that, the toxicity of ZnO nano materials and its block material to *c. elegans* is similar, with LC_{50} 2.3 mg L^{-1}, while the LC_{50} of Al_2O_3 and TiO_2 block materials to *c. elegans* is twice with its corresponding nanomaterials (Wang et al., 2009). The toxicity experiment of Ag nanoparticles to *c. elegans* found that Ag nanoparticles showed significant inhibition on the growth of *c. elegans*, and Ag nanoparticles have internalized as the body composition of *c. elegans*, suggesting that nanomaterials may also have other potential toxicity to organism effects (Meyer et al., 2010).

5 CONCLUSIONS

According to the above studies, man-made nanomaterials are indeed toxic to organisms, and different nanomaterials have different biological toxicity to different kinds of organisms. At present, researches about the ecological toxicity of synthetic nanomaterials are still in its infancy, the test method and toxicity evaluation method are not standard, and there is no uniform standard reference (Peralta-Videa et al., 2011). The acute toxicity experiment is still the commonly used evaluation method. Therefore, in order to obtain comprehensive understanding of the toxic effects and health risks for man-made nanomaterials to organisms, it's urgent to explore new toxicity evaluation indexes, and to study in-depth the toxic mechanisms of these materials.

ACKNOWLEDGMENTS

The research work was supported by Science and Technology Development Plan projects of Weifang under Grant No. 2014ZJ1055 and Startup Fund at Weifang University of Science & Technology under Grant No. sdlgy2013w002.

REFERENCES

[1] Baan, R.A. (2007). Carcinogenic Hazards from Inhaled Carbon Black, Titanium Dioxide, and Talc not Containing Asbestos or Asbestiform Fibers: Recent Evaluations by an IARC Monographs Working Group. *Inhalation Toxicology, 19*(s1), 213–228.

[2] Baun, A., Hartmann, N.B., Grieger, K., & Kusk, K.O. (2008). Ecotoxicity of engineered nanoparticles to aquatic invertebrates: a brief review and recommendations for future toxicity testing. *Ecotoxicology, 17*(5), 387–395.

[3] Dabrunz, A., Duester, L., Prasse, C., Seitz, F., Rosenfeldt, R., Schilde, C., Schulz, R. (2011). Biological Surface Coating and Molting Inhibition as Mechanisms of TiO_2 Nanoparticle Toxicity in *Daphnia magna*. *PLoS ONE, 6*(5), 109–112.

[4] Gratzel, M. (2001). Photoelectrochemical cells. [10.1038/35104607]. *Nature, 414*(6861), 338–344.

[5] Jose Ruben, M., Jose Luis, E., Alejandra, C., Katherine, H., Juan, B.K., Jose Tapia, R., & Miguel Jose, Y. (2005). The bactericidal effect of silver nanoparticles. *Nanotechnology, 16*(10), 2346.

[6] Lee, S.W., Kim Sm Fau—Choi, J., & Choi, J. (2009). Genotoxicity and ecotoxicity assays using the freshwater crustacean Daphnia magna and the larva of the aquatic midge Chironomus riparius to screen the ecological risks of nanoparticle exposure. *Environmental Toxicology and Chemistry, 28*(1), 86–91.

[7] Lee, W.-M., & An, Y.-J. (2013). Effects of zinc oxide and titanium dioxide nanoparticles on green algae under visible, UVA, and UVB irradiations: No evidence of enhanced algal toxicity under UV pre-irradiation. *Chemosphere, 91*(4), 536–544.

[8] Lee, W.-M., An, Y.-J., Yoon, H., & Kweon, H.-S. (2008). Toxicity and bioavailability of copper nanoparticles to the terrestrial plants mung bean (Phaseolus radiatus) and wheat (Triticum aestivum): Plant agar test for water-insoluble nanoparticles. *Environmental Toxicology and Chemistry, 27*(9), 1915–1921.

[9] Lin, D., & Xing, B. (2008). Root Uptake and Phytotoxicity of ZnO Nanoparticles. *Environmental Science & Technology, 42*(15), 5580–5585.

[10] Liu, Y., Tourbin, M., Lachaize, S., & Guiraud, P. (2014). Nanoparticles in wastewaters: Hazards, fate and remediation. *Powder Technology, 255*(0), 149–156.

[11] Lok, C.-N., Ho, C.-M., Chen, R., He, Q.-Y., Yu, W.-Y., Sun, H., Che, C.-M. (2006). Proteomic Analysis of the Mode of Antibacterial Action of Silver Nanoparticles. *Journal of Proteome Research, 5*(4), 916–924.

[12] Lovern, S.B., & Klaper, R. (2006). Daphnia magna mortality when exposed to titanium dioxide and fullerene (C60) nanoparticles. *Environmental Toxicology and Chemistry, 25*(4), 1132–1137. doi: 10.1897/05-278r.1.

[13] Lyon, D.Y., Fortner, J.D., Sayes, C.M., Colvin, V.L., & Hughes, J.B. (2005). Bacterial cell association and antimicrobial activity of a C60 water suspension. *Environmental Toxicology and Chemistry, 24*(11), 2757–2762.

[14] Meyer, J.N., Lord, C.A., Yang, X.Y., Turner, E.A., Badireddy, A.R., Marinakos, S.M., Auffan, M. (2010). Intracellular uptake and associated toxicity of silver nanoparticles in Caenorhabditis elegans. *Aquatic Toxicology, 100*(2), 140–150.

[15] Nel, A. (2006). Toxic Potential of Materials at the Nanolevel. *Science, 311*(5761), 622–627.

[16] Nowack, B., & Bucheli, T.D. (2007). Occurrence, behavior and effects of nanoparticles in the environment. *Environmental Pollution, 150*(1), 5–22.

[17] Oberdörster, E., Zhu, S., Blickley, T.M., McClellan-Green, P., & Haasch, M.L. (2006). Ecotoxicology of carbon-based engineered nanoparticles: Effects of fullerene (C60) on aquatic organisms. *Carbon, 44*(6), 1112–1120.

[18] Peralta-Videa, J.R., Zhao, L., Lopez-Moreno, M.L., de la Rosa, G., Hong, J., & Gardea-Torresdey, J.L. (2011). Nanomaterials and the environment: A review for the biennium 2008–2010. *Journal of Hazardous Materials, 186*(1), 1–15.

[19] Salvador, A., Pascual-Martí, M.C., Adell, J.R., Requeni, A., & March, J.G. (2000). Analytical methodologies for atomic spectrometric determination of metallic oxides in UV sunscreen creams. *Journal of Pharmaceutical and Biomedical Analysis, 22*(2), 301–306.

[20] Scheringer, M. (2008). Nanoecotoxicology: Environmental risks of nanomaterials. [10.1038/nnano.2008.145]. *Nat Nano, 3*(6), 322–323.

[21] Smith, C.J., Shaw, B.J., & Handy, R.D. (2007). Toxicity of single walled carbon nanotubes to rainbow trout, (Oncorhynchus mykiss): Respiratory toxicity, organ pathologies, and other physiological effects. *Aquatic Toxicology, 82*(2), 94–109.

[22] Templeton, R.C., Ferguson, P.L., Washburn, K.M., Scrivens, W.A., & Chandler, G.T. (2006). Life-Cycle Effects of Single-Walled Carbon Nanotubes (SWNTs) on an Estuarine Meiobenthic Copepod†. *Environmental Science & Technology, 40*(23), 7387–7393.

[23] Wang, H., Wick, R.L., & Xing, B. (2009). Toxicity of nanoparticulate and bulk ZnO, Al_2O_3 and TiO_2 to the nematode Caenorhabditis elegans. *Environmental Pollution, 157*(4), 1171–1177.

[24] Wiesner, M.R., Lowry, G.V., Jones, K.L., Hochella, J.M.F., Di Giulio, R.T., Casman, E., & Bernhardt, E.S. (2009). Decreasing Uncertainties in Assessing Environmental Exposure, Risk, and Ecological Implications of Nanomaterials†‡. *Environmental Science & Technology, 43*(17), 6458–6462.

[25] Ying, Z., Tiecheng, R., Yuguo, L., Jinxue, G., & Wenxin, L. (2006). Dependence of the cytotoxicity of multi-walled carbon nanotubes on the culture medium. *Nanotechnology, 17*(18), 4668.

[26] Zhu, Y., Zhao, Q., Li, Y., Cai, X., & Li, W. (2006). The Interaction and Toxicity of Multi-Walled Carbon Nanotubes with *Stylonychia mytilus*. *Journal of Nanoscience and Nanotechnology, 6*(5), 1357–1364.

Material Science and Environmental Engineering – Chen (Ed.)
© 2016 Taylor & Francis Group, London, ISBN 978-1-138-02938-5

Research on the preparation and activity of BiOBr-TiO_2 nanobelt heterostructured photocatalyst with high visible light responsive activity

Y.H. Ao, J.L. Xu, P.F. Wang & C. Wang
Key Laboratory of Integrated Regulation and Resource Development on Shallow Lakes, College of Environment, Ministry of Education, Hohai University, Nanjing, China

ABSTRACT: A new type of heterostructured photocatalyst (BiOBr-TiO_2 nanobelt) has been successfully synthesized to increase the photocatalytic properties of TiO_2 nanobelt. The phase structure and morphology of as-prepared photocatalysts were characterized by X-Ray Diffraction (XRD) and Scanning Electron Microscopy (SEM), respectively. The photocatalytic activity of the samples was assessed by degradation of dye X-3B (C.I.Reactive Red 2) under visible light irradiation. Results showed that the heterostructured photocatalysts exhibited higher performance than the pure titania nanobelt. Furthermore, the dosages of BiOBr on the activity of the heterostructures were also investigated. Results showed that the optimal dosage was 5% (Bi/Ti). The enhanced photocatalytic performances of BiOBr-TiO_2 nanobelt heterostructure will provide an approach for potential practical application in various organic wastewater treatments.

Keywords: nanocomposite; semiconductor; heterostructure; bismuth oxybromide; photocatalysis

1 INTRODUCTION

Environmental pollution is one of the most serious problems facing mankind today. Human kind has made great efforts using green and eco-friendly methods to deal with the problem (Hu et al. 2013; Yu et al. 2013). Since the discovery of photocatalytic splitting of water on TiO_2 electrodes by Fujishima and Honda in 1972 (Fujishima et al. 1972), TiO_2 has attracted considerable attention as photocatalyst in solving environmental and energy issues. But there are still some problems to restrict the extensive application of titania due to its wide band structure, low quantum efficiencies, slow reaction rate and poor solar energy utilization (Li et al. 2011; Wang et al. 2013). For the resolving of these problems, many efforts have been made to the improvement of performance of titania. On one hand, a variety of nanostructured TiO_2, such as nanobelts (Liu et al. 2012), nanowires (Inoue et al. 2010), nanorods (Gao et al. 2012), nanotubes (Macak et al. 2007), nanoflakes (Wang et al. 2010), have been explored to improve its photocatalytic efficiency. On the other hand, doping or coupling to titania with other semiconductors can enlarge the absorption of titania to the visible light region, thus increase the utilization of solar energy (Lin et al. 2011(a); Murcia et al. 2011).

Among mostly reported bismuth-based nanostructures, bismuth oxyhalide (BiOX, X = Cl, Br, I) are considered to be one of the most promising photocatalysts in pollution control owing to their unique layered structure. BiOBr is a lamellar-structured p-type semiconductor with an indirect band gap, which was demonstrated to have effective removal capacity for heavy metal ions and organic dyes (Li et al. 2013; Liu et al. 2012). However, the absorbtion of visible light for pure BiOBr is very limited to practical applications, which make it necessary to develop effective ways to change this situation (Cao et al. 2012). Wei et al. (Wei et al. 2013) reported hybrid BiOBr-TiO_2 nanocomposites with high visible light photocatalytic activity for water treatment. However, there is no paper focused on the TiO_2 nanobelt-BiOBr heterostructured photocatalyst.

In the present work, we found a facile, rapid, and mild method to prepare BiOBr-TiO_2 nanobelt heterostructure, which exhibited remarkable enhancement of photocatalytic activity in the degradation of dye X-3B. The coupling with BiOBr can effectively extend the photo-response of TiO_2 to the visible light region. It can also form p-n heterojunction between BiOBr nanoparticles and TiO_2 nanobelt, thus inhibit the rate of hole-electron recombination and improve the photcatalytic activity.

2 MATERIALS AND METHODS

All chemicals were of analytical reagent grade and solutions were prepared with deionized water.

The TiO_2 nanobelts were synthesized by hydrothermal method and subsequent heat treatment, as reported by Wang et al. previously (Wang et al. 2010). In a typical synthesis procedure, 0.2 g of the as-prepared TiO_2 nanobelts was dispersed in 50 ml deionized water by ultrasonic treatment for 15 min to form suspension A. In addition, definite amount of Bi $(NO_3)_3 \cdot 5H_2O$ and KBr were dissolved into 15 mL of ethylene glycol respectively. The two ethylene glycol solutions were mixed to form transparent solution B. Afterwards, the solution B was added to suspension A drop by drop and kept at room temperature for 4 h with violently stirring. The precipitate was collected by centrifugation and washed with ethanol and deionized water for several times. At last, the products were dried at 80 °C in an oven. Therefore, we obtained different catalyst samples with atomic ratio of Bi/Ti = 1%, 2%, 5%, 8%, respectively.

The morphology and phase structure of the samples was carried out by X-Ray Diffraction technique (XRD), Scanning Electron Microscopy (SEM) and Transmission Electron Microscopy (TEM). The photocatalytic activity of the various catalysts was assessed by degradation of reactive brilliant red (X-3B) under visible light irradiation. 30 mg of photocatalysts was dispersed into 100 mLX-3B solution with an initial concentration of 20 mg/L, and maintained in dark for 30 min. The concentration of X-3B was analyzed by monitoring the absorbance at 535 nm on UV–vis spectrophotometer (Hitachi UV-3600).

3 RESULTS AND DISCUSSIONS

The XRD patterns of the BiOBr-TiO_2 nanobelts heterostructured catalyst are given in Figure 1. The diffraction peaks of TiO_2 can be indexed as mixed anatase (JCPDS No. 21-1272) and rutile phase (JCPDS No. 21-1276). The peaks marked with stars indicate the successful coupling of BiOBr (JCPDS No. 09-0393) nanoparticles onto the TiO_2 nanobelt. Furthermore, the diffraction peaks corresponding to BiOBr became sharper when the atomic ratio of Bi/Ti increases, implying higher crystallization degree (Lin et al. 2011(b); Cao et al. 2012).

Figure 2 (a) exhibited the SEM micrograph of sample with atomic ratio of 5% Bi/Ti. It can be observed that the TiO_2 with belt-like structure was decorated by granular BiOBr. Samples with higher content of BiOBr nanoparticles progressively show an increased density on the surface of the TiO_2 nanobelts (the data not shown here). The pure TiO_2 nanobelts have clean and smooth surfaces, which provide a good platform for loading of BiOBr nanoparticles. TEM image for sample with 5% Bi/Ti is shown in Figure 2 (b), which also confirms the formation of the heterostructure between BiOBr nanoparticles and TiO_2 nanobelts. The results show that BiOBr nanoparticles sized in the range of 50–200 nm are randomly dispersed onto the surface of the TiO_2 nanobelts.

In order to evaluate the photocatalytic performance, the degradation of X-3B over various heterostructured photocatalysts is performed under visible light irradiation. The results are depicted in Figure 3. The adsorbtion and degradation rate improve with the increase of Bi/Ti in a certain range. However, it is obvious that the adsorbtion ability and photocatalytic efficiency of heterostructured

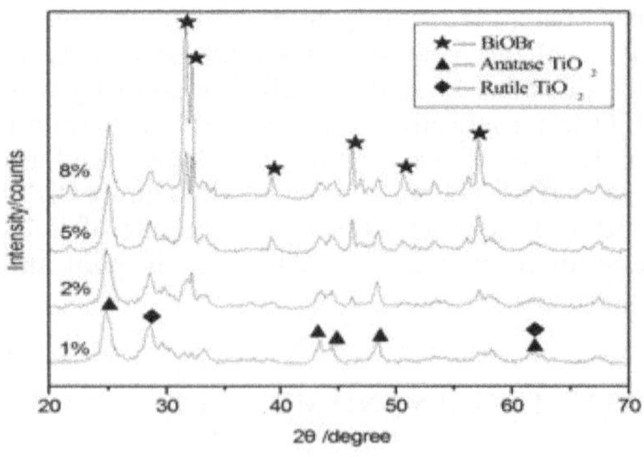

Figure 1. XRD patterns of different atomic ratio of Bi/Ti.

Figure 2. SEM images (a) and TEM images (b) of BiOBr/TiO_2 sample with atomic ratio of 5% Bi/Ti.

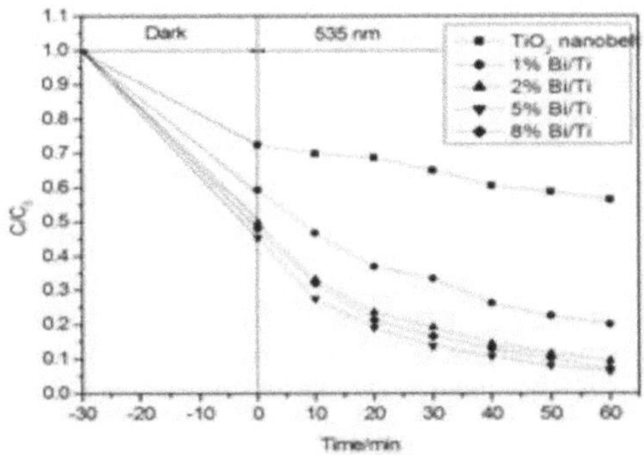

Figure 3. Dark absorption and visible light-induced photocatalytic degradation of X-3B over TiO_2 nanobelt and BiOBr-TiO_2 nanobelt heterostructures.

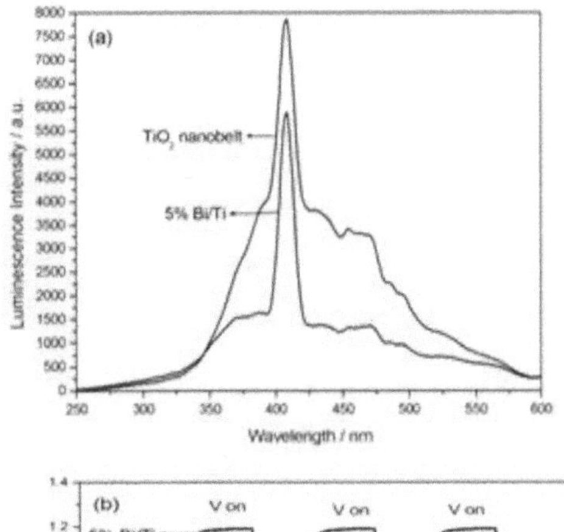

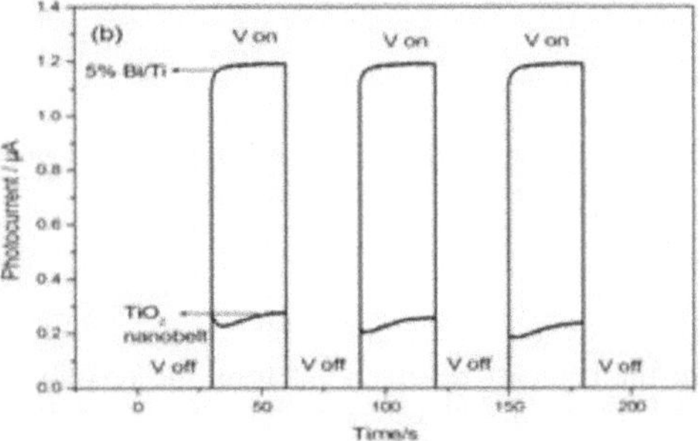

Figure 4. (a) PL spectra and (b) Photocurrent of BiOBr/TiO_2 with 5% Bi/Ti and TiO_2 nanobelt.

catalyst with 5% Bi/Ti is the highest. The degradation rate over X-3B is nearly 95% after 1 h visible light irradiation, while the corresponding photodegradation rate of the pure TiO_2 nanobelts is no more than 40%. But the adsorbtion rate and degradation rate decreased when the amount of BiOBr nanoparticles increases. The reason may be that the photoinduced electrons and holes produced in BiOBr cannot efficiently transfer to the TiO_2 and they will recombine inevitably, thus decreasing the photocatalytic activity of the composites (Huang et al. 2010). Therefore, the loading amount of BiOBr nanoparticles plays a significant impact on the photocatalytic performance of the BiOBr-TiO_2 nanobelt heterostructure (Zhang et al. 2011; Liu et al. 2011).

The Photoluminescence (PL) spectrum reflects the transfer of photo-induced electron-hole pairs between the two semiconductors, which is one of the key factors for the photocatalytic activity (He et al. 2015). The lower the PL intensity, the lower recombination rate of photo-generated electrons and holes is, thus the higher the photocatalytic activity (Liu et al. 2015). As shown in Figure 4 (a), the pure TiO_2 samples exhibited a strong emission peak at about 420 nm, which is attributed to the recombination process of self-trapped excitations. The PL intensity of BiOBr/TiO_2 hybirds (with 5% Bi/Ti) is much lower than that of pure TiO_2, indicating the charge transfer between the TiO_2 and BiOBr through the formed heterojunctions can effectively enhance separation efficiency of photo-generated carriers.

The charge transfer of BiOBr/TiO_2 heterojunctions was also examined by the experiments of transient photocurrent responses. The photocurrent responses of as-prepared samples were obtained by the intermittent visible light irradiation of 30 s in 50 mL of Na_2SO_4 solution with a 0 V applied potential bias versus Ag/AgCl reference electrode. As shown in Figure 4 (b), the rise and fall of the photocurrent response are evidently observed for each switch-on and switch-off event. For comparison, the photocurrents of pure TiO_2 are also displayed in the figure. It can be seen that typical BiOBr/TiO_2 composites show much higher photocurrent than pure TiO_2, which also confirms that the introduction of BiOBr into TiO_2 can hinder the recombination of electron-hole pairs. This result is well correspond to the PL experiments.

Based on the results above, a synergetic photoinduced mechanism can be proposed (as shown in Fig. 5). The p-n junctions will be generated at the interfaces when BiOBr nanoparicles are assembled onto the TiO_2 nanobelts. Under visible light irradiation, on one hand, the electrons in the X-3B are excited, and then transferred to the CB of TiO_2. The electron can also be transferred to the CB of BiOBr. All these electrons can participate in the reduction reaction to yield superoxide radical anions (O2–). This is a photo-sensitization process. On the other hand, the BiOBr can also be excited by the visible light and generates electron-hole pairs. The holes can be transferred to the VB of TiO_2. And these holes can oxidize H_2O/OH– to •OH. The holes and •OH can oxidize X-3B molecular to CO_2 and H_2O with the O_2– corporately. Appropriate atomic ratio of Bi/Ti can facilitate

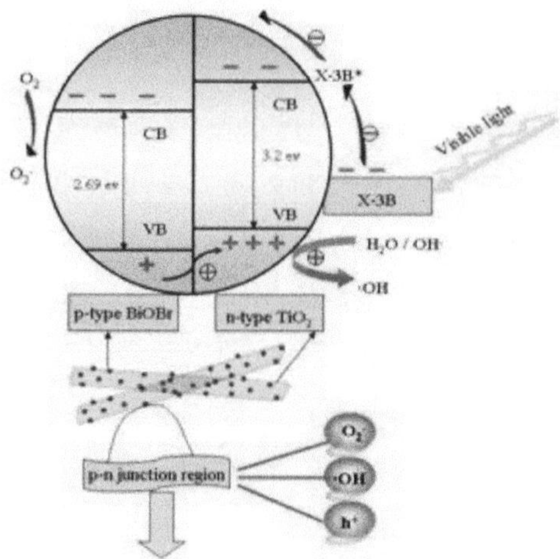

Figure 5. p-n junction band structure and schematic diagram of electron–hole separation process.

the effective transfer of the photoinduced carriers in the heterostructure. Therefore, the recombination of photoinduced carriers can be effectively suppressed. The heterostructured catalyst with high electron–hole separation efficiency provides a promising way to modify low quantum efficiency of TiO_2 nanobelt, and thereby extend its visible light respond range (Lin et al. 2011).

4 CONCLUSIONS

In summary, we successfully synthesized a new photocatalyst with high crystallinity and good photodegradation properties based on BiOBr nanoparticles and TiO_2 nanobelts. Results showed that BiOBr-TiO_2 nanobelt heterostructure have higher photocatalytic activity than pure TiO_2 nanobelt under visible light irradiation. The optimal atomic ratio of Bi/Ti to synthesis BiOBr-TiO_2 nanobelt heterostructure is 5%. This research provides a basis for developing new type of visible light driven type II heterostructured catalysts for environmental remediation.

ACKNOWLEDGEMENTS

We are grateful for grants from National Science Funds for Creative Research Groups of China (No. 51421006), the National Science Fundation of China for Excellent Young Scholars (No. 51422902), Program for Changjiang Scholars and Innovative Research Team in University (No. IRT13061), the Key Program of National Natural Science Foundation of China (No. 41430751), the National Natural Science Foundation of China (No. 51479065), Fundamental Research Funds for the Central Universities (2013B32114, 2013B14114), and PAPD.

REFERENCES

[1] Cao J., Xu B.Y., Lin H.L., et al. 2012. Chemical etching preparation of BiOI/BiOBr heterostructures with enhanced photocatalytic properties for organic dye removal. *Chemical Engineering Journal* 185–186: 91–99.

[2] Cheng H.F., Huang B.B., Dai Y., et al. 2010. One-Step Synthesis of the Nanostructured AgI/BiOI Composites with Highly Enhanced Visible-Light Photocatalytic Performances. *Langmuir* 26(9): 6618–6624.

[3] Fujishima A., Honda K. 1972. Electrochemical photolysis of water at a semiconductor electrode. Nature 238: 37–38.

[4] Gao M.J., Li Y.X., Guo M., et al. 2012. Effect of Substrate Pretreatment on Controllable Growth of TiO_2 Nanorod Arrays. Journal of Materials Science Technology 28(7): 577–586.

[5] He Y.M., Zhang L.H., Fan M.H., et al. 2015. Z-scheme SnO2-x/g-C3 N4 composite as an efficient photocatalyst for dye degradation and photocatalytic CO2 reduction. Solar Energy Materials and Solar Cells 137: 175–184.

[6] Hu A., Liang R., Kurdi S., et al. 2013. Enhanced photocatalytic degradation of dyes by TiO_2 nanobelts with hierarchical structures. Journal of Photochemistry and Photobiology A: Chemistry 256: 7–15.

[7] Inoue Y., Noda I., Torikai T., et al. 2010. TiO_2 nanotube, nanowire, and rhomboid-shaped particle thin films fixed on a titanium metal plate. Journal of Solid State Chemistry 183(1): 57–64.

[8] Li G.F., Qin F., Wang R.M., et al. 2013. BiOX (X = Cl, Br, I) nanostructures: Mannitol-mediated microwave synthesis, visible light photocatalytic performance, and Cr (VI) removal capacity. Journal of Colloid and Interface Science 409: 43–51.

[9] Li N., Zhu L., Zhang W.D., et al. 2011. Modification of TiO_2 nanorods by Bi2MoO6 nanoparticles for high performance visible-light photocatalysis. Journal of Alloys Compounds 509(41): 9770–9775.

[10] Lin J.J., Shen J.X., Wang R.G., et al. 2011(a). Nano-p-n junctions on surface-coarsened TiO_2 nanobelts with enhanced photocatalytic activity. Journal of Materials Chemistry 21(13): 5106–5113.

[11] Lin J.J., Shen J.X., Wang T.L., et al. 2011(b). Enhancement of photocatalytic properties of TiO_2 nanobelts through surface-coarsening and surface nanoheterostructure construction. Materials Science & Engineering: B 176(12): 921–925.

[12] Liu R.Y., Hu P.G., Chen S.W. 2012. Photocatalytic activity of Ag3PO4 nanoparticle/TiO_2 nanobelt heterostructures. Applied Surface Science 258(24): 9805–9809.

[13] Liu Y.L., Zhong L., Peng Z.Y., et al. 2011. Self-assembly of Pt nanocrystals/one-dimensional titanate nanobelts heterojunctions and their great enhancement of photocatalytic activities. CrystEngComm 13(17): 5467–5473.

[14] Liu Z.S., Wu B.T., Zhu Y.B., et al. 2012. Fe-Ions Modified BiOBr Mesoporous Microspheres with Excellent Photocatalytic Property. Catalysis Letters 142(12): 1489–1497.

[15] Liu Z.S., Wu B.T., Niu J.N., et al. 2015. BiPO4/BiOBr p-n junction photocatalysts: One-pot synthesis and dramatic visible light photocatalytic activity. Materials Research Bulletin 63: 187–193.

[16] Macak J.M., Tsuchiya H., Ghicov A., et al. 2007. TiO_2 nanotubes: Self-organized electrochemical formation, properties and applications. Current Opinion in Solid State & Materials Science 11(1–2): 3–18.

[17] Murcia L.S., Hidalgo M.C., Navio J.A., et al. 2011. Novel Bi2 WO6–TiO_2 heterostructures for Rhodamine B degradation under sunlike irradiation. Journal of Hazardous Materials 185(2–3): 1425–1434.

[18] Wang C.X., Yin L.W., Zhang L.Y., et al. 2010. Large Scale Synthesis and Gas-Sensing Properties of Anatase TiO_2 Three-Dimensional Hierarchical Nanostructures. Langmuir 26(15): 12841–12848.

[19] Wei X.X, Cui H.T., Zhao L.F., et al. 2013. Hybrid BiOBr–TiO_2 nanocomposites with high visible light photocatalytic activity for water treatment. Journal of Hazardous Materials 263(2): 650–658.

[20] Wang W.Z., Ao L. 2010. A soft chemical synthesis of TiO_2 nanobelts. Materials Letters 64(8): 912–914.

[21] Wang Y.J., Wang Q.S., Zhan X.Y., et al. 2013. Visible light driven type II heterostructures and their enhanced photocatalysis properties: a review. Nanoscale 5(18): 8326–8339.

[22] Yu H.G., Zhu Z.F., Zhou J.H., et al. 2013. Self-assembly and enhanced visible-light-driven photocatalytic activities of Bi2MoO6 by tungsten substitution. Applied Surface Science 265: 424–430.

[23] Zhang H., Fan X.F., Quan X., et al. 2011. Graphene Sheets Grafted Ag@AgCl Hybrid with Enhanced Plasmonic Photocatalytic Activity under Visible Light. Environmental Science & Technology 45(13): 5731–5736.

Material Science and Environmental Engineering – Chen (Ed.)
 ISBN 978-1-138-02938-5

Biosynthesis of cadmium sulfide nanoparticles using *Desulfovibrio desulfuricans*

J. Gong
Department of Environmental Economics, Shanxi University of Finance and Economics, Taiyuan, P.R. China

X.M. Song & Y. Gao
College of Chemical and Environment Engineering, North University of China, Taiyuan, P.R. China

S.Y. Gong
The First Clinical Medical College, Shanxi Medical University, Taiyuan, P.R. China

Y.F. Wang & J.X. Han
Department of Environmental Economics, Shanxi University of Finance and Economics, Taiyuan, P.R. China

ABSTRACT: CdS nanoparticles were synthesized using *Desulfovibrio desulfuricans* at room temperature and atmospheric pressure in a single-step process. The kinetics of the biosynthesis of cadmium sulfide and the cellular distribution of cadmium were also studied. The microstructure of the CdS nanoparticles was characterized by means of SEM, HRTEM, XRD and EDS. The particles are essentially granular, and the particles size is in the range of approximately 8–10 nm. The transformation rate of CdS nanoparticles was 94.6% after 96 h. The mechanism study showed that CdS nanoparticles can be produced inside the cells and released into solution. In the cells, cadmium was most highly concentrated in the membrane, followed by the cell wall concentration and the cytoplasm.

Keywords: cadmium sulfide; nanoparticles; microbial method; *Desulfovibrio desulfurican*

1 INTRODUCTION

Much attention has been placed on the potential applications of CdS nanoparticles due to their small size effect, quantum effect and surface effect (Sanchez Lopez J.C. et al. 1999; Wang Z.H. et al. 2011). Therefore, many physical chemistry methods have been developed in recent decades. CdS nanoparticles can be synthesized by precipitation of substances containing sulfur and cadmium ions under different conditions (Zheng J. et al. 2008; Yang Z.X. et al. 2012; Seoudi R. et al. 2010; Wang X.L. et al. 2010; Ma X.C. et al. 2005). Researchers have used biological systems to synthesize biominerals in recent years owing to some unique advantages of these systems: the synthesis process is controlled by microorganisms instead of an artificial construct, it can be performed at room temperature and normal pressure, and the nanophase materials generated are byproducts of bioremediation. The microorganisms used for the preparation of sulfide nanoparticles in these systems include bacteria, fungi and yeast (S. Rajeshkumar et al. 2014). In an early study, CdS nanoparticles were synthesized in *Candida glabrata* cells (Reese R.N. & Winge D.R. 1988). The fiber structure of CdS nanoparticles can be formed in *E. coli* cells cultured in medium containing $CdCl_2$ and Na_2S (Sweeney R.Y. et al. 2004). CdS nanoparticles can be synthesized intracellularly by the bacteria (Mandal D. et al. 2006; Hongjuan Bai et al. 2009). Yeast was also used for microbial synthesis of semiconductor CdS nanoparticles. Their characterization and their use in the fabrication of an ideal diode were studied (Kowshik M. et al. 2002). The extracellular enzyme secreted by the fungus *fusarium oxysporum* can mediate extracellular synthesis of CdS nanoparticles (Absar Ahmad et al. 2002).

In this paper, we report an investigation of the synthesis of cadmium sulfide nanoparticles by *Desulfovibrio desulfuricans*, which commonly exists in the natural environment under room temperature and atmospheric pressure, using a single step process. The kinetics of the biosynthesis of cadmium sulfide and the cellular distribution of cadmium were also studied by the determination of cadmium ion and cadmium sulfide.

2 EXPERIMENTAL MATERIALS AND METHODS

2.1 *Bacterial strain and cultivation*

Desulfovibrio desulfuricans, which had been isolated from the sewage treatment plant in Taiyuan city, was obtained from the College of Life Science, Shanxi University, Taiyuan, China. Cultures of the strain were prepared by growing the organism aseptically at neutral pH and room temperature in anaerobic culture medium consisting of the following (Gong Jun et al. 2007): 0.5 $g \cdot L^{-1}$ K_2HPO_4; 1.0 $g \cdot L^{-1}$ NH_4Cl; 1.0 $g \cdot L^{-1}$ Na_2SO_4; 0.1$CaCl_2 \cdot 2H_2O$; 2.0 $g \cdot L^{-1}$ $MgSO_4 \cdot 7H_2O$; 0.1 $g \cdot L^{-1}$ ascorbic acid; 3.5 ml of 70% Na-lactate; and 1.0 $g \cdot L^{-1}$ yeast extract.

2.2 *Microbiological synthesis of cadmium sulfide nanoparticles*

After neutralizing and sterilizing 5 L culture medium, 1500 mg $CdCl_2$ dissolved in 100 mL of distilled water was aseptically added to the medium. Then 10% cultures of *Desulfovibrio desulfuricans* at 3×10^8 cells $\cdot$ mL^{-1} were inoculated, and the solution was cultured anaerobically at room temperature for 96 h. After yellow sediments appeared and settled down naturally, the supernatants were discarded, and the products were washed three times with distilled water. The separating products were dried 3 h at 50°C, and cadmium sulfide nanoparticles powders were obtained.

X-Ray Diffraction (XRD) was performed in an X-ray diffractometer (D8 Advance Bruker) using Cu Kα radiation. The diffracted intensities were recorded at θ angles from 0° to 80°. Energy Dispersive Spectrdmeter (Link-Inca) was used for the elemental analysis. Microstructure observations were performed using a high-resolution transmission electron microscope (JEM-2010).

2.3 *The transformation of cadmium during cultivation*

Eight aliquots of 500 mL anaerobic culture mediums of 1.31 mM $CdCl_2$ were sterilized. After inoculation with 10% *Desulfovibrio desulfuricans* cultures, the solutions were incubated at 30°C in the dark under aerobic conditions. Then, samples were taken at different points (0 h, 12 h, 24 h, 36 h, 48 h, 60 h, 72 h, 84 h and 96 h). The bacterial suspensions were centrifuged at 1800× g for 10 min, and the cadmium concentrations in the supernatants and precipitates were measured. Yellow sediments that settled naturally were washed several times with distilled water and dried until a constant weight was obtained. The cadmium concentration was analyzed by atomic absorption spectrophotometry, and the weight of the white sediment of cadmium sulfide was measured using the gravimetric method.

2.4 *The distribution of cadmium in the cells*

Two bacterial suspensions, one without cadmium and one with cadmium were also centrifuged at 1800× g for 10 min, and the supernatants were discarded. The precipitates were fixed and made into embedded sections. Then, microstructure observations were performed using a Transmission Electron Microscope (TEM, HITACHI H-600-2).

The incubation of a 5 L culture of 1.31 mM $CdCl_2$ was performed as previously described. After 48 h cultivation, the cells were harvested by centrifugation (1800× g, 10 min), washed twice with 50 mM Tris buffer containing 2.5 mM EDTA, pH 8.1 and then resuspended in the same buffer. Lysozyme was added to a concentration of 200 $mg \cdot mL^{-1}$ and the cells were incubated for 30 min at 25°C. All subsequent steps were carried out at 0–4°C. Spheroplasts were collected by centrifugation at 15,000 g for 15 min and resuspended in 30 mM Tris buffer containing 3 mM EDTA, pH 8.1. The supernatant obtained was the periplasmic fluid and consisted of a peptidoglycan layer. The spheroplasts were then disrupted by ten 10-s bursts with a Vibronic Ultrasonic processor and centrifuged at 16000 g for 10 min to remove debris and unbroken cells. The resulting supernatant consisted of membranous and cytoplasmic fractions and was centrifuged at 50,000 g for 60 min. The resulting supernatant was the cytoplasmic fraction, and the precipitate was the membranous fraction. Then the solutions from different time pionts were used for the metal analysis.

Twenty milliliters of bacterial suspension was centrifuged at 1800× g for 10 min, and the supernatants were discarded. The precipitates that had been fixed with glutaraldehyde fixation liquid were incubated at 4°C for 1 h and then centrifuged at 1800× g for 10 min. The precipitate was resuspended in deionized water, and the solution was treated with five 10-s ultrasonic bursts with the Vibronic Ultrasonic processor. The prepared sample was used for observation under a High-Resolution Transmission Electron Microscope (HRTEM, JEM-2010), and an Energy Dispersive Spectrometer (Link-Inca) was used for the elemental analysis.

3 RESULTS AND DISCUSSION

3.1 *Preparation and characterization of nanoparticles CdS*

The X-ray diffraction patterns of the sample cultured for 96 h are shown in Figure 1. Three peaks can be indexed as cubic CdS, (1 1 1), (2 2 0), and (3 1 1) faces, respectively, by comparison with the

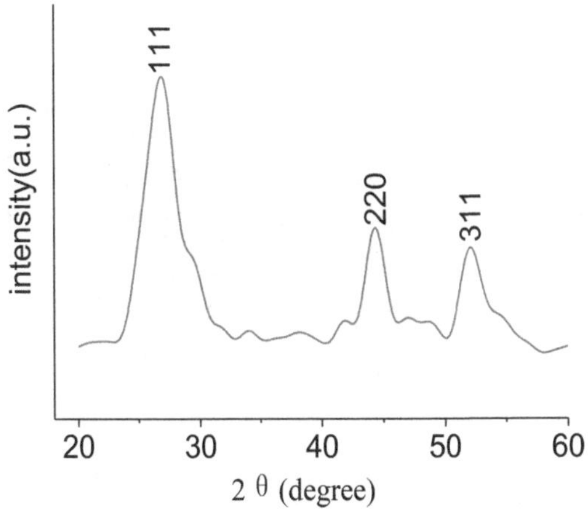

Figure 1. X-ray diffraction pattern of the particles synthesized.

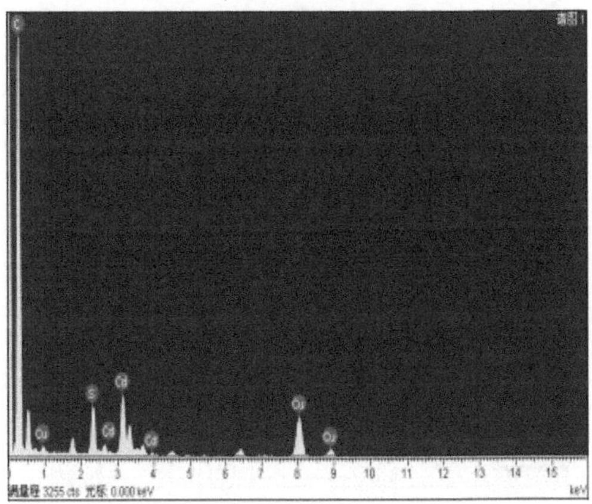

Figure 2. Energy dispersive spectrometer graph of the particles synthesized.

data from the JCPDS file. It is clear from the figure that the CdS particle is highly crystallized and is of high purity based on the lack of additional peaks. All of the diffraction peaks can be indexed to the cubic crystals. The EDX pattern shows that the CdS nanoparticles are composed of the elements Cd and S. SEM and HRTEM images of CdS particles are showed in Figures 3 and 4, respectively. The particles are essentially granular, and the particles size is in the approximate range of 8–10 nm.

3.2 *The transformation of cadmium during cultivation*

The content curve of cadmium in solution in Figure 5 shows that the cadmium content in the solution declines rapidly within 60 h and

Figure 3. SEM pattern of the particles synthesized.

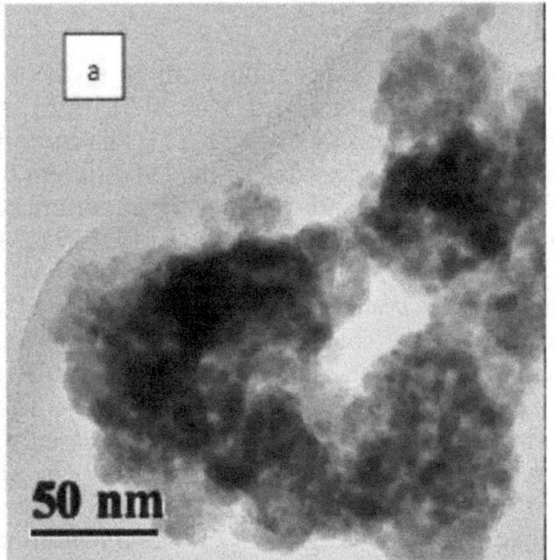

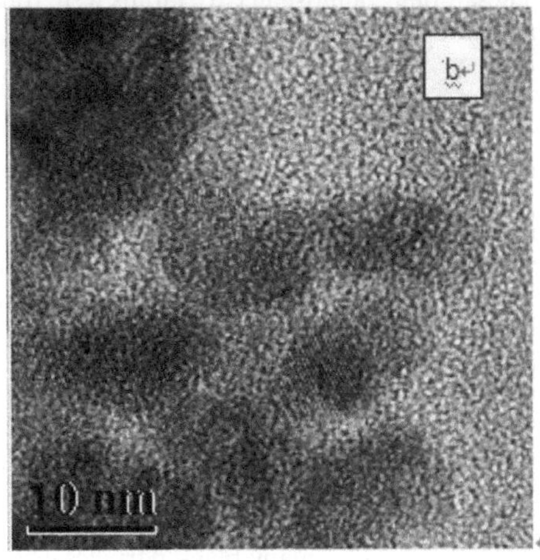

Figure 4. HRTEM micrographs of prepared CdS particles. (a) HRTEM image at low amplificatory times, (b) HRTEM image at high amplificatory times.

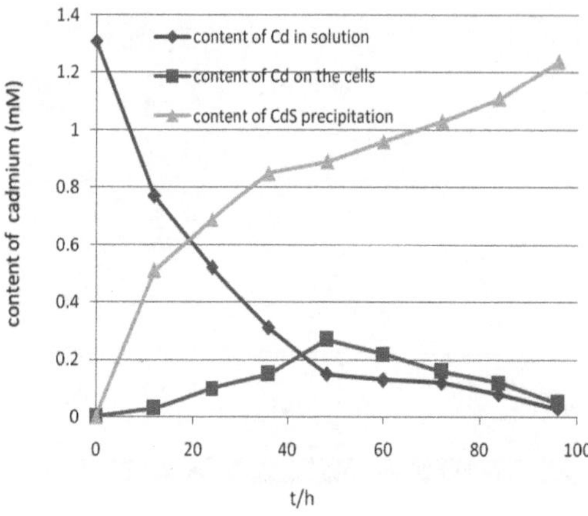

Figure 5. The transformation of cadmium during cultivation.

then stabilizes. The cadmium content in the solution decreased from 1.31 mM to 0.015 mM, and the transformation rate was 98.5%. The cadmium content in the cells was determined at different time points. The results indicate that the cadmium concent in the cells gradually increased to 0.15 mM at 48 h and then began to decrease. This is because the *Desulfovibrio desulfuricans* cells transformed the cadmium in the solution into CdS nanoparticles. The transformation rate of nanoparticles CdS (the molar concentration of CdS products × 100%/ the molar concentration of $CdCl_2$) increased with increasing culture time and reached 94.6% after 96 h (Fig. 5). Therefore, CdS nanoparticles were produced during the growth of *Desulfovibrio desulfuricans*, and considering the amount of runoff during sampling, almost all Cd^{2+} atoms were transformed into CdS after 96 h.

3.3 *The distribution of cadmium in cells*

Desulfovibrio desulfuricans can utilize sulfate as a terminal electron acceptor when performing anaerobic oxidation of organic substrates (H.D. Peck et al. 1982). As a consequence, they produce and accumulate large amounts of S^{2-} by dissimilatory sulfate reduction. Therefore, S^{2-} produced by the strain can be used as a sulfur source for the production of CdS nanoparticles. TEM analysis of the cell cultures with and without cadmium was performed to determine the distribution of cadmium in the cells. The HRTEM micrographs of the strain cultured in the culture medium containing Cd^{2+} showed that many black particles existed in the cells in comparison to the strain cultured in the culture medium without Cd^{2+}. HRTEM and energy dispersive spectrometer analysis were also

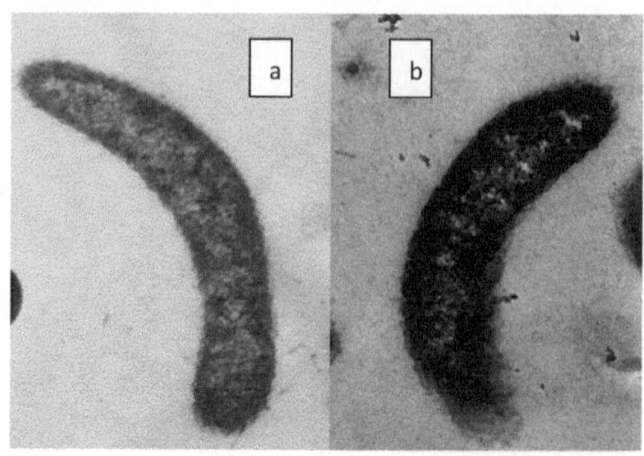

Figure 6. TEM micrographs of *Desulfovibrio desulfuricans*: (a) cell structure of the strain cultured in culture medium without Cd^{2+}, (b) cell structure of the strain cultured in culture medium containing Cd^{2+}.

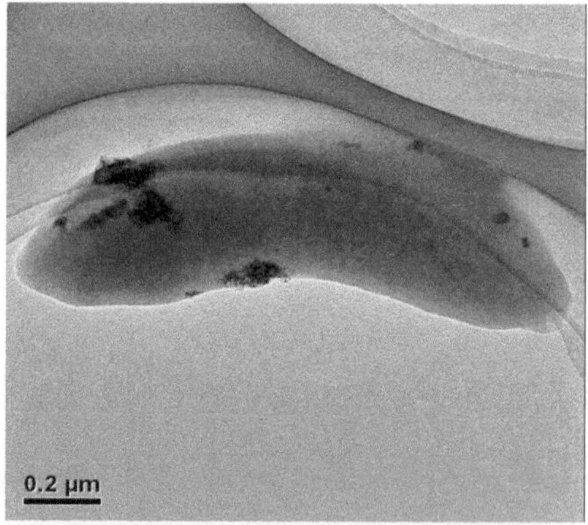

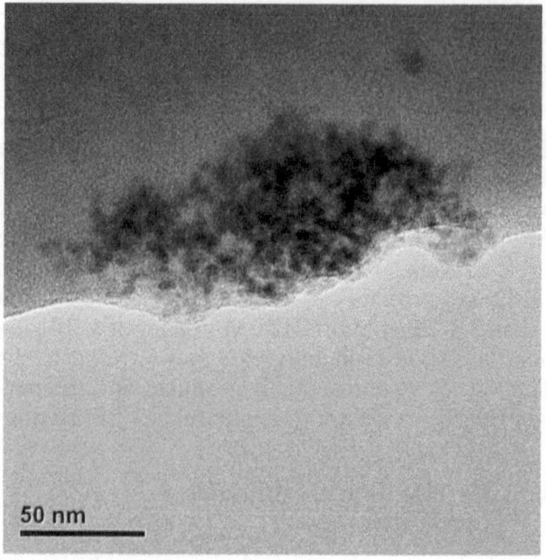

Figure 7. HRTEM micrographs of *Desulfovibrio desulfuricans*.

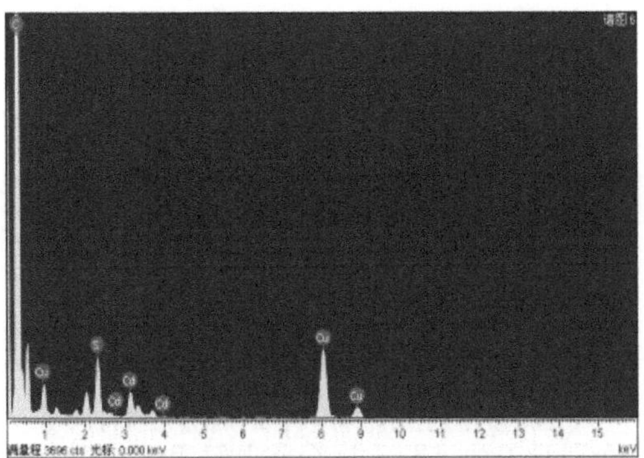

Figure 8. Energy dispersive spectrometer image of the particles in cells.

Table 1. Distribution and uptake of Cd based on subcellular fractionation of *Desulfovibrio desulfuricans*.

Subcellular location	Cadmium content (μg/g wet cells)	Percentage (%)
Cell wall	4.92	17.8
Membrane	18.4	66.7
Cytoplasm	4.27	15.5

performed to verify the elemental composition of the particles in the cells. Figures 7 and 8 clearly show that the particles in the cells are mainly composed of Cd and S and that size of the particles is in the nanoscale. Therefore, it can be deduced that CdS nanoparticles can be produced inside *Desulfovibrio desulfuricans* cells.

In develop a better understanding of the distribution of cadmium in the cells; the concentrations of cadmium in different subcellular locations were determined. The cadmium contents of the cell wall, membrane and cytoplasm were 4.92 μg/g wet cells, 18.4 μg/g wet cells and 4.27 μg/g wet cells, respectively. In the cells, cadmium was most highly concentrated in the membrane, followed by the cell wall concentration and then the cytoplasm. The maximum proportion of the cadmium content in the membrane was 66.7%. Various adsorption proteins exist in membranes and participate in solute transportation; therefore, cadmium accumulates in membranes.

4 CONCLUSIONS

The microbial method for the synthesis of CdS nanoparticles uses the live microorganism *Desulfovibrio desulfuricans* and can automatically-control the size of the CdS nanoparticles without using strict external conditions. The formation of the homogeneous phase of CdS nanoparticles was confirmed by XRD analysis and HRTEM study. Considering the amount of runoff during sampling after 96 h, almost all Cd^{2+} atoms were transformed into CdS nanoparticles. The mechanistic study showed that CdS nanoparticles can be produced inside cells and released into solution and that the cadmium content in cells is the highest in the membrane. Further understanding of the mechanism of CdS nanoparticles synthesis requires further investigations.

ACKNOWLEDGEMENTS

We acknowledge the services provided by the Sophisticated Analytical Instrumentation Facility, Institute of Coal Chemistry, CAS, Taiyuan, China and Shanxi Medical University. We received financial support from the National Natural Science Foundation of China (31200049) and the Natural Science Foundation for Young Scientists of Shanxi Province, China (2012011009-1).

REFERENCES

[1] Sanchez Lopez J.C., Reddy E.P., Rojas T.C., et al. 1999. Preparation and characterization of CdS and ZnS nanosized particles obtained by the inert gas evaporation method. *Nanostructured Materials* 12:459–462.

[2] Wang Z.H., Pan L., Wang L.L., et al. 2011. Urchin-like CdS microspheres self-assembled from CdS nanorods and their photocatalytic properties. *Solid State Sciences* 13(5):970–975.

[3] Zheng J., Song X., Chen N., et al. 2008. Highly symmetrical cds tetrahedral nanocrystals prepared by low-temperature chemical vapor deposition using polysulfide as the sulfur source. *Crystal Growth and Design* 8(5):1760–1765.

[4] Yang Z.X., Zhong W., Zhang P., et al. 2012. Controllable synthesis, characterization and photoluminescence properties of morphplogy-tunable CdS nanomaterials generated in thermal evaporation processes. *Appl Surf Sci* 258(19):7343–7347.

[5] Seoudi R., Shabaka A., Eisa W.H., et al. 2010. Effect of the prepared temperature on the size of CdS and ZnS nanoparticles. *Physica* 405:919–924.

[6] Wang X.L., Feng Z.C., Fan D.Y., et al. 2010. Shape-Controlled Synthesis of CdS Nanostructures via a Solvothermal Method. *Crystal Growth Design* 10:5312–5318.

[7] Ma X.C., Xu F., Liu Y.K., Liu X.M., et al. 2005. Double-dentate solvent-directed growth of multi-armed CdS nanorod-based semiconductors. *Mater Res Bull* 40:2180–2188.

[8] S. Rajeshkumar, M. Ponnanikajamideen, C. Malarkodi, M. Malini, G. Annadurai. 2014. Microbe-mediated synthesis of antimicrobial semiconductor nanoparticles by marine bacteria. *J Nanostruct Chem* 4:96–102.

[9] Reese R.N., Winge D.R. 1988. Sulfide stabilization of the cadmium–γ-glutamyl peptide complex of Schizosacch-aromyces pombe. *J Biol Chem* 263:12832–12835.

[10] Sweeney R.Y., Mao C., Gao X., Burt J.L., Belcher A.M., Georgiou G., Iverson B.L. 2004. Bacterial biosynthesis of cadmium sulfide nanocrystals. *Chem Biol* 11:1553–1559.

[11] Mandal D., Bolander M.E., Mukhopadhyay D., et al. 2006. The use of microorganisms for the formation of metal nanoparticles and their application. *Appl Microb Biotech* 69(5):485–492.

[12] Hongjuan Bai, Zhaoming Zhang, Yu Guo, Wanli Jia. 2009. Biological Synthesis of Size-Controlled Cadmium Sulfide Nanoparticles Using Immobilized Rhodobacter sphaeroides. *Nanoscale Res Lett* 4:717–723.

[13] Kowshik M., Deshmukh N., Vogel W., et al. 2002. Microbial synthesis of semiconductor CdS nanoparticles, their characterization, and their use in the fabrication of an ideal diode. *Biotechnol Bioeng* 78(5):583–585.

[14] Absar Ahmad, Priyabrata Mukherjee, Deendayal Mandal, et al. 2002. Enzyme mediated extracellular synthesis of CdS naoparticles by the fungus, fusarium oxysporum. *J. Am. Chem. Soc* 124(41):12108–12109.

[15] Gong Jun, Zhang ZhaoMing, Bai HongJuan, Yang GuanE. 2007. Microbiological synthesis of nanophase PbS by Desulfotomaculum sp. *Science in China Series E: Technological Sciences* 50(1):1–6.

[16] H.D. Peck, J.R., J. Le Gall. 1982. Biochemistry of Dissimilatory Sulphate Reduction. *Phil. Trans. R. Soc. Lond. B* 298:443–466.

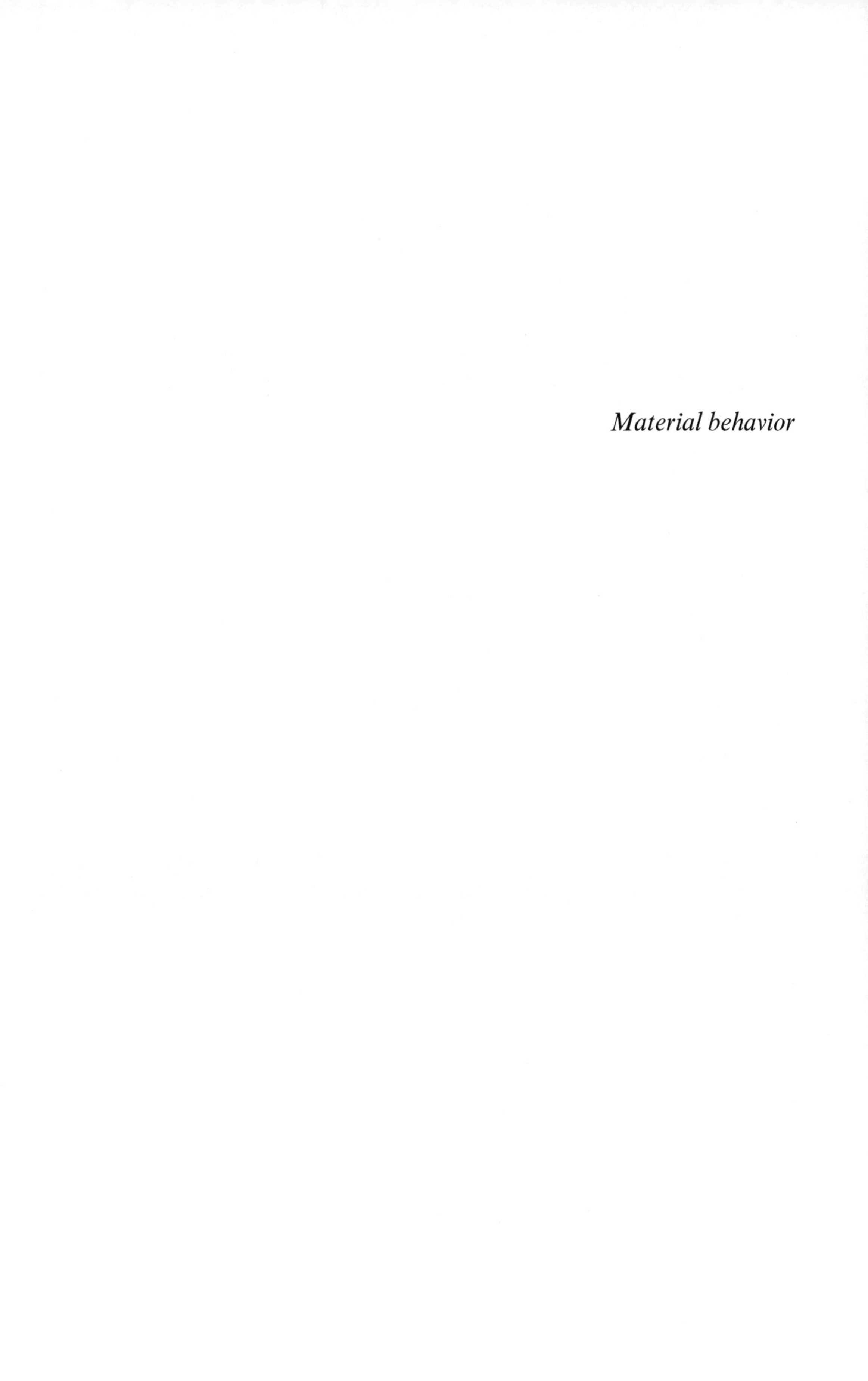

Material behavior

Material Science and Environmental Engineering – Chen (Ed.)
 ISBN 978-1-138-02938-5

Structural dynamic analysis on idealized joint model in portal crane

D.Q. Zhou, X.H. Cao & P. Yu
Wuhan University of Technology, Wuhan, China

ABSTRACT: In order to properly evaluate the errors resulted from the idealized joint constraint model which is used to simulate the hinge function of joint bearing; this paper presents an effort to quantitate the errors through observation of the changes in case the idealized joint constraint model is replaced by a type of more accurate revolute joint model. The dynamic analysis of the luffing process of a gantry crane four-bar linkage arm system is taken as the example to demonstrate the application of the developed models for comparison. The developed models are different from the idealized joint constraint model. The effects of joint clearance, contact deformation, and friction are comprehensively taken into account in the new model and the system dynamic responses with these two joint models in the same typical working conditions were obtained and compared. The comparison clearly reveals some limits of the idealized joint constraint model, which may provide essential assistance for engineering machinery designers.

Keywords: idealized joint; dynamic characteristics; limits; portal crane

1 INTRODUCTION

In the process of structural dynamics simulation analysis, the metal structure of engineering machinery is generally considered as a multibody system which is composed of multiple rigid body or flexible body, linked by living hinge. In practical application, living hinge may be a sliding bearing or a rolling bearing. For the convenience of analysis, the traditional way is to consider the bearing as an ideal plane hinge. But, the link of sliding bearing, which is close to ideal flat hinge, also exists clearance, contact deformation and friction, which can lead to the deviations between the simulation results and actual stress and running state of the structure [1, 2, 3]. In practical, the conditions of the rolling bearings is similar to that, and the ignored factors such as clearance, contact deformation, and friction is very complex. Therefore, the dynamic simulation analysis result of ideal hinge constraint model is more difficult to judge.

The purpose of this paper is to reveal the possible deviation in dynamic simulation calculation of ideal hinge constraint model, using a typical the four-bar combination jib of a portal crane as an example. The method is to use a more accurate sliding bearing model with clearance instead of the ideal hinge constraint model to analyze articulation joint in the system. Through comparing the calculation results, the limitations of ideal hinge constraint model are demonstrated. In the simulation of sliding bearing with clearance, the force and movement form does not conform to the reality, although the actual structure uses rolling bearing. But the two bearings both meet the Hertz contact collision dynamics, and have the same modeling accuracy, thus, the calculation deviation between the sliding bearing model with clearance and ideal hinge constraint model helps to infer the calculation deviation aroused by ideal hinge constraint model. Then the limits of the Idealized Joint Constraint Model in Structural Dynamic Analysis are verified.

The reason to apply above calculation method is that it is easier to establish sliding bearing model with clearance than a commonly used spherical roller bearing at present. In recent years, in order to analyze how hinge gap effects dynamic performance and motion precision of the mechanical multibody system better [4, 5] domestic and international numerous scholars put forward many kinds of modeling method of sliding bearing with clearance and they have been successfully applied to high speed and high precision space agency and the defense automation equipment, etc. Such as Peng-fei Guo [6] analyzed the influence of clearance error on output based on collision Dubowsky hinge model and the Monte Carlo probabilistic method. Zheng-feng Bai [7] carried on a dynamic simulation of joint with clearance based on ADAMS for aerospace equipment, such as solar panels and soft lunar lander. Lankarani [8, 9] put forward a nonlinear spring damping contact collision force model, and analyzed dynamic characteristics of mechanism with clearance. Taking factors, such as wear and lubrication into consideration, Flores [10] analyzed dynamic characteristics of slider-crank and other mechanisms with clearance through numerical simulation based on nonlinear spring damping contact collision force and the friction model.

The paper focused on portal crane arm frame system, and the suitable sliding bearings model with clearance is established according to its mechanical and movement characteristics. Rigid-flexible coupling model of arm frame system is built based on the ADAMS, and dynamic characteristic curve of arm frame system is simulated under ranged conditions. Comparing with the corresponding simulation results of ideal hinge constraint model, the limits of the Idealized Joint Constraint Model in Structural Dynamic Analysis provides the evidence of its applicability.

2 MODELING OF CLEARANCE REVOLUTION JOINT

2.1 *Vector model of motion pair with clearance*

Ideally, the center distance between shaft and shaft sleeve is zero, and the existence of the clearance makes the center distance is not zero, resulting in a collision and friction. Model of motion pair with clearance generally contains three state motion model, two state motion models and continuous contact model [11, 12]. Portal crane belongs to the continuous contact model. Because the boom system has low peed and four connecting rods is the double rocker mechanism. The schematic figure of shaft and shaft sleeve of is shown in Figure 1.

The difference between radius of shaft sleeve and shaft is clearance $c = R_1 - R_2$; Eccentric vector of shaft sleeve to axle is $e = \sqrt{e_x^2 + e_y^2}$. In the process of contact between sleeve and the shaft will cause deformation, and the penetration depth of the contact point is $\delta = e - c$. When $\delta < 0$, the shaft sleeve is in a state of freedom of movement, not contacting with shaft. When $\delta = 0$, the shaft sleeve contacts shaft or starts shedding. When $\delta > 0$, the shaft sleeve contacts shaft and elastic deformation occurs.

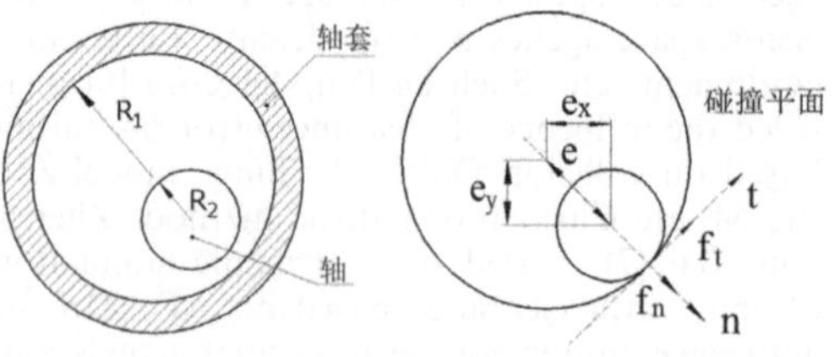

Figure 1. The schematic figure of clearance revolution joint.

2.2 *Contact collision force model of the clearance*

According to Hertz contact theory, it is assumed that damping is the source of the energy loss in the process of collision in mechanism with clearance. The collision between collision hinge shaft and bearing is low-speed collision, and the contact force formula is [13]:

$$F_n = \begin{cases} K_n\delta^n + C\dot{\delta} & \delta \geq 0 \\ 0 & \delta < 0 \end{cases} \tag{1}$$

In this formula:

K_n is contact stiffness coefficient of collision body;
C is damping coefficient in the process of collision;
δ^n is the elastic deformation in the process of collision;
$\dot{\delta}$ is relative impact velocity.

In order to improve the calculation accuracy and make the model more applicable to the analysis of overloading and the big gap of the portal crane, the contact stiffness coefficient in formula (1) was improved according to the method in literature [4].

The improved nonlinear stiffness coefficient [14, 15]:

$$K_n = \frac{1}{8}\pi E^* \sqrt{\frac{2\delta(3c + 2\delta)^2}{(c + \delta)^3}} \tag{2}$$

Damping coefficient is:

$$C = \frac{3K_n(1 - c_e^{\,2})e^{2(1-c_e)}\delta^n}{4\dot{\delta}^{(-)}} \tag{3}$$

$\dot{\delta}^{(-)}$ in formula (3) is the initial relative velocity for the impact point. And the recovery coefficient is c_e. The nonlinear contact coefficient is n, taking $n = 1.5$.

2.3 *The friction model of joint clearance*

Considering the influence of the dry friction and ignoring lubrication, modified Coulomb friction model is used to establish friction between rotating hinge mechanism with clearance and bearing. Then accurate friction value can be concluded [4]:

$$\mu(v_\tau)=\begin{cases}-\mu_d sign(v_\tau) & |v_\tau|>v_d \\ -\left\{\mu_s+(\mu_d-\mu_s)\left(\frac{|v_\tau|-v_s}{v_d-v_s}\right)^2\left[3-2\left(\frac{|v_\tau|-v_s}{v_d-v_s}\right)\right]\right\}sign(v_\tau) & v_s\le|v_\tau|\le v_d \\ \mu_s-2\mu_s\left(\frac{v_\tau+v_s}{2v_s}\right)^2\left(3-\frac{v_t+v_s}{v_s}\right) & |v_\tau|<v_s\end{cases} \tag{4}$$

In this formula, v_τ is relative sliding velocity of the collision point between shaft and bearing; μ_d is sliding friction coefficient; μ_s is static friction coefficient; v_s is static friction critical speed; v_d is the maximum friction critical speed.

3 RIGID-FLEXIBLE COUPLING MODEL OF PORTAL CRANE BOOM SYSTEM BASED ON ADAMS

3.1 *Rigid-flexible coupling model of portal crane boom system based on ADAMS*

On the basis of MQ2536 portal crane drawings from a port, the rigid arm frame system model is established. Using Rigid to flex function in ADAMS, rigid body is replaced with the flexible body, and the modal neutral file is generated by ANSYS. After replacement, the conditions are deleted, which do not conform to the actual deformation by observing the modal deformation condition mode of each component of.

3.2 *The definition of sliding bearing model contact with clearance in ADMAS*

Four-bar combination boom system contains four hinge points. The boom, the turntable hinged point 1, large rod, and a-frame hinge point 2 are under larger impact because they are responsible for the luffing mechanism and link to the bas. And the deviating from the ideal state of hinge constraint model is most. So, the analysis focuses on the influence of the hinge point 1 and 2.

In ADAMS, the degrees of freedom of the constraint model are added, and CONTACT is to simulate actual collision of hinge point 1 and 2. In order to get accurate simulation results, according to the formula (1) to (4), CNFSUB subroutine is written by C language to modify the CONTACT force function. C program is compiled and the target files, DLL and LIB library files is generated in order. Then the link of ADAMS/View and the objective function is linked. The library file z is loaded in the Solver Setting, and subroutine is called in the CONTACT.

4 DYNAMIC SIMULATION ANALYSIS ON TWO HINGE MODELS

4.1 *The simulation condition assumptions and analysis*

The variable amplitude conditions of portal crane are defined according to Table 1. The application of Rotational Joint Motion in ADAMS drives motor. After editing Motion, motor accelerates slowly from static to a constant speed rotation, then slows down and then starts reversely from the original position. The whole operation process, which contains load running and back in situ after acceleration and down, is referred to a cycle, as shown in Figure 2.

4.2 *The dynamic simulation results and analysis on two hinged models*

4.2.1 *The maximum stress of main structure*

Figure 3 shows the process of stress time courses in variable amplitude of corresponding node in flange plate in structure section of central boom. According to the figure, it can be seen that the

Table 1. Portal crane d luffing system performance parameters.

Lifting weight/t	Max radius/m	Min radius/m	Average velocity/ mmin(−1)
25	36	12	50

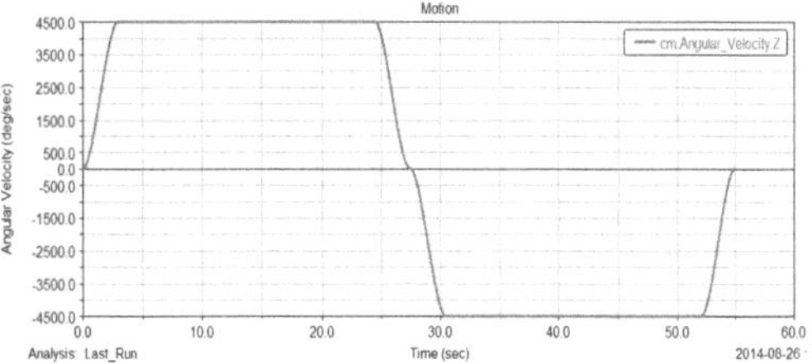

Figure 2. The rotating angular velocity curve of the motor in one cycle.

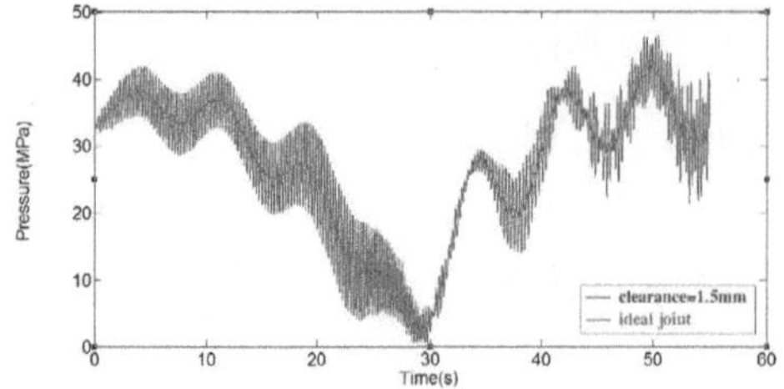

Figure 3. Stress curve of boom in the process of variable amplitude.

stress changes of ideal smooth hinge model has no fluctuation during the luffing, but to hinge model with clearance, although the trend has not changed, but some part has high frequency fluctuations. The difference between those maximum stress is 10.1% and the difference between their stress amplitude is 15.3%. Similar results can be concluded after taking the structure simulation for other position. Therefore, there is more than 10% deviation if the stress of main structure is calculated by ideal hinge constraint model.

4.2.2 *The time history of displacement, velocity and acceleration of forward endpoint in truck bridge*

Figure 4 to 5 are time history of front horizontal displacement, horizontal velocity and horizontal acceleration of two models under the bridge of truck bridge. As you can see, the curve of displacement and velocity were similar, but the acceleration under the clearance between hinge models has an obvious local jitter. Therefore, there will be a very big deviation under the usage of ideal hinge model to calculate the acceleration characteristics of the predicted structure.

4.2.3 *Changes of local stress around hinge point*

According to Figure 6, the stress curve near the pivotal point under two kinds of model structure infers that in the ideal hinge model stress changes smoothly. To the model with clearance, the general change trend is similar, but with high frequency vibration. From the numerical as shown in Table 2, the difference between two maximum stresses is up to 56.9% during the whole process. And it's far greater than the deviation about 10% to 20% of hinged point stress in main body structure, which means the ideal hinge constraint model will complete failure when analysis is focused on local stress near the hinged point.

4.3 *Discuss*

By comparing ideal hinge constraint model with the rotating hinge with clearance which considering more comprehensive factors, it can be found that the main cause of the simulation deviation results between the former from the latter are:

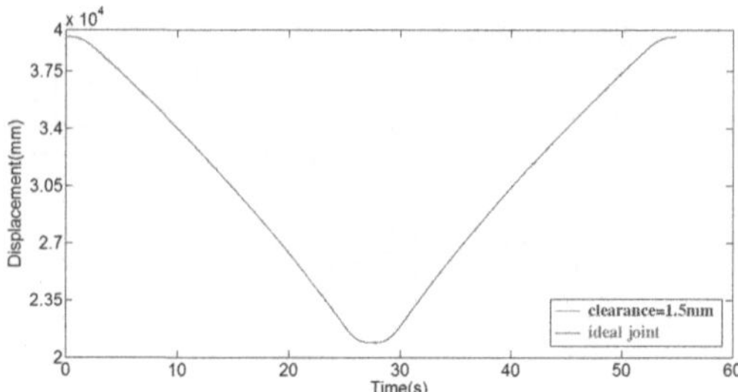

Figure 4. The curve of horizontal displacement of forward endpoint in truck bridge.

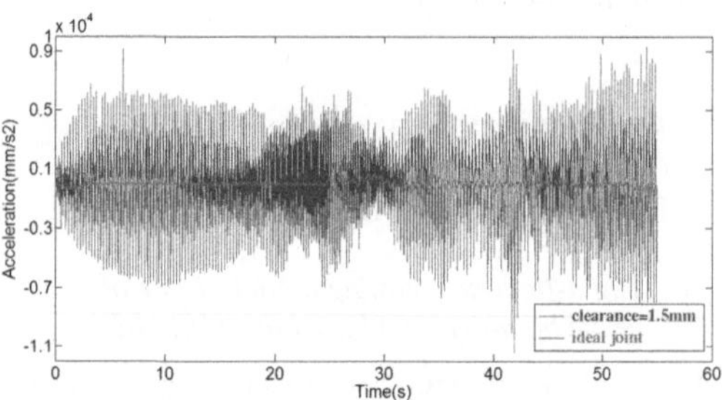

Figure 5. The curve of horizontal acceleration of forward endpoint in truck bridge.

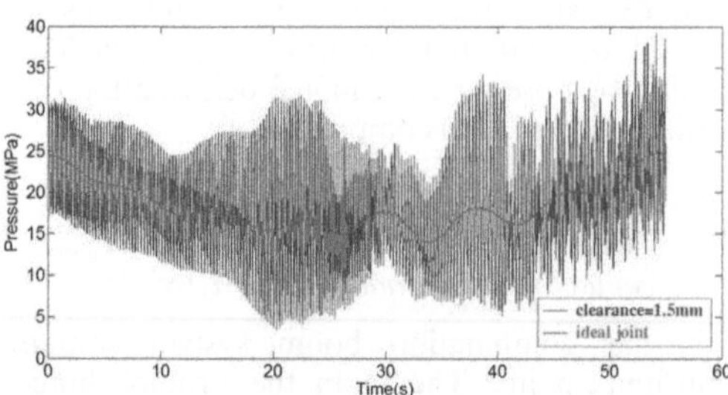

Figure 6. Local stress curve of hinged point in big rod.

Table 2. The local maximum stress value near hinged point.

Hinged point in big rod	Maximum stress/MPa
Ideal hinge	24.94304
Clearance = 1.5 mm	39.13851
Range of variation	56.90%

1. Under ideal hinge constraint model, the opposite force of supports works on one point on

projected area. Meanwhile, according to the principle of contact, the opposite force of supports works on an arc to the hinge model with clearance. Obviously, the latter is closer to the actual;

2. The clearance causes the structure deformation, and thus, produces additional stress and deformation.
3. The secondary factors such as friction. The stress deviation away from the hinged point on the structure is mainly determined by (2) and (3). And the stress near the hinged point is determined by (1) with the largest deviation.

Because of the combination of (1), (2), (3), with the deviation of stress, there will be a large transient wave, and strain, velocity and acceleration can also fluctuate accordingly. But displacement response is almost unaffected because the displacement change caused by clearance and strain can be ignored to the size of the structure.

5 CONCLUSION

Based on comparison of the simulation results of various kinetic parameters from four connecting rod combination portal crane arm frame system under the ideal hinge model and hinge model with clearance, it is shown that ideal hinge model has the following limitations:

1. For the local stress calculation of hinged point attachment, it is invalid to adopt simplified ideal hinge constraint model. More precise and complex hinged model should be established. Or, it should be calculated by traditional method based on statics analysis.
2. When using a simplified ideal hinge constraint model to calculate stress and strain and the corresponding properties such as strength, stiffness, and fatigue life to the main structure far from hinged point, the error is around 15%. Though not precise enough, it still can be used to a certain extent.
3. When evaluating the dynamic characteristics of the mechanism kinematics, if displacement and velocity are cared about, they will achieve satisfactory results by adopting simplified ideal hinge model. But when acceleration is considerate, hinge clearance should be considered to establish accurate model for analysis.

ACKNOWLEDGMENTS

This paper is supported by the basic research project of the Ministry of Communications (2015329811290).

REFERENCES

[1] Megahed, S.M., Haroun, A.F. Analysis of the dynamic behavioral performance of mechanical systems with multibody clearance joints [J]. J. Compute. Nonlinear Dyn. 7, 011002 (2012).

[2] Muvengei, O., Kihiu, J., Ikua, B. Numerical study of parametric effects on the dynamic response of planar multibody systems with differently located frictionless revolute clearance joints [J]. Mech. Mach. Theory 53, 30–49 (2012).

[3] Erkaya, S., Uzmay, I. Experimental investigation of joint clearance effects on the dynamics of a slider-crank mechanism. Multibody Syst. Dyn. 24, 81–102 (2010).

[4] Schwab A.L. A comparison of revolute joint clearance models in the dynamic analysis of rigid and elastic mechanical systems [J]. Mechanical and Machine Theory, 2002, 37(9):895–913.

[5] Muvengei O., Kihiu J., Ikua B. Dynamic analysis of planar rigid-body mechanical systems with two-clearance revolute joints [J]. Original Paper, 2013, Nonlinear Dyn (2013) 73:259–273.

[6] Guo Pengfei, Yan shaoze. MonteCarlo simulation of motion error for four-bar mechanism with clearance[J]. Journal of Tsinghua University, 2007, 47(11):1989–1993.

[7] Bai Zhengfeng. Research on Mechanism Dynamic Characteristics with joint clearance [D]. Harbin: Harbin Institute of Technology, 2011.

[8] Lankarani H.M., Nikravesh P.E. A Contact Force Model With Hysteresis Damping for Impact Analysis of Multibody Systems [J]. Journal of Mechanical Design.1990, 112:369–376.

[9] Koshy C., PFlores, HMLankarani. Study of the effect of contact force model on the dynamic response of mechanical systems with dry clearance joints: computational and experimental approaches [J]. Original Paper, 2013.01, Nonlinear Dyn (2013) 73:325–338.

[10] Flores P. Dynamic Analysis of Mechanical Systems with Imperfect Kinematic Joints [D]. Portugal: Universidade Do Minho Ph.D thesis, 2004 Computers and Structures. 2004, 82:1359–1369.

[11] Guo Xinglin, Zhao Zikun. Flexible dynamic analysis of crank-rocker mechanism with clearance [J]. Mechanical degree. 2010, 32(6):905–909.

[12] Luo Xiaoming. Joint friction analysis of multibody system [D]. 2011, 07.

[13] Koshy C., PFlores, HMLankarani. Study of the effect of contact force model on the dynamic response of mechanical systems with dry clearance joints: computational and experimental approaches [J]. Original Paper, 2013.01, Nonlinear Dyn (2013) 73:325–338.

[14] Liu C.S., Zhang K., Yang R. The FEM analysis and approximate model for cylindrical joints with clearances [J]. Mechanism and Machine Theory. 2007, 42:183–197.

[15] Pereira, C.M., Ramalho, A.L., Ambrósio, J.A. A critical overview of internal and external cylinder contact force models [J]. Nonlinear Dyn. 63, 681–697 (2011).

Material Science and Environmental Engineering – Chen (Ed.)

Stress concentration factors for the multi-planar tubular Y-joints subjected to axial loadings

B. Wang
Powerchina Huadong Engineering Corporation Limited, Hangzhou, China

N. Li & X. Li
State Key Laboratory of Coastal and Offshore Engineering, Dalian University of Technology, Dalian, China

Y. Li
China United Engineering Corporation, Hangzhou, China

ABSTRACT: The support structure for an offshore wind turbine is subjected to combined hydrodynamic loads and aerodynamic loads. The tubular joints are the weakest component leading to fatigue failure of the whole structure. Considering the load combinations under fatigue limit state, two finite element models of offshore wind turbines composed of Tripod and Pentapod substructures are analyzed, respectively. Then, the load types subjected to axial loadings of three-planar and five-planar tubular Y-joints are determined, respectively. The finite element models of three-planar and five-planar tubular Y-joints are established and used to calculate hot spot stresses, respectively. The stress concentration factors along the weld of the three-planar and the five-planar tubular Y-joints under the axial forces are obtained. The effects of geometrical parameters on SCFs are studied. Moreover, the numerical SCF results are compared with the empirical ones from DNV-RP-C203 for simple tubular Y joints.

Keywords: Y-joints; axial loading; stress concentration factor; finite element analysis; offshore wind turbine

1 INTRODUCTION

Offshore wind power development is rapidly growing driven the growing demand for sustainable energy. Fatigue is the significant factor resulting in the failure of support structure of offshore wind turbine subjected to combined aerodynamic and hydrodynamic loadings. Multi-pod piled substructure is widely used in China. The connection between diagonal braces and central column as the major loadbearing components is defined as the tubular joint for joint type multi-planar Y. The weld itself and geometrical discontinuity around tubular joint cause stress concentration which is a detrimental impact on fatigue performance of offshore structures.

The fatigue design is based on use of S-N curves, which are obtained from fatigue tests. In S-N curves, the number of cycles that a tubular joint can sustain before fatigue failure is determined by the hot spot stress range. The stress concentration due to the weld itself is included in the S-N curve to be used, while the stress concentration due to the geometry effect of the actual detail is determined by means of calculation of hot spot stress. Generally, the hot spot stress is calculated from the nominal stress and Stress Concentration Factor (SCF). The SCF has been a issue for tubular structures in the past decades. At present, there are many published parametric SCF equations for tubular joints subjected to different basic loadings. The SCFs for simple tubular joints are listed in the fatigue design codes such as API [1] and DNV-RP-C203 [2]. Hot spot stresses at failure-critical locations for 4 different tubular joints (DK, DKT, X-type) are derived, and the effects of planar and non-planar braces are considered by Dong [3]. Woghiren [4] studied a parametric stress analysis of various configuration of rack plate stiffened multi-planar KK-joint using the finite element method. Considering bending effects, Karamanos [5] computed SCFs in DT-joint. Ahmadi [6] proposed a set of parametric SCFs equations for the three-planar tubular KT-joint under three different axial loading conditions. Ahmadi [7] also established a set of SCF parametric equations for the fatigue design of internally ring-stiffened KT-joint.

Using solid element, the finite element models of three-planar and five-planar tubular Y-joints are established, respectively. The stress concentration factors along the weld of the three-planar and the five-planar tubular Y-joints under the axial loads are obtained. The effects of geometrical parameters on SCFs are studied.

2 BASIC INFORMATION

2.1 *Selection of geometrical parameters*

Typical tripod and pentapod substructures of offshore wind turbine are shown in Figure 1. For tripod substructure, the joint connected with three diagonal braces and a central column forms a three-planar tubular joint with the angle of 120° among the diagonal braces by horizontal projection. For pentapod substructure, the joint linked with five diagonal braces and a central column forms a five-planar tubular joint with the angle of 72° among the diagonal braces by horizontal projection. The numbering of the diagonal braces for three-planar and five-planar tubular Y-joints are displayed in Figure 2.

From the extensive studies by the researchers, the geometrical parameters that affect the SCFs of tubular joints include the ratio of chord length to chord radius, α, the ratio of brace diameter to chord diameter, β, the ratio of chord radius to chord wall thickness, γ, the ratio of brace wall thickness to chord wall thickness, τ, the angle between the chord axis and brace axis, θ, and the angle among diagonal braces, ϕ.

Based on substructure designs of practical projects in China, the ranges of geometrical parameters are determined. Then, the values of geometrical parameters are selected and listed in Table 1. The SCFs of these cases are calculated.

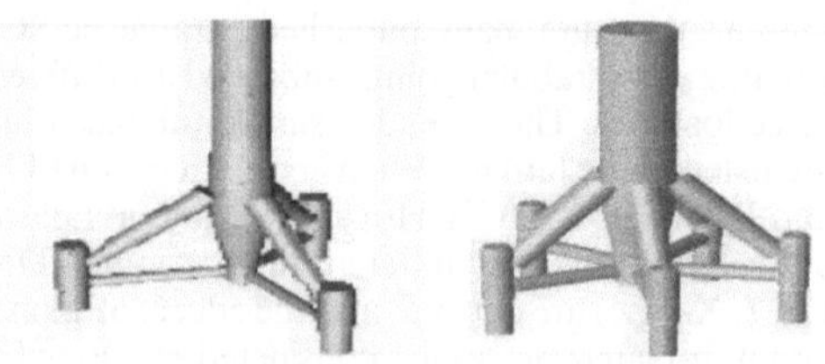

Figure 1. Substructures of offshore wind turbine.

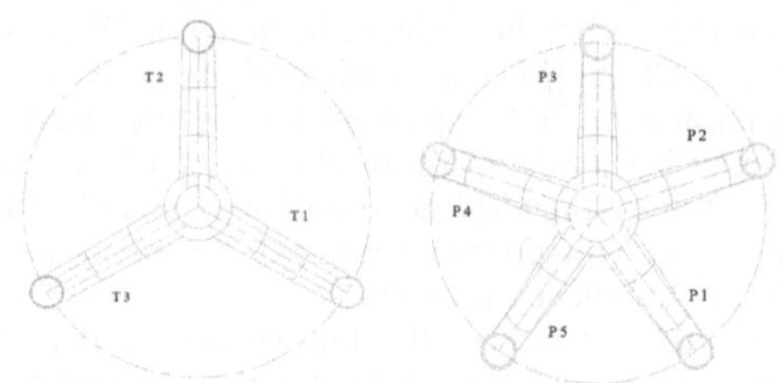

Figure 2. Numbering of diagonal braces.

Table 1. Geometrical parameters of multi-planar tubular Y-joints.

Parameters	Three-planar Y-joint	Five-planar Y-joint
α	12, 13, 14	9, 11, 13
β	0.50, 0.58, 0.66	0.35, 0.38, 0.43
γ	20, 24, 25, 28	30, 33, 37
τ	0.60, 0.70, 0.80	0.55, 0.65, 0.70
θ	40°, 42.5°, 45°	35°, 40°, 45°
ϕ	120°	72°

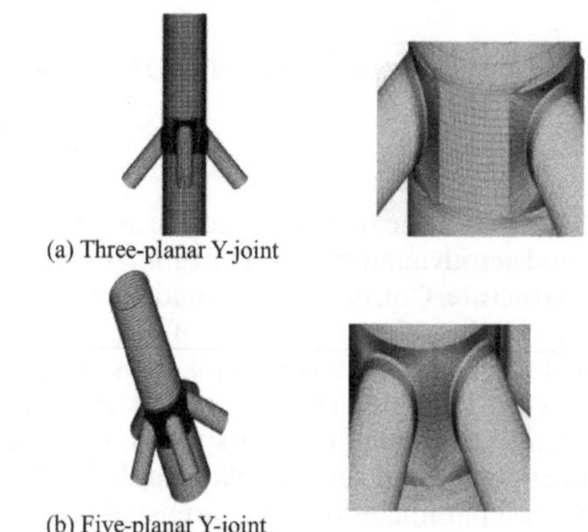

(a) Three-planar Y-joint

(b) Five-planar Y-joint

Figure 3. FE model of multi-planar tubular Y-joints.

2.2 *Finite element modeling*

Accurate modeling of the weld profile is one the most critical factors affecting the accuracy of SCF results. In the present study, the welding size along the brace/chord intersection satisfies the AWS D1.1 specification [8].

Using ANSYS, the SOLID95 element type is used in the study to model the chord, brace and the weld profile. This method will produce more accurate and detailed stress distribution near the intersection in comparison with a simple shell analysis [9]. The material is steel with elastic modulus, E = 206 GPa, Poisson's ratio $\nu = 0.3$.

To guarantee the mesh quality, a sub-zone mesh generation method is used during the FE modeling. In this method, the entire joint is divided into several different zones according to the computational requirements. This method can easily control the mesh quantity and quality and avoid badly distorted elements. To verify the convergence of EF results, convergence test with different mesh densities is conducted. The number of elements for both thickness of the chord, and thickness of the brace is 3. 56 elements are used along the curve of the weld toe. The FE models of three-planar and five-planar tubular Y-joints are displayed in Figure 3.

Both chord ends are assumed to be fixed, thus, the degrees of freedom, $U_x, U_y, U_z, \theta_x, \theta_y, \theta_z$, of the corresponding end nodes are restrained. The axial loading is applied to the nodes in the brace end sections according to different axial loading conditions.

2.3 *Determination of loading conditions*

Using SACS, some practical structures of offshore wind turbine with tripod and pentapod substructures are analyzed subjected to the aerodynamic and hydrodynamic loadings for fatigue limit state. From the results, the loading conditions for three-planar and five-planar Y-joints under in-plane bending moment are obtained and shown in Figure 4 and Figure 5, respectively.

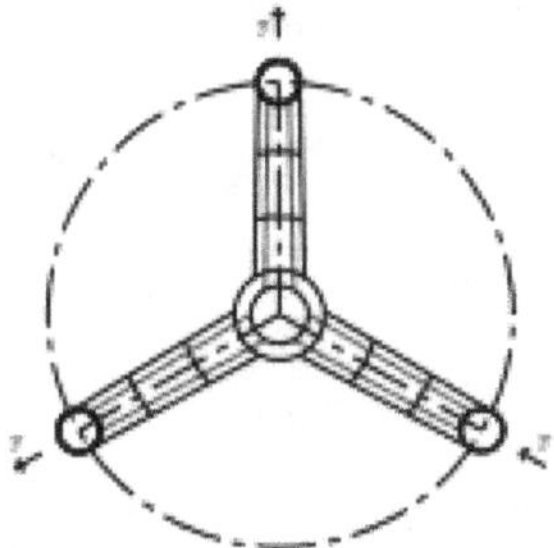

(a) Axial force on three braces

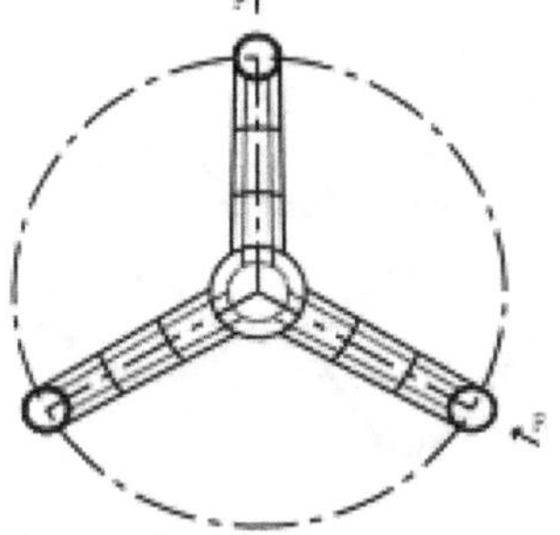

(b) Axial force on two braces

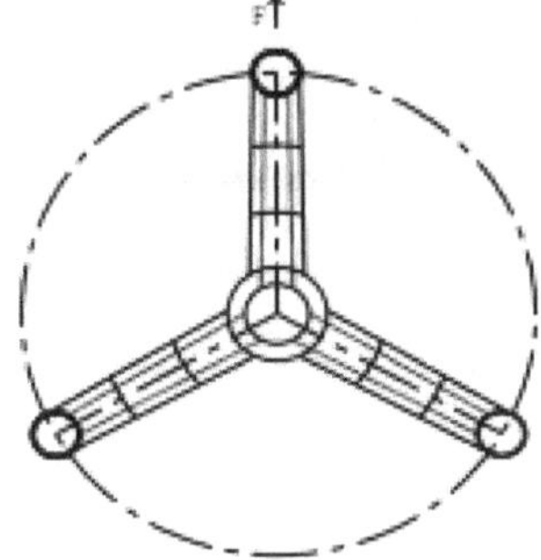

(c) Axial force on one brace

Figure 4. Axial loadings of three-planar tubular Y-joints.

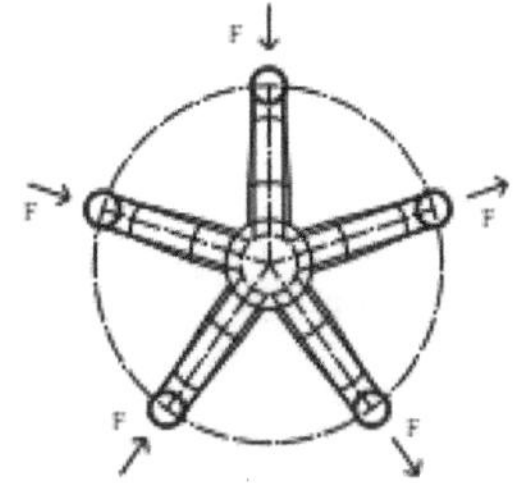

(a) Uniform axial force on five braces

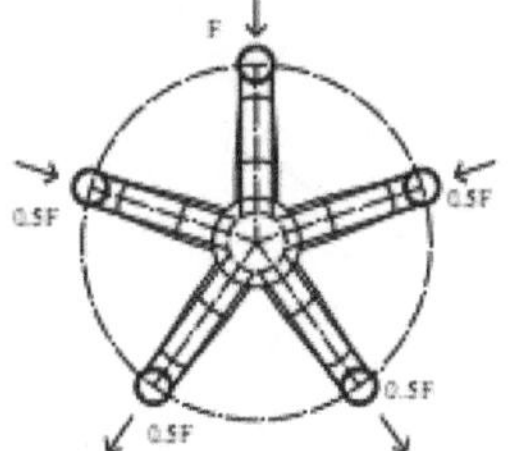

(b) Different axial force on five braces

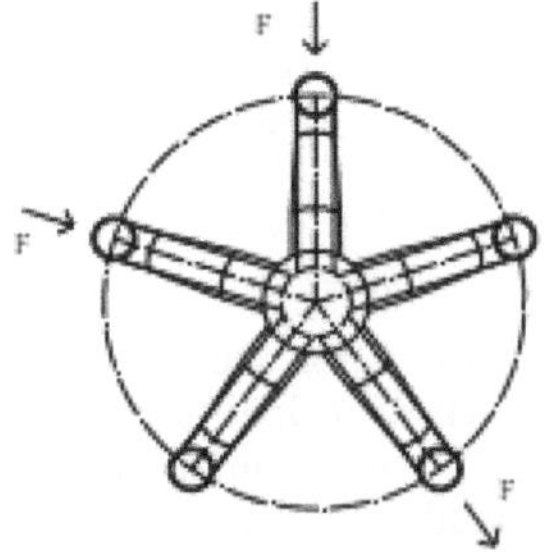

(c) Axial force on three braces

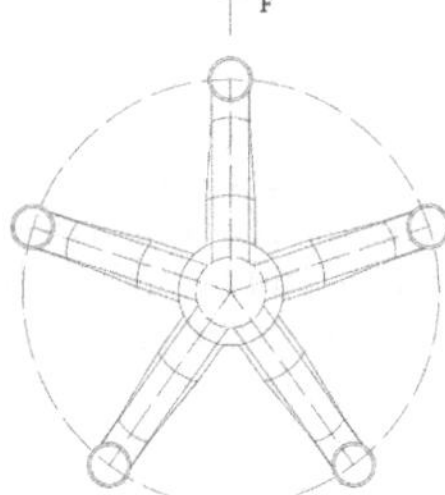

(d) Axial force on one brace only

Figure 5. Axial loadings of five-planar tubular Y-joints.

3 DISTRIBUTION OF SCFS ALONG THE WELD TOE

For each uni-planar Y-joint composed of the multi-planar Y-joint, the numbering of the diagonal braces shown in Figure 2 is used to represent each uni-planar Y-joint. The definition of polar angle Φ along the spatial curve of the weld toe for each uni-planar Y-joint is shown in Figure 6.

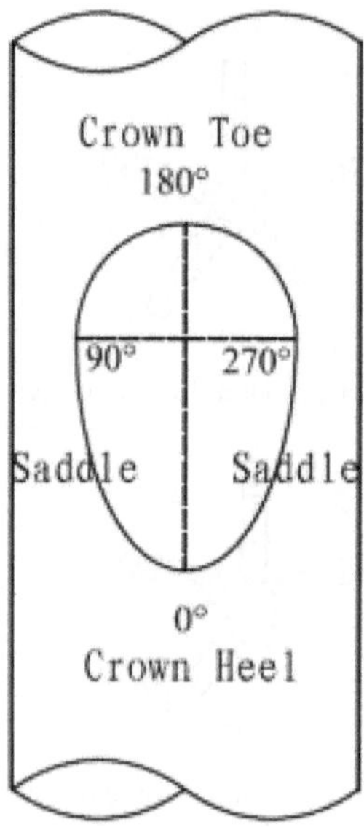

Figure 6. Polar angle Φ definition along the spatial curve of the weld toe for uni-planar Y-joint.

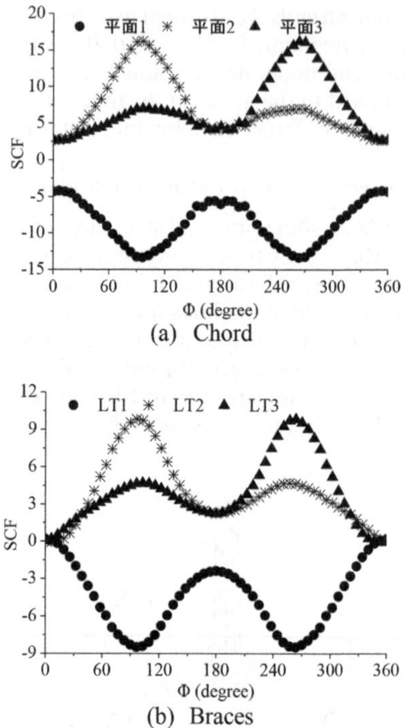

Figure 7. Distribution of SCFs along the weld of three-planar Y-joint for LT1.

3.1 *Three-planar Y-joint*

The SCFs distribution along the weld toe of each planar Y-joint subjected to load condition LT1 is displayed in Figure 7. The maximum SCF locates in the saddle of the weld toe for Y-joints subjected to axial loadings. For LT1, the applied axial loadings are symmetrical to the plane from the diagonal brace T1 and the central column, thus, the SCFs distribution of Y-joint T1 is symmetrical to the crown of the weld in chord and brace. Although the SCFs distributions for Y-joint T2 and T3 are unsymmetrical to the weld crown, the trend of SCFs distribution for Y-joint T2 is the opposite of that for Y-joint T3. The maximum SCF for Y-joint T2 occurs in $\Phi = 90°$, but that for Y-joint T3 in $\Phi = 270°$. Due to the interaction between T2 and T3, the maximum SCFs from T2 and T3 are larger than those from T1.

3.2 *Five-planar Y-joint*

The SCFs distribution along the weld toe of each planar Y-joint subjected to load condition LP1 is displayed in Figure 8. Similar results from five-planar Y-joint can be found corresponding to three-planar Y-joint. The maximum SCF locates in the saddle of the weld toe for each planar Y-joint. The SCFs distribution of Y-joint P4 is symmetrical to the crown of the weld in chord and brace due to the symmetry of the applied axial loadings. The interaction among the planar Y-joints P3, P4 and P5 applied to compressive forces cause that the maximum SCFs from P3 and P5 are much larger than those from P4.

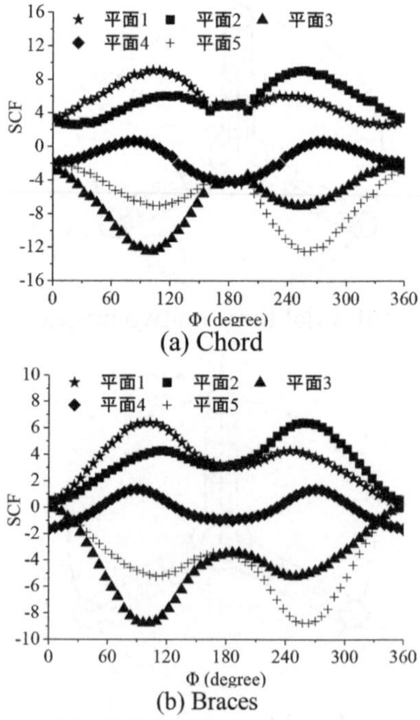

Figure 8. Distribution of SCFs along the weld of five-planar Y-joint for LT1.

4 GEOMETRICAL EFFECTS ON THE SCFS OF THREE-PLANAR Y-JOINT

Referred to the geometrical parameters of three-planar tubular Y-joint listed in Table 1, the corresponding FE models are established, respectively. The SCFs of three-planar tubular Y-joint subjected to axial loadings are obtained according to the loading conditions displayed in Figure 4. The comparison of FE results and results calculated by DNV-RP-C203 for simple tubular Y-joints are shown in Figure 9–Figure 13.

It can be seen from Figure 9–Figure 13 that the SCFs in chord are larger than those in braces. The SCFs under the loading condition LT1 are larger than the other loading conditions.

As shown in Figure 9–Figure 13, the SCFs in both chord and braces from DNV are smaller than those from FE in any cases.

4.1 *Effect of the α on the SCFs*

From Figure 9(a), the SCFs in chord increase with the increasing α in the range of 12 and 14, and the SCFs from FE analysis for loading condition LT1 and LT4 are larger than those from DNV. From Figure 9(b), the SCFs in braces increase with the increase of α as well, and the SCFs from DNV are smaller than those from FE.

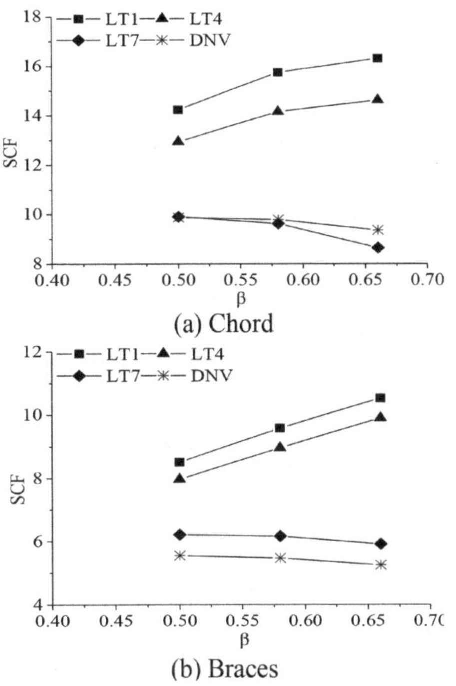

Figure 10. Effect of β on SCF for three-planar tubular Y-joints ($\alpha = 13$, $\gamma = 20$, $\tau = 0.8$, $\theta = 45°$).

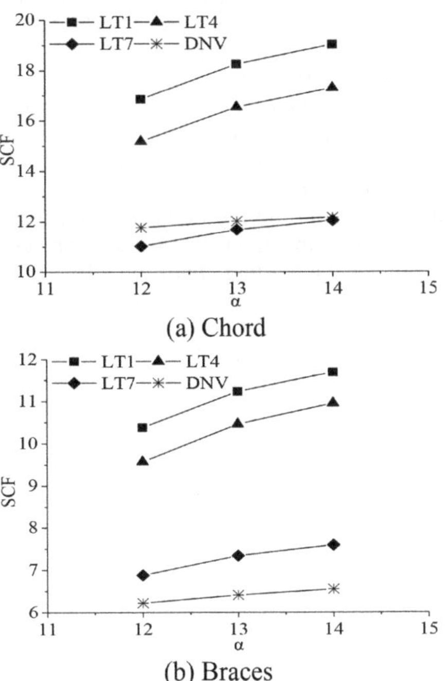

Figure 9. Effect of α on SCF for three-planar tubular Y-joints ($\beta = 0.58$, $\gamma = 25$, $\tau = 0.8$, $\theta = 45°$).

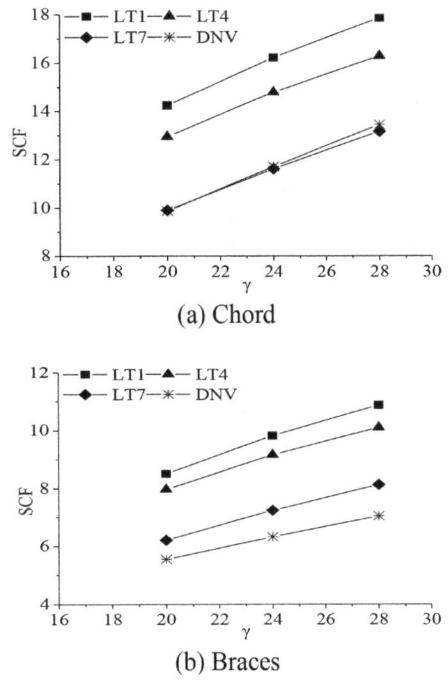

Figure 11. Effect of γ on SCF for three-planar tubular Y-joints ($\alpha = 13$, $\beta = 0.5$, $\tau = 0.8$, $\theta = 45°$).

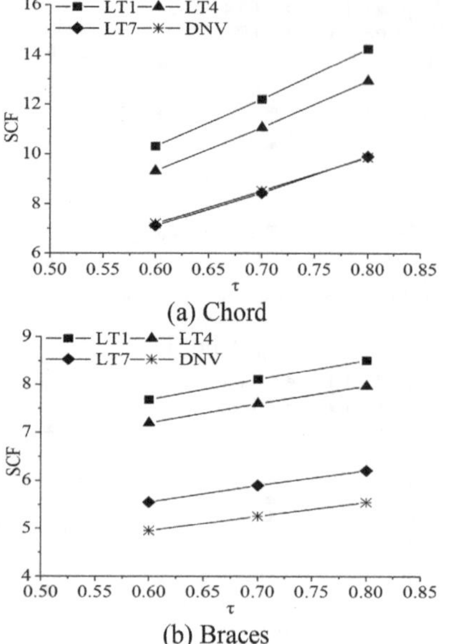

(a) Chord

(b) Braces

Figure 12. Effect of τ on SCF for three-planar tubular Y-joints ($\alpha = 13$, $\beta = 0.5$, $\gamma = 20$, $\theta = 45°$).

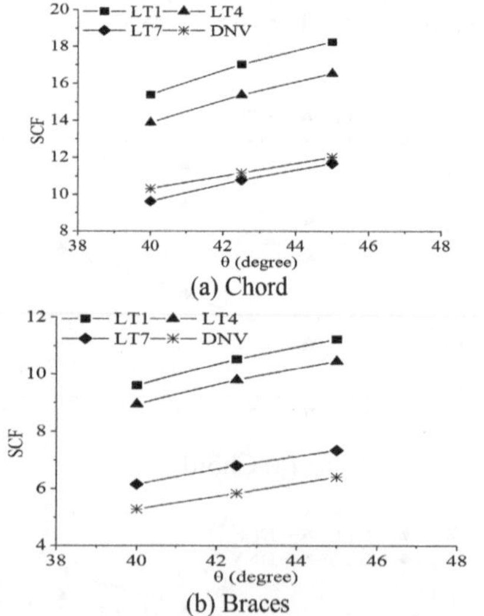

(a) Chord

(b) Braces

Figure 13. Effect of θ on SCF for three-planar tubular Y-joints ($\alpha = 13$, $\beta = 0.58$, $\gamma = 25$, $\tau = 0.8$).

4.2 *Effect of the β on the SCFs*

From Figure 10, when more than one diagonal braces are applied to axial loadings, the SCFs from FE analysis increase with the increasing β in the range of 0.50 and 0.66, while the SCFs from DNV decrease with the increase of β. Thus, the trend of SCFs from FE results is different of those from DNV with the change of β for the axial forces on multi-braces.

4.3 *Effect of the γ on the SCFs*

From Figure 11, the SCFs in chord and braces increase with the increasing γ in the range of 20 and 28. The SCFs from DNV are smaller than those from FE subjected to the three loading conditions.

4.4 *Effect of the τ on the SCFs*

From Figure 12, the SCFs from DNV are smaller than those from FE subjected to the three loading conditions. The SCFs in chord obviously increase, while the SCFs in braces slowly increase, with the increasing τ in the range of 0.6 and 0.8.

4.5 *Effect of the θ on the SCFs*

From Figure 13, the SCFs increase with the increasing θ in the range of 40 and 45. The SCFs from FE analysis are larger than those from DNV.

5 GEOMETRICAL EFFECTS ON THE SCFS OF FIVE-PLANAR Y-JOINT

Based on the geometrical parameters of five-planar tubular Y-joint listed in Table 1, the corresponding FE models are established, respectively. The SCFs of five-planar tubular Y-joint subjected to axial loadings are obtained according to the loading conditions displayed in Figure 5. The comparison of FE results and results calculated by DNV-RP-C203 for simple tubular Y-joints are shown in Figure 14–Figure 18.

It can be seen from Figure 14–Figure 18 that the SCFs in chord are larger than those in braces. The SCFs under the loading condition LP1 are larger than the other loading conditions.

As shown in Figure 14–Figure 18, the SCFs in chord and braces from DNV are smaller than those from FE under loading condition LP1 in any cases.

5.1 *Effect of the α on the SCFs*

Figure 14 shows that the SCFs increase with the increasing α, indicating that the parameter α has significant effect on SCFs.

5.2 *Effect of the β on the SCFs*

From Figure 15, the effect of the parameter β on SCFs in chord is complex. The other geometrical parameters should be considered to determine SCFs.

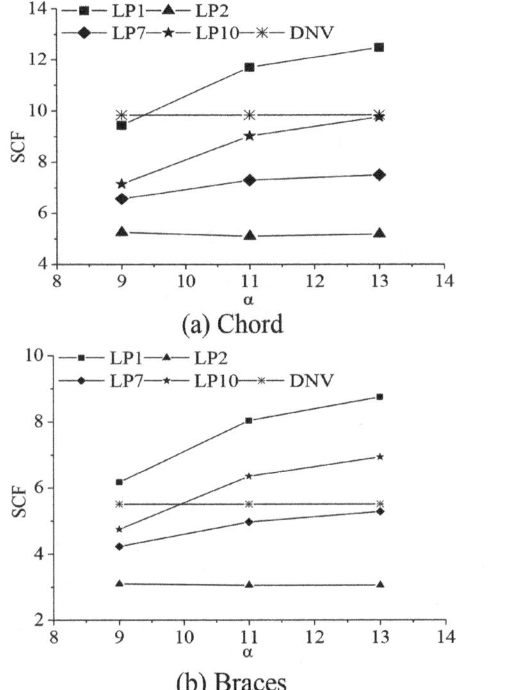

Figure 14. Effect of α on SCF for five-planar tubular Y-joints ($\beta = 0.35$, $\gamma = 30$, $\tau = 0.7$, $\theta = 40°$).

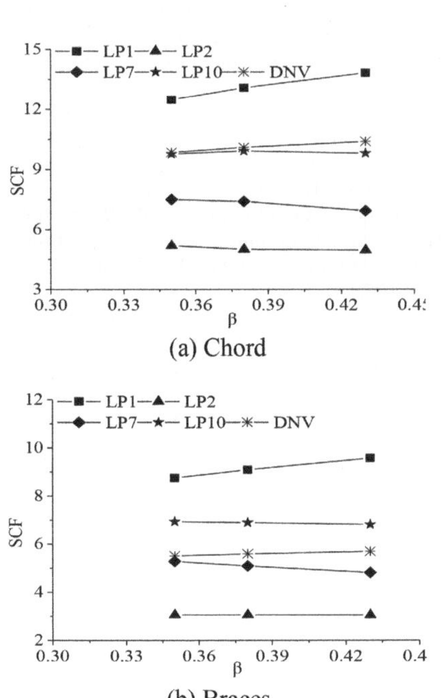

Figure 15. Effect of β on SCF for five-planar tubular Y-joints ($\alpha = 13$, $\gamma = 30$, $\tau = 0.7$, $\theta = 40°$).

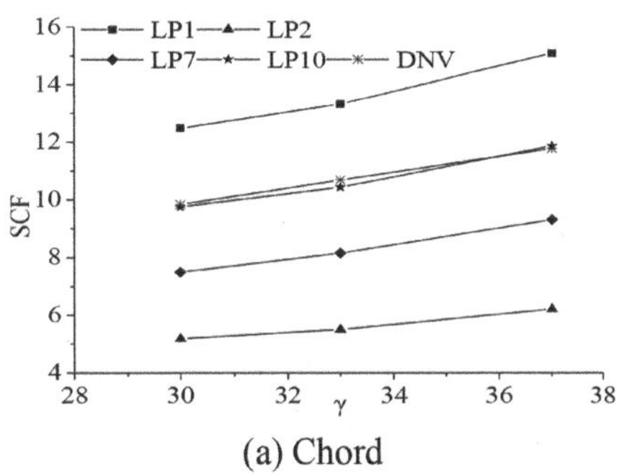

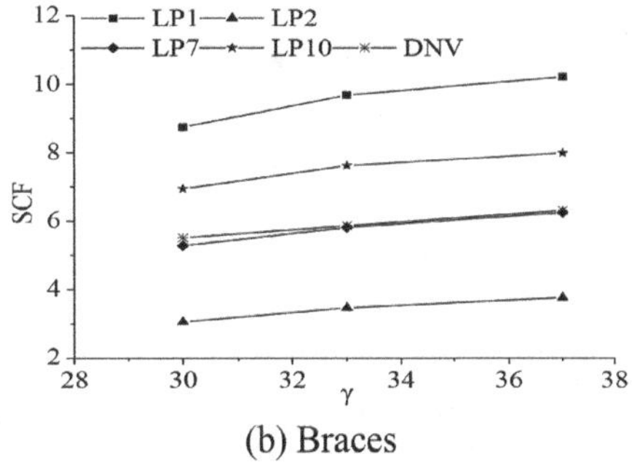

Figure 16. Effect of γ on SCF for five-planar tubular Y-joints ($\alpha = 13$, $\beta = 0.35$, $\tau = 0.7$, $\theta = 40°$).

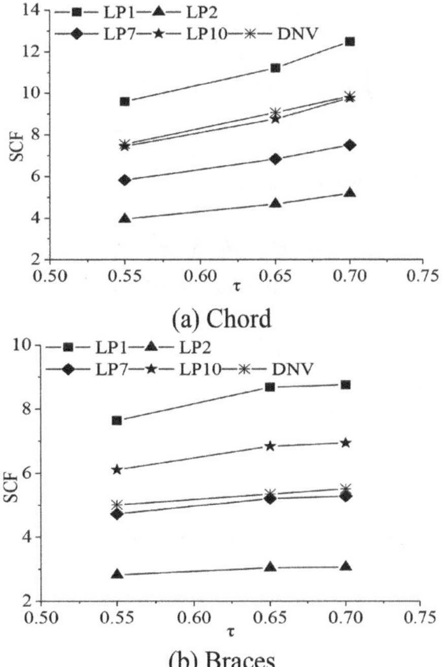

Figure 17. Effect of τ on SCF for five-planar tubular Y-joints ($\alpha = 13$, $\beta = 0.35$, $\gamma = 30$, $\theta = 40°$).

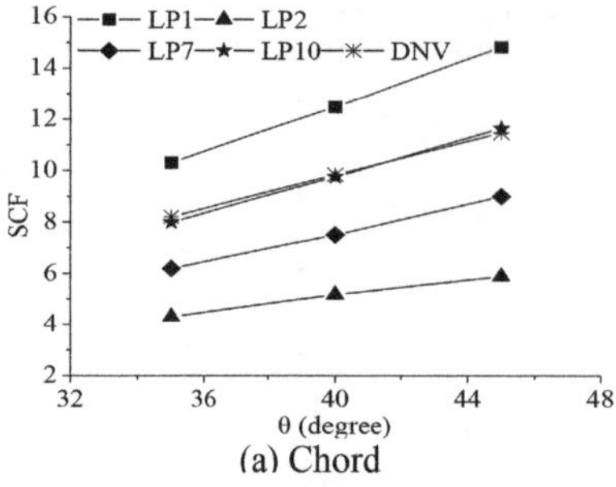

(a) Chord

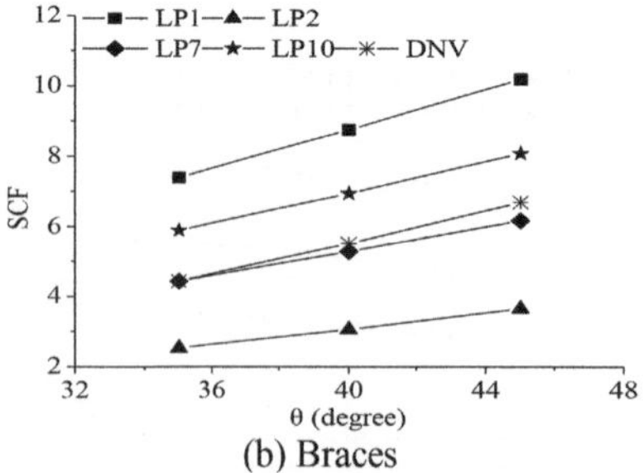

(b) Braces

Figure 18. Effect of θ on SCF for five-planar tubular Y-joints ($\alpha = 13$, $\beta = 0.35$, $\gamma = 30$, $\tau = 0.7$).

5.3 *Effect of the γ on the SCFs*

As displayed in Figure 16, the parameter γ has significant influence on SCFs in chord.

5.4 *Effect of the τ on the SCFs*

The SCFs in both chord and braces increase with the increasing parameter τ in the range of 0.55 and 0.70 as shown in Figure 17.

5.5 *Effect of the θ on the SCFs*

From Figure 18, the parameter θ has obvious impact on SCFs on both chord and braces. And the SCFs increase with the increase of parameter θ in the range of 35 and 45.

6 CONCLUSIONS

Using FE simulation, the effects of geometrical parameters on SCFs of three-planar and five-planar tubular Y-joints subjected to axial loadings are analyzed. The following conclusions can be drawn.

1. For both three-planar and five-planar tubular Y-joints subjected to axial loadings, SCFs in chord are larger than those in braces. SCFs are maximum when all the diagonal braces are applied to axial loadings, indicating that the interaction among each planar Y-joints composed of multi-planar Y-joints occurs.
2. For both three-planar and five-planar tubular Y-joints subjected to axial loadings, the parameter α, β, γ, τ, θ have significant influence on SCFs.
3. For the SCFs in chord and braces under the condition that all diagonal braces are applied to axial loadings, the results from DNV are smaller than FE results. Thus, if the SCFs in chord from DNV suitable for simple tubular Y-joints are used to carry out fatigue check of multi-planar tubular Y-joints, the assessment will not be conservative.

ACKNOWLEDGEMENTS

This work is funded by the National Natural Science Foundation of China (Grant No. 51121005), and supported by the Science Foundations of Powerchina Huadong Engineering Corporation Limited (No. KY120228-03-07 and KY2014-02-41) and the Open Fund Project of State Key Lab of Coastal and Offshore Engineering in Dalian University of Technology (No. LP1413). Their financial supports are gratefully acknowledged.

REFERENCES

[1] API. Recommended Practice for Planning, Designing and Constructing Fixed Offshore Platforms—Load and Resistance Factor Design. 1997.

[2] DNV-RP-C203. Fatigue Design of Offshore Steel Structures. 2011.

[3] Dong W.B., Moan T., Gao Z. Long-term fatigue analysis of multi-planar tubular joints for jacket-type offshore wind turbine in time domain. Engineering Structures, 2011, 33: 2001–2014.

[4] Woghiren C.O., Brennan F.P. Weld toe stress concentrations in multi-planar stiffened tubular KK joints. International Journal of Fatigue. 2009, 31(1): 164–172.

[5] Karamanos S.A., Romeijn A., Wardenier J. SCF equations in multi-planar welded tubular DT-joints including bending effects. Marine structures. 2002, 15(2): 157–173.

[6] Ahmadi H., Mohammad A.L., Mohammad H.A. The development of fatigue design formulas for the outer brace SCFs in offshorue three-planar tubular KT-joints. Thin-Walled Structures, 2012, 58(67–78).

[7] Ahmadi H., Mohammad A.L., Shao Y.B., Mohammad H.A. Parametric study and formulation of outer-brace geometric stress concentration factors in internally ring-stiffened tubular KT-joints of offshore structures. Applied Ocean Research, 2012, 38: 74–91.

[8] American Welding Society (AWS). Structural Welding Code, AWS D1.1. Miami, USA, 2009.

[9] Chiew S.P., Soh C.K., Wu N.W. General SCF design equations for steel multiplanar tubular XX-joints. International Journal of Fatigue. 2000, 22(4): 283–293.

Material Science and Environmental Engineering – Chen (Ed.)

Effect of dry friction structures on stator blade

J.H. Xie & H. Wu
Wuhan Second Ship and Research Institute, Wuhan, Hubei, China

R.S. Yuan & Z.Y. Zhang
School of Energy and Power Engineering, Xi'an Jiaotong University, Xi'an, Shaanxi, China

ABSTRACT: Stator blades generate large amplitude vibration under the effect of alternating airflow load. By using dry friction damping, vibration of stator blades can be restrained effectively. Nevertheless, the contact surfaces structure of a dry friction damper is extremely complicated. In this paper, the stator blade is simplified by 3-*DOF* beam model with lumped masses, and the damper is described by one-bar microslip model. Then the vibration characteristics of a stator blade are analyzed. The results show that dampers could not only dissipate energy, but also change stiffness of blade. Moreover, the optimum positive pressure exists in the computation, under which stator blade has the minimum vibration response.

Keywords: stator blade; dry friction; microslip; damping; vibration

1 INTRODUCTION

Dry friction structures are sufficiently applied in turbine stator blades vibration control technique, because of their simple structure and excellent damping effect. Stator blades are installed into carrier-rim by shroud and blade root structures and their vibration are restrained effectively by dry friction phenomena appeared between carrier-rim and shroud or blade root structure. When stator blade installed into carrier-rim, the contact pairs come into contact and compressed due to gravity and structure. On the other hand, it can be loaded by bolts to insure the contact pairs are compressed tight enough.

When stator blades vibration under alternating steam excitation force, dry friction can be generated by relative displacement that occur between shroud or blade root and carrier-rim, and stator blades vibration are restrained. The dry friction structures, on the one hand, dissipate vibration energy. On the other hand, the structure not only changes the mass of the blade, but also changes the stiffness of the blade so that the inherent frequency and the exciting force frequency of the blade could avoid each other well. Thus, the purpose of reducing the vibration level of the blade is achieved.

Using dry friction structure to inhibit vibration is a very effective method, a large number of scholars have conducted fruitful research on it. After a large number of researches on friction phenomena, Coulomb proposed the coulomb's friction law. Based on coulomb's friction model, a single degree of freedom system with dry friction damping structure is analyzed by Den Hartog [1], and analytic solution was obtained. Goodman connected spring and coulomb's friction model in series, and named it as macro slip hysteresis model. Based on macro slip hysteresis model, Griffin [2–4] investigated blade vibration properties under different positive pressure, and optimal positive pressure is obtained. After study of the interface coupling movement sufficiently, a three-dimensional model that could be used to determine Viscous-sliding-separation is established by Yang and Menq [5–8].

Since the macro slip models is less an estimation than an accurately description of dry friction, the micro slip models are proposed by Mindlin, Metherell et al [9–11]. Based on micro slip model, blade vibration properties are obtained by Gabor, meanwhile, the damper is optimized. A new model that can be used to investigate dry friction under different roughness and pressure is developed by Yalin Liu et al [12].

2 THE ONE-BAR MICROSLIP MODEL

The damper, as shown in Figure 1, is modeled by a rectangular bar pressed against a rigid surface with a normal load q, and subjected it to a tangential force F. The normal load on the bar is assumed to be constant over the width of the bar, and defined by a quadratic normal load function in the lengthwise direction.

$$q(x) = q_0 + q_2 \frac{4\left(xl - x^2\right)}{l^2} \tag{1}$$

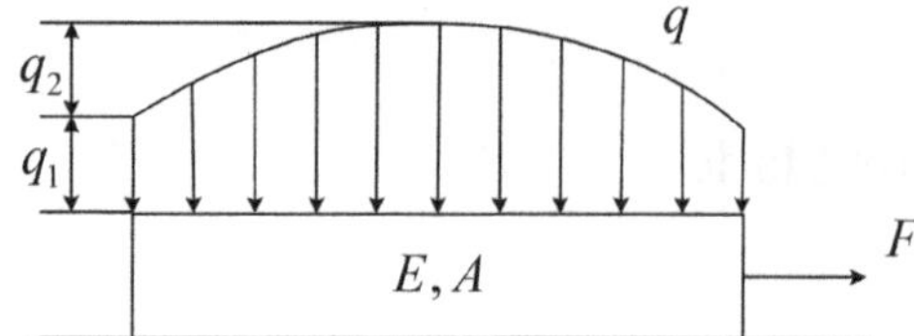

Figure 1. One-bar microslip model.

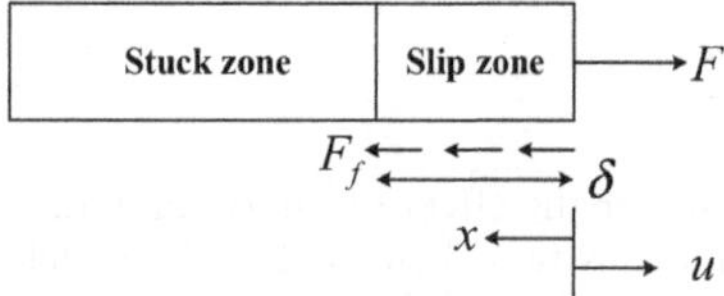

(a) Stuck and slip zones when initially loading the bar

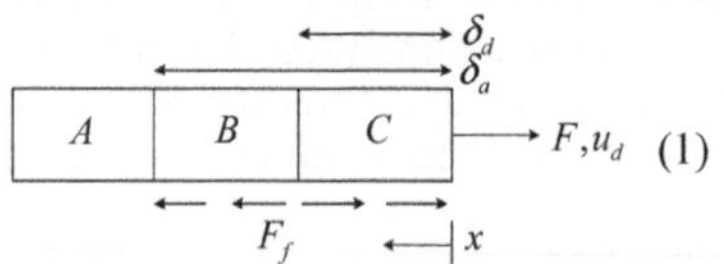

(b) Illustrative plot of force and displacement when F is decreasing (2)

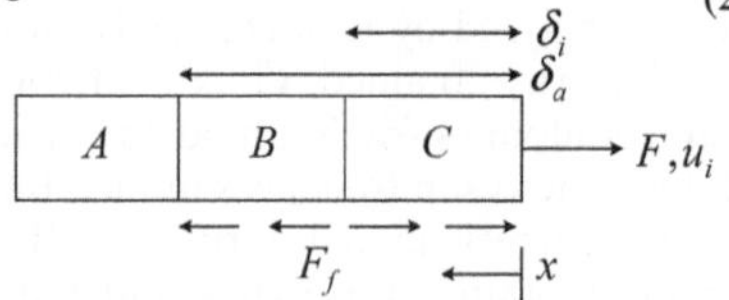

(c) Illustrative plot of force and displacement when F is increasing

Figure 2. Microslip model for damper.

The tangential force F acted on the rectangular bar varies sinusoidally, and it can be expressed as

$$F = F_a \sin \omega t \tag{2}$$

The bar has a modulus of elasticity E, and a cross-section area A. The length of the bar is l, the coefficient of friction μ is assumed to be constant across the contact zone and independent, whether the bar is sliding or not.

As the force is applied, the bar starts to slip. Then the bar may be divided into two zones, one is slipping and the other is stuck (as shown in Fig. 2(*a*)). The length of the slip zone is defined as the slip length δ and the slip length corresponding to the maximum force F_{amp} is δ_a.

The zone that is slipping is stretched after the initial loading with $F = F_{\text{amp}}$. As the force decreases from F_{amp} to $-F_{\text{amp}}$, the bar may be divided into three zones in this analysis, which is shown in Figure 2(*b*)

Zone *A* is stuck and has zero strain
Zone *B* is stuck and stretched
Zone *C* is slipping and compressed.

The length of the compressed zone is denoted by the slip length δ_d. As F decreases, zone *C* will increase and zone *B* will decrease. This will continue until $F = -F_{\text{amp}}$. Zone *B* is then eliminated and δ_d equals δ_a.

Here we have the opposite situation as in the previous part. When the force increases from $-F_{\text{amp}}$ to F_{amp}, the bar may be divided into three zones in this analysis, which is shown in Figure 2(*c*)

Zone *A* is stuck and has zero strain
Zone *B* is stuck and compressed
Zone *C* is slipping and stretched.

The length of stretched zone is denoted as the slip length δ_i. As F increases, zone *C* will increase and zone B will decrease. This will continue until $F = F_{\text{amp}}$. Zone *B* is then eliminated and δ_d equals δ_a.

The length of the slip zone is defined as the slip length *δ*. The slip length corresponding to the maximum force F_{amp} is δ_a. The displacement of the bar is defined as u. According to the mechanics analysis, it can be seen that force *F* and displacement u are the function of slip length *δ*.

As the force decreases from F_{amp} to $-F_{\text{amp}}$, that is when $\partial F/\partial t < 0$

$$\begin{cases} F = F_d(\delta_d) \\ u = u_d(\delta_d) \end{cases} \qquad 0 \le \delta_d \le \delta_a \tag{3}$$

And

$$\begin{cases} F_d(\delta_a,\delta_d) = \mu q_0(\delta_a - 2\delta_d) + \dfrac{2\mu q_2}{3l^2}(3l\delta_a^2 - 2\delta_a^3 + 4\delta_d^3 - 6l\delta_d^2) \\ u_d(\delta_a,\delta_d) = \dfrac{\mu}{EA}\left(q_0\dfrac{\delta_a^2 - 2\delta_d^2}{2} + q_2\left(\dfrac{4(\delta_a^3 - 2\delta_d^3)}{3l} - \dfrac{\delta_a^4 - 2\delta_d^4}{l^4}\right)\right) \end{cases} \tag{4}$$

As the force decreases from $-F_{\text{amp}}$ to F_{amp}, that is when $\partial F/\partial t > 0$

$$\begin{cases} F = F_i(\delta_i) \\ u = u_i(\delta_i) \end{cases} \qquad 0 \le \delta_i \le \delta_a \tag{5}$$

And

$$\begin{cases} F_i(\delta_a,\delta_i) = \mu q_0(2\delta_i - \delta_a) + \dfrac{2\mu q_2}{3l^2}(2\delta_a^3 - 3l\delta_a^2 + 6l\delta_i^2 - 4\delta_i^3) \\ u_i(\delta_a,\delta_d) = \dfrac{\mu}{EA}\left(q_0\dfrac{2\delta_i^2 - \delta_d^2}{2} + q_2\left(\dfrac{4(2\delta_i^3 - \delta_a^3)}{3l} - \dfrac{2\delta_i^4 - \delta_a^4}{l^2}\right)\right) \end{cases} \tag{6}$$

The linearization techniques are discussed to transform the nonlinear properties of friction into equivalent damping and stiffness. Two criteria are established in Lazan's linearization method, for equivalence of the two loops

a. The same value of loop area which is the same as damping energy per cycle W.
b. The amplitude of force F_{amp} and displacement u_{amp} should be the same.

The first criterion gives the equivalent viscous damping c_{eq}, while the second criterion together with c_{eq} gives the equivalent stiffness k_{eq}.

$$w(\delta_a) = \frac{2\mu^2\delta_a^3}{3EA}\left[q_0^2 + \left(\frac{4\delta_a}{l} - \frac{14\delta_a^2}{5l^2}\right)q_0 q_2 + \left(\frac{8\delta_a^4}{7l^4} - \frac{4\delta_a^3}{l^3} + \frac{16\delta_a^2}{5l^2}\right)q_2^2\right] \tag{7}$$

$$c_{eq} = \frac{2\mu^2\delta_a^3}{3EA\pi\omega u_{amp}^2}\left[q_0^2 + \left(\frac{4\delta_a}{l} - \frac{14\delta_a^2}{5l^2}\right)q_0 q_2 + \left(\frac{8\delta_a^4}{7l^4} - \frac{4\delta_a^3}{l^3} + \frac{16\delta_a^2}{5l^2}\right)q_2^2\right] \tag{8}$$

$$k_{eq} = \sqrt{\left(\frac{F_{amp}}{u_{amp}}\right)^2 - \left(\omega c_{eq}\right)^2} \tag{9}$$

3 FORCED VIBRATION ANALYSIS MODEL OF STATOR BLADE

In this article, the stator blade is simplified by 3-*DOF* beam model with lumped masses, and the damper is described by one-bar microslip model. Then the vibration characteristics of a stator blade are analyzed. The lumped masses are indicated by m_1, m_2 and m_3, respectively; the stiffness of beams are indicated by k_1 and k_2.

The innovation points in this paper are embodied in the application Field. The stator blade is simplified by 3-*DOF* beam model with lumped masses for the first time, and the shroud and blade root dry friction structures are simulated by the dampers that located at m_1 and m_2. As is shown in Figure 3, the linearized equivalent damping and stiffness of dampers are represented by C_{eq1}, k_{eq1} and C_{eq2}, k_{eq2}, respectively.

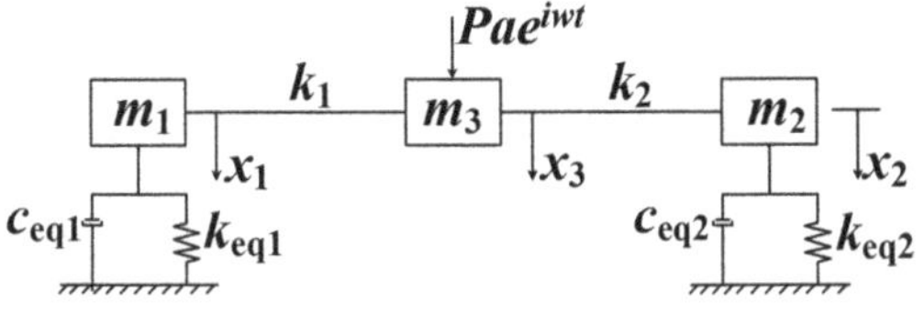

Figure 3. Forced vibration analysis model of stator blade.

Based on the forced vibration analysis model of stator blade mentioned above, the kinetic equation is as follows

$$\begin{cases} m_1\ddot{x}_{1(t)} + C_{eq1}(\delta_{a1})\dot{x}_{1(t)} + \left(k_{eq1}(\delta_{a1}) + k_1\right)x_{1(t)} - k_1 x_{3(t)} = 0 \\ m_2\ddot{x}_{2(t)} + C_{eq2}(\delta_{a2})\dot{x}_{2(t)} + \left(k_{eq2}(\delta_{a2}) + k_2\right)x_{2(t)} - k_2 x_{3(t)} = 0 \\ m_3\ddot{x}_{3(t)} + (k_1 + k_2)x_{3(t)} - k_1 x_{1(t)} - k_2 x_{2(t)} = P_a e^{iwt} \end{cases} \tag{10}$$

From equation (10), it can be easily seen that, the vibration property of m_1 and m_2 relate to each other due to m_3.

Assuming harmonic motion yields

$$x_{1(t)} = x_{1a}e^{iwt},\quad x_{2(t)} = x_{2a}e^{iwt},\quad x_{3(t)} = x_{3a}e^{iwt} \tag{11}$$

We define the complex stiffness as

$$\begin{cases} K_1(\delta_{a1}) = k_{eq1}(\delta_{a1}) + iwC_{eq1}(\delta_{a1}) \\ K_2(\delta_{a2}) = k_{eq2}(\delta_{a2}) + iwC_{eq2}(\delta_{a2}) \end{cases} \tag{12}$$

Eqs. (12) and (13) in Eq. (11) gives the following algebraic equations

$$\begin{cases} -w^2 m_1 x_{1a} + K_1(\delta_{a1})x_{1a} + k_1(x_{1a} - x_{3a}) = 0 \\ -w^2 m_2 x_{2a} + K_2(\delta_{a2})x_{2a} + k_2(x_{2a} - x_{3a}) = 0 \\ -w^2 m_3 x_{3a} + (k_1 + k_2)x_{3a} - k_1 x_{1a} - k_2 x_{2a} = Pa \end{cases} \tag{13}$$

Solving Eq. (13) for x_1, x_2 and x_3 yields

$$\begin{cases} x_{1a} = \dfrac{k_1}{K_1(\delta_{a1}) + k_1 - w^2 m_1}x_{3a} \\ x_{2a} = \dfrac{k_2}{K_2(\delta_{a2}) + k_2 - w^2 m_2}x_{3a} \\ x_{3a} = \dfrac{P_a}{\left(k_1 + k_2 - w^2 m_3\right) - \dfrac{k_1^2}{K_1(\delta_{a1}) + k_1 - w^2 m_1} - \dfrac{k_2^2}{K_2(\delta_{a2}) + k_2 - w^2 m_2}} \end{cases} \tag{14}$$

The beam and damper data are

$k_1 = 1.0 \times 10^7$ N/m, $k_2 = 1.0 \times 10^7$ N/m, $m_1 = 0.05$ kg, $m_2 = 0.05$ kg

$E_1A_1 = 40000$ N, $l = 0.2$ m, $Pa = 1$, $Q_0 = \mu q_0 l/\text{Pa}$, $Q_2 = Q_0/2$.

4 RESULTS AND DISCUSSION

Through blades vibration analysis with lacing in Figure 4, amplitude frequency response curves of m_1, m_2, and m_3 in the range of 0~3000 Hz are achieved. The results are shown in Figures 4, 5, and 6. Different positive pressure and damper parameters under different resonance states are shown in Table 1.

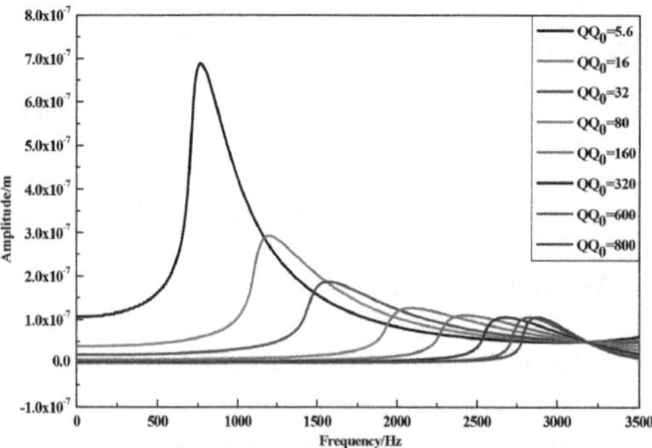

Figure 4. Vibration response of m_1 and m_2.

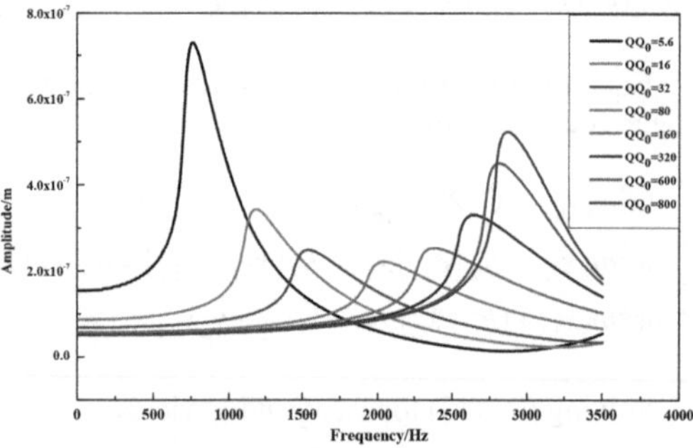

Figure 5. Vibration response of m_3.

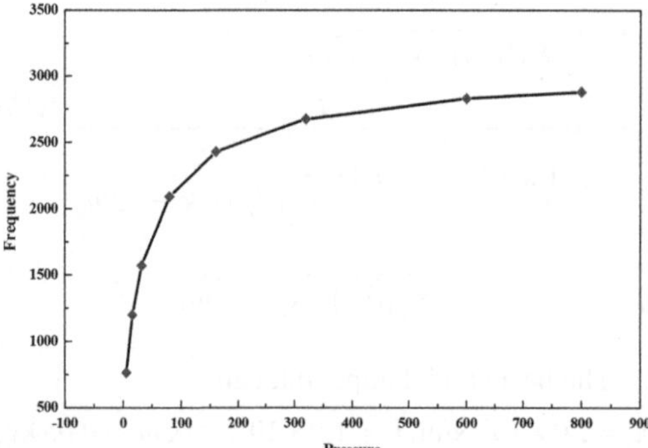

Figure 6. The resonance frequencies change with positive pressure.

Table 1. Vibration parameters under resonant conditions.

Q_0	ω_{res}/Hz	δ_a/m	k_{eq}/N/m $\times 10^6$	c_{eq}/N/ (m/s)	w/ J $\times 10^{-7}$
5.6	765	0.040	2	160	11.5
16	1200	0.016	4	264	5.3
32	1570	0.009	8	357	3.8
80	2090	0.005	15	514	3.4
160	2430	0.003	22	669	3.9
320	2675	0.002	32	880	5.2
600	2830	0.002	43	1140	7.1
800	2880	0.001	50	1290	8.2

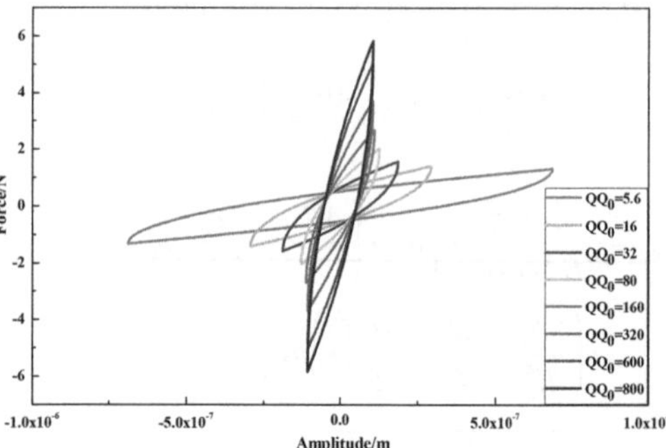

Figure 7. Hysteresis loop of damper under resonance state.

Since the mass distributions and damping parameters of shroud and blade root are identical, and k_1 equals to k_2. The vibration responses of m_1 and m_2 equal to each other. Figures 4 and 5 show amplitude-frequency response curves in the frequency range of 0~3500 Hz of m_1, m_2 and m_3.

It can be seen that, the vibration amplitude of m_3 is larger that of m_1 and m_2 under any circumstances. Moreover, there are optimal normal pressures that make vibration response of m_1, m_2, and m_3 minimum, respectively. The optimal normal pressure, under which m_1 and m_2 have minimum response, is 600, for m_3, which is 80. As a stator blade, reduction of m_3 is more meaningful than that of m_1 and m_2, since the m_3 has maximum deflection and can reflect vibration property of stator blade more accurately.

The Figure 7 shows the resonance frequencies of stator blade change with positive pressure. It is obvious that there is a positive correlation between the resonant frequency and positive pressure, but as the positive pressure rises, the increase amplitude decreases, and tending to flat finally. Based on vibration theory knowledge, the resonant frequency is relevant to stiffness regardless of damping.

Damper not only changes damping property, but also stiffness of system, and the tangential stiffness of system a positive correlation with normal pressure. On one hand, the tangential stiffness changes sharply with the increase of normal pressure, on the other hand, the smaller the initial tangential stiffness, the larger the response of system, as tangential stiffness changes and vice versa. So the resonant frequency changes sharply at first and flatly later.

In addition, it can be found that the amplitude response of both m_1 and m_2 change sharply with the increase of positive pressure when positive pressure in a lower level, and which change flatly when pressure in a higher level, but compared with m_1 and m_2, amplitude response of m_3 changes sharper.

Figure 7 shows the hysteresis loop in a vibration cycle of damper under resonance state in different positive pressures. Table 1 gives out the vibration parameters under resonant conditions.

The area surrounded by a hysteresis loop represented the energy dissipation amount in a vibration cycle of the damper. From Figure 7 and Table 1, it could be seen that the energy dissipation amount of the damper increases first and then decreases along with the increase of positive pressure. Based on Formula (7), the energy dissipation amount in a vibration cycle of the damper were relevant with the positive pressure and the sliding part length δ_a. The energy dissipation amount is positively correlated with the positive pressure and negatively correlated with the sliding part length. There is a strong negative correlation between sliding part length δ_a and positive pressure, and there is a positive pressure that makes damper has minimal energy consumption in a vibration cycle. From Figure 1, it can be seen that the positive pressure, that makes damper consumes minimum energy, is 80, which is same as that of m_3.

5 CONCLUSIONS

In this article, the stator blade is simplified by 3-DOF beam model with lumped masses, and the damper is described by one-bar microslip model. Then the vibration characteristics of a stator blade with two dampers under different positive pressure and frequency are analyzed. To sum up, following conclusions can be drawn.

1. There are optimal normal pressures that make vibration response of m_1, m_2 and m_3 minimum, respectively. The optimal normal pressure, under which m_1 and m_2 have minimum response, is 600, for m_3, which is 80.
2. The vibration response amplitude of lumped mass m_3 is greater than that of m_1 and m_2.
3. The energy dissipation amount of the damper increases first and then decreases along with the increase of positive pressure.
4. Damper not only changes damping property but also stiffness of system, and there is a positive correlation between the resonant frequency and positive pressure.

REFERENCES

[1] Den Hartog J.P. Forced vibration with combined and viscous friction[J]. Transactions of the ASME, 1931, 53(9): 107–115.

[2] J.H. Griffin. Friction Damping of Resonant Stresses in Gas Turbine Engine Airfoils[J]. Journal of Engineering for Gas Turbines and Power, 1980, 102: 329–333.

[3] C.H. Menq, J.H. Griffin. A Comparison of Transient and Steady State Finite Element Analyses of the Forced Response of a Frictionally Damped Beam[J]. Journal of Vibration, Acoustics, Stress, and Reliability in Design, 1985, 107: 19–25.

[4] J.H. Griffin. A Review of Friction Damping of Turbine Blade Vibration[J]. International Journal of Turbo and Jet Engines, 1990, 7: 297–307.

[5] B.D. Yang, C.H. Menq. Characterization of 3D Contact Kinematics and Prediction of Resonant Response of Structures Having 3D Frictional Constraint[J]. Journal of Sound and Vibration, 1998, 217(5): 909–925.

[6] J.J. Chen, B.D. Yang, C.H. Menq. Periodic Response of Blades Having Three-Dimensional Nonlinear Shroud Constraints[J]. Journal of Engineering for Gas Turbines and Power, 2001, 123: 901–909.

[7] Menq C.H., Griffin J.H., Bielak J. The influence of a variable normal load on the forced vibration of a frictionally damped structure[J]. Journal of Engineering for Gas Turbines and Power,1986, 108: 300–305.

[8] Yang B.D., Chu M.L., Menq C.H. Stick-slip-separation analysis and non-linear stiffness and damping characterization of friction contacts having variable normal load[J]. Journal of Sound and Vibration, 1998, 210(4): 461–481.

[9] Metherell A.F., Diller S.V. Instantaneous energy disspation rate in a lap joint-uniform clamping pressure[J]. Journal of Applied Mechanics, 1967, 34(3): 612–617.

[10] Mindlin R.D., Mason W.P., Osmer T.F., et al. Effect of an oscillating tangential force on the contact surfaces of elastic spheres[C]. 1st US National Congress of Applied Mechanics. Chicago: ASME, 1951: 203–208.

[11] Gabor Csaba. Modeling micro slip friction damping and its influence on turbine blade vibration[D]. Department of Mechanical Engineering, Linöping University, 1998.

[12] Liu Y., Shangguan B., Xu Z. A friction contact stiffness model of fractal geometry in forced response analysis of a shrouded blade[J]. Nonlinear Dynamics, 2012, 70(3): 2247–2257.

Material Science and Environmental Engineering – Chen (Ed.)
© 2016 Taylor & Francis Group, London, ISBN 978-1-138-02938-5

Microstructure and shear strength of low-silver SAC/Cu solder joints during the thermal cycle

H.S. Wang, G.S. Gan, D.H. Yang, G.Q. Meng & H. Luo
Chongqing Municipal Engineering Research Center, Institutions of Higher Education for Special Welding Materials and Technology (Chongqing University of Technology), Chongqing, China

ABSTRACT: A new type of low-silver hypoeutectic SAC lead-free solder was prepared, and then hot dip soldering was used to form the joints of copper. The microstructure evolution of IMCs and the variation of shear strength of solder joints were investigated during the thermal cycle. The results have shown that the thickness of the IMC layer increased over cycles, and the shape of the IMC changed from fine scalloped to wavy, eventually to smooth. However, the IMCs in solders rapidly gathered up and grew up from the uniform distribution of fine particles. With the increase of cycles, the shear strength of the solder joints had a sharp decline, then decreased slowly and eventually stabilized. All solder joints were broken within the solder, and the fracture appearance was belonging to ductile fracture.

Keywords: low-silver hypoeutectic SAC lead-free; hot dip soldering; thermal cycle; shear strength

1 INTRODUCTION

The cost of Sn-Ag-Cu (SAC, Ag ≥ 3.0%) solder is two times higher than the Sn-Pb eutectic solder, which has been concerned by the industry (Illés et al. 2014; Shnawah et al. 2012). In order to reduce cost, low silver SAC solders came into existence (Am et al. 2012). such as SAC105 (Shnawah et al. 2012), SAC0705 (Liu et al. 2014), SAC0507 (Hammad. 2013) and SAC0307 (Mookam et al. 2012). However, the low-silver SAC solder will lead to a series of problems, like declined the wettability, coarsened the matrix microstructure and reduced the mechanical properties (Han, et al. 2014; Moshrefi-Torbati et al. 2011). The solder joint would fail in advance during their service time due to the coarsening of matrix microstructure and the growth behavior of IMC (Mookam et al. 2012). The Cu/ (SAC-X-Y)/Cu solder joint was prepared by hot dip soldering with a new low-silver SAC lead-free solder in this paper. The evolution of the IMC and its influence on the mechanical properties during the thermal cycle were also researched.

2 EXPERIMENTAL

Tin with 99.95% (in mass, similarly hereinafter), copper and silver with 99.99% were weighed as the mass ratio of 98.87: 0.68: 0.45. Then the heating and melting of all metals were carried out under the protection of the molten salt (KCl: LiCl = 1.3:1) in a crucible resistance furnace. Finally, the low-silver SAC-X-Y solder alloy was obtained by pouring into a mold after adding a certain mass ratio of X and Y elements at 673 K for 1 h.

The copper with rosin-base fluxed (it was composed of 75 g isopropanol, 25 g foral, and 0.4 g diethylamine hydrochloride) were put in a particular mold, and then the mold were immersed into the molten of low-silver SAC-X-Y solder at 543 K for 10 s to get the Cu/SAC-X-Y/Cu solder joints.

The samples were put into the high and low temperature test chamber. Thermal cycle is a temperature interval of −40~125°C. The dwell time at each temperature extreme was 0.5 h, and the ramp rate was about 1.83°C/min. Different samples were

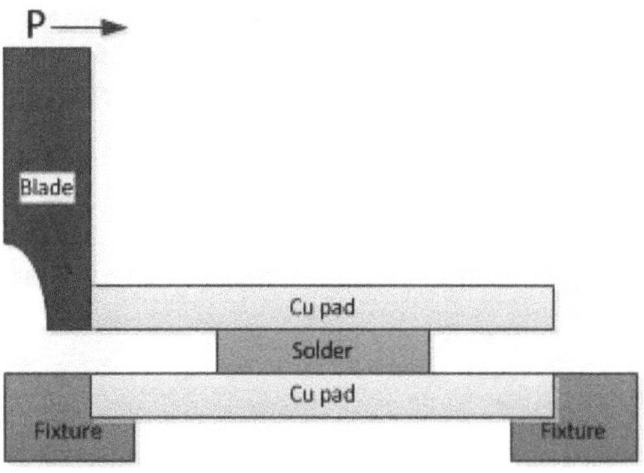

Figure 1. The shear process.

taken out after 30, 60, 90, and 120 cycles, respectively. Optical metallographic microscope and SEM were used to analyze the interfacial microstructures. The thickness of the IMC layer was calculated by the Image-pro plus software. The average shear strength of several solder joints (shear velocity is 10 mm/min) was identified by the PTR-1101 bonding strength tester. The shear process is shown in Figure 1.

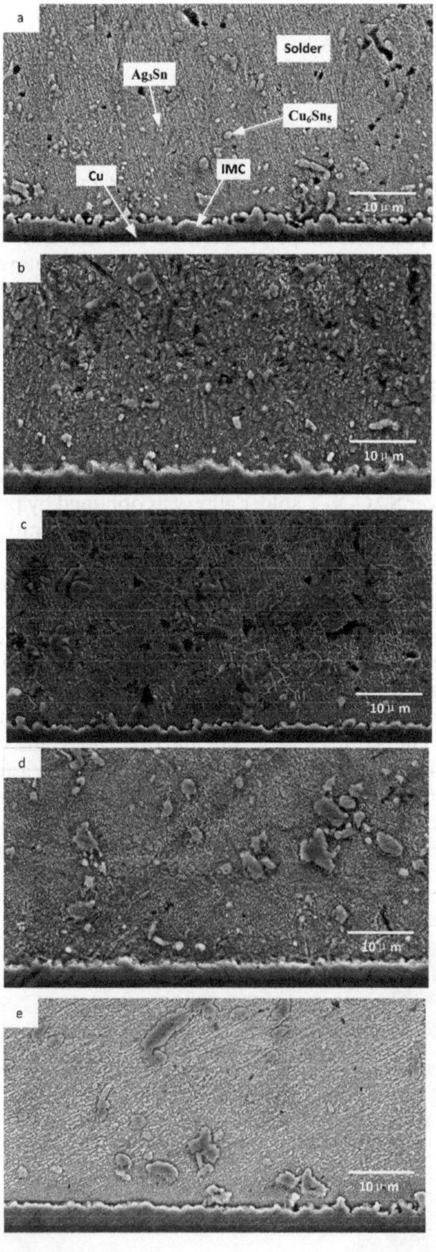

Figure 2. Effect of thermal cycles on the morphology of IMC. (a) 0 cycle; (b) 30 cycles; (c) 60 cycles; (d) 90 cycles; (e) 120 cycles.

3 RESULTS AND DISCUSSION

3.1 *Effect of thermal cycles on the IMCs*

Effect of thermal cycles on the morphology of IMC of Sn-0.45 Ag-0.68Cu/Cu is shown in Figure 2. The solder joint was consisted of the Cu matrix, IMC layer and the solder. Typical scallop-type IMCs are observed near the solder of as-reflowed joint, and its components were Cu_6Sn_5 phase which is verified through EDS and earlier research. Tiny second phase diffusing in the solder were Cu_6Sn_5 and a small amount of Ag_3Sn. The thickness of IMC was significantly thicker than as-reflowed joint, and its average thickness varied from 2.86 μm to 3.25 μm after 30 cycles. Relationship between the IMC thickness and thermal cycles is shown in Figure 2(a).

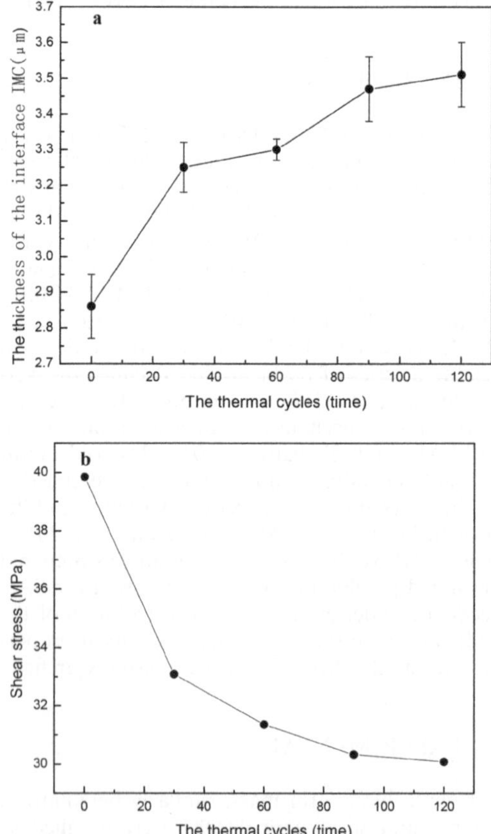

Figure 3. Effect of thermal cycles on the thickness of IMC (a) and the shear strength of solders (b) during the thermal cycle.

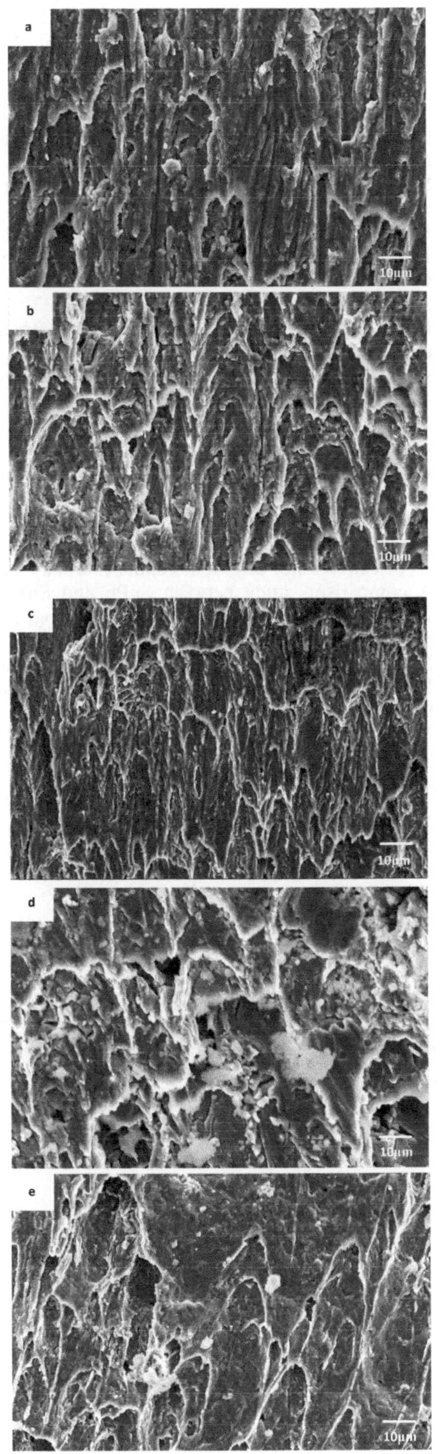

Figure 4. The fracture morphology of joints during the thermal cycle. (a) 0 cycle; (b) 30 cycles; (c) 60 cycles; (d) 90 cycles; (e) 120 cycles.

The atomic diffusion speed was fast in the interface due to high soldering temperature, however, the soldering time was short and the atomic diffusion was not sufficient, so the interface IMC was uneven and became wavy. During the thermal cycle, the diffusion rate was slower, but the inter-diffusion of elements between Sn atoms in the solder and Cu atoms in the copper substrate generated Cu_6Sn_5 to make the interfacial IMC smooth with the increase of cycles. At the same time, the IMCs in solders rapidly gathered up and grew up from the uniform distribution of fine particles.

3.2 *Effect of thermal cycles on the mechanical properties of solder joints*

The shear strength of solder joint during thermal cycles is illustrated in Figure 3(b). It can be seen that the shear strength of as-reflowed solder joint is 39.84 MPa, and then the shear strength of solder joint declined with the increasing of cycles. After 30 cycles, the shear strength of the solder joints was 33.08 MPa which decreased by about 16.97%. With the increase of cycles, the residual stress in the solder joints was to be released, which caused the toughness of the joints to become good and made the downturn of the shear strength to decrease. Brittle second phase (Cu_6Sn_5) in solders grew up during the thermal cycle and small Ag_3Sn particles reduced (Fig. 2), which led to decreased ability of anti grain boundary sliding, there is one other reason for the decreasing of the shear strength. The minimum shear strength of solder joints was 30.09 MPa after 120 cycles, which decreased by about 19%.

The fracture morphology of solder joint during the thermal cycle is shown in Figure 4. All solder joints were broken within the solder, the fractures present the parabolic dimple, and its orientation was the same with the shear orientation. With the increase of cycles, the Cu_6Sn_5 phase increased rapidly (Fig. 2), but it was not enough to make the joints brittle fracture. Therefore, the dimples of the fracture became larger and the number was less, showing a feature of ductile fracture.

4 CONCLUSIONS

1. With the increase of cycles, the interfacial IMCs of Sn-0.45 Ag-0.68Cu/Cu joints were thickening, and their shape changed from the scallop into the wavy shape, eventually to smooth. The IMC in solder rapidly gathered up and grew up from the uniform distribution of fine particles.
2. The shear strength of the solder joint had a sharp decline, then decreased slowly and eventually stabilized with the increase of cycles. The shear strength of as-reflowed solder joint was

39.84 MPa, and the minimum shear strength of solder joints was 30.09 MPa after 120 cycles. All solder joints were broken within the solder, and the fracture appearance was belonging to ductile fracture.

ACKNOWLEDGMENTS

This work was supported by the Scientific Research Staring Foundation of Chongqing University of Technology (NO. 2012ZD12), the Innovation Foundation of Graduate of Chongqing University of Technology (NO. YCX2013101 and YCX2014215) and Chongqing Municipal Engineering Research Center of Institutions of Higher Education for Special Welding Materials and Technology (NO. SWMT201502, SWMT201503 and SWMT201505) respectively.

REFERENCES

[1] Illés, B., & Horváth, B. 2014. Tin whisker growth from micro-alloyed SAC solders in corrosive climate. Journal of Alloys and Compounds 616(31): 116–121.

[2] Shnawah, D.A.A., Sabri, M.F.B.M., Badruddin, I.A. 2012. A review on thermal cycling and drop impact reliability of SAC solder joint in portable electronic products. Microelectronics Reliability 52: 90–99.

[3] Am, Y., Jw, J., Jh, L., Jk, K., & Ms., K. 2012. Tensile properties and thermal shock reliability of Sn-Ag-Cu solder joint with indium addition. Journal of Nanoscience and Nanotechnology 12(4): 3655–3657.

[4] Shnawah, D.A., Sabri, M.F.M., Badruddin, I.A., Said, S.B.M., & Che, F.X. 2012. The bulk alloy microstructure and mechanical properties of Sn-1Ag-0.5Cu-xAl solders (x = 0, 0.1 and 0.2 wt.%). J Mater Sci: Mater Electron 23(11): 1988–1997.

[5] Liu, Y., Sun, F., Liu, Y., & Li, X. 2014. Effect of Ni, Bi concentration on the microstructure and shear behavior of low-Ag SAC-Bi-Ni/Cu solder joints. J Mater Sci: Mater Electron 25(6): 2627–2633.

[6] Hammad, A.E. 2013. Investigation of microstructure and mechanical properties of novel Sn-0.5 Ag-0.7Cu solders containing small amount of Ni. Materials and Design 50(17):108–116.

[7] Mookam, N., Kanlayasiri, K. 2012. Evolution of intermetallic compounds between Sn-0.3 Ag-0.7 Cu low-silver lead-free solder and Cu substrate during thermal aging. Journal of Materials Science & Technology 28(1): 53–59.

[8] Han, C., & Han, B. 2014. Board level reliability analysis of chip resistor assemblies under thermal cycling: A comparison study between SnPb and SnAgCu. Journal of Mechanical Science and Technology 28(3): 879–886.

[9] Moshrefi-Torbati, M., & Swingler, J.M. 2011. Reliability of printed circuit boards containing lead-free solder in aggressive environments. J Mater Sci: Mater Electron 22(4): 400–411.

Material Science and Environmental Engineering – Chen (Ed.)
© 2016 Taylor & Francis Group, London, ISBN 978-1-138-02938-5

Stress Concentration Factors for the multi-planar tubular Y-joints subjected to in-plane bending loadings

B. Wang
Powerchina Huadong Engineering Corporation Limited, Hangzhou, China

N. Li & X. Li
State Key Laboratory of Coastal and Offshore Engineering, Dalian University of Technology, Dalian, China

J.Q. Xu & J.W. Zhang
Powerchina Huadong Engineering Corporation Limited, Hangzhou, China

ABSTRACT: The support structure for an offshore wind turbine is subjected to hydrodynamic loads and aerodynamic loads. The tubular joints are the weakest component leading to fatigue failure of the whole structure. Considering the load cases and their combination under fatigue limit state, two offshore wind turbines composed of Tripod and Pentapod substructures are analyzed, respectively. Then, the load types subjected to in-plane bending loadings of three-planar and five-planar tubular Y-joints are determined, respectively. Using solid element, the finite element models of three-planar and five-planar tubular Y-joints are established, respectively. The effects of geometrical parameters on SCFs are studied. Moreover, the numerical SCF results are compared with the empirical ones from DNV-RP-C203 for simple tubular Y joints.

Keywords: Y-joints; in-plane bending loading; stress concentration factor; finite element analysis; offshore wind turbine

1 INTRODUCTION

Multi-pod piled substructures are widely used in development of offshore wind farm due to the limits of fabrication, transportation, and construction in China. Fatigue is the significant factor resulting in the failure of support structure of offshore wind turbine subjected to combined aerodynamic and hydrodynamic loadings. The connection between diagonal braces and central column as the major loadbearing components is defined as the tubular joint for joint type multi-planar Y. The weld itself and geometric discontinuity around tubular joint causes stress concentration, which is a detrimental impact on fatigue performance of offshore structures.

The fatigue design is based on use of S–N curves, which are obtained from fatigue tests. S–N curve graphically presents the dependence of fatigue life (N) on fatigue strength (S). In S–N curves, the number of cycles that a tubular joint can sustain before fatigue failure is determined by the hot spot stress range. The stress concentration due to the weld itself is included in the S–N curve to be used, while the stress concentration due to the geometry effect of the actual detail is determined by means of calculation of hot spot stress. Generally, the hot spot stress is calculated from the nominal stress and Stress Concentration Factor (SCF). The SCF has been a issue for tubular structures in the past decades. At present, there are many published parametric SCF equations for tubular joints subjected to different basic loadings. The SCFs for uni-planar tubular joints are listed in the fatigue design codes, such as API[1] and DNV-RP-C203[2]. Hot spot stresses at failure-critical locations for 4 different tubular joints (DK, DKT, X-type) are derived, and the effects of planar and non-planar braces are considered by Dong[3]. Woghiren[4] studied a parametric stress analysis of various configuration of rack plate stiffened multi-planar KK-joint using the finite element method. Considering bending effects, Karamanos[5] computed SCFs in DT-joint. Ahmadi[6] proposed a set of parametric SCFs equations for the three-planar tubular KT-joint under three different axial loading conditions. Ahmadi[7] also established a set of SCF parametric equations for the fatigue design of internally ring-stiffened KT-joint.

Based on the analysis of practical projects, the load types subjected to in-plane bending loadings of three-planar and five-planar tubular Y-joints are determined. Using solid element, the finite element models of three-planar and five-planar tubular Y-joints are established, respectively. The effects of geometrical parameters on SCFs are studied.

2 BASIC INFORMATION

2.1 *Selection of geometric parameters*

Typical tripod and pentapod substructures of offshore wind turbine are shown in Figure 1. For tripod substructure, the joint connected with three diagonal braces and a central column forms a three-planar tubular joint with the angle of 120° among the diagonal braces from horizontal projection. For pentapod substructure, the joint linked with five diagonal braces and a central column forms a five-planar tubular joint with the angle of 72° among the diagonal braces from horizontal projection.

From the extensive studies by the researchers, the geometric parameters that affect the SCFs of tubular joints include the ratio of chord length to chord radius, α, the ratio of brace diameter to chord diameter, β, the ratio of chord radius to chord wall thickness, γ, the ratio of brace wall thickness to chord wall thickness, τ, the angle between the chord axis and brace axis, θ, and the angle among diagonal braces, ϕ.

Based on substructure designs of practical projects in China, the ranges of geometric parameters are determined. Then, the values of geometric parameters are selected and listed in Table 1. The SCFs of these cases are calculated.

2.2 *Finite element modeling*

Accurate modeling of the weld profile is one the most critical factors affecting the accuracy of

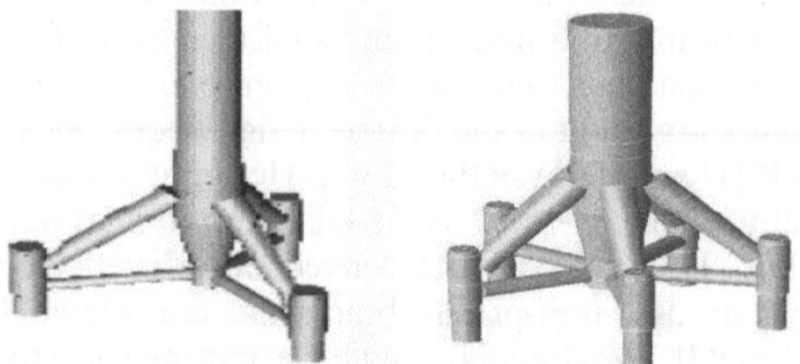

Figure 1. Substructures of offshore wind turbine.

Table 1. Geometric parameters of multi-planar tubular Y-joints.

Parameters	Three-planar Y-joint	Five-planar Y-joint
α	12, 13, 14	9, 11, 13
β	0.50, 0.58, 0.66	0.35, 0.38, 0.43
γ	20, 24, 25, 28	30, 33, 37
τ	0.60, 0.70, 0.80	0.55, 0.65, 0.70
θ	40°, 42.5°, 45°	35°, 40°, 45°
ϕ	120°	72°

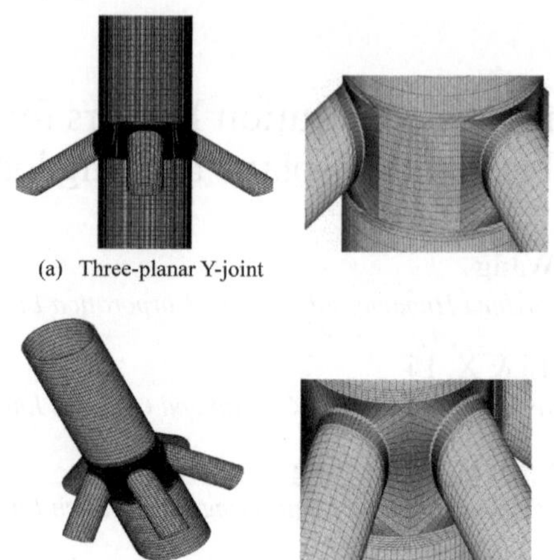

(a) Three-planar Y-joint

(b) Five-planar Y-joint

Figure 2. FE model of multi-planar tubular Y-joints.

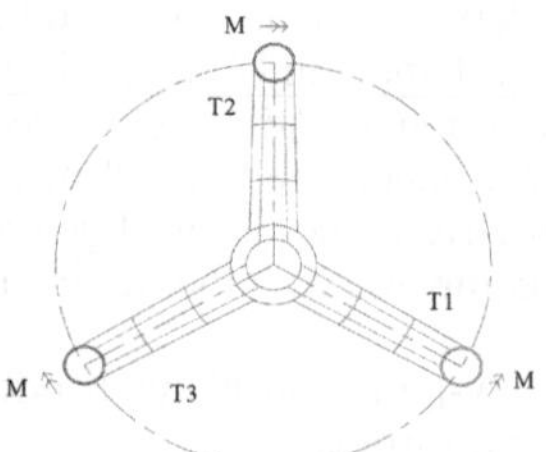

(a) In-plane bending on three braces

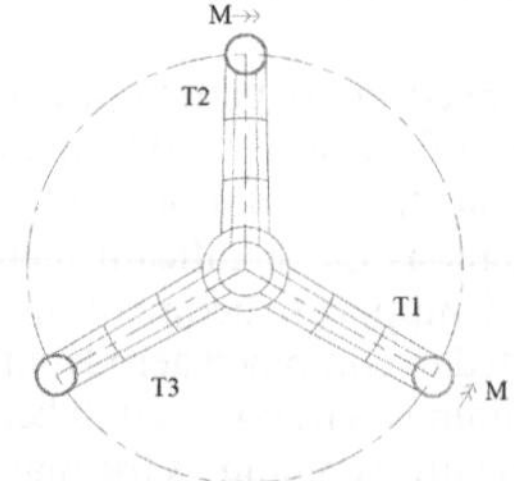

(b) In-plane bending on two braces

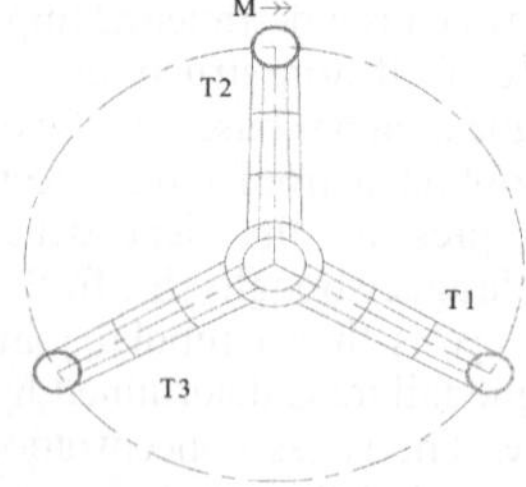

(c) In-plane bending on one brace only

Figure 3. In-plane bending loads of three-planar tubular Y-joints.

SCF results. In the present study, the welding size along the brace/chord intersection satisfies the AWS D1.1 specification[8].

Using ANSYS, the SOLID95 element type is used in the study to model the chord, brace, and the weld profile. This method will produce more accurate and detailed stress distribution near the intersection in comparison with a simple shell analysis[9]. The material is steel with elastic modulus, $E = 206$ GPa, Poisson's ratio $\nu = 0.3$.

To guarantee the mesh quality, a sub-zone mesh generation method is used during the FE modeling. In this method, the entire joint is divided into several different zones according to the computational requirements. This method can easily control the mesh quantity and quality and avoid badly distorted elements. To verify the convergence of EF results, convergence test with different mesh densities is conducted. The number of elements for both thickness of the chord, and thickness of the brace is 3. 56 elements that are used along the curve of the weld toe. The FE models of three-planar and five-planar tubular Y-joints are displayed in Figure 2.

Both chord ends are assumed to be fixed, thus, the degrees of freedom, U_x, U_y, U_z, θ_x, θ_y, θ_z, of the corresponding end nodes are restrained. The nodes in the brace end section are rigidly coupled to a node in the center of the section. The in-plane bending moment is applied to the node.

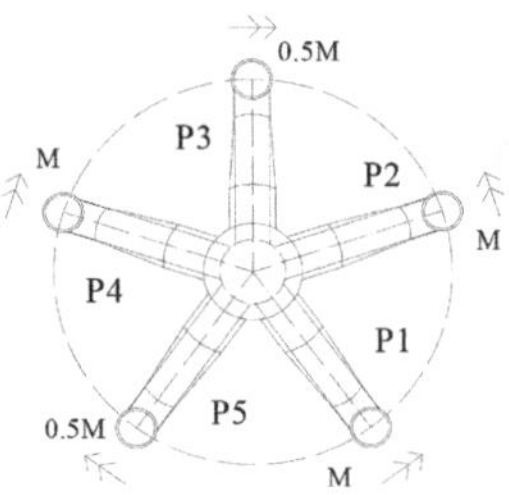

(a) In-plane bending on five braces

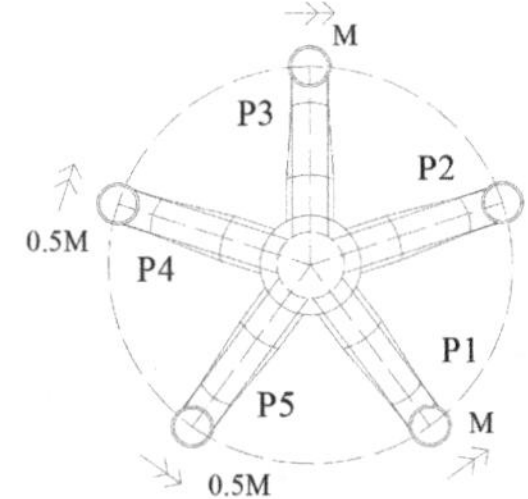

(b) In-plane bending on four braces

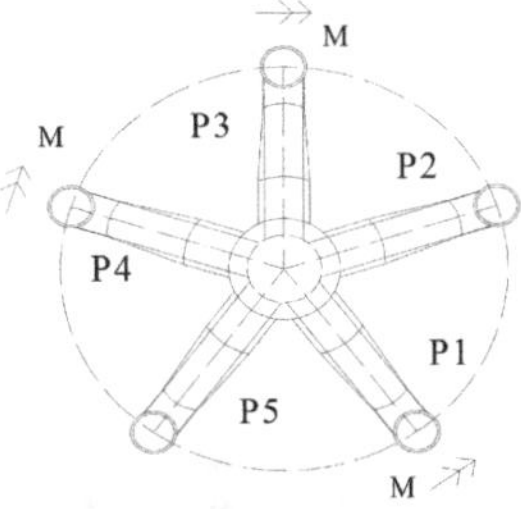

(c) In-plane bending on three braces

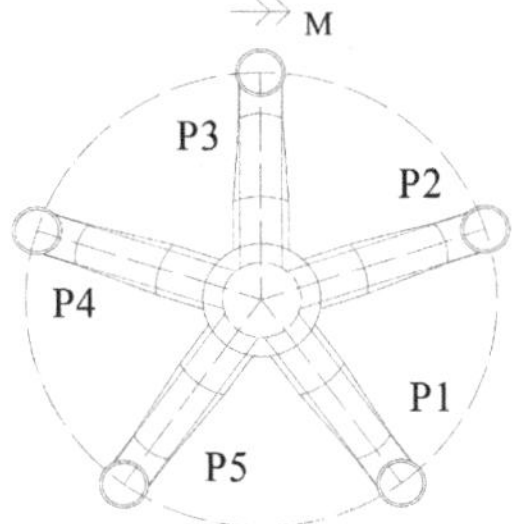

(d) In-plane bending on one brace

Figure 4. In-plane bending loads of five-planar tubular Y-joints.

2.3 *Determination of loading conditions*

Using SACS, some practical structures of offshore wind turbine with tripod and pentapod substructures are analyzed subjected to the aerodynamic and hydrodynamic loadings for fatigue limit state. From the results, the loading conditions for three-planar and five-planar Y-joints under in-plane bending moment are obtained and shown in Figure 3 and Figure 4, respectively.

3 GEOMETRICAL EFFECTS ON THE SCFS OF THREE-PLANAR Y-JOINT

Referred to the geometric parameters of three-planar tubular Y-joint listed in Table 1, the corresponding FE models are established, respectively. The SCFs of three-planar tubular Y-joint subjected to in-plane bending moment are obtained according to the loading conditions displayed in Figure 3. The comparison of FE results and results calculated by DNV-RP-C203 for simple tubular Y-joints are shown in Figure 5–Figure 9. The equations of SCFs of uni-planar tubular Y-joints are given in DNV-RP-C203.

It can be seen from Figure 5–Figure 9 that the SCFs in chord are larger than those in braces. The SCFs under the loading condition LT2 are larger than the other loading conditions.

As shown in Figure 5–Figure 9, the SCFs in chord from DNV are smaller than those from FE in some cases. The maximum deviation is about 9.0%. The SCFs in braces from DNV are smaller than FE results in any cases.

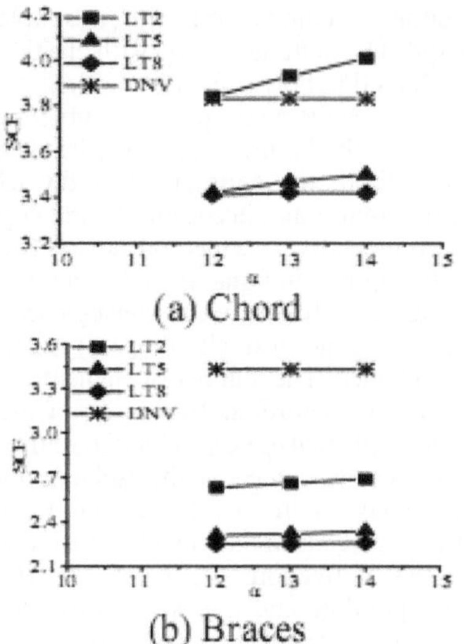

Figure 5. Effect of α on SCF for three-planar tubular Y-joints ($\beta = 0.58$, $\gamma = 25$, $\tau = 0.8$, $\theta = 45°$).

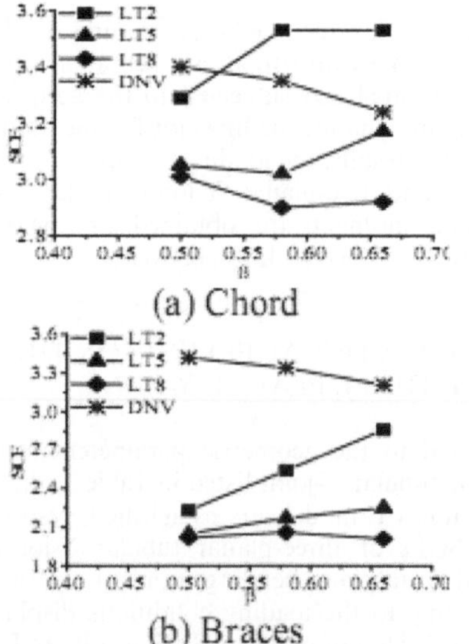

Figure 6. Effect of β on SCF for three-planar tubular Y-joints ($\alpha = 13$, $\gamma = 20$, $\tau = 0.8$, $\theta = 45°$).

3.1 *Effect of the α on the SCFs*

From Figure 5(a), the SCFs in chord increase with the increasing α in the range of 12 and 14, and the SCFs from FE analysis for loading condition LT2 are larger than those from DNV. From Figure 5(b), the SCFs in braces almost keep constant with the increase of α, and the SCFs from DNV are larger than those from FE.

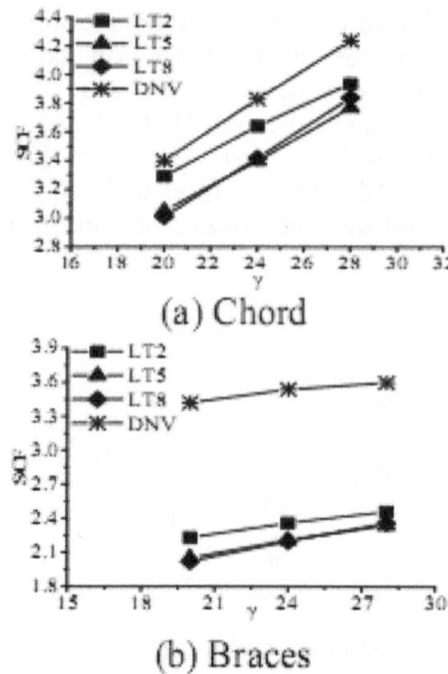

Figure 7. Effect of γ on SCF for three-planar tubular Y-joints ($\alpha = 13$, $\beta = 0.5$, $\tau = 0.8$, $\theta = 45°$).

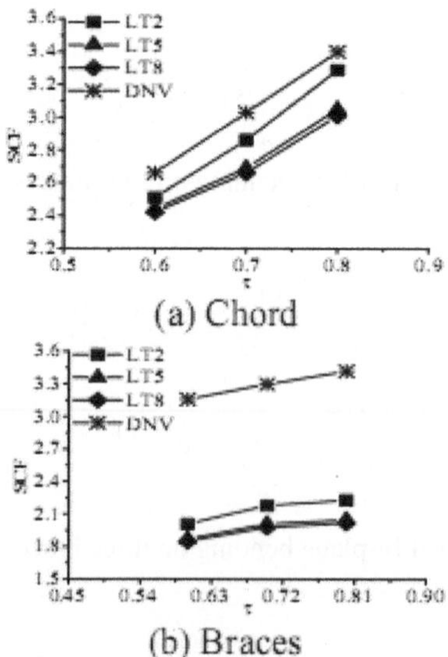

Figure 8. Effect of τ on SCF for three-planar tubular Y-joints ($\alpha = 13$, $\beta = 0.5$, $\gamma = 20$, $\theta = 45°$).

3.2 *Effect of the β on the SCFs*

From Figure 6, the SCFs increase with the increasing β in the range of 0.50 and 0.66. The SCFs in chord from FE analysis for loading condition LT2 are larger than those from DNV, while the SCFs of braces from DNV are larger than those from FE subjected to the three loading conditions.

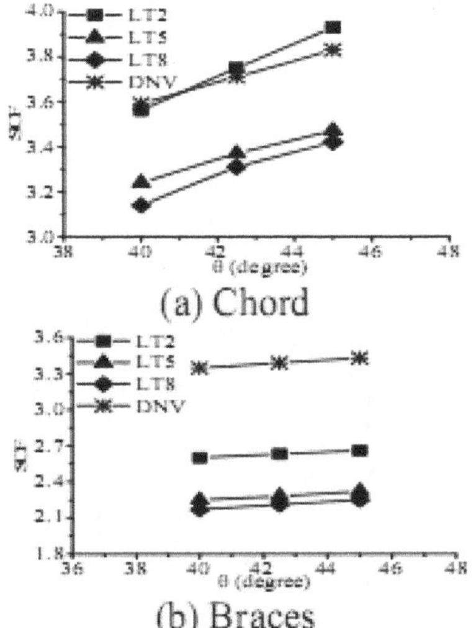

Figure 9. Effect of θ on SCF for three-planar tubular Y-joints ($\alpha = 13$, $\beta = 0.58$, $\gamma = 25$, $\tau = 0.8$).

3.3 *Effect of the γ on the SCFs*

From Figure 7, the SCFs from DNV are larger than those from FE subjected to the three loading conditions. The SCFs of chord increases quickly, while the SCFs of brace slowly increase, with the increasing γ in the range of 20 and 28.

3.4 *Effect of the τ on the SCFs*

From Figure 8, the SCFs from DNV are larger than those from FE subjected to the three loading conditions. The SCFs in chord obviously increase, while the SCFs of braces slowly increase, with the increasing τ in the range of 0.6 and 0.8.

3.5 *Effect of the θ on the SCFs*

From Figure 9, the SCFs increase with the increasing θ in the range of 40 and 45. The SCFs in chord from FE analysis for loading condition LT2 are larger than those from DNV, while the SCFs in braces from DNV are larger than those from FE subjected to the three loading conditions.

4 GEOMETRICAL EFFECTS ON THE SCFS OF FIVE-PLANAR Y-JOINT

Based on the geometric parameters of five-planar tubular Y-joint listed in Table 1, the corresponding FE models are established, respectively. The SCFs of five-planar tubular Y-joint subjected to in-plane bending moment are obtained according to the loading conditions displayed in Figure 4. The comparison of FE results and results calculated by DNV-RP-C203 for simple tubular Y-joints are shown in Figure 10 – Figure 14.

It can be seen from Figure 10 – Figure 14 that the SCFs in chord are larger than those in braces.

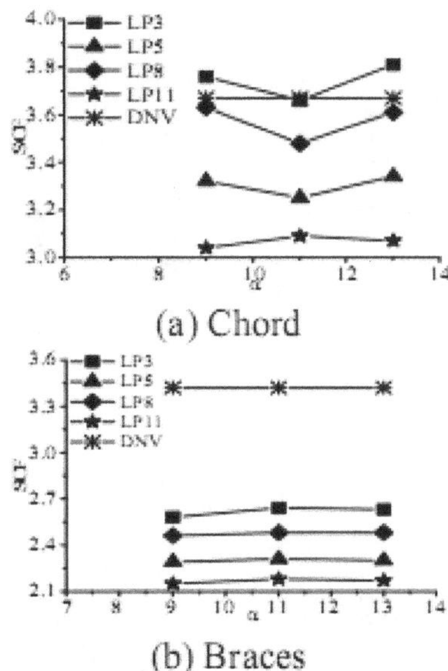

Figure 10. Effect of α on SCF for five-planar tubular Y-joints ($\beta = 0.35$, $\gamma = 30$, $\tau = 0.7$, $\theta = 40°$).

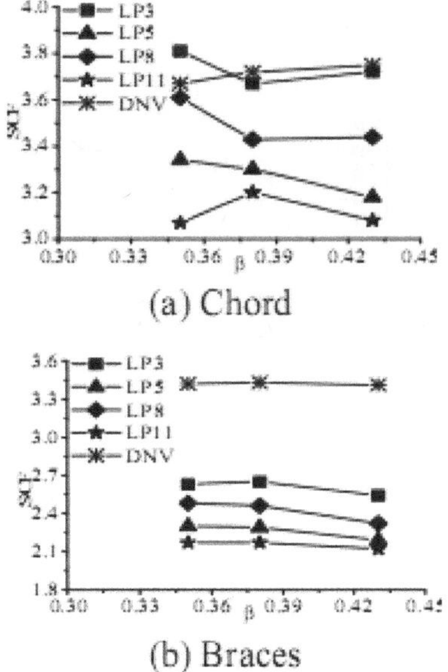

Figure 11. Effect of β on SCF for five-planar tubular Y-joints ($\alpha = 13$, $\gamma = 30$, $\tau = 0.7$, $\theta = 40°$).

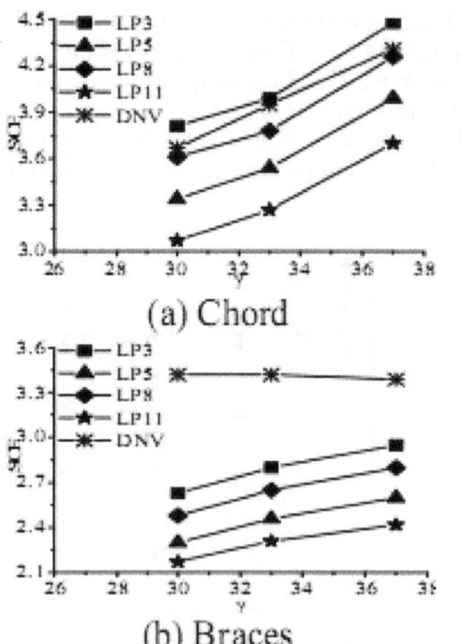

Figure 12. Effect of γ on SCF for five-planar tubular Y-joints ($\alpha = 13$, $\beta = 0.35$, $\tau = 0.7$, $\theta = 40°$).

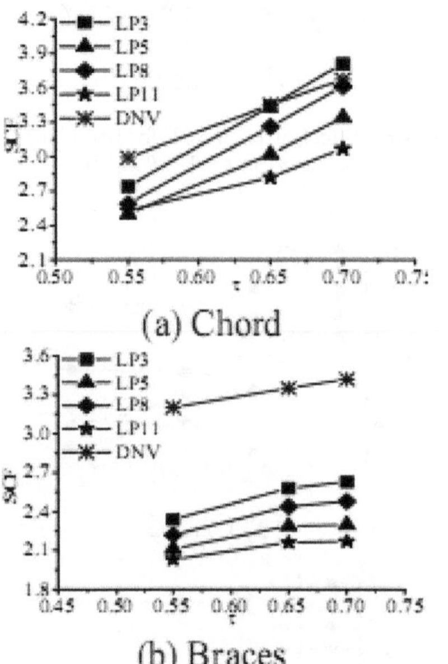

Figure 13. Effect of τ on SCF for five-planar tubular Y-joints ($\alpha = 13$, $\beta = 0.35$, $\gamma = 30$, $\theta = 40°$).

The SCFs under the loading condition LP3 are larger than the other loading conditions.

As shown in Figure 10–Figure 14, the SCFs in chord from DNV are smaller than those from FE in some cases. The maximum deviation is about 5.7%.

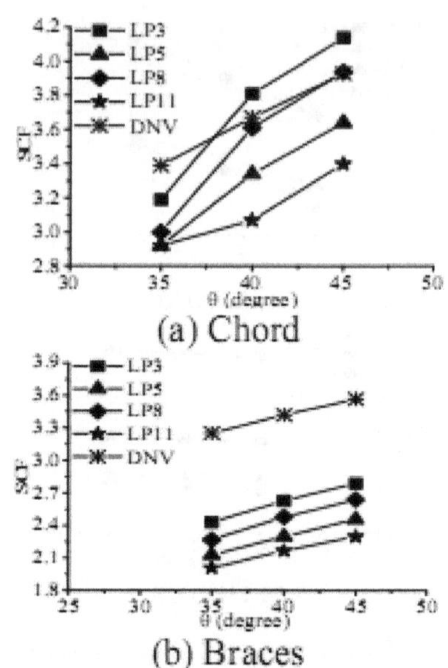

Figure 14. Effect of θ on SCF for five-planar tubular Y-joints ($\alpha = 13$, $\beta = 0.35$, $\gamma = 30$, $\tau = 0.7$).

The SCFs in braces from DNV are smaller than FE results in any cases.

4.1 *Effect of the α on the SCFs*

Figure 10 shows that the parameter α has effect on SCFs in chord, while has little effect on SCFs in braces.

4.2 *Effect of the β on the SCFs*

From Figure 11, the effect of the parameter β on SCFs in chord is complex. The other geometric parameters should be considered to determine SCFs.

4.3 *Effect of the γ on the SCFs*

As displayed in Figure 12, the parameter γ has significant influence on SCFs in chord.

4.4 *Effect of the τ on the SCFs*

The SCFs in both chord and braces increase with the increasing parameter τ in the range of 0.55 and 0.70 as shown in Figure 13.

4.5 *Effect of the θ on the SCFs*

From Figure 14, the parameter θ has obvious impact on SCFs on both chord and braces. And the

SCFs increase with the increase of parameter θ in the range of 35 and 45.

5 CONCLUSIONS

Using FE simulation, the effects of geometric parameters on SCFs of three-planar and five-planar tubular Y-joints subjected to in-plane bending moment are analyzed. The following conclusions can be drawn.

1. For both three-planar and five-planar tubular Y-joints subjected to in-plane bending moment, SCFs in chord are larger than those in braces. SCFs are maximum when all the diagonal braces are applied to in-plane bending moment, indicating that the interaction among each planar Y-joints composed of multi-planar Y-joints occurs.
2. For both three-planar and five-planar tubular Y-joints subjected to in-plane bending moment, the parameter α has little effect on SCFs, while the parameters γ, τ, θ have significant influence on SCFs.
3. For the SCFs in braces, the results from DNV are larger than numerical results; while for the SCFs in chord, the results from DNV are smaller than numerical results in some cases. Thus, if the SCFs in chord from DNV suitable for simple tubular Y-joints are used to carry out fatigue assessment of multi-planar tubular Y-joints, the check will not be conservative.

ACKNOWLEDGMENTS

This work is funded by the National Natural Science Foundation of China (Grant No. 51179026), and supported by the Science Foundations of Powerchina Huadong Engineering Corporation Limited (No. KY120228-03-07 and KY2014-02-41) and the Open Fund Project of State Key Lab of Coastal and Offshore Engineering in Dalian University of Technology (No. LP1413). Their financial supports are gratefully acknowledged.

REFERENCES

[1] API. Recommended Practice for Planning, Designing and Constructing Fixed Offshore Platforms—Load and Resistance Factor Design. 1997.

[2] DNV-RP-C203. Fatigue Design of Offshore Steel Structures. 2011.

[3] Dong W.B., Moan T., Gao Z. Long-term fatigue analysis of multi-planar tubular joints for jacket-type offshore wind turbine in time domain. Engineering Structures, 2011, 33: 2001–2014.

[4] Woghiren C.O., Brennan F.P. Weld toe stress concentrations in multi-planar stiffened tubular KK joints. International Journal of Fatigue. 2009, 31(1): 164–172.

[5] Karamanos S.A., Romeijn A., Wardenier J. SCF equations in multi-planar welded tubular DT-joints including bending effects. Marine structures. 2002, 15(2): 157–173.

[6] Ahmadi H., Mohammad A.L., Mohammad H.A. The development of fatigue design formulas for the outer brace SCFs in offshorue three-planar tubular KT-joints. Thin-Walled Structures, 2012, 58(67–78).

[7] Ahmadi H., Mohammad A.L., Shao Y.B., Mohammad H.A. Parametric study and formulation of outer-brace geometric stress concentration factors in internally ring-stiffened tubular KT-joints of offshore structures. Applied Ocean Research, 2012, 38: 74–91.

[8] American Welding Society (AWS). Structural Welding Code, AWS D1.1. Miami, USA, 2009.

[9] Chiew S.P., Soh C.K., Wu N.W. General SCF design equations for steel multiplanar tubular XX-joints. International Journal of Fatigue. 2000, 22(4): 283–293.

Material Science and Environmental Engineering – Chen (Ed.)
 ISBN 978-1-138-02938-5

Influence of bearing materials on elastohydrodynamic lubrication of angular contact ball bearing

B.M. Wang, X.Y. Liao & Z.X. Wu
School of Mechanical Electronical Engineering, Lanzhou University of Technology, Gansu, Lanzhou, China

X.S. Mei
School of Mechanical Engineering, Xi'an Jiaotong University, Shanxi, Xi'an, China

ABSTRACT: Based on the elastohydrodynamic lubrication (EHL) theory, the EHL mathematical model of point contact for angular contact ball bearing is established. The comprehensive calculating method and its corresponding program were given for calculating the maximum pressure and minimum thickness of oil film in angular contact ball bearing. Taking 7006 bearing as an example, the results show that, compared with full-steel and full ceramic angular contact ball bearing, oil film is the thickest in hybrid ceramic angular contact ball bearing, oil film pressure is the least in hybrid ceramic angular contact ball bearing, temperature rise is the lowest in full ceramic angular contact ball bearing; comprehensive analysis shows that lubrication property of hybrid ceramic angular contact ball bearing is best in these three kinds of angular contact ball bearings. In order to investigate the effect of different ceramic materials parameters on bearing EHL, six kinds of full ceramic bearings are further analyzed. Oil film is the thickest in bearing with SiC ball and ZrO_2 ring; oil film pressure is minimum in bearing with SiC ball and ZrO_2 ring, temperature rise is the lowest in bearing with Si_3N_4 ball and ring. Analysis shows that, the ball density has great influence on bearing EHL because of speed effect; synthetic elasticity modulus is also an important factor, the bigger synthetic elastic modulus is, the thicker film thickness is, however, too high elasticity modulus leads to stress concentration, which also make the pressure rise; it is also found that specific heat capacity has the most influence on bearing temperature rise.

Keywords: angular contact ball bearing; point contact; EHL; contrastive analysis; material

1 INTRODUCTION

Speed machining has become the mainstream of metal cutting. To achieve this goal, the rotational speed of the spindle and feed system must be increased. So, angular contact ball bearings get wide use in this field, because of its high speed, high temperature resistant, and good rigidity [1]. Lubrication plays an important role in the proper functioning of a bearing, preventing ball-to-ring contact which may damage the bearing and lead to failure. This also reduces the energy consumption of a machine since friction forces become less important when surface separation is ensured by a complete lubricant film. In many operating conditions, the pressure generated in the lubricant film is high enough to induce considerable elastic deformation of the contacting bodies. This lubrication regime is known as ElastoHydroDynamic (EHD) [2]. In this sense, it is essential to understand the EHL in angular contact ball bearings.

In the past decade, a great deal of work has been done in understanding bearing EHD. Literature [3] introduces the influence of surface waviness and lubrication of ball bearing on bearing stiffness by experimental verification. Literature [4, 5] expounds the advantages of mix ceramic ball bearing. The lubrication film thickness and hydrodynamic contact pressure are calculated, and the influences of working load, speed, initial contact angle, and ball diameter on the lubrication state are taken into consideration in literature [6,7,8]. Literature [9–11] describe the effects of oil-supply and working speed on the resistance of the oil film and the temperature of the bearing. It should be pointed out that these literatures just discuss one kind material bearing. With the rapid development of material industry, more and more new materials, especially ceramic materials, are used in angular contact ball bearing. So the effect of materials on bearing EHL cannot be ignored.

In this paper, based on the elastohydrodynamic lubrication (EHL) theory, the EHL mathematical model of point contact for angular contact ball bearing is established. The elastohydrodynamic lubricating property of angular contact ball bearing with different materials is contrasted and analyzed. That can help us to find the best material to make the best property of bearing. [12,13]

2 THE ELASTOHYDRODYNAMIC LUBRICATION MATHEMATICAL MODEL OF POINT CONTACT FOR ANGULAR CONTACT BALL BEARING

Most of angular contact ball bearings are lubricated by oil/air. The principle of oil/air lubrication is that a drop of oil is split by air flow, dispersed in streaks, and transported by air flow. The lubrication is compressed in the nozzle and fed to the friction point in droplet from with the help of air. The oil remains at the friction point and the air can escape unimpeded into the open [14].

For ease of analysis, point contact elastohydrodynamic lubrication pair can be simplified as an ellipsoid and an infinite plane (Fig. 1). The ellipsoid rotates around X axis, the tangential velocity of the contact point is u_2; the infinite plane moves along the fixed direction. Its speed is u_1 and is parallel to u_2. [15–17]

2.1 *Basic equations*

Assuming that the lubricant is Newtonian fluid, the basic equation of isothermal steady state lubrication between rolling elements and ring is given by [18–20]

$$\frac{\partial}{\partial x}\left[\left(\frac{\rho}{\eta}\right)_e h^3 \frac{\partial p}{\partial x}\right]+\frac{\partial}{\partial y}\left[\left(\frac{\rho}{\eta}\right)_e h^3 \frac{\partial p}{\partial y}\right]=12\frac{\partial}{\partial x}(\rho^* u_e h) \tag{1}$$

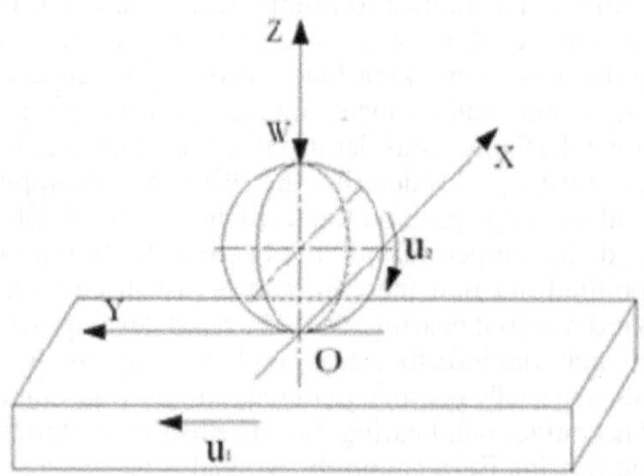

Figure 1. Schematic diagram of point contact elastohydrodynamic lubrication.

where variables are defined as follows:

$$\left(\frac{\rho}{\eta}\right)_e = 12(\eta_e \rho'_e / \eta'_e - E');$$

$$\rho^* = \frac{\rho'_e \eta_e (u_b - u_a) + \rho_e u_e}{u_e};$$

$$\rho_e = \frac{1}{h}\int_0^h \rho dz; \quad \rho'_e = \frac{1}{h^2}\int_0^h \rho \int_0^z \frac{dz'}{\eta^*} dz;$$

$$\frac{1}{\eta_e} = \frac{1}{h}\int_0^h \frac{1}{\eta^*} dz; \quad \frac{1}{\eta'} = \frac{1}{h^2}\int_0^h \frac{z}{\eta^*} dz;$$

$$\eta^* = \frac{\eta \cdot \tau}{\tau_0 \sinh(\tau/\tau_0)}.$$

where ρ is lubrication oil density, p represents lubrication oil pressure, h is film thickness, u_e is entrainment velocity of lubrication oil, η is dynamic viscosity of lubrication oil, η^* is equivalent viscosity of lubrication oil, τ is shear stress of lubrication oil, τ_0 is characteristics shear stress, x, y, z represent coordinate variable.

The boundary conditions of Reynolds equation are

$$\begin{cases} p(x_{in}, y) = p(x_{out}, y) = p(x, y_{in}) = p(x, y_{out}) = 0 \\ p(x, y) \geq 0 \\ x_{in} < x < x_{out} \\ y_{in} < y < y_{out} \end{cases} \tag{2}$$

2.2 *Film thickness equation*

When the elastic deformation between two contact surfaces is considered, oil film thickness of bearing is given by follows [21]

$$h(x,y) = h_{00} + \frac{x^2}{2R_x} + \frac{y^2}{2R_y} + \frac{2}{\pi E'}\iint \frac{p(x',y')}{\sqrt{(x-x')^2+(y-y')^2}} dx'dy' \tag{3}$$

where h_{00} is rigid body displacement of geometric, E' is synthetical elastic modulus, R_x, R_y represent equivalent curvature radius.

2.3 *Force equilibrium equation*

In the state of full film lubrication, oil film undertakes the load completely, so the resultant force of film pressure in the contact area must keep balance with external load. It can be written as

$$\iint p(x,y)\, dxdy = W \tag{4}$$

where W is normal external load.

2.4 *Viscosity-pressure equation*

Lubrication oil viscosity can be reflected by Reynolds viscosity-pressure relation, which expresses the relationship between viscosity and pressure exactly

$$\eta = \eta_0 \exp\left\{A_1 + \left[-1 + (1 + A_2 p)^{Z_0} (A_3 t - A_4)^{-S_0}\right]\right\} \quad (5)$$

where η_0 is lubrication oil environment viscosity, $A_1 = \ln\eta_0 + 9.67$, $A_2 = 5.1 \times 10^{-9}$ Pa^{-1}, $A_3 = 1/(t_0 - 138)$ K^{-1}, $A_4 = 138/(t_0 - 138)$, $Z_0 = \alpha/(A_1 A_2)$, $S_0 = \beta/(A_1 A_3)$, α is viscosity-pressure coefficient, β is viscosity-temperature coefficient, Z_0 and S_0 represent dimensional constant, t is lubrication oil temperature, t_0 is environment temperature.

2.5 *Density-pressure equation*

In the studies of elastohydrodynamic lubrication, the relationship between lubrication oil density and pressure is expressed as follows

$$\rho = \rho_0 \left[1 + C_1 p/(1 + C_2 p) - C_3 (t - t_0)\right] \quad (6)$$

where ρ_0 is lubrication oil environment density, $C_1 = 0.6 \times 10^{-9}$ Pa^{-1}, $C_2 = 1.7 \times 10^{-9}$ Pa^{-1}, $C_3 = 7.5 \times 10^{-4}$ K^{-1}.

In order to reduce the parameters and improve the stability of the calculation, dimensionless processing should be carried in each equation. Define the dimensional parameters as follows

$$P = p/p_H, \bar{\eta} = \eta/\eta_0, \bar{\rho} = \rho/\rho_0, W = w/(E'R),$$
$$U = \eta_0 u/(E'R), H = hr/b^2$$

The mathematical model of the equation can be described with normalized dimensionless parameters.

3 CALCULATION METHOD AND CORRESPONDING PROGRAM

Film pressure is solved by multigrid method, film thickness is calculated with multigrid integral method. Multigrid method means iterate on dense grid and sparse grid by turns for one problem, so that the components of high frequency deviation and low frequency deviation can be eliminated quickly. The grid divided into four layers, the highest layer has 257 nodes. In order to make the pressure and load the iterative convergence, two successive iterative relative errors must keep less than 10^{-3} respectively. The calculation flow chart is shown in Figure 2.

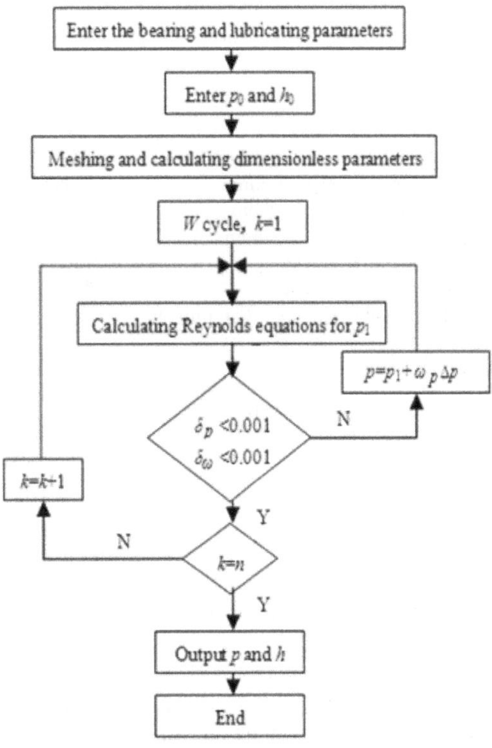

Figure 2. Computation chart of process.

4 EXAMPLE ANALYSIS

4.1 *Lubrication property of outer ring of ceramic angular contact ball bearing*

By taking 7006C angular contact ball bearing as an example, its structural parameters are shown in Table 1. The material of rings is $GCr_{15,}$ the material of ball is Si_3N_4. The material parameters of GCr_{15} and Si_3N_4 are shown in Table 2.

Bearing axial preload is 60 N and rotational speed is 8000 rpm. Numerical solution of maximum film pressure p minimum oil film thickness h and maximum temperature rise, which reflect lubrication between outer ring and ball in ceramic angular contact ball bearing. It is shown in Figure 3.

Figure 3 shows that oil film necking phenomena appears in the export zone, this is due to the lubrication oil appear divergent in this area, which makes a sharp drop in the oil film pressure. When the oil film pressure of export zone is lower than Hertz pressure, it will cause necking phenomena. Because of necking, the flowing oil gets huge compression, which causes pressure increases dramatically and narrow secondary pressure peak appears in the corresponding position. Especially, both sides of contact area, oil is divergent along the gap

Table 1. Structural parameters of a ceramic angular contact ball bearing.

Parameters	Value
The number of ball, Z	16
The initial contact angle, $\alpha/°$	15
The ball diameter, D_b/mm	5.998
The inner diameter, d/mm	30
The outer diameter, D/mm	55
The material of rings	GCr_{15}
The material of ball	Si_3N_4
Rotational speed, $n/(\times 10^4 r \cdot min^{-1})$	0~5

Table 2. Material parameters of GCr_{15} and Si_3N_4.

Parameters	GCr_{15}	Si_3N_4
Elasticity modulus, E/GPa	206	310
Poisson ration υ	0.29	0.26
Density $\rho/kg \cdot m^{-3}$	7850	3200
Specific heat capacity, $C/J \cdot (kg \cdot K)^{-1}$	470	840
Thermal conductivity, $\lambda/W \cdot (m \cdot K)^{-1}$	38	25

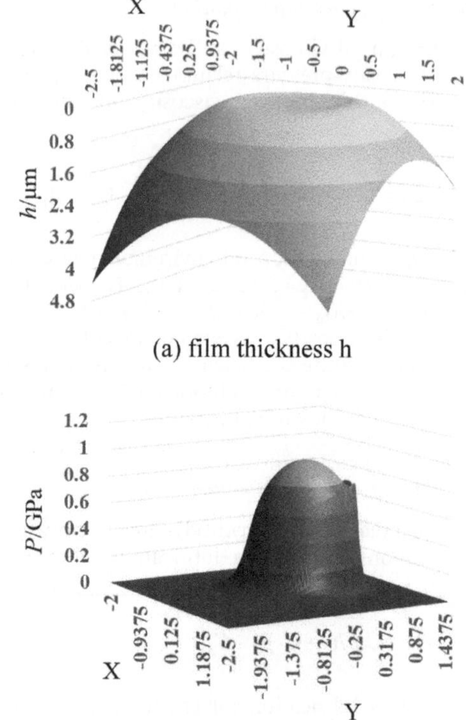

(a) film thickness h

(b) film pressure p

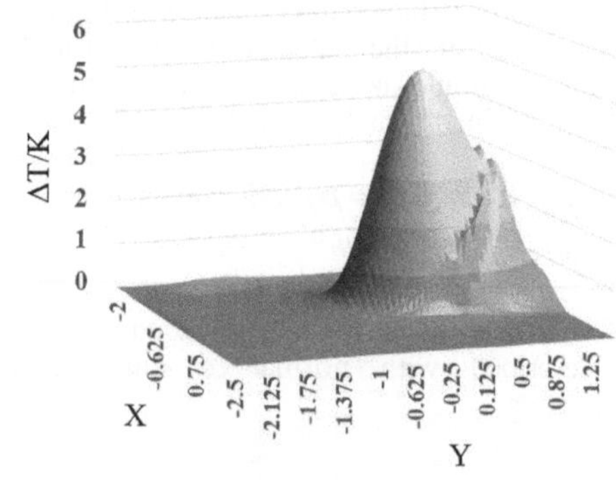

(c) temperature rise ΔT

Figure 3. Numerical solution of h, p, and ΔT of ceramic angular contact ball bearing.

of side leakage flow direction. A sharp drop in pressure makes the contact face raised. The elastohydrodynamic oil film displays as a horseshoe, the minimum film thickness exists in the both sides of export zone. The oil film pressure closes to Hertz pressure distribution in the middle contact area, and film are approximately parallel. At the secondary pressure peak, due to the shrinkage oil film, the necking phenomenon forms in the export zone, where the film thickness is the minimum.

4.2 *Lubrication property contrast of ceramic angular contact ball bearing with full steel, mix ceramic and full ceramic*

In Figure 4 oil film pressure, thickness, and temperature rise in three different kinds of angular contact; ball bearings are compared when outer ring contact with ball under different speed and load. The three kinds of bearings are full steel angular contact ball bearing with GCr_{15} rings and balls, mix ceramic angular contact ball with GCr_{15} rings are and Si_3N_4 balls, full ceramic angular contact ball bearing with Si_3N_4 rings and balls.

Figure 4 shows that lubrication oil film thickness is thicker in mix ceramic angular contact ball bearing than in common full steel and full ceramic angular contact ball bearing, and film pressure is less in mix ceramic angular contact ball bearing than full steel and full ceramic angular contact ball bearing under the same conditions. Compared with thermal conductivity, specific heat capacity has the greater influence on bearing temperature rise. Because of its biggest specific heat capacity, temperature rise in full ceramic angular contact ball bearing is the lowest. Comprehensive analysis shows that lubrication property of mix ceramic angular contact ball bearing is best in these three kinds of angular contact ball bearings. It is also found that the ball density has great influence on bearing EHL because of speed effect.

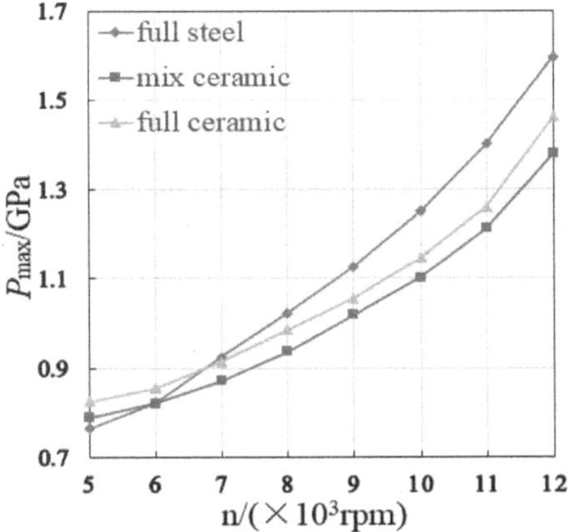

(a) maximum oil film pressure p

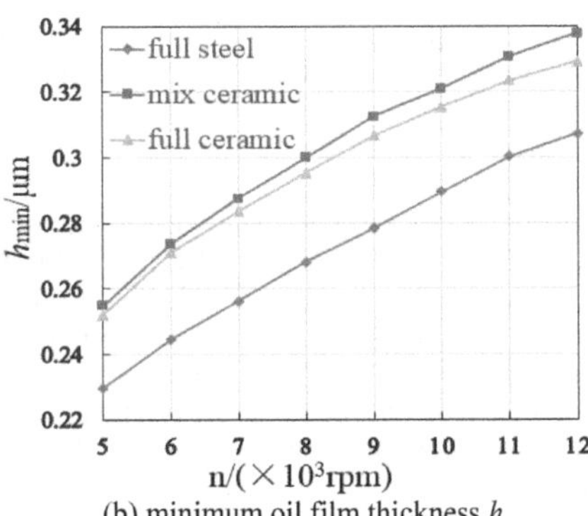

(b) minimum oil film thickness h

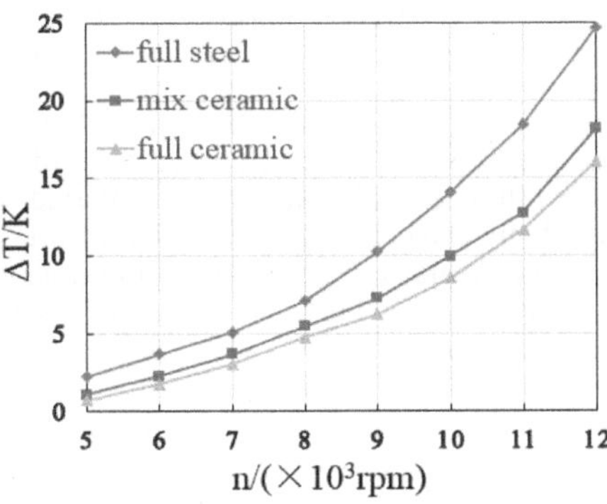

(c) maximum temperature rise ΔT

Figure 4. Numerical solution of p, h and ΔT of full steel, mix ceramic and full ceramic angular contact ball bearing under different speed (F = 60 N).

Figure 4(a) shows when speed rise up to 6000 rpm, film pressure of full steel angular contact ball bearing rise sharply. It is because the centrifugal force of rolling element that cannot be ignored when speed is high enough.

The comparison of mix ceramic bearing and full ceramic bearing show that different materials have different elastic modulus that can lead to larger elastic deformation and bigger contact area, so the oil film pressure is relatively small. Meanwhile, too big elasticity modulus leads to stress concentration, which also makes the pressure rise. Compared with synthetic elasticity modulus, oil film pressure has the greater influence on oil film thickness. Therefore, mix ceramic angular contact ball bearing has the thickest oil film.

In order to investigate the effect of other material parameters on bearing EHL, six kinds of full ceramic bearings are further analyzed.

4.3 *Lubrication property contrast of different material full ceramic angular contact ball bearings*

At present, besides Si_3N_4 ceramic materials used in bearings are SiC and ZrO_2. Material parameters of SiC and ZrO_2 are shown in Table 3, respectively, choose SiC, ZrO_2 and Si_3N_4 as ring and ball material of ceramic angular contact ball bearing, six different kinds of bearings are shown in Table 4.

Figure 5 present six kinds of full ceramic angular contact ball bearing film pressure and thickness contrast diagrams under different speed. Figure 5(a) shows film pressure is less in bearing with SiC ball and ZrO_2 ring than other ceramic angular contact ball bearing under the same conditions. The reason lies on SiC and ZrO_2 have the largest difference modulus of elasticity, which lead to largest elastic deformation and biggest contact area, the oil film pressure is relatively smaller.

In Figure 5(b) it can be found that the oil pressure is the biggest influence factor for oil film thickness,

Table 3. Material parameters of SiC and ZrO_2.

Parameters	SiC	ZrO_2
Elasticity modulus, E/GPa	410	210
Poisson ration υ	0.14	0.3
Density ρ/kg·m^{-3}	3220	5600
Specific heat capacity, C/J·(kg·K)$^{-1}$	450	400
Thermal conductivity, λ/W·(m·K)$^{-1}$	83.6	2.5

Table 4. Material of ball and rings for full ceramic angular contact ball bearing.

	1	2	3	4	5	6
Ball	Si_3N_4	Si_3N_4	Si_3N_4	SiC	SiC	SiC
Rings	Si_3N_4	ZrO_2	SiC	ZrO_2	SiC	Si_3N_4
E′/(GPa)	332.5	272.4	370.4	297.4	418.2	370.4

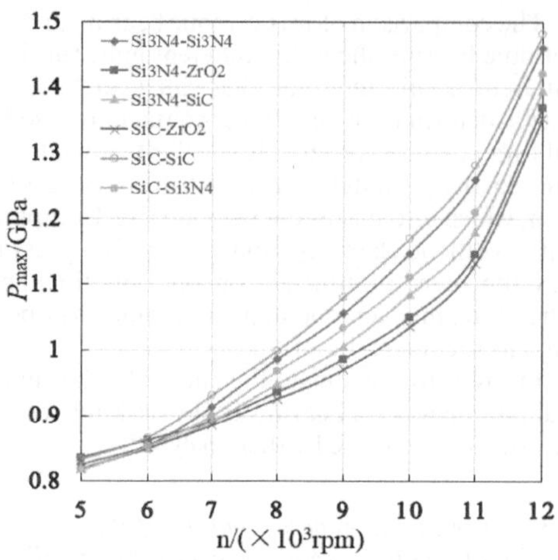

(a) maximum oil film pressure *p*

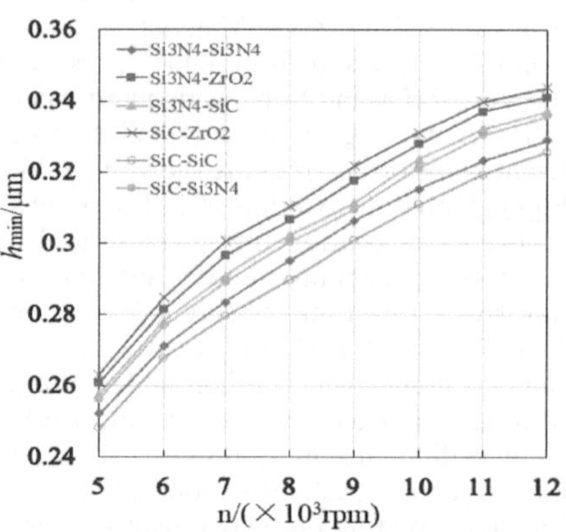

(b) minimum oil film thickness *h*

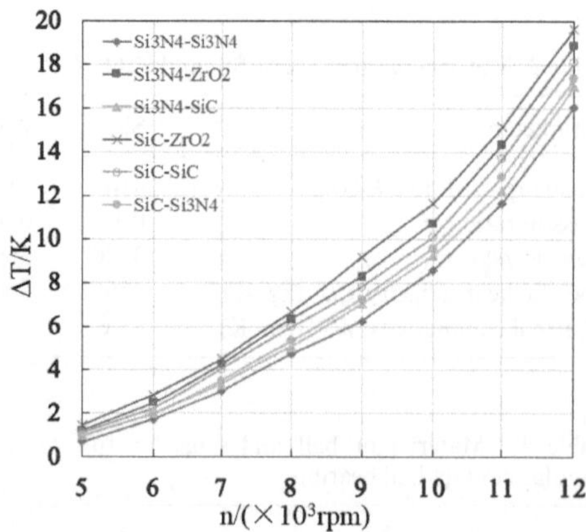

(c) maximum temperature rise ΔT

Figure 5. Six kind of full ceramic angular contact ball bearing lubrication property contrasts under different speed.

as well. Oil film thickness is thicker in bearing with SiC ball and ZrO_2 ring than other ceramic angular contact ball bearing. Similarly, compared with synthetic elasticity modulus, oil film pressure has the greater influence on oil film thickness. Smaller oil pressure leads to thicker oil film thickness.

Figure 5(c) shows bearing with Si_3N_4 ball and ring have the lowest temperature rise, which is also because of its biggest specific heat capacity.

4 CONCLUSIONS

1. Among full steel, mix ceramic and full ceramic angular contact ball bearings, mix ceramic angular contact ball bearing has the thickest oil film and the least oil film pressure, full ceramic angular contact ball bearing has the lowest temperature rise. Comprehensive analysis shows that lubrication property of mix ceramic angular contact ball bearing is best in these three kinds of angular contact ball bearings.
2. In order to investigate the effect of other material parameters on bearing EHL, six kinds of full ceramic bearings are further analyzed. Oil film is the thickest in bearing with SiC ball and ZrO_2 ring, while oil film pressure is minimum in bearing with SiC ball and ZrO_2 ring, temperature rise is lowest in bearing with Si_3N_4 ball and ring.
3. Analysis shows that, the ball density has greater influence on bearing EHL because of speed effect. Both oil film pressure and synthetic elasticity modulus has great influence on bearing, synthetic elasticity modulus which is determined by elasticity modulus and poisson ration of ball and ring. Compared with synthetic elasticity modulus, oil film pressure has the greater influence on oil film thickness. Meanwhile, too big elasticity modulus leads to stress concentration, which also makes the pressure rise.
4. It is also found that compared with thermal conductivity, specific heat capacity has the greater influence on bearing temperature rise.

ACKNOWLEDGMENT

The authors gratefully wish to acknowledge the supports provided by the National Science Foundation (51165024), the Gansu Province Science Foundation (1208RJZA131), and the Project supported by the Development Program for Outstanding Young Teachers in Lanzhou University of Technology (1002ZCX004).

REFERENCES

[1] Wu, Y.H. 2006. Motorized spindle unit technology in CNC machine tool [M]. Beijing: China Machine Press.

[2] Wen, S.Z. 1990. Tribological principle [M]. Beijing: Tsinghua University Press.

[3] Du, Q.H. & Yang S.N. 2007. Radial stiffness of ball bearings considering influences of surface waviness and lubrication [J]. Journal of Vibration and Shock, 2007, 26(10):152–156.

[4] Wang, J. 2004. Experimental research on oil-air lubrication and preload for hybrid ceramic ball bearing [J]. Journal of Yanshan University, 2004, 28(1): 92–94.

[5] Xue, J.R. et al. 2002. Present status and development tendency on hybrid ceramic bearing [J]. Bulletin of the Chinese Ceramic Society, 2002,21(6):53–57.

[6] Almqvist, T. et al. 2004. A comparison between computational fluid dynamic and Reynolds app roaches for simulating transient EHL line contacts [J]. Triblolgy International, 2004, 37(1): 61–69.

[7] MA, J.J. et al. 2000. EHL properties of roller contacts and its application part: A numerical analysis of EHL for engineering logarithmic roller [J]. Tribology, 2000, 20(1): 63–66.

[8] Li, S.S. et al. 2011. Experimental study of the lubrication performances in ultra high speed lubricated by oil-air[J]. Lubrication Engineering, 2011, 36(10):25–28.

[9] Kim, S.J. et al. 2007. Complementary effects of solid lubricants in the automotive brake lining [J]. Tribology International, 2007, 40: 15–20.

[10] Eriksson, M. & Jacobson, S. 2000. Tribology surfaces of organic brake pads [J]. Tribology International, 2000, 33: 817–827.

[11] Yang, P.R. 1998. Numrical analysis of fluid lubrication [M]. Beijing: National Defence Industry Press.

[12] Briggs, W.L. et al. 2011. A multigrid tutorial. Tsinghua University Press.

[13] Harris, T.A. et al. 2001. A method to calculate frictional effects in oil-lubricated ball bearings, Tribol. Trans. 44(4)(2001)704–708.

[14] Hamrock, B.J. & Dowson, D. 1977. Isothermal elastohydrodynamic lubrication of point contacts, part IV-starvation results. ASME J Lubr Technol 1977; 99:15–33.

[15] Venner. C.H. 1991. Multilevel solutions of the EHL line and point contact problems. PhD thesis, University of Twente, Netherland.

[16] Venner, C.H. et al. 2004. Waviness deformation in starved EHL circular contacts. ASME J Tribol 2004; 126:248–57.

[17] Yang, P. & Wen, S. 1990. A generalized Reynolds equation for non-Newtonian thermal elastohydrodynamic lubrication. ASME J. Tribol. 112, 631–636 (1990).

[18] Liu, X. et al. 2001. Non-Newtonian thermal analyses of point EHL contacts using the eyring model. ASME J Tribol.123, 816–821 (2001).

[19] Cheng, H.S. & Sternlicht, B.1965. A numerical solution for the pressure, temperature and film thickness between two infinitely long, lubricated rolling and sliding cylinders, under heavy loads. ASME J. Basic Eng. 87, 695–707(1965).

[20] Chaomleffel, J-P. et al. 2007. Experimental Results and Analytical Film Thickness Predictions in EHD Rolling Point Contacts [J] 1 Tribio Int (2007), doi:10. 1016/J. Tribo Int. 2007. 02. 005.

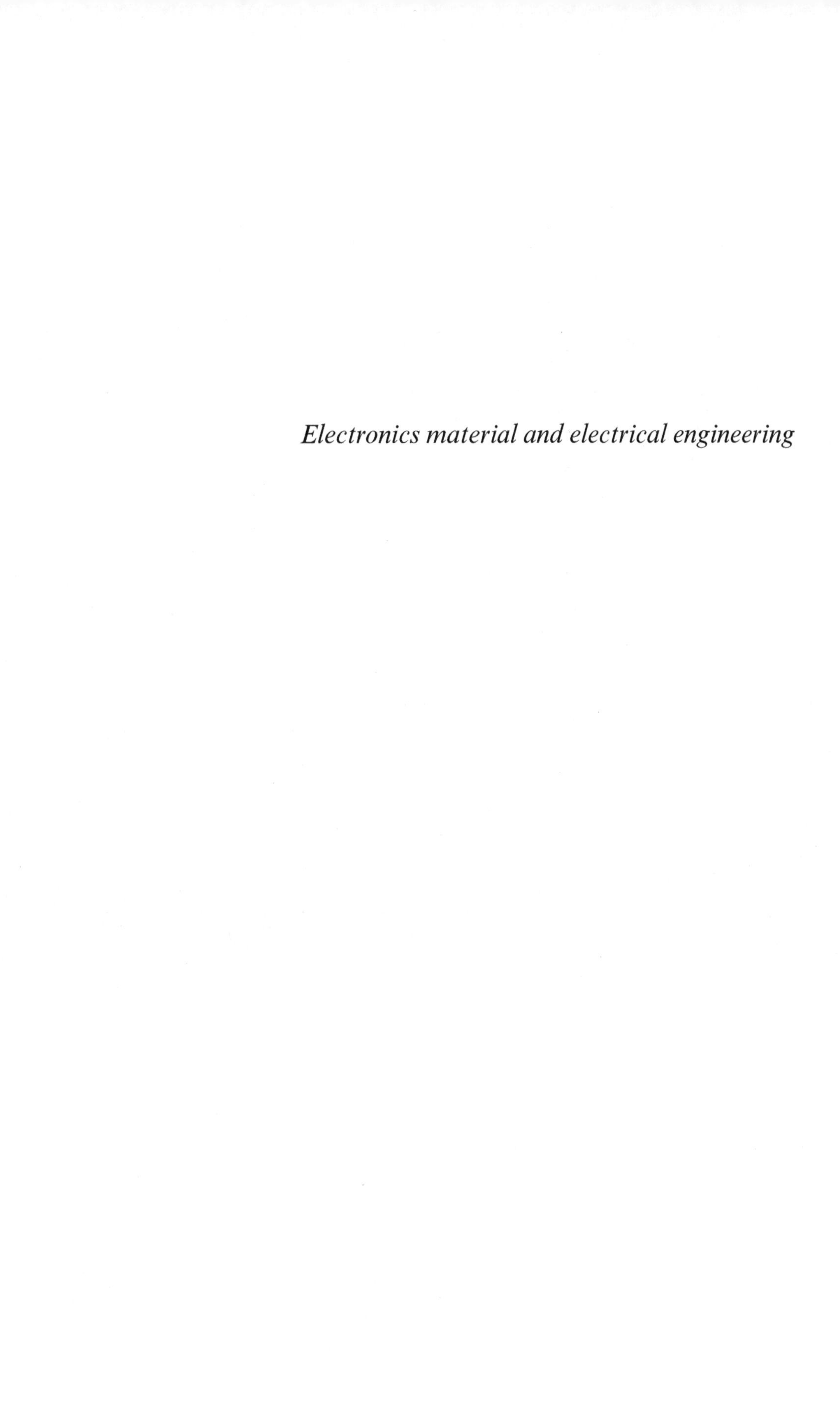

Electronics material and electrical engineering

Material Science and Environmental Engineering – Chen (Ed.)

Electrochemical reduction of CO_2 by a Cu/graphene electrode

X. Liu, L.S. Zhu & H. Wang
College of Environmental Science and Engineering, Beijing Forestry University, Beijing, China

Z.Y. Bian
College of Water Sciences, Beijing Normal University, Beijing, China

ABSTRACT: Cu/graphene gas diffusion electrode was prepared from the synthesized catalyst. The cyclic voltammogram analyses indicated Cu/graphene nanocomposite catalysts posed high catalytic activity for the CO_2 reduction and the rate-determining step was the CO_2 diffusion process from bulk solution to electrode surface. The electrochemical reduction of CO_2 was investigated in a diaphragm electrolysis device, using Cu/graphene gas-diffusion electrode as a cathode and a Ti/RuO_2/IrO_2 net anode. The reaction parameters were explored and the optimum performance conditions for the electrocatalytic reduction of CO_2 were achieved. Formic acid concentration showed highest production rate under a condition of 4 V, 0.5 mol/L of $KHCO_3$, and with 40 min electrolytic time. In the present system, the formic acid concentration reached to 16.6 mg/L and the current efficiency was 25.3% after 40 min electrolysis.

Keywords: CO_2; electrochemical reduction; Cu/graphene electrode; formic acid

1 INTRODUCTION

Currently, global warming caused by excess CO_2 emission has received much attention. Developing the low-power CO_2 reduction technology has a profound influence on protecting our environment and promoting the sustainable development of our society and economy. Several efforts have been made on the conversion of CO_2 to carbonaceous fuels (Jiang et al. 2010).

The electrochemical reduction of CO_2 is an important and convenient method, which can easily determine the reduction product distribution, current efficiency and selectivity through setting different voltages. A variety of electrode materials have been explored for the CO_2 reduction, such as metal plate electrode (Ertem et al. 2013), gas diffusion electrode (Lu et al. 2013), and the modified electrode (Prakash et al. 2013). An electrode with lower over potential of CO_2 reduction and higher over potential of hydrogen evolution should be designed for the CO_2 conversion in aqueous solutions. In recent years, graphene was widely used as catalyst because of its good conductivity, chemical stability, and high mechanical strength (Ebbesen et al. 1996).

In this paper, Cu nanoparticle were decorated on graphene by reduction of graphene oxide, and then characterized by Cyclic Voltammetry (CV) technologies. An electrocatalytic system with Cu/graphene gas-diffusion cathode was investigated for the CO_2 reduction. The product formation was measured and considered according to the operation conditions.

2 DEVICE AND EXPERIMENT

The graphene oxide can be prepared by the improved Hummers method (Hummers & Offeman 1958), and Cu/graphene catalyst for the electrochemical reduction of CO_2 was prepared by means of sodium borohydride reduction in a graphene oxide suspension.

The experiment device for the electrochemical reduction of CO_2 was constructed and described as the previous report (Lu et al. 2013). Electrolysis was conducted in the terylene diaphragm cell of 100 mL. The whole system consists of DC power supply, an aeration device, electrolytic cell, electrode composition; diaphragm electrolysis system with 729 polyester filter cloths as a diaphragm, with Ti/RuO_2/IrO_2 electrode as anode, Cu/graphene gas-diffusion electrode as cathode, electrolytic cell made of organic glass.

Cyclic Voltammetry (CV) spectra were recorded using a potentiostat/galvanostat (EG&G Model 273A) with a standard three-compartment cell. A Pt wire was used as a counter electrode, an Ag/AgCl electrode as a reference electrode, and the Cu/graphene catalyst modified electrode as a working electrode. 0.5 mol/L $KHCO_3$ was used as electrolyte, which was saturated with CO_2 (N_2) by feeding CO_2 (N_2) to the cell for 30 min before the electrochemical measurements, and gas was continued throughout the electrolysis. The temperature was kept constant at 25 °C during the temperature. The scan rate was 100 mV/s.

Formic acid was determined using an ion chromatograph (DIONEX ICS3000) analyses by comparing the retention time of the standard reference compounds. The separation was performed using an AS-11 column at the flow rate of 1.2 mL/min, with 250 mmol/L NaOH as the mobile phase at 30 °C. The total run time was 15 min and an injection volume is 10 μL.

3 RESULTS AND DISCUSSION

3.1 *Electrochemical analysis of Cu/graphene catalyst*

The catalytic activity of Cu/graphene composites for the reduction of CO_2 is elucidated by cyclic voltammogram of Cu/graphene modified glassy carbon electrode in 0.5 mol/L $KHCO_3$ solution (Fig. 1).

It can be seen from Figure 1 that there was an intense reduction peak at about −1.3 V in the presence of CO_2. It indicated that the two-electron reduction of CO_2 to $HCOO^-$ lead to the formation of reduction current peak using catalysts modified electrode.

$$CO_2 + H_2O + 2e^- \rightarrow HCOO^- + OH^- \quad (1)$$

Researchers assumed that the first intermediate of CO_2 electro-reduction was CO_2^- anion radical, which was formed by the initial single electron transfer. The CO_2^- is nucleophilic substance, it will take a proton at the nucleophilic carbon atom from hydrogen donor, forming HCOO, subsequently. HCOO· is then reduced to $HCOO^-$. Hydrogen donor can be O-H, N-H, or metal-H.

By changing operation parameters (voltage, electrolysis time, and concentration of electrolyte) for the electrochemical reduction of CO_2, the optimum electrolysis conditions to produce formic acid were achieved.

3.2 *Influence of voltage*

The measured concentrations of formic acid with supplied voltage variation during the electrochemical reduction of CO_2 are shown in Figure 2 under the conditions of 0.5 mol/L $KHCO_3$ electrolyte and reaction time of 30 min.

It can be seen from Figure 2, the concentration of formic acid increased with the increase of supplied voltage till 4 V, after that, the concentration of formic acid decreased deeply with the increase of voltage. The formic acid concentration reached 16.5 mg/L at 4 V. Therefore, 4 V was the best voltage for electrolysis in present system. In the process of electrochemical reduction of CO_2, HCO_3^- is directly involved in the reaction, the voltage is low, and the reduction of CO_2 was controlled by the charge transfer. Therefore, with the increase in the voltage, the amount of generated acid also increases. When the voltage exceeds the optimum voltage and increases gradually when formic acid was further reduced into other products, so the production amount of formic acid also decreased.

3.3 *Study on electrolysis time*

The measured concentrations of formic acid changes with the electrolysis time in 0.5 mol/L $KHCO_3$ solution at 4 V are shown in Figure 3.

It can be seen from Figure 3, in the electrolysis time, concentration of formic acid increased with the increase of the electrolysis time. The concentration of formatted formic acid reached a high

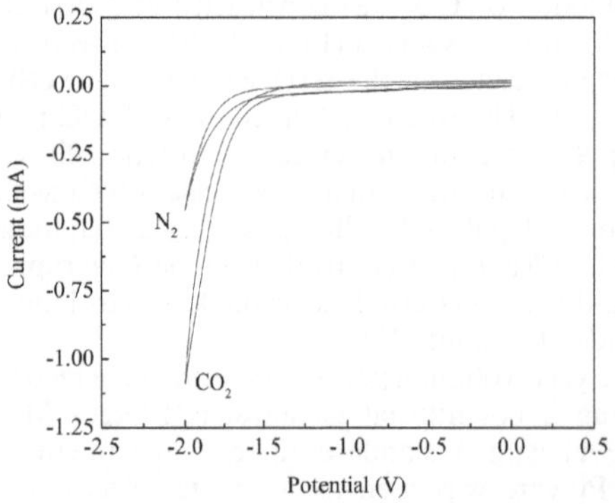

Figure 1. Cyclic voltammograms for Cu/graphene by catalyst modified glassy carbon electrode in 0.5 mol/L $KHCO_3$ solution.

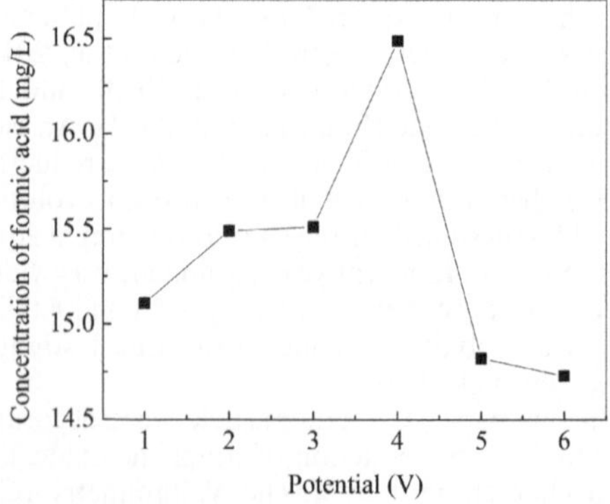

Figure 2. Concentration variation of formic acid with different voltages.

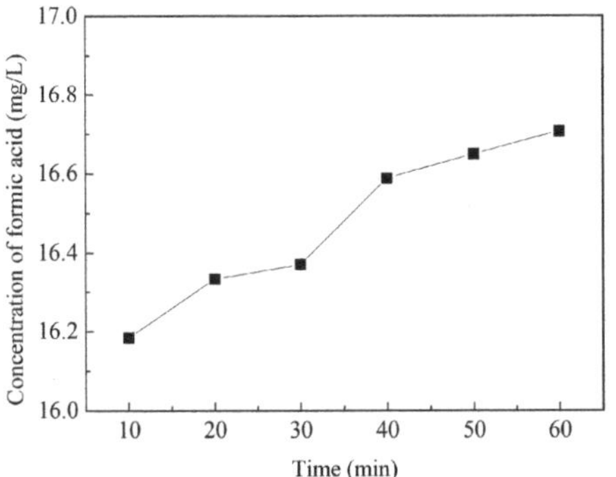

Figure 3. Concentration variation of formic acid with different electrolysis time.

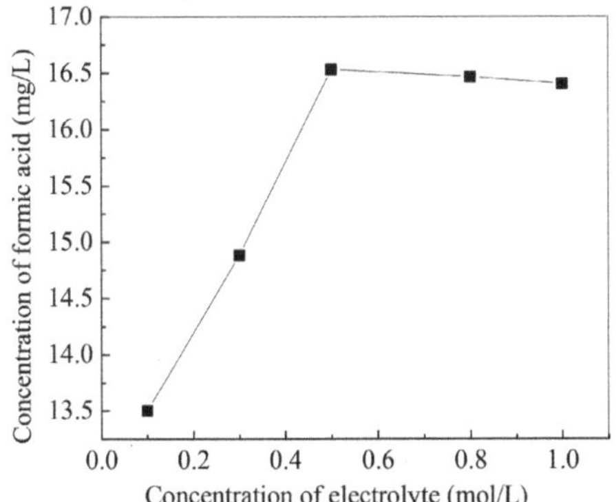

Figure 4. Concentration variation of formic acid with different concentration of $KHCO_3$.

level of 16.6 mg/L when the electrolysis time was 40 min. And the increase trend of the formic acid concentration was slow after 40 min electrolysis.

3.4 *Influence of concentration of electrolyte*

Electrolyte $KHCO_3$ concentrations were tested with 0.1, 0.2, 0.3, 0.4, 0.5, 0.6, 0.7, 0.8, 0.9, and 1 mol/L at the voltage 4 V for 40 min, respectively. The concentration variation of formic acid with electrolyte solution concentration is shown in Figure 4.

As seen in Figure 4, the amount of formic acid increased with increasing of electrolyte concentration within a certain range. In 0.5 mol/L $KHCO_3$ solution, the concentration of formic acid produced a maximum of 16.5 mg/L. HCO_3^- was involved in the reduction process of CO_2. So the reduction of CO_2 could be controlled with low concentration of $KHCO_3$. When the $KHCO_3$ concentration increased to a certain value, formic acid formation from CO_2 could be controlled by charge transfer and the concentration of $KHCO_3$ did not improve the amount of formatted formic acid. In addition, with the concentration of HCO_3^- increasing, the pH of solution continued to rise and high pH increased the potential of formic acid formation.

According to the calculation formula for the current efficiency, the current efficiency refers to electrolysis, the actual amount of deposition or dissolution ratio of the amount of substance and precipitation or dissolution by calculated according to the theory, usually denoted by the symbol η. The experimental data into the formula to calculate the current efficiency of the best conditions was 25.3%.

4 CONCLUSIONS

Cu/graphene could electrocatalytic reduction of CO_2 to formic acid. The CV behaviors showed that the Cu/graphene catalysts promoted the reduction by lowering the reduction potential and increasing the reduction current. The optimum operation conditions for the CO_2 electrocatalytic reduction was achieved in present Cu/graphene system: the electrolyte concentration of $KHCO_3$ was 0.5 mol/L, the voltage was 4 V, electrolysis time was 40 min. And the concentration of formic acid produced was 16.6 mg/L with current efficiency of 25.3%.

ACKNOWLEDGMENTS

This work was supported by the Fundamental Research Funds for the Central Universities (No. TD2013-2 and 2012LYB33), the National Natural Science Foundation of China (No. 51278053 and 21373032), and grant-in-aid from Kochi University of Technology and China Scholarship Council.

REFERENCES

[1] Ebbesen, T.W., Lezec, H.J. & Hiura, H. 1996. Electrical conductivity of individual carbon nanotubes. *Nature* 382: 54–56.

[2] Ertem, M.Z., Konezny, S.J., Araujo, C.M. & Batista, V.S. 2013. Functional role of pyridinium during aqueous electrochemical reduction of CO_2 on Pt(111). *The Journal of Physical Chemistry Letters* 4(5): 745–748.

[3] Hummers, W.S. & Offeman, R.E. 1958. Preparation of graphitic oxide. *Journal of the American Chemical Society* 80(6): 1339–1339.

[4] Jiang, Z., Xiao, T., Kuznetsov, V.L. & Edwards, P.P. 2010. Turning carbon dioxide into fuel. *Philosophical Transactions of the Royal Society A* 368: 3343–3364.

[5] Lu, G., Wang, H., Bian, Z.Y. & Liu, X. 2013. Electrochemical reduction of CO_2 to organic acids by a Pd-MWNTs gas-diffusion electrode in aqueous medium. *The Scientific World Journal* 2013: 1–8.

[6] Prakash, G.K.S., Viva, F.A. & Olah, G.A. 2013. Electrochemical reduction of CO_2 over Sn-Nafion® coated electrode for a fuel-cell-like device. *Journal of Power Sources* 223: 68–73.

Material Science and Environmental Engineering – Chen (Ed.)
© 2016 Taylor & Francis Group, London, ISBN 978-1-138-02938-5

First principle study on the electronic structure of NbB and ZrB under pressure

Y.Q. Chai
Department of Physics and Electronic Information, Langfang Teachers University, Langfang, P.R. China

ABSTRACT: This study utilizes accurate first principle method to conduct a comparative study on the electronic structure of NbB and ZrB under pressure. According to the results, product compression effect on the density of states near the Fermi level and volume compression effect on the energy band structure show differences exists between the two borides. The study shall be able to provide a significant theoretical reference for further experimental research to understand superconducting materials under high pressure.

Keywords: boride; electronic structure; density of state; fermi energy

1 INTRODUCTION

The discovery of new superconductor MgB_2, which superconducting transition temperature (Tc) of 39 K (Nagamatsu et al., 2001), has caused great attention among the physics community. In recent years, many researchers have studied the new superconductor of Borides. Current findings can be classified as computational materials synthesis, physical testing, and the theoretical computations (Buzea & Yamashita, 2001; Singh, 2001; Chai et al., 2003) including the influence on pressure. It has been reported NbB and ZrB both have superconducting properties, and their superconducting transition temperature (Tc) are relatively small.

Pressures can be used as a very effective way to adjust the physical characteristics of regulating materials, especially, for providing very important information on the in situ effects of superconductivity for the type of pairing mechanism. For further analysis the superconductivity of a boride superconductors which change with pressure, or the first principle based on crystal energy band calculation procedures is used to analysis and calculation of electronic structure properties of a boride which has bigger to the transition temperatures (NbB and ZrB) under high pressure volume after compression to study the influence of super pressure conductor on superconductivity in a boride. Different volume compression ratio on the comparison of two kinds of superconductor electronic structure calculation of the volume of compressed.

2 COMPUTATIONAL METHODOLOGY AND INPUT PARAMETERS

In order to systematically analyze these borides, WIEN2 K program is used for all calculations in this study. Recommended potential options are selected as "GGA of Perdew-Burke-Ernzerh of 96"; separation energy is selected as –6.0 Ry; $RMT \times Kmax = 8$; and Brillouin zone K number is 6000.

According to Table 1, NbB and ZrB have different space groups. Nb and Zr are located in the fifth period in the periodic table of the elements and have relatively closed valence electron numbers. However, due to different space groups and different cell volumes. According to the theory of BCS, these differences are related to the superconducting transition temperature Tc.

Under high pressure, the changes of original cell volume are bound to change their superconductivity. After shrinkage in volume in the ratio of –5%, –10%, –15%, –20%, Table 2 shows that both total energy and Femi Energy (E_F) change. It is also found that the changes of total energy changes much smaller than, the changes of E_F. Moreover, the changes of NbB are closed to linear. Figure 1 shows the E_F changes of NbB and ZrB before and after shrinking.

Table 1. Lattice parameters and Tc.

	a (a.u)	b (a.u.)	c (a.u.)	Tc (K)
NbB	6.232	16.486	5.9828	8.25
ZrB	8.787	8.7872	8.7872	3.40

Table 2. The total energy and E_F changes under pressure.

	Total Energy (E)		Femi Energy (EF)	
V/V_0	NbB	ZrB	NbB	ZrB
1.00	−15381.53	−7248.15	0.9399	0.76052
0.95	−15381.52	−7248.12	1.00397	0.80509
0.90	−15381.49	−7248.09	1.06324	0.85548
0.85	−15381.44	−7248.04	1.12606	0.91314
0.80	−15381.35	−7247.98	1.18223	1.07998

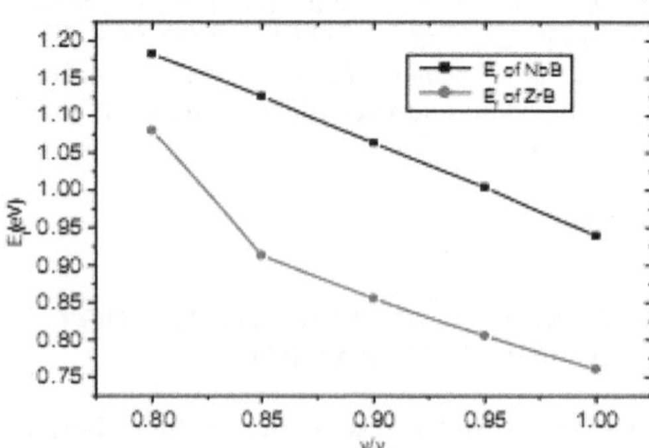

Figure 1. E_F changes in contraction of volume before and after shrinking.

3 PRODUCT COMPRESSION EFFECT ON THE DENSITY OF STATES NEAR THE FERMI LEVEL

Table 3 shows the changes of Fermi level density of states after volume compression. Figure 2 shows the two a boride superconductor in volume before and after compression 20% variation in DOS diagram.

From Table 3 and Figures 2 and 3, after volume shrinkage, DOS near the Fermi level for both NbB and ZrB are changed with volume. And the difference of variation amplitude is very large, as total DOS of NbB and decreases with the volume reduction and the changes are relatively larger. It is also found that when the volume compression is 10%, the total DOS of NbB decreases obviously. On the other hand, for ZrB, the DOS decreases at the 5% volume compression, and increases at the 10% volume compression, and then with more volume shrinkage, the total DOS has increased again.

For NbB and ZrB, near the Fermi surface, DOS of B atom compared to DOS of metal atoms and Interstitial DOS is very small. In NbB, the B DOS accounts for total 5.6% of total DOS; in ZrB, the B DOS accounts for total DOS 20.9%.

Table 3. DOS of ZrB changes under pressure.

V/V_0	ZrB total	Zr total	B total	Zr-B interstitial
1.00	1.03395	0.42264	0.21651	0.3948
0.95	0.96315	0.40522	0.20948	0.34846
0.90	1.01735	0.42398	0.23145	0.36192
0.85	1.00968	0.43375	0.23571	0.34022
0.80	1.00424	0.33506	0.20649	0.46269

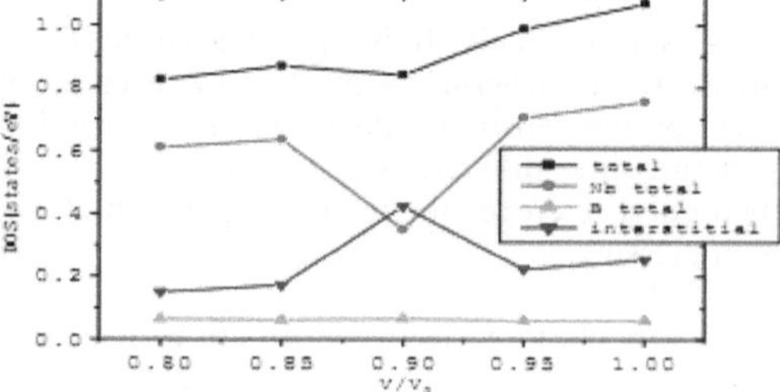

Figure 2. The changes of NbB DOS before and after volume compression.

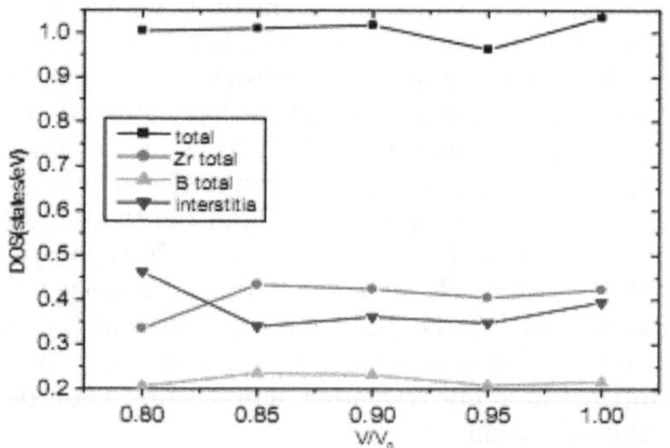

Figure 3. The changes of ZrB DOS before and after volume compression.

Changes of volume shrinkage after near the Fermi level B DOS complex. Overall volume contraction, the changing range of B DOS is relatively small.

For NbB, the B DOS with the volume contraction increases linearly (6.0%~r 7.9% of total DOS); while for ZrB, B DOS with the volume contraction also increased (20.9%~23.3%). The increase reaches the maximum at 15% volume compression. In both types of superconductors, DOS of metal atom occupy relatively larger proportion, especially, for NbB. Nb DOS accounted for 70.7% of the total DOS without volume shrinkage. Nb DOS decreases after volume shrinkage and reaches its minimum at 10% volume shrinkage. However, in ZrB, the metal atoms changes oscillatory.

4 VOLUME COMPRESSION EFFECT ON THE ENERGY BAND STRUCTURE

Figure 4 is the energy band diagram of the NbB and ZrB at atmospheric pressure, and Figure 5 is the volume shrinkage after 20% in the energy band diagram. By contrast, under pressure influences volume shrinkage of two superconductor energy band structures. Although changes in the basic distribution band are very small, there is still a certain degree of broadening. However, in the two kinds of superconductors with almost no change near the f point. Moreover, near the Fermi level, ZrB had a band at X more to rice.

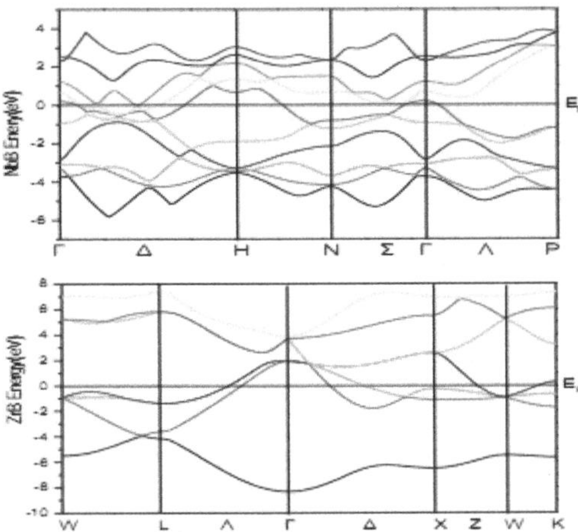

Figure 4. The energy band diagram of NbB and ZrB.

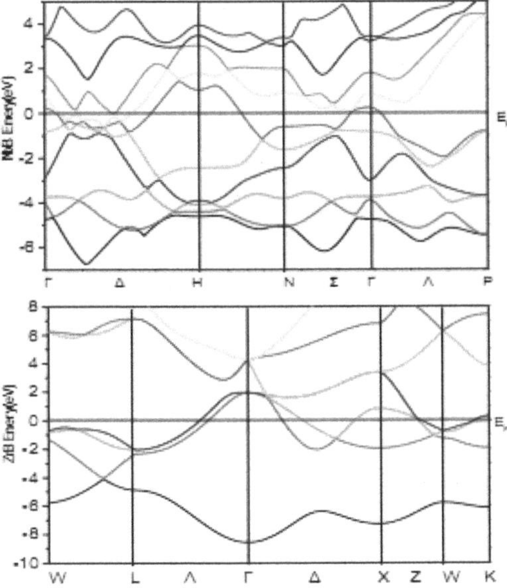

Figure 5. Compressed NbB and ZrB energy band diagram.

5 CONCLUSIONS

At present, among all the discovered borides, only NbB is found to have a higher superconducting transition temperature Tc. However, the other borides are found to have very small Tc even though some are with superconducting properties. Therefore, we can conclude that some boride structures may result in different properties. Current researchers explain this phenomenon, as a result of different lattice constants and valence electrons.

For the study of the electronic structure of MgB_2, it is recently pointed out that the superconductivity of MgB_2 mainly depends on the B DOS and MgB_2 B atomic characteristic band situation. Moreover, from the study of the electronic structure of NbB, it was found that since the B DOS account for a very small proportion of total DOS, the superconducting property should depend on the metal atom. Under high pressure, it is found that B DOS increased with the volume shrinkage.

For this type of boride superconductors, if one of the B atoms has the similar effect of the B atoms in MgB_2, then the boride superconductor transition temperature Tc, shall be increased with increasing pressure. The analysis of the electronic structure of these two kinds of boride volume shrinkage in this study provide some theoretical help and can be served as the basis of understanding superconducting materials under high pressure.

REFERENCES

[1] Buzea C., Yamashita T. 2001. Review of the superconducting properties of MgB2, Superconductors, Science & Technology 14(11), 115–146.

[2] Chai Y.Q., Jin C.Q., Liu B.Q. 2003. Comparative study on electronic structures of MgB2—like borides. Acta Physicasinica 52(11).

[3] Inorganic Crystal Structure Database RETRIEVE Version 2.01 February 1997.

[4] Nagamatsu J., Nakagawa, N., Muranaka, T., Zenitani, Y., Akmitsu, J. 2015. Superconductivity at 39 K in magnesium diboride. *Nature*, 410(6824), 63–64.

[5] Singh, P.P. 2001. Role of boron p-electrons and holes in superconducting MgB_2, and other diborides: A fully relaxed, full-potential electronic Structure Study, Physical Review Letters, 87(8): 087004.

Material Science and Environmental Engineering – Chen (Ed.)

Facial synthesis hollow hemisphere $LiFePO_4$ cathode for lithium-ion batteries

Y.H. Luo
School of Chemical Engineering, Nanjing University of Science and Technology Nanjing, Jiangsu, China
Sinosteel Anhui Tianyuan Technology Co. Ltd., Maanshan, Anhui, China

N. He
Sinosteel Maanshan Institute of Mining Research Co. Ltd., Maanshan, Anhui, China

Y.C. Wang
Sinosteel Anhui Tianyuan Technology Co. Ltd., Maanshan, Anhui, China

W.G. Cao
School of Chemical Engineering, Nanjing University of Science and Technology Nanjing, Jiangsu, China

F. Pan
School of Chemical Engineering, Nanjing University of Science and Technology Nanjing, Jiangsu, China
National Quality Supervision and Inspection Center for Industrial Explosive Materials Jiangsu, Nanjing, Jiangsu, China

ABSTRACT: In our work, the facial synthesis method of $FePO_4$ with different crystal styles was put forward, and its effect on the preparation and electrochemical performance of $LiFePO_4$ was investigated. The results showed that the crystal form of $FePO_4$ was amorphous and monoclinic structured at the treatment temperature of 30°C and 80°C, respectively. $LiFePO_4$ prepared from both precursors were both hollow hemisphere shaped and olivine structured. $LiFePO_4$ produced from the monoclinic structured precursor exhibited smaller size and better electrochemical performance, its discharge capacities was 155.9 mAh · g^{-1} and 141.8 mAh · g^{-1} at the rates of 0.1 C and 1 C, respectively. After 150 cycles, its capacity retention was about 97.8% and 95.1% at 0.1 C and 1 C, respectively.

Keywords: $FePO_4$; $LiFePO_4$; lithium-ion battery; composite materials; energy storage and conversion

1 INTRODUCTION

Since $LiFePO_4$ was reported as a cathode material for lithium-ion battery by Good enough in 1990s[1], it has been considered as a promising cathode material for Li-ion battery for its advantages, such as low cost, no toxicity, high thermal stability, a relatively high theoretical capacity of 170 mA h g^{-1} and excellent cycling performance[2–5]. $LiFePO_4$ composite has been successfully put into practice by carbon coating, which is a good solution to enhance the conductivity of pristine $LiFePO_4$ [6–8]. Aiming to improve the electrochemical properties, different precursors, such as $FePO_4$, Fe_2O_3, and FeC_2O_4 have been used as precursors to prepare $LiFePO_4$ composite. Among them, $FePO_4$ was attached much more attention as a good precursor, and different methods were studied to synthesis $FePO_4$ [9–14]. However, they have complicated synthetic routes and high synthetic cost, besides, there is a report about the effect of the crystal style of $FePO_4$ on the preparation and electrochemical performance of $LiFePO_4$ composite.

Herein, we report a facial synthesis method of $FePO_4$ with different crystal form to investigate its effect on the electrochemical performance of $LiFePO_4$ composite.

2 EXPERIMENTAL

The precursor was prepared using low-cost $FeSO_4$ and H_3PO_4 as raw materials. 200 ml 0.13 mol/L$FeSO_4$ and 9 ml 85% H_3PO_4 were mixed together in the reaction still, stirred for 1 h, then 200 ml ethanol was added into the solution in condition of vigorous stirring for 3 h. Subsequently, the temperature of the slurry was kept at 30°C and 80°C for 5 h, respectively, and was filtrated, washed, and dried. The precursor prepared at 30°C was labeled as S1; the precursor prepared at 80°C was labeled as S2. Then, the as-prepared precursor and $LiOH \cdot H_2O$ were mixed using a mill. Simultaneously, a proper amount of $C_6H_{12}O_6$ was added into the mixture as carbon source. After drying treatment, the mixture was calcined at 650 °C for 12 h and cooled down under a N_2 atmosphere. Finally, the $LiFePO_4$ composite sample was obtained. They were labeled as L1 (prepared from S1) and L2 (prepared from S2), respectively.

The crystalline phases of synthesized precursor and $LiFePO_4$ composite were identified with X-Ray Diffraction (XRD, X'Pert Pro MPD, Cu Kα radiation, $\lambda = 1.54178$ Å). The particle morphologies and sizes of the samples were examined by Field-Emission Scanning Electron Microscopy (FE-SEM, SIRION). The Brunauer–Emmett–Teller (BET) surface area of as-synthesized samples was determined by using an instrument of (FBT-5″). The elemental result was analyzed by X-ray fluorescence spectrometer (Axios).

The electrochemical performances were investigated in CR2025 coin-type lithium half-cells. NMP was employed as the solvent, blended slurry consisting of active material, acetylene black, and PVDF binder in a weight ratio of 75:15:10 was coated on Al sheet (Φ = 12 mm) and dried at 110 °C in a vacuum for 12 hours. Lithium foil served as the reference and counter electrode and Celgard 2300 membrane was used as the separator. The electrolyte is consisted of a solution of 1 M LiPF6 in Ethylene Carbonate (EC)/Dimethyl Carbonate (DMC) (1:1 by volume). Coin cells were assembled in a dry glove box filled with high-purity argon. The charge and discharge performance was determined by an automatic NEWARE battery cycler (Neware BTS-610) in a voltage range of 2.5–4.3 V at room temperature (25 ± 2°C).

3 RESULTS AND DISCUSSION

Table 1 shows the elemental analysis of $FePO_4$ prepared at different temperatures. For both samples, the mole ratio of P to Fe is 1.005, indicating the high purity of the prepared $FePO_4$.

Figure 1 shows the XRD patterns of the as-prepared $FePO_4$ and $LiFePO_4$. The XRD patterns of S1 (Fig. 1a) and S2 (Fig. 1b) can be indexed to amorphous and monoclinic structure, respectively. All the diffraction peaks of L1 (Fig. 1c) and L2 (Fig. 1d) are in agreement with the olivine structured $LiFePO_4$, which indicates that the crystal form of $FePO_4$ has no significant effect of the structure of $LiFePO_4$.

Table 1. Elemental analysis of $FePO_4$ prepared at different temperatures (%).

Temperature	Fe	P	S	Ca	Mg	Si	Na
S1	29.18	16.36	0.033	0	0.022	0	0.014
S2	29.30	16.34	0.041	0	0.019	0	0.023

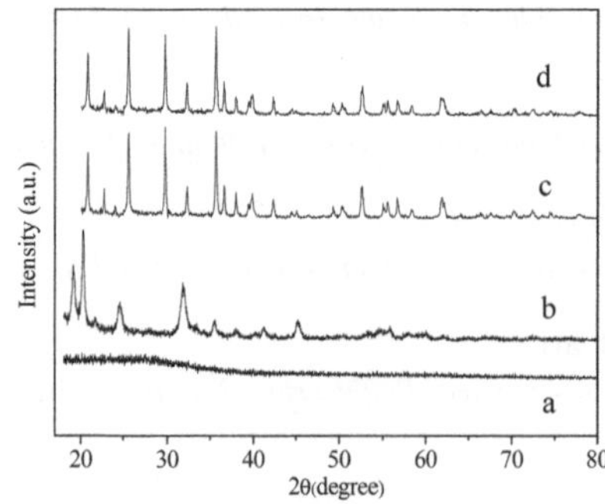

a - amorphous $FePO_4$; b - monoclinic $FePO_4$; c - $LiFePO_4$ prepared with a; d - $LiFePO_4$ prepared with b

Figure 1. XRD of the as-prepared $FePO_4$ and $LiFePO_4$.

Figure 1 shows the XRD patterns of the as-prepared $FePO_4$ and $LiFePO_4$. The XRD patterns of S1 (Fig. 1a) and S2 (Fig. 1b) can be indexed to amorphous and monoclinic structure, respectively. All the diffraction peaks of L1 (Fig. 1c) and L2 (Fig. 1d) are in agreement with the olivine structured $LiFePO_4$, which indicates that the crystal form of $FePO_4$ has no significant effect of the structure of $LiFePO_4$.

The SEM images of the as-prepared $FePO_4$ and $LiFePO_4$ are presented in Figure 2. From which we can see that S1(Fig. 2a) and S2 (Fig. 2b) both show the irregular shape, and porous structure can be seen on their surface, different from the spherical and spheroidal shaped $FePO_4$ reported in literatures[15], which may be due to template role played by ethanol. With close observation, the particle size of S2 is a bit smaller that that of S1, such result can be explained by the formula of crystal nucleation form rate proposed by A.E. Nielsen[16] and equation of crystal growth rate proposed by Dieksen, etc[17]. According to their research, the crystal nucleation form rate is faster at higher temperature, thus the particle size of $FePO_4$ prepared at 80°C is relatively smaller.

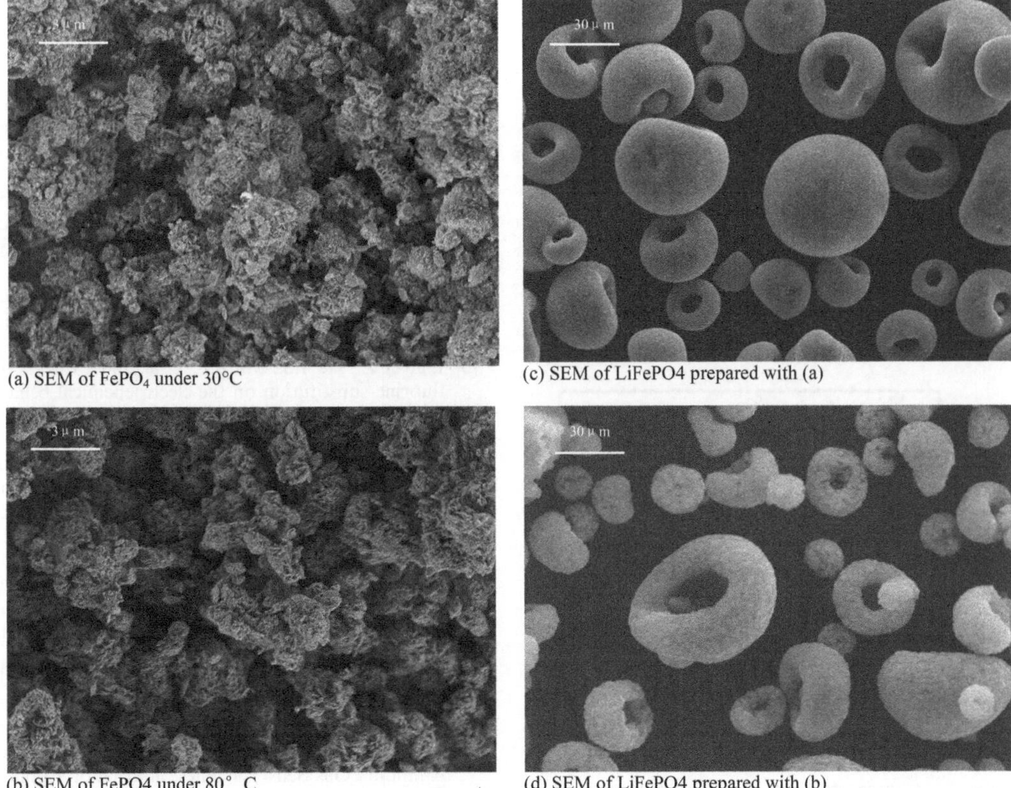

(a) SEM of $FePO_4$ under 30°C

(b) SEM of FePO4 under 80° C

(c) SEM of LiFePO4 prepared with (a)

(d) SEM of LiFePO4 prepared with (b)

Figure 2. SEM of the as-prepared $FePO_4$ and $LiFePO_4$.

The surface area of S2 (68 $m^2 \cdot g^{-1}$) is bigger than that of S1 (45 $m^2 \cdot g^{-1}$), which may have an effect on the morphology of the prepared $LiFePO_4$. Figure 2(c) and Figure 2 (d) show the SEM images of L1 and L2, respectively. As is shown in the images, they are all hollow-hemisphere shaped, it should be noted that such appearance was first synthesized to our best knowledge. Obviously, L2 is smaller than L1, and the surface area of L2 (15 $m^2 \cdot g^{-1}$) is bigger than that of L1 (10 $m^2 \cdot g^{-1}$) can also be seen, which may indicate the difference between their electrochemical performance.

Figure 3 shows the charge and discharge curves and cycling performance of $LiFePO_4$ composite at different rates. As displayed in Figure 3(A), each charge–discharge curve exhibits a flat and long potential plateau around 3.4 V. The discharge capacities of L1 are 150.7 $mAh \cdot g^{-1}$ and 135.2 $mAh \cdot g^{-1}$ at the rates of 0.1 C and 1 C, respectively. The discharge capacities of L2 are 155.9 $mAh \cdot g^{-1}$ and 141.8 $mAh \cdot g^{-1}$ at the rates of 0.1 C and 1 C, respectively. Obviously, L2 exhibits better electrochemical performance.

Table 2. The BET of phosphate iron and lithium iron phosphate ($m^2 \cdot g^{-1}$).

Sample	S1	S2	L1	L2
BET	45	68	10	15

Additionally, L2 also demonstrates better cycle stability. After 150 cycles, the capacity retention of L2 is about 97.8% and 95.1% at 0.1 C and 1 C, respectively, and the capacity retention of L1 is about 94.6% and 90.3% at 0.1 C and 1 C, respectively, as illustrated in Figure 3 (B). The better electrochemical performance of L2 can be explained by its smaller particle size. The smaller the particle size is, the smaller the diffusion resistance of Li ion is, thus the utilization of the active materials improves, meanwhile, the volume change caused by the embedded and out behavior of Li ion is smaller, which benefits the structural stability, thus slower capacity fade can be seen.

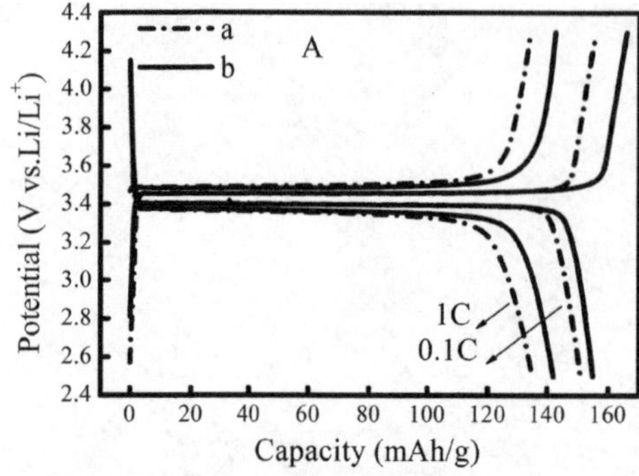

(a) Charge and discharge curves of $LiFePO_4$

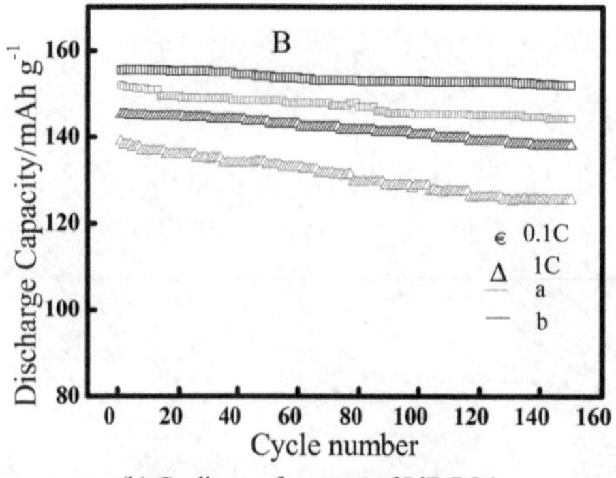

(b) Cycling performance of LiFePO4

Figure 3. Charge and discharge curves and cycling performance of $LiFePO_4$.

4 CONCLUSION

This work has prepared $FePO_4$ with amorphous and monoclinic crystal form via facial synthesis method and studied its effect on the electrochemical performance of $LiFePO_4$. The results showed that $LiFePO_4$ prepared from both precursors were both hollow hemisphere shaped and olivine structured. $LiFePO_4$ produced from the monoclinic structured precursor exhibited smaller size and better electrochemical performance, its discharge capacities was 155.9 $mAh \cdot g^{-1}$ and 141.8 $mAh \cdot g^{-1}$ at the rates of 0.1 C and 1 C, respectively. After 150 cycles, its capacity retention was about 97.8% and 95.1% at 0.1 C and 1 C, respectively.

REFERENCES

[1] Padhi A.K., Nanjundaswamy K.S., Goodenough J.B. 1997. Phospho-olivines as positive—electrode materials for rechargeable lithium batteries. J Electrochem Soc. 144(4): 1188–1194.

[2] Gaberscek M., Dominko R., Jamnik J. 2007. Is small particle size more important than carbon coating? An example study on $LiFePO_4$ cathodes. electrochem Commun. 9 (12): 2778–2783.

[3] Chen W.M., Qie L., Yuan L.X. 2011. Insight into the improvement of rate capability and cyclability in $LiFePO_4$/polyaniline composite cathode. Electrochimica Acta. 56(6): 2689–2695.

[4] Delacourt C., Laffont L., Bouchet R., et al. 2005. Toward understanding of electrical limitations (electronic, ionic) in $LiMPO_4$ (M = Fe, Mn). J Electrochem Soc. 152(5): A913–A921.

[5] Sun C.S., Zhang Y., Zhang X.J., et al. 2010. Structural and electrochemical properties of Cl-doped $LiFePO_4$. J Power Sources. 195: 3680–3683.

[6] Liao X.Z., He Y.S., Ma Z.F., et al. 2007. Effects of fluorine-substitution on the electrochemical behavior of $LiFePO_4$ cathode materials. J Power Sources. 174(2): 720–725.

[7] Huang Y.H., Park K.S., Goodenough J.B. 2006. Improving lithium batteries by tethering carbon-coated $LiFePO_4$ to polypyrrole. J Electrochem Soc. 153(12): A 2282–A 2286.

[8] Guo X.K., Fan Q., Yu L., et al. 2013. Sandwich-like $LiFePO_4$/graphene hybrid nanosheets: in situ catalytic graphitization and their high-rate performance for lithium ion batteries. J. Mater. Chem. 1: 11534–11538.

[9] Dilbar Ahmat, Erkin Tursun, Xawkat Abliz, et al. 2010. Fabrication of Composite Optical wave—guides Based onThin Films Consisted of Iron Phosphate Nanoparticles and Their Applications as Ammonia Gas Sensor. Chinese Journal of Applied Chemistry. 27: 965–969.

[10] Wang Ye, Otsuka K. Partial. 1997. Oxidation of ethane by reductively activated oxygen over iron phosphate catalyst. Journal of Catalysis. 171: 106–114.

[11] Huang Y.Y., Rena H.B., Peng Z.H., et al. 2009. Synthesis of $LiFePO_4$arbon composite from nano-FePO4 by a novel stearic acid assisted rheological phase method. Electrochimica Acta. 55: 311–315.

[12] Dirksen J.A. 1991. Fundamentals of crystallization: Kinetic effects on particle size distributions and morphology. Chem Eng Sci. 46: 2389–2427.

[13] Shi Z.C., Li Y.X., et al. 20065. Mesoporous FePO4 with Enhanced Electrochemical Performance as Cathode Materials of Rechargeable Lithium Batteries. Electrochem. Solid-State. Lett. 8: 396–399.

[14] Xua Y.B., Lua Y.J., Yin Ping, et al. 2008. A versatile method for preparing FePO4 and study on its electrode performance in lithium ion batteries. Journal of Materials Processing Technology. 204: 513–519.

[15] Qian L.C., Xia Y., Zhang W.K., et al. 2012. Electrochemical synthesis of meso-porous $FePO_4$ nanoparticles for fabricating high performance $LiFePO_4$ cathode materials. Microporous and Mesoporous Materials. 152: 128–133.

[16] Nielsen A.E. 1964. Kinetics of Precipitation. Oxford: Pergamon Press.

[17] Dirksen J.A. 1991. Fundamentals of crystallization: Kinetic effects on particle size distributions and morphology. Chem Eng Sci. 46: 2389–2427.

Material Science and Environmental Engineering – Chen (Ed.)

Synthesis of Ce-doped zircon ceramics designed to immobilize tetravalent actinides

X.N. Liu
Research Center of New Energy Materials, Southwest University of Science and Technology, Mianyang, P.R. China

Y. Ding & X.R. Lu
Key Subject Laboratory of National Defense for Radioactive Waste and Environmental Security, Southwest University of Science and Technology, Mianyang, P.R. China

ABSTRACT: $Zr_{1-x}Ce_xSiO_4$ ($0 \le x \le 0.1$) ceramics were prepared by solid-state reaction method. The effects of Ce content on the phase structure of the ceramics were investigated. The results show that the ceramics with $x < 0.04$ show homogeneous single zircon phase, while the samples with $x \ge 0.04$ exhibit coexistence of two crystal phases (zircon and $Ce_2Si_2O_7$). The bulk density of the Ce-doped zircon ceramics increases with the increase in neodymium content. These results indicate that the obtained zircon ceramics waste forms are promising candidate for immobilization of tetravalent actinides.

Keywords: Zircon; ceramics; tetravalent actinides; nuclear waste immobilization

1 INTRODUCTION

Immobilization of actinides from high level nuclear wastes is one of the major concerns of nuclear industry due to their higher radiotoxicity and long half-life [1]. The most promising mineral host for actinides includes pyrochlore, zirconolite, monazite and zircon [2–5]. Zircon has long been recognized as a natural occurring and highly durable phase and thus it's suitable to immobilize actinides [5, 6].

Zircon containing uranium and thorium concentrations up to 5000 ppm occurs in nature. Zircon is an extremely durable mineral [7], often found as a heavy mineral in stream sediments. And the heavy mineral still shows limited chemical alteration or physical abrasion after transport over great distances [8]. Indeed, zircon has been extensively studied, because it is a phase commonly used in the U/Pb dating of minerals that may be as old as several billion years [9, 10]. The structure of zircon is well known [11]; it consists of triangular dodecahedral ZrO_8 groups which form edge-sharing chains parallel to the *a*- and *b*-axis, and SiO_4 tetrahedral monomers which form edge-sharing chains with alternating ZrO_8 groups parallel to the *c*-axis [12]. In natural zircon, U and Th replace the Zr site in low concentrations. Moreover, compositions of $ASiO_4$, in which A^{4+} = Zr, Hf, Th, Pa, U, Np, Pu, and Am, have been synthesized. The regular increase in the unit cell volume with the increasing ionic radius of the A-site cation confirms the homologous topologies of these structures. Four of these compositions, hafnon ($HfSiO_4$), zircon, coffinite ($USiO_4$), and thorite ($ThSiO_4$), occur naturally. The results of present structure refinements [13] and structural analysis [14, 15] suggest complete miscibility between $ZrSiO_4$ and $HfSiO_4$, but there are miscibility gaps between the $ZrSiO_4$-$USiO_4$-$ThSiO_4$ joins [16]. Zircon with 9.2 at.% plutonium substituting for Zr has been synthesized [17, 18]. This is equal to a waste loading of 10 wt.% Pu, but the maximum extent of the solubility of Pu in zircon and influence on the phase and structure have not been investigated. The fact that a pure, end member composition, $PuSiO_4$, has been synthesized suggests the possibility of extensive substitution of Pu for Zr [13].

In this paper, cerium was used as surrogate for tetravalent actinides. A series of Ce-doped zircon waste form $Zr_{1-x}Ce_xSiO_4$ ($0 \le x \le 0.1$) have been synthesized by solid state reaction method. The obtained ceramics were characterized by powder X-Ray Diffraction (XRD) and Scanning Electron Microscopy (SEM) in order to investigate the effect of Ce loading on the phase and structure of Ce-doped zircon waste form.

2 EXPERIMENTIAL

General formula design: According to the theory of isomorphism, we speculate the Zr in $ZrSiO_4$ structure can be substituted by Ce (cerium, surrogates for tetravalent actinides). The chemical general formula $Zr_{1-x}Ce_xSiO_4$ is employed. The detailed compound designs are listed in Table 1.

Zircon waste forms preparation: CeO_2 (Shanghai Aladdin Co. Ltd., purity of 99.99%), ZrO_2 (Shanghai Aladdin Co. Ltd., purity of 99.99%) and SiO_2 (Chengdu KESHI Chemical, purity of 99.9%) were used as raw materials. Series of compositions with general stoichiometry as $Zr_{1-x}Ce_xSiO_4$ ($0 \le x \le 0.1$) were synthesized by solid state reaction method. Before use, all raw materials were heated at 120 °C overnight to remove adsorptive water. Then stoichiometric amounts of the powders in designed ratios were weighed, and ground/mixed in ethyl alcohol medium (purity of 99.7%, the mass ratio of ethyl alcohol to powder is about 3:1) by ball milling. The powders were dried again and then pressed into pellets of 12 mm in diameter and 0.5 mm in thickness at a pressure of 10 MPa. The pellets were sintered at 1550 °C for 72 h to fabricate dense bulk ceramics in air atmosphere. The temperature increase rates were about 5 °C per minute.

Characterization: The obtained zircon waste forms were characterized by powder X-Ray Diffraction (XRD). The XRD patterns of the samples were recorded on a PANalytical X-Pert Pro X-ray diffractometer using Cu-Kα (λ = 1.5406 and 1.5444 Å) radiation, over the two theta range of 10–90° with step width and time of 0.02° and 1.2 s, respectively. The XRD patterns were analyzed by Rietveld refinement and Le Bail refinement methods using Fullprof-2k software package. The microstructure of the sintered samples was observed by Scanning Electron Microscope (SEM, Ultra 55, and Germany). The densities of obtained ceramics were measured by Archimedes method.

Table 1. Detailed composition of $Zr_{1-x}(Ce_x)SiO_4$ ($0 \le x \le 0.1$).

			Amount of raw materials (g)		
Sample	*x*	Composition	CeO_2	SiO_2	ZrO_2
ZS-1	0	$ZrSiO_4$	0.0000	0.6008	1.2322
ZCS-1	0.01	$Zr_{0.99}Ce_{0.01}SiO_4$	0.0172	0.6008	1.2199
ZCS-2	0.02	$Zr_{0.98}Ce_{0.02}SiO_4$	0.0344	0.6008	1.2076
ZCS-3	0.03	$Zr_{0.97}Ce_{0.03}SiO_4$	0.0516	0.6008	1.1952
ZCS-4	0.04	$Zr_{0.96}Ce_{0.04}SiO_4$	0.0688	0.6008	1.1829
ZCS-5	0.05	$Zr_{0.95}Ce_{0.05}SiO_4$	0.0861	0.6008	1.1706
ZCS-6	0.06	$Zr_{0.94}Ce_{0.06}SiO_4$	0.1033	0.6008	1.1583
ZCS-7	0.07	$Zr_{0.93}Ce_{0.07}SiO_4$	0.1205	0.6008	1.1459
ZCS-8	0.08	$Zr_{0.92}Ce_{0.08}SiO_4$	0.1377	0.6008	1.1336
ZCS-9	0.09	$Zr_{0.91}Ce_{0.09}SiO_4$	0.1549	0.6008	1.1213
ZCS-10	0.10	$Zr_{0.9}Ce_{0.1}SiO_4$	0.1721	0.6008	1.1090

3 RESULTS AND DISCUSSION

The effect of cerium loading on the structure of Ce-doped zircon ceramics waste form have been Comprehensive studied. The progress of the structure formation in $Zr_{1-x}Ce_xSiO_4$ ($0 \le x \le 0.1$) compositions was followed by analyzing their XRD patterns recorded after sintered at 1550 °C for 72 h in air. Figure 1 shows XRD patterns of Ce-doped zircon samples. As can be seen in Figure 1, it was found that the primary crystalline phase of all samples is zircon. Thus, the formation of Ce-doped zircon ceramic can be considered to completing at 1550 °C for 72 h. Furthermore, it can be noted that the Ce-doped zircon ceramics with $x < 0.04$ show homogeneous single zircon phase, while the samples with $x \ge 0.04$ shows the presence of the secondary phases cerium silicate ($Ce_2Si_2O_7$). This indicates that compositions with $x \ge 0.04$ exhibit coexistence of two phases (zircon and $Ce_2Si_2O_7$). Furthermore, it can be observed that the peak intensities of silicate neodymium ($Nd_2Si_2O_7$) increase with raising neodymium loading.

In order to investigate the effect of cerium loading on the microstructure of the obtained Ce-doped zircon ceramics, Scanning Electron Microscope (SEM) analysis was employed. Figure 2 shows the SEM images of Ce-doped zircon ceramics with various cerium loading. It can be observed that all samples exhibit round-shaped grains, the average grain size is about 1–3 μm and the grain boundary is clear. However, the grain size becomes smaller and more uniform with increasing cerium loading, because the doping neodymium results in uniformly grain growth. Figure 2(a) and (b) show that the particles have only one shape (a single phase), which is consistent with the results of XRD. However, as can be seen in Figure 2(c), the

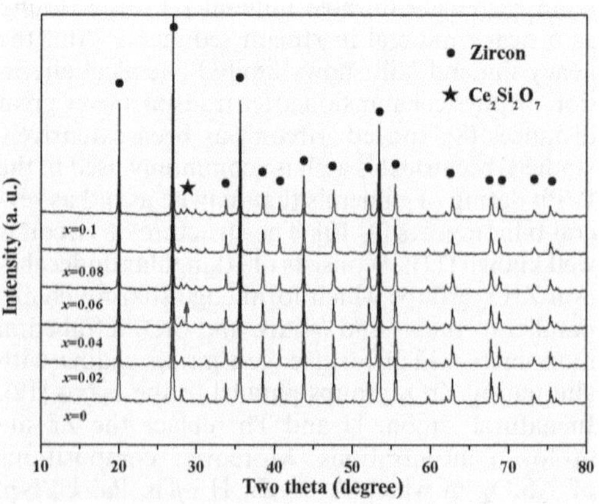

Figure 1. Representative XRD patterns of $Zr_{1-x}Ce_xSiO_4$ ($0 \le x, y \le 0.1$) compositions.

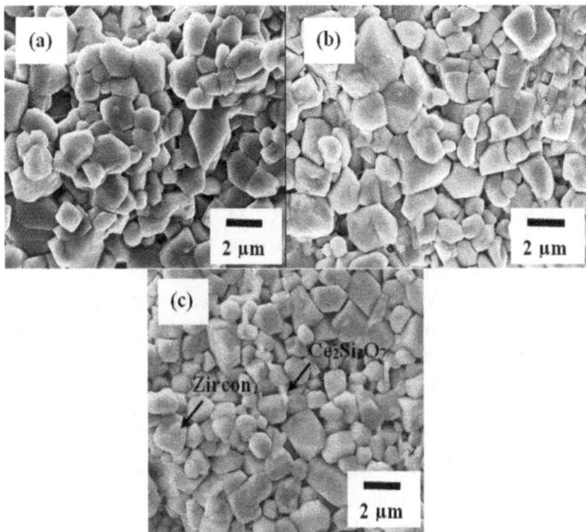

Figure 2. Representative SEM images of $Zr_{1-x}Ce_xSiO_4$ compositions. (a) x = 0, (b) x = 0.2 and (c) x = 0.4.

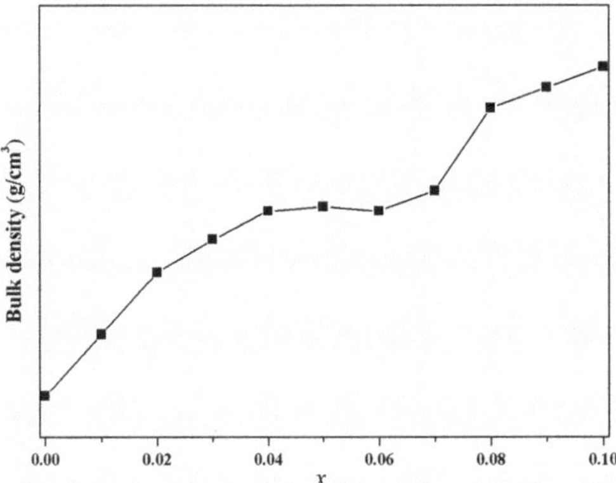

Figure 3. Bulk density of $Zr_{1-x}Ce_xSiO_4$.

two shape particles are present (round and flake morphology). This indicates that two phases are formed, which is also consistent with the XRD result of sample with x = 0.4.

In order to investigate the effect of cerium loading on the density of the obtained Ce-doped zircon ceramics, the densities of ceramics with the variation of cerium loading were measured and compared. Figure 3 shows the bulk density of the prepared Ce-doped zircon ceramics. It is found that all samples show high density values ranging from 4.41 to 4.56 g cm^{-3} which reaches over 95% of the theoretical density. Moreover, the bulk densities of the samples increase with the increase in neodymium loading. The composition with $x = 0.1$ ($Zr_{0.9}Nd_{0.1}SiO_{3.95}$) shows the highest density of about 4.56 g/cm^3 which is 98% of the theoretical values. This may be due to the formed low melting $Ce_2Si_2O_7$ phase appears as liquid phase which in turn improve the densification of the materials [19].

4 CONCLUSIONS

A series composition with general stoichiometry as $Zr_{1-x}Ce_xSiO_4$ ($0 \leq x \leq 0.1$) has been prepared by solid state reaction method at 1550 °C for 72 h in air. The results show that the Ce-doped zircon ceramics with $x < 0.04$ show homogeneous single zircon phase, while the samples with $x \geq 0.04$ exhibit coexistence of two phases (zircon and $Ce_2Si_2O_7$). The bulk densities of the Nd-doped zircon ceramics increase with cerium loading. These results suggest that zircon ceramics are promising candidate for the immobilization of tetravalent actinides.

ACKNOWLEDGEMENT

The authors are grateful for financial support from A Project Supported by Scientific Research Fund of SiChuan Provincial Education Department (13ZB0177) and the Scientific Research Fund of Mianyang city (14S-01-2).

REFERENCES

[1] Loiseau P., Majerus O., Aubin-Chevaldonnet V., Bardez I., Quintas A. Glasses, glass-ceramics and ceramics for immobilization of highly radioactive nuclear wastes [M]. New York: Nova Science Publishers, 2009.

[2] Lu X., Ding Y., Dan H., Wen M., Mao X., Wu Y., Wang X. High capacity immobilization of TRPO waste by $Gd_2Zr_2O_7$ pyrochlore [J]. Mater Lett 136(2014)1–3.

[3] Jafar M., Sengupta P., Achary S.N., Tyagi A.K. Phase evolution and microstructural studies in $CaZrTi_2O_7$ (zirconolite)–$Sm_2Ti_2O_7$ (pyrochlore) system [J]. J Eur Ceram Soc 34(2014)4373–4381.

[4] Yang H., Teng Y., Ren X., Wu L., Liu H., Wang S., Xu L. Synthesis and crystalline phase of monazite-type $Ce_{1-x}Gd_xPO_4$ solid solutions for immobilization of minor actinide curium [J]. J Nucl Mater 444(2014)39–42.

[5] Ewing R.C., Lutze W., Weber W.J. Zircon: A host-phase for the disposal of weapons plutonium [J]. J Mater Res 10(1995)243–246.

[6] Harker A.B, Flintoff J.F. Polyphase Ceramic for Consolidating Nuclear Waste Compositions with High Zr-Cd-Na Content [J]. J Am Ceram Soc 73(1990)1901–1906.

[7] Hanchar J.M., Miller C.F. Zircon zonation patterns as revealed by cathodoluminescence and backscattered electron images: implications for interpretation of complex crustal histories [J]. Chem Geol 110(1993)1–13.

[8] Hutton C.O. Studies of heavy detrital minerals [J]. Geol Soc Am Bull 61(1950)635–710.
[9] Mueller P.A., Heatherington A.L., Wooden J.L., Shuster R.D., Nutman A.P., Williams I.S. Precambrian zircons from the Florida basement: A Gondwanan connection [J]. Geology 22(1994)119–122.
[10] Barton E.S., Altermann W., Williams I.S., Smith C.B. U-Pb zircon age for a tuff in the Campbell Group, Griqualand West Sequence, South Africa: Implications for Early Proterozoic rock accumulation rates [J]. Geology 22(1994)343–346.
[11] Robinson K., Gibbs G.V., Ribbe P.H. The structure of zircon: a comparison with garnet [J]. Am Miner 56(1971)782–789.
[12] Taylor M., Ewing R.C. The crystal structures of the $ThSiO_4$ polymorphs: huttonite and thorite [J]. Acta Crystallogr B 34(1978)1074–1079.
[13] Keller C. Untersuchungen ueber die germanate und silicate des typs ABO_4 der vierwertigen elemente Thorium bis Americium [J]. Nukleonik 5(1963)41–48.
[14] Spear J.A. The actinide orthosilicates [J]. Reviews in Mineralogy, Mineralogical Society of America 5(1982)113–135.
[15] Speer J.A., Cooper B.J. Crystal structure of synthetic hafnon, $HfSi0_4$, comparison with zircon and the actinide orthosilicates [J]. Am Mineral 67(1982)804–808.
[16] Mumpton F.A., Roy R. Hydrothermal stability studies of the zircon-thorite group [J]. Geochim Cosmochim Acta 21(1961)217–238.
[17] Weber W.J. Radiation-induced defects and amorphization in zircon [J]. J Mater Res 5(1990)2687–2697.
[18] Weber W.J. Self-radiation damage and recovery in Pu-doped zircon [J]. Radiat Eff Defect S 115(1991)341–349.
[19] Thomas S., Sayoojyam B., Sebastian M.T. Microwave dielectric properties of novel rare earth based silicates: $RE_2Ti_2SiO_9$ [RE = La, Pr and Nd] [J]. J Mater Sci Mater-El 22(2011)1340–1345.

Material Science and Environmental Engineering – Chen (Ed.)

A light and broadband absorber utilizing resistive FSS and metamaterial layer

Y. Shen, Z.B. Pei, H.Y. Yuan & S.B. Qu
College of Science, Air Force Engineering University, Xi'an, Shaanxi, China

Y.Q. Pang
Electronic Materials Research Laboratory, Key Laboratory of the Ministry of Education, Xi'an Jiaotong University, Xi'an, Shaanxi, China
College of Science, Air Force Engineering University, Xi'an, Shaanxi, China

ABSTRACT: This paper presents the design, analysis and measurement of a light and broadband absorber utilizing a resistive FSS and a metamaterial layer, which are spaced by a honeycomb panel. The equivalent circuit theory is employed to analyze the working mechanism of the absorber with the assistance of Smith chart, and the validation has been performed by fitting the simulated S-parameters with the circuit model. It is shown through the simulation that the fractional bandwidth of 89.6% is realized for the reflectivity below −10 dB under the normal incidence. Furthermore, the total thickness of the proposed absorber is only 0.105 λ at the low operating frequency. We also fabricated an absorber sample and the measured result is well consistent with the simulation.

Keywords: absorbers; resistive FSS; metamaterials; broadband; lightweight

1 INTRODUCTION

Absorbing materials are often used to reduce the reflection of Electromagnetic (EM) waves on a surface. Conventional design methods consist in the insertion of losses on the surface of the material. As a typical example, the Salisbury screen was proposed decades ago because of its structural simplicity, which is comprised of a continuous resistive sheet placed a quarter-wavelength above the conducting plate. However, the bandwidth of Salisbury screens is narrow. Jaumann absorbers of the multilayer configuration can be employed to broaden absorption bandwidth, but they result in large thickness. To earn the wider bandwidth and the thinner thickness, Munk proposed resistive Frequency Selective Surface (FSS) in the design of absorbing materials. Since the resistive FSS not only contain a resistive component, but are reactive as well; the impedance matching can be obtained over a wide frequency range, and thus, a broadband absorber can be achieved. Because of the broadband absorption, simple configuration and easy fabrication, the FSS absorbers have been widely studied including theoretical analysis and applications in the past several years [1–3].

In 2008, a metamaterial absorber was first proposed to gain perfect absorption by Landy et al [4]. Due to the advantages of miniaturization, thin thickness, and high-efficiency absorbance, metamaterial absorbers have attracted significant attention. However, the metamaterial absorbers are almost of narrowband due to the strong resonant characteristic. As a result, many studies focused on the broadening of absorption bandwidth by using the multi-resonance [5, 6] or multi-interference [7], but the achieved bandwidth is limited all the same. Quite recently, Yoo et al [8] proposed a resistor-loaded FSS absorber to achieve broadband absorption with the incorporation of a metamaterial layer.

In this paper, we design an absorber which mainly consists of resistive FSS and metamaterial layer. The metamaterial layer used here extended the bandwidth of total system on the basis of the broadband absorption performance for resistive FSS absorber. Meanwhile, the resistive FSS can avoid the complex fabrication and lower the cost comparing with the resistor-loaded FSS for broadband absorption. Simulation indicates that the proposed absorber has broadband absorption more than 90% in the frequency of 6.0 to 19.0 GHz. The overall thickness of the absorber is 5.2 mm, being equal to 0.105 λ at the low operating working frequency. The density is only 0.58 g/cm^3 which satisfied the demand of lightweight. Finally, the experimental verification was also presented that the measured result is well consistent with the simulated result.

2 UNIT CELL AND SIMULATED RESULTS

The proposed absorber is composed of a resistive FSS and a metamaterial layer spaced by a honeycomb panel, as shown in Figure 1. The resistive FSS is made of the hexagonal resistive patches, which are placed on a thin substrate with the thickness of d_1. The radius and surface resistance of the hexagonal patches are r_1 and R_s, respectively. The parameter s_1 is the space between the adjacent patches. The hexagonal resistive patches are made up of resistive film, whose thickness can be ignored. The unit cell of the metamaterial layer is given in Figure 1(b), which consists of metal hexagonal-loops printed on a substrate backed with a metal plate. The inner radius and width of the hexagonal-loop are r_2 and w, respectively. The space between the adjacent loops and the thickness of the substrate are s_2 and d_2. All metals are made by copper films with an electric conductivity $\sigma = 5.8 \times 10^7$ S/m and a thickness of 0.02 mm. The substrates are the commercial FR4 with the relative permittivity of 4.3 and loss tangent of 0.025. The thickness of the honeycomb panel used as spacer is d and the relative permittivity is assumed to be 1.1.

The absorptive efficiency of an absorber is defined as $A(\omega) = 1 - R(\omega) - T(\omega) = 1 - |S_{21}|^2 - |S_{11}|^2$, where $A(\omega)$, $|S_{11}|^2$ and $|S_{21}|^2$ are the absorbance, reflectivity and transmissivity, respectively. Because of the metal back plate, the transmission (S_{21}) of the proposed system is zero. Thus, the absorbance is calculated by $A(\omega) = 1 - |S_{11}|^2$ in this paper. Absorbance can be maximized by minimization of reflection from the proposed structure. The layout of the unit cell structure is optimized by Genetic Algorithm and the geometric parameters are shown as follows: $R_s = 106\ \Omega$/sq, $r_1 = 6.4$ mm, $s_1 = 5.2$ mm, $d_1 = 0.4$ mm, $r_2 = 8.6$ mm, $s_2 = 1.4$ mm, $d_2 = 0.8$ mm, $w = 0.4$ mm, $d = 4.0$ mm. Figure 2 indicates the simulated reflectivity spectrum of the absorber for normal incidence. There is a deep dip at 7.7 GHz, and the reflectivity lower than −10 dB ($|\Gamma| < -10$ dB) of a bandwidth in 6.0~19.0 GHz. At the same time, we also analyzed the structure at oblique incidence for TE and TM wave, and the simulated results are given in Figure 3.

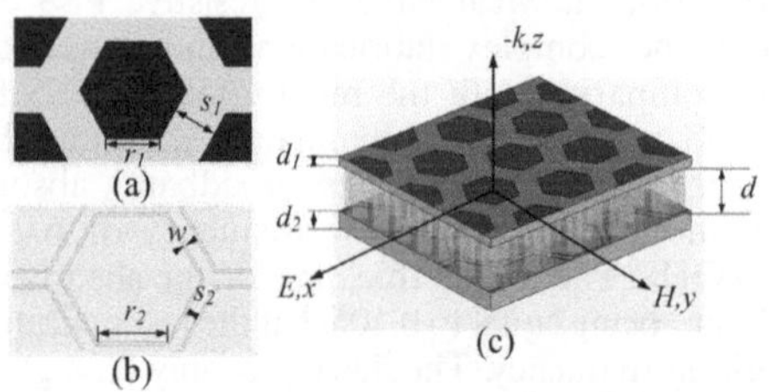

Figure 1. Geometry of the absorber. (a) Top view of the unit cell of the resistive FSS, (b) Top view of the unit cell of the metamaterial layer, (c) Perspective view of the system.

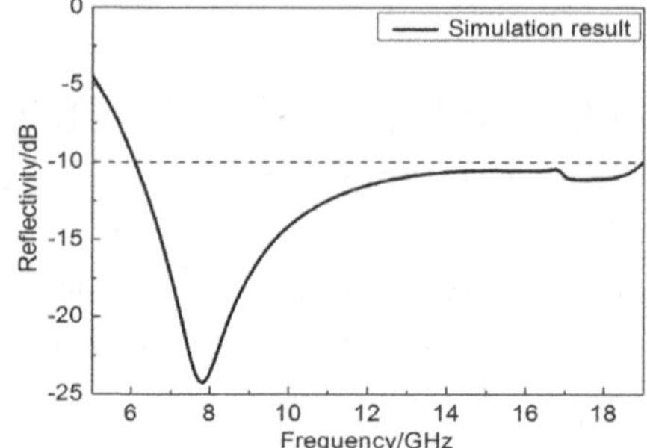

Figure 2. Simulated reflectivity of the absorber under the normal incidence.

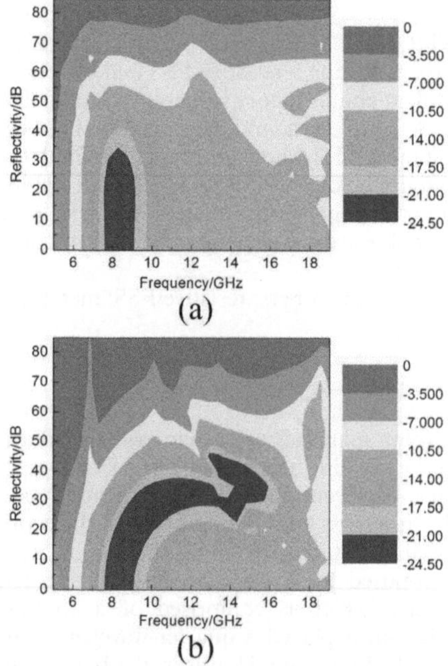

Figure 3. Reflectivity from the full-wave simulation at different angle of incidence for (a) TE wave, (b) TM wave.

The absorber has a steady absorptive efficiency more than 80% in the band of 6.0~19.0 GHz is limited to an angular of 60° for TE wave, and 50° for TM wave. With the increasing of incidence angle, not only the absorption level begins to decrease for two waves, but the absorption bandwidth also shifts to the high frequency region for TM wave.

3 THE OPTIMIZATION EQUIVALENT CIRCUIT ANALYSIS

In this part, we will analyze the absorbing principle of the proposed absorber base on the approach of equivalent circuit model. Figure 4 shows the circuit model of the proposed absorber with a combination of three sections. Section A is a resistive FSS, which is a conventional circuit analog absorber when placed in a distance front of the metal back plate. [5–7] The equivalent circuit model of the resistive FSS made of square resistive film was developed in Ref.4, which consists of a series *RLC* circuits shown in Figure 4. Section B is the dielectric substrates of the FR4 and the honeycomb panel used as interconnection Transmission-Line (TL) here to enhance the absorber mechanical performance. Section C is a metamaterial layer backed with metal plate, which can improve the absorbance in the high frequency region. To absorbers, the reflection coefficient is

$$\Gamma = \frac{Y_0 - Y_{in}}{Y_0 + Y_{in}} \tag{1}$$

where Y_0 and Y_{in} are the characteristic admittance of air and the input admittance of absorbers, respectively. In order to attain a reflection condition of zero, Y_0 must be equal to Y_{in}. From Figure 4, Y_{in} is given by

$$Y_{in} = Y_{RLC} + Y_1 \tag{2}$$

$$Y_{RLC} = \frac{1}{R + j\omega L + \frac{1}{j\omega C}} = G_0 + jB_0 \tag{3}$$

$$G_0 = \frac{\omega^2 RC^2}{\left(1 - \omega^2 RC\right)^2 + \omega^2 R^2 C^2} \tag{4}$$

$$B_0 = \frac{\omega C - \omega^3 RC^2}{\left(1 - \omega^2 RC\right)^2 + \omega^2 R^2 C^2} \tag{5}$$

where Y_{RLC} is the admittance toward resistive FSS, Y_1 is the admittance toward the interconnection TL and the metamaterial layer. From the equation for the TL, Y_1 is

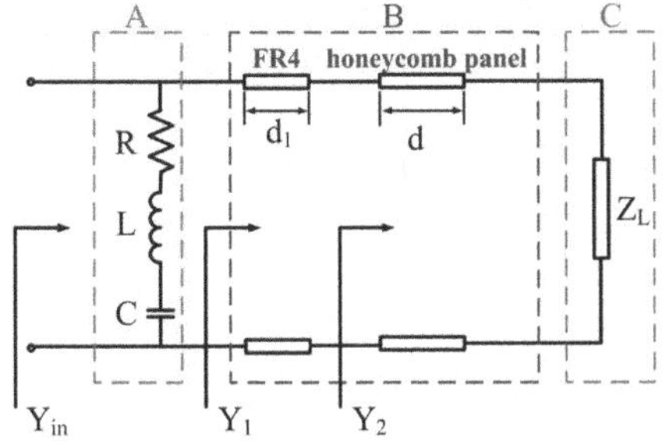

Figure 4. The circuit model of the absorber, which is a combination of Section A, Section B and Section C.

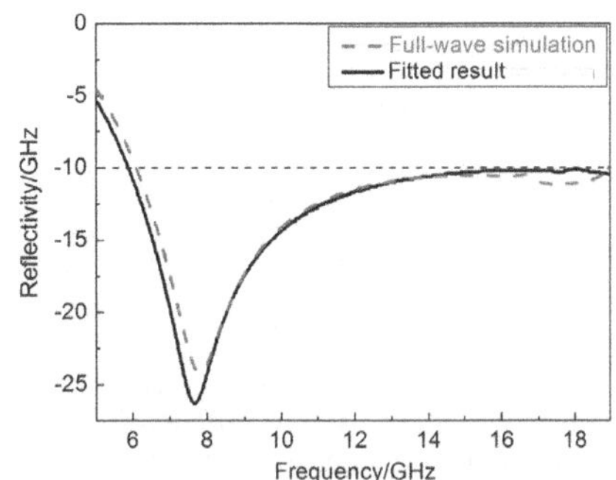

Figure 5. Comparison of the reflectivity between full-wave simulation and fitted result.

$$Y_1 = Y_{01} \frac{Y_2 + jY_{01}\tan(\beta_1 d_1)}{Y_{01} + jY_2\tan(\beta_1 d_1)} = G_d + jB_d \tag{6}$$

$$Y_2 = Y_{02} \frac{Y_L + jY_{02}\tan(\beta d)}{Y_{02} + jY_L\tan(\beta d)} \tag{7}$$

where Y_{01} and Y_{02} are the characteristic admittances of the FR4 substrate and the honeycomb panel, respectively. β_1 and β are the propagation constants for the FR4 substrate and the honeycomb panel. Y_2 is the admittance toward the honeycomb panel and the metamaterial layer. Y_L is the characteristic admittance of the metamaterial layer.

Since the parameters of the proposed absorber optimized by CST are identifiable, the value of the equivalent parameters *R*, *L*, *C* of the resistive FSS in the circuit model is also destined. For the resistive FSS, the equivalent resistance value *R* in the microwave frequency region can be approximated as [9]:

$$R = \frac{S_{unitcell}}{S_{patch}} R_s \tag{8}$$

where $S_{unitcell}$ and S_{patch} are the areas of unit cell and resistive patches. According to the parameters mentioned above, the value of *R* is 234.8 Ω. The values of *L* and *C* can be determined by fitting the S-parameters simulated by CST with the circuit model shown in Figure 4 in Advanced Design System. In the fitting process, the S-parameters of the part of Sections B and C were imported from the CST for simplicity. The determined values are $L = 2.1$ nH and $C = 0.07$ pF, respectively. Figure 5 present the reflectivity of the full-wave simulation, as well as fitted result by the equivalent circuit model shown in Figure 4.

4 EXPERIMENT MEASUREMENT

To investigate the absorptive properties of the proposed absorber experimentally, we fabricated the sample of the proposed absorber, and the prototype is shown in Figure 6. The overall size is 310×310 mm^2, which consists of 19×22 hexagonal unit cell. The resistive FSS is made of resistive film, which is printed on the PET utilizing the carbon slurry by mean of silk screen printed technology. The PET is of 12 μm-thickness, which will not affect the EM performance of the system. The metamaterial layer is fabricated by the technology of printed circuit board. The unit cell of the resistive FSS and the metamaterial layer based on FR4 substrate are shown in Figure 6(b) and (c), and the honeycomb panel of 4 mm is shown in Figure 6(d).

The absorption performance was measured in a microwave anechoic chamber. An Agilent E8363B vector analyzer and two broadband double-ridged horn antennas are used to emit and receive the EM wave. The measured reflection coefficient comparing with the simulated result is shown in Figure 7. On the whole, the trend of the measurement is in a good agreement with the simulation when taking into the errors during the fabrication and measurement.

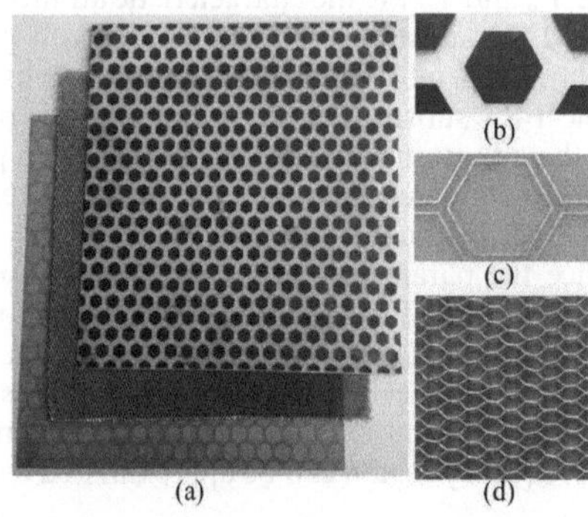

Figure 6. Pictures of the fabricated absorber. (a) 310 × 310 mm prototype, (b) unit cell of the resistive FSS, (c) unit cell of the metamaterial layer, and (d) the honeycomb panel of 4 mm.

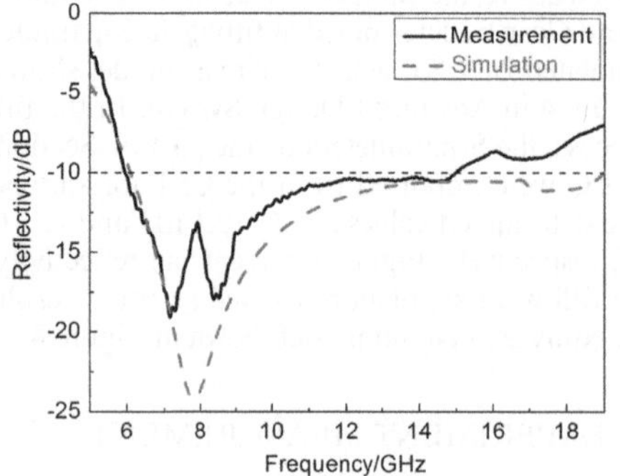

Figure 7. Comparison of the reflectivity between the measurement and full-wave simulation.

5 CONCLUSIONS

In summary, we have proposed a lightweight and broadband absorber made of a resistive FSS, a honeycomb panel and a metamaterial layer. The simulated result shows that this proposed absorber has broadband absorption more than 90% over the range from 6.0 to 19.0 GHz. The absorber is also possessed of polarization-independent and incident-angle-insensitive. The absorbing principle of this proposed absorber has been analyzed by means of equivalent circuit theory with assistance of Smith chart. The lumped parameters of L and C were attained by fitting the simulated S-parameters with the circuit model, which supports the validity of the equivalent circuit theory effectively. The good agreement between simulation and measure implied the validity of our design. Moreover, the electromagnetic configuration of the proposed absorber is similar to that of the sandwich structure composites, which have been widely used in the engineering fields. As a result, this thin and broadband absorber may find potential applications in the design of advanced composites with absorbing, as well as mechanism properties.

REFERENCES

[1] Pinto Y., Sarrazin J., Lepage A.C., and Begaud X., Capet N., 2014. Design and measurement of a thin and light absorbing material for space applications. *Appl. Phys. A* 115: 541–545.

[2] Sun L.K., Cheng H.F., Zhou Y.J., and Wang J., 2012. Design of broadband microwave absorber utilizing FSS screen constructed with coupling configurations. *Appl. Phys. A* 109: 873–875.

[3] Pang Y.Q., Cheng H.F., Zhou Y.J., and Wang J., 2012. Analysis and enhancement of the bandwidth of ultrathin absorbers based on high-impedance surfaces. *J. Phys. D: Appl. Phys.* 45: 104–108.

[4] Landy N.I., Sajuyigbe S., Mock J.J., Smith D.R., and Padilla W.J., 2008. Perfect metamaterial absorber. *Phys. Rev. Lett.* 100: 402–405.

[5] Pang Y.Q., Cheng H.F., Zhou Y.J., and Wang J., 2013. Double-corrugated metamaterial surfaces for broadband microwave absorption. *J. Appl. Phys.* 113: 4907–49011.

[6] Ding F., Cui Y.X., Ge X.C., Jin Y., and He S.L., 2012. Ultra-broadband microwave metamaterial absorber. *J. Appl. Phys.* 100: 506–509.

[7] Sun J.B., Liu L.Y., Dong G.Y., and Zhou J., 2011. An extremely broad band metamaterial absorber based on destructive interference. *Opt. Express* 19: 21155–21162.

[8] Yoo M.Y., and Lim S.J., 2014. Polarization-independent and ultrawideband metamaterial absorber using a hexagonal artificial impedance surface and a resistor-capacitor layer. *IEEE Trans. Antennas Propag.* 62: 2652–2658.

[9] Costa F., Monorchio A., and Manara G., 2010. Analysis and design of ultra thin electromagnetic absorbers comprising resistively loaded high impedance surfaces. *IEEE Trans. Antennas Propag.* 58: 1551–1558.

Material Science and Environmental Engineering – Chen (Ed.)
 ISBN 978-1-138-02938-5

Effect of irradiation on the microstructures, ferroelectric and dielectric properties of P (VDF-TrFE) copolymer films

L. Tian, Y.P. Li, J. Sun & X.H. Zhang
Hunan Institute of Enginnering, Xiangtan, Hunan, China

ABSTRACT: Poly (vinylidene fluoride-trifluoroethylene) [P (VDF-TrFE)] copolymer were irradiated with different electron dose. After irradiated with 84 Mrad and 112 Mrad dose, the normal P (VDF-TrFE) copolymer were changed into relaxor. The temperature dependence of dielectric properties for 84 and 112 Mrad irradiated specimens were analyzed by the modified Curie-Weiss law and the infinite range model. It was found that compared with the 84 Mrad irradiated specimen, the relaxor behavior was enhanced for 112 irradiated one. Besides, the local order parameter decreased with increasing temperature and the sharper temperature dependence of the local order parameter for 84 Mrad irradiated sample was observed.

Keywords: P (VDF-TrFE) films; irradiation; relaxor; infinite range model

1 INTRODUCTION

Outstanding electrostrictive, piezoelectric and electro caloric have been observed in inorganic ferroelectric relaxor [1–2]. Recently, experimental results have confirmed that some organic polymer, e.g. the electron-irradiated poly (vinylidene fluoride-trifluoroethylene) [P (VDF-TrFE)] and the poly (vinyliene fluoride-trifluoroethylene-chlorofluoroethylene [P (VDF-TrFE-CFE)] also exhibit relaxor behavior [3–6]. Up to now, several theories have been proposed to explain the peculiar dielectric properties of relaxor and give reasonable good results when applied to many inorganic relaxors [7–10]. For example, Yang et al have used modified Curie-Weiss law to describe the dielectric behaviors of $Pb_{0.5}Sr_{0.5}Ti_3$ relaxor [7]. Usman et al have described the dielectric constant for $BaZr_xTi_{1-x}O_3$ relaxor using an infinite range model [8]. However, those theories are scarcely applied to organic relaxors. From previous intensive studies from many investigators, it is known organic ferroelectric relaxors have many features that resemble those of inorganic relaxors [6, 12–14]: (1) They exhibit diffused phase transition. (2) The temperature corresponding to dielectric constant peak increases with increasing frequency. (3) They have similar characteristic temperatures e.g. Burn temperature, the intermediate temperature, phase transition temperature and freezing temperature. (4) The relaxor behavior is related to the presence of polar nanoregions. Those similar features for both kinds of relaxors suggest that the theories on inorganic relaxors can also be applied to organic relaxors. Thus, in the present paper the P (VDF-TrFE) films were irradiated by different dose. Firstly, the electron irradiation on microstructure, polarization loops and dielectric properties are studied. Then the relaxor behavior of irradiated P (VDF-TrFE) has been investigated using modified Curie-Weiss law and infinite range model.

2 EXPERIMENTS

Copolymer of P (VDF-TrFE) with VDF/TrFE molar ratio of 70/30 was first dissolved in dimethyl formamide to form a dilute solution. The P (VDF-TrFE) films were fabricated on an aluminum (Al) coated polyimide substrate. The thickness of the films investigated was about 180 nm. The as-grown films were annealed at 135°C for 4 hours. The electron irradiation was carried out in air atmosphere with irradiation dose of 0, 42, 84, 112 Mrad respectively. To characterize crystalline structure, X-ray diffraction measurements were performed with nickel filter Cu Kα irradiation (Bruker analytical x-ray system). Finally, the top electrodes were prepared by thermal evaporation method to form capacitor structure for electrical characterization. The polarization versus electric filed (*P-E*) hysteresis loops were measured using a Radiant Precision LC system. The dielectric properties were tested using an E4980 A precision LCR meter with an ac drive voltage of 0.02 V. The temperature was maintained using a computer controlled cryostat (MMR Tech., Inc.).

3 RESULTS AND DISCUSSION

3.1 *Electron irradiation on the microstructure of P (VDF-TrFE) thin film*

Figure 1 demonstrates X-Ray Diffraction (XRD) results of P (VDF-TrFE) copolymer with different irradiation dose. As seen from the figure, a reflection peak of 19.8° was obviously seen for 42 Mrad irradiated sample, which keeps the same position with the reflection peak of non-irradiated P (VDF-TrFE) film. The result indicates that 42 Mrad dose irradiation cannot change the normal ferroelectric P(VDF-TrFE) into relaxor because the reflection peak at 19.8° corresponds to the all-trans ferroelectric phase (polar phase)[15]. However, after 84 Mrad and 112 Mrad irradiation, the reflection peak shifts to a lower angle of 18.4°, which has the same position with the reflection peak of P (VDF-TrFE-CFE) terpolymer [6]. Thus, we can infer that higher irradiation dose of 84 Mrad and 112 Mrad can change the normal ferroelectric into relaxor.

3.2 *Electron irradiation effect on ferroelectric properties of P (VDF-TrFE) films*

Figure 2 shows the polarization-electric field (*P-E*) loops of all specimens measured at room temperature. A normal ferroelectric hysteresis loop is found for the 0 Mrad irradiated specimen. After 42 Mrad irradiation, the shape of hysteresis loop keeps similar with the non-irradiated sample although its remanent polarization decreases. It is worth noting that after 84 Mrad and 112 Mrad irradiation, the *P-E* loops become slim, which have similar shape with that of P (VDF-TrFE-CFE) terpolymer films [6]. These experimental results support that the normal ferroelectric P (VDF-TrFE) thin films

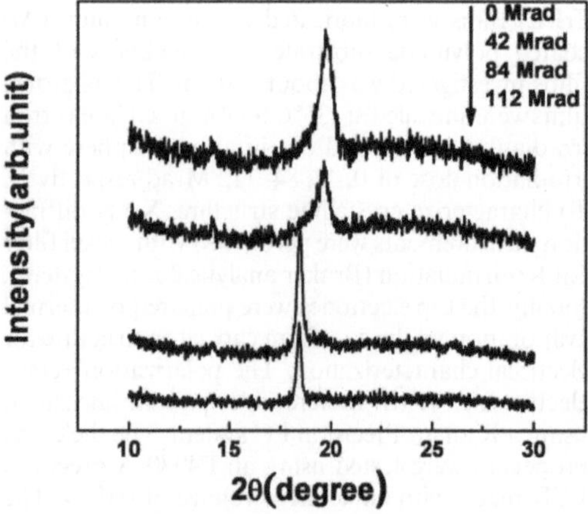

Figure 1. XRD results of P (VDF-TrFE) films irradiated with 0, 42, 84 and 112 Mrad dose.

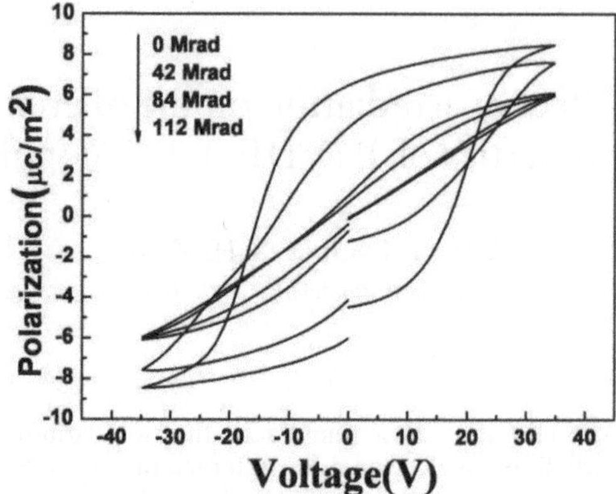

Figure 2. *P-E* loops of P (VDF-TrFE) films irradiated with 0, 42, 84 and 112 Mrad dose.

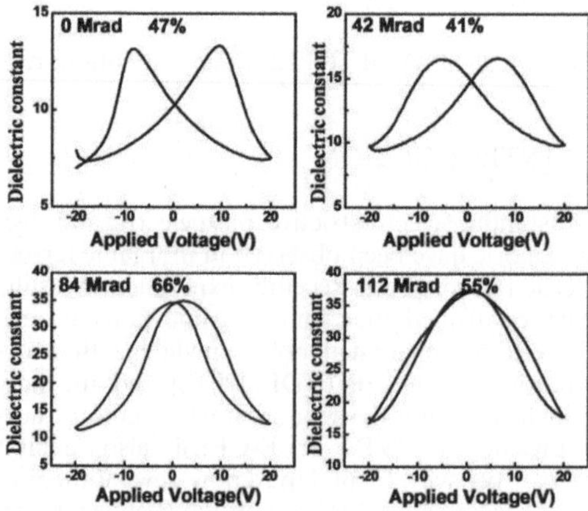

Figure 3. Dielectric constant-voltage dependence and the tunability values of P (VDF-TrFE) films irradiated with 0, 42, 84 and 112 Mrad dose.

have been changed into relaxor after 84 Mrad and 112 Mrad irradiation.

3.3 *Electron irradiation effect on tunability of P (VDF-TrFE) thin films*

Figure 3 presents the dielectric constant under dc bias voltage (ε-v) at room temperature for all samples. The present tunability is determined from the dielectric constant of the down curves of the ε-v. The tunability is 47%, 41%, 66%, and 55% for 0, 42, 84 and 112 Mrad irradiated samples, respectively. It is obvious that the 84 Mrad and 112 Mrad irradiated samples exhibit larger values of tunability. This is mostly due to the relaxor nature of both samples. Generally, the relaxor behavior is related to the polar nanoregions in ferroelectric films [10]. Those polar nanoregions exist related kinetic processes such as reversal under the applied electrical field, and then contribute a lot to the tenability [6–7]. Besides, note that compared with 84 Mrad irradiated sample, the 112 Mrad irradiated one has larger value of tunability. This phenomenon can be explained from the XRD results of both samples. On the one hand, the Full Width at Half Maximum (FWHM) of the diffraction peak of 84 Mrad is smaller, which suggests the sample has larger grain size. On the other hand, the intensity of reflection peak is larger for 84 Mrad-irradiated samples. Thus, the better microstucture for 84 Mrad sample maybe a key factor for the better tunability.

3.4 *The relaxor behavior of irradiated P(VDF-TrFE) film*

As mentioned above, the 84 Mad and 112 Mrad irradiated samples have been changed into relaxor. In order to discuss the relaxor behavior of irradiated P (VDF-TrFE) films, the temperature dependence of dielectric constant for both samples in heating process at different frequencies is measured and the results are shown in Figure 4. As can be seen, both samples exhibit relaxor behavior: the maximum dielectric constant (ε_m) undergoes a decrease with increasing frequency and the peak temperature (T_m) shifts to a higher temperature with increasing frequency. In generally, the parameters of ΔT_{relax} and $\Delta T_{diffuse}$ are introduced to characterize the relaxation degree and the diffuseness degree. They are defined as follows:

$$\Delta T_{relax} = T_{m(100KHz)} - T_{m(1KHz)} \tag{1}$$

$$\Delta T_{diffuse} = T_{0.9m(100KHz)} - T_{m(100KHz)} \tag{2}$$

The values of ΔT_{relax} ($\Delta T_{diffuse}$) extracted from Figure 4 are 2 K (18 K) and 3 K (21 K) for 84 Mrad and 112 Mrad irradiated samples, indicating higher relaxation degree and the diffuseness degree for 112 Mrad irradiated sample. For explaining the behavior of diffused phase transition of relaxors, a modified Curie-Weiss law is usually used [16]:

$$1/\varepsilon - 1/\varepsilon_m = (T - T_m)^r/C \quad (1 \le r \le 2) \tag{3}$$

where r and C are constant, and the parameter r is an diffusive parameter ranging between 1 and 2. For $r = 2$, it implies a complete diffused phase transition for an ideal ferroelectric relaxor. For $r = 1$, it refers a normal ferroelectric. The dielectric data at 1 MHz is fitted with equation 3 and the values of r are 1.90 and 1.57, as shown in insets of Figure 4.

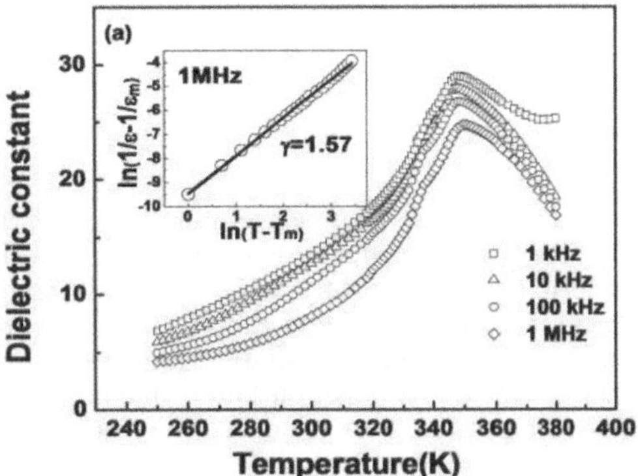

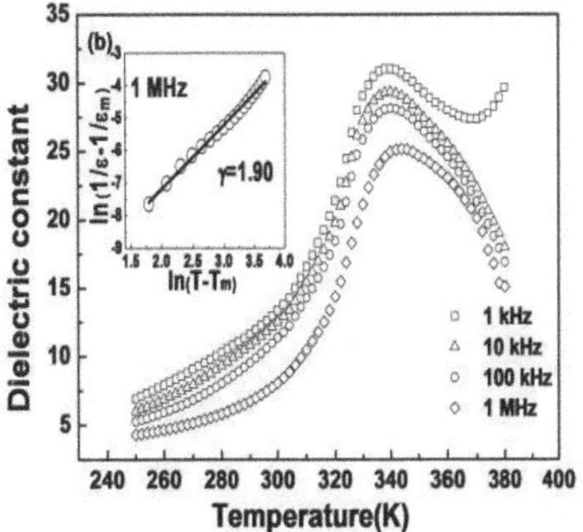

Figure 4. (a)–(b) The dielectric constant as a function of temperature measured at different frequencies for 84 and 112 Mrad irradiated samples. The insets are the $\ln(1/\varepsilon - 1/\varepsilon_m)$ as a function of $\ln(T - T_m)$ and the parameter fitted with the modified Curie-Weiss Law.

The larger value of r of 1.90 indicates the enhanced relaxor behavior of 112 Mrad irradiated P (VDF-TrFE).

Generally, the diffused phase characteristics of relaxors are related to the correlative behaviors of neighboring polar region. To characterize the degree of correlation with the neighboring polar region, the local order parameter of q is used. The q is given in the following equation, which is derived from the infinite range model developed by Sherrington and Kirkpatrick [17].

$$q(T) = 1 - \frac{\chi T}{C + \theta\chi} \tag{4}$$

where χ, C, θ are the reciprocal susceptibility, the Curie-Weiss constant and the Curie-Weiss temperature respectively. The values of the Curie-Weiss constant C and the Curie-Weiss temperature θ in equation 4 can be obtained by fitting the dielectric data to the Curie-Weiss law. The Curie-Weiss law can be expressed as

$$\chi = \frac{T-\theta}{C} \tag{5}$$

Figure 5 shows the temperature dependence of reciprocal susceptibility and fitting curve with curie-Weiss law for both samples. As can be seen, the reciprocal susceptibility can be well fitted by equation 5 at higher temperature range, and the values of the Curie-Weiss constant C and the Curie-Weiss temperature θ are obtained. The C (θ) values are 1265 (317) and 1052 (336) for 84 Mrad and 112 Mrad irradiated samples. After substitute the values of C and θ into equation (4), the parameter q can be calculated. Figure 6 presents the temperature dependence of the parameter q at 1 MHz for 84 Mrad and 112 Mrad irradiated samples. For both samples, the local order parameter decreases with increasing temperature. Similar dependence q on temperature was found in our previous study on P(VDF-TrFE-CFE) terpolymer [18]. Besides, the slope dq/dT in figure is sharper when using 84 Mrad irradiation, which suppress the relaxor and diffused phase transition [19–20].

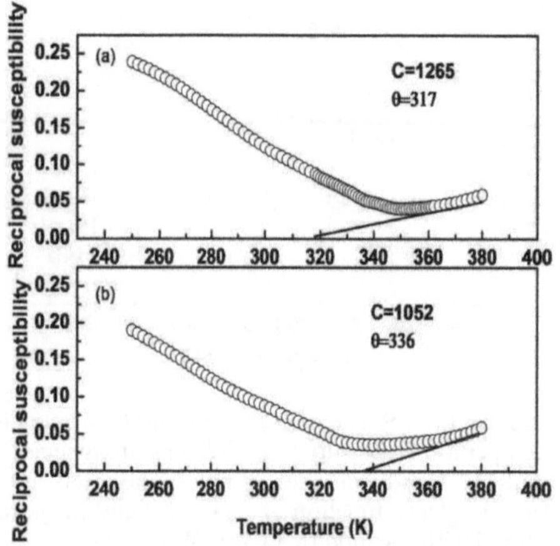

Figure 5. Temperature dependence of reciprocal susceptibility and fitting parameters with the Curie-Weiss law for (a) 84 Mrad and (b) 112 Mrad irradiated samples.

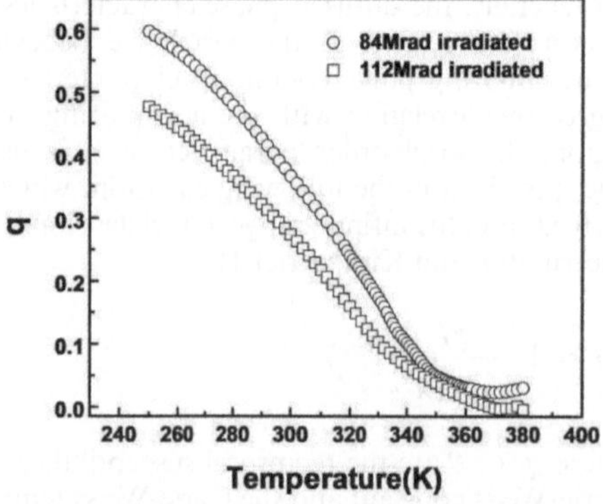

Figure 6. Local order parameter as a function of temperature for 84 Mrad and 112 Mrad irradiated samples according to infinite range model.

4 CONCLUSION

The P(VDF-TrFE) copolymer films were fabricated by sol-gel method and then were irradiated by various electron doses. From the XRD and *P-E* results, it is found after 84 Mrad and 112 Mrad irradiation, the normal ferroelectric P (VDF-TrFE) films can be changed into relaxor. Then the relaxor behavior of 84 Mrad and 112 Mrad irradiated samples were investigated by the modified Curie-Weiss law and the infinite range model. Compared with the 84 Mrad irradiated specimen, the relaxor behavior is enhanced for 112 irradiated one. Besides, we have found the local order parameter for both samples decreases with increasing temperature. Furthermore, the temperature dependence of local order parameter becomes sharper for 84 Mrad irradiated sample.

ACKNOWLEDGMENTS

A project supported by Scientific research fund of Hunan Provincial Education Department (Grant No. 14B041) and by the Natural Science Foundation of Hunan Province (Grant No. 2015 JJ6024 and No. 14 JJ6040).

REFERENCES

[1] Mischenko, A.S. et al. 2006. Giant electrocaloric effect in the thin film relaxor ferroelectric $0.9PbMg_{1/3}Nb_{2/3}O_3$-$0.1PbTiO_3$ near room temperature. *Applied Physics Letters* 89:242912.
[2] Park, S.E. et al. 1997. Ultrahigh strain and piezoelectric behavior in relaxor based ferroelectric single crystals. *Journal of Applied Physics* 82:1804–1811.
[3] Q.M. Zhang, et al. 2000. Relaxor ferroelectric behavior in high-energy electron-irradiated poly(vinylidene fluoride-trifluoroethylene) copolymers. *Applied Physics a-Materials Science & Processing* 70(3):307–312.
[4] Bharti, V. & Zhang, Q.M. 2001. Dielectric study of the relaxor ferroelectric poly(vinylidene fluoride-trifluoroethylene) copolymer system. *Physical Review B* 63(18):184103.
[5] Bao, H.M. et al. 2007. Phase transitions and ferroelectric relaxor behavior in P(VDF-TrFE-CFE) terpolymers. *Macromolecules* 40(7):2371–2379.

[6] Wang, J.L. et al. 2008. High electric tunability of relaxor ferroelectric Langmuir-Blodgett terpolymer films. *Applied Physics Letters* 93:192905.
[7] Yang, J. & Chu, J.H. 2007. Analysis of diffuse phase transition and relaxorlike behaviors in $Pb_{0.5}Sr_{0.5}TiO_3$ films through dc dielectric-field dependence of dielectric response. *Applied Physics Letters* 90:242908.
[8] Usman, M. et al. 2013. Order parameter and scaling behavior in $BaZr_xTi_xO_3$ relaxor ferroelectrics. *Applied Physics Letters* 103:262905.
[9] Liu, Y. et al. 2007. Structurally frustrated polar nanoregions in $BaTiO_3$-based relaxor ferroelectric systems. *Applied Physics Letters* 91:152907.
[10] Tenne, D.A. et al. 2003. Raman study of $Ba_xSr_{1-x}TiO_3$ films: Evidence for the existence of polar nanoregions. *Physical Review B* 67(1):012302.
[11] Ang, C. & Yu, Z. 2004. dc electric-field dependence of the dielectric constant in polar dielectrics: Multipolarization mechanism model. *Physical Review B* 69(17):1–8.
[12] Liu, B.L. et al. 2014. The intermediate temperature T^* revealed in relaxor polymer. *Applied Physics Letters* 104:222907.
[13] Viehland, D. et al. 1990. Freezing of the polarization fluctuations in lead magnesium niobate relaxors. *Journal of Applied Physics* 68:2916.
[14] Jeong, I.K. et al. 2005. Direct observation of the formation of pola nanoregions in $Pb(Mg_{1/3}Nb_{2/3})O_3$ using neutron pair distribution function analysis. *Physical Review Letters* 94(14):147602.
[15] Fernandez, M.V. et al. 1987. Study of annealing effects on the structure of Vinylidene Fluoride-Trifluoroethylene copolymer using WAXS and SAXS. *Macromolecules* 20(8):1806–1811.
[16] Martirena, H.T. et al. 1974. Grain-size and pressure effects on the dielectric and piezoelectric properties of hot-pressed PZT-5. *Ferroelectrics* 7(1):151–152.
[17] Sherrington, D. & Kirkpatrick, S. 1975. Solvable model of a spin-glass. *Physical review letters* 35(26): 1792–1796.
[18] Tian, L. et al. 2015. Effects of Electron Irradiation on the Dielectric Behavior of Langmuir-Blodgett Terpolymer Films. *Ferroelectrics* 487(1):81–87.
[19] Viehland, D. et al. 1992. Deviation from Curie-Weiss behavior in relaxor ferroelectrics. *Physical Review B* 46(13):8003–8006.
[20] Viehland, D. et al. 1991. Dipolar-glass model for lead magnesium niobate. *Physical Review B* 43(10):8316–8320.

Material Science and Environmental Engineering – Chen (Ed.)
 ISBN 978-1-138-02938-5

Analysis of stop-band FSS with all dielectric metamaterial

F. Yu, S.B. Qu, J. Wang, H.L. Du & J.F. Wang
College of Science, Air Force Engineering University, Xi'an, Shaanxi, China

L. Lu
Air Force Xi'an Flight Academy, Shaanxi, Xi'an, China

ABSTRACT: We present a new method for designing, a stop-band Frequency Selective Surface (FSS) with rectangular bore elements using dielectric. The stop-band FSS were designed to obtain resonant frequency at 38.5 GHz and 42 GHz, with about 4 GHz bandwidth at −3 dB. Thus, stop-band response absorption is realized. From two aspects of equivalent electromagnetic parameters and the surface current, the mechanism of the response is analyzed. Theoretical analysis shows that the stop-band response arises from impedance mismatch by the negative permeability and negative permittivity. Design of the stop band FSS is simple and is easy to be implemented, so such FSS may have application values in designing novel stop band FSS.

Keywords: stop-band; FSS; all dielectric

1 INTRODUCTION

Frequency Selective Surface (FSS), as a kind of spatial filter, is planar or curved surfaces composed of periodically arranged scatter arrays. FSS can exhibit one or many pass-bands or stop-bands. Conventional FSSs are usually two-dimensional structures, which are composed of periodically arranged metallic patches or slits. Metamaterials are artificial composite structures or composite materials that exhibit u-natural physical properties [1–4]. The concept of metamaterial arisen not only gives us a new material form, but also provides us a new material design concept, which brings revolutionary changes to the world view and methodology of human beings [5].

All-dielectric metamaterials are now receiving more and more attentions in the academic world. This kind of metamaterial includes no metallic part and thus has the advantages of low loss, simple structure, better homogeneity, and isotropy at high frequencies (in the millimeter waves, submillimeter waves, infrared waves, etc).

This paper shows that a stop-band FSS can be obtained by single-layer FSS screen with simple elements, such as rectangular bore, a stop-band response is obtained. The mechanism of the response is analyzed in both equivalent electromagnetic parameters and the surface current.

2 SIMULATION AND DESIGN

The unit of stop-band analyzed structure is shown in Figure 1. It is composed by an array of rectangular bore elements in a dielectric layer. The dielectric grid and the air are the two different kind of material structure. The dielectric layer with relative permittivity equal to 9.7 was used. The bore has periodicity extension in the X and Y directions, respectively. The rectangular bore has width w and

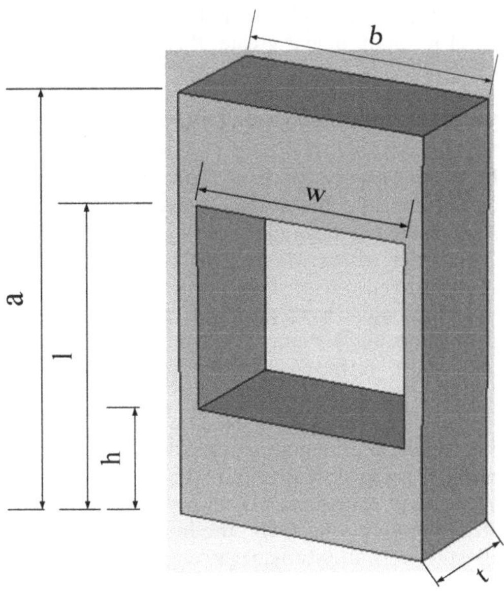

Figure 1. The unit geometry of FSS structures. The dimensions from the optimization are: a = 5.009, b = 3.018, l = 2.716, h = 1.3, w = 2.568, t = 1.474.

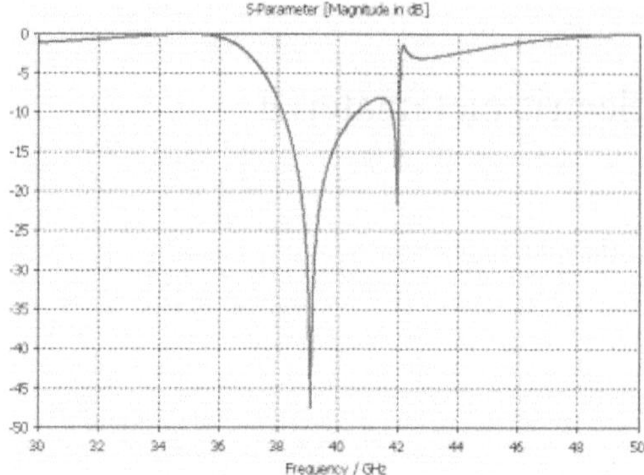

Figure 2. Transmittance of all-dielectric FSS showing TE modes at normal angle of incidence.

length (l – h). The FSS layer material adapt Al_2O_3, whose dielectric constant is $\varepsilon = 4.9$ and loss angle tangent $\tan\delta = 0.003$.

The effect of stop-band can be produced by this structure. In order to testify our predictions, computer simulations were performed using CST Microwave Studio.

Figure 2 gives the simulated magnitudes of S_{11} parameters. The TE modes were independently simulated at normal incidence. This design was predicted to provide over 10 dB of suppression over a bandwidth of 4 GHZ. As can be seen from the graph, the stop band is a single band, the pole is 39.13 GHz and 42 GHz, 3 dB stop band is 38~42 GHz, the bandwidth is 4 GHz, The reflectivity of resonant pole is greater than –45 dB.

3 THEORETICAL ANALYSIS

From two aspects of equivalent electromagnetic parameters and the surface current, the mechanism of the two responses is analyzed.

In order to see whether the two requirements of absorbers are satisfied, the effective constitutive parameters were retrieved from simulated S parameters [6, 7]. Figure 3(a) (b) and (c), show the impedance, retrieved effective permeability and permittivity, respectively.

As shown in Figure 3(b), at the frequency 39.13 GHz, the imaginary parts of effective permeability is negative. The unit of the FSS can be seen a magnetic resonance. As shown in Figure 3(c), at the frequency 42 GHz, the imaginary parts of effective permittivity is negative, respectively. The unit of the FSS can be seen a electric resonance.

In Figure 3(a), over the frequency range 30–36 GHz and 42–50 GHz, the relative and imaginary impedance is matched, which means $Z_E \approx Z_0$

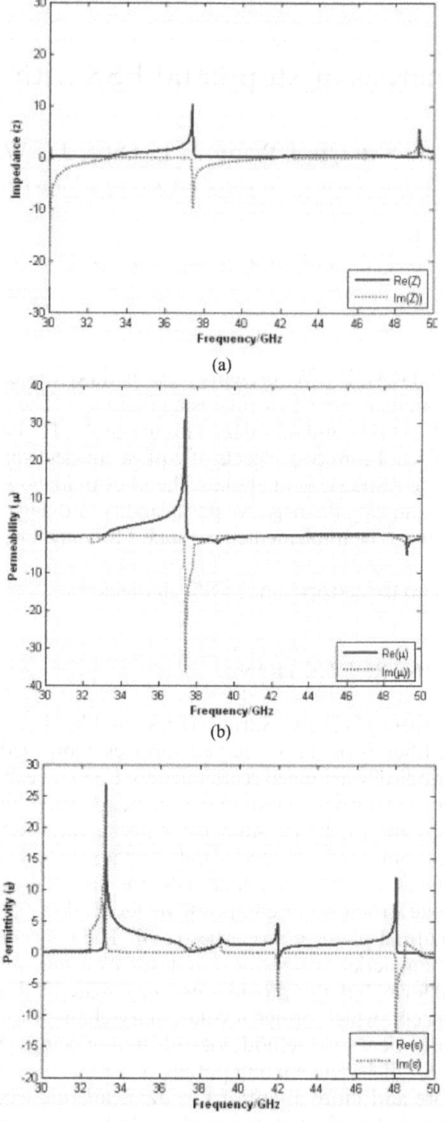

Figure 3. Real parts and imaginary parts of effective permeability and permittivity of the FSS from the simulation scattering parameters (a) impedance, (b) effective permeability, (c) effective permittivity.

(where Z_0 is the impedance of the free space). And over the frequency range 36–42 GHz, the relative impedance mismatched, which caused the appearance of the stop-band.

We also monitored the field distributions at 39.13 GHz and 42 GHz, to analyze the physical mechanism of realizing strong absorption by the FSS in Figure 1. Figure 4(a)–(d) show the electric

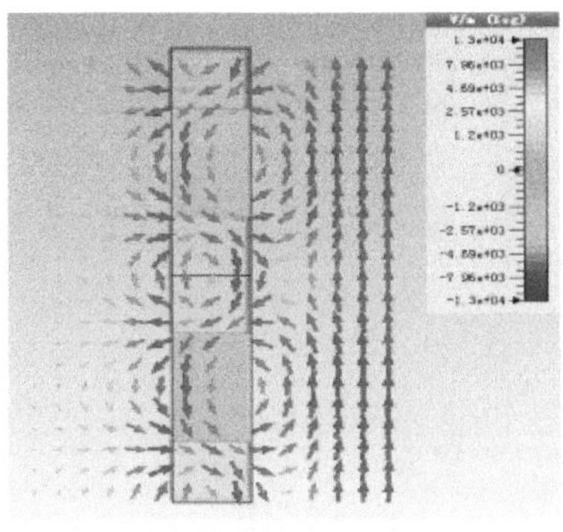

(a)

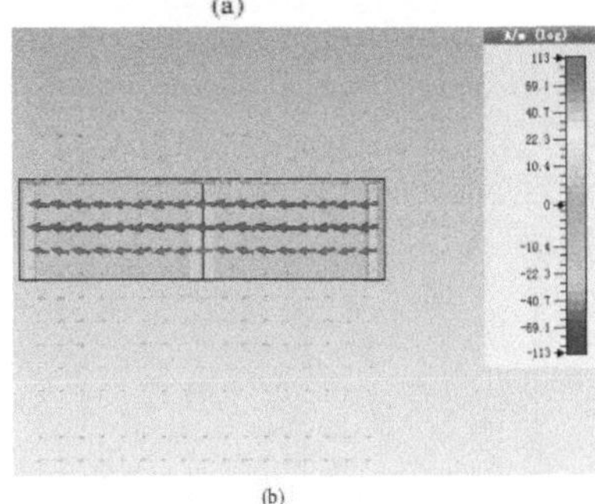

(b)

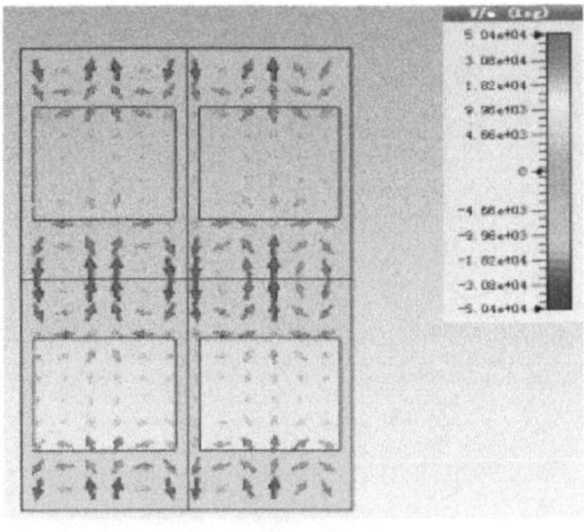

(c)

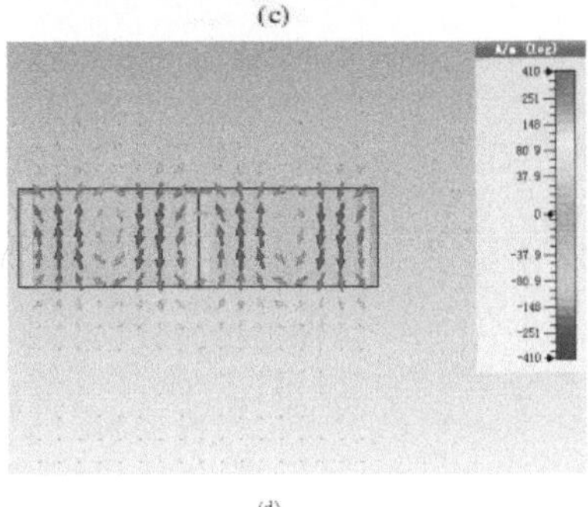

(d)

Figure 4. Field distributions of the unit cell at 39.13 GHz and 42 GHz: (a) electric field distribution at 39.13 GHz (left view), (b) magnetic field distribution at 39.13 GHz (top view), (c) electric field distribution at 42 GHz (front view), (d) magnetic field distribution at 42 GHz (top view).

field, magnetic field at 39.13 GHz and 42 GHz, respectively.

By Figure 4(a) and (b), the electric field distribution is annular and magnetic field distribution is parallel. It is obvious that the electric resonator behaves like a magnetic resonance. By Figure 4(c) and (d), both the electric field distribution and magnetic field distribution are annular, so it is obvious that the electric resonator behaves like an electric resonance.

4 CONCLUSION

In this paper, by numerical simulations and theoretical analysis, we proposed a stop-band FSS using the all dielectric metamaterial. The resonant frequency at 38.5 GHz and 42 GHz, with about 4 GHz bandwidth at −3 dB. It can be also used in EM waves with selective stop-band frequency.

REFERENCES

[1] Munk B.A. Frequency Selective Surfaces: Theory and Design. [M]. New York: Wiley-Inter science, 2000. 1~4.

[2] Cui Tie Jun, Smith David R, Liu Ruopeng. Metamaterials theory, design, and applications [M]. New York: Springer, 2009. 1~2.

[3] V.G. Veselago. The electrodynamics of substances with simultaneously negative values of ε and μ [J]. Soviet Physics Uspekhi, 1968, 10: 509~514.

[4] J.B. Pendry, A.J. Holden, W.J. Stewart et al. Extremely low frequency plasmons in metallic mesostructures [J]. Physical Review Letters, 1996, 76: 4773~4776.

[5] R.A. Shelby, D.R. Smith, S. Schultz. Experimental verification of a negative index of refraction [J]. Science, 2001, 292: 77~79.

[6] X.D. Chen, T.M. Grzegorczyk, B.-I. Wu, J. Pacheco Jr., J.A. Kong, Phys. Rev. E 70 (2004) 016608.

[7] D.R. Smith, D.C. Vier, Th. Koschny, C.M. Soukoulis, Phys. Rev. E. 71 (2005) 036617.

[8] Fei Yu, Shaobo Qu, Zhuo Xu, Jiafu Wang. Investigations on the Design of All-dielectric Frequency Selective Surfaces [J]. PIERS Proceedings, Suzhou, China, 2011: 472~475.

Material Science and Environmental Engineering – Chen (Ed.)
© 2016 Taylor & Francis Group, London, ISBN 978-1-138-02938-5

Properties of fly ash used in ceramsite which from the power plant in Luanping county, Hebei province, China

M.H. You, F. Zhou, H.T. Peng, X.P. Zhang, T. Wang, D. Wang, Z.B. Yu & Y.D. Zhu
College of Water Conservancy and Civil Engineering, China Agricultural University, Beijing, China

ABSTRACT: The recycling of fly ash is very important from an environmental point of view. The aim of this study is to discuss the feasibility of fly ash used for ceramsite production. The properties of the fly ash from a power plant located at Luanping county of Hebei province in China have been studied in this paper. The chemical composition of the fly ash was ascertained by Wavelength Dispersive X-Ray Fluorescence Spectrometry (WDXRF). The morphology of the fly ash was studied by Scanning Electron Microscope (SEM). The fineness and water demand ratio of the fly ash were studied by material physical properties test. The results show that the main chemical components are SiO_2 and Al_2O_3. The microscopic morphology showed the main phases are a glass phase, together with quartz, mullite etc. Fineness and water demand ratio of the fly ash are lower than the average of fly ash in China respectively. The lower fineness and water demand ratio, the less energy cost when produce ceramsite. It is a viable approach for making ceramsite with this fly ash.

Keywords: ceramsite; fly ash; chemical composition; morphology

1 INTRODUCTION

Fly ash is one of the residues generated in combustion, and comprises the fine particles that rise with the flue gases. The large amount of fly ash emissions will cause serious environmental pollution. At present, fly ash has been widely used in the building materials, backfill, road building, agriculture, chemical industry, high-performance ceramic materials, etc. all over the world. The utilization of fly ash is 46.2% in England, 65% in Germany and 75% in France. This paper, analyzed the fly ash from the Power plant at Luanping county of Hebei province in China. Therefore, the objectives of this study are:

1. to judge whether this fly ash can be used for ceramsite,
2. to improve the application and additional value of fly ash in China.

Produnce ceramsite is one of the major use of fly ash. The main ingredients that used for ceramsite are SiO_2 and Al_2O_3. Form glass melt at high temperature. The main chemical reactions as follows.

The decomposition of organic matter (1)–(3):

$$C + O_2 \rightarrow CO_2 \tag{1}$$

$$2C + O_2 \rightarrow 2CO \tag{2}$$

$$CO_2 + C - \rightarrow 2CO \tag{3}$$

The decomposition of carbonate (4)–(6)

$$CaMg(CO_3)_2 \xrightarrow{730\sim760°C} MgO + CaCO_3 + CO_2 \uparrow \tag{4}$$

$$MgCO_3 \xrightarrow{500\sim750°C} MgO + CO_2 \uparrow \tag{5}$$

$$CaCO_3 \xrightarrow{550\sim1000°C} CaO + CO_2 \uparrow \tag{6}$$

Iron oxide decomposition and reduction (7)–(10)

$$6Fe_2O_3 \xrightarrow{>1150°C} 4Fe_3O_4 + O_2 \uparrow \tag{7}$$

$$6Fe_2O_3 \xrightarrow{>1150°C} 4Fe_3O_4 + O_2 \uparrow \tag{8}$$

$$Fe_2O_3 + CO \rightarrow 2FeO + CO_2 \uparrow \tag{9}$$

$$Fe_2O_3 + CO \rightarrow 2FeO + CO_2 \uparrow \tag{10}$$

The materials used for ceramsite should accord with the following requirements.

1. The main chemical compounds of material are SiO_2 and Al_2O_3. And the material should contain flux oxide like Na_2O, K_2O, CaO, MgO and foaming agent.
2. The chemical composition of the material used for ceramsite should be controlled in a certain range: SiO_2 (53%–79%), Al_2O_3 (12%–16%), flux oxide (8%–24%).

3. In terms of physical properties, fineness and water demand ratio of fly ash should be as low as possible. Because low fineness and water demand ratio can reduce the demand for heat energy when manufacturing ceramsite.

2 MATERIALS AND METHODS

This fly ash was collected from the Power plant at Luanping county of Hebei province in China.

This research adopted Scanning Electron Microscope (SEM), Wavelength Dispersive X-Ray Fluorescence Spectrometry (WDXRF) and material physical properties test (fineness, water content). Through these methods, we obtained some aspects of performance of this fly ash. Analyzing whether this fly ash can be applied to manufacture ceramsite.

3 WAVELENGTH DISPERSIVE X-RAY FLUORESCENCE SPECTROMETRY (WDXRF)

WDXRF is a convenient and sensitive method for determining the chemical constituents. Through the WDXRF, we analyzed the chemical compounds of this fly ash and ascertained the proportion of various chemical elements. Table 1 shows the main chemical elements and their proportion. Table 2 shows the main chemical compounds and their proportion.

Table 1. Main chemical elements and proportion.

Chemical elements	Proportion (%)
O	46.5
Al	12.6
Si	25.8
Fe	3.98
K	2.40
Ca	3.10
C	2.68
Na	0.866
Mg	0.589

Table 2. Main compounds and proportion.

Compounds	Proportion (%)
SiO_2	51.7
Al_2O_3	22.7
CO_2	9.34
Fe_2O_3	5.13
CaO	3.97
K_2O	2.65
Na_2O	1.13
MgO	0.936

Table 1 show that the most chemical elements of fly ash from Power plant at Luanping county of Hebei province in China are O, Si, Al. Table 2 shows that the main chemical compounds of fly ash are SiO_2 and Al_2O_3. Table 2 shows content of SiO_2 is 51.7%, content of Al_2O_3 is 22.7% and content of oxides is 8.7% in this fly ash. The chemical compounds of this fly ash accord with requirement of used for ceramsite.

4 SCANNING ELECTRON MICROSCOPE (SEM)

Scanning Electron Microscope (SEM) is a type of electron microscope that materials images of a sample by scanning it with a focused beam of electrons. In this research, in order to observe the appearance morphology of the fly ash, we analyzed the fly ash with SEM. In order to observe more clearly, the fly ash was scanned by SEM with different multiple. Figure 1 shows the image of the fly ash at 100 μm. Figure 2 shows the image of the fly ash at 10.0 μm. Figure 3 shows the image of the fly ash at 2.00 μm. Figure 4 shows the image of the fly ash at 500 nm.

Through the use of Scanning Electron Microscope (SEM), we can draw a conclusion that the majority of the particles are regular sphere and also some particles are irregular shape like banding, columnar, etc.

The majority of the particles ranged in size from approximately 1 to 50 μm and consisted of solid spheres, we can see agglomerated particles and the majority of the particles are in regular sphere (Fig. 1). From Figure 2 we can see

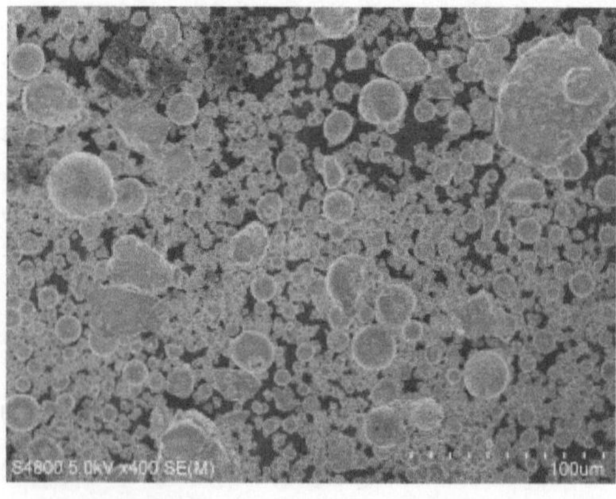

Figure 1. SEM image of fly ash at 100 μm.

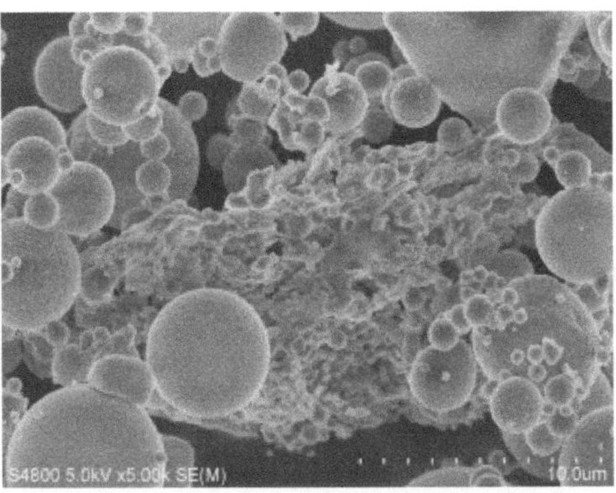

Figure 2. SEM image of fly ash at 10.0 μm.

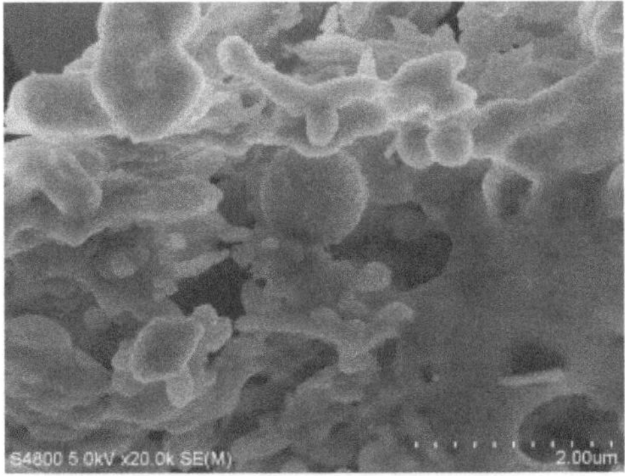

Figure 3. SEM image of fly ash at 2.00 μm.

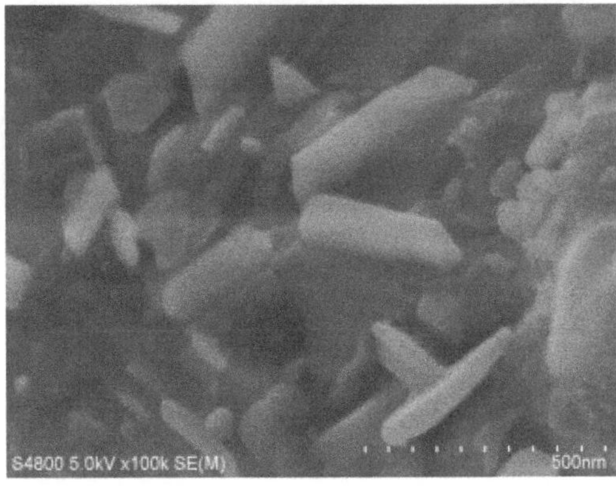

Figure 4. SEM image of fly ash at 500 nm.

particles and particle aggregates. Some particles of size 1 to 0.5 μm in diameter are rounded, with a globular surface. Some particles with irregularly shaped amorphous may be due to interparticle contact each other or rapid cooling in Figure 3.

Table 3. The material physical properties.

Items	Experimental data
Fineness (%)	21.7
Ratio of water demand (%)	88
Ignition loss (%)	1.17
Content of sulphur trioxide (%)	0.22
Water content (%)	0.1

The surface of particle adsorbs many irregularly shaped crystals (Fig. 4).

5 MATERIAL PHYSICAL PROPERTIES TEST

In order to determine the physical properties such as fineness, water content of the fly ash from Power plant at Luanping county of Hebei province in China, the material physical properties of the fly ash were tested. The result of material physical properties test was showed in Table 3.

Table 3 shows that the fineness of this fly ash is 21.7%. The average fineness of fly ash is 58% in China. The fineness of the fly ash from Power plant at Luanping county of Hebei province in China is lower than the average fineness in China, and also lower than 45% criterion of manufacture ceramsite. The ratio of water demand of this fly ash is 88%. The average ratio of water demand of fly ash is 102% in China. The ratio of water demand of this fly ash is smaller than the average value in China. Lower fineness and water demand ratio can reduce the demand for heat energy when manufacturing ceramsite. So the physical properties of the fly ash from Power plant at Luanping county of Hebei province in China is suitable for manufacture of ceramsite.

6 CONCLUSIONS

The chemical composition and physical properties of the fly ash were analyzed. The results show that the most chemical elements of fly ash from Power plant at Luanping county of Hebei province in China are O, Si, Al and the main chemical compounds of fly ash are SiO_2 and Al_2O_3. The chemical compounds of this fly ash accord with requirement of used for ceramsite. The fineness of this fly ash is 21.7%. The ratio of water demand of this fly ash is 88%. Lower fineness and water demand ratio can reduce the demand for heat energy when manufacturing ceramsite. In terms of surface morphology, the fly ash contains many crystals and pores. The fly ash from Power plant at Luanping county of

Hebei province in China is suitable for manufacturing ceramsite. But, there is also a disadvantage that gas forming materials like C, Fe_2O_3, $CaCO_3$ are less. Expansion rate is small when manufacturing ceramsite. So, we suggest adding some gas forming materials to use this fly ash to manufacture ceramsite.

ACKNOWLEDGMENTS

This work was supported by National Undergraduate Training Programs for Innovation and Entrepreneurship in China (No. 15100003).

REFERENCES

[1] Fei Yang, Dachuan Zhao: Functional materials. Vol. 41 (2010) p518.

[2] Xiaoyong Lu, Xiaoyan Zhu: Journal of Liaoning Technical University. Vol. 24 (2) (2005) p295.

[3] Shiyuan Yang, Fangjie Yang: Tile world. (7) (2010) p45.

[4] Shiyuan Yang, Fangjie Yang: Tile world. (8) (2010) p42.

Material Science and Environmental Engineering – Chen (Ed.)
 ISBN 978-1-138-02938-5

Effect of anionic surfactants with different carbon chain lengths on CO_2 hydrate formation

Z.Z. Jia & Y. An
Institute of Chemistry and Chemical Engineering, Guizhou University, Guiyang, China

F. Wang
University of Chinese Academy of Science, Beijing, China
Shandong Industrial Engineering Laboratory of Biogas Production and Utilization, Key Laboratory of Biofuels, Qingdao Institute of Bioenergy and Bioprocess Technology, Chinese Academy of Sciences, Qingdao, Shandong, China

S.J. Luo & R.B. Guo
Shandong Industrial Engineering Laboratory of Biogas Production and Utilization, Key Laboratory of Biofuels, Qingdao Institute of Bioenergy and Bioprocess Technology, Chinese Academy of Sciences, Qingdao, Shandong, China

ABSTRACT: In this work, Sodium Dodecyl Sulfate (SDS) and octadecyl sodium sulfate (SOS) were used as promoters respectively to study the effect of surfactants with different carbon chain lengths on CO2 hydrate formation. Under unstirred condition, SDS showed efficient promotion effect on hydrate formation and the increase in SDS concentration (0–1 CMC) lead to better performance, while no obvious promotion was observed when SOS was used. At stirring condition of 800 rpm, both SDS and SOS could enhance the CO_2 hydrate formation efficiently and SDS showed better performance than SOS at the same CMC concentration. However, SOS could enhance the hydrate formation efficiently at much lower quantity concentration in comparison to SDS, and fewer promoters would be used in potential CO_2 capture and sequestration process.

Keywords: CO_2 hydrates; surfactant; carbon chain length

1 INTRODUCTION

Carbon dioxide has been identified as the major contributor to global warming [1], therefore, capture and sequestration of CO_2 has attracted great attention during the past several decades [2].

Gas hydrates are clathrate crystalline compounds formed by water and gas molecules at suitable temperature and pressure through the occupation of gas molecules in the cavities formed by water molecules under hydrogen-bond [3]. However, the slow hydrate formation is the main problem in hydrate-base CO_2 capture and sequestration.

Surfactants have been confirmed to promote hydrate formation greatly and much work has been conducted on the hydrate formation with surfactants as promoter [4–9]. However, the promotion effect of surfactants on hydrate formation was influenced by the surfactant concentration greatly and a certain concentration was necessary to obtain efficient promotion effect [10, 11]. To improve the promotion effect of surfactants and ensure the surfactants with good promotion effect at low dosage, other additives were also used together with surfactants in hydrate formation [2, 12, 13]. For example, Abolfazl Mohammadi et al [2] conducted the CO_2 hydrate formation in the presence of silver nanoparticles and SDS and found that both SDS and silver nanoparticles single did not showed significant effect on decreasing the induction time and increasing the storage capacity of CO_2 hydrates while the mixture of SDS and silver nanoparticles significantly increased the storage capacity of CO_2.

In addition, in the studies carried out by Jeffry Yoslim et al [14], sodium alkyl sulfate surfactants with different carbon chain lengths (C12, C14, C16) were used as promoter in methane-propane hydrate formation. Similarly efficient promotion effect was obtained at C12 2200 ppm, C14 300 ppm and C16 40 ppm respectively, which meant that the increase in the carbon chain length could reduce the surfactant amount used in hydrate formation. Similar study was also conducted by Kazunori

Okutani et al [4] in methane hydrate formation. However, studies on surfactants with different carbon chain lengths to reduce the surfactant amount in CO_2 hydrate formation have rarely been reported. In this work, surfactants with the same hydrophilic group but different carbon chains, SDS (C12) and SOS (C18), were used as promoter in CO_2 hydrate formation to study the effect of different surfactants on CO_2 hydrate formation.

2 EXPERIMENTS

2.1 *Materials*

Sodium Dodecyl Sulfate (SDS, AR, CMC 8.6 mmol/L) and octadecyl Sodium Sulfate (SOS, AR, CMC 0.17 mmol/L) were provided by Xiya Reagent Co., Ltd. CO_2 (99.99%) was purchased from Heli Gas Co., Ltd (Qingdao, China). The deionized water used in this work was laboratory-made. The structures of the anionic surfactants are shown in Figure 1.

2.2 *CO_2 hydrate formation experiments*

CO_2 hydrate formation was conducted in a 350-mL stainless steel reactor and the schematic diagram of experimental apparatus is shown in Figure 2. Firstly, reaction solution with volume of 50 mL was added into the reactor and then reactor was cleaned with CO_2 three times and cooling started. After the temperature reached 275.15 K, the reactor was pressurized with CO_2 to 3 MPa. The temperature and pressure of the hydrate formation system were recorded on the computer during the whole process.

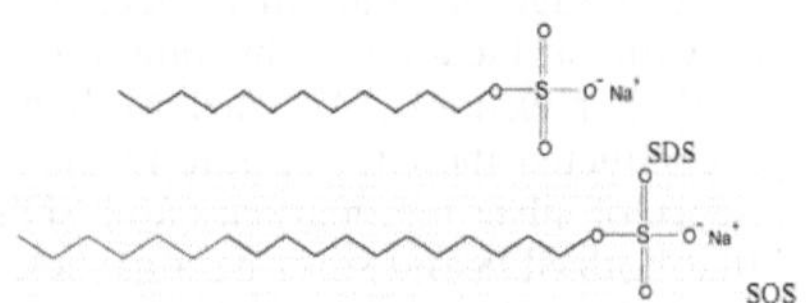

Figure 1. Molecular structures of the anionic surfactants.

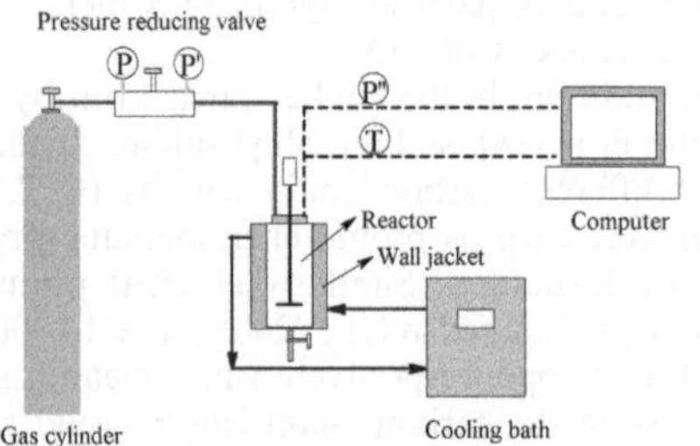

Figure 2. Schematic diagram of the hydrate formation apparatus.

2.3 *Calculation of CO_2 consumption*

CO_2 consumption (n) at time t was calculated as follow.

Where P_0 and P_t are the pressure of the reactor at time 0 and t, V is the volume of the gas phase in the reactor, R is the universal gas constant, T_0 and T_t are the temperature in the reactor at time 0 and t, z_0 and z_t are the compressibility factors at time 0 and t.

z_t and z_0 are calculated through the Pitzer Correlations for the compressibility factor [15]:

Where T_c is 304.2 K, P_c is 7.338 MPa, ω is 0.224.

3 RESULTS AND DISCUSSION

3.1 *CO_2 hydrate formation at unstirred condition*

Figure 3 shows the pictures of surfactant solutions with different concentrations at hydrate formation temperature (275.15 K). At 0.5–1 CMC, SLS precipitated slightly, while at 2 CMC it precipitated seriously. For SOS, at 0.5–2 CMC, no serious precipitation was observed. Therefore, both SDS and SOS were used at the concentration of 0.5, 0.75, 1 and 2 CMC in this work.

Figure 4 (A) shows the evolutions of gas consumption during the CO_2 hydrate formation with SLS as promoter and the hydrate formation process could be divided as three stages: CO_2 dissolution period, induction period and hydrate growth period. The CO_2 dissolution period was usually finished within the initial 30 min and in these experiments the CO_2 consumption below 0.04 mol could be viewed as the CO_2 dissolution into the solution. The induction period was regarded as the time from charging CO_2 into the reactor to the time obvious pressure decrease was observed.

As shown in Figure 4, the existence of SDS could enhance the hydrate formation and the increase in the SDS concentration within a certain range

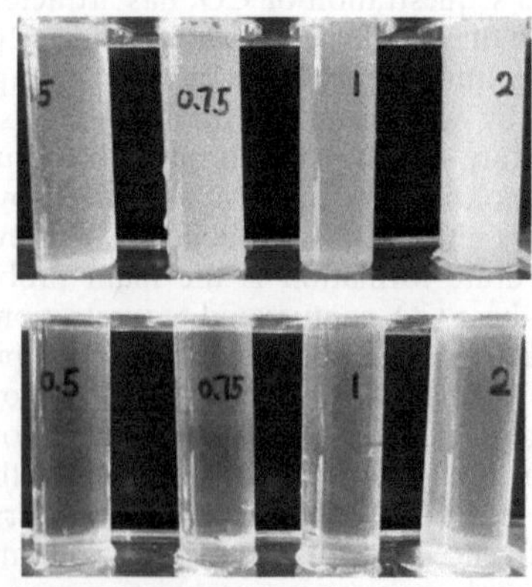

Figure 3. Surfactant solutions with different concentrations at hydrate formation temperature (275.15 K) (left-SDS, right-SOS).

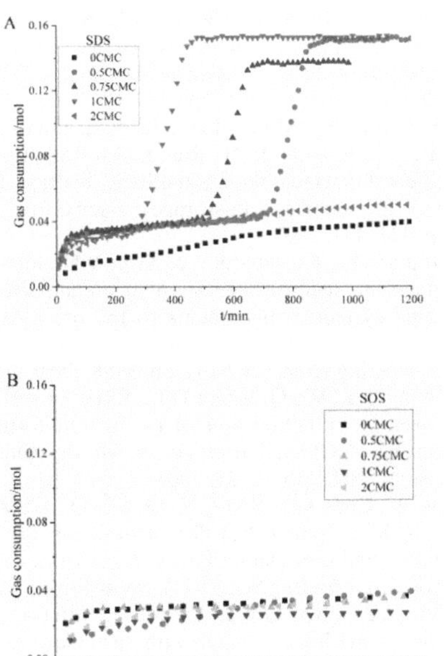

Figure 4. Evolutions of gas consumption during CO_2 hydrate formation with different SDS (A) and SOS (B) concentrations at unstirred condition. (Initial pressure-3 MPa; temperature-275.15 K).

(0–1 CMC) resulted in shorter induction period and higher hydrate formation rate. For example, as the concentration of the SLS increased from 0.25 CMC to 1 CMC, the induced period decreased from 700 min to 215 min. On one hand, surfactants could reduce the gas-liquid interfacial energy and contribute the diffusion of gas molecules into the liquid phase and higher surfactant concentration made more contribution. On the other hand, as discussed in previous reports [16–19], surfactant molecules could be adsorbed on the hydrate surface and contribute to the reaction between gas and water molecules by reducing the hydrate-liquid interfacial tension. However, at SDS concentration of 2 CMC, no obvious promotion effect on hydrate formation was observed. On one hand, higher SDS concentration resulted in the increase in the ion concentration in the liquid phase, which could inhibit the hydrate formation. On the other hand, as shown Figure 3, at hydrate formation temperature SDS precipitated seriously at 2 CMC and a lot of SDS crystals could be precipitated from the liquid phase, which might not be conducive to the hydrate formation.

Figure 4(B) shows the evolutions of gas consumption during the CO_2 hydrate formation with SOS as promoter and no obvious promotion effect on hydrate formation was observed, indicating that SDS showed much better promotion effect on CO_2 hydrate formation in comparison to SOS.

3.2 *CO_2 hydrate formation at 800 rpm*

Figure 5 shows the gas consumption during CO_2 hydrate formation with different surfactant concentrations at 800 rpm and much faster hydrate formation and higher gas consumption were obtained compared with the hydrate formation at unstirred condition, which indicated that stirring could promoted hydrate formation efficiently, especially when SOS was used as promoter. In the hydrate formation at unstirred condition, hydrates formed initially at the gas-liquid interface and then would cover the interface, which consequently retarded the diffusion of gas into the liquid phase, resulting in slow hydrate formation. While at stirring condition, the hydrates formed at the gas-liquid interface could be removed promptly and the diffusion of gas into the liquid phase could be enhanced, therefore, much faster hydrate formation was achieved.

In the CO_2 hydrate formation with SDS as promoter, the increase in the SDS concentration within a certain range lead to faster hydrate formation and excessive SDS concentration would inhibit the hydrate formation, which was consistent with that at unstirred condition. While in CO_2 hydrate formation with SOS as promoter, similar promotion effect was observed at all the SOS concentration except 2 CMC, which meant that low concentration of SOS was enough to promote the hydrate formation. As shown in Figure 3, no obvi-

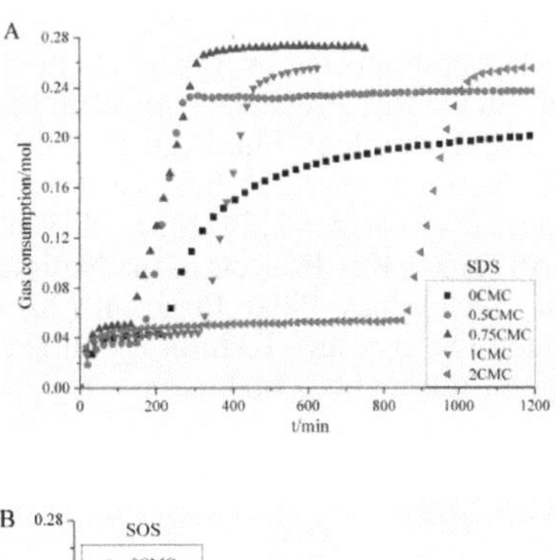

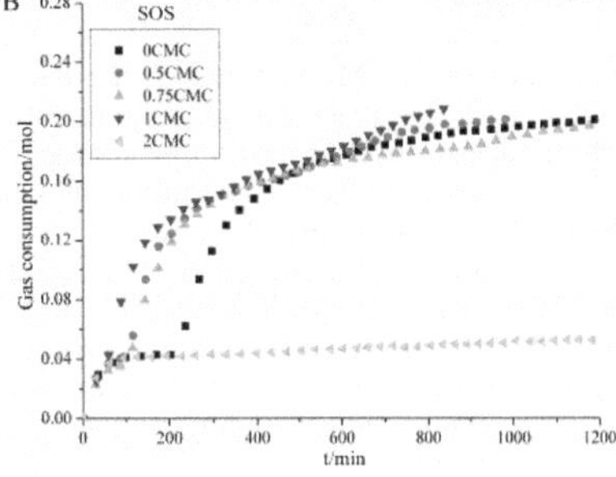

Figure 5. Evolutions of gas consumption during CO_2 hydrate formation with different SDS (A) and SOS (B) concentrations at 800 rpm. (Initial pressure-3 MPa; temperature-275.15 K).

ous precipitation of SOS was observed at 2 CMC, the poor promotion effect of SOS at 2 CMC might be caused by the inhibition effect of high ion concentration in the liquid phase.

Figure 5 also shows that at stirring condition SOS produced much shorter induction period than SDS, especially at low concentration, while SDS showed higher hydrate growth rate and gas consumption than SOS, which meant higher storage capacity of CO_2 hydrates. However, as the CMC of SDS and SOS were 8.6 mmol/L and 0.17 mmol/L, it could be imaged that SOS could produce well promotion on CO_2 hydrate formation at much lower molar amount in comparison to SDS.

4 CONCLUSIONS

CO_2 hydrate formation was carried out with SDS and SOS as promoters at unstirred and stirring conditions respectively. At unstirred condition, SDS showed much better promotion effect on hydrate formation than SOS. While at 800 rpm, both SDS and SOS could enhance the CO_2 hydrate formation efficiently. Even SDS showed better performance than SOS at the same CMC concentration, SOS could enhance the hydrate formation efficiently at much molar concentration.

ACKNOWLEDGEMENT

This work was supported by Taishan Scholar Program of Shandong Province, Guizhou Province Science and Technology Fund (Qian. 2013_2131) and the National Natural Science Foundation of China (31101918, 21307143), 863 Project (2012AA052103), Key Projects in the National Science and Technology Pillar Program (20140015) and Qingdao Science and Technology and People's Livelihood Project (14-2-3-69-nsh).

REFERENCES

[1] Eslamimanesh A., Mohammadi A.H., Richon D., Naidoo P. & Ramjugernath D. Application of gas hydrate formation in separation processes: a review of experimental studies. The Journal of Chemical Thermodynamics, 46, pp. 62–71, 2012.

[2] Abolfazl Mohammadi, Mehrdad Manteghian a, Ali Haghtalab, Amir H. Mohammadi & Mahboubeh Rahmati-Abkenar, Kinetic study of carbon dioxide hydrate formation in presence of silver nanoparticles and SDS. Chemical Engineering Journal, 237, pp. 387–395, 2014.

[3] Walsh M.R., Koh C.A., Sloan E.D., Sum A.K. & Wu D.T., Microsecond Simulations of Spontaneous Methane Hydrate Nucleation and Growth Science, 326(5956), pp. 1095–1098, 2009.

[4] Okutani K., Kuwabara Y. & Mori Y.H., Surfactant effects on hydrate formation in an unstirred gas/liquid system: An experimental study using methane and sodium alkyl sulfates. Chem Eng Sci, 63, pp. 183–194, 2008.

[5] H. Ganji, M. Manteghian, K. Sadaghiani zadeh, M.R. Omidkhah & H. Rahimi Mofrad, Effect of different surfactants on methane hydrate formation rate, stability and storage capacity. Fuel, 86(3), pp. 434–441, 2007.

[6] Okutani K., Kuwabara Y. & Mori Y.H., Surfactant effects on hydrate formation in an unstirred gas/liquid system: Amendments to the previous study using HFC-32 and sodium dodecyl sulfate. Chemical engineering science, 62(14), pp. 3858–3860, 2007.

[7] Zhang B.Y., Wu Q. & Sun D.L., Effect of surfactant Tween on induction time of gas hydrate formation. Journal of China University of Mining and Technology, 18(1), pp. 18–21, 2008.

[8] Lin W., Chen G.J., Sun C.Y., Guo X.Q., Wu Z.K. & Liang M.Y., Yang L.Y., Effect of surfactant on the formation and dissociation kinetic behavior of methane hydrate. Chem Eng Sci, 59(21), pp. 4449–4455, 2004.

[9] Sun C.Y., Chen G.J. & Yang L.Y., Interfacial tension of methane + water with surfactant near the hydrate formation conditions. J Chem Eng Data, 49, pp. 1023–1025, 2004.

[10] Zhong Y. & Rogers R.E., Surfactant effects on gas hydrate formation. Chem Eng Sci, 55(19), pp. 4175–4187, 2000.

[11] Jonathan V. & Phillip S., Evaluating Surfactants and Their Effect on Methane Mole Fraction during Hydrate Growth. Ind. Eng. Chem. Res., 51(40), pp. 13144–13149, 2012.

[12] C.F.d.S. Lirio, F.L.P. Pessoa & A.M.C. Uller, Storage capacity of carbon dioxide hydrates in the presence of Sodium Dodecyl Sulfate (SDS) and Tetrahydrofuran (THF). Chemical Engineering Science, 96, pp. 118–123, 2013.

[13] A. Kumar, T. Sakpal, P. Linga & R. Kumar, Influence of contact medium and surfactants on carbon dioxide clathrate hydrate kinetics. Fuel, 105, pp. 664–671, 2013.

[14] Yoslim J., Linga P. & Englezos P., Enhanced growth of methane-propane clathrate hydrate crystals with sodium dodecyl sulfate, sodium tetradecyl sulfate, and sodium hexadecyl sulfate surfactants. J Cryst Growth, 313(1), pp. 68–80, 2010.

[15] Smith J.M., Van Ness H.C. & Abbott M.M. Introduction to Chemical Engineering Thermodynamics, McGraw-Hill Education, Singapore, pp. 96–98, 2001.

[16] Zhang J.S., Lee S. & Lee J.W. Kinetics of methane hydrate formation from SDS solution. Ind Eng Chem Res, 46(19), pp. 6353–6359, 2007.

[17] Zhang J.S., Lee S. & Lee J.W. Does SDS micellize under methane hydrate-forming conditions below the normal Krafft point?. Journal of colloid and interface science, 315(1), pp. 313–318, 2007.

[18] Zhang J.S., Lo C., Somasundaran P., Lu S., Couzis A. & Lee J.W. Adsorption of sodium dodecyl sulfate at THF hydrate/liquid interface. J. Phys. Chem. C, 112(32), pp. 12381–12385, 2008.

[19] Kashchiev D. & Firoozabadi A. Induction time in crystallization of gas hydrates. Journal of crystal growth, 250(3), pp. 499–515, 2003.

Material Science and Environmental Engineering – Chen (Ed.)
© 2016 Taylor & Francis Group, London, ISBN 978-1-138-02938-5

First-principles calculation of thermoelectric $NaCoO_2$ bulk

P.X. Lu
College of Materials Science and Engineering, Henan University of Technology, Zhengzhou, China

X.M. Wang
College of Information Science and Engineering, Henan University of Technology, Zhengzhou, China

ABSTRACT: In order to investigate the relationship among the electronic structure, the lattice dynamics and the thermoelectric properties of $NaCoO_2$ bulk, the first-principles calculation was conducted by using density functional theory and semi-classical Boltzmann theory in this paper. Our results suggest that $NaCoO_2$ bulk has a typical semiconductor behavior with a band gap of 0.40 eV near the Fermi energy level. Its small electrical conductivity and large Seebeck coefficient are attributed to its small concentration of carriers. In addition, its thermal conductivity is mainly contributed from the phonon thermal conductivity while its electron thermal conductivity is very small. Its small phonon thermal conductivity is resulted from its small phonon density of states. So our work provides a complete understanding on the mechanism of the electron and phonon transport in $NaCoO_2$ bulk.

Keywords: electronic structure; thermoelectric properties; $NaCoO_2$; first-principles calculation

1 INTRODUCTION

Thermoelectric materials, which can be used as solid-state Peltier cooler or power generator from waste heat, will play an important role in solving global fossil energy crisis and minimizing environmental pollution (L.E. 2008). A material with an excellent performance needs a high electrical conductivity, a large Seebeck coefficient and a low thermal conductivity simultaneously (G.J. & E.S. 2008). Recently, sodium cobalt oxide $NaCoO_2$ has been recognized as encouraging thermoelectric materials due to their lower cost, nonpoisonous, simple synthesis procedure and the anti-oxidization in oxidized atmosphere for long time (K. et al. 2003 & M. et al. 2002). Many efforts in improving the thermoelectric properties of the thermoelectric materials were made on either improving the Seebeck coefficient, or increasing the electrical conductivity or decreasing the thermal conductivity. The experimental results indicated that $NaCoO_2$ bulk had a large Seebeck coefficient, a small electrical conductivity and meanwhile a high thermal conductivity (T. et al. 2006). Meanwhile, in recent years theoretical calculations have also been conducted widely to reveal the mechanism of the electron and phonon transport in $NaCoO_2$ bulk. The lattice thermoelectric properties of Na_xCoO_2 bulk vary with its stoichiometry, structural defects, temperature and nanosheet geometry (D.O. & D.B. 2014). Its thermal conductivity can be suppressed significantly by an Einstein-like rattling mode at low energy (D.J. et al. 2013). Moreover, atoms doping can lead to a metal-insulator transition at low temperature in $Na_{0.8}Co_{1-x}Sm_xO_2$ (J. & Z.P. 2012). The formation energies of defects and the resultant changes in electronic structure of $NaCoO_2$ and $Na_{0.5}CoO_2$ have been evaluated by using first-principles calculation (M. et al. 2010). In addition, the thermoelectric properties of the $NaCoO_2$ compound have also been investigated by using simple lattice dynamical model or Boltzman theory (P.K. et al. 2005 & S. et al. 2010). To the best of our knowledge, however, there have been no reports demonstrating the relationship among the electronic structure, the dynamics and the thermoelectric properties of $NaCoO_2$ bulk up to now. Therefore, our work is aimed to provide a complete understanding on the mechanism of the electron and phonon transport in $NaCoO_2$ bulk by using first-principles calculation.

2 CALCULATION METHODS

$NaCoO_2$ bulk has a rhombohedral symmetry (R-3 mH, Group No. 166) with lattice parameters of $a = b = 0.2888$ nm, $c = 1.5606$ nm, $\alpha = \beta = 90°$, $\gamma = 120°$, in which Na atoms occupy (0, 0, 0.5) site, Co atoms (0, 0, 0) site while O atoms (0, 0, 0.275) (Y. et al. 2003). Calculations were performed using CASTEP (Cambridge Serial Total Energy Package), a first-principle pseudopotential method based on the Density-Functional Theory (DFT)

(M.D. et al. 2002). The used pseudopotential was the Norm-conserving pseudopotential. As for the approximation of the exchange correlation term of the DFT, the Generalized Gradient Approximation (GGA) of Perdew was adopted with Perdew-Burke-Ernzerhof parameters. The cutoff energy of atomic wave functions was set at 830 eV. Sampling of the irreducible wedge of Brillouin zone was performed with a regular Monkhorst-Pack grid of special k-points, which was $12 \times 12 \times 4$. The high symmetry points in Brillouin zone were selected as G (0.000, 0.000, 0.000), A (0.000, 0.000, 0.500), H (–0.333, 0.667, 0.500), K (–0.333, 0.667, 0.000), G (0.000, 0.000, 0.000), M (0.000, 0.500, 0.000), L (0.000, 0.500, 0.500), H (–0.333, 0.667, 0.500). All atomic positions in our models had been relaxed according to the total energy and force using BFGS scheme, based on the cell optimization criterion (RMs force of 0.01 eV/nm, displacement of 0.00005 nm). The calculation of total energy and electronic structure was followed by cell optimization with SCF tolerance of 5.0×10^{-6} eV. The Partial Density of States (PDOS) for electrons and phonons was obtained by broadening the discrete energy levels using a smearing function of 0.05 eV Full-Width at Half-Maximum (FWHM) on a grid of k-points generated by the Monkhorst–Pack scheme (H.J. & J.D. 1976).

The temperature dependence of heat capacity C and Debye temperature Θ_D is obtained by calculating the phonon dispersion. The phonon mean free path λ is calculated in terms of Eq. 1:

$$\lambda = 10^{-8} \exp\left(\frac{\Theta_D}{\eta T}\right) \tag{1}$$

where Θ_D is the Debye temperature, T is the absolute temperature and η is a constant equal to 1.62, which depends on the scattering limitation of phonons (M. et al. 2009). The effective mass of carriers m^* can be obtained by differentiating the energy at Fermi energy level twice with respect to the wave vector of the high symmetric points in Brillouin zone. The temperature dependence of the Fermi energy is calculated by Eq. 2:

$$E_F = E_F^0\left[1 - \frac{\pi^2}{12}\left(\frac{k_B T}{E_F^0}\right)^2\right] \tag{2}$$

where E_F^0 is the Fermi energy at 0 K, T is the absolute temperature and k_B is the Boltzmann constant. The Fermi-Dirac distribution function $(-\partial f/\partial E)$ is written as Eq. 3:

$$\left(-\frac{\partial f}{\partial E}\right) = \frac{1}{k_B T}\frac{1}{\left(e^{(E-E_F)/k_B T}+1\right)}\cdot\frac{1}{\left(e^{-(E-E_F)/k_B T}+1\right)} \tag{3}$$

where k_B is the Boltzmann constant, E is the energy band, E_F is the Fermi energy and T is the absolute temperature. The group velocity of carriers υ_e(E) is calculated by differentiating the energy with respect to the wave vector in Brioullin zone (C. et al. 2005). Consequently, the electrical conductivity and the Seebeck coefficient are defined as Eq. 4 and Eq. 5, respectively (Y. et al. 2010):

$$\sigma = \frac{e^2}{3}\int dE\left(-\frac{\partial f}{\partial E}\right) g(E)\tau(E)\upsilon^2(E) \tag{4}$$

$$\alpha = \frac{e}{3T\sigma}\int dE\left(-\frac{\partial f}{\partial E}\right) g(E)\tau(E)\upsilon^2(E)(E-E_F) \tag{5}$$

where e is the electron charge, $(-\partial f/\partial E)$ is the Fermi distribution function, $g(E)$ is the Density of States (DOS), υ_e(E) is the group velocity of electrons, E and E_F are the energy band and the Fermi energy, respectively. $\tau(E)$ is the relaxation time, which can be simplified as an energy-independent constant equal to 6.4×10^{-25} s using the experimental electrical conductivity. The electron thermal conductivity κ_e is estimated using the Wiedemann–Franz law (N.W. & N.D. 1976), and the phonon thermal conductivity κ_p is given by Eq. 6 (T. et al. 1996):

$$\kappa = \frac{1}{3}\rho\upsilon_p C\lambda \tag{6}$$

where ρ is the theoretical density equal to 4367 kg/m^3? C is the heat capacity and υ_p represents the average velocity of phonons equal to 1360 m/s. The average velocity of phonons is calculated from the phonon dispersion by differentiating the frequency of phonons with respect to the wave factor of the high symmetry points in Brillouin zone. Therefore, the figure of merit ZT is determined by Eq. 7:

$$ZT = \frac{\alpha^2\sigma}{\kappa_e + \kappa_p}T \tag{7}$$

where α is the Seebeck coefficient σ is the electrical conductivity, κ_e and κ_p are the thermal conductivity contributed from electrons and phonons, respectively. T is the absolute temperature.

3 RESULTS AND DISCUSSION

3.1 *Electronic structure*

The Partial Density of States (PDOS) of $NaCoO_2$ bulk is presented in Figure 1. A typical semiconductor characteristic for $NaCoO_2$ bulk can be observed clearly due to the band gap of 0.40 eV existing near the Fermi energy level. In addition, it is clear that

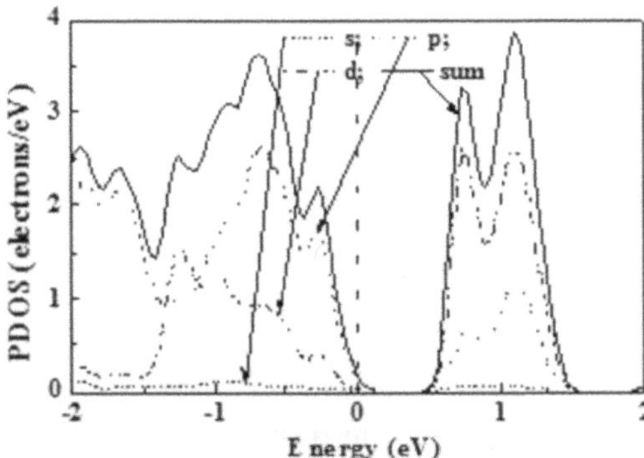

Figure 1. PDOS of $NaCoO_2$ bulk.

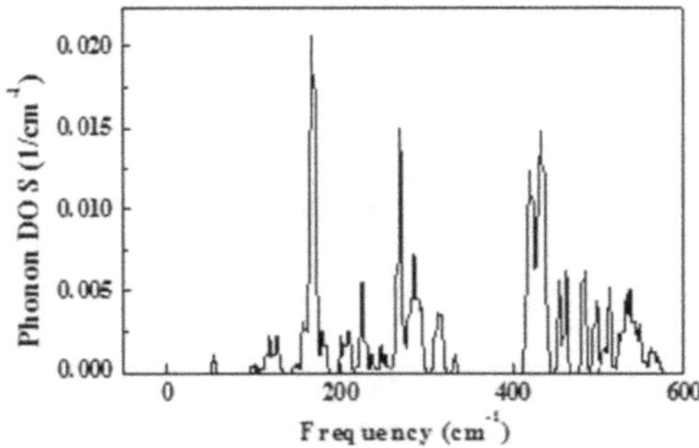

Figure 2. Phonon DOS of $NaCoO_2$ bulk.

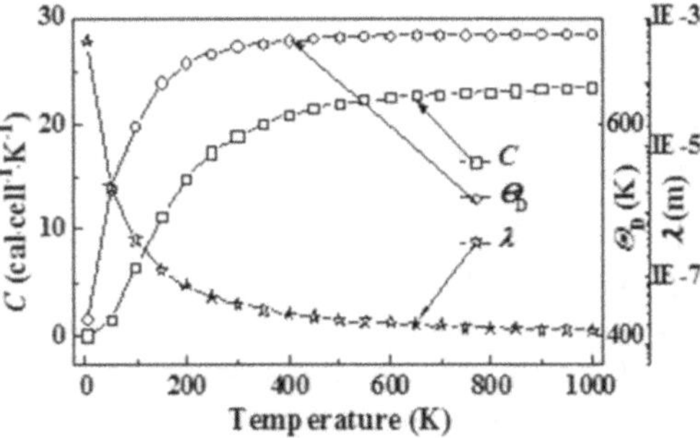

Figure 3. Heat capacity, Debye temperature and mean free path of $NaCoO_2$ bulk.

the electrons at the top of Valance Band (VB) (from −2 eV to 0 eV) in $NaCoO_2$ bulk mainly come from the p-orbital and d-orbital as well as small s-orbital. Meanwhile, the electrons at the bottom of Conduction Band (CB) (from 0 eV to +2 eV) are contributed from both the p-orbital and the d-orbital. Hence, a typical p-d orbital hybridizing behavior between the parent atoms exists in $NaCoO_2$ bulk. In general, the electrical conductivity is mainly determined by the Density of States (DOS) near Fermi energy level (J.M. 1972). The small DOS of 2.15 electrons/eV for $NaCoO_2$ bulk may be a main reason for its high Seebeck coefficient and small electrical condcutivity. Pseudogap energy, an energy gap between the nearest two peaks near Fermi energy level in the DOS curve, can reflect the covalent bonding strength between the parent atoms. The pseudogap energy of $NaCoO_2$ bulk is merely 1.05 eV, which indicates the relatively weak covalent bonding would result in a low phonon vibration frequency and a small phonon thermal conductivity (according to the formula $\omega = \sqrt{k/m}$ (ω is the phonon vibration frequency, k is the force constant, and m is the effective mass).

3.2 *Lattice dynamics*

The phonon density of states for $NaCoO_2$ bulk is shown in Figure 2. Two gaps at about 200 cm^{-1} and 350 cm^{-1} can be observed obviously. Such gaps can be explained by the distribution of the bond lengths in the unit cell (M. et al. 1997). The existing three different covalent bonding behaviors (s-orbital, p-orbital and d-orbital) between the parent atoms show effectively three different force constants, which leads to the frequency gaps and the discontinuous spectra in the phonon DOS curves.

The temperature dependent of the heat capacity, Debye temperature and mean free path for $NaCoO_2$ bulk are shown in Figure 3. The calculated heat capacity at room temperature is 19 cal/cell · K. The total heat capacity includes two parts, one from the phonon contribution (C_p) and the other from the electron contribution (C_e). In a low temperature (below 300 K), the heat capacity coincides with the formula $C = AT^3$, which corresponds to the C_p. In a high temperature (above 300 K), however, the heat capacity is directly proportional to AT, which belongs to the C_e. The calculated Debye temperatures at room temperature are 670 K, which will result in a small mean free path and a small phonon thermal conductivity. The mean free path of phonons for $NaCoO_2$ bulk is small, which would lead to a small phonon thermal conduction.

3.3 *Thermoelectric properties*

The electrical conductivity and Seebeck coefficient for $NaCoO_2$ bulk are presented in Figure 4. The electrical conductivity at 300 K is 53000 $\Omega^{-1} \cdot m^{-1}$, which is well consistent with the experimental data (S. et al. 2006). The small electrical conductivity can be attributed to the small density of states near the Fermi level. The negative sign of the Seebeck coefficients for $NaCoO_2$ bulk indicates that the major carriers are electrons in the whole temperature. Moreover, the Seebeck coefficient at

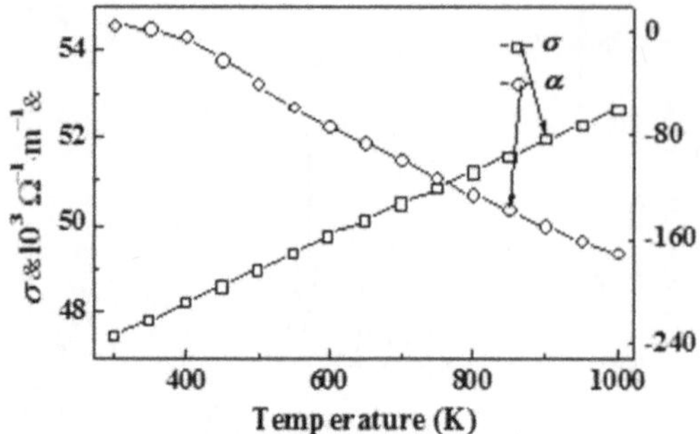

Figure 4. Electrical conductivity and Seebeck coefficient of $NaCoO_2$ bulk.

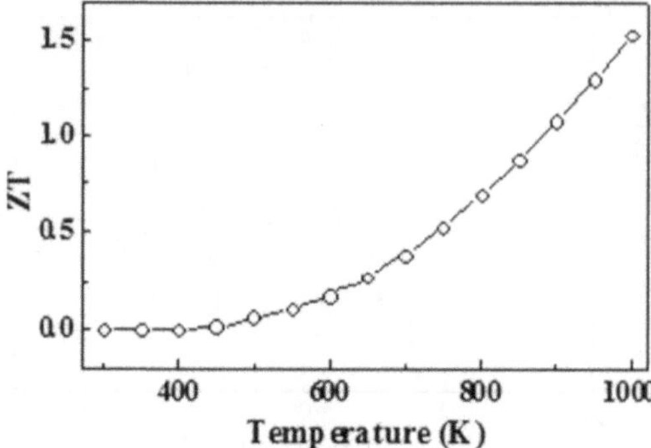

Figure 6. Figure of merit of $NaCoO_2$ bulk.

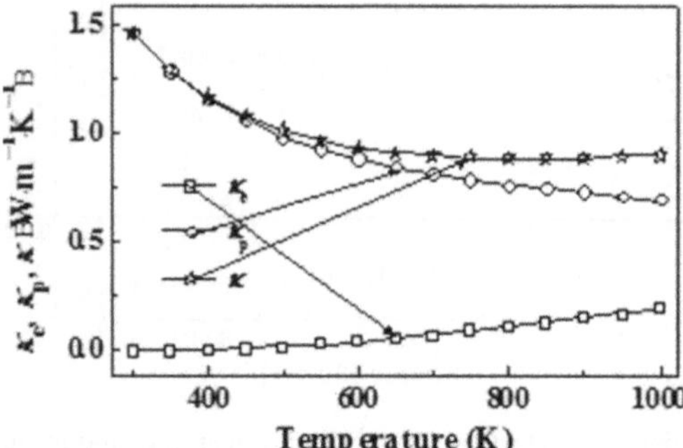

Figure 5. Total, electron and phonon thermal conductivity of $NaCoO_2$ bulk.

300 K is −170 μV · K^{-1}, which is close to the experimental result (J. et al. 1983). Such large Seebeck coefficient may be resulted from its small concentration of carriers.

Figure 5 plots the total, electron and phonon thermal conductivity of $NaCoO_2$ bulk as a function of temperature. The κ is divided into two parts: the electron thermal conductivity κ_e and the phonon thermal conductivity κ_p. It can be seen that the κ_e and κ_p at 1000 K of $NaCoO_2$ bulk is 0.21 W·m^{-1}·K^{-1} and 0.7 W·m^{-1}·K^{-1}, respectively. So, the total thermal conductivity is mainly resulted from the phonon thermal conductivity. Such small thermal conductivity is originated from the small electrical conductivity caused by its small concentration of carriers and the small phonon thermal conductivity resulted from small phonon density of state.

Figure 6 shows the figure of merit for $NaCoO_2$ bulk. The ZT value at 1000 K is close to 1.51, which can be attributed to its large Seebeck coefficient and small thermal conductivity. Such ZT value would be improved in further if its electrical conductivity can be increased through a transition from semicondcutor to semimetal behavior via nanostructuring engineering (G. et al. 2011).

4 CONCLUSIONS

Summing up, the electronic structures, the lattice dynamics and the thermoelectric properties of $NaCoO_2$ bulk have been calculated successfully by using density functional theory and Boltzmann transport theory. $NaCoO_2$ bulk has a typical semiconductor characteristic with a band gap of 0.40 eV near the Fermi energy level. Its small DOS of 2.15 electrons/eV may be a main reason for its small electrical conductivity and large Seebeck coefficient. Its small thermal conductivity is resulted from the small phonon thermal conductivity caused by its small phonon density of states. It can be predicted that the ZT value would be improved if its electrical conductivity can be increased in further through a transition from semiconductor to semimetal via nanostructuring engineering.

ACKNOWLEDGEMENTS

This project is supported financially by Science Foundation of Henan University of Technology (Grant Nos 2011BS056 and 11JCYJ12).

REFERENCES

[1] Bell, L.E. 2008. Cooling, heating, generating power, and recovering waste heat with thermoelectric systems. *Sci.* 321: 1457–1461.

[2] Snyder, G.J. & Toberer, E.S. 2008. Complex thermoelectric materials. *Nat. Mater.* 7: 105–114.

[3] Takada, K. Sakurai, H. Takayama-Muromachi, E. Izumi, F. Dilanian, R.H. & Sasaki, T. 2003. Superconductivity in two-dimensional CoO_2 layers. *Nat. (Lond.)* 422: 53–55.

[4] Ito, M. Nagira, T. Oda, Y. Katsuyama, S. Majima, K. & Nagai, H. 2002. Effect of partial substitution of 3d transition metals for Co on the thermoelectric properties of $Na_xCo_2O_4$. *Mater. Trans.* 43: 601–604.

[5] Seetawan, T. Amornkitbamrung, V. Burinprakhon, T. Maensiri, S. Tongbai, P. Kurosaki, K. Muta, H. Uno, M. & Yamanaka, S. 2006. Effect of sintering temperature on the thermoelectric properties of $Na_xCo_2O_4$. *J. Alloys Compds.* 416: 291–295.

[6] Demchenko, D.O. & Ameen, D.B. 2014. Lattice thermal conductivity in bulk and nanosheet Na_xCoO_2. *Computational Materials Science* 82: 219–225.

[7] Voneshen, D.J. Refson, K. Borissenko, E. Krisch, M, Bosak, A. Piovano, A. Cemal, E. Enderle, M. Gutmann & M.J. Hoesch, M. 2013. Suppression of thermal conductivity by rattling modes in thermoelectric sodium cobaltate. Nature materials 12 (11): 1028–1032.

[8] Sun, J. & Guo, Z.P. 2012. The metal-insulator transition of $Na_{0.8}Co_{1-x}Sm_xO_2$. *Intergrated Ferroelectrics* 137: 120–125.

[9] Yoshiya, M. Okabayashi, T. Tada, M. & Fisher, C.A.J. 2010. A first-principles study of the role of Na vacancies in the thermoelectricity of Na_xCoO_2, *J. Electr. Mater.* 39(9): 1681–1686.

[10] Jha, P.K. Troper, A. da Cunha Lima, I.C. Talatic, M. & Sanyald, S.P. 2005. Phonon properties of intrinsic insulating phase of the cobalt oxide superconductor $NaCoO_2$. *Physica B* 366: 153–161.

[11] Tosawat, S. Athorn, V.U. Prasarn, C. Chanchana, T. & Vittaya, A. 2010. Evaluating Seebeck coefficient of Na_xCoO_2 from molecular orbital calculations. *Computational Materials Science* 49: S225–S230.

[12] Takahashi, Y. Gotoh, Y. & Akimoto, J. 2003. Single-crystal growth, crystal and electronic structure of $NaCoO_2$. *Journal of Solid State Chemistry* 172: 22–26.

[13] Segall, M.D. Lindan, P.J.D. Probert, M.J. Pickard, C.J. Hasnip, P.J. Clark S.J. & Payne, M.C. 2002. First-principles simulation: ideas, illustrations and the CASTEP code. *J. Phys.: Cond. Matt.* 14: 2717–2744.

[14] Monkhorst, H.J. & Pack, J.D. 1976. Special points for Brillouin-zone integrations. *Phys. Rev. B* 13: 5188–5192.

[15] Murata, M. Nakamura, D. Hasegawa, Y. Komine, T. Taguchi, T. Nakamura, S. Jaworski, C.M. Jovovic, V. & Heremans, J.P. 2009. Mean free path limitation of thermoelectric properties of bismuth nanowire. *J. Appl. Phys.* 105: 113706.

[16] Stiewe, C. Bertini, L. Toprak, M. Christensen, M. Platzek, D. Williams, S. Gatti, C. Müller, E. Iversen, B.B. Muhammed M. & Rowe, M. 2005. Nanostructured $Co_{1-x}Ni_x(Sb_{1-y}Tey)_3$ skutterudites: Theoretical modeling, synthesis and thermoelectric properties. *J. Appl. Phys.* 97: 044317.

[17] Kono, Y. Ohya, N. Taguchi, T. Suekuni, K. Takabatake, T. Yamamoto, S. & Akai, K. 2010. First-principles study of type-I and type-VIII $Ba_8Ga_{16}Sn_{30}$ clathrates. *J. Appl. Phys.* 107: 123720.

[18] Ashcroft, N.W. & Mermin, N.D. 1976 *Solid State Physics.* New York: Harcourt Brace Press.

[19] Caillat, T. Kulleck, J. Borshchevsky, A. & Fleurial, J.P. 1996. Preparation and thermoelectric properties of the skutterudite-related phase $Ru_{0.5}Pd_{0.5}Sb_3$. *J. Appl. Phys.* 79: 8419–8426.

[20] Ziman, J.M. 1972. *Principles of the Theory of Solid.* Cambridge: Cambridge University Press.

[21] Menon, M. Richter, E. & Subbaswamy K.R. 1997. Structural and vibrational properties of Si clathrates in a generalized tight-binding molecular-dynamics scheme. *Phys. Rev. B* 56: 12290.

[22] Tosawat, S. Vittaya, A. Thanusit, B. Santi, M. Prasit, T. Ken, K. Hiroaki, M. Masayoshi, U. & Shinsuke, Y. 2006. Effect of sintering temperature on the thermoelectric properties of $Na_xCo_2O_4$. *J. Alloys Compds.* 416: 291–295.

[23] Molenda, J. Delmas, C. & Hagenmuller, P. 1983. Electronic and electrochemical properties of Na_xCoO_{2-y} cathode, *Solid State Ionics* 9–10: 431–435.

[24] Zhou, G. Li, L. & Li, G.H. 2011. Semimetal to semiconductor transition and thermoelectric properties of bismuth nanotubes. *J. Appl. Phys.* 109: 114311.

Material Science and Environmental Engineering – Chen (Ed.)

A new fault-distance algorithm using single-ended quantity for distance relaying

Z.Y. Xu, J.W. Meng, S.P. Li, Y. Wang & Y.C. Qiao
School of Electrical and Electronic Engineering, North China Electric Power University, Beijing, China

T.B. Zong
College of Electronic Information and Control Engineering, Beijing University of Technology, Beijing, China

ABSTRACT: Transition resistance adversely affects the accuracy of the measured impedance on a transmission line, which may result in tripping, missing, or maloperation of distance relays. This paper proposes a new algorithm to calculate the distance between the relay point and the fault point on a transmission line. The algorithm is based on the phase angle relationship of the fault point voltage and its variation before and after the fault moment, featuring robustness to the uncertainties in load current and transition resistance. The equations of fault distance measurement were given for phase–ground fault and phase–phase fault using single-ended fundamental frequency electrical quantities. The simulation results show that the principle of the algorithm is correct and is of a high accuracy.

Keywords: transition resistance; distance protection; fault distance; load current; transmission line

1 INTRODUCTION

Microcomputer distance protection is widely used in high-voltage transmission line protection. Microcomputer protection uses the fault impedance calculated from local measurements of voltage and current to reflect the fault distance with flexible implementation. When a high-impedance fault occurs on a transmission line, the transition resistance will weaken the fault feature of the power grid, adversely affecting the right action of protection. Long time delay removal and even operation rejection caused by high impedance fault occurred sometimes, and what's more, it's likely to cause further serious power system accident. How to eliminate the impact of the transition resistance on impedance measurement is a very realistic issue facing the distance protection.

Some previous studies strengthened the tolerance to the transition resistance by adjusting the boundary of the protection action. However, when a fault occurs outside the intended protection zone, with the increment of the transition resistance, the compensating voltage will commutate resulting in over-reach. And some others added an additional impedance compensation section in the measuring impedance, so that the measured impedance can accurately reflect the actual fault distance. Although this algorithm can effectively eliminate the protection rejection problem, but due to the impact of load levels it may cause over-reach resulting in non-selective action of the relay. In addition, some references presented a technique of fault resistance compensation in the phase coordinate. The estimation of the line fault impedance was obtained in an iterative manner with improved accuracy. This technique is only suitable for single-sourced transmission lines which are rare in practice.

The algorithm proposed in this paper is based on the phase angle relationship of the fault point voltage and its variation before and after the fault moment to calculate the fault distance. When a fault occurs along the transmission line, the phase angle difference between the fault point voltage and its variation before and after the fault moment varies slightly and will not get affected by transition resistance and load levels. As a result, with corresponding reasonable setting compensation angles for the faults of different types the algorithm can ensure a high accuracy and avoid the effects of the uncertain load current and fault path resistance. Since only single-ended fundamental frequency electrical quantity is used in this new algorithm, it is easy to implement it in practice. The simulation results show that the algorithm has a high accuracy and high tolerance to transition resistance and load current, which can provide effective and reliable protection for high-voltage transmission line.

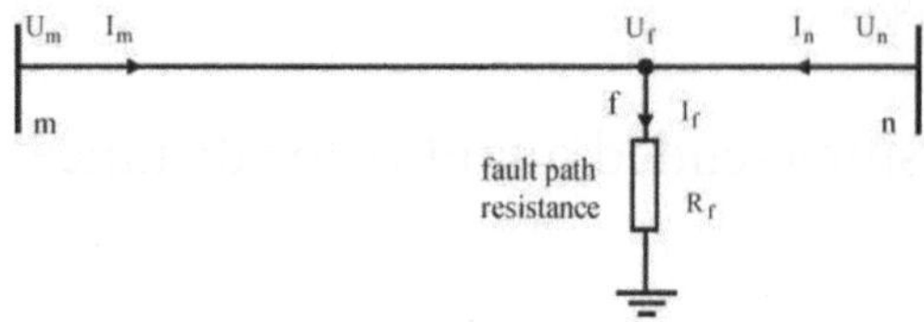

Figure 1. Transmission-line fault with transition resistance.

2 IMPACT OF TRANSITION RESISTANCE

A transmission line with bilateral power is shown in Figure 1. In this paper, single-phase grounding fault through the transition resistance R_f is taken as an example to analyze the impact of the transition resistance on distance protection.

The measure impedance in a traditional distance relay at end m is expressed as:

$$Z_m = Z_{mf} + \frac{I_f}{I_m} R_f \tag{1}$$

Due to the existence of the transition resistance at the fault point, the measured impedance consists of Z_{mf} excepted to reflect the actual distance from m to the fault point as well as an additional value $(I_f/I_m)R_f$ co-produced by the fault current, load current, and transition resistance.

The additional impedance mostly presents resistive-capacitive at sending end, which causes over-reach of the relay. Conversely, the additional impedance mostly presents resistive-inductive at receiving end, causing under-reach of the relay. Affected by power angle on both sides of the transmission line, impedance of power system, line length, fault position and many other factors, the value of additional impedance is uncertain and will vary over a wide range. Therefore, the transition resistance affects the distance protection greatly.

3 FAULT-DISTANCE ALGORITHM

The measured impedance in traditional distance relay couldn't reflect the actual fault distance when transition resistance exists. The new algorithm proposed in this paper for fault impedance measurement will be free from the impact of transition resistance and load current. It is assumed that the sampled data of the end voltage and line current (instantaneous values) are available while the fault occurrence has already been detected with appropriate classification. Once this is confirmed, the algorithm described in this paper will be applied to calculate the accurate distance from the distance relay to the fault point by using the fundamental components of the voltages and currents measured at end m or n.

3.1 *Phase–ground fault*

A phase–ground fault with transition resistance on a transmission line is shown in Figure 2. For any fault on an overhead transmission line in a power system, the fault path is predominantly resistive. As a result, the phase angles of the fundamental fault point voltage and fault path current are always equal.

Fault point voltage can be expressed by the fundamental phase voltage, line current, and zero-sequence current compensation at the relay point as follow

$$U_f = U_\varphi - xZ_1(I_\varphi + 3kI_0) \tag{2}$$

Similarly, the variation of the fault point voltage before and after the fault moment can also be expressed as

$$\Delta U_f = \Delta U_\varphi - xZ_1(\Delta I_\varphi + 3kI_0) \tag{3}$$

U_φ, I_φ and I_0 represent the fundamental phase voltage, line current, and zero-sequence current compensation measured at the relay point after fault, respectively. ΔU_φ and ΔI_φ represent the variation of the phase voltage and line current before and after the fault moment at the relay point. Z_1 is positive-sequence series impedance of the protected line, per phase and per unit length (Ω/km). $k = (Z_0 - Z_1)/3Z_1$ is the zero-sequence current compensation factor and x is the distance from the relay to the fault point, which is to be determined.

The following part will analyze the relationship of the phase angle between U_f and ΔU_f. Corresponding to Figure 2, for a phase–ground fault the fault current I_f can be obtained as follow:

$$I_f = \frac{U_{f|0|}}{\left(Z_{\Sigma(1)} + Z_{\Sigma(2)} + Z_{\Sigma(0)}\right)/3 + R_f} \tag{4}$$

$U_{f|0|}$ is the fault point voltage before the fault moment, $Z_{\Sigma(1)}$, $Z_{\Sigma(2)}$, and $Z_{\Sigma(0)}$ represent the

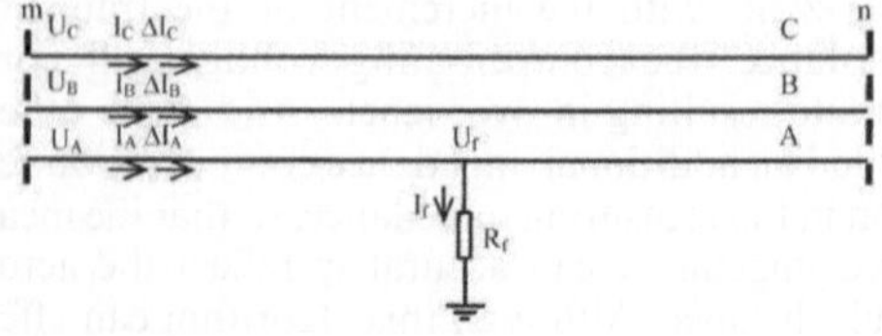

Figure 2. Phase–ground fault.

equivalent impedance of positive/negative/zero sequence networks looking inward from the fault point, respectively. So we got the simplified equivalent fault component network shown in Figure 3. $Z_\Sigma = (Z_{\Sigma(1)} + Z_{\Sigma(2)} + Z_{\Sigma(0)})/3$ can be taken as the equivalent impedance outside the fault point.

From Figure 3, it's easy to obtain the variation of the fault voltage as follow

$$\Delta U_f = U_f - U_{f|0|} = -I_f Z_\Sigma \tag{5}$$

And the fault point voltage can be expressed as

$$U_f = R_f I_f \tag{6}$$

According to (5) and (6), we obtain

$$U_f = -\left(R_f / Z_\Sigma\right)\Delta U_f \tag{7}$$

The impedance angle of Z_Σ is assumed as 90° – σ and σ is taken as compensation angle in this algorithm. And we obtained the phase angle relationship between U_f and ΔU_f as follow, where Arg (A) means the phase angle of A.

$$U_f = -R_f \frac{\Delta U_f}{|Z_\Sigma| e^{j(90°-\sigma)}} = \frac{R_f}{|Z_\Sigma|}\Delta U_f e^{j(90°+\sigma)} \tag{8}$$

$$Arg(U_f) = Arg\left(\Delta U_f e^{j(90°+\sigma)}\right) \tag{9}$$

The impedance angle of Z_Σ is independent of transition resistance and load current and only related to the location of the fault. However, with the change of the fault location, it changes slightly. By calculation and simulation verification in the next section we have reason to believe that a reasonable set of compensation angle can meet considerable accuracy of the measured distance from fault point to relay point when a single-phase grounding fault occurs throughout the range of protection on the long-distance transmission line.

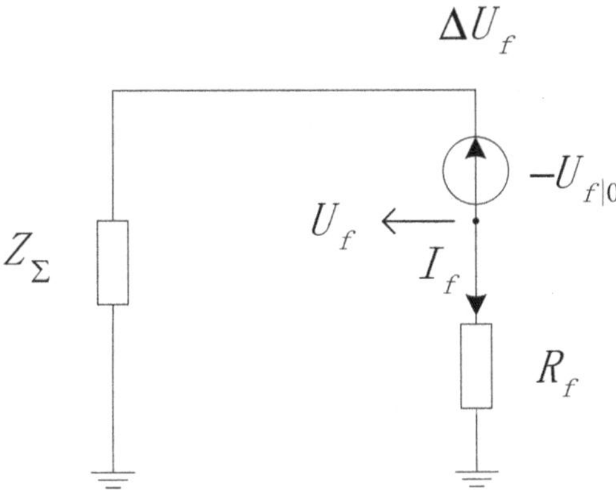

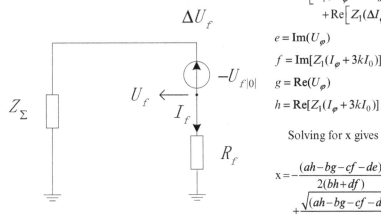

Figure 3. Equivalent network of the fault component.

U_f and ΔU_f and $\Delta U_f e^{j(90°+\sigma)}$ can be expressed as follows in real and imaginary parts:

$$\begin{aligned} U_f = &\mathbf{Re}\left(U_\varphi - xZ_1(I_\varphi + 3kI_0)\right) \\ &+ j\mathrm{Im}\left(U_\varphi - xZ_1(I_\varphi + 3kI_0)\right) \end{aligned} \tag{10}$$

$$\begin{aligned} \Delta U_f = &\mathbf{Re}\left(\Delta U_\varphi - xZ_1(\Delta I_\varphi + 3kI_0)\right) \\ &+ j\mathrm{Im}\left(\Delta U_\varphi - xZ_1(\Delta I_\varphi + 3kI_0)\right) \end{aligned} \tag{11}$$

$$\begin{aligned} \Delta U_f e^{j(90°+\sigma)} &= j\Delta U_f e^{j\sigma} \\ &= -\mathbf{Im}\left(\Delta U_f e^{j\sigma}\right) + j\mathbf{Re}\left(\Delta U_f e^{j\sigma}\right) \end{aligned} \tag{12}$$

According to (10), (11) and (12), (9) can be expressed as

$$\frac{\mathrm{Im}\left(U_\varphi - xZ_1(I_\varphi + 3kI_0)\right)}{\mathrm{Re}\left(U_\varphi - xZ_1(I_\varphi + 3kI_0)\right)} = -\frac{\mathrm{Re}\left(\Delta U_f e^{j\sigma}\right)}{\mathrm{Im}\left(\Delta U_f e^{j\sigma}\right)} \tag{13}$$

which can be rearranged as

$$\frac{a + bx}{c - dx} = \frac{e - fx}{g - hx} \tag{14}$$

where

$$a = \mathbf{Im}(\Delta U_\varphi)\sin\sigma - \mathbf{Re}(\Delta U_\varphi)\cos\sigma$$

$$\begin{aligned} b = &\mathbf{Im}\left[Z_1(\Delta I_\varphi + 3kI_0)\right]\sin\sigma \\ &- \mathrm{Re}\left[Z_1(\Delta I_\varphi + 3kI_0)\right]\cos\sigma \end{aligned}$$

$$c = \mathbf{Re}(\Delta U_\varphi)\sin\sigma + \mathbf{Im}(\Delta U_\varphi)\cos\sigma$$

$$\begin{aligned} d = &\mathbf{Im}\left[Z_1(\Delta I_\varphi + 3kI_0)\right]\cos\sigma \\ &+ \mathrm{Re}\left[Z_1(\Delta I_\varphi + 3kI_0)\right]\sin\sigma \end{aligned}$$

$$e = \mathbf{Im}(U_\varphi)$$

$$f = \mathbf{Im}[Z_1(I_\varphi + 3kI_0)]$$

$$g = \mathbf{Re}(U_\varphi)$$

$$h = \mathbf{Re}[Z_1(I_\varphi + 3kI_0)]$$

Solving for x gives (15) as follow.

$$\begin{aligned} x = &-\frac{(ah - bg - cf - de)}{2(bh + df)} \\ &+ \frac{\sqrt{(ah - bg - cf - de)^2 - 4(bh + df)(ce - ag)}}{2(bh + df)} \end{aligned} \tag{15}$$

The per phase fault impedance from the relay to the fault point for a phase–ground fault can be expressed as

$$Z_{\varphi} = \mathrm{x}Z_1 \tag{16}$$

Note that φ represents phase a, b, or c. Equations (15) and (16) are suitable for a phase–ground fault.

The phase can be selected by the fault classification procedure that is to apply the equations.

3.2 *Phase–phase fault*

Similarly, for phase–phase fault shown in Figure 4, the following relation can be obtained:

$$U_{\varphi\varphi\mathrm{f}} = -2R_{\mathrm{f}}\frac{\Delta U_{\varphi\varphi\mathrm{f}}}{Z_{\Sigma}} \tag{17}$$

Note that here $Z_{\Sigma} = Z_{\Sigma(1)}+Z_{\Sigma(2)}$. The impedance angle of Z_{Σ} is assumed as $90° - \sigma$. So we obtained the phase angle relationship between $U_{\varphi\varphi\mathrm{f}}$ and $\Delta U_{\varphi\varphi\mathrm{f}}$ as follow.

$$Arg\left(U_{\varphi\varphi\mathrm{f}}\right) = Arg\left(\Delta U_{\varphi\varphi\mathrm{f}}e^{j(90°+\sigma)}\right) \tag{18}$$

Fault point voltage and its variation after the fault moment can be expressed as

$$U_{\varphi\varphi\mathrm{f}} = U_{\varphi\varphi} - \mathrm{x}Z_1 I_{\varphi\varphi} \tag{19}$$

$$\Delta U_{\varphi\varphi\mathrm{f}} = \Delta U_{\varphi\varphi} - \mathrm{x}Z_1 \Delta I_{\varphi\varphi} \tag{20}$$

(18) Can be expressed as

$$\frac{\mathrm{Im}(U_{\varphi\varphi} - \mathrm{x}Z_1 I_{\varphi\varphi})}{\mathrm{Re}(U_{\varphi\varphi} - \mathrm{x}Z_1 I_{\varphi\varphi})} = -\frac{\mathrm{Re}(\Delta U_{\varphi\varphi\mathrm{f}}e^{j\sigma})}{\mathrm{Im}(\Delta U_{\varphi\varphi\mathrm{f}}e^{j\sigma})} \tag{21}$$

which can be rearranged as

$$\frac{a+b\mathrm{x}}{c-d\mathrm{x}} = \frac{e-f\mathrm{x}}{g-h\mathrm{x}} \tag{22}$$

where

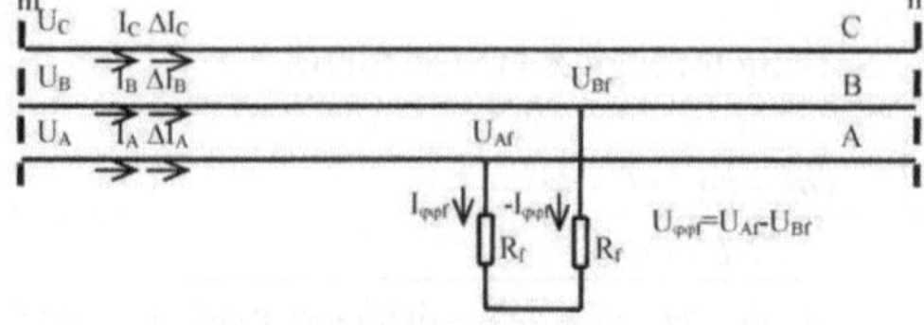

Figure 4. Phase–phase fault.

$$a = \mathrm{Im}(\Delta U_{\varphi\varphi})\sin\sigma - \mathrm{Re}(\Delta U_{\varphi\varphi})\cos\sigma$$

$$b = \mathrm{Im}(Z_1\Delta I_{\varphi\varphi})\sin\sigma - \mathrm{Re}(Z_1\Delta I_{\varphi\varphi})\cos\sigma$$

$$c = \mathrm{Re}(\Delta U_{\varphi\varphi})\sin\sigma + \mathrm{Im}(\Delta U_{\varphi\varphi})\cos\sigma$$

$$d = \mathrm{Im}(Z_1\Delta I_{\varphi\varphi})\cos\sigma + \mathrm{Re}(Z_1\Delta I_{\varphi\varphi})\sin\sigma$$

$$e = \mathrm{Im}(U_{\varphi\varphi})$$

$$f = \mathrm{Im}(Z_1 I_{\varphi\varphi})$$

$$g = \mathrm{Re}(U_{\varphi\varphi})$$

$$h = \mathrm{Re}(Z_1 I_{\varphi\varphi})$$

Solving for x gives (23) as follow.

$$\mathrm{x} = -\frac{(ah-bg-cf-de)}{2(bh+df)} + \frac{\sqrt{(ah-bg-cf-de)^2 - 4(bh+df)(ce-ag)}}{2(bh+df)} \tag{23}$$

Similar to (16), the per phase fault impedance at the relay point for a phase–phase fault is defined as

$$Z_{\varphi\varphi} = \mathrm{x}Z_1 \tag{24}$$

where Z_1 is again the per phase positive-sequence impedance for a unity length of the line. Note that subscript $\varphi\varphi$ represents ab, bc, or ca. Equations (23) and (24) can be applied to any fault involving two phases, including phase–phase fault, phase–phase–ground fault, and three-phase symmetrical fault.

4 SIMULATION VERIFICATION

The simulation of a series of faults of different types with transition resistance at several locations along the transmission line are carried out by using the PSCAD/EMTDC package to verify and evaluate the proposed fault impedance algorithm for distance relaying. The pre-fault power level is also varied while the fault inception angle is randomly set. The following fault types are included: single phase–ground fault, phase–phase fault, and three-phase symmetrical fault. The full-cycle Fourier correlation analysis is used as the filtering algorithm to extract the fundamental frequency components.

A 500-kV, 400-km transmission line with two end systems S1 and S2 is shown in Figure 5. Line model uses distributed parameter. The parameters of these end systems and the transmission line is given as follow. The studied zone-one

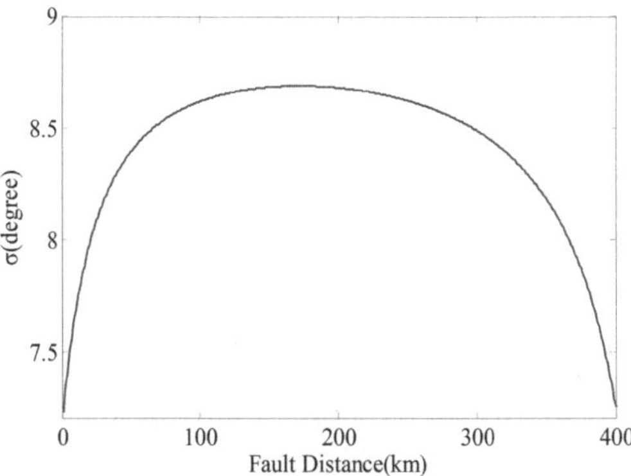

Figure 5. Compensation angle σ (A-g fault).

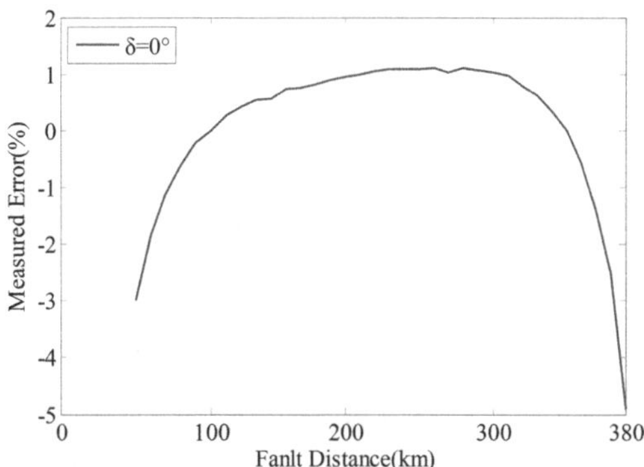

Figure 6. Fault distance errors with compensation angle σ of 8.27° (A-g fault, 300 Ω transition resistance).

distance relay is set to protect 85% of the 400-km line viewed from the relay point.

Positive-/negative-sequence parameters of the transmission line: R = 0.1839 × 10^{-4} [Ω/m]; X_L = 0.2630 × 10^{-3} [Ω/m]; X_C = 500 [MΩ · m]. Zero-sequence parameters of the transmission line: R = 0.1417 × 10^{-3} [Ω/m]; X_L = 0.6027 × 10^{-3} [Ω/m]; X_C = 800.5217 [MΩ · m].

S1 system parameters: Z_1 = 0.69+j9.97 Ω, S2 system parameters: Z_1 = 1.38+j19.94 Ω.

4.1 *Setting compensation angle* σ

For the condition of single phase–ground fault, σ is the complementary angle of the impedance angle of $Z_\Sigma = (Z_{\Sigma(1)}+Z_{\Sigma(2)}+ Z_{\Sigma(0)})/3$. Figure 5 shows the relation curve between σ and the distance from the distance relay to the fault point. As can be seen from Figure 6, with the change of fault location on the transmission line, varies from 7.2° to 8.7°.

The zone-one distance relay in practical projects for 400 km transmission line concerns mainly whether fault distance can be calculated accurately when the fault location is within a range of 50 km to 380 km away from the relay point, especially near the boundary of the protection zone (i.e., 340 km from the relay point). Ensuring the accuracy around the boundary of the protection zone can prevent under-reach and over-reach. So the compensation angle σ of 8.27° at 340 km is taken as the setting angle for phase–ground fault.

In order to verify the reasonableness of the setting angle, phase–ground faults with 300 Ω transition resistance at different location along the 400-km line were simulated. The percentage errors shown in the following figures are calculated by using the obtained fault distance with respect to the exact fault distance. Figure 6 presents the estimated fault distance errors along the 400-km line from end m with a power angle δ of 0°. It is obvious that the maximum estimation error is less than 5% on this tough condition with 300 Ω transition resistance, which can meet the engineering requirement. To conclude, adopting the compensation angle at 340 km as the setting angle is reasonable and feasible with a high accuracy.

As for phase–phase fault, σ is the complementary angle of the impedance angle of $Z_\Sigma = Z_{\Sigma(1)}+Z_{\Sigma(2)}$. Similarly, the corresponding compensation angle 4.1° when the fault point is 340 km away from the relay point is adopted as the setting angle. It likewise has a considerable accuracy and the simulation for verification is in next section.

4.2 *Evaluation of fault-distance steady-state error*

Steady-state errors of the measured fault distance are studied in this section for different load levels (power angle) and transition resistance values. Figure 7 shows the error for solid A-g faults whose location is within a range of 50 km to 380 km away from the relay point with a power angle of 0°. It is apparent that the maximum measurement error is −1.8% and the error at the end of the protection zone is nearly 0.

Figure 8 presents the estimated fault distance errors for phase–ground faults whose location is within a range of 50 km to 380 km away from the relay point with 300 Ω transition resistance; the system and load conditions are the same as for Figure 8. The maximum error is less than −2.3% for a heavy-load line and −4.8% for a no-load line. The measurement error at the end of the line under a heavy load condition is less than that under a no load condition.

The actual fault point voltage U_f and its variation ΔU_f calculated based on a lumped parameter line model as shown in (2) and (3) is affected by the capacitive current in a distributed nature, which contributes to the errors. On a no-load and high

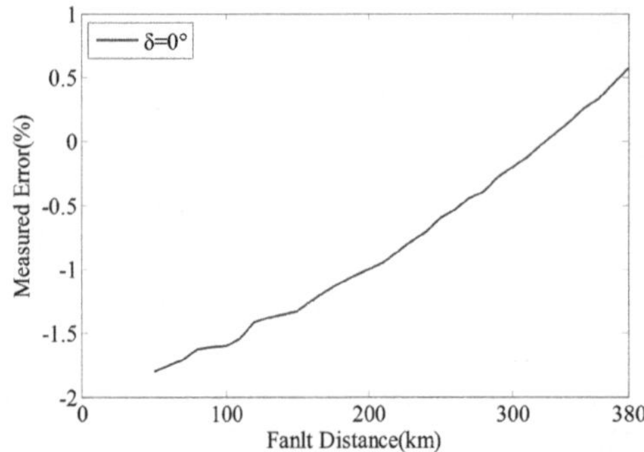

Figure 7. Fault distance errors (A-g solid fault, δ = 0°).

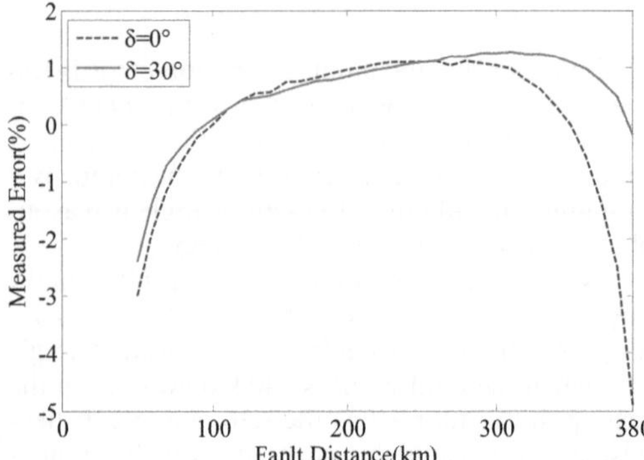

Figure 8. Fault distance errors (A-g fault, 300 Ω transition resistance).

transition resistance condition, the relatively small variation of current caused by the fault is comparable to or even less than the distributed capacitive current along the line resulting in larger error; The longer the fault distance is, the greater influence the distributed capacitive current will have. The same effect still exists in a heavy-load condition, but U_f and ΔU_f can be more accurately calculated from the voltage and current measured at the relay point by using a lumped parameter line model because the line voltage drop is dominated by the series load current. It is worth pointing out that the errors shown in Figure 9 are already acceptable in most applications of fault distance measurement for distance relaying.

Figure 9 shows the errors of fault distance for BC faults along the line. The maximum error is less than –4.5% for a heavy-load line and –4.9% for a no-load line. As can be seen from the Figure 10, the maximum error appears at the location of 50 km away from the relay point, and the measurement error near the boundary of the protection zone (i.e., 340 km from the relay point) is substantially

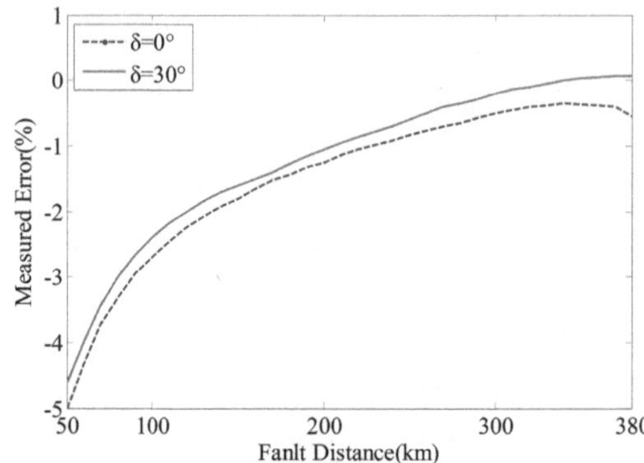

Figure 9. Fault distance errors for BC faults with 10 Ω per phase transition resistance.

Table 1. Single-phase–ground fault with different source impendence angles.

Equivalent source impedance/(°)	Error/(%)
87	–1.47
88	–1.62
89	–1.73

equal to zero. So the algorithm proposed in this paper can maintain high accuracy at the end of the protection zone to distinguish whether the fault is within or outside the intended protection zone, correctly. The errors shown in Figure 10 are acceptable in most applications.

The angle of the equivalent source impedance is to be affected by the shunt load connected at the ends of the transmission line. The shunt load impedance at the ends of the transmission line will typically result in a variation of the equivalent source impedance by no more than 3° in the phase angle. With the system it becomes stronger, and the effect will reduce.

Table 1 presents the calculated fault distance errors with different equivalent source impedance phase angles at end m for a A-g fault under the same general conditions as for Figure 9 with equivalent source impedance at end n unchanged. The fault point is 360 km from end m with 300 Ω transition resistance, while the power angle is 30°. The simulation results show that the maximum error for fault distance is only –1.73%, which is acceptable in practical applications. It can be concluded that the accuracy of the proposed algorithm is not significantly affected by the shunt load at the end of the protected line. However, the aforementioned analysis implies that the algorithm would be of a higher accuracy if the end systems are stronger.

5 SUMMARY

This paper proposes a new algorithm to calculate the fault distance on a transmission line based on the phase angle relationship of the fault point voltage and its variation before and after the fault moment with only single-ended fundamental frequency electrical quantity is used. With corresponding reasonable setting compensation angles for the faults of different types, the algorithm can ensure a high accuracy and avoid the effects of the uncertain load current and fault path resistance.

The algorithm has been tested by using PSCAD/EMTDC simulation for the faults of different types with different load levels. The simulation results show that the proposed algorithm is of a high accuracy and indeed immune to the uncertainties in the load current and transition resistance, as well as insensitive to the variations of the source impedance, which can provide effective and reliable protection for high-voltage transmission line.

REFERENCES

[1] Wang, X.G. & Huang S.F. 2010. Impact of fault resistance for fault transient component. *Power System Protection and Control* 38 (2): 18–21.

[2] Huang, S.F. & Jiang, Q.K. 2011. Analysis on impact of transition resistance on multiphase compensation impedance element and its countermeasures. *Power System Technology* 35 (8): 202–206.

[3] Jie, L. & Shihong, M. 2011. A Protection Criterion for High Resistance Grounding of Transmission Line Based on Phase-Segregated Active Power Differential Principle. *Power System Technology* 35 (8): 197–201.

[4] Bin, L. & Cai, J.X. 2002. Refusal-to-move Analysis for Zero Sequence Protection of Model 11 Microcomputer. *Heilongjiang Electric Power 24 (5): 381–382,386.*

[5] Liu, F.Y. & Chen, T.B. 2002. Analyzing the reason of the RAZFE relay's refusing in the fault of Geshuang line. *Relay* 30 (2): 68–69.

[6] Yang, L. & Yang T.F. 2010. Study of adaptive grounding distance relay. *Transactions of China Electrotechnical Society* 25 (4): 77–81.

[7] Zhang, H.Z. & Wang, W.Q. & Zhu, L.N. 2008. Ground distance relay based on fault resistance calculation. *Power System Protection and Control* 36 (18): 37–42.

[8] Daros, A. & Salim, R.H.M. 2008. Ground distance relaying with fault-resistance compensation for unbalanced system. *IEEE Trans. Power Del* 23 (3): 1319–1326.

[9] Xu, Z.Y. & Xu, G. & Ran, L. 2010. A new fault-impedance algorithm for distance relaying on a transmission line. *Power Delivery, IEEE Transactions on* 25(3): 1384–1392.

[10] Eriksson, L. & Saha, M.M. 1985. An accurate fault locator with compensation for apparent reactance in the fault resistance resulting from remote-end in feed. *IEEE Trans. Power App. Syst* 104 (2): 424–436.

Material Science and Environmental Engineering – Chen (Ed.)
© 2016 Taylor & Francis Group, London, ISBN 978-1-138-02938-5

Ten meters S-band antenna gain measurement based on sun method

L.J. Zhou, X.W. Su, Y.M. Nie & X.G. Zhai
China Satellite Maritime Tracking and Controlling Department, Jiangyin, China

ABSTRACT: A method for ten meters s-band antenna gain measurement is proposed in detail. The theoretical and simulating results indicate that the sun method have the same precision as the comparison method. However, antenna gain measurement based on sun method is simpler and much more feasible, which is experimentally proved to be applied widely in actual engineering measurement.

Keywords: sun method; antenna gain; *Y* factor method

1 INTRODUCTION

Reflector antenna is widely used in satellite communication, telemetry, remote control, and radio astronomy equipment. As an important index of the antenna of the antenna gain widespread attention, research of the measurement method is also carried out widely and deeply [1–5].

The current commonly used Comparative method for measuring antenna gain test results meets the requirements. But the test distance is directly proportional to the square of Antenna diameter and inversely proportional to wavelength. For large reflector antenna with high working frequency point, testing distance requires several kilometers, needing to set up nearly 100 meters beacon towers, costly capital, the beacon erection accuracy cannot be guaranteed, and the test site existing ground reflection. These defects will seriously affect the antenna gain test [6–10].

Measurement of large diameter antenna by using the sun method undoubtedly can fully satisfy the far field test conditions of distance. And it can also overcome the reflection on the ground and the environmental effects witches exist in the comparison method. The sun is the strongest radio source and signal noise ratio. Although the solar radiation is not stable, but the solar flux density can be found on the Internet. It can greatly improve the measurement accuracy.

2 PRINCIPLE AND METHOD

The basic principle of sun method for measuring antenna gain is that: first, measuring the noise power level of the antenna by pointing to the sun and the cold air background, respectively. Second, calculating the difference between the two to get the *Y* factor, and calculating the G/T_{sys} value of the system. Third, measuring the system noise temperature by using *Y* factor method. Integrated the above information can be used to obtain the antenna gain *G*. As shown in Figure 1.

2.1 *The calculation formula is as follows*

$$\frac{G}{T_{sys}} = \frac{8\pi k(Y-1)K_1K_2}{\lambda^2 S} \tag{1}$$

where k is Boltzmann constant, $k = 1.38 \times 10^{-23}$ (J/K);
λ is wavelength (m);
S is the solar flux density ($W/m^2 \cdot Hz$), A solar flux unit is 10–22 $W/m^2 \cdot Hz$;
K_1 is absorbing attenuation factor of atmosphere;
K_2 is Beam correction factor.

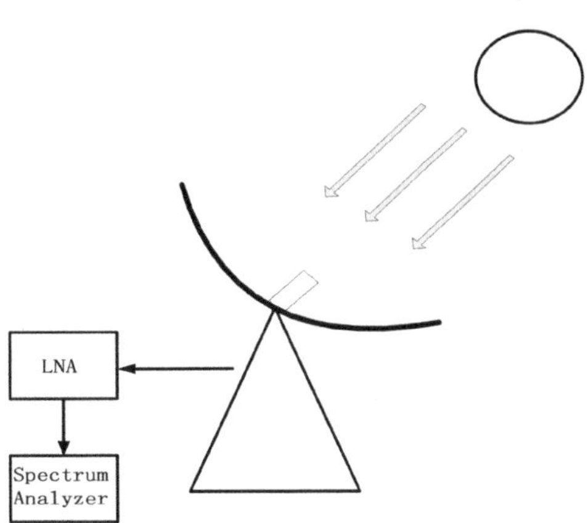

Figure 1. Principle of measurement of antenna gain of sun law.

By formula (1) can be obtained the antenna gain which measured in decibels:

$$G = 10 \times \log\left(\frac{8\pi k(Y-1)T_{sys}K_2}{\lambda^2 S}\right) + K_1 \tag{2}$$

Formula (2) shows that, it only measures the magnitude and the system noise temperature of *Y* factor, and can calculate the antenna gain. We use *Y* factor method to measure system noise temperature. The specific method is that we measure the noise temperature by using spectrum analyzer when field discharge (LNA) is respectively connected with the normal standard load and reception antenna. The antenna should point cold air in the whole testing process.

$$T_{sys} = \frac{(T_0 + T_{LAN})}{Y_0} \tag{3}$$

where

T_0 is the physical temperature of the antenna, the unit is K;

T_{LAN} is the noise temperature of low noise amplifier, the unit is K;

Y_0 is the noise power difference of low noise amplifier.

The formula (2) can be simplified:

$$G = 15..68 + 10\log(Y-1) + 20\log f - 10\log S + K_1 + K_2 + 10\log T_{sys} \tag{4}$$

where

f is the test frequency, the unit is GHz;

$Y = 10^{\frac{Y_E}{10}}$;

$Y_E = Y_{antenna} - Y_{coldair}$.

2.2 *Modifying factor*

2.2.1 *Atmosphere absorbing attenuation factor K_1*

Atmosphere absorbing attenuation factor is a function of frequency, antenna elevation and atmosphere which increases with the increase of frequency, and decreases with the increase of the antenna elevation angle. Calculation of atmosphere absorbing attenuation factor is more complex, we can get it by the follow formula.

$$K_1 = \frac{0.02}{\sin(EL)} \tag{5}$$

where

v is a constant which can be obtained from the Rec.ITU R P.676 5 [11].

El is the elevation when the antenna is pointing to the sun.

2.2.2 *Beam correction factor K_2*

Beam correction factor have something to do with the brightness temperature distribution of radio source and direction of the antenna shape. The calculation formula is as follows.

$$K_2 = \frac{\iint_\Omega B(\theta,\phi)d\Omega}{\iint_\Omega B(\theta,\phi)P(\theta,\phi)d\Omega} \tag{6}$$

where

$B(\theta, \phi)$ is the brightness temperature distribution of radio source.

$P(\theta, \phi)$ is the Antenna normalized power pattern.

When the half power beam width of antenna is larger than the maximum angular diameter of the sun beam correction factor and can be calculated from the formula as follows.

$$\mathrm{K}_2 = 10 \times \log\frac{\chi}{1-e^{\chi}} \tag{7}$$

And

$$\chi = \frac{0.2106}{\mathrm{HPBW}^2} \tag{8}$$

3 TEST AND ANALYSIS

3.1 *The test result of sun method*

The gain test results of a type of 10 meters diameter reflector antenna for S band is shown in Table 1.

Table 1. The gain results measured by sun method.

Parameters	Test results
Frequency (GHz)	2.25
Instrument noise power (dBm)	–93.7
Field discharge load (dBm)	–66.83
EL = 450 cold air (dBm)	–72.02
Antenna noise temperature T_A (K)	66.8
Field discharge noise temperature (K)	65
Point to the sun (dBm)	–55.03
Solar flux (dBm)	91.906

Table 2. The gain results measured by comparison method.

Parameters	Test results
Frequency (GHz)	2.25
Standard horn gain (dB)	16.4
The horizontal direction standard horn received power (dBm)	–29.09
The vertical direction standard horn received power (dBm)	–29.11
The horizontal direction Antenna receiving power (dBm)	–6.62
The vertical direction Antenna receiving power (dBm)	–6.83

By formula (4):

$$G = 15.86 + 10\log(50-1) + 20\log 2.25 - 10\log 91.906 + 0.05 + 0.14 + 21.20 = 41.56\ dB \tag{9}$$

3.2 *The test result of comparison method*

The gain test results of the antenna are shown in Table 2.

The calculation formula is as follows:

$$G = G_0 + [A_2 - A_1] \tag{10}$$

where

G_0 is a standard gain horn gain;

A_1 is the received level when frequency spectrum instrument and standard gain horn connected;

A_2 is the received level when spectrum analyzer and the measured antenna connected;

For the circular polarization antenna, put the standard gain horn antenna in vertically polarized state to measure the $A_\perp$ first. Then change the location in horizontal polarization state to measure the $A_{//}$. The total circular polarization energy for the antenna is:

$$A = \log[10^{0.1A//} + 10^{0.1A_\perp}] \tag{11}$$

By the formula (10) and formula (11) have: G = 41.78 dB.

4 THE RESULT ANALYSIS

From the above test we know that the antenna gain results of the sun method and the comparison method is the same. It means that the sun method is right and the test result is more accurate.

5 CONCLUSION

Compared to the comparison method, measuring the antenna gain, which D/λ is equal to 100 by the sun method can meet the far field test conditions and overcome the test site ground reflection effect. At the same time, the test result is more accurate. It is of practical value and has a broad application prospect.

REFERENCES

[1] Lu Jun, Zhao Hui. Scale model method tests the influence of aircraft on radar direction map [A]. The 2011 National Conference of microwave and millimeter wave [C], 2011.

[2] Yao Demiao, Cai Jianming, Study on the wideband and high-gain microstrip antenna element [J], Journal of electronics, 1996, 526–531.

[3] Chen I.J., Huang C.S., Hsu P. Cirularly polarized patch antenna array fed by coplanar waveguide [J]. IEEE Trans. AP, 2004, 29(6); 1607–1609.

[4] Zio Ikowski R.W. and Kipple A.D. Application of double negative materials to increase the power radiated by electrically small antennas. IEEE Trans Antennas Propagat, Vol51: 2626–2640, Oct. 2003.

[5] Qin Shunyou, Xu Deseng. The satellite communication earth station antenna engineering measurement technology [M]. People's Posts and Telecommunications Press. 2006.

[6] Li Xiangzhou, Study on the gain of antenna is determined by using the measured antenna pattern [D]. Shandong Normal University, 2001.

[7] Zhang Lianhong, Yuan Chengli, Liu Guangyan; An effective way to improve the gain of antenna [A]. The 2009 Annual Conference of the national antenna [C], 2009.

[8] Wu Xiang, The high gain of millimeter wave radar antenna development [A]. The 2011 National Conference of microwave and millimeter wave [C], 2011.

[9] Li Hongbin, Fang Shaojun, Ding Weiping; Design of low cost and high gain microstrip antenna [A]. The 2009 National Conference of microwave and millimeter wave [C], 2009.

[10] Ye Sheng, Jin Ronghong, Gen Junping. Design of broadbandandhighgainmicrostripantennaarray[A]. The 2009 Annual Conference of the national antenna [C], 2009.

[11] Rec.ITU-R P.676-5. Attenuation Atmospheric Gases [S].

Material Science and Environmental Engineering – Chen (Ed.)

A circular polarization conversion based on multilayer split square ring Frequency Selective Surface

Y. Fan, Y.F. Li, D.Y. Feng, S.B. Qu & J.B. Yu
College of Science, Air Force Engineering University, Xi'an, Shanxi, China

M.D. Feng
Electronic Materials Research Laboratory, Key Laboratory of the Ministry of Education, Xi'an Jiaotong University, Xi'an, Shanxi, China

ABSTRACT: In this letter, we propose to achieve a broadband circular polarization conversion based on multilayer split square ring frequency selective surfaces. It can convert linear polarization into circular polarization within a broadband of 19.3 GHz (39.5–58.8 GHz); a three layer prototype to operate in U-band has been designed and simulated. The low symmetry of unit cells provides the advantages of both low cross-polarization and low sensibility of the incidence angle. As simulated in CST, it has been shown that axial ratios lower than 3 dB are obtained for angles of incidence as high as 25° over the frequency range of 39.5–58.8 GHz.

Keywords: circular polarization; polarization conversion; Frequency Selective Surface; split square ring

1 INTRODUCTION

Circular Polarization Wave plays (CPW) an important role in antenna system and communication system. CPW provides the advantages of both low error rate and strong anti-interference. Conventional polarization manipulations are always realized using the wave-plate, which is made of birefringent material such as crystalline solids and liquid crystals [1]. However, narrow bandwidth and low transformation efficiency limit its use in the micro-optical systems. Currently, polarization manipulation can be achieved by the anisotropic or chiral materials [2], yet still with thickness limitations and bulky configurations.

In former works, several methods have been used to convert LPW into CPW. However, the need for broadband polarization conversion in high gain multibeam antenna systems limits the use of narrow band polarizer based on circular wave guides [3–4]. Thus, transmission-type circular polarizer based on open planar structures can be an attractive solution to avoid the band limitation due to the cut-off properties of the circular waveguide. This conversion can be built with several Frequencies Selective Surfaces (FSS) that provide different transmission characteristics for the vertical and horizontal components of an incident LPW (linear polarization wave) with a 45° polarization angle. The FSS can be capacitive for one component and inductive for the other component. Thanks to the FSS, the first component is phase advanced while the second one is phase retarded. Therefore, a phase difference appears between the two components of the transmitted wave. If the magnitude of both components is equal and the phase difference is equal to or close to 90°, the incident LPW is transformed into a transmitted CPW. By cascading FSS, one can increase both the transmission coefficient and the bandwidth of the polarizer allowing broader bandwidths compared with the ones provided by the circular polarizer based on circular waveguides.

In this letter, we achieve wide-band transmission-type circular polarization conversion using three-layer split square ring frequency selective surfaces. The simulated results indicates that polarization conversion transmission is greater than 90% over a wide frequency range from 39.5 to 58.8 GHz, and the simulated axis ratios of the transmitted waves are less than –3dB in the frequency range 35–58.8 GHz. The proposed circular polarization conversion may find potential applications in the design of wide-band circular polarization antennas [5–6].

2 WORKING PRINCIPLE AND DESIGN

The proposed polarization conversion is composed of three-layer multiple FSS separated by air gaps, as shown in Figure 1. The optima geometry of the

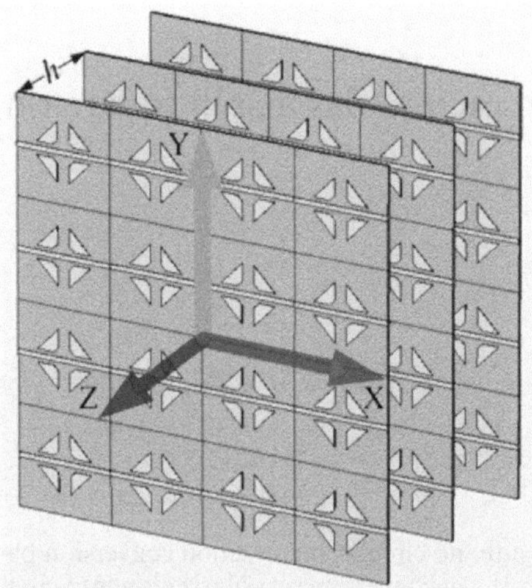

Figure 1. Schematic view of circular polarization conversion.

FSS unit cell is shown in Figure 2(a). It contains a split square ring resonator bisected by a metal strip. The geometrical parameters are described as follows: the periodicity of the FSS unit cell is $a = 4$ mm; $d_1 = 2$ mm, and $d_2 = 1$ mm are the inner and the outer length of the square ring, respectively; $w_1 = 0.1$ mm is the width of the metal strip, $w_2 = 0.5$ mm is the length of the vertical and the horizontal gaps, and $t_1 = 0.1$ mm is the thickness of one FSS layer. All the elements are etched on a F4B dielectric substrate ($\varepsilon_r = 2.65$). The polarizer is then built by cascading different FSS layers separated by a distance of $h = 1$ mm, as indicated in Figure 1. It is important to note that the FSS elements of the different layers have the same orientation and are perfectly aligned with respect to their centers.

The main property of the ring resonator is the resonant behavior of its reflection coefficient. This resonant reflection occurs when the circumference of the ring is approximately equal to the wavelength. The addition of the gaps and the metal strip to the ring allows different transmission characteristics for the *x*- and *y*-polarizations. Now consider that an incident LPW, whose electric field vector E_{inc} is oriented at $\theta = 45°$ to the *x*-axis (Fig. 2(b)), travels toward the $-z$ direction. This field vector can be decomposed into two orthogonal linearly polarized components of equal magnitude, as equation (1) below:

$$E_{inc} = E_{\perp} + E_{||} = E_0(m_y + m_x)e^{jkz} \tag{1}$$

where E_0 is the magnitude of the incident electric field, m_x and m_y are the unit vectors in *x*- and

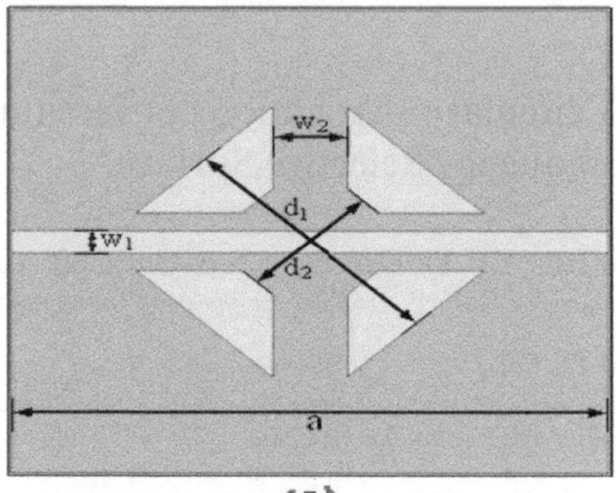

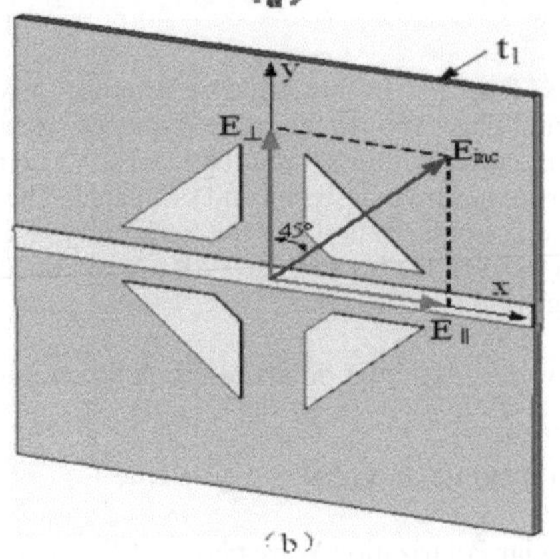

Figure 2. (a) The front view of one FSS layer; and (b) the perspective view of one FSS layer.

y-directions, respectively, and *k* is the wave vector. The field vectors $E_{\perp}$ and $E_{||}$ are, respectively, perpendicular and parallel to the metal strip. For $E_{\perp,}$ the FSS exhibits a capacitive behavior mainly due to the horizontal gaps; however, for $E_{||}$ the FSS becomes primarily inductive due to the metal strip. At the output of the polarization conversion, the transmitted electric field can also be presented as a sum of two orthogonal linearly polarized components:

$$E_t = E_0(\Gamma_{\perp} m_y + \Gamma_{||} m_x)e^{jkz} \tag{2}$$

where $\Gamma_{\perp} = |\Gamma_{\perp}|e^{j\varphi_{\perp}}$ and $\Gamma_{||} = |\Gamma_{||}|e^{j\varphi_{||}}$ are the transmission coefficients for the y- and x-polarizations, respectively. As a result, a phase difference $\Delta\varphi = \varphi_{\perp} - \varphi_{||}$ appears between the two components of the transmitted wave. If both the spacing between layers and the element's geometry are adjusted to obtain a $\Delta\varphi$ of $-90°$ and $|\Gamma_{\perp}| = |\Gamma_{||}|$, the incident LPW can be converted into a transmitted CPW.

3 SIMULATION RESULTS

To verify the designed three-layer circular polarization, the full-wave numerical simulation was performed using CST Microwave Studio and simulated results are given in Figures 2 and 3. To eliminate the effect of absorption, lossless substrate and Perfect Electric Conductor (PEC) are adopted for the FSS unit cell. The four boundaries along x- and y-direction are set to be periodic boundaries. The conversion is first simulated under normal x- and y-polarization incident waves illuminating from z-direction, respectively. The separation between the layers is 1 mm. The simulated magnitudes of the transmission coefficients $|r_\perp|$ and $|r_\parallel|$ corresponding to the scattering of $E_\perp$ and $E_\parallel$, respectively, are shown in Figure 3, as well as the reflection coefficient $|\Gamma_\perp|$ and $|\Gamma_\parallel|$. The magnitude of the reflection coefficients is less than −10dB in the frequency band from 39.5 to 60GHz. Figure 4 indicates the phase difference Δφ. A Δφ of −90°±5°is obtained in the frequency band from 39.5 to 58.8 GHz.

According to the reference, the Axis Ratio (AR) of the transmitted wave can be calculated as equations (3) and (4):

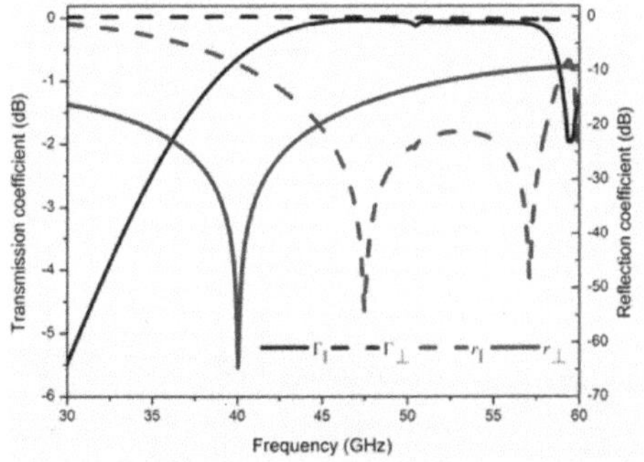

Figure 3. Transmission coefficient and reflection coefficient corresponding to $E_\perp$ and $E_\parallel$.

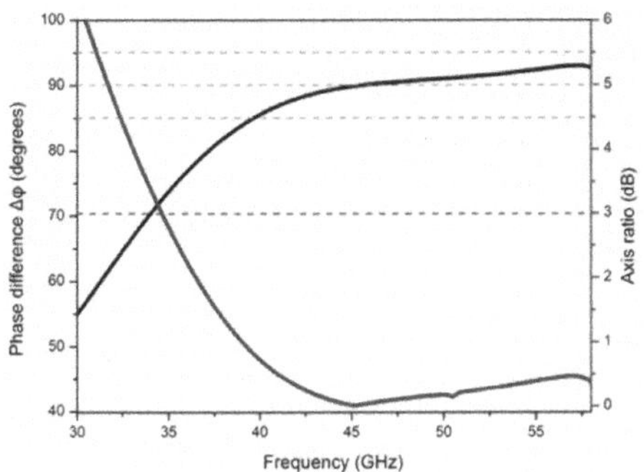

Figure 4. Phase difference $\Delta\varphi$ and the axis ratio AR.

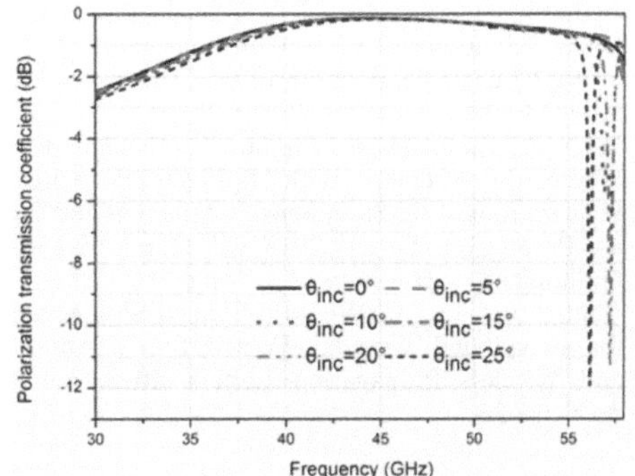

Figure 5. The incidence angle of the polarization conversion transmission versus the frequency.

$$AR = \left(\frac{|E_\parallel|^2 + |E_\perp|^2 + \sqrt{n}}{|E_\parallel|^2 + |E_\perp|^2 - \sqrt{n}} \right)^{\frac{1}{2}} \quad (3)$$

$$n = |E_\parallel|^4 + |E_\perp|^4 + 2|E_\parallel|^2 |E_\perp|^2 |\cos(2\Delta\varphi) \quad (4)$$

The simulated AR of the transmitted wave is shown in Figure 4. The value of AR is lower than 3 dB in the frequency band from 35 to 58.8 GHz.

The linear to circular polarization transmission coefficient under y-polarized wave from +z direction with different incident angles is shown in Figure 5. Obviously, the conversion remains high polarization transmission when θ_{inc} varies from 0° to 25°. The peak transmission is at 45 GHz. And the valley value is caused by the resonance of the square split ring, which is changed with the variation of the incident angle.

4 CONCLUSION

In summary, a circular polarization conversion based on three-layer square split ring FSS has been achieved and investigated in this letter. Numerical simulations indicate that the circular polarization conversion transmission is over 90% in 39.5–58.8 GHz and the axis ratio for the transmitted wave is lower than 3dB in the frequency range 35–58.8 GHz. The polarization transmission keeps as high as θ_{inc} 25°.

REFERENCES

[1] Hungshan. C., Yihsin. L., Chiaming. C., Yujen. W., Abhishek. K.S., Jiatong. S., Vladimir G.C., "A polarized bifocal switch based on liquid crystals operated electrically and optically", Journal of Applied Physics, 2015, vol. 117, no. 4, pp. 044502.

[2] Cadusch. J.J., James. T.D., Djalalian-Assl. A., Davis. T.J., Roberts. A., "A Chiral Plasmonic Metasurface Circular Polarization Filter", Photonics Technology Letters, 2014, vol. 26, no. 23, pp. 2357–2360.
[3] Yang. Z., Kunpeng. W., Zhijun. Z., Zhenghe. F., "A Waveguide Antenna With Bidirectional Circular Polarizations of the Same Sense", Antennas and Wireless Propagation Letters, 2013, vol. 12, pp. 559–562.
[4] Guihong. L., Huiqing. Z., Tong. L., Long. L., Changhong. L., "CPW-Fed S-Shaped Slot Antenna for Broadband Circular Polarization", Antennas and Wireless Propagation Letters, 2013, vol. 12, pp. 619–622.
[5] Chong. Z., Junhong. W., Meie. C., Zhan. Z., Zheng. L., "A New Kind of Circular Polarization Leaky-Wave Antenna Based on Substrate Integrated Waveguide", International Journal of Antennas and Propagation, vol. 2015.
[6] Jayashree P.S., Raj. K. Mahadev. D.," Circular Polarization in Defected Hexagonal Shaped Microstrip Antenna", Wireless Personal Communications, 2014, vol. 75, no. 2, pp. 843–856.

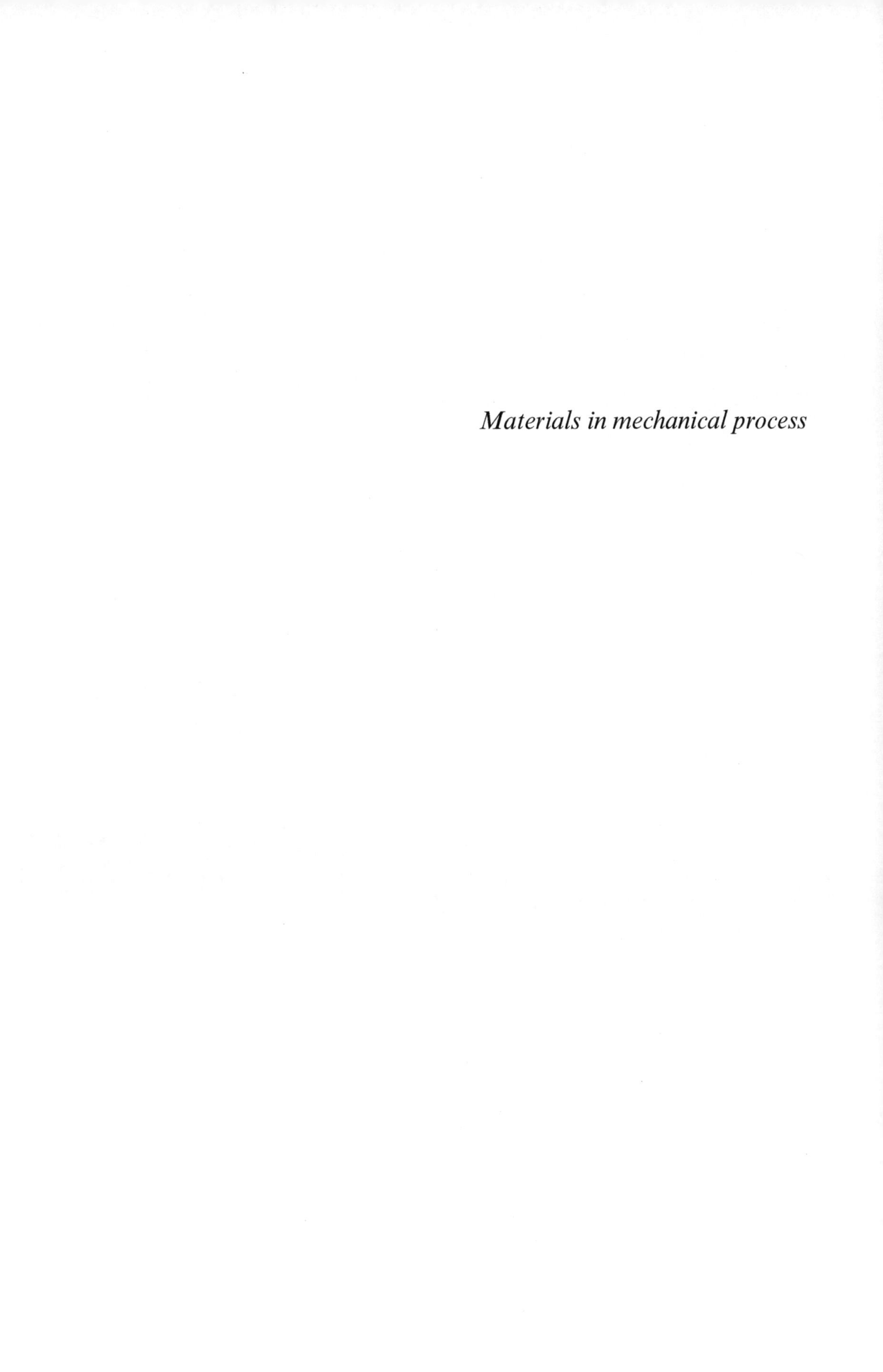

Materials in mechanical process

Material Science and Environmental Engineering – Chen (Ed.)
© 2016 Taylor & Francis Group, London, ISBN 978-1-138-02938-5

Research on the large gear detecting system based on LabVIEW

L.M. Hao, X.J. Zhang, B.X. Li, S.F. Li, Y.H. He & R.J. Jia
School of Mechanical Engineering and Automation, University of Science and Technology Liaoning, Anshan, Liaoning, China

ABSTRACT: From the analysis of characteristics for large gear, large gear tooth profile error detection scheme based on the coordinate method was put forward. Through coordinate transformation, realized the unity of the work piece coordinate system and measuring coordinate system, the mathematical model of standard involute tooth profile was established. Applying LabVIEW software to develop large gear tooth profile measuring system software, the entire system adopted modular design and can carry on the tooth profile of automatic measurement, data acquisition, error evaluation, and the measurement result display and other functions. The experimental results show that the measured gear involute tooth profile curve and the theoretical calculation of the tooth profile curve are in the basic agreement, the tooth profile error meets national standard, verifies the correctness of this measuring principle and the feasibility of related methods, and provides a theoretical basis for the improvement and perfection of the subsequent detection system.

Keywords: large gear; tooth profile error; mathematical model; detection

1 INTRODUCTION

As the key or the important basic parts of large complete sets of equipment, large gears are more and more widely applied in ships, mines, metallurgy, power generation, aerospace, and other equipment; its quality, performance, service life directly affects the whole technical and economic index. And the large gear has the characteristics of large size, heavy weight, leads to exist many difficulties of its tooth profile measurement, so the research on the large gear tooth profile error measurement is becoming more and more urgent and important.

2 GEAR MEASURING PRINCIPLES

Coordinate method is a kind of high feasibility scheme in the large gear measurement. It is mainly used to measure the gear geometry error, the measurement principle takes the gear as a complex shaped geometric solid, based on the measurement coordinate system, and the measurement of gear tooth surface geometry deviation is according to design parameters.

In mechanics, the involute equation of gear tooth profile is usually derived of the Cartesian coordinates as shown in Figure 1. It can be seen from Figure 1; the set of coordinate system (i.e. the workpiece coordinate system) is taking the rotation center of gear as the original point and make line between the contact point A on base circle of involute and the gear rotary center as the X axis. The involute equation is given as:

$$\begin{cases} x = r_b \cos\beta + r_b\beta\sin\beta \\ y = r_b \sin\beta - r_b\beta\cos\beta \end{cases}, \text{ and}$$

$$\beta = \frac{\overset{\frown}{AB}}{r_b} = \frac{\overline{BK}}{r_b} = \tan\alpha_i$$

where r_b represents the radius of base circle and α_i is the pressure angle of any point on involute.

But in actual measurement, the measurement tool is clamped in the middle of two teeth. Then, it needs to establish measurement coordinate system

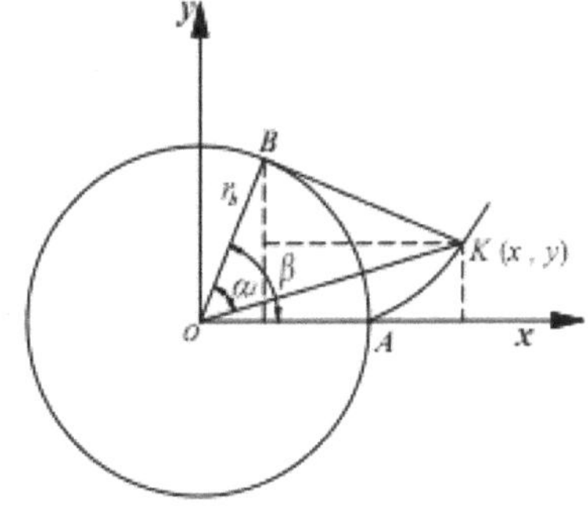

Figure 1. Gear involute in the workpiece coordinates.

XOY, the alveolar center line is selected as X axis; the line passing through the gear center and vertical with X axis is Y axis to, as shown in Figure 2. So it is necessary to rotate the workpiece coordinate system to a certain angle. The relationship between the coordinate after rotation transform and the original coordinate is as followed:

$$\begin{cases} X = x\cos\varphi - y\sin\varphi \\ Y = x\sin\varphi + y\cos\varphi \end{cases}$$

where φ represents alveolar half angle on base circle.

$$\varphi = \frac{e_b}{2r_b} = \frac{1}{2}\frac{\cos\alpha\left(\frac{\pi m}{2} - mzinv\alpha\right)}{\frac{mz\cos\alpha}{2}} = \frac{\pi}{2z} - inv\alpha$$

where e_b represents alveolar width on base circle, α is pressure angle on pitch circle.

For standard gear $\alpha = 20^\circ$ then $inv\alpha = \tan\alpha - \alpha = 0.0149$. So, the involute tooth profile curve equation in the actual measurement coordinate system is defined as:

$$\begin{cases} X = \dfrac{mz\cos\alpha}{2}\Big\{[\cos(\tan\alpha_i) + \tan\alpha_i \sin(\tan\alpha_i)] \\ \quad \times\cos\left(\dfrac{\pi}{2z} - 0.0149\right) - [\sin(\tan\alpha_i) \\ \quad - \tan\alpha_i\cos(\tan\alpha_i)] \times \sin\left(\dfrac{\pi}{2z} - 0.0149\right)\Big\} \\ Y = \dfrac{mz\cos\alpha}{2}\Big\{[\cos(\tan\alpha_i) + \tan\alpha_i \sin(\tan\alpha_i)] \\ \quad \times\sin\left(\dfrac{\pi}{2z} - 0.0149\right) + [\sin(\tan\alpha_i) \\ \quad - \tan\alpha_i\cos(\tan\alpha_i)] \times \cos\left(\dfrac{\pi}{2z} - 0.0149\right)\Big\} \end{cases}$$

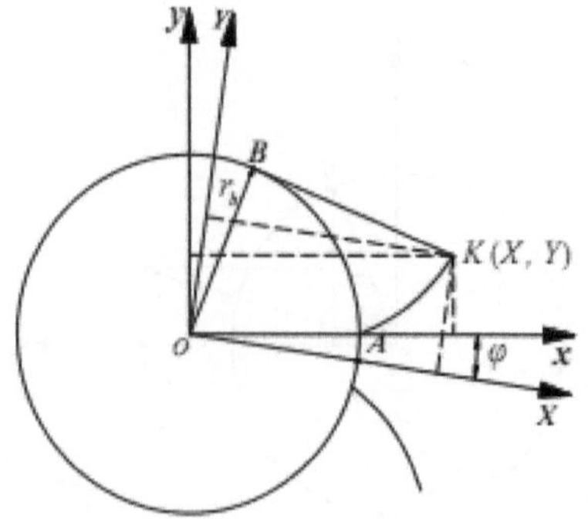

Figure 2. Gear involute in the measure coordinates.

The mathematical model of the standard involute gear tooth profile can be established, according to the above equation.

3 DETECTION SYSTEM PROGRAM DESIGN

LabVIEW is a kind of virtual instrument software development tool based on G language that is invented by the USA NI company. It is one of the most widely used virtual instrument software platform in the world at present, mainly applied to the instrument control, data acquisition, data display, and so on. Different from traditional programming languages, LabVIEW uses the graphical language, powerful programming convenient, intuitive and friendly man-machine interface. It is largely used

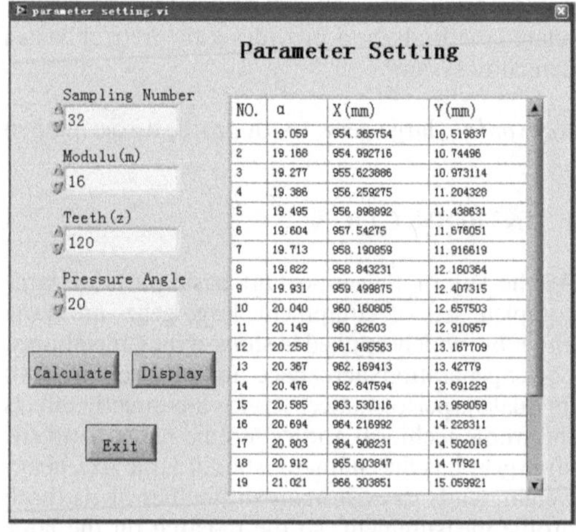

parameter setting.vi

Parameter Setting

Sampling Number: 32

Modulu(m): 16

Teeth(z): 120

Pressure Angle: 20

NO.	α	X(mm)	Y(mm)
1	19.059	954.365754	10.519837
2	19.168	954.992716	10.74496
3	19.277	955.623886	10.973114
4	19.386	956.259275	11.204328
5	19.495	956.898892	11.438631
6	19.604	957.54275	11.676051
7	19.713	958.190859	11.916619
8	19.822	958.843231	12.160364
9	19.931	959.499875	12.407315
10	20.040	960.160805	12.657503
11	20.149	960.82603	12.910957
12	20.258	961.495563	13.167709
13	20.367	962.169413	13.42779
14	20.476	962.847594	13.691229
15	20.585	963.530116	13.958059
16	20.694	964.216992	14.228311
17	20.803	964.908231	14.502018
18	20.912	965.603847	14.77921
19	21.021	966.303851	15.059921

Calculate Display Exit

Figure 3. Parameter setting module front panel.

measuring.vi

Measuring Data

序号	X(mm)	Y(mm)
1	954.525754	10.357837
2	955.220716	10.59596
3	955.919886	10.864114
4	956.519275	11.108328
5	957.209892	11.502631
6	957.94675	11.722051
7	958.610859	11.939619
8	959.353231	12.163364
9	959.911875	12.424315
10	960.648805	12.690503
11	961.36403	12.967957
12	962.021563	13.235709
13	962.675413	13.50579
14	963.428594	13.783229

Measure Save Exit

Figure 4. Tooth profile measurement module front pane.

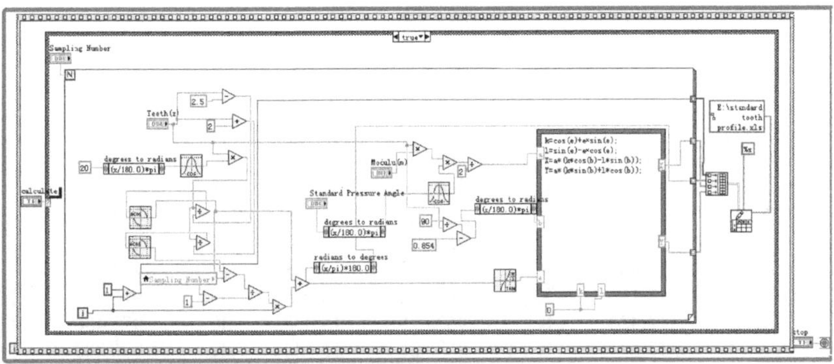

Figure 5. Parameters setting and calculation module block diagram.

Table 1. Involute gear tooth profile calculation value.

	X	Y		X	Y		X	Y
K_1	954.366	10.520	K_{12}	961.496	13.168	K_{23}	969.148	16.219
K_2	954.993	10.745	K_{13}	962.169	13.427	K_{24}	969.870	16.517
K_3	955.624	10.973	K_{14}	962.848	13.691	K_{25}	970.597	16.819
K_4	956.259	11.204	K_{15}	963.530	13.958	K_{26}	971.328	17.126
K_5	956.899	11.439	K_{16}	964.217	14.228	K_{27}	972.063	17.436
K_6	957.543	11.676	K_{17}	964.908	14.502	K_{28}	972.803	17.750
K_7	958.191	11.917	K_{18}	965.604	14.779	K_{29}	973.548	18.068
K_8	958.843	12.160	K_{19}	966.304	15.060	K_{30}	974.297	18.389
K_9	959.499	12.407	K_{20}	967.008	15.344	K_{31}	975.050	18.715
K_{10}	960.161	12.658	K_{21}	967.717	15.632	K_{32}	975.809	19.045
K_{11}	960.826	12.911	K_{22}	968.430	15.923			

for research, principle design, testing and implement the instrument system, in order to improve the work efficiency. Therefore, LabVIEW software is used to develop the large gear detection system.

The system is composed of parameters setting module, measurement module and tooth profile curve display module. Parameters setting module can realize the basic parameters of gears, computation of standard tooth profile and data save functions, and so on. The front panel of parameter setting module as shown in Figure 3, the standard involute tooth profile mathematic model through entering gear modulus, the number of teeth and the standard pressure angle can be established, according to the set of sample points, click the "calculate" button to calculate corresponding coordinate value of each point on the standard involute. Click "display" button, the calculation results are shown in the list on the right side of the panel, at the same time, the calculated data is saved in the excel file form. On the basis of the above function developed the parameter setting module block diagram as shown in Figure 5. Measurement module is shown in Figure 4 can realize data acquisition, real-time display and the preservation of measurement data. Tooth curve display module can realize displaying the standard tooth profile and the actual tooth profile at the same time, the measured gear tooth error can be determined by comparison.

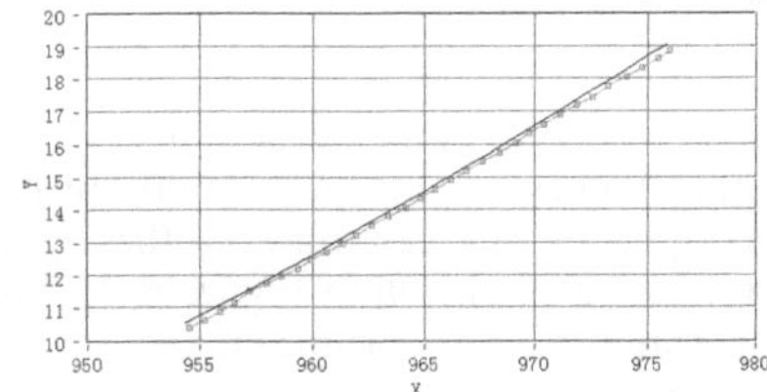

* Theoretical tooth profile curve   the measured tooth profile curve

Figure 6. Tooth profile error curves.

4 EXPERIMENTAL ANALYSES

With the modulus m = 16 mm, number of teeth z = 120, precision grade level 8 standard involute

Table 2. Involute gear tooth profile measuring value.

	X	Y		X	Y		X	Y
K_1	954.526	10.458	K_{12}	962.022	13.236	K_{23}	969.740	16.314
K_2	955.221	10.696	K_{13}	962.675	13.506	K_{24}	970.418	16.598
K_3	955.920	10.964	K_{14}	963.429	13.783	K_{25}	971.125	16.887
K_4	956.519	11.108	K_{15}	964.169	13.961	K_{26}	971.841	17.174
K_5	957.210	11.503	K_{16}	964.827	14.236	K_{27}	972.519	17.407
K_6	957.947	11.722	K_{17}	965.516	14.507	K_{28}	973.214	17.749
K_7	958.611	11.940	K_{18}	966.243	14.798	K_{29}	974.045	18.037
K_8	959.353	12.163	K_{19}	966.924	15.075	K_{30}	974.728	18.325
K_9	959.912	12.424	K_{20}	967.628	15.352	K_{31}	975.486	18.712
K_{10}	960.649	12.691	K_{21}	968.425	15.644	K_{32}	976.000	18.956
K_{11}	961.364	12.968	K_{22}	969.130	15.929			

Table 3. Difference in values in Y-direction.

i	ΔY	i	ΔY	i	ΔY	i	ΔY
1	–0.062	9	0.017	17	0.005	25	0.067
2	–0.049	10	0.033	18	0.019	26	0.048
3	–0.009	11	0.057	19	0.015	27	–0.029
4	–0.096	12	0.068	20	0.008	28	–0.001
5	0.064	13	0.078	21	0.012	29	–0.031
6	0.046	14	0.092	22	0.006	30	–0.065
7	0.023	15	0.003	23	0.095	31	–0.003
8	0.003	16	0.008	24	0.081	32	–0.089

spur gear as the detection object, select 32 measuring points, the detection system can calculate theoretical value of the standard involute tooth profile, the coordinates of each point value as shown in Table 1.

Based on the sampling measurements obtained measuring values of the gear tooth profile as shown display module can draw out the tooth profile error curves as shown in Figure 6. In Table 2, the application of tooth profile curve.

It can be seen from Figure 6, measured master gear's involute tooth profile curve and theoretical calculation involute tooth profile curve are in basic agreement. The X direction coordinate values is greater, and difference value of two curves in X direction is relatively small, so the tooth profile error is mainly caused by difference value in Y direction (see Table 3). In order to determine the range of variation of the error, it needs to determine the mean $\overline{\Delta Y}$ and standard deviation σ.

$$\overline{\Delta Y} = \frac{1}{n}\sum_{i=1}^{n} \Delta Y_i = 0.0129$$

$$\sigma = \sqrt{\sum_{i=1}^{n} (\Delta Y_i - \overline{\Delta Y})^2 \Big/ (n-1)} = 0.050161$$

According to estimation of t distribution, t = 3, the uncertainty of the error $E = t\sigma/\sqrt{n} = 0.026602$, finally the error of data is 0.0129 ± 0.026602 mm. The national standard GB/T 10095.1-2001 specified tooth profile deviation of this precision gear should be less than 0.048 mm, the results accord with the national standard requirements.

5 CONCLUSIONS

Measurement of gear tooth profile error using the coordinate method, through coordinate transformation realized the unity of the workpiece coordinate system and measurement coordinate system, and established the mathematical model of standard involute tooth profile. Applied LabVIEW software to develop large gear tooth profile measuring system software, the entire system adopt modular design, may carry on the tooth profile of automatic measurement, data acquisition, error evaluation, and the measurement result display and other functions. Through the experiment, the results show that the measured tooth profile curve are well with theoretical calculation involute tooth profile curve, tooth profile error accords with national standard, verifies the correctness of measurement principle and the feasibility of related method and technology, and provides a theoretical basis for the improvement and perfection of the subsequent detection system.

REFERENCES

[1] W.L. Li. 2007. Integrated Measuring Technology on Tooth Profile Error and Longitudinal Form Error of Large Gears, J. Tool Engineering 41 (4).

[2] H.K. Xie. 2004. Development of gear measuring technology and instrument in recent years. Tool Engineering 24 (9).

[3] Y.L. Wang. 2002. Research on involute of gear tooth profile practical rectangular coordinate equation. Journal of Mechanical Transmission 26 (3).

[4] Q. Tan & X.S. Zeng. 2009. Precise Solution and Parametric Modeling on Tooth Profile of Involute Gear. Machine Tool and Hydraulics 37 (9).

[5] J.X. Liu. 2013. LabVIEW 2012 Chinese version of virtual instrument from entry to the master, China Machine Press, Beijing.

[6] C. Wang. 2013. LabVIEW programming and cases analysis, Beihang University press, Beijing.

Material Science and Environmental Engineering – Chen (Ed.)
 ISBN 978-1-138-02938-5

Analyzing the improved model of the Bourdon tube using a uniform strength cantilever beam FBG pressure sensor

J. Shao
School of Mechanical Engineering, Xi'an Shiyou University, Xi'an, China

Z.A. Jia
Key Laboratory of Optical Fiber Sensing, School of Science, Xi'an Shiyou University, Xi'an, China

J.H. Liu
School of Electrical Engineering, Xi'an Jiaotong University, Xi'an, China

X.X. Bai
School of Mechanical Engineering, Xi'an Shiyou University, Xi'an, China

ABSTRACT: Because of the limit of the existing theoretical model of the Bourdon tube with uniform strength cantilever beam Fiber Bragg Grating (FBG) pressure sensor, the new theoretical model was built with taking the effect of the displacement of the beam into account in the paper. We assumed that the beam and the Bourdon tube were connected in parallel and they had the same displacement, then we got the new theoretical model. Based on the new model, the relationship between the pressure sensitivity and structural parameters were analyzed. The new model shows that there are the optimal structure parameters for getting the maximal pressure sensitivity. Meanwhile, Comparing with experimental results, the new theoretical models are more precise than the existing theoretical models. The new model had solved the problem of the existing model. It can be used to design elastic element and optimize parameters. This work can provide scientific basis to choose elastic element conveniently and properly.

Keywords: fiber grating; pressure sensor; Bourdon tube with uniform strength cantilever beam; theoretical modeling

1 INTRODUCTION

As sensing element, fiber Bragg gratings have the advantage of light, small, immune to electromagnetic interference, and intrinsic flame proof. It adapted to be used in the foul and hazardous environment, such as oil and gas pipelines [1], stations and warehouses for transportation and storage of oil and gas [2], storage of hazardous gas [3, 4].

The Bourdon tube is widely used in pressure instrument [5]. The sensitivity of the optical Fiber grating sensor based on the Bourdon tube is 2.2×10^{-5} MPa^{-1} [6]. If combining the Bourdon tube and the cantilever beam [7], it could reach 1.8×10^{-4} MPa^{-1}. In order to avoid chirp, the beam would be made as uniform strength beam [8, 9].

In the literatures available, the goal, that the stiffness of the uniform strength beam is far less than that of the Bourdon tube for ignoring the effect of the uniform strength beam to the displacement of the Bourdon tube, can be getting through choosing different materials [1, 7, 10]. However, the stiffness of elastic element not only depends on material, but also structure. Therefore, the old model has limits and it is difficult for optimizing the design of sensors.

In this paper, the new model of the Bourdon tube with uniform strength cantilever beam FBG sensor will be built. This model can be used for designing elastic element and the more important is for optimizing the structure of elastic element.

2 THEORETICAL MODEL

The fiber Bragg grating pressure sensor composed of the Bourdon tube elastic element and the uniform strength cantilever beam with fiber Bragg grating, composition block diagram shown in Figure 1 and the schematic diagram shown in Figure 2. The free end of the beam is linked with the free end of the Bourdon tube. The other end of the beam is fixed. The grating is glued near the fixed end of the beam. Under the pressure p, the free end will move w and the strain of the fiber grating changed to ε, that makes the wavelength of fiber grating shifted $\Delta\lambda_B$.

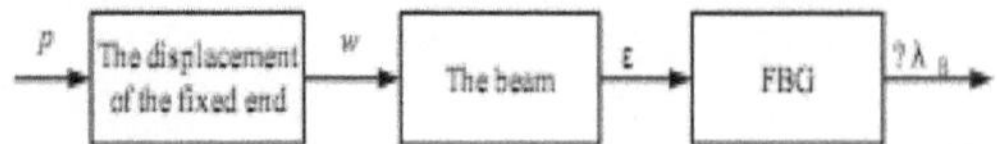

Figure 1. The block diagram of FBG Bourdon tube pressure sensor.

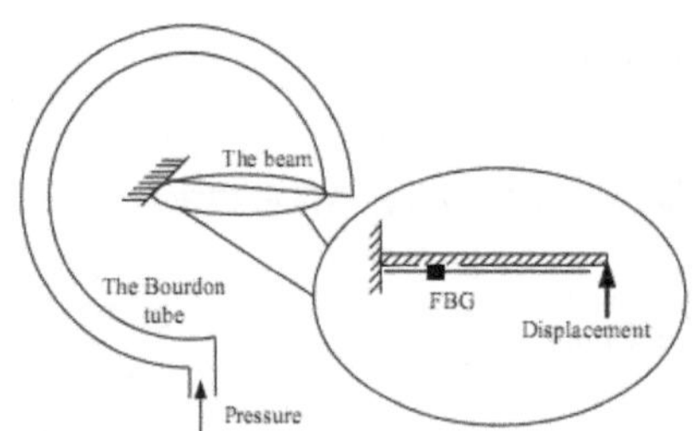

Figure 2. Schematic diagram of FBG Bourdon tube pressure sensor.

There are three steps to build the theoretical model. They are: ① the relationship between pressure p and the displacement of the free end w; ② the displacement w and the strain ε of the beam; ③ ε and $\Delta\lambda_B$.

2.1 *The relationship between pressure p and the displacement of the free end w*

There are also three steps for the relationship between pressure p and the displacement of the free end w, as shown in Figure 3. These steps are ① the relationship between the traction force F_q and pressure p; ② the stiffness of the bourdon tube; ③ the stiffness of the beam.

Based on the steps and referred to the article [5], the relation between the input pressure p and the displacement w is

$$p = \frac{k_t + k_{lt}}{\sqrt{k_q^2 + k_j^2}} w \tag{1}$$

where k_q is tangential coefficient of the Bourdon tube, k_j is radial coefficient of the Bourdon tube, k_{tt} is stiffness of the Bourdon tube, $k_{tt} = \sqrt{k_q^2+k_j^2}/k_1$, k_{lt} is stiffness of the beam, k_1 is a Coefficient of the Bourdon tube. All the parameters above are concerned with the structure of the Bourdon tube and the beam.

2.2 *The relationship between the displacement of the free end w and the strain ε of the beam*

Under force F, the relationship between the displacement of the free end w and the strain ε of the beam is

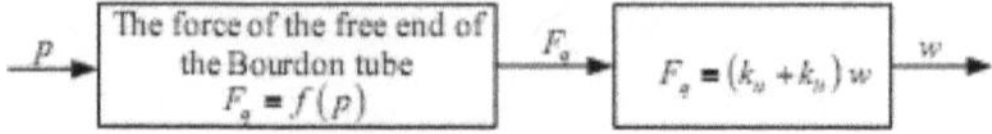

Figure 3. The block diagram of the pressure and the displacement of free end of the beam.

$$\varepsilon = \frac{h_l}{L_l^2} w \tag{2}$$

where h_l is thickness of the beam, L_l is length of the beam.

2.3 *The relationship between the strain ε of the beam and the shift $\Delta\lambda_B$ of wavelength of FBG*

With the coupled-mode theory of FBG, the relationship between the wavelength shift $\Delta\lambda_B$ of FBG and the strain can be expressed as [11]

$$\Delta\lambda_B = \lambda_{B0}(1 - p_e)\varepsilon \tag{3}$$

where, $\Delta\lambda_B$ is the wavelength shift of FBG, $\Delta\lambda_B = \lambda_B - \lambda_{B0}$; λ_{B0} is the free wavelength of the FBG at ambient temperature, p_e is the photoelastic coefficient of this optical fiber, ε is the strain of the FBG.

Substituting Eq. (1) and Eq. (2) into Eq. (3), the static model of the sensor can be expressed as

$$\begin{aligned}\frac{\Delta\lambda_B}{\lambda_{B0}} &= (1 - p_e)\cdot\frac{\sqrt{k_q^2 + k_j^2}}{k_{tt} + k_{lt}} k_2 p \\ &= k_t p\end{aligned} \tag{4}$$

where k_t is the sensitivity of the pressure sensor, it can be expressed as

$$k_t = (1 - p_e)\cdot\frac{\sqrt{k_q^2 + k_j^2}}{k_{tt} + k_{lt}}\cdot k_2 \tag{5}$$

If neglected the effect of the beam to the Bourdon tube, then the model is

$$\frac{\Delta\lambda_B}{\lambda_{B0}} = (1 - p_e)\cdot k_1 k_2 p \tag{6}$$

Eq. (6) is the old model in the existing articles [7–9].

3 RESULTS AND DISCUSSION

Based on the new model of this paper Eq. (4), taken a Bourdon tube as example, the relationship between thickness h_l, length L_l of the beam and the sensitivity k_t will be discussed.

3.1 *The relationship between the structure of the beam and the sensitivity*

The elastic modulus of the Bourdon tube E_t is 115 GPa and that of the beam E_l is 135 GPa and the relationship between thickness h_l or length L_l and the sensitivity k_t is shown in Figure 4.

From Figure 4, we can see that (1) given a fixed length, there is the best thickness for the maximum sensitivity. And this is the most important for the new model. (2) Given different lengths and thicknesses, there are sensitivity contours, as shown in Figure 5.

3.2 *Discussion about the pressure sensitivity*

Comparing the two theoretical models, we can get that the new model is closer to the experimental result, as shown on Table 1.

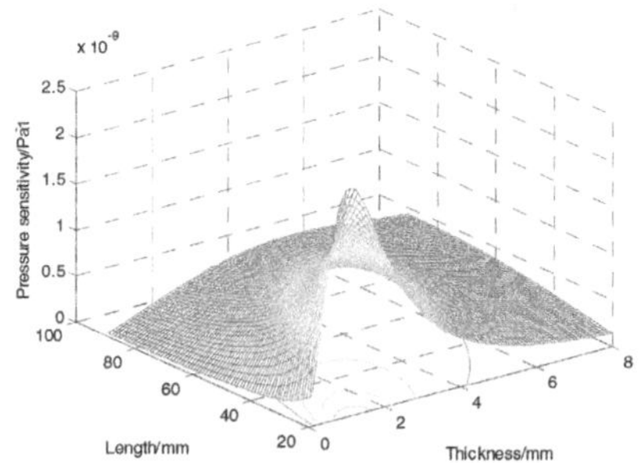

Figure 4. The relationship between the pressure sensitivity and length or thickness of the beam.

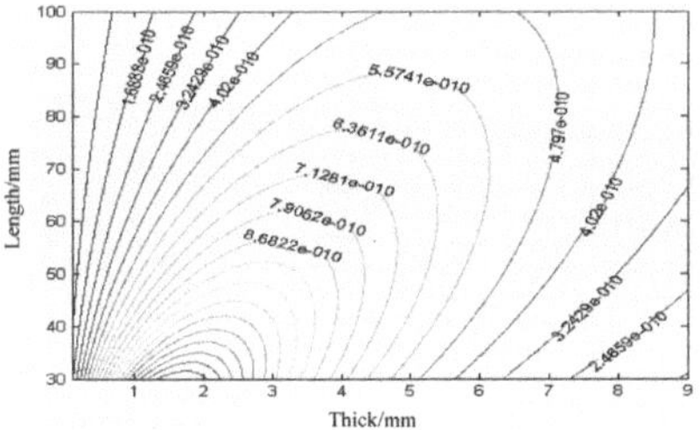

Figure 5. Pressure sensitivity contours.

Table 1. Comparing theoretical models.

Parameters	New model	Existing model
Modulus of elasticity E_l/GPa	135	
Length L_l/mm	75.54	
Thick h_l/mm	1.66	
Theoretical pressure sensitivity/Pa^{-1}	3.85×10^{-10}	3.95×10^{-10}
Experimental pressure sensitivity/Pa^{-1}	2.80×10^{-10} [9]	

However, the theoretical pressure sensitivity 3.85×10^{-4} MPa^{-1} is a little farther than the experimental pressure sensitivity. There are two reasons for this phenomenon 2.80×10^{-4} MPa^{-1}: one is that the strain is different as the different attachment process and the other is not the best setting angle when the beam and the Bourdon tube.

4 CONCLUSION

The static model of the Bourdon tube with uniform strength beam FBG sensor is built in this paper. Based on this model, the relationship between the structure of the beam and the pressure sensitivity is analyzed. Comparing with the old model, the new model can reflect the relation among the parameters more comprehensively than the old one and give the better basis for designing the elastic element.

ACKNOWLEDGMENTS

This work is financially supported by the National Natural Science Foundation of China (No. 61240028), the National Natural Science Fund of Shaanxi Province (No. 2011JM8028), the Scientific Research Plan Projects of Shaanxi Education Department (No. 11JK0952), the Scientific Research Foundation for the Doctoral Program of Xi'an Shiyou University (No. 2011BS001), and Students Research Training Program of Xi'an Shiyou University (No. 1196).

REFERENCES

[1] Wang Hongliang, Qiao Xueguang, FU Haiwei, et al. FBG sensor of measuring pressure with lag of stress active compensation and with temperature active compensation [J]. Journal of Optoelectronics ·Laser. 2008, 19(1): 1–5. (In Chinese).

[2] Fang Laihua, WU Zongzhi, Liu Ji, et al. Development of Safety Monitoring and Management Information System for Oil gas Storage and Transportation Station [J]. China Safety Science Journal. 2008, 18(2): 111–118. (In Chinese).

[3] Yin Caihong, Sun Jinhua. Fire and explosion hazard analysis of the principal process in methanol production from natural gas [J]. Journal of Safety and Environment. 2012, 12(5): 169–173. (In Chinese).

[4] Wang Feiyue, Xu Zhisheng. Risk assessment on the store system of inflammable and easy explosion toxic gas [J]. Industrial Safety and Environmental Protection. 2002, 28(8): 33–36. (In Chinese).

[5] Pang zhenji, Zhang biguang. Instrument parts and structure [M]. Tianjin: Tianjing University Press, 1990. (In Chinese).

[6] Zhou Hong, Qiao Xueguang, Wang Hongliang, et al. Measurement of pressure and displacement by using the optical Fiber grating sensor based on the C-shaped elastic tube [J]. Journal of Xidian University. 2005, 32(1): 142–145. (In Chinese).
[7] Liu Yunqi, Liu Zhiguo, et al. Pressure and temperature properties of fiber grating bourdon tube pressure sensor [J]. Acta Photonica Sinica, 1998, 27 (12): 111121114. (In Chinese).
[8] YU Youlong, Liu Zhiguo, Dong Xiaoyi, et al. No-chirped linearly tuning technique for fiber Bragg grating with a cantilever beam [J]. Acta Optica Sinica, 1999, 19 (7):8732876. (In Chinese).
[9] Shao Jun, Liu Junhua, Qiao Xueguang, et.al. A FBG Pressure Sensor Based on Bourdon and Cantilever Beam of Uniform Strength [J]. Journal of Optoelectronics · Laser. 2006, 17(7): 807–809. (In Chinese).
[10] Yang Yang, Development of edge interrogation technology-based FBG Bourdon tube pressure sensor. [J]. Optical Technique. 2009, 35(1): 53–56. (In Chinese).
[11] LIAO Yanbiao. Fiber Optics [M]. Beijing: Tsinghua University Press, 2000. 196–202. (In Chinese).

Material Science and Environmental Engineering – Chen (Ed.)
© 2016 Taylor & Francis Group, London, ISBN 978-1-138-02938-5

Chaos synchronization of two Light-Emitting Diode systems with complex dynamics via adaptive H-infinity control

Y. Che, X. Cui, B. Liu & C. Han
Tianjin Key Laboratory of Information Sensing and Intelligent Control, School of Automation and Electrical Engineering, Tianjin University of Technology and Education, Tianjin, China

M. Lu
School of Information Technology and Engineering, Tianjin University of Technology and Education, Tianjin, China

ABSTRACT: A Light-emitting diode with ac-coupled nonlinear optoelectronic feedback can exhibit complex sequences of periodic oscillations and chaotic spiking. In this paper, we proposed an adaptive H-infinity control to achieve the chaos synchronization of two light-emitting diode systems. The unknown nonlinear part in the synchronization error dynamics is first approximated by radial basis function neural networks, then the effects of the approximate errors and disturbances are attenuated by H-infinity control. The simulation results demonstrate the efficiency of the proposed method.

Keywords: chaos synchronization; Light-Emitting Diode; complex dynamics; adaptive H-infinity control

1 INTRODUCTION

Chaos is a universal phenomenon in non-linear system, which may be responsible for many regular regimes of operation (Rabinovich 1998). Meanwhile, synchronous oscillations underlie many critical processes (Gray et al. 1989), this makes synchronization an attractive field which has drawn a lot of attentions. The synchronization of chaotic dynamical systems has been studied widely for the last few years since Pecora and Carroll's seminal work (Pecora & Carroll 1990). The potential of synchronized coupled chaotic lasers was soon recognized and first experimental demonstrations appeared for CO_2 and solid-state lasers (Roy & Thornburg 2009, Sugawara et al. 1994); and later for semiconductor lasers with optical feedback (Sivaprakasam & Shore 1999). Meanwhile, many control methods for chaos synchronization have been developed (Lian et al. 2002).

Recently, a GaAs Light-Emitting Diode (LED) with ac-coupled nonlinear optoelectronic feedback has been shown to exhibit complex dynamics including mixed mode oscillations and chaos (Marino et al. 2011). In this paper, we study the synchronization of two light-emitting diode systems with complex dynamics via adaptive H-infinity control, which is robust to the system uncertainty sources such as unmodelled dynamics. Because of the ability to approximate uniformly continuous functions to arbitrary accuracy (Sanner & Slotine 1992), we first use the Radial Basis Function Neural Network (RBFNN) to deal with the uncertainty and derive the update laws based on the Lyapunov stability theory. Then we adopt H-infinity control method to attenuate the effects caused by approximate errors and disturbances. Hence, the controller can not only ensure closed-loop stability, but also achieve an H-infinity tracking performance (Chen et al. 1996) for the synchronization error dynamics of two chaotic light-emitting diode systems.

2 LED SYSTEM MODEL AND ITS COMPLEX DYNAMICS

For numerical and analytical purposes, the LED system dynamics is written in dimensionless form (see (Marino et al. 2011) for details):

$$\begin{aligned} \dot{x}_1 &= x_1(x_2 - 1) \\ \dot{x}_2 &= \gamma(\delta_0 - x_2 + \alpha(x_3 + x_1)/(1 + s(x_3 + x_1)) - x_1 x_2) \\ \dot{x}_3 &= -\varepsilon(x_3 + x_1) \end{aligned} \tag{1}$$

where δ_0, γ, ε, α and s are system parameters.

Figure 1 shows a detailed bifurcation diagram computed from Eq. (1) by varying δ_0 over a small interval contiguous to the initial Hopf bifurcation. The system passes through a cascade of period doubled and chaotic attractors. A period doubling route to chaos occurs.

The (a, c) panels in Figure 2 show some typical patterns with different values of δ_0 obtained by numerical integration of Eq. (1). The (b, d) panels are the corresponding phase portraits in $x_1 - x_2 - x_3$. We observe chaotic spiking with $\delta_0 = 1.15$ and periodic oscillation with $\delta_0 = 1.1$.

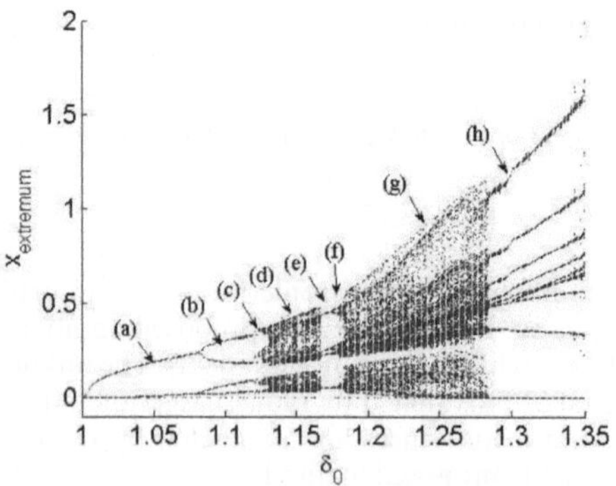

Figure 1. Bifurcation diagram for the peak values of x_1 as the parameter δ_0 is varied. Other parameters are $s = 0.2$, $\alpha = 1.002$, $\gamma = 3.3 \times 10^{-3}$, and $\varepsilon = 4 \times 10^{-5}$.

3 CHAOS SYNCHRONIZATION OF LED SYSTEMS VIA H-INFINITY CONTROL

Consider a pair of master-slave unidirectional coupled chaotic LED system as following:

$$\begin{aligned}\dot{x}_m &= Ax_m + f(x_m) + d_m \\ \dot{x}_s &= Ax_s + f(x_s) + u(t) + d_s\end{aligned} \tag{2}$$

where $x_m(t) = (x_{1m}, x_{2m}, x_{3m})^T$, $x_s(t) = (x_{1s}, x_{2s}, x_{3s})^T$ are state vectors of the system, $A \in R^{n\times n}$ is a constant parameter matrix for the linear part of the system, and $u(t) = (u_1, u_2, u_3)^T$ is the control input. d_m and d_s are added to simulate disturbances in the system. The subscripts m and s stand for the master and the slave systems, respectively.

Assumption1. Assume that A is known and the states x_m and x_s are measurable.

Define the error vector as $e = x_s - x_m$, the dynamics of synchronization error can be expressed as:

$$\dot{e} = Ae + F(x_m, x_s) + d + u(t) \tag{3}$$

where $F(x_m, x_s) = f(x_s) - f(x_m)$ is the nonlinear part of the error dynamics.

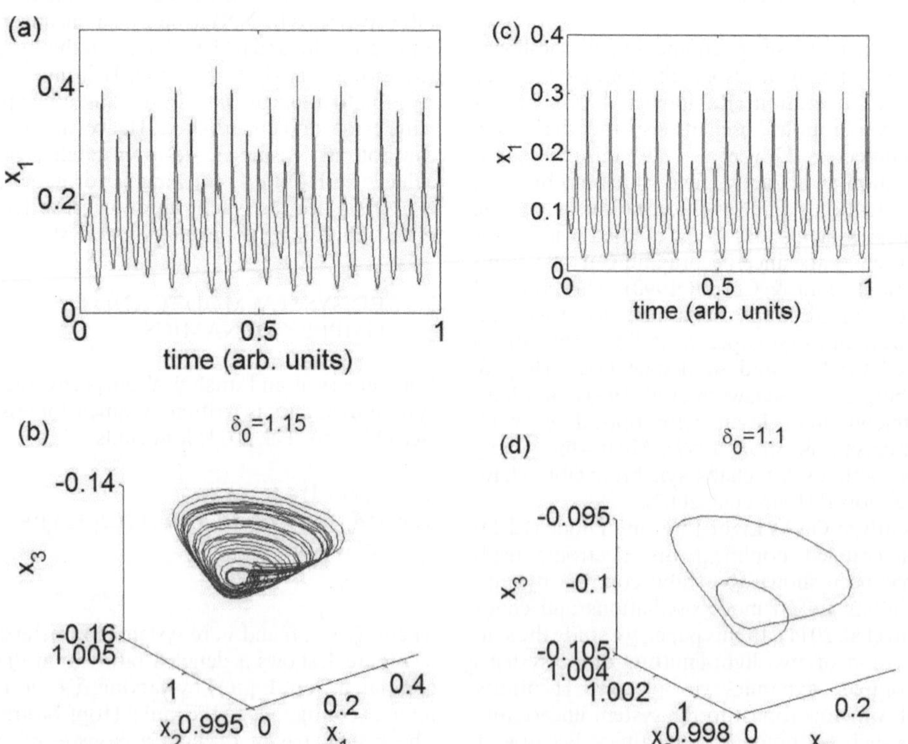

Figure 2. Responses of the LED system with different values of parameter δ_0: (a, c) Time series of the variable x_1; (b, d) the corresponding phase portraits in $x_1 - x_2 - x_3$.

The problem of synchronization between the master and the slave now can be transformed into a problem of how to design a controller $u(t)$ to realize the stabilization of the error system (3) at the origin.

If $F(x_m, x_s)$ is exactly known, one can design the ideal active control function $u^*(t)$ as:

$$u^*(t) = -Be - F(x_m, x_s) - d \tag{4}$$

where $B \in R^{n\times n}$ is a constant feedback gain matrix.

Then the error dynamical system (3) can be rewritten as $\dot{e} = -Qe$ with $Q = B - A$ and $Q \in R^{n\times n}$.

Under *Assumption 1*, by properly selecting the feedback gain matrix B such that the eigenvalues of the matrix Q is positive definite, one can make the error dynamical system $\dot{e} = -Qe$ globally asymptotically stable at the origin, thus implying that the systems (2) is globally asymptotically synchronized.

However, the nonlinear function $F(x_m, x_s)$ and the disturbance d can't be always available in real system. Thus the ideal controller (4) can't be abtained. In this paper, we use RBFNNs to approximate the nonlinear function $F(x_m, x_s)$ in the following form:

$$F(x_s, x_m) = \Phi(x_s, x_m)\Theta^* + \varepsilon \tag{5}$$

Then the error dynamics Equation (3) becomes:

$$\dot{e} = Ae + \Phi(x_s, x_m)\Theta^* + \omega + u(t) \tag{6}$$

where $\omega = \varepsilon + d$ is the lumped uncertainty.

Assumption 2. The lumped uncertainty is assumed such that $\omega \in L[0,T], \forall T \in [0,\infty]$.

Then, we obtain a controller in Theorem 1 below which guarantees H-infinity tracking performance [10] for the overall system with uncertain non-linear functions.

Theorem 1. Consider the synchronization error dynamics (3) with uncertain non-linear function $F(x_m, x_s)$, which is approximated as Eq. (5). Suppose Assumptions 1 and 2 are satisfied and choose the control input as $u(t) = -Be - \Phi(x_s, x_m)\hat{\Theta} - e/2\rho^2$. Choose the adaptive law of $\hat{\Theta}$ as $\dot{\hat{\Theta}} = \Gamma\Phi^T e$ where Γ is a constant matrix. Then the H-infinity tracking performance (Chen et al. 1996) for the overall system satisfies the following relationship:

$$\frac{1}{2}\int_0^T e^T Qe dt \le \frac{1}{2}e^T(0)e(0) + \frac{1}{2}\tilde{\Theta}^T(0)\Gamma^{-1}\tilde{\Theta}(0) + \frac{1}{2}\rho^2\int_0^T \omega^2 dt \tag{7}$$

Proof. Consider the following Lyapunov function candicate $2V = e^T e + \tilde{\Theta}^T\Gamma^{-1}\tilde{\Theta}$. Differentiating it with respect to time, one can obtain

$$\begin{aligned}\dot{V} &= e^T\dot{e} + \tilde{\Theta}^T\Gamma^{-1}\dot{\hat{\Theta}} = -e^TQe - \frac{e^2}{2\rho} + e\omega \\ &\le -e^TQe + \frac{1}{2}\rho^2\omega^2 - \frac{1}{2}\left(\frac{e}{\rho} - \rho\omega\right)^2 \\ &\le -e^TQe + \frac{1}{2}\rho^2\omega^2\end{aligned} \tag{8}$$

By assumption 2, and integrating both sides of (8) from time $t = 0$ to $t = T$, we obtain

$$V(T) - V(0) \le -\frac{1}{2}\int_0^T e^TQedt + \frac{1}{2}\rho^2\int_0^T \omega^2 dt \tag{9}$$

Since $V(T) \ge 0$, then

$$\begin{aligned}\frac{1}{2}\int_0^T e^TQedt &\le V(0) + \frac{1}{2}\rho^2\int_0^T \omega^2 dt \\ &\le \frac{1}{2}e^T(0)e(0) + \frac{1}{2\beta}\tilde{\Theta}^T(0)\Gamma^{-1}\tilde{\Theta}^T(0) \\ &+ \frac{1}{2}\rho^2\int_0^T \omega^2 dt\end{aligned} \tag{10}$$

This complete the proof.

Thus the H-infinity control performance is achieved for a prescribed attenuation level ρ, and the synchronization of coupled system (2) can be obtained.

4 SIMULATION RESULTS

In this section, numerical simulations are showed out for the synchronization of the uniderectional coupled LED systems via the proposed H-infinity control.

The control parameters are chosen as $\boldsymbol{B} = diag\{1,1,1\}$, $\beta = 0.005$. We switch on the controller at time $t = 1$. Then the simulation results are shown in Figure 3. In *Case 1*, We first illustrate the synchronization behaviors of two identical chaotic LED systems with different initial conditions as shown in Figure 3a and b. In *Case 2*, we control the periodic LED to follow a chaotic LED as shown in Figure 3c and d. Before the control is implemented, the master and slave LED systems exhibit their own original complex dynamical behaviors. After the controller is applied, the synchronization error converges to a very small neighborhood of the origin point and chaos synchronization is obtained.

5 SUMMARY

In this paper, chaos synchronization of two light-emitting diode systems with complex dynamics via H-infinity control has been studied. We use

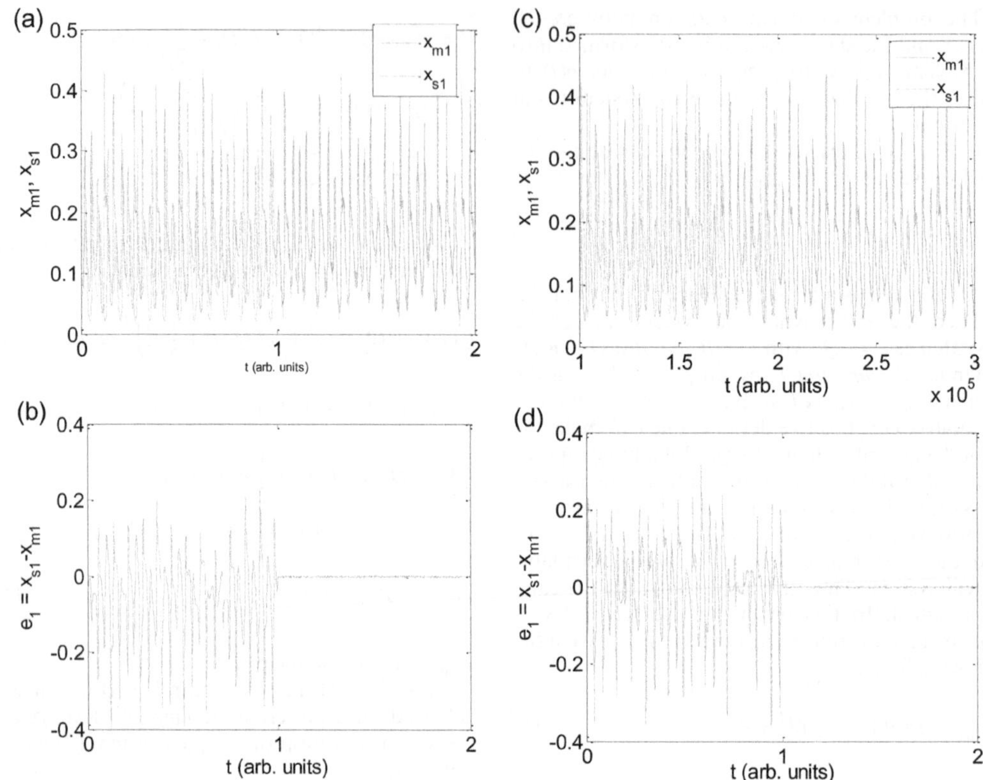

Figure 3. Responses of waveforms of x_{m1}, x_{s1} (a, c) and the corresponding errors $e_1 = x_{s1} - x_{m1}$ (b, d) for two cases: (a, b) Case 1 and (c, d) Case 2, whereas the control signal is switched on at time $t = 1$.

RBFNN to approximate the uncertain non-linear part of synchronization error system, and H-infinity control to eliminate the noise or disturbances. On the basis of Lyapunov stability theorem, stable synchronization between the master and slave LED can be ensured. The simulation results demonstrate the effectiveness of the method.

ACKNOWLEDGMENTS

This work is supported by The National Natural Science Foundation of China (Grants No. 61178081, and No. 61104032).

REFERENCES

[1] Rabinovich, M.I. & Abarbanel, H.D.I. 1998. The role of chaos in neural systems. Neuroscience 87(1): 5–14.

[2] Gray, C.M., König, P., Engel, A.K., et al. 1989. Oscillatory responses in cat visual cortex exhibit inter-columnar synchronization which reflects global stimulus properties. Nature 338(6213): 334–337.

[3] Pecora, L.M., & Carroll, T.L. 1990. Synchronization in chaotic systems. Physical review letters 64(8): 821.

[4] Roy, R., & Thornburg Jr, K.S. 1994. Experimental synchronization of chaotic lasers. Physical Review Letters 72(13): 2009.

[5] Sauer, M., & Kaiser, F. 1998. On-off intermittency and bubbling in the synchronization break-down of coupled lasers. Physics Letters A 243(1): 38–46.

[6] Sivaprakasam, S., & Shore, K.A. 1999. Signal masking for chaotic optical communication using external-cavity diode lasers. Optics letters 24(17): 1200–1202.

[7] Lian, K.Y., Liu, P., Chiang, T.S., & Chiu, C.S. 2002. Adaptive synchronization design for chaotic systems via a scalar driving signal. IEEE Trans. Circ. Syst. I. 49: 17–25.

[8] Marino, F., Ciszak, M., Abdalah, S.F., Al-Naimee, K., Meucci, R. & Arecchi, F.T. 2011. Mixed mode oscillations via canard explosions in light-emitting diodes with optoelectronic feedback. Phys. Rev. E 84: 047201(1–5).

[9] Sanner, R.M., & Slotine, J.J. 1992. Gaussian networks for direct adaptive control. Neural Networks, IEEE Transactions on 3(6): 837–863.

[10] Chen, B.S., Lee, C.H. & Chang, Y.C. 1996. H-infinity tracking design of uncertain nonlinear SISO systems: Adaptive fuzzy approach, IEEE Trans. Fuzzy Syst., 4: 32–43.

Material Science and Environmental Engineering – Chen (Ed.)
 ISBN 978-1-138-02938-5

Evaluation of low measurement uncertainty for efficiency of three-phase induction motor

F. Li
China Electric Power Research Institute, Beijing, China

X.B. Yang
Yunnan Province Power Electric Design Institute, Kunming, China

M.J. Shi, Y.X. Zhang & J. Li
China Electric Power Research Institute, Beijing, China

ABSTRACT: Uncertainty established on error theory is of quantitative, and a parameter of measurement. With wide promotion and application of high efficiency and ultra-efficiency motor, the measurement method and test system accuracy for motor efficiency has been enhanced. More and more researchers dedicate to uncertainty theory for measurement of motor efficiency. In this paper, the analysis and calculation on various losses of internal motor along with influent factors for uncertainty has been made based on B method evaluation. Expanded uncertainty for efficiency measurement is proposed, the evaluation method will provide vital reference value for uncertainty theory research of efficiency measurement of induction motor.

Keywords: three-phase induction motor; measurement; uncertainty; loss; efficiency

1 INTRODUCTION

Uncertainty established on error theory is quantitative, and a parameter of measurement. In the early 1970s, more and more international measurement scholars prefer to use uncertainty, rather than error, and the term of uncertainty is widely used in measurement filed.

In 2012, China promulgated a new version of Evaluation and Expression of Uncertainty in Measurement. The uncertainty measurement has become one of the essential contents in laboratory [1]. Meanwhile, a new version of the Testing Methods of Three-phase Induction Motor was issued. Compared with the version of 2005, new testing methods and uncertainty measurement concerning efficiency of induction motor has been put forward, including B method evaluation, which brings out the lowest uncertainty. Under permit tent condition, B method evaluation is recommended for efficiency measurement of induction motor [2]. In addition, in recent years, measurement uncertainty for efficiency of three-phase induction motor (especially IE2/IE3 motor) is increasingly getting attention in corresponding industries.

2 MEASUREMENT UNCERTAINTY AND STANDARD UNCERTAINTY

According to the definition of JJF 1001-2011, *General Measurement Terms and Definitions*, uncertainty, a non-negative parameter, is characteristics of the dispersion of the testing parameter value. An integrated testing value should include measurement uncertainty represented by the standard deviation, which is the standard uncertainty. Under specified testing condition, the method that measurement uncertainty obtained by statistical analysis is a method evaluation, and the uncertainty can be represented with u_A; the measurement uncertainty obtained by the different from A method evaluation is evaluation of B type, which can be represented with u_B. B method evaluation is evaluated by testing and other relative information, which comprises subjective identification component [3].

The standard uncertainty of each input parameter in one testing model can be composed, and represents with u_C, which is called composite standard uncertainty. Composite standard uncertainty is still of standard deviation, and be characteristic of the dispersion of measurement results.

An expanded uncertainty U is equivalent to coverage coefficient k multiply composite standard uncertainty u_C, namely, $U = ku_C$. In most cases, $k = 2$, thus the confidence interval of U is approximately half width of 95% [1].

3 MATHEMATICAL MODEL OF EFFICIENCY MEASUREMENT FOR MOTOR

According to the formula of efficiency

$$\eta = \frac{P_2}{P_1} \times 100\% \quad (1)$$

In formula:
P_1—input power
P_2—output power
$P_2 = P_1 - P_T$
$P_T = P_{fw} + P_{Fe} + P_s + P_{cu1s} + P_{cu2s}$

Therefore, the formula of efficiency can be expressed as follows:

$$\eta = \frac{1}{P_1}\left(P_1 - P_{cu1s} - P_{Fe} - P_{fw} - P_{cu2s} - P_s\right) \times 100\% \quad (2)$$

As can be seen above the equation, there exists 6 input items, namely, P_1, P_{cu1}, P_{Fe}, P_{fw}, $P_{cu2\,s}$, P_s. The effect on measurement uncertainty which may bring out by other influent factors, such as equipments, researchers, and environment, should be analyzed in the measurement progress of P_{cu1}, P_{Fe}, P_{fw}, $P_{cu2\,s}$, P_s. The factors which affect uncertainty of loss can be evaluated with B method evaluation.

According to analysis result, impact on uncertainty caused by thermal terminal resistance, wind abrasion, iron loss, and stray losses, which are obtained through least square method, can be ignored.

4 REPETITION WITH A METHOD EVALUATION FOR MEASUREMENT UNCERTAINTY

Under the same measurement conditions, i.e., the same measurement procedure, operator, operating conditions, and the same locations, 6 times efficiency testing have been made respectively on the same motor, and the calculation result is shown in Table 1.

According to the Bessel equation, the standard deviation of the measured values of efficiency as follows:

$$s(\eta) = \sqrt{\frac{1}{n-1}\sum_{i=1}^{n}(\eta_i - \bar{\eta})^2} = 0.14\% \quad (3)$$

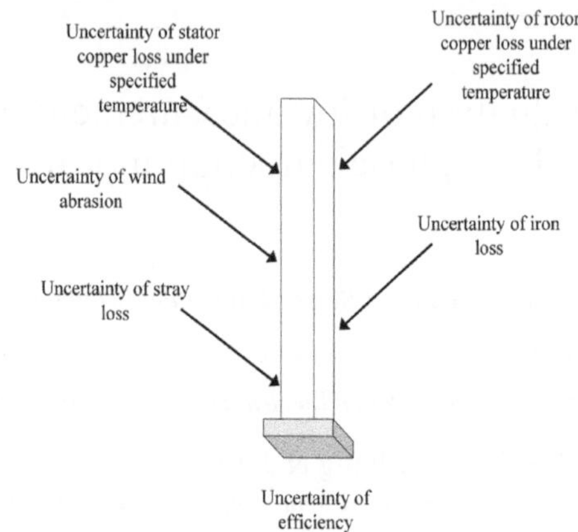

Figure 1. Fish-bone diagram of influence factors of efficiency uncertainty.

Table 1. The calculation result of 6 times testing.

	η (%)
1	92.54
2	92.53
3	92.87
4	92.67
5	92.82
6	92.68
Avg.	92.69

The uncertainty through repetition with A method evaluation is [4]

$$u_A(\eta) = \frac{1}{\sqrt{n}} s(\eta) \cdot 100\% = 0.057\% \quad (4)$$

5 B METHOD EVALUATION FOR MEASUREMENT UNCERTAINTY

5.1 *Uncertainty of each direct measurement parameter*

5.1.1 *Temperature logging device*

Temperature θ: when $U = 0.02$ °C, $k = 2$, and the measurement temperature is 120°C, the error is $\Delta\theta = 0.42$°C, thus the composite uncertainty is $u(\theta) = 0.2427$°C.

5.1.2 *Torque and rotate speed sensor*

Rotate speed n: the relative uncertainty is 0.005 r/min, when measurement speed is 2000 r/min, its measurement error is $\Delta n = 0.1$ r/min, and the composite uncertainty is $u(n) = 0.0578$ r/min.

Torque T: when measurement torque is 400 N·m, and the reading deviation of indicator is $\Delta T = 0.7$ N·m, thus the composite uncertainty is $u(T) = 0.4041$ N·m.

5.1.3 *Power analyzer*

1. Current uncertainty
   Measurement uncertainty of current I from the calibration certificate of the power meter: $U_{rel} = 0.03\%$, $k = 2$; If automatic measurement range is used for current that the expectation value is 30 A, the error is $\Delta I_1 = 0.0087$ A and the uncertainty should be $u(I_1) = 0.0068$ A.
   Measurement uncertainty of voltage current I from current transformer: the expectation value is 110 A, and the error is $\Delta I_2 = 0.032$ A, thus the uncertainty is $u(I_2) = 0.01848$ A.
   Eventually, the composite uncertainty of current is $u(I) = 0.0197$ A.
2. Voltage uncertainty
   Measurement uncertainty of voltage U from the calibration certificate of the power meter: $U_{rel} = 0.01\%$, $k = 2$; If automatic measurement range is used for voltage that the expectation value is 380 V, the error is $\Delta U = 0.0257$ V, and the composite uncertainty should be $u(U) = 0.0241$ V.
3. Power uncertainty
   Measurement uncertainty of AC active power from the calibration certificate of the power meter: $U_{rel} = 0.03\%$, $k = 2$; If automatic measurement range is used for active power that the expectation value is 34200 W, the error is $\Delta P = 10.6$ W, and the composite uncertainty should be $u(P) = 7.9856$ W.

5.1.4 *Resistance uncertainty*

Measurement uncertainty of resistance R from the calibration certificate of the resistance meter: when 200 mΩ measurement range is used for resistance that the expectation value is 100 mΩ, the error is $\Delta R = 0.009$ mΩ and the composite uncertainty should be $u(R) = 5.2915 \times 10^{-6}$.

5.2 *Uncertainty evaluation of loss*

When y is determined by linear measurement function of $x_1, x_2 \ldots x_N$, the estimated value of y can be expressed as follow:

$$y = f(x_1, x_2, \ldots, x_N) \tag{5}$$

When the correlation coefficient of each input items is zero, the composite uncertainty of y can be represented with $u_c(y)$.

$$u_c^2(y) = \sum_{i=1}^{N}\left[\frac{\partial f}{\partial x_i}\right]^2 u^2(x_i) \tag{6}$$

5.2.1 *Uncertainty of stator copper loss under specified temperature*

Uncertainty of stator copper loss under Specified temperature is shown in Figure 2.

1. Uncertainty of initial cold terminal resistance of stator winding
   Cold stator winding resistance is calculated as follows:

$$R_1 = \left[R_{UV} + R_{VW} + R_{WU}\right]/3 \tag{7}$$

In the progress of testing, take cold resistance of winding as $R_C = R_1$, cold temperature of winding $\theta_c = \theta_1$, θ_1 should be the maximum of three test points. Thus, the square of uncertainty for stator cold resistance and wind temperature are as follows:

$$u(R_1) = 3.05 \times 10^{-6}\,\Omega \tag{8}$$

$$u(\theta_1) = 0.2427°\mathrm{C} \tag{9}$$

2. Uncertainty of thermal terminal resistance
   Due to R_w is obtained by fitting with least square method; the uncertainty of curve fitting can be ignored. Therefore

$$U(R_w) = U(R) = 5.2915 \times 10^{-6}\ \Omega \tag{10}$$

Winding cold resistance is $R_c = R_1$, winding cold temperature is $\theta_c = \theta_1$.

3. Uncertainty of winding operating temperature
   Winding temperature θ_w is determined by thermal terminal resistance R_w, and the formula is

$$\theta_w = \frac{R_w}{R_c} \times (K_1 + \theta_c) - K_1 \tag{11}$$

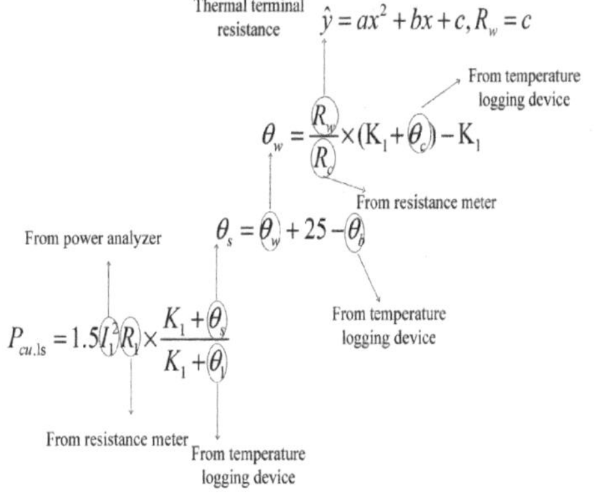

Figure 2. Influence factors of stator copper loss uncertainty.

Therefore, the squared of operating temperature uncertainty of stator winding is $u(\theta_w) = 0.3291°C$.

4. Uncertainty of specified temperature of stator
   The specified temperature θ_s can be calculated as

$$\theta_s = \theta_w + 25 - \theta_b \quad (12)$$

The squared of specified temperature uncertainty is $u(\theta_s) = 0.2657°C$.

5. Stator cooper loss uncertainty under specified temperature
   The formula of winding resistance under specified temperature is

$$R_s = R_1 \times \frac{K_1 + \theta_s}{K_1 + \theta_1} \quad (13)$$

The squared of winding resistance uncertainty under specified temperature is $u(R_s) = 1.1137 \times 10^{-4}\ \Omega$.

The stator copper loss under specified temperature can be expressed as

$$P_{cu1s} = 1.5 I_1^2 R_s \quad (14)$$

Therefore, the uncertainty of stator copper loss under specified temperature is $u(P_{cu1s}) = 1.8236$ W.

5.2.2 *The uncertainty of other parameters*

1. Under the condition that ignoring the impact on uncertainty of curve fitting for wind abrasion, whose uncertainty is comprised by the uncertainty of constant loss and winding resistance.

$$u(P_{fw}) = u(P_{con}) = 7.9874\ W \quad (15)$$

2. Ignoring the impact on uncertainty of curve fitting for iron loss, whose uncertainty is comprised by the uncertainty of constant loss and wind abrasion.

$$u(P_{fe}) = \sqrt{u^2(P_{con}) + u^2(P_{fw})} = 11.2958\ \text{W} \quad (16)$$

3. According to the same analysis above, the uncertainty of other parameters are as follows:
   a. Rotor copper uncertainty under specified temperature

$$u(P_{cu2s}) = 1.5767\ \text{W} \quad (17)$$

   b. Ignoring the impact on uncertainty of curve fitting for residual loss, stray loss uncertainty is [5]

$$u(P_s) = \sqrt{4a^2T^2u^2(T)} = 1.6514\ \text{W} \quad (18)$$

5.2.3 *Uncertainty of apparent total loss*

Apparent total loss is calculated as follows:

$$P_T = P_{fw} + P_{Fe} + P_s + P_{cu1s} + P_{cu2s} \quad (19)$$

Thus, the squared of Apparent total loss uncertainty is

$$u^2(P_T) = u^2(P_{fw}) + u^2(P_{Fe}) + u^2(P_s) + u^2(P_{cu1s}) + u^2(P_{cu2s}) = 14.2774\ \text{W} \quad (20)$$

5.2.4 *Uncertainty of efficiency*

Efficiency is calculated as

$$\eta = \frac{P_2}{P_1} = \frac{P_1 - P_T}{P_1} \times 100\% \quad (21)$$

The uncertainty of efficiency evaluated by B method evaluation is

$$u_B(\eta) = \frac{1}{P_1}\sqrt{u^2(P_T) + \left(\frac{P_T}{P1}u(P_1)\right)^2} \times 100\% = 0.02414\% \quad (22)$$

6 COMPOSITE STANDARD UNCERTAINTY

Refer to the formula of composite uncertainty

$$u_C(\eta) = \sqrt{u_A(\eta)^2 + u_B(\eta)^2} = 0.062\% \quad (23)$$

The system expanded uncertainty can be obtained $U = k.u_C(\eta) = 0.124\%$, where $k = 2$.

The expanded uncertainty of motor efficiency is small. The efficiency difference of 55 kW, 4 pole motor is 1.3% in GB 18613-2012. From the analysis result in this paper, the expanded uncertainty obtained under the condition that ignoring curve fitting uncertainty is slightly larger than the value specified by GB 18613-2012, but still less than 1/3 of the specified difference.

7 CONCLUSION

Analysis of each influence factors for motor loss has been made. In order to get expanded uncertainty, the uncertainties and composite uncertainties of each factor have been calculated. From the analysis result, we can learn that.

There is a great impact on efficiency uncertainty caused by various loss uncertainties, and the uncertainty of stray loss is more complicated.

The condition of testing and analysis is complied with the requirement of GB 1032-2012 in this paper, and efficiency uncertainty is less than 1/3 of specified difference.

The evaluation above of motor efficiency uncertainty can offer important reference value for theory research in this field.

REFERENCES

[1] National Technical Committee of Legal Metrology metering management. JJF 1059.1-2012, Evaluation and Expression of Uncertainty in Measurement [S] Beijing: China Quality Inspection Press.

[2] National Standardization Technical Committee of rotary motor. GB/T1032-2012, Three-phase Asynchronous Motor Test Methods [S] Beijing: China Standardization Publish press.

[3] National Technical Committee of Legal Metrology metering management JJF 1001-2011, General Terms and Definitions of Measurement [S]. Beijing: China Standard Press.

[4] Wang Zhuanjun, Chin Wei Wei, Chen Gen, ect. 2010. Research and Design of Testing System Concerning Low uncertainty for high and Ultra-efficiency motor [J] Machines & Control Application, 37 (4): 1–5.

[5] Wang Zhuanjun. 2011. Evaluation of Uncertainty of Stray Loss for Three-phase Asynchronous Motor Under Loading State [J]. Machines & Control Application, 38 (9): 62–65.

Material Science and Environmental Engineering – Chen (Ed.)
© 2016 Taylor & Francis Group, London, ISBN 978-1-138-02938-5

Study on multi-cylindrical Magneto-Rheological Fluid Clutch

D.M. Chen, H. Zhang & Q.G. Cai
Department of Mechanical Engineering, Academy of Armored Force Engineering, Beijing, China

ABSTRACT: The existed Magneto-Rheological Fluid Clutches (MFC) have a low utilization rate of the magnetic field, they are bulky and easily effected by the centrifugal force in high speed rotating situation. And the temperature rise causes the leakage of Magneto-Rheological Fluid (MRF). In order to overcome those shortcomings, design a multi-cylindrical MFC with a pressure balance mechanism and the coil placed beside. Use Ansys Workbench to analyze the temperature field of the clutch, and simulation the magnetic field in Ansoft. From the results, the MFC can transmit a torque of 375 N · m when the current is between 2.5~3 A, and the MFC can work in a reasonable temperature.

Keywords: magneto-rheological fluid; clutch; cylindrical; simulation

1 INTRODUCTION

The clutch has been widely used in various fields of engineering to transmit power. The main types of clutch are friction disc clutch, magnetic clutch, jaw clutch and so on. With the development of industry, the clutch has a higher demand on intelligence control. The researchers turn to MRF. Combine the magneto-rheological effect of MRF with the structure of clutch; they design a new type of clutch—MFC. MFC has many advantages, such as response rapidly and can be controlled easily.

The currently existing MFC can't transmit a large torque. Although by increasing the size or the number of coil may improve the torque transmission capacity, it will bring other problems such as the leakage of the fluid, overweight, and temperature rising [1–3]. In this paper we design a MFC through optimizing the structure and magnetic circuit to overcome the above problems.

2 THE THEORIES OF MFC

2.1 *The structure forms of MFC*

According to the place of the relative plates in shear model, MFC can be divided into two kinds of structures, cylindrical and disc.

MFC which adopt a single cylinder or a pair of disc can only transmit a small torque and can't meet the demand of large torque. By increasing the number of the cylinder or disc can multiply the working area to increase the transmission capacity.

There are three different places of the coil: inside the clutch, outside the clutch, and beside the clutch, as shown in Figure 1.

On the basis of predecessors' research results, we found that the MFC with a multi-cylindrical structure and the coil beside the clutch can solve the problems caused by centrifugal force, size, and weight.

2.2 *The principle of cylindrical MFC*

Cylindrical MFC use the magneto rheological effect to transmit torque, as shown in Figure 2. The clutch has input and output cylinder, input

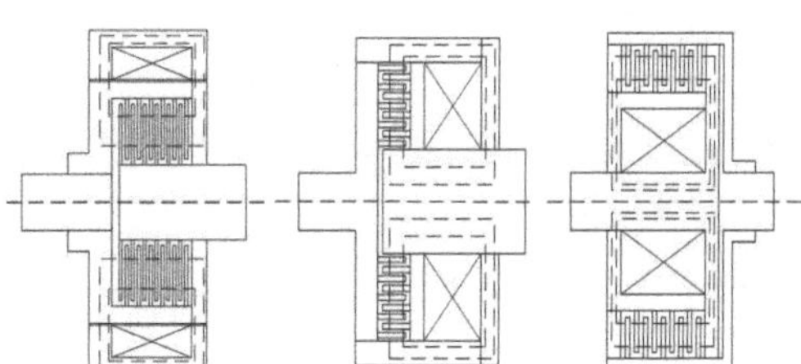

Figure 1. The different places of coil.

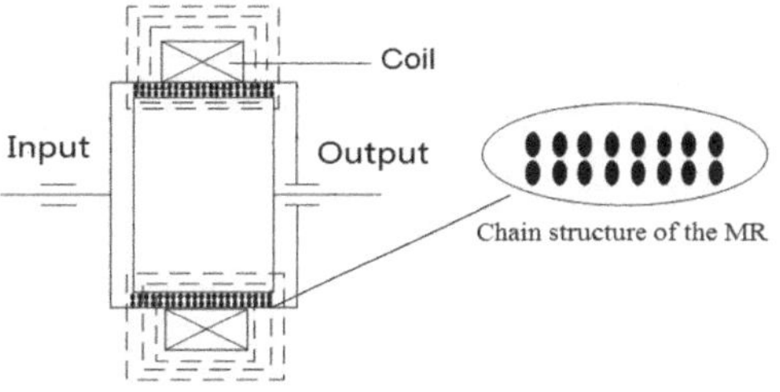

Figure 2. The principle of cylindrical MFC.

and output shaft, coil, MRF, and bearings. The clearance of clutch is filled up with MRF. Use sealing devices to prevent the leakage of the fluid.

When there is no current in the coil, no magnetic field in the clutch is created, and the MRF flows freely like liquid. When the current is switched on, the magnetic field is created in the clutch. The magnetic particles in the MRF begin to form chain structures and connect the two cylinders by shearing. The transmission capacity is related with the magnetic field strength. When the magnetic saturation is reached, no more chains will be formed, and the clutch transmits the maximum torque.

2.3 *The torque formula of cylindrical MFC*

Simplify the relevant conditions and establish a model of a cylindrical MFC with two cylinders, as shown in Figure 3. In the cylindrical coordinate system, the radius of the input cylinder is r_2, the radius of the output cylinder is r_1, and the corresponding angular velocity is ω_2 and ω_1.

Then, the torque is:

$$T = \frac{4\pi L r_1^2 r_2^2 \ln(r_2/r_1)}{r_2^2 - r_1^2}\tau_B + \frac{4\pi L r_1^2 r_2^2(\omega_2 - \omega_1)}{r_2^2 - r_1^2}\eta \quad (1)$$

where L is the actual working length between the cylinders.

When the clutch is multi-cylindrical, the torque is:

$$T_n = nT = \sum_1^n \frac{4\pi L r_{i1}^2 r_{i2}^2 \ln(r_{i2}/r_{i1})}{r_{i2}^2 - r_{i1}^2}\tau_{Bi} + \frac{4\pi L r_{i1}^2 r_{i2}^2(\omega_{i2} - \omega_{i1})}{r_{i2}^2 - r_{i1}^2}\eta \quad (2)$$

There are two parts: magnetic torque and viscosity torque.

Magnetic torque:

$$T_{nB} = \sum_1^n \frac{4\pi L r_{i1}^2 r_{i2}^2 \ln(r_{i2}/r_{i1})}{r_{i2}^2 - r_{i1}^2}\tau_{Bi} \quad (3)$$

Viscosity torque:

$$T_{n\eta} = \sum_1^n \frac{4\pi L r_{i1}^2 r_{i2}^2(\omega_{i2} - \omega_{i1})}{r_{i2}^2 - r_{i1}^2}\eta \quad (4)$$

The viscosity torque is very small when compared to the magnetic torque in reality. Usually, we only consider about the magnetic torque.

There are four variables in the magnetic torque: the axial length L, inner radius r_1, and outer radius r_2, and the shear yield strength τ_B. They are based on the size, electric current and other factors.

3 THE DESIGN OF THE MFC

3.1 *The magneto-rheological fluid*

The MRF used in this paper is ZGY-8, which was developed by our team. The compositions are showed in Table 1.

Through experiment testing, we got the B-H and characteristic curve of the MRF, as shown in Figure 4 and Figure 5.

After data fitting in MATLAB, got the relationship between magnetic flux density B and shear yield stress τ_0:

$$\tau_0 = -160.8B^3 - 54.05B^2 + 130.8B - 5.7 \quad (5)$$

Table 1. The compositions of the MRF.

Magnetic particles	Particle size	Volume fraction	Additives
Hydroxyl iron powder	2.5 (μm)	24.5%	Oleic acid

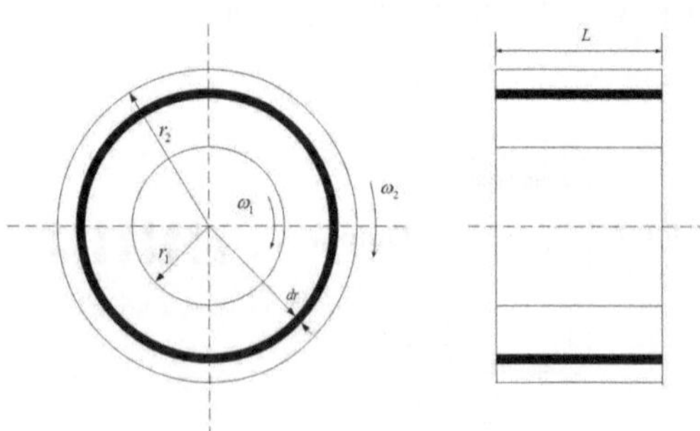

Figure 3. The model of cylindrical MFC.

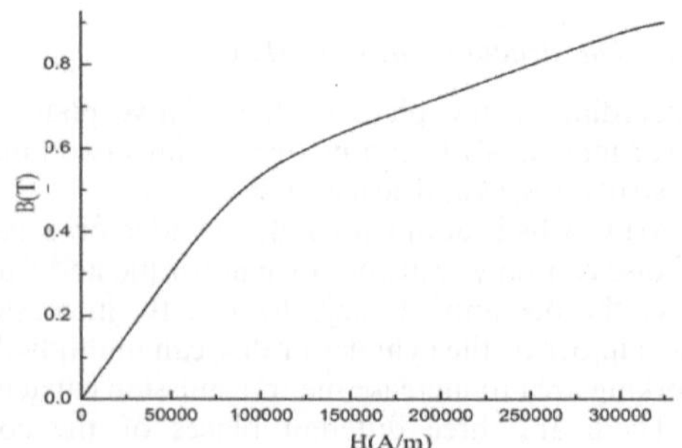

Figure 4. The B-H curve of the MRF.

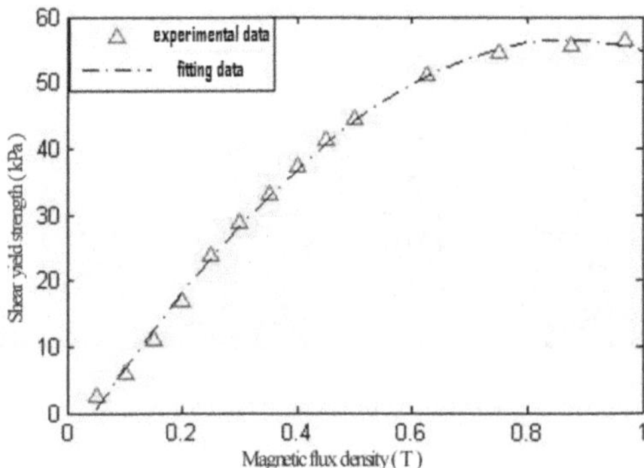

Figure 5. The characteristic curve of the MRF.

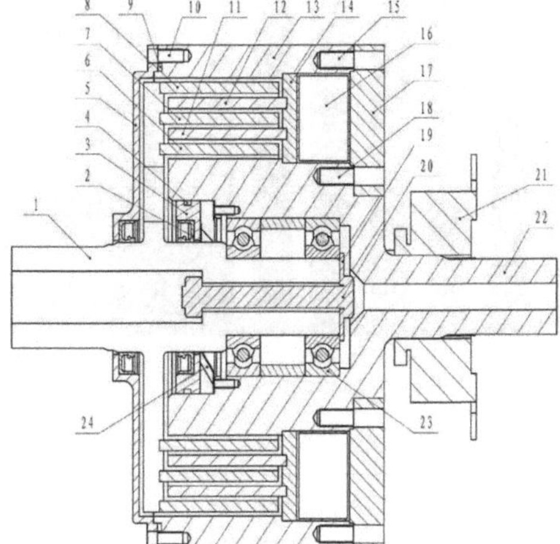

1-the input shaft, 2-hole spring collar, 3-the pistons, 4,9-seal, 5-cover, 6,7,8 - input cylinder, 10, 15, 18 - hexagon socket screw, 11, 12 - output cylinder, 13 - outside hub, 14 - the transition plate, 16 - coil, 17 - coil cover, 19 - gaskets, 20 - bolts, 21 - conductive ring, 22 - output shaft, 23 - deep groove ball bearings, 24 - spring .

Figure 6. The assembly drawing.

3.2 *The assembly drawing*

Based on the structure characteristic of cylindrical MFC and the torque transmission target and the space of the transmission system, we design a multi-cylindrical MFC, as shown in Figure 6.

The power transmission process is as follows.

Engine—the input shaft 1—the input cylinders 6, 7, 8—output cylinders 11, 12, outside hub 13—the transition plate 14, end cover 17—output shaft 22.

The control process is as follows.

Input current—conductive ring 21—coil 16 (creating a magnetic field)—magnetic flux density.

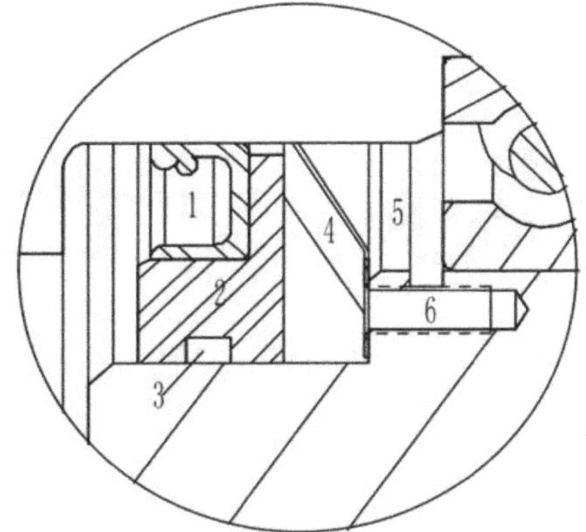

1 - lip-type packing, 2 -piston, 3 - o-rings, 4 - spring , 5 - hole spring collar, 6 - pan head screw

Figure 7. Pressure balance mechanism.

3.3 *Pressure balance mechanism*

The coil and shear friction of MRF produce a large amount of heat when the coil is electrified, which would rise the temperature of the clutch. When it stops working, the temperature will gradually fall back. A pressure balance mechanism is designed to eliminate the influence of the expansion of MRF caused by the temperature rising. It consists of 6 parts. The sectional view of the mechanism is shown in Figure 7.

When working, the expansion liquid pushes the piston 2 outward. The spring 4 is placed behind the piston, converting heat energy into elastic potential energy. When the temperature is dropped, the piston recovers to the original position.

4 SIMULATION

4.1 *The temperature simulation*

Build a 3D model of the MFC in Ansys Workbench, as shown in Figure 8. Ignore the effects of bolts and bearings to facilitate the calculation.

The clutch produces the largest heat when starting to connect. In this paper, the maximum current in the coil is 3 A.

From the simulation result in Figure 9, we can see that: after reaching steady state, the highest temperature in the MFC is 131.81°C, the lowest is 105.69°C, and the maximum temperature difference is 26.12°C. The highest temperature point is in the MRF work areas, while the lowest point is at the end of the output shaft. The inner temperature is a little higher than the surface.

The results show that the MFC works in a reasonable temperature.

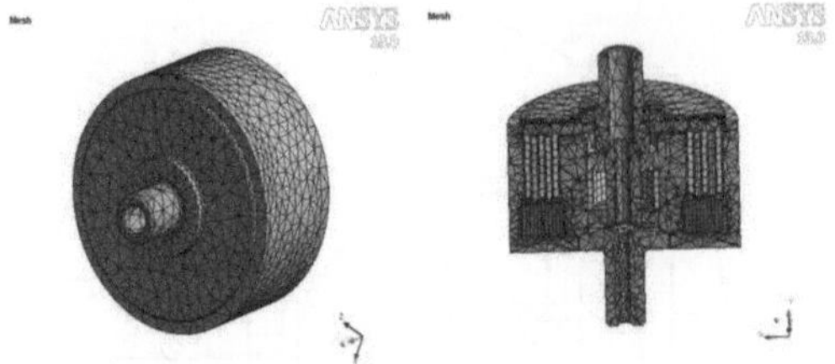

Figure 8. The meshing of the model.

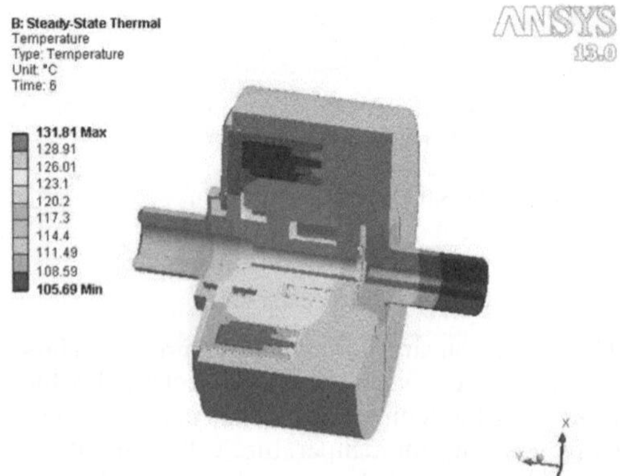

Figure 9. The temperature field of the MFC.

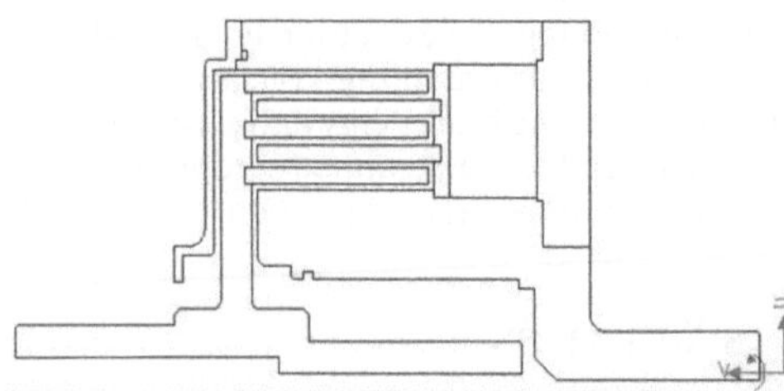

Figure 10. The finite element model.

4.2 *The magnetic field simulation*

Build a finite element model of the MFC in Ansoft [4], as shown in Figure 10, simulation the magnetic flux density in work areas and the distribution of the flux lines.

Figure 11 is the distribution of the flux lines when the current is 2.5 A and 3 A. As can be seen from the figure, the flux lines formed a closed loop along the designed path, and pass through the work areas, vertically. And with the increase of the current, the magnetic flux density becomes larger. It proved that the design of the magnetic circuit is reasonable.

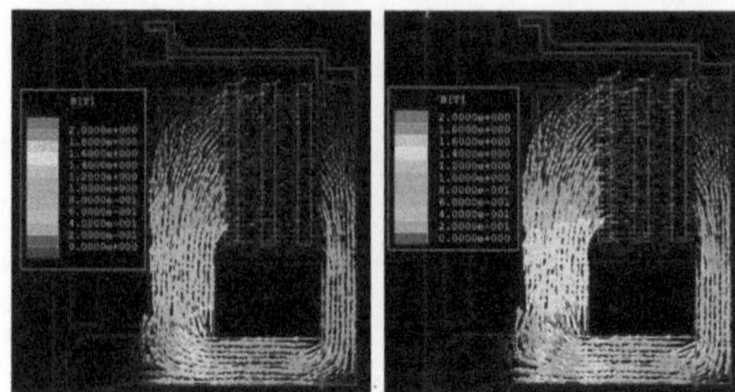

Figure 11. The distribution of the flux lines.

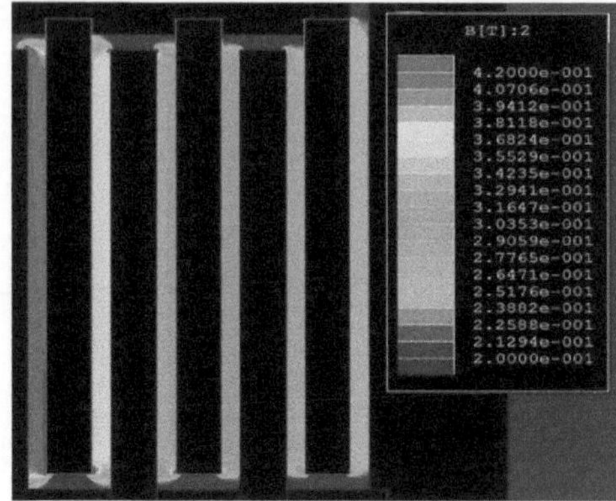

Figure 12. The magnetic flux density.

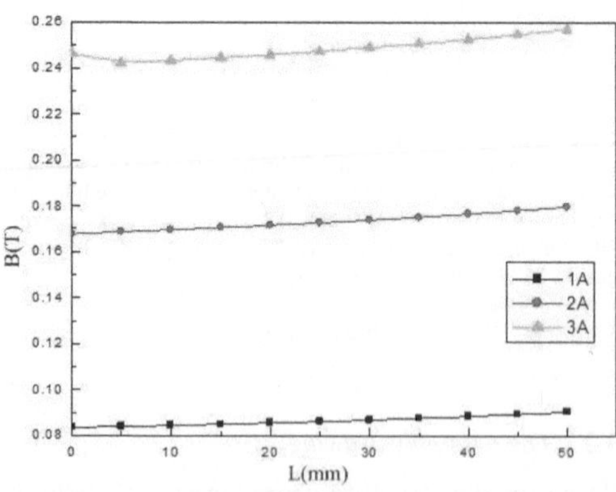

Figure 13. The flux density of the outermost work area under different current.

The magnetic flux density is shown in Figure 12. The distribution is not uniform. For different work areas, the density in the inside areas is larger than the outside.

The magnetic flux density of the outermost work area under different current is shown in Figure 13. The base point of *L* is at the side away from the coil. The result shows that the closer the coil, the larger the density, but the fluctuation is less than 0.2T.

From equation (2) and the simulation results, when the current is between 2.5~3 A, the MFC can transmit a torque of 375 N · m.

5 CONCLUSIONS

1. Introduced the structure forms of MFC and the principle of cylindrical MFC. Got the torque transmission formula of multi-cylindrical MFC.
2. Got the B-H and characteristic curve of the MRF through experiment and the relationship between magnetic flux density and shear yield stress.
3. Based on the structure of cylindrical MFC and the torque transmission target and the space of the transmission system designed a multi-cylindrical MFC and introduced the work and control process.
4. Simulated the temperature field, the results show that the highest temperature is 131.81°C, the MFC can work in a reasonable temperature.
5. Simulated the magnetic field in Ansoft. When the current is between 2.5~3 A, the MFC can transmit a torque of 375 N · m.

REFERENCES

[1] Kavlicoglu B.M., Gordaninejad Evrensel et al. Heating of a High-torque Magneto-rheological Fluid Limited Slip Differential Clutch [J]. *Smart Material Structure*. 2008, 19: 235–241.

[2] Lampe D., Thess A., Dotzauer C. MRF Clutch Design Considerations and Performance [C]. *The 6th International Conference on New Actuators Bremen*. 1998: 81–84.

[3] Joško Deur, Senior Member. Modeling of Electro-magnetic Circuit of a Magnetorheological Fluid [C]. *2009 IEEE Multi-conference*, Saint Petersburg, Russia. 2009, 113–118.

[4] Demin Chen, Qingge Cai, Hong Zhang. Simulation and Test Study of multidisc Magneto-rheological Fluid Clutch [C]. *Frontier of Manufacturing Science and Measuring Technology IV*. 2014, 464–467.

Material Science and Environmental Engineering – Chen (Ed.)
© 2016 Taylor & Francis Group, London, ISBN 978-1-138-02938-5

Pressure-driven flow considering contact angle effect

C. Yang
Heilongjiang Institute of Technology, Harbin, Heilongjiang, China

ABSTRACT: Understanding pressure-driven flow into capillary is important in microcasting technology. We establish experimentally a certain liquid–solid system, or mercury–mica system. The contact angle formed by a drop of liquid on the solid surface is measured using the contact angle meter. It is found that the contact angle decreases by about 20° as the radius of droplet decreases from 0.325 mm to 0.03 mm. The theories for pressure-driven flow are generalized to a nonlinear second-order differential equation, which takes the effects of the external pressure force, the inertial force, and the contact angle into account. The critical external pressure difference is obtained via the differential equation. The influence of contact angle on the critical external pressure difference is significant due to the change of contact angle with drop size, and the difference between the theoretical result and the experimental result at $r_0 = 0.02$ mm is up to 12.5 kPa.

Keywords: pressure-driven flow; contact angle; surface tension; capillary

1 INTRODUCTION

The microcasting technology becomes more important in micro-system industries because of economic and mass production of micro-components for most applications. It has been recently reported that micro-products were produced by microcasting technology (Baumeister et al. 2002, 2004a, b). In the characterization of micro-casting by the external pressure difference, the dynamic properties of the liquid metal front and the contact angle effects will have important consequences. In order to produce precision micro-components using microcasting technology, one should understand pressure-driven flow in cylindrical capillary.

Research on capillary flows, dating its origin back to more than one century ago, is still a subject of interest due to the widespread applications in materials manufacturing, and modern engineering fields (Barraza et al. 2002). For a fully developed capillary flow, many researchers (Xiao et al. 2006, Hamraoui et al. 2002, Zhmud et al. 2000, and Kim et al. 2002) consider the surface tension effect.

Washburn (Washburn 1921) derived the flow equation to describe capillary phenomenon, and liquid rise in a cylindrical capillary within a reservoir to verify the results. Washburn's equation predicts that the capillary rise should be proportional to the square root of time:

$$l = \sqrt{\frac{\gamma r_0 \cos\theta}{2\mu}} \cdot \sqrt{t} \tag{1}$$

where γ is the surface tension of the liquid having viscosity μ, θ is the contact angle between the liquid and the capillary wall, and r_0 is the tube radius. Several modifications to Washburn equation have been reported to include the viscosity, inertial and entrance pressure loss effects (Barraza et al. 2002, Duarte et al. 1996). The simplest relevant model that has been studied is the cylindrical capillary in contact with an infinite liquid reservoir. Many have focused on the study of the conventional liquid flow, such as the surface tension effect (Xiao et al. 2006, Hamraoui et al. 2002, Zhmud et al. 2000, Kim et al. 2002), gravity effect (Jong et al. 2007), viscosity effect (Xiao et al. 2006, Hamraoui et al. 2002, Zhmud et al. 2000), and the influence of contact angle (Xiao et al. 2006, Hamraoui et al. 2002, Martic et al. 2002). The viscous force for capillary rise is $8\pi\mu l\, dl/dt$ (Mises et al. 1971), and inertia force is $\rho\pi r_0^2 d\,(ldl/dt)/dt$ (Bosanquet, 1923). In general, as applied to a viscous noncompressible liquid in a long cylindrical capillary, the overall balance of forces may be expressed as

$$2\pi r_0 \gamma \cos\theta = 8\pi\mu l \frac{dl}{dt} + \rho\pi r_0^2 \frac{d}{dt}\left(l\frac{dl}{dt}\right). \tag{2}$$

Contact angle between the liquid and the wall plays an important role in cylindrical capillary flow. Contact angle phenomena have been studied for almost two centuries. Some of the important foundations in this area were established in the early 19th century, such as the remarkable work of Thomas Young (Young, 1805). Generally, the classical Young equation expresses the mechanical equilibrium for the three-phase contact line. In other words, a sessile drop on an ideal solid surface is in equilibrium with a vapor phase. The Young equation is

$$\gamma_{lv}\cos\theta_e = \gamma_{sv} - \gamma_{sl} \tag{3}$$

where γ_{lv} is the liquid–vapor interfacial tension, γ_{sv} is the solid-vapor interfacial tension, γ_{sl} is the solid-liquid interfacial tension and θ_e is the equilibrium contact angle. In 1977, Neumann (Boruvka et al. 1977) proposed the second type of multiple-value contact angle phenomenon called the drop size dependence of contact angle. It has been observed that for a sessile drop on an ideal solid surface in equilibrium with the liquid's vapor, the contact angle varies with the drop size from a few degrees up to 10–20°. The mechanical equilibrium condition for any point at the three-phase contact line is given by the modified Young equation (Boruvka et al. 1977)

$$\gamma_{lv}\cos\theta = \gamma_{sv} - \gamma_{sl} - \sigma k_{gs} \tag{4}$$

in which the new variable σ and k_{gs} are respectively the line tension and the local curvature of the three-phase line in the plane of solid surface. Only if the solid surface is horizontal, smooth, homogeneous and rigid is the drop truly axisymmetric and the three-phase contact line a smooth circle. Eq. (4) becomes

$$\gamma_{lv}\cos\theta = \gamma_{sv} - \gamma_{sl} - \frac{\sigma}{R}, \tag{5}$$

where R is the radius of the three-phase contact circle. When $R \gg \sigma$, Eq. (5) reduces to the classical Young equation. Combining Eq. (3) with Eq. (5) yields

$$\cos\theta = \cos\theta_\infty - \frac{\sigma}{\gamma_{lv}}\frac{1}{R}. \tag{6}$$

We can see that the contact angle will vary with drop size. The researches on the effect of drop size on contact angle have been carried out by many researchers (Good et al. 1979, Gaydos et al. 1987, Li et al. 1991, Drelich et al. & Lin et al. 1993, Duncan et al. 1995, Amirfazli et al. 1998). A review was made by D. Li (Li, 1996) in 1996.

In this paper, the effect of drop size on contact angle was observed with a solid-liquid-vapor system, i.e. mica-mercury-air system. The influence of the variable contact angle on liquid flow driven by the external pressure in horizontal capillary was discussed. The critical external pressure difference driving liquid flow in horizontal capillary was obtained by the nonlinear second-order differential equation for the pressure-driving flow. Finally, the validation of the theory was demonstrated via comparison of the numerical results and experimental data obtained in this study.

2 MATERIALS AND METHODS

In this study, the sessile drop technique for measuring the contact angles was employed and the three-phase contact line is supposed to be the smooth circle (Boruvka et al. 1977). Producing such drops requires horizontal, smooth, homogeneous and rigid solid surface. The solid surface used in the measurements was new cleaved mica sheets because of its sufficient smoothness, homogeneity and rigidness. The first step in the preparation procedure was to produce clean mica sheets. The mica piece was split into sheets and then used in this experiment at once. The procedure was completed as soon as possible to prevent adsorption on the solid surface of dust though the whole procedure was accomplished in the super-clean room.

The liquid used was mercury (99.999%) obtained from Beijing Chemical Reagent Co., Inc. No further purification was attempted. The mercury drops were dripped on the different mica sheets separately. In other words, each mercury drop was measured onto a new cleaved mica sheet. The size of mercury drop varied from 0.1 mm to 1 mm. Each sessile drop was photographed three times at room temperature.

The experimental setup was the contact angle meter (OCA20) made in Germany. After focusing the microscope, we acquired digital images from the drop. In each case the camera's lenses were deliberately adjusted out of focus and then brought back into focus before re-photographing the same drop. All measurements were performed at ambient temperature of 22 ± 1°C. The data required were obtained from the contact angle meter directly.

3 RESULTS AND DISCUSSION

The contact angles of mercury on clean mica were measured in air at 22 ± 1°C. Figure 1 shows the experimental data for mercury on the clean mica. At the top part of Figure 1 the value of contact angles are shown and the bottom frame shows the cosine of the contact angles. It can be seen from the top part in Figure 1 that the contact angle of mercury on the mica decrease as the radius of contact circle decreases seriously below about 0.325 mm, above which the contact angle is essentially constant, at 147.2 ± 1.3°. The smallest value of contact angle,

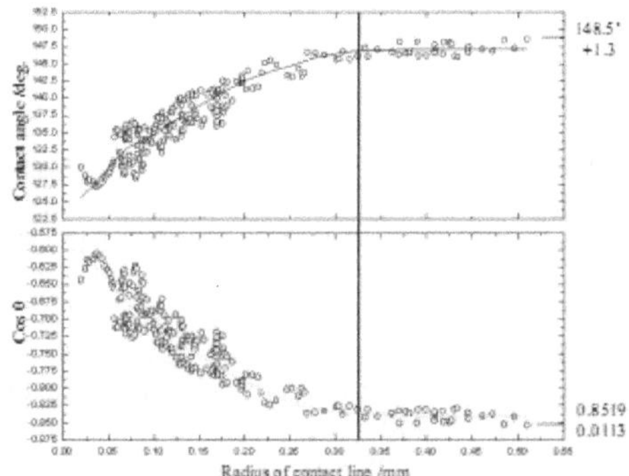

Figure 1. Contact angle for mercury on clean mica. Solid lines fitted to sessile drop.

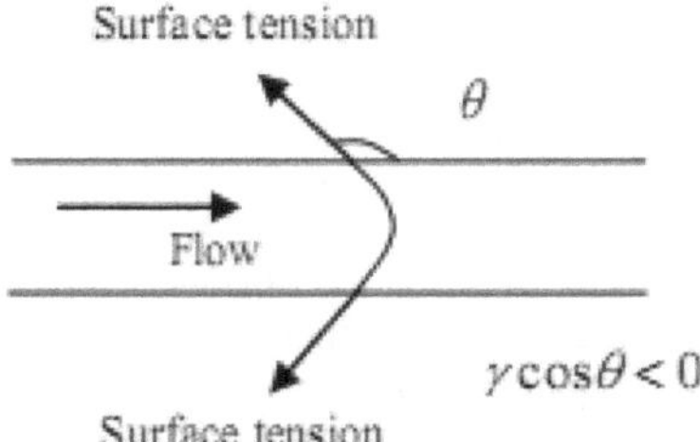

Figure 2. Pressure-driven flow: $\theta > 90°$.

127.1°, was obtained at about 0.02 mm radius of contact circle. The bottom frame in Figure 1 illustrates that the cosine of the contact angle increases with the decrease of the radius of contact circle, and the constant result is -0.8406 ± 0.0113 and the largest value is -0.6032. As can be seen, the contact angle is a function of the radius of contact circle, which is in agreement with the results of Ref. (Good et al. 1979).

Let us assume that the capillaries are made of mica and the dynamic contact angle of liquid in the capillaries equals to the static contact angle which is determined only by the characteristics of the capillaries wall and the working fluid (Hamraoui et al. 2002). For the fluid of mercury in the capillaries of different radius, the external pressure must be applied because of non-wettability with mica surface, which is defined as pressure-driven flow. Figure 2 shows a pressure-driven flow with a negative effect, or flow hindrance. Depending on the negative effect, the range of θ varies: $90° < \theta \le 180°$.

According to the above analysis, Eq. (2) becomes

$$\Delta p \pi r_0^2 + 2\pi r_0 \gamma \cos\theta = 8\pi\mu l \frac{dl}{dt} + \rho \pi r_0^2 \left[l \frac{d^2 l}{dt^2} + \left(\frac{dl}{dt}\right)^2 \right] \tag{7}$$

where l is the flow length, t is the flow time, and Δp is the external pressure difference, or the difference between the external pressure and the atmospheric pressure. The other symbols are the same with those in Eq. (2). The following approach was reported in Ref. (Xiao et al. 2006) using dimensionless time, $t^* = t/t_0$, and flow length, $l^* = l/l_0$, where the characteristic time t_0 and length h_0 are defined as: $t_0 = \rho r_0^2/8\mu$, $l_0 = r_0$. Using these definitions, Eq. (7) becomes a generalized dimensionless equation

$$\Delta p \pi r_0^2 \frac{\rho r_0 (\Delta p r_0 + 2\gamma\cos\theta)}{64\mu^2} = \frac{Bo}{64Oh^2} + \frac{\cos\theta}{32Oh^2}, = l^* \frac{dl^*}{dt^*} + l^* \frac{d^2 l^*}{dt^{*2}} + \left(\frac{dl^*}{dt^*}\right) \tag{8}$$

where the Oh is the Ohnesorge number representing the radio of viscous force to surface tension force $Oh = \mu/(r_0\rho\gamma)^{1/2}$, Bo is the Bond number denoting the radio of external pressure force to surface tension force $Bo = \Delta p r_0/\gamma$.

It can be seen from Eq. (8) that the left side must be positive. Clearly there is an equilibrium point between the force induced by the external pressure and the surface tension. Let us assume that the left side of Eq. (8) is zero and the critical pressure difference values are obtained for the different radius of capillary. The critical pressure difference is expressed as

$$\Delta p_c = \frac{2\gamma\cos\theta}{r_0}. \tag{9}$$

The format of Eq. (9) is the same as that of the equation of capillary phenomenon (Adam et al. 1941), but actually, they are not identical. The Δp_c in Eq. (9) indicates the critical pressure difference and is not the pressure of the equilibrium equation of capillary phenomenon. Only when the pressure difference is greater than the critical pressure difference the liquid flow in capillary may occur.

Figure 3 shows the relationship between the radius of capillaries and the critical pressure difference for different contact angles of the mercury on the mica. The left-hand curve is the practical experimental results for the critical pressure difference versus the radius of capillary, and the right-hand curve is the theoretical ones. The critical pressure difference decreases with the increase of the radius of capillary when the contact angle is supposed to be constant. It can be observed in Figure 3 that when the constant contact angle is $\theta = 147.2°$, which is θ_∞ in Newmann equation (Boruvka et al. 1977) for the mercury on mica system, and the radius of capillary is $r_0 = 0.02$ mm the critical pressure difference value is up to 40.8 kPa.

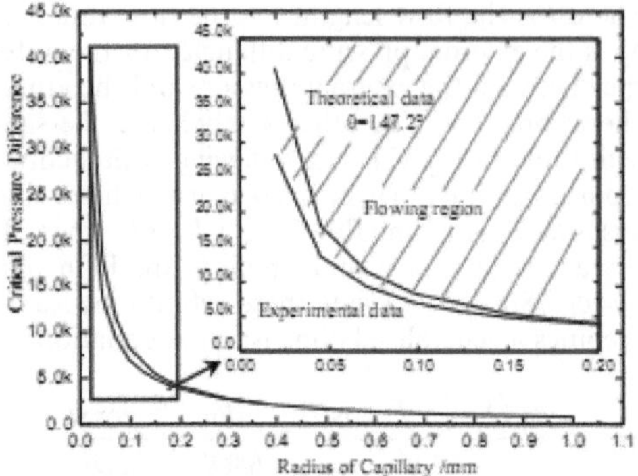

Figure 3. Relationship between critical pressure difference and radius of capillary: theoretical data denoting the critical pressure difference obtained by the theoretical contact angle, and experimental data denoting the critical pressure difference obtained by the contact angle measured in this experiment, and shaded part is flow region.

Actually, the situation discussed above is ideal. As shown in Figure 1, the contact angle of the mercury on mica decreases as the drop size decreases when $r_0 < 0.3$ mm, and the trend is in agreement with the results of ref. (Yekta-Fard et al. 1988). It can be shown in Figure 3 that for the mercury on mica system the influence of the contact angle on the critical pressure is significant, and the critical pressure difference value is only 28.3 kPa, much less than the theoretical result, 40.8 kPa, for $r_0 = 0.02$ mm. That is to say, the flow of mercury for $r_0 = 0.02$ mm may be obtained in theory only when $\Delta p > 40.8$ kPa, but actually, it can be completed as long as $\Delta p > 28.3$ kPa, and the difference between two situations is up to 12.5 kPa. In Figure 3, the two curves are coincidence when $r_0 > 0.35$ mm. This may be attributed to that the contact angle of mercury is constant above the value of radius of the contact circle. For the fluid of mercury in capillary of different radius, it can be seen in Figure 3 that the right side of experimental data curve denoting the relationship for critical pressure difference versus radius of capillary is the flow region, and the left side of it is the non-flowing region. In other words, the liquid flow may be seen only in the flowing region.

From the dimensional analysis, Eq. (8) can be expressed by the following dimensionless function form:

$$l^* \frac{d^2 l^*}{dt^{*2}} + \left(\frac{dl^*}{dt^*}\right)^2 + l^* \frac{dl^*}{dt^*} = f\left(Bo, Oh, \cos\theta\right) \qquad (10)$$

Eq. 6 shows $\cos\theta$ as a function of R, but we can not find their numerical relation because of uncertainty of the line tension. Then, we extensively perform numerical solution of Eq. (10) via the experimental results.

Figure 4 presents the flow length, l^*, as a function of time, t^*. With the validation of analysis on pressure-driven flow with different contact angles at different capillary radius, parametric studies on the pressure-driven flow are presented in Figure 4a–4d, which show the numerical solutions; experimental data represent the numerical solutions in the case of the contact angles measured in experiments and theoretical data denote the numerical predictions with the equilibrium contact angle. A principal purpose of the parametric studies is to compare the experimental results and the theoretical results in a wider parameter range. In Figure 4, we can see that the difference the flow length between the experimental results and the theoretical results is obtained at the same external pressure and fixed time. The difference increases as the capillaries radius decreases above about 0.1 mm and it decreases below about 0.1 mm. We recognize that these could be explained by the negative effect of the surface tension. In other words, the variational contact angle with the different capillary radii plays an important role in the surface tension when $r_0 > 0.1$ mm, and the capillary radius will become the dominant factor for the surface tension with further decreasing it.

The external pressure difference driving liquid flow increases as the capillaries radii decrease from 0.3 mm to 0.02 mm while the Bond number (*Bo*) decreases from 1.74 to 1.68 and the Ohnesorge number increases from 0.0012 to 0.0045. The effects of Ohnesorge number (*Oh*) and Bond number on the flow process are shown in Figure 4a–4d, respectively. Since the Ohnesorge number is defined as the ratio of viscous force (resistance to flow) to surface tension (resistance to flow), Ohnesorge number increases with decreasing capillary radius due to increased viscous resistance relative to the negative surface tension. The effect of Ohnesorge number on pressure-driving flow in this study may not be obvious because viscous force and surface tension are resistance. Bond number is defined as the ratio of external pressure difference to the surface tension and decreases with decreasing capillary radii due to increased negative surface tension relative to the driving pressure force. This may be explained that the surface tension increases rapidly as the capillary radius decreases.

It is seen in Figure 4a–4d that for theoretical results the flow length decreases as the capillary radius decreases at a fixed time. This may be due to the surface tension that plays an important role in the flow process. For the experimental results the flow length increases as the capillaries radius decrease above about 0.1 mm and it decreases below 0.1 mm. This may be due to the external pressure plays the principal role when $r_0 > 0.1$ mm, and the surface tension is the dominant resistance with further decreasing the capillary radius, or $r_0 < 0.1$ mm.

The effects of inertia forces are usually significant only in the early stages of flow or when r_0 is large or μ small. For very small radius, viscous forces are dominant, the inertial terms can be neglected, and one obtain

$$\Delta p + 2\pi r_0 \gamma \cos\theta = 8\pi\mu l \frac{dl}{dt}. \tag{11}$$

Integral to Eq. (11) leads to the equation

$$l^2 = \frac{\Delta p + 2\pi r \gamma \cos\theta}{4\pi\mu} t. \tag{12}$$

Early studies of Newman (Newman, 1968) on capillary flow of Newtonian liquids indicate the above relationship between time and the flow length. But for the critical pressure difference the result is the same with Eq. (9) and will no larger be discussed in this paper.

4 SUMMARY

For the mercury on mica system we have found that the contact angle may decrease with decreasing drop size, when the radius of curvature of the three-phase line falls below a threshold value. The differential equation for the pressure-driving flow in the literature are generalized into a nonlinear second-order differential equation which includes the effects of the external pressure, the inertial force, the surface tension, and the contact angle as a function of the capillary radius. For the pressure-driving flow there is a critical pressure difference above which the liquid flow may be seen and below which the flow is impossible. The critical pressure in this experiment is smaller than that in theory for the very small capillary radius due to the decreasing contact angle with decreasing capillary radius. The external pressure plays an important role in pressure-driving flow when the capillary radius is above about 0.1 mm; the surface tension increases sharply and is dominant when the capillary radius is below about 0.1 mm.

ACKNOWLEDGMENTS

The author is grateful to the youth cadreman item (1253G050) for funding through Education Department of Heilongjiang Province.

REFERENCES

[1] Baumeister, G. Mueller, K. Ruprecht, R. & Hausselt, J. 2002. *Microsyst. Technol* 8(2–3): 105–108.
[2] Baumeister, G. Ruprecht, R. & Hausselt, J. 2004. *Microsyst. Technol* 10(3): 261–264.
[3] Baumeister, G. Ruprecht, R. & Hausselt, J. 2004. *Microsyst. Technol* 10(6–7): 484–488.
[4] Barraza, H.J. Kunapuli, S. & O'Rear, E.A. 2002. *J. Phys. Chem. B* 106(19): 4979–4987.
[5] Xiao, Y. Yang, F. & Pitchumani, R. 2006. *J. Colloid Interface Sci* 298(2): 880–888.
[6] Hamraoui, A. & Nylander, T. 2002. *J. Colloid Interface Sci* 250(2): 415–421.
[7] Zhmud, B.V. Tiberg, F. & Hallstensson, K. 2000. *J. Colloid Interface Sci* 228(2): 263–269.
[8] Kim, D.S. Lee, K.C. Kwon, T.H. & Lee, S.S. 2002. *J. Micromech. Microeng* 12(): 236–246.
[9] Washburn, E.W. 1921. *Phys. Rev* 17(3): 273–283.
[10] Duarte, A.A. Strier, D.E. & Zanette, D.H. 1996. *Am. J. Phys* 64(4): 413–417.
[11] Jong, W.R. Kuo, T.H. Ho, S.W. Chiu, H.H. & Peng, S.H. 2007. *Int. Commun. Heat Mass Transf* 34(2): 186–196.
[12] Martic, G. Gentner, F. Seveno, D. Coulon, D. Coninck, J.D. & Blake, T.D. 2002. *Langmuir*. 18(21): 7971–7976.
[13] Mises, R.V. & Friedrichs, K.O. (ed.) 1971. *Fluid dynamics*. New York: Springer-Verlag New York Inc.
[14] Bosanquet, C.H. 1923. *Philos. Mag* 45(6): 525–531.
[15] Young, T. 1805. *Philos. Trans. R. Soc. London* 95(12): 65–87.
[16] Boruvka, L. & Neumann, A.W. 1977. *J. Che. Phys* 66(12): 5464–5471.
[17] Good, R.J. & Koo, M.N. 1979. *J. Colloid Interface Sci* 71(2): 283–292.
[18] Gaydos, J. & Neumann, A.W. 1987. *J. Colloid Interface Sci* 120(1): 76–86.
[19] Li, D. Lin, F.Y.H. & Neumann, A.W. 1991. *J. Colloid Interface Sci* 142(1): 224–231.
[20] Drelich, J. Miller, J.D. & Hupka, J. 1993. *J. Colloid Interface Sci* 155(2): 379–385.
[21] Lin, F.Y.H. Li, D. Nenmann, A.W. 1993. *J. Colloid Interface Sci* 159(1): 86–95.
[22] Duncan, D. Li, D. Gaydos, J. Neumann, A.W. 1995. *J. Colloid Interface Sci* 169(2): 256–261.
[23] Amirfazli, A. Kwok, D.Y. Gaydos, J. & Neumann, A.W. 1998. *J. Colloid Interface Sci* 205(1): 1–11.
[24] Li, D. 1996. *Colloids Surf* 116(1–2): 1–23.
[25] Adam, N.K. (ed.) 1941. *The Physics and Chemistry of Surfaces*. London: Oxford University Press.
[26] Yekta-Fard, & M. Ponter, A.B. 1988. *J Colloid Interface Sci* 126(1): 134–140.
[27] Newman, S. 1968. *J. Colloid Interface Sci* 26(2): 209–213.

Material Science and Environmental Engineering – Chen (Ed.)
© 2016 Taylor & Francis Group, London, ISBN 978-1-138-02938-5

Error analysis of cam mechanism of translating roller follower

H.W. Zhou
College of Mechanic and Electronic Engineering, Changchun University of Science and Technology Changchun, Jilin, P.R. China

ABSTRACT: By analyzing the structure of cam mechanism of translating roller follower, the factors influencing the position precision of the cam mechanism were obtained. Then a mathematical model was established to study the influences of each position error on the displacement, speed, and accelerated speed of cam mechanism. Furthermore, the relationship between position error and kinematic error was revealed.

Keywords: roller follower; cam mechanism; position error

1 INTRODUCTION

Cam mechanism has simple and compact structure and can be designed easily. Merely with properly designed cam contour, the follower will move arbitrarily as expected. For production equipment of high efficiency, the efficiency can be improved by taking advantages of the slight parts and simple structure of cam. Therefore, cam mechanism has been widely applied in automation equipment. As roller follower can effectively reduce the friction between cam and follower, it is extensively used in cam mechanism. To guarantee the operation precision of cam mechanism, the position error of the mechanism has to be analyzed so as to provide basis for the machining and assembling of the mechanism.

2 SOURCE OF ERROR

The cam mechanism of translating roller follower is generally composed of a push rod, a roller, and a cam. Considering this, the error of the cam mechanism comes from the errors of assembling precision of roller, contour precision of cam, and assembling precision of cam.

3 POSITION ERROR

3.1 *Error coming from the roller*

The error contributed by the roller is directly associated with the structure of the roller. Roller is commonly assembled on the axis of the push rod, so the error ΔR coming from the roller consists of the outer diameter error ΔR_1 of the roller and the gap ΔR_2 between the roller and the axis. That is,

$$\Delta R = \Delta R_1 - \Delta R_2 \tag{1}$$

The position error of the push rod is

$$\Delta h_1 = \frac{\Delta R}{\cos\alpha} \tag{2}$$

α is the pressure angle of the roller on the contour line of the cam.

3.2 *Error induced by the contour line of the cam*

The error Δh_2 of the contour of the cam contains the machining error Δh_r in the processing of the cam and the gap Δd between the cam and the

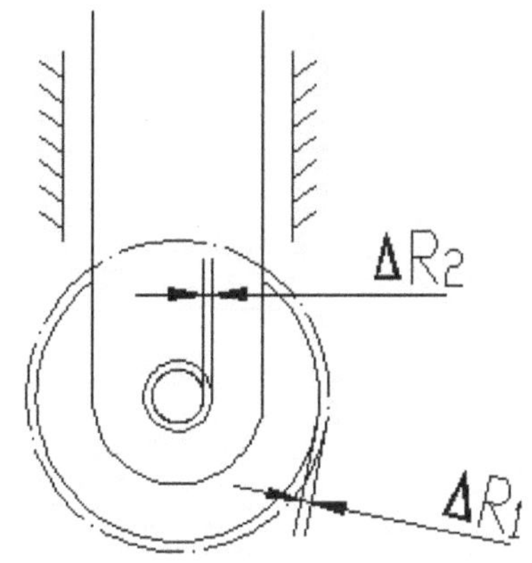

Figure 1. Error coming from the roller.

assembling axis. In general working environment, the cam and the assembling axis present interference fit, that is, $\Delta d = 0$. therefore

$$\Delta h_2 = \Delta h_r \tag{3}$$

3.3 *Error caused by eccentricity of the cam*

The position error Δh_3 of the push rod contributed by the eccentricity of the cam is mainly composed of the position error Δe of the hole of the cam. The position error of the push rod caused by Δe is

$$\Delta h_3 = \frac{\Delta e \cos\beta}{\cos\alpha} \tag{4}$$

β is the angle between Δe and the common normal.

3.4 *Position error of cam mechanism*

The above analysis reveals that the position error Δh of the cam mechanism is the sum of all the errors above, that is,

$$\Delta h = \Delta h_1 + \Delta h_2 + \Delta h_3 \tag{5}$$

4 INFLUENCES OF POSITION ERROR ON THE KINEMATIC ERROR OF CAM MECHANISM

The influences of the position error of the cam on the movement of the follower can be analyzed by constructing mathematical model. By establishing a coordinate system illustrated in Figure 2, the kinematic equation of follower is

$$S = S(\varphi) \tag{6}$$

Correspondingly, the polar coordinate equation of the contour line of the cam is:

$$r = \left(\left(\sqrt{(s+s_0)^2 + \left(\frac{ds}{d\varphi} - e\right)^2} - r_0 \right)^2 + \left(\frac{ds}{d\varphi}\right)^2 - 2\left(\sqrt{(s+s_0)^2 + \left(\frac{ds}{d\varphi} - e\right)^2} - r_0 \right) \frac{ds}{d\varphi} \cos\theta_s \right)^{\frac{1}{2}} \tag{7}$$

$$\theta = \varphi + \varphi_0 - \theta_r$$

The displacement of the follower is

$$S_b = r\sin\theta_{rb} + r_0 \sin(\theta_{rb} + \gamma) - S_0 \tag{8}$$

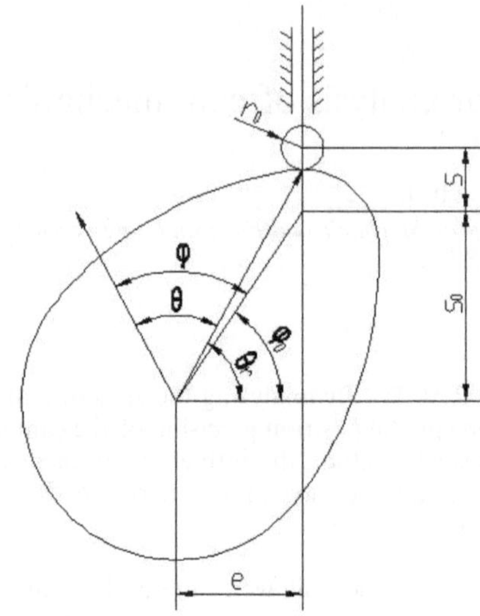

Figure 2. The coordinate system of cam mechanism.

here,

$$\theta_{rb} = 2\tan^{-1}\frac{-r_0 \sin\gamma \pm \sqrt{r_0^2 - 2r_0 r\cos\gamma + r^2 - e^2}}{e + r_0\cos\gamma + r}$$

$$\theta_{rb} = \varphi + \varphi_0 - \theta_{rb}$$

$$\gamma = \tan^{-1}\frac{\dfrac{dr}{d\varphi}}{\gamma\left(1 - \dfrac{d\theta_r}{d\varphi}\right)}$$

By calculating partial derivatives of different variables using formula (8), the following influencing coefficients are obtained:

The influencing coefficient of eccentricity error on the displacement of the follower is

$$K_{s1} = \frac{\partial s_b}{\partial e} = -\frac{k_1}{k_2} + \frac{e}{s_0} \tag{9}$$

The influencing coefficient of eccentricity error on the moving speed of the follower is

$$K_{v1} = \frac{\partial^2 s_b}{\partial e \partial\varphi} = -\frac{1}{k_2}\frac{dk_1}{d\varphi} + \frac{k_1}{k_2}\frac{dk_2}{d\varphi} \tag{10}$$

The influencing coefficient of eccentricity error on the accelerated speed of the follower is

$$K_{a1} = \frac{\partial^3 s_b}{\partial e \partial\varphi^2}$$

$$K_{a1} = -\frac{1}{k_2}\frac{d^2k_1}{d\varphi^2} + \frac{1}{k_2^2}\left(2\frac{dk_1}{d\varphi}\frac{dk_2}{d\varphi} + k_1\frac{d^2k_1}{d\varphi^2}\right) - 2\left(\frac{dk_2}{d\varphi}\right)^2\frac{k_1}{k_2^3} \quad (11)$$

The influencing coefficient of radius error of the roller on the displacement of the follower is

$$K_{s2} = \frac{\partial s_b}{\partial r_0}$$

$$K_{S2} = \frac{k_1}{k_2}\cos(\theta_{rb}+\gamma) - \frac{r_0+r_b}{s_0} + \sin(\theta_{rb}+\gamma) \quad (12)$$

The influencing coefficient of radius error of the roller on the moving speed of the follower is

$$K_{v2} = \frac{\partial^2 s_b}{\partial r_0 \partial\varphi}$$

$$K_{v2} = \cos(\theta_{rb}+\gamma)\left(\frac{1}{k_2}\frac{dk_1}{d\varphi} + \frac{k_1}{k_2^2}\frac{dk_2}{d\varphi}\right) + \left(\frac{\partial\theta_{rb}}{\partial\varphi} + \frac{d\gamma}{d\varphi}\right)\left((\cos(\theta_{rb}+\gamma) + \frac{k_1}{k_2}\sin(\theta_{rb}+\gamma)\right) \quad (13)$$

The influencing coefficient of eccentricity error on the accelerated speed of the follower is

$$K_{a2} = \frac{\partial^3 s_b}{\partial r_0 \partial\varphi^2}$$

$$K_{a2} = \cos(\theta_{rb}+\gamma)\left(\frac{1}{k_2}\frac{d^2k_1}{d\varphi^2} - \frac{k_1}{k_2^2}\frac{d^2k_2}{d\varphi^2} - \frac{2}{k_2^2}\frac{dk_1}{d\varphi}\frac{dk_2}{d\varphi} + \frac{2k_1}{k_2^3}\left(\frac{dx_2}{d\varphi}\right)^2\right) + 2\sin(\theta_{rb}+\gamma)\left(\frac{\partial\theta_{rb}}{\partial\varphi} + \frac{d\gamma}{d\varphi}\right) \times\left(\frac{k_1}{k_2^2}\frac{dk_2}{d\varphi} - \frac{1}{k_2}\frac{dk_1}{d\varphi}\right) + \left(\frac{\partial^2\theta_{rb}}{\partial\varphi^2} + \frac{d^2\gamma}{d\varphi^2}\right) \times\left(\cos(\theta_{rb}+\gamma) - \frac{k_1}{k_2}\sin(\theta_{rb}+\gamma)\right) - \left(\frac{\partial\theta_{rb}}{\partial\varphi} + \frac{d\gamma}{d\varphi}\right)\left(\frac{k_1}{k_2}\cos(\theta_{rb}+\gamma) + \sin(\theta_{rb}+\gamma)\right) \quad (14)$$

The influencing coefficient of radius of base circle on the displacement of the follower is

$$K_{s3} = \frac{\partial s_b}{\partial r_b} = -\frac{r_b+r_0}{s_{042}} \quad (15)$$

where

$$k_1 = r\cos\theta_{rb} + r_0\cos(\theta_{rb}+\gamma)$$

$$k_1 = r\sin\theta_{rb} + r_0\sin(\theta_{rb}+\gamma)$$

$$\frac{d\theta_r}{d\varphi} = \left(\frac{dr}{d\varphi}\cos\theta_r + r_0\sin\theta_s\frac{d\theta_s}{d\theta}\right)\frac{1}{r\sin\theta_r}$$

5 CONCLUSION

The following conclusions were drawn from the analysis above:

1. The eccentricity error significantly influences each error coefficient. Therefore, large eccentricity should be avoided and the eccentricity error has to be controlled in the machining and assembling of the cam mechanism.
2. The eccentricity error exhibits large influences on the overall mechanism; while the radius error of the roller significantly impacts the speed error and accelerated speed error of the mechanism.
3. As the radius of base circle increases, all the influencing coefficients reduce.

REFERENCES

[1] Zeshao Fu, Mechanism Design, Chengdu: Chengdu University of Science and Technology Press, 1988.

[2] Heng Sun. Zuomo Chen. 1996. Mechanical Principle. Beijing: Higher Education Press.

[3] Hongzhong Li. 2005. Analysis of position error of secondary cam. 64–65. Journal of Mechanical Transmission.

[4] Weimin Zhang. Yunpeng Gong. 1995. Influences of cam error of translating follower. 424–428. Jounal of Northeastern University.

[5] Weimin Zhang et al. 1995. Precision of parallel indexing cam. 338–343. Journal of Tianjin University.

[6] Chaoying Liu. 1995. Analysis and calculation of position error in the design of disk cam. 19–24. Journal of Jilin Institute of Architecture and Civil Engineering.

[7] Zhenyu Hong et al. 2003. Analysis of position error of planetary indexing cam. 129–131. Mechanical Science and Technology for Aerospace Engineering.

Material Science and Environmental Engineering – Chen (Ed.)

Error analysis and precision estimation of the locating device of welding fixture of rear axle bracket

H.W. Zhou

College of Mechanic and Electronic Engineering, Changchun University of Science and Technology, Changchun, Jilin, P.R. China

ABSTRACT: Based on the mathematical model for the linkage of the welding fixture of rear axle bracket, the influences of tolerance of each linkage on the overall error of the mechanism were calculated through differential methods. The reasonable tolerances of each linkage were therefore obtained. According to the computation results, the precision of the processed fixture was estimated.

Keywords: error analysis; locating device; precision estimation

1 INTRODUCTION

In the machining of rear axle bracket of A4 car, all the components of the bracket have to be welded integrally. Owing to the influences of internal stress of the welded workpieces, the workpieces are likely to deform. To ensure that the shape and size of the rear axle bracket can meet design requirements, the workpieces have to be detected and calibrated. Considering the large production, the welding fixture of rear axle bracket is expected to meet the requirements of accurate measurement, stability, and rapid assembling and disassembling. Characterized by accurate motion, convenient assembling, etc., planar linkage was applied as the locating device in the welding fixture designed for the rear axle bracket.

As planar linkage shows more complex structure comparing with mechanism with higher pairs, there are more factors causing error. To ensure that the influence of error of planar linkage on the detection precision of rear axle bracket is within the tolerance, the error of planar four-bar linkage needs to be investigated. The error of linkage is mainly caused by the length error of linkage and the gap between kinematic pairs. Therefore, by studying the influence of length error of linkages on the locating precision, especially position precision, of welding fixture, the manufacturing tolerances were distributed reasonably for each linkage according to the precision requirements of the fixture.

2 ERROR OF KINEMATIC MECHANISM OF GAUGE

The output displacement s (or output angle ψ) of planar linkage is the function of geometric size of the linkage and input angle ϕ. To calculate the displacement s (or output angle ψ), the length l of the linkage and the input angle ϕ are regarded as a constant and an independent variable, respectively. While, in the analysis of error, the length l of the linkage is applied as an independent variable. When the degree of freedom of the workpieces is f, the numbers of workpieces with two, three, and four kinematic pairs are n_2, n_3, and n_4. Then, the number Z of independent variables is calculated.

$$Z = f + n_2 + 3n_3 + 5n_4 \tag{1}$$

If l_1, l_2, l_3 and l_4 are the lengths of the linkages, the transmission function is

$$s = s(\phi, l_1, l_2, l_3, l_4) \tag{2}$$

With little errors of Δs and Δl_i, by developing function (2) based on Taylor series and ignoring the terms higher than second order, the following formula for calculating errors is obtained:

$$\Delta s = \frac{\partial s}{\partial \phi}\Delta\phi + \frac{\partial s}{\partial l_1}\Delta l_1 + \frac{\partial s}{\partial l_2}\Delta l_2 + \frac{\partial s}{\partial l_3}\Delta l_3 + \frac{\partial s}{\partial l_4}\Delta l_4 \tag{3}$$

In the proposed planar linkage (Fig. 1), linkage l_1 is the driving link, while linkage l_3 is the follower, and the complex number vector equation is

$$l_1 e^{i\phi_1} + l_2 e^{i\phi_2} = l_3 e^{i\phi_3} + l_4 \tag{4}$$

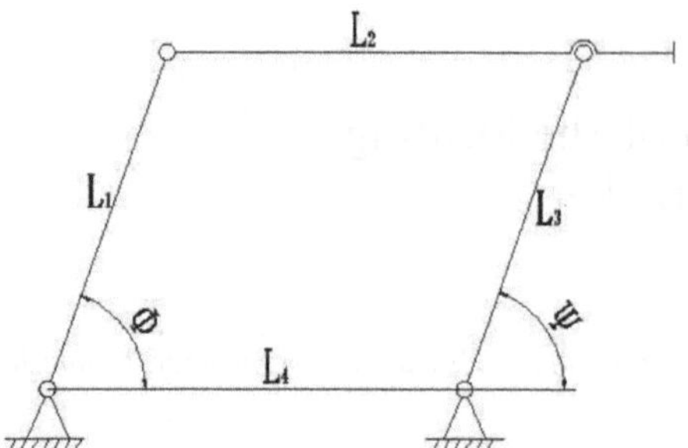

Figure 1. Linkage mechanism of the welding fixture of rear axle bracket of A4.

Then, the error of output angle ϕ_3 is

$$\Delta\phi_3 = \frac{\partial\phi_3}{\partial\phi_1}\Delta\phi_1 + \frac{\partial\phi_3}{\partial l_1}\Delta l_1 + \frac{\partial\phi_3}{\partial l_2}\Delta l_2 + \frac{\partial\phi_3}{\partial l_3}\Delta l_3 + \frac{\partial\phi_3}{\partial l_4}\Delta l_4 \tag{5}$$

In the analysis of error $\Delta\ l_l$ which induces the error of output angle ϕ_3, ϕ, l_1, l_2, l_3 and l_4 are constants. Meanwhile, $\Delta\ l_1$ causes errors of ϕ_3 and ϕ_1 simultaneously. By computing partial derivative for formula (4), we obtain

$$e^{i\phi_1} + il_2\frac{\partial\phi_2}{\partial l_1}e^{i\phi_2} = il_3\frac{\partial\phi_2}{\partial l_1}e^{i\phi_3} \tag{6}$$

By dividing the both sides of formula (6) with $e^{i\phi_3}$ and acquire the real part, we obtain

$$\cos(\phi_1 - \phi_2) = -l_3\frac{\partial\phi_3}{\partial l_1}\sin(\phi_3 - \phi_2) \tag{7}$$

$$\frac{\partial\phi_3}{\partial l_1} = -\frac{\cos(\phi_1 - \phi_2)}{l_3\sin(\phi_3 - \phi_2)} \tag{8}$$

Likewise, it is obtained that

$$\frac{\partial\phi_3}{\partial\phi_1} = \frac{l_1\sin(\phi_1 - \phi_3)}{l_3\sin(\phi_3 - \phi_2)} \tag{9}$$

$$\frac{\partial\phi_3}{\partial l_2} = -\frac{1}{l_3\sin(\phi_3 - \phi_2)} \tag{10}$$

$$\frac{\partial\phi_3}{\partial l_3} = \frac{\cos(\phi_3 - \phi_2)}{l_3\sin(\phi_3 - \phi_2)} \tag{11}$$

$$\frac{\partial\phi_3}{\partial l_4} = \frac{\cos(\phi_2)}{l_3\sin(\phi_3 - \phi_2)} \tag{12}$$

By substituting the error and corresponding partial derivative in formula (5), the error of putout angle Ψ and thereby the error of displacement s are obtained.

3 DISTRIBUTION OF TOLERANCES FOR EACH LINKAGE

The above analysis shows that $\frac{\partial\phi_3}{\partial l_1}$ 和 $\frac{\partial\phi_3}{\partial l_3}$ vary with ϕ_1 most significantly. Therefore, to avoid the output angle Ψ (displacement s) fluctuates with ϕ_1, the tolerances $\Delta\ l_1$ and $\Delta\ l_3$ of l_1 and l_3 need to be little. While, as $\frac{\partial\phi_3}{\partial l_2}$ 和 $\frac{\partial\phi_3}{\partial l_4}$ changes slightly, the tolerances $\Delta\ l_2$ and $\Delta\ l_4$ of l_2 and l_4 can be large values.

Suppose that the length tolerance of each linkage is $\Delta\ l_i$, the limit deviation of output angle Ψ caused by the length error is

$$\Delta\psi = \sqrt{\sum_{i=1}^{n}\left(\frac{\partial\psi}{\partial l_i}\right)^2\left(\frac{\Delta l_i}{2}\right)^2} \tag{13}$$

In the design of linkage, same precision degree is generally expected to be adopted for the length tolerance, that is, the tolerance increases with the increase of the length of linkage. In the case that $\Delta\ l_i = e$, $\Delta l_i = l_i/l_1\ e$. $\Delta\Psi$ should be less than the design requirement $[\Delta\Psi]$ of the fixture, that is,

$$\Delta\psi = \frac{e}{2}\sqrt{\sum_{i=1}^{n}\left(\frac{\partial\psi}{\partial l_i}\right)^2\left(\frac{\Delta l_i}{2}\right)^2} \le [\Delta\psi] \tag{14}$$

that is

$$e \le \frac{2[\Delta\psi]}{\sqrt{\sum_{i=1}^{n}\left(\frac{\partial\psi}{\partial l_i}\right)^2\left(\frac{\Delta l_i}{2}\right)^2}} \tag{15}$$

As the lengths of each linkage are known, their tolerances are calculated according to the lengths.

4 ESTIMATION OF LOCATING PRECISION

In the design of fixture, the sizes and tolerances of each linkage can be designed according to the calculation results. However, the exact precision of the processed workpieces is difficult to be determinated. Aiming at this, probability statistics method was adopted to estimate the possible range of the position error of linkages.

Suppose that the mean square error of error Δl_i is $\sigma_i(i = 1, 2, 3, 4)$. As all the errors Δl_i are independent, according to the operation law of variance, the mean square error of the motion position error Δs of working point is

$$\sigma_s = \sqrt{\sum_{i=1}^{4}\left(\frac{\partial s}{\partial l_i}\sigma_i\right)^2} \tag{16}$$

The limit values of errors Δl_i and Δs are $\pm\delta_i(i=1,2,3,4)$ and $\pm\delta s$. By substituting the ratio of limit error to the mean square error in formula (16), we obtain

$$\delta_s = k_s\sqrt{\sum_{i=1}^{4}\left(\frac{\partial s}{\partial l_i}\frac{\delta_i}{k_i}\right)^2} \tag{17}$$

As error Δl_i is independent, the quantitative relation between the limit error δ_s and original limit error δ_i can be estimated using formula (17). $\partial s/\partial l_i$ is the error coefficient. k_s and k_i (i = 1, 2, 3, 4) are determinated based on the proportion distributions and confidence probabilities of each error through probability statistics. Owing to the original error Δl_i is subject to normal distribution, the overall error Δs complies to normal distribution as well. When the confidence probabilities are the same, $k_s = k_i$, then formula (17) is rewritten as

$$\delta_s = \sqrt{\sum_{i=1}^{4}\left(\frac{\partial s}{\partial l_i}\delta_i\right)^2} \tag{18}$$

5 CONCLUSION

In the design of planar linkages of the welding fixture of rear axle bracket, mathematical model can be applied to discover the leading factors influencing the error. Therefore, the lengths and tolerances of each linkage can be designed reasonably. On this basis, after calculating the tolerances of each linkage, the precision of the overall linkage mechanism can be computed through the method of probability statistics.

REFERENCES

[1] Danian Hua. Zhihong Hua. Jingping Lv. 1995. Design of linkage mechanism. Shanghai: Shanghai Scientific and Technical Publishers.

[2] Hongzhong Li. 2005. Error analysis of planar linkages, Machine Building & Automation. 24–26.

[3] Binghui Fan. Zhenxu Feng. Li Yunjiang. 1998. Locating error of four-bar boom movement. 17–19. Mining & Processing Equipment.

[4] Zongren Qian. 2001. Error analysis of planar linkage mechanism. 12–13. Guangxi Machinery.

[5] Xiuye Wang. Mingqin Zhang. 2002. Analysis of position error of four-bar linkage. 81–83. Journal of Shandong Jianzhu University.

[6] Jiyuan Zhang. 1996. Mechanism Computation. Beijing: National defence industry press.

[7] Zeshao Fu. 1988. Mechanism Design. Sichuan: Chengdu University of Science and Technology Press.

Material Science and Environmental Engineering – Chen (Ed.)
© 2016 Taylor & Francis Group, London, ISBN 978-1-138-02938-5

Study on welding repair process of turbine blade

W.J. Gou
Beijing Polytechnic, Beijing, China
Tianjing University of Technology and Education, Tianjing, China

D.J. Meng, L.H. Wang & W. Hu
Beijing Polytechnic, Beijing, China

ABSTRACT: Martensitic stainless steel (0Cr13Ni5Mo) is widely used in the manufacture of large turbine blades. When the turbine is running, sand abrasion and cavitation in the high-speed flow of water, seriously damage turbine blades, at the same time cavitated blade units can generate severe vibration and noise. Cavitation wear and blade vibration reduce the efficiency of power generation, and also seriously threaten the safe operation of hydropower stations. Welding cavitated wear parts are an important means of turbine blade repair, but in order to shorten the period of power plant's shut down for maintenance, developing the technology of impeller in-situ repair based on the robotic automated job is of great importance; and using martensitic stainless steel as the base material for welding of turbine blades, the welding process parameters and welding specification is critical in situ repair.

Keywords: repair of turbine blade; welding process; welding experiment

1 FEATURES OF WELDING JOB AND MODE OF METAL TRANSFER

1.1 *Characteristics of turbine cavitation and program of welding repair*

There are mainly two means of repair welding operations of large turbine blade: inside-pit and outside-pit repair. Inside-pit repair is in-situ repair, and it repairs without dismantling and moving turbine blades; because of short maintenance cycle and high efficiency, this method is widely used. Repair process is divided into three stages: cleaning surface before welding—welding—grinding and painting after welding towards cavitated parts [1, 2]. While impellers of large francis turbine are mostly in a vertical position when repairing, the type of welded line is vertical or horizontal welding; combining the forming direction of cavitation stripe and direction of robot's welding, vertical welding repairing is adopted during welding process of turbine, which is helpful to repairing welding and grinding modification after welding [3, 4]. Figure 1 shows the prototype of turbine cavitation.

1.2 *Welding form and globular transfer*

Because base material for the turbine blade is ultra-low carbon stainless steel, welding process is achieved by cavitation site welding, and surfacing to form is mainly used during welding process, the welding process requires a fast and efficient deposition rate and MIG welding is used; because formation trends and characteristics of Francis turbine cavitation, vertical welding is mainly adopted. The high welding deposition rate of MIG/MAG welding is suitable for welding of non-ferrous metals—martensitic stainless steel, and it has three kinds of transferring forms: spray transfer, drop-like transfer, and short circuiting transfer, Figure 2 shows the welding parameters relationship of MIG, MIG/MAG welding [5, 6].

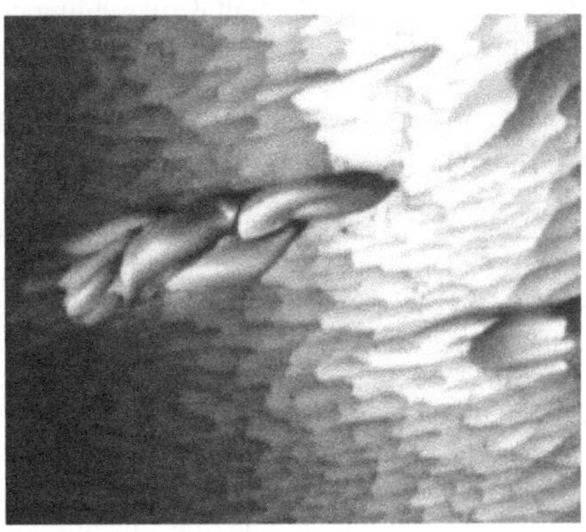

Figure 1. Prototype of turbine cavitation.

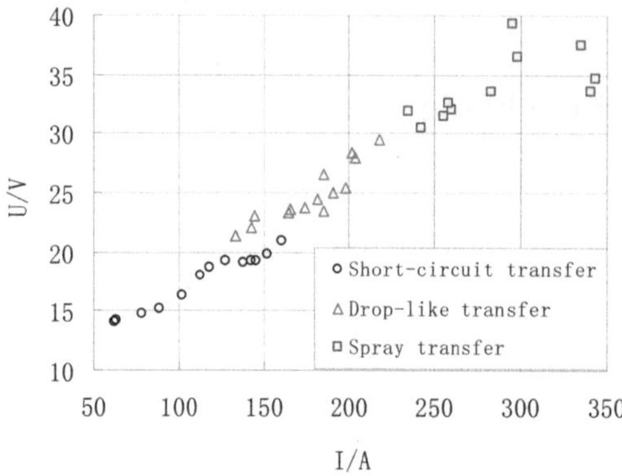

Figure 2. Welding parameters relations of MIG, MIG/MAG.

2 STUDY ON WELDING PARAMETERS OF MARTENSITIC STAINLESS STEEL

2.1 *Research methods and contents*

Compared by experiment method, martensitic stainless steel vertical welding and mode of metal transfer forms are selected; the impact of different modes of metal transfer on the welding effect is tested; better welding fabrication is achieved by changing the welding parameters.

Design observation equipment for globular transfer and the morphology of molten bath to know the situation of droplet transfer and characteristics of the morphology of molten bath; design gouging groove of bar type cavitation and surface cavitation to make surfacing experiments, and change the welding process parameters for surfacing, to determine welding process parameters of martensitic stainless steel under the vertical posture [7, 8].

2.2 *Equipments and conditions of the experiment*

1. Panasonic-YD500GR3 digital arc welding power source; (2) KUKA industrial robots, for holding a welding torch, to change torch inclination and achieve automatic welding; (3) high-speed camera to observe and record droplet transfer; (4) shielding gas: Ar and Ar + O_2; (5) the wire for welding is 0Cr13Ni5Mo of 1.2 mm diameter; materials of test specimens is 1Cr13, welding parts analog test specimens of bar cavitation gouging groove. In order to meet the vertical posture of the turbine blade under actual working conditions, tests were carried out by using the mold of vertical welding, in welding experiments the steel plates are fixed to the vertical mold which is at 90 degree angle with horizontal plane, and the imaging system is used to shoot and record the state of melting drop and the molten bath.

2.3 *Effect of process parameters on welding fabrication*

2.3.1 *Effect of modes of metal transfer on welding fabrication*

During the test vertical welding position is simulated in welding, so the specimen is perpendicular to the horizontal table, and make experiments for three transitional forms in MIG/MAG welding. High-speed cameras are used to record droplet shedding process and shape of molten bath.

2.3.1.1 Experiment of spraying transfer

During the process of spraying the set voltage is 260 A; set current is 30 V; extension of welding wire stem is 18 mm, welding speed v = 12.5 mm/s; protective gas is argon; welding torch is up at 15 degrees angle with horizontal (Fig. 7), welding direction is from top to bottom, and during welding process arc of welding is stable. Figure 3 shows the high-speed camera shooting of top-down welding spray transfer in vertical welding posture. Since stable arc welding during the spray transfer makes welding wire and base material form a continuous molten "metal flow" near the weld pool, high welding heat is released during welding, molten metal cannot be cooled in a relatively short period of time, and continuously flows downwardly under the force of gravity, forming a discontinuous weld and melt tumors, the effect of forming surfacing cannot be achieved.

With the same parameters, when the welding direction is from bottom to top, the welding quality is worse, the molten metal flowing down, basically cannot form a weld, only forming non-continuous tumors or flow-like metal-cover plate.

2.3.1.2 Droplet transition experiment

During droplet transition experiments set voltage is 180 A; set current is 30 V, during welding machine displays current of 167–173 A, voltage of 29 V; extension of welding wire stem is 16 mm, welding speed v = 10 mm/s; shielding gas is argon. Figure 4 shows high-speed photography of droplet transition in the vertical welding posture. Droplet transfer forms a large number of welding tumors, but not a continuous weld, further more droplet transition is of low efficiency, and is not suitable for large-scale welding work, therefore, droplet

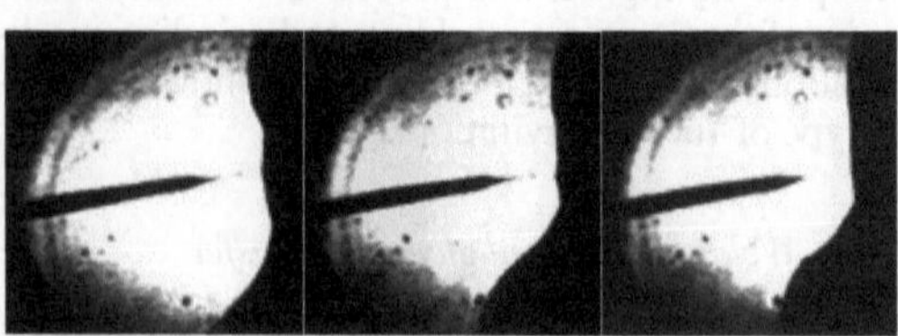

Figure 3. High-speed camera shooting of welding spray transfer in vertical posture.

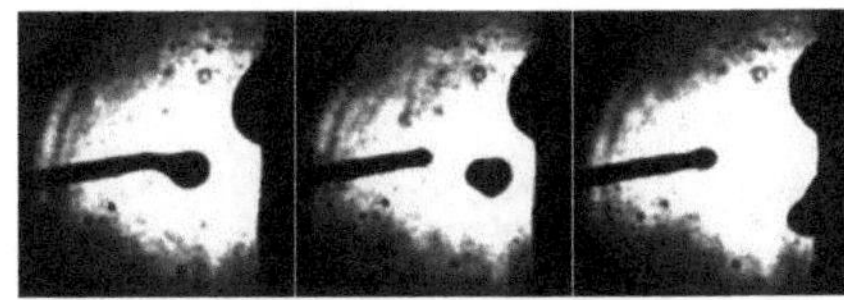

Figure 4. High-speed photography of droplet transition in vertical welding posture.

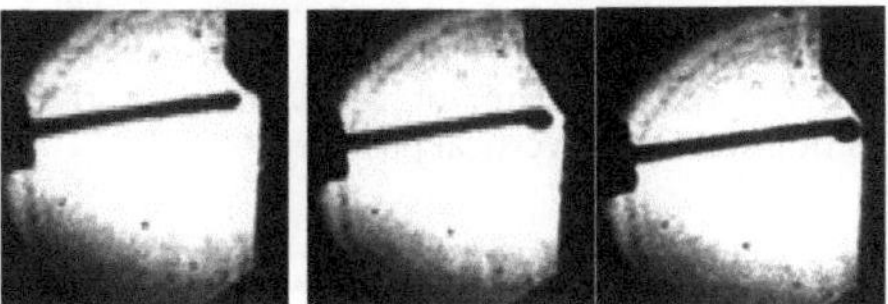

Figure 5. High-speed camera shooting of droplet transfer form in vertical welding posture.

transition can not complete the welding process of bar shape cavitation and surface type cavitation.

2.3.1.3 Short-circuit transfer experiments

During short-circuit experiment set voltage is 150 A; set current is 18 V, during welding machine displays current of 135–148 A, shows voltage of 17.5 V; extension of welding wire stem 12 mm, welding speed v = 5 mm/s; shielding gas is argon; welding is in downward direction, welding torch is at an angle of 15 degrees. Figure 5 shows short circuit in vertical welding posture, which recorded a welding cycle of short circuit transition, from droplet generation process of growing up, to short-circuit process of droplet's contact with the molten bath. During downward welding short circuit transition forms is a continuous weld bead, the effect of weld bead is better, you can make it droplet transfer form in vertical welding posture, and it can be used in welding processes of bar type Cavitation and surface Cavitation.

2.3.2 *The influence of the shielding gas on vertical welding molding*

The shielding gas will affect on welding molding, metal wettability, droplet flowing ability, air holes and excess weld metal reinforcement in weld line. In the experiment for the welding characteristics that turbine blades are stainless steel, pure Ar and mixed gases of 90% Ar + 10% O_2 are used as the shielding gas in the welding experiments. Welding effects are shown in Figure 6: comparisons of welding effects of different shielding gas. In the left two weld lines pure Ar is used as the shielding gas, while the first weld on the right gas mixture of 90% Ar + 10% O_2 is used as shielding gas.

Analyzing from the weld appearance: when pure Ar is used as shielding gas, due to the large viscosity and surface tension of the liquid metal, its fluidity

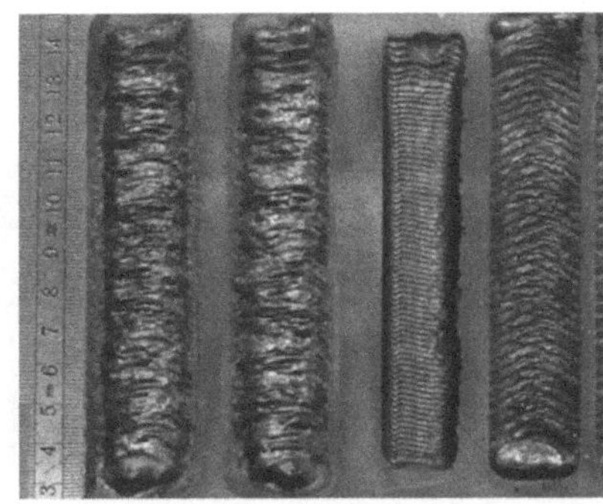

Figure 6. Comparing the welding effect of different shielding gas.

is bad, and there are obvious signs of swinging, big undulation and rough texture on the surface of the weld. Height of welding bead is not easy to achieve the welding requirement of less than 1 mm. After using 90% Ar + 10% O_2 gas mixture, due to the addition of O_2, welding undulation gets smaller, welding texture has been greatly improved, the experiment showed that using gas mixture of 90% Ar + 10% O_2 as protective gas can effectively control the height of welds, can get a steady welding state by optimizing the welding parameters to achieve the requirements of welding process.

2.3.3 *Swing amplitude, swing frequency and welding speed*

The width of MIG/MAG short-circuit transition must be greater than the width of the turbine blade cavitation, therefore, the use of the swing in the welding process is necessary, and the test specimens of swing amplitude, swing frequency, welding speed are shown as in Figure 7. According to the experimental data, the relationship between swing, frequency, and welding speed can be drawn from the combination. The width of welding bead: swing amplitude × 2 + 1; swing amplitude should ensure two weld beads overlapped 1/3–1/2, i.e. 1 Hz welding speed is 2.5–3.5 (mm/s), as shown in the below Figure 7: bead 5 and 9 match better, while bead 6 and 10 match worse; when other conditions remain unchanged, swing frequency is increased, undulation of bead can be reduced, so that the weld is more smooth.

2.3.4 *Influence of welding direction, inclination of the welding torch on welding molding*

In welding experiment the two kinds of welding—from top to bottom and from bottom to top welding in vertical welding posture—are compared. Welding process of turbine blades is automatic welding. In the welding process the welding parameters cannot change with metal solidification speed in weld bead, during bottom-up welding, before the bead beneath

Figure 7. Matching of swing amplitude, swing frequency and welding speed.

weld pool the has been cooled, there is new more molten metal in weld pool, it will not cool down and flow. During process, since the metal below the weld pool is cool, the molten metal in weld pool can be rapidly cooled. It is verified in the experiment that in vertical welding, top-down welding is better.

In the welding experiment $Ar + O_2$ is used as shielding gas, the metal surface tension is reduced, and the flowing ability of liquid metal is increased, which is not conducive to vertical welding. To mitigate this disadvantage, during welding the welding torch is made at an angle of 105 degrees with weld bead, to make the welding torch incline up to 15 degrees. In Figure 7 weld lines of upper and lower parts are compared, in welding experiments top-down welding in vertical welding position is applied. When the torch, arc welding process by means of droplet transfer thrust upward component of the liquid metal flows under the influence of gravity offset, as shown in Figure 7 comparing higher half with the corresponding lower half of the weld points: the tilting of torch can effectively control the flow of liquid metal.

3 SHORT-CIRCUIT TRANSITION WELDING PARAMETERS OF MIG/MAG WELDING IN VERTICAL POSTURE

The mode of metal transfer of MIG welding is related to the factors such as types of welded metal and welding wire, extension of wire diameter and stem, composition of shielding gas, welding current and voltage of arc welding, etc. In the experiments top-down welding in vertical welding posture is used, torch tilts up to 15 degrees, droplet transfer is short-circuit transition, wire is 0Cr13Ni5Mo of 1.2 mm in diameter, extension of wire stem is 12 mm, shielding gas is 90% Ar + 10% O_2, swing frequency is 0.45 Hz, swing amplitude is 8.5 mm, and the average welding speed is 75 mm/min. The weld appearance when average voltage is 18.5 V and current is 14.5 A is shown in Figure 8.

Figure 8. Welding seam of MIG/MAG short-circuit transition welding in vertical posture.

4 CONCLUSION

Experiments show that adjusting the welding parameters can achieve stable MIG/MAG short-circuits transfer welding process parameters. Matching of the shielding gas, welding direction, inclination of welding torch, torch swing amplitude, swing frequency, welding speed, and bead sorting, provides a foundation for the realization of automatic welding process of large turbine blades Cavitation. The process based on this experiment can make reference for automatic welding process issues like martensitic stainless steel or high carbon alloy.

REFERENCES

[1] Xue Wei, Chen Zhaoyun. The theoretical study on abrasion and cavitation in hydraulic machinery. *Electric Machine and Hydraulic Turbine*. Vol. 44 (1999) No. 6, p. 44–48.

[2] Jin Xiaoqiang, Ji Linhong, etc. Machining method research of the on site working robot for turbine blades maintaining. *Machinery Design and Manufacture.* Vol. 45 (2008) No. 1, p. 186–188.

[3] Fang Chaoqun. Cavitation and restoration of 700 MW water turbine blades in Three Gorges Power Plant. *Hydropower and New Energy*. Vol. 92 (2010) No. 6, p. 1–2.

[4] Tu Yangwen, ChenGang, etc. Turbine automation research status in maintenance equipment. *Welding Technology*. Vol. 35 (2006) No. 4, p. 9–12.

[5] Yu Jianrong, Jiang Lipei, Shi Yaowu. CO_2 arc welding Droplet transfer Charact eristic parameters quantitative evaluation. *Chinese Journal of Mechanical Engineering.* Vol. 38 (2002) No. 2, p. 137–140.

[6] Peng Yajie, Yu Jianfeng, LI Yun, etc. Discussion on large-area repair welding technology control of hydraulic turbine. *Mechanical and Electrical Technique of Hydropower Station*. Vol. 34 (2011) No. 4, p. 35–39.

[7] BaoYefeng, Zhou Yun, Wu Yixiong, etc. Study on metal transfer of MIG/MAG. *Electric Welding Machine.* Vol. 36 (2006) No. 3, p. 55–58.

[8] Yuan Minzhe, Ye Jianlin, Mao Hui, etc. Stainless steel plate welding research and application. *Metal Materials and Metallurgy Engineering*. Vol. 39 (2011) No. 5, p. 36–39.

Material Science and Environmental Engineering – Chen (Ed.)

Requirement analysis research of a new type armored anti-riot vehicle based on QFD and AHP

C.Z. Ma, Y.X. Ling & W.X. Zhang
Officers College of Chinese Armed Police Force, Chengdu, China

ABSTRACT: Aiming at the problem how to solve the demand importance of the new armored anti-riot vehicles in requirement analysis stage, a new method based on the combination of QFD and AHP was proposed and applied in the research. The method regarded the troops' actual requirement as the starting point, determined the force demand importance ultimately through cluster analysis on the requirements of the army, and showed the analysis results vividly through the house of quality. The results show that, this method can effectively overcome the shortcomings of qualitative analysis of traditional QFD method and the troops demand importance is quantified. In addition, the conviction of requirement analysis results of the new armored anti-riot vehicle is improved

Keywords: QFD; armored anti-riot vehicle; requirement analysis; quality house; AHP

1 INTRODUCTION

Requirement analysis is the starting point of the equipment construction. The requirement analysis work of the new armored anti-riot vehicle should adhere to the task requirements [1], and give full consideration to the actual needs of the troops. The main contents include gaining the real requirements of the armored anti-riot vehicles, organizing and analyzing the requirements, calculating the demand importance with the method of AHP, carrying out the quality plan, and establishing the house of quality.

2 THE QFD METHOD

Quality Function Deployment, or QFD, is a method of product development which driven by the customer demands and can maximize meet the needs of customers [2]. The basic process mainly includes four stages, namely the customer requirements acquisition, the requirements hierarchy analysis, the confirming for the importance of customer requirements, and the quality planning. The core of QFD is to build the house of quality. House of quality, or HOQ, is a visual image of the two element matrix expansion graph [3], the relation matrix is the most important tool of HOQ. Figure 1 shows the basic structure of HOQ.

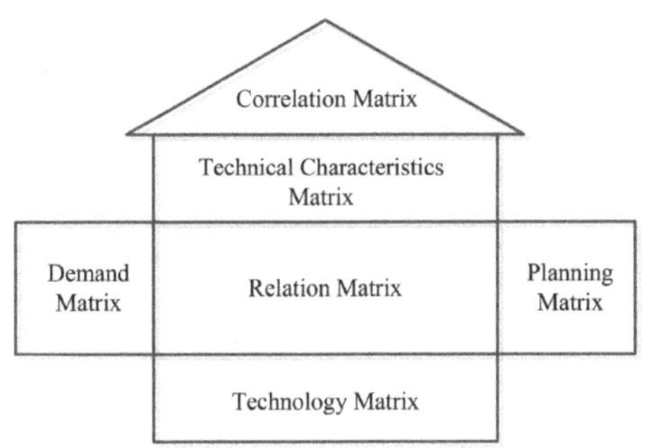

Figure 1. Fundamental structure of HOQ.

3 THE AHP METHOD

AHP is a multi-objective decision analysis method of a combination of qualitative analysis and quantitative analysis [4]. Through establishing the judgment matrix, the complex system can be hierarchical, and qualitative and quantitative factors can be combined in the process of decision analysis, which largely eliminates the limitations of human [5]. Because AHP method has obvious advantages, it is suitable for the application in the QFD to determine the demand importance.

The implementation of AHP method consists of the following steps:

1. Analyzing all kind of requirements' relationship, establish the hierarchical structure;
2. According to the scoring criteria, the requirements must be compared with each other, then estimate the quantitative judgment matrix of the demand importance;

3. Aiming at customer requirements for each project, calculate the geometric average weights, and carry on the consistency check.

4 INTEGRATED APPLICATIONS OF QFD AND AHP

4.1 *Requirements elicitation*

Requirements elicitation is the successful implementation assurance of QFD; requirements can be obtained through questionnaires, face-to-face panels, telephone surveys and other market research methods. This paper mainly adopts checking information, interviewing experts, researching troops and other research units to acquire the original requirements of the new armored anti-riot vehicles.

4.2 *Hierarchical analysis of requirements*

The raw data acquired must be trans-formed, translated and arranged, which is converted into something for a quality house. Affinity diagram method [6] can streamline and cluster the demand information effectively, so as to get the hierarchical customer demand. The paper uses this method to cluster and analyze the new armored anti-riot vehicles' original demand.

Firstly, the obtained army needs are recorded in the cards for the first affinity classification, as to get the third demand level.

Secondly, the similar content cards are organized together for the second affinity classification. Then similar needs in third level belong to the same demand, as to get the second demand level.

Finally, the hierarchical structure of the force requirements in accordance with the "first level", "second levels", "third levels" are to fill in the corresponding position, as shown in Table 1.

4.3 *The demand importance*

The demand importance refers to the quantitative score of the requirements from customers, it is used to display the various demand importance for customers, which is the essential evaluation index in QFD. Combined with the actual needs of the troops, the research uses the AHP method to analyze and calculate the customer demand

Table 1. Troops demand deployment table of the new type armored anti-riot vehicle.

The first level	The second level	The third level	Number
The new type armored anti-riot vehicle with satisfactory for the army	Good motor performance	Good rapidity;	C_1
		Good trafficability;	C_2
		Good transportability	C_3
	Good thermal performance	Capture the target quickly;	C_4
		Good reactivity of weapons;	C_5
		Hit the target accurately;	C_6
		Destroy the target effectively	C_7
	Good protection performance	Bullet-proof;	C_8
		Anti-violence;	C_9
		Anti-explosion;	C_{10}
		Anti-collision;	C_{11}
		Good protection in-car	C_{12}
	Good communication and command performance	Good control performance of information management;	C_{13}
		Good confidentiality of information transmission	C_{14}
	Excellent human-machine environment	Convenient operation;	C_{15}
		Comfortable ride;	C_{16}
		Low noise in-car	C_{17}
		Small vibration	C_{18}
		Suitable temperature in-car	C_{19}
	Good reliability, maintainability and supportability	Low failure rate;	C_{20}
		Long life length;	C_{21}
		Complete supporting facilities;	C_{22}
		Easy to pair;	C_{23}
		Powerful guarantee	C_{24}
	Economic rationality	Low price;	C_{25}
		Gas saving;	C_{26}
		Low maintenance costs	C_{27}

importance. An AHP questionnaire survey has been conducted previously among many drivers, captains, car repair personals and industry experts of the armored anti-riot vehicles. Use comparison methods by contrasting each other, and implement AHP in the second level and the third level.

4.3.1 *Demand importance calculation in the second level*

1. List the demand importance evaluation matrix, as shown in Figure 2.
   According to the score standard in Table 2, each demand in second levels should be compared with each other, and then constructs the quantitative demand importance evaluation matrix.
   Judgment matrix denoted as $A=(a_{ij})_{n\times n}$, a_{ij} refers to the importance, which the *i* factor compared to the *j* factor, $a_{ij}=1/a_{ji}$. A total of 10 experts are invited to judge, getting ten judgment matrixes.

$$A_1=\begin{bmatrix} 1 & 1/3 & 1/3 & 3 & 5 & 1 & 5 \\ 3 & 1 & 2 & 5 & 7 & 4 & 7 \\ 3 & 1/2 & 1 & 5 & 7 & 3 & 7 \\ 1/3 & 1/5 & 1/5 & 1 & 1 & 1/3 & 3 \\ 1/5 & 1/7 & 1/7 & 1 & 1 & 1/5 & 1 \\ 1 & 1/4 & 1/3 & 3 & 5 & 1 & 5 \\ 1/5 & 1/7 & 1/7 & 1/3 & 1 & 1/5 & 1 \end{bmatrix}\cdots$$

$$A_{10}=\begin{bmatrix} 1 & 1/3 & 1/5 & 3 & 3 & 1/3 & 5 \\ 3 & 1 & 1/3 & 5 & 4 & 1 & 9 \\ 5 & 3 & 1 & 7 & 8 & 3 & 9 \\ 1/3 & 1/5 & 1/7 & 1 & 2 & 1/5 & 3 \\ 1/3 & 1/4 & 1/8 & 1/2 & 1 & 1/5 & 3 \\ 3 & 1 & 1/3 & 5 & 5 & 1 & 5 \\ 1/5 & 1/9 & 1/9 & 1/3 & 1/3 & 1/5 & 1 \end{bmatrix}$$

2. Geometric average weights and the consistency check
   Calculate the vector $W=(w_1, w_2, ..., w_i, ..., w_n)^T$, and the formula is as follows:

$$\overline{w_i}=n\sqrt{\prod_{j=1}^{n} a_{ij}} \quad (1)$$

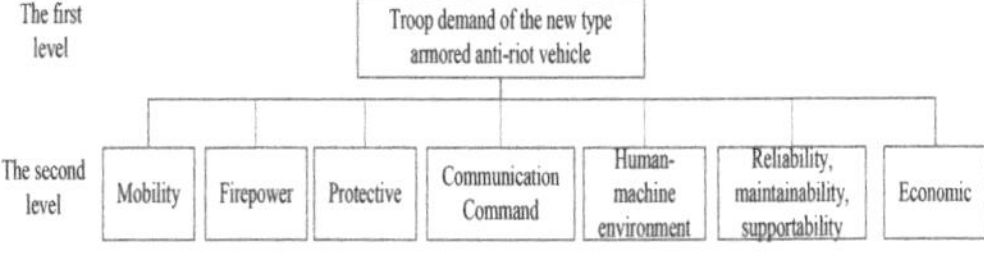

Figure 2. Troops demand second-level structure of the new type armored anti-riot vehicle.

Table 2. Contrast score standard of demand importance.

The value	Explanation
1:1	Both equally important
3:1	The former is slightly more important than the latter
5:1	The former is more important than the latter
7:1	The former is very more important than the latter
9:1	The former is definitely more important than the latter
2, 4, 6, 8	Intermediate value of the above situations
Reciprocal value of above ratio	For the latter than the former

Table 3. *RI* value table.

The order of matrix	*RI*
1	0
2	0
3	0.58
4	0.9
5	1.12
6	1.24
7	1.32
8	1.41
9	1.45

$$w_i=\overline{w_i}\Big/\sum_{i=1}^{n}\overline{w_i} \quad (2)$$

In the formula, w_i is the importance value of the *i* factor?

Consistency test is determined by calculating the consistency ratio *CR*. The calculation formula is as follows:

$$CR=CI/RI \quad (3)$$

$$CI=\frac{\lambda_{\max}-n}{n-1} \quad (4)$$

In the formula, *CR* is the consistency ratio, *CI* is the consistency index, and *RI* is the average random consistency index. $\lambda_{\max}$ is the maximum Eigen-value of the judging matrix A, and n means the order of matrix A. The value of *RI* is shown in Table 3.

When $CR<0.1$, the evaluation matrix's consistency is acceptable; when $CR\geq 0.1$, the evaluation matrix needs to be modified.

By using the analysis software MATLAB to complete data analysis, make a demand judgment matrix consistency test table of the new armored anti-riot vehicles. After the inspection, all matrixes satisfy $CR < 0.1$, it's indicated that the judgment matrix is consistency, as shown in Table 4.

3. The demand importance calculation
Through taking the averagely weighting of the ten judgment matrixes given by experts, the comprehensive judgment matrix is obtained:

$$A = \begin{bmatrix} 1.000 & 0.367 & 0.379 & 2.500 & 1.453 & 1.333 & 3.521 \\ 2.850 & 1.000 & 2.083 & 3.750 & 2.733 & 3.353 & 4.750 \\ 2.833 & 0.963 & 1.000 & 3.750 & 2.250 & 3.423 & 4.500 \\ 0.333 & 0.213 & 0.177 & 1.000 & 0.825 & 0.342 & 3.000 \\ 0.767 & 0.459 & 0.552 & 1.125 & 1.000 & 0.879 & 2.364 \\ 0.778 & 0.243 & 0.375 & 2.250 & 1.500 & 1.000 & 4.250 \\ 0.345 & 0.149 & 0.183 & 0.333 & 0.517 & 0.375 & 1.000 \end{bmatrix}$$

Though calculating A, it's known that $\lambda_{max} = 7.403$; $CR = 0.051 < 0.1$, which states clearly that the comprehensive judgment matrix A is consistent. By the formula (1), (2), it may have the importance vector:

$$\begin{aligned} \overline{W} &= (\overline{w}_1, \ldots, \overline{w}_7)^T \\ &= (1.131, 2.670, 2.310, 0.522, 0.893, 1.002, 0.347)^T, \end{aligned}$$

normalized to obtain the demand importance of the second level:

$$\begin{aligned} W &= (w_1, \ldots, w_7) \\ &= (0.127, 0.301, 0.260, 0.059, 0.101, 0.113, 0.039). \end{aligned}$$

Table 4. Consistency check table of troops demand judgment matrix.

Number	1	2	3	4	5
λ_{max}	7.279	7.468	7.315	7.423	7.434
CR	0.035	0.059	0.040	0.053	0.055
Number	6	7	8	9	10
λ_{max}	7.265	7.362	7.184	7.446	7.358
CR	0.033	0.046	0.023	0.056	0.045

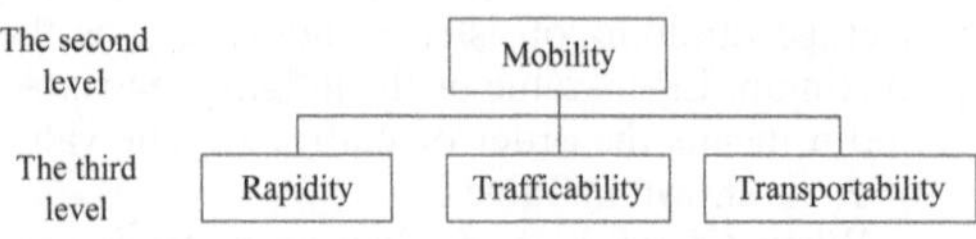

Figure 3. Motility demand three-level structure of the new type armored anti-riot vehicle.

Sorted the importance degree from small to large, the second level priority is as follows: firepower, protection, mobility, reliability, maintainability and supportability, human-machine environment, communication command, economic.

4.3.2 *Demand importance calculation of the third level*

Using the demand of the second level for the comparison criteria, it can get the demand of the third level in a layer of criterion, in accordance with the above method.

It's known that $(w_{11}, w_{12}, w_{13}) = (0.582, 0.309, 0.109)$, the other demand importance can be obtained in the same way. Because of limited space, there is no statistics.

4.4 *Quality planning*

The ultimate goal of quality planning is to establish the quality house [7]. In this research, the troops requirements is the left wall input in quality house (as space is limited, it only lists the troops demand of the second level); the roof of quality house gives the relationship between the equipment's characteristics (▲ means "positive correlation", and ▼ means "negative correlation"), it mainly depends on the experience of experts; in the relationship matrix between the force requirements and equipment characteristics, • says the strong correlation, ◎ said the middling correlation, ○ says the week correlation, and blank says not

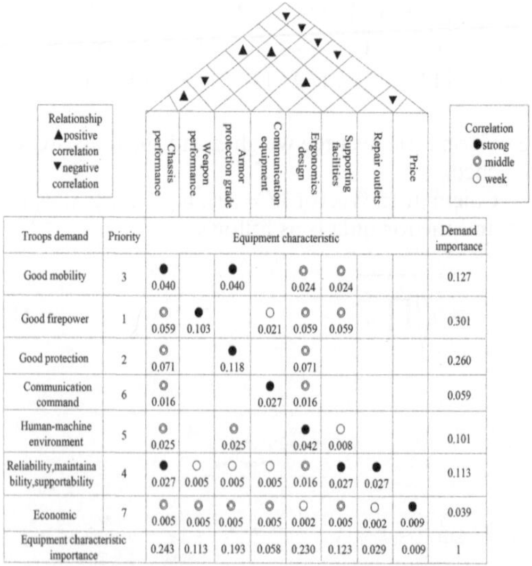

Figure 4. Requirement HOQ of the new type armored anti-riot vehicle.

related; according to the proportional distribution method [8], equipment characteristic importance can be obtained, for example the firepower demand importance 0.301 is distributed according to the proportion of 5:3:1 to the symbols •, ◎, ○, • is 0.103, ◎ is 0.059, ○ is 0.021, so the total is 0.059 + 0.103 + 0.021 + 0.059 + 0.059 = 0.301. The house of quality is shown in Figure 4.

5 SUMMARY

In this research, the QFD and AHP method is integrated, to identify and determine the real army needs of the new armored anti-riot vehicles, and analysis to determine the force demand importance. This method is not only quantifying in the whole requirement analysis process, to overcome the shortcomings of the traditional QFD method's qualitative analysis, but also making the research visual, to ensure the credibility and comprehensive of analysis results. The results show that the method can effectively solve this problem.

REFERENCES

[1] Wang Kai, Sun Wanguo. 2008. Military requirement demonstration for weapon equipment [M]. Beijing: National Defence Industry Press.

[2] Hao Yongjing. 2001. Study on the theory and method of quality function deployment [D]. Hebei University of Technology.

[3] Fan Bin. 2007. Several theoretical methods and applications for quality function deployment [D]. University of Qingdao.

[4] Dong Sihui, Su Bo. 2012. Improved analytic hierarchy process and its application in coal mine safety assessment [J]. Journal of Liaoning Technical University, 31(5):690–694.

[5] Cai Wei, Wu Bing. 2005. Application of analytic hierarchical process in mine ventilation evaluation [J]. Journal of Liaoning Technical University, 24(2):149–152.

[6] Xiong Wei. 2012. Quality function deployment: theory and methodology [M]. Beijing: Science Press.

[7] Che Ada, Yang Mingshun. 2008. Methods and applications for quality function deployment [M]. Beijing: Publishing House of electronics industry.

[8] Prasad B. 1998. Review of QFD and Related deployment techniques [J]. Journal of Manufacture System, 17(3):221–234.

Material Science and Environmental Engineering – Chen (Ed.)
 ISBN 978-1-138-02938-5

Experimental study of flow boiling heat transfer deterioration in helical-coiled tube

C.L. Ji
School of Energy and Power Engineering, Shandong University, Jinan, Shandong, China
Department of Municipal and Environmental Engineering, Shandong Urban Construction Vocational College, Jinan, Shandong, China

J.T. Han
School of Energy and Power Engineering, Shandong University, Jinan, Shandong, China

X.P. Liu
Department of Municipal and Environmental Engineering, Shandong Urban Construction Vocational College, Jinan, Shandong, China

Y.M. Feng
Shandong Poyry Engineering Consulting Company Limited, Jinan, Shandong, China

ABSTRACT: An investigation was experimentally carried out with R134a to explore flow boiling heat transfer deterioration characteristics and analyze their mechanism of heat transfer deterioration in horizontal helical-coiled tubes. In high quality area, in low mass flow rate, the medial wall temperature soars first generally, then the lateral wall temperature rises, but their take-off point spacing is small and take-off point is near the quality for 0.73; in high mass flow rate, the medial and lateral wall temperature soars nearly simultaneously, and rising point is near the quality for 0.78.

Keywords: heat transfer deterioration; horizontal helical-coiled tube; wall temperature rising

1 EXISTING RESULTS

Helically-coiled tube is extensively used in the nuclear, petrochemical, power generation, cryogenic and aerospace industries, as well as in refrigeration and heat pump systems, due to its high efficiency in heat transfer and compactness in volume. In the operation process of the equipment, if the heat load increases to a certain value, the wall temperature will increase rapidly in the case of constant pressure and the mass flow rate. This leads to the local heat transfer coefficient decreasing sharply in helically-coiled tube, so the heat exchange equipment burns. This phenomenon is called the heat transfer deterioration, and corresponding to the heat load is called the critical heat load. How to prevent or predict the flow boiling heat transfer deterioration has been the hotspot of scientific research [1–7]. Guo Liejin [8] studied flow boiling heat transfer deterioration in the high pressure water/steam loop in helically-coiled tube, and analyzed the condition and mechanism of the heat transfer deterioration. They pointed out the wall temperature increased dramatically when the heat transfer deterioration occurred, and the anterior and posterior wall temperature was greater than the other two. Chen [9] was studied experimentally flow boiling heat transfer characteristics of R134a in horizontal helically-coiled tube. The results showed that the wall temperature increased with the increasing of the heat load under the constant conditions. Crain [10] studied experimentally flow boiling heat transfer characteristics in vertical helically-coiled tube. The results showed that the heat transfer deterioration of the upper side and the lower side of the same section occurred first, and then the lateral, but the medial had maintained high heat transfer levels. Further pointed out the existence of the secondary flow is helpful to the stability of liquid membrane. Cumo [11] conducted the experiment on Dry-out characteristics of R12 in the vertical helically-coiled tube and the same diameter circle tube. The results showed that the critical heat flux in helically-coiled tube is higher than the circle tube, and rising rate of the wall temperature was relatively small when the heat transfer deterioration occurred.

In summary, many scholars had studied the flow boiling heat transfer characteristics in helical-coiled tube, but the characteristics and mechanism research of flow boiling heat transfer deterioration is still not deep enough. Therefore, the characteristics of the flow boiling heat transfer were studied experimentally in a wide quality range in the helical-coiled tube, and analysis the wall temperature distribution of the different sections.

2 EXPERIMENTAL DEVICE AND HEAT TRANSFER DETERIORATION STANDARD

2.1 *Experimental device*

The experiment was performed in two-phase flow test bench at the Institute of Refrigeration and Cryogenics in Shandong University. The experimental structure as shown in Figure 1. Mainly included the refrigerant loop, cooling water circuit, measurement system and data acquisition system. Preheating section and experimental section were used for direct heating regulated DC power supply. The pump provided power for the whole cycle. Refrigerant R134a was outputted by metering pump. Mass flow rate of R134a was measured by mass flow meter, and heated to the required conditions in the preheating, then continued to be heated and measured into the test section, into

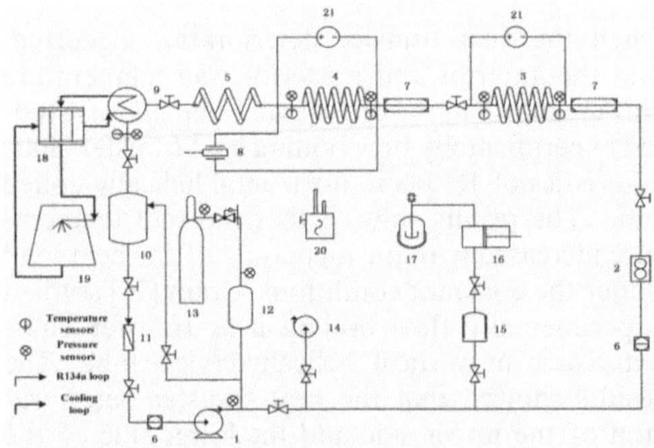

Figure 1. Schematic diagram of experimental system. 1-Plunger type metering pump; 2-Mass flowmeter; 3-Experimental section the helically-coiled tube; 4-Balancesection the helically-coiled tube; 5-The flow pattern observation section; 6-The observation mirror; 7-Entrance observation mirror; 8-Differential pressure transmitter; 9-Condenser; 10-The liquid storage tank; 11-Dry filter; 12-Pressure regulator; 13-Airbag; 14-Vacuum pump; 15-The buffer tank; 16-The refrigerant machine; 17-The refrigerant tank; 18-Water chilling unit; 19-Cooling tower; 20-Halogen leak detector; 21-Regulated DC power supply.

tube-in-tube condenser inverse heat transfer with coolant in the water chiller. Rl34a was cooled to liquid and stored in the liquid storage tank for continuous cycle.

The experimental section for $\phi 10 \times 0.84$ was bent using stainless steel 06Cr19Ni10, as shown in Figure 2a. Total Length 3200 mm; Effective Heating Length 2853.2 mm; Screw Diameter 300 mm; Pitch 45 mm. Along the helically-coiled circumferential every 45° spaced evenly 8 groups of T type thermocouples, as shown in Figure 2b; along the diameter each 45° arrangement 4 pair T type thermocouples, as shown in Figure 2c. The pressure sensors measured the pressure value in different positions in the experimental system. The data of the temperature, pressure, flow rate and the output signal in the experimental system was collected and pretreated by the Agilent34980 software. Along the flow direction, divided into the left side ($\beta = 90°$), right side ($\beta = 270°$), lateral ($\beta = 270°$) and medial ($\beta = 180°$).

2.2 *Judgement method of heat transfer deterioration*

The helical-coiled tube as experimental section. Its import and export were equipped with pressure transmitter and armoured thermocouple respectively. First, increasing heat load of the preheating section (the same type of material with experimental section); then gradually increased the heat load of the experimental section until the pressure, flow and the inlet quality reached the desired value; it was considered to begin the heat transfer deterioration when export wall temperature occurred pulsation or rising; then continued to increase the heat load, while observing change of the wall temperature near the exit. If the wall temperature was more than 20°C, it was thought to occur the heat transfer deterioration; repeated the above steps, until to obtain enough and repetitive data. Based on programmable function of acquisition software Agilent Bench Link Data Logger Pro., we established a method to determine the heat transfer deterioration.

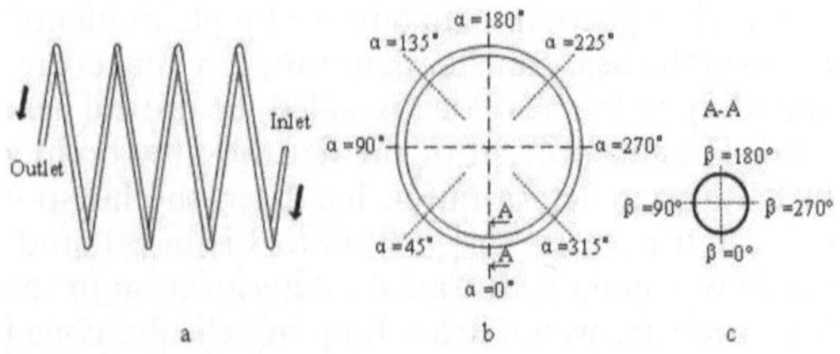

Figure 2. Experimental section and thermocouple arrangement.

The experimental parameters range: pressure P = 0.5~1.25 MPa; mass flow rate, G = 50~500 $kgm^{-2}s^{-1}$, heat flux density 0~70 kWm^{-2}, the entrance heat balance quality: X = –0.15–1.0.

3 EXPERIMENTAL RESULTS AND ANALYSIS

The wall temperature distribution curves in the helical-coiled tube is shown in Figure 3. In the low quality area (x < 0.65), the lateral (β = 0°) velocity is larger than the medial (β = 180°) because of the action of the centrifugal force. So the outside heat transfer ability is stronger than the medial, and the inside wall temperature is relatively lower. With the quality increasing, up to the 0.65–0.75 range, the lateral wall temperature is higher than the medial. The reason is the interaction of the gas phase entrainment and centrifugal force. The liquid film of the lateral point is quickly evaporated, while the medial point to maintain a certain thickness liquid layer because of the two flow, so the the medial heat transfer is better than that of medial lateral point. Under the experimental conditions, wall temperature rises rapidly and heat transfer deterioration occurs in the quality for 0.75–0.85 rang. In the high quality area (x > 0.65), the medial wall temperature rises first general in low mass velocity, then lateral wall temperature rises, but takeoff point spacing is smaller and the quality of soaring point is about 0.73. As shown in Figure3(a); the wall temperature of the medial and lateral rises almost at the same time in high mass velocity, and the quality of soaring point is about 0.73. As shown in Figure 3(b).

As Figure 4 shows the critical heat flux changes with the quality changing in the range of experimental temperature. In the low quality area (x < 0.3), the critical heat flux is very higher. The reason is the interaction of the macroscopic convection in the nucleate boiling region and the micro violent convection near the tube wall. The wall surface bubbles form, grow up and out, not only to take its latent heat, and the near wall superheat liquid flowing to the mainstream center. The mainstream center cold fluid flows to the position of the bubble departure and absorbs the heat of the boundary layer, so that the critical heat flux is relatively high. With the quality increasing, up until about 0.5, the critical heat flux density drops to the lowest point, there are prone to dry-out, resulting in the deterioration of heat transfer. The quality for 0.5–0.8 range, vapor velocity increasing, in favor of vapor-liquid interfacial shear stress increasing, so the film is thin. This intensifies the evaporation process from the liquid to the vapor, and enhances the heat transfer between the liquid and tube wall, and makes the critical heat flux density value larger. The quality up to 0.8, the critical heat flux density begins to reduce sharply. The main reason is that the annular flow liquid film near the wall becomes thin with the quality increasing, and torn eventually by the main steam, so the wall local can not covered by the film, resulting in wall temperature rising, eventually leads to the annular flow to be mist flow, dry-out again.

As Figure 5 shows the effect of heat flux density on the average heat transfer coefficient. For the

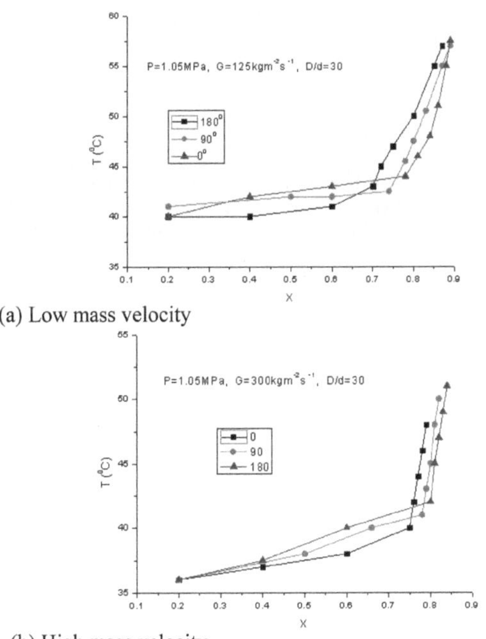

Figure 3. Curve of each point wall temperature change along the circumferential.

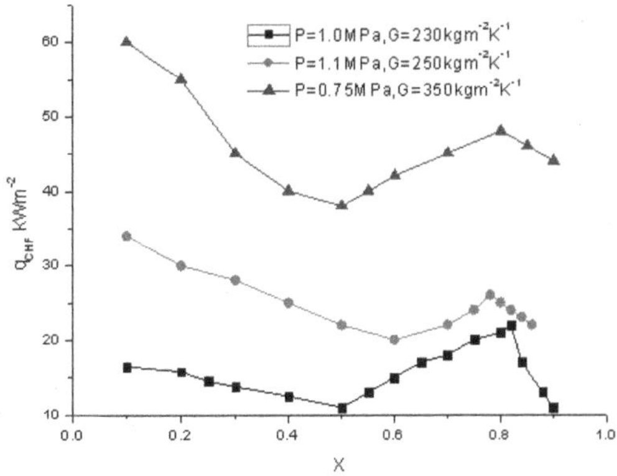

Figure 4. Critical heat flux vs X.

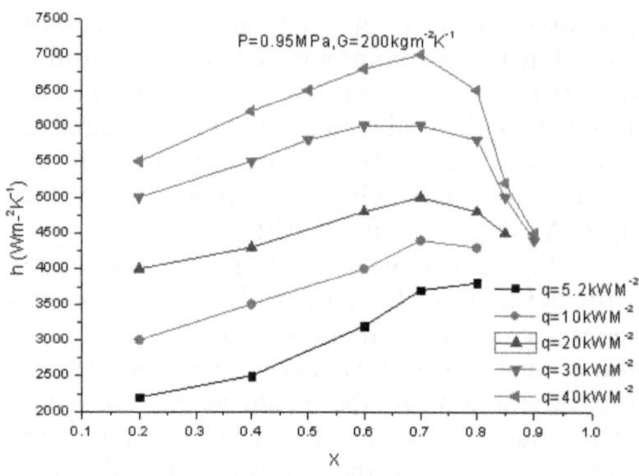

Figure 5. Average heat transfer coefficient vs X.

lower heat flux density, boiling heat transfer coefficient in the low quality increases with the increasing of the quality. The boiling heat transfer coefficient decreases with the increase of the quality when the quality reaches the critical value. The heat transfer coefficient begins to decrease quickly in the quality for 0.72 when heat flux density increased to 30 kW m^{-2}. This shows that high heat load is prone to appear the dry-out phenomenon, and the heat transfer deterioration occurs.

4 CONCLUSION

In this paper, using R134a as refrigerant, the experimental studies flow boiling heat transfer deterioration characteristics and the influence factors for the horizontal helical-coiled tube, the results of the study show, the quality for 0.75–0.85, the wall temperature rapidly soaring, the heat transfer deterioration occurs. The heat transfer coefficient is also in constant increase with the increase of heat flux, but the quality about 0.72, the heat transfer coefficient begins to decrease.

ACKNOWLEDGEMENT

This paper was supported by the National Natural Science Foundation of China (51076084).

REFERENCES

[1] Owhadi A. Bell k.J. Crain J. 1968. Forced convection boiling inside helically coiled tubes. International Journal of Mass Transfer 11(11): 1179–1793.

[2] Chen C.N. Han J.T. Shao L. 2010. Study on dry-out CHF characteristics of R134a flow boiling inhorizontal helically-coiled tubes. Nuclear Power Engineering 31(5): 76–80. (in Chinese).

[3] Ahmed M. Elsayed C. Raya K. 2012. Investigation of flow boiling heat transfer inside small diameter helically coiled tubes. International Journal of Refrigeration 35(8): 2179–2187.

[4] Shao L. Han J.T. 2013. Experimental investigation on flow patterns and pressure drop of R134a flow boiling in a horizontal helically-coiled tube. Journal of Enhanced Heat Transfer 20(3): 225–233.

[5] Chen C.N. Han J.T. Jen T.C. 2011. Dry-out CHF correlation for R134a flow boiling in a horizontal helically-coiled tube. International Journal of Heat and Mass Transfer 54(1–3): 739–745.

[6] Shao L. Wang M.X. Han J.T. 2013. Experimental investigation on two-phase flow patterns and wall temperatures of flow boiling in horizontal helically-coiled pipe. Proceedings of the CSEE 33(26): 81–86. (In Chinese).

[7] Ji C.L. Han J.T. Yin J. 2014. Fitting heat transfer coefficient correlation on flow boiling and error analysis for the helically-coiled tube. Journal of Shandong Uuniversity (engineering science) 44(5): 83–87. (In Chinese).

[8] Guo L.J. Chen X.J. 1994. An experimental research on the boiling heat transfer deterieration in horizontal helically steam generating tubes at high pressure. Journal of Nuclear Science and engineering 14(4): 289–295.

[9] Chen C.N. Han J.T. Jen T.C. 2011. Experimental study on critical heat flux characteristics of R134a flow boiling in horizontal helically coiled tubes. International Journal of Thermal Science 50(1): 169–177.

[10] Crain B.J. Bell K.J. 1973. Forced convection heat transfer to a two-phase mixture of water and steam in helical coiled tubes. AIChE Symp series 69(131): 30–36.

[11] Cumo M. Farello G.E. Ferrari G. 1972. The influence of curvature in post dry-out heat transfer. International Journal of Heat and Mass Transfer 15(11): 2045–2062.

Material Science and Environmental Engineering – Chen (Ed.)

Performance study on a novel design of a jet-inspired horizontal axis tidal stream turbine by a CFD numerical tool

M. Orji, S.Q. Hu & Z. Liang
College of Shipbuilding Engineering, Harbin Engineering University, Harbin, China

J. Yan
College of Hydraulic and Electric Engineering, Harbin Engineering University, Harbin, China

ABSTRACT: This study presents three-dimensional steady-state numerical simulations of a jet-inspired horizontal axis tidal stream turbine. The turbine was designed to be less bulky, light-weight, and could spin at a very low tidal speed while still maintaining a high performance coefficient. The objective of this work is to explore the potential of a small or micro horizontal axis tidal turbine in generating electrical power at a lower tidal current speed. Small tidal turbines of this nature are the predominate tidal power extracting devices in rural or urban areas where the traditional tidal turbines could not be installed due to low head current velocity or space constraints. This study is also intended to draw the attention of scholars that there is a possibility of extracting tidal energy from some of the available marine sites which might seem not to be a suitable tidal energy resource due to low head current flow. The performance analysis of the design is carried out using Ansys CFX-CFD numerical tool. At a design flow velocity of 2 m/s, the maximum power coefficient of 0.455 was obtained at the tip speed velocity ratio of 1.0. The influence of flow phenomenon such as pressure and velocity streamline distributions around the blade were carefully analyzed and presented. The strong tip vortices moving in a helical path behind the rotor axis were captured and presented. The tip vortex is responsible for inducing velocities. The optimum tip speed ratio of 1.0 at 150 rpm which gives the optimal power coefficient is relatively lower than the optimum tip speed ratio of 5.0 for a standard three-bladed tidal turbine. The good news about employing this jet-inspired horizontal axis tidal turbine is that it can be used to generate power in areas or tidal sites starting up at a lower current velocities.

Keywords: jet-inspired tidal turbine; tidal currents; tip speed ratio; grid resolution and power coefficient

1 INTRODUCTION

Reversing climate change is the dominant concern about modern age. Transport fuels, heat and power generation creates green house gases that pollutes and destroys the entire universe. Nature has given us an answer. The oceans have free energy in abundance and this power can be harnessed to protect the environment. Invisible, sustainable and practical ways to exploit the power of the tides exists and have to be implemented. Underwater tidal turbine is unique and it is an in-stream generator. A decade of research and development has resulted in the design that is economically and ecological sound. Harnessing tidal energy would make a massive difference supplying 15% of the world's energy. Tidal power provides a predictable and reliable source of electricity. Tidal energy is renewable and clean, which assures it will be able to meet the energy needs of future generations. Tidal current or tidal stream technologies convert the kinetic energy into useable energy. Tidal energy is created through the conversion of tidal kinetic energy into electrical energy, which can be used to replace unclean energy production, such as the burning fossil fuels (Rourke et al., 2009). The technology developments are comparable to the development of wind turbines; however a number of fundamental differences in their design and operation exist. Their optimal performance still remains critical as the quest for clean energy increases. One of the biggest differences between air and water, of course, is the density. Water density is about 1,025 kg/m^3. By comparison, the air, depending on pressure and temperature, ranges from about 1.1 to 1.4 kg/m^3. The difference is about a factor of 800 which represents a substantial physical load for a blade. The complexity of the maritime environment is critical and demands expertises in the design that can survive the extreme loadings imposed by surface waves and sea and yet remain cost-effective in extracting tidal energy. Tidal current technologies have made enormous strides in development toward commercialization in the past five to seven years.

Based on an overview of existing tidal current projects, 76% of all turbines are horizontal axis turbines and 12% are vertical axis turbines according to International Renewable Energy Agency (IRENA., 2014). The innovative tidal-turbine technology can be used all over the world. This design incorporates features which results in minimal cost and optimal performance. Low cost power generation guarantees a fast pay back of capital cost. Rotor blade is the key component of a turbine system, it extracts energy from the tide and primarily dictates the performance, loads and dynamics of the whole turbine system (Bir et al., 2011). Many researchers have studied tidal current power systems. (Jo et al., 2012) studied the performance of a horizontal axis tidal turbine by blade configuration using CFD method. They designed a 0.5 m diameter turbine using S814 NREL airfoil. The maximum power coefficient was found to be 0.51 at a tip speed ratio of 5.0, with an inflow velocity of 1 m/s. Performance study on a counter-rotating tidal current turbine by CFD and model experimentation was conducted by Lee et al. (2014). They found that dual rotor turbine set up produced more power than a single rotor blade configuration with the inflow velocity set at 1 m/s and higher. Other studies on numerical CFD analysis of a horizontal axis tidal turbine system includes (Yavus et al., 2015; Yao et al., 2012; Faudot et al., 2011; Bahaj et al., 2007b).

The present paper focused on the development of a novel design of a small tidal turbine, whose rated power might seem negligible when compared with the traditional horizontal axis tidal turbine. This turbine has a slightly complex configuration with light weight, easy to mount and less space constraints. It relies on torque produced by the tidal currents acting on the blades to generate electrical power. A CFD numerical tool was implemented to predict the performance analysis of this turbine to improve system reliability. Such tools are critical to the design and operation of new technologies which are needed to maximize energy output.

2 ROTOR DESIGN

The objective is to design a less bulky tidal rotor blade system that could spin at a very low tidal speed while still maintaining a reasonable lift force. The blades are shorter when compared to the traditional tidal turbines. It spins effectively at lower tides, but can also sustain higher tidal currents in which other turbines would stall or break. This rotor blade shape is different from the traditional horizontal axis tidal turbine. Prior to designing a rotor blade a preliminary assessment of the marine site is necessary to determine the hydrographic/meta ocean, seabed conditions, environmental and other factors. For determining the design current speed and rotor diameter, hydrographic elements should be considered. In our study, a 0.25 m diameter turbine was designed and the design current speed was set to 2 m/s. To estimate the power output of our blade, we substituted the following parameters in the given equation (1). The estimated output power coefficient (Cp) is 0.4, the power train efficiency coefficient (η) is 0.9, the velocity (U) is 2 m/s, the diameter of the turbine (D) is 0.25 m, seawater density (ρ) is 1025 [kg/m^3]. The power output of the turbine can be obtained from the expression below.

Table 1. Turbine blade design specifications.

Design parameters	Designation
Blade Aerofoil	Self constructed
Turbine diameter [m]	0.25
Number of blades	10
Density [kg/m^3]	1025
Current velocity [m/s]	2
Blade length [m]	0.08
Hub diameter [m]	0.01

Figures 1–2. CAD Models of jet-inspired horizontal axis tidal turbine.

$$P_{expect} = \eta C_p \left(\frac{\rho \pi D^2 U^3}{8} \right) \quad (1)$$

The optimum tip speed ratio of 5.0 is the standard dimensionless value designated for a three bladed tidal or wind turbine system. Performance analysis using TSR is widely used in tidal turbine blade design (Jo et al., 2013). This quantity is related to the rotational speed for a tidal turbine which can be obtained from the expression in equation (2) below.

$$N_{rpm} = 60\left(\frac{U_D}{\pi D} \right)\lambda \quad (2)$$

To verify the turbine model, based on the specifications provided, three-dimensional CAD model of the blade was designed using Solidworks 2014 design software package.

3 CFD SIMULATION

The three-dimensional model of the turbine was employed in the CFD simulation. The finite volume method is one of the numerical techniques applied in well-established commercial CFD codes to solve the governing equations of the fluid. Current practices for CFD-based workflows utilize steady Reynolds-Average Navier Stokes (RANS) with one or two equation turbulence models (Wilcox, D.C., 2015; Menter, F., 1994).

In this study, the commercial software ANSYS WORKBENCH CFX (V14.5) was implemented to perform the steady-state simulation. The CFX code solver solves the RANS equations using finite volume discretization to obtain the hydrodynamic loads on around the blade surfaces. Basic assumptions such as regarding the flow as stationary around the blade, incompressible, three-dimensional and in a steady-state were made in the analysis field to enable

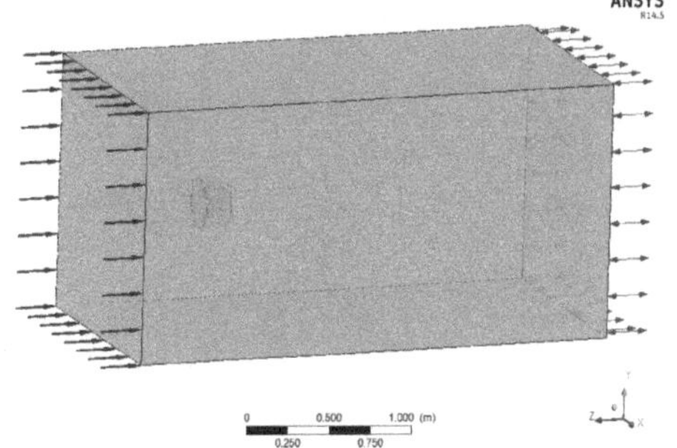

Figure 3. Computational domain of the turbine.

us solve this problem. Designing a reasonable flow field domain is significant to the CFD analysis. The analysis field consists of two domains, the internal rotational flow field domain which houses the blade and outer stationary flow field domain. The outer domain is a rectangular shape of length 3 m, width 1.5 m and height 1.5 m. This domain extends 3D in the upstream direction and 9D in the downstream direction, where D is the diameter of the turbine. The internal rotational domain is a cylinder having a diameter of 0.28 m and a height of 0.16 m. The origin of the coordinate was located at the center of the hub, the internal domain is meant to rotate along z-axis of the coordinate system. The Boundary Conditions (BC) of the flow field domains are defined as follows: At the inlet section of the external domain a normal velocity was defined using the CFX Expression Language (CEL), this is the design velocity. An opening boundary condition was defined at the outlet region, followed by specifying an outlet boundary pressure of 0 (Pa). The no-slip wall condition was applied at the bottom wall and side walls of the flow field external domain as specified in (Jo et al., 2012).

Free-slip wall condition was applied at the top section. A cylindrical domain housing the blade which is connected to the external domain serves as the interface. Multiple Rotating Frames (MRF) method is effectively applied to consider rotating motion of the blade. The blade was modeled as a wall with no-slip condition. The frozen rotor condition and the pitch change option "None" are chosen to conduct this simulation, (Lee et al., 2014) also implemented that. The frozen rotor model is widely used in the axial compressors and turbines because it can change the frame of reference while still maintaining the relative position of the components. The RANS model with two-equation k-ω turbulence model was applied to solve the fluid domain. This model gives accurate predictions on the onset and amount of flow separation under adverse pressure gradients (Lee et al., 2014). Different runs were performed for the rotational speed values of 50 rpm, 100 rpm and 150 rpm with the inflow current velocity kept constant. The convergence rate of the steady-state simulation was monitored during the iteration process by means of residuals target set to 1E-05.

4 GRID RESOLUTION

Mesh generation continue to be significant bottle neck in CFD workflow. Studies have shown that adequate grid resolution is required to capture the full range of turbulence structures in models. Significant human intervention is often required in the mesh generation process due to lack of robustness.

This is evidenced by the inability of the existing mesh generation softwares to consistently produce valid high quality meshes of the desired resolution about complex configurations on the first attempt. Hence, conducting grid convergence study is necessary and significant.

The 3D model was meshed using Ansys workbench mesher, the external domain was meshed using structured mesh grid algorithm while the mesh around the rotating domain and the blade were meshed using hexa grid. By this approach, it was possible to get rid of the difficulty in mesh generation around complex geometry. Non-matching interfaces, between the internal rotating domain encompassing the turbine blade and outer sub-domains, were defined and solved using the General Grid Interface (GGI) interpolation method. Consequently, this connection requires more memory. Generally, boundary layers require a special treatment in a CFD model. The mesh near the blade was refined to ensure that sufficient number nodes near the wall can resolve the boundary layer flow and accurately predict the fluid characteristics.

In the interest of conducting a mesh convergence study, three meshes of different resolution have been created. This is to ensure that the solution of the problem is independent of the mesh resolution. Fine meshes were distributed mainly on the blade surfaces to increase solution accuracy. A more refined mesh was concentrated more near the root as this is the critical point on the blade. The table below shows the number of nodes and elements for the cases under consideration.

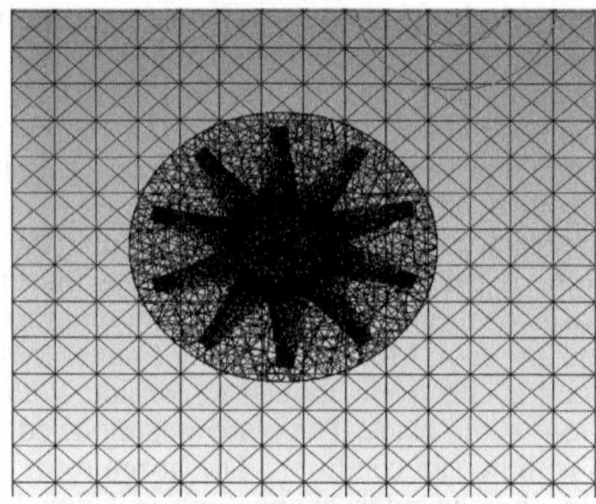

Figures 4–5. Grid representation of the rotor blade.

Table 2. Mesh modeling information.

	Total grids	
Cases	Nodes	Elements
Case_1	137872	709696
Case_2	341619	1821822
Case_3	1087984	5880633

5 CFD ANALYSIS AND DISCUSSION

The hydrodynamic loading around the turbine blades were carefully studied with the essential data from the flow analysis. Figures 6–11 below

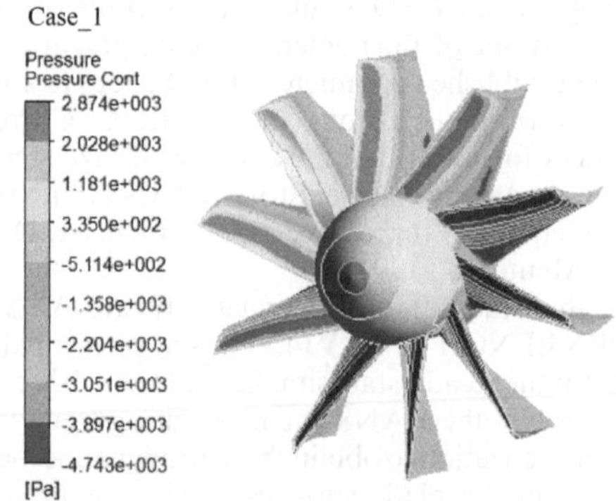

Figure 6. Pressure contour at 100 rpm.

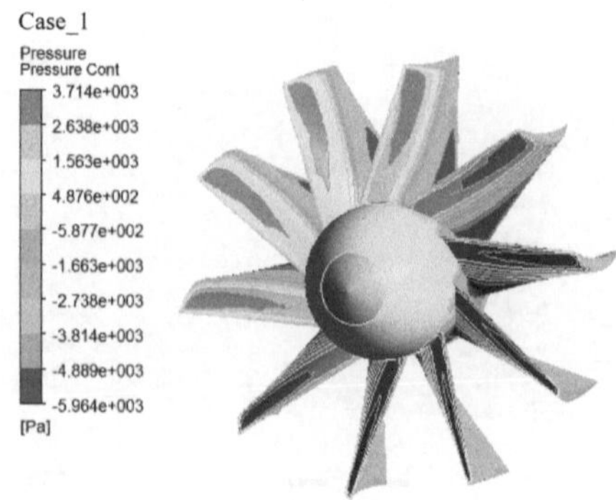

Figure 7. Pressure contour at 150 rpm.

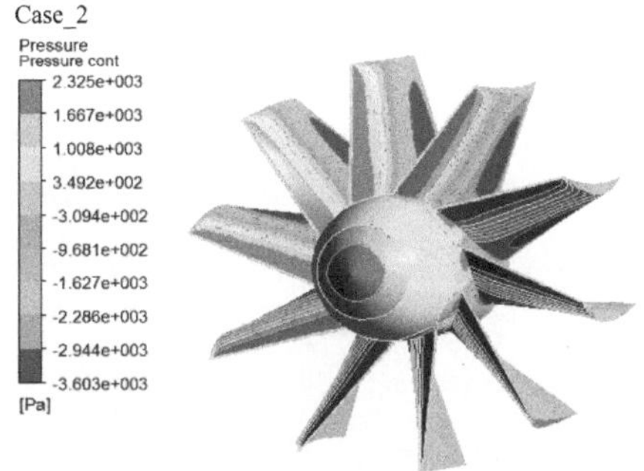

Figure 8. Pressure contour at 50 rpm.

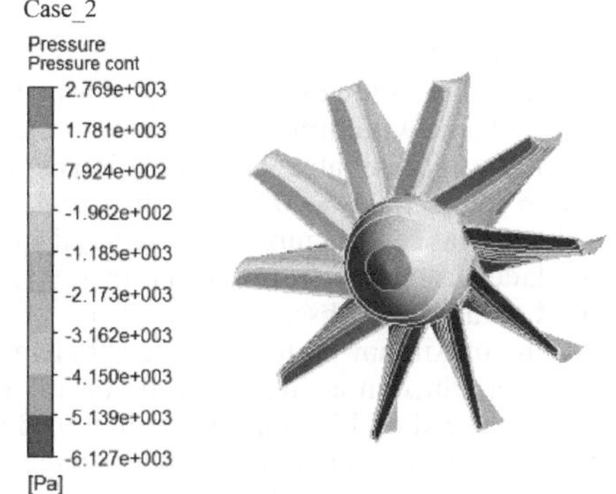

Figure 9. Pressure contour at 100 rpm.

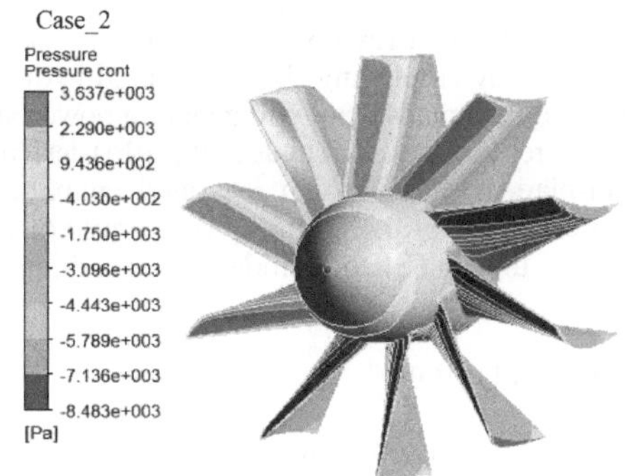

Figure 10. Pressure contour at 150 rpm.

shows the entire pressure distribution around the blades. Three cases with three different grid resolutions were studied for the rotational speed values selected. The pressure differences in both case_1 and case_2 were carefully examined. It was observed that case_2 pressure distribution results were mesh independent. The flow field around the

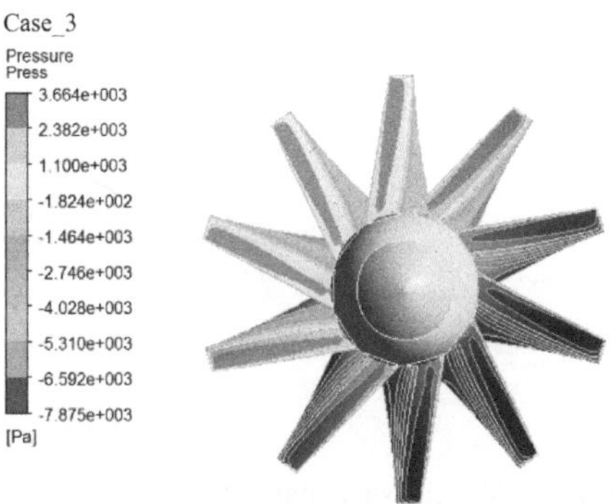

Figure 11. Pressure contour at 150 rpm.

blade is quite complex, this is because as the blade rotates, regions of high and low pressure appear. This pressure difference creates lift. To accurately analyze the complexities involved, it is vital to understand the behavior of boundary layer under the influence adverse pressure gradient. The pressure is found to be on increase at the pressure side and on the decrease at the suction side of the blade. The pressure reduction on the rotor blade suction side near the trailing edge is explained by higher magnitude of the velocity at the trailing edge.

In case 3, with a rotational speed of 150 rpm and a more refined mesh this time, the pressure on the leading edge region decreased slightly while that at the tip region increased. The increase in pressure downstream can drastically alter the flow field. There is always danger of separation in the region where the pressure increases, reported by (Gersten, K. et al., 2005). We would not go into details of boundary layer, as it is beyond the scope of this research. However, more information about boundary layers could be referenced from (Gersten, K. et al., 2005). Throughout the simulation, the tidal loads pressured the rotor blade structure more at the leading edge and the hub region.

The velocity streamlines distribution is shown above. Streamline and plane velocity representations are presented and analyzed. In this simulation it can be observed that the flow is strongly accelerated toward the turbine blade. As the rotational speed increases, the streamline expands to maximum at tip speed ratio where maximum power coefficient appears. Wake formation caused by the turbine develops and becomes more complicated if the rotational speed is increased further. Behind the turbine at 150 rpm, rings of vortices are formed. The vortices move in a helical path behind the rotor. There is a sharp increase in the magnitude of the velocity close to the trailing edge of the blade while the velocity behind the rotating turbine blade decreases rapidly as shown in Figures 12–15.

Figure 12. Velocity streamline distribution at 50 rpm.

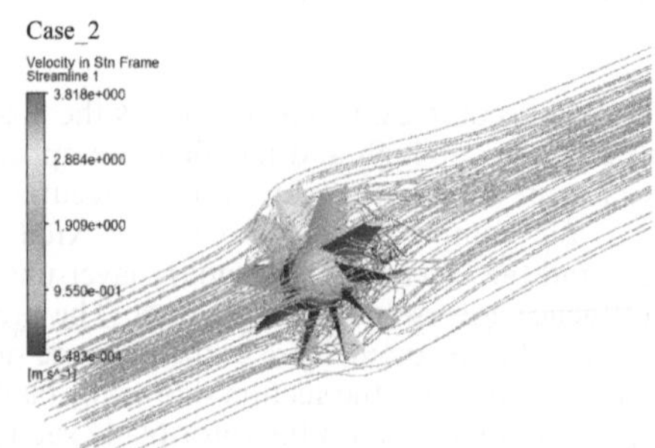

Figure 13. Velocity streamline distribution at 100 rpm.

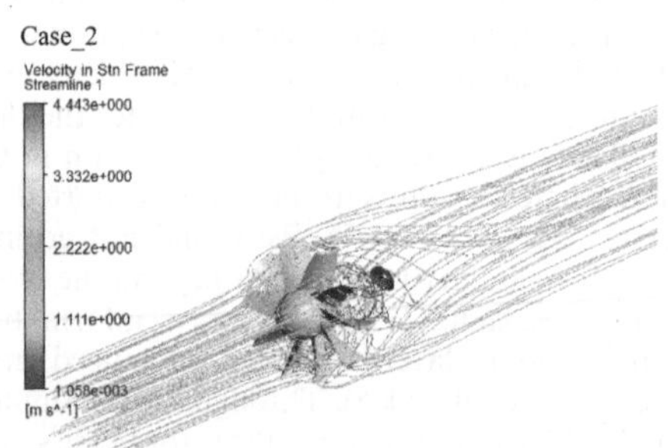

Figure 14. Velocity streamline distribution at 150 rpm.

Performance coefficient, also regarded as the efficiency defines the performance of a tidal turbine. This efficiency is usually termed as the power coefficient. Thus, the power coefficient of the rotor can be defined as the ratio of actual power developed by the rotor to the theoretical power available in the tidal current. The power coefficient of a turbine depends on many factors such as the profile of the rotor blades, blade arrangement and setting etc. It is important for the blade designer to try to fix these parameters at its optimum level so as to attain maximum efficiency at a wide range of tidal current velocities.

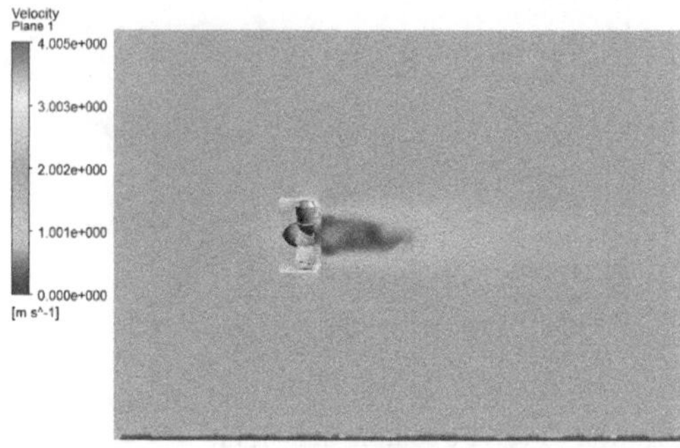

Figure 15. Velocity plane distribution at 150 rpm.

In this paper three cases were currently studied, in each case three simulations were performed on the turbine blade with different grid resolutions to determine the optimal value. The simulation parameters were set as follows: the incoming tidal current velocity was set at 2 m/s. The rotational speed of the blade was varied from 50 rpm to 150 rpm. The TSR values were observed in the range of 0.3 to 1.0. The maximum performance coefficient of 0.455 was obtained at a tip speed ratio of 1.0 and a rotational speed of 150 rpm. According to Betz, a limit is set on the maximum performance coefficient for a tidal stream turbine.

Our results indicated high efficiency in comparison with other designs. The optimum tip speed ratio of 1.0 is quite low when compared to a standard three bladed tidal turbine with an optimum tip speed ratio of 5.0 (Yavus et al., 2015). This simply means that by employing the jet-inspired hydrokinetic tidal turbine, we can still generate power at a lower current velocities. (Yavus et al., 2015) designed a twin-blade hydrofoil which generated a maximum power efficiency of 0.457 at a tip speed ratio of 3.5, still lower than a TSR of standard tidal turbine.

6 CONCLUSION

This research is focused on a performance analysis of a novel design of a jet-inspired horizontal axis tidal turbine, a less bulky turbine for the first time. The computational procedure based on CFD was developed for the efficient performance of this type of tidal blade configuration and the following conclusions were drawn:

1. A performance analysis of the turbine is important and was verified by CFD. The reliability of the new blade design was estimated using CFD numerical tool by presenting the streamlines and hydrodynamic pressure fields around the blade.

2. We observed the straight interaction between the mean flow at the rotor and the formation of free vortices which is oriented in a helical path behind the rotor. In a tidal turbine system, velocity is induced due to strong tip vortices. This is an important phenomenon in modeling tidal turbine using blade element method.
3. The maximum efficiency of 0.455 was obtained at the tip speed ratio of 1.0 in this three-dimensional steady-state simulation. The optimum tip speed ratio of 1.0 for the jet-inspired turbine is too low when compared with the optimum tip speed ratio of 5.0 for a standard tidal turbine application. This is an indication that the jet-inspired tidal turbine can operate in a lower tidal current velocities.
4. Lastly to prove this simulation results beyond reasonable doubt, an experiment should be conducted to improve the accuracy and properly validate the numerical results.

ACKNOWLEDGMENT

This work was supported by the National Project of Special Fund for Marine Renewable Energy (No. GHME2010CYO1, GHME2013ZCO1) and S&T project of Heilongjiang (No. 12531527) and "111 project" foundation (No. B07019) from State Administration of Foreign Experts Affairs of China and Ministry of Education of China.

REFERENCES

[1] Bahaj, A.S., Batten, W.M.J., McCann, G., 2007b. Experimental verifications of numerical predictions for the hydrodynamic performance of horizontal axis marine current turbines. Renew. Energy 32, 2479–2490.

[2] Bir, G.S., Lawson, M.J. & Li, Y., 2011. Structural design of a horizontal-axis tidal current turbine composite blade.

[3] Chul Hee Jo, Jun Ho Lee, Yu Ho Rho & Kang Hee Lee., 2013. Performance analysis of a HAT tidal current turbine and wake flow characteristics. Renewable Energy, 65, pp. 175–182

[4] Faudot, C. & Dahlhaug, O.G., 2011. Tidal turbine blades: design and dynamic loads estimation using CFD and blade element momentum theory. In: Proceedings of the 30th International Conference on Ocean, Offshore and Arctic Engineering.

[5] Gersten, K. & Schlichting, H., 2005. Boundary layer theory, 8th edition, pp. 40, Springer: Berlin, Germany.

[6] IRENA (International Renewable Energy Agency). 2014. "Ocean energy technology: Innovation, Patents, Market Status and Trends." June 2014, IRENA, Abu Dhabi.

[7] Jo, C.H., Yim, J.Y., Lee, K.H., Rho, & Y.H., 2012. Performance of horizontal axis tidal current turbine by blade configuration. Renew. Energy, 42. pp. 195–206.

[8] Lee,N.J,Kim,I.C.,Kim,C.G.,Hyun,B.S.,&Lee,Y.H., 2014. Performance study on a counter-rotating tidal current turbine by CFD and model experimentation Renewable Energy, 79. pp. 122–126.

[9] Menter, F., 1994. "Two-equation eddy-viscosity turbulence Models for engineering applications", AIAA Journal Vol. 32, pp. 1598–1605.

[10] Rourke O., Fergal., Boyle, Fergal., & Reynolds, Anthony., 2009. "Tidal Energy Update." Applied Energy 87.2 (2010): pp. 398–409., Academic Search Premier, EBSCO. Web, 13 Jan, 2010.

[11] Wilcox, D.C., 2006. Turbulence modeling for CFD, DCW industries, 3rd edition.

[12] Yao,J.,Yuan,W.,Wang,J.,Xie,J.,Zhou,H.,Peng,M.& Sun, Y., 2012. Numerical simulation of aerodynamic performance for two dimensional wind turbine airfoils. Procedia Eng. 31, pp. 80–86.

[13] Yavuz, T., Koc, E., Kilkis, B., Erol, O., Can Balas & Aydemir, T., 2015. Numerical and experimental analysis of the twin—blade hydrofoil for hydro and wind turbine applications. Ocean Engineering 97, pp. 12–20.

[14] Yavuz, T., Koc, E., Kilkis, B., Erol, O., Can Balas & Aydemir, T., 2015. Performance analysis of the airfoil-slat arrangements for hydro and wind turbines applications. Renewable Energy, 74, pp. 414–421.

Material Science and Environmental Engineering – Chen (Ed.)
© 2016 Taylor & Francis Group, London, ISBN 978-1-138-02938-5

Hydrolysis of BNPP catalyzed by the copper (II) complex

W. Hu & S.X. Li
College of Chemistry and Pharmaceutical Engineering, Sichuan University of Science and Engineering, Zigong Sichuan, P.R. China
University Key Laboratory of Green Chemistry of Sichuan Institutes of Higher Education, Zigong, Sichuan, P.R. China

W. Ying
University Key Laboratory of Green Chemistry of Sichuan Institutes of Higher Education, Zigong, Sichuan, P.R. China

J. He
College of Chemistry and Pharmaceutical Engineering, Sichuan University of Science and Engineering, Zigong Sichuan, P.R. China

ABSTRACT: The kinetics of hydrolysis of Bis (4-Nitrophenyl) Phosphate (BNPP) in the catalytic system containing surfactant ligand of N, N, N-tricarboxyethyl-laurylamine (L) and Cu (II) was investigated. The analysis of specific absorption spectrums of the hydrolytic reaction systems indicated that key intermediates made up of BNPP and Cu (II) complexes are formed in the reaction process of BNPP catalytic hydrolysis. The kinetic parameter of BNPP catalytic hydrolysis has been calculated and the activation energy for the catalytic hydrolysis is 38.89 kJ mol^{-1}.

Keywords: Cu (II) complex; mimic hydrolase; catalytic kinetics; phosphate diester

1 INTRODUCTION

In recent years, many researchers have already concentrated on the study of artificial enzymes, and many mimetic model of hydrolase have been widely applied in the study of the ester hydrolysis, such as the hydrolysis of carboxylic acid esters [1–4], amino acid esters [5,6], and phosphate esters [7–12]. These mimetic hydrolases or artificial hydrolases have similar catalytic function to natural enzymes, but they are structurally less complex and more stable than natural enzymes, and they can provide information on mechanistic aspects of enzyme action.

In this paper, the kinetics of hydrolysis of BNPP in the catalytic system made up of containing N, N, N-tricarboxyethyl-laurylamine and bound Cu (II) have been studied as artificial hydrolytic metalloenzymes. The results showed that the Cu (II) complexes as hydrolase mimics exhibit good catalytic activity and similar catalytic character to natural enzyme.

2 EXPERIMENTAL

2.1 *Materials*

Bis(4-Nitrophenyl) Phosphate (BNPP) and the buffering agent [tri(hydroxymethyl) aminomethane] were purchased from Sigma Chemical Co. The ionic strength of the buffer solution was maintained at 0.1 M KNO3. The pH of the buffer solution was measured at 25°C using a Radiometer PHM 26 pH meter fitted with G202C glass and K4122 calomel electrodes. The H_2O used in the experiment was redistilled and deionized H_2O. The BNPP stock solution for the experiment was prepared in redistilled deionized H_2O. Other reagents, unless they indicated, were all of analytical grade and were used without further purification.

2.2 *Kinetic method*

Kinetic experiment was carried out by UV-vis methods with a GBC 916 UV-vis spectrophotometer equipped with a thermostatic cell holder. Cu (NO)3 · $6H_2O$ (5×10^{-2} mol dm^{-3}) and the ligand (5×10^{-2} mol dm^{-3}) were dissolved in deionized water (500 mL) and anhydrous ethanol (500 mL), respectively. Then, the two solutions were mixed in proportion. A 3 mL mixed solution was injected into a 1 cm cuvette in the spectrophotometer with a temperature control system. A solution of BNPP of the desired concentration was injected into the curvette, the absorption changes of BNPP hydrolysis product (p-nitrophenol) at wavelength of 400 nm were recorded. The pseudo-first–order rate constant (k_{obsd}) for the hydrolysis of the substrate (BNPP) was obtained by initial rate method under the conditions of more than 10-fold excess of substrate concentration over the concentration of the catalyst. Each value was the average of three determinations, and its average relative standard deviation was lower than 3%.

3 RESULTS AND DISCUSSION

3.1 *Apparent first order rate constants of the BNPP hydrolysis at 25°C*

Experimental results show that a metal complex ([Cu (II)]/[L] = 1:1) is formed in the solution, and it is the active species for the BNPP catalytic hydrolysis. The structure of metal complex maybe shows as Figure 1. Apparent first order rate constants (kobsd) of the hydrolysis of BNPP catalyzed by the Cu (II) complexes were shown in Table 1, The data of Table 1 show, compared with the spontaneous hydrolysis of BNPP in water [13], the rate of catalytic hydrolysis of BNPP increases by factors of ca. 10^7 in the Cu (II) + L system. The results show that the Cu (II) complexes are efficient catalysts for BNPP hydrolysis. The reason is that the ligands are surfactant and it with metal ions to form metallomicelles.

3.2 *Mechanism of BNPP catalytic hydrolysis*

The characteristic spectra of BNPP observed were as following: the maximum absorbance wavelengths were 282 nm in aqueous solution, 296 nm in CuL aqueous solution. The red shift of the maximum absorbance wavelength of BNPP in the presence of the complex means that the energy of light wave absorbed is higher for BNPP in H_2O than that in the solution containing the complex, which implies an intermediate has been made of the complex and BNPP [14]. Moreover, the spectrum test has also showed that one final products of BNPP hydrolysis is p-nitrophenol and the other is phosphoric acid.

The complex is a hydrated complex in aqueous solution. The oxygen atom of BNPP molecular has strong coordinate ability with metallic ion, thus when BNPP add the aqueous solution, it is easy to coordinate with the metal ion of the hydrated complex through the oxygen atom. This will promote form of the metastable intermediate containing BNPP molecule and the hydroxy complex. The spectral analysis can also prove its existence.

On the base of the above analysis, we assume that the mechanism of BNPP hydrolysis catalyzed by the CuL is similar to that by hydrolytic metalloenzyme. We can see that: firstly, the intermediate is generated due to the coordination of active species and BNPP; secondly, H_2O of the intermediate is activated by metal ion and attacks the positive P atom of the BNPP molecules as a nucleophile to promote the departure of the p-nitrophenol group; thirdly, the second p-nitrophenol is released and the catalyst recovers quickly.

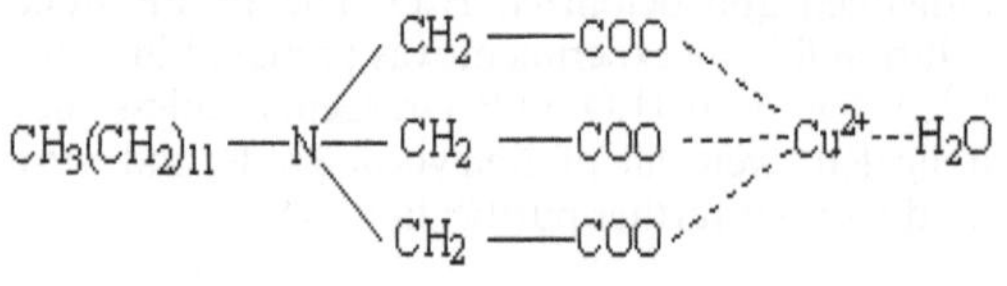

$$CH_3(CH_2)_{11}-N(CH_2-COO)_3\cdots Cu^{2+}\cdots H_2O$$

Figure 1. The structure of the complex.

3.3 *Kinetics on catalytic hydrolysis of BNPP*

The rate of BNPP spontaneous hydrolysis is much lower than that of BNPP catalytic hydrolysis, so the products of BNPP spontaneous hydrolysis can be neglected in kinetic calculation. Hence, the reaction rate equation can be operated as [15]:

$$\frac{1}{k_{obsd}} = \frac{1}{k_1} + K\frac{1}{[S]} \tag{1}$$

On the basis of the equation (1), the straight lines for 1/kobsd versus 1/[S] are gained (shown in Fig. 2). From Figure 2, it can be seen that the plots show a good linear relationship, which indicates the mechanism proposed above for the BNPP

Table 1. 10^3 kobsd (s^{-1}) of BNPP hydrolysis catalyzed by the complex.

104[S] (mol dm^{-3})	pH = 8.5	pH = 9.0	pH = 9.5	pH = 10.0	pH = 10.5
0.33	2.92	3.56	2.53	2.35	2.17
0.50	3.25	3.98	2.81	2.62	2.39
0.67	3.55	4.32	3.05	2.92	2.75
0.83	3.81	4.48	3.35	3.10	2.93
1.00	3.87	4.61	3.48	3.23	3.01

Condition: 25°C, [complex] = 5×10^{-3} mol dm^{-3}.

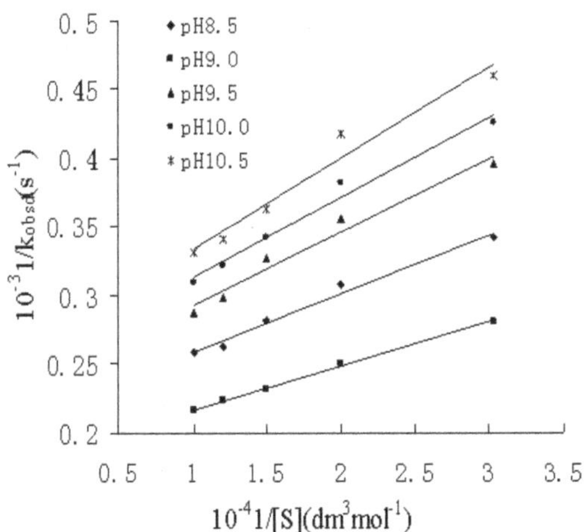

Figure 2. The plots of $1/k_{obsd}$ versus1/[S] for BNPP hydrolysis catalyzed by the complex in different pH. 25°C.

Table 2. 10^3 k1 of BNPP hydrolysis catalyzed by the complex.

pH	8.52	9.0	9.5	10.0	10.5
$10^3 k_1$	4.63	5.40	4.16	3.39	3.74

catalytic hydrolysis being reasonable. On the basis of Figure 2, k_1 (the first order rate constant for the release of the first p-nitrophenol of BNPP) of BNPP hydrolysis catalyzed by the complex can be evaluated via the linear fit by the least-square method, which is shown in Table 2.

3.4 *Apparent activity energy of the BNPP hydrolysis*

In order to investigate the effect of reaction temperature on the rate of BNPP catalytic hydrolysis, the experiments were performed at five different temperatures.

According to Arrhenius equation:

$$k_{\text{obsd}} = Z\, e^{-Ea/RT} \tag{2}$$

Rearranging Eq. (2), we can get

$$-\ln k_{\text{obsd}} = \frac{E_a}{RT} - \ln Z \tag{3}$$

In order to calculate the apparent activation energy of BNPP catalytic hydrolysis, a plot of $-\ln_{k\text{obsd}}$ versus 1/T was drawn as shown in Figure 3. The slopes of the straight lines, expressed as I, are 4680.1 for CuL solution system.

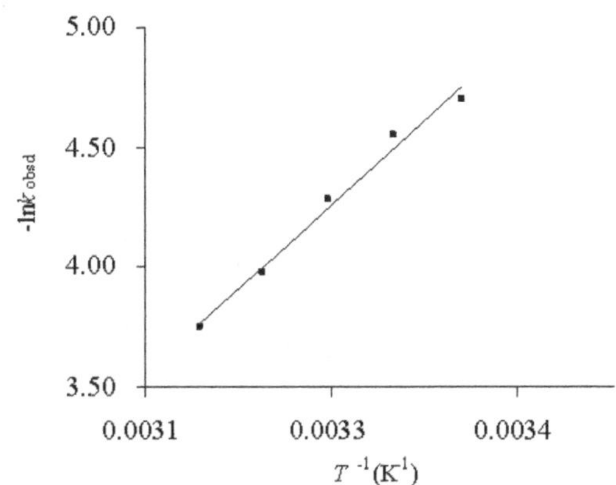

Figure 3. The plots of k_{obsd} vs. T.

According to Eq. (3), we have:

$$Ea = I \times R\ (8.314\ J\ K^{-1} mol^{-1}) \tag{4}$$

The apparent activity energy of BNPP catalytic hydrolysis is calculated based on Eq. (4), and the results is 38.89 kJ mol^{-1}.

Conditions: pH = 9.0, [ML] = 5×10^{-3} mol dm^{-3}, [BNPP] = 5×10^{-5} mol dm^{-3}

ACKNOWLEDGEMENTS

This work was supported by the Key Project of the Science and Technology Department of Zigong (No. 2014HX09, No. 2013H11), the Opening Project of Key Laboratory of Green Catalysis of Sichuan Institutes of High Education (No. LYJ14202), the Key Project Supported by Sichuan Provincial Education Department (13ZA0117), the Specialized Research Fund for the Doctoral Program of Sichuan University of Science and Engineering (No. 2012RC18) and the Cultivation Fund of National Natural Science Foundation of China, Sichuan University of Science and Engineering (No. 2013PY04).

REFERENCES

[1] Paolo, S., Paolo, T., Umberto, T., (1998) Colloid. Surface. A. 144, 71.

[2] Jairam, R., Lau, M.L., Adorante, J., (2001) J. Inorg. Biochem. 84, 113.

[3] Jiang, B., Zang, R., Xie, J., Du, J., Meng, X., Zeng, X., (2005) J. Disper. Sci. Technol. 26, 105.

[4] Li, J., Li, H., Feng, F., Xie, J., Li, S., Zhou, B., Qin, S., (2005) Chin. J. Chem. 23, 678.

[5] You, J., Yu, X., Liu, K., T (1999) etrahedron-Asymmetr. 10, 243.

[6] You, J., Yu, X., Su, X., J. (2003) Mol. Catal. A Chem. 202, 17.
[7] Negi, S., Schneider, H.J., (2002) Tetrahedron. Lett. 43, 411.
[8] Hartshorn, C.M., Deschamps, J.R., Singh, A., R. (2003) eact. Funct. Polym. 55, 219.
[9] Shu L.C., Wei H., Jian Z.L., Bin X., Lei J.H., Xiao L,Z., (2013) Prog. React. Kinet. Mech. 38, 240.
[10] Wei H., Shen X.L., Jian Z.L., Ying W., Jin Z., (2011) Prog. React. Kinet. Mech. 36, 227.
[11] Jiang, F., Jiang, B., Cao, Y., Meng, X., Yu, X., Zeng, X., (2005) Colloid. Surface. A. 254, 91.
[12] Mong, Y., Jiang, F., Hu, W., Meng, X., Yu, X., Zeng, X., (2006) J. Disper. Sci. Technol. 27, 15.
[13] Young, M.J., Wahnon, D., Hynes, R.C., Chin, J. (1995) J. Am. Chem. Soc. 117, 9441.
[14] Wanger, M.R., Walker, F.A. (1983) Inorg. Chim. Acta. 22, 3021.
[15] Wei H., Ying W., Jie Y., Jian Z.L., Xing G.M., Hu C.W., Zeng X.C., (2006) J. Dispersion Sci. Technol. 8, 1085.

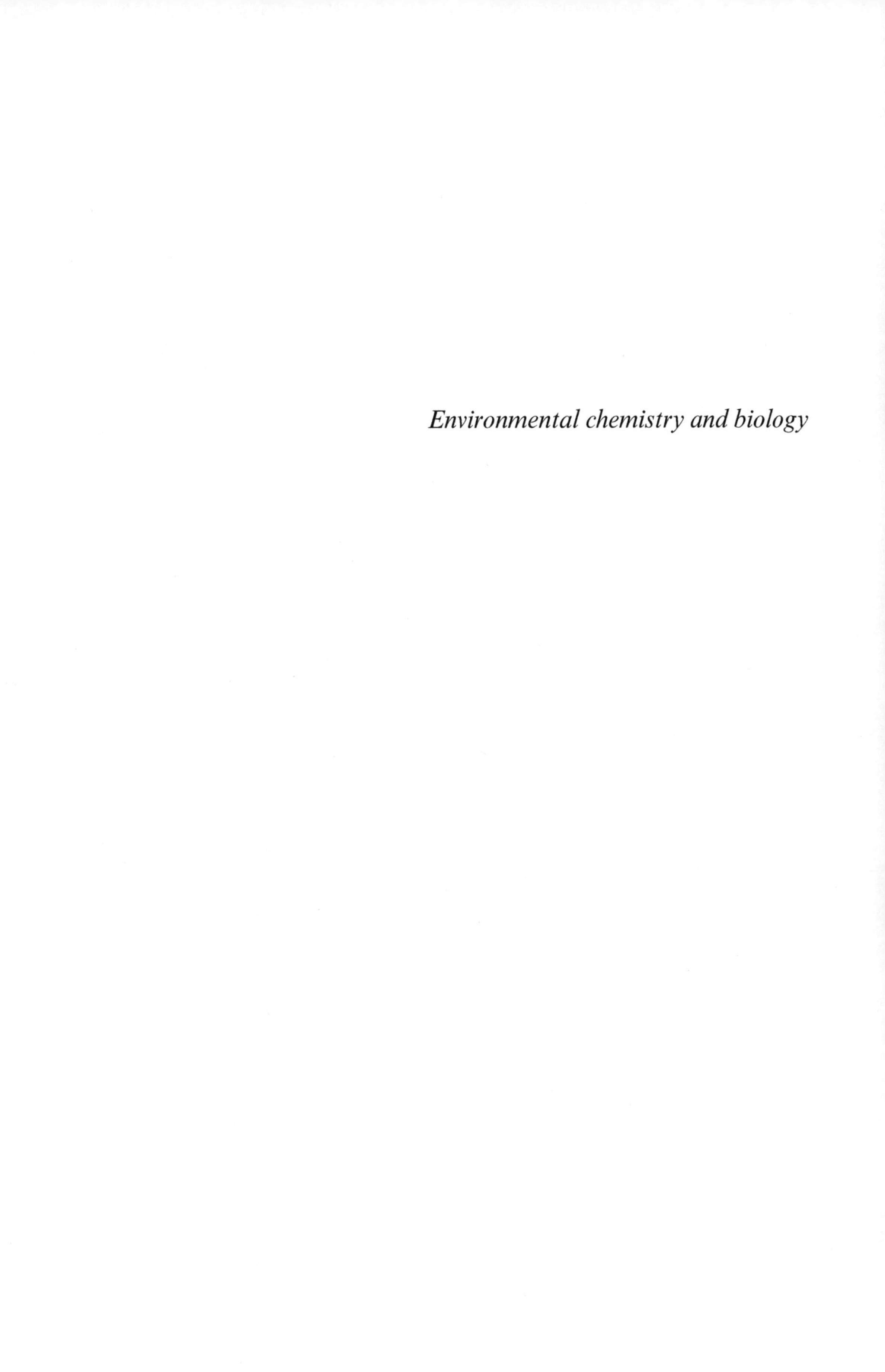

Environmental chemistry and biology

Material Science and Environmental Engineering – Chen (Ed.)
© 2016 Taylor & Francis Group, London, ISBN 978-1-138-02938-5

Research of removing nitrate by sulfur-based autotrophic denitrification technology

W.W. Ma, H. Wang & D.Y. Jiao
College of Water Resources and Environment Engineering, China University of Geosciences (Beijing), Haidian District, Beijing, China

M.L. Chen
China Northeast Design and Research Institute, Municipal Engineering Co. Ltd., Chaoyang District, Changchun City, Jilin Province, China

ABSTRACT: The nitrate pollution of groundwater and surface water is an increasingly serious problem, so the research on this issue becomes a hotspot. This paper introduces the nitrate pollution situation of water in recent years, expounds the main principles of sulfur-based autotrophic denitrification technology and summarizes the physical and chemical factors, the applications and research process of it, and points the research prospect of sulfur-based autotrophic denitrification technology out.

Keywords: sulfur-based autotrophic denitrification; *thiobacilla denitrificanss*; nitrate

1 INTRODUCTION

Nowadays, a lot of groundwater and surface drinking water have serious nitrate pollution problem in our country. The content of NO_3^--N in groundwater has been up to 100 mg/L in some area. Nitrate pollution mainly comes from the excessive use of chemical fertilizers in agricultural production, industrial and domestic wastewater discharge [1]. The high nitrate concentration will cause water to produce some of the carcinogens nitrosamines, which has certain threat to human health. The conventional treatment methods of nitrates in groundwater are ion exchange method, reverse osmosis, electrodialysis and so on; however they all have the problem of high cost or secondary pollution risk. Therefore, more economic and efficient methods for removing nitrate in both groundwater and surface drinking water are very necessary.

Biological denitrification technology is considered to be the most efficient, economical and widely used technology of denitrification. The common autotrophic denitrification technology-include hydrogen autotrophic denitrification and sulfur autotrophic denitrification technology [2]. This paper mainly introduces the disposing nitrogen technology of autotrophic denitrification with sulfur as electron donor.

2 THE PRINCIPLE OF SULFUR AUTOTROPHIC DENITRIFICATION TECHNOLOGY

Some sulfur-oxidizing bacteria of inorganic chemical nutrient and light nutritional type, like *Thiobacilla denitrificanss* can use square of reduction (S^{2-}, S, SO_3^{2-}, etc.) as electron donor under anoxic and anaerobic conditions. By sulfur of reduction to gain energy, at the same time using nitrate as electron acceptor to the reduction of nitrogen to complete autotrophic denitrification process. The principle of this process shown in the following formula (as an example of S):

$$55S + 50NO_3^- + 38H_2O + 20CO_2 + 4NH_4^+ \rightarrow 4C_5H_7O_2N + 25N_2 + 55SO_4^{2-} + 64H^+ \quad (1)$$

3 THE INFLUENCE FACTORS OF SULFUR-BASED AUTOTROPHIC DENITRIFICATION TECHNOLOGY

Researchers have studied the various factors which affects the effects of sulfur autotrophic denitrification process. The activity status of *thiobacillus denitrifications* has a great influence on the effect of sulfur autotrophic denitrification treatment. So keep thiobacillus denitrifications in its best condition could achieve better sulfur autotrophic denitrification treatment effect. The study of ChenXuan etc. showed that the bese pH for thiobacillus denitrifications is 7.0 [3]. Other studies by ChenXuan, etc. [4] have shown that the temperature have obvious effect on thiobacillus denitrifications, growth and denitrification rate. But the optimum growth temperature of thiobacillus denitrifications is inconsistent with its optimum denitrification temperature. The optimum temperature for thiobacillus denitrifications growth is 29. 5°C, lower than its optimal temperature of denitrifying which are 32. 8°C. Strain is sensitive to pH changes, and its optimum growth pH (about 6. 90) is almost the same as the optimum denitrifying pH (6. 82). ChenXuan, etc. [4] also found that *thiobacillus denitrificans* have higher denitrification activity in salinity of 0%~2.0%, but its denitrification activity in salinity which is higher than 2. 0% showed obvious inhibitory effect. The study of Campos, etc., found that nitrite and sulfate have obvious inhibitory effect on sulfur autotrophic denitrification [5]. Koenig, etc. [6] found that the smaller sulfur particles are, the shorter minimum hydraulic retention time of outflow water nitrate required. When the hydraulic retention time is constant, the highest nitrate concentration which can handle by denitrification is related to the size of sulfur particle. HuangLiren, etc. [7] found that the nitrate concentration of output from the sulfur autotrophic fixed bed reactor is related to Hydraulic Retention Time (HRT). Wang, etc. [8] confirms that the best S/N proportion is 5:3 by intermittent test.

4 THE RESEARCH PROGRESS OF SULFUR AUTOTROPHIC DENITRIFICATION TECHNOLOGY

It can be seen by the formula (1) that the autotrophic denitrification process produces H^+ to consumption basicity and lower the pH of the environment. However the best pH for thiobacillus denitrificans is neutral (7.0). The reaction need to be added alkaline in order to ensure the smooth progress. Using limestone as the source of the alkali to establishment sulfur/limestone autotrophic denitrification system is an economic and effective method to solve this problem. Moon, etc. [9] found when sulfur/limestone volume ratio is 1:1; the nitrate conversion rate is the highest. LiuLinhua, etc. [9] stabilized the pH at 7 by using *thiobacillus denitrificans* to get inoculation and using upflow sulfur/ limestone filter column to remove nitrates in groundwater. Besides, there are also a process called Sulfur—Limestone Autotrophic Denitrification (SLAD) process which use elemental sulfur as electron donor to make autotrophic denitrification process [11~13]. In this process, the limestone is not only used to neutralize acid which formed in denitrification, but also provided the carbon source of the synthesis of denitrifying bacteria cells. Because the autotrophic denitrification bacteria widely exists in soil and sediment deposition and SLAD process doesn't need carbon source, so SLAD can replace these process such as artificial wetland or the alternative process of oxidation pond process [14]. Except the elemental sulfur, S^{2-} and $S_2O_2^{3-}$ as electron donor denitrification is also found in various kinds of research. Sierra-Alvarez, etc. [15] using sulfur/limestone autotrophic denitrification process to remove nitrate nitrogen in the groundwater, when the NO_3^--N of inflow is 1.3 mmol·L^{-1}, the highest inflow water load can reach 18. 1 mmol·$(L·d)^{-1}$, the average removal rate of NO_3^-N is 95. 9%, and outflow water does not have the accumulation of NO_2^--N and S^{2-}. WangHui, etc. [16] treat nitrate wastewater by using $Na_2S_2O_3$ as electron donor and combined it with activated carbon, under the load of 0. 96 kg·$(m^3·d)^{-1}$ it can maintain the removal rate of NO_3^--N above 90%, and no accumulation of NO_2^--N.

The engineering application of sulfur autotrophic denitrification process usually use a certain volume of sulfur/limestone as reaction medium, and use the vaccination denitrification sulfur bacillus as biological ingredients to build Permeable Reactive Barrier (PRB). Moon, etc. [17] found that using sulfur autotrophic denitrification column inoculate thiobacillus denitrificans analogue PRB to remove nitrate in groundwater can completely transform NO_3^--N (60 mg·L^{-1}) to nitrogen. It is important to note that the thickness of permeable reactive barrier should be more than 50 cm. Sulfur autotrophic denitrification column simulation experiment of LiuLinhua, etc. [18] found that the nitrate removal rate of groundwater is higher than 98% when the NO_3^--N of inflow is 25 mg·L^- during the operating time.

Using limestone to improve the basicity in sulfur autotrophic denitrification process is economic and effective, but it increased the amount of Total Dissolved Solids (TDS) of outflow. For a high concentration nitrate wastewater, it will affect denitrification efficiency because of the lack of alkalinity caused by low dissolution rate of limestone. So it is more and more important to find other materials except limestone as bacteria cells to compound inorganic carbon source to improve the basicity t. For example RuanYun jie, etc. [19] using corallite instead of limestone as reaction filler.

WangAijie, etc. [20, 21] put forward the concept of simultaneous desulfurization and denitrification first in 2004. By controlling the technology conditions, make the electron balanced in sulfide oxidation and nitrate reduction process. Thus realized removal sulfide and nitrate in one reaction at the same time. A. Wang [22] study on the feasibility and the key factors of simultaneous desulfurization and denitrification process by using a denitrification sulfur bacillus strains screened by himself. Sulfide concentration and sulfur nitrogen ratio are the most important key factors determined by the batch test, and it also showed that the sulfide concentration upper bound is 300 mg/L, the best ratio of sulfur and nitrogen is 5/3. It is confirmed that when the sulfur nitrogen ratio is 5/3 the running effect is the best by continuous experiment. When the concentration of sulfide is 200 mg/L in inflow, the elemental sulfur conversion rate and denitrification rate were about 90% and 80% respectively. Their research also shows that in the suitable ratio of substrates and operating conditions, it almost can remove all the S^{2-}, NO_3^-N in organic wastewater contained simultaneously. S^{2-}, NO_3^-N and almost completely converted them to elemental sulfur and nitrogen. By microbial ecosystem which use sulfur autotrophic denitrifying bacteria and heterotrophic denitrifying bacteria of synergistic effect. In addition to make an innovation of startup mode of synchronous desulfurization denitrification system, namely build autotrophic denitrification system first, then add organic to promote the breeding of heterotrophic denitrifying bacteria. Then, it is effectively avoided imbalance between populations because of the autotrophic bacteria slow growth.

5 CONCLUSION

At present, the use of sulfur autotrophic denitrification process of thiobacillus denitrifications is only at the laboratory research stage. There are many technical problems need to be solved in the large-scale application. But the process has the advantages of easy operation, free maintenance, no secondary pollution, low energy consumption, low cost, high efficiency, and so on, so it achieved more and more attention from both domestic and foreign research scholars. Based on its existing problems, such as bacteria does not have a strong tolerance of temperature and salinity, the technology of immobilized bacteria, etc. In the future we should do more research on the breeding, domestication of strains and the optimization of autotrophic denitrification process conditions. In short, the use of sulfur autotrophic denitrification technology by thiobacillus denitrifications has opened up a new shortcut for biological denitrification and it will have a wider application prospect.

ACKNOWLEDGMENT

The research was financially supported by the National Science and Technology Pillar Program (NO2012BAJ25B00).

REFERENCES

[1] Shrimali M., Singh K.P., New methods of nitrate removal from water. Environ Pollut, 2001, 112(3): 351–359.

[2] Yanhao Zhang, Ning Yang, Kang Xie, et al. Autotrophic denitrification technology is reviewed [J]. Chemical industry and environmental protection, 2010, 30(3):225–228.

[3] Xuan Che, Jiamin Wu, Hongxin Tan, Guozhi Luo, Jvlong Qi. Autotrophic denitrification research progress and application in the circulating water aquaculture system. Fishery modernization, 2007, (1):15.

[4] Xuan Che, Guozhi Luo, Hongxin Tan, et al. The separation and identification of denitrification thiobacillus denitrification characteristics research [J]. Environmental science, 2008, 29(10):2931–2937.

[5] Campos J.L., Carvalho S., Portela R., et al. Kinetics of denitrification using sulphur compounds: Effects of S/N ratio, endogenous and exogenous compounds [J]. Bioresource Technology, 2008, 99:1293–1299.

[6] Koenig A., Liu L.H., Autotrophic denitrification of landfill leachate using elemental sulfur [J]. Water Science Technology, 1996, 34(5):469–476.

[7] Liren Huang, Linghua Liu. Denitrification thiobacillus landfills infiltration sewage processing research [J]. Environmental science, 1997, 18(5):51–54.

[8] Wang A., Du D., Ren N. An Innovative Process of Simultaneous Desulfurization and Denitrification by Thiobacillus Denitrifican [J]. Journal of Environmental Science and Health, 2005, 40(10): 1939–1950.

[9] Moon H.S., Chang S.W., Nam K. et al. Effect of reactive media composition and co-contaminants on sulfur-based autotrophic denitrification. Environ Pollut, 2006, 144(3):802–807.

[10] Linghua Liu, Zhishi Wang, Zhansheng Wang. Sulfur/limestone autotrophic denitrification dynamics model of the filter column [J]. Environmental science, 1994, 13(5):439–447.

[11] Zhang Tianc, Zeng Hui. Development of a response surface for prediction of nitrate removal in sulfur-limestone autotrophic denitrification fixed-bed reactors. Jnviron Engin-asce, 2006, 132(9):1068–1072.

[12] ZengHui, Zhang Tianc. Evaluation of kinetic parameters of a sulfur-limestone autotrophic denitrification biofilm process. WaterRes, 2005, 39(20): 4941–4952.

[13] Flere J.M., Zhang Tianc. Nitrate removal with sulfurlimestone autotrophic denitrification processes. J Environ Eng-asce, 1999, 125(8):721–729.

[14] Zhang Tianc, Lampe D.G. Sulfur: Limestone autotrophic denitrification processes for treatment of nitrate-contam ina-tedwater: Batch experiments. WaterRes, 1999, 33(3):599–608.

[15] Sierra-Alvarez R., Beristain-Cardoso R., Salazar M., et al. Chemolithotrophic denitrification with elemental sulfur for groundwater treatment. [J]. Water Research, 2007, 41(6):1253–1262.
[16] Hui Wang, Weili Zhou, Lihua Ouyang, et al. Sulfur autotrophic denitrification combined with nitrate wastewater biological activated carbon treatment [J]. China's water supply and drainage, 2011, 27(9):29–32.
[17] Moon H.S., Shin D.Y., Nam K., et al. A long-term performance test on an autotrophic denitrification column for application as a permeable reactive barrier [J]. Chemosphere, 2008, 73:723–728.
[18] Linghua Liu, Zhishi Wang, Zhansheng Wang. Sulfur/limestone filter column to remove the research of nitrate [J]. Environmental engineering, 1995, 13(3):11–15.
[19] Yunjie Ruan, Guozhi Luo, Hongxin Tan, et al. Sulfur/corallite packing bed autotrophic denitrification reactor [J]. Journal of fujian agriculture and forestry university (Natural science edition), 2009, 38(2):198.
[20] Aijie Wang, Dazhong Du, Nanqi Ren. Denitrification thiobacillus desulphurization and denitrification in wastewater treatment technology, the application of [J]. Journal of Harbin institute of Technology University, 2004, 36(4):423–425.
[21] Dazhong Du. Denitrification thiobacillus simultaneous desulfurization and denitrification technology of the key factors and running effect research [D]. Harbin: Harbin institute of technology, 2004.
[22] Wang A., Du D., Ren N. An Innovative Process of Simultaneous Desulfurization and Denitrification Thiobacillus Denitnficans [J]. Journal of Environmental Science and Health, Part A, 2005, 40 (10):1939–1950.

Material Science and Environmental Engineering – Chen (Ed.)
 ISBN 978-1-138-02938-5

Transmission model of Ebola in the dynamic small-world network

C.C. Liao
North China Electric Power University (Baoding), Hebei, China

ABSTRACT: Since many real networks generally have characteristics of small-world networks, this paper proposes a revised SIR model on a dynamical small-world network model on the basis of the mean-field theory. The comparison of the result of simulation and reality shows that this method can be the basis of forecasting and preventing the spread of infectious virus. It truly reflects the situation and transmission of Ebola.

Keywords: infectious diseases model; dynamic small-world network; Ebola

1 INTRODUCTION

1.1 *Background*

Throughout human history, infectious diseases have come and gone, with some killing thousands of people and others taking less of a human toll. To reduce death toll and pecuniary loss, many scientists have attributed to the study of epidemics [1].

In 2014, Ebola Virus Disease (EVD or "Ebola"), a severe and often fatal illness, outbroke in Western Africa and spread at a high rate of speed. The escaling scale, duration and mortality arouse the world's attention. The World Health Organization and its partners assisted the affected countries to scale-up the control measures.

Meanwhile, many researchers have modeled the transmissions of the infectious disease to forecast and prevent its spread according to the collected data and its mechanism. Alessandro Vespignani, a physical scientist from Northeastern University, predicted the situation of the spreading virus by develop a worldwide model on the basis of population movement mode and data.

1.2 *Our work*

From the viewpoint of traditional theories, the connection between people in social as the rule network and the main prediction model is the reaction-diffusion model. Considering movement contact in reality of the social network during the process of the spread of the virus, the above model could not reflect the actual transmission. Therefore, we could regard one person as a node and think of the intimate contact relationship between two people as the branch between two nodes, and then introduce a dynamic small-world network into the traditional transmission model. We first consider the various stages of Ebola development after a person is infected, especially the latent period, and then propose a modified SIRL model differing from the traditional SIR model [2]. To get closer to reality, we use a combination of the SIRL model and the dynamic small-world network.

2 THE SPREADING MODEL

2.1 *Small-world network*

Small-world network model was a network model developed by Watts and Strogatz in 1998 based on the human social network, which could convert the rule network into the stochastic network [3]. In general, real world is dynamically changing. In real life, people always have closely linked objects (e.g. relatives, colleagues, friends) and occasionally contact with others. That is, there exist a long-term close contact and a short-term movement contact simultaneously in the network. To represent the above-mentioned phenomenon, the Dynamic Small-World Network (DSW [4–5]) model contains a fixed short-range connection and a time-varying long-range connection.

To better show the real network, we combine the revised SIR model with the dynamic small-world network to propose a transmission model of Ebola. Considering the vast population and complex social network, we view a crowd of people contacting with each other closely as a corporation, and these corporations satisfy the demands of the dynamic small-world network.

2.2 *The revised SIRL model on the network*

According to the propagation law of Ebola, one infectious cycle could be divided into three periods: latent period, infective period and isolation period.

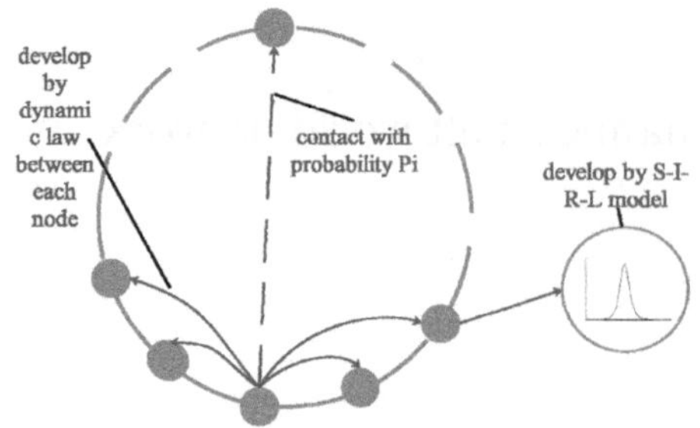

Figure 1. The schematic diagram of Ebola transmission.

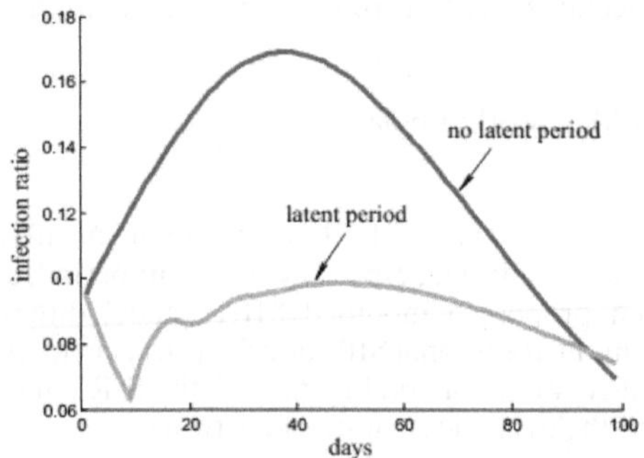

Figure 2. Tends of infection ratio with or without the latent status.

That is to say, there are three basic statuses, susceptible status (S), infection status (I) and removed status (R). Particularly, removed persons includes both healed persons and dead persons.

2.2.1 *Whether to consider latent period*

Using MATLAB to simulate the model with the latent period and without the latent period could show the difference between the two situations.

From the above figure, whether to consider the latent status could make a difference. At first, the infection ratio increases rapidly in the traditional SIR model, while it reduces if people in the latent period is taken into consideration. In the traditional SIR model, the latent period is ignored, which is equal to one person moving into the infection period as soon as he/she is infected. However, the latent period seems to be a butter period. During previous days, the situation that some infected people are not detected leads to the reduction in presentation. Besides, the maximum of the SIR model in infection is higher than the maximum of the SIRL model.

In view of the difference between whether to consider the latent period, we add the latent period to the SIR model to cater the actual fact. Thus, there are four basic statuses: susceptible status (S), latent status (L), infection status (I) and removed status (R).

- S is a conventional letter representing the number of susceptible people.
- I is a conventional letter representing the number of infected people.
- R is a conventional letter representing the number of removed people, including both healed persons and dead persons.
- L is a conventional letter representing the number of latent people.

Similar to the traditional SIR model, each node (person) changes according to the following rules.

1. One infection node could infect the other nodes in one corporation that is vulnerable with probability α.
2. Some infected nodes recovering with probability β would be removed from the total infected nodes and do not take part in the spread of the disease any more.
3. Some infected nodes cannot be healed until death and would be removed from the network and thus do not take part in the spread any more.

On the basis of equivalent relationships among the amounts of four nodes, the following real-time formulas could be easily derived:

$$\frac{\partial S(t)}{\partial t} = L(t-t_i) \tag{1}$$

$$\frac{\partial I(t)}{\partial t} = L(t-t_i) - I(t-t_i) \tag{2}$$

$$\frac{\partial R(t)}{\partial t} = \beta I(t) \tag{3}$$

$$\frac{\partial L}{\partial t} = \partial S(t) I(t) - L(t-t_i) \tag{4}$$

2.3 *Dynamic small-world model*

From the perspective of the overall situation, the characteristics of Ebola, high gregariousness, uneven distribution and central node structure are clear. Besides these characteristics, this model is also inspired by the ideology of network science. Obviously, the influence of one infected person on other susceptible people is different. To simplify the analysis, one corporation consisting of persons with close contact is taken into consideration as a node in this model.

Before clarifying this model, we take the state of m = 2 into consideration to introduce the structure of the dynamical small world, namely every node in the network has two fixed nearest nodes.

- Each node in the network represents a corporation in the social network. Arrange the N given nodes (corporation) in the network in a ring.

- Join every node to its closest m nodes.
- There are fixed Short-Range links (SR links) and time-varying and Long-Range links (LR links). SR links do not change with time, while LR links occur randomly and instantly when the disease spreads.
- LR links are established between the node i and the node j with probability γ.

In the circuit theory, the current of node i to node j is equal to the quotient of voltage difference divided by resistance. Similarly, between two nodes with short-range links, the influence of node i on node j could be quantified in the following equation:

$$S_{SR} = \frac{(I_i - I_j)}{\lambda} \tag{5}$$

Once two nodes with long-range links link with each other, infected persons in node i and node j may be infected partly:

$$S_{SL} = \frac{(I_i - I_j)}{\lambda} \tag{6}$$

3 RESULT ANALYSIS

3.1 *The simulation of the model*

To demonstrate the above-mentioned veracity, we make its dynamic and numerical simulation using the MATLAB language and compare the simulated result with real data obtained.

As shown in Figure 3, the simulation achieves high conformity with the development in the chosen country, which verifies the availability of introducing the dynamic small-world network into simulating the human social network, and study and predict the Ebola's transmission situation.

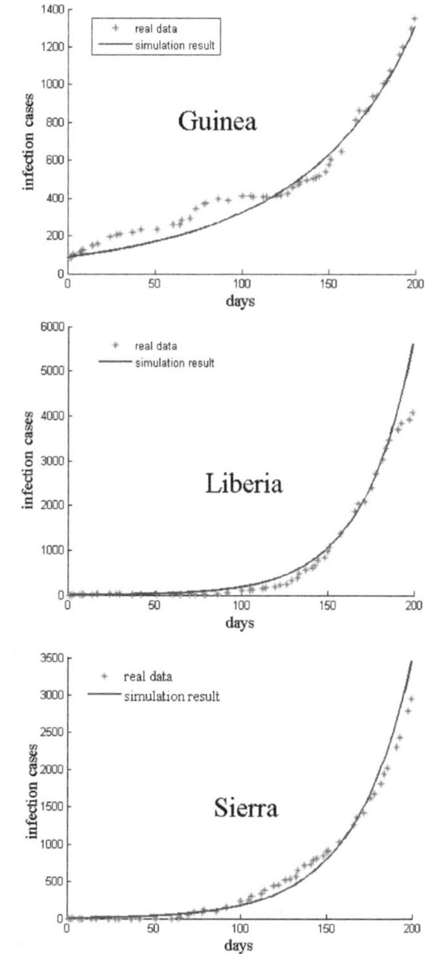

Figure 3. Simulated result (three representative countries).

Table 1. Ebola transmission in some countries.

Index	α	β
Liberia	0.3	0.265
Sierra leone	0.3	0.270
Guinea	0.3	0.287
Cote d'Ivoire	0.3	0.244
Ghana	0.3	0.250
Mali	0.3	0.239
Mauritania	0.3	0.236

3.2 *Control intervention is effective*

At the beginning, rescuers could only start with isolating and nursing, which may make a difference.

Though there are only control strategies, such as isolating the infected people, strengthening of the public health infrastructure and other response interventions, these strategies do work and slow the rise of Ebola epidemic to some extent.

To analyze the impact of response interventions more intuitively and further, we choose Liberia as an example.

From Figure 4, it is easy to draw some findings as the basis of the medication transport system.

- The quantity of infected people tends to zero after a long time, which means labor control, such as isolating the infected people, making a difference to stop Ebola.
- Even labor control is an effective measure; the effect is still not satisfactory enough until Ebola is under control. The amount of infected people

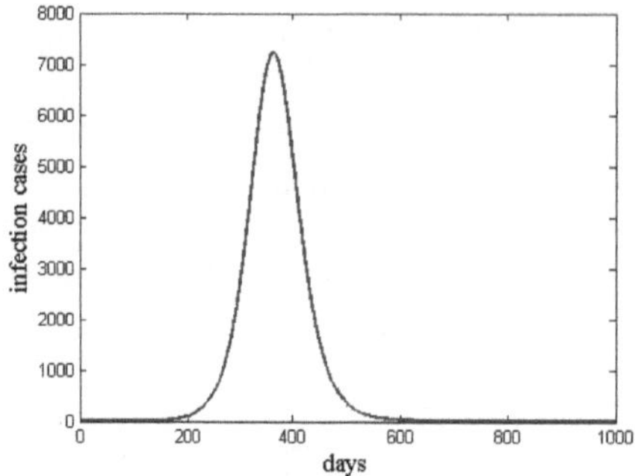

Figure 4. Trends of the quantity of infected people with or without the latent status.

Table 2. Ebola transmission in some countries.

Index	I	H
Liberia	3065	5
Sierra leone	2027	2
Guinea	989	2
Cote d'Ivoire	89	0
Ghana	3	0
Mali	35	0
Mauritania	3	0

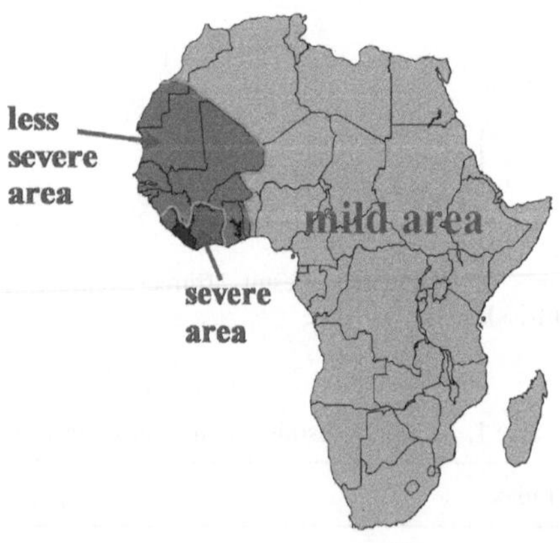

Figure 5. Trends of infection ratio with or without latent status.

reaches to a peak, about 7000. The long-lasting time and the huge number of infected people would lead to enormous death tolls.

- Concrete values of these curves could be calculated by the MATLAB language to figure up the ratio of isolated people to population.
- Though there are no vaccines or drugs to cure the infected people completely, labor control's effort makes the ratio of cure rise to 0.265.

Obviously, the ratio of cure is still tiny to eradicate Ebola or its current strain with too great sacrifice. More effective measures are a crying need, particularly to provide new vaccine or medication with pertinence.

3.3 *Dividing the epidemic area*

According to the above data, it is easy to map out areas with similar infected conditions.

We divide Africa into three areas: severe infected area (Liberia, Sierra Leone, Guinea), moderate infected area (Cote d'Ivoire, Ghana, Mali, Mauritania) and mild infected area (Morocco, Algeria, Niger, Nigeria, Sudan, Angola).

4 CONCLUSION

Now Ebola epidemic is still not optimistic. Modeling one more reasonable and effective transmission model is the basis of understanding the mechanism of epidemic, forecasting and preventing its spread. Either the SI model or the SIR model could reflect some characteristics of the transmission of the virus, but does not link to reality very well.

This paper proposes the transmission model on the dynamical small-world network, in consideration of the influence of the latent period. On the basis of the above model and collected data, the result is reasonable. Besides, dividing the epidemic area has current significance for controlling Ebola optimally.

REFERENCES

[1] Grassly N.C., Fraser C. 2008. Mathematical models of infectious disease transmission. *Nature Reviews Microbiology*: 477–487.

[2] Kermack W.O., McKendrick A.G. 1927. Contributions to the mathematical theory of epidemics. *Proceedings of the Royal Society of London A* 115: 700–721.

[3] Lei Ning. 2014. Dynamical Small World network model and stability analysis.

[4] Stone T.E. et al. 2010. Comparative effects of avoidance and vaccination in disease spread on dynamic small-world network.

[5] Li Chanchan et al. 2014. Epidemic Spreading in Dynamic Small-World Networks with Community Structure. *Complex Systems and Complexity Science* 11: 23–39.

Material Science and Environmental Engineering – Chen (Ed.)

Enhanced removal of nutrient by Bio-Film floating bed

Y.H. Shui, Y.M. Jin, C. Su, C.L. Yang, Y.Y. Jin, R. Zhang & J.H. Shen
Chengdu Textile College, Innovation Team for Wastewater Treatment, Chengdu, China

ABSTRACT: The main reason for eutrophication is that nitrogen and phosphorus discharged to a water body cannot be removed effectively. In order to improve removal efficiency by plant floating bed. Bio-film floating bed was to be constructed with non-woven fabric as filler. The natural formation of bio-film on fabric and purification ability of the bio-film floating bed was investigated for two years. Control Hydraulic Loading (HL) was 1 $m^3(m^2d)^{-1}$ by intermittent water. Results showed that the bio-film can be formation on the non-woven fabric rapidly. The biomass and microbial diversity can be increased significantly under the natural condition and form good micro food chain. Removal of nutrient can be enhanced by bio-film floating bed. Total Phosphorus (TP) and Total Nitrogen (TN) were further more reduced 0.69 mg/L and 4.72 mg/L and removal rates were enhanced by 17.9%, 20.5% than plant floating bed, respectively. Constructed biological membrane can be used in the plant floating bed in the purification of eutrophic water body.

Keywords: eutrophication; plant floating bed; bio-film; removal rate

1 INTRODUCTION

It was in 1919 that the concept of oligotrophy and eutrophy had been first put into the field of lake research by Nauman and the problem of water eutrophication had been put forward. But it has not been paid much attention to until the 1960s. At present, with the rapid growth of world economy and urbanization process, water eutrophication has already been the focus on in the worldwide environmental problem. In China, over 80% of drinking water comes from surface waters. Due to high nutrient load mainly from human living and agriculture, the lakes, rivers and reservoirs are often highly eutrophic. Although various kinds of restoration technologies have emerged for sake of controlling the water eutrophication, the phytoremediation has been paid much attention to by the government, scientific academia, and the business circles in the world [1,2] because of its better accommodation to the environment requirement. Especially the technology of plant floating bed, combing the water treatment and water ecological restoration, has plenty of advantages, such as, low cost, no more spaced to be taken, simple management, commercial value of floating plant, and much landscape value, etc [3]. Plant floating bed is also known as ecological floating bed or artificial floating island. It was used as spawning grounds for fish in 1950s; Schwimmkampen floating island founded in Germany promoted the development of plant floating bed technology in 1979 [4]. Artificial floating island was widely accepted by scientists in the field of environmental and lake study in the mid-term of 1990s [5] over 7000 m^2 of waters has been installed ecological floating bed in Japan 1995 [6], meanwhile, it has also spread in China, Australia, Canada, India, Spain, England and America in succession. However, there were still problems in this technology, for instance, limited purification ability; consequent pollution of the plant dead body to the water. Therefore, how to raise ecological remediation in plant floating bed system was vital to its further application and popularization.

The biological membrane method has been widely used in varieties of sewage treatment process, as this technology has many characteristics, such as, less sludge, no sludge bulking, high adaptability to the quality and quantity of the sewage, simple management, etc [7]. The core part of this technology was the biological filler, on which the microbes grew and the bio-film formed. It affected the growth, breed, exfoliation, and population succession of the microbes, and also functioned as a filter to intercept the suspended matter in the water. Therefore, the system of bio-film intensified plant float bed was constructed to study how the bio-film acted in this technology.

2 MATERIAL AND METHODS

2.1 *Construction of the system*

As Figure1 and Figure 2 shown, the system (1000 mm long, 30 mm wide and 1000 mm deep) was fabricated with 10 mm thick PVC planks.

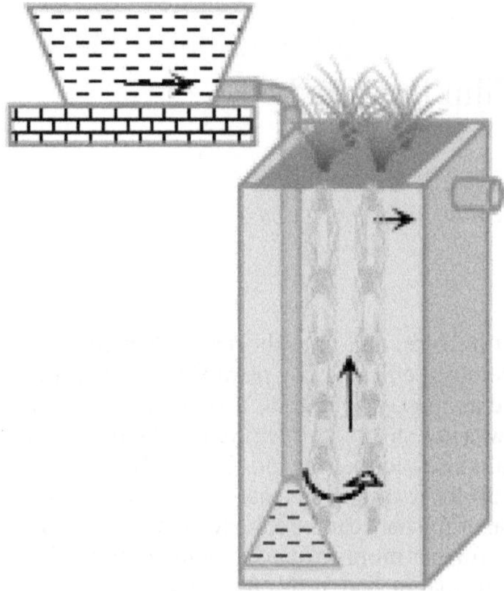

Figure 1. Schematic diagram of the system.

Figure 2. The real picture of non-woven bio-film carrier.

The floated bed was made of waste foam board, on which the small herbaceous plants such as *Myriophyllum verticillatum* L. and *Trifolium repens* L. were cultivated. Bio-film, which was made of non-woven fabrics (0.2 cm thick, 0.028 $g \cdot cm^{-2}$ specific gravity and 2 cm wide), was hung in one group constructed Bio-Film Floating Bed (BFB) and another one is Floating Bed (FB).

2.2 *Mode experiments*

The system was carried out outdoor from September in 2012 to May in 2014, and the wastewater was treated in an intermittent way. Through gravity, 280–300 L entropic water flow to the bottom of the reactor with a current velocity varied from 12 to 15 $L \cdot h^{-1}$ by adjusting the stop valve. Then the water flow from top to bottom through the pipe, and flew out from the overflow pipe, 15 cm to the top of the reactor. Finally, the treated water flows to the next unit.

The tested water was taken from serious eutrophication water body, and the average index values of inflow quality were shown in Table 1. Samples taken from inflow, 15 to 20 cm deep from the bottom, and the outflow was analysed promptly.

3 RESULTS AND DISCUSSION

3.1 *The formation of bio-film*

Gravel, slag, coke, cork, wood, aluminum, synthetic organic plate were used as bio-filler in wastewater treatment by K.lmhoff, Mahr, Sierp and Waring etal from 1920s to 1930s [8,9]. Then varieties of new bio-fillers were developed in sewage treating, such as porous Polyurethane (PU) foam, foamglass, activated carbon fiber [10]. Owing to the high price limited these materials application. Besides, as non-woven fabric was an industrial material with a low price and easy attachment for the microbes with its rough surface, it could be used in large specific areas [11].

It was succeeded when the filler turned from white to dark brown after the system started. From April to May in 2012, 4 phylum (*Cyanophyta, Chlorophyta, Bacillariophyceae, Euglenophyta*), 32 genus, 166 species and 18 kinds of Zooplankton (*Vorticellidae*, *Elegans*, *Rotifer*, *Daphnia*, *Chironomid larvae*, et al) were observed through microscopy. It turned out that the non-woven fabric bio-film could be a perfect space for microbes' growth and habitation, for its powerful role in ameliorating the system, increasing the quantity and species of microbes, and contribution to the micro-food chain's formation and stabilization.

3.2 *Purification effect on TP*

The purification effects of the constructed system on TP were shown in Figure 3 and Figure 4.

As showed in Figure 3, the TP variation trend of the inflow and outflow was consistent compared

Table 1. Averages and standard deviation statistical values of domestic sewage.

Index	TP (mg/L)	DP (mg/L)	TN (mg/L)	NH_3-N (mg/L)	NO_3-N (mg/L)	NO_2-N (mg/L)	COD_{Mn} (mg/L)
Mean value	3.82	2.67	23.12	14.41	1.10	0.54	23.38
Min	1.97	0.40	12.67	4.16	0.01	0.01	6.98
Max	6.60	6.00	43.02	32.37	5.71	2.46	62.33
St.D	1.28	1.31	7.39	8.13	1.85	0.78	17.59

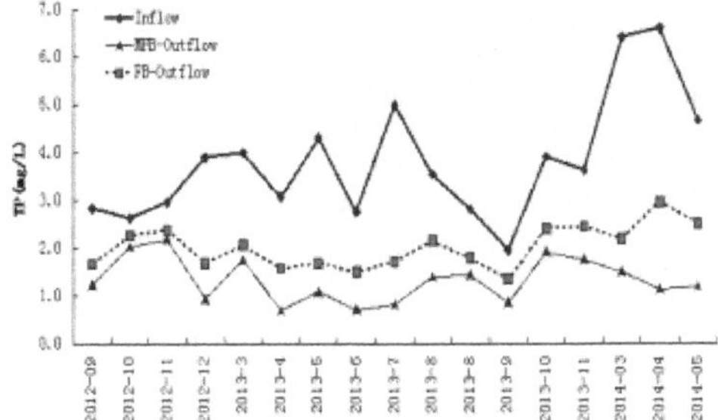

Figure 3. TP change curves of inflow and outflow.

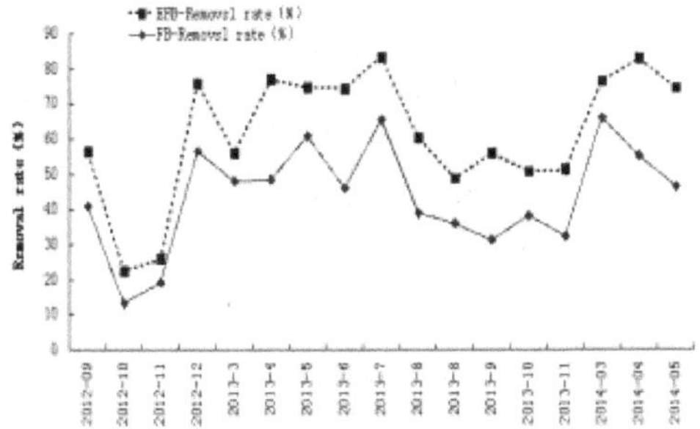

Figure 4 Change curves of TP removal rate.

BFB and FB. TP concentration in the outflow of the FB was 2.03 ± 0.44 mg/L, while the BFB was 1.34 ± 0.47 mg/L, so it turned out that the bio-film could remove the phosphorus with an effectively average removal rate of 34%.

Furthermore, it was clearly shown in Figure 4 that the existence of bio-film contributed to the phosphorus removal greatly, because the average TP removal rate increased by 17.9% as the bio-film system has a 61.6% TP removal rate while the control group, 43.7%. Maximum value of increased TP removal rate in BFB even reached at 83.4%. It might be the reasons that the phosphorus was the essential nutrient element in plant growth, the transformation of some dissolved inorganic phosphorus by the plants or the root zone of the plant, and also the increased adsorption, filtration and precipitation of the granular phosphate by the bio-film.

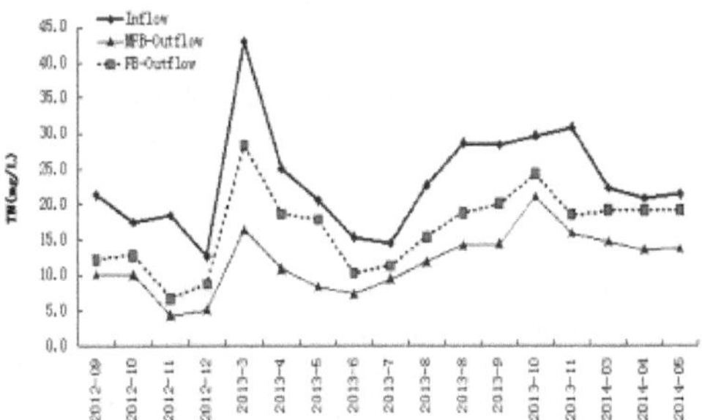

Figure 5. TN change curves of inflow and outflow.

3.3 *Purification effect on TN*

The purification effects of the constructed system on TN were shown in Figure 5 and Figure 6.

As showed in Figure 5, the TN variation trend of the inflow were consistent similar to TP's curves. TN concentration of FB outflow was 16.56 ± 5.60 mg/L, while the BFB was 11.84 ± 4.31 mg/L, so it became apparent that the bio-film could remove the nitrogen with an average removal rate of 20%, effectively. Meanwhile, it was also pointed out in Figure 6 that the average TN removal rate increased by 28.1% as the bio-film system has a 48.6% TN removal rate while the control group, 28.1%. The maximum value of increased TN removal rate in the bio-film group even arrived at 76.8%. It stated that the bio-film could also enhance the nitrogen removal in the system.

According to the outflow concentrations and the removal rate, nitrogen removal was weaker than phosphorus; on the other hand, the increased removal rate of nitrogen was higher than phosphorus. It might be the reason for different removal mechanism between nitrogen and phosphorus. Nitrogen degradation mainly depended on nitrifying and denitrifying, while phosphorus depended on the absorption and precipitation. As for the bio-film's appearance, biomass and species of the microbes increased a lot, so that the adsorption, filtration and precipitation of the bio-film or the bio-filler itself raised greatly.

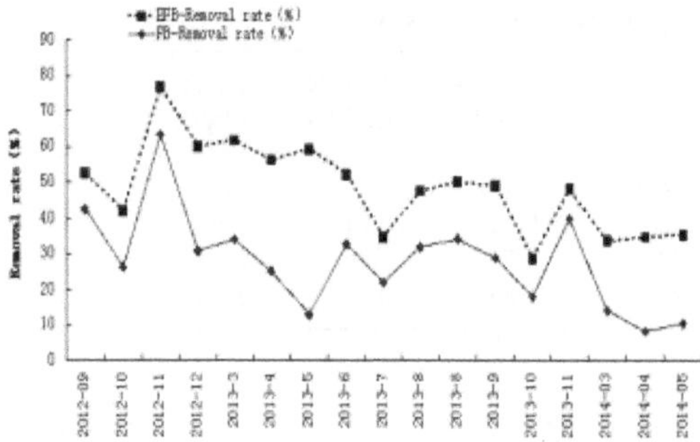

Figure 6. Change curves of TN removal rate.

4 CONCLUSION

As bio-film was formed by the introduction of bio-filler in the plant floated bed system, the biomass and species of microbes increased a lot during the treatment period, and then, a fine miniature food chain has come into being. Thus, the pollutants in the eutrophic water were degraded in the biological metabolism way. It was verified in the study that the bio-film could not only intensify the phosphorus removal power of the plant floated system, but also the nitrogen, as clearly data showed that the average TP concentration of outflow decreased 0.69 mg/L with a increased average removal rate by 17.9%, while TN, 4.72 mg/L with a increased average removal rate by 20.5%. Besides, the remediation effect and stability of the plant floated bed system was enhanced greatly, as the existence of bio-film, reduced the influence of the system by the changes of inflow or seasons. It is therefore recommended that bio-film floating bed systems be widely used for the treatment of pollutant river water.

ACKNOWLEDGEMENTS

The research work was supported by technology and program project of Science and Technology Department of Sichuan Province (2014RZ0034). Innovation team for wastewater treatment of Chengdu Textile College.

REFERENCES

[1] Hui Zhua, Baixing Yana, Yingying Xua. Removal of nitrogen and COD in horizontal subsurface flow constructed wetlands under different influent C/N ratios. Ecological Engineering, 63(2014):58–63.

[2] Hsiao-Ling Lua, Chen-Ruei Kub, Yuan-Hsiou Chang. Water quality improvement with artificial floating islands. Ecological Engineering, 74(2015): 371–375.

[3] Wenping Cao, Yanqiu Zhang. Removal of nitrogen (N) from hypereutrophic waters by Ecological Floating Beds (EFBs) with various substrates. Ecological Engineering, 62(2014):148–152.

[4] Hoeger S. Schwimimkampen Germany's artificial floating islands. Journal of soil and Water conservation, 1988, 43(4).

[5] Xiuping Yue, Jian Yuan.Water treatment filter material and packing [M]. Beijing: Chemical industry press, 2010:21. (in Chinese).

[6] Yaming Liu, Haibo Cui, Zhuoli Hao. Research progress of packing application in biological treatment of wastewater. Chemical Engineering and Equipment. 2008(11):115–116. (in Chinese).

[7] Fei Qu, Yiping Ma. The research of new type biological packing in sewage treatment. Synthetic Fiber in China. 2006, 35(3):34–37. (in Chinese with English abstract).

[8] Lisha Chen, Reti Hai. Application of Ecological Carbon Fiber Filler in Treatment of Intensive Aquaculture Wastewater. Environmental Science and Technology, 2012, 35(5):93–97. (in Chinese with English abstract).

[9] Mengchun Gao, Zonglian She, Min Yang. Study on the pretrement of water in Yellow River by bio-non-woven fabric reactor. Industrial Water Treatment. 2005, 25(12):37–39. (in Chinese with English abstract).

[10] Shaoxiong Yuan, Wenying Chen. Basic Research on the Influence of Non-woven Packing on Domestic Sewage Self-purification. Journal of Anhui Agri. Sci. 2010, 38(25):13986–13988. (in Chinese with English abstract).

[11] Lin Xinbin, Dafang Fu. Study on treating municipal wastewater by nonwoven fabrics membrane bioreactor. China Water & Waste water. 2007. 23(13): 66–69. (in Chinese with English abstract).

Material Science and Environmental Engineering – Chen (Ed.)
 ISBN 978-1-138-02938-5

Studies on the diversity of Sulfate-Reducing Bacteria in sediments from Baihua Lake with molecular biological techniques

L.Y. Fu

Guizhou Provincial Key Laboratory for Information Systems of Mountainous Areas and Protection of Ecological Environment, Guizhou Normal University, Guiyang, P.R. China
School of Environment, Tsinghua University, Beijing, China
Research Center of Water Pollution Control Technology, Chinese Research Academy of Environment Sciences, Beijing, China

L.Y. Li, J.W. Hu, X.F. Huang, P.H. Xia, J. Luo, M. Jia, Y.M. Lin, X.D. Shi & J.L. Li

Guizhou Provincial Key Laboratory for Information Systems of Mountainous Areas and Protection of Ecological Environment, Guizhou Normal University, Guiyang, P.R. China

ABSTRACT: Sulfate-Reducing Bacteria (SRB) inhabiting the sediment were studied in Baihua Lake located in a typical karstic area of the Yunnan-Guizhou Plateau, China. The molecular biological techniques of Polymerase Chain Reaction (PCR)-amplified ribosomal DNA (rDNA) followed by sequence analysis of the resulting clones was performed to analyze SRB subgroups. The 16S rDNA genes from mixed bacterial DNAs were amplified by PCR with primer sets which comprise six main phylogenetic groups. The phylogenetic tree constructed demonstrated the following subgroups of SRB: Desulfobulbus, Desulfonema, Desulfovibrio, Desulfobacca, Desulfuromonas and Desulfovirgula.

Keywords: bacteria; Baihua Lake; karst; sediment; 16S rDNA

1 INTRODUCTION

In a lake ecosystem, various microorganisms participate in material cycles. Sulfate-Reducing Bacteria (SRB) are a diverse group of anaerobic bacteria that have the ability to use sulfate as a terminal electron acceptor in the consumption of organic matter, with the concomitant production of H_2S (Daly et al., 2000; Shen et al., 2012a, b). They are ubiquitous in the environment and have pivotal roles in the biogeochemical cycling of carbon and sulfur.

During recent decades, biological treatment with SRB has gradually been considered as the most promising alternative for Acid Mining Drainage (AMD) decontamination. This kind of wastewater normally contains high concentrations of sulfate and heavy metals, therefore the search for isolation culture of some unique species of SRB having highly resistant ability to metals is extremely important for the development of a bioremediation technology (Martinsa et al., 2009). However, only an estimated 20% of the naturally occurring bacteria have hitherto been isolated and characterized, since the selective enrichment cultures fail to mimic the conditions of their natural habits under which particular microorganisms can proliferate (Muyzer et al., 1993). Nowadays, the application of molecular biological methods to investigate the occurrence and distribution of bacteria in the environment has the advantage of providing direct information on the community structures of the microbes. These methods constitute the cloning of ribosomal copy DNA or Polymerase Chain Reaction (PCR)-amplified ribosomal DNA (rDNA) followed by sequence analysis of the resulting clones (Ward et al., 1990). In earlier studies, 16S rDNA primers were designed to detect SRB subgroups and used successfully to identify the presence of SRB (Daly et al., 2000). Consequently, the molecular biological techniques can offer new opportunities for determining the genetic diversity of microbial communities and identifying several uncultured microorganisms.

Baihua Lake (E 106°27′-106°34′, N 26°35′-26°42′) is a karst man-made reservoir situated in Qingzhen County, approximately 22 km northwest of Guiyang City which is the capital of Guizhou Province located on the Yunnan-Guizhou Plateau in southwest China. The lake is one of the five key drinking-water sources for Guiyang City with a population of approximately 3 million. With the development of various industries and agriculture in surrounding areas, Baihua Lake has been polluted seriously by nutrients and heavy metals, especially by mercury (Huang et al., 2009a, b). The average concentration of mercury and sulphur in coal in Guizhou Province is substantially higher than coal produced in any other provinces in China (Feng et al., 2002). Compeau and Bartha (1985) indicated that SRB were the principal methylators of mercury in anoxic estuarine sediments. The ecological and health effects of mercury pollution are greatly exacerbated by environmental transformation of the less hazardous forms of this metal to extremely toxic and biomagnification-prone methylmercury compounds (Effries, 1982; Compeau and Bartha, 1985; Hu et al., 2011).

Previous studies have examined the correlations between SRB content and geochemical factors in sediments from Baihua Lake (Deng et al., 2009; Shen et al., 2012a, b). To provide the relevant data that could contribute to establishing bacteriological indicator systems for environmental monitoring and offering theoretical guidance for the management in sediments of this karstic artificial lake, the main objectives of this study were to apply the phylogenetic analysis based on the 16S rDNA sequence comparison to the DNA extracted from sediments to obtain useful information on the occurrence of SRB taxa.

Figure 1. Location of the sampling stations on Baihua Lake.

2 MATERIALS AND METHODS

1.1 *Sample collection and analysis*

1.1.1 *Sediment sampling*

Surface sediment samples used in this study were collected by a grab sampler during July 2012. According to the size and shape of Baihua Lake and the direction of water flowing (from south to north) and hydrogeological conditions, six sampling sites as shown in Figure 1 were precisely selected in situ with the assistance of Global Position System (GPS-72, Garmin Corporation made in Taiwan, China), namely Dachong (E106°29.351′, N26°37.111′), Tangchong (E106°29.605′, N26°37.401′), Pingpu (E106°30.953′, N26°39.219′), Yapengzhai (E106°31.388′, N26°40.131′), Baifan (E106°32.121′, N26°39.928′), Guanyinshanzhuang (E106°31.766′, N26°39.337′).

The collected sediment samples were rapidly sealed and stored in sterile plastic bags with ice. Upon collection, all samples were transported to the laboratory and maintained in the refrigerator at 4 °C for immediate processing. In addition, the surface water were collected by application of an organic glass water sampler and then placed in 500-mL polyethylene sampling bottles.

1.1.2 *Physicochemical characteristics of samples and SRB population analysis*

The analyzed indicators of the water samples included Dissolved Oxygen (DO), pH, conductivity, Suspended Solids (SS), five-day Biochemical Oxygen Demand (BOD_5), the water transparency were analyzed according to the standard methods (Commission of China EPA, 2002). DO, pH and conductivity of the surface water were determined directly by a multifunctional meter instrument (DZB-718, Shanghai Precision & Scientific Instrument Co., Ltd., China). The moisture content of the sediment samples were determined by measuring the weight loss after drying at 105 °C for 24 h.

The SRB were analyzed with the three-tube Most Probable Number (MPN) method (Deng et al., 2009; Republic of China Ministry of Chemical Industry, 1993). For the selective cultivation of the SRB, a liquid medium B (Soil Microbe Research Association of Japan, 1983; Postgate, 1984) was applied. In this experiment, 1.0000 g of fresh sediment of each sample was weighed into 100-mL beakers and tenfold dilution series were prepared with physiological saline for SRB. 0.4 g of ascorbic acid as a reducing agent and 1.2 g of ferrous ammonium sulfate as the indicating agent which had been separately sterilized irradiated with ultraviolet-visible light for 30 min, were then solved in 40 mL sterilized water separately. One mL of each of the above two solutions was introduced into 100 mL sterilized medium in the super-clean bench. Then 1 mL of dilute solution was injected into each test tube, and then the test tubes were filled completely with the medium for the aim of oxygen tolerance. The blackening of the medium was regarded as a positive result. The number of the SRB was obtained from a most probable number table (McCaig et al., 1999).

1.1.3 *Identification of SRB subgroups by molecular cloning*

Total DNA extraction. A subsample of the sediment (approx. 2.5 g, wet weight) was homogenized

Table 1. 16S rDNA-targeted PCR primer sequences specific for SRB subgroups from Baihua Lake.

Primer	Target site[a]	Sequence 5′-3′[b]	Specificity	Annealing temp. (°C)	Expected size of product (bp)
DFM140	140-158	TAGMCYGGGATAACRSYKG	Group 1	58	700
DFM842	842-823	ATACCCSCWWCWCCTAGCAC			
DBB121	121-141	CGCGTAGATAACCTGTCYTCATG	Group 2	66	1120
DBB1237	1237-1215	GTAGKACGTGTGTAGCCCTGGTC			
DBM169	169-183	CTAATRCCGGATRAAGTCAG	Group 3	64	840
DBM1006	1006-986	ATTCTCARGATGTCAAGTCTG			
DSB127	127-148	GATAATCTGCCTTCAAGCCTGG	Group 4	60	1150
DSB1273	1273-1252	CYYYYYGCRRAGTCGSTGCCCT			
DCC305	305-327	GATCAGCCACACTGGRACTGACA	Group 5	65	860
DCC1165	1165-1144	GGGGCAGTATCTTYAGAGTYC			
DSV230	230-248	GRGYCYGCGTYYCATTAGC	Group 6	61	610
DSV838	838-818	SYCCGRCAYCTAGYRTYCATC			

[a]16S rDNA positions; *E. coli* numbering; [b]Ambiguities: R (G or A); Y (C or T); K: (G or T); M (A or C); S (G or C); W (A or T).

and then subjected to the isolation of genomic DNA. DNA was extracted from the sediment sample taken from site Pingpu using a Fast DNA Spin Kit SK8233 (Sangon Biotech., Shanghai, China), according to the manufacturer's protocol.

PCR amplification. The 16S rDNA genes from mixed bacterial DNAs were amplified by PCR with primer sets which comprise six main phylogenetic groups listed in Table 1 (Daly et al., 2000; Gerasimchuk et al., 2010). Primers were synthesized by Sangon Biotech (Shanghai) Co., Ltd. The PCR amplification were applied with a thermocycler (BBI, Canada) using Taq PCR Master Mix (BS9297). PCR products were electrophoresed through a 1% (w/v) agarose gel in 1 × Tris/Acetate/EDTA (TAE) running buffer containing ethidium bromide (0.2 μg/mL) to ensure correct PCR products. Under the exposure to UV light (U-3010 UV-Vis spectrophotometer from Hitachi Ltd., Japan), the objective bands of 700 bp (for Group 1), 1120 bp (for Group 2), 840 bp (for Group 3), 1150 bp (for Group 4), 860 bp (for Group 5) and 610 bp (for Group 6) were eluted, respectively, and then purified from the agarose using TIANgel Midi Purification Kit SK1131 (Sangon Biotech., China). Marker SM0332 (Sangon Biotech., China) was included to enable estimation of the molecular mass of the DNA bands amplified.

Cloning and sequencing of 16S rDNA. The PCR products were ligated into the pUCm-T vector cloning system (SK2211, Sangon Biotech., China) and then transformed into *E. coli* DH5a competent cells (SK2301, Sangon Biotech., China). The transformation was selected on agar LB medium with Xgal/IPTG and ampicillin. The extraction of plasmids was performed with an SK1191 kit (Sangon Biotech., China). Sequence of 16S rDNA was determined by an ABI-3730 automatic capillary sequencer (Applied Biosystem Int., USA).

Phylogenetic analysis. All DNA sequences analysis was inspected with Chromas software Version 1.62 (32-bit, Technelysium). They were then compared with the GenBank database using BLAST (Basic Local Alignment Search Tool) and manually aligned with respective sequences from SRB strains and related taxa, according to the similarities in secondary structure indentified by the ClustalX program (Martinsa et al., 2009). An original phylogenetic tree was constructed by the neighbour-joining algorithm using MEGA software package, version 5.1 program. Confidence limits for the tree topologies were estimated using bootstrap analysis of 1000 replications.

2 RESULTS AND DISCUSSION

2.1 *Population and distribution of SRB from Baihua Lake*

The physicochemical characteristics of samples collected and the amount of SRB in sediments from Baihua Lake is shown in Table 2. The population of SRB in the sediments from Baihua Lake was generally slightly higher than that in the year of 2010 (Shen et al., 2012a). The population of SRB in the sediment sample at site Dachong was significantly higher than that at the other sites, indicating the speed of sulfur cycle as well as the carbon cycle at location Dachong was relatively faster.

The population of the SRB varied significantly in the different sediments and presented the largest number in the sediment sample from site Dachong. The previous studies reported that the concentrations of the Simultaneously Extracted Metals (SEM) near sites Tangchong and Dachong were significantly higher than that at the other sites (Zhang et al., 2013; Fu et al., 2013). Since AVS is

Table 2. Physicochemical characteristics of sediment samples and surface water samples from Baihua Lake.

Sampling sites	Surface water samples						Sediment samples			
	DO	pH	Transparency (m)	Conductivity (μS/cm)	SS (mg/L)	BOD_5 (mg/L)	Depth (m)	Color	Moisture content (%)	SRB ($\times 10^4$ MPN/g, dry weight)
Dachong	12.37	8.12	2.8	380	30	9.46	9	Gray	69.84	3.65
Tangchong	11.86	8.38	2.5	375	52	7.16	10	Gray	79.44	0.97
Pingpu	14.29	8.74	1.9	379	58	8.86	10	Gray	83.45	1.81
Yapengzhai	12.29	8.75	2.3	371	52	8.42	21	Black	82.66	0.55
Baifan	11.86	8.84	2	369	52	8.06	8	Gray	78.72	0.07
Guanyin-shanzhuang	14.2	8.79	1.9	369	91	8.59	13	Black	70.87	0.26

Table 3. BLAST results of the SRB based on comparative sequence analysis of 16S rDNA gene retrieved from the surface sediments from Baihua Lake.

Clone No.	Phylogenetic group	Relative description	GenBank accession No.	Max identity
SRB-G2-1	*Deltaproteobacteria*	Uncultured *Desulfobulbus* sp. clone EB41	DQ831531	98%
SRB-G2-2	*Deltaproteobacteria*	Uncultured *Desulfobulbus* sp. clone EB41	DQ831531	98%
SRB-G2-3	*Deltaproteobacteria*	Uncultured *Desulfobulbus* sp. clone WB25	DQ831537	97%
SRB-G2-4	*Deltaproteobacteria*	Uncultured *Desulfobulbus* sp. clone EB41	DQ831531	99%
SRB-G2-5	*Deltaproteobacteria*	Uncultured *Desulfobulbaceae* bacterium clone TDNP_Wbc97_92_1_234	FJ517134	98%
SRB-G2-6	*Deltaproteobacteria*	Uncultured *delta proteobacterium* clone AKYH946	AY921871	98%
SRB-G2-7	*Deltaproteobacteria*	*sulfate-reducing bacterium* R-PropA1	AJ012591	94%
SRB-G2-8	*Deltaproteobacteria*	*sulfate-reducing bacterium* R-PropA1	AJ012591	98%
SRB-G2-9	*Deltaproteobacteria*	*Cystobacter ferrugineus 16S rRNA gene*, strain Cb fe13	AJ233900	93%
SRB-G2-10	*Deltaproteobacteria*	Uncultured *Desulfobulbus* sp. clone 2_32_D5_b	JQ086856	98%
SRB-G2-11	*Deltaproteobacteria*	Uncultured *Desulfobulbus* sp. clone OTU-X1-25	JQ668541	99%
SRB-G2-12	*Deltaproteobacteria*	Uncultured *Desulfobulbus* sp. clone WB25	DQ831537	98%
SRB-G2-13	*Deltaproteobacteria*	Uncultured *Desulfobulbus* sp. clone WB25	DQ831537	98%
SRB-G2-14	*Deltaproteobacteria*	Uncultured *Desulfobulbaceae* bacterium clone TDNP_Wbc97_92_1_234	FJ517134	99%
SRB-G2-15	*Deltaproteobacteria*	Uncultured *Desulfobulbaceae* bacterium clone TDNP_Wbc97_92_1_234	FJ517134	99%
SRB-G2-16	*Deltaproteobacteria*	Uncultured *Desulfobulbus* sp. clone A280	EU283456	94%
SRB-G5	*Deltaproteobacteria*	Uncultured *Syntrophobacterales* bacterium partial 16S rRNA gene, clone AMGC11	AM935385	98%
SRB-G6-1	*Deltaproteobacteria*	Uncultured *Syntrophobacterales* bacterium partial 16S rRNA gene, clone AMGC11	AM935385	98%
SRB-G6-2	*Firmicutes*	Uncultured *Peptococcaceae* bacterium clone Alk2-2A	EU522657	100%
SRB-G6-3	*Deltaproteobacteria*	Uncultured *Desulfobacca* sp. clone De63C12	GU472643	96%
SRB-G6-4	*Deltaproteobacteria*	*Desulfovibrio giganteus* DSM 4370	AF418170	92%
SRB-G6-5	*Chlorobi*	Uncultured *Firmicutes bacterium* clone LMC52	JN868216	99%
SRB-G6-6	*Deltaproteobacteria*	*Desulfobacca acetoxidans* strain ASRB2	NR_028662	97%
SRB-G6-7	*Firmicutes*	Uncultured *Pelotomaculum* sp. clone B01 16S ribosomal RNA gene, partial sequence	EU888820	97%
SRB-G6-8	*Deltaproteobacteria*	Uncultured *Desulfuromonadales* bacterium clone ZL44	JF733677	99%
SRB-G6-9	*Deltaproteobacteria*	Uncultured *Desulfobacca* sp. clone De63C12	GU472643	99%
SRB-G6-10	*Unclassified Bacteria*	Uncultured *soil bacterium* PBS-III-27	AJ390456	92%
SRB-G6-11	*Deltaproteobacteria*	Uncultured *Desulfobacca* sp. clone De63C12	GU472643	97%
SRB-G6-12	*Chlorobi*	Uncultured *Firmicutes bacterium* clone LMC52	JN868216	99%
SRB-G6-13	*Deltaproteobacteria*	Uncultured *delta proteobacterium* clone Z195MB47	FJ484762	98%
SRB-G6-14	*Firmicutes*	Uncultured *Clostridia bacterium* clone EV818FW062101BH4MD9	DQ079649	94%
SRB-G6-15	*Deltaproteobacteria*	Uncultured *delta proteobacterium* clone LQH296	JN868184	98%
SRB-G6-16	*Deltaproteobacteria*	Uncultured *Desulfobacca* sp. clone De63C12	GU472643	97%

produced by anaerobic respiration of SRB in sediments when organic matter becomes oxidized while sulfate becomes reduced typically, the high population of the SRB at the northern part of the lake was affected by the pollution level of the water at the inlet of Baihua Lake where both sites Tangchong and Dachong are close to the inlet. The water source of Baihua Lake is mainly from the overflow of Hongfeng Lake which is located upstream of

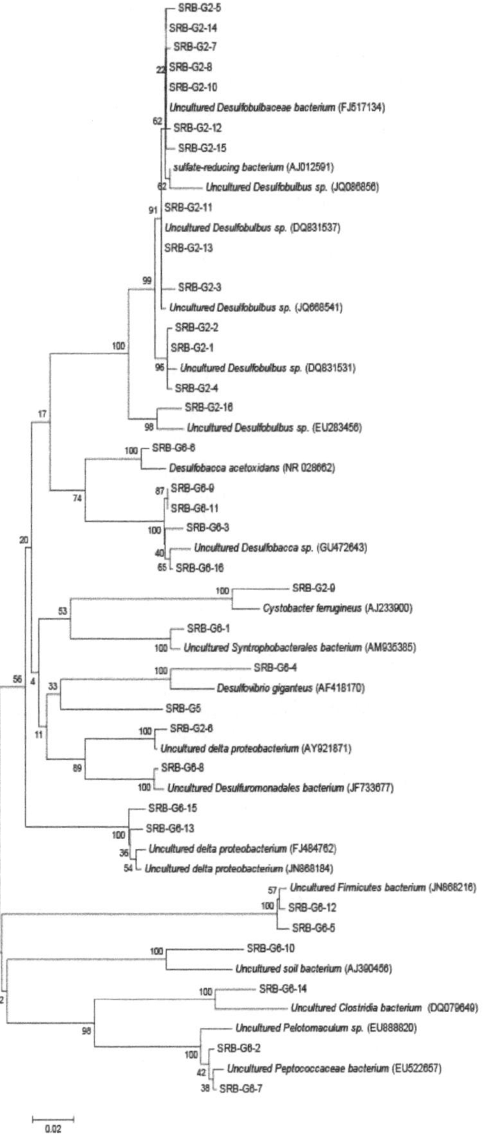

Figure 2. Phylogenetic relationship based on 16S rDNA gene sequences of clones extracted from surface sediments from Baihua Lake with closely related sequences from the GenBank database.

Baihua Lake and also polluted seriously. In addition, another possible reason for the high concentrations at sites Tangchong and Dachong are that these sites are adjacent to Qingzhen County, where the urban sewage was discharged into the lake.

2.2 *Analysis of bacterial biodiversity of SRB based on 16S rDNA gene sequence from Baihua Lake*

Since biological treatment with SRB was a promising alternative for the bioremediation technology and SRB were the principal methylators of mercury in anoxic estuarine sediments, the analysis of bacterial biodiversity of SRB in the sediments from Baihua Lake was conducted based on 16S rDNA gene sequences. The molecular method performed was also expected to give some valuable information for the work of the isolation culture of certain functional SRB species in the future. The PCR amplification of 16S rDNA by application of the primer sets of Group 2, Group 5 and Group 6 resulted in products of the predicted size of 11250 bp, 860 bp and 610 bp, respectively. The results obtained from Chroms software showed that the PCR products amplified by primer set of Group 5 was pure, and thus the sequence can be directly acquired from the analysis via this software. However, the PCR products separately amplified by primer sets of Group 2 and Group 6 did not give the pure single sequence, and thus 32 clones were randomly selected. Compared with the GenBank database, the BLAST results of the SRB clones based on the 16S rDNA extracted from the surface sediments from Baihua Lake were analyzed as shown in Table 3.

To obtain a reliable description of the phylogenetic relationship of the SRB population in the surface sediment sample from Baihua Lake, the characterized sequence of pure SRB cultures together with the clones were included in the analysis. As shown in Figure 2, the phylogenetic tree constructed demonstrated the following subgroups of SRB: *Desulfobulbus*, *Desulfonema*, *Desulfovibrio*, *Desulfobacca*, *Desulfuromonas*, and *Desulfovirgula*. The results of the identification and diversity of SRB with the molecular methods seem satisfactory.

3 CONCLUSIONS

In summary, the enumeration by cultivation of SRB presented significant changes with spatial variations in sediments from a Chinese karst lake of Baihua. A decrease in the population of the five types of bacteria under study revealed the improvement of the eutrophic level of this lake. In addition, a preliminary phylogenetic analysis of the 16S rDNA sequences indicated the existence of SRB and presented a complex SRB diversity in this karst lake. Given the rich microorganism resources in Baihua Lake, a further study is required to culture and isolate some SRB strains for the practical application of environmental remediation, such as the AMD treatment.

ACKNOWLEDGEMENT

This paper was supported by the National Natural Science Foundation of China (Grants No. 21367009 and 20967003).

REFERENCES

[1] Baumgartner, L.K., Reid, R.P., Dupraz, C., Decho, A.W., Buckley, D.H., Spear, J.R., Przekop, K.M. & Visscher, P.T. 2006. Sulfate reducing bacteria in microbial mats: Changing paradigms, new discoveries. *Sedimentary Geology* 185(3–4): 131–145.

[2] Blanco, I., de Serres, F.J., Cárcaba, V., Lara, B. & Fernández-Bustillo, E. 2012. Alpha-1 antitrypsin deficiency PI*Z and PI*S gene frequency distribution using on maps of the world by an Inverse Distance Weighting (IDW) multivariate interpolation method. *Hepatitis Monthly* 12(10 HCC): e6024.

[3] Chen, F.W. & Liu, C.W. 2012. Estimation of the spatial rainfall distribution using Inverse Distance Weighting (IDW) in the middle of Taiwan. *Paddy Water Environment* 10(3): 209–222.

[4] Commission of China EPA, 2002. Monitoring and analyses methods for water and wastewater (4th Edition). China Environmental Science Press, Beijing, China.

[5] Compeau, G.C. & Bartha, R. 1985. Sulfate-reducing bacteria: Principal methylators of mercury in anoxic estuarine sediments. *Applied and Environment Microbiology* 50(2): 498–502.

[6] Daly, K., Sharp, R.J. & McCarthy, A.J. 2000. Development of oligonu-cleotide probes and PCR primers for detecting phylogenetic subgroups of sulfate-reducing bacteria. *Microbiology* 146(7): 1693–1705.

[7] David, C.G., Bruno, D., Philippe, P., Guillemette, J. & Philippe, D. 2005. Structure of sediment-associated microbial communities along a heavy-metal contamination gradient in the marine environment. *Applied and Environment Microbiology* 71(2): 679–690.

[8] Deng, J.J., Huang, X.F., Hu, J.W., Li, C.X., Yi, Y. & Long J. 2009. Distribution of several microorganisms and activity of alkaline phosphatase in sediments from Baihua Lake. *Asia-Pacific Journal Chemical Engineering* 4(5): 711–716.

[9] Devereus, R., Delaney, M., Widdel, F. & Stahl, D.A. 1989. Natural relationships among sulfate-reducing enbacteria. *Journal Bacteriology* 171(12): 6689–6695.

[10] DiToro, D.M., Mahony, J.D., Hansen, D.J., Scott, K.J., Hicks, M.B., Mayr, S.M. & Redmond, M.S. 1990. Toxicity of cadmium in sediments: the role of acid volatile sulfide. *Environmental Toxicology Chemistry* 9(12): 1487–1502.

[11] Effries, T.W. 1982. The microbiology of mercury. Progr. Ind. Microbiol., 16: 23–75.

[12] Feng, X.B., Sommar, J., Lindqvist, O. & Hong, Y.T. 2002. Occurrence, emissions and deposition of mercury during coal combustion in the province Guizhou, China. *Water, Air, Soil Pollution* 139(1–4): 311–324.

[13] Finster, K., Liesack, W. & Thamdrup, B. 1998. Elemental sulfur and thiosulfate disproportionation by Desulfocapsa sulfoexigens sp. nov., a new anaerobic bacterium isolated from marine surface sediment. *Applied and Environment Microbiology* 64(1): 119–125.

[14] Fu, L.Y., Hu, J.W., Shen, W., Huang, X.F., Luo, J., Jia, M. & Zhang, J.P. 2014. Occurrence and implications of SEM-AVS for surface sediments from Baihua Lake, China. *Soil and Sediment Contamination: An International Journal* 23(3): 287–312.

[15] Fu, L.Y., Luo, J., Hu, J.W., Wu, Q., Huang, X.F. & Tian, L.F. 2010. Estimation of organochlorine pesticides (DDT and HCH) in surface sediments from Baihua Lake. *Selected Proceedings of the Fifth International Conference on Waste Management and Technology* (ISBN 978–1–935068–68–6), 534–538. Scientific Research Publishing, Inc, USA.

[16] Gerasimchuk, A.L., Shatalov, A.A., Novikov, A.L., Butorova, O.P., Pimenov, N.V., Lein, A.Y., Yanenko, A.S. & Karnachuk, O.V. 2010. The search for sulfate-reducing bacteria in mat samples from the lost city hydrothermal field by molecular cloning. *Microbiology* 79(1): 96–105.

[17] Hu, H.Y., Feng, X.B., Zeng, Y.P. & Qiu, G.L. 2011. Progress in research on microbial methylation of mercury. *Chinese Journal of Ecology* 30(5): 874–882.

[18] Huang, X.F., Hu, J.W., Deng, J.J., Li, C.X. & Qin, F.X. 2009a. Speciation of heavy metals in sediments from Baihua Lake and Aha Lake. *Asia-Pacific Journal Chemical Engineering* 4(5): 635–642.

[19] Huang, X.F., Hu, J.W., Li, C.X., Deng, J.J., Long, J. & Qin, F.X. 2009b. Heavy metals pollution and potential ecological risk assessment of sediments from Baihua Lake. *International Journal Environmental Health Research* 19(6): 405–419.

[20] Janssen, P.H., Schuhmann, A., Bak, F. & Liesack, W. 1996. Disproportionation of inorganic sulfur compounds by the sulfate-reducing bacterium Desulfocapsa thiozymogenes gen. nov., sp. nov. *Archives of Microbiology* 166(3): 184–192.

[21] Kondo, R. & Butani, J. 2007. Comparison of the diversity of sulfate-reducing bacterial communities in the water column and the surface sediments of a Japanese meromictic lake. *Limnology* 8(2): 131–141.

[22] Laanbroek, H.J., Abee, T. & Voogd, I.L. 1982. Alcohol conversions by Desulfobulbus propionicus Lindhurst in the presence and absence of sulfate and hydrogen. *Arch Microbiol* 133: 178–184.

[23] Leloup, J., Fossing, H., Kohls, K., Holmkvist, L., Borowski, C. & Jørgensen, B.B. 2009. Sulfate-reducing bacteria in marine sediment (Aarhus Bay, Denmark): abundance and diversity related to geochemical zonation. *Environmental Microbiology* 11(5): 1278–1291.

[24] Li, Q.H., Liu, S.P., Lin, T., Chen, L.L., Chen, F.F., Gu, J.G. & Yang, K. 2011. Design and analysis of

ecotechnological engineering for improving water quality at an estuary into reservoir in Guizhou Province. *China Water & Wastewater* 27(4): 47–53 (in Chinese with English abstract).

[25] Matulewich, V.A. & Finsten, M.S. 1978. Distribution of autotrophic nitrifying bacteria in a polluted river (the Passaic). *Applied and Environmental Microbiology* 35(1): 67–71.

[26] McCaig, A.E., Phillips, C.J., Stephen, J.R., Kowalchuk, G.A., Harvey, S.M., Herbert, R.A., Embley, T.M. & Prosser, J.I. 1999. Nitrogen cycling and community structure of proteobacterial β-subgroup ammonia-oxidizing bacteria within polluted marine fish farm sediments. *Applied and Environmental Microbiology* 65(1): 213–220.

[27] H., Natasa, N., Vesna, F., Vesna, J., Martina, L., Sonja, L., Radojko, J., Ingrid, F., Qu, L.Y., Jadran, F. & Damjana, D. 2003. Total mercury, methylmercury and selenium in mercury polluted areas in the province Guizhou, China. *Science of The Total Environment* 304(1–3): 231–256.

[28] Muyzer, G., De Waal, E.C. & Uitierlinden, A.G. 1993. Profiling of complex microbial populations by denaturing gradient gel electrophoresis analysis of polymerase chain reaction-amplified genes coding for 16S rDNA. *Applied and Environmental Microbiology* 59(3): 695–700.

[29] Mattled, M.U. & Terry, J.B. 1993. Remobilization of heavy metals retained as oxyhydroxides or silicates by Bacillus subtilis Cells. *Applied and Environmental Microbiology* 59(12): 4323–4329.

[30] Northup, R.R., Yu, Z.S., Dahlgren, R.A. & Vogt, A.A. 1995. Polyphenol control of nitrogen release from pine litter. *Nature* 337: 227–229.

[31] Postgate, J.R. 1984. The Sulphate-reducing Bacteria. Cambridge: Cambridge University Press.

[32] Republic of China Ministry of Chemical Industry. 1993. GB/T14643.5–93 Industrial circulating cooling water-Sulfate-reducing bacteria-MPN test (in Chinese). Beijing: Standards Press of China.

[33] Shen, W., Hu, J.W., Fu, L.Y. & Jin, M. 2012a. Distribution of several microorganisms in sediments from Baihua Lake. Advanced Materials Research 396–398: 1923–1927.

[34] Shen, W., Hu, J.W., Huang, X.F., Jin, M., Fu, L.Y., Zhang, J.P. & Luo, J. 2012b. Study on correlations between sulfate-reducing bacteria and geochemical factors in sediments from Baihua Lake. Advanced Materials Research 599: 76–80.

[35] Soil Microbe Research Association of Japan. 1983. Ye, W.Q., Zhang, W.Q., Zhou, P.L., Xue, J.Z. & Zhang, Y.L. trans. Analysis method of soil microorganism (in Chinese). Beijing: Science Press.

[36] Ward, D.M., Weller, R., & Bateson, M.M.. 1990. 16S rDNA sequences reveal numerous uncultured microorganisms in a natural community. Nature (London) 345: 63–65.

[37] Widdel, F. & Bak, F. 1992. Gram-negative mesophilic sulfate-reducing bacteria (2nd edn). In: Balows A, Trüper & H.G, Dworkin & M, Harder & W, Schleifer & K.H (eds.), The prokaryotes: 3352–3378. New York: Springer.

Material Science and Environmental Engineering – Chen (Ed.)
 ISBN 978-1-138-02938-5

The inhibitory factors of the anammox process

Z.J. Yin, L.H. Zang & D.D. Ji
College of Environmental Science and Engineering, Qilu University of Technology, Ji'nan, Shandong, China

ABSTRACT: Anaerobic ammonium oxidation (anammox) is a novel and perspective technology for nitrogen removal compared with conventional biological treatment methods. The objective of this paper was to investigate the inhibitory effects of anammox salinity, organic matter, heavy metal and other conditions on anammox microbial activity. Higher salinity inhibits anammox activity and salinity destroys anammox. Organic inhibition is dependent on the level of organic and the species. Most heavy metals inhibited anammox growth and activity. The mechanism of inhibition on anammox will help to prevent improvements on the process and the application of the anammox process.

Keywords: anammox; heavy metals; salinity; organic matter

1 INTRODUCTION

Anaerobic ammonium oxidation (anammox) is a biological process in which ammonium is directly converted to dinitrogen gas under anoxic conditions with nitrite as the electron acceptor (Equation (1)) [1,2]:

$$NH_4 + 1.32\ NO_2 + 0.066HCO_3 + 0.13H^+ \rightarrow 1.02\ N_2 + 0.26NO_3^- + 0.066CH_2O_{0.5}N_{0.15} + 2.03H_2O \quad (1)$$

When treating the wastewater at a low C/N ratio, the anammox process is more environmental friendly and cost-effective due to the lower oxygen demand, the lack of a requirement for external carbon sources and the reduction in the final sludge quantities. The mechanism is shown in Figure 1 [3, 4].

One of the main drawbacks of the anammox process is a long start-up period mainly due to the slow growth rates of anammox bacteria.

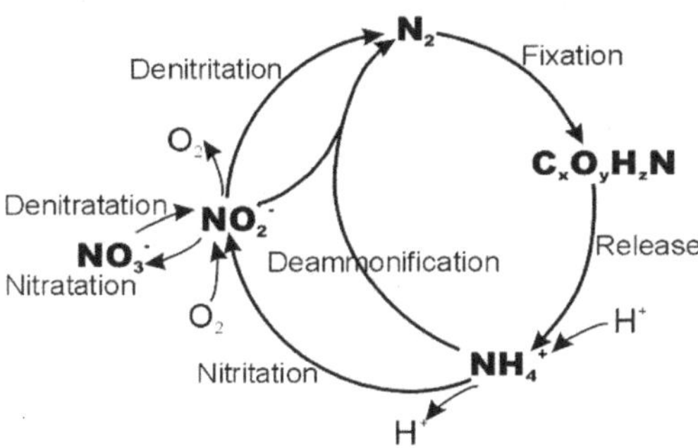

Figure 1. Nitrogen cycle.

Anammox bacteria are strictly anaerobes and autotrophs, they are very difficult to be cultured, and the doubling time has been reported to be approximately 11 days [5]. Additionally, wastewater commonly contains compounds that might induce inhibitory effects on Anammox Activity (AA). Also, the impact of a severe toxic substance would be particularly problematic for anammox due to the long recovery periods needed, which could kill the biomass. It is essential to understand influencing the activity of anammox bacteria that can improve its applicability.

This paper aims to give a summary of previous and current research on the inhibition of the anammox process by various substances, focusing on the inhibitory effects on the anammox process.

2 SALINITY INHIBITION ON ANAMMOX

Salinity is an important parameter for wastewater treatment because many kinds of industrial wastewater are rich in ammonium and contain high salt concentrations (total salinity >1%) [6]. The high salt concentration in saline wastewater originating from many industries can induce cell plasmolysis or death due to the dramatic increase in osmotic pressure, and thus may alter the biochemical properties of activated sludge or the microbial community structure [7]. Liu et al. reported that even with long-term domestication, freshwater anammox sludge cannot tolerate salinity higher than 30 g L^{-1}. The relevant research results and the corresponding anammox under different operating conditions are summarized in Table 1. There are massive swings in salinity for many kinds of industrial wastewater,

Table 1. Research on the salinity inhibition of anammox.

Reactor	Salts	Concentration g/L	Operation mode	Effect	Reference
Up-flow column reactor	NaCl	20	Long-term test	Significant inhibition	[9]
Up-flow column reactor	NaCl	30	Long-term test	NRR of 4.5 kgN/m^3d	[9]
Up-flow fixed bed column reactor	NaCl	30	Long-term test	Stable NRR of 1.7 kgN/m^3d	[8]
Up-flow fixed bed column reactor	NaCl	>30	Long-term test	NRR sharply declined	[8]
UASB	NaCl	>10	Long-term test	Losing much biomass	[7]
SBR	NaCl	<7	Long-term test	Anammox activity not affected	[6]

Table 2. The heavy metal inhibition of anammox.

Inhibitor	Operation mode	Concentration (mg/L)	Effect	Reference
Cu_2^+	Batch	12.9	IC_{50}	[14]
Cu_2^+	Long-term	5.0	94% loss in activity	[14]
Cd_2^+	Batch	11.16 ± 0.42	IC_{50}	[16]
Ag^+	Batch	11.52 ± 0.49	IC_{50}	[16]
Hg_2^+	Batch	60.35 ± 2.47	IC_{50}	[16]
Zn_2^+	Batch	7.6	IC_{50}	[17]
Ni_2^+	Batch	48.6	IC_{50}	[17]

while a gradual and long exposure to salinity is essential for achieving higher anammox activity in saline wastewater [8]. With long-term operation at high salt concentrations and the concentration of salinity less than 7 g/L, anammox bacteria may be successfully improved by acclimatization. Further research needs to be undertaken.

3 ORGANIC MATTER INHIBITION ON ANAMMOX

Anammox bacteria are slow-growing, strictly anoxic autotrophic microorganisms, which use CO_2 as the sole carbon source [9]. The application of anammox to domestic wastewater has been limited because of its high concentrations of organic matter, which inhibit the activity of anammox bacteria [10]. The research on the inhibition of anammox by organic matter would help to treat both nitrogen and organic matter containing wastewater through the applications of the anammox process.

Generally, there are two different organic matter inhibition mechanisms in anammox. In the first proposed mechanism, biodegradable organic carbon fosters the growth of heterotrophic bacteria (HET). More heterotrophic growth will lead to a higher sludge production, which in turn often leads to a higher sludge loss, and thus to a decrease in the Sludge Retention Time (SRT). Sludge loss is the total amount of sludge that leaves the system. A low SRT is particularly critical for the slowly growing anammox [11]. In the second proposed mechanism, alternative external organic or inorganic electron donors could be of use to replenish these electrons. Indeed, it has been established that these organisms are able to convert organic and inorganic compounds to sustain their metabolism, most notably formate, acetate and propionate [12, 13].

4 HEAVY METAL INHIBITION ON ANAMMOX

Some nitrogen-rich wastewater streams, such as landfill leachate and metal refinery wastewater, contain high levels of heavy metals [14]. The effect of heavy metals on microbial growth and activity is dependent on the concentration and the type of sludge and the operation time. Trace amounts of some metals, which are the components of many enzymes or coenzymes, can improve microbial activity. However, excessive concentrations can produce inhibitory effects and can even be toxic [15]. The relevant research results and the corresponding anammox are summarized in Table 2, showing that heavy metals are not easily biodegradable and can accumulate in organisms, causing biological accumulation toxicity. While there are only a few studies on the heavy metal inhibition of anammox, it is necessary to undertake further studies.

5 CONCLUSIONS

This paper reviews the impact of inhibition on the anammox process, and discusses the inhibition of anammox. Improving the performance of the

Anammox process under inhibitory conditions can lead to better popularization and application of anammox for biological nitrogen removal. Anammox can tolerate the inhibition, while the sensitivity to inhibition depends on the functional microbial community, the species of inhibition, time of operation and concentrations. The following aspects require the attention of researchers in the future:

a. Reducing the inhibition of the anammox process through continuous feeding or decreased feed rates can greatly reduce the inhibition, However, the anammox process utilizes ammonium and nitrite as the substrate and electron acceptor, respectively, thus reducing feeding rates could result in limited reaction kinetics. So, it should be maintain the balance between maximizing loading rate and minimizing inhibition on the anammox process.
b. The inhibition mechanisms are inconclusive for most situations. Further study on anammox is needed to help researchers to understand, apply and reduce its effect on wastewater treatment processes. For this purpose, pure or highly enriched cultures with identified metabolic pathways and functional genes would be required.
c. Further study needs to determine the key control strategy for preventing Anammox bacteria inhibition and improving the efficiency of nitrogen removal by the anammox process.

ACKNOWLEDGMENT

This paper was supported by the National Science and Technology Support Program of China (grant no. 2014BAC25B01) and the National Science and Technology Support Program of China (grant no. 2014BAC28B01).

REFERENCES

[1] Jetten, M.S. et al. 1999. The anaerobic oxidation of ammonium. *FEMS Microbiol. Rev* 22(5): 421–437.
[2] Strous, M. et al. 1998. The sequencing batch reactor as a powerful tool for the study of slowly growing anaerobic ammonium-oxidizing microorganisms. *Appl. Microbiol. Biotechnol* 50(5): 589–596.
[3] MC, V.L. & S, S. 2006. Biological treatment of sludge digester liquids. *Water Sci. Technol* 53(12): 11–20.
[4] B, W. et al. 2007. Development and implementation of a robust deammonification process. *Water Sci. Technol* 56(7): 81–8.
[5] Egli, K. et al. 2001. Enrichment and characterization of an anammox bacterium from a rotating biological contactor treating ammonium-rich leachate. *Arch. Microbiol* 175(3): 198–207.
[6] Yi, Y. et al. 2011. Effect of Salt on Anammox Process. *Procedia Environmental Sciences* 10: 2036–2041.
[7] Ma, C. et al. 2012. Impacts of transient salinity shock loads on anammox process performance. *Bioresour. Technol* 112: 124–13.
[8] Liu, C. et al. 2009. Effect of salt concentration in anammox treatment using nonwoven biomass carrier. *Journal of Bioscience and Bioengineering* 107: 519–523.
[9] Boran, K. et al. 2013. How to make a living from anaerobic ammonium oxidation. *FEMS Microbiol. Rev* 37(3): 428–61.
[10] Dapena-Mora. et al. 2007. Evaluation of activity and inhibition effects on anammox process by batch tests based on the nitrogen gas production. *Enzyme and Microbial Technology* 40(4): 859–865.
[11] Sarina, J. et al. 2014. Successful application of nitritation/anammox to wastewater with elevated organic carbon to ammonia ratios. *Water research* 49(1): 316–326.
[12] Strous, M. et al. 2006. Deciphering the evolution and metabolism of an anammox bacterium from a community genome. *Nature* 440: 790–794.
[13] B, K. et al. 2007. Candidatus "Anammoxoglobus propionicus" a new propionate oxidizing species of anaerobic ammonium oxidizing bacteria. *Syst Appl Microbiol* 30: 39–49.
[14] Yang, G.F. et al. 2013. The effect of Cu(II) stress on the activity, performance and recovery on the Anaerobic Ammonium-Oxidizing (Anammox) process. *Chemical Engineering Journal* 226: 39–45.
[15] C, L. et al. 2007. Inhibition of heavy metals on fermentative hydrogen production by granular sludge, Chemosphere 67: 668–673.
[16] Bi, Z. et al. 2014. Inhibition and recovery of Anammox biomass subjected to short-term exposure of Cd, Ag, Hg and Pb. *Chemical Engineering Journal* 244: 89–96.
[17] Li, G.B. et al. 2015. Inhibition of anaerobic ammonium oxidation by heavy metals. *Journal of Chemical Technology and Biotechnology* 90(5): 830–837.

Material Science and Environmental Engineering – Chen (Ed.)
© 2016 Taylor & Francis Group, London, ISBN 978-1-138-02938-5

Effects of geophagous earthworms *Eisenia fetida* on sorption of Phthalate Acid Esters to soils

L.L. Ji & L.P. Deng
Key Laboratory of Integrated Regulation and Resource Development on Shallow Lakes, College of Environment, Ministry of Education, Hohai University, Nanjing, Jiangsu, China

ABSTRACT: Effects of geophagous earthworms on the fate of Phthalate Acid Esters (PAEs) in soil are unclear. In this study, the batch technique was used to study the sorption behavior of three typical phthalate acid esters, Dimethyl Phthalate (DMP), Diethyl Phthalate (DEP) and Dibutyl Phthalate (DBP), on soil and earthworm casts with different aging times. The tested soil and casts were characterized with elemental composition and size fraction to assess the role of the diverse characteristics in sorption. Due to the higher organic carbon content and fine particles by digestion of earthworms, the casts showed stronger sorption capacity than soil. For a given sorbent, the sorption affinity correlated well with hydrophobicity of PAEs. The normalized sorption of PAEs indicates that π-π electron-donor-acceptor interaction was also important for sorption. The results suggest that geophagous earthworms may change sorption behavior and thus the bioavailability and transport of PAEs in soil.

Keywords: Phthalate Acid Esters; earthworm casts; sorption; aging time

1 INTRODUCTION

Phthalic Acid Esters (PAEs) are commonly used to increase the flexibility, pliability and elasticity of plastic products and widely used in general plastic products, cosmetics, personal care products, food packaging and medical products (Wang et al. 2013, Cai et al. 2008, Xia et al. 2011). Since the PAEs are physically bound to the polymer chains, they are easily discharged into the environment during the manufacturing, using and disposal process (Roslev et al. 2007, Liang et al. 2008). It has been reported that PAEs were distributed all over the world and frequently detected in air (Wang et al. 2012), water (Xie et al. 2007), soil (Kong et al. 2012), food and marine ecosystems (Zhao et al. 2004, Lu et al. 2009). PAEs has attracted wide attention because of the teratogenicity, mutagenicity, carcinogenicity and reproductive toxicity (Li et al. 2012, Daiem et al. 2012), therefore the environmental behavior of PAEs has been become a hot area of research.

Soil sorption is considered to be a significant environmental process for organic compounds. This process plays an important role in the migration, degradation and bioavailability of organic compounds. A few studies have shown that the sorption mechanism of soil to organic pollutants depends on the content and structure of soil mineral and soil organic matter. In temperate soils, earthworms are keystone soil organisms which affect significantly both the microbial compartment and the physico-chemical properties of the soil they inhabit. They greatly modify soil properties either directly through consumption or indirectly by disturbing the soil structure by casting and burrowing activities (Hickman & Reid 2008, Binet et al. 1998). It has been estimated that the endogeic species of the geophagous earthworms can produce about 20–200 tones dry weight casts per hectare per year in soil, and after several years, all the surface soil in their habitat may be completely transformed into casts. In addition, earthworms enhance the decomposition and transformation of organic matter in soil, promoting the formation of soil humus and organic-rich soil particles from organic matter in the soil mixed with minerals, which contributes earthworm casts to enriching with organic carbon, clay particles, and specific compounds (protein, amino sugars, polyphenols) (Bolan & Baskaran, 1996, Mora et al. 2003). Jongmans and his teams have found that sorption of organic pollutants to earthworm casts was significantly stronger than the surrounding soil (Jongmans et al. 2003). Earthworms enhanced the degradation of many organic pollutants, including pesticides, polycyclic aromatic compounds, chlorinated aromatics and hydrocarbons in soil (Kinney et al. 2008, Shan et al. 2010), by aerating and improving the nutritional status of the soil coupled with distributing microorganisms and promoting their activities. Therefore, the fate of organic pollutants in soil can be intensively affected by earthworms and casts (Hickman &

Reid 2008). However, there is little study on the effects of geophagous earthworms on sorption behavior of organic pollutants to soil, especially different aging times of earthworm casts.

Here we selected three representative phthalates, Dimethyl Phthalate (DMP), Diethyl Phthalate (DEP) and Dibutyl Phthalate (DBP), mainly investigated sorption behavior of phthalates to the parent soil and earthworm casts, also their respective OC contents, size fractions and aging times were monitored to assess the role of the diverse soil and cast characteristics in sorption.

2 MATERIAL AND METHODS

2.1 *Sorbates and sorbents*

Parent soil was collected from one agricultural field outside Rugao Jiangsu Province China, earthworms *Eisenia fetida* were collected by digging in an agricultural field in Jurong, Jiangsu Province. The soil and the earthworms were brought to the laboratory in a nylon bag. Earthworm casts were prepared in the laboratory by incubating 100 earthworms in 15 kg of the field soil in plastic box (40 cm × 25 cm × 20 cm) at 20 °C. The parent soil was incubated under the same conditions as a control. Fresh casts produced in the box were collected daily over a period of 7 d. Aging casts were prepared by incubating the fresh cast in new vessels at 20 °C for 7 d or 28 d. The soil and earthworm casts were freeze-dried, sieved (< 2 mm), and stored < 4 °C for use. Seleted soil and earthworm casts properties are given in Table 1. Total organic carbon, total N and total H were determined using an element analyzer (Elementar Vario EL III, Germany). Particle size distribution was analyzed using the low energy sonication fraction method as described by Stemmer et al. (Stemmer et al. 1998).

Sorbates include Dimethyl Phthalate (DMP), Diethyl Phthalate (DEP) and Dibutyl Phthalate (DBP) were purchased from ITC (Shanghai, China) with purities of > 97%. Chemical structures of three phthalates are given in Figure 1. Solute aqueous

Table 1. Selected physical and chemical properties of soil and earthworm casts with different aging times.

Parameter	Parent soil	Fresh cast	Aging 7d cast	Aging 28d cast
Clay (%)	1.71	2.34	2.39	2.48
Silt (%)	28.5	36.4	36.0	35.8
Sand (%)	70.8	61.3	61.5	61.8
C (%)	3.86	7.81	7.54	5.80
H (%)	0.76	1.30	1.24	0.82
N (%)	0.33	0.84	0.82	0.61

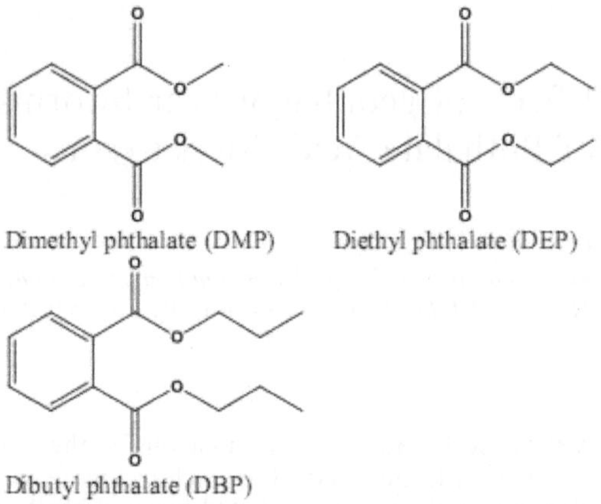

Dimethyl phthalate (DMP) Diethyl phthalate (DEP)

Dibutyl phthalate (DBP)

Figure 1. Chemical structures of three selected phthalates.

solubility (S_w) and *n*-Octanol-Water Partition Coefficient (K_{ow}) are listed in Table 2.

2.2 *Batch sorption experiments*

Sorption experiments were carried out in polytetrafluoroethylene-lined screw cap glass vials of capacity 22 mL receiving 25–300 mg sorbents (soil and earthworm casts) and sufficient volume of background solution containing 0.02 M NaCl and 200 mg/L NaN_3 (as the bio-inhibitor). The amount of sorbents and the volume of solution were selected for each sorbate to assure data compatibility while maintaining analytical accuracy. The pH of sorbent suspension was adjusted to about 6.0 by 0.1 M HCl and 0.1 M NaOH. PAEs solute was added in carrier of methanol that was kept below 0.1% by volume to minimize co-solvent effects. The vials were shaken on a rotary mixer at room temperature for 5 days in the dark. After still standing for 12 h, the vials were centrifuged at 3500 r/min for 10 min and the concentrations of the phthalate acid esters in the supernatants (1 mL) were analyzed by High Performance Liquid Chromatography (HPLC) using Bridge—C18 column (4.6 × 150 mm) (Waters). Isocratic elution was performed with a UV detector under the following conditions: 30% water/70% acetonitrile (v: v) with a wavelength of 226 nm for PAEs.

2.3 *Statistical analysis*

The sorbate concentration in the solid-phase was calculated by Equation 1 below:

$$q = \frac{\left(C_o - C_w\right) \times V}{m} \quad (1)$$

The solid liquid distribution coefficient (K_d) in balance is ratio of equilibrium concentration of

Table 2. Sorbate water solubility (S_w) and *n*-octanol-water partition coefficient (K_{ow}) of phthalates.

Compound	S_w (mmol/L)	Log K_{OW}
Dimethyl phthalate	21.88	1.53
Diethyl phthalate	3.63	2.39
Dibutyl phthalate	0.044	4.61

solid phase and liquid phase concentration, it was calculated by Equation 2 below:

$$K_d = \frac{q}{C_W} = \left(\frac{C_o}{C_W} - 1\right) \times \frac{V}{m} \quad (2)$$

where q = the concentration of PAEs adsorbed on the sorbent (soil or earthworm casts) at equilibrium (mmol/kg); C_o and C_w = the concentration of the phthalic acid ester in the solution at initial state and equilibrium (mmol/L), respectively; V = the volume of solution (L); m = the weight of solid phase sorbent (kg); and K_d = the distribution coefficient (L/kg). The data analysis and drawing were described by a plotting software program (Sigmaplot 10.0).

3 RESULT AND DISCUSSION

3.1 *Characteristics of soil and earthworm casts*

The properties of the soil and the different casts were summarized in Table 1. In the experiments for the cast preparation, the soil was homogeneously mixed and no extra organic material or food was supplied to the earthworms, thus selective feeding on soil particles with higher organic matter by the earthworms was avoided. The earthworm casts were higher than parent soil in terms of the contents of clay, silt, organic carbon content and nitrogen, which mainly attributed to the grinding action of soil particles in the earthworm gut (Hickman & Reid 2008, Shan et al. 2011). Earthworm could mixed mineral soil with soil organic matter forming aggregate structure, thus providing physical protection for organic matter and improve the potential ability of soil organic carbon sequestration, meanwhile, earthworms can increase soil nitrogen fixation, increase the loss of some inorganic nutrients in the soil, thereby increasing the loss of some inorganic nutrients in the soil (Gao et al. 2013). The pH values of the casts did not change over the aging time and were about 0.3 units lower than that of the soil. The decrease of pH in the casts might be owing to oxidation of organic substances and producing humic in the earthworm gut.

3.2 *Sorption model of PAEs on soil and earthworm casts*

The sorption data are fitted with the Freundlich sorption model: $q_e = K_F C_e^n$ (weighed on $1/q$), where q (mmol/kg) = the solid-phase, C_e (mmol/L) = aqueous-phase concentrations at sorption equilibrium; K_F (mmol^{1-n} L^n/kg) = the Freundlich affinity coefficient; and n (unitless) = the Freundlich linearity index. The model fitting parameters are summarized in Table 3. The Freundlich model fits the sorption data reasonably, and for all sorbate/sorbent combinations, sorption is highly nonlinear (departures of n from 1 imply nonlinear sorption) within the tested concentration ranges. It indicated that the sorption was a continuous energy distribution of sorption sites (Sander & Pignatello 2005) and the surface sorbent properties determined the sorption mechanisms (Jordão et al. 2010). All the nonlinearity n values of PAEs by these sorbents tested in this study were less than 1, which was consistent with previous works that the sorption isotherm of organic chemicals by soils and casts are nonlinear (Hickman & Reid 2008, Sun et al. 2012, Ying et al. 2003). The K_{oc} for different sorbate-sorbent combinations was also calculated (values at 0.01 S_w and 0.1 S_w). The following discussion was based on the sorption parameters of the FM fitting.

3.3 *Sorption isotherms of PAEs on soil and earthworm casts*

Sorption of DMP, DEP and DBP on the parent soil and casts with various aging times (0, 7 and 28d) were obtained in Figure 2. At the same pH about 7, the sorption of the PAEs on the casts was always higher than on the soil, and the sorption capacity of earthworm casts with different aging times is almost the same, Because the pH values in the casts were almost the same and the contents of clay, sand, silt and organic carbon are not changing with increasing aging time, while, obviously higher than parent soil, the higher sorption capacity of the casts might be mainly attributed to the higher contents of the fine soil particles (clay and silt) and modification of the organic carbon through the gut passage. It's reported that activity of the earthworms is the most important and potential factors determining the sorption capacity of the casts, such as surrounding condition and feeding activities (Alekseeva et al. 2006). Studies suggested that geophagous soil fauna (including earthworms) may lead to an accumulation of aromatic components in the process of ingesting soil by selective digestion of the proteinaceous components of soil humic substances (Li & Brune 2005, Don et al. 2008), on the condition, such selective digestion behavior would inevitably modify the composition and surface properties of soils, and increase the possibility

Table 3. Freundlich model parameters K_F and $n \pm$ standard deviation for isotherms measured for sorption, along with OC-normalized soil-to-water distribution coefficient (K_{oc}).

Compound	Sorbent	K_F (mmol^{1-n}L/kg)	n	R^2	K_{oc} (L/kg) 0.01S$_w$/0.1S$_w$
DMP	Soil	14±4	0.72±0.04	0.96	540/280
	Fresh cast	6.9±0.4	0.60±0.01	0.99	160/60
	Aging 7 d cast	5.5±0.3	0.52±0.01	0.99	150/500
	Aging 28 d cast	6.4±0.5	0.53±0.02	0.99	220/70
DEP	Soil	3±1	0.51±0.05	0.93	440/145
	Fresh cast	13±1	0.62±0.01	0.99	610/250
	Aging 7 d cast	14±1	0.63±0.01	0.99	670/280
	Aging 28 d cast	13±1	0.54±0.02	0.99	1030/360
DBP	Soil	16±2	0.37±0.02	0.98	53800/12600
	Fresh cast	250±26	0.73±0.02	0.99	25600/13800
	Aging 7 d cast	260±34	0.71±0.03	0.99	32000/16400
	Aging 28 d cast	91±7	0.60±0.02	0.99	34600/13800

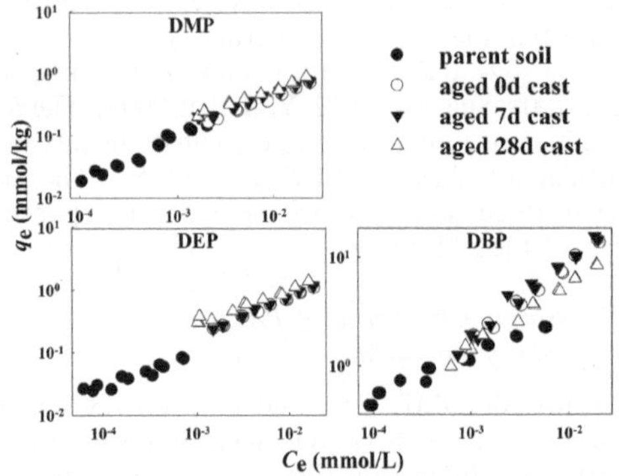

Figure 2. Sorption isotherms of DMP, DEP and DBP on soil and earthworm casts with different aging times.

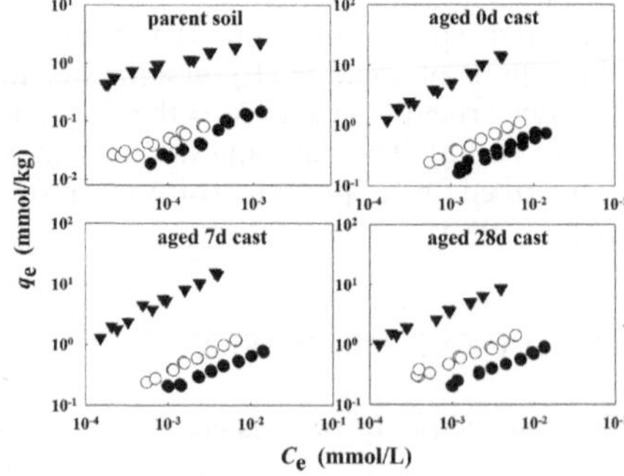

Figure 3. Comparison of sorption data between DMP, DEP and DBP with different sorbents (• = DMP, ○ = DEP, ▼ = DBP).

of exposure of hydrophobic compounds to the aromatic moiety of the soil organic carbon, and therefore enhancing the sorption capacity of the casts. Besides, feeding activity of the earthworms, by grinding soil aggregates in their gut, could modify the structure of casts, increase the particle surface and form soil aggregation, this was in agreement with previous studies of Larink (Larink et al. 2001) who had found that the swelling value and porosity values of earthworm casts were 50.9% and 20.0% higher than parent soil, respectively, which may also increase the sorption of the PAEs on the casts.

Although the organic carbon and particle size (clay, silt) may contribute to PAEs sorption on soil, but for organic chemical contaminants, there are still other properties are also a great influence on the sorption, e.g. polarity, hydrophobicity, and molecular configuration (Pan et al. 2008). The sorption of the three PAEs on the soil and all the casts are shown in Figure 3, it could be clearly seen that their sorption affinity followed the order: DBP > DEP > DMP, the difference of sorption capacity was mainly caused by the different properties of sorbates. At the pH values (6.61–7.07) of the soil and the casts in our study, more than 99% of the PAEs were present in the molecular form, PAEs with the strongest hydrophobicity had the highest sorption on the soil and the casts (Fang et al. 2014). Phthalates in the water solubility were contrary to the hydrophobic property, one has the minimum solubility in water, minimal affinity for water, the maximum hydrophobicity, therefore, the maximum sorption capacity. Generally, the sorption can be predicted using the hydrophobic parameters of organic chemical, such as K_{OW} (octane-water distribution coefficient) or logK_{OW}. The hydrophobicity of DBP is higher than that of DEP and the DMP, DBP is more easily adsorbed by soil than DEP and DMP.

3.4 *The potential sorption mechanisms*

Since aged casts contains various organic components with various structures and properties and the sorption point of organics is not uniform, several mechanisms may exist during the process of sorption (Liao et al. 2013). Earthworm casts with different aging time contained more organic carbon and, coupled with smaller particle size (clay, silt), thus having a higher specific surface area, and their sorption capacity for PAEs were greater than that of parent soil. The nonlinear characteristic sorption isotherms of PAEs were also attributed to the organic matter heterogeneity, which has been confirmed as the most important factor in nonlinear sorption (Li et al. 2013). Three intermolecular interactions including a hydrophobic interaction (e.g. van der Waals, dipole–dipole, dipole-induced dipole), hydrogen bonding, and π–π bonding are involved in organic matter sorption mechanisms (Li et al. 2013). However, PAEs with apolar alkyls are nonionic molecules that do not contain hydrogen atoms attached to electronegative atoms, electrostatic attraction and hydrogen bonding can therefore be eliminated as major binding mechanisms in a neutral environment. In addition, the previously mentioned casts contained a certain amount of humic acids, and studies have shown that interaction of Humic Acids (HA) and PAEs could increase the sorption of PAEs, at the meantime, competitive sorption between humic acid and PAEs occurred. Although the hydrophobic interactions are the dominant factor mediating the sorption of PAEs to HA and soil, we evaluated a potential cause of the sorption, the occurrence of π–π electron donor–acceptor interactions with π-donor sites in organic matter may be enhanced the sorption of PAEs to casts (Sullivan et al. 1982, Wen et al. 2013).

The contribution of potential specific sorption mechanisms other than hydrophobic effects can be better understood with normalization of solute hydrophobicity, Figure 4 shows the results of hydrophobicity-normalized sorption isotherms

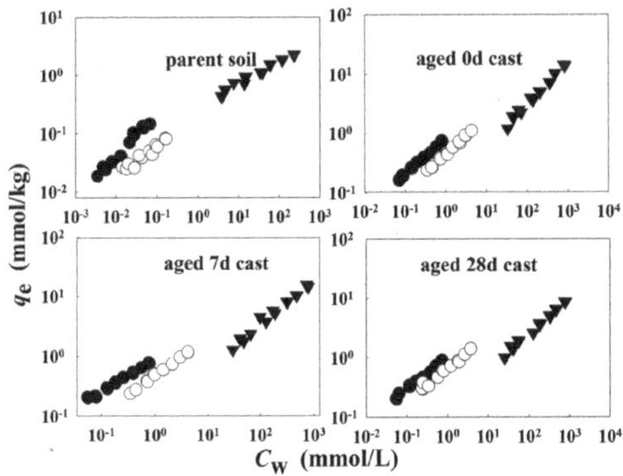

Figure 4. Normalized sorption isotherms of DMP, DEP and DBP for different sorbents using solute octanol–water partition coefficient (K_{OW}) (• = DMP, o = DEP, ▼ = DBP).

to different soils plotted as q_e against C_w, which represents the solute concentration in octanol that would be in equilibrium with the observed aqueous concentration (C_e), i.e. $C_w = C_e \cdot K_{ow}$. The hydrophobicity of DBP is higher than that of DEP and DMP and the sorption affinity of the DBP was higher than that of the DEP and DMP on aged casts, while with normalization of solute hydrophobicity, the sorption affinity of PAEs decreased in the order of DMP> DEP > DBP, indicating that other interaction mechanisms may operate in addition to the hydrophobic effect. If not, the sorption isotherms of tested PAEs should be similar. We propose that π–π EDA interactions were another primary mechanism for the enhanced sorption of PAEs on the soil and casts. DMP, DEP and DBP may act as π-electron acceptors due to the strong electron-withdrawing ability of the ester functional group, and hence interact strongly with graphene structure of the organic matters in soil and casts. Therefore, hydrophobic interactions and π-π bonding were postulated as the major mechanism for the sorption of PAEs on aging casts.

4 CONCLUSION

The structures of soil were changed through the gut passage of the geophagous earthworms so that the sorption of PAEs by soils was enhanced. Sorption isotherms of PAEs on soil and earthworm casts were fitted the FM model well. The higher sorption capacity of the earthworm casts was probably attributed to the higher fine fraction particles, clay and the soil organic carbon in the earthworm casts. For a given sorbent, the sorption affinity follows an order of DBP > DEP > DMP, which correlates well with their hydrophobicity and alkyl chain length. The results revealed that earthworms would change the bioavailability and the environmental fate of PAEs, especially where earthworms are abundant.

ACKNOWLEDGEMENT

This work was supported by the National Science Foundation of China (Grant 41101479), the China Postdoctoral Science Foundation (2011M500849, 2012T50461), and the Fundamental Research Funds for Central Universities (2011B08814) and a Project Funded by the Priority Academic Program Development of Jiangsu Higher Education Institutions (PAPD).

REFERENCES

[1] Binet, F. 1998. Significance of earthworms in stimulating soil microbial activity. *Biology and Fertility of Soils* 27(1): 79–84.

[2] Bolan, N.S. & Baskaran, S. 1996. Characteristics of earthworm casts affecting herbicide sorption and movement. *Biology and fertility of soils* 22(4): 367–372.

[3] Butenschoen, O. 2009. Carbon and nitrogen mobilisation by earthworms of different functional groups as affected by soil sand content. *Pedobiologia* 52(4): 263–272.

[4] Cai, Q.Y. 2008. The status of soil contamination by Semivolatile Organic Chemicals (SVOCs) in China: a review. *Science of the Total Environment* 389(2): 209–224.

[5] Daiem, M.M.A. 2012. Environmental impact of phthalic acid esters and their removal from water and sediments by different technologies–A review. *Journal of environmental management* 109: 164–178.

[6] Don, A. 2008. Organic carbon sequestration in earthworm burrows. *Soil Biology and Biochemistry*, 40(7): 1803–1812.

[7] Fang, C.R. 2014. Sorption behavior of dibutyl phthalate and dioctyl phthalate by aged refuse. *Environmental Science and Pollution Research*: 1–9.

[8] Gao, B. 2013. Sorption of phthalic acid esters in two kinds of landfill leachates by the carbonaceous sorbents. *Bioresource technology*, 136, pp. 295–301.

[9] Hickman, Z.A. & Reid, B.J. 2008. Earthworm assisted bioremediation of organic contaminants. *Environment International* 34(7): 1072–1081.

[10] Jongmans, A.G. 2003. Soil structure and characteristics of organic matter in two orchards differing in earthworm activity. *Applied Soil Ecology* 24(3): 219–232.

[11] Jordão, C.P. 2010. A study on Al (III) and Fe (II) ions sorption by cattle manure vermicompost. *Water, Air, & Soil Pollution* 210(1–4): 51–61.

[12] Kinney, C.A. 2008. Bioaccumulation of pharmaceuticals and other anthropogenic waste indicators in earthworms from agricultural soil amended with biosolid or swine manure. *Environmental Science & Technology* 42(6): 1863–1870.

[13] Kong, S. 2012. Diversities of phthalate esters in suburban agricultural soils and wasteland soil appeared with urbanization in China. *Environmental Pollution* 170: 161–168.

[14] Larink, O. 2001. Regeneration of compacted soil aggregates by earthworm activity. *Biology and Fertility of soils* 33(5): 395–401.

[15] Liang, P. 2008. Application of dispersive liquid–liquid microextraction and high-performance liquid chromatography for the determination of three phthalate esters in water samples. *Analytica chimica acta* 609(1): 53–58.

[16] Liao, Y. 2013. Effect of deposit age on adsorption and desorption behaviors of ammonia nitrogen on municipal solid waste. *Environmental Science and Pollution Research* 20(3): 1546–1555.

[17] Li, C. 2012. Biodegradation of an endocrine-disrupting chemical di-n-butyl phthalate by Serratia marcescens C9 isolated from activated sludge. *African Journal of Microbiology Research* 6(11): 2686–2693.

[18] Li, J. 2013. Kinetics, equilibrium, and mechanisms of sorption and desorption of 17α-ethinyl estradiol in two natural soils and their organic fractions. *Science of The Total Environment* 452: 404–410.

[19] Li, X. & Brune, A. 2005. Selective digestion of the peptide and polysaccharide components of synthetic humic acids by the humivorous larva of *Pachnoda ephippiata* (Coleoptera: Scarabaeidae). *Soil Biology and Biochemistry* 37(8): 1476–1483.

[20] Lu, Y. 2009. Biodegradation of dimethyl phthalate, diethyl phthalate and di-butyl phthalate by *Rhodococcus* sp. L4 isolated from activated sludge. *Journal of hazardous materials* 168(2): 938–943.

[21] Mora, P. 2003. Physico-chemical typology of the biogenic structures of termites and earthworms: a comparative analysis. *Biology and Fertility of Soils* 37(4): 245–249.

[22] Pan, B. & Xing, B. 2008. Adsorption mechanisms of organic chemicals on carbon nanotubes. *Environmental Science & Technology* 42(24): 9005–9013.

[23] Roslev, P. 2007. Degradation of phthalate esters in an activated sludge wastewater treatment plant. *Water Research* 41(5): 969–976.

[24] Sander, M. & Pignatello J.J. 2005. Characterization of charcoal adsorption sites for aromatic compounds: insights drawn from single-solute and bi-solute competitive experiments. *Environmental science & technology* 39(6): 1606–1615.

[25] Shan, J. 2010. Selective digestion of the proteinaceous component of humic substances by the geophagous earthworms *Metaphire guillelmi* and *Amynthas corrugatus*. *Soil Biology and Biochemistry* 42(9): 1455–1462.

[26] Shan, J. 2011. Enhancement of chlorophenol sorption on soil by geophagous earthworms (*Metaphire guillelmi*). *Chemosphere* 82(2): 156–162.

[27] Stemmer, M. 1998. Organic matter and enzyme activity in particle-size fractions of soils obtained after low-energy sonication. *Soil Biol Biochem* 30: 9–17.

[28] Sullivan, K.F. 1982. Adsorption of phthalic acid esters from seawater. *Environmental science & technology* 16(7): 428–43.

[29] Sun, K. 2012. Polar and aliphatic domains regulate sorption of Phthalic Acid Esters (PAEs) to biochars. *Bioresource technology* 118: 120–127.

[30] Wang, J. 2013. Soil contamination by phthalate esters in Chinese intensive vegetable production systems with different modes of use of plastic film. *Environmental Pollution* 180: 265–273.

[31] Wang, W. 2012. Distributions of phthalic esters carried by total suspended particulates in Nanjing, China. *Environmental monitoring and assessment* 184(11): 6789–6798.

[32] Wen, Z.D. 2013. Effects of humic acid on phthalate adsorption to vermiculite. *Chemical Engineering Journal* 223: 298–303, 2013.

[33] Xia, X. 2011. Levels, distribution, and health risk of phthalate esters in urban soils of Beijing, China. *Journal of environmental quality* 40(5): 1643–1651.

[34] Xie, Z. 2007. Occurrence and air-sea exchange of phthalates in the Arctic. *Environmental science & technology* 41(13): 4555–4560.

[35] Ying, G.G. 2003. Sorption and degradation of selected five endocrine disrupting chemicals in aquifer material. *Water research* 37(15): 3785–3791.

[36] Zhao, X.K. 2004. Adsorption of dimethyl phthalate on marine sediments. *Water, air, and soil pollution* 157(1–4): 179–192.

Material Science and Environmental Engineering – Chen (Ed.)
© 2016 Taylor & Francis Group, London, ISBN 978-1-138-02938-5

Inhibitory effects of nisin on *Staphylococcus aureus* and *Xanthomonas*

B.G. Zhang, Y.E. Qing, X.T. Wang, G.Q. Zhao & G.H. Hu
School of Life Science and Engineering, Lanzhou University of Technology, Lanzhou, China

ABSTRACT: Nisin was produced with whey, which is a by-product of cheese. The bacteriostatic effect of nisin was investigated at various temperatures, pH and concentrations of zinc. The results showed that nisin was stable at the pH level below 4. The antimicrobial activities of nisin enhanced as the pH level decreased. The temperature had more effect on the antimicrobial activities of nisin at the pH level above 4. The antimicrobial activities of nisin decreased as the pH level rose. The activity of nisin almost lost at alkaline condition. The antibacterial effect of nisin on Gram-positive bacteria such as *Staphylococcus aureus* was better than that on Gram-negative bacteria such as *Xanthomonas campestris*. In addition, divalent metal ions had an obvious effect on the bacteriostatic effect of nisin.

Keywords: Nisin; *Staphylococcus aureus*; *Xanthomonas campestris*; Zinc oxide

1 INTRODUCTION

Nisin is a bacteriocin produced by several strains of *Lactococcus lactis*, and recognized as GRAS [1]. Nisin is an antimicrobial peptide with 34 amino acids and 3.5 kDa. Due to its antibacterial activity, attempts have been made to use it as a natural preservative in foods used in more than 40 countries [2]. Nisin could reduce the levels of *L. monocytogenes* in sliced fermented sausages with no added sodium salt [3]. The antibacterial properties of nisin have been well studied and applied in a variety of foods including vegetables, dairy products and meats. Nisin has the potential to be used as an inducing agent in *in situ* delivery systems of bioactive peptides and proteins by genetically modified bacteria in the intestine [4]. That means nisin could survive in the intestinal environment. In many cases, nisin can be combined with other sterilization methods to improve the effect of sterilization [5–7]. The specific antibacterial mechanism of nisin remains controversial, although it has obtained a very good application. Based on this above consideration, we studied the inhibitory effect and mechanism of nisin on two kinds of bacteria, and the effect of metal ions on the bacteriostatic effect of nisin.

2 MATERIALS AND METHODS

2.1 *Bacterial strains and growth conditions*

S. aureus (ATCC25923) was propagated in a beef extract peptone medium at 37°C for 24 h. *Xanthomonas campestris* was propagated in a beef lactose agar medium at 28°C for 24 h. Bacteria were stored at −80°C as stock cultures in their corresponding culture media supplemented with 5% glycerol, and sub-cultured twice before their use in experiments.

2.2 *Nisin preparation*

A stock solution of nisin (10^6 IU/mL) was prepared by suspending 160 mg pure nisin (WeiRi, China) in 10 mL of 0.02 mol/L HCl. This solution was subsequently centrifuged at 1500× g for 20 min, sterilized by filtration through a 0.22 mm filter and stored at 4°C for up to 4 days.

2.3 *Antibacterial activity*

Antibacterial activity was assessed against *S. aureus* and *Xanthomonas*. The cultures were inoculated in nutrient broth and incubated for 24 h at 37°C and 28°C, respectively, on a shaker incubator. The cells were collected by centrifugation at 10,000 rpm at 4°C for 20 min. After collection, the cells were washed with Phosphate Buffer Saline (PBS, pH 7.2) twice. The final concentration was fixed at 10^5 cells/ml of PBS. Initial counts were estimated by the spread plating method on nutrient agar. Different concentrations of nisin were added to the tubes containing the bacterial cells. Tubes were kept at room temperature for 24 h, and final counts were estimated by the plating method.

2.4 *Detection of pH stability of nisin*

Cultures of *S. aureus* and *Xanthomonas* were inoculated at the level of 1% for 24 h in broth medium and Lysogeny broth as the control group.

The inoculated group contained 16 mg/mL of nisin. Then, the medium was cultured at 37°C, 150 r/min for 0, 12, 18 and 32 h to detect the pH value.

2.5 *Detection of thermostability of nisin*

The antimicrobial of nisin was studied at different temperatures. A sterile filter paper was soaked in nisin solution at 10, 25, 30, 40, 60, 80, 100, 110 and 120°C for 20 min. Then, the inhibitory zone diameter was measured according to the paper-disc agar-diffusion method.

2.6 *Influence of zinc ion on nisin*

A 0.0081 g sample of ZnO was dissolved in 0.02 mol/L of HCL to formulate 0.001 mol/L of ZnO solution. Bacteria were cultured for 24 h. This was followed by 1% of *S. aureus* and *Xanthomonas* cultured for 24 h in broth medium and Lysogeny broth as the control group that contained 0.001 mol/L of ZnO. The inoculated group contained 16 mg/mL of nisin and 0.001 mol/L of ZnO. Then, the medium was cultured at 37°C, 150 r/min for 3, 18 and 32 h to detect the OD value at 650 nm.

2.7 *Measurement of the protein content of the culture medium*

Coomassie brilliant blue assay was selected to measure the protein content of the medium. Briefly, 0.1 g of bovine serum albumin was dissolved in 100 mL of distilled water to prepare a 1.0 mg/mL of standard protein solution. A solution of 100 g Coomassie Brilliant Blue G-250 was dissolved in 50 mL of 95% ethanol, and then added 120 mL of 85% phosphoric acid to dilute with water to 1 L. The solution was stored in a brown vial separately. Then, seven test tubes were taken, with one for blank, samples water and reagents were added. The standard protein solution at concentrations of 0, 0.01, 0.02, 0.04, 0.06, 0.08 and 0.1 mL was added to each test tube and filled with distilled water to 0.1 mL. Then, 5.0 mL Coomassie Blue G-250 reagent was added to each test tube, and finally added 4.9 mL distilled water. The test tubes were immediately shaken. After 2–5 minutes, each sample was measured with a spectrum colorimeter at 595 nm, to record the data and draw the standard curve. Cultures of *S. aureus* and *Xanthomonas* were inoculated at the level of 1% for 24 h in broth medium and Lysogeny broth as the control group. The inoculated group contained 16 mg/mL of nisin. Then, the medium was cultured at 37°C, 150 r/min for 0, 12, 24 and 36 h to detect the OD value at 595 nm. The protein content of each group was calculated according to the standard curve.

2.8 *Statistical analysis*

All experiments were performed in triplicate, and data are presented as means ± Standard Errors (SE). Significant differences between measurements for the control and treated samples were analyzed using the one-way factorial analysis of variance (ANOVA), followed by Duncan's *post hoc* test (SPSS 16.0).

3 RESULTS

3.1 *Curves of bacteriostasis of nisin*

The antimicrobial properties of nisin are shown in Figure 1. We can see that the lag phase of *Staphylococcus aureus* and *Xanthomonas* was significantly delayed compared with *Staphylococcus aureus*-supplemented nisin. The growth of the blank group was greater than that of the experimental group within the first 20 hours. While after 24 hours, the growth of the experimental group became greater than that of the blank group. This proved that nisin has antibacterial effects. The inhibitory effect of nisin on *Staphylococcus aureus* was stronger than that on *Xanthomonas*. Nisin can significantly delay the lag phase of bacteria.

3.2 *pH stability of nisin*

As shown in Figure 2, with the continuous growth of *Staphylococcus aureus*, the pH of the culture medium was increasing, and the increase in the experimental group was significantly higher than that in the control group. This change was found to be opposite for *Xanthomonas*. That means nisin can change the pH of the culture to change its own antibacterial activity.

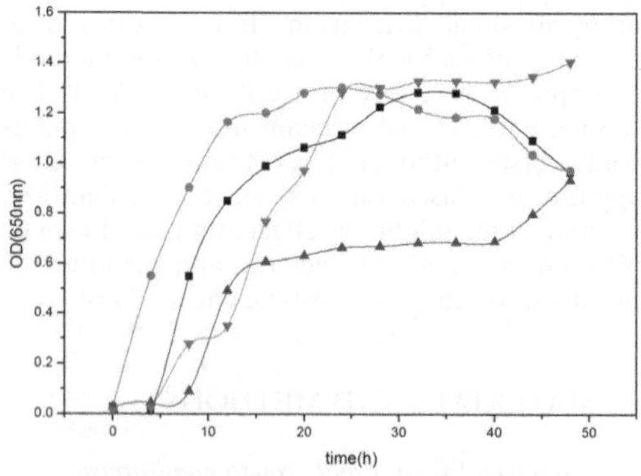

(■, nisin+*S. aureus*; ●, *S. aureus*; ▲, nisin+*Xanthomonas*; ▼, *Xanthomonas*.

Figure 1. The growth curve of two kinds of bacteria.

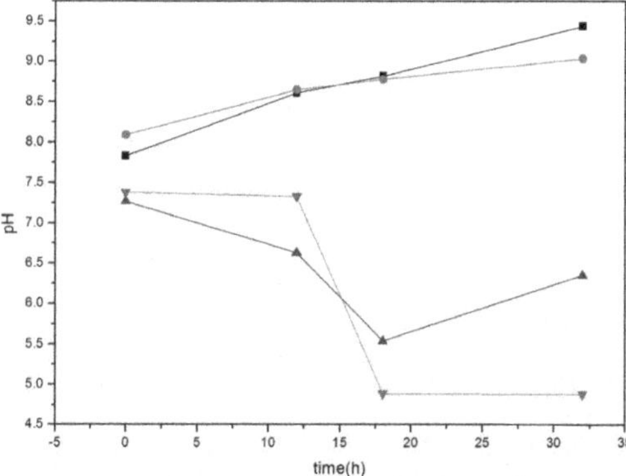

■, nisin+*S. aureus*; ●, *S. aureus*;
▲, nisin+*Xanthomonas*; ▼, *Xanthomonas*.

Figure 2. The change in the pH level along with incubation time.

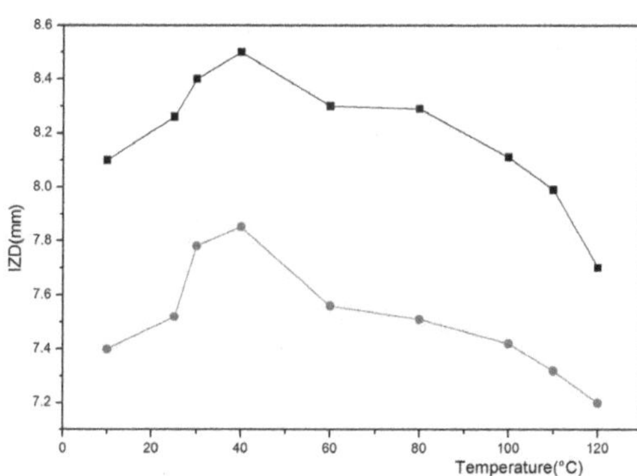

■, *Staphylococcus aureus*; ●, *Xanthomonas Campestris*.

Figure 3. The influence of the temperature on the bacteriostatic effect of nisin.

3.3 *Thermostability of nisin*

As shown in Figure 3, the inhibition zone diameter decreased as the temperature increased. The IZD of nisin treated at 110°C and 120°C for 20 min decreased little compared with the control group. That means nisin has a good heat resistance.

3.4 *Influence of zinc ion on nisin*

As shown in Figure 4, the OD value of the two bacteria was significantly higher than that of the experimental group. The OD value of *Staphylococcus aureus* broth was significantly greater than that of *Xanthomonas* broth. This means the bacteriostatic effect of nisin has a close relationship with teichoic acid.

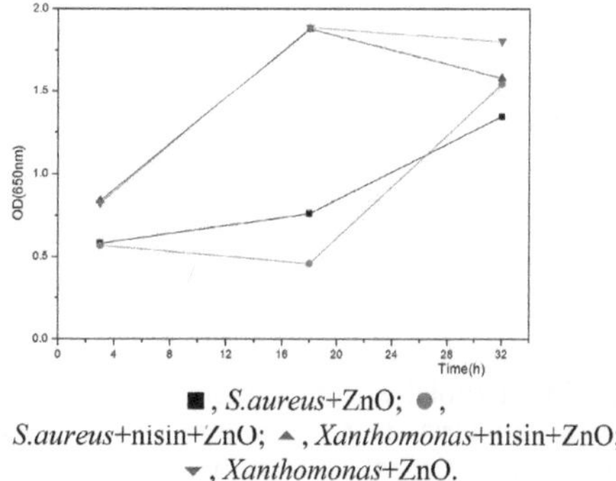

■, *S.aureus*+ZnO; ●, *S.aureus*+nisin+ZnO; ▲, *Xanthomonas*+nisin+ZnO; ▼, *Xanthomonas*+ZnO.

Figure 4. The influence of zinc ions on the bacteriostatic effect of nisin.

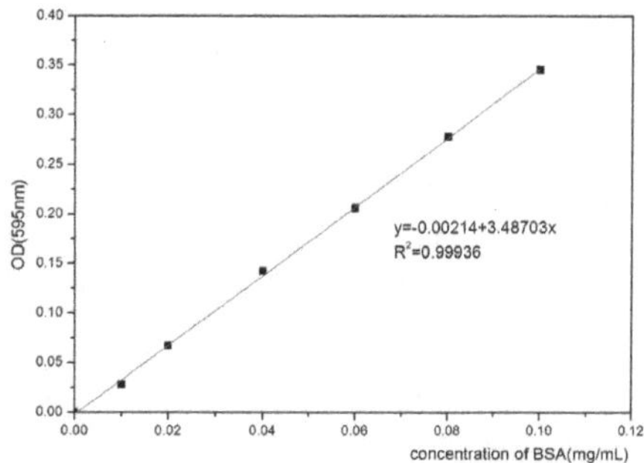

Figure 5. Standard curve of the protein.

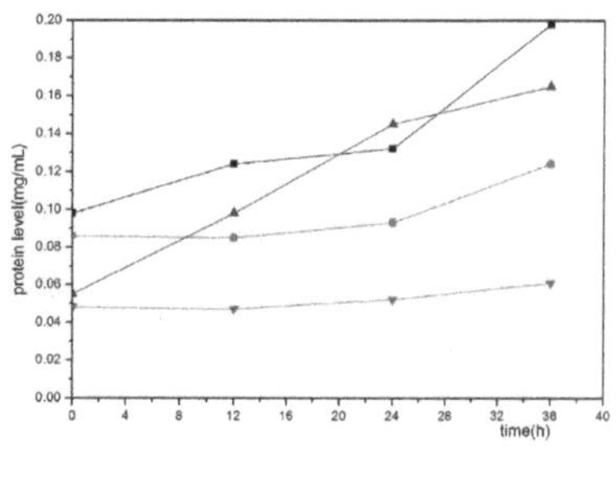

■, nisin+*S. aureus*; ●, *S. aureus*; ▲, nisin+*Xanthomonas*; ▼, *Xanthomonas*.

Figure 6. The change in protein content.

3.5 *Change in the protein content of culture medium*

As shown in Figure 5 and Figure 6, the protein content of the culture medium increased significantly.

Nisin may consume proton motive force-sensitive cells, inhibiting the biosynthesis of peptidoglycan. The synthesis of cell membrane and phospholipids was hindered. This led to the leakage of intracellular materials such as protein and nucleic acids, and caused sensitive cell lysis.

4 CONCLUSION

The antimicrobial effect of nisin against Gram-positive bacteria (e.g. *S. aureus*) was greater than that against Gram-negative bacteria (e.g. *Xanthomonas*). At a certain range, the antibacterial activity of nisin increased with the pH level decreasing. Divalent metal ions hindered the antimicrobial effect of nisin. The antimicrobial mechanism of Nisin might be related to teichoic acid of bacteria. Nisin consumed proton motive force-sensitive cells, inhibiting the biosynthesis of peptidoglycan. The synthesis of cell membrane and phospholipids was hindered. This led to the leakage of intracellular materials such as protein and nucleic acids, and caused sensitive cell lysis. Thus, the cell growth and proliferation were inhibited.

ACKNOWLEDGMENT

This work was financially supported by the National Undergraduate Training Programs for Innovation and Entrepreneurship (No. 201410731012) and the Foundation of Lanzhou University of Technology.

REFERENCES

[1] Muppalla, S.R., R. Sonavale: Radiation Physics and Chemistry, Vol. 81 (2012) No. 12, p. 1917.

[2] García, P., B. Martínez: International Journal of Food Microbiology, Vol. 141 (2010) No. 3, p. 151.

[3] Marcos, B., T. Aymerich: Food Control, Vol. 30 (2013) No. 1, p.325.

[4] J. Reunanen, P.E.J. Saris: Biotechnology Letters, Vol. 31 (2009). No. 8, p. 1229.

[5] Bi, X., Y. Wang: Food Control, Vol. 41 (2014), No. 1, p. 139.

[6] Razavi Rohani, S.M. LWT-Food Science and Technology, Vol. 44 (2011), No. 10, p. 2260.

[7] Tong, Z., L. Zhou: Peptides, Vol.32 (2011), No. 10, p. 2021.

Material Science and Environmental Engineering – Chen (Ed.)
© 2016 Taylor & Francis Group, London, ISBN 978-1-138-02938-5

Toxicological effects of perfluorooctanoic acid (PFOA) on *Daphnia magna*

G.H. Lu
Key Laboratory of Integrated Regulation and Resources Development, College of Environment, Hohai University, Nanjing, China
College of Hydraulic and Civil Engineering, Agricultural and Animal Husbandry College of Tibet University, Linzhi, China

B.N. Ma, S. Li & L.S. Sun
Key Laboratory of Integrated Regulation and Resources Development, College of Environment, Hohai University, Nanjing, China

ABSTRACT: A suite of comprehensive toxicity tests were conducted including 48 h acute toxicity test, 7-day sub-acute toxicity test, 21-day chronic test, as well as feeding experiment. Daphnias were exposed to aqueous solutions of PFOA at concentrations ranged from 0.032 to 20 mg/L based on the results of acute toxicity, the biomarkers including Acetylcholinesterase (AChE), Superoxide Dismutase (SOD) and Catalase (CAT) were determined after 7-day of exposure, and the survival, growth and reproduction of the daphnias were monitored in 21-day life cycle. AChE, SOD and CAT activities as well as the filtration rate and ingestion rate were inhibited at concentrations equal to or higher than 0.16 mg/L. PFOA inhibited both growth and reproduction of *D. magna* during the testing period, and concentration dependence was apparent for the number of first brood per female, body length and intrinsic of natural increase.

Keywords: PFCs; reproduction; biomarker; *Daphnia magna*

1 INTRODUCTION

Perfluorinated Compounds (PFCs) with different functions have been extensively used in many technological applications either individually or as composites in manufactured products [1]. For example, PFCs are used as water-, oil- and grease-repellents or as surfactants [2]. In recent years, PFCs have become a focus of public health concern due to their ubiquitous presence in environmental media, food and drinking water as well as in wildlife [3, 4, 5]. PFCs break down very slowly in the environment, and have been shown to accumulate in animals including humans [6, 7, 8].

PFOA has been detected in different aquatic environments throughout the world. The contents of PFOA were <0.02~0.52 ng/g dry weight and 0.02~0.18 ng/g dry weight in the sediments of Taihu Lake and Liaohe River in China [9]. PFOA was detected at concentrations of 60~174 ng/L in the River Po watershed in N-Italy [10]. PFOA was detected in 65% of New Jersey Public Drinking Water Systems at the concentrations ranged from 5 to 39 ng/L [11]. The concentrations of PFOA are also generally found in the ng/L range in natural waters in China. For example, the mean concentrations of PFOA were detected at 55.0 ng/L in Donghu Lake in Wuhan, at 21.7 ng/L in Taihu Lake, and at 82 ng/L in surface waters from Northern China [12, 13, 14].

Recent studies have indicated unintended biological activity of PFCs on non-target organisms. The goals of this study are to investigate the acute toxicity, sub-acute toxicity and reproductive toxicity of PFOA on *Daphnia magna*. We conducted a comprehensive toxicity test including a 48 h acute toxicity test, a 7-day sub-acute toxicity test as well as a 21-day chronic test. Besides, a feeding experiment was investigated to explore whether the target compound could interfere with food intake and ultimately cause toxicity on *D. magna*. All of above are important for ecotoxicological risk assessments of PFCs.

2 MATERIALS AND METHODS

2.1 *Chemicals*

PFOA (98% purity) was purchased from Tokyo Chemical Industry Co., Ltd. (Tokyo, Japan). The test solution was prepared immediately prior to use by diluting the stock solution with daphnia culture medium (consisting of 64.75 mg/L $NaHCO_3$, 5.75 mg/L KCl, 123.25 mg/L $MgSO_4 \cdot 7H_2O$, and 294 mg/L $CaCl_2 \cdot 2H_2O$) reconstituted according to standard OECD guideline [15].

Acetylthiocholine iodide, 5, 5-dithiobis (2-nitrobenzoic acid) were purchased from Sinopharm Chemical Reagent Co., Ltd. (Shanghai, China). Bovine serum albumin was purchased from Shanghai Huixing Biochemistry Reagent Co., Ltd. (Shanghai, China) and the purity was >98%. All other chemicals were of analytical grade and were obtained from Shanghai Chemical Reagent Co., Ltd. (Shanghai, China).

2.2 *Animals*

D. magna were originally obtained from the Chinese Center for Disease Control and Prevention (Beijing, China). Culture medium for daphnias was renewed three times weekly The daphnias were fed with the green algae *Scenedesmus obliquus* daily, which was supplied by Wuhan Institute of Hydrobiology, Chinese Academy of Sciences. Cultures were maintained and tested in an environmental chamber having a light/dark photoperiod of 16 h/8 h at a constant temperature (20°C). Offspring of *D. magna* were separated at regular intervals, from cultures that were 3–5 weeks old and the test animals were ≤ 24 h juveniles.

2.3 *Acute toxicity test*

A 48-h acute toxicity test for *D. magna* was performed according to the modified OECD standard procedure [15]. Five concentration treatments (3, 10, 30, 100, and 300 mg/L) plus a blank control were conducted. Ten neonates (<24 h old) from a designated brood were placed in a 100 mL glass beaker containing 45 mL test solution for each test concentration and control. Test daphnias were not fed during the testing period. Each treatment was replicated three times simultaneously. The status of immobilization and mortality were checked at 48 h, and results were recorded. The daphnias unable to swim within 15 s after gentle agitation of the test container are considered to be immobile. Those animals whose heartbeats have stopped are considered dead. The heartbeats were watched under a stereomicroscope (4 × magnification).

2.4 *Sub-acute toxicity test*

Sixty neonates (<24 h old) from a designated brood were placed in a 1 L glass beaker for each test concentration and control, and the test solution was renewed daily. All experiments were performed in triplicate. Cultured daphnias aged 7-days were used the determination of the activities of Acetylcholinesterase (AChE), Superoxide Dismutase (SOD) and Catalase (CAT).

After exposure, daphnias were homogenized in 100 mM phosphate buffer at pH 7.4 and containing 100 mM KCl and 1 mM EDTA with 1:4 (wet wt./buffer volume ratio) using a glass homogenizer (Kimble Kontes, Vineland, NJ, USA). After centrifugation at 10,000 × g for 10 min at 4°C, the supernatants were used as the enzyme extract. AChE activity was determined at 405 nm by the method of Guilhermino et al. [16], which was expressed as nmol/mg protein/min. SOD activity was determined at 420 nm by the method of Marklund [17]. SOD activities were expressed as U/mg protein. One U was defined as the amount of enzyme required to cause 50% inhibition of pyrogallol auto-oxidation. CAT activity was determined using ammonium molybdate [18], which was expressed μmol/mg protein/min. Protein concentrations were determined at 595 nm using a method developed by Bradford [19], with bovine serum albumin as the standard.

2.5 *Chronic toxicity test*

The effects of PFOA on the reproductive output were assessed by a semi-static test according to the standard procedure for *D. magna* reproduction test [20]. Based on the results of acute toxicity, neonates (<24 h old) were exposed to various concentrations of PFOA (0.032, 0.16, 0.8, 4 and 20 mg/L). Single daphnia was cultured in a glass beakers of 100 mL filled with 45 mL of test solution for 21-day at 20±1°C. The testing was performed in 20 replicates for each exposure concentration. The daphnias were fed 1×10^6 cells of algae per animal per day. The test solution was renewed every other day. Survival, growth and reproduction (fecundity) of the daphnias were monitored for each of the 20 replicates. Growth of the surviving adults of each treatment was determined after 21-day of exposure. The length of each surviving animal was measured from the apex of the helmet to the base of the tail spine. Days to the first pregnancy, days to the first brood, and the number of first brood per female and the average offspring in each brood were the criteria used to evaluate the fecundity. Neonates were counted daily and were removed from the beakers.

The intrinsic rate of population growth (r) was also examined in the present study. This endpoint integrates the measures of age-specific survival and fecundity to estimate the effect of toxicant exposures on population growth [21]. The intrinsic rate of natural increase (r) was calculated using the following formula: $\Sigma l_x m_x e^{-rx} = 1$; where l_x is the proportion of individuals surviving to age x, m_x is the age-specific fecundity (number of neonates produced per surviving female at age x), and x in days. As r calculated in *D. magna* after 21-days is indistinguishable from r estimated for the entire lifespan, due to the great importance of early reproduction [22], all calculations were based on 21-day experiments.

2.6 *Feeding experiment*

The feeding experiment was run according to a method described by Zhu et al. [23]. Filtration and ingestion rates were used as measures of the feeding experiment. Ten neonates (<24 h) were placed in a 100 mL glass beakers containing 50 mL of test solution with a dark photoperiod for 5 h at 20±1°C. During the exposed period the daphnias were fed with 1×10^6 cell/mL *Scenedesmus obliquus*. Then final food concentration was measured with a hemocytometer under an electron microscope (400 × magnification). Filtration rate (F) is defined as the volume of medium swept clear by a daphnia in a unit of time and the ingestion rate (I) as the number of cells consumed by a daphnia during a specific time interval. Equations (1)–(3) from Gauld [24] were used to calculate the average F [μL/(ind·h)] and I [cells/(ind·h)]:

$$F = \frac{V}{n} \times \frac{\ln C_0 - \ln C_t}{t} - A \tag{1}$$

$$A = \frac{\ln C_0 - \ln C_t'}{t} \tag{2}$$

$$I = F \times \sqrt{C_0 \cdot C_t} \tag{3}$$

where C_0 and C_t are initial and final food concentrations (cell/μL), t is time (duration of the experiment in hours), and n is the number of daphnias in volume V (μL); and A is a correction factor for changes in the control with final concentration C_t' after time t. The expression $\sqrt{C_0 \cdot C_t}$ represents the geometric mean of food concentration during time t.

2.7 *Statistical analysis*

All data were tested for normal distribution by Shapiro-Wilk's test and for homogeneity of variances by Levene's tests. The results were expressed as the mean with Standard Deviation (SD). To analyze reproduction, ingestion and biomarker data, one-way Analysis Of Variance (ANOVA) and t tests with Dunnett's test performed the SPSS statistical package (SPSS Co., Chicago, IL, USA). For other types of toxicity data were also performed using SPSS. All differences were considered significant at $P < 0.05$.

3 RESULTS AND DISCUSSION

3.1 *Acute toxicity of PFOA*

No mortality occurred in the control group in the acute toxicity test. The immobilization and mortality of daphnias increased continuously with elevated PFOA concentrations (see Fig. 1). The median Effective Concentration (EC_{50}) and median Lethal Concentration (LC_{50}) were calculated by one variable linear regression analyses of the negative logarithm of compound concentrations with the immobilization or mortality rates as the relative toxic potency. The EC_{50} and LC_{50} values of PFOA on *D. magna* were 110.7 mg/L and 139.0 mg/L, respectively. Our results are comparable with previous studies by Zheng et al. [25] and Li et al. [26], which reported the 48-h EC_{50} and 48 h-LC_{50} values of PFOA on *D. magna* were 114.3 and 181.0 mg/L, respectively.

3.2 *Biomarker responses*

AChE, SOD and CAT activities in daphnias are shown in Figure 2. No mortality occurred during experiments. AChE activity in *D. magna* treated with 0.032 mg/L of PFOA was not significantly different from the control value. However, AChE activity was significantly inhibited by higher concentrations of PFOA (≥0.16 mg/L, $P < 0.05$). The

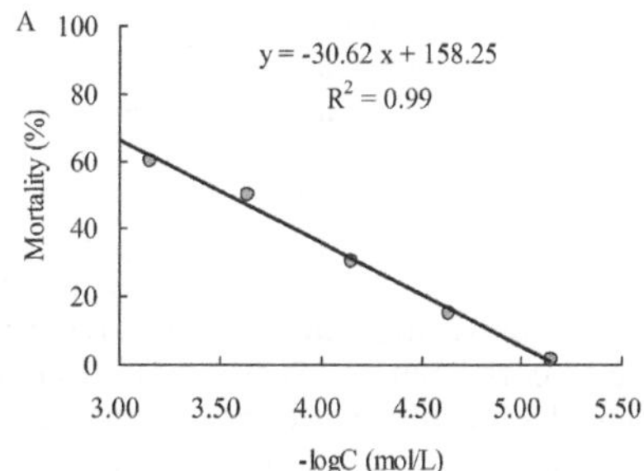

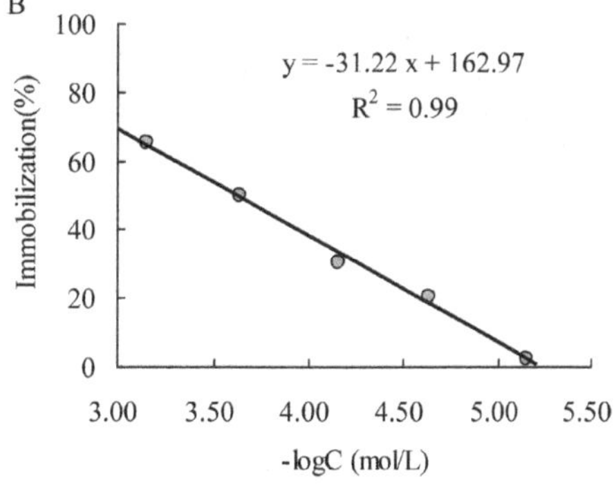

Figure 1. Concentration-response plots of PFOA on *D. magna*.

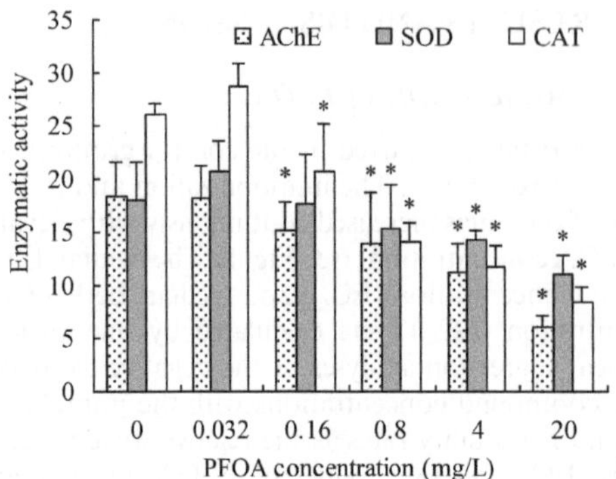

Figure 2. Biomarker responses in *D. magna* exposed to PFOA.

inhibition rate of AChE activity matches the concentration increase, and the maximal inhibition rate was 67%, produced by the highest concentration of PFOA. SOD activity decreased significantly at concentrations equal to or higher than 0.8 mg/L ($P < 0.05$). However, the two lowest concentrations of PFOA did not significantly change SOD activity. CAT activity was significantly inhibited by all tested concentrations except for the lowest concentration. Furthermore, the change of enzyme activities exhibited obvious concentration dependence.

AChE activity plays an important role in many physiological functions, when the activity of the enzyme decreases, neuronal and muscle injury may occur [27]. A significant AChE activity decrease by 20% or mare can be considered as a clear toxicological effect of xenobiotic exposure, such as reductions in the feeding activity and swimming rate [28]. Xuereb et al. [29] also found that the feeding rate and locomotor impairment in *Gammarus fossarum* were directly correlated to AChE inhibitions of higher than 50% for chlorpyrifos and methomyl. In the present study, the inhibition of AChE activity being up to 67% at the highest exposure concentration may cause disruption of the nervous system.

CAT plays a critical role in dismutation of the hydrogen peroxide, whereas SOD dismutates the superoxide anion radical [30]. The SOD-CAT system provides the first defense line against oxygen toxicity and usually used as a biomarker indicating Reactive Oxygen Species (ROS) production [31]. Pollution induces the expression of antioxidant enzymes that allow organisms to partially or totally overcome stress resulting from exposure an unsafe environment. However, the decrease of SOD and CAT activities may have been due to excessive ROS production [32]. Our result indicated that treatment with PFOA resulted in an increase of ROS. ROS, in turn, stimulated the response of antioxidant defenses in *D. magna* and resulted in an impaired its physiological functions.

3.3 *Chronic toxicity of PFOA*

The survival, body length, reproduction and population parameters of adult daphnias were assessed after 21-day of exposure in the present study and the results are shown in Table 1. Any above parameters did not change significantly at the lowest concentration of PFOA. However, the number of first brood per female, body length and intrinsic of natural increase (r) of daphnias decreased significantly at concentrations ≥0.16 mg/L ($P < 0.05$). The inhibition of growth and reproduction matches the concentration increase of PFOA. In addition, the highest concentration of PFOA significantly prolonged the time to first pregnancy and the time to first brood of daphnias and decreased the number of offsprings.

Reproductive success, and in particular the intrinsic rate of natural increase (r), has been recommended as a superior laboratory toxicological endpoint compared to the acute mortality, because it combines lethal and sublethal effects into one meaningful measure [21, 33]. In the present study, the reproduction parameter "number of first brood per female", the population parameter "intrinsic of natural increase" as well as the growth parameter "length", are confirmed to be more sensitive endpoints in the 21-day chronic toxicity test. This result is consistent with our previous studies, in which *Daphnia magna* was exposed to other PFCs or nanoscale metal oxides [21, 34].

3.4 *Feeding behavior of D. magna exposed to PFOA*

The effects of PFOA on the feeding behavior of *D. magna* are shown in Figure 3. The filtration rate (F) and ingestion rate (I) were significantly inhibited by PFOA at concentrations ≥0.8 mg/L ($P < 0.05$). Concentration dependence was apparent for both F and I, and the highest inhibition rates were 61% and 51% compared with control values, respectively. The median Inhibitory Concentrations of filtration (IC_{50F}) and median inhibitory concentrations of ingestion (IC_{50I}) were estimated through one variable linear regression analyses, and were 1.54 and 3.0 mg/L, respectively. These values were much lower than the corresponding EC_{50} for immobilization after 48 h of exposure.

It has been reported that the food availability and food acquisition efficiency may impair cladoceran fitness [35]. Toxicants that affect the movement of appendages and the coordination of the nervous system may reduce ingestion rates [36]. In addition, PFCs could accumulate within the gastrointestinal tract [37], which may interfere with normal food intake and cause the toxicity observed in *D. magna*. These findings suggest that the sensitive feeding response is required to evaluate the effect of low-level PFCs.

Table 1. Size and fecundity of *D. magna* exposed to PFOA in a 21-day life study.

Concentration/ (mg/L)	Time to first pregnancy/d	Time to first brood/d	Number of first brood per female/ind	Number of off springs per brood per female/ind	Length/mm	Intrinsic rate of natural increase (r)
Control	6.56±0.73	9.17±1.34	7.10±1.10	7.46±1.01	2.78±0.09	0.288±0.010
0.032	6.83±0.83	10.07±1.00	7.01±1.83	7.17±2.15	2.66±0.16	0.277±0.027
0.16	7.29±0.91	10.25±1.18	5.07±1.59*	7.16±1.81	2.45±0.24*	0.244±0.027*
0.80	7.46±1.90	10.47±1.85	4.79±2.52*	7.08±1.55	2.32±0.27*	0.228±0.036*
4.0	7.88±1.50*	11.07±2.79*	3.81±2.17*	5.89±1.86*	2.28±0.30*	0.219±0.039*
20	8.23±2.92*	11.92±2.66*	2.46±2.12*	3.98±1.87*	2.18±0.36*	0.203±0.029*

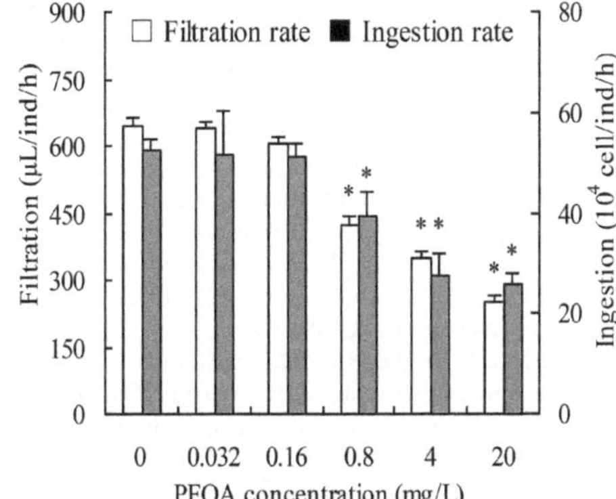

Figure 3. Filtration and ingestion rates of *D. magna*.

4 CONCLUSIONS

This study demonstrated the toxicological effects of PFOA on *Daphnia magna*. The EC_{50} and LC_{50} values were 110.7 mg/L and 139.0 mg/L after 24 h of acute exposure, respectively. AChE, SOD and CAT activities in *D. magna* exposure to PFOA for 7-day was significantly inhibited in a concentration-dependent manner. Parental exposure in *D. magna* transferred adverse effects to offspring. The number of first brood per female, body length and intrinsic of natural increase of *D. magna* are confirmed to be more sensitive parameters for the 21-day chronic exposure to PFOA. In addition, the exposure of *D. magna* to PFOA resulted in the reduction of food intake, which could affect their growth and reproduction. The results of the present study suggested that physiological and biochemical endpoint need to be incorporated into the risk assessment of PFCs.

ACKNOWLEDGMENTS

This study was supported by the National Natural Science Foundation of China (Grant 51279061) and the Fundamental Research Funds for the Central Universities (2014B07514).

REFERENCES

[1] Henrik, V. & Eriksson, P., Perfluorooctane Sulfonate (PFOS) and Perfluorooctanoic Acid (PFOA). Reproductive and Developmental Toxicology, pp. 623–635, 2011.

[2] Renner, R., Growing concern over perfluorinated chemicals. Environmental Science & Technology, 35(7), pp. 154A–160A, 2001.

[3] Mommaerts, V., Hagenaars, A., Meyer, J., DeCoen, W., Swevers, L., Mosallanejad, H. & Smagghe, G., Impact of a perfluorinated organic compound PFOS on the terrestrial pollinator *Bombus terrestris* (Insecta, Hymenoptera). Ecotoxicology, 20(2), pp. 447–456, 2011.

[4] Murakami, M., Adachi, N., Saha, M., Morita, C., Takada, H. 2011. Levels, temporal trends, and tissue distribution of perfluorinated surfactants in freshwater fish from asian countries. Archives of Environmental Contamination Toxicology, 61(4), pp. 631–641.

[5] Rudel, H., Muller, J., Jurling, H., Bartel-Steinbach, M. & Koschorreck, J., Survey of patterns, levels, and trends of perfluorinated compounds in aquatic organisms and bird eggs from representative German ecosystems. Environmental science and pollution research international, 18(9), pp. 1457–1470, 2011.

[6] Powley, C.R., George, S.W., Russell, M.H., Hoke, R.A. & Buck, R.C., Polyfluorinated chemicals in a spatially and temporally integrated food web in the Western Arctic. Chemosphere, 70(4), pp. 664–672, 2008.

[7] Cai, Y.Q., Wang, J.M., Shi, Y.L. & Pan, Y.Y., Perfluorooctane Sulfonate (PFOS) and other fluorchemicals in viscera and muscle of farmed pigs and chickens in Beijing, Chinese Science Bulletin, 55(31), pp. 3550–3555, 2010.

[8] Leondiadis, L., Vassiliadou, I., Costopoulou, D. & Ferderigou, A., Levels of Perfluorooctane Sufonate (PFOS) and Perfluorooctanoate (PFOA) in blood samples from different groups of adults living in Greece. Chemosphere, 8(1), pp. 1199–1206, 2010.

[9] Zhu, L.Y., Yang, L.P. & Liu, Z.T., Occurrence and partition of perfluorinated compounds in water and sediment from Liao River and Taihu Lake, China. Chemosphere, 83(6), pp. 806–814, 2011.

[10] Loos, R., Locoro, G., Huber, T., Wollgast, J., Christoph, E.H., De Jager A., Gawlik B.M, Hanke G, Umlauf G. & Zaldivar J.M., Analysis of Perfluorooctanoate (PFOA) and other perflurinaed compounds (PFCs) in the River Po watershed in N-Italy. Chemosphere, 71(2), pp. 206–313, 2008.

[11] Post, G.B., Louis, J.B., Cooper, K.R., Boros-Russo, B.J. & Lippincott, R.L., Occurrence and Potential Significance of Perfluorooctanoic Acid (PFOA) Detected in New Jersey Public Drinking Water Systems[J]. Environmental Science & Technology, 43(12), pp. 4547–4554, 2009.

[12] Chen, J., Wang, L.L., Zhu, H.D., Wang, B.B., Liu, H.C., Cao, M.H., Miao, Z., Hu, L., Lu, X.H. & Liu, G.H., Spatial Distribution of Perfluorooctanoic Acids and Perfluorinate Sulphonates in Surface Water of East Lake. Environmental Science, 33(8), pp. 2586–2591, 2012.

[13] Yang, L.P., Zhu, L.Y. & Liu, Z.T., Occurrence and partition of perfluorinated compounds in water and sediment from Liao River and Taihu Lake, China. Chemosphere, 83(6), pp. 806–814, 2011.

[14] Wang, T.Y., Khim, J.S., Chen, C.L., Naile, J.E., Lu, Y.L., Kannan, K., Park, J., Luo, W., Jiao, W.T., Hu, W.Y. & Giesy, J.P., Perfluorinated compounds in surface waters from Northern China: Comparison to level of industrialization. Environment International, 42, pp. 37–46, 2012.

[15] Organization for Economic Cooperation and Development (OECD) 2004. Guideline for Testing of Chemicals No.202, OECD Publishing, Paris.

[16] Guilhermino, L., Lopes, M.C., Carvalho, A.P. & Soared, A.M.V.M., Inhibition of acetylcholinesterase activity as effect criterion in acute tests with juvenile, *Daphnia magna*. Chemosphere, 32, pp. 727–738, 1996.

[17] Marklund, S. & Marklund, G., Involvement of the superoxide anion radical in the autoxidation of pyrogallol and a convenient assay for superoxide dismutase. European Journal of Biochemistry, 47, pp. 469–474, 1974.

[18] Góth, L., A simple method for determination of serum catalase activity and revision of reference range. Clinca Chimica Acta, 196, pp. 143–151, 1991.

[19] Bradford, M.M., A rapid and sensitive method for the quantitation of microgram quantities of protein utilizing the principle of protein-dye binding. Analytical Biochemistry, 72, pp. 248–254, 1976.

[20] Organization for Economic Cooperation and Development (OECD) 1998. Guidelines for testing chemicals NO.211, OECD Publishing, Paris.

[21] Zhao, H.Z., Lu, G.H., Xia, J., and Jin, S.G., Toxicity of Nanoscale CuO and ZnO to *Daphnia magna*. Chemical Research in Chinese Universities, 28(2), pp. 209–213, 2012.

[22] Van, Leeuwen. C.J., Luttmer, W.J. & Griffieon, P.S., The use of cohorts and populations in chronic toxicity studies with *Daphnia magna*: a cadmium example. Ecotoxicology and Environmental Safe, 9(1), pp. 26–39, 1985.

[23] Zhu, X.S., Chang, Y. & Chen, Y.S., Toxicity and bioaccumulation of TiO_2 nanoparticle aggregates in *Daphnia magna*. Chemosphere, 78(3), pp. 209–215, 2010.

[24] Gauld, T., The grazing rate of marine copepods. Journal of the Marine Biological Association of the United Kingdom, 29(3), pp. 695–706, 1951.

[25] Zheng, X.M., Feng, Z. & Liu, H.L., China POPs Forum 2010 & 5th National Symposium on Persistent Organic Pollutants, 17 May 2010, Toxicity of typical perfluorinated compounds to *Danhnia Magna* and zebrefish (brachydanio rerio) embryos. Nanjing University, Nanjing China.

[26] Li, M.H., Toxicity of perfluorooctane sulfonate and perfluorooctanoic acid to plants and aquatic invertebrates. Environmental Toxicology, 24(1), pp. 95–101, 2009.

[27] Dettbarn, W.D., Milatovic, D. & Gupta, R.C., Oxidative stress in anticholinesterase-induced excitotoxicity. In: Gupta RC (ed) Toxicology of organophosphate and carbamate compounds. Academic Press, Burlington, pp. 511–532, 2006.

[28] United States Environmental Protection Agency (USEPA). SCE policy issues related to the food quality protection act. Office of pesticide programs science policy on the use of cholinesterase inhibition for risk assessment of organophos-phate and carbamate pesticides. Federal register 63, 1998.

[29] Xuereb, B., Lefe`vre, E., Garric, J. & Geffard, O., 2009. Acetylcholinesterase activity in Gammarus fossarum (Crustacea Amphipoda): linking AChE inhibition and behavioural alteration. Aquatic Toxicology, 94, pp. 114–122, 2009.

[30] Kim, S., Kim, W., Chounlamany, V., Seo, J., Yoo, J., Jo, H.J. & Jung, J., Identification of multi-level toxicity of liquid crystal display wastewater toward *Daphnia magna* and *Moina macrocopa*. Journal of Hazardous Materials, 227, pp. 327–333, 2012.

[31] Li, Z.H., Zlabek, V., Velisek, J., Grabic, R., Machova, J., Kolarova, J., Li, P. & Randak, T., Acute toxicity of carbamazepine to juvenile rainbow trout (*Oncorhynchus mykiss*): effects on antioxidant responses, hematological parameters and hepatic EROD. Ecotoxicology Environmental Safe, 74, pp. 319–327, 2011.

[32] Xu, D., Li, C., Wen, Y. & Liu, W., Antioxidant defense system responses and DNA damage of earthworms exposed to Perfluorooctane Sulfonate (PFOS). Environmental Pollution, 174, pp. 121–127, 2013.

[33] Pestana, J.L.T., Loureiro, S., Baird, D.J. & Soares, A.M.M., Pesticide exposure and inducible antipredator responses in the zooplankton grazer, *Daphnia magna* Straus. Chemosphere, 78, pp. 41–48, 2010.

[34] Lu GH, Liu JC, Sun LS, Yuan LJ. Toxicity of Perfluorononanoic acid and perfluorooctane sulfonate to *Daphnia magna*. Water Science and Engineering, 2015.

[35] Pereira J.L., Mendes C.D. & Gonçalves F., Short- and long-term responses of *Daphnia* spp. to propanil exposures in distinct food supply scenarios. Ecotoxicology Environmental Safe, 68, pp. 386–396, 2007.

[36] Yi, X., Kang, S.W. & Jung, J., Long-term evaluation of lethal and sublethal toxicity of industrial effluents using *Daphnia magna* and *Moina macrocopa*. Journal of Hazardous Materials, 178(1–3), pp. 982–987, 2010.

[37] Kim, K.T., Klaine, S.J., Cho, J., Kim, S.H. & Kim, S.D., Oxidative stress responses of *Daphnia magna* exposed to TiO_2 nanoparticles according to size fraction. Science of the Total Environment, 408(10), pp. 2268–2272, 2010.

Material Science and Environmental Engineering – Chen (Ed.)

Adsorption of rhodamine B onto TiO_2-loaded Activated Carbon Fibers

P. Chen, D. Zhu, L. Jiang, J.Y. Chen, R.L. Liu & J.W. Feng
School of Civil Engineering, Hefei University of Technology, Hefei, China

ABSTRACT: Activated carbon fibers loaded with TiO_2 composite materials (TiO_2-ACF) were prepared by the sol-gel method. The results showed that the pseudo-second-order kinetic model and the Langmuir model were more suitable to describe the adsorption of rhodamine B onto TiO_2-ACF. Monolayer adsorption and heterogeneous polymolecular layer adsorption existed during the adsorption of rhodamine B onto TiO_2-ACF. The adsorption capacity of rhodamine B onto TiO_2-ACF increased significantly with increasing temperature, which was an endothermic process.

Keywords: TiO_2-ACF; rhodamine B; adsorption

1 INTRODUCTION

TiO_2 was a kind of promising photocatalytic material, but it was easy to reunite and not easy to recycle. This problem could be solved by loading TiO_2 on fiber materials, such as Activated Carbon Fibers (ACF). ACF loaded with TiO_2 composite materials (TiO_2-ACF) is a carbon-based adsorption material with the characteristic of highly developed pore structure, strong adsorption ability and recycling performance, which is widely used in pollutant removal and water purification processes (Hou et al. 2009, Shi et al. 2012). However, the specific surface area, total pore volume and average pore diameter of TiO_2-ACF are different with ACF; therefore, the adsorption characteristics of pollutants onto TiO_2-ACF may be different from ACF.

Rhodamine B is a kind of synthetic dye; it is used widely in the dye industry. When dye wastewater is discharged into the water body, water transparency is reduced and the aquatic organisms and microbial growth are influenced, meanwhile, the water self-purification and water ecological balance are destroyed (Chang & Lin 2005, Garcia-Segura et al. 2013).

In the present paper, rhodamine B was selected as the target pollutant, and the adsorption characteristics of aqueous rhodamine B onto TiO_2-ACF were examined.

2 EXPERIMENTAL PROCEDURE

2.1 *Materials*

Rhodamine B used in this study was of analytical grade and solutions were prepared with deionized water. The sol-gel method was used to prepare the TiO_2-ACF: 10 mL tetrabutyl titanate was dissolved in 15 mL anhydrous ethanol, 4 mL acetylacetonate and 6 mL nitric acid, and were added to the solution (with magnetic stirring), and then 25 mL anhydrous ethanol and 5 mL deionized water were added, and ultrasonic stirring was continued for 3 h, and then the samples were stored for 72 h. ACF was added to the sol solvent, evaporating was conduced, and finally it was put into the tube furnace under the shielding gas N_2, and the furnace temperature was set to 600 °C.

2.2 *Calculation of adsorption capacity*

The adsorption capacity of TiO_2-ACF for rhodamine B was calculated by using the following equation:

$$q = \frac{C_o - C_e}{M} V \tag{1}$$

where q = desorption capacity (mg/g); C_o = the initial concentration of rhodamine B (mg/L); C_e = the liquid phase concentration of rhodamine B at equilibrium (mg/L); V = solution volume (L); and M = quantity of TiO_2-ACF (g).

2.3 *Adsorption kinetics experiments*

A 0.03 g sample of TiO_2-ACF was added to 150 mL rhodamine B solution (C_o = 30.00 mg/L, pH = 4.30), which was then shaken at a certain temperature (288 K, 298 K and 308 K). At given intervals, small aliquots were extracted and centrifuged. The concentration of the supernatant was measured by a UV/visible photometer.

2.4 *Adsorption isotherm experiments*

A 0.03 g sample of TiO_2-ACF was added into 150 mL rhodamine B solutions at different concentrations (pH = 4.30), which was then shaken at a certain temperature (288 K, 298 K and 308 K). At given intervals, small aliquots were extracted and centrifuged. The concentration of the supernatant was measured by a UV/visible photometer.

3 RESULTS AND DISCUSSION

3.1 *Adsorption kinetics*

Adsorption kinetics of TiO_2-ACF is of great importance in understanding the adsorption mechanism and adsorption reactor design. Herein, the adsorption kinetics of rhodamine B was studied at different temperatures (288 K, 298 K and 308 K). It could be seen from Figure 1 that the amount of adsorbed rhodamine B increased gradually with increasing adsorption time, and the final equilibrium was achieved within approximately 24 h.

The adsorption kinetics was simulated by the pseudo-first-order model and the pseudo-second-order model (Equations 2 and 3):

$$\lg(q_e - q_t) = \lg q_e - \frac{k_1 t}{2.303} \quad (2)$$

$$\frac{t}{q_t} = \frac{1}{k_2 q_e^2} + \frac{t}{q_e} \quad (3)$$

where q_e = adsorbed amounts of rhodamine B onto TiO_2-ACF at equilibrium (mg/g); q_t = adsorbed amounts of rhodamine B onto TiO_2-ACF at time t (h); k_1 = corresponding rate constants of adsorption (1/min); and k_2 = corresponding rate constants of adsorption (g/mg/min).

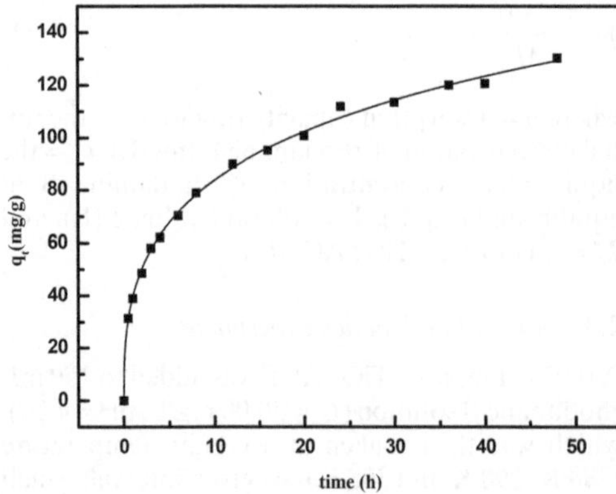

Figure 1. Adsorption kinetics of rhodamine B onto TiO_2-ACF (C_0 = 30.00 mg/L, V = 150.00 mL, t = 24 h, pH = 4.30, $m_{ACF\text{-}TiO2}$ = 30.00 mg).

The parameters of the pseudo-first-order model and the pseudo-second-order model are summarized in Table 1. The pseudo-first-order rate constant k_1 and the equilibrium adsorption amount q_e were obtained though non-linear regression fitting, and the related kinetic parameters are summarized in Table 1. And the calculated q_e values were 54.59, 93.76 and 94.06 mg/g for 288 K, 298 K and 308 K, respectively. For the pseudo-first-order kinetic model, the calculated q_e values were far lower than the corresponding experimental q_e values, i.e. 70.21, 130.28 and 149.44 mg/g, respectively. These results indicated that the pseudo-first-order kinetic model was not appropriate to describe the adsorption process. While for the pseudo-second-order kinetic model, the calculated correlation coefficients (R^2) for the equations were 0.961, 0.985 and 0.997, respectively. The pseudo-second-order rate constant (k_2) decreased with increasing temperatures, and the calculated q_e values agreed very well with the experimental values. These results indicated that the pseudo-second-order kinetic model was appropriate to describe the adsorption process.

3.2 *Adsorption isotherm*

The effect of the temperature on the adsorption was analyzed by adsorption isotherm, and the results are shown in Figure 2. As shown in Figure 2, the adsorption of rhodamine B onto TiO_2-ACF was studied at different temperatures (288 K, 298 K and 308 K); the adsorption amount of rhodamine B increased gradually with increasing temperature, and the equilibrium absorption capacity also increased gradually with increasing initial concentrations of rhodamine B.

Different adsorption mechanisms were analyzed by different adsorption isotherm equation parameters, and the surface property of the adsorbent

Table 1. Parameters of kinetic models for the adsorption of rhodamine B onto TiO_2-ACF.

T (K)	288	298	308
$q_{e,exp}$ (mg/g)	70.21	130.28	149.44
Pseudo-first-order model			
$q_{e,cal}$ (mg/g)	54.59	93.76	94.06
k_1 (× 10^{-3} 1/min)	1.02	1.01	1.60
R^2	0.938	0.974	0.956
Pseudo-second-order model			
$q_{e,cal}$ (mg/g)	70.92	131.93	154.80
k_2 (× 10^{-5} g/(mg/min))	5.12	3.28	4.62
R^2	0.961	0.985	0.997

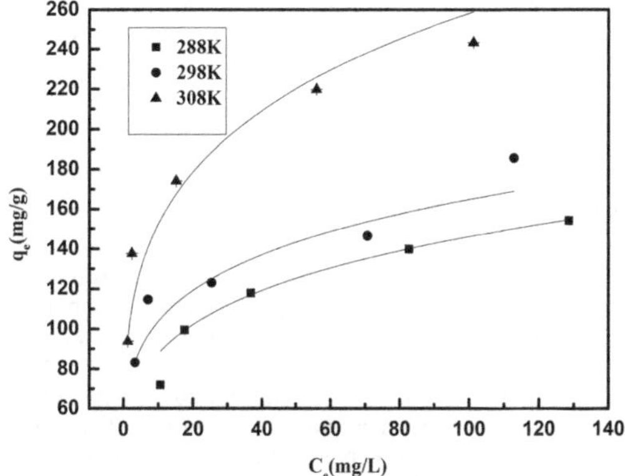

Figure 2. Effect of the temperature on the adsorption of rhodamine B onto TiO_2-ACF (V = 150.00 mL, t = 24 h, pH = 4.30, $m_{ACF\text{-}TiO2}$ = 30.00 mg).

Table 2. Parameters of isotherm equations for the adsorption of rhodamine B onto TiO_2-ACF.

	Langmuir			Freundlich		
T(K)	q_m (mg/g)	b	R^2	n	a	R^2
288	118.483	0.047	0.970	3.340	23.936	0.946
298	186.916	0.119	0.965	5.242	70.137	0.882
308	247.524	0.259	0.995	5.120	100.970	0.926

was determined by isotherm equation parameters. Langmuir, Freundlich and Dubinin-Radushkevich (D-R) isothermal equations were the commonly used isotherm equations (Jakubov & Mainwaring 2002, Li et al. 2009, Park et al. 2003, Xiu & Li 2000). Herein, the data on the adsorption of rhodamine B onto TiO_2-ACF were simulated by the Langmuir isothermal equation and Freundlich isothermal equations. Langmuir and Freundlich isothermal equations are expressed in the following equations (Equations (4) and (5)):

$$\frac{C_e}{q_e} = \frac{1}{bq_m} + \frac{C_e}{q_m} \tag{4}$$

$$\lg q_e = \lg a + \frac{1}{n}\lg C_e \tag{5}$$

Table 2 summarizes all the constants and R^2 values obtained from the two isotherm models (Equations 4 and 5). It was observed that the Langmuir model gave higher R^2 values, indicating that the adsorption of rhodamine B onto TiO_2-ACF was monolayer adsorption and non-uniform adsorption. The reason might be that the distribution of pore size and the structure of pore changed after loading TiO_2 onto the surface of ACF, and a large number of mesoporous materials produced during this process. And the calculated $1/n$ was less than 1, indicating that the adsorption process of rhodamine B onto TiO_2-ACF was conducive for the adsorption process.

3.3 *Adsorption thermodynamics*

To investigate the adsorption processes, the changes in thermodynamic parameters, such as standard enthalpy (ΔH^0), standard entropy (ΔS^0) and standard free energy (ΔG^0), must be determined, which can be calculated according to the following equations (Equations (6)–(8)) (Fan et al. 2011, Sun et al. 2013):

$$\ln K_d = \frac{\Delta S^0}{R} - \frac{\Delta H^0}{RT} \tag{6}$$

$$\Delta G^\circ = -RT\ln K_d \tag{7}$$

$$K_d = \frac{C_{Ae}}{C_e} \tag{8}$$

where R = 8.314 J/mol/K; T = absolute temperature (K); C_{Ae} = adsorbed concentration of rhodamine B onto TiO_2-ACF at equilibrium (mg/g); and C_e = concentration of aqueous rhodamine B at equilibrium (mg/g).

The calculated ΔH^0, ΔS^0 and ΔG^0 for the adsorption of rhodamine B onto TiO_2-ACF are listed in Table 3 and shown in Figure 3.

From Table 3, we can see that the ΔH^0 and ΔS^0 values were positive. The positive ΔH^0 value indicates that the adsorption process was endothermic. The reason for this was that the molar volume of water was much smaller than that of rhodamine B, and a layer of water film was easily formed on the surface of TiO_2-ACF; only water molecules were desorbed, the rhodamine B molecules were absorbed, thus the desorption of water molecules was an endothermic process, and the adsorption of rhodamine B was and exothermic process. However, the molecular weight of rhodamine B was equal to 26.6 times of water molecules, and the desorption of a number of water molecules was needed

Table 3. Thermodynamic parameters for the adsorption of rhodamine B onto TiO_2-ACF.

q_e	ΔH^0	ΔS^0	ΔG^0 (kJ/mol)		
(mg/g)	(kJ/mol)	(kJ/(kmol))	288 K	298 K	308 K
20	104.63	0.36	0.30	−3.97	−6.93
30	102.10	0.35	0.84	−2.91	−6.18

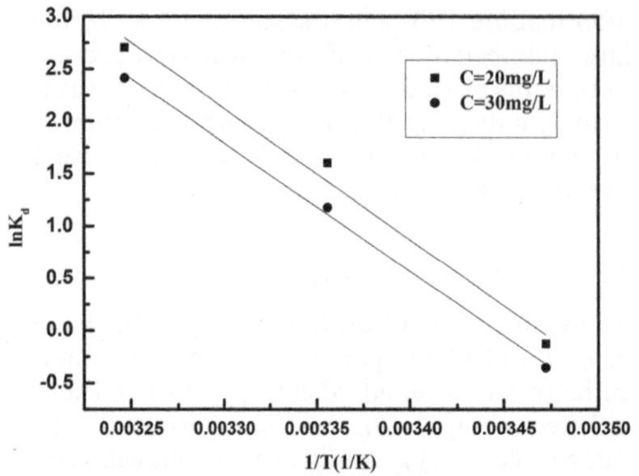

Figure 3. Fit plot for $\ln_{kd}$ versus 1/T for the adsorption of rhodamine B on the TiO_2-ACF ($V = 150.00$ mL, $t = 24$ h, pH = 4.30, $m_{ACF\text{-}TiO2} = 30.00$ mg).

for the adsorption of a rhodamine B molecule. And desorption heat was greater than the liberated heat of the adsorption, and the whole process of adsorption was and endothermic process. The value of ΔS^0 was also positive, which indicated that the adsorption process was a random adsorption process in the solid-liquid interface, and the randomness of adsorption increased with increasing temperature. The value of ΔG^0 was negative, and was more negative with increasing temperature, which showed that the adsorption was spontaneous, and higher temperature was conducive to the adsorption process.

4 CONCLUSION

Removal of aqueous rhodamine B by adsorption onto TiO_2-ACF was conducted, and the experimental results showed that the equilibrium time of adsorption was approximately 24 h; the pseudo-second-order kinetic model was appropriate to describe the adsorption process; adsorption of rhodamine B onto TiO_2-ACF was monolayer and non-uniform adsorption; meanwhile, the adsorption process was endothermic, random and spontaneous.

ACKNOWLEDGMENTS

The authors gratefully acknowledge the support from the Natural Science Foundation of China under Grant No. 51208163 and the Open Fund of State Key Laboratory of Hydrology-Water Resources and Hydraulic Engineering, Hohai University (2013491211).

REFERENCES

[1] Chang, C.L. & Lin, T.S. 2005. Decomposition of toluene and acetone in packed dielectric barrier discharge reactors. *Plasma Chemistry and Plasma Processing* 25(3): 227–243.

[2] Fan, J.J. & Cai, W.Q. & Yu, J.G. 2011. Adsorption of N719 dye on anatase TiO_2 nanoparticles and nanosheets with exposed (001) facets: equilibrium, kinetic, and thermodynamic studies. *Chemistry-An Asian Journal* 6(9): 2481–2490.

[3] Garcia-Segura, S. & Dosta, S. & Guilemany, J.M. & Brillas, E. 2013. Solar photoelectrocatalytic degradation of Acid Orange 7 azo dye using a highly stable TiO_2 photoanode synthesized by atmospheric plasma spray. *Applied Catalysis B: Environmental* 132(1): 142–150.

[4] Hou, Y. & Qu, J. & Zhao, X. & Lei, P. & Wan, D. & Huang, C.P. 2009. Electro-photocatalytic degradation of acid orange II using a novel TiO_2/ACF photoanode. *Science of the Total Environment* 407(5): 2431–2439.

[5] Jakubov, T. & Mainwaring, D. 2002. Modified dubinin radushkevich/dubinin astakhov adsorption equations. *Journal of Colloid Interface Science* 252(2): 263–268.

[6] Li, K.Q. & Zheng, Z. & Feng, J.W. & Zhang, J.B. & Luo, X.Z. & Zhao, G.H. & Huang, X.F. 2009. Adsorption of *p*-nitroaniline from aqueous solutions onto activated carbon fiber prepared from cotton stalk. *Journal of Hazardous Materials* 166: 1180–1185.

[7] Park, S.J. & Jang, Y.S. & Shim, J.W. & Ryu, S.K. 2003. Studies on pore structures and surface functional groups of pitch-based activated carbon fibers. *Journal of Colloid and Interface Science* 260(2): 259–264.

[8] Shi, J.W. & Cui, H.J. & Chen, J.W. & Fu, M.L. & Xu, B. & Luo, H.Y. 2012. TiO_2/activated carbon fibers photocatalyst: Effects of coating procedures on the microstructure, adhesion property, and photocatalytic ability. *Journal of Colloid and Interface Science* 388(1): 201–208.

[9] Sun, Z.M. & Yu, Y.C. & Pang, S.Y. & Du, D.Y. 2013. Manganese-Modified Activated Carbon Fiber (Mn-ACF): Novel efficient adsorbent for Arsenic. *Applied Surface Science* 284(11): 100–106.

[10] Xiu, G.H. & Li, P. 2000. Prediction of breakthrough curves for adsorption of lead (II) on activated carbon fibers in a fixed bed. *Carbon* 38(7): 975–981.

Pollution control

Material Science and Environmental Engineering – Chen (Ed.)
 ISBN 978-1-138-02938-5

A complete solution of the optimal pollution path

X.X. Yan
Xi'an University of Science and Technology, Shaanxi, Xi'an, China

J.S. Zhang
Yan'an University, Shaanxi, Yan'an, China

S.H. Zou
Xi'an University of Science and Technology, Shaanxi, Xi'an, China

ABSTRACT: We use the DHSS model as a benchmark, cite pollution in utility function and constraints, construct an optimal control model and solve it in detail, and analyze the optimal pollution path. We find that when considering pollution, the utility is increased with discount; and the optimal pollution path is increased and then decreased, and tends to zero gradually.

Keywords: exhaustible resources; optimal control; optimal pollution path; pollution constraints

1 INTRODUCTION

In the current situation of China, planning exhaustible resource extraction path is reasonably one of the important issues under the dual requirements to keep the effective supply of society and strong pollution control. In this paper, we revisit a question posed by Benchekroun and Withagen (2011): what patterns of the optimal extraction of exhaustible resources will be when we think about pollution? Benchekroun and Withagen (2011) provided the closed form solution to the Dasgupta-Heal-Solow-Stiglitz (DHSS) model (Dasgupta and Heal, 1974; Solow, 1974; Stiglitz, 1974) and Cobb-Douglas production function.

We choose the objective function to maximize the minimum rate of consumption throughout the time horizon (Solow, 1974) because there exists a sustainable constant positive rate of consumption. We also suggest that the production function is Cobb-Douglas and instantaneous utility is logarithmic for Stiglitz, given this special case in the literature. Pollution is thought as a detrimental effect on utility and production. Previously, we use the Lucas-Uzawa model (Robert & Lucas, 1988) of exponential integral function defined utility similar to that used in the article of H. Benchekroun and C. Withagen (2011). Boucekkine and Ruiz-Tamarit (2008) and Boucekkine et al. (2008) showed that the model can be expressed as a hypergeometric function. We use utility function for pollution as used by Stokey (1998). For the fixed potential output, pollution is an increasing and convex function of the actual output. With exponential integral function and pollution function all strictly increasing and strictly convex, the utility function must be concave and increasing toward the southeast.

In the constraint conditions, we assume pollution as a stock variable and can be accumulated. We suggested that the flow of pollution is affected by damage from extraction. The flow of pollution is directly proportional to the energy used in consumption. For the production function, we consider two ways: without and within.

The rest of the paper is organized as follows. In Section 2, we introduce the model and present preliminary results. In Section 3, we solve the problem. In Section 4, we analyze the optimal pollution path. In Section 5, we end by offering concluding remarks.

2 THE MODEL

Let $K(t)$ and $S(t)$ denote the stock of man-made capital and the stock of exhaustible resource at instant of time t. The variables $C(t)$ and $R(t)$ are consumption and resource extraction at instant of time t and are assumed to be non-negative. The variables $P(t)$ are the stock of pollution at instant of time t and are assumed to be non-negative. Let α be the production elasticity of man-made capital $0 < \alpha < 1$. The rate of pure time preference (discounted value) is ρ. We assume $\rho > 0$ to be strictly positive. Let ω be the emission of the use of exhaustible resource extracted ($\omega > 0$). For any variable $x(t)$, we adopt the convention $x(t) = dx/dt$. Consider the following optimal control problem of the DHSS economy, which we refer to as the DHSS problem:

$$Max\int_0^{\infty} e^{-\rho t}U(C,P)dt \tag{1}$$

Subject to

$$\dot{K}(t) = K(t)^a R(t)^{1-a} - C(t) \tag{2}$$

$$\dot{S}(t) = -R(t) \tag{3}$$

$$\dot{P}(t) = \omega R(t) \tag{4}$$

$$K(0) = K_0 > 0;\ S(0) = S_0 > 0;\ P(0) = P_0 > 0 \tag{5}$$

With

$$U(C,P) = \ln C - \ln P$$

Let $\lambda(t)$, $\mu(t)$ and $\phi(t)$ be denoted as the co-state variables associated with the stock of capital, the stock of exhaustible resource, and the stock of pollution, respectively. The current value Hamiltonian is given by

$$H_c = U(C,P) + \lambda(K^{\alpha}R^{1-\alpha} - C) + \mu(-R) + \phi(\omega R) \tag{6}$$

The maximum principle yields

$$\frac{\partial H_c}{\partial R} = 0 \Rightarrow \mu - \omega\phi - \lambda(1-\alpha)K^{\alpha}R^{-\alpha} = 0 \tag{7}$$

$$\frac{\partial H_c}{\partial C} = 0 \Rightarrow \lambda = C^{-1} \tag{8}$$

$$\dot{\lambda} = -\frac{\partial H_c}{\partial K} + \rho\lambda \Rightarrow \dot{\lambda} = \rho\lambda - \lambda\alpha R^{1-\alpha}K^{\alpha-1} \tag{9}$$

$$\dot{\mu} = -\frac{\partial H_c}{\partial S} + \rho\mu \Rightarrow \dot{\mu} = \rho\mu \tag{10}$$

$$\dot{\phi} = -\frac{\partial H_c}{\partial P} + \rho\phi \Rightarrow \dot{\phi} = \rho\phi + P^{-1} \tag{11}$$

Any the solution that satisfies the above system along with the following transversality conditions:

$$\lim_{t\to\infty} e^{-\rho t}\lambda(t)K(t) = 0 \tag{12}$$

$$\lim_{t\to\infty} e^{-\rho t}\mu(t)S(t) = 0 \tag{13}$$

$$\lim_{t\to\infty} e^{-\rho t}\phi(t)P(t) = 0 \tag{14}$$

3 SOLVING THE PROBLEM

Remark 1. Here we cite Benchekroun and Withagen (2011, P. 615–616, remark 1), and define the exponential integral:

$$E_a(z) = \int_1^{\infty} e^{-zu}u^{-a}\,\mathrm{d}u,\ a \in \mathbb{R},\ z > 0\ \lim_{z\to\infty} E_1(z) = 0$$

and $\lim_{z\to 0^+} E_1(z) = \infty$

And define

$$\varphi = (1-\alpha)\left(\frac{\mu_0}{1-\alpha}\right)^{\frac{\alpha-1}{\alpha}},\ \pi(t) = \lambda_0^{\frac{\alpha-1}{\alpha}} + \varphi t,\ x(t) = \frac{\rho\pi(t)}{\varphi},$$

the variable $x(t)$ is an affine function of time with $x(t) = x_0 + \rho t$.

So, we have

$$\mu_0 = -\frac{\Psi(x_0)}{\rho S_0}e^{x_0}$$

$$C(t) = e^{-\rho t}\pi(t)^{(\alpha/(1-\alpha))}$$

$$S(t) - S_0 = S_0\left(\frac{\Psi(x(t))}{\Psi(x_0)} - 1\right)$$

$$P(t) = P_0 - \frac{\omega S_0}{\Psi(x_0)}\left(E_1(x(t)) - E_1(x_0)\right)$$

The solution of x_0 is the same as that used in the literature (Benchekroun and Withagen, 2011).

Similar to $S(t)$ and S_0, we can express $P(t)$ and P_0 as follows:

$$P(t) = \frac{\omega}{\rho\,\mu_0}e^{-x_0}E_1(x(t))$$

$$P_0 = \frac{\omega}{\rho\,\mu_0}e^{-x_0}E_1(x_0)$$

4 ANALYZING THE OPTIMAL POLLUTION PATH

We maximum utility, i.e. $\frac{dU(C,P)}{dx_0} = \frac{\sigma}{1-\sigma}\frac{1}{x_0+\rho t} > 0$, so when x_0 increases, $U(C,P)$ increases. For $\frac{dx_0}{d\rho} = \frac{\frac{dX}{d\rho}}{\frac{dh_1(x_0)}{dx_0}}$, $\frac{dX}{d\rho} = \frac{S_0}{K_0}(1-\alpha)^{\alpha/(1-\alpha)}(-\alpha)\rho^{1/(\alpha-1)} < 0$,

for all $x_0 > 0$, $\frac{dh_1(x_0)}{dx_0} < 0$, so $\frac{dx_0}{d\rho} > 0$. The proof is the same as that given in H. Benchekroun and C. Withagen's paper (see P634).

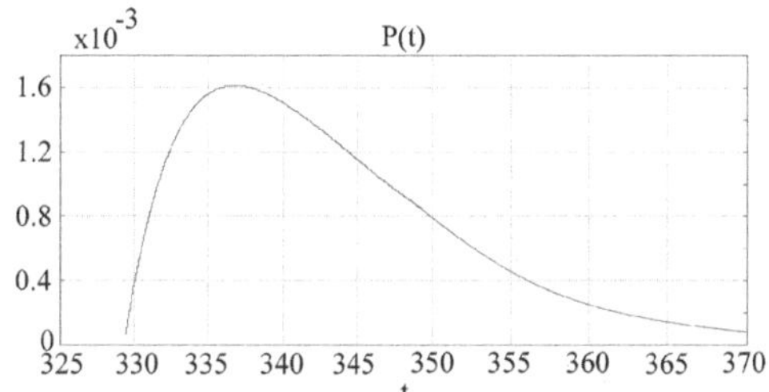

Figure 1. The pollution path.

So, when considering pollution, we obtain the following result: x_0 is the increasing function of ρ. So, $U(C,P)$ is the increasing function of ρ too, which means if the maximum of ρ is $\tilde{\rho}$, the maximum of $U(C,P)$ is $\tilde{U}$, if we assume $\rho = 0.03$, $\alpha = 0.6$, $\omega = 0.05$, $P_0 = 1$, $P_1^* = 0.6P_0$, $K_0 = 1$ and $S_0 = 0.5$, through the simulation, as shown in Figure 1.

5 CONCLUSION

This paper assumes a perfectly competitive market environment and cite pollution constraint, and maximizes the social welfare as the goal to construct the optimal control model of pollution constraints. By solving the model, we can draw the variable solution, and then conduct the analysis of the optimal path of pollution. The result shows that when considering pollution, the utility is increased with discount. The optimal pollution path is increased and then decreased, and tends to zero gradually.

ACKNOWLEDGMENT

This paper was supported by the Shaanxi Province Soft Science Research Project (2015KRM112), the Philosophy and Social Sciences Prosperous Development Project of Xi'an University of Science and Technology (2014SY03), and the Cultivation Fund of Xi'an University of Science and Technology (201364).

REFERENCES

[1] Benchekroun H., Withagen C. The optimal depletion of exhaustible resources: A complete characterization [J]. Resource and Energy Economics, 2011, 33(3):612–636.

[2] Dasgupta P., Heal G. The Optimal Depletion of Exhaustible Resources [J]. The Review of Economic Studies, 1974, 41:3–28.

[3] Solow R.M. Intergenerational Equity and Exhaustible Resources [J]. The Review of Economic Studies, 1974, 41:29–45.

[4] Stiglitz J. Growth with Exhaustible Natural Resources: Efficient and Optimal Growth Paths [J]. The Review of Economic Studies, 1974, 41:123–137.

[5] Robert E. Lucas, Jr. On the Mechanics of Economic Development [J]. Journal of Monetary Economics, 1988, 22:3–42.

[6] Boucekkine R., Ruiz-Tamarit J. Special functions of the study of economics dynamics: The case of the Lucas-Uzawa model [J]. Journal of Mathematical Economics, 2008, 44:33–54.

[7] Boucekkine R., Martinez B., Ruiz-Tamarit J. Global dynamics and imbalance effects in the Lucas-Uzawa model [J]. International Journal of Economic Theory, 2008, 4(4):503–518.

[8] Stokey. N.L. Are There Limits to Growth? [J]. International Economic Review, 1998, 39:1–31.

Material Science and Environmental Engineering – Chen (Ed.)

Pollution characterization of water-soluble anions in TSP and PM_{10} in Wuhan Economic and Technological Development Zone, China

A.H. Li

Hubei Key Laboratory of Industrial Fume and Dust Pollution Control, Jianghan University, Wuhan, P.R. China
School of Chemistry Environmental Engineering, Jianghan University, Wuhan, P.R. China

J.T. Jiang & Q.G. Liu

Hubei Key Laboratory of Industrial Fume and Dust Pollution Control, Jianghan University, Wuhan, P.R. China

ABSTRACT: The contents of TSP and PM_{10} were determined from March to April in 2014 in Wuhan Economic and Technological Development Zone (WEDZ). Meanwhile, four kinds of water-soluble anions that comprised of fluoride (F^-), chloride (Cl^-), sulfate (SO_4^{2-}), and nitrate (NO_3^-) in TSP and PM_{10} were detected by IC. TSP and PM_{10} daily mass concentrations ranged from 161.6~446.3 μg/m^3 and 110.1~241.1 μg/m^3, with average concentrations of 324.6 μg/m^3 and 175.3 μg/m^3, respectively. In TSP, the contributions of F^-, Cl^-, NO_3^-, SO_4^{2-} were 0.29%, 2.97%, 13.47% and 12.67%, and followed the order of $NO_3^- > SO_4^{2-} > Cl^- > F^-$. In the case of PM_{10}, the contributions were 0.22%, 1.39%, 17.70% and 17.26%, and followed the same order with TSP. The mean ratios of NO_3^- and SO_4^{2-} in TSP and PM_{10} were 1.06 and 1.03, indicating that stationary sources and other industrial emissions and vehicle emissions were both important sources to the different sizes particles in WEDZ during the monitoring period.

Keywords: TSP; PM_{10}; water-soluble anions

1 INTRODUCTION

In the recent years, China's dramatic economic rise, rapid industrial development, population growth, construction and demolition projects, and the increase in traffic flow critically affected the atmospheric environment, especially with regards to contamination of atmospheric particulate matter (Wang, J. et al. 2013). Atmospheric PM is a complex mixture of elemental and organic carbon, ammonium, nitrates, sulfates, mineral dust, trace elements and water. It may be of natural or anthropogenic origin and can be emitted directly into the atmosphere, or formed in the atmosphere from gaseous precursors. An increasing number of metropolises, such as Beijing, Guangzhou, Shenzhen (Zhang K. et al. 2007, Han et al. 2014, and Wu et al. 2013) were greatly influenced by atmospheric haze which was mainly caused by atmospheric Particulate Matter (PM). The study of atmospheric particle concentrations, sizes and chemical composition at the receptors is essential to apportion the sources of the aerosols and the processes associated with their formation (Fang et al. 2006, Kong et al. 2011, and Jorge et al. 2013).

Water-soluble ions are major components of the atmospheric particulates. They can compose up to 60–70% of the total mass of suspended particulate matter (Ali-Mohamed 1991). Therefore, as an important part of chemical composition in TSP and PM_{10}, the research involving in water-soluble ions became the most concern in the field of atmospheric pollution (Zhao et al. 2011, Galindo et al. 2011, Li et al. 2014).

In this study, TSP and PM_{10} have been monitored from March 11th to April 8th; the pollution characterizations of four water-soluble anions of TSP and PM_{10} in Wuhan Economic and Technological Development Zone were investigated.

Wuhan Economic and Technological Development Zone is on the Yangtze River III terrace which is mainly composed of clay soil about 18~30 meters thick. The average ground bearing is over 30 t/m^2. Hydrology and engineering geology condition is appropriate for various industrial uses. Wuhan Economic and Technological Development Zone is of a subtropical wet monsoon climate with four distinct seasons. Its annual average temperature is 16.8°C and the yearly average rainfall is 1093.3 mm. The annual average sunshine is 208.9 days. Northeast wind dominates throughout the year, while Southeast wind dominates in summer.

2 EXPERIMENTAL

2.1 *Sampling*

The sampling site was located in Jianghan University in WEDZ, the apparatuses used in this study were installed on the rooftop of a building (25 m above ground) and the sampling port was 1.5 m above the ground. During the whole monitoring, 34 TSP and PM_{10} samples were collected on the glass fiber filters, using the middle-volume air sampler (made in China) at the flow rate of 100 L/min. Before and after sampling, the filters were firstly equilibrated in a HWS-150 thermostatic and humidistatic chamber (made in China) for 24 h and then weighted with Sarorius CPA225 electronic balance (made in Germany, which has low limit detection of 10 μg). The samples were stored at −18°C for later analysis.

2.2 *Analytical methods*

After sampling, each aerosol sample was leached using 10 mL ultrapure water (specific resistance: 18.3 MΩ•cm) to extract the water-soluble anions for 1h in KQ-250B mode ultrasonic cleaning instrument (made in China). The extracts were filtered through a syringe membrane (0.45 μm) and stored at 4°C for later analysis. An ion chromatography (Dionex Corp., ICS-2500 USA) was used to measure the concentrations of the four water-soluble anions including F^-, Cl^-, NO_3^- and SO_4^{2-}, which detected with ASRS-4 suppressor and an AS14-HC column (Dionex Corp.), using 12 mM NaOH as the eluent at the flow rate of 0.8 mL/min.

2.3 *Standard curves of the four water-soluble anions*

The mixed standard solutions of F^-, Cl^-, NO_3^- and SO_4^{2-}, were prepared for the ion chromatographic analysis. Then the four kinds of water-soluble anions were quantitatively analyzed by internal standard method using peak area, respectively. The results showed that the correlation coefficients of standard curve were greater than 0.99 for F^-, Cl^-, NO_3^- and SO_4^{2-} (see Table 1).

Table 1. Linear regression equation and correlation coefficient of ion chromatographic analysis for F^-, Cl^-, NO_3^- and SO_4^{2-}.

Water-soluble anions	Linear regression equation	Correlation coefficient
F^-	y = 1.4914x + 0.4574	0.9984
Cl^-	y = 1.2447x − 2.1188	0.9992
NO_3^-	y = 0.5403x − 1.4233	0.9979
SO_4^{2-}	y = 0.2656x − 1.3540	0.9970

3 RESULTS AND DISCUSSION

3.1 *Daily mean concentration of TSP and PM_{10} in WEDZ during the monitoring period*

The daily mean concentrations of TSP and PM_{10} were measured for a total of 34 samples in Wuhan Economic and Technological Development Zone during March to April 2014. The results were showed in Figure 1.

As shown in Figure 1, the average concentrations of TSP and PM_{10} were 324.6 μg/m³ and 175.3 μg/m³, respectively, in WEDZ during the sampling period. There were 64.7% of the daily mean concentrations of TSP exceeded the prescribed limit of China Second-class Ambient Air Quality Standard (300 μg/m³). As for PM_{10}, there were 76.5% of the daily mean concentrations of TSP exceeded the prescribed limit of China Second-class Ambient Air Quality Standard (150 μg/m³). The results indicated that the urban air quality in Wuhan was relatively polluted by suspended atmospheric particles.

However, compared to the main cities in China, such as Beijing (Zhang K. et al. 2007), Xi'an (Shen et al. 2009), the concentrations of TSP and PM_{10} in WEDZ were relatively low. The result of correlation analysis revealed that the correlation coefficient between TSP and PM_{10} was equal to 0.69, which indicated that a certain linear correlation

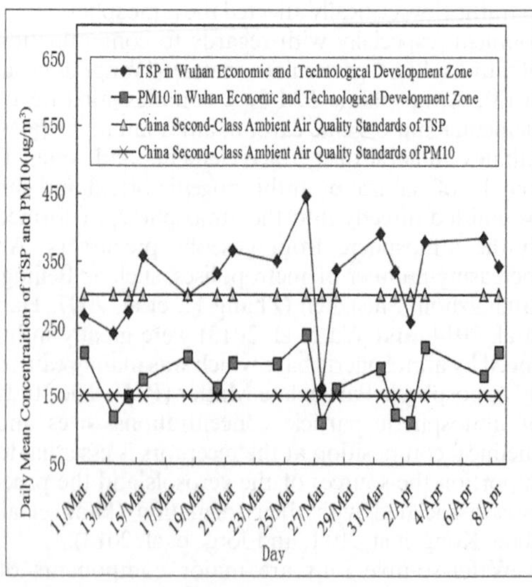

Figure 1. The daily mean concentrations of TSP and PM_{10} in WEDZ during the sampling period.

between TSP and PM_{10} were observed. However, the average of the PM_{10} accounting for TSP was 54.0% to demonstrate that PM_{10} was key component of atmospheric particulate in WEDZ during the monitoring period.

3.2 *Characterization of water-soluble anions in TSP and PM_{10}*

The daily mean concentrations of the water-soluble ions (F^-, Cl^-, NO_3^- and SO_4^{2-}) were measured for a total of 34 samples in Wuhan Economic and Technological Development Zone during March to Apirl 2014. The concentration range and daily mean concentration of four kinds of water-soluble anions in TSP and PM_{10} are summarized in Table 2.

As shown in Table 2, the mean concentrations of F^-, Cl^-, NO_3^- and SO_4^{2-} were 0.93 μg/m³, 9.64 μg/m³, 43.71 μg/m³ and 41.13 μg/m³ in TSP and 0.38 μg/m³, 2.44 μg/m³, 31.03 μg/m³ and 30.26 μg/m³ in PM_{10}, respectively. The concentrations of four kinds of water-soluble anions in TSP and PM_{10} followed the same order of $NO_3^- > SO_4^{2-} > Cl^- > F^-$. NO_3^- and SO_4^{2-} were the dominant anions in TSP and PM_{10} in WEDZ, and contributed 88.9% and 95.6% to the total measured water-soluble anions. The ratios of F^-, Cl^-, NO_3^- and SO_4^{2-} in PM_{10} and TSP were 0.41, 0.25, 0.71 and 0.74, respectively. The results demonstrated that the F^- and Cl^- were mainly dominant in TSP, while NO_3^- and SO_4^{2-} were primarily dominant in PM_{10}.

The concentrations of water-soluble anions in TSP and PM_{10} in the other regions around the world were listed in Table 3.

As shown in Table 3, compared to those regions, such as Xi'an, Qinghai Lake, Shanghai, and Chongqing (Shen et al. 2009, Zhang N.N. et al. 2014, Wang Y. et al. 2006), the concentrations of water-soluble anions in TSP and PM_{10} in WEDZ were relatively high. The concentrations of SO_4^{2-} in TSP in Xi'an, Shanghai and Qinghai Lake were 34.0 μg/m³, 17.83 μg/m³ and 5.04 μg/m³ respectively, and followed the order of WEDZ > Xi'an > Shanghai > Qinghai Lake. The concentrations of SO_4^{2-} in PM_{10} in Beijing, Guangzhou, Delhi (Sharma et al. 2014) were 53.7 μg/m³, 22.5 μg/m³ and 9.86 μg/m³ respectively, and followed the order of Beijing > WEDZ > Guangzhou > Delhi.

3.3 *Mass ratio of NO_3^-/SO_4^{2-}*

The most important sources of SO_2 and NO_x are coal combustion and oil-fired power plants, with some contribution from transportation (Larssen et al. 2006). SO_4^{2-} originates primarily from stationary sources, while NO_3^- is mainly derived from the mobile sources. Thus, the mass ratio of $[NO_3^-]/[SO_4^{2-}]$ has been used as an indicator of the relative importance of stationary versus mobile sources of sulfur and nitrogen in the atmosphere (Yao et al. 2002, Kim et al. 2000). Mass ratio of $[NO_3^-]/[SO_4^{2-}]$ in some regions were shown in Table 4.

Table 2. Concentrations and ratios of water-soluble anions in TSP and PM_{10} in WEDZ (μg/m³).

	TSP		PM_{10}		
Species	Range	Mean	Range	Mean	PM_{10}/TSP
F^-	0.36~2.16	0.93	0.15~0.96	0.38	0.41
Cl^-	1.47~15.98	9.64	0.86~4.24	2.44	0.25
NO_3^-	13.59~98.19	43.71	9.06~60.43	31.03	0.71
SO_4^{2-}	17.67~63.00	41.13	20.37~38.11	30.26	0.74

Table 3. Concentrations of water-soluble anions in TSP and PM_{10} in WEDZ and other some regions (μg/m³).

Species sites	Sampling period	Size fraction	SO_4^{2-}	NO_3^-	Cl^-	F^-
Qinghai Lake [13]	Jun–Sep 2010	TSP	5.04	1.3	0.39	–
Xi'an [12]	Oct 2006–Sep 2007	TSP	34.0	16.1	4.6	–
Shanghai [14]	Sep 2003–Jan 2005	TSP	17.83	14.19	8.06	0.94
Beijing [2]	October 2004	PM_{10}	53.7	50.2	–	–
Guangzhou [3]	July 2006	PM_{10}	22.5	3.8	1.9	–
Delhi, India [15]	Jan 2010–Dec 2011	PM_{10}	9.86	8.89	3.36	–
Wuhan Economic and Technological	March–April 2014	TSP	41.13	43.71	9.64	0.93
Development Ozone (This study)		PM_{10}	30.26	31.03	2.44	0.38

Table 4. Mass ratio of $[NO_3^-]/[SO_4^{2-}]$ in some regions.

Species sites	$[NO_3^-]/[SO_4^{2-}]$ in TSP	$[NO_3^-]/[SO_4^{2-}]$ in PM_{10}
Xi'an	0.47	–
Shanghai	0.8	–
Beijing	–	0.93
Guangzhou	–	0.17
Delhi	–	0.90
Wuhan Economic and Technological Development Ozone (This study)	1.06	1.03

As shown in Table 4, the daily averaged mass ratio of $[NO_3^-]/[SO_4^{2-}]$ in TSP in WEDZ varied between 0.49 and 1.96 with an average of 1.06, which was significantly higher than those reported in other regions, e.g. 0.47 in Xi'an (Shen et al. 2009), 0.80 in Shanghai (Wang Y. et al. 2006). High $[NO_3^-]/[SO_4^{2-}]$ mass ratio in TSP in WEDZ indicated that the mobile source of the air pollution becomes more and more predominant.

In case of PM_{10}, that ranged from 0.40 to 1.72 with an average of 1.03 and was also higher than those reported regions, e.g. 0.93 in Beijing (Zhang K. et al. 2007), 0.17 in Guangzhou (Han et al. 2014) and 0.90 in Delhi (Sharma et al. 2014). The NO_3^- mass concentration in WEDZ was slightly higher than SO_4^{2-}, which implies that the mobile sources were with the same importance as the stationary sources compared to the regions with lower $[NO_3^-]/[SO_4^{2-}]$.

4 CONCLUSION

In summary, the urban air quality in WEDZ was polluted by the atmospheric particulate matter TSP and PM_{10} to a certain extent. Four kinds of water-soluble anions in TSP and PM_{10} were investigated. The results showed that SO_4^{2-} and NO_3^- were the dominant anions in atmospheric PM in WEDZ. The mass ratio of $[NO_3^-]/[SO_4^{2-}]$ has been calculated, which indicated that the stationary sources and mobile sources were both the major sources of air pollution in WEDZ.

ACKNOWLEDGEMENTS

This work was financially supported by Scientific Research Fund of Hubei Provincial Education Department (No. B2014074), Technologies R&D Program Fund of (No. 201250499145-7) and High-level Professionals Research Start-up Fund of Jianghan University (No. 2012020).

REFERENCES

[1] Ali-Mohamed, A.Y. 1991. Estimation of inorganic particulate matter in the atmosphere of Isa Town, Bahrain, by dry deposition. *Atmos. Environ*, Part B 25: 397–405.

[2] Fang G.C., Wu Y.S., Chen J.C., Rau J.Y., Huang S.H. & Lin C.K. 2006. Characterization of chemical species in PM_{10} and $PM_{2.5}$ aerosols in suburban and rural sites of central Taiwan. *Journal of Hazardous Materials* 132: 269–276.

[3] Galindo N., Yubero E., Nicolás J.F., Crespo J., Pastor C., Carratalá A. & Santacatalina M. 2011. Water-soluble ions measured in fine particulate matter next to cement works. *Atmospheric Research* 45: 2043–2049.

[4] Han T.T., Liu X.G., Zhang Y.H., Gu J.W., Tian H.Z., Zeng L.M., Chang S.Y., Cheng Y.F., Lu K.D. & Hu M. 2014. Chemical characteristics of PM_{10} during the summer in the mega-city Guangzhou, China. *Atmospheric Research* 137: 25–34.

[5] Jorge P., Andres A. & Xavier Q. 2013. PM_{10} and $PM_{2.5}$ sources at an insular location in the western Mediterranean by using source apportionment techniques. *Science of Total Environment* 456–457: 267–277.

[6] Kim, B.M., Teffera, S., Zeldin, M.D. 2000. Characterization of $PM_{2.5}$ and PM_{10} in the South Coast Air Basin of Southern California: part 1 espatial variations. *Journal of the Air and Waste Management Association* 50 (12): 2034–2044.

[7] Kong S.F, Ji Y.Q., Lu B., Chen L., Han B., Li Z.Y. & Bai Z.P. 2011. Characterization of PM_{10} source profiles for fugitive dust in Fushun-a city famous for coal. *Atmospheric Environment* 45: 5351–5365.

[8] Larssen, T., Lydersen, E., Tang, D., He, Y., Gao, J., Liu, H., Duan, L., Seip, H.M., Vogt, R.D., Mulder, J., Shao, M., Wang, Y., Shang, H., Zhang, X., Solberg, S., Aas, W., Okland, T., Eilertsen, O., Angell, V., Liu, Q., Zhao, D., Xiang, R., Xiao, J., Luo, J. 2006. Acid rain in China: rapid industrialization has put citizens and ecosystems at risk. *Environ. Sci. Technol.*: 418–425.

[9] Li L., Yin Y., Kong S.F., Wen B., Chen K., Yuan L., Li Q. 2014. Altitudinal effect to the size distribution of water soluble inorganic ions in PM at Huangshan, China. *Atmospheric Environment* 98: 242–252.

[10] Sharma, S.K. Mandal, T.K. Mohit Saxena, Rashmi, Rohtash, Sharma, A. & Gautam, R. 2014. Source apportionment of PM_{10} by using positive matrix factorization at an urban site of Delhi, India. *Urban Climate* 10: 656–670.

[11] Shen Z.X, Cao J.J., Arimoto R., Han Z.W., Zhang R.J., Han Y.M., Liu S.X., Okuda T., Nakao S. & Tanaka S. 2009. Ionic composition of TSP and $PM_{2.5}$ during dust storms and air pollution episodes at Xi'an, China. *Atmospheric Environment* 43: 2911–2918.

[12] Wang, J., Hu, Z.M., Chen, Y.Y., Chen, Z.L. & Xu, S.Y. 2013. Contamination characteristics and possible sources of PM_{10} and $PM_{2.5}$ in different functional areas of Shanghai. *China Atmospheric Research* 68: 221–229.

[13] Wang Y., Zhuang G., Zhang X., Huang K., Xu C., Tang A., Chen J. & An Z. 2006. The ion chemistry, seasonal cycle, and sources of $PM_{2.5}$ and TSP aerosol in Shanghai. *Atmospheric Environment* 40: 2935–2952.

[14] Wu G., Du X., Wu X.F, Fu X., Kong S.F, Chen J.H., Wang Z.S. & Bai Z.P. 2013. Chemical composition, mass closure and sources of atmospheric PM_{10} from industrial sites in Shenzhen, China. *Journal of Environmental Sciences* 25: 1626–1635.

[15] Yao, X.H., Chan, C.K., Fang, M., Cadle, S., Chan, T., Mulawa, P., He, K.B., Ye, B. 2002. The water-soluble ionic composition of $PM_{2.5}$ in Shanghai and Beijing, China. *Atmospheric Environment* 36 (26): 4223–4234.

[16] Zhang K., Wang Y.S., Wen T.X, Meslmani Y. & Murray F. 2007. Properties of nitrate, sulfate and ammonium in typical polluted atmospheric aerosols (PM_{10}) in Beijing. *Atmospheric Research* 84: 67–77.

[17] Zhang N.N., Cao J.J., Liu S.X., Zhao Z.Z, Xu H.M. & Xiao S. 2014. Chemical composition and sources of $PM_{2.5}$ and TSP collected at Qinghai Lake during summertime. *Atmospheric Research* 138: 213–222.

[18] Zhao J.P., Zhang F.W., Xu Y. & Chen J.S. 2011. Characterization of water-soluble inorganic ions in size-segregated aerosols in coastal city, Xiamen. *Atmospheric Research* 99: 546–562.

Material Science and Environmental Engineering – Chen (Ed.)
 ISBN 978-1-138-02938-5

Heavy metal distribution features and pollution assessment in different grain-size surface sediments at the Yellow River Estuary

S. Lin & M. Cong
College of Environment Sciences and Engineering, Ocean University of China, Qingdao, Shandong, China

X.X. Luo
College of Environment Sciences and Engineering, Ocean University of China, Qingdao, Shandong, China
Key Laboratory of Marine Environmental Science and Ecology, Ministry of Education, Ocean University of China, Qingdao, Shandong, China

ABSTRACT: Heavy metals are important pollutants in sediments, which cause a great potential hazard. In this paper, surface sediment samples collected from the Yellow River Estuary were categorized by their grain sizes, and then concentrations of Cu, Pb, Zn and Cd in unsized sediments and in sediments with grain size < 63 μm were measured. The results showed that concentrations of the four heavy metals in sediments with grain size < 63 μm were higher than those in unsized sediments. This indicated that concentrations of heavy metals were higher in sediments with a smaller grain size and higher organic content. On the whole, heavy metal concentrations were relatively low at the Yellow River Estuary, and high concentrations of heavy metals appeared around Lijin pontoon. Single-factor pollution statuses of heavy metals belonged to the 'low' or 'moderate' level, which was relatively low overall and led to the 'low' potential ecological risk of the study area.

Keywords: Yellow River Estuary; surface sediment; heavy metals; ecological risk assessment

1 INTRODUCTION

Yellow River Estuary and its adjacent sea, where the National special Marine reserve of Yellow River Estuary is located, is a major spawning, breeding and feeding ground of Bohai Sea. In recent years, with the rapid development of economy and the swell in the population of the Yellow River basin, sewage discharged to this area is increasing day by day (Shen et al. 1989; Luo et al. 2011). Marine sediment is the source and sink of environmental pollutants such as heavy metals. When the sediment is disturbed, heavy metal elements could be released to the environment and then cause secondary pollution. Heavy metals are important pollutants that cause potential hazard. Unlike other contaminants, they cannot be degraded by microorganisms. Conversely, they may concentrate in organisms and become persistent pollutants, which could have a great impact on aquatic organisms. In this paper, the Yellow River Estuary was taken as the study area; concentrations of Cu, Pb, Zn and Cd in unsized sediments and sediments with grain size less than 63 μm were analyzed, and then distribution characteristics and pollution status of this area are discussed to provide the basis for ecological environmental protection and contamination control.

2 MATERIALS AND METHODS

2.1 *Sampling*

Surface sediment samples of 7 stations along the direction of the stream channel to the Yellow River Estuary were collected. The sampling sites are shown in Figure 1.

Sampling and pretreatment were carried out according to the 'The Specification for Marine Monitoring' (GB 17378.5-2007). Station P1 was

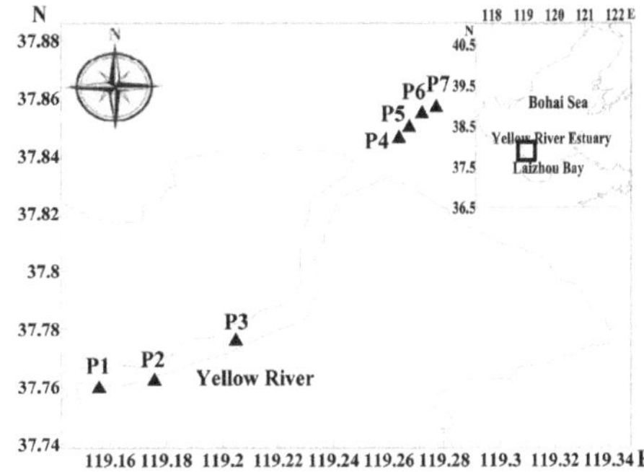

Figure 1. Sampling sites of the Yellow River Estuary.

located near Lijin pontoon where the salinity was 0; whereas the salinity of station P7 reached up to 28.8. The salinity presented a notable increasing trend from P1 to P7.

2.2 *Sample analysis and data processing*

After dispersion by sonic oscillation, parts of sediment samples were divided into two groups, namely samples with grain size less than 63 μm and those with grain size greater than 63 μm, by sieving through a 250-mesh sieve, and then the samples were dried for later use.

Some sediment samples were naturally air-dried and foreign matters were removed, and then the samples were fully mixed, ground and sieved through an 80-mesh nylon sieve. These samples were unsized.

The storage and analysis of sediment samples were carried out according to 'The Specification for Oceanographic Survey' (GB/T12763.3-2007). Concentrations of Cu, Pb and Zn in the samples were measured by flame photometry using an atomic absorption spectrophotometer; graphite furnace system was used to determinate Cd concentration. The standard samples (GBW07314) were used as internal samples for quality assurance and quality control. The recoveries of heavy metal concentrations ranged from 90% to 110%, and the standard deviations of parallel samples were less than 10% (Luo et al. 2011).

3 RESULTS AND DISCUSSION

3.1 *Grain size distribution of surface sediments*

Along the direction of the stream channel to the sea, the proportion of small grain-sized components in the sediment samples presented an increasing trend. In the 7 samples, mass percent of components with grain size less than 63 μm, such as silt and clay particles, ranged from 20.17% to 82.20% and their the average mass percent was 53.44%, while that of components with grain size greater than 63 μm, such as sand, ranged from 17.80% to 79.83% and the average was 46.56%.

3.2 *Concentrations and distributions of heavy metals in surface sediments*

We could find out from Figures 2 and 3 that heavy metal concentrations in unsized sediments and in sediments with grain size less than 63 μm were higher at stations P1 and P7 than at the other stations. By and large, heavy metal concentrations in surface sediments were relatively low in the study area, with almost all heavy metals conforming to

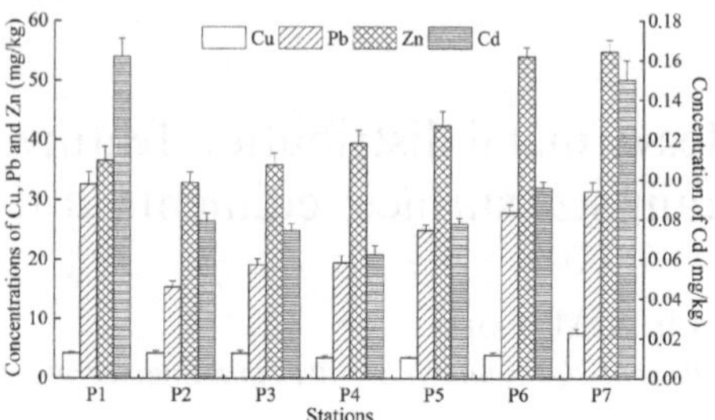

Figure 2. Heavy metal concentrations of unsized surface sediments.

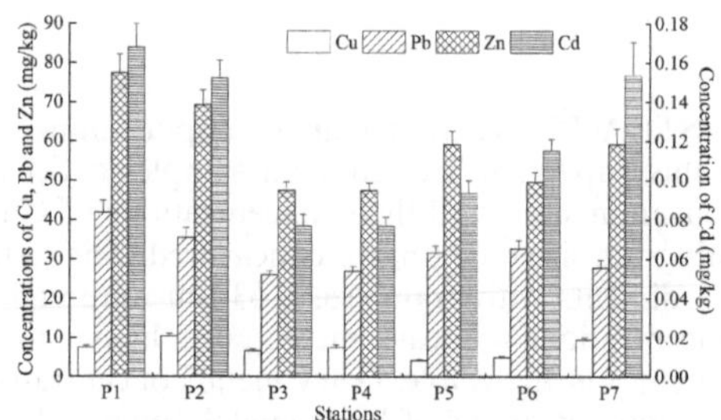

Figure 3. Heavy metal concentrations of surface sediments with grain size less than 63 μm.

the first criterion of the marine sediment quality standard. There was a good correlation between heavy metal concentrations and sediment grain sizes. Along the direction of the stream channel to the sea, heavy metal concentrations increased as the proportion of fine particles in sediments increased. Mass percent of the four heavy metals tended to be higher in sediments with grain size less than 63 μm. Along the stream channel to the sea, as the content of fine particle matters increased, mass percent of heavy metals in sediments with grain size less than 63 μm increased gradually. The finer the sediment particulates were, the higher the organic content was, and heavy metals would be absorbed or deposited in sediments more easily.

3.3 *Pollution assessment of heavy metals in surface sediments*

The single-factor pollution index (SI) and the comprehensive pollution index (CI) were employed to evaluate the heavy metal pollution status of sediment at the Yellow River Estuary. Since there was no single standard in selecting the background values of heavy metals and the values could be more suitable when considering the research status of the study area, referring to the studies of the Yellow River Estuary (Hanson et al. 1993;

Luo et al. 2010), the background values of Cu, Pb, Zn and Cd were chosen as 24.68, 17.50, 73.01 and 0.12 mg/kg respectively. The assessment results are shown in Figure 4. The SI values of Pb were above 1 in almost all the samples. However, most SI values were below 1 for the other heavy metals, indicating their 'low' single-factor pollution level in the study area. Concentrations of Pb and Cd were the highest at station P1. CI values were all below 5 for the 7 stations, which showed the 'low' pollution level of the heavy metals.

3.4 *Potential ecological risk assessment of heavy metals in surface sediments*

The Hakanson potential ecological risk index was used to assess the risk of heavy metals in surface sediments at the Yellow River Estuary. The results are shown in Figure 5. Except that the single potential ecological risk index (SRI) value of Cd was above 40 at station P1, the SRI values of the four heavy metals were below 40 for all the stations. The average SRI value ranged from 0.45 to 40.36 for the unsized samples, while the value ranged from 0.65 to 42.13 for the samples with grain size less than 63 µm, indicating the 'low' single potential ecological risk level of heavy metals. Potential ecological risk factors were Cd, Pb, Cu and Zn in the order of their SRI values. The SRI value of Cd was the highest, so that Cd was the primary potential ecological risk factor. In the case of the comprehensive potential ecological Risk Index (RI), all the stations had RI values far below 150, which meant that the comprehensive potential ecological risk of heavy metals was at 'low' level.

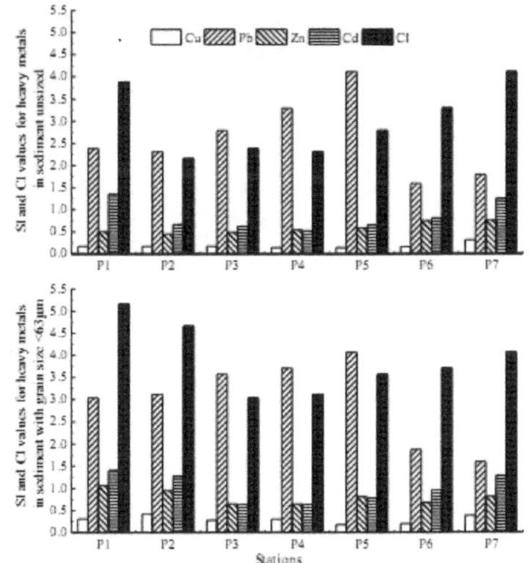

Figure 4. SI and CI values for heavy metals in unsized sediments and sediments with grain size <63 µm.

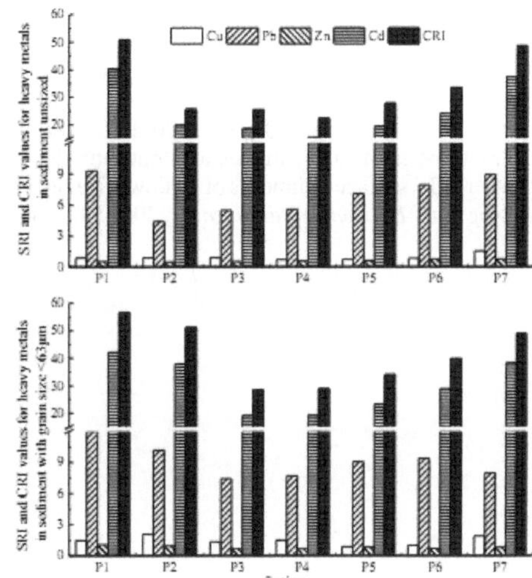

Figure 5. SRI and CRI values for heavy metals in unsized sediments and sediments with grain size <63 µm.

4 CONCLUSIONS

The concentrations of Cu, Pb, Zn and Cd were relatively low at the Yellow River Estuary. Their distributions were related to sediment grain sizes to a certain extent, that is heavy metal concentrations increased as fine particulate content increased. Single-factor pollution level was relatively low; concentrations of Pb and Cd reached the 'moderate' level at 2 stations (P1 and P7). Comprehensive potential ecological risk level was 'low' at the study area; and among the 4 heavy metals, Cd had the highest potential risk.

ACKNOWLEDGMENTS

This study was supported by the NSFC-Shandong Joint Fund for Marine Science Research Centers (Grant No. U1406403) and the National Marine Public Welfare Research Project of China (Grant No. 201405007).

REFERENCES

China National Standardization Management Committee. 2007. GB/T12763.3-2007. The Specification of Oceanographic Survey—Part 3: Marine meteorological observations. Beijing: *Standards Press of China*.

General Administration of Quality Supervision, Inspection and Quarantine of the People's Republic of China, Standardization Administration of the People's Republic of China. 2007. GB 17378.5-2007. The Specification for Marine Monitoring—Part 5: Sediment analysis. Beijing: *Standards Press of China*.

Hanson P.J., Evans D.W., Collby D.R. 1993. Assessment of elemental contamination in estuarine and coastal environments based on geochemical and statical modeling of sediments. *Mar Environ Res.*, 36: 237–266.

Luo X.X., Tian J., Yang J.Q., et al. 2011. Distribution characteristics of heavy metals and nutrient elements in inter-tidal surface sediments of Yellow River estuary. *Ecology and Environmental Sciences*, 20(5): 892–897.

Luo X.X., Zhang R., Yang J.Q., et al. 2010. Distribution and pollution assessment of heavy metals in surface sediment in Laizhou Bay. *Ecology and Environmental Sciences*, 19(2): 262–269.

Shen Z.L., Lu J.P., Liu X.J. 1989. Inorganic nitrogen and phosphate in estuary of the Huanghe River and its near waters. *Studia Mar. Sinica*, 30: 51–79.

Material Science and Environmental Engineering – Chen (Ed.)

Assessment of heavy metal pollution and ecological risks for sediments from Baihua Lake, China

M.Y. Fan
Guizhou Provincial Key Laboratory of Information System of Mountainous Areas and Protection of Ecological Environment, Guizhou Normal University, Guiyang, Guizhou, China

H. Yang
Guizhou Provincial Key Laboratory of Information System of Mountainous Areas and Protection of Ecological Environment, Guizhou Normal University, Guiyang, Guizhou, China
Institute of South China Karst, Guizhou Normal University, Guiyang, Guizhou, China

Y.M. Lin & X.D. Shi
Guizhou Provincial Key Laboratory of Information System of Mountainous Areas and Protection of Ecological Environment, Guizhou Normal University, Guiyang, Guizhou, China

J.L. Li
Guizhou Provincial Key Laboratory of Information System of Mountainous Areas and Protection of Ecological Environment, Guizhou Normal University, Guiyang, Guizhou, China
Institute of South China Karst, Guizhou Normal University, Guiyang, Guizhou, China

L.Y. Li
Guizhou Provincial Key Laboratory of Information System of Mountainous Areas and Protection of Ecological Environment, Guizhou Normal University, Guiyang, Guizhou, China

J.W. Hu
Guizhou Provincial Key Laboratory of Information System of Mountainous Areas and Protection of Ecological Environment, Guizhou Normal University, Guiyang, Guizhou, China
Institute of South China Karst, Guizhou Normal University, Guiyang, Guizhou, China

ABSTRACT: In order to assess the pollution level of heavy metals (Pb, Cr, Cd, Hg, As, Cu, Fe and Zn) in the sediments from Baihua Lake, the sediment samples were collected at 15 sampling locations were analyzed in this investigation. The results showed that the variations in the concentrations of Fe, Zn, Hg, As and Pb were large, and that Fe, Zn, Hg and As were all high for the sampling location Yueliangwan. According to the results of the individual pollution index, the major part of Baihua Lake was polluted to different degrees by Hg, As, Cd and Zn. The integrated pollution index of each sampling location ranged from 1.13 to 41.66, and Yueliangwan, Jinyinshan and Laojiutu were the most polluted among all the sampling locations. Their integrated pollution indices were 41.66, 18.27 and 9.61, respectively. Since the Hg and As pollution was the heaviest in sediments, the control of these two heavy metals pollution became particularly important for this lake.

Keywords: Baihua Lake; heavy metals; sediments; nemerow integrated pollution index

1 INTRODUCTION

Sediment is the "sink" and "source" of nutrients, heavy metals and persistent organic pollutants in the ecosystems of lakes (Zoumis et al., 2001; Zhong et al., 2007). Among these pollutants, heavy metals have wide sources, and are difficult to decompose and easy to accumulate in the environment and along the food chain. Thus, heavy metals are considered as one of the most important pollutants because of causing toxicity to organisms in the environment (Dai, 2006; Huang et al., 2009). Baihua Lake, as one of the five drinking water resources for approximately 3 million people, is located in Qingzhen County (E 106°27′–106°34′, N 26°35′–26° 42′) and only 16 km west of Guiyang, the capital of Guizhou Province (Huang, 2008). As a deep lake on Yunnan-Guizhou plateau, Baihua has more inlet tributaries than outlets, leading to a long cycle of water exchange and is hence not conducive to the discharge of pollutants. In recent years, the concentrations, speciation, distribution characteristics of heavy metals and the methods of potential ecological risk index have been studied in sediments from Baihua Lake (Huang et al., 2009; Huang, 2008; Gao et al. 2012; Gao et al. 2012; Tian et al. 2012). However, this cannot objectively reflect the overall pollution level for the lake. In order to fully reflect the overall pollution level, the methods of the individual pollution index and the Nemerow integrated pollution index were conducted in the present work to study the pollution of heavy metals in sediments. This is significant to assess the current environmental quality of the lake (Yu et al., 2010; Tian et al., 2011; Singh et al., 2005; Qiao et al., 2010).

2 MATERIALS AND METHODS

2.1 *Sample collection*

Sediments were sampled during October 2014 in Baihua Lake. The sampling locations (Fig. 1) were selected based on the water flow direction, water depth and the surrounding environment of the lake. The sampling locations covered the entire lake with fifteen sampling locations. The sediment samples were taken by the grab sampler, and put into polyethylene plastic bags after drying at room temperature, and then treatment of the samples were carried out after labeling the samples. After pretreatment, the original samples were analyzed.

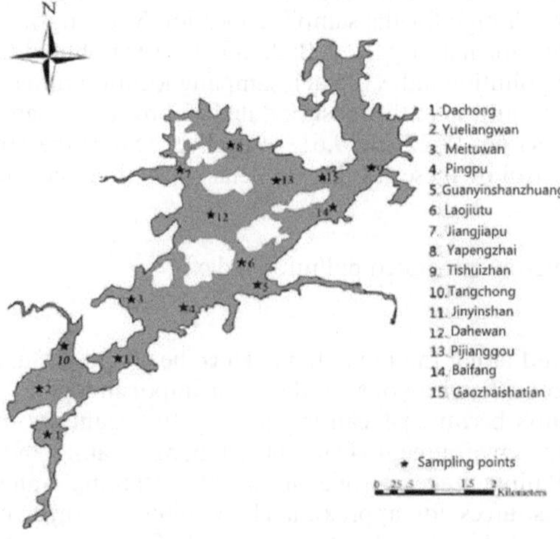

Figure 1. Location of the sampling points.

The collected samples were transferred from the polyethylene plastic bag to a cleaned glass dish treated with HNO_3. The samples were ground into powder for examination after drying at room temperature.

2.2 *Instruments and reagents*

The following instruments were used in the investigation: (1) atomic fluorescence spectrometer (AFS-933) was obtained from Titan Instruments Corporation, Beijing, China; (2) atomic absorption spectrometer (ZEEnit 700P) from Analytic Jena AG, German; (3) atomic absorption spectrophotometer (WFX-210) from Ray Leigh Corporation, Beijing, China; and (4) water purification system (Nex Power 2000) from Hu man Corporation, Korea. All reagents used in this study were made in China. The Hydrochloric Acid (HCL), nitric acid (HNO_3), and perchloric acid ($HCLO_4$) were guaranteed regents, and the other regents were of analar grade. Deionized water was prepared using the above-listed water purification system.

2.3 *Sample analysis*

Analysis of Pb, Cr, Cd, Cu, Fe and Zn: 0.1500 g of each sediment sample were weighed accurately into a 250 ml beaker, and 20 ml of digesting mixture was made up of nitric acid and perchloric acid (nitric acid:perchloric acid = 4:1 (v/v)). The beakers were placed on an adjustable electric heating plate, and heated at low temperature for 60 minutes, then the electric heating plate was adjust to the highest temperature, and heating was continued for a while [4]. The left solution and sediment were transferred into a 50 ml volumetric flask and diluted to the full volume with deionized water. The prepared solutions were determined later.

Analysis of Hg and As (Li, 2004): 0.1000~0.1200 g of each sediment sample were weighed into a 50 ml tube, and 10 ml of digesting solution was made up of hydrochloric acid, nitric acid and deionized water hydrochloric acid:nitric acid:water = 3:1:4 (v/v). The tubes were placed in a boiling water bath for 2 h. Then, 5 ml of reducing agent solution (10% thiourea solution) were added into each tube, and tubes were diluted to the desired volume with deionized water. The prepared solutions were determined later.

3 RESULTS AND DISCUSSION

3.1 *The analysis of the concentrations of heavy metals*

The results of the heavy metal analysis are shown in Figure 2 and Table 1. As shown in Figure 2, the average concentrations of heavy metals in sediments from the lake were as follow: Fe > Zn > As > Pb > Cu > Cr > Hg > Cd, and variations of Fe, Zn, Hg, As and Pb were large.

From Table 1, it can be seen that the concentration of Fe ranged from 281.21 to 1121.23 mg/kg in sediments, with the maximum value for the sampling location Guanyinshanzhuang and the minimum value for the sampling location Jiangjiapu. The concentration of Zn ranged from 69.77 to 385.64 mg/kg, with the maximum value for the sampling location Jinyinshan and the minimum value for the sampling location Jiangjiapu. And the concentration of Cr ranged from 13.57 to 48.27 mg/kg, with the maximum value for the sampling location Baifan and the minimum value for the sampling location Dachong.

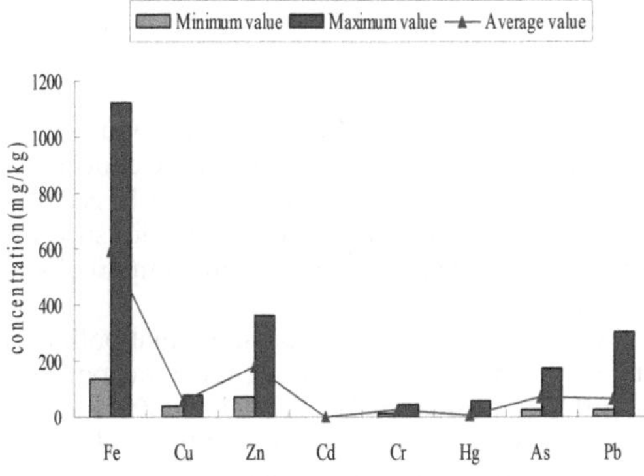

Figure 2. Average concentrations of the heavy metals.

Interestingly, the concentrations of As and Hg reached to a high level in sediments from Baihua Lake. The sampling location Yueliangwan was the most seriously polluted location from an overall perspective because the concentrations of Fe, Zn, Hg and As were all high in this location.

3.2 *The methods of the individual pollution index and the Nemerow integrated pollution index*

In order to further determine whether the heavy metals enrichment can cause the problems associated with heavy metal pollution, data were analyzed by using the methods of the individual pollution index (Yao et al., 2005) and the Nemerow integrated pollution index (Nemerow, 1985).

The formula for the individual pollution index is as follows:

$$I_i = \frac{C_i}{S_i} \tag{1}$$

where C_i is the measured heavy metal concentration and S_i is the limited value of the heavy metal in the evaluation criterion: I_i is the individual pollution index.

To evaluate the overall environmental quality, we also used the Nemerow index, which employs both maximum and average values of pollutant concentrations reflecting their polluted levels:

$$P_N = \sqrt{\frac{MaxI_i^2 + AvgI_i^2}{2}} \tag{2}$$

Table 1. Concentrations of the heavy metals in sediments from Baihua Lake (dry weight [dw]).

Name	Fe (mg/kg)	Cu (mg/kg)	Zn (mg/kg)	Cd (mg/kg)	Cr (mg/kg)	Hg (mg/kg)	As (mg/kg)	Pb (mg/kg)
Dachong	702.77	56.63	89.27	0.00	13.57	0.00	38.93	25.77
Yueliangwan	847.52	57.76	335.58	0.54	13.91	58.11	174.86	55.81
Meituwan	601.91	63.80	133.56	0.42	21.96	1.32	89.63	44.33
Pingpu	590.89	79.84	205.46	1.20	21.50	4.52	41.27	63.22
Guanyinshanzhuang	1121.23	56.68	174.06	1.60	39.40	1.45	79.10	51.40
Laojiutu	601.20	43.09	254.39	2.77	27.76	13.17	103.59	51.42
Jiangjiapu	281.21	39.73	69.77	1.08	20.35	1.00	29.54	33.41
Yapengzhai	512.55	62.06	129.11	1.70	32.87	1.30	71.53	303.80
Tishuizhan	544.08	67.64	130.95	1.17	24.87	5.45	105.39	46.26
Tangchong	133.99	37.49	79.07	0.27	15.36	3.85	27.02	42.76
Jinyinshan	664.28	64.27	358.64	0.75	20.59	17.94	73.44	27.89
Dahewan	461.61	58.90	162.50	1.44	24.00	3.74	75.64	60.59
Pijianggou	392.74	67.62	199.86	0.83	16.73	1.28	53.46	20.74
Baifan	409.54	71.54	157.75	2.44	48.27	1.46	45.00	56.18
Gaozhaishatian	1029.98	69.93	265.37	0.87	38.67	1.17	74.29	54.98

Table 2. Evaluated indices of heavy metal pollution in sediments.

Name	Cu	Zn	Cd	Cr	Hg	As	Pb	P_N
Dachong	0.57	0.30	0.00	0.05	0.00	1.56	0.07	1.13
Yueliangwan	0.58	1.12	0.90	0.06	58.11	6.99	0.16	41.66
Meituwan	0.64	0.45	0.70	0.09	1.32	3.59	0.13	2.63
Pingpu	0.80	0.68	2.00	0.09	4.52	1.65	0.18	3.35
Guanyinshanzhuang	0.57	0.58	2.67	0.16	1.45	3.16	0.15	2.69
Laojiutu	0.43	0.85	4.62	0.11	13.17	4.14	0.15	9.61
Jiangjiapu	0.40	0.23	1.80	0.08	1.00	1.18	0.10	1.36
Yapengzhai	0.62	0.43	2.83	0.13	1.30	2.86	0.87	2.22
Tishuizhan	0.68	0.44	1.95	0.10	5.45	4.22	0.13	4.07
Tangchong	0.37	0.26	0.45	0.06	3.85	1.08	0.12	2.79
Jinyinshan	0.64	1.20	1.25	0.08	17.94	2.94	0.08	18.27
Dahewan	0.59	0.54	2.40	0.10	3.74	3.03	0.17	2.85
Pijianggou	0.68	0.67	1.38	0.07	1.28	2.14	0.06	1.64
Baifan	0.72	0.53	4.07	0.19	1.46	1.80	0.16	3.01
Gaozhaishatian	0.70	0.88	1.45	0.15	1.17	2.97	0.16	2.23

a: The individual pollution index is graded as follows: $I_i \leq 1.0$, clean; $I_i > 1.0$, polluted.
b: The Nemerow integrated pollution index is graded as follows: $P_N \leq 0.7$, clean; $0.7 < P_N \leq 1.0$, still clean; $1.0 < P_N \leq 2.0$, slight pollution; $2.0 < P_N \leq 3.0$, moderate pollution; $P_N > 3.0$, serious pollution (HJ/T166-2004).

where P_N is the integrated pollution index; $MaxI_i$ is the maximum value of all indices; and $AvgI_i$ is the average of all indices.

The evaluated indices of heavy metal pollution are given in Table 2.

The results from Table 2 show that the maximum individual pollution indices were 58.11, 6.99, 4.62 and 1.20; thus, the total Hg, As, Cd and Zn in sediments reached the risky level ($I_i > 1.0$), while the other heavy metals were under the permissible level ($I_i \leq 1.0$). The individual index of Hg and As was high, indicating that Hg and As pollution was the most severe among these heavy metals.

According to the standards for soil pollution (HJ/T166-2004), P_N was greater than or equal to 3.0 for the different sampling locations of the whole area, indicating serious pollution of heavy metals. Obviously, the Nemerow integrated pollution indices for these sampling locations Yueliangwan, JinYinshan and Laojiutu were 41.66, 18.27 and 9.61, respectively, which were significantly higher than those for the other location and at the serious pollution level. The reasons for their high integrated pollution indices were mainly from the Hg and As pollution, which might be related to the discharge from the surrounding coal mines (Tian et al., 2012).

4 CONCLUSIONS

The results showed that the average concentrations of heavy metals from Baihua Lake in decreasing order as follows: Fe > Zn > As > Pb > Cu > Cr > Hg > Cd. In addition, variations in the concentrations of Fe, Zn, Hg, As and Pb were large. It is important to note that the concentrations of Fe, Zn, Hg and As were all high for the sampling location Yueliangwan.

According to the results of individual pollution index, the major part of Baihua Lake was polluted to different degrees by Hg, As, Cd and Zn. In addition, the Nemerow integrated pollution indices for the sampling locations Yueliangwan, JinYinshan and Laojiutu were significantly higher than those for the other sampling locations. And the results obtained from the Nemerow integrated pollution indices imply a dangerous pollution level for this water body. Since Hg and As were the most serious pollutants, the control of these two metal pollution was particularly important for this lake.

ACKNOWLEDGMENT

This study was supported by the National Natural Science Foundation of China (Grant No. 21367009) and by the Government of Guizhou Province (Project No. LKS [2013] 09).

REFERENCES

[1] Zoumis, T., Schmidt, A., Grigorva, L. & Calmano, W. 2001. Contamination in sediments: remobilization and demobilization. *The Science of Total Environment* 266: 195–202.

[2] Zhong, J.H. & Fan, C.X. 2007. Advance in the study on the effectiveness and environmental impact of sediment dredging. Journal of Lake Science 19(1): 1–10. (in Chinese)

[3] Dai, S.G. 2006. *Environmental Chemistry*. Higher Education Press, Beijing, 30–45. (in Chinese)

[4] Huang, X.F., Hu, J.W., Li, C.X., Deng J.J., Long J. & Qin F.X. 2009. Heavy-metal pollution and potential ecological risk assessment of sediments from Baihua Lake, Guizhou, P.R. China. *International Journal of Environmental Health Research* 19(6): 405–419.

[5] Huang, X.F., 2008. Studies on Characteristics of Pollution in Sediments from Baihua Lake. *Guizhou: Guizhou normal university*: 15–16.

[6] Gao, J., Dong, X., Liang, L.C., Zhang, Y. & Chen, Z. 2012. The morphology Distribution of Heavy Metals in the Section of Maixi Estuary Sediments in Baihua Lake. *Journal of Anhui Agricultural Sciences* 40(12): 7303–7305.

[7] Gao, J., Dong, X., Liang, L.C., Zhang, Y. & Chen, Z. 2012. Vertical distribution of the heavy metal in Maixi River sediments and its potential ecological risk. *Guizhou Agricultural Sciences* 40(3): 207–210.

[8] Yu, Y., Zhang, M., Qian, S.Q., Li, D.M. & Kong, F.X. 2010. Current status and development of water quality of lakes in Yunnan-Guizhou Plateau. *Journal of Lake Science* 22(6): 820–828. (in Chinese)

[9] Tian, L.F., Hu, J.W., Qin F.X., Huang, X.F., Liu, F., Luo, G.L. & Jin, M. 2011. Contamination characteristics of heavy metal elements in the water body from Lake Hongfeng. *China Environment Science* 31(3): 481–489. (in Chinese)

[10] Singh, K.P., Mohan, D., Singh, V.K. & Malik, A. 2005. Studies on distribution and fractionation of heavy metals in Gomti river sediments-a tributary of the Ganges, India. *Journal of Hydrology* 312: 14–27.

[11] Qiao, Y.M., Gu, J.G., Yang, Y. & Huang, C.J. 2010. The distribution, enrichment and pollution assessment of heavy metals in surface sediments of sea areas around the Nanao Island. *Journal of Tropical Oceanography* 29(1): 77–84. (in Chinese)

[12] Li, Y. 2004. Dual channel hydride generation atomic fluorescence spectrometry method for the simultaneous determination of arsenic and mercury in soil. *Agricultural Environment Development*. 22(1): 41–42.

[13] Yao, C.X., Chen, Z.L., Zhang, J. & Hou J. 2005. Heavy Metal Pollution Assessment of Vegetables in Pudong Zone of Shanghai. Journal of Agro-Environment Science. 2005, 24, 761–765.

[14] Nemerow, N.L. 1985. Stream, *Lake, Estuary, and Ocean Pollution*. Van Nostrand Reinhold, New York.

[15] Tian, L.F., Hu, J.W., Luo, C.L., Ma J.J., Huang X.F. & Qin F.X. 2012. Ecological risk and stability of heavy metals in sediments from Lake Baihua in Guizhou province. *Acta Scientiae Circumstantiae* 32(4): 885–894.

Material Science and Environmental Engineering – Chen (Ed.)

Combustion of biomass and waste materials in the Czech Republic: Determination of mixed emission limits

T. Výtisk & R. Janalík
Department of Power Engineering, Technical University of Ostrava, Ostrava, Czech Republic

ABSTRACT: This article describes the problems with combustion and co-combustion of biomass in the Czech Republic and also the transition from combustion of biomass to combustion and co-combustion of waste materials. For the determination of mixed emission limits, which must be kept at co-combustion, the Department of Power Engineering at the Technical University of Ostrava has created a software. The use of this special software will enable the operators of the sources of pollution to make faster decision when selecting a new fuel mix, to ensure a minimal amount of emissions.

Keywords: combustion; co-combustion; biomass; waste material; mixed emission limit; software

1 INTRODUCTION

Biomass, as the primary renewable source of energy, has an irreplaceable place in the power supply of the Czech Republic, and its role has been increasing over time. Although the amount of energy produced from biomass for energy purposes cannot significantly compete with other primary energy sources, biomass could become an alternative to fossil fuels, whose stocks have been shrinking. In the Czech Republic, biomass is one of the few renewable energy sources, and the production of which can rely on natural conditions. This means that biomass can be expected to be the main source of meeting the objective to produce 1% of energy in the Czech Republic from renewable energy sources by 2020.

2 CO-COMBUSTION OF BIOMASS

Combustion of biomass represents an alternative to conventional fossil fuels, and it is a renewable energy source that is environmental friendly and its incineration produces less emission of carbon dioxide into the atmosphere. The addition of crushed wood waste will save up to one-fifth of the consumption of coal in some heating plants, but despite that, these plants have decided not to use biomass in the future because "The Energy Regulation Authority" (ERA), which annually allocates the support for all renewable energy sources, has decided to significantly reduce the support due to concerns of biomass boom, similar to the photovoltaic one.

Since 1 January 2013, there has been a change in the legislation regarding the reduction of the support of co-combustion of biomass with coal in the Czech Republic, which has had the most profound impact on the area of combined production of electricity and heat with the most efficient use of biomass in heating plants. There is a reduction in the support of co-combustion of biomass with coal, which is also the result of ill-considered support of the production of electricity in photovoltaic power plants and nearly three times higher support of separate incineration of wood chips for the production of electricity, causing a significant reduction in the amount of biomass incinerated in heating plants. A decrease in the amount of wood chips required for the production of heat gets the producers of wood chips into trouble, because of considerable funds invested into production and transport to consumers and also workers employed for this purpose, usually in rural regions with a high unemployment rate.

3 COMBUSTION AND CO-COMBUSTION OF WASTE MATERIALS

3.1 *Legislative framework of combustion of waste materials and the technical potential context and indenting*

The Czech legislation [2, 3] introduced the term of "heat waste treatment", which includes incineration as well as co-incineration of waste in a law focused on air. According to this law, the heat treatment of waste is the oxidation of waste or its treatment by means of another thermic process, including the incineration of the produced substances if this could lead to higher levels of pollution compared with incineration of the same amount of natural gas with the same energy content. Incineration plants should not include plants that fundamentally differ from conventional incineration plants technically and in terms of their impact on the environment, e.g., some plants operating on the principle of pyrolysis or plasma processes.

The Air Protection Act therefore also defines waste incineration plant as a "stationary source designed for the heat treatment of waste, whose main purpose is not the production of energy or other products, and any stationary source in which more than 40% of the heat is produced by means of heat treatment of hazardous waste or which is focused on heat processing of untreated mixed municipal waste".

A common condition, however, is that waste, regardless of whether it is combusted in an incineration plant or co-combusted in another energy source and regardless of whether it is removal or energy use, must be thermally treated only in such plants that have the required permits issued by the regional authority, as well as in plants where such treatment is explicitly permitted.

Waste is anything that is waste according to the Waste Act and, at the same time, the law dealing with air protection does not define the term of alternative fuel, which is why this term is, in practice, most often used to define waste or fuel derived from waste, which is mechanically, biologically or otherwise adjusted for subsequent incineration using its energy potential [4]. It is therefore clear that air pollution sources that incinerate waste fuels are still sources that thermally treat waste and must have the required permissions.

There is a shift of opinion in the Czech Republic in the field of support of energy utilization of otherwise unusable residual mixed municipal waste, admitting the necessity of energy utilization of waste, which has been a long established practice in the EU countries. However, the important parameters necessary for successful energy utilization of waste must be known and defined, and they are listed below [5]:

- Fuels from waste must be clearly defined in terms of their quality.
- The production of fuels from waste must not use hazardous waste.
- Untreated mixed municipal waste cannot be designated as fuel from waste.
- Production of fuels from waste must be carried out under clearly defined conditions and only in facilities intended for that purpose, which are in possession of a decision issued by the relevant state administration authority necessary for this activity according to the Waste Act (Regional Authority).
- Use of fuels from waste should be possible only in the sources defined for this purpose, always with the IPPC permit, which lays down the specific conditions.
- Fuels from waste must not be co-incinerated anywhere else than in the facilities of the defined type.
- Fuel produced from waste during co-combustion should always meet the limit for emissions (always using incineration tests on a specific source). The emission limit guarantees the quality of fuel and the process of energy utilization of fuel produced from waste itself.
- The emission limits for co-incineration of fuels produced from waste can be as strict as for the co-incineration of waste, always using the mixing equation defined by the Decree on the Air Protection Act.

The principle of co-combustion conditions is given at the European level, and the defined conditions are set in such a way to ensure the highest possible degree of environment protection.

In the view of the above-presented facts, it can be concluded that the co-incineration of waste in the Czech Republic can be carried out in most types of power station coal boilers, in incineration plants with an output exceeding 50 MW, including in particular:

- Power station facilities with fluid and granulation (powder) boilers and, occasionally, with gasification.
- Heating plant sources with fluid, granulation (powder) or grate boilers.

Czech databases [6] were used in order to identify potentially suitable incineration sources with the rated thermal output exceeding 50 MW, in compliance with the fact that the original list was eliminated with regard to the technology used for incineration. The sources incinerating liquid or gaseous fuels, such as fuel oil and natural gas, were excluded. The selection was aimed especially at large sources of pollution, where the co-combustion of waste is legally possible. The total installed rated thermal output in the above-identified potentially suitable existing incineration plants is about 32.5 GW (figure including cement furnaces).

Since the beginning of 2013, the Czech Republic has slowly been shifting from the combustion and co-combustion of biomass to the combustion or co-combustion of waste.

3.2 *Software used to determine the emission limits*

The above-presented facts can raise interest of operators of stationary sources of air pollution in the co-combustion of waste, but they also cause some problems. The complex legislative system provides a complicated description and explanation of the necessity of setting emission limits, the amount of which is influenced by the amount and composition of the fuel and co-incinerated waste.

A suitable choice of this fuel mix can enable the operators of the sources of pollution to achieve more favorable emission limits than with separate incineration of basic fuel, which is certainly a motivating factor in the decision-making process regarding the use of this type of alternative fuel.

For a better orientation in this field and for a simple balance of the incineration process with the resulting value of the mixed emission limit, the Department of Power Engineering at the Technical University of Ostrava has created a software that is designed to calculate the specific emission limits for stationary sources engaged in the heat treatment of waste or fuel produced from waste (Alternative Solid Fuel according to Czech version of the European Standard EN 15357) together with the basic fuel, which, at the same time, are not defined as waste incineration plants or cement furnaces. The use of the software by operators of the sources of pollution is relatively simple and is based on the knowledge of the type of incineration plant, its rated heat input, the type and composition of the main incinerated fuel, including its calorific value, the type and composition of waste intended for incineration, including its minimum calorific value and its quantity, in order to determine the heat input supplied to the stationary source in this waste (the ratio of the source heat input). The actual amount of produced flue gases, which is crucial for determining the mixed concentrations and the resulting emission limits, is determined on the basis of stoichiometric (theoretical) calculations of the amount of flue gases resulting from the combustion of waste and fuel and their quantities (heat inputs in the fuel and waste).

The software establishes a new procedure to calculate the daily specific emission limits for HSP (Harmful Solid Particles), NO_x, SO_2, CO, HCl pollutants and the concentration of O_2 in flue gases at the output from stationary sources that are focused on the heat treatment of waste together with the main fuel.

4 CONCLUSION

The method of calculating the emission limits according to the above-presented formulas, representing the average values between emission limits determined for waste incineration plants and the values weighted according to the volume of flue gas produced during the co-combustion of fuel and waste, adheres to the existing legal regulations of the Czech Republic and the EU. It is based on Act no. 201/2012 of the Coll. "On the Air Protection" of the Ministry of Environment and the Decree of the Ministry of Environment of the Czech Republic no. 415/2012 of the Coll. "The Permissible Level of Pollution and its Detection".

The use of this special software will enable the operators of the sources of pollution to make faster decision when selecting a new fuel mix, and it will definitely contribute to a more rapid expansion of co-incineration of waste materials in practice, which will significantly reduce the amount of emissions and the amount of waste stored in landfills at the moment.

REFERENCES

[1] Council Directive no. 75/442/EHS on waste.

[2] MoE of the CR no. 201/2012 of the Coll.—Air Protection Act.

[3] MoE of the CR no. 415/2012 of the Coll.—Decree on the Permissible Level of Pollution and its Determination.

[4] Act no. 165/2012 of the Coll.—Act on Supported Energy Sources and on the Change of Several Acts.

[5] MoA of the CR—Action Plan for Biomass in the CR for the period of 2012–2020.

[6] ČHM—Register of Emissions and Air Pollution Sources REZZO (Register of Emissions and Air Pollution Sources).

Material Science and Environmental Engineering – Chen (Ed.)
 ISBN 978-1-138-02938-5

Identifying sources of heavy metal using the multivariate geostatistical method

D.X. Chen
Jinling College, Nanjing University, Nanjing, China

X.Y. Sui
Jiangsu Land Consolidation and Rehabilitation Center, Nanjing, China

S.L. Zhou
School of Geographic and Oceanographic Sciences, Nanjing University, Nanjing, China

ABSTRACT: The development of urbanization, industrialization and agricultural intensification has exerted increasing influences on the quality of soil environment. With Kunshan as the researched area, and by combining multivariate statistics and geostatistics, this paper studies the sources of 8 heavy metals in the soil of Kunshan. The result showed that the sewage irrigation in chemical industry is the major source of Hg, Cd and Pb pollutions, printing and dyeing industry is that of Cr and Ni pollutions, and electronic industry is that of Zn and Cu pollutions in the soil of Kunshan respectively; metallurgy and electroplating industries have influences on the Hg, Cd, Pb, Cr, Ni, Zn, and Cu pollutions in the soil of Kunshan; and the source of As is unknown, but is greatly related to the distribution of lake and water system.

Keywords: pollution sources; heavy metals; geostatistic; multivariate

1 INTRODUCTION

In the past several decades, heavy metal elements in the soil have attracted broad attention for their damages to ecosystem (Chai et al., 2004). Plentiful researches have been carried out on heavy metal pollutions in soil mostly with geostatistical method or classical statistical method. Traditional geostatistics could only treat single variable, but neglects the statistical dependence among multiple variables; while classical statistical method could only analyze the statistical dependence among multiple variables, but could not analyze the spatial correlation of multiple variables. Therefore, carrying out statistical dependence and spatial correlation analysis on the correlation of heavy metals in soil by combining classical statistics and geostatistics could disclose the rule of spatial distribution of heavy metals in soil, dig out the major elements affecting the spatial distribution of heavy metals, determine the sources of heavy metal pollutions in soil (Lin, 2002; Lv et al., 2014; Nazzal et al. 2014; Xie et al., 2011), and provide important data for risk evaluation of heavy metal pollutions in soil.

With Kunshan featured by the rapid development of urbanization, industrialization and agricultural intensification as the researched area, and by combining multivariate statistics and geostatistics, this paper studies the distribution characteristics and sources of heavy metals in the soil of Kunshan, determines their natural or artificial influential factors, and provides a scientific evidence for the sustained utilization and reasonable management of the soil resources in areas with rapid industrialization, urbanization and high agricultural intensification.

2 MATERIALS AND METHODS

2.1 *Soils sampling*

This paper selects Kunshan in the Yangtze River Delta Region with rapid economic development as the researched area, samples from 5 functional areas (chemical industry, printing & dyeing industry, smelting & electroplating industry, aquatic breeding industry, and pot garden industry) reflecting the urbanization, industrialization and agricultural intensification process as well as the major industrial pollution characteristics in Kunshan and having big possibility of potential pollutions in soil, and collects 124 soil sample points in total, as shown in Figure 1.

Figure 1. Distribution of sampling sites.

2.2 *Sample analysis*

Hg and As are tested with reduction gasification—Atomic Fluorescence Spectrometry; Cr, Cu, Ni, Pb and Zn are tested with three-acid (HF-HNO_3-$HClO_4$) digestion and inductively coupled plasma atomic emission spectroscopy (ICP); and Cd is tested with three-acid digestion and graphite furnace atomic absorption. The results of the above-mentioned tests all reflect the full quantity of heavy metals. Here, national geochemical standard sample is adopted for quality control.

2.3 *Multivariate statistics mapping*

This paper adopts multivariate statistical method (principal component analysis and cluster) to identify the homology and discrepancy of the major pollutions of heavy metals in the soil of Kunshan, and then on this basis, carries out multivariate statistics spatial mapping. Here, multivariate statistics spatial mapping is an important method for multi-factor mapping, and its basic thought is firstly to carry out principal component analysis on the content of heavy metal elements, extract the major factors with big contribution rate, and calculate the score of principal components; then to carry out spatial mapping with the data of principal component score and geostatistical Kriging interpolation method; and finally to analyze the sources and causes of the heavy metal pollutions according to the mapping information.

3 RESULTS AND DISCUSSIONS

3.1 *Heavy metal content statistical characteristics*

The result of statistical analysis on the content of heavy metals in the soil of Kunshan is as shown in Table 1. With local element background as evaluation standard, except that Cr has 48.39% sample points out of standard, all the other heavy metal elements have over 70% sample points out of standard, and where, the out-of-standard rate of Cd and As is up to 97.10% and 95.16% respectively. It shows that, the content of heavy metals in the soil of Kunshan is greatly disturbed by human activities; and with extraneous sources entering into the soil, the gathering degree of heavy metals is obvious. The above analysis proves that, the degree of Hg and Cd pollutions is the highest, and that of As, Pb and Zn pollutions is relatively serious; Cr is relatively close to soil background value, and has small out-of-standard rate; it could be considered that basically no extraneous source enters into the soil except for some point-source pollutions in local areas.

3.2 *Multivariate statistical analysis*

3.2.1 *Principal component analysis*

The accumulative variance contribution rate of 5 Factors has reached 85.39%, and could explain over 85% total variance. The characteristic root of the 5 factors extracted is over 1. Therefore, the 5 principal component factors could well represent all the heavy metal elements.

Through factor loading analysis (Table 2), we could get the major influential factors of the heavy metals in the soil. The content of Pb and Cd is mainly affected by factor 1, that of Cr and Ni by 2, that of Cu and Zn by 3, that of Hg by 4, and that of as by 5.

3.2.2 *Cluster analysis*

For further inspecting the extraction result of influential factors, cluster is adopted to analyze the

Table 1. Heavy metal content statistical characteristics.

Metals	Range ($mg\ kg^{-1}$)	Average value ($mg\ kg^{-1}$)	Background value ($mg\ kg^{-1}$)	Out of standard (%)
HG	0.01~0.92	0.26	0.163	83.87
As	6.07~24.22	12.86	8.80	95.16
Cr	39.6~136.9	68.67	65.72	48.39
Cu	13.7~72.50	28.53	22.78	70.97
Ni	19.4~74.30	35.88	29.12	90.32
Pb	12.14~83.8	28.22	20.39	76.61
Zn	47.8~332.4	103.33	73.03	88.71
Cd	0.07~2.72	0.22	0.116	97.10

Table 2. Factor loading matrix.

	Rotation factors				
Metals	Factor 1	Factor 2	Factor 3	Factor 4	Factor 5
HG	0.311	0.056	0.194	0.849	−0.109
As	0.061	0.077	−0.028	−0.065	0.982
Cr	−0.057	0.842	0.065	0.403	0.053
Cu	0.325	0.168	0.654	0.374	0.166
Ni	0.202	0.879	0.177	−0.200	0.062
Pb	0.854	0.122	0.289	0.044	0.045
Zn	0.095	0.108	0.909	0.045	−0.110
Cd	0.828	0.020	0.026	0.317	0.041

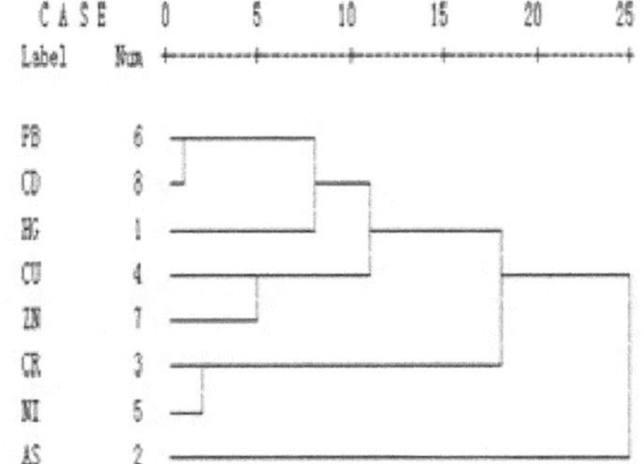

Figure 2. Cluster of heavy metals.

mutual relationship of heavy metal elements. As shown in Figure 2, for strong mutual correlation, Pb, Cd and Hg are formed into one type, so are Ni and Cr, and Zn and Cu respectively; whilst for weak correlation of As and other elements, they form one type respectively. The result of cluster analysis and that of principal component analysis is roughly coherent, except for slight discrepancy in the classification of Hg.

3.2.3 *Principal component score mapping*

Principal component is usually the reflection of dominant influential factor in some space. Ordinary kriging interpolation in geostatistical methods is applied to generate the spatial distribution chart of each component score within Kunshan (Fig. 3). Also, GIS tool is applied to carry out superposition analysis on the distribution chart of major rivers, lakes, and key polluting enterprises in Kunshan.

The high-value area distribution of Factor 1 basically has close relationship with the distribution of key industrial pollution-discharging enterprises from chemical industry and metallurgy industry in Kunshan. According to related materials, chemical production is easy to produce Hg and Pb pollutions, whilst metallurgy industry is easy to generate Cd, Pb and Cr pollutions. Therefore, it could be considered that, chemical and metallurgy enterprises are the major influential factors of Factor 1.

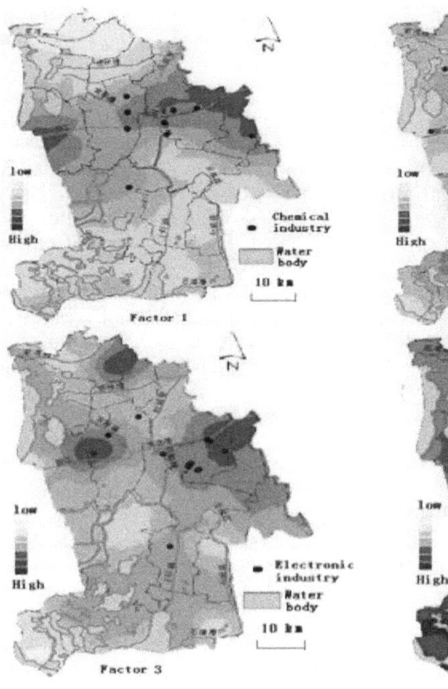
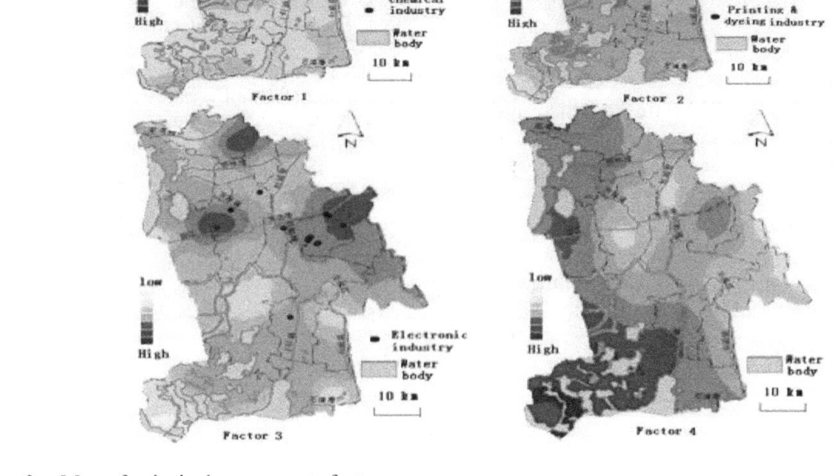

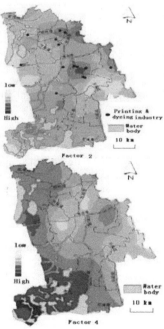

Figure 3. Map of principal components factors.

The score of factor 2 presents island-shaped distribution after interpolation. By superposing the distribution chart of printing & dyeing, electroplating, and metallurgy enterprises among key polluting enterprises in Kunshan and the spatial distribution chart of Factor 2, as shown in Figure 3, we find that the high-value area of Factor 2 has close relationship with the distribution of the abovementioned 3 types of enterprises. By combining with the field investigation in Kunshan, we find that the island-shaped high-value area tallies with the former special electroplating area in Kunshan. Therefore, it could be considered that, Factor 2 is mainly affected by the point-source pollutions of printing & dyeing, electroplating and metallurgy enterprises.

The score of factor 3 presents island-shaped distribution. By superposing the distribution chart of electronic, electroplating, and metallurgy enterprises among key polluting enterprises in Kunshan and the spatial distribution chart of factor 3, as shown in Figure 3, we find that the high-value area of factor 3 has close relationship with the distribution of the abovementioned 3 types of enterprises. The field investigation in Kunshan shows that, the content of Cu and Zn in the soil nearby partial aquatic breeding enterprises is relatively high. Therefore, factor 3 is mainly affected by the point-source pollutions of electronic, electroplating and metallurgy enterprises in Kunshan, and is affected by the livestock breeding in agriculture in partial areas.

Factor 4 is as shown in Figure 3. The high-value area of As is distributed along the compact district of water network, like Yangcheng Lake, Kuilei Lake, Bailian Lake, and Baixian Lake, etc. Spatial distribution has extremely big spatial correlation with the distribution of lake and water system, and it's because that As is easy to move with water, and to gather in the soil along with agricultural irrigation.

4 CONCLUSIONS

With combination of multivariate statistics and geostatistics, this paper analyzes the spatial distribution pattern and sources of heavy metal elements in the soil of Kunshan, and draws the following conclusion:

1. The sewage irrigation in chemical industry is the major source of Hg, Cd and Pb pollutions, printing and dyeing industry is that of Cr and Ni pollutions, and electronic industry is that of Zn and Cu pollutions in the soil of Kunshan respectively; metallurgy and electroplating industries have influences on the Hg, Cd, Pb, Cr, Ni, Zn and Cu pollutions in the soil of Kunshan.
2. The source of As pollution in the soil of Kunshan is unknown, but its distribution is affected by the distribution of river and lake, and generally, the content of As in compact districts of lake and river is relatively high.
3. The combination of multivariate statistics and geostatistics, with its high-efficiency information integration ability, is an effective method for studying the distribution pattern of soil elements and analyzing pollution sources.

ACKNOWLEDGMENT

This work is supported by the National Basic Research Program of China (973 Program) (No. 2002CB410810) and National Science Fund for Fostering Talents in Basic Science, Grant No. J0630535.

REFERENCES

[1] Chai, S., Wen, Y., Wei, X., Zhang, Y., Dong, H. & Chen, Y. 2004. Heavy metal content characteristics of agricultural soils in the Pearl River Delta. *Acta Scientiarum Naturalium Universitatis Sunyatseni* 43: 90–94.

[2] Lin, Y. 2002. Multivariate geostatistical methods to identify and map spatial variations of soil heavy metals. *Environmental Geology* 42: 1–10.

[3] Lv, J., Liu, Y., Zhang, Z. & Dai, B. 2014. Multivariate geostatistical analyses of heavy metals in soils: Spatial multi-scale variations in Wulian, Eastern China. *Ecotoxicology and environmental safety* 107: 140–147.

[4] Nazzal, Y., Ghrefat, H. & Rosen, M.A. 2014. Application of multivariate geostatistics in the investigation of heavy metal contamination of roadside dusts from selected highways of the Greater Toronto Area. Environmental earth sciences 71(3): 1409–1419.

[5] Xie, Y., Chen, T.B., Lei, M., Yang, J., Guo, Q.J., Song, B. & Zhou, X.Y. 2011. Spatial distribution of soil heavy metal pollution estimated by different interpolation methods. Accuracy and uncertainty analysis. *Chemosphere* 82(3): 468–476.

Material Science and Environmental Engineering – Chen (Ed.)

Cd removal in laboratory scale of unvegetated vertical subsurface flow constructed wetland using mixed substrates

S.X. Gao, J. Ren, J.X. Hao, L. Dai & L. Tao
School of Environmental and Municipal Engineering, Lanzhou Jiaotong University, Lanzhou, Gansu, China

ABSTRACT: In this study, five mixed substrates namely SSFGF, FSSGF, FSSFF, FSSFG, and FSFGF, consisting of fly ash, sludge, soil, fine cinder, gravels, and fine sand, was made in order to test the removal effect of cadmium at five initial concentrations. Multiple comparisons and analysis of variance were used to analyze the removal efficiency of cadmium. The test results show that for five substrates, the effluent concentrations are significantly different among five initial concentrations. Moreover, when the decomposition time, the removal efficiency of Cd from the solution is better for substrate SSFGF-based wetland than those at three initial concentrations (5 mg/L, 20 mg/L, 30 mg/L, and 50 mg/L), and then it is slightly better for FSSFF and FSSGF. At an initial concentration of 10 mg/L, the removal efficiency is better for substrate FSSFF-based wetland than those and then it is slightly better for SSGFF.

Keywords: cadmium; constructed wetland; initial concentration; mixed substrate; wastewater

1 INTRODUCTION

Heavy metals present a serious threat to our environment as hazardous pollutants because of their toxicity, persistence, and bioaccumulation problems. Cadmium is a non-essential element and one of the most hazardous trace elements, being considered a "priority metal" from the standpoint of potential hazard to human health (Ayuso & Sanchez 2007).

Recently, the use of constructed wetlands systems to remove metals is an area of interest to many researchers. Currently, numerous studies have investigated the removal efficiency of constructed wetlands in varieties of wastewater (Bragato et al. 2006, Carleton et al. 2000, Jan 2005, Khan et al. 2009, Lan et al. 1992, Mays & Edwards 2001, Mitsch & Wise 1997). The results have shown that constructed wetlands are very effective in removing heavy metals from polluted wastewaters including aluminum, zinc and chromium, lead, cadmium, iron, nickel, chromium, and copper.

Several studies have shown many materials were used as filter media to treat wastewater polluted by heavy metal. It is reported that many materials such as laterite, coke and gravel, cocopeat, zeolite, and limestone were efficient to remove heavy metals (Allende et al. 2011, Chen et al. 2009, Wood & McAtamney 1996). Other studies proved that the plants did not contribute to total heavy metal removal to a higher extent (Galletti et al. 2010, Scholz & Xu 2002). The overall performance of a CW is, therefore, primarily dependent on the selection of suitable substrates.

The use of mixed substrates in constructed wetlands has been suggested by different researchers, with the aim of improving removal performance. For example, it shows that six different packing orders of filter media including gravel, sand, granular activated carbon, charcoal, and Filtralite (light expanded clay) were applied to removal of lead and copper of simulating pretreated mine wastewater in vertical-flow wetlands (Scholz & Xu 2002). Nevertheless, little research in this area has been reported. This study aims at exploring the performance of five mixed substrates: fly ash, sludge, soil, gravel, fine cinder, and fine sand in the removal of Cd using vertical-flow constructed wetland systems and identifying some substrate configuration and optimal combination.

2 MATERIALS AND METHODS

2.1 *The lab-scale unvegetated constructed wetland systems*

The lab-scale constructed wetlands, located in Lanzhou Jiaotong University of Lanzhou China, were constructed. Five wetlands were filled with five different substrates (mixed substrates) with an external packed bed size of 0.8 m × 1 m × 1 m (length × width × height). Each unit had a drainage layer of stones at the base, which was 0.1 m deep. The drainage layer was topped with a layer of mixed substrates that was 0.7 m deep. In the bottom of each unit, the systems were installed polyvinylchloride pipe (PVC) of 0.019 m in diameter to effluent. The constructed wetlands were all fed with cadmium wastewater from the same source through flexible tubing of 0.013 m in diameter from the top of the wetlands. The mean Hydraulic Loading Rate (HLR) was kept at 0.5 $m^3/m^2 \cdot d$ across the five systems. After being treated by the CWs, the heavy metal wastewater flowed out of the system through the effluent pipe.

2.2 *Materials*

Six substrate materials, namely, fly ash, sludge, soil, gravel, fine cinder and fine sand were selected for this study. Soil was obtained from the campus of school of environmental and municipal engineering of Lanzhou Jiaotong University, Lanzhou City, China. Fine sand was obtained from Shapotou district, Ningxia Province, China. Gravel was obtained from construction site, Lanzhou city, China. Fine cinder was obtained from boiler room of Lanzhou Jiaotong University, Lanzhou City, China. Sludge was obtained from sewage treatment plant of Lanzhou, Lanzhou City, China. Fly ash was purchased. Five different packing configuration of filter substrate were applied to the five vertical-flow wetland systems (Table 1).

2.3 *The artificial wastewater and Instruments*

Aqueous solutions containing cadmium ions (Cd^{2+}) at various concentrations (5 mg/L, 10 mg/L, 20 mg/L, 30 mg/L and 50 mg/L) were prepared by dissolving cadmium sulfate ($Cd(NO_3)_2 \cdot 5H_2O$) powders in deionized water. $Cd(NO_3)_2 \cdot 5H_2O$ was AR grade.

Water samples were collected at the following time intervals: 0.25 h, 0.5 h, 1 h, 2 h, 4 h, 8 h, 12 h, 24 h and 48 h. The concentration of the metals was detected by inductively-coupled plasma atomic emission spectrometry (Varian, IRIS Intrepid II XSP, USA).

Table 1. Packing configuration of vertical-flow constructed wetlands (kg).

Substrate	Fly ash	Sludge	Soil	Fine cinder	Gravel	Fine sand
SSFGF	0	130	230	160	310	286
FSSGF	150	130	230	0	310	286
FSSFF	150	130	230	160	0	286
FSSFG	150	130	230	160	310	0
FSFGF	150	130	0	160	310	286

2.4 *Statistical analyses*

Analysis of variance (ANOVA) for the data was performed on all data sets. Duncan's test was used to differentiate means where appropriate. A level of $P < 0.05$ was used in all comparisons. Statistical analysis was performed using the STATISTICA software (Statsoft 1993).

3 RESULTS AND DISCUSSION

3.1 *Cd removal efficiency of mixed substrates*

Result indicated effluent concentration of Cd from the five substrates were affected, however, the effects varied with substrates for Cd level at five initial concentration and after 48 h operation (Table 2). It was found that for five substrates there was significant difference among five initial concentrations ($P < 0.001$).

At the Cd initial concentration of 5 mg/L, the Cd concentration was not significantly different between substrate SSFGF, FSSGF, and FSSFF, and was significantly higher than that in the other two substrates. The Cd concentration was not significantly different between substrate FSSFG and FSFGF, and it was significantly lower than that in the other three substrates.

At the Cd initial concentration of 10 mg/L, there was significant difference among the five substrates. The Cd concentration was significantly

Table 2. Cd concentration of effluent from unvegetated vertical subsurface flow constructed wetland using mixed substrates.

Initial concentration (mg/L)	SSFGF	FSSGF	FSSFF	FSSFG	FSFGF	F-value
5	1.77a	1.68a	1.80a	1.37b	1.40b	11.138***
10	1.34a	5.56b	1.60c	2.60d	0.95e	9959.812***
20	1.60a	3.35b	1.93c	4.70d	5.21e	7103.887***
30	0.46a	1.29b	1.40c	1.55d	1.99e	6923.267***
50	0.42a	1.33b	3.04c	3.22d	2.96c	556.210***

Data for this table are the average of six repetitions. Data in the same line followed by the same letter are not significantly different at 5% level of probability (Duncan's multiple comparisons test). F-ratios indicating statistical significance at ***$P < 0.001$, **$P < 0.01$ and *$P < 0.05$.

greater in FSSGF than that in the other four substrates. The Cd concentration was significantly lower in FSFGF than that in the other four substrates. The Cd concentration was lower in substrate SSFGF, FSSFF and FSSFG than that in substrate FSSGF and greater than that in FSFGF. The amounts of Cd decreased in the sequence of substrate FSSGF > substrate FSSFG > substrate FSSFF > substrate SSFGF > substrate FSFGF after 48 h operation.

At the Cd initial concentration of 20 mg/L, there was significant difference among the five substrates. The Cd concentration was significantly greater in FSFGF than that in the other four substrates. The Cd concentration was significantly lower in SSFGF than that in the other four substrates. The Cd concentration was lower in substrate FSSGF, FSSFF and FSSFG than that in substrate FSFGF and greater than that in SSFGF. The Cd concentration was lower in substrate FSSFF than that in substrate FSSGF, FSSFG and FSFGF and greater than that in SSFGF. The amounts of Cd decreased in the sequence of substrate FSFGF > substrate FSSFG > substrate FSSGF > substrate FSSFF > substrate SSFGF after 48 h operation.

At the Cd initial concentration of 30 mg/L, there was significant difference among the five substrates. The Cd concentration was significantly greater in FSFGF than that in the other four substrates. The Cd concentration was significantly lower in SSFGF than that in the other four substrates. The Cd concentration was lower in substrate FSSGF, FSSFF and FSSFG than that in substrate FSFGF and greater than that in SSFGF. The amounts of Cd decreased in the sequence of substrate FSFGF > substrate FSSFG > substrate FSSFF > substrate FSSGF > substrate SSFGF after 48 h operation.

At the Cd initial concentration of 50 mg/L, there was significant difference among the five substrates. The Cd concentration was greater in FSSFG than that in the other four substrates. The Cd concentration was lower in SSFGF than that in the other four substrates. The Cd concentration was greater in FSSGF than that in SSFGF and lower than that in the other three substrates. The amounts of Cd decreased in the sequence of substrate FSSFG > substrate FSSFF > substrate FSFGF > substrate FSSGF > substrate SSFGF after 48 h operation.

3.2 *Cd removal process of mixed substrates*

The percentage removal of Cd from solution by the five substrates at five different initial Cd concentrations varied with operation time (Fig. 1). At the beginning of 4 h, all the systems exhibited very greater fluctuations.

During the first 18 h, the removal efficiency of substrate SSFGF-based wetland was higher than those of the other substrates-based wetlands, and the slightly better removal efficiency was found in the substrate FSSFF-based wetland. Then until the ultimate time it was better in substrate FSFGF-based wetland than those. During the whole decomposition time, all the removal efficiencies of five substrates differently decreased (Fig. 1a).

Before the first 17 h, the higher removal efficiency of Cd was found in the substrate FSSFF-based wetland, it reached the maximum 98.41% at 8 h operation time, subsequently the removal efficiency decreased reaching the minimum on time 24 h and then increased in the following time. Afterwards, the slightly better removal efficiency was found in the substrate FSFGF-based wetland before 17 h and the removal efficiency differently increased and reached the maximum values on time 48 h during the whole decomposition time. The removal efficiency of substrate SSFGF-based wetland differently increased and reached the maximum values on time 24 h, and it was higher than those after 22 h (Fig. 1b).

During the decomposition period, all the constructed wetlands performed well and the removal efficiencies were almost all above 80%. And the

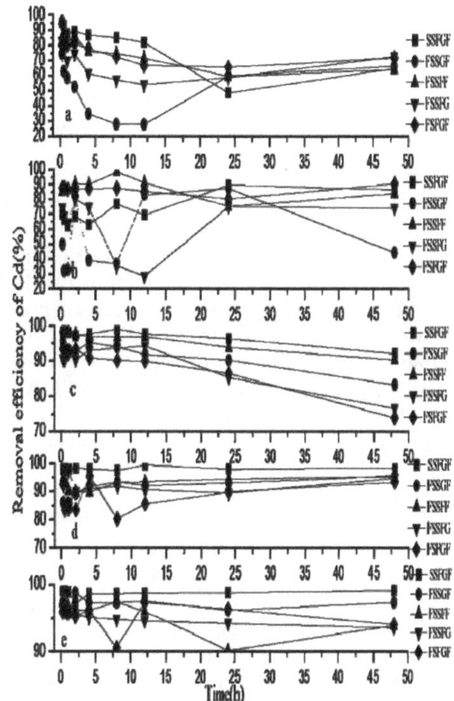

Figure 1. Removal of Cd in 5 mg/L (a), 10 mg/L (b), 20 mg/L (c), 30 mg/L (d) and 50 mg/L (e) initial concentrations under unvegetated vertical subsurface flow constructed wetland using mixed substrates.

efficiency decreased differently and reached the minimum values on time 48 h. Before 2 h, the removal efficiency of the substrate SSFGF-based wetland was lower than that of the FSSFF-based wetlands, in the following time it was better in the latter (Fig. 1c).

During the decomposition period, the removal efficiency of substrate SSFGF-based wetland was obviously higher than those of the other substrates-based wetlands. At the beginning of 1 h, the removal efficiency was greater in the substrate FSSFG-based wetland, and then it decreased greatly and was lower than that in SSFGF and FSSFF after 8h. After 8 h, the slightly better removal efficiency w as found in the substrate FSSFF-based wetland and it continually increased (Fig. 1d).

During the whole 48 h decomposition periods, all the constructed wetlands performed well and the removal efficiencies were almost all above 90%. The removal efficiency of substrate SSFGF-based wetland was better than those of the other substrates-based wetlands. Before 2 h, it decreased differently; subsequently it increased continually and reached the maximum values on 48 h. Next, the greater removal efficiency was found in substrate FSSGF-based wetland, and it increased differently and reached the maximum values on 2 h (Fig. 1e).

4 CONCLUSIONS

For five substrates, all the effluent concentrations were significantly different among five initial concentrations (5 mg/L, 10 mg/L, 20 mg/L, 30 mg/L and 50 mg/L) from five substrates.

The removal efficiency of Cd from the solution was better for substrate SSFGF-based wetland than those of the other substrates-based wetlands at three initial concentrations (5 mg/L, 20 mg/L and 30 mg/L), and then it was slightly better for substrate FSSFF-based wetland. At initial concentration of 10 mg/L, the removal efficiency was better for substrate FSSFF-based wetland than those of the other substrates-based wetlands, and then it was slightly better for substrate FSFGF-based wetland. At relatively high initial concentration of 50 mg/L, it was better for substrate SSFGF-based wetland than those of the other substrates-based wetlands and then it was slightly better for substrate FSSGF-based wetland.

ACKNOWLEDGMENTS

This project is supported by the Program for Changjiang Scholars and Innovative Research Team in the University of Ministry of Education of China (No. IRT0966).

REFERENCES

[1] Allende, L.K., Fletcher, T.D. & Sun, G. 2011. Enhancing the removal of arsenic, boron and heavy metals in subsurface flow constructed wetlands using different supporting media. *Water Science and Technology* 63(11): 2612–2618.

[2] Ayuso, A.E. & Sanchez, G.A. 2007. Removal of cadmium from aqueous solutions by palygorskite. *Journal of Hazardous Materials* 147: 594–600.

[3] Bragato, C., Brix, H. & Malagoli, M. 2006. Accumulation of nutrients and heavy metals in Phragmitesaustralis (Cav.) Trin. ex Steudel and Bolboschoenus maritimus (L.) Palla in a constructed wetland of the Venice lagoon watershed. *Environmental Pollution* 114(3): 967–975.

[4] Carleton, J.N., Grizzard, T.J., Godrej, A.N., Post, E.H., Lampe L. & Kenel, P.P. 2000. Performance of a constructed wetlands in treating urban stormwater runoff. *Water Environmental Research* 72(3): 295–304.

[5] Chen, M.Z., TANG, Y.Y., LI, X.P. & YU, Zh.X. 2009. Study on the Heavy Metals Removal Efficiencies of Constructed Wetlands with Different Substrates. *Journal of water Resource and Protection* 1: 1–57.

[6] Galletti, A., Verlicchi, P. & Ranieri, E. 2010. Removal and accumulation of Cu, Ni and Zn in horizontal subsurface flow constructed wetlands: Contribution of vegetation and filling medium. *Science of the Total Environment* 408: 5097–5105.

[7] Jan, V. 2005. Removal of Heavy Metals in a Horizontal Sub-Surface Flow Constructed Wetland. *Journal of Environmental Science and Health* 40: 1369–1379.

[8] Khan, S., Ahmad, I., Shah, T.M., Rehman, S. & Khaliq, A. 2009. Use of constructed wetland for the removal of heavy metals from industrial wastewater. *Journal of Environmental Management* 90: 3451–3457.

[9] Lan, C., Chen, G., Li, L. & Wong, M.H. 1992. Use of cattails in treating wastewater from a Pb/Zn mine. *Journal of Environmental Management* 16(1): 75–80.

[10] Mays, P.A. & Edwards, G.S. 2001. Comparison of heavy metal accumulation in a natural wetland and constructed wetland receiving acid mine drainage. *Ecological Engineering* 16(4): 487–500.

[11] Mitsch, W.J. & Wise, K.M. 1997. Water quality, fate of metals, and predictive model validation of a constructed wetland treating acid mine drainage. *Water Research* 6(32): 1888–1900,

[12] Scholz, M. & Xu, J. 2002. Performance comparison of experimental constructed wetlands with different filter media and macrophytes treating industrial wastewater contaminated with lead and copper. *Bioresource Technology* 83: 71–79.

[13] Wood, R.B. &McAtamney, C.F. 1996. Constructed wetlands for waste water treatment: the use of laterite in the bed medium in phosphorus and heavy metal removal. *Hydrobiologia* 340: 323–331.

Material Science and Environmental Engineering – Chen (Ed.)
ISBN 978-1-138-02938-5

Experimental research on fine particles removal from flue gas by foaming agglomeration

X. Jiang, C.P. Wang, G. Li & H.Y. Li
Energy Engineering Research Institution, Qingdao University, Qingdao, China

ABSTRACT: Fine particles emission to the atmosphere from coal combustion needs to be deeply removed by higher efficient way. The chemical agglomeration technology was used in this paper to aggregate the fine particles by foam layer, and the aggregation and removal efficiency was the main target. The factors, such as foam agent type, solution concentration, flue gas temperature and stabilizer affected the efficiency of particles aggregation were investigated. The removal efficiency changing trends of the SDBS and SDS foaming agent are different with the flue gas temperature increasing, and the solution concentration influences more than the flue gas temperature. The adding of stabilizer is beneficial to the improvement of the removal efficiency. The highest aggregation and removal efficiency of fine particles can reach 59% in the research.

Keywords: fine particle; aggregation removal; removal efficiency; foaming agent; stabilizers

1 INTRODUCTION

The fly ash produced by burning coal has become one of the main sources of atmospheric particulates. The surface of inhalable particles (PM_{10}) and fine particulate matters ($PM_{2.5}$) from burning coal enriches trace heavy metal elements. After these particulate matters enter the atmosphere, it will cause serious damage to the environment and human health (Shi et al. 2014). At present, the removal efficiency of larger dust particles can reach 99% with the application of Electrostatic Precipitator (ESP) or water film dust catcher in power plant, but the capture efficiency of PM_{10} and $PM_{2.5}$ is much lower. At the front of dust removal equipment in coal-fired power plant, the mass percentage of PM_{10} is about 8%. However, at the back of dust removal equipment, almost all particulate matters are PM_{10}, and the mass percentage of $PM_{2.5}$ accounts for more than 50% (Lu et al. 2014).

Physical aggregation method includes acoustic agglomeration, light aggregation, electrocoagulation, phase transformation, etc. (Dai et al. 2011). Chemical agglomeration technology is able to effectively promote agglomeration of fine particles from coal combustion and achieve synergistic removal of various pollutants. The key of this technology is making agglomeration agent inject at the front of final dust removal equipment, which promotes fine particles aggregate and grow up to large particles. The technology of spray flocculation for removing larger particles has been early applied in the mining industry, namely spray dedusting (Heidenreich et al. 1995). To increase the contact area between solid and liquid, aggregation solution is needed to form foam layer, or adding flocculants to promote fine particles attach to the surface of foams. Some of the scientific research units in world have carried out preliminary research in this field and determined the commonly used agglomeration solutions, flocculants, etc. Usually these reagents are polymers, such as Sodium Dodecyl Sulfate (SDS), nonionic polyacrylamide, etc., (Jiang et al. 2002).

For coal-fired boilers, the flue gas flow is huge, physical agglomeration method is not suitable (Rozpondek et al. 2009). Chemical agglomeration technology has the advantages of low operating costs and high removal efficiency, and may be used for large flue gas flow. In this paper, the characteristics of collision and agglomeration between fine particles from coal firing and foam layers were studied, the improvement of aggregation and removal efficiency was the main target. The factors, such as foam agent type, solution concentration, flue gas temperature and stabilizer et al. affected the efficiency of particles aggregation were investigated, in order to develop a new control technology of particulate matters with desulfurization and provide the reference for the industrial application of chemical agglomeration technology.

2 EXPERIMENTAL RESEARCH

2.1 *Particle size analysis of sample*

Fine particulate matters used in experiment were derived from fly ash, which was collected after ESP in power plant. The particle size distribution is shown in Figure 1, the maximum particle size of sample is lesser than 10 μm, and the mass percentage of $PM_{2.5}$ is about 33.45%. The particles whose sizes are about 5 μm account for more than a half, the percentage is about 53%. Therefore, particles used in experiment belong to the fine particles that can not be captured efficiently in traditional dust removal devices; the following experimental research about fine particles removal from flue gas by foaming agglomeration is highly significant to reduction of air pollution.

2.2 *Foaming agent and experimental system*

In the experiment, Sodium Dodecyl Sulfate (SDS) Solution and Dodecyl Benzene Sulfonate (SDBS) solution were used as foaming agent, these solutions could produce abundant foams, and the foam layers played a key role to the aggregation of fine particles. Coconut diethanolamide and triethanolamine were used as stabilizers, these agents were conducive to the stability of foams.

According to the reference (Zhao et al. 2007), foam solutions were prepared as follows: the reagents used in experiment were respectively weighed and dissolved in a certain amount of distilled water, dodecyl benzene sulfonate (SDBS) solution and Sodium Dodecyl Sulfate (SDS) solution were respectively prepared with the concentration of 0.3%, 0.4%, 0.5%, 0.6% and 0.7%. Three kinds of composite solutions were respectively composed of 0.3% SDS solution and 0.1%, 0.2%, 0.3% coconut diethanolamide solution, and another three composite solutions were respectively composed of 0.3% SDS solution and 0.1%, 0.2%, 0.3% triethanolamine solution.

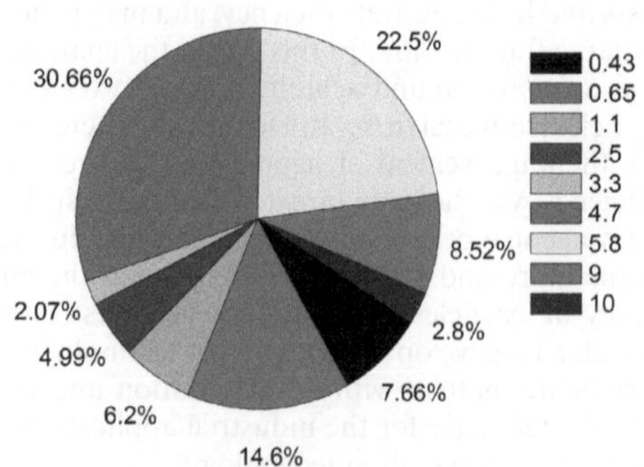

Figure 1. Particle size distribution of sample.

The experimental schematic diagram is shown in Figure 2. The experimental system mainly includes 1-Water tank, 2-Pump, 3-Nozzle, 4-Flue gas channel, 5-Air distributor, 6-Fly ash conveying equipment, 7-blower, 8-collection bottle, etc. In the experimental process, foaming agents are injected into the flue gas channel by the nozzle, and contacted with ash particles from the channel bottom, and then fine particles are aggregated and removed.

2.3 *Experimental procedure*

(1) Filter papers that had been marked respectively were transferred into draught drying cabinet. After heated and dried, the weight of each filter paper was weighed and recorded. (2) Foaming solution was introduced into water tank, sufficient fly ash particles were put into the conveying equipment, and an appropriate amount of distilled water was added in collection bottle. (3) After the flue gas channel was heated to the designed temperature, the pump was opened to make foaming agent inject into the gas channel, then the blower was started to blow ash particles into gas channel. (4) When ash particles went through the gas channel, they were contacted with foam layer, part of the particles were aggregated with each other and trapped in the collection bottle, the residual fine particles overflowed from the water with flue gas. The operation was stopped after 10 minutes, the rubber tube was taken out from collection bottle, and the blower was stopped. (5) A piece of filter paper was used and put in buchner funnel, the circulating water vacuum pump, buchner flask and buchner funnel were connected well, the ash particles in collection bottle were collected by leaching. (6) After leaching collection, the filter paper was heated and dried in draught drying cabinet for

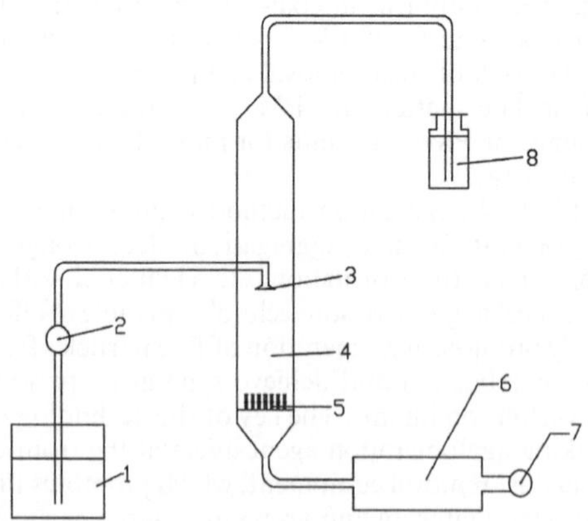

Figure 2. Schematic diagram of experimental system.

1 h, and the weight of filter paper was weighed and recorded, then the recorded data was used to minus the original weight of filter paper, the calculated difference was the weight of collected ash particles. (7) On other conditions, the above process was repeated, different kinds of foaming agents were injected into the gas channel or the channel was set at different temperatures. Without any foaming agent, distilled water in collection bottle was only used to capture ash particles, because more particles could overflow from the water with flue gas, the weight of particles collected in bottle is lower.

For the convenience of explaining particle removal efficiency, the weight of particles collected in bottle without using any foaming agent is defined as m_1, on the contrary, with using foaming agent, the weight of particles collected in bottle is defined as m_2, so the increase in weight which is caused by agglomeration effect of foaming agent is $m_2 - m_1$. Therefore, the removal efficiency of foaming agent can be calculated by Equation 1, where η = the removal efficiency.

$$\eta = \frac{m_2 - m_1}{m_1} \times 100\% \tag{1}$$

3 RESULTS AND ANALYSIS

3.1 *The influence of flue gas temperature*

Because the flue gas temperature is about 100°C after wet desulfurization, and after dry or semi dry desulfurization the gas temperature is a little higher, the gas temperatures in the experiment were chosen as 100°C, 110°C, 120°C, 130°C, and 140°C, respectively. The removal efficiency of fine particles, which is varied with foaming agent concentrations, is showed in Figures 3 and 4.

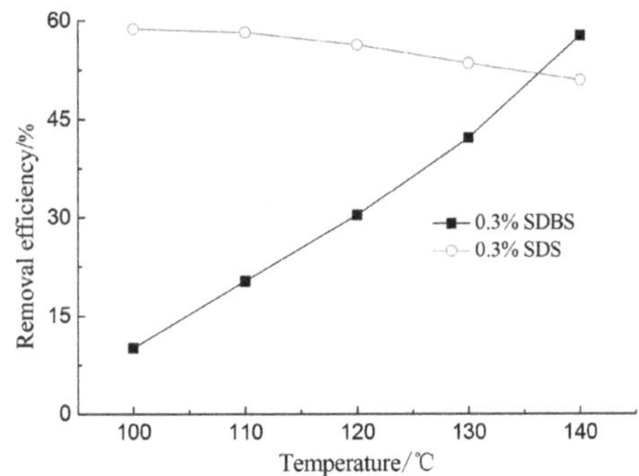

Figure 3. Removal efficiency of 0.3% foaming solutions.

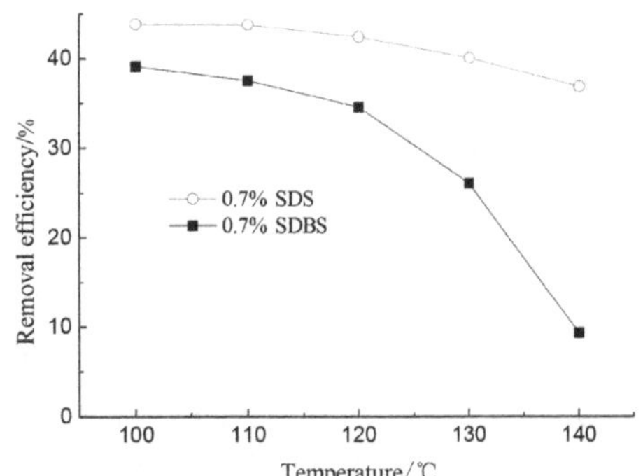

Figure 4. Removal efficiency of 0.7% foaming solutions.

As shown in Figure 3, the removal efficiency of SDS solution is slightly influenced by the flue gas temperature, and it decreases a little with the temperature increasing and is always about 55%. However, the removal efficiency of SDBS solution is seriously affected by gas temperature. In the experimental temperature range, the foaming rate of SDBS dilute solution increases with temperature rising, so the removal efficiency of this foaming agent increases rapidly as temperature rise. Even when the temperature reaches 140°C, the removal efficiency of SDBS solution is higher than the one of SDS solution. It can therefore be argued that the SDBS solution with 0.3% concentration is more suitable to aggregate and remove fine particles at higher temperature.

When the solution concentration was increased to 0.7%, the removal efficiency is shown in Figure 4. Through comparison between Figure 3 and Figure 4, it is showed that with the concentration increasing to 0.7%, the removal efficiency of SDS solution is significantly lower than the one in Figure 3. And unlike the features in Figure 3, the removal efficiencies of two solutions present the same trend with the temperature changing. The removal efficiency of SDS solution still decreases slightly with the temperature increasing and is about 40%, which decreases by 15% than the one in Figure 3. This may be related to the foaming rate of SDS solution that decreases with increasing concentration.

The removal efficiency of SDBS solution whose concentration is 0.7% decreases obviously with the rise of temperature, which is opposite to the efficiency changing trend of SDBS solution that is showed in Figure 3. And in Figure 4 the highest removal efficiency of SDBS solution is only about 40%, it is lower than the one of SDBS solution in Figure 3. Based on the above analysis, the concentration of two foaming solutions has more influence on the removal efficiency than the temperature. At higher temperature, the removal efficiency of lower concentration solution is greater.

3.2 *The influence of solution concentration*

The flue gas temperature is set at 100°C, the removal efficiencies under different solution concentrations are compared in Figure 5.

In concentration range from 0.3% to 0.7%, the removal efficiency changing trends of two solutions are opposite. The removal efficiency of SDS solution rapidly decreases with concentration increasing, and the change of removal efficiency is no longer apparent when the concentration is greater than 0.5%. On the contrary, the efficiency of SDBS solution rapidly increases with the concentration rising, but the highest efficiency of this solution is still lower than the lowest efficient of SDS solution. In other words, at low gas temperature, the SDS solution is more effective to aggregate and remove particles than the SDBS solution. But the highest removal efficiency of SDS solution is only 58%. Hence, it is necessary to study the effect of stabilizers.

3.3 *The influence of stabilizer concentration*

In order to further improve the removal efficiency of foaming solution, two commonly used stabilizers, namely coconut diethanolamide and triethanolamine, were added into the 0.3% SDS solution, and when the gas temperature was set at 140°C, the effect of their concentrations on removal efficiency was investigated. As shown in Figure 6, two kinds of stabilizers both play a role in improving the removal efficiency of fine particles. The efficiency of SDS solution improves with the concentration of two stabilizers increasing, however the increasing rates are different. With concentration of coconut diethanolamide increasing, the removal efficiency grows slowly and steadily, the increment is small, only about 3.5%. However, the efficiency is improved observably with the increasing concentration of triethanolamine. Especially when 0.3% triethanolamine stabilizer is used, the increasing rate reaches the highest value, and the biggest increment of efficiency is about 10%. The highest removal efficiency of the fine particles reaches 59% in the experiment.

The above experimental results show that, for the unit that has reached the dust emission standard of 50 mg/m^3, it is feasible to further reduce the emissions of fine particles and reach the new environmental standard of 20 mg/m^3 when chemical agglomeration technology is applied.

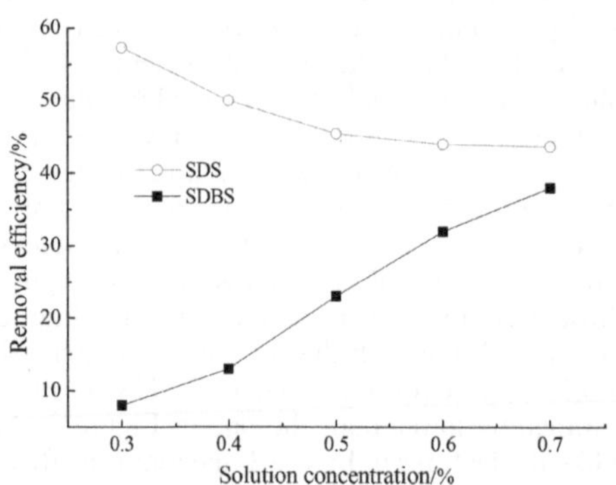

Figure 5. The influence of foaming agent concentration.

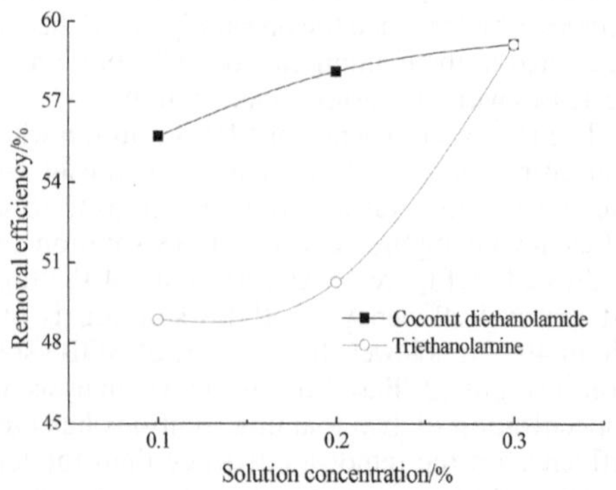

Figure 6. The influence of stabilizer concentration.

4 CONCLUSIONS

1. The removal efficiencies of two commonly used foaming solutions are differently affected by flue gas temperature. The removal efficiency of SDBS solution is significantly affected by gas temperature. The removal efficiency of 0.3% SDBS dilute solution increases with gas temperature rising, but the efficiency changing trend of 0.7% SDBS solution is opposite to it. The removal efficiency of SDS solution is slightly influenced by the gas temperature. In the experimental temperature range, the removal efficiency of lower concentration solution is higher.
2. The concentration of two foaming solutions has more influence on the removal efficiency than the gas temperature. At lower gas temperatures (such as 100°C), the removal efficiency of SDBS solution rapidly increases with the concentration rising, however, the efficiency changing trends of two solutions are opposite, the efficiency of SDS solution decreases with concentration increasing.
3. Two kinds of stabilizers both play a role in improving the removal efficiency of SDS solution. With the concentration of coconut diethanolamide increasing, the removal efficiency grows slowly and steadily, but the efficiency is improved observably with the increasing

concentration of triethanolamine. When the concentration of triethanolamine is 0.3%, the highest removal efficiency of fine particles reaches 59% in the experiment.

REFERENCES

[1] Dai, X.D., Xu, X.L., Liao, M.F. 2011. Emission status and control technology of ultra fine particles in coal-fired power plants. *Energy Environmental Protection* 25(6): 1–4.

[2] Heidenreich, S. & Ebert, F. 1995. Condensational droplet growth as a preconditioning technique for the separation of submicron particles from gases. *Chemical Engineering and Processing* 34(3): 235–244.

[3] Jiang, Z.G., Du, C.F., Sun, J. 2002. Experimental study and analysis on performance and dedusting efficiency of frothing generator. *Journal of University of Science and Technology Beijing* 9(1): 5–8.

[4] Lu, J.Y. & Ren, X.D. 2014. Analysis and discussion on formation and control of primary particulate matter generated from coal-fired power plants. *Journal of the Air & Waste Management Association* 64(12): 1342–1351.

[5] Rozpondek, M. & Siudek, M. 2009. Pollution control technologies applied to coal-fired power plant operation. *Acta Montanistica Slovaca* 14(2): 156–160.

[6] Shi, Y.T., Do, Q., Gao, J.M. 2014. Hazard evaluation modeling of particulate matters emitted by coal-fired boilers and case analysis. *Environmental Science* 35(2): 470–474.

[7] Zhao, Y.C., Zhang, J.Y., Wei, F. 2007. Experimental study on agglomeration of submicron particles from coal combustion. *Journal of Chemical Industry and Engineering* 58(11): 2876–2881.

Material Science and Environmental Engineering – Chen (Ed.)
© 2016 Taylor & Francis Group, London, ISBN 978-1-138-02938-5

Elimination of chaos in a light-emitting diode system with optoelectronic feedback

Y. Qin, B. Liu, X. Wang, C. Han & Y. Che
Tianjin Key Laboratory of Information Sensing and Intelligent Control, School of Automation and Electrical Engineering, Tianjin University of Technology and Education, Tianjin, China

ABSTRACT: We present numerical simulations of a three-dimensional model of a light-emitting diode with ac-coupled nonlinear optoelectronic feedback, where chaos can occur in the original model within a wide range of the voltage threshold parameter. We propose two simple methods to eliminate chaos. One is to introduce a time delay in the feedback loop, and the other one is to increase the feedback gain without delay. The effectiveness of the proposed methods is demonstrated using bifurcation diagrams and the underlying mechanisms are discussed.

Keywords: chaos elimination; light-emitting diode; time delay; optoelectronic feedback

1 INTRODUCTION

Chaos, as a complex phenomenon in non-linear system, has attracted a great deal of attention in the last few decades. Chaos has been detected not only in many dynamical model systems, but also in various physical natures, such as neural system (Rabinovich 1998), chemical systems (Koper et al. 1992) and recently in a semiconductor laser with optoelectronic feedback (Al-Naimee et al. 2009, Al-Naimee et al. 2010). Due to their critical dependence on the initial conditions, chaotic systems are intrinsically unpredictable, which is commonly undesirable in real applications. Thus, chaos control becomes an important topic in nonlinear science, and has been widely investigated in many science and engineering fields. The most common goal of control for chaotic system is to eliminate chaotic behavior, that is, to convert chaotic motion to a stable periodic or a constant one. Both feedback (Ott et al. 1990, Myneni et al. 1999, Corron et al. 2010) and non-feedback (Mettin et al., 1995, Braiman & Goldhirsch 1991) control methods have been proposed and realized.

Recently, a GaAs Light-Emitting Diode (LED) with ac-coupled nonlinear optoelectronic feedback has been shown to exhibit complex dynamics including mixed mode oscillations and chaos (Marino et al. 2011). Compared to the semiconductor laser, LEDs display the same dynamics and are much more easily controllable. In this paper, we study the elimination of chaotic spiking in the above LED system. We propose two methods, that is, increasing the feedback delay time with constant feedback strength and increasing the feedback strength without time delay. The chaotic oscillations of the original LED system can be converted to periodic oscillations in both ways.

2 CHAOTIC LED SYSTEM MODEL

For numerical and analytical purposes, the LED system dynamics is written in dimensionless form (see (Marino et al. 2011) for details):

$$\begin{aligned}\dot{x}_1 &= x_1(x_2 - 1)\\ \dot{x}_2 &= \gamma(\delta_0 - x_2 + \alpha(x_3 + x_1)/(1 + s(x_3 + x_1)) - x_1x_2)\\ \dot{x}_3 &= -\varepsilon(x_3 + x_1)\end{aligned} \tag{1}$$

where x_1 and x_2 are suitably normalized photon and population-inversion densities and x_3 describes the nonlinear ac feedback loop. γ is the ratio between photon and carrier lifetimes. δ_0, ε and s are system parameters. α is the feedback strength. Throughout this paper, the fixed parameters values are set as $s = 0.2$, $\gamma = 3.3 \times 10^{-3}$ and $\varepsilon = 4 \times 10^{-5}$.

Since $x_3 + x_1$ is proportional to the feedback voltage (Marino et al. 2011). When we introduce a delayed feedback voltage in the physical system, the corresponding dimensionless form dynamical system (1) becomes

$$\begin{aligned}\dot{x}_1 &= x_1(x_2 - 1)\\ \dot{x}_2 &= \gamma\left(\delta_0 - x_2 + \alpha\frac{x_3(t-\tau) + x_1(t-\tau)}{1 + s(x_3(t-\tau) + x_1(t-\tau))} - x_1x_2\right)\\ \dot{x}_3 &= -\varepsilon(x_3 + x_1)\end{aligned} \tag{2}$$

where τ is the feedback time delay parameter.

3 CHAOS ELIMINATION BY INCREASING THE FEEDBACK TIME DELAY

In this case, we fix the feedback strength at $\alpha = 1.002$. Figure 1 shows a detailed bifurcation diagram of the peak values of computed from Eq. (2) by varying δ_0 over a small interval with different feedback time delays. As shown in Figure 1(a), which is obtained from the original model, i.e. $\tau = 0$, the large ranges of the extremum distribution indicate chaotic behavior of the system for all values of δ_0. As the feedback time delay increases to $\tau = 1$ (Fig. 1(b)) and $\tau = 5$ (Fig. 1(c)), the ranges of the extremum distribution become smaller and the maximum and the minimum separate dramatically. The LED system exhibits quasi-periodic oscillations for large feedback delay time τ. Figure 2 gives the bifurcation diagrams as a function of τ with different values of δ_0, which systematically reveal the effects of increasing τ on elimination of Chaos. Some typical patterns with different values of δ_0 and τ are shown in Figure 3. The corresponding 3D phase portraits are given in Figure 4.

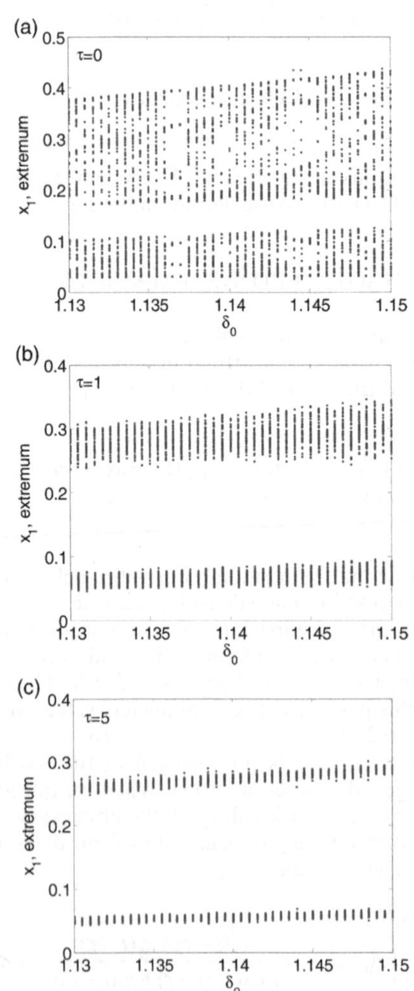

Figure 1. Bifurcation diagrams for the peak values of x_1 as the parameter δ_0 is varied, with different feedback delays. (a) $\tau = 0$, (b) $\tau = 1$, (c) $\tau = 5$.

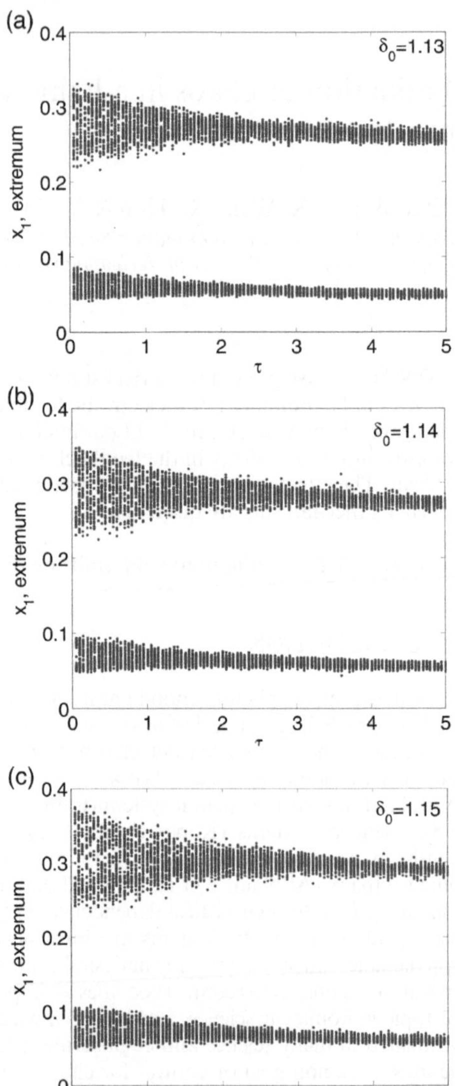

Figure 2. Bifurcation diagram for the peak values of x_1 as the parameter τ is varied with different δ_0. (a) $\delta_0 = 1.13$, (b) $\delta_0 = 1.14$, (c) $\delta_0 = 1.15$.

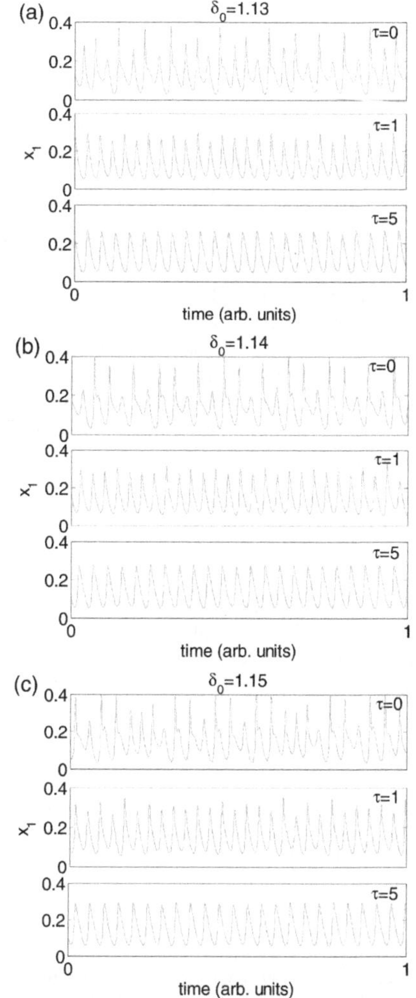

Figure 3. Responses of the LED system with different values of parameter δ_0 and $\tau = 0, 1, 5$: (a) $\delta_0 = 1.13$, (b) $\delta_0 = 1.14$, and (c) $\delta_0 = 1.15$.

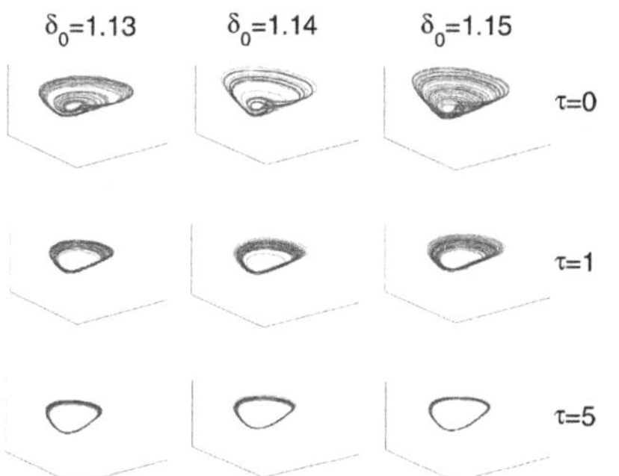

Figure 4. 3D phase portraits of the LED system with different values of parameters δ_0 and τ. With τ large enough, the system becomes quasi-periodic for all values of δ_0.

4 CHAOS ELIMINATION BY INCREASING THE FEEDBACK STRENGTH

In this case, we use the original model (1), that is, no time delay is considered. The system exhibits chaotic motions for a range of δ_0 as shown in Figure 1(a). We investigate how the feedback strength α affects the LED dynamics. Figure 5 gives the bifurcation diagrams for the peak values of x_1 as a function of α with different values of δ_0. With increase of α, the LED system transits from chaotic regime to periodic regime via a cascade of inverse period-double bifurcations. In Figure 6, we show effects of different values of α on the bifurcation diagrams as a function of δ_0. As long as the feedback strength α is big enough, the LED system exhibits periodic oscillations for all values of parameter δ_0. The elimination of chaos is obtained. Figure 7 show examples of 3D phase portraits of the responses of the LED system

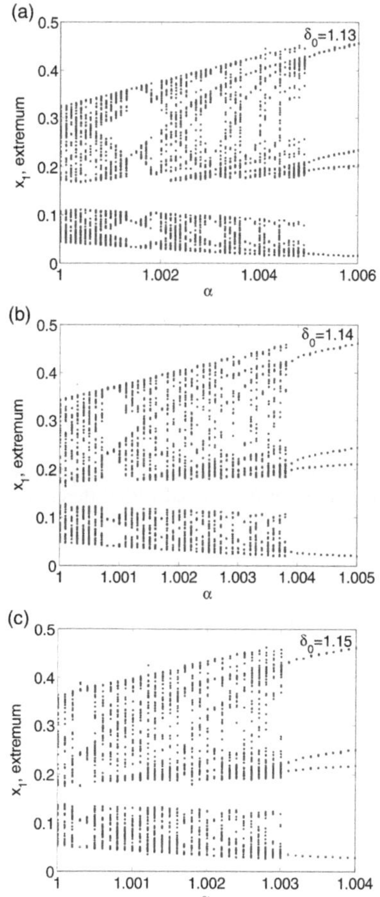

Figure 5. Bifurcation diagram for the peak values of x_1 as the parameter α is varied with different δ_0. (a) $\delta_0 = 1.13$, (b) $\delta_0 = 1.14$ and (c) $\delta_0 = 1.15$.

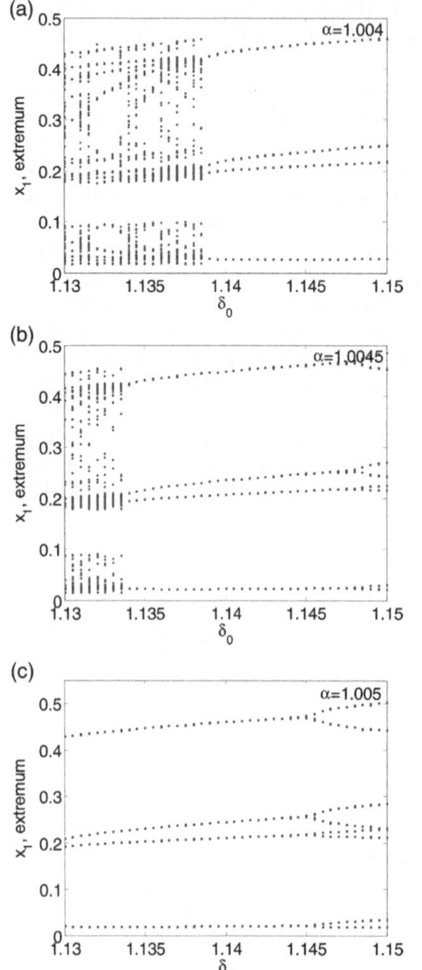

Figure 6. Bifurcation diagram for the peak values of x_1 as the parameter δ_0 is varied with different α. (a) $\alpha = 1.004$, (b) $\alpha = 1.0045$ and (c) $\alpha = 1.005$.

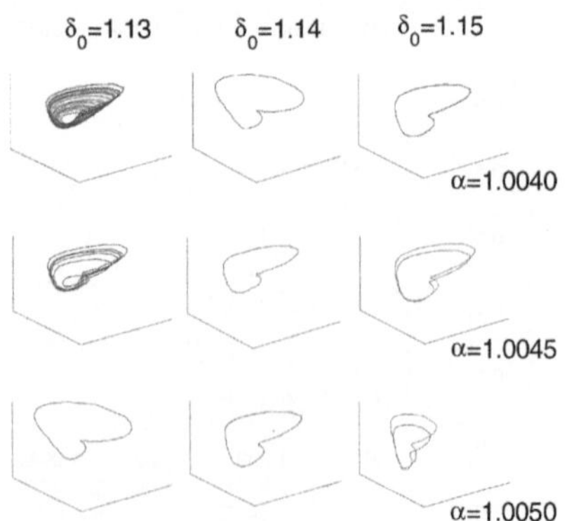

Figure 7. 3D phase portraits of the LED system with different values of parameters δ_0 and α. With α large enough, the system becomes periodic for all values of δ_0.

with different parameters δ_0 and α, which illustrates the effectiveness of method.

5 SUMMARY

In this paper, two methods for chaos elimination of a light-emitting diode system have been proposed. By increasing the feedback delay time, one can make the behavior the LED system less chaotic and more periodic, although exact periotic oscillations can't be obtained. By increasing the feedback strength without delay, one can convert the chaotic LED system to a periodic one with desired periods. The main advantage of these methods is that no extra control elements are needed, which facilitates their applications. The simulation results demonstrate the effectiveness of the proposed methods.

ACKNOWLEDGMENTS

This work is supported by the National Natural Science Foundation of China (Grants No. 61178081, and No. 61104032). We would also acknowledge the support of Tianjin University of Technology and Education (Grants Nos: RC14-09, RC 14-49 and RC14-59).

REFERENCES

[1] Rabinovich, M.I. & Abarbanel, H.D.I. 1998. The role of chaos in neural systems. *Neuroscience* 87(1): 5–14.

[2] Koper, M.T.M., Gaspard, P. & Sluyters, J.H. 1992. Mixed-mode oscillations and incomplete homoclinic scenarios to a saddle focus in the indium/thiocyanate electrochemical oscillator. *J. Chem. Phys.* 97(11): 8250–8260.

[3] Al-Naimee, K. Marino, F. Ciszak, M. Meucci, R. & Arecchi, F.T. 2009. Chaotic spiking and incomplete homoclinic scenarios in semiconductor lasers with optoelectronic feedback. *New J. Phys.* 11: 1–11.

[4] Al-Naimee, K., Marino, F., Ciszak, M., Abdalah, S.F. Meucci, R. & Arecchi, F.T. 2010. Excitability of periodic and chaotic attractors in semiconductor lasers with optoelectronic feedback. *Eur. Phys. J. D*, 58(2): 187–189.

[5] Ott, E., Grebogi, C., & Yorke, J.A. 1990. Controlling chaos. *Phys Rev Lett* 64: 1196.

[6] Myneni, K., Barr, T.A., Corron, N.J., & Pethel, S.D. 1999. New method for the control of fast chaotic oscillations. *Phys Rev Lett*. 83: 2175–2178.

[7] Corron, N.J., Pethel, S.D., & Hopper, B.A. 2000. Controlling chaos with simple limiters. *Phys Rev Lett*. 84: 3835.

[8] Braiman Y., Goldhirsch I. 1991. Taming chaotic dynamics with weak periodic perturbations. *Phys Rev Lett* 66: 2545.

[9] Mettin, R., Kurz, T. 1995. Optimized periodic control of chaotic systems. *Phys Lett A* 206: 331–339.

[10] Marino, F., Ciszak, M., Abdalah, S.F., Al-Naimee, K., Meucci, R. & Arecchi, F.T. 2011. Mixed mode oscillations via canard explosions in light-emitting diodes with optoelectronic feedback. *Phys. Rev. E* 84: 047201(1–5).

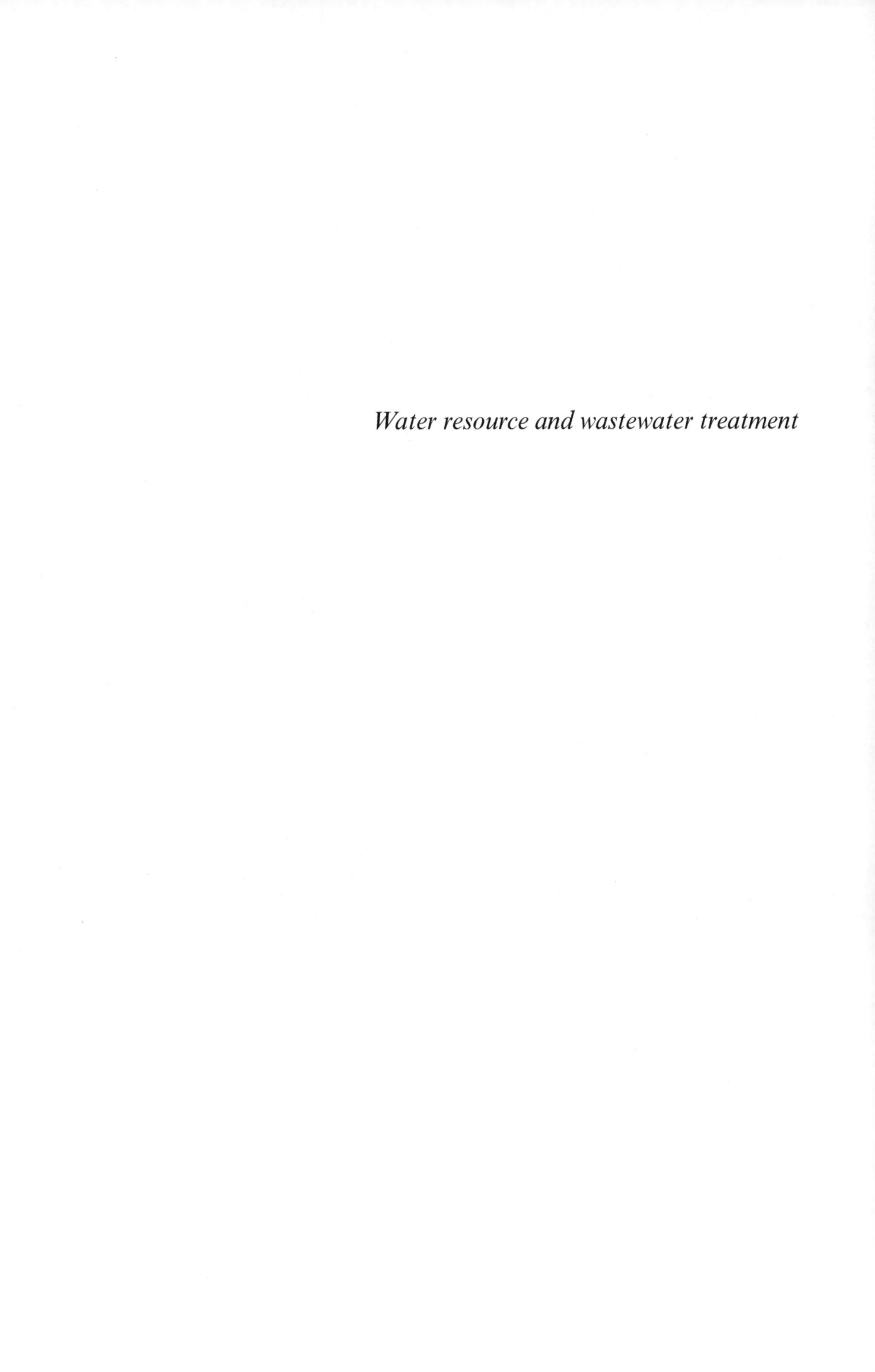

Water resource and wastewater treatment

Material Science and Environmental Engineering – Chen (Ed.)
© 2016 Taylor & Francis Group, London, ISBN 978-1-138-02938-5

Research on removal of residual antibiotics in water

Y. Yu
College of Environmental Science and Engineering, Guilin University of Technology, Guilin, China

Y.Q. Lu
College of Environmental Science and Engineering, Guilin University of Technology, Guilin, China
Guangxi Scientific Experiment Center of Mining, Metallurgy and Environment, Guilin, China

ABSTRACT: Antibiotics are widely used for preventing diseases including human diseases and animal diseases. Antibiotics were overused seriously in China, Vietnam, and other developing countries. Recently, with the rapid development of livestock agriculture and aquiculture, the types and dosage of antibiotics are keeping on increasing. Antibiotics have been detected in sewage and drinking water at trace concentrations. They pose potential harms to ecosystem and human health, and become a concerned environmental problem. How to remove the residual antibiotics from water is our current primary research direction.

Keywords: water; antibiotics; residual; removal

1 INTRODUCTION

The discovery and application of antibiotics in the prevention and treatment of livestock and aquaculture diseases in humans and animals as an additive to accelerate animal growth and increase production areas have played a very important role [1]. After people and animals taking antibiotics drugs, metabolites can produce residual antibiotics which are not absorbed so into our municipal water, may lead to unexpected dangers. In the sewage treatment plant, inlet and outlet of the water, surface water, even the groundwater can be found antibiotic residues. These antibiotics are gradually affecting our health and the environment. It is very important to establish a scientific method of analysis for knowing the environmental pollution of antibiotics.

2 ANTIBIOTICS

2.1 *The definition of antibiotics*

Antibiotics, refers to the very low concentrations of all living matter suppression and the drug which has killing effect. Such as bacteria, viruses, parasites, and even anti-tumor drugs are all areas of antibiotics.

However, the daily life and medical among antibiotics referred mainly directed against bacteria, viruses, and microbial drugs [2]. It is considerably more species. Without rigorous distinction, the antibiotic may also be referred to antibiotics [3]. Some antibiotics may be a substance produced during microbial growth and reproduction, or may be entirely synthetic or partially synthetic chemicals. Now people commonly used antibiotics in clinical microbiological culture medium liquid extracts and chemically synthesized or semi-synthetic compounds. Currently known kinds of natural antibiotics are more than ten thousands [4].

2.2 *The source of the antibiotic*

There are three major sources of water and antibiotics: "The first is from people." the expert explained. Antibiotics in the body cannot be completely absorbed. Nearly half will be discharged with the urine. The part is utilized by the body, which is only converted to metabolites. Antibiotics and its metabolites, after being expelled from the body, go into the wastewater system.

The second source is feed, animal feed, and aquaculture. Animal feed often add antibiotics to help prevent animal disease and promote growth, increase production. There are local aquacultures, antibiotics directly poured into the water, so that contaminated water directly.

Third are pharmaceutical and hospital medical waste. It is understood that the use of antibiotics in human medical drug accounted for more than 6% of the total prescription drugs and more than 70% of veterinary medicine. According to incomplete statistics, amount the top 15 drugs in China's current use and sales, 10 kinds are antimicrobial drugs.

2.3 *Status of the domestic use of antibiotics*

In China, antibiotic usage is very high. According to statistics, the use of antibiotics in the hospital accounted for 30–50% of total prescription drug. Parts of them are being used. Such as the antibiotics, which are used of treatment of viral infections, bacterial infections, parasitic infections, mycoplasma, chlamydia and other microbial infections. There is a considerable part of the irrational use. At present, China has a lot of unreasonable using problems. This requires a more rigorous and more scientific guidance management in the use of antibiotics [5]. Where overtime, excess, not symptomatic use or strictly regulate the use of antibiotics, all belong to the abuse of antibiotics. In Europe and other developed countries, the use of antibiotics in the hospital is roughly accounted for about 10% of all drugs. However, this figure is the amount of antibiotic use in hospitals of the lowest 30%, even up to 50% in primary hospital [5]. China has become one of the most serious problems of abuse of antibiotics in the world. Antibiotic abuse in our country has the following reasons [6–8]: (1) Drug competition amount manufacturers. (2) Unreasonable patient relationship. (3) Drug supervision is weak. (4) Irregular use of antibiotics in animal husbandry.

For its part, the State Food and Drug Administration currently have issued a lot files about the rational use of antimicrobial agents to carry out. Doctors call on that ordinary people and patients should have a sense of fair use. You should under the guidance of doctors to use antibiotics. Do not blindly buy the antibiotics. People should be rational drug use, in order to reduce the adverse effects of antibiotics on the environment and human.

2.4 *Contamination of antibiotics at home and abroad*

Antibiotics can do some extent metabolism in the human or animal body. However, the extent of absorption of antibiotics in humans or animals is relatively low. After the use of antibiotics, only about 30 percent of the antibiotics can be absorbed via the human or animal metabolism. Nearly 70% of the activity of antibiotics drugs goes through excrement (or feces) and discharge into the environment [9–10]. At the moment, most domestic and abroad sewage treatment plant's process design did not consider the removal of antibiotics and other trace organics. Thus the antibiotic removal efficiency in sewage treatment plant in the process is very low [9]. Thus, the urban sewage residues of antibiotics pass through the sewage treatment system, and ultimately discharged into the environment. In the countryside, people or animals, taking antibiotics, quite a number of them excrete with the pool and its droppings. Through the sewer network, precipitation and surface runoff and other ways to eventually released into the environment, including surface water, groundwater, sediment and soil.

2.5 *Water pollution levels of antibiotics*

In other countries, there are many studies on antibiotic residues in the environment. Many studies have shown that although different countries and regions, in the same body of water (such as water, surface water, etc.). The residual content of antibiotic magnitude is consistent. Overall, the total concentration of hospital wastewater antibiotic residues is greater than 1 $\mu g \cdot L^{-1}$ the antibiotic total concentration of urban wastewater is less than 1 μgL^{-1}. The total concentration of antibiotics in the surface water, groundwater, and seawater are mostly smaller than 1 $\mu g \cdot L^{-1}$. Of course, parts of places the total concentration of the antibiotic are higher than 1 $\mu g \cdot L^{-1}$. The difference is that the concentration of different studies chose a different antibiotic object. For different environmental media Water is the environmental media that has more antibiotic pollution researches on it. The study involved a variety of antibiotics categories: Including macrolides, aminoglycosides, tetracyclines (such as tetracycline, chlortetracycline, oxytetracycline, etc.), sulfonamides, and quinolones (including the first to third generation quinolone antibiotic), etc. The United States was the first country to carry out investigations of antibiotic contamination in a wide range. 1999–2000, the United States Bureau of Land Quality Survey (USGS) monitored antibiotic contamination of domestic 139 rivers. The target compounds are quinolones, tetracyclines, sulfonamides and four macrolide antibiotics. Overall most of the content of various antibiotics is 10 $ng \cdot L^{-1}$. And the highest detection rate is TMP which is up to 27.4%. The content of chlortetracycline has the highest value. It is up to 420 $ng \cdot L^{-1}$ [11]. Murata and etc. also investigated antibiotic pollution in 37 major rivers in Japan. They studied the seven kinds of sulfa antibiotics and levels and distribution of four kinds of macrolide antibiotics and Trimethoprim and other 12 kinds of antibiotics. The total concentration of antibiotics were between 0 to 626 $ng \cdot L^{-1}$ [12]. In China, XuWeihai [13], etc. investigated the antibiotic pollution in the Pearl River Delta and Hong Kong's Victoria Harbour. They found that the degree of pollution in Hong Kong Victoria Harbour is relatively low. In addition to amoxicillin, other antibiotics were lower than the detection limit of quantification. The Pearl River was detected in high concentrations of antibiotic residues. In the wet and dry seasons, the residual concentration of Pearl River median range of antibiotics were 11–67 $ng \cdot L^{-1}$ and 66–460 $ng \cdot L^{-1}$. ZhangRuijie and others did a large scale of antibiotic pollution in Bohai and Yellow sea. Survey results show that erythromycin, sulfamethoxazole and trimethoprim are the strengths of pollutants region. Its concentration is in the range of 0.10–16.6 ng·L^{-1} [14]. Overall, the study of antibiotics in the water pollution is mainly in developing countries or developed areas. It has very little study in underdeveloped areas.

3 DETERMINATION OF ANTIBIOTICS

Instrumental analysis methods for the detection of antibiotics are mainly by gas chromatography—mass spectrometry, high performance liquid chromatography and high performance liquid chromatography—mass spectrometry. Because food and environmental media have a wide range of antibiotic residues, low levels (ppt ~ ppm level) and complex substrate, it requires analysis equipment not only with a wide detection range, high sensitivity, but also have a strong anti-jamming capability. Antibiotics are class of compounds, which are strong water-soluble and less volatile. Using gas chromatography–mass spectrometry conduct must be doing complex derivative reaction analysis. Uncontrollable reaction process reproducibility has great impact on the analysis results. Currently it has little use. Liquid chromatography detection methods of antibiotics become mainstream technology now.

3.1 *Comparison of several test method detection limit*

1. HuGuanjiu et al. [15] analyzed the five samples of tetracycline in water environmental with liquid chromatography—UV detection method. When 1000 times concentrated water sample is under the conditions, the method detection limit is 0.05–0.14 μg·L^{-1}.
2. Prat et al. [16] analyzed water samples in 10 kinds of quinolone antibiotics with liquid chromatography—fluorescence detection method. When water sample is 500 times concentrated, the method detection limit is 0.1–2.0 ng·L^{-1}.
3. McArdell et al [17] applied HPLC-ESI-MS-MS technology, which could successfully tested inlet and outlet water of macrolide antibiotic residues in the sewage treatment plant. When the sample is 5000 times concentration, method detection limit is 2.0–35.0 ng·L^{-1}.
4. Reverte et al. [18] applied HPLC-MS to detect tetracyclines and quinolones residues in sewage treatment plant inlet and outlet water. When the sample is 1000 times concentrated, the method detection limit is 4.0–6.0 ng·L^{-1}.
5. Lindsey et al. [19] applied HPLC-ESI (+)/APCI (+)—MS technique to detect the tetracyclines and sulfa antibiotic residues in groundwater and surface water.

Overall HPLC-ESI (+)/APCI (+)—MS's detection limit is the lowest. It can accurately detect the content of antibiotic residues in water. Of course, there are many advanced methods being studied now. Some are broad applicability of the method, but some are not.

4 GENERAL TREATMENTS OF ANTIBIOTICS

According to GB (21903-2008) (fermentative pharmaceutical industrial water pollutant discharge standard) wastewater treatment must meet the following criteria: COD ≤120 mg·L^{-1}; BOD ≤ 40 mg·L^{-1}; NH_3-N ≤ 35 mg·L^{-1}; SS ≤ 60 mg·L^{-1}. High concentration of antibiotic wastewater is quite tricky. It is also our main researches now. The widely used treatment includes biological and physical chemical treatment.

4.1 *Biological treatment*

At present, the application of control technology at home and abroad is sparse and immature. Main engineering is aerobic process. The costs of investment and processing are high. But the actual wastewater treatment rate is low. Europe and Japan and other countries had already treated its wastewater during the 1940s when they began produced penicillin. Due to the processing techniques limit. Most of them in 1970s still used the activated sludge, biological filter, etc. From the beginning of the 1970s, they would transfer this kind of conventional bulk API manufacturing to developing countries. Instead of developing high-tech, high value-added new drug, one of the reasons is that pollution problems. So to some extent the application level of aerobic, anaerobic biological treatment processes currently used in the country on behalf of such actual technical level of wastewater treatment. Some researchers also used a combination of 2 aerobic treatments and 2 anaerobic treatments. The purpose is to ensure that treatment discharge standards when the effluent is treated by aerobic treatment of anaerobic (COD for 1000–4000 mg·L^{-1}). Meanwhile, high COD wastewater nitrogen with an aerobic and anaerobic process can also achieve the purpose of denitrification.

4.2 *Physical and chemical method*

1. Coagulation and sedimentation method
   It is a method that quantitative agent is put in raw sewage in order to make pollutant in water coagulated and sedimentation after bridging and destabilization. It could quickly decrease the content of concentration of COD, menstruum, suspended matter and mycelium, which could

reduce inhibition and poisoning of menstruum in microbe during further processing. It is a method with low cost and broad application prospect.

2. Advanced oxidation method

   It could destroy activity of antibiotics because we could transfer refractory material to small molecular substances degraded easily, which means that inhibition in antibiotics to microbes is decreased and the further processing is not difficult as before. Disadvantage is that if we use Fe/C Micro electrolysis-Fenton oxidative degradation method, large quantity of residual Fe^{2+} and Fe^{3+} will have negative effect on further processing.

5 PROBLEMS IN ANTIBIOTIC REMOVAL

1. Some antibiotics are discharged to environment as biological metabolites not original medicines. So it is necessary to calculate residual of antibiotic metabolites in samples (water, soil, and sediment) to evaluate antibiotics distributing and transferring in environment comprehensively.
2. Raw sewage is discharged by lots of factories in order to decrease cost. So factories should change the sewage discharge modes to reduce antibiotics in water.
3. Developing countries and regions tend to use inexpensive antibiotics while developed countries prefer the high-priced antibiotics. We should establish the plans according to the reality, not just keep pace with the methods in developed countries because difference exists between developing and developed countries as the result of statistics.

ACKNOWLEDGMENTS

This work was financially supported by the Guangxi Scientific Experiment Center of Mining, Metallurgy and Environment (KH2012ZD004) and Guangxi Talent Highland for Hazardous Waste Disposal Industrialization.

REFERENCES

[1] Schlüsener M.P., Bester K. Persistence of antibiotics such as macrolides, tiamulin and salinomycin in soil [J]. Environmental Pollution, 2006, 143(3): 565–571.

[2] ZhuLi. Clinical rational use of antibiotics. Chinese community physicians, medical professionals. 2008, 24: 15.

[3] DongBirong. Rational use of antibiotics [J] Chengdu medicine: 2002, 28 (4): 247–248.

[4] ZhangRuijie. Yellow Sea and the East River area typical antibiotic contamination of the environment. [D] Shandong: Yantai Institute of Coastal Zone Research, 2011.

[5] XiaoYonghong, Experts talk about the rational use of antibiotics. http://health.sohu.eom/7/0404/75/column219747516.shtml.

[6] Zong Yi, Li Jing abuse situation and harm of antibiotics [J] Family Medicine, 2009, (4): 11.

[7] ShiSong, Situations and Countermeasures about the abuse of antibiotics [J] Chinese Modern Drug Application, 2009, 3 (7): 181–182.

[8] Xiao Yonghong, How many Chinese people misuse of antibiotics. http://health.msn.com.cn/Info/20100819/17091123890_2.shtml.

[9] Kummerer, K., Henninger, A., Promoting resistance by the emission of antibiotics from hospitals and households into effluents [J]. Clinical Microbiology and Infection 2003, 9: 1203–1214.

[10] Yang, J.F., Ying, G.G., Zhao, J.L., et al. Spatial and seasonal distribution of selected antibiotics in surface waters of the Pearl Rivers, China [J]. J. Environ. Sci. Heal. B. 2011, 46: 272–280.

[11] Kolpin D.W., Furlong E.T., Meyer M.T., et al. Pharmaceuticals, hormones, and other organic wastewater contaminants in US streams, 1999–2000: A national reconnaissance [J]. Environmental Science & Technology, 2002, 36(6): 1202–1211.

[12] Murata, A., Takada, H., Mutoh, K., et al. Nationwide monitoring of selected antibiotics: Distribution and sources of sulfonamides, trimethoprim, and macrolides in Japanese rivers [J]. Sci. Total Environ. 2011, 409: 5305–5312.

[13] Xu W.H., Zhang G., Zou S.C; et al. Determination of selected antibiotics in the Victoria Harbour and the Pearl River, South China using high-performance liquid chromatography-electrospray ionization tandem mass spectrometry [J]. Environmental Pollution, 2007, 145(3): 672–679.

[14] Jiang L., Hu X.L., Yin D.Q., et al. Occurrence, distribution and seasonal variation of antibiotics in the Huangpu River, Shanghai, China [J]. Chemosphere, 2011, 82: 822–828.

[15] HuGuanjiu, WangBin, SunCheng. High Performance Liquid Chromatography in environmental water samples five kinds of tetracyclines residues [J] Environmental Chemistry, 2007, 26 (l): 106–107.

[16] Determination of quinolones in water samples by solid-phase extraction and liquid chromatography with fluorimetric detetion [J]. Journal of Chromatography A, 2004, 1041: 27–33.

[17] C.S. McArdell, E. Molnar, M.J-F. Suter, et al. Occurrence and fate of macrolide antibiotics in wastewater treatment plants and in the Glatt valley watershed, Switzerland [J]. Environmental Science and Technology, 2003, 37: 5479–5486.

[18] S. Reverte, F. Borrull, E. Pocurull, et al. Determination of antibiotic compounds in water by solid-phase extraction high performance liquid chromatography-electrospray mass spectrometry. Journal of Chromatography A, 2003, 1010: 225–232.

[19] M.E. Lindsey, M. Meyer, E.M. Thurman. Analysis of trace levels of sulfonamide and tetracycline antimicrobials in groundwater and surface water using solid phase extraction and liquid chromatography/mass spectrometry [J]. Analytical Chemistry, 2001, 73: 4640–4646.

Material Science and Environmental Engineering – Chen (Ed.)

Optimization of multiple-tube fermentation for determination of Fecal coliform in medical wastewater

W.Y. Zhao
College of Environment Science and Engineering, Guilin University of Technology, Guilin, China
Guangxi Scientific Experiment Center of Mining, Metallurgy and Environment, Guilin University of Technology, Guilin, China

Z. Wang, J. Chen & C.W. Zhao
College of Environment Science and Engineering, Guilin University of Technology, Guilin, China

L.W. Xu
College of Environment Science and Engineering, Guilin University of Technology, Guilin, China
Guangxi Scientific Experiment Center of Mining, Metallurgy and Environment, Guilin University of Technology, Guilin, China

ABSTRACT: The multiple-tube fermentation method is one of the national standard testing Fecal coliform bacteria, which exist problems including operation cumbersome and time consuming. Using the 5-tube zymotechnics, the 10-tube zymotechnics, peptone culture method, EC cultivation, etc., combined with the national standard multiple-tube fermentation method separately testing Fecal coliform bacteria in medical wastewater, the results showed that, compared with the national standard multiple-tube fermentation method, peptone culture method can save 24–48 hours, better data reliability, which could be a optimization method for testing Fecal coliform bacteria in medical wastewater, which can saving time, manpower, material resources, having good popularization value.

Keywords: medical wastewater; Fecal coliform; multiple-tube fermentation technique; peptone culture method

1 INTRODUCTION

Fecal coliform is one of important standards in testing the water pollution, which also has much scientific significance especially in quickly and accurately testing the waste water from densely populated hospitals. Fecal coliform as a standard of Fecal pollution is mainly come from the excreta of human and animals, and it's one of subspecies of total coliform group. Through monitoring the Fecal coliform in the medical waste water, Fecal coliform is the technical basis in evaluating the disinfection results of medical waste water. At present the frequently used method in testing the Fecal coliform in waste water is multiple-tube fermentation method and filter membrane method. Dai Qiaorong, from the environmental monitor station of Long Gang, Shen Zhen, made experiments to test the Fecal coliform from 30 different samples of water through these two methods and made the conclusion that the multiple-tube fermentation method is more available to the waste water while the filter membrane method is more available to the clean surface water. According to the extensive application of the multiple-tube fermentation method, this experiment aimed to doing optimization study.

2 EXPERIMENTAL MATERIAL AND EQUIPMENT

2.1 *The water samples*

All of the water samples are unsterilized medical waste water from different hospitals of Gui Lin, Guang Xi. Using the sterilization container, the water samples are refrigerated in 2–4 degrees Celsius and be conducted within 2 hours.

2.2 *Material main reagents and equipment*

Material Main reagents: Lactose peptone culture, Three times Lactose peptone culture and EC culture.

The equipment: Vertical pressure steam sterilizer. Clean bench. SHZ-type bath thermostat. SPX-250B-D type constant temperature incubator. Glass tubes of 18×180 mm and 25×200 mm. Inoculating loop of 3 mm.

3 EXPERIMENTAL METHODS

3.1 *The national standard method*

Multiple-tube fermentation method is one of the most conventional methods to test Fecal coliform. The experimental procedure:

a. To take 5 tubes of water samples from each 3 different volumes of water samples (the dilution ratios are different according to the pollution level).
b. To inoculate the water samples to Lactose peptone culture to doing fermentation. And then to cultivate the samples in the degree of 37C±0.5°C for about 24±2 hours and then to observe the samples. If produce acid and gas, it proved that the test was positive.
c. To inoculate the positive water samples to EC culture to cultivate in the degree of 44.5°C±0.5°C for about 24±2 hours. (The liquid level of bath thermostat is higher than the medium level in the tubes.) If produce acid and gas, it proved that the Fecal coliform is positive.
d. To check MPN table to calculate the result.

3.2 *The five tube zymotechnics*

The experimental procedure:
a. To mix the water samples and then dilute it.
b. To take one tube of water samples in each size of 100 ml, 10 ml, 1 ml, 0.1 ml, and 0.01 ml. (When the size is 100 ml, then the Three times Lactose peptone cultureis 50 ml. When the size is 10 ml, then the Three times Lactose peptone cultureis 5 ml. When the size is 1 ml or the size is less than 1 ml, then the Lactose peptone cultureis 10 ml).
c. The experimental procedure of primary fermentation and secondary fermentation is same as the National standard method.
d. To check MPN table to calculate the result.

3.3 *The ten tube zymotechnics*

The experimental procedure:
a. To inoculate every water sample in 2 different sizes and then the number of each sample is 5 tubes. (When the size is 100 ml, then the Three times Lactose peptone culture is 5 ml. When the size is 1 ml or the size is less than 1 ml, then the Lactose peptone culture is 10 ml.)
b. The experimental procedure of primary fermentation and secondary fermentation is same as the National standard method.
c. To check MPN table to calculate the result.

3.4 *The EC cultivation method*

The EC cultivation makes use of the character of Fecal coliform that it will produce acid and gas even in the degree of 44.5°C. The experimental procedure:
a. To directly inoculate the sample to EC culture without primary fermentation. (The measure of inoculation is same as the national standard method.)
b. After cultivating the sample in the water in degree of 44.5°C±0.5°C for about 24±2 hours, and then to check MPN table to calculate the result.

3.5 *The peptone culture method*

Fecal coliform as aerobic and facultative anaerobic and Gram-negative bacteria can ferment the lactose and produce acid and gas. The peptone culture method makes use of the character of Fecal coliform that it will produce acid and gas even in the degree of 44.5°C. The experimental procedure:
a. To inoculate the sample to Lactose peptone culture. (The measure of inoculation is same as the national standard method.)
b. To cultivate the sample in the water in degree of 44.5°C±0.5°C for about 24±2 hours, and then to observe. If produce acid and gas, it proved that the Fecal coliform is positive.
c. To check MPN table to calculate the result.

4 EXPERIMENTAL RESULTS AND ANALYSIS

4.1 *Experimental results*

The testing results are from the MPN table. The results of 26 medical wastewater samples are in Table 1. According to Table 1, the content of Fecal coliform by using optimization methods is mostly different from the result come from the national standard method. Both results by using 10-tube zymotechnics method and peptone culture method are inside the national standard method's 95% confidence interval. By using 5-tube zymotechnics method and EC cultivation, the result of simple sample proportion which are inside the national standard method's 95% fiducial range is 61.54% and 92.31%.

4.2 *Analysis of experimental results*

All the data in the experimental come from Correlation analysis and t tests by using SPSS 18.0.

4.2.1 *Correlation analysis of optimization method and national standard method*

From Table 2, the optimization method and national standard method of test results have a certain correlation, in addition to the five pipe method is with the national standard method of correlation coefficient was significant positive correlation, only three other method and national standard method of correlation coefficient reached

Table 1. The MPN results of Fecal coliforms by improved and national standard methods respectively (A/L).

Sample	National standard method (10^5)	5-tube zymotechnics (10^5)	10-tube zymotechnics (10^5)	EC cultivation (10^5)	Peptone culture method (10^5)	95% confidence interval (10^5)
1	94	9.5	70	33	79	28~220
2	27	180	63	11	14	9~80
3	17	95	15	9	34	5~46
4	22	28	34	7	12	7~67
5	110	90	79	79	110	31~250
6	49	90	34	8	33	17~130
7	12	28	17	4	12	3~28
8	22	9.5	24	9	33	7~67
9	94	180	130	70	110	28~220
10	21	22	43	14	11	7~63
11	14	95	30	17	8	4~34
12	9	19	10	5	17	2~21
13	11	9	16	7	8	2~25
14	33	91	21	14	26	11~93
15	26	22	34	22	27	9~78
16	79	1800	170	34	94	25~190
17	63	190	49	17	79	21~150
18	46	920	24	27	43	16~120
19	12	9	12	9	11	3~28
20	5	<9	8	2	9	<0.5~13
21	17	9	15	14	17	5~46
22	26	28	63	33	33	9~78
23	34	90	33	26	23	11~89
24	70	950	170	27	70	23~170
25	9	9	9	7	7	2~21
26	43	200	79	23	46	15~110

Table 2. The correlation analysis of improved and national standard.

Sample	Degrees of freedom	Pearson correlation coefficient	Correlation coefficient P
5-tube zymotechnics	26	0.422*	0.032
10-tube zymotechnics	26	0.742**	0.000
EC cultivation method	26	0.839**	0.000
Peptone culture method	26	0.955**	0.000

Note: * *indicates the relationship was extremely significant level ($P < 0.01$), *indicates the relationship was significant positive correlation ($P < 0.05$).

extremely significant level, including peptone culture method and even reached 95.5%. Prove that the four kinds of optimization test data with the national standard method have a significant positive correlation between test data.

4.2.2 *T-test of optimization method and national standard method*

1. T-test of 5-tube zymotechnics and National standard method

According to Table 3, compare with national standard method, the statistic F's observed value of 5-tube zymotechnics is 44.100, which shows that the P value is 0.000. The significant level a is 0.05, so that P is less than a. Assuming that variance is not the same, the statistic T's observed value is 5.605, corresponding to the two tailed opening rate P value of 0.000 which is still less than 0.05. Therefore, the average value of two overall has remarkable difference, that the two results of determination are statistically significant different. Using the 5-tube zymotechnics to detect medical waste water sample is unreliable.

2. T-test of 10-tube zymotechnics and National standard method
   According to Table 4, compare with national standard method, the statistic F's observed value of 10-tube zymotechnics (Table 4). The T test analysis of 10-tube zymotechnics and National Standardchinics is 2.021, which shows that the P value is 0.161. The significant level a is 0.05, so that P is large than a. Assuming that variance is the same, the statistic T's observed value is –1.028, corresponding to the two tailed opening rate P value of 0.309 which is large than 0.05. Therefore, the average value of two overall doesn't have remarkable difference, that the two results of determination are not statistically significant different. Using the 10-tube zymotechnics to detect medical waste water sample is reliable.
3. T-test of EC cultivation method and National standard method
   According to Table 5, compare with national standard method, the statistic F's observed value of EC cultivation method is 6.806, which shows that the P value is 0.012. The significant level a is 0.05, so that P is large than a. Assuming that variance is the same, the statistic T's observed value is 2.245, corresponding to the two tailed opening rate P value of 0.019 which is less than 0.05.

Table 3. The T test analysis of 5-tube zymotechnics and national standard.

	Same variance	Different variance
Leven examine of variance equation		
F	44.100	
Sig.	0.000	
T test of mean equation		
t	5.494	5.605
Df	49.000	25.142
Sig. (two side)	0.000	0.000
Average difference	33.050	33.050
Standard errors	6.015	5.897
95% confidence interval		
Minimum	20.962	20.909
Highest	45.138	45.191

Table 4. The T test analysis of 10-tube zymotechnics and national standard.

	Leven examine of variance equation		T test of mean equation					95% confidence interval	
	F	Sig.	t	Df	Sig. (two side)	Average difference	Standard errors	Minimum	Highest
Same variance	2.021	0.161	–1.028	50	0.309	–11.038	10.738	–32.607	10.530
Different variance			–1.028	43.144	0.310	–11.038	10.738	–32.692	10.605

Table 5. The T test analysis of EC cultivation and national standard.

	Leven examine of variance equation		T test of mean equation					95% confidence interval	
	F	Sig.	t	Df	Sig. (two side)	Average difference	Standard errors	Minimum	Highest
Same variance	6.806	0.012	2.425	50	0.019	16.808	6.931	2.887	30.728
Different variance			2.425	41.778	0.020	16.808	6.931	2.819	30.797

Table 6. The T test analysis of peptone culture method and national standard.

	Leven examine of variance equation		T test of mean equation					95% confidence interval	
	F	Sig.	t	Df	Sig. (two side)	Average difference	Standard errors	Minimum	Highest
Same variance	0.091	0.765	–0.004	50	0.996	–0.038	8.678	–17.468	17.391
Different variance			–0.004	49.689	0.996	–0.038	8.678	–17.471	17.394

Therefore, the average value of two overall has remarkable difference, that the two results of determination are statistically significant different. Using EC cultivate on method to detect medical waste water sample is unreliable.

4. T-test of Peptone culture method and National standard method
According to Table 6, compare with national standard method, the statistic F's observed value of peptone culture method is 0.091, which shows that the P value is 0.765. The significant level a is 0.05, so that P is large than a. Assuming that variance is the same, the statistic T's observed value is -0.004, corresponding to the two tailed opening rate P value of 0.996 which is large than 0.05. Therefore, the average value of two overall does not have remarkable difference, that the two results of determination are not statistically significant different. Using peptone culture method to detect medical waste water sample is reliable.

5 CONCLUSION

1. In the four optimization method, the 5-tube zymotechnics not only have the worst relativity compared to the national standard method, but its data's difference in statistical significance is unreliable. This is the result of the uncontrollably accidental errors made by the only sample of every kind of water sample in the experiment. Although the 5-tube zymotechnics can reduce some workload, it cannot reduce the fermentation time.
2. The data of the 10-tube zymotechnics is, to some degree, reliable and can reduce the workload; however, it still needs the first and secondary fermentation which consume too long time. Comparing to the national standard method, it greatly increases the uncertainty causing by the systematic effect, such as diluting and adding sample during the experiment, which has some influence on the experiment effect. Therefore, although all its data are inside the national standard method's 95% fiducially range, its relativity and reliability are not the best.
3. The EC cultivation reduces the method of first fermentation and conducts the secondary fermentation directly, which greatly reduce the fermentation time. It also gets 83.9%, an apparently positive correlation to the national standard method, but the data in the national standard method's fiducial range is only 92.31%. Moreover, its t test proves that the data has unreliability. According to observing the test data of EC cultivation, it is found that the data of EC cultivation is mostly smaller than the national standard method, effecting by the unobviously positive effect causing by the too low or too high content of cholate in the culture solution.
4. Peptone culture method has the best popularization value in the four optimization methods. Firstly, peptone culture method reduces the first fermentation which cuts 24 hours for the test; secondly, its data is average load in the national standard method's fiducially range and has apparently positive correlativity; finally, its data does not have obvious difference with the data tested by the national standard method in statistics. So the result shows that the peptone culture method has the advantage of the accuracy, speed and saving manpower and material resources to test Fecal coliform in waste medical water.

REFERENCES

[1] State Environmental Protection Administration. Water and wastewater monitoring analysis method [M].4. Beijing: China Environmental Press, 2002:701–706.

[2] Lee Yong. Evaluation of uncertainty in determination of sewage Fecal coliform in manifold zymotechnics [J]. www.gdchem.com, 2013,21 (40):146–147.

[3] Dai Qiaorong, Liu Yu, Li Yan. Determination of water to explore the of Fecal coliforms multiple tube fermentation method and membrane filtration method, [J]. technology entrepreneur, 2012,7 (on):213.

[4] State Environmental Protection Administration. Water quality-Determination of Fecal coliform-manifold zymotechnics and filter membrane (HJ/T347–2007) [M] Beijing: Chinese Environment Science Press, 2007.3.10.

[5] Xie Rong. Study on Determining Fecal Coliform by 5 Tube Zymotechnics [J]. Arid Environmental Monitoring 2004,18 (3):191–192.

[6] Xie Rong. Study on Determining Fecal Coliform by 5 Tube Zymotechnics [J]. Heilongjiang Environmental Joumal, 2009,33 (4):37–39.

[7] Zhang Jilong, He Jiming, Yuan Quan, Ou sensitivity. Comparison multiple tube fermentation method and its improved using in water Fecal coliform assay detection [J]. Sichuan environment, 2012,10,31 (5): 8–11.

[8] Zhang Shaofeng, Liu Guojiang, Wei Chunlei and Fecal coliform detection methods and research progress of [J]. Marine Science Bulletin, 2008,27 (3):102–105.

[9] Feng Qingying, Liu Juan, Yang Dawei. Improvement of Detection Method of Fecal Coliforms in Surface Water [J] Environmental Monitoring in China. 2009.25 (6):43–46.

Material Science and Environmental Engineering – Chen (Ed.)
© 2016 Taylor & Francis Group, London, ISBN 978-1-138-02938-5

Research progress in High-strength Rare earth Ammonia Wastewater treatment

W.B. Hou, L.H. Wu & W.J. Zhang
College of Environmental Science and Engineering, Guilin University of Technology, Guilin, P.R. China

Y. Jin
College of Civil Engineering and Architecture, Guilin University of Technology, Guilin, P.R. China

Y. Ye
Ministry of Environmental Protection Water Project Management Office, Beijing, P.R. China

ABSTRACT: Large amounts of HRAW are discharged from rare earth smelting process, and it has caused serious pollution on the receiving water in China. In this paper, research progress in High-strength Rare earth Ammonia Wastewater (HRAW) was summarized in detail. Although a lot of technologies were introduced in literature, it is still a hard work to treat HRAW with a reasonable cost due to the current treatment process. Single method was proved to be difficult to meet discharge standards. It is feasible to adopt combination processes to reduce costs and enhance treatment effects. In addition, the recycling of ammonia by-product should be preferred at the same time in order to offset the cost of wastewater treatment.

Keywords: rare earth; high-strength; ammonia wastewater; treatment

1 INTRODUCTION

As a country with large scale of rare earth mining and manufacturing, the proven reserves in China are approximately 65.88 million tons, which are mainly distributed in Inner Mongolia, Jiangxi and other regions. The distribution of rare earth mining is characterized by unbalanced distribution between the northern and southern areas. Since the 1960s, China started the industrial production of rare earth, and gradually formed a set of rare earth process (as shown in Fig. 1) (Tao & Ding 2013). By the time, the corresponding environmental problems arisen following the development of the rare earth industry. For an example, High-strength Rare earth Ammonia Wastewater (HRAW) generated from rare earth manufacturing, and has a large amount of complex components with high hardness, high ammonia content and difficult to treat. The indiscriminate discharge of this kind of wastewater caused serious pollution to the water.

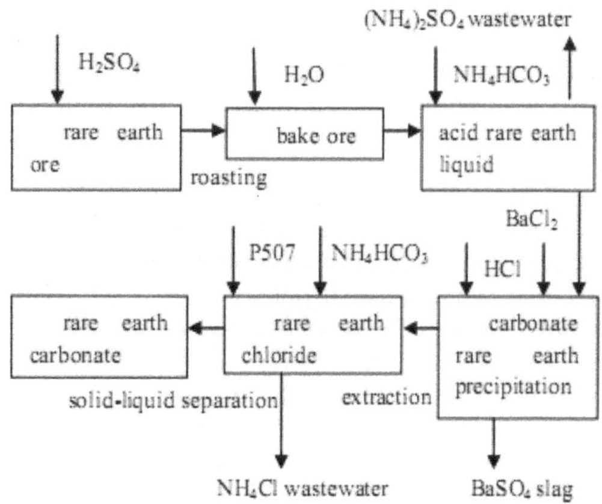

Figure 1. Rare earth manufacturing process.

2 HRAW

2.1 *Emission sources of HRAW*

Two kinds of major wastewater are discharged in the rare earth manufacturing: ammonium sulfate wastewater and ammonium chloride wastewater. Ammonium sulfate wastewater generate in the process of rare earth carbonate manufacturing, where ammonium sulfate is used as raw materials in the production of rare earth ore roasting. The ammonium sulfate concentration in wastewater is up to 5000 mg/L as N, and contains small amounts of calcium and magnesium ions. Ammonium chloride wastewater is discharged after ammonia saponification, or from secondary carbon sink. The chloride concentration in wastewater is 10000–20000 mg/L.

2.2 *The treatment status of HRAW*

In China, smelting method still dominates in rare earth manufacturing. Usually, large amounts of HRAW produce in the smelting process. However, due to the high treatment cost, the HRAW is treated not well and even untreated before discharge. The discharge of HRAW has caused serious environmental problems such as eutrophication. High levels of ammonia also enter water body, which also increased the cost of drinking water treatment. Therefore, it is urgent to make efforts of developing technology for treatment of HRAW, and thereafter support the sustainable development of rare earth industry.

3 THE TREATMENT OF HRAW

3.1 *Stripper method*

The basic principle of ammonia stripper method is gas-liquid mass transfer. The pH of wastewater is first adjusted in order to make the ammonium ions in wastewater conversion into the free ammonia, and then the NH3 is transferred into the air by a large amount of air to achieve the purpose of removing NH4+ from wastewater. The pH, temperature and gas-liquid ratio are the key factors in ammonia stripper method. The ammonia concentration and other impurities concentration in wastewater affect less. This method is mainly used in ammonium sulfate wastewater treatment, which has more complex component. Chen et al (Chen et al. 2007) has used orthogonal experiment method in ammonia stripping test, and the results showed that the primary and secondary order of factors affecting the efficiency of ammonia stripping is pH > temperature > stripping time. The preferred level conditions were pH ≥ 12, temperature = 30°C, time = 1h, gas-liquid ratio = 2000. Under this condition, the ammonia removal rate could reach above 90%. Ammonia stripper method has been widely used in pretreatment of high concentration ammonia wastewater (Le et al. 2007, K.C. 1997, Hu 2003) with relative good treatment performance, meanwhile the processing cost is relatively low. Huang et al (Huang et al. 2008) has achieved the nitrogen removal rate above 94% with residual ammonia concentration in effluent below 100 mg/L when the pH of the wastewater was adjusted to 12, gas-liquid ratio was 3000 to 4000, and temperature was in the range of 35–45 °C. Zhou et al (Zhou & Qi 2010) concluded that at the condition of pH = 11, temperature 40 °C, the gas-liquid ratio 600: 1, the stripping time 60 min, the ammonia removal rate could reach 94.5%, and the ammonia concentration in wastewater changed from 2897 mg/L to 159 mg/L after stripping. The biggest cost in the running process of ammonia stripper method is the consumption of alkali to adjust pH value. By using lime, the cost could be reduced, but it was difficult to clean up the sedimentation, which was often found in the tank. On the other hand, the running cost will turn high if using soda ash or solid alkali (Wang et al. 2012).

3.2 *Chemical precipitation methods*

Magnesium Ammonium Phosphate (MAP) precipitation method removes ammonia mainly by adding Mg^{2+} and PO_4^{3-} in ammonia wastewater, then MgNH4PO4 (MAP) precipitation generates with NH4+, thus the ammonia in wastewater is removed. The biggest characteristic of this method is that the ammonia can be recycled as generated MAP compound fertilizer. The advantages of chemical precipitation method are the high nitrogen removal efficiency, simple process, precipitation reaction unrestricted from temperature, toxins in water or other factors, easy to control, and the magnesium ammonium phosphate precipitation formed contains N, P, Mg, which can be used as fertilizer recycling (Zou et al. 2007). Huang et al (Huang et al. 2008) achieved high removal rate of ammonia up to 98% under the conditions of pH = 12, n(Mg):n(N):n(P) = 1.2:1:1.3, reacted time 20 min. The residual ammonia was only 100–200 mg/L comparing to initial concentration of 2333 mg/L. The treated effluent water could reach the national wastewater discharge standards. Although a lot of cases are published up to now, this method is not suggested due to the high running cost. This is partly due to the fact that it is difficult to find a lot of magnesium ore in the market. MAP as a production is still not popular in China.

3.3 *Electrodialysis*

Electrodialysis method is a kind of membrane separation technology. This method arranges anion and cation exchange membrane alternately between the positive and negative electrodes, and separates them with a special partition. It is composed of the desalination system and concentration system. Then connected, the ammonia in the wastewater is concentrated (Chen 1998) by using the permeability of ion exchange membrane. The method of electrodialysis can concentrate the ammonium in the wastewater, and the wastewater can be directly reused after treated. By the further evaporation, concentrate getting from dialysis can be recycled as ammonium salt. This method has been applied in large-scale. Pan et al. (Pan & Lu 2002) have used the secondary electrodialysis method to increase the NH_4Cl solution concentration from 6% to 13%, and this technology has been successfully applied in the Baotou Hefa rare earth company with a treatment capacity of 20 $t{\cdot}h^{-1}$. Tang et al. (Tang & Ling 2008) under the conditions of electrodialysis voltage was 55 v, influent flow was 24 L/h and ammonia wastewater influent conductivity was 2920 us/cm, the influent ammonia which concentration is 534.59 mg/L, after treated, the concentrated water and fresh water each took up 19% and 81% in effluent room, ammonia content was 2700 mg/L and 13 mg/L respectively.

However, the application of electrodialysis has some limitations. This method must be operated in careful, and many factors should be considered such as current density, flow rate, temperature, concentration difference and etc. The footprint is very high due to the complicated process and auxiliary equipments. And it does not fit to ammonium sulfate wastewater, where calcium and magnesium impurities is high, and calcium sulfate scaling will affect the stable operation of the device. The rare earth smelting process produces large amount of HRAW and the water quality changes frequently, but electrodialysis method requires very strict conditions for water quality, therefore, this method is difficult to control, and the treatment performance is difficult to guarantee (Xue et al. 2000).

3.4 *Biological nitrogen removal method*

Biological Nitrogen Removal (BNR) processes is recommended to remove nitrogen from wastewater. HRAW belongs to the low C/N ratio wastewater, and traditional BNR is inhibited by the lack of carbon source. Han et al (Han et al. 2000) adopted UASB process to treat high HRAW wastewater. After75d experiments, short nitrification and denitrification occurred in the reactor, and the removal rate reached 40%. The results showed that BNR is possible in treatment of the low C/N HRAW.

BNR can handle HRAW with strong adaptability. However, there are some disadvantages exists at the same time: ①domesticated species need long start-up period; ②Training strains requirements are strict, difficult to control, and it can cause nitrite nitrogen increases in effluent if operation is not carried out in proper; ③It is difficult to deal with the microorganism in effluent after denitrification.

4 SUGGESTIONS

Due to the difference of production technology, raw materials and product structure, the discharged HRAW is different in rare earth industry. Therefore, there is no uniform pattern of wastewater treatment. Different companies should have different treatment processes to solve wastewater problems. At present, although there are many related research and sophisticated treatment process, most rare earth companies only treated it partially. Rare earth plant generally recovery ammonia by steam stripping method or recycling chloride by evaporation, condensation, crystallization. While stripping ammonia emissions into the atmosphere will cause secondary pollution; the market sale prospects of ammonium chloride recycled by crystallization also was not very optimistic, resulting in ammonium chloride recovered from plants deposited large accumulation, the concentration of ammonia in the effluent is still high (Wang et al. 2013).

Among the treatment method of ammonia wastewater (Table 1), ion exchange, liquid membrane, biological, soil irrigation method and circulating cooling water denitrification method are specific in wastewater quality, which only can deal with low concentration of ammonia wastewater. Biological methods can treat high ammonia wastewater by combination with other methods, first, adopted other methods to remove ammonia, in order to reach the conditions that biological method can treat, then using the biological method for further processing.

Although ammonia stripper method can remove ammonia effectively, but there are some shortcomings such as temperature, chemical and power consumption. Chemical precipitation method is a common in treating high concentration ammonia wastewater, but it is difficult to find a lot of magnesium ore on the market at present. Electrodialysis

Table 1.

Method	Influent concentration as N (mg/L)	Effluent concentration as N (mg/L)	Cost	Reference
Stripper method	2897	159	Average	[7]
Chemical precipitation methods	2333	<50	High	[9][10]
Electrodialysis	535	13	High	[12][13][14]
Biological nitrogen removal method	500	300	Low	[15]

method can undertake the concentration of ammonia wastewater under certain conditions; however, due to its complex process and cumbersome controlled conditions, the industrial practice is difficult at present.

Stripper method could deal with high concentration ammonia wastewater, but the wastewater after treated can't meet the discharge standards. Ammonium comprehensive recovery method and vacuum distillation recovery chloride method could bring some economic benefits to the enterprise, but neither of these methods can deal with ammonium sulfate wastewater. Therefore, for the HRAW produced in rare earth production, choosing treatment process should be based on the specific conditions and the actual situation, and in conjunction with other technology for co-processing.

5 CONCLUSION

Due to the difference of production technology, raw material and product structure, the HRAW produced in rare earth industry is different; therefore it is impossible to adopt a uniform wastewater treatment mode. Single method is difficult to make the effluent meet emission standards after treatment. It is feasible to adopt combination processes to reduce costs and enhance treatment effects, such as chemical precipitation + liquid membrane method, stripper + adsorption method or chemical + biological method. In addition, the recycling of ammonia by-product should be considered at the same time, in order to offset the cost of wastewater treatment, and reach the goal of uniting the economic, environmental and social benefits.

ACKNOWLEDGEMENTS

This research was supported by Guangxi Natural Science Foundation (2013GXNSFCA019018; 2014GXNSFBA118265), the Scientific Research Fund of Guangxi Education Department (2013ZD031; 2013ZL076; ZL2014051), Guangxi Science and Technology Development Project (1347004-15), the Guangxi Talent Highland for Hazardous Waste Disposal Industrialization.

REFERENCES

[1] Chen Dongsheng. (1998) Waste water treatment by membrane processes [J]. Membrane science and technology, 18(5):32–34.

[2] Chen Lirong, Dai Baocheng, Wu Wenfei, Zheng Kuncan. (22007). Study on air stripping of high concentration ammoniac wastewater in rare-earth industry. Technology & Development of chemical industry, 36 (9):38–40.

[3] Cheung K.C. (1997). Ammonia stripping as a pretreatment for landfill leachate. Water, Airand Soil Pollution, 94:209~ 221.

[4] Huang Haiming, Xiao Xianming, Yanbo. (2008). Experimental research on treatment of ammonia nitrogen wastewater by ammonia stripping in a rare earths separation. Factory. Chinese journal of environmental engineering, 2(8):1062–1065.

[5] Huang Haiming, YanBo, Chen Qihua, Fu Zhong, Xiao Xianming. (2008). Removal of ammonia-nitrogen from rare earths wastewater by chemical precipitation. Environmental chemistry, 27(6):775–778.

[6] Hu Jifeng. (2003). Treatment technology for waste water containing ammonia and industrial design. Technology of water treatment, 29(4):244~ 246.

[7] Han Jianhong, Gao Lingqin, Fan Buhe. (2010). Study on treatment of rare-earth wastewater by UASB processes. Water & Wastewater Engineering, vol.36:251–253. (in Chinese)

[8] Pan Qi, Lu Xiaohua. (2002). Study on treatment of NH4Cl wastewater by eletrodialysis. Hubei Chemical, 6:15–16.

[9] Tao Xiaoming, Ding Zhonghao. (2013). Research on treatment of rare earth ammonia nitrogen wastewater. Journal of green science and technology, 6:168–170.

[10] Tang Yan, Ling Yun. (2008). Separation of ammonia and water by electrodialysis [J]. China resources comprehensive utilization, (3):27–29.

[11] Wang Chunmei, Zhang Yongqi, Huang Xiaowei, Long Zhiqi. (2012). The technical status of rare earth smelting ammonia-nitrogen wastewater treatment. Energy Saving of Nonferrous Metallurgy, 2(1):11–15.

[12] Wang Meirong, Jia Huaijie, Hao Qian. (2013). Studies on the processing and recycling of ammonia-nitrogen wastewater from earths factories. Chinese rare earths, 31(6):46–51.

[13] Xue Deming, Hong Gongwei, Wu Guofeng, Li Xiaoyan. (2000). Concentration and recovery of ammonium salts in the waste from rare-earth industry by electrodialysis. Membrane science and technology, 20(2):61–65.

[14] Xiaohui Le, i Norio Sugiura, Chuanping Feng, et al. (2007). Pretreatment of anaerobic digestion effluent with ammonia stripping and biogas purification. Journal of Hazardous Materials, 145: 391~397.

[15] Zou anhua, Sun Tichang, Song Cunyi, Xing Yi, Guo Suhong. (2007). Recovery of magnesium ammonium phosphate from ammonia nitrogen wastewater by chemical precipitation. [J]. Journal of university of science and technology beijing, 29(6):562–566.

[16] Zhou Youxin, Qi Jianxiang. (2010). Blow-off experimentation on treating high ammonia nitrogen wastewater in rare-earth smelting. Coal technology, 29(12):207–209.

Material Science and Environmental Engineering – Chen (Ed.)
 ISBN 978-1-138-02938-5

Water resources analysis and assessment of recent ten years in Tianjin district

M.C. Li, Q. Si & Y. Wang
Laboratory of Environmental Protection in Water Transport Engineering, Tianjin Research Institute of Water Transport Engineering, Tianjin, China

ABSTRACT: The shortage of water resources is an important problem for Tianjin district. Following the development of regional economy, especially the construction of Tianjin Binhai new district, this phenomenon is aggravated continuously. Hence, many water diversion projects have been considered for Tianjin district, such as the south-to-north water diversion project. The water resources is very important for the survival of human being, so it is very important for analyzing and evaluating the water resources for guiding the management and control. In this paper, the water resources of recent ten years in Tianjin are analyzed. The results of analysis and assessment show the water resources are hugely shortage in recent years.

Keywords: water resources; analysis; assessment; Tianjin

1 INTRODUCTION

Following the development of national and regional economy, the demand of water resources is aggravated continuously in Tianjin. At the same time, the absence of water resources is very serious in Tianjin. So the analysis and assessment of the water resources is very important for guiding the management and control.

Water resources are the very important factor for the survival of human being. Especially, the absence of water resources is increasing obviously in the North China. So the research on the water resources have been studied widely to obtain the optimal management scheme in recent years, such as the fuzzy comprehensive evaluation for water resources carrying capacity [1], AHP method [2], set pair analysis method [3], SD model method [4], normal cloud model method [5], etc.

In this paper, the water resources index in Tianjin is analyzed. Coupling the advantages of set pair analysis and cloud theory, a synthetic assessment method and systematic procedure [6] are developed and applied to evaluate the water resource level.

2 RESEARCH DISTRICT

Tianjin is a municipality direct under the Central Government, as well as an opening city. It's situated in the eastern part of the North China Plain, covering an area of 11,300 km^2 (Fig. 1). As one of China's biggest industrial centers, Tianjin has built up an all-round industrial system with machinery, electronics, textiles, chemicals, metallurgy, foodstuff, etc. Following the development of regional economy, the demand of water resources has been become more and more hugely.

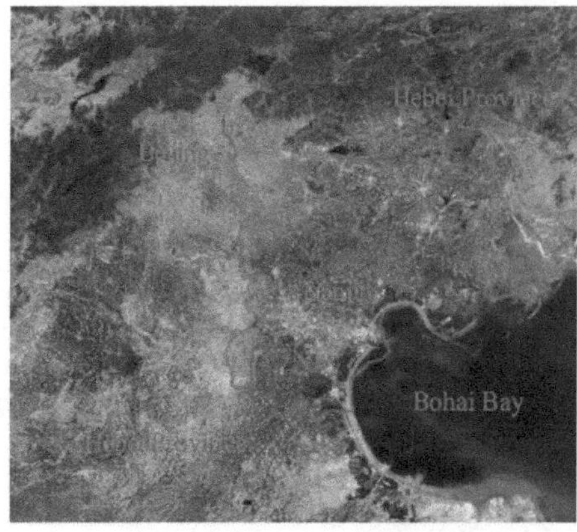

Figure 1. The position of research district.

3 ASSESSMENT METHOD

The characteristic of assessment is a combined process between assessment factors and assessment standard.

3.1 *Cloud theory*

Uncertainty is the basic characteristic of phenomena and things in the world. Uncertainty of quantity can be determined by the concept method and it has general significance. The randomness and ambiguity of concept can react in cloud theory by the mathematical expectation, entropy and hyper entropy. So the uncertainty transformation is achieved between concept and quantity.

Cloud theory [7] is developed based on probability and fuzzy mathematic. Its basic algorithm is to build an uncertainty transformation model for the exchange between concept and quantity. The randomness and ambiguity can reacted in this model, which has been applied to system evaluation, algorithm improvement, decision support, intelligent control, data mining, knowledge discovery and network security [8].

3.2 *Set pair analysis theory*

Uncertainty set pair analysis theory [9] can effectively deal with the high dimensional nonlinear problem owing to the combination of certainty and uncertainty. So the set pair analysis method is used to many fields such as mathematics, economy, resources and environment, etc.

3.3 *Coupled method*

Coupling the advantages of set pair analysis and cloud theory, a synthetic assessment method and systematic procedure [6] are developed and applied to evaluate the water resource level. The procedure of coupled method is same as the carrying capacity assessment [6].

4 WATER RESOURCE ANALYSIS AND ASSESSMENT

The Average per capita water resources in recent ten years of Tianjin is analyzed in this paper.

Coupled with the assessment standard, the water resource level is evaluated by the coupled method.

4.1 *Analysis*

The Average per capita water resources in recent ten years of Tianjin is shown in Figure 2.

Figure 2 shows the average per capita water resources time series in the research district. The biggest value of average per capita water resources occurs in 2012, which is the rainstorm's influence around the North China district.

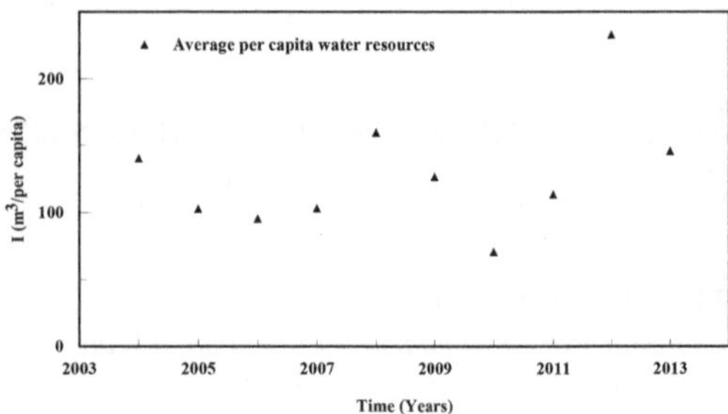

Figure 2. The average per capita water resources in recent ten years of Tianjin.

Table 1. Level standard of assessment index.

	Level of indexes			
Index	Good (I)	General (II)	Poor (III)	Very poor (IV)
Apcwr	≥1700	1700–1000	1000–500	≤500

Note: Apcwr is average per capita water resources.

4.2 *Assessment*

Evaluation index of water resources is selected for testing the coupled method. Figure 2 shows the average per capita water resources time series. The assessment standard [10] is used to evaluate water resources level listed in Table 1.

Following the coupled assessment procedure, the water resource level of Tianjin in recent ten years is evaluated. The water resources levels are all very poor (IV) in recent ten years.

5 CONCLUSIONS

In this paper, the average per capita water resources of recent years in Tianjin district is analyzed. Coupling the advantages of set pair analysis and cloud theory, a synthetic assessment method and systematic procedure are applied to evaluate the water resource level. The assessment results show that the water resources levels are all very poor (IV) in recent ten years. So the water resources protection and optimal management is an important work for Tianjin district.

ACKNOWLEDGMENT

This work was supported by the National Natural Science Foundation of China (No. 51209110), the project of Science and Technology for Development of Ocean in Tianjin (KJXH2011-17) and the National Nonprofit Institute Research Grants of TIWTE (TKS130215 and KJFZJJ2011-01).

REFERENCES

[1] M. Liu, Z.L. Nie, J.Z. Wang, L.F. Wang, Fuzzy comprehensive evaluation of groundwater resources carrying capacity in North China plain. Bulletin of Soil and Water Conservation, vol.34, No.6, pp. 311–315, 2014.

[2] Z.Z. Song, Y.L. Gu, To evaluate water resources carrying capacity in all districts of Ningxia based on AHP, Natural Science Journal of Harbin Normal University. Vol.31, No.1, pp. 60–62, 2015.

[3] L.L. Chen, Regional carrying capacity evaluation of water resources based on set pair analysis, Hydropower and New Energy. No.127, pp.1–4, 10, 2015.

[4] Y. Luo, B.S. Yao, The water resources carrying capacity based on SD model in Changsha, China Rural Water and Hydropower. No.1, pp.42–46, 2015.

[5] G.H. Wei, L. Ma, Evaluation of regional water resources carrying capacity based on normal cloud model, Water Saving and Irrigation. No.1 pp.68–71, 2015.

[6] Li Ming-chang, Zhang Guang-yu, Si Qi, Research on district and assessment method of inorganic nitrogen in multi-waterway marine waters. Asia Navigation Conference, 2013.

[7] Li De-yi, Du Yi. Artificial intelligence with uncertainty. Beijing: National Defence Industry Press, 2005. (in Chinese).

[8] Fu Bin, Li Dao-guo, Wang Mu-kuai. Review and prospect on research of cloud model. Application Research of Computers, 2011, 28(2): 420–426. (in Chinese).

[9] Zhao Ke-qin. Set pair analysis and its application. Hangzhou: Zhejiang Science and Technology Press, 2000. (in Chinese).

[10] Wan Xiao-ming. Study on standard system of water resource sustainable utilization. Hohai University. 2005. (in Chinese).

Material Science and Environmental Engineering – Chen (Ed.)
© 2016 Taylor & Francis Group, London, ISBN 978-1-138-02938-5

Study on total amount control of water pollutants in county areas based on pollution receiving red-line

X. Zhang, Z.C. Dong & G.D. Wu
State Key Laboratory of Hydrology-Water Resources and Hydraulic Engineering, Hohai University, Nanjing, China

Y.X. Ding
Jiangsu Surveying and Design Institute of Water Resources, Yangzhou, China

ABSTRACT: Water pollution is widely distributed in China, imposing a huge threat on the economic development of local areas. To address this problem, in this study, the total water pollution control system is established. Pollution receiving red-line theory is applied in this system. The Sihong County is used as a case, with the control on the total amount of water pollutants in 2020. This study can not only improve local water environment but also provide a reference of water pollution control for other counties alike.

Keywords: the total water pollutant control; pollution receiving red-line; county areas

1 INTRODUCTION

The concept "total water pollutant control" was first proposed in the late 1960s by the Japanese scholars, and the purpose was to improve the water environment by taking effective measures to control the total amount of water pollutants within a given area (Helmer and Hespanhol 1997). In 1997, US Environmental Protection Agency proposed the Total Maximum Daily Loads Plan (USEPA 1999). This plan requires the calculation of maximum pollutants loads under the condition of meeting water quality standard. China attempted to control the total amount of water pollutants in the late 1970s, during which total BOD control standards in the Songhua River was worked out initially (Bao et al. 2000). In 1998, the National Environmental Protection Agency proposed that the pollutants control made a change from the concentration of pollutants to the total pollutant amounts. After that, the control on the total amount of water pollutants experienced three stages: total control of goals, total control of capacities and the combined control of goals and capacities.

In recent decade, with the growth of the county economy, increasing water environmental issues have become an important factor restricting the county's economic, social modernization and the development of water conservancy (Wu et al. 2006). In order to address this issue, we should pay more attention on protecting water environments. Therefore, how to control pollutants in water bodies has become a hot topic. In this paper, combining the current water management system with the total amount of water pollution receiving red-line theory, from a macro-perspective, we restricted the total emissions to improve the water environment.

2 THE TOTAL WATER POLLUTION CONTROL SYSTEM

National documents in 2011 proposed the concept of "three red lines"; the third red line is to establish water function areas restricting water pollutants receiving red-line, and to control the total amount of water pollutants discharging into water bodies. In consideration of the characteristics for a county, a total water pollution control system is established as shown in Figure 1.

Pollution source analysis as a basis for the control on water pollutants, aims to find out the sources of pollutants in a river; thereby the amount of pollutants emitted can be reasonably calculated and the source control can be strengthened. Usually, pollution sources can include point sources and non-point sources and sometimes endogenous and foreign sources exist in some areas.

Amphibious relationship analysis mainly aims to determine the corresponding relationship between the water function area and the control area of the land surface confluence so that pollutant amount discharging into rivers in individual water function areas can be summed up. This corresponding

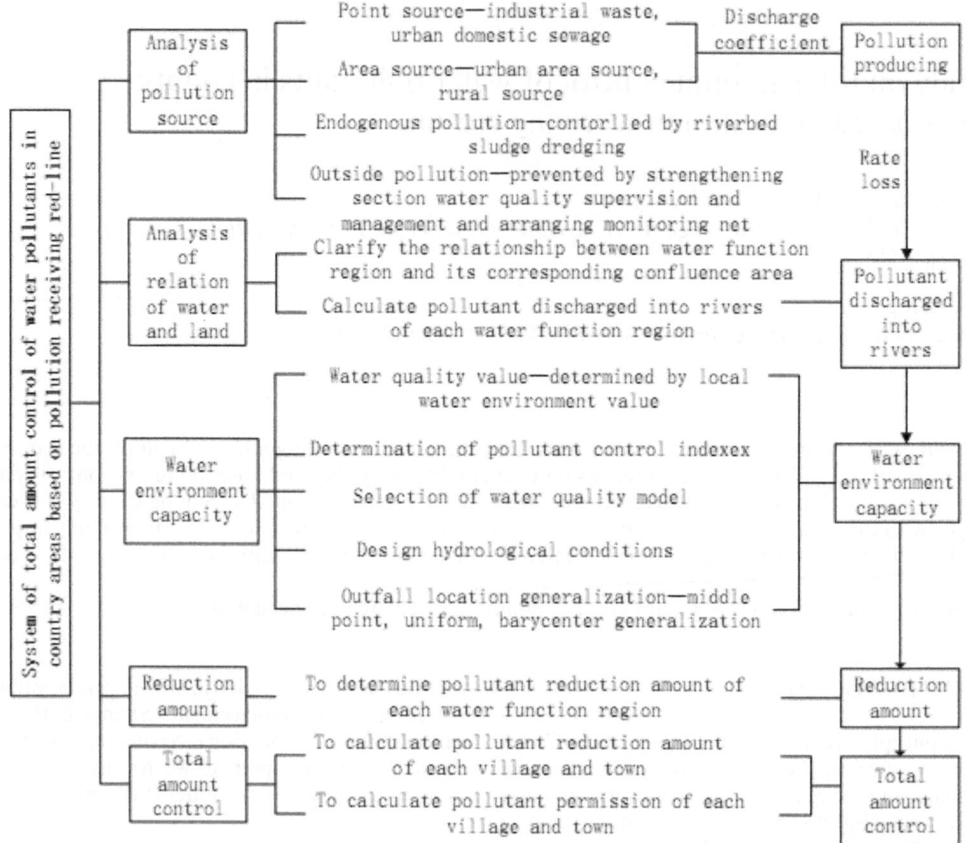

Figure 1. System of total amount control of water pollutants in county areas based on pollution receiving red-line.

relationship is based on town administrative divisions, river systems, distribution of sewage outfall into rivers, topography.

Water environmental capacity analysis is the core of the whole control system, providing a scientific basis for total amount controls. In this paper, water environmental capacity is ideal, which is generalized by selecting appropriate water quality models and sewage outfalls according to local environmental goals and water pollution features.

Reduction amount analysis is to determine the amount of water pollutants reduced in individual function areas according to the amount of pollutants discharging into rivers and pollution receiving capacity in corresponding functional areas. This relationship is shown in Equation 1.

$$W_R = W_I - W_C \tag{1}$$

where W_R is the amount of water pollutants reduced, W_I is the amount of pollutants pouring into rivers and W_P is pollution receiving capacity.

In this paper, CODcr and NH_3-N are selected as pollutants (Wei et al. 2006; Mao and Wang 2003). These pollutants comprised point sources and non-point sources. Therefore there is Equation 2 below:

$$W_I = \alpha_P W_P + \alpha_A W_A \tag{2}$$

where W_P, W_A is the amount of point sources and non-point sources respectively and α_P, α_A are the rate loss in rivers for point sources and non-point sources respectively.

In addition, allowable emission amounts in individual towns are made, ensuring the sustainable use of water resources. The weight of pollutants emission in a town is determined by its corresponding function area and this weight is then assigned to amounts of pollutants reduction in this town. The towns as the basic statistical units (see Equation 3) determine the reduction targets. Allowable emissions are the difference between emissions and reduction targets (Equation 4).

$$W_{R2} = \sum_{i=1}^{n} \alpha_i * W_{R1i} \quad (3)$$

$$W_{AL} = W_E - W_{R2} \quad (4)$$

where W_{R2}, W_{AL}, W_E are amounts of reduction, allowable emissions and emissions in a town, respectively. α_i is the weight of pollutants emission in a town on a functional area, n is the number of water functional areas for a town.

3 CASE STUDY

2010 is chosen as the base year and the control on total mounts of water pollutants in 2020 of Sihong County is studied based on established control system.

3.1 *Regional overview*

Sihong County is located in the western part of Jiangsu Province, middle reaches of the Huaihe River, and the western side of Hongze Lake. It consists of 14 towns and 9 villages and the total area is 2,731.4 km^2. The design sizes of three sewage treatment plants (south, north, and Shuanggou) are 50,000 t/d, 50,000 t/d and 15,000 t/d respectively. In recent years, the Sihong County has accelerated the process of industrialization, urbanization and agricultural industrialization. However, the discharge of waste water has imposed a serious influence on water ecological environment. Most rivers to some extent are threatened by various pollutants. In 2010, the number of sections which did not meet the standard was 12, which accounted for 80% of the total 15 water quality sections in major rivers of the county. The water environmental degradation not only affects economic development but also poses a significant threat to human health.

3.2 *Calculation of water environment capacity*

The calculation area (comprising 20 major rivers and one lake) consists of 26 function regions—2 conservancy regions (Hongze Lake and Xuhong River), 5 buffer regions, 2 reservation regions and

Table 1. Predictions of water environment capacity of each water function region in 2020.

River	Water function region	Q (m^3/s)	u (m/s)	Water environment capacity (t/a)	
				CODcr	NH_3-N
Hongze lake	C	–	–	0	0
Huaihe river	B	31.5	0.02	3181	178
Huaihongxinhe	B	2.5	0.03	386	21
river	R	2.5	0.05	462	24
Xuhonghe river	C	–	–	0	0
Xinbianhe river	B	3	0.1	78	4
	R	3	0.12	385	19
Xinsuihe river	B	5	0.05	738	39
	A	5.02	0.06	388	20
Laosuihe river	B	0.2	0.05	46	2
	A	0.21	0.03	25	1
Ximinbianhe river	A	0.35	0.003	162	22
Tonghe river	A	2	0.1	83	4
Xishahe river	A	0.25	0.01	122	9
Suihe river	P	0.32	0.003	301	33
	T	0.42	0.003	106	24
Lanshanhe river	A	0.15	0.01	75	10
Liminhe river	A	0.18	0.003	84	11
Laobianhe river	T	0.45	0.005	119	26
Andonghe river	A	0.3	0.003	141	19
Wanggouhe river	A	0.05	0.001	22	3
Douhuaixinhe river	A	0.02	0.001	9	1
Gaohuaixinhe river	A	0.03	0.001	14	2
Suibeihe river	A	0.07	0.003	33	4
Zaochenhe river	A	0.1	0.004	47	6
Lugouhe river	A	0.07	0.002	31	4

17 development regions (14 agricultural water consumption regions, 2 transitional regions and 1 pollutant discharge control region). According to sewage water emission and water quality conditions, CODcr and NH_3-N are used as the control indexes of total pollutants. One-dimensional water quality model is selected to calculate water environment capacity according to river characteristics and the formula is as follows:

$$M = \left[C_S \exp\left(\frac{kL}{2u}\right) - C_0 \exp\left(\frac{-kL}{2u}\right) \right] Q \tag{5}$$

where M is water environmental capacity of function regions, t/a; Q is design flow in computation reaches, m^3/s; L is the length of computation reaches, m.

1. Determination of the Model Parameters
   Water quality objectives of water function regions C_s is selected according to "Surface Water Quality Standards" (GB3838-2002) and "Surface Water (Environment) Function Regionalization of Jiangsu Province". The initial concentration C_0 is assumed to be target value of upstream water quality. If the computation reach was headwater, C_0 is assumed to be headwater quality.

Integrated attenuation coefficient of pollutants k is determined according to results from "Comments on Water Environment Capacity and Limiting Sewage Amount of Huaihe River Basin": $k_{CODcr} = 0.05 + 0.68u$ (l/d); $k_{NH3\text{-}N} = 0.061 + 0.551u$ (l/d), where u is the design flow rate, m/s.

In terms of design flow Q, two occasions were consideration: ① To rivers with enough runoff data, we choose the most withered month average discharge in 90% guarantee rate to be design flow Q; ② To rivers with no runoff data, we do rainfall frequency analysis based on 1956–2010 rainfall data of Sihong Station and select a typical year (the annual precipitation is close to annual precipitation in 90% guarantee rate) to do rainfall-runoff analysis. The average runoff is calculated based on the statistics that average annual rainfall during the dry season accounts for 40% of annual rainfall, and design flow Q equals the average runoff multiplied by the river catchment area. If there are not flow rate data, design flow rate u is the ratio of design flow and wetted area.

2. Results
   According to the target value and boundary conditions above, water environment capacity of each function region is calculated by using

Table 2. Pollutant discharge of each village and town in 2020 (the part treated by sewage treatment plant was deducted.

	CODcr (t/a)		NH_3-N (t/a)	
Village and town	Point source	Area source	Point source	Area source
Qingyang town	6523.46	6.16	905.08	177.46
Shuanggou town	1309.58	3.21	205.75	16.12
Sihe village	235.25	2.52	44.4	13.07
Fengshan village	149.41	1.49	27.39	8.7
Tian'ganghu village	140	1.82	24.36	10.81
Shangtang town	332.51	3.5	55.01	21.27
Weiying town	298.2	2.12	47.12	11.24
Chemen village	270.62	1.52	50.95	11.52
Yaogou village	581.71	1.27	120.31	9.08
Shiji village	122.11	1.76	21.4	10.02
Chengtou village	215.78	1.58	40.38	9.55
Linhuai town	251.53	0.42	39.45	2.41
Chenwei village	237.14	2.09	42.74	14.25
Bancheng town	231.65	0.54	37.91	3.44
Sunyuan town	314.35	2.37	51.99	15.16
Meihua town	263.71	1.69	46.98	12.59
Guiren town	404.73	2.21	64.24	18.95
Jinsuo town	279.68	1.25	46.09	9.81
Caomiao village	141	2.27	23.12	13.02
Zhuhu town	357.35	1.71	64.08	12.5
Jieji town	236.45	3.41	35.16	17.91
Taiping town	271.66	3.05	43.75	17.13
Longji town	283.1	2.06	43.08	13.45

selected water quality model. The conservancy regions and drinking water source area are prohibited to discharge pollutants into, so water environment capacity of the two types of areas is zero. The total water environment capacities of CODcr and NH_3-N of all function regions in 2020 were 7038 t/a and 486 t/a separately (Table 1).

3.3 *Calculation of total pollutant discharge*

According to pollution source monitoring data provided by Sihong Hydrology Bureau and Environmental Monitoring Station, pollution sources in Sihong County consist of point source, area source, and endogenous pollution and outside pollution. In this study, we take point source (industrial waste, urban domestic sewage) and area source (rural domestic sewage, livestock and poultry manure, chemical fertilizer and pesticide pollution) into consideration. Endogenous pollution is controlled by riverbed sludge dredging and outside pollution is prevented by strengthening section water quality supervision and management and arranging water quality monitoring network. The calculation formulas of discharge of point pollution and area pollution are as follows:

$$Q_{ind} = Q_s \times c \tag{6}$$

$$Q_u = P_u \times d_u \tag{7}$$

$$Q_r = P_r \times d_r \tag{8}$$

$$Q_l = P_l \times d_l \tag{9}$$

$$Q_c = S_c \times d_c \tag{10}$$

where Q_{ind} is pollutant discharge from industrial wastewater; Q_s is discharge of industrial wastewater; c is concentration of pollutants; Q_u is pollutant discharge from urban domestic sewage; P_u is urban population; Q_r is pollutant discharge from rural domestic sewage; P_r is rural population; Q_l is pollutant discharge from livestock and poultry manure; P_l is the number of livestock and poultry; Q_c is pollutant discharge from chemical fertilizer and pesticide pollution; S_c is the area of cultivated land; d_u, d_r, d_l and d_c are all discharge coefficients.

Table 3. Predictions of pollutant discharged to rivers and reduction amount of each water function region in 2020.

River	Water function region	Pollutant discharged to rivers (t/a)		Reduction amount (t/a)	
		CODcr	NH3-N	CODcr	NH3-N
Hongze lake	C	0	0	0	0
Huaihe river	B	235.8	46	0	0
Huaihongxinhe	B	9.2	1.6	0	0
river	R	1211.5	179.7	749.5	155.7
Xuhonghe river	C	0	0	0	0
Xinbianhe river	B	2.2	0.3	0	0
	R	914.3	190.1	529.3	171.1
Xinsuihe river	B	6.9	0.9	0	0
	A	6	0.3	0	0
Laosuihe river	B	11	0.6	0	0
	A	20.8	0.7	0	0
Ximinbianhe river	A	6.2	1.2	0	0
Tonghe river	A	2.8	0.3	0	0
Xishahe river	A	7.9	0.9	0	0
Suihe river	P	1830.5	206.4	1529.5	173.4
	T	278.2	46	172.2	22
Lanshanhe river	A	265.9	47.9	190.9	37.9
Liminhe river	A	787.5	133.4	703.5	122.4
Laobianhe river	T	2519	329.3	2400	303.3
Andonghe river	A	364	59.7	223	40.7
Wanggouhe river	A	11.5	1.6	0	0
Douhuaixinhe river	A	2.2	0.3	0	0
Gaohuaixinhe river	A	6.9	0.9	0	0
Suibeihe river	A	6	0.3	0	0
Zaochenhe river	A	42.2	1.9	0	0
Lugouhe river	A	7.1	1.2	0	0

Under the condition of the unchanged industrial structure of Sihong County, pollutant concentration of industrial wastewater in 2020 is estimated to be present value (the concentrations of CODcr and NH_3-N are 116 mg/L and 25 mg/L respectively). Other related parameters were determined according to "Twelfth Five-Year Plan of Sihong County", "Total urban planning of Sihong County during 2011–2030" and "Integrated Water Resources Planning of Suqian City, Jiangsu Province". Based on formulas above, we estimated pollutant discharge of each village and town of Sihong county in 2020 (Table 2).

3.4 *Amounts of pollutants discharged to rivers and reduction amount*

Based on water and land relation, we estimate pollutant discharge to each water function region. Point pollutant discharged into which function region is determined by the outfall location. The amount of area pollutant discharged into water function region is estimated based on river catchment area. The amount of total pollutant discharged into each function region equals the amount of pollutant multiplied by the rate loss of pollutant, and the reduction amount equals the quantity of pollutant discharged into rivers subtracting water environment capacity. The amounts of CODcr and NH_3-Nin all function regions discharged into rivers are 8556 t/a and 1252 t/a respectively, and the reduction amounts of CODcr and NH_3-N are 6498 t/a and 1026 t/a separately (Table 3).

3.5 *Countermeasure*

Amount of point pollutants could be reduced by constructing sewage treatment plants to strengthen sewage treatment and improving water reuse rate, etc. The polluted area could be reduced by accelerating the construction of ecological agriculture, optimizing the development of animal husbandry and improving the rural clean facilities, etc. We should combine point-pollutants reduction with area-pollutants reduction to control pollutant discharge effectively, and area-pollutants reduction should be complementary while point-pollutant should be primary.

4 CONCLUSIONS

The sustainable utilization of water resources in the county is an important support for the sustainable development of economy and society. Faced with the current situation of water resources protection, pollution receiving red-line is introduced to the total control on water pollutants. The total water pollution control system is established, refining the approved process of water environmental capacity and inversely calculating allowable emissions. The Sihong County is used as a case. The total water environmental capacities of CODcr and NH_3-N of all function regions in 2020 are 7038 t/a and 486 t/a respectively. The amounts of CODcr and NH_3-Nin of all function regions discharged into rivers are 8556 t/a and 1252 t/a respectively, and the reduction amounts of CODcr and NH_3-N are 6498 t/a and 1026 t/a respectively. The rational calculation process and results in this study provide an important basis for preventing water pollution and reducing pollutant discharge in local villages and towns. Besides that, it could also provide a reference for other counties to control total water pollution and to improve local water environments.

REFERENCES

[1] Helmer, R. and Hespanhol, I. 1997. Water pollution control: a guide to the use of water quality management principles, St Edmundsbury Press, Suffolk.

[2] USEPA. 1999. Protocol for developing nutrient TMDLs Washington DC: USEPA.

[3] Bao, C.K., Zhang, M., Shang, J.C. 2000. Research on Total Quantity Control of Pollutants Emission in the Songhua River Basin of Jilin Province, Scientia Geographica Sinica. 20(1): 60–64 (in Chinese).

[4] Wu, Y.Y., Li, Y.S., Liu, W.J. 2006. Study on Gini coefficient method of total pollutant load allocation for water bodies. Research of Environmental Sciences. 19(2) (in Chinese).

[5] Wei, C.H., Huang, X., Wen, X.H. 2006. Pilot study on municipal wastewater treatment by a modified submerged membrane bioreactor, Water Science & Technology. 53(9): 103–110.

[6] Mao, Y.Y., Wang, X.D. 2003. Pollutants bearing capacity of Huaihe River basin in Jiangsu Province, Water Resources Protection. 19(3) (in Chinese).

Material Science and Environmental Engineering – Chen (Ed.)
 ISBN 978-1-138-02938-5

Study on Ion Selective Electrode in ammonia nitrogen monitoring application of water and soil loss

Y. Zhao
Water Resources Research Institute of Shandong Province, Ji'an, China
Shandong Provincial Key Laboratory of Water Resources and Environment, Ji'an, China

ABSTRACT: Runoff nutrient monitoring of water and soil loss usually adopts field acquisition followed by laboratory analysis which is very complex. To verify the reliability of the rapid nutrient measurement method of Ion Selective Electrode (ISE) in water and soil loss monitoring, the nitrogen ammonia ISE as an example was compared with traditional Nessler reagent spectrophotometry by simulating rainfall. Studies were conducted with regard to accuracy, anti-ion interference and migration rule in slope runoff. Experimental results showed that: With confidence level of 95%, results of the two methods are no significant difference; the maximum value of measurement range of ISE is 1000 mg/L and that of spectrophotometric method is 2 mg/L; ISE can automatically resist ion interference; when the power function model is used to simulate ammonia nitrogen migration in slope runoff, the fitting coefficient of measurement value of electrode method is 0.9537 and that of spectrophotometric method is 0.9305.

Keywords: Ion Selective Electrode; nutrient monitoring of water and soil loss; ammonia nitrogen

1 INTRODUCTION

Owing to unreasonable fertilization in quantity, nitrogen and other nutrients in slope farmland running off with runoff and sediment have become one of important sources of non-point source pollution of agriculture [1]. Nitrogen loss not only causes Lake Eutrophication, but also influences human health. Ammonia nitrogen has been included in human health risk evaluation indices in developed countries as non-cancer gens [2].

Chinese scholars have researched a lot about nitrogen loss rule in farmland and explored related rainfall runoff, soil characteristics, land utilization mode, fertilization conditions and other influence factors [3–4]. Monitoring for nitrogen substance is indispensable in these research processes. Runoff samples collected in the field are usually processed and divided to mud sample and water sample in regular water and soil loss monitoring. In addition, they are analyzed in the laboratory and summation of them represents total nutrient loss quantity caused by erosion. Taking measurement of ammonia nitrogen as an example, the water sample often adopts Nessler reagent spectrophotometry, gaseous phase molecular absorption spectroscopy, flow injection analysis method, etc [5–6]. Ammonia nitrogen in mud sample adopts soil physicochemical method for pretreatment before measuring filtering fluid [7]. Although these methods are of high accuracy and strong universality, instruments are expensive and cannot be carried in the field, with high requirements for operation skill, which is tedious and complicated for experimenters. At present, there are many methods for nutrient monitoring of water and soil loss in the country and automation monitoring equipment of water and soil loss with high integration level and rapid monitoring method is still lacked.

Ion selective electrode method is characterized with simple operation, rapid measurement, convenient carrying, which has obtained good effect in environmental protection industry of online sewage monitoring. However, it is rarely reported that ion selective electrode is applied to nutrient monitoring of water and soil loss. To verify the reliability of the rapid nutrient measurement method in water and soil loss monitoring, the paper compared electrode method with the mostly frequently-used Nessler reagent spectrophotometry in actual nutrient monitoring of water and soil loss (taking ammonia nitrogen as an example) and conducted experiment research in the hope of providing a rapid and practical monitoring method for nutrient monitoring of water and soil loss.

2 EXPERIMENTAL

2.1 *Reagent and instrument*

1. Ammonium standard solution. Weigh 3.819g dried ammonium chloride (NH_4Cl) at 100 °C

to be dissolved in water, move it to 1000 mL volumetric flask, dilute to marking line, shake uniformly and obtain standard stock solution of ammonium. The solution contains 1 mg ammonia nitrogen per ml; put standard stock solution of 5 mL ammonium to 500 mL volumetric flask, dilute to marking line with non-ammonia water, shake uniformly and obtain ammonium standard working solution with concentration of 0.01 mg/mL. The preparation methods of standard solutions with other concentrations are the same as above.

2. Sodium potassium tartrate solution. Weigh 50 g $KNaC_4H_6O_6 \cdot 4H_2O$ to be dissolved in 100 mL water. Heat and boil to remove ammonia, cool it and fix volume to 100 mL.
3. Nessler reagent. Weigh 16 g sodium hydroxide, dissolve it in 50 mL water, fully cool to room temperature. Additionally weigh 7 g potassium iodide and 10g mercury iodide and dissolve them to water, slowly inject the solution to sodium hydroxide solution while agitation, dilute to 100 mL with water, store in polyethylene bottle and keep airtightly.
4. ISA solution (ion strength adjustment) 1 mol/L sodium sulfate is used to adjust measurement conditions of approaching electrical conductivity of calibration solution. The electrode is not sensitive to composition of ISA solution.
5. Non-ammonia water Prepare according to national standard HJ 636-2012 [8].
6. Tu-1810 ultraviolet and visible spectrophotometer (Made in Beijing).
7. MPS-K-16 water quality monitoring system and analysis software (Made in Germany).

2.2 *Preparation of runoff sample*

Manual rainfall simulator and mobile slope-variation steel tank (length × width × depth 200 cm × 100 cm × 20 cm) are used for test under unsaturated soil of brown soil in northern earth-rock mountain region in National Water and Soil Conservation Demonstration Park of Qilong Bay, Laiwu City, Shandong Province. Clay fertilizer is allocated artificially and urea is applied as 1000 kg/hm^2; rainfall intensity is set up as a constant, 50 mm/h, with gradient of 20°; plastic drum is used to obtain runoff sample once every other 5min and filter paper is used for filtration and spare use.

2.3 *Measurement by Nessler reagent spectrophotometric method (method 1)*

1. Test principle: Ammonia nitrogen in the form of free ammonia or ammonium ion reacts with Nessler reagent to generate light red brown complex and the absorbance is in direct proportion to ammonia nitrogen content. Absorbance is measured at wavelength 420 nm.
2. Calibration curve: Respectively put 0, 0.5, 1, 3, 5, 7 and 10 mL ammonium standard solution in a series of 50 mL colorimetric tube, dilute to the marking line with non-ammonia water, prepare standard series solution, add 1 mL potassium sodium tartrate solution to standard series solution and add 1.5 mL Nessler reagent, shake uniformly, put for 10min, measure absorbance in 20 mm cuvette with distilled water as reference under wavelength 420 nm.
3. Blank test
   Replace water sample with water and carry out pretreatment and measurement with the same step as the sample.
4. Result calculation
   Calculate mass concentration of ammonia nitrogen with regression equation (1) of standard curve:

$$y = bx + a \tag{1}$$

where y = calibration absorbance, namely, difference between absorbance of standard solution and blank absorbance of reagent; x = ammonia nitrogen content (calculate with NH_4^+) mg; b = slope of regression equation; a = intercept of regression equation.

2.4 *Measurement by Ion selective electrode method (method 2)*

1. Test principle
   The top of the ion selective electrode in contact with water sample is a layer of sensitive membrane which is selective towards specific ions. When the selective membrane of electrode contacts ion solution, certain electric potential is generated in and out of the membrane. Size of such electric potential depends on activity of free ion in the solution. The parameter electric potential is converted to concentration of corresponding ion through Nernst equation and calculation equation (2) is

$$U = U_0 \pm K\log(\alpha_i) \tag{2}$$

where: U = measurement potential value, mV; U_0 = potential value of reference electrode, mV; α_i = ion activity, mol/L; K = electrode slope. When the ion in the solution is positive ion, "+" is used for expression in the equation; when the ion in the solution is negative ion, "–" is used for expression in the equation.

NH_4^+ ion selective electrode adopts Teflon membrane in MPS-K-16 water quality monitoring system. As for other interference ions, special test electrode is used for compensation and analysis software is used to eliminate impact of interference ion.

2. Electrode calibration
   Analysis software supports calibration from 2 points to 6 points at highest. As the detection limit of Nessler reagent spectrophotometric method in standard HJ 535-2009 is 0.025 mg/L and measurement scope is 0.1–2 mg/L, it is sufficient to adopt 2 points for calibration. Prepare NH_4^+ standard solution with concentration of 0.2 and 2 mg/L respectively, add 5 mL and ISA to adjust scope of electric conductivity. The calibration starts from minimum concentration and proceeds successively. Before the electrode is used, it shall be cleaned with distilled water and dried in the air. At the time of calibration, the electrode shall be put in the solution for about 3 min and values can be read after it becomes stable. Calibration concentration and corresponding measurement value shall be input in software.
3. Measurement
   Put the electrode in the solution to be tested for about 3min and directly read measurement value on the interface of Analysis software after it becomes stable; set up time interval, such as 5min before measuring continuously and finally output results from the computer.

3 RESULTS AND DISCUSSION

3.1 *Data accuracy with different methods*

To verify whether the Nessler reagent spectrophotometric method (method 1) and ion selective electrode method (method 2) has the same accuracy to measure ammonia nitrogen with the same level, two methods are respectively used to measure ammonia nitrogen in the same runoff sample in parallel. 6 measurements are required. On this basis, consistency of the two groups of data is judged with mathematical statistic method (t inspection). Measurement results of the two methods are shown in Table 1.

According to mathematical statistic method (t inspection), $x_1 = 1.65$, $x_2 = 1.634$ (x-sample average), $s_1 = 0.01169$, $s_2 = 0.00737$ (s-sample standard deviation), degree of freedom $f = n_1 + n_2 - 2 = 10$ (n-sample size), statistics $t = 1.614$. Corresponding significance level is 0.05, $t_{0.05}(10) = 2.228$ with degree of freedom of 10. As $t < t_{0.05}(10)$, the measurement results of two experiment analysis methods have no significant differences in measurement of ammonia nitrogen of blind sample at a confidence level of 95%. Namely, they have the same accuracy.

3.2 *Comparison test of anti interference ability*

Take ammonium solution with 3 different concentration levels, high, medium and low in ion interference experiment, respectively add common Ca^{2+} interference ion with 3 different concentration levels for orthogonal experiment. The experiment condition is that: masking reagent potassium sodium tartrate solution masking metal ion is not added in Nessler reagent spectrophotometric method (method 1); at the same time, as concentration part of ammonium solution exceeds measurement scope of Nessler reagent spectrophotometric method, dilution method is used to make it within the measurement scope of the method.

As seen from measurement result of ammonia nitrogen content of runoff sample in Table 2, with existence of interference ion Ca^{2+}, the result produces drastic positive interference if Nessler reagent spectrophotometric method (method 1) is used to measure NH^{4+} value. Namely, the interference degree increases with increasing of concentration of interference ion. With low-concentration ammonium solution (concentration of ammonium solution is 1.00 mg/L), impact on measurement of ammonia nitrogen is significant. Ca^{2+} increases from 40 mg/L to 160 mg/L and

Table 1. Measurement results of ammonia nitrogen content in the runoff.

Measurement methods	Method 1	Method 2
Parallel measurement results of ammonia nitrogen content ($mg \cdot L^{-1}$)		
1	1.63	1.628
2	1.64	1.633
3	1.65	1.641
4	1.63	1.63
5	1.64	1.645
6	1.66	1.627
Average x	1.65	1.634
Standard deviation s	0.01169	0.00737

Table 2. Measurement results of interference experiment.

Monitoring methods	Concentration of interference ion Ca^{2+} ($mg \cdot L^{-1}$)	NH_4^+ measurement value ($mg \cdot L^{-1}$)		
Standard value		1.00	10.00	30.00
Method 1	40	1.02	10.22	30.15
	80	1.18	10.92	31.36
	160	1.25	12.10	32.02
	Average	1.15	11.08	31.18
Method 2	40	0.988	10.082	30.098
	80	1.022	10.013	30.067
	160	1.031	10.151	30.204
	Average	1.014	10.082	30.123

interference degree increases from 2% to 25%; with increasing of concentration of ammonium solution, the interference degree of interference ion for measurement result reduces. Namely, when the concentration of ammonium solution is 30.00 mg/L, interference degree only increases to 6.73% from 0.5%. Reason for above result is that Ca^{2+} and other metal ions react with Nessler reagent to generate yellowish-brown complex and produce heterochrome or turbidity, resulting in increase of absorbance and result increase. The larger concentration of interference ion is, the larger interference caused will be. Generally, needle is added according to concentration of interference ion to eliminate masking agent of such metal ion. Hence, at the time of measuring blind sample, especially when interference ion has large concentration and no masking agent is added, the impact of Nessler reagent spectrophotometric method on measurement cannot be ignored. Masking agent must be added to eliminate error and standard curve shall be drawn again. If the interferent is other types of substances, pretreatment and other processes are required [9].

As for measurement result of ion selective electrode method (method 2), as MPS-K-16 has multiple probes, it can be used to measure a dozen common water quality parameters such as K^+, Ca^{2+}, Na^+, Cl^-, electric conductivity, pH value and the impact of interference ion can be eliminated if certain substance parameter is input on the interface of software. Hence, the measurement result of electrode method does not have enormous impact and is closer to standard value compared with Nessler reagent spectrophotometric method, without the need of pretreatment and addition of any substance.

3.3 *Ammonia nitrogen migration rule experiment of slope runoff*

Two methods, Nessler reagent spectrophotometric method and ion selective electrode method are respectively used to measure NH_4^+ solute concentration value in the runoff on the basis of preparation of runoff sample. NH_4^+ concentration values of runoff in different rainfall durations in the process of 30min rainfall after runoff generation are shown in Table 3.

Power function is adopted to fit change process of ammonia nitrogen concentration in the runoff sample with rainfall duration. Results show that: the change relationship between ammonia nitrogen concentration and rainfall duration in runoff sample measured by two methods can be fitted well with power function model and fitting coefficient R^2 is up to 0.9. It is consistent with deep model rule of mixing layer studied by Gang Kong [10], Quanjiu Wang [11], etc. In addition, the correlation coefficient of method 2 is larger. After above reasons are analyzed, it may be caused by different test principles of the two methods. Free ammonia and ammonia nitrogen existing in the form of NH_4^+ react with Nessler reagent to generate light red brown complex in spectrophotometric method and measurement results include trace free ammonia; however, the selective membrane of ion selective electrode generates certain electric potential in and out of the membrane, with regard to relationship with activity of free ion in the solution and measurement result is only NH_4^+ (ionic form). Hence, the fitting result is better. There may be other reasons and a lot of experiments are to be conducted with a view to exploring.

Table 3. Relationship between solute concentration of power function fitting runoff and change of rainfall duration.

Monitoring methods	Rainfall duration (min)	Ammonia nitrogen concentration in the runoff ($mg \cdot L^{-1}$)	Power function fitting equation	Correlation coefficient
Method 1	10	0.73	$Ck(t) = 1.102\,t^{-0.1918}$	0.9305
	15	0.63		
	20	0.61		
	25	0.60		
	30	0.58		
	35	0.56		
Method 2	10	0.723	$Ck(t) = 1.111\,t^{-0.1916}$	0.9537
	15	0.644		
	20	0.628		
	25	0.603		
	30	0.594		
	35	0.551		

Table 4. Analysis method comparison table.

	Ion selective electrode	Spectrophotometric method
Measurement scope	0.01–1000 mg/L, wide measurement scope	0.01–2 mg/L, dilution required for part out of range
Analysis speed	Measurement + result display = 5 min	Pretreatment + colorimetric determination = 1 h (standard curve takes more time)
Pretreatment	Pretreatment is not applicable to interference substance containing suspended matter, colored water quality, metals and interference shall be eliminated in the software interface, with short measurement period.	If there is residual chlorine in the sample, appropriate quantity of sodium thiosulfate solution can be added for removal and starch-potassium iodide test paper shall be used to detect whether residual chloride has been removed. Appropriate quantity of potassium sodium tartrate can be added at the time of developing to eliminate interference of calcium, magnesium and other metal ions. If the water sample is turbid or colored, it can be treated with pre-distillation method or flocculation and sedimentation method. The measurement period is long.
Experiment process	Zero point shall be calibrated automatically and working curve shall be created automatically to automatically display and store measurement date. All functions can be completed through computer and uninterrupted work in the field is allowed for a long time.	Analysts make calibration curve artificially, carry out colorimetric determination in spectrophotometer and input measured luminosity value to the computer. It can only be conducted in the laboratory.
Use cost	Directly measure, without addition of other reagents.	Chemical reagents are required.

The experiment result of ammonia nitrogen migration rule of slope runoff shows that: ion selective electrode method can be adequately applied and is of high reliability with regard to measuring ammonia nitrogen in specific field experiment.

3.4 *Comparison of two methods in other aspects*

Apart from above experiment, the paper has compared comprehensively in other aspects and obtained the comparison results of ion selective electrode and traditional spectrophotometric method in measurement of ammonia nitrogen (Table 4).

Application examples in actual field nutrient monitoring show that measurement data of MPS-K-16 monitor (taking ammonia nitrogen ion selective electrode as an example) are reliable, with extensive measurement scope and simple and rapid operation. At the same time, the monitor integrates multi-parameter data acquisition function and can carry out uninterrupted work in the field for a long time. In addition, the system can realize remote data wireless transmission. The instrument design is suitable for field working conditions and can be applied to severe environment conditions, such as sewage, strong acid and strong base occasion or high hot and arid regions. The probe can maintain long-term stability without the need of maintenance and can be widely applied to water and soil loss monitoring.

4 CONCLUSIONS

From the present investigation, the following conclusions may be drawn:

1. Ammonia nitrogen of parallel runoff sample measured by two methods is provided with *t* inspection and results have no significant difference at a confidence level of 95%.
2. The measurement scope of ion selective electrode method is 0.01–1000 mg/L, which is far more than that of spectrophotometric method, 0.01–2 mg/L and can be applied to water quality of various runoffs, including high-concentration irrigation water and seriously polluted river water. The electrode method has strong capacity of resisting ion interference. With existence of other iron interference, the measurement result is closer to standard value; however, spectrophotometric method needs to add masking agent to eliminate interference and draw standard curve again.
3. Power function simulation model can better describe the change process of solute concentration of slope runoff with rainfall duration in field simulation rainfall experiment, which is consistent with migration rule result researched by predecessors; the fitting coefficient ($R^2 = 0.9537$) of electrode method of is greater than that of spectrophotometric method ($R^2 = 0.9305$).

ACKNOWLEDGMENTS

This work was financially supported jointly by the Promotion of Advanced Water Science and Technology Program of Shandong (SDSLKY201317, SDSLKY201504).

REFERENCES

[1] Ling, Chen. & De-fu, Liu. et al. 2012. Soil Nitrogen Loss from Sloping Farmland under Artificial Rainfall in Xiangxi River Valley. Journal of Ecology and Rural Environment 28(6):616–621 (in Chinese).

[2] Junting, Yu. Shuai, Zhang. Zhibin, Zhang. 2010. Effects of environmental factors on release of sediment nitrogen in shallow lake. Journal of Shandong Jianzhu University 25(1):58–61 (in Chinese).

[3] Guanghua, Jing. Xingxiu, Yu. et al. 2012. Characteristics of soil nitrogen loss under different intense rainfalls in Yimeng mountainous area. Transactions of the Chinese Society of Agricultural Engineering 28(6):120–125 (in Chinese).

[4] Kun, Li. Tao, Jiang. Jianyao, Chen. et al. 2011. A comparative experiment of nitrogen loss from the humid hillside areas of the south China. Chinese Journal of ecology 20(3):447–451 (in Chinese).

[5] Zhiqiang, Huang. Niannian, Yuan. 2012. Drainage and nitrogen loss in dry soils with controlled drainage. Journal of Hohai University (Natural Sciences) 40(4): 412–419 (in Chinese).

[6] Yali, Zhang. Xing, Chang. Mingan, Shao. et al. 2004. Impact of rainfall intensity on soil mineral nitrogen loss by runoff on loess slope. Transactions of the Chinese Society of Agricultural Engineering 20(3):55–58 (in Chinese).

[7] Rushen, Lu. 2000. The analysis method of Soil Agricultural Chemistry. Beijing: Chinese Agricultural Science Press.

[8] The people's Republic of China Ministry of environmental protection. 2012. Water quality-Determination of total nitrogen-Alkaline potassium persulfate digestion UV spectrophotometric method. Beijing: Chinese Environment Science Press.

[9] The people's Republic of China Ministry of environmental protection. 2009. Water quality-Determination of ammonia nitrogen-Nessler's reagent spectrophotometry. Beijing: Chinese Environment Science Press.

[10] Gang, Kong. Quanjiu, Wang. Jun, Fan. 2007. Research on Nutrient Loss from Loessial Soil under Different Slope. Journal of Soil and Water Conservation 21(3):14–18 (in Chinese).

[11] Quanjiu, Wang. Tian-liang, Mu. & Hui, Wang. 2009. The effects of slope on solute concentration in runoff on the loess slope. Agricultural Research in the Arid Areas 27(4):176–179 (in Chinese).

Material Science and Environmental Engineering – Chen (Ed.)

Identifying desired groundwater remediation strategies by using PROMETHEE and GAIA methods

L.X. Ren, H.W. Lu, L. He & Z. Wang
Canada-Sino Research Academy for Resources and Environment, North China Electric Power University, Beijing, China

ABSTRACT: This study applied the PROMETHEE-GAIA method to identify compromised groundwater remediation strategies. The PROMETHEE can capture various preferences of the decision makers, while GAIA can provide a global view of the problem. The approach is applied to a petroleum-contaminated aquifer in southeastern China, and the results showed that option A_{11} is the most desired groundwater remediation strategy for 5-year remediation duration, A_{22} for 10-year, A_{31} for 15-year and A_{41} for 20-year.

Keywords: PROMETHEE; GAIA; groundwater remediation

1 INTRODUCTION

In recent decades, groundwater quality has been deteriorated by the growing industrialization, irrigation and waste disposal activities (Park et al. 2007). This process is imperceptible and irreversible, which has already caused environmental problems and threatened human health. Therefore, it is urgent to make considerable efforts to remedy groundwater pollution and ensure the sustainable use of groundwater. The Pump And Treat (PAT) method, as a well-known and often-applied technique, has been used for more than 20 years in the remediation of groundwater contamination (Singh et al. 2011; Hsiao et al. 2002).

Many optimization methods have been proposed to deal with groundwater remediation problems. For example, Singh et al. (2011) introduced a nondominated sorting genetic algorithm to solve a groundwater remediation problem; Hsiao et al. (2002) integrated a genetic algorithm with constrained differential dynamic programming to calculate optimal solutions; He et al. (2009) advanced an integrated simulation, inference, and optimization method for optimizing groundwater remediation systems. However, most of the above approaches aim at single objective problems. In this paper, five actions in four remediation periods are demonstrated by using an integrated PROMETHEE (preference ranking organization method for enrichment evaluation) and GAIA (geometrical analysis for interactive aid) methods, in order to identify an optimized remediation strategy under different scenarios.

2 METHODOLOGY

PROMETHEE is a simple outranking assessment method, which has been widely applied in many decision-making situations (Brans and Vincke 1985; Elevli 2014). The approach includes the steps as follows:

1. Consider the problem of groundwater remediation with 5 actions $A = \{A_{t1}, A_{t2}, A_{t5}\}$ and 11 evaluation criteria $C = \{P_{t1} \dots P_{t5}, C_{t1} \dots C_{t5}, TC_t\}$ (i.e. pumping rate in five injection/extraction wells, pollutant concentrations in five monitoring wells and costs) for t remediation periods, respectively.
2. Confirm a specific preference function P (linear function is chosen in this paper) to compute preference degree $P_{tk}(A_{ti}, A_{tj})$. The value of $P_{tk}(A_{ti}, A_{tj})$ ranges from 0 to 1, which reflects the preference degree between two actions.
3. Assign weight $w = \{w_{t1}, w_{t2}, \dots, w_{t11}\}\}$ to the corresponding criteria, which reflect the relative importance level of each criterion.
4. Compute outranking degree $\pi_t(A_{ti}, A_{tj})$, positive outranking flow ($\phi^+(A_{ti})$) and negative outranking flow as follows:

$$\pi_t(A_{ti},A_{tj}) = \sum_{k=1}^{n} w_{tk} P_{tk}(A_{ti},A_{tj}) \tag{1}$$

$$\phi_t^+(A_{ti}) = \sum_{i=1}^{m} \pi_t(A_{ti},A_{tj}) \tag{2}$$

$$\phi_t^-(A_{ti}) = \sum_{i=1}^{m} \pi_t(A_{tj},A_{ti}) \tag{3}$$

5. Complete the net outranking flow and obtain the ranking of alternatives (see Eq. 4).

$$\phi_t(A_i) = \phi_t^+(A_{ti}) - \phi_t^-(A_{ti}) \tag{4}$$

3 RESULTS

A real-world petroleum-contaminated aquifer is studied to demonstrate the performance of the proposed method. The PAT system is selected as an in-situ treating technique during the groundwater remediation process. In the simulation domain (Fig. 1), two injection, three extraction, and five monitoring wells were selected. Four remediation durations, i.e. 5, 10, 15, and 20 years, were suggested for investigating the impact of different remediation schemes in different periods. It should be noted that the contaminant concentration in each monitoring well for five actions in each period has to meet the environmental standard (i.e. 0.3 mg/L) issued by Quality Standard for Ground Water.

3.1 *Scenario 1*

The criteria weights for scenario 1 are calculated as shown in Table 1. From Figure 2(A) and Figure 3(A), A_{11} is determined as the recommended choice, followed by A_{13}, A_{12}, A_{14} and A_{15}. Figure 2(A) shows the contribution of each criterion to the net flow. Criteria P_{13} and P_{14}, CM_{12} and CM_{13} are close to each other respectively. Criteria P_{12} and P_{13}, P_{14} and P_{15}, P_{12} and CM_{14} are not related to each other and express independent feature. Criteria P_{11} and P_{14}, P_{13} and CM_{14} are conflicted with each other as they are oriented totally in opposite directions. Based on this set of parameters, these actions present different features in preference. A_{11} performs excellent in criteria P_{13} and TC_1; A_{12} is better with respect to criterion P_{12}; A_{13} is superior with respect to criterion P_{11}; A_{14} has outstanding strength on criteria

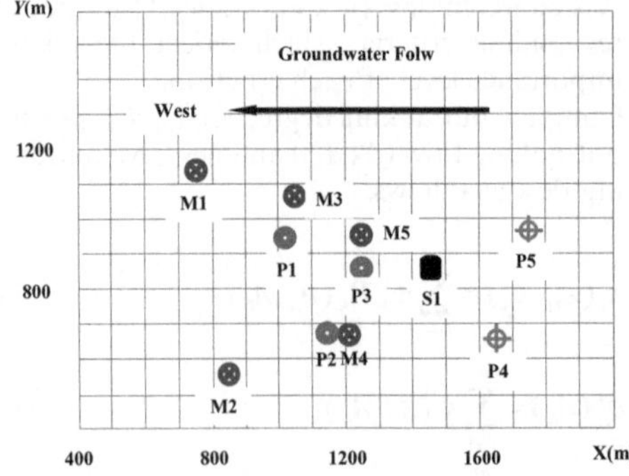

Figure 1. Caption of simulation domain.

Table 1. Criteria weights for each remediation duration.

Weight	5-year	10-year	15-year	20-year
P_{t1}	0.018	0.024	0.029	0.052
P_{t2}	0.023	0.027	0.029	0.045
P_{t3}	0.033	0.038	0.040	0.032
P_{t4}	0.033	0.039	0.040	0.027
P_{t5}	0.033	0.040	0.040	0.027
CM_{t1}	0.079	0.072	0.142	0.075
CM_{t2}	0.079	0.072	0.142	0.075
CM_{t3}	0.129	0.118	0.105	0.131
CM_{t4}	0.129	0.118	0.114	0.131
CM_{t5}	0.230	0.213	0.066	0.126
TC_t	0.214	0.242	0.253	0.280

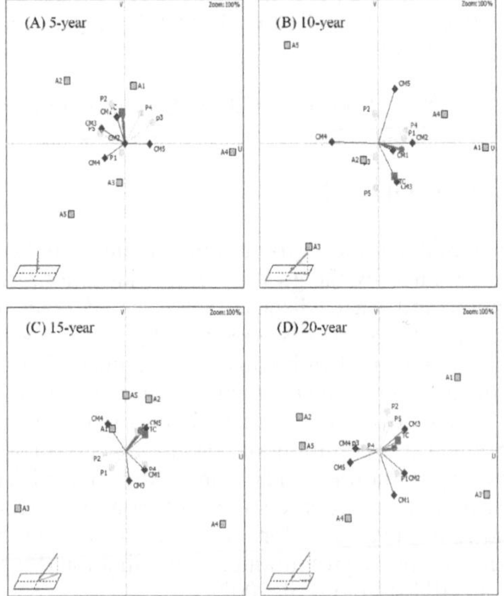

Figure 2. Caption of GAIA plane with action performances for each scenario.

P_{13} and CM_{15}; A_{15} shows negative with respect to criteria P_{14} and CM_{15}.

3.2 *Scenario 2*

Compared with that of 5-year periods, the weight of CM_{23} is decreased from 0.129 to 0.118, and the weight of TC_2 is adjusted from 0.214 to 0.242 when the remediation duration is 10-year. The feasible alternatives to be ranked are A_{22}, A_{12}, A_{24}, A_{23} and A_{25}, as shown in Figure 2(B) and Figure 3(B). P_{21} and P_{24}, TC_2 and CM_{23} are almost correlated, respectively; P_{21} and P_{22}, CM_{21} and CM_{25}, CM_{22} and P_{25} are not related with each other, respectively; CM_{25} and P_{25}, CM_{22} and CM_{24} are highly conflicted with each other, respectively. In the

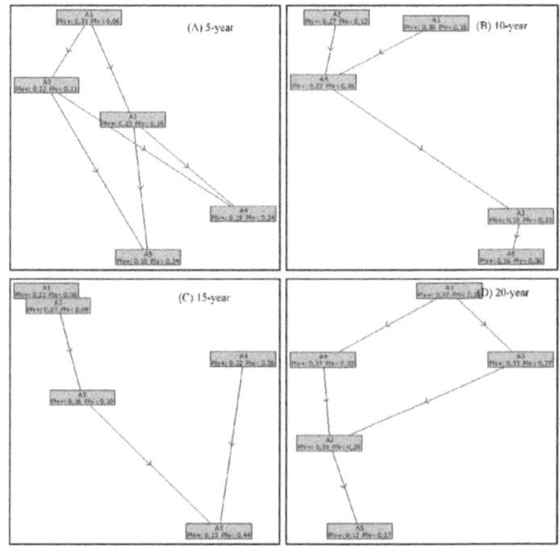

Figure 3. Caption of complete ranking of the five alternatives for each scenario.

GAIA plane, it can be observed that the decision axis is short, which means that the PROMETHEE decision axis has weak decision power. Because all the considered criteria are highly conflicting, this problem has a poor solution accuracy of 67.6%.

3.3 *Scenario 3*

For the 15-year remediation strategy, the weights of criteria have been increased corresponding to the growth of remediation durations except CM_{33}, CM_{34} and CM_{35}. By using the PROMETHEE-GAIA method, A_{31} is observed to be the most appropriate remediation scheme for this application, as shown in Figure 2(C) and Figure 3(C). A_{32} has the second preference and A_{33} is the least favored remediation process. Four criteria (i.e. P_{33}, P_{35}, CM_{35} and TC_3) are superimposed on each other, which indicate that they have the same effect. It can be observed that the value of A_{31} (i.e. 0.23) in ϕ^+ is similar to A_{32}, that of A_{31} in ϕ^- is a little more than A_{32}. It is also found that this selection has a good solution accuracy of 85.4%.

3.4 *Scenario 4*

Because remediation of large aquifer systems with high concentration distribution is a time-consuming and costly process, the weight of cost is increased to 0.280. Among all five considered remediation schemes, A_{41} is found to be the best process, followed by A_{43}, A_{42}, A_{44} and A_{45} (Fig. 2(D) and Fig. 3(D)). Compared with the other alternatives, A_{41} is better with respect to criteria P_{43} and TC_4; A_{42} is superior with respect to criterion P_{43}; A_{23} is better with respect to P_{41}, CM_{41} and CM_{43}; CM_{41} and CM_{45}, P_{45} and CM_{42}, and CM_{41} and CM_{43} are independent to each other as they are oriented in orthogonal axes. When the pumping duration is 20 years, the pumping rate at each well for the optimized scheme (i.e. A_{41}) is 3.402, 2.333, 3.985, 3.305 and 0.292 m^3/h, respectively, and the contamination concentrations in the area can meet the environmental standard.

4 CONCLUSIONS

In this paper, the PROMETHEE and GAIA methods are integrated to identify compromised groundwater remediation strategies for a petroleum-contaminated aquifer in southeastern China. Five alternatives and eleven criteria in four remediation periods were chosen, respectively. According to the net flow of PROMETHEE and the visualization of GAIA, A_{11} is determined as the recommended choice in 5-year remediation period, A_{22} for 10-year, A_{31} for 15-year and A_{41} for 20-year remediation periods. In future studies, more uncertain parameters may be introduced in order to provide systematic and quantitative analysis comprehensively.

ACKNOWLEDGEMENTS

This research was supported by the China National Funds for Excellent Young Scientists (51222906), National Natural Science Foundation of China (41271540), Program for New Century Excellent Talents in Universities of China (NCET-13-0791), and Fundamental Research Funds for the Central Universities.

REFERENCES

[1] Park, D.K. & Ko, N.Y. & Lee, K.K. 2007. Optimal groundwater remediation design considering effects of natural attenuation processes: pumping strategy with enhanced-natural-attenuation. Geosciences Journal 11(4): 377–385.

[2] Singh, T.S. & Chakrabarty, D. 2011. Multiobjective optimization of pump-and-treat-based optimal multilayer aquifer remediation design with flexible remediation time. Journal of Hydrologic Engineering 16(5): 413–420.

[3] Hsiao, C.T. & Chang, L.C. 2002. Dynamic Optimal Groundwater Management with Inclusion of Fixed Costs. Journal of Water Resources Planning and Management. 128: 57–65.

[4] He, L. & Huang, G.H. & Lu, H.W. 2009. A coupled simulation-optimization approach for groundwater remediation design under uncertainty: An application to a petroleum-contaminated site. Environment Pollution 157(8–9): 2485–2492.

[5] Brans, J.P. & Vincke. P.H. 1985. A preference ranking organization method. Management Science. 31: 647–656.

[6] Elevli, B. 2014. Logistics freight center locations decision by using Fuzzy-PROMETHEE. Transport 29(4): 412–418.

Material Science and Environmental Engineering – Chen (Ed.)

The water quality and its pollution sources in Poyang Lake wetlands: A Ramsar site in China

D.M. Xie
Jiangxi Science and Technology Normal University, Nanchang, Jiangxi, China

Y.M. Zhou
School of Geography and Environment, Jiangxi Normal University, Nanchang, Jiangxi, China
Key Laboratory of Poyang Lake Wetland and Watershed Research, Ministry of Education, Jiangxi Normal University, Nanchang, Jiangxi, China

G.H. Jin
Jiangxi Province Meteorological Information Center, Nanchang, Jiangxi, China

ABSTRACT: Lakes are important wetlands, and they are highly contaminated because of stagnation of their water bodies. Consequently, this study examined Poyang Lake as an example. The characteristics of the water quality and its pollution sources from 2008 to 2011 were analyzed. The results revealed that the water quality has changed markedly in recent years, and that the quality has increasingly worsened. Furthermore, in the dry season in Poyang Lake (generally from October to March in the following year), the water quality was more than 70% poorer than the Class III standard. The contaminants mainly comprised total phosphorus and ammonia nitrogen, and they mostly converged from the industrial effluent and sanitary waste around Poyang Lake. The water quality has deteriorated badly, and the structure and function of the ecosystem is likely to encounter serious challenges in the Poyang Lake wetlands.

Keywords: water quality; pollutant sources; environmental protection; Poyang Lake wetlands

1 INTRODUCTION

Lakes are widespread, huge in number and diverse in type in China. Ma et al. (2010) discovered that, in China, there are 2693 natural lakes with an area greater than 1.0 km^2. However, as a result of a range of factors, the total area comprising natural lakes has decreased rapidly in recent years. Over the last 30 years, the area comprising natural lakes has decreased by more than 0.12×10^4 km^2 in the Yangtze River watershed; only 34.16% of the total area covered by lakes now comprises natural lakes (The Forestry Bureau of P.R. China, 2000). Natural lakes play an important and diverse role in ecosystems; however, they are highly contaminated because of stagnation of their water bodies and sustained increasing external pressure (Gurluk and Rehber, 2006; Wersal et al, 2006). According to the *China Environmental Condition Bulletint*, issued by the Ministry of Environmental Protection of the P.R. China (2012), the water quality was clean in only 8 of the 64 large lakes.

Poyang Lake is the largest freshwater lake in China. It is one of the first national wetland reserves, one of the first sites to be listed for China in the Ramsar List of Wetlands of International Importance, and the most abundant biological resource in the wetlands ecosystem in China. Consequently, it is internationally the most significant wetland reserve (Wang et al, 2010). Some studies have confirmed that Poyang Lake is the least contaminated lake in China (Mao et al, 2011). In recent years, more focus has been directed towards the ecology and environment of Poyang Lake (Hu et al, 2010), especially when the Chinese Government approved *the planning of the Poyang Lake Eco-Economic Region in Jiangxi province*, in which they proposed a dam-building project in the lake. Poyang Lake has subsequently attracted worldwide attention (Wang et al, 2014).The ecosystem of the lake is exceedingly complex because it is affected by the Yangtze River, the Five River system (Ganjiang River, Fuhe River, Xinjiang River, Raohe River and Xiuhe River, in the territory of Jiangxi Province) and circumjacent rivers (Xie et al, 2011). Recent research has demonstrated that marked changes have occurred within the water quality of Poyang Lake (Wang et al, 2007). However, most research has focused on heavy metal contamination (Luo et al, 2008), hydrological characteristics, the relationship between river and lake (Hu et al, 2007; Guo et al, 2008), freshwater fish resources (Wang et al, 2003) and schistosomiasis control (Hu et al, 2005). There is a lack of research on the characteristics of the water quality and changes within it, and the reasons underlying such changes. Our study has examined these characteristics and the emerging trends within the water quality—and the underlying reasons—based on the data issued by *the Jiangxi provincial water resources bureau* in Poyang Lake in order to protect Poyang Lake using scientific and effective methods.

2 DATA AND METHOD

2.1 *Study area*

Poyang Lake is located in northern Jiangxi Province, China, on the southern bank of the Yangtze River (115°47′–116°45′E, 28°22′–29°45′N). The climate shows subtropical humid monsoon characteristics, influenced mainly by the cold Siberian temperatures and subtropical high pressure. The daytime temperature is cold in spring, with frequent precipitation in summer, but is dry in autumn and winter (Compilation Committee of Studies on Poyang Lake, 1988; Jiangxi Water Bureau, 2007). The total precipitation is more than 1640 mm and precipitation is concentrated between April and June in the Poyang Lake watershed (Xie et al, 2013). Water feeds into the lake from the Ganjiang River, Fuhe River, Xinjiang River, Raohe River and Xiuhe River in Jiangxi Province and flows out again into the Yangtze River. The basin of these waterways covers 97% of the land area of the province. Poyang Lake has previously been known as Pengli Lake, Penglize Lake, Pengze Lake, Peng Lake, Yanglan Lake and Gongting Lake. After a long evolutionary period, it was formed around 1600 years ago. Historically, its area was more than 5000 km^2 (Liu and Ye, 2000; Wang et al, 2006). The significant changes in water levels and, subsequently, the total area covered by water have led to significant dynamic changes among the different types of wetland, as well as to changes in the ecological landscape caused by the transition between water and land phases. Consequently, the ecosystem of Poyang Lake is subject to regular fluctuations over time. It is regarded as a lake during times of high water levels in the wet season (usually from April to September) and as a river during times of low water levels in the dry season (usually from October to March in the following year). Spatially, therefore, it is planar during periods of flooding, but linear during dry periods (Min and Zhan, 2012).

Poyang Lake is an important reservoir of the Yangtze River. The volume of water resources feeding into the Yangtze River is more than 1450×10^8 m^3, which is equivalent to 15.6% of the annual total volume of water resources volume in the Yangtze River in every year. Around 95% of the global population of white cranes (*Grus leucogeranus*) winter at Poyang Lake every year, and the lake provides a habitat for more than 360 species of birds—half of them migratory—and many endangered fish. The lake has great natural value and is a provider of many ecological services, for example: water conservation, flood management, soil protection, recreation, scientific research, species habitats, purification, CO_2 fixation, O_2 release, and organic matter production (Yan, 2004).

2.2 *Research method*

The six hydrological stations of Waizhou, Lijiadu, Meigang, Bofengkeng, Shizhenjie and Yongxiu respectively corresponded to the monitoring sites of the Five River system, which comprises Ganjiang River, Fuhe River, Xinjiang River, Raohe River (Changjiang River and Le'an River join in Yaogongdu Village, Poyang County, where the river then becomes Raohe River, Bofengkeng in Changjiang River and Shizhejie in Le'an River) and Xiuhe River. The eight hydrological stations of Ganjiang south inlet, Ganjiang main inlet, Fuhe inlet, Xinjiang east inlet, Xinjiang west inlet, Changjiang inlet, Le'an inlet and Xiuhe inlet respectively corresponded to the monitoring sites of the Five Rivers where they entered Poyang Lake. The seven hydrological stations of Kangshan, Longkou, Tangyin, Poyang, Duchang, Xingzi and Hukou (located from south to north) represented the monitoring sites in Poyang Lake itself. The quality of the water quality was analyzed from 2008 to 2011. When the water quality was poorer than a Class III standard, the water quality was identified as being poorer than the normal water safety standard according to the *Environmental Quality Standard for Surface Water GB3838-2002* (The Ministry of Environmental Protection of P.R. China, 2002). Subsequently, we performed analyses on the monitoring site data using bivariate correlations and the *Pearson* correlation coefficient test, based on SPSS 21 software.

2.3 *Data and interpretation*

Data were gathered from the *Bulletin of Water Resources Dynamic Monitoring in Poyang Lake* issued by the *Hydrological Bureau of Jiangxi Province* (2007~2012). The standard of water quality was classified according to the *Environmental Quality Standard for Surface Water GB3838-2002*

(The Ministry of Environmental Protection of P.R. China, 2002). Furthermore, although the water quality is the result of a complex ecosystem, analysis was performed only on the level of contamination and water volume in Poyang Lake. In this study, the Arabic numerals 1, 2, 3, 4, 5 and 6 correspond to the water quality standards I, II, III, IV, V and poorer than V, respectively. The water quality was analyzed once every month for factors such as transparency, dissolved oxygen, pH, total nitrogen, total phosphorus, and ammonia nitrogen. The analysis results represented the water quality in each month.

3 RESULTS AND ANALYSIS

3.1 *Characteristics of the water quality in Poyang Lake basin*

3.1.1 *The Five River system*

Result showed that the water quality was poorest in the Le'an River—usually standard V or poorer. In 2008, 2009, 2010 and 2011, the water quality was poorer than standard III for 9 months, 11 months, 9 months and 12 months, respectively. In Fuhe River and Xinjiang River, the water quality was poorer than standard III for only one month in every year. In Ganjiang River and Xiuhe River, the water quality was clean—standard II or III in every year. The water quality was comparatively stable in the Five River system except for the Le'an River. The ratio of times when the quality was poorer than standard III was almost 36% in every year. In 2008, 2009, 2010 and 2011, the quality was poorer than standard III 26 times, 28 times, 24 times and 27 times, respectively.

3.1.2 *The Five River system feeding into Poyang Lake*

Result showed that the water quality was frequently poorer than standard III in the Five River system feeding into Poyang Lake from 2008 to 2011. The water quality was poorest in the Le'an River, and the quality was poorer than standard III throughout 2008–2011. The water quality was poorer in Xinjiang East, and almost poorer than standard III between 2008 and 2011. The water quality was comparatively clean in the inlet of Xiuhe River; however, the water quality increasingly worsened—no poorer than standard III in 2008, only once poorer than standard III in 2009 and 2010, and twice poorer than standard III in 2011. The water quality increasingly worsened in the inlet of the Five River system feeding into Poyang Lake between 2008 and 2011. In 2008, 2009, 2010 and 2011, the ratio of times when the quality was poorer than standard III was, respectively, 45.8%, 56.3%, 50% and 46.3%. The water quality was poorer than standard III more frequently in the dry season, and the ratio of times when the quality was poorer than standard III was more than 70% between October and March in the following year.

3.1.3 *Poyang Lake*

Result showed that the water quality increasingly worsened in Poyang Lake between 2008 and 2011. In 2008, the quality was poorer than standard III 38 times—a ratio of 45%. In 2009, the quality was poorer than standard III 55 times—a ratio of 65%. In 2010, the quality was poorer than standard III 54 times—a ratio of 64%. In 2011, the quality was poorer than standard III 65 times—a ratio of 77%. The water quality was poorer than standard III more frequently in the dry season, and the ratio of times when the quality was poorer than standard III was more than 70% between October and April in the following year. However, this ratio was comparatively poorer in the wet season. The ratio of times when the quality was poorer than standard III was nearly 30% between June and August. In July and August, the ratio of times when the quality was poorer than standard III was, respectively, 43% and 71%. In the **Poyang** monitoring sites, in 2008, 2009, 2010 and 2011, the quality was poorer than standard III, respectively, 6, 10, 8 and 11 times. In the **Longkou** monitoring sites, in 2008, 2009, 2010 and 2011, the quality was poorer than standard III, respectively, 7, 10, 7 and 11 times. In the **Kangshan** monitoring sites, in 2008, 2009, 2010 and 2011 the quality was poorer than standard III, respectively, 6, 10, 11 and 8 times. In the **Tangyin** monitoring sites, in 2008, 2009, 2010 and 2011 the quality was poorer than standard III, respectively, 4, 8, 6 and 9 times. In the **Duchang** monitoring sites, in 2008, 2009, 2010 and 2011 the quality was poorer than standard III, respectively, 4, 7, 8 and 10 times. In the **Xingzi** monitoring sites, in 2008, 2009, 2010 and 2011 of the quality was poorer than standard III, respectively, 6, 5, 7 and 9 times. In the **Hukou** monitoring sites, in 2008, 2009, 2010 and 2011 of the quality was poorer than standard III, respectively, 5, 7, 7 and 7 times.

3.2 *The correlation between different monitoring sites and the water quality*

3.2.1 *The Five River system and the inlet of Five River system feeding into Poyang Lake*

Result showed that the correlation was weak between the Five River system and the inlet of Five River system feeding into Poyang Lake; there were significant correlations only between Xinjiang East inlet and Xinjiang River ($R = 0.602$, $P = 0.01$), and Le'an inlet and Le'an River ($R = 0.545$, $P = 0.01$). The result explained the limited effect of the upstream Five River system on the water quality in the monitoring sites of the Five River system feeding into Poyang Lake.

3.2.2 *The Five River system and Poyang Lake*

Result showed that the correlation of monitoring sites was weak between the Five River system and Poyang Lake; there were only significant correlations of monitoring sites between Le'an River and Poyang (R = 0.48, P = 0.01), Longkou (R = 0.589, P = 0.01) and Tangyin (R = 0.499, P = 0.01) stations. The results illustrated that the upstream Five River system had a slight effect on the water quality in the monitoring sites of Poyang Lake.

3.2.3 *The inlets of the Five River system and Poyang Lake*

Result showed that the correlation of the monitoring sites was significant in the inlets of the Five River system and Poyang Lake. There were significant correlations of monitoring sites in Poyang and Changjiang inlet (R = 0.62, P = 0.01) and Le'an inlet (R = 0.769, P = 0.01), and then general correlations of monitoring sites in Poyang and Xinjiang East inlet and Xinjiang West inlet.

There were significant correlations of monitoring sites in Longkou and Changjiang inlet (R = 0.551, P = 0.01), and Le'an inlet (R = 0.605, P = 0.01) and Xinjiang West inlet (R = 0.503, P = 0.01), and then general correlations of monitoring sites in Longkou and Xinjiang East inlet.

There were significant correlations of monitoring sites in Kangshan and Changjiang inlet (R = 0.643, P = 0.01), Xinjiang East inlet (R = 0.513, P = 0.01), Fuhe inlet (R = 0.653, P = 0.01) and Xinjiang west inlet (R = 0.77, P = 0.01), and then general correlations of monitoring sites in Kangshan and Ganjiang Main inlet.

There were significant correlations of monitoring sites in Tangyin and Changjiang inlet (R = 0.543, P = 0.01), and Xinjiang East inlet (R = 0.571, P = 0.01) and Xiuhe inlet (R = 0.592, P = 0.01), and then general correlations of monitoring sites in Tangyin and Le'an inlet, and Xinjiang East inlet and Ganjiang Main inlet.

There were significant correlations of monitoring sites in Duchang and Changjiang inlet (R = 0.625, P = 0.01), Le'an inlet (R = 0.599, P = 0.01), Xinjiang West inlet (R = 0.578, P = 0.01) and Xiuhe inlet (R = 0.6, P = 0.01), and then general correlations of monitoring sites in Duchang and Xinjiang East inlet and Ganjiang Main inlet.

There were significant correlations of monitoring sites in Xingzi and Xinjiang West inlet (R = 0.557, P = 0.01) and Xiuhe inlet (R = 0.774, P = 0.01), and then general correlations of monitoring sites in Xingzi and Xinjiang East inlet and Ganjiang Main inlet.

There were significant correlations of monitoring sites in Hukou and Changjiang inlet (R = 0.539, P = 0.01), and Xinjiang West inlet (R = 0.624, P = 0.01) and Ganjiang Main inlet (R = 0.708, P = 0.01), and then general correlations of monitoring sites in Hukou and Fuhe inlet and Xiuhe inlet.

The results illustrated that the water quality in the inlet of the Five River system directly affected the water quality in Poyang Lake, and that the water quality in the inlet of the Five River system had an overwhelming effect on the water quality in Poyang Lake.

4 DISCUSSION AND CONCLUSIONS

Our study illustrated the water quality at selected regions; most regions were contaminated with industrial wastewater and domestic sewage. The contamination in Poyang Lake originated from the downstream of Waizhou station, Lijiadu station, Meigang station, Bofengkeng station, Yongxiu station, and total Le'an River.

The water quality at Poyang station in Poyang Lake was mostly contaminated from Le'an River, Changjiang River and Poyang County. The water quality at Longkou station in Poyang Lake was mostly contaminated from Poyang County. The water quality at Kangshan station in Poyang Lake was mostly contaminated from Xinjiang West, Ganjiang South and Fuhe River. The water quality at Tangyin station in Poyang Lake was mostly controlled by the water quality in Kangshan and Longkou. The water quality at Duchang station in Poyang Lake was mostly contaminated from rivers and Duchang County. The water quality at Xingzi station in Poyang Lake was mostly contaminated from rivers and Xingzi County. The water quality at Hukou station in Poyang Lake was mostly contaminated from the Poyang Lake region.

According to the investigation and analyses, we can draw the following conclusions:

i. There were marked differences in the water quality between 2008 and 2011 in Poyang Lake; the water quality had increasingly worsened. Both scientists and environmentalists were concerned that the ecosystem structure and function had declined in the Poyang Lake wetland.
ii. The water quality was poorer in southern Poyang Lake than in northern Poyang Lake.
iii. The water quality in the inlet of the Five River system directly affected the water quality in Poyang Lake. The most common contaminants were total phosphorus and ammonia nitrogen according to the *Bulletin of Water Resources Dynamic Monitoring in Poyang Lake*; these contaminants originated from industrial wastewater

and domestic sewage from the Poyang Lake region.

iv. The water quality was poorer than Class III standard mostly between October and April of the following year, and the water quality was mostly good between May and September. The change in the water quality was influenced by the change in water volume, and the contaminants were deliquated in the wet season. However, the seasonal change in the water quality does not explain the level of contamination in Poyang Lake.

There is a large population and developed industry and agriculture in the Poyang Lake region in Jiangxi Province—the ratios of population and GDP are, respectively, 50% and 60% of Jiangxi Province—but the ratio of land area is only 30% of Jiangxi Province. In 2009, the Chinese Government approved the *Poyang Lake Eco-Economic Region Plan* for Jiangxi province (The Development and Reform Commission of P.R. China, 2010). From then on, substantial investment has led to a rapid increase in GDP (The Development and Reform Commission of P.R. China, 2011): in recent years the value of investment has been almost 2000 billion RMB in the Poyang Lake region. However, the natural resources and eco-environment were disturbed and damaged when the GDP increased rapidly without strict measures of environmental and regulatory protection. We investigated these social and economic developments and the agricultural and service industry in the Poyang Lake region, and concluded that rapid development has occurred in recent years. However, the infrastructure construction of environmental protection and ecological conservation lagged behind the economic development. More and more contaminants, such as industrial waste, domestic sewage, and poultry and livestock manure, were scoured by runoff into Poyang Lake, leading to contamination of the water quality. The degeneration and loss of the natural resource and eco-environment are irreversible; for natural wetland especially, the degeneration and loss are irreparable (de Groot et al, 2006).

Strict countermeasures are necessary to strengthen and protect the water quality in Poyang Lake, as follows:

i. Industrial wastewater and domestic sewage must be cleaned before feeding into the runoff, so that the diffusion of containment is controlled from the source of pollutants.
ii. An ecological zone should be built round Poyang Lake, in which the construction of factories, reclaiming of mines, and building of infrastructure for service industry should be prohibited.
iii. An organization should be founded to manage and coordinate a consistent approach to the conservation of Poyang Lake. The responsibility and obligations of such an organization would be to formulate protection planning, laws and regulations, and strengthen the governance of Poyang Lake.

ACKNOWLEDGEMENTS

This research was financially supported by the National Natural Science Foundation of China (Grant No. 31360120), the Natural Science Foundation of Jiangxi Province (Grant No. 20132BAB203030), the Science and Technology Supporting Programme (Grant No. 20151BBG70014), and the Science Project of the Education Department in Jiangxi Province (Grant No. GL1318, and Grant No. GJJ13561).

REFERENCES

[1] Compilation Committee of Studies on Poyang Lake (1988) Studies on Poyang Lake; Shanghai Science and Technology Press: Shanghai, 60–71 (in Chinese).

[2] De Groot, R.; Stuip, M.; Finlayson, M.; Davidson, N. (2006) Valuing wetlands: guidance for valuing the benefits derived from wetland ecosystem services. Ramsar Convention Secretariat.

[3] Development and Reform Commission of P.R. China (2011) 70% of major construction projects were in process in the Poyang Lake Eco-Economic Region. http://dqs.ndrc.gov.cn/zbjq/201112/t20111228_453243.html. Accessed April 10, 2014 (in Chinese).

[4] Development and Reform Commission of P.R. China (2009) the planning of the Poyang Lake Eco-Economic Region. http://www.ndrc.gov.cn/fzgggz/dqjj/qyzc/200912/t20091228_322189.html. Accessed April 10, 2014 (in Chinese).

[5] Dongming, X.; Guohua, J.; Yangming, Z.; Gengying, J.; Lingguang, H.; Yuping, Y. et al. (2013) Study on Ecological Function Zoning for Poyang Lake Wetland: a RAMSAR site in China. Water Policy, 15, 922–935.

[6] Forestry Bureau of P.R. China (2000) China National Wetland Conservation Action Plan; China Forestry Press: Beijing, 9 (in Chinese).

[7] Guo, H.; Hu, Q.; Jiang, T. (2008) Annual and seasonal streamflow responses to climate and land-cover changes in the Poyang Lake basin, China. J. Hydrol., 355, 106–122.

[8] Gurluk, S.; Rehber, E. (2006) Evaluation of an Integrated Wetland Management Plan: Case of Uluabat (Apollonia) Lake, Turkey. Wetlands, 26(1), 258–264.

[9] Hu, G.H.; Hu, J.; Song, K.Y.; Lin, D.D.; Zhang, J.; Cao, C.L. et al. (2005) The role of health education and health promotion in the control of schistosomiasis: experiences from a 12-year intervention study in the Poyang Lake area. Acta Trop., 96, 232–241.

[10] Hu, Q.; Feng, S.; Guo, H.; Chen, G.Y; Jiang, T. (2007) Interactions of the Yangtze River flow and hydrologic processes of the Poyang Lake, China. J. Hydrol., 347, 90–100.

[11] Hu, Z.P.; Ge, G.; Liu, C.L.; Chen, F.S.; Li, S. (2010) Structure of Poyang Lake wetland plants ecosystem and influence of lake water level for the structure. Resources and Environment in the Yangtze Basin, 19(6), 597–605 (in Chinese).

[12] Jiangxi Water Bureau (2007) Jiangxi Drainage; Yangtze Press: Wushan, 90–91 (in Chinese).

[13] Liu, X.Z.; Ye, J.X. (2000) Wetland of Jiangxi; China Forestry Press: Beijing, 26–27 (in Chinese).

[14] Luo, M.B.; Li, J.Q.; Cao, W.P.; Wang, M.L. (2008) Study of heavy metal speciation in branch sediments of Poyang Lake. J. Environ. Sci, 20, 161–166.

[15] Ma, R.H.; Yang, G.H.; Duan, H.T.; Jiang, J.H.; Wang, S.M.; Feng, X.Z. et al. (2010) China's lakes at present: Number, area and spatial distribution. Sci. China Earth Sci., doi: 10.1007/s11430–010–4052–6 (in Chinese).

[16] Mao, Z.P.; Zhou, H.D.; Wang, S.Y.; Liu, C. (2011) Characteristics of water environment in Poyang Lake. Journal of China Institute of Water Resources and Hydropower Research, 4, 267–273 (in Chinese).

[17] Min, Q.; Zhan, L.S. (2012) Characteristics of low-water level changes in Lake Poyang during 1952–2011. J. Lake Sci., 24(5), 675–678 (in Chinese).

[18] Ministry of Environmental Protection of P.R. China (2002) Surface water environmental quality standard (GB 3838–2002). http://kjs.mep.gov.cn/hjbhbz/bzwb/shjbh/shjzlbz/200206/t20020601_66497.htm. Accessed March 31, 2014 (in Chinese).

[19] Ministry of Environmental Protection of P.R. China (2012) Bulletin China environment. http://jcs.mep.gov.cn/hjzl/zkgb/2012zkgb/. Accessed March 24, 2014 (in Chinese).

[20] Wang, H.Z.; Xu, Q.Q.; Cui, Y.D.; Liang, Y.L. (2007) Macrozoobenthic community of Poyang Lake, the largest freshwater lake of China, in the Yangtze floodplain. Limnology, 8, 65–71.

[21] Wang, P.; Lai, G.Y.; Huang, X.L. (2014) Simulation of the impact of Lake Poyang project on the dynamic of lake water level. J. Lake Sci., 26(1), 29–36 (in Chinese).

[22] Wang, X.H., Yan, B.Y., Wu, G.C. (2006) Mountain-River-Lake Program; Science Press: Beijing, 9 (in Chinese).

[23] Wang, X.L., Xu, L.G., Yao, X., Yu, L., Zhang, Q. (2010) Analysis on the soil microbial biomass in typical hygrophilous vegetation of Poyang Lake. Acta Ecol. Sin., 30(18), 5033–5042 (in Chinese).

[24] Wang, Z.W.; Wu, Q.J.; Zhou, J.F.; Ye, Y.Z.; Tong, J.G. (2003) Silver carp, Hypophthalmichthys molitrix, in the Poyang Lake belong to the Ganjiang River population rather than the Changjiang River population. Environ. Biol. Fishes, 68, 261–267.

[25] Wersal, R.M.; Madsen, J.D.; McMillan, B.R.; Gerard, P.D. (2006) Environmental factors affecting biomass and distribution of Stuckenia pectinata in the Heron Lake System, Minnesota, USA. Wetlands, 26(2), 313–321.

[26] Xie, D.M.; Zheng, P.; Deng, H.B.; Zhao, J.Z.; Fan, Z.W.; Fang, Y. (2011) Landscape responses to changes in water levels at Poyang Lake wetlands. Acta Ecol. Sin., 31(5), 1269–1276 (in Chinese).

[27] Yan, B. (2004) Evaluation of the wetland ecosystem services of Poyang Lake. Resources Science 26(3), 61–68 (in Chinese).

Material Science and Environmental Engineering – Chen (Ed.)

Effect of sudden expansion flow on mass transfer performance of tubular electrochemical reactor used for organic wastewater treatment

J. Su, Y.X. Wen & Y.F. Long
School of Chemistry and Chemical Engineering, Guangxi University, Nanning, Guangxi, China

H.B. Lin
College of Chemistry, Jilin University, Changchun, Jilin, China

ABSTRACT: The design of a tubular electrochemical reactor is attractive for electrochemical process because of its flexibility to scale-up and space saving. In this paper, the effect of sudden expansion entrance on the rate of mass transfer in a tubular electrochemical reactor with parallel electrodes is studied. Overall mass transfer coefficients were measured using the limiting current method in different expansion ratios, and the equation of $\gamma = (s/S)^{-0.188}$ is the experiential correlation of mass transfer enhancement factor.

Keywords: mass transfer; sudden expansion flow; tubular electrochemical reactor

1 INTRODUCTION

Electrochemical oxidation method is a perspective technology for organic wastewater treatment because its advantages such as free secondary pollution, environmental compatibility, mild reaction conditions, easy automation, etc. However, the practical application of this technology still faces some urgent problems such as low current efficiency and weak mass transfer performance, which affects the pollutant transport from the electrolyte to the electrode surface. An obvious strategy for electrochemical oxidation process optimization is to improve the mass transfer performance to obtain the maximum degradation rate and current efficiency. The enhancement of mass transfer means destruct the hydraulic boundary layer by intensifying the flow turbulence (such as stirring) or adjusting the operation parameter (such as flow rate).

The treatment of mass transfer can be described precisely by mathematical equations consisted of dimensionless parameters. Since some of the dimensionless parameters, such as Sherwood number and Reynolds number, are related to the reactor configuration. The research of mass transfer for industrial scale highlights the desired effect within a diverse range of electrode-electrolyte geometry which has significant effect on solution and current path (Griffiths et al. 2005, Stankovic 1994, Storck & Hutin 1980, Wragg & Leontaritis 2000). For examples, entrance region has an important influence on the performance of such reactors. It is characterized by the presence of flow downstream the inlet. Much work has been devoted to the study of mass/heat transfer in abrupt enlargement inlet and significant enhancement of mass/heat transfer is observed. Rizk et al. (1996) used this method based on reduction of ferricyanide species on nickel electrodes to study the effect of an axisymmetric sudden flow expansion on mass transfer rates under highly turbulent flow conditions. The mass transfer distribution in the recirculation zone in a function of dimensionless factor has been investigated by Tagg et al. (1979). Parameter of expansion ratio in a filter-press type electrochemical reactor has been used to express the peak mass transfer in axisymmetric sudden expansion. Djati et al. (2001) have researched on the influence of different fluid inlet types, slits or tubes, on mass transfer in the rectangular reactor. Mass transfer increased with decrease in the expansion ratio, that corresponded to the recirculation flow downstream the sudden expansion.

The research object of this paper is extended to a tubular electrochemical reactor in the present work. Because this design is flexible to scale-up and space saving. In some new applications for electrochemical technology such as metal removal (Martínez-Delgadillo et al. 2010, Rodriguez & Martinez 2005), organic degradation (Körbahti & Tanyolaç 2009, Liang et al. 2008) and electrochemical disinfection, tubular reactor is also selected. Our study is focus on the effect of sudden expansion inlet on the rate of mass transfer in a tubular electrochemical reactor with parallel electrodes. And the mass transfer characteristic can be optimized by reactor configuration adjustment and hydrodynamic improvement. The established mass transfer equations with inlet expansion ratio as parameter were verified based on the experimental results.

2 EXPERIMENTAL SECTION

The limiting current technique is a rapid and precise method for mass transfer rate measurement (Gostick et al. 2003, Ross & Wragg 1965, Tagg et al. 1979). Ions in the bulk solution move towards the electrode surface by diffusion. The limiting-current is observed when this transport process becomes the rate-limiting factor. A frequently used reaction system for this technique is the reduction of ferricyanide in cathode:

$$Fe(CN)_6^{3-} + e^- \rightarrow Fe(CN)_6^{4-} \tag{1}$$

For the measurement of mass transfer coefficient, a vertical tubular cell made of transparent polymethyl methacrylate (PMMA) was installed. The tube was 0.56 m tall and 0.03 m for its diameter. The diameters of inlet and outlet for the reactor were both 0.03 m. The working titanium plate cathode which placed in one side of the channel was 0.50 m long and 0.02 m wide. And the Ti/RuO_2-TiO_2-SnO_2 anode with the same geometry as cathode was fixed opposite the cathode with an inter-electrode spacing of 0.02 m. Variety of baffles was installed in the inlet of cell for studying their effect on mass transfer.

The arrangement of holes in the baffles is shown in Figure 1. Liquid flowed enter the reactor through the circular arranged holes during the experiments. And expansion ratio s/S for different inlets is defined by the ratio of entrance cross-section area

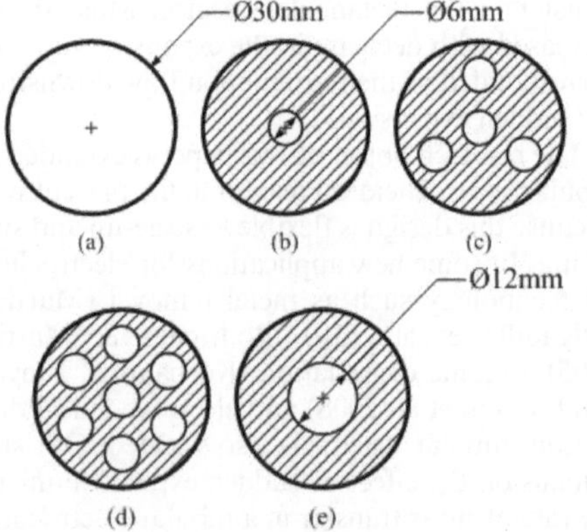

Figure 1. Schematic diagram for holes arrangement of sudden expansion inlet. (a) s/S=1; (b) s/S=0.04; (c) s/S=0.16; (d) s/S=0.28; (e) s/S=0.16.

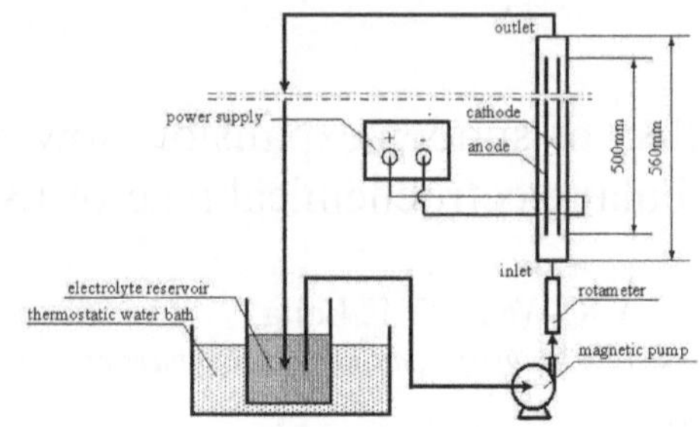

Figure 2. Flow cell arrangement.

Table 1. Physical property of the solution at 25°C.

Property	Value
Viscosity, μ	1.1042×10^{-3} kg·m^{-1}·s^{-1}
Density, ρ	1.046×10^{3} kg·m^{-3}
Diffusion coefficient, D	6.693×10^{-10} m^{2}·s^{-1}
Schmidt number, Sc	1126

to that of the vertical cell. It is mentioned that the expansion ratio *s*/*S* of baffle (c), which equal to 0.16, was the same as that of baffle (e).

The cell arrangement for mass transfer coefficient measure is presented in Figure 2. It is consisted of a 2×10^{-3} m^3 reservoir equipped with a thermostatic water bath, a magnetic pump, a rotameter covering a flow rate range of 0.02 m^3 per hour to 0.16 m^3 per hour, and the tubular electrochemical reactor. Before each measurement, nitrogen was provided into the electrolyte for 30 minutes to remove oxygen. The temperature in the reservoir was maintained to 25°C by thermostatic water bath. The electrolyte solution was consisted of 0.01mol/L $K_3Fe(CN)_6$, 0.01 mol/L $K_4Fe(CN)_6$, and 1 mol/L NaOH was used as supporting electrolyte. The physical properties for electrolyte solution are given in Table 1. Polarization curves with a limiting current plateau were obtained for different expansion ratios and hydrodynamic conditions. The limiting current obtained from these curves was used to calculate the mass transfer coefficient according to the equation:

$$k = \frac{I_{\lim}}{nAFC_0} \tag{2}$$

3 RESULTS AND DISCUSSION

3.1 *Average mass transfer coefficient varied with flow rate*

Values of average mass transport coefficient were characterized by measuring the limiting current over a range of electrolyte velocities. The limiting current plateau appeared approximately in the same region of 1.70V were monitored. The effect of increasing the volume flow rate on the limiting current for the absence of abrupt larger inlet is shown in Figure 3. An enhancement of limiting current appeared with the flow rate increasing. Average mass transfer is expressed in terms of Sh_{av} as a function of the Re_{av} and Sc_{av} defined using the characteristics of the cell. The data were well related by the equation based on the hydraulic diameter of 0.03m and volume flow rate ranged from 20L/h to 160L/h ($Re = 313–2502$):

$$Sh = 0.112 Re_{av}^{0.76} Sc_{av}^{1/3} \tag{3}$$

This experimentally determined dimensionless correlations differ by a few percent (20–35%) compared with the literature correlation for the similar parallel plate cell (Oduoza et al. 1997). This difference can be caused by (1) the difference in the preparation of electrolyte and experimental temperature; (2) the difference of hydrodynamic condition.

3.2 *Average mass transfer coefficient varied with expansion ratio*

Average mass transport was characterized by measuring the limiting current over a range of expansion ratio (Fig. 4). The limiting current increased with the decrease of expansion ratio. Furthermore, for the liquid flow through the circular arranged holes of expansion inlet, the limiting current of the cell with expansion entrance (e) differed by a maximum of 15% compared with that of (c) when they had the equivalent value of s/S (Fig. 5). The experimental data indicated that, the average mass transfer coefficient was significantly associated with expansion ratio but little related to distribution and amount of circular arranged holes in baffles. This phenomenon is evident when flow rate is less than 100L/h.

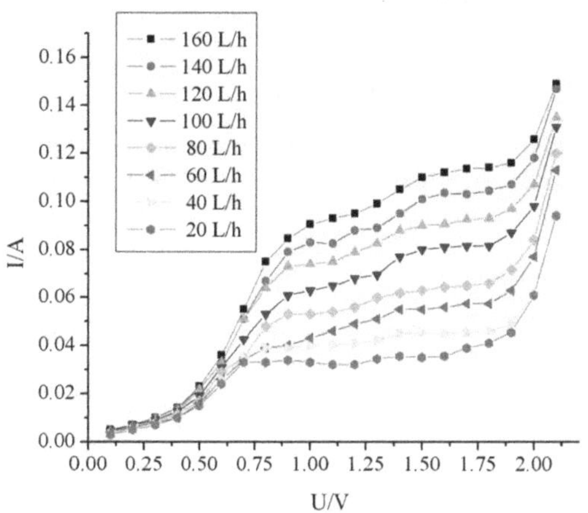

Figure 3. Current varied with cell potential for different flow rate in the absence of sudden expansion entrance.

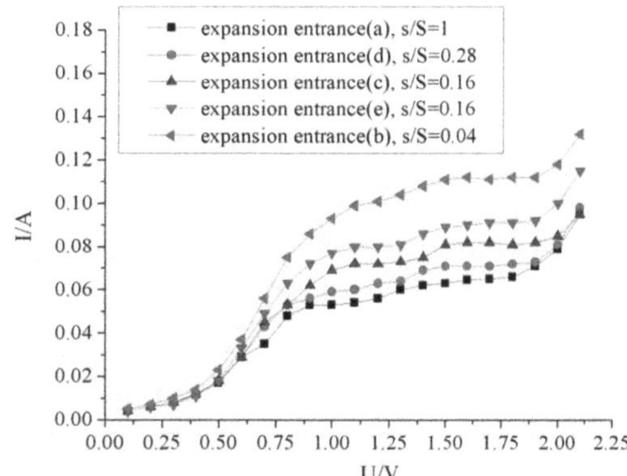

Figure 4. Current varied with cell potential for different expansion ratio for the flow rate equal to 80L/h.

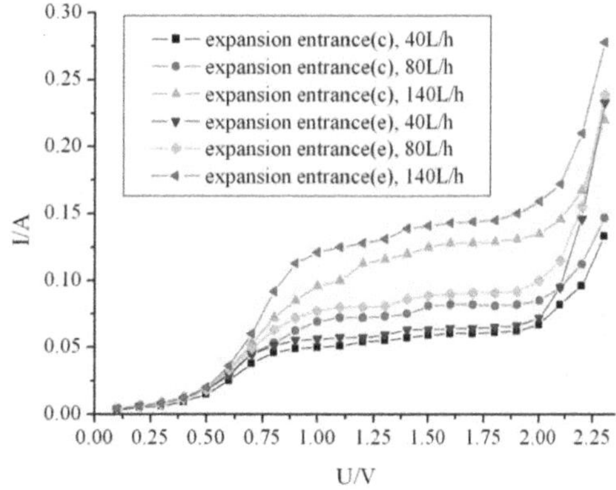

Figure 5. Current varied with cell potential for the same expansion ratio equal to 0.16 with different flow rates.

When the expansion factor s/S is used as a modified parameter to predict average mass transfer performances in the electrochemical reactor with sudden expansion inlet, the experimental data for expansion ratio values between 0.04 and 1 are correlated by

$$Sh = 0.112 Re_{av}^{0.76} Sc_{av}^{1/3} (s/S)^{-0.188} \tag{4}$$

Figure 6 shown the comparison between experimental data to the equation (4) when the condition of flow rate is equal to 80L/h. The mass transfer

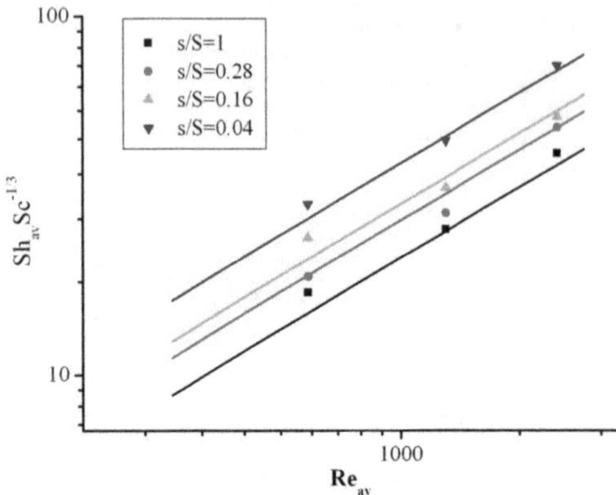

Figure 6. Comparison between experimental data with the equation. (flow rate = 80L/h; dot-experimental data; line-calculated values of dimensionless equation).

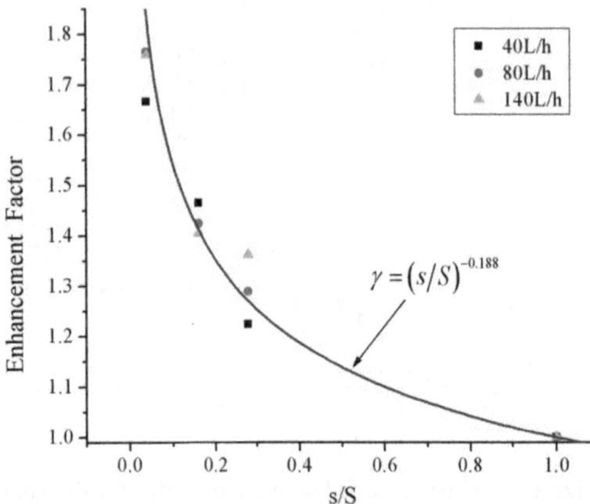

Figure 7. Mass transfer enhancement factor vs. sudden expansion factor at different flow rate. (dot-experimental data; line-calculated values of dimensionless equation).

characteristic increases with the expansion factor decrease. This is corresponding to an increase of the recirculation flow downstream of the sudden expansion. A useful parameter is the mass transfer enhancement factor, γ, which can be defined as:

$$\gamma=\frac{k_{av,s/S}}{k_{av,s/S=1}}=\frac{I_{\lim,s/S}}{I_{\lim,s/S=1}} \tag{5}$$

Deriving from equation (4), $\gamma = (s/S)^{-0.188}$ is its expression for this experimental tubular electrochemical reactor. The plot of mass transfer enhancement factor *vs.* abrupt expansion factor was presented in Figure 7 for flow rate between 20L/h and 160L/h.

4 CONCLUSIONS

The tubular electrochemical reactor is a promising equipment used for the electrochemical degradation of organic pollutant. In order to improve the mass transfer performance, which affects the pollutant transport from the electrolyte to the electrode surface, determination of mass transfer performance for abrupt enlarger entrance has been studied. It is shown that the expansion ratio s/S is a pertinent parameter to describe mass transfer enhancement. However, the average mass transfer coefficient was little related to distribution and amount of circular arranged holes in baffles. And $\gamma = (s/S)^{-0.188}$ is the experiential correlation of mass transfer enhancement factor.

NOMENCLATURE

A = electrode area, m^2
D = diffusion coefficient, $m^2 \cdot s^{-1}$
Sc = Schmidt number
k = mass transfer coefficient, $m \cdot s^{-1}$
I = current, A
n = number of electrons
F = Faraday constant, $A \cdot s \cdot mol^{-1}$
Sh = Sherwood number
Re = Reynolds number
s = sum area of holes, m^3
S = area of tubular section, m^3

Greek symbols
μ = viscosity, $kg \cdot m^{-1} \cdot s^{-1}$
ρ = density, $kg \cdot m^{-3}$
γ = ratio of mass transfer enhancement

Subscripts
lim = limiting
av = average

ACKNOWLEDGMENTS

This research was supported by the Guangxi Natural Science Foundation (No. 2012GXNSFBA053019).

REFERENCES

[1] Djati, A., Brahimi, M., Legrand, J. & Saidani, B. 2001. Entrance effect on mass transfer in a parallel plate electrochemical reactor. *Journal of Applied Electrochemistry* 31(8): 833–837.

[2] Gostick, J., Doan H.D., Lohi A. & Pritzker, M.D. 2003. Investigation of Local Mass Transfer in a Packed Bed of Pall Rings Using a Limiting Current Technique. *Industrial Engineering and Chemistry Research* 42(15): 3626–3634.

[3] Griffiths, M., Leon, C.P. & Walsh, F.C. 2005. Mass transport in the rectangular channel of a filter-press electrolyzer (the FM01-LC reactor). *AIChE. Journal* 51(2): 682–687.

[4] Körbahti, B.K. & Tanyolaç, A. 2009. Electrochemical treatment of simulated industrial paint wastewater in a continuous tubular reactor. *Chemical Engineering Journal* 148(2–3): 444–451.

[5] Liang, W., Qu, J., Wang, K., Wang, J., Liu, H. & Lei, P. 2008. Electrochemical degradation of cyanobacterial toxin microcystin-LR using Ti/RuO_2 electrodes in a continuous tubular reactor. *Environmental Engineering Science* 25(5): 635–642.

[6] Martínez-Delgadillo, S.A., Mollinedo P.H.R., Gutiérrez, M.A., Barceló, I.D. & Méndez, J.M. 2010. Performance of a tubular electrochemical reactor, operated with different inlets, to remove Cr(VI) from wastewater. *Computers & Chemical Engineering* 34(4–5): 491–499.

[7] Oduoza, C.F., Wragg, A.A. & Patrick, M.A. 1997. The effects of a variety of wall obstructions on local mass transfer in a parallel plate electrochemical flow cell. *Chemical Engineering. Journal* 68(2–3), 145–155.

[8] Rizk, T.Y., Thompson, G.E. & Dawson, J.L. 1996. Mass transfer enhancement associated with sudden flow expansion. *Corrosion Science,* 38(10): 1801–1814.

[9] Rodriguez, M.G. & Martinez, S.A. 2005. Removal of Cr(VI) from wastewaters in a tubular electrochemical reactor. *Journal of Environmental Science and Health. Part A: Environmental Science and Engineering and Toxicology* 40(12): 2215–2225.

[10] Ross, T.K. & Wragg, A.A. 1965. Electrochemical mass transfer studies in annuli. *Electrochimica Acta* 10(11), 1093–1106.

[11] Stankovic, V.D. 1994. Limiting current density and specific energy consumption in electrochemical cells with inert turbulence promoters. Journal of Applied Electrochemistry 24(6), 525–530.

[12] Storck, A. & Hutin, D. 1980. Mass transfer and pressure drop performance of turbulence promoters in electrochemical cells. *Electrochim. Acta. 26,* 127–137.

[13] Tagg, D.J., Patrick, M.A. & Wragg, A.A. 1979. Heat and mass transfer downstream of abrupt nozzle expansions in turbulent flow. Transactions of the Institution of Chemical Engineers 57, 176–181.

[14] Wragg, A.A. & Leontaritis, A.A. 2000. Effect of baffle length on mass transfer in a parallel plate rectangular electrochemical cell. *Journal of Applied Electrochemistry* 30(12), 1439–1444.

Material Science and Environmental Engineering – Chen (Ed.)

Pretreatment of ABS resin wastewater by coagulation-dissolved air flotation apparatus

Y. Li, Q.B. Wang & X.W. He
School of Chemical and Environmental Engineering, China University of Mining and Technology (Beijing), Beijing, China

ABSTRACT: In order to improve the effectiveness of the treatment of colloidal suspensions in Acrylonitrile-Butadiene-Styrene (ABS) resin wastewater, this study optimized the parameters of coagulation-dissolved air flotation for the pretreatment of the wastewater. The results show that the Suspended Solids (SS) and Chemical Oxygen Demand (COD) removal efficiency maintain in the range of 95%~98% and 55%~70% at the pH level of 5, the coagulant dosage of 100 $mg \cdot L^{-1}$, the cationic Polyacrylamide (PAM) dosage of 2.5 $mg \cdot L^{-1}$, the residence time of 15 min and the reflux ratio of 40%. The optimized coagulation-dissolved air flotation device can be operated stably, and its effluent meets the demands for biochemical treatment (SS <80 $mg \cdot L^{-1}$ and COD <1500 $mg \cdot L^{-1}$). Therefore, this device has a broad application for the pretreatment of latex wastewater.

Keywords: optimization; ABS resin wastewater; coagulation; dissolved air flotation

1 INTRODUCTION

Wastewaters from Acrylonitrile-Butadiene-Styrene (ABS) resin manufacturing plants contain relatively high concentrations of toxic and refractory organic matter, which produced three sections as follows. (1) Latex wastewater from the butadiene polymerization step, which is intermittently discharged; the water quality varies greatly, with suspended matter ranging from 0.3 to 10 μm at an average size and 65~1200 $mg \cdot L^{-1}$ with concentration and COD content ranging from 1000 to 20,000 $mg \cdot L^{-1}$. (2) Intermittently discharged latex wastewater from the graft polymerization step, which also has varying water quality: the average size of suspended matter is 0.3~10 μm, the suspended matter concentration ranges from 250 to 3000 $mg \cdot L^{-1}$, and the COD content varies from 1500 to 15,000 $mg \cdot L^{-1}$. (3) Wastewater from the gelatinous drying step, which is continuously discharged; the average size of suspended matter in this waste ranges from 10 to 100 μm, its concentration ranges from 200 to 800 $mg \cdot L^{-1}$, and the COD content varies from 1000 to 1500 $mg \cdot L^{-1}$ (Li et al. 2013).

The study of ABS resinproducing industrial wastewater is limited at present. Coagulation, internal electrolysis, and electrochemical and advanced oxidation processes were used to pretreat ABS resin wastewater in the study by Guo L.X., Lai B., Srinivasa R. Popuri, Chang Cheng-Nan Xie C.S., and Wang B. (Guo et al. 2005; Lai et al. 2012; Popuri, Chang and Xu. 2011; Chang and Lin. 1997; Xie. 2009; Wang. 2010). These studies have demonstrated the effective degradation of organic compounds and improved wastewater biodegradability; however, these technologies are in the stage of laboratory research, and their applications are limited by their immature level of development, high cost and high energy requirements. In China, ABS resin wastewaters are mixed and then pretreated with coagulation-dissolved air flotation (Chen et al. 2007; Liu et al. 2008; Zhao, Liu and Liu. 2003). Because of the fluctuant wastewater quality and small particle size, the suspended solids are difficult to remove. In addition, there are many problems associated with the traditional dissolved air flotation process, which can lead to unstable operation, frequent jamming of the dissolved air releaser and the drainage does not meet the standards (Hanafy and Nabih. 2007; Zhang et al. 2014; Santo et al. 2012; Lakghomi, Lawryshyn and Hofmann. 2015). Therefore, it is urgent to optimize the coagulation-dissolved air flotation device for ABS resin wastewater. The objective of this paper is to improve the coagulation-dissolved air flotation device, and investigate the pretreatment of ABS resin wastewater.

2 MATERIALS AND METHODS

2.1 *Chemicals*

Coagulant XR powder was supplied by Yixing Tongsheng Chemical Products Co., Ltd. The applied solutions were prepared at the rate of 0.4%. Flocculant FO4440SSH powder, a kind of high-molecular-weight cationic Polyacrylamides (PAM), was supplied by SNF (China) Flocculant Co., Ltd. The applied solutions were prepared at the rate of 0.04%.

The wastewater was obtained from an ABS synthetic resin plant in north China. Table 1 presents the pH, turbidity, Suspended Solids (SS), Chemical Oxygen Demand (COD), and Total Organic Carbon (TOC) value of the wastewater samples.

2.2 *Analytical methods*

The methods and equipment used to determine the pH, turbidity, Suspended Solids (SS), Chemical Oxygen Demand (COD) and Total Organic Carbon (TOC) are presented in Table 2. Measurements were performed in accordance with standard methods for the examination of pollutants in water and wastewater (Li. 2012).

2.3 *Test equipment*

The dissolved air system and the release system used in traditional coagulation-dissolved air flotation methods have been improved by implementing advanced dissolved air flotation technology developed domestically and abroad using theoretical research and pilot field tests. The test equipment and technological process are shown in Figure 1.

The coagulation-dissolved air flotation apparatus used in this study is improved, and has the following characteristics in comparison with conventional apparatuses: first, the air saturator, with an operating pressure of 0.5–0.7 MPa, is oriented in the horizontal direction and a jet device is situated at the inlet of the air saturator. This design improves the dissolved air efficiency by increasing

Table 1. The main water quality criteria.

pH	Turbidity (NTU)	SS ($mg \cdot L^{-1}$)	COD ($mg \cdot L^{-1}$)	TOC ($mg \cdot L^{-1}$)
6~9	300~4000	100~5000	1000~5000	400~2000

Table 2. Analytical methods and equipment used in this study.

Parameter	Method	Apparatus
pH	Electrometry	pH Meter
Turbidity	Nephelometry	Turbidimeter
SS	Gravimetry	Electronic Balance
COD	Titration	Digester
TOC	Spectrophotometry	Spectrophotometer

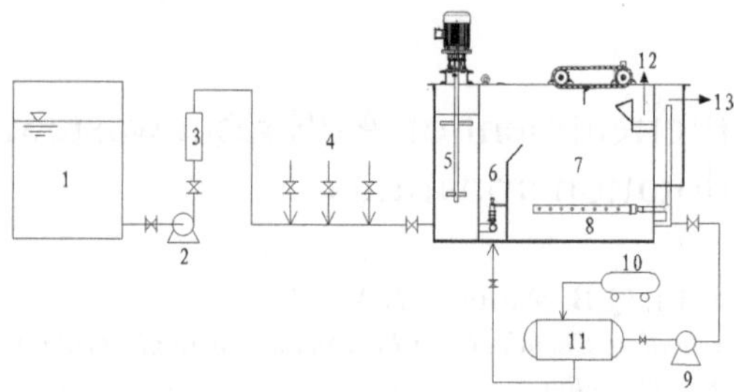

1. wastewater tank, 2. pipeline pump, 3. flowmeter, 4. dosing system, 5. reaction tank, 6. contact tank, 7. separating pond, 8. catch-pipe, 9. reflux pump, 10. air compressor, 11. air saturator, 12. scum, 13. effluent

Figure 1. Test apparatus and process flow diagram.

the dissolved air pressure and the area of the aqueous-vapor interface, thus reducing the residence time. As a result, the volume of the air saturator is reduced to one-sixth that of a conventional air saturator. Second, based on the principle of energy dissipation by hose resistance, a new type of anti-blocking releaser is used, allowing a wide flow path at the leaser outflow. In addition, the high flow speed resulting from the high-pressure dissolved air contributes to the reduction of blockage.

2.4 *Test methods*

The experiments were performed at a petrochemical plant in northern China; the designed handling capacity of the test device is 1.0 m^3/h, and the dissolved air operating pressure is 0.6 MPa. Raw water was poured into the original water tank, mixed with de-emulsifiers, coagulated through pipes and then transferred into the reaction cell. Water and chemicals flowed into the flotation contact tank to be separated after completion of the reaction. Scum was scraped into the scum groove from the water surface to the outflow. The clean water was transferred into the water tank through the water collection tube at the bottom. Using SS in the effluent as an evaluation index, the effects of chemical type, dosage, influent pH, residence time and reflux ratio were studied.

3 RESULTS AND DISCUSSION

3.1 *Effect of coagulant XR dosage on process efficiency*

The improved coagulation-dissolved air flotation unit was used to remove suspended solids and COD using 10 $mg \cdot L^{-1}$ cationic PAM together with 25, 50, 75, 100, 150 and 200 $mg \cdot L^{-1}$, respectively, of coagulant XR reagent, and its effectiveness was monitored in 2 hours later.

As shown in Figure 2(a), given a constant cationic PAM dosage of 10 $mg \cdot L^{-1}$, the SS removal efficiency

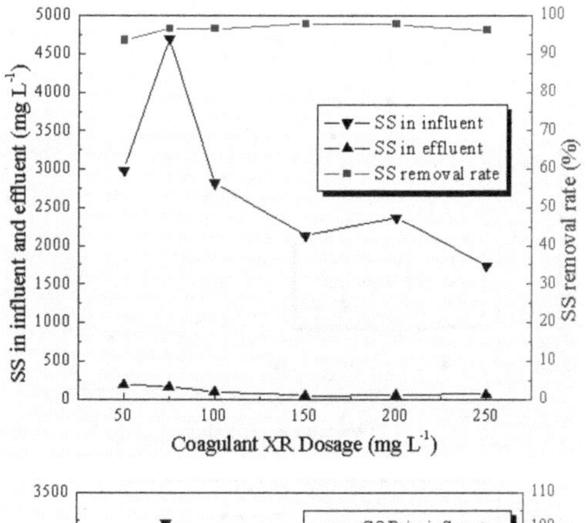

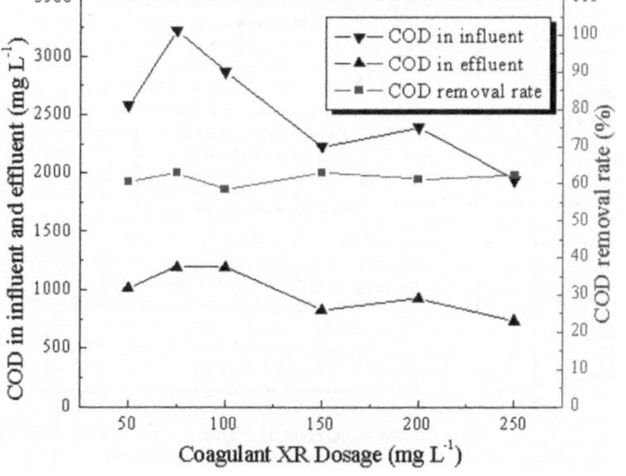

Figure 2. (a) Effect of coagulant XR dosage on SS removal. (b) Effect of coagulant XR dosage on COD removal.

increased with increasing coagulant XR dosage. A dosage of 100 mg·L^{-1} achieved the maximum SS removal efficiency, yielding a removal rate of 97.8% and effluent SS concentration of 46 mg·L^{-1}. However, upon continuing to add the coagulant XR, the SS removal efficiency did not increase further, but declined. Because of the high concentration of SS in ABS resin wastewater, the colloidal particles did not destabilize when the coagulant XR was insufficient. With the increasing coagulant XR dosage, the electric double layer was compressed, and the colloidal particles were agglomerated rapidly and removed. With increasing dosage of the coagulant XR, the electric charge of the colloidal particle inversed, and the colloidal particle stabilized again. The removal of suspended solids declined.

The removal of COD should follow the law theoretically (Guo, Xu and Lv. 1996; Wu et al. 2007), but it did not conform to the law in the study. As shown in Figure 2(b), the concentration of COD in the effluent fluctuated greatly, but the removal of COD was almost unchanged and maintained in the range of 55%~65%. Most of the suspended COD and parts of the soluble COD could be removed by the coagulation-dissolved air flotation process. Because the wastewater quality fluctuated greatly, and the ratio of suspended COD and soluble COD was fluctuant, the trend of the COD removal was not obvious. In other words, the effect of the coagulant XR dosage on the removal of the suspended solid was larger than removal of COD; therefore, the dosage of 100 mg/L was optimal.

3.2 *Effect of cationic PAM dosage on process efficiency*

The improved coagulation-dissolved air flotation unit was used to remove suspended solids and COD using the coagulant XR dosage of 150 mg·L^{-1} together with the cationic PAM dosage of 2.5, 5.0, 7.5, 10, 15 and 20 mg·L^{-1}, respectively, and its effectiveness was monitored in 2 hours later.

As shown in Figure 3(a), at a constant coagulant XR dosage of 100 mg·L^{-1}, the SS removal efficiency increased with an increased dosage of cationic PAM. A cationic PAM dosage of 7.5 mg·L^{-1}

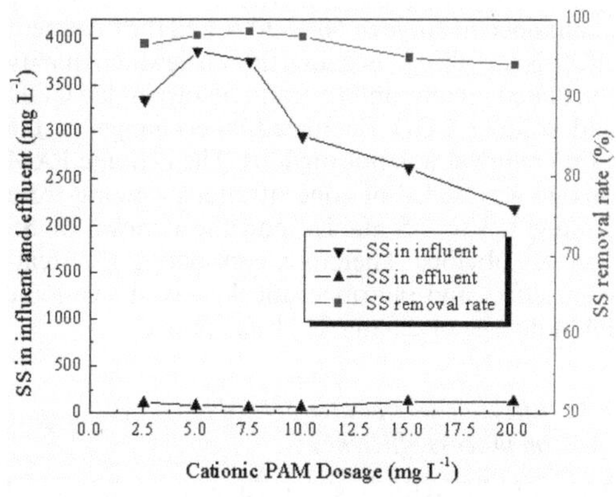

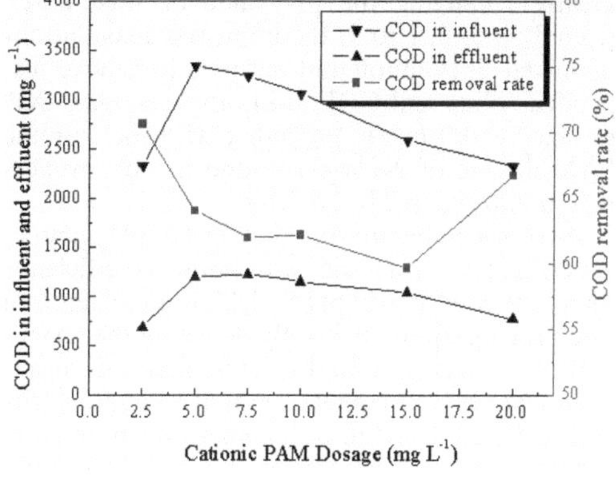

Figure 3. (a) Effect of cationic PAM dosage on SS removal. (b) Effect of cationic PAM dosage on COD removal.

achieved the maximum SS removal efficiency, yielding a removal rate of 98.3% and an effluent SS concentration of 63 mg·L^{-1}. With the increasing dosage of cationic PAM, the SS removal efficiency declined slightly. The cationic PAM has the effects of electrical neutralization and adsorption bridging, when the PAM was added, the negative charge of the colloidal particle surface was neutralized further and the distance between the particles became short. At the same time, the active groups, consisting of long chains of cationic PAM, extended into the water. The particle size of flocs increased gradually due to specific adsorption on the surface of colloidal particles and bridging between particles, eventually reaching equilibrium between the PAM concentration and floc particle size. When the PAM concentration continued to increase, coverage of the particle surface approached 100%. At this point, the particle surface no longer had available adsorption sites and a bridging effect could not be achieved, resulting in the role of space compression. Particles disperse because of the steric effect, and the particle size of flocs decreases (Zhu et al. 2007; Yan, Wang and Chen. 2013).

As shown in Figure 3(b), the removal of COD was maintained in range of 60%~70%, and the treatment effect is stabilized. Because the wastewater quality fluctuated greatly, and the ratio of suspended COD and soluble COD fluctuated, the change in the COD removal was not obvious. The cationic PAM dosage was added at concentrations ranging from 2.5 mg·L^{-1} to 7.5 mg·L^{-1} and the removal of SS was not obvious. Therefore, considering the treatment effect and pharmaceutical costs, the cationic PAM dosage of 2.5 mg·L^{-1} was optimal.

3.3 *Effect of wastewater pH value on process efficiency*

The pH of ABS resin wastewaters fluctuated greatly. Changing the pH value of wastewater, from 3, 4, 5, 6, 7 to 9, the improved coagulation-dissolved air flotation unit was used to remove suspended solids and COD using the coagulant XR dosage of 100 mg·L^{-1} together with the cationic PAM dosage of 2.5 mg·L^{-1}, and its effectiveness was monitored in 2 hours later.

As shown in Figure 4(a), the effect of pH on process efficiency was strong, with improved efficiency under the condition at pH 5, yielding an SS removal rate ranging up to 95.7%. By detecting the coagulant XR, which is a kind of aluminum salt, it has been reported in the literature that aluminum salts can be transformed into oligomers, medium-sized polymers and other polymers through hydrolysis, and that oligomers and medium-sized polymers are major active materials that facilitate flocculation, although the formation of these active materials depends on the pH value (Yang et al. 2013; Huang, et al. 2012). As shown in Figure 4(b), the treatment effect of COD was similar to that of SS, with improved efficiency under acidic conditions. Under the acidic condition, the dissociation degree and hydrophilicity of the dissolved organic matter are changed, and the dissolved organic matter become adsorbed easily and are attached to the floc co-precipitates. Therefore, the effect of wastewater treatment was best at pH 5~6.

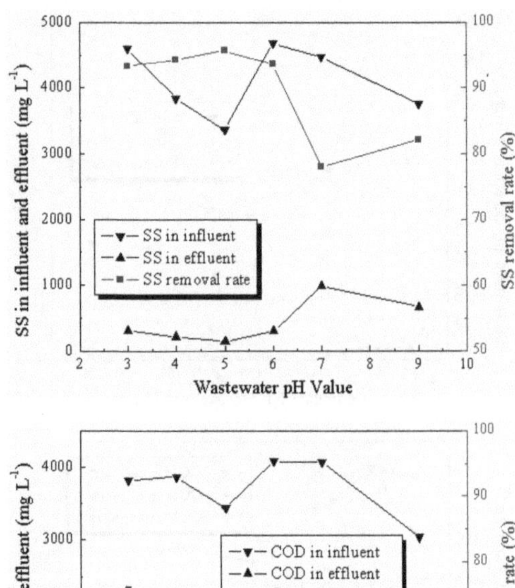

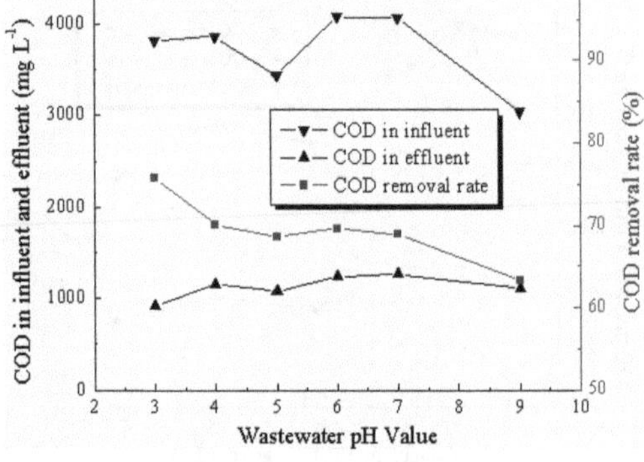

Figure 4. (a) Effect of pH value on SS removal. (b) Effect of pH value on COD removal.

3.4 *Effect of residence time on process efficiency*

Residence time has a strong influence on the coagulation/flocculation process. Changing the residence time from 12.5 min, 15 min, 19 min, 25 min to 37.5 min (inflow 0.4 m^3, 0.6 m^3, 0.8 m^3, 1.0 m^3 and 1.4 m^3, respectively), the improved coagulation-dissolved air flotation unit was used to remove suspended solids and COD using the coagulant XR dosage of 100 mg·L^{-1} together with the cationic PAM dosage of 2.5 mg·L^{-1}, and its effectiveness was monitored in 2 hours later.

Figure 5(a) and 5(b) shows the SS removal rate as a function of residence time. The concentration

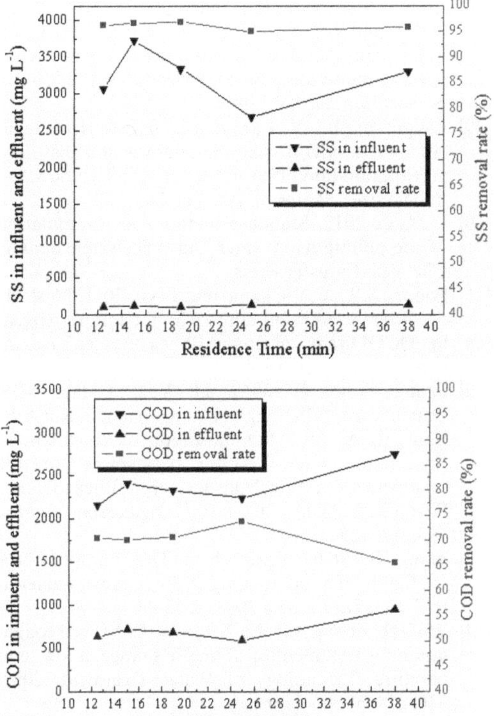

Figure 5. (a) Effect of residence time on SS removal. (b) Effect of residence time on COD removal.

of SS and COD in the effluent had a small fluctuant with the different residence time, and the removal of SS and COD was high, and the concentration of SS and COD in the effluent were below 140 mg · L^{-1} and 1000 mg · L^{-1}, respectively, and the removal rates of SS and COD were above 95% and 60%, respectively. The concentration of SS in the effluent was 104 mg · L^{-1} and the removal rate reached up to 96.9% at the residence time of 19 min. The floc formed in the process of coagulation collided and adhered to tiny bubbles, and then formed floc with a bigger volume and lower density, so that the floc floated up speedily and removed. Considering the wide variation in raw water quality, a 15 min residence time was optimum for achieving maximum SS removal efficiency.

3.5 *Effect of reflux ratio on process efficiency*

Changing the reflux ratio from 20%, 40%, 60%, 80% to 100%, the improved coagulation-dissolved air flotation unit was used to remove suspended solids and COD using the coagulant XR dosage of 100 mg · L^{-1} together with the cationic PAM dosage of 2.5 mg · L^{-1}, and its effectiveness was monitored in 2 hours later.

The reflux ratio affects the efficiency of the coagulation/flocculation unit, thus also affecting equipment investment costs and daily operation expenditures. As a result, the reflux ratio should be minimized as much as possible while maximizing the removal efficiency. As shown in Figure 6(a) and 6(b), the concentration of SS and COD in the effluent declined and the removal ratio increased with the increasing reflux ratio. The ratio of dissolved-air content to SS in the raw water (A/S), which is important for the air-flotation technology, varies with a varying reflux ratio. Improving the A/S value, which is equivalent to increasing the reflux ratio, helps improve efficiency. The amount of bubbles and flocs is balanced at a certain reflux ratio: a relative excess of bubbles yields no significant improvement in efficiency because of the low utilization of bubbles, but energy dissipation is increased with the increasing reflux ratio. Figure 6(a) shows that the effect of the reflux ratio on SS removal was small when the reflux ratio was exceeded 40%. Figure 6(b) shows that the concentration of COD in the effluent fluctuated largely under the condition of different reflux ratios, but the COD removal maintained in the range of 50%~60%, which did not affect the subsequent

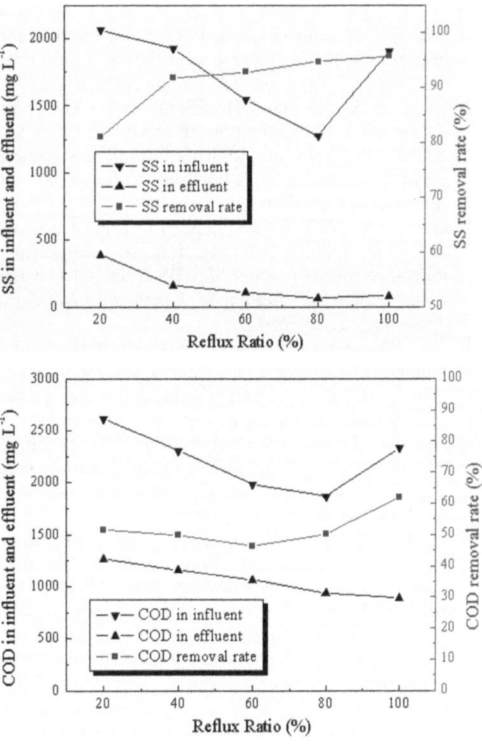

Figure 6. (a) Effect of reflux ratio on SS removal. (b) Effect of reflux ratio on COD removal.

biological treatment. Therefore, when comprehensively considering equipment investment and energy dissipation, a 40% reflux ratio was determined to be the optimum for achieving maximum SS and COD removal efficiency.

4 CONCLUSIONS

1. The optimum coagulation-air flotation conditions for the treatment of ABS resin wastewater were selected: the coagulant XR and cationic PAM dosages were 100 mg · L^{-1} and 2.5 mg · L^{-1}, respectively. The optimum pH was 5~6.
2. The optimum operating parameters for the treatment of ABS resin wastewater by the improved coagulation-dissolved air flotation device were determined. The residence time was 15 min, and the optimum reflux ratio was 40%.

ACKNOWLEDGMENT

This research was supported by the Special S&T Project on Treatment and Control of Water Pollution (No. 2012ZX07201-005-02-03).

REFERENCES

[1] Chang, C.N., and J.G. Lin. 1997. "The Pretreatment of acrylonitrile and styrene with the ozonation process." *Water Science and Technology* 36(2):263–270.

[2] Chen, X.Y., J. Gao, H. Zhou, G.T. Ying, and Z. Xu. 2007. "Modification on treatment process of wastewater from acrylonitrile-butadiene-styreneter polymer production." *Petrochemical Technology and Application* 25(6):544–546.

[3] Guo, L.X., X.T. Zhao, G.X. Li, L.H. Cao, and F.Q. Liu. 2005. "Investigation of coagulation-microelectrolysis process of ABS-resin laden wastewater treatment." *Journal of Lanzhou University of Technology* 31(5):36–42.

[4] Guo P.Y., Xu Y.L., Lv P.C. "Reduction of COD in biologically and chemically treated water with coagulating process." Fuel and Chemical Processes, 1996, 27(2):93–96. (in Chinese).

[5] Hanafy, M., and H.I. Nabih. 2007. "Treatment of oily wastewater using dissolved air flotation technique." *Toxicological and Environmental Chemistry* 29(2):143–159.

[6] Huang, J., Z.H. Yang, G.M. Zeng, M. Ruan, H.Y. Xu, W.C. Gao, Y.L. Luo, and H.M. Xie. 2012. "Influence of composite flocculant of PAC and MBFGA1 on residual aluminum species distribution." *Chemical Engineering Journal* 191:269–277.

[7] Li, Y., F.Q. Li, Y.D. Song, and X.W. He. 2013. "Research progress on treatment technology of ABS resin wastewater." *Contemporary Chemical Industry* 42(12):1668–1670. (in Chinese).

[8] Lai, B., Y.X. Zhou, H.K. Qin, C.Y. Wu, C.C. Pang, L. Yu, and J.X. Xu. 2012. "Pretreatment of wastewater from Acrylonitrile-Butadiene-Styrene (ABS) resin manufacturing by microelectrolysis." *Chemical Engineering Journal* 179:1–7.

[9] Lakghomi, B., Y. Lawryshyn, and R. Hofmann. 2015. "A model of particle removal in a dissolved air flotation tank: Importance of stratified flow and bubble size." *Water Research* 68:262–272.

[10] Li, G.G. 2012. Standard methods for examination of the pollutants in water and wastewater. Beijing: Chemical Industry Press.

[11] Popuri, S.R., C.Y. Chang, and J. Xu. 2011. "A study on different addition approach of Fenton's reagent for DCOD removal from ABS wastewater." *Desalination* 277(1–3):141–146.

[12] Santo, Carlos E., Vitor J.P. Vilar, Cidalia M.S. Botelho, Amit Bhatnagar, Eva Kumar, and Rui A.R. Boaventura. 2012. "Optimization of coagulation–flocculation and flotation parameters for the treatment of a petroleum refinery effluent from a Portuguese plant." *Chemical Engineering Journal* 183:117–123.

[13] Wang, B. 2010. "Research of TiO_2 photocatalysis oxidation method treating ABS organic wastewater." *Ningxia Electric Power* 2:52–56.

[14] Wu Y.S., Zhang B.P., Lu X.J., et al. Enhanced coagulation for treatment of domestic sewage at low temperature. Technology of Water Treatment, 2007, 33(1):54–58. (in Chinese).

[15] Xie, C.S. 2009. "Study on treatment of ABS wastewater by photocatalysis." phD diss., Shandong University.

[16] Yan, W.L, Y.L. Wang, and Y.J. Chen. 2013. "Effect of conditioning by PAM polymers with different charges on the structural and characteristic evolutions of water treatment residuals." *Water Research* 17(47):6445–6456.

[17] Yang, Z.L., B. Liu, B.Y. Gao, Y. Wang, and Q.Y. Yue. 2013. "Effect of Al species in polyaluminum silicate chloride (PASiC) on its coagulation performance in humic acid–kaolin synthetic water." *Separation and Purification Technology* 111:119–124.

[18] Zhao, D.F., H.H. Liu, and G.D. Liu. 2003. "Biochemical treatment technology of wastewater produced from acrylonitrile-butadiene-styrene resin." *Journal of the University of Petroleum, China (Edition of Natural Science)* 27(5):113–115.

[19] Zhang, Q.D., S.J. Liu, C.P. Yang, F.M. Chen and S.L. Lu. 2014. "Bioreactor consisting of pressurized aeration and dissolved air flotation for domestic wastewater treatment." *Separation and Purification Technology* 138:186–190.

[20] Zhu, Z., T. Li, D.S. Wang, Z.H. Yao, and H.X. Tang. 2007. "Effect of cationic PAM dosage on the micro-properties of flocs." *Environmental Chemistry* 26(2):175–179.

Material Science and Environmental Engineering – Chen (Ed.)
 ISBN 978-1-138-02938-5

Experimental research on the disposition of overflowing rain sewage in Chongqing through artificial wetland of free aeration waterfall

X.H. Peng, W.B. Wang & Y.Q. Li
Logistics Engineering University of CPLA, China

ABSTRACT: The aim of this article is to introduce a technology of artificial wetland through Free Aeration Waterfall and Classified Water Puff, and its parameter set. Through the disposition of overflowing rain sewage in Chongqing, it was discovered that the proposed technology has advantages in terms of small land occupation, high level of water aeration, good effects on disposition, few blind corner or jam. It is particularly suitable for a particular terrain with some slant.

Keywords: free aeration waterfall; classified water puff; parameter set; overflowing rain sewage; jam

1 INTRODUCTION

In China, most of the cities retained Converged Sewage Channels nowadays. With the development of urbanization, the cement area of urban land is increasing day by day, and the pollution from overflowing rain sewage in cities is increasingly devastating, especially in these southern cities, which abounds in rainfall. Chongqing is typically a mountainous city, and there is a big difference in terms of geographical height from the banks alongside the two rivers. The possibility of building storage pools by utilizing the unexplored area or green land is small. It ought to take into account full utilization of the natural terrain for the disposition of overflowing sewage right on the spot. This paper conducted an experiment on the disposition of overflowing sewage in Chongqing through the artificial wetland of free aeration waterfall, and it avoided the defects of common artificial wetland in terms of slow water flow, long stay of sewage, long submergence of flora roots and construction stuff in water, and low level of material aeration. On the other hand, the technology has advantages in terms of no necessity of air exposure, low energy-consumption, and switch from detriments into benefits.

2 CRAFTS

2.1 *The quality of sewage water in Chongqing*

Currently, there are 143 overflowing channels within the sewage networks in the urban center of Chongqing Municipal. Due to the geographical conditions of those outlets, there are 28 channels that can be able to accomplish disposition independently, accounting for 19.6% of the whole overflow networks. There are 9 channels that can be able to accomplish disposition on the basis of low standard and high cooperation, accounting for 6.3% of the whole overflow networks. There are 106 channels that cannot be able to accomplish disposition, but release only, accounting for 74.1%. If it would be able to accomplish disposition of the sewage water from the channels that have the disposition capability, it could decrease the disposition workload of pollution sources as much as possible under the current conditions.

Table 1 presents the index on the quality of overflowing rain sewage in Chongqing Municipal.

Table 2 presents the quality of overflowing sewage after the first disposition (average index).

Table 1. The quality of overflowing rain sewage in Chongqing Municipal.

Index of overflowing rain sewage in CQ Municipal	Density (mg/L)
TSS	363.8
COD	124.2
BOD	71.5

Table 2. Quality of puff water from artificial wetland of free aeration waterfall.

Quality of puff water from artificial wetland of FAW	Density (mg/L)
TSS	75.5
COD	98.7
BOD	55.9
TN	7.33
TP	0.93
NH3-N	2.95

2.2 *The choice on the type of artificial wetland and its crafts' process*

2.2.1 *Choice on the type of artificial wetland*

According to the concrete terrain in Chongqing, it ought to choose the artificial wetland of interconnected artificial wetland of vertical waterfall with three levels, which is shown as a shape of stairs. Each level consists of three parts, namely, wetland bed, stuff, and flora of wetland. The slant degree on each level of the bottom of the wetland bed is 2%. It is classified into Level One, Level Two, and Level Three from the top to the bottom, respectively. Figure 1 shows the specific apparatus.

2.2.2 *The choice on the crafts' process*

Due to the high mountains and stiff slopes in Chongqing, the volume of flowing materials in the rain sewage water is relatively high, and the change of water volume is huge. Furthermore, as the flowing continues, the water quality changes to a large degree. Consequently, at first, the sewage water ought to go through the primary disposition pool, before the disposition of artificial wetland. Figure 2 shows the crafts' process.

2.3 *Choice of flora*

One of the keys on artificial wetland is the choice of flora. There is a huge difference among different flora in terms of disposition capability. Those floras with strong and advanced root system and thriving capability of survival and reproduction are better in terms of the disposition effects of sewage water. According to the climate in Chongqing and its special geographical environment, it is suitable to choose Calamus, Cordate, and Desmodium that are adaptable to thrive in the South.

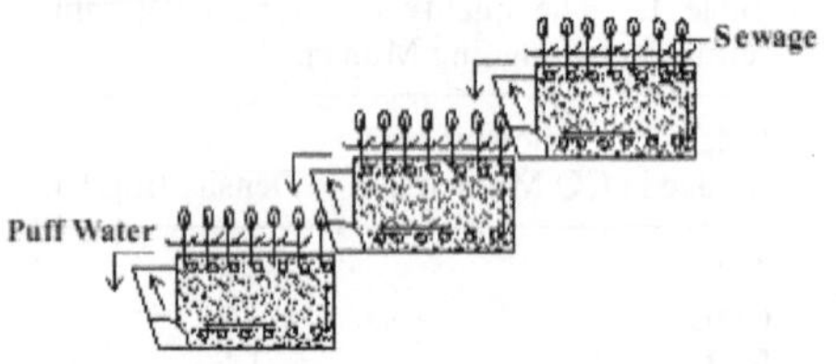

Figure 1. Artificial wetland of free aeration waterfall.

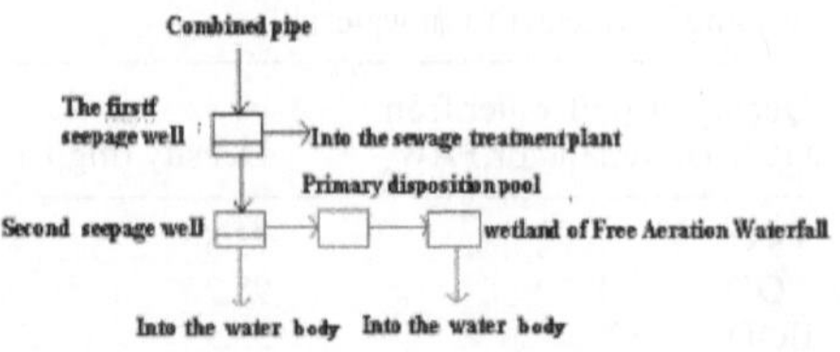

Figure 2. Crafts' process picture of disposition of rain sewage water.

2.4 *Choice of stuff*

It is necessary to choose the stuff of multi-function and recombination, which is capable to deprive ammonia, nitrogen, and phosphorus. Artificial wetland is classified into interconnected three levels, and it ought to use dust-size stuff only for the wetland of each level, and the size of the stuff for the wetlands of each level becomes smaller and smaller in order, following the water flow direction (which means the biggest size at the puff point and the smallest size at the outlet point). There shall be laid with the mixture of fly ash and soil, as shown in Figure 3.

Setting several layers for the water flow could not only make the sewage water to go through layers of various dust sizes, decreasing jams, but also bring great convenience to clean and repair the pool. Meanwhile, setting a channel system of multi-level water puff, using the switch of valves, could make other wetlands of various levels operate as usual, when any level or several levels were under inspection and repair, and it will elevate the operational reliability of artificial wetland apparently.

2.5 *Choice of terrain*

Almost all the areas alongside the banks of Yangtze River and Jialing River in the Chongqing Urban Centre are allocated lands for urban construction, and the density of buildings is quite high. According to the realistic terrain, it would be appropriate to choose marsh land or wasteland of low commercial value with the natural slant degree at around 3%, and the lands chosen ought to be readily prepared for any geological disasters and shall be above the position of anti-flood level.

Figure 3. Filler particle size chart.

Table 3. Order of porous rate.

Level number (from right to left)	d
1	0.4
2	0.38
3	0.35

3 CONCLUSION

3.1 *Confirming the superficial area of wetland layer*

The superficial area of the wetland layer can be calculated by using the following formula:

$$A_S = 5.2Q\,(\ln S_0 - \ln S_e) \tag{1}$$

S_O, S_e—puff density of BOD and outlet density of BOD, mg/L;
A_s—superficial Area of wetland layer, m^2; and
Q—flux set on average.

The density at the puff and outlet of the three units of wetland layers is presented in Table 4.

According to Table 4 and formula (1), the superficial area of the three units of wetland layers is listed in Table 5.

3.2 *Hydraulic workload*

The average hydraulic workload of wetland is

$$q = \frac{Q}{A_s} \tag{2}$$

According to Formula (2), the calculated hydraulic workload of the three units is presented in Table 6.

3.3 *Stay time of hydraulics*

The stay time of hydraulics is given in the following formula:

$$HRT = \frac{A_S \times D \times n}{Q} \tag{3}$$

The calculated stay time of the interconnected three-level hydraulics is presented in Table 8.

The hydraulic stay time of the whole apparatus of artificial wetland is 2.5 d.

3.4 *The calculation formula of the cross-sectional area of wetland*

$$A_C = \frac{Q}{K_S \times S}$$

A_C—area, m^2;
K_S—hydraulic transmitting velocity; and
S—hydraulic slant degree, m/m.

The cross-sectional area of wetland layers is presented in Table 9.

According to the formula: $W = A_c/D$, which could be calculated by the width of wetland units.
W—width of the wetland, unit: m.

According to the formula: $L = A_s/W$, which could be calculated by the length of wetland layers.

The size of wetland layers is given in Table 10.

Table 4. Puff density of BOD.

Level no.	Puff density of BOD (mg/L)
1	55.9~28
2	28~11.6
3	11.6~2.7

Table 5. Superficial area of the wetland layer.

Level no.	Superficial area of the wetland layer
1	360
2	560
3	750

Table 6. Hydraulic workload of wetland layers.

Level no.	m^3/m$^2\cdot$d
1	0.28
2	0.18
3	0.13

Table 7. Technological parameters of wetland.

Level no.	Height of stuff	Porous rate
1	0.6	0.4
2	0.4	0.38
3	0.3	0.35

Table 8. Hydraulic stay time of wetland layers.

Level no.	d
1	0.86
2	0.85
3	0.79

Table 9. Cross-sectional area of wetland layers.

Level no.	Area, m^2
1	10.8
2	8.0
3	7.5

Table 10. The size of length and width of wetland layers.

Level no.	Length	Width
1	20	18
2	28	20
3	30	25

3.5 *Apparatus of water puff and outlet*

The water puff system of wetland shall guarantee the distribution of water as even as possible. The experiment adopts porous channels, which are set below the wetland layers, and the water puff of the system is modulated by valves. In order to control the water level of wetland effectively, it is set at the bottom of stuff layers with transfixed converged channels, rotated curve unit and controlling valves. The converged channels at the bottom are set with vent channels, in which the aperture size is the same as the converged channels. The angle between the back-side layer and the bottom of layer is 100–120°. The top of the back-side layer is set with the ushering plank vertically. The middle and back parts of wetland layers are set with partitions, in which the width is the same as that of wetland layers. The partition will divide wetland layers into stuff pool and converged water pool, and there remains a water flow channel between the bottom of partitions and the bottom of layers, and the top of the back-side layer is lower than the top of partitions.

3.6 *Partitions and anti-infiltration apparatus*

It is set with anti-infiltration layers, with the infiltration index being smaller than 10^{-8} m/s, and partitions at the bottom and side of wetland layers, in order to prevent sewage water from downward infiltration.

Throughout the 3-year operation, the water quality at the outlets complied with the requirements of the design, with the outlet water COD of 10 mg/L on average, SS of around 8 mg/L on average, and NH3-N of 1.5 mg/L on average.

4 CONCLUSION

The artificial wetland has the following advantages:

1. To erase blind corners quite well, and full utilization of the pool accommodation.
2. The coming water flows into it from high grounds. By using the natural potential energy, zero-consumption of energy could be realized, and the advantage of energy-saving is apparent.
3. The outlet water flows freely, and then it passes through the effects of paralleled porous partitions outside the pool wall, and consequently, trivial water drops are fully exposed to air, which elevates the effects of aeration.
4. After the free aeration waterfall of sewage, the water would enter the following artificial wetland, which makes the bacteria in the stuff pool more active. And the life activities, once again, would accelerate the metabolism of bacteria, which would lead to better effects of sewage disposition than any traditional artificial wetland.
5. It is set with vent channels at the bottom of stuff, which would create positive conditions for nitrification and denitrification.
6. It has strong adaptability to water quality and quantity. When requirements of outlet water quality elevates, we could increase the levels of artificial wetland. When the water volume is increased, we could parallel several single-level wetlands, or broaden the layer's size of the single-level wetland.

The artificial wetland has a high level of aeration effects, simple structure, apparent land-saving effectiveness, and it would never generate and blind corners, unlikely for any jams, especially suitable for a particular terrain with some slant.

REFERENCES

[1] RISN-TG006-2009. Technological Guidance on Sewage Disposition through Artificial Wetland.

[2] HJ2005-2010. Technological Norms on Sewage Disposition Projects through Artificial Wetland.

[3] Research and Discussion of Relevant Norms on Artificial Wetland Design. Jinpeng Su, Supplement 2011.

Material Science and Environmental Engineering – Chen (Ed.)
© 2016 Taylor & Francis Group, London, ISBN 978-1-138-02938-5

Preliminary study on desorption of cadmium from soil using pretreated citric acid industrial wastewater

Y.Y. Gu, G.D. Liu, L.H. Yan, H. An & Y. Wang
College of Chemical Engineering, China University of Petroleum (East China), Qingdao, China

ABSTRACT: Electrokinetic remediation is an efficient in situ soil remediation technology. In the process of electrochemical remediation, soil environment such as soil pH will generally change dramatically, affecting the sorption/desorption of heavy metals in soil. Meanwhile, the use of enhancement agents may modify the sorption/desorption characteristics of heavy metals onto/from soil particle surfaces, influencing the electrokinetic remediation efficiency. Citric acid industrial wastewater which contains large amounts of citric acid and other organic acids is a potential enhancement agent for electrokinetic remediation of heavy metal-contaminated soils. In this work, the effects of pretreated citric acid industrial wastewater on cadmium desorption from soil are investigated. The results showed that the wastewater could considerably enhance cadmium desorption into solution in a wide pH range. The proportions of cadmium extracted were 7%–67% higher by the wastewater than those by DI water at pH 3–9. Citric acid in the wastewater is more efficient in desorbing cadmium than acetic acid. Cadmium in exchangeable fraction decreased while those in carbonate, organic, and residual fractions increased after desorption using the wastewater at pHs 5, 7, and 10.

Keywords: cadmium; desorption; citric acid industrial wastewater; electrokinetic remediation

1 INTRODUCTION

In recent years, heavy metal pollution of soil is becoming more and more serious in China and the hazard of "cadmium rise" resulting from cadmium pollution of soil has been identified and reported. Electrokinetic remediation is a promising method for soil decontamination, especially for soils that have low hydraulic permeability (Yeung & Gu 2011; Gao et al. 2013; Suzuki et al. 2004). In electrokinetic remediation, a direct current electric field is applied across the contaminated soil and the contaminants are removed in-situ by electromigration, electroosmosis, and/or electrophoresis (Yeung 2009). During electrokinetic remediation process, the sorption and desorption characteristics on the soil particle surfaces are very important factors which decide the migration and transformation, bioavailability and toxicity, affecting the efficiency of soil remediation directly.

The sorption/desorption of heavy metals depend on soil pH as well as the chemical composition of soil solution. At the electrodes, H^+ or OH^- ions are generated due to the electrolysis of water, resulting in pH variation of soil across the soil specimen with treatment time (Lysenko & Mishchuk 2009). The soil pH affects On the other hand, enhancement agents, such as EDTA, citric acids, and acetic acid, are usually introduced to promote contaminant removal from soil in electrokinetic remediation (Yeung & Gu 2011; Gomes et al. 2012). Citric acid industrial wastewater which contains large amounts of organic acids is supposed to be a potential alternative for the commercial enhancement agents widely used in previous research (Gu & Yeung 2011, 2012; Gu et al. 2014).

In this study, the desorption of cadmium from soil particle surfaces using pretreated citric acid industrial wastewater at different soil pHs are investigated and sequential extraction of cadmium before and after desorption at different pHs are conducted to analysis the effects of the wastewater on cadmium desorption from the soil.

2 MATERIALS AND METHODS

2.1 *Soil*

The soil used in this study was collected from the campus of China University of Petroleum (East China). It is air-dried, pulverized, and passed through a 2 mm sieve before tested. The physicochemical properties of the soil are listed in Table 1.

2.2 *Citric acid industrial wastewater*

The citric acid industrial wastewater after acidification pretreatment from Shandong Jinhe citric

Table 1. Physicochemical properties of the soil.

Property	Value
pH (1:2.5 by weight)	7.36
Organic content [g/kg]	2.96
Electric conductivity (1:5) [dS/cm]	0.089
Particle size distribution [%]	
0.2–2 [mm]	67.5
0.02–0.2 [mm]	22.4
<0.02 [mm]	10.1

Table 2. Physicochemical properties of the wastewater.

Property	Value
pH	4.28
COD/(mg/L)	12760
Cl^{-1}/(mg/L)	3.77
Ca^{2+}/(mg/L)	1481.18
Mg^{2+}/(mg/L)	920.74
Acetic acid/(mg/L)	4500
Glucose/(mg/L)	3.35
Citric acid/(mg/L)	4875.4

acid factory was used in the study. The pH of the wastewater is 4.28. The physicochemical properties of the wastewater are tabulated in Table 2.

2.3 *Acid/base buffer capacity of soil*

1.5 g air-dried soil was mixed with 30 mL deionized water (DI water) in a 50 mL centrifuge tubes and the soil mixtures were shaken for 24 h at 25°C. Then, a measured volume of 1 M H_2SO_4 or NaOH was added to the centrifuge tubes and the soil mixtures were shaken for another 24 h until the soil pH stabilized. Soil pH was measured using a pH meter. Each test was conducted in duplicate.

2.4 *Desorption tests*

The soil mixtures were prepared by adding 1 g of air-dried soil and 0.5 mM $Cd(NO_3)_2$ to 50 mL centrifuge tubes. They were shaken for 24 h at 25°C before centrifuged. The cadmium concentration in the supernatant was measured by Atomic Absorption Spectrometry (AAS). In order to measure the cadmium concentration sorbed onto the soil particle surfaces accurately, the centrifuge tubes were weighed before and after the supernatant was decanted. Afterwards, 10 mL of the wastewater or DI water was added to the centrifuge tube and pH was adjusted using 0.1/1 M HNO_3 or NaOH to 1–12. The soil mixtures were shaken for 24 h and centrifuged before the final pH and cadmium concentration of the supernatant was measured.

Table 3. Sequential extraction procedures.

Cadmium fractions	Extractants and conditions
EX*	8 mL of 1 M MgCl2, pH 7.0, 1 h
CAR*	8 mL of 1 M NaOAc adjusted to pH 5.0 with HOAc, 5 h
Fe/Mn*	20 mL of 0.04 M NH2OH-HCl in 25% (v/v) HOAc, 96±3°C, 5 h
OR*	3 mL of 0.02 M HNO3 and 5 mL of 30% H2O2, pH 2 (adjusted with HNO3), 85±2°C for 2 h; 3-mL aliquot of 30% H2O2 (pH 2 with HNO3), 85 ± 2°C for 3 h; 5 mL of 3.2 M NH4OAc in 20% (v/v) HNO3, diluted to 20 mL, 30 min
RE*	Concentrated HF-HClO4 mixture

*EX—exchangeable; CAR—carbonates; Fe/Mn—Fe/Mn oxides; OR—organic; RE—residual.

After the soil was spiked with cadmium following the above sorption procedures, 10 mL 4875.4 mg/L of citric acid or 4500 mg/L of acetic acid were added to the soil spiked with the initial cadmium concentration of 0.538 mg/g as the purging solutions wastewater in order to determine the effects of different constituents of the pretreated wastewater on cadmium desorption.

At the end of the batch sorption/desorption test at pH 5, 7, and 10, sequential extraction of cadmium were conducted on the soil samples after the supernatants were decanted. The initial cadmium sorbed on soil particle surfaces is 1.764 mg/g. The procedures of sequential extraction experiments are listed in Table 3. Cadmium concentrations of each extraction were measured by AAS. Each test was performed in duplicate to ensure the precision and repeatability of the experiment.

3 RESULTS AND DISCUSSION

3.1 *Acid/base buffer capacity of soil*

The acid/base buffer capacity of the soil varies with the soil pH as shown in Figure 1. The initial pH of the soil suspension is 6.66. The slope of the curve is flatter at pH 3–9, indicating a lower buffer capacity of soil in this pH range. It means that the soil pH could decrease to 3 or increased to 9 relatively easily with the addition of acid or alkaline. On the other hand, the buffer capacity of the soil increases sharply at pH 2–3 and 9–10, consuming much more acid or alkaline when soil pH decreases from 3 to 2 or increases from 9 to 10.

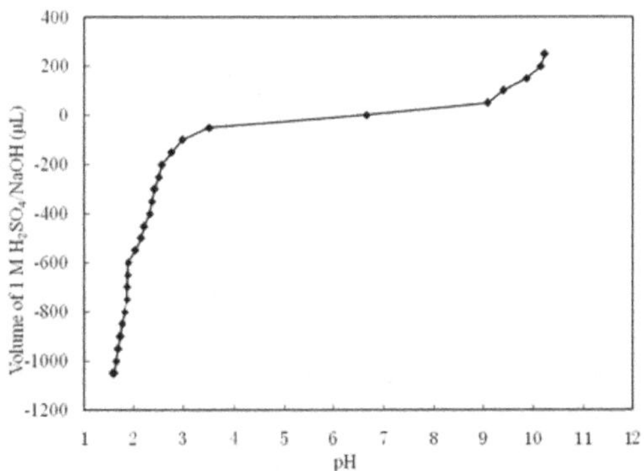

Figure 1. Acid/base buffer capacity of the soil.

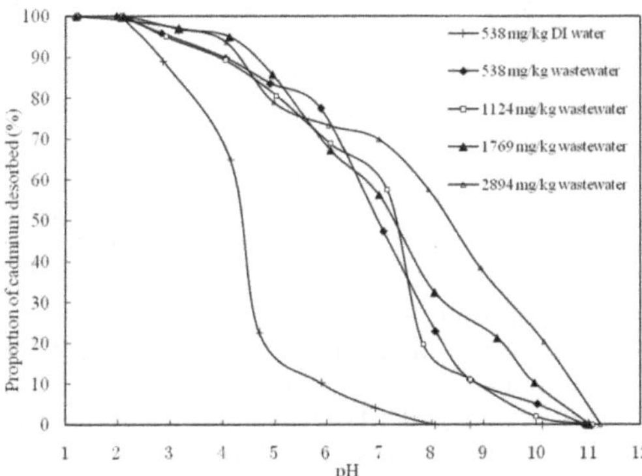

Figure 2. Cadmium desorption using the wastewater in comparison DI water.

3.2 *Cadmium desorption from soil*

Cadmium desorption curves in the presence of citric acid industrial wastewater of different cadmium concentrations at pH 1–12 are depicted in Figure 2, in comparison those with DI water. It can be observed that, at pH 7, only 4% of cadmium was desorbed using DI water, while 47 of cadmium could be desorbed into soil solution with the enhancement of the wastewater when the initial cadmium concentration was 0.538 mg/g. When the soil pH increased to 8, cadmium on the soil particle surfaces could barely be desorbed using DI water. However, with the addition of the wastewater, the upper bound of soil pH limit for cadmium desorption is extended to 10. At pHs 3–9, the wastewater increased the proportion of cadmium desorption by 7%–67% more than those by DI water at the same soil pH. The wastewater significantly changed the desorption characteristics of cadmium from soil particle surfaces in a wide pH range.

The proportion of cadmium desorbed from soil decreased sharply with the increase in soil pH. It can also observed from the figure that cadmium desorption increased with the cadmium concentration initially sorbed on the soil particle surfaces. Similar results were also obtained using synthetic citric acid industrial wastewater as the purging solution to desorb cadmium from a natural clayey soil of high acid/base buffer capacity in our previous work (Gu & Yeung 2011).

3.3 *Effects of wastewater constituents on cadmium desorption from soil*

Citric acid and acetic acid in the pretreated citric acid industrial wastewater were found to be the most efficient constituents in desorbing cadmium from soil due to complexion (Gu & Yeung 2011). In this study, the purging solutions of citric acid and acetic acid were prepared to evaluate the effects of the two constituents on cadmium desorption at different pHs. Citric acid was more efficient in desorbing cadmium than acetic acid as shown in Figure 3, which may due to the stronger complexing ability of citrate than acetate. The proportion of cadmium desorbed from soil in the presence of citric acid and acetic acid decreased with the increase in soil pH. Similar results were reported in some previous studies (Mustafa et al. 2006; Yuan et al. 2007). The enhancement of cadmium desorption from soil using the wastewater was the synergetic desoption by citric acid, acetic acid, and other constituents in the wastewater including other constituents such as low molecular weight organic acids, cations, NH_4^+, Cl^-, etc.

3.4 *Sequential extraction of cadmium*

Cadmium distribution before and after desorption using the wastewater is shown in Figure 4(a) and 4(b), respectively. Before desorption, approximately

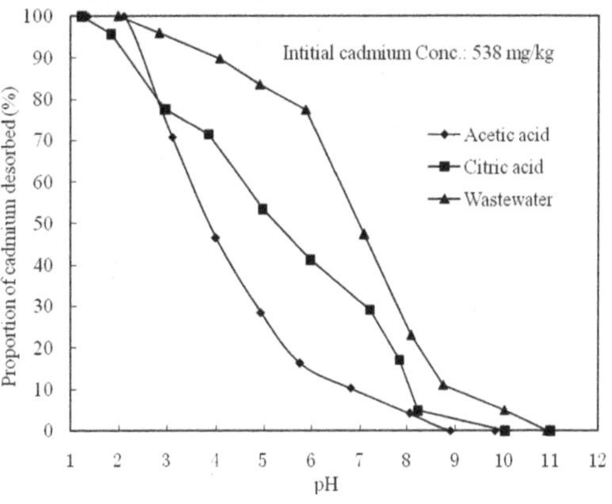

Figure 3. Cadmium desorption using different constituents of the wastewater.

90% of the cadmium spiked into the soil existed in exchangeable and carbonate fractions as observed in Figure 4(a). With the increase in soil pH from 5 to 10, cadmium in exchangeable fraction decreased and that associated with carbonate fraction increased due to greater metal hydrolysis. Cadmium in other factions barely changed with soil pH.

It can be observed in Figure 4(b) that the exchangeable fraction decreased while those in carbonate, organic, and residual fractions increased after desorption using the wastewater at pHs 5, 7, and 10. The large organic content in the wastewater may increases cadmium associated with organic matters. As the exchangeable fraction has higher bioavailability than carbonate fraction (Pociecha & Lestan 2009), the result indicates that the cadmium bioavailability decreased after desorption using the wastewater. Similar results were obtained using synthesized wastewater (Gu et al. 2013).

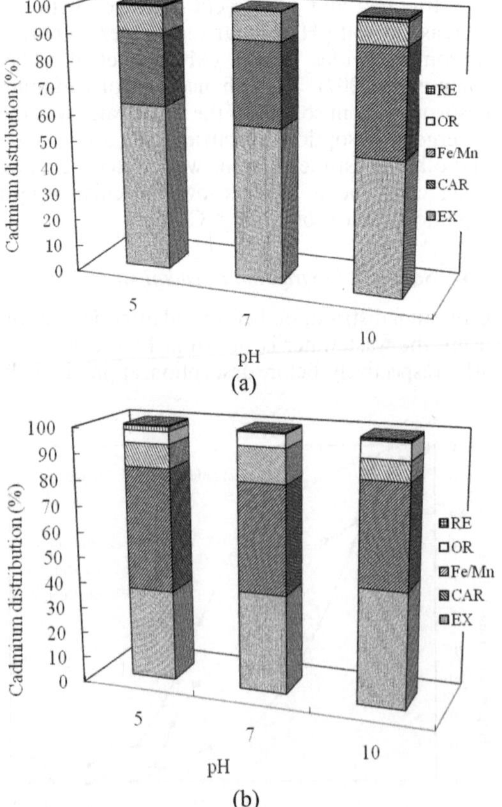

Figure 4. Cadmium distribution in soil at different pHs before and after desorption using the wastewater. (a) Before desorption; (b) after desorption.

After cadmium desorption using the wastewater, cadmium remained in soil in exchangeable, carbonate, and Fe/Mn oxide factions decreased. The exchangeable fraction was desorbed the most from soil. At a lower pH, more cadmium in these fractions can be extracted by the wastewater.

4 CONCLUSIONS

From this work, the following conclusions can be drawn:

1. The acid/base buffer capacity of the soil varies with the soil pH. It is possible to generate a pH gradient from 3 to 9 from the anode to the cathode during the electrokinetic remediation process due to electrolysis of water.
2. Cadmium desorption can be considerably enhanced with the addition of citric acid industrial wastewater in comparison with DI water. The proportion of cadmium desorbed from soil decreased with the increase in soil pH and the decreased in initial cadmium concentration spiked in the soil.
3. Citric acid in the wastewater is a more effective purging solution compared to acetic acid.
4. Cadmium distribution in exchangeable fraction shifted to carbonate fraction with the increase in soil pH. After cadmium desorption using the wastewater, the proportion of cadmium in exchangeable fraction decreased while that in carbonate and oxide fractions increased.
5. Citric acid industrial wastewater after acidification pretreatment is a promising enhancement agent for cadmium desorption and electrokinetic soil remediation process.

ACKNOWLEDGEMENTS

This work is supported by the National Nature Science Foundation of China (41201303); the Fundamental Research for the Central Universities (14CX02052A, 14CX02191A); Qingdao Science and Technology Program for young scientists (14-2-4-86-jch) and State Key Laboratory of Pollution Control and Resource Reuse Foundation (PCRRF13023).

REFERENCES

[1] Gao, J., Luo, Q.S., Zhu, J., Zhang, C.B. & Li, B.Z. 2013. Effects of electrokinetic treatment of contaminated sludge on migration and transformation of Cd, Ni and Zn in various bonding states. *Chemosphere* 93: 2869–2876.

[2] Gomes, H.I., Dias-Ferreira, C. & Ribeiro, A.B. 2012. Electrokinetic remediation of organochlorines in soil: Enhancement techniques and integration with other remediation technologies. *Chemosphere* 87: 1077–1090.

[3] Gu, Y.-Y. & Yeung, A.T. 2011. Desorption of cadmium from natural Shanghai clay using citric acid industrial wastewater. *Journal of Hazardous Materials* 191: 144–149.

[4] Gu, Y.-Y. & Yeung, A.T. 2012. Use of citric acid industrial wastewater to enhance electrochemical remediation of cadmium-contaminated natural clay. In R.D. Hryciw, A. Athanasopoulos-Zekkos & N. Yesiller (eds) *Geotechnical Special Publication No. 225: GeoCongress 2012, Virginia, 2012*. American Society of Civil Engineers. 3995–4004.

[5] Gu, Y.Y., Fu, R.B. & Li, H.J. 2014. Electrochemical remediation of cadmium-contaminated soil enhanced by citric acid industrial wastewater. *Journal of Chemical Industry and Engineering (China)* 65(8): 3170–3177. (in Chinese).

[6] Gu, Y.-Y., Yeung, A.T., Tsang, D.C.W. & Fu, R.B. 2013. Applications of citric acid industrial wastewater and phosphonates for soil remediation: effects on temporal change of cadmium distribution. Soil & Sediment Contamination 22(8): 876–889.

[7] Lysenko, L.L. & Mishchuk, N.A. 2004. Electrohydrodynamic method of pH regulation at soil decontamination. *Colloids and Surfaces A: Physicochemical and Engineering Aspects* 333: 59–66.

[8] Mustafa, G., Kookana, R.S. & Singh, B. 2006. Desorption of cadmium from goethite: effects of pH, temperature and aging. *Chemosphere* 64:856–865.

[9] Pociecha, M. & Lestan, D. 2009. EDTA leaching of Cu contaminated soil using electrochemical treatment of the washing solution. *Journal of Hazardous Materials* 165: 533–539.

[10] Suzukia, T., Niinaea, M., Kogaa, T., Akitab, T., Ohta, M. & Chosoba, T. 2004. EDDS-enhanced electrokinetic remediation of heavy metal-contaminated claysoils under neutral pH conditions. *Colloids and Surfaces A: Physicochemical and Engineering Aspects* 440: 145–150.

[11] Yeung, A.T. & Gu, Y.-Y. 2011. A review on techniques to enhance electrochemical remediation of contaminated soils. *Journal of Hazardous Materials* 195: 11–29.

[12] Yeung, A.T. 2009. Geochemical processes affecting electrochemical remediation. In K.R. Reddy & C. Cameselle (eds), Electrochemical remediation technologies for polluted soils, sediments and groundwater. John Wiley & Sons, Hoboken, N.J., 65–94.

[13] Yuan, S.H., Xi, Z.M., Jiang, Y., Wan, J.Z., Wu, C., Zheng, Z.H. & Lu, X.H. Desorption of copper and cadmium from soils enhanced by organic acids. *Chemosphere* 68: 1289–1297.

Material Science and Environmental Engineering – Chen (Ed.)
 ISBN 978-1-138-02938-5

Study on preparing ultra-5A molecular sieve and adsorbing phenol in wastewater

J.L. Yang, M.L. Du, C. Chen & C.X. Yu
School of Chemistry and Chemical Engineering, Xi'an University of Science and Technology, Xi'an, China

ABSTRACT: Molecular sieve is a porous material with a large specific surface area and pore volume, which is also an ideal adsorbent in wastewater treatment. In this study, a 5A molecular sieve is synthesized in the traditional method, and then modified by the guide agent method to prepare the ultra-5A molecular sieve. The results are characterized by XRD, SEM, IR and other methods. We suggested that the ultra-5A molecular sieve absorbs phenol in the wastewater by changing the time, temperature, pH of adsorption and other conditions, and found the best adsorption conditions as follows: adsorption time is 10 min; the adsorption effect of ultra-5A molecular sieve is better when pH = 3 and is up to 97%; as the adsorption temperature increases, the adsorption efficiency of phenol increases but the increase is little, so the most optimum adsorption temperature of phenol is at room temperature. In addition, this has some reference value for increasing the treatment efficiency of phenol.

Keywords: ultra-5A molecular sieve; wastewater; phenol; adsorption

1 INTRODUCTION

Zeolite is an important class of catalysts, adsorbents and ion exchangers, which has a uniform pore size, large surface area, pore volume and excellent ion exchange quality, is widely used in detergent industry, petrochemical industry, and fine chemical, environmental protection, development of new functional materials and many other fields, and plays an important role in national economy.

With the iron and steel metallurgy, coking, oil refining, petrochemical and other industries developing rapidly, the type and quantity of wastewater containing phenol is increasing day and day. Although there more number of studies on zeolites absorbing heavy metal ions and other inorganic non-metallic ions in water [1–4], only a few studies are available on the adsorption of organic phenols, and no report on ultra-molecular sieve adsorption of phenol. In this study, we use the directing agent method to prepare an ultra-5A molecular sieve, explore its performance of treating phenol in wastewater, probe the optimal conditions for adsorption, and search new ways for treating phenol in wastewater.

2 EXPERIMENTAL STUDY

2.1 *Experimental procedures for zeolite synthesis*

In this study, we use the directing agent method to prepare the ultra-5A molecular sieve, in which the aluminum source is $NaAlO_2$ and the silicon source is Na_2SiO_3 as the raw material. The ideal ratio of raw materials to prepare the ultra-5A molecular sieve is $Na_2O:Al_2O_3:SiO_2:H_2O = 3:1:2:185$, according to the literature [5].

Sucrose percentage content can be calculated by using the following formula: sucrose content/$(NaAlO_2+ Na_2SiO_3) = 1.67\%$.

Samples of 8.197 g $NaAlO_2$ and 28.422 g $Na_2SiO_3 \cdot 9H_2O$ were dissolved in 60 mL and 80 mL water, respectively, mixed and stirred for 10 minutes at 50°C in a water bath. In addition, a certain amount of sucrose was dissolved in 20 mL water as the directing agent. And two configuration solutions were mixed under the conditions of stirring continuously at 45°C to produce a white gel, and then kept at room temperature for 12 h. Sucrose solution was added and heated to 90°C after continuous stirring for 10 min, and reacted for 3 h at a constant temperature. The product was leached and washed with water at pH levels of 9–10. The ultra-4A molecular sieve was obtained under the conditions of drying at 110°C and after which it was ground. The prepared ultra-4A molecular sieve was dissolved in 85 mL $CaCl_2$ solution whose concentration was 200 g/L. It was then heated to boiling with continuous stirring for 30 minutes, leached and washed again. The sample was placed in an oven at 110 °C to dry. The ultra-5A molecular sieve obtained after crushing.

2.2 *Structure characterization of the molecular sieve*

2.2.1 *XRD analysis*

As shown in Figure 1, the X-ray diffraction data of ultra-5A molecular sieve that we prepared are in good agreement with the standard A-type zeolite. And seen from the whole diffraction pattern, the number of its peak is more, the diffraction peak intensity is strong, and there is no mixed crystal.

2.2.2 *Scanning electron microscope*

From the above SEM images, we can see that, the shape of the ultra 5A molecular sieve that we prepared with uniform size is regular; it has a very positive quartet and the particle size is about 600 nm. But the diameter of molecular sieve 5A prepared by the conventional hydrothermal method is about 2–3 μm.

2.2.3 *Infrared spectroscopy*

Figure 3 shows the IR spectra of ultra-5A molecular sieve. The figure shows that the ultra-5A molecular sieve synthesized by the directing agent method has very similar infrared skeletal vibrational bands to that of the standard samples. There is OH^- vibration of adsorbed water at 3428 cm^{-1}; characteristic peaks of adsorbed water at 1649 cm^{-1}; near 999 cm^{-1}, the absorption peak corresponds to the internal tetrahedral Si-O-Si bond asymmetric stretching vibration; near 723 cm^{-1}, the absorption peak corresponds to the symmetric stretching vibration of the Si-O-Si bond; double ring vibration of the absorption peak is near 592 cm^{-1}; near 443 cm^{-1}, the Si-O bending vibration of the absorption peak exists.

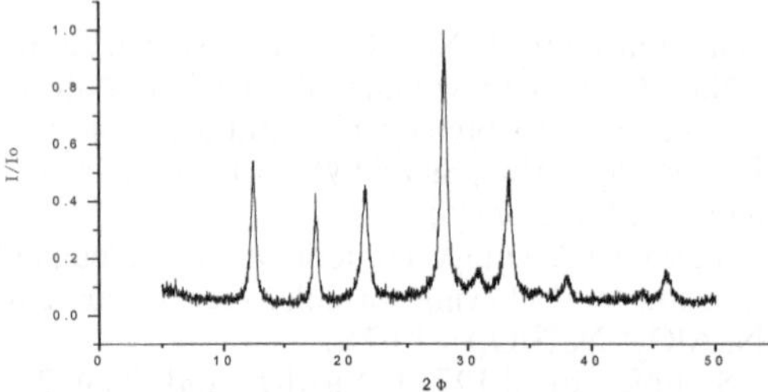

Figure 1. XRD diagram of the ultramicro 5A molecular sieve.

Figure 2. SEM photograph of the 5A molecular sieve.

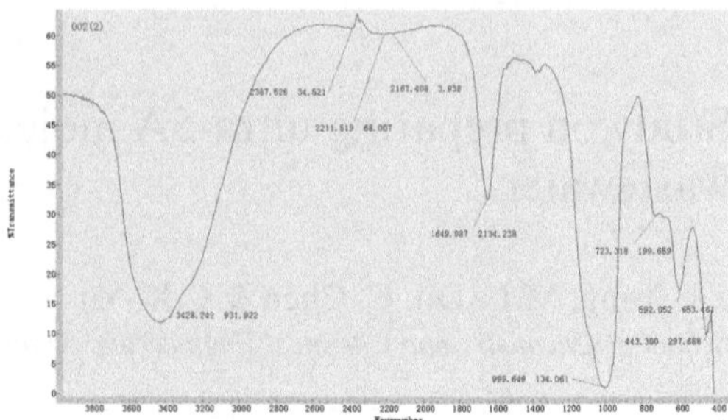

Figure 3. IR diagram of the ultra-5A molecular sieve.

2.3 *Experimental design and study of phenol adsorption*

2.3.1 *Determination standard curve of absorption phenol*

1. Determination of the best absorption peak of phenol. According to the literature [6], absorption peaks occur at 288.5 nm after adding NaOH or KOH in the aqueous solution containing phenol. So we take the phenol solution after alkalization as the determination sample and deionized water as the blank sample, and determine best absorption peaks of phenol under the conditions of alkaline conditions in the present experiment. We found that the best absorption of phenol occurs at 296 nm under the present experimental conditions, so in the follow-up study we take 296 nm as the best absorption peak.
2. Determination of the absorption of the standard curve.

2.3.2 *Phenol adsorption*

1. Drawing the adsorption efficiency curve of phenol. We took 0.1 g conventional 5A molecular sieve and 20 ml 500 mg/L phenol solution in a 1–6 Erlenmeyer flask, respectively. We then took the multi-speed oscillator for adsorption. We added 0.1 g sample and 20 mL deionized water in 7 flasks as a blank experiment.

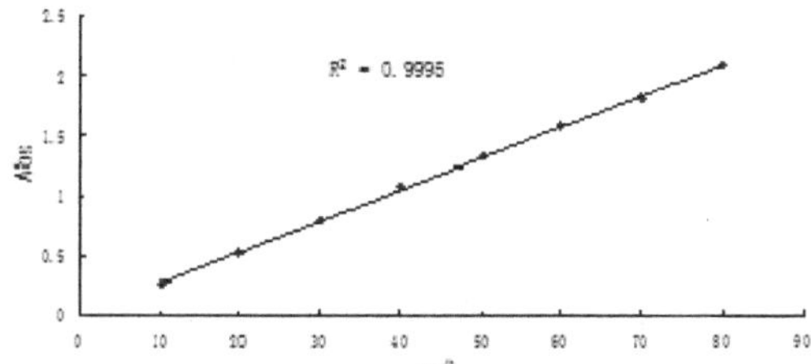

Figure 4. Determination of the phenol absorption standard curve under the basic condition.

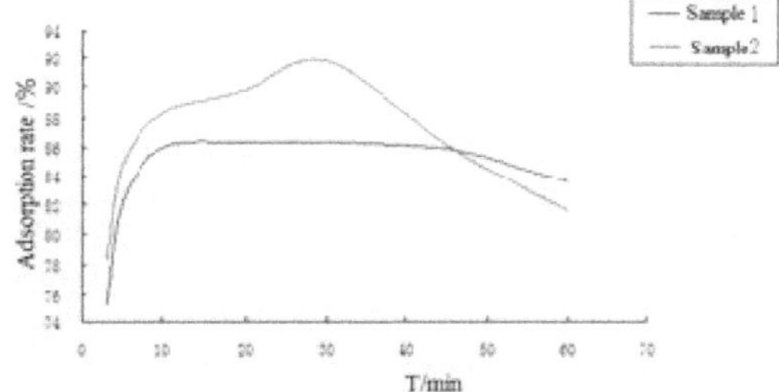

Figure 5. Adsorption efficient curve of phenol.

After 3 minutes, we removed 1 flask and filtered the mixture, and took the amount of 10 mL clear liquid, and then added a drop of 1 mol/L NaOH solution, and measured the absorbance. We removed 2–6 flasks at 5, 10, 20, 30, 45 and 60 min for operation 2. Then, we took the ultra-molecular sieve 5A, and repeated the above experiment. The results are shown in Figure 5.

Seen from Figure 5, adsorption efficiency of ultra-5A molecular sieve (sample 2) is better, adsorption rate is faster at first 10 min, then turn slowly, appear slow significantly after 30 min. By the comparison of sample 1, the adsorption efficiency of ultra-5A molecular sieve is 5% higher than the conventional 5A molecular sieve, and the adsorption rate at the first 10 min is significantly higher than the conventional 5A molecular sieve, and the result of the experiment is as anticipated. The ultra-5A molecular sieve has a special pore structure and large specific surface area, and is a polar adsorbent. Strong adsorption force of molecular sieve is produced by dispersion force and electrostatic force. Because Ca^{2+} ions in 5A molecular sieve cavities generate positive electric fields partly, this static electricity makes molecular sieve play a significant choice adsorption role in polar, unsaturated and easily polarized molecules. Non-polar groups $C_6H_5^-$ present in phenol has a strong adsorption pole with molecular sieve surface, and because the molecular diameter is moderate, these compounds can enter molecular sieve pores and get adsorbed.

2.3.3 *pH and temperature effect on adsorption efficiency*

The above adsorption isotherm curves are measured at room temperature (25°C) and pH is under the neutral condition. In order to understand how pH and temperature affect phenol adsorption, we select the adsorption time as 10 min (with significant changes in the adsorption rate) and measure the adsorption efficiency of molecular sieve at different pH values and temperatures.

1. Study the effect of pH on the adsorption efficiency at room temperature. As can be seen from the above experiment, adsorption efficiency of phenol under acidic conditions is much higher than that under alkaline conditions. The main reason for this is as follows: phenol is a weak acid ion compounds, is partially ionized within a certain pH range, and exists in ions and non-ionic forms in the aqueous solution. When the pH lowers, H^+ concentration increases in the solution; non-ionic form of compounds is in a relatively larger proportion. Non-ionic form is highly hydrophobic, and more easily absorbed by the hydrophobic surface of the molecular sieve in comparison with ionic forms; when the pH reaches 12, phenol is almost in the ionic form, resulting in much lower adsorption rate. Therefore, the pH level of 3 is favorable for phenol adsorption, which fully meets the national emission standards.
2. Temperature influence of the adsorption efficiency under pH 3. Setting the pH level as 3 in the solution and the adsorption time as 10 minutes, we measure the adsorption efficiency of

Table 1. Study of the pH effect on adsorption efficient under 25°C and 10 min.

	Abs adsorption efficiency/%					
Serial no	pH = 3		pH = 7		pH = 12	
Sample 1	0.94	92.92	1.85	85.94	4.18	68.18
Sample 2	0.43	96.82	1.54	88.29	4.10	68.78

Table 2. Study of the temperature effect on adsorption efficient under conditions of pH = 3 and 10 min.

	Abs adsorption efficiency/%					
Serial no	25°C		50°C		75°C	
Sample 1	0.94	92.92	0.85	93.62	0.78	94.10
Sample 2	0.43	96.82	0.37	97.29	0.35	97.42

molecular sieve by changing the adsorption temperature. The results are summarized in Table 2.

From Table 2, it can be seen that the adsorption rate at which ultra-5A zeolite adsorbs phenol is increased with temperature, but with little effect. In addition, the adsorption rate at which the ultra-5A zeolite molecular sieve absorbs phenol can reach more than 95% even at room temperature. So 5A molecular sieve processing wastewater with phenol at room temperature can achieve very good results, fully meeting the national emission standards.

3 CONCLUSIONS

In this study, ultra-5A molecular sieve was successfully synthesized by adding the directing agent on the basis of the traditional hydrothermal method. Characterized by SEM, XRD and IR, the results show that the ultra-5A molecular sieve that we prepared whose crystalline is better and particle size is more uniform, had a size reaching up to 700 nm, a regular shape, and very positive quartets.

We also measured the adsorption efficiency curve of phenol under the conditions of room temperature and neutral pH value, and obtained variation of adsorption efficiency with time under these conditions. The best adsorption time of ultra-molecular sieve 5A was 10 min; the effect of ultra-5A molecular sieve on the adsorption of phenol was better at pH = 3 and up to 97%. As the temperature increased, the adsorption efficiency of phenol increased but to a little extent, so the most optimum adsorption temperature of phenol was at room temperature.

REFERENCES

[1] Dalian Institute of Chemical Physics Research zeolite group, Science in China. Zeolite, Beijing: Science Press, 1978.

[2] H. Tao, G. X. Xu, H.Y. Xie, K.Y. Wang. The study of 13 X zeolite treatment wastewater containing phenol, China Water & Wastewater [J], 2002,18 (4): 50–52.

[3] D.S. Kong. The experimental study of NaX zeolite adsorption phenol in wastewater. Guizhou, Liupanshui Teachers College [J], 2009,21 (3): 24–26.

[4] D.Z. Zhang. The study of nano-zeolite composite removal of BTEX in water, Tianjin Institute of Urban Construction Master Thesis, 2007.

[5] H.M. Chen. The study of NaY zeolite synthesis, Master degree thesis of Dalian University of Technology, 2008.

[6] J.L. Huang, Z.Y. Bao. UV absorption spectroscopy and its application [M]. Beijing: China Science and Technology Press, 1992.

Material Science and Environmental Engineering – Chen (Ed.)
© 2016 Taylor & Francis Group, London, ISBN 978-1-138-02938-5

Photo-Electro-Fenton process for wastewater treatment

X.Q. Wang, L.H. Zang & D.D. Ji
College of Environmental Science and Engineering, Qilu University of Technology, Ji'nan, Shandong, China

ABSTRACT: Photo-electro-Fenton method is a novel wastewater treatment technology, and it is attracting more and more attention. The reaction mechanisms, advantages and disadvantages of the traditional Fenton, Photo-Fenton, Electro-Fenton and Photo-Electro-Fenton processes are described. The existing main problems and development tendency are put forward.

Keywords: Photo-Electro-Fenton; AOPs; wastewater

1 INTRODUCTION

The organic contaminants in wastewater, surface water and ground water may come from agricultural runoff, contaminated soil, storage leakage of hazardous compounds and industrial wastewater. These organic compounds cause serious threat to public health due to their toxicity, endocrine disrupting, and mutagenesis, even potential carcinogen to humans, animals and aquatic life in general. Many organic pollutants are considered as toxic and harmful even when existing at very low concentrations. Therefore, it is important to remove them from the contaminated water. However, in certain cases, conventional treatment methods such as biological processes are not effective for the recalcitrant contaminants present [1, 2].

Consequently, oxidation processes are preferred to degrade such organics in wastewater. The processes including combined conventional with non-conventional AOPs such as photo-Fenton, electro-Fenton, photo-electro-Fenton are put forward. In addition, solar-irradiated processes have been studied, which were associated with the use of light from non-natural sources in order to decrease the costs [3]. However, the solar energy-based processes have restricted applications due to receiving less solar radiation in some countries.

2 METHODS

2.1 *Fenton process*

Fenton process is an advanced oxidation process that can be effectively used in the destruction of hazardous organic pollutants in water via generation of hydroxyl radicals by catalytic decomposition of hydrogen peroxide using ferrous ion as the catalyst. The Fenton reaction was discovered by H.J.H. Fenton in 1894 [4]. The advantage of the Fenton method is mineralization of the organic substances when compared with separation and concentration of toxic chemicals by physical and mechanical methods such as flocculation, precipitation, adsorption on activated carbon, and filtration, which can generate secondary pollution [5]. In recent years, the Fenton reaction has been efficiently utilized for the removal of many hazardous organics from wastewater [6, 7].

Fenton process relies on the pH of the solution mainly due to the presence of iron and hydrogen peroxide. The optimum pH for the Fenton reaction was discovered to be around 3, regardless of the target substrate [8, 9, 10, 11]. Thus, the efficiency of the Fenton process to degrade organic compounds is reduced at high or low pH. Usually, the rate of degradation increases with an increase in the concentration of ferrous ion [12]. However, the extent of increase is sometimes found to be limited above a certain concentration of ferrous ion as reported by Lin [13] and Kang and Hwang [14] et al. Also, an enormous increase in the ferrous ions will lead to an increase in the unutilized quantity of iron salts, which will contribute to an increase in the total dissolved solids content of the effluent stream, and this is not permitted. Thus, laboratory-scale studies are required to establish the optimum loading of ferrous ions to mineralize the organics.

2.2 *Electro-Fenton process*

There is a greater interest in the development of effective electrochemical treatments for the destruction of toxic and biorefractory organics [15]. In the Electro-Fenton (EF) process, pollutants are degraded not only by the action of

Fenton's reagent in the bulk, but also with anodic oxidation at the anode surface. The electro-Fenton process is classified into four categories depending on Fenton's reagent addition or formation. In type 1, hydrogen peroxide and ferrous ion are electro-generated using a sacrificial anode and an oxygen-sparging cathode, respectively [16]. In type 2, hydrogen peroxide is externally added while ferrous ion is produced from the sacrificial anode, as given in Equation (1) [17]:

$$Fe \rightarrow Fe^{2+} + 2e^- \quad (1)$$

In type 3, ferrous ion is externally added and hydrogen peroxide is generated using an oxygen-sparging cathode [18,19]. In type 4, hydroxyl radical is produced using Fenton reagent in an electrolytic cell, and ferrous ion is regenerated through the reduction of ferric ions on the cathode, and hydrogen peroxide is produced using hydrogen, and oxygen via catalysis of Pd (Fig. 1).

2.3 *Photo-Fenton processes*

A combination of hydrogen peroxide and UV radiation with Fe^{2+} or Fe^{3+} oxalate ion (Photo-Fenton (PF) process) produces more hydroxyl radicals compared with the conventional Fenton method or photolysis, which, in turn, increases the rate of degradation of organic pollutants [21]. Fenton reaction accumulates Fe^{3+} ions in the system, and the reaction does not proceed once all Fe^{2+} ions are consumed. The photochemical regeneration of ferrous ions (Fe^{2+}) by photo-reduction (Equation (2)) of ferric ions (Fe^{3+}) occurs in the photo-Fenton reaction [22]:

$$Fe(OH)^{2+} + hv \rightarrow Fe^{2+} + \cdot OH \quad (2)$$

Direct photolysis of H_2O_2 (Equation (3)) produces hydroxyl radicals, which can be used for the

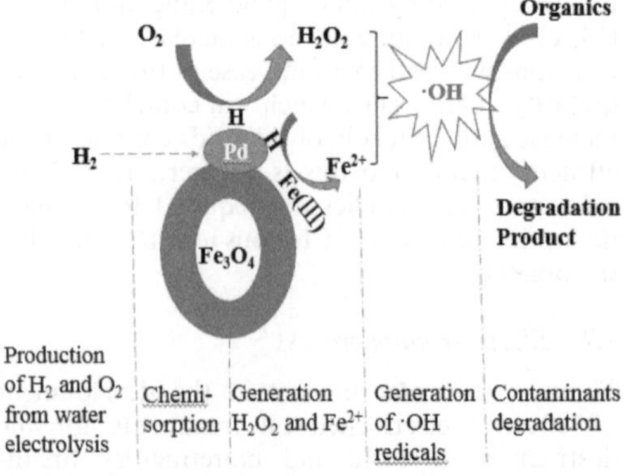

Figure 1. Proposed mechanisms for Pd/MNPs catalytic E-Fenton degradation of phenol [20].

degradation of organic compounds. Under UV-A illumination, while a H_2O_2-assisted Photoelectrocatalytic Oxidation (PECO) reaction occurs on the surface of the TiO_2 photo anode, an E-Fenton reaction takes place in the solution [23]. The experimental results demonstrated that 2,4-DCP degradation in the aqueous solution was greatly enhanced because of the interaction between the two types of reactions:

$$H_2O_2 + hv \rightarrow 2 \cdot OH \quad (3)$$

2.4 *Photo-Electro-Fenton process*

The catalytic effect of Fe^{2+} in the electro-Fenton process can be enhanced by irradiating the contents with UV light. Therefore, the combination of electrochemical and photochemical processes with the Fenton process is called the Photo-Electro-Fenton (PEF), which generates a greater quantity of free radicals due to the combination effect [24]. The direct photolysis of the acid solution containing peroxide generates hydroxyl radicals through the hemolytic breakdown of the peroxide molecule according to Equation (3). This reaction increased the oxidative capability of the process due to the additional production of hydroxyl radicals. Therefore, the degradation of the target organic substrate can be enhanced when the solution is irradiated with UV light, in addition to the application of the electro-Fenton process. Photochemical regeneration of Fe^{2+} by the photoreduction of Fe^{3+} ions and photo-activation of complexes renders the photo-electro-Fenton systems more efficient [25, 26].

3 CONCLUSIONS

Advanced oxidation processes are environmental-friendly processes for the degradation of refractory compounds. Different AOPs have been the most appropriate technique for the specific treatment. Hybrid methods are not economically viable techniques to degrade a large quantum of effluents from the industries. Hence it is advisable to use these methods as pretreatment to reduce the toxicity, which is significant for biological treatment. These combined methods are expected to reduce the reactor size and decrease the operating cost.

ACKNOWLEDGMENT

This paper was supported by the National Science and Technology Support Program of China (grant no. 2014BAC25B01) and the National Science and Technology Support Program of China (grant no. 2014BAC28B01).

REFERENCES

[1] Garg, A. et al. 2010. Oxidative phenol degradation using non-noble metal based catalysts. *Clean-soil, air, water*, 38: 27–34.

[2] Bernal-Martinez L.A. et al. 2010. Synergy of electrochemical and ozonation processes in industrial wastewater treatment. *Chemical Engineering Journal*, 165: 71–77.

[3] Anderson J.V. et al. 1991. Development of solar detoxification technology in the USA: an introduction. *Solar Energy Materials*, 24: 538–549.

[4] Fenton H.J.H. 1894. Oxidation of tartaric acid in the presence of iron. *Journal of the Chemical Society*, Transactions, 65: 899–910.

[5] Hartmann M. et al. 2010. Wastewater treatment with heterogeneous Fenton-type catalysts based on porous materials. *Journal of Materials Chemistry*, 20: 9002–9017.

[6] Neyens E. & Baeyens J. 2003. A review of classic Fenton's peroxidation as an advanced oxidation technique. *Journal of Hazardous Materials*, 98: 33–50.

[7] Bautista P. et al. 2008. An overview of the application of Fenton oxidation to industrial wastewaters treatment. *Journal of Chemical Technology and Biotechnology*, 83: 1323–1338.

[8] Babuponnusami A. & Muthukumar K. 2011. Degradation of phenol in aqueous solution by Fenton, sono-Fenton, Sono-photo-Fenton methods. *Clean-Soil, Air, Water*, 39: 142–147.

[9] Cesar P. & John K. 1996. Overview on photocatalytic and electrocatalytic pretreatment of industrial non-biodegradable pollutants and pesticides. *CHIMIA: International Journal for Chemistry*, 50: 50–55.

[10] Ting W.P. et al. 2008. The reactor design and comparison of Fenton, electro-Fenton and photoelectron-Fenton processes for mineralization of Benzene Sulfonic Acid (BSA). *Journal of Hazardous Materials*, 156: 421–427.

[11] Kurt U. et al. 2007. Reduction of COD in wastewater from an organized tannery industrial region by electro-Fenton process. *Journal of Hazardous Materials*, 143: 33–40.

[12] Lin S.H.& Lo C.C. 1997. Fenton process for treatment of desizing wastewater. *Water Research*, 31: 2050–2056.

[13] Lin S.H. et al. 2000. Operating characteristics and kinetics studies of surfactant wastewater treatment by Fenton oxidation. *Water Research*, 33: 1735–1741.

[14] Kang Y.W. & Hwang K.Y. 2000. Effects of reaction conditions on the oxidation efficiency in the Fenton process. *Water Research*, 34: 2786–2790.

[15] Brillas E. & Casado J. 2002. Aniline degradation by electro-Fenton and peroxicoagulation processes using a flow reactor for waste water treatment. *Chemosphere*, 47: 241–248.

[16] Badellino C. et al. 2006 Oxidation of pesticides by in situ electrogenerated hydrogen peroxide: study for the degradation of 2,4-dichlorophenoxyacetic acid. *Journal of Hazardous Materials*, 137: 856–864.

[17] Luo M.S. et al. 2014. An integrated catalyst of Pd supported on magnetic Fe_3O_4 nanoparticles: Simultaneous production of H_2O_2 and Fe^{2+} for efficient electro-Fenton degradation of organic contaminants. *Journal of water research*, 48: 190–199.

[18] Gogate P.R. & Pandit A.B. 2004. A review of imperative technologies for wastewater treatment II: hybrid methods. *Advances in Environmental Research*, 8:553–597.

[19] Faust B.C. & Hoigne J. 1990. Photolysis of Fe (III)-hydroxy complexes as sources of OH radicals in clouds, fog and rain, Atmos. *Atmospheric Environment*, 24A: 79–89.

[20] Zhao B.X. et al. 2, 4-Dichlorophenol Degradation by an Integrated Process: Photoelectrocatalytic Oxidation and E-Fenton Oxidation. *Journal of Hazardous Materials*, 83: 642–646.

[21] Babuponnusami A. & Muthukumar K. 2012. Advanced oxidation of phenol: a comparison between Fenton, electro-Fenton, sono-electro-Fenton and photo-electro-Fenton processes. *Chemical Engineering Journal*, 183: 1–9.

[22] Brillas E. et al. 2003. Mineralization of herbicide 3,6-dichloro-2-methoxybenzoic acid in aqueous medium by anodic oxidation, electro-Fenton and photoelectro-Fenton. *Electrochimica*, 48: 1697–1705.

[23] Boye B. et al. 2003. Anodic oxidation, electro-Fenton and photoelectro-Fenton treatments of 2, 4, 5-trichlorophenoxyacetic acid. *Journal of Electroanalytical Chemistry*, 557: 135–146.

Material Science and Environmental Engineering – Chen (Ed.)

Up-flow hydrolysis reactor as a pretreatment process for treating domestic sewage

J. Li
School of Chemical and Environmental Engineering, China University of Mining and Technology (Beijing), Beijing, China

J. Zhou
Beijing Service Bureau for Diplomatic Missions, Beijing, China

ABSTRACT: Hydrolysis reactor adopting the up-flow granular sludge bed configuration as a pretreatment process was investigated for its feasibility and process performance. The combined process of the hydrolysis reactor and A/O was adopted for treating the domestic sewage operated at a total HRT of 23 h. The results showed that the average COD, TN, SS, and TP concentrations measured in the final effluent of the hydrolysis–A/O process amounted to 40.26 mg/L, 13.68 mg/L, 12.00 mg/L and 0.63 mg/L. This corresponded to the average removal rates of COD, TN, SS, and TP that could reach 90.40%, 66.36%, 95.38% and 89.15%, respectively. This showed that the combined system represented an effective sewage treatment process.

Keywords: up-flow; hydrolysis; pretreatment

1 INTRODUCTION

Hydrolysis process as a major pretreatment technique of toxic and refractory industrial wastewater has become a viable and most commonly used technology. In recent years, the successful operation in some municipal sewage plants has suggested that the hydrolysis process has made an attractive candidate for application such as domestic sewage pretreatment process. Consequently, if being put into use as a field scale, the process would offer various advantages, such as easy construction, less oxygen, less nitrite and nitrate production, and no need for organic carbon and maintenance and ability to withstand fluctuations in pH, temperature, and influent substrate concentrations.

Hydrolysis reactor keeps the anaerobic reaction in the hydrolysis–acidification stage by controlling the reaction time. The aggregation of biomass helps in excellent settleability and good activity, which allowed a high throughput of effluent through the reactor. The success of hydrolysis reactor has been attributed to its biomass retention capability through effective separation of solid retention time from HRT, allowing for high loads and short HRTs. The characteristics of the hydrolysis reactor are that it combines long sludge age and short HRT (3.8 h), and provides small footprints for the bioreactor. Essential anaerobic bacteria consortia are immobilized within the granulae; the diversity and distribution of bacteria are determined by various chemical, physical and biological factors.

This research was based on the upgrade project of the Wuxi Chengbei municipal sewage plant in Jiangsu Province, China. In this study, we put forward the hydrolysis–A/O co-treatment process for treating domestic sewage, which adopted a hydrolysis reactor before the A/O process, and refluxed some effluents of the secondary sedimentation tank back to the hydrolysis reactor.

2 MATERIALS AND METHODS

2.1 *Experimental set-up*

A start-up experiment was performed in a laboratory scale. The 33.6 L maximum working volume hydrolysis reactor consisted of a glass column with an internal diameter of 15 cm and a height of 210 cm, and an inverted cone to retain the granular sludge and outlet ports for gas. The influent was delivered from the center of the column reactor bottom to ensure an even distribution of the feed, and the effluent from the hydrolysis reactor outlet flowed into the A/O system.

2.2 *Seed sludge*

The development of the anaerobic granular sludge in the hydrolysis reactor generally required

7–10 days, and did not require additional sludge. An extremely short period was adopted for an easy startup of the hydrolysis reactor. A/O using seed sludge was obtained from the sludge thickening tank of the Wuxi Chengbei municipal sewage plant, where the Sludge Volume Index (SVI) was 28 mg/L and the ratio of Volatile Suspended Substances to Total Suspended Substances (VSS/TSS) in the sludge was 0.53.

2.3 *Raw water*

Raw water was obtained from the effluent of the rotational flow grit chamber of the Wuxi Chengbei municipal sewage plant, and the designed inflow was 180 L/d.

2.4 *Analytical methods*

The close reflux, titrimetric method (APHA, 1995) was used to determine COD concentration. Nitrate and nitrite concentrations were determined using the chromotropic acid method and the colorimetric method, respectively (APHA, 1995). TSS and biomass (expressed in VSS), TN and NH4+–N concentrations were determined in accordance with the Standard Methods (APHA, 1995). pH was monitored by using a PHS–3D acidometer (Shanghai Sanxin Instrumentation Inc., China). DO and temperature were monitored by using a JPBJ-608 portable quick analyzer (Shanghai Precision & Scientific Instrument Co., Ltd, China).

3 RESULTS AND DISCUSSION

3.1 *COD and SS removal*

Time course for the respective COD and SS in the hydrolysis–A/O system at different phases is shown in Figure 1 (a), (b). During the experimental periods, the influent COD and SS were 161.98–1008.00 mg/L and 94.00–596.00 mg/L, and the average COD and SS concentrations were 487.54 mg/L and 312.00 mg/L. After the hydrolysis reactor, the average COD and SS concentrations were 104.52 mg/L and 32.00 mg/L, the contributions of the hydrolysis reactor to the removal rates of COD, SS were 76.84% and 88.53% by calculation.

The residual COD and SS concentrations remaining in the hydrolysis effluent were mainly biodegradable organics, which would be further eliminated in the A/O process. The average COD and SS concentrations in the final effluent of A/O were 40.26 mg/L and 12.00 mg/L. The total removal rates of COD and SS reached 90.40% and 95.38%, respectively. This indicated the achievement of advanced removal rates of COD and SS in this system.

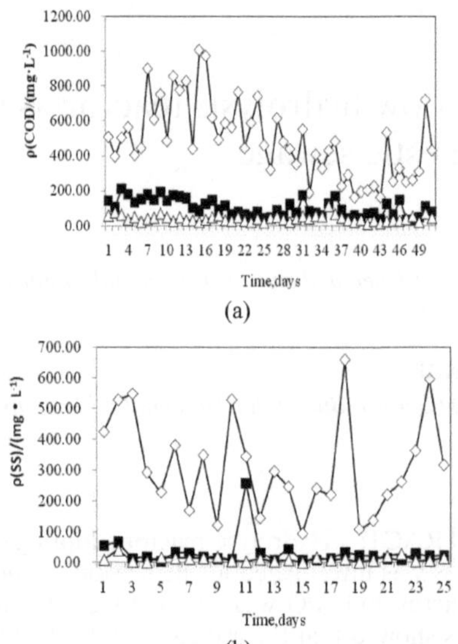

Figure 1. Time course for respective COD and SS concentrations in the hydrolysis–A/O system at different phases.

The Sludge Retention Time (SRT) was longer than the Hydraulic Retention Time (HRT) in the hydrolysis reactor, which would lead to the transformation from macromolecular refractory organics into micromolecular labile organic compounds by a large number of hydrolytic bacteria. As given in Table 1, the average SCOD of influent in the hydrolysis reactor was 44.99%. After hydrolysis treated, the average SCOD concentration was 71.72%, which was increased by 26.73% compared with influent concentration. At the same time, Volatile Fatty Acids (VFA) of the influent and the effluent in the hydrolysis reactor were tested, which made it clear that the VFA concentration increased from 51 mg/L of influent to 72 mg/L of effluent.

Under the same conditions of dissolved oxygen and sludge concentration (sludge being taken from oxidation ditch of the Chengbei municipal sewage plant), we tested the denitrification rate of "raw water + sludge" and "hydrolysis effluent + sludge", respectively. The largest denitrification rates of "raw water + sludge", "hydrolysis effluent + sludge" were, respectively, 16.86 mg/g MLSS·d and 13.57 mg/g MLSS·d, which increased by 24%. Therefore, the increasing concentrations of SCOD and VFA would have a positive influence on carbon sources and denitrification in the follow-up process.

3.2 *Nitrogen removal*

Time course for respective TN concentrations in the hydrolysis–A/O system at different phases is shown

Table 1. Comparison of the influent and the effluent in the hydrolysis reactor.

No.	COD/(mg·L^{-1}) Influent	COD/(mg·L^{-1}) Effluent	SCOD/(mg·L^{-1}) Influent	SCOD/(mg·L^{-1}) Effluent	SCOD ratio of influent	SCOD ratio of effluent
1	318.09	242.55	139.17	123.26	43.75%	50.82%
2	479.04	207.59	192.42	143.71	40.17%	69.23%
3	184.31	80.39	94.12	76.47	51.06%	95.12%

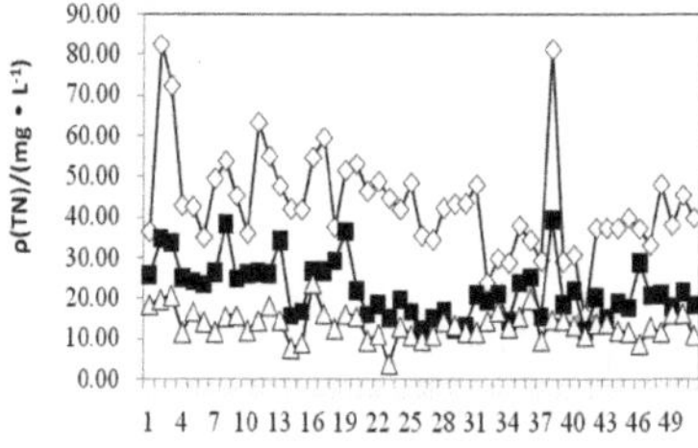

(◇) Influent, (■) hydrolysis effluent and (△) final effluent

Figure 2. Time course for respective TN concentrations in the hydrolysis–A/O system at different phases.

in Figure 2. The influent TN concentration of the hydrolysis–A/O system was 15.54–82.50 mg/L, with the average TN concentration being 43.33 mg/L. After the hydrolysis reactor, the average TN concentration was 22.28 mg/L, the effluent TN concentration of the hydrolysis reactor was 21.05 mg/L less than the influent concentration, and the contribution of the hydrolysis reactor to the removal rate of TN was 46.88% by calculation. The residual TN concentration remaining in the hydrolysis effluent would be further eliminated in the A/O process, with the average TN concentration in the final effluent of A/O being 13.68 mg/L, and the total TN removal rate reaching up to 66.36%.

As can be seen in Table 2 and Table 3, the solid-state nitrogen concentration decreased, while the ratio of dissolved nitrogen to TN increased. The micromolecular organic compounds were absorbed and degraded easily through the cell membrane, and they were treated by microorganisms in a short time; thus, nitrogen removal in the follow-up process would be increased.

Under the same conditions of the dissolved oxygen amount and the influent quantity, with sludge taken from oxidation ditch, and using a respiration rate measurement apparatus, the nitrification rates of "sludge + raw water" and "sludge + hydrolysis effluent", were monitored respectively (Table 4). The result showed that the nitrification rate of "sludge + hydrolysis effluent" was higher than that of "sludge + raw water", with the average being

Table 2. Concentration of various nitrogen forms of influent and hydrolysis effluent.

Analysis items	Influent	Hydrolysis effluent
TN/(mg·L^{-1})	63.61	41.43
Dissolved nitrogen, 0.45 μm/(mg·L^{-1})	35.71	25.85
Solid-state nitrogen/(mg·L^{-1})	27.90	15.58
NH_4^+–N/(mg·L^{-1})	30.02	22.01
NO_3^-–N/(mg·L^{-1})	0.53	0.49
NO_2^-–N/(mg·L^{-1})	0.02	0.05
Org–N/(mg·L^{-1})	33.02	18.88
Solubility Org-N/(mg·L^{-1})	4.69	2.69

Table 3. Ratio of various nitrogen forms of influent and hydrolysis effluent.

Analysis items	Influent	Hydrolysis effluent
TN/(mg·L^{-1})	63.61	41.43
Dissolved nitrogen /TN/%	56.13	62.39
NH_4^+–N/TN/%	47.19	53.13
NO_2^-–N/TN/%	0.03	0.12
NO_3^-–N/TN/%	0.83	1.18
Org–N/TN/%	51.91	45.57

0.44 mg/g·h, increasing by about 15%. Therefore, Org–N was hydrolyzed into NH_4^+–N in the hydrolysis reactor, which would promote the nitrification and be helpful for the biological treatment in the follow-up process.

To further study the denitrification capability of the hydrolysis reactor, an independent hydrolysis reactor was established. The designed inflow was 100 L/d and adding NO_3^- (form of $NaNO_3$) to the inflow. In Figure 3, experimental data showed that influent NO_3^-–N concentration of the hydrolysis reactor was 1.71–27.61 mg/L, the average NO_3^-–N concentration of the hydrolysis effluent was 6.73 mg/L, and its average removal rate was 48.07%.

In order to clarify the existence of the denitrification role in the hydrolysis reactor, we

Table 4. Comparison test of nitrification rate with hydrolysis effluent and raw water.

Analysis items	The first group		The second group	
	Sludge + raw water	Sludge + hydrolysis effluent	Sludge + raw water	Sludge + hydrolysis effluent
MLSS in reactor/($mg \cdot L^{-1}$)	3238	3238	2732	2732
Gross OUR/($mg \cdot L^{-1} \cdot h^{-1}$)	33.97	33.80	26.89	22.37
Post-inhibition OUR/($mg \cdot L^{-1} \cdot h^{-1}$)	24.75	23.46	18.87	12.87
OUR of Nitration/($mg \cdot L^{-1} \cdot h^{-1}$)	9.22	10.34	8.02	9.50
SAUR*/($mg \cdot g^{-1} \cdot h^{-1}$)	2.85	3.19	2.95	3.48

*SAUR specific nitrification rate.

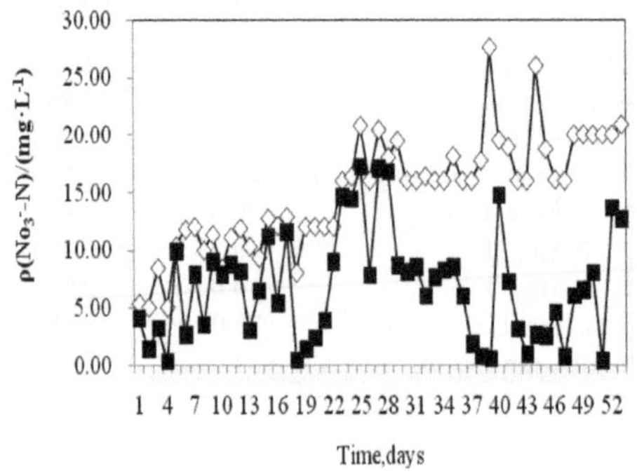

(◇) Influent, (■) hydrolysis effluent

Figure 3. Time course for respective NO_3^-–N concentrations in the independent hydrolysis reactor at different phases.

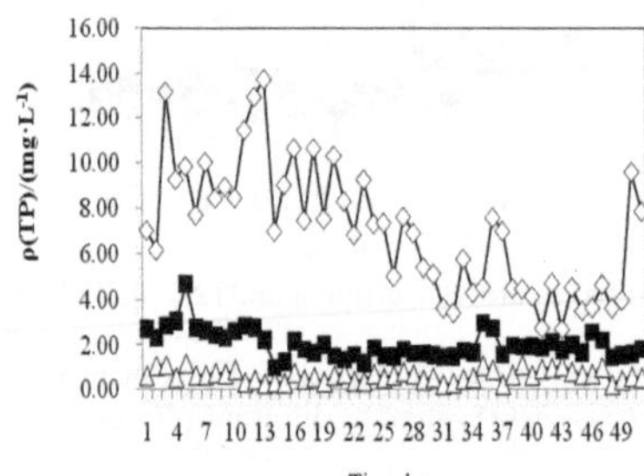

(◇) Influent, (■) hydrolysis effluent and (Δ) final effluent

Figure 4. Time course for respective TP concentrations in the hydrolysis–A/O system at different phases.

tested the sludge taken from the hydrolysis reactor under the conditions of adding 40 g $NaNO_3$ and excess alcohol. MLSS was 1100 mg/L and MLVSS was 664 mg/L, and the denitrification rate of hydrolysis sludge was 26.02 mg NO_3^-–N/(g MLVSS · h).

3.3 *TP removal*

Through the quantitative discharge of sludge in the hydrolysis reactor every day, the results clearly demonstrated that the achieved PO_4^- removal rates in the hydrolysis reactor were high. Time course for respective TP concentrations in the hydrolysis–A/O system at different phases is shown in Figure 4. Under the condition of no additional chemical treatment, experimental data showed that the influent TP concentration of the hydrolysis–A/O system was 2.71–13.75 mg/L, and the average TP concentration was 7.08 mg/L. After the hydrolysis reactor, the average TP concentration was 2.04 mg/L, and the contribution of the hydrolysis reactor to the removal rate of TP was 67.21% by calculation. This may be due to retention and denitrification causing phosphorus uptake. The residual TP concentration remaining in the hydrolysis effluent would be further eliminated in the A/O process. The average TP concentration in the final effluent of A/O was 0.63 mg/L, and the total TP removal rate reached to 89.15%.

3.4 *Hydrolysis reactor stabilization in the follow-up A/O process*

The wastewater quality discharged was unstable, and the pollutant concentrations had relatively large fluctuations. High concentrations of the pollutants impacted repeatedly on the treatment system in this trial, but the removal rates of COD and TN remained above 95.73% and 56.32%. As summarized in Table 5, the system had a stable effluent when COD and TN concentrations were less than 1008 mg/L and 63 mg/L, respectively. As a result, it meant that the hydrolysis reactor could be resistant to a higher shock, thus making the follow-up process more stable.

Table 5. Pollutants' load impact test on COD and TN concentrations.

Analysis items	1	2	3	4
COD of influent/ (mg·L^{-1})	898.40	856.00	828.00	1008.00
COD of hydrolysis/ (mg·L^{-1})	179.20	176.00	160.00	88.00
COD of effluent/ (mg·L^{-1})	38.40	26.00	30.00	20.00
Hydrolysis removal rate/%	80.05	79.44	80.68	91.27
System removal rate/%	95.73	96.96	96.38	98.02
TN of influent/ (mg·L^{-1})	49.45	63.26	47.56	41.77
TN of hydrolysis/ (mg·L^{-1})	26.42	26.50	34.27	16.75
TN of effluent/ (mg·L^{-1})	11.77	14.46	14.63	9.07
Hydrolysis removal rate/%	46.57	58.11	27.94	59.90
System removal rate/%	76.20	77.14	69.23	78.29

4 CONCLUSIONS

The experiments reveal that the hydrolysis reactor as a pretreatment unit is hardly influenced by the influent load. The removal rates of COD, SS, TN and TP are significantly high as long as the anaerobic sludge layer is not lost, and the following process withstands high load impacts. A complete aerobic denitrification function can take place in the hydrolysis reactor; the TN removal rate of the effluent increases through refluxing nitrification liquid of the effluent back to it.

The investigation of this combined process will have an actual significance in the engineering application further. To be applied as a promising new technology, additional research on the critical factors such as sludge stratification and lost is required in the future.

ACKNOWLEDGMENTS

This research was funded by the National S&T Pillar Program (2006038059002) and the National S&T Major Project in the 11th Five year Plan of China (2009ZX07313-003). The authors gratefully acknowledge the Wuxi Municipal Engineering Administrative Office and the Wuxi Chengbei municipal sewage plant.

REFERENCES

[1] Zhang Yaobin, Ma Yongguang, Quan Xie, et al. 2009. Rapid startup of a hybrid UASB–AFF reactor using bi-circulation. *Chemical Engineering Journal*, 155: 266–271.

[2] Li Xiaochen, Pan Juncheng. 2004. Feasibility study of hydrolysis as biological denitrification pretreatment process for wastewater. *JiangSu Environmental Science and Technology*, 17(1): 1–3. (in Chinese).

[3] Luostarinen A Sari, Rintala A Jukka. 2005. Anaerobic on-site treatment of black water and dairy parlour wastewater in UASB-septic tanks at low temperatures. *Water Research*, 39: 436–448.

[4] Jiang Lanlan, Liang Ting, Liu Xuehong. 2008. Study of the up-flow sludge bad hydrolytic tank in urban comprehensive sewage treatment. *Technology of Water Treatment*, 34(5): 84–87. (in Chinese).

[5] Zheng Jun, Zhang Gang, Wang Jian, et al. 2007. Application of UASB/hydrolytic acidification/ biological aerated filter process in treatment of high-concentration polyester wastewater. *China Water & Wastewater*, 23(10): 51–54. (in Chinese).

[6] Tawfika A, Ohashi A, Harada H. 2006. Sewage treatment in a combined Up-flow Anaerobic Sludge Blanket (UASB)-Down-flow Hanging Sponge (DHS) system. *Biochemical Engineering Journal*, 29: 210–219.

[7] Liu Jun, et al. 2000. Study on anaerobic hydrolysis biological treatment of urban wastewater. *Water & Wastewater Engineering*, 26(7): 10–13. (in Chinese).

[8] Elmitwalli A Tarek, Otterpohl Ralf. 2007. Anaerobic biodegradability and treatment of grey water in Up flow Anaerobic Sludge Blanket (UASB) reactor. *Water Research*, 41: 1379–1387.

[9] Apha, Awwa, Wpcf. 2000. Standard Methods for the Examination of Water and Wastewater. Washington, D.C: *American Public Health Association, American Water Works Association and Water Pollution Control Federation.*

[10] He Shuying, Li Jixiang, Xu Yatong, et al. 2008. The microbial community structure of brewery wastewater treatment system by hydrolytic acidification and sbr technics. *Technology of Water Treatment*, 34(3): 35–38. (in Chinese).

Material Science and Environmental Engineering – Chen (Ed.)
 ISBN 978-1-138-02938-5

The study of geothermal reservoir coupling systems in dual-porosity rock

B. Cui, Y. Duan & M.T. Xu
School of Mining Engineering, Guizhou Institute of Technology, Guiyang, China

ABSTRACT: In this article, we build coupled T-H-M models to solve the problem of thermal-mechanical-hydraulic interaction in fracture-porous rock. The model that has been developed to illustrate the coupled process of thermal, hydraulic and matrix deformation. It includes the physical coupling between the fracture and the reservoir matrix when dealing with both thermal and water transport. Three modules have been assigned for representing this coupled behavior: the mechanical stress-strain module, Darcy's fluid flow modules, and thermal conduction-convection modules. The flow and heat transport in the porous and the fractured phase are simulated by fluid flow and thermal modules while the single mechanical module represents the entire mechanical response of the dual porosity rock mass.

Keywords: dual porosity medium; T-H-M coupling; thermal conductivity; pore pressure

1 INTRODUCTION

Fractured porous media is one of many complex geological attributes that is of high interest. The concept of "Double-Porosity" has been introduced in the early 1960's, which considers the fractured rock mass consisting of two porous systems [1–2], which are the matrix, having high porosity and low permeability, and the fracture, having low porosity and high permeability. It is well known that fractured rocks exhibit changes in mechanical compliance and hydraulic conductivity when subjected to thermal, hydraulic, mechanical, and chemical forces. In many engineering applications it is important to be able to predict the direction and magnitude of these changes.

However, the interplay between temperature, effective stress, chemical potential, and fracture response is complex: it is not only influenced by anisotropic and spatially varying fracture properties, but also by fracture properties that are dynamic, and evolve with the dynamic nature of the applied forces [3]. The behavior of the system that constitutes the three main interacting processes: thermal, hydrological and mechanical, is generally referred to as the Thermo-Hydro-Mechanical (THM) coupled processes.

Research in THM behavior of fractured porous media covers a very wide range of fields and applications, from underground nuclear waste disposal [4–6], geothermal energy exploitation [7–9], rock mechanics and mining [10], geology and geotectonic [11–12], transport of oil and gas in deep reservoirs [13–14], engineering/construction material [15], to engineering geological barriers [16–17]. Yet, despite the amount of research being done, there is still insufficient understanding of the processes to be able to construct a feasible design or prediction regarding its behavior.

Early mathematical models developed to describe such behavior approached the problem by treating the fractured rock and the matrix as a continuous medium [18] and representing the fractured rock mass by an equivalent porous medium [19]. However, general interaction theories are required to describe the effect of the motion of fluid on the motion of matrix and vice versa. Much of this has been satisfied through introduction of the general theory of consolidation [20] which relates the influence of pore pressure due to the existence of fluid in a porous rock. This study made it possible for engineers and scientists to solve problems in a much wider scope. The addition of the thermal component coupled later in the mid 1980's [21] formed a basis to a fully coupled THM solution.

Although, there has been extensive research and computer modeling in this area in the past two decades, the porosity within the fracture and the matrix has long been considered as one entity as originally formulated for a continuous body of porous medium [20]. However, this does not truly reflect the influences of fluid pressures and rock stresses on the fracture and the matrix due to the porosity within the fracture being more sensitive than that of the matrix [22].

In order to describe the behavior of a dual-porosity medium in a geothermal environment, an equation governing the thermal behavior must be coupled to the original dual-porosity model. A set of Mechanical-Hydraulic-Thermal (T-H-M) coupled equations have been formulated and discredited to form a finite element statement for the behavior of a dual-porosity medium. The developed finite element model can be numerically solved with any finite element software.

2 GOVERNING EQUATIONS

2.1 *Solid deformation in the rock*

The strain-displacement relationship is defined as:

$$\varepsilon_{ij} = \frac{1}{2}(u_{i,j} + u_{j,i}) \tag{1}$$

where ε_{ij} is the component of the total strain tensor and u_i is the component of displacement. The equilibrium equation with self weight and neglecting inertial effects is given as

$$\sigma_{ij,j} + f_i = 0 \tag{2}$$

where σ_{ij} is the component of the total stress tensor and f_i the component of body force. In the paper, the subscript P denote pore, and F is fracture, with thermoelastic response, and utilizing two distinct pore fluid pressures as[3]

$$Gu_{i,kk} + \frac{G}{1-2\nu}u_{k,ki} = (\alpha_P p_i + \alpha_F p_i) + \alpha_T T_i - f_i, \tag{3}$$

where u is displacement; υ is Poisson ratio; f_i is body force per unit volume; α_1 and α_2 are Biot's coefficients in pore and fracture; α_T is thermal coefficient; G is elastic modulus of the dual porosity media.

2.2 *Conservation of mass*

The equilibrium equation for fluid transport in an isotropic medium according to Darcy's law is

$$\beta\dot{\mathbf{P}} - \nabla^{\mathrm{T}}\left[\frac{k}{\mu}\nabla(\mathbf{p} + \gamma z)\right] + Q_f = 0 \tag{4}$$

The term P on the LHS of equation (4) refers to the rate of change of pressure with time t, due to grain compressibility β. The second term is Darcy's flux, which is function of the permeability k; dynamic viscosity μ; specific weight of the fluid γ; elevation z; and fluid pressure p. The third term Q_f, is the source term.

However, accounting for the thermal effects and the fluid mass transfer between the porous and fractured phases, the equilibrium equation for the porous phase may be rewritten as

$$-\nabla^{\mathrm{T}}\left[\frac{k_P}{\mu_P}\nabla(\mathbf{p}_P + \gamma z)\right] - \beta_P\dot{\mathbf{P}}_P - \varphi(\mathbf{p}_P - \mathbf{p}_F) + \dot{\boldsymbol{\varepsilon}} + (Q_f)_P = 0 \tag{5}$$

where the added component in the first term on the LHS is a positive component referring to the flow of fluid mass driven by the changes of the volume due to strain $\dot{\boldsymbol{\varepsilon}}$, and the second term is the fluid mass transfer between the porous and the fractured phase, where the coefficient φ which governs the quasi-steady response [2] is a function of the geometry of the porous block and the flow characteristics (permeability and dynamic viscosity) within, as

$$\varphi = \alpha^{\#}\frac{k_P}{\mu} \tag{6}$$

here, $\alpha^{\#} = 60/(s^*)^2$, and s^* is the space of the fracture.

Similarly, the equilibrium equation for the fractured phase may be expressed as

$$-\nabla^{\mathrm{T}}\left[\frac{k_F}{\mu_F}\nabla(\mathbf{p}_F + \gamma z)\right] - \beta_F\dot{\mathbf{P}}_F + \varphi(\mathbf{p}_P - \mathbf{p}_F) + \dot{\boldsymbol{\varepsilon}} + (Q_f)_F = 0 \tag{7}$$

The equation (6) differs from that of the equation 错误！未找到引用源。(porous phase), in the negative component of the mass transfer between the porous and fractured phase.

2.3 *Thermal behavior*

The governing equation for fluid flow which relates the heat flux q_T, and its potential, which is the temperature, is defined by Fick's law which is expressed in matrix form as

$$\mathbf{q}_T = -\boldsymbol{\lambda}\nabla T \tag{8}$$

where λ is the thermal conductivity.

The general thermal transport equilibrium equation for an isotropic medium is defined by

$$\rho C\dot{T} - \nabla^{\mathrm{T}}\lambda\nabla T - \rho C\nabla\mathbf{q}_f + Q_T = 0 \tag{9}$$

where ρ is the density, C is the heat capacity, and Q_T are is the heat source.

3 NUMERICAL SIMULATION

Check the margin setting (Page Setup dialog box in File menu) and column settings (see Table 1 for correct settings).

Table 1. Parameters for the fracture-matrix rock.

Parameter	Symbol	Value
Initial fracture aperture	$2b$	200 μm
Reservoir matrix porosity	ϕ_P	0.05
Fluid velocity in the fracture	V_0	1 m/day
Thermal dispersivity	β_T	0.05 m
Reservoir matrix diffusion coefficient	D_m	$1.0 \times 10^{-5}\ m^2\ day^{-1}$
Reservoir matrix tortuosity	T	0.1
Reservoir matrix specific heat capacity	C_P	800 $Jkg^{-1}K^{-1}$
Rock density	ρ_s	2600 kgm^{-3}
Fracture fluid density	ρ_f	1000 kgm^{-3}
Thermal conductive of fracture fluid	λ_f	0.5 $Wm^{-1}K^{-1}$
Specific heat capacity of fracture fluid	C_f	5000 $Jkg^{-1}K^{-1}$
Initial temperature (matrix and fracture)	T_0	423 K (150°C)
Constant injection well temperature	T_i	300 K (27°C)

3.1 *Rock seam model and parameters*

The process of geothermal exchange in nature may be simply shown as Figure 1. In order to simulate this coupling process in computer, a calculation model should be built.

The process includes mass deformation, pore pressure and temperature changes. The parameters of the rock are shown in Table 1.

There is a rectangular dual porosity rock mass (Fig. 2). The cold water pass through the fracture-matrix composed rock seam and outlet from the other side.

3.2 *Boundary and initial conditions*

The mechanical boundaries are fixed in the direction normal to the surface for all except the top face where a compressive stress of −10 N/m^2 in the z direction is applied. Reasonable displacements were given. For the case with an elastic modulus and Poisson's ratio of the porous phase at 100 GPa and 0.25 respectively, the normal and shear stiffness of the rock mass of 5 GPa and 0.1 GPa respectively, and the fracture spacing in the z direction of 1 m. Water at room temperature is injected in to the fracture in a hot porous matrix at 150°C. Normal stress in the Z direction acts on the top and bottom surface of the rectangular block at 10 Pa. Displacements are fixed in the y direction at the inlet and outlet boundaries and fixed in the x direction at the sides of the block. The modeling of the heat transport can be completed by creating a heat source at the inlet boundary, and assigning

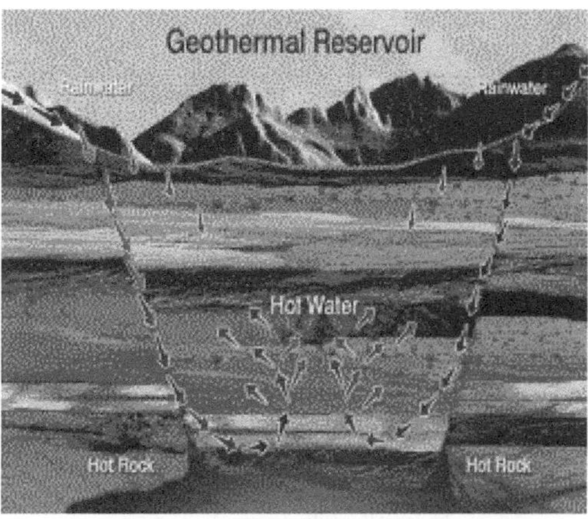

Figure 1. Coupled process in geothermal reservoir [23].

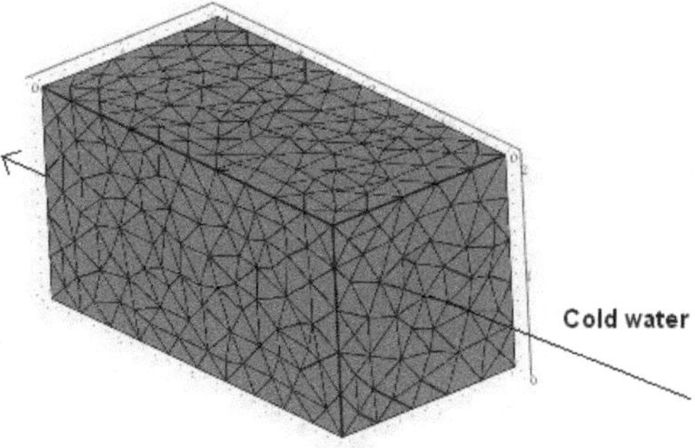

Figure 2. A rectangular 4 × 2 × 2 m block is assigned for the geometry of the domain.

different thermal conduction in the porous phase and the fractured phase. Heat flux at 1000 MW/m^2 is developed at the outlet.

3.3 *Results analysis*

Figure 3 shows the pressure [Pa] distribution in the reservoir and the arrow is the velocity field [m/s]. It is clear that the pressure at the position of the inlet surface is 1.014×10^5 Pa, and it is much more greater than that of other positions and from Figure 4 we can find that when cold water goes through the rock mass, the displacements (5.742×10^{-3} m) at the loading surface is close to that obtained from manual calculation (5.720×10^{-3} m). The flow velocities of the porous phase and the fracture phase is found to be 1.25×10^3 m/s and 1.25×10^6 m/s which is the exact solution obtained from calculation. The heat flux given by the model is exactly as calculated manually.

Table 2. List of specifically defined functions for material properties.

Material	Name	Expression
Porosity phase	Elastic modulus, E_P	$3(1+\nu_P)K_d$
	Thermal expansion coefficient, α_P	$\phi_P\alpha_f+(1-\phi)\alpha_s$
	Thermal conductivity, λ_P	$\phi_P\lambda_f+(1-\phi)\lambda_s$
	Density, ρ_P	$\phi_P\rho_f+(1-\phi)\rho_s$
Fractured phase	Heat capacity, C_F	$\phi_F C_f$
	Elastic modulus, E_F	K_f
	Density, ρ_F	$\phi_F\rho_f$
Thermal	Thermal expansion coefficient α_F	$\phi_F\alpha_f$
	Thermal conductivity, λ_F	$\phi_F\lambda_f$

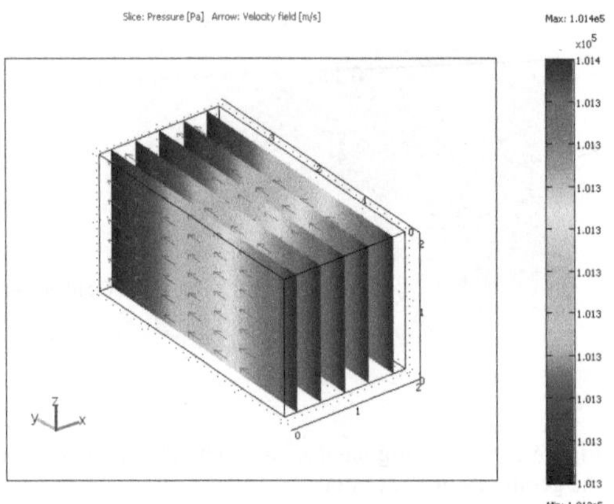

Figure 3. Pressure and velocity field.

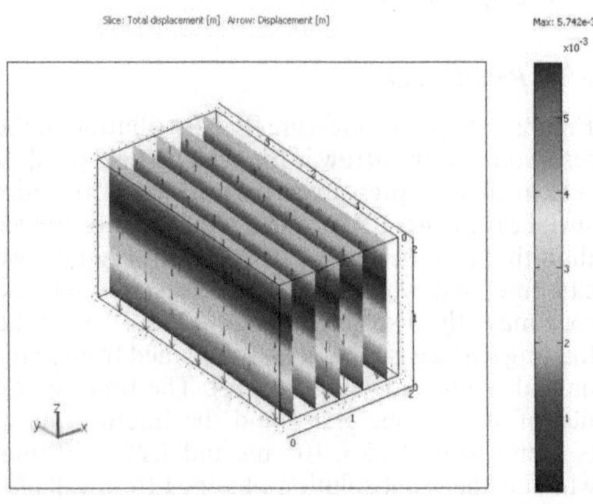

Figure 4. Displacements when cold water passing by.

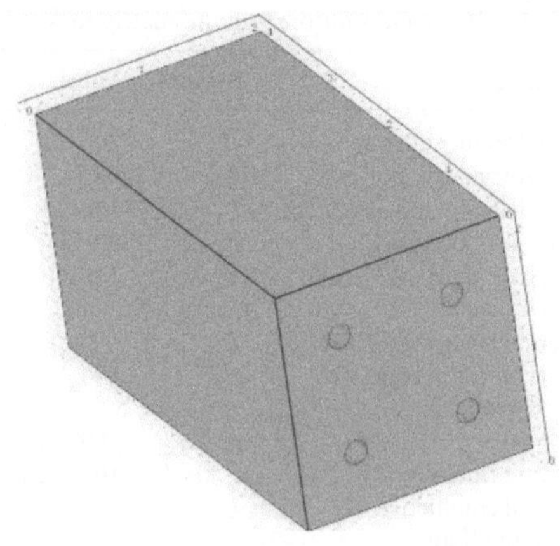

Figure 5. Model of hot water across solid matrix.

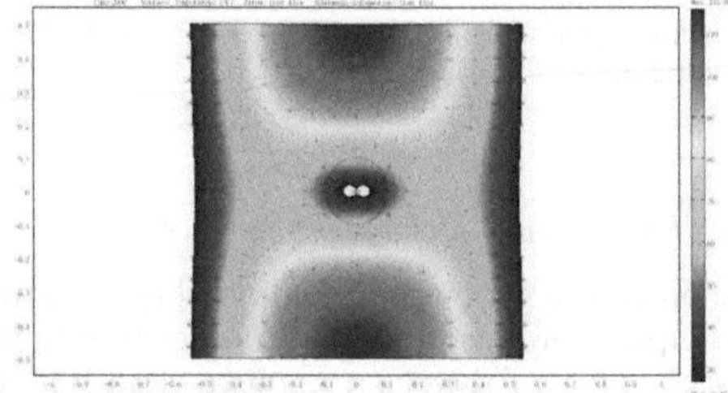

Figure 6. Quarter of the model coupled between thermal and mechanical.

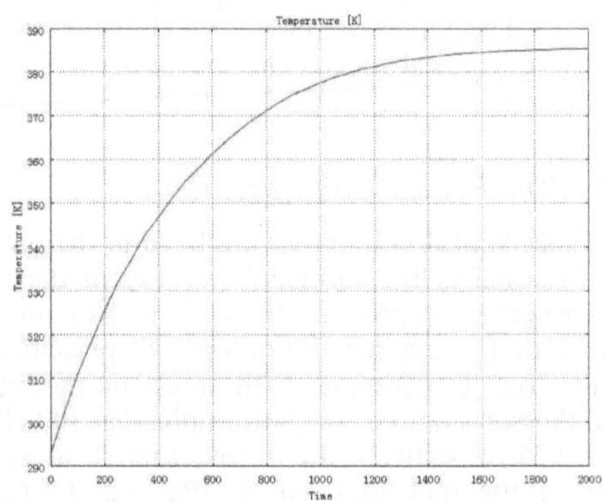

Figure 7. The evolution of temperature with time at the point of x = 0, y = 0.4.

If the hot water conduct thermal to cold, dual porosity rock, there will be a model as follows. The top and bottom of the rock are fracture and full of hot water and the left and right sides of the rock are thermal isolated. Because the model is symmetry, we can study a quarter of it.

There is deformation at both left and right sides. The pore volume shrinks greatly and the heat flux appears much around the deformed pore. Figure 7 shows the temperature versus time plot. It is easily seen the temperature is close to the steady-state value at the end of the modeling calculation.

4 CONCLUSIONS

1. A new coupled T-H-M model has been developed to solve the problem of thermal-mechanical-hydraulic interaction in fractured rock mass. The model that has been developed couples simultaneously thermal and matrix deformation. It includes the physical coupling between the fracture and the reservoir matrix when dealing with both thermal and water transport.
2. The model is capable of providing correct solutions for static analysis. Although the time dependent solutions have not been verified and the model is not completed with the capability to update permeability values due to effects of stress and temperature, some results are shown to present the capability of the model to date.
3. Future study of the coupling process for dual porosity medium need to consider the following items: ① The dynamic changes of the cleat aperture and permeability of the reservoir matrix due to the effects from stresses and temperature; ② When the applied stress produces a change in hydraulic properties or the change in fluid pressure produces a change in mechanical properties (such as changes of compressibility or elastic modulus, etc.), the T-H-M coupling mechanism; ③ Thermal-Hydrological-Mechanical-Chemical coupling processes and numerical simulator which used to examine THMC processes in a construct that is applicable to the broad variety of engineering applications.

REFERENCES

[1] Barenblatt, G.E., I.P. Zheltov, et al. (1960). "Basic concepts in the theory of seepage of homogeneous liquids in fissured rocks." Journal of Applied Mathematics and Mechanics 24: 1286–1303.

[2] Warren, J.E. and P.J. Root (1963). "The behavior of naturally fractured reservoirs." Society of Petroleum Engineers Journal 3: 245–255.

[3] Joshua Taron, Derek Elsworth, Ki-Bok Min. Numerical simulation of thermal-hydrologic-mechanical-chemical processes. International Journal of Rock Mechanics & Mining Sciences 46 (2009) 842–854.

[4] Wang, J.S.Y., C.F. Tsang, et al. (1981). "A study of regional temperature and thermohydrologic effects of an underground repository for nuclear wastes in hard rock." Journal of Geophysical Research 86(B5): 3759–3770.

[5] Tsang, C.F., O. Stephanson, et al. (2000). "A discussion of thermo-hydro-mechanical (THM) processes associated with nuclear waste re-positories." International Journal of Rock Mechanics and Mining Sciences 37(1–2): 397–402.

[6] Yow, J.L. and J.R. Hunt (2002). "Coupled processes in rock mass performance with emphasis on nuclear waste isolation." International Journal of Rock Mechanics and Mining Sciences 39(2): 143–150.

[7] Hicks, T.W., R.J. Pine, et al. (1996). "A hydro-thermo-mechanical numerical model for HDR geothermal reservoir evaluation." International Journal of Rock Mechanics and Mining Sciences & Geomechanics Abstracts 33(5): 499–511.

[8] Germanovich, L.N., R.P. Lowell, et al. (2001). "Temperature-dependent permeability and bifurcations in hydrothermal flow." Journal of Geophysical Research-Solid Earth 106(B1): 473–495.

[9] Rutqvist, J. and C.F. Tsang (2003). "Analysis of thermal-hydrologic-mechanical behavior near an emplacement drift at Yucca Mountain." Journal of Contaminant Hydrology 62–3: 637–652.

[10] Neaupane, K.M., T. Yamabe, et al. (1999). "Simulation of a fully coupled thermo-hydro-mechanical system in freezing and thawing rock." International Journal of Rock Mechanics and Mining Sciences 36(5): 563–580.

[11] Faulkner, D.R. and E.H. Rutter (2003). "The effect of temperature, the nature of the pore fluid, and subyield differential stress on the permeability of phyllosilicate-rich fault gouge." Journal of Geophysical Research-Solid Earth 108(B5).

[12] Neuzil, C.E. (2003). Hydromechanical coupling in geologic processes. Hydrogeology Journal 11(1): 41–83.

[13] Koutsabeloulis, N. and S.A. Hope (1998). Coupled stress/fluid/thermal multiphase reservoir simulation studies incorporating rock mechanics. SPE/ISRM EUROCK-98 Symposium, Norway.

[14] Pao, W.K.S., R.W. Lewis, et al. (2001). "A fully coupled hydro-thermo-poro-mechanical model for black oil reservoir simulation." International Journal for Numerical and Analytical Methods in Geomechanics 25(12): 1229–1256.

[15] Schrefler, B.A., G.A. Khoury, et al. (2002). "Thermo-hydro-mechanical modelling of high performance concrete at high temperatures." Engineering Computations 19(7–8): 787–819.

[16] Thomas, H.R., Y. He, et al. (1998). "An examination of the validation of a model of the hydro/thermo/mechanical behaviour of engineered clay barriers." International Journal for Numerical and Analytical Methods in Geomechanics 22(1): 49–71.

[17] Collin, F., X.L. Li, et al. (2002). "Thermo-hydro-mechanical coupling in clay barriers." Engineering Geology 64(2–3): 179–193.

[18] Gray, W.G., K. O'Neill, et al. (1976). Simulation of heat transport in fractured, single-phase geothermal reservoirs. Summaries Second Workshop Geothermal Reservoir Engineering. Stanford, California, Stanford University: 222–228.

[19] Pritchett, J.W., S.K. Garg, et al. (1976). Geohydrological Environmental Effects of Geo-thermal Power Production—Phase IIA. La Jolla, California, Systems, Science and Software.

[20] Biot, M.A. (1941). "General theory of three-dimensional consolidation." Journal of Applied Physics 12: 155–164.

[21] Noorishad, J., C.F. Tsang, et al. (1984). Coupled Thermal-Hydraulic-Mechanical Phenomena in Saturated Fractured Porous Rocks: Numerical Approach. Journal of Geophysical Research 89 (B 12): 10, 365–10, 373.

[22] Wittke, W. (1973). Percolation through fissured rock. International Association of Engineering Geology Bulletin 7: 3–28.

[23] Swenson, D.V., R. DuTeau, et al. (1997). "A coupled model of fluid flow in jointed rock applied to simulation of a hot dry rock reservoir." International Journal of Rock Mechanics and Mining Sciences & Geomechanics Abstracts 34: 308.

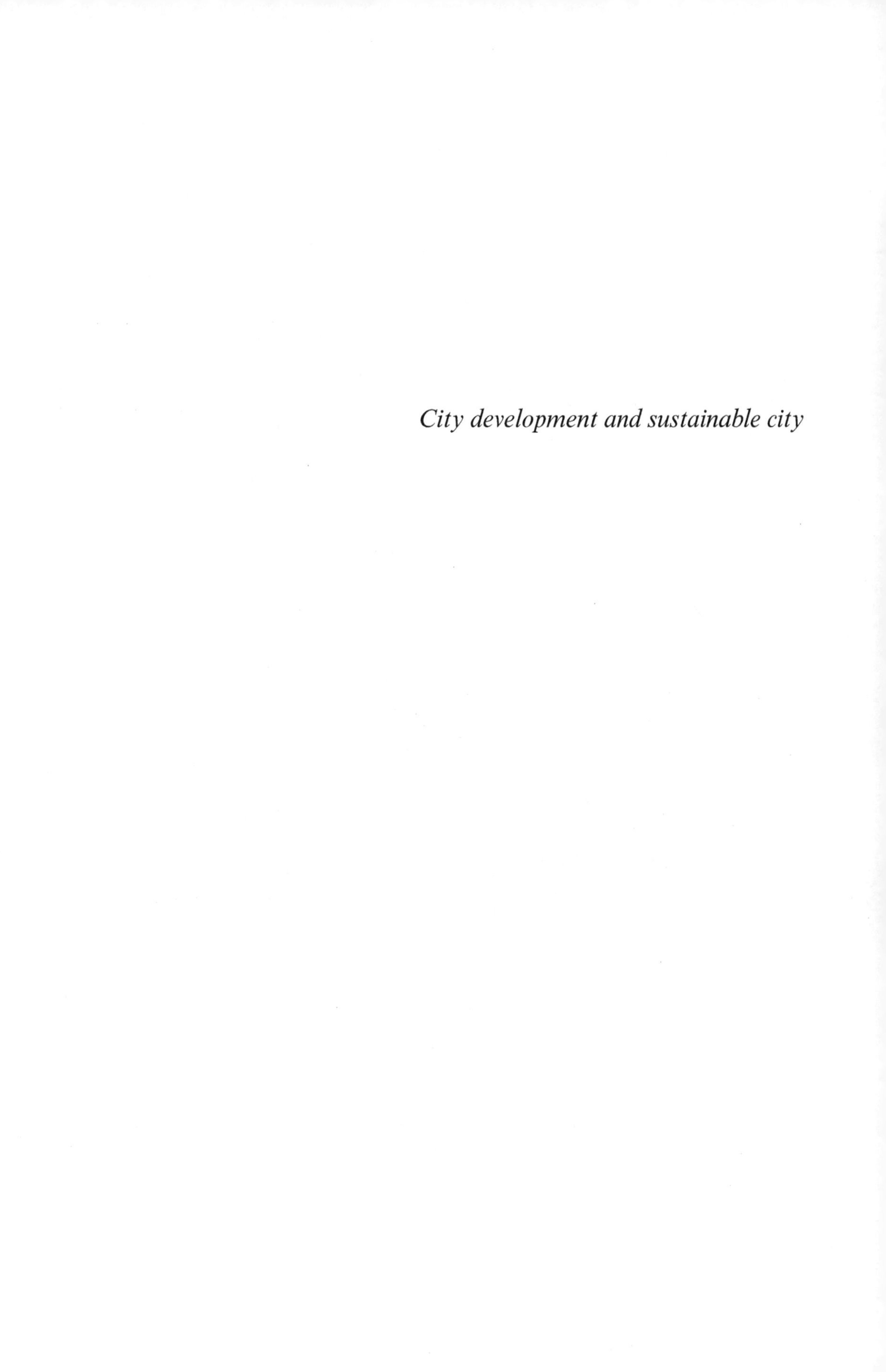

City development and sustainable city

Material Science and Environmental Engineering – Chen (Ed.)
© 2016 Taylor & Francis Group, London, ISBN 978-1-138-02938-5

Primary investigation on the planning and design of urban river landscape

Y. Chen & Y. Wang
Institute of Art and Design, Nanchang Hangkong University, Nanchang, China

ABSTRACT: As the important carrier of urban ecological environment, urban river landscape plays an important role in socioeconomic development with a significant value in promoting urban environment quality, enriching urban landscape content and prompting urban socioeconomic development. This paper explores the components and features of urban river landscape. From the perspective of ecology and region, this paper proposes the design principles of river landscape for the harmony and unification of river landscape, ecological environment, and urban economic and social development.

Keywords: river landscape design; ecology; regional culture

1 INTRODUCTION

Since China's reform and opening up, the accelerated urbanization process has led to the deterioration of urban environment gradually. The convergence of a mass of industrial and domestic sewage and wastes in urban rivers has resulted in the constant interruption and vanishing of urban rivers; some rivers have even turned into waste rivers that jeopardize the health of the residents living around tremendously. Since 1990s, many Chinese cities have started to realize the importance of urban environment. Urban river landscape design has become the hot point of urban landscape planning and design at present. As the open space in urban area, river landscape has entered into various levels of a city, in the wake of the increasingly enhanced spiritual demands of residents and tourists in culture, leisure and entertainment.

2 OVERVIEW ON URBAN RIVER LANDSCAPE

Urban rivers normally indicate natural or artificial river sections flowing through urban regions. Urban rivers are closely related to urban development, providing urban areas with the services, including water source, drainage, transportation, trade communication, and leisure and entertainment. Urban river landscape is the important content in modernized urban landscape design and the closest river network and system with the natural environment in urban areas.

2.1 *Components of urban river landscape*

In the wake of urban development, river space has shown a trend of diversification and complication, different from the ordinary landscape design of urban public space. Urban space is the juncture of water area and land area, which shall not only meet the landscape features of ordinary public space, but also will take the influence on the urban space ecosystem into consideration of designing. Compared with the forms of other urban spaces, there are diversified components for urban river space landscape, including water body, river bank, transportation factor, and landscape factor. Each component has its own features.

2.1.1 *Water body*

Water body is the main body of urban river landscape. The hydrological environment of a river, which is under the influence of its surrounding geographic environment and natural ecology, can determine the basic orientation of river landscape design. Water body can affect multiple senses (vision, hearing, smelling, kinesthesia and touch) and impact on men's judgment and comment on this special environment. The existence of water can make men's communication place become more intelligential and attractive; the natural properties of water flow, flexibility and volatility, can add diversity and enjoyment to urban landscape, while the ecosystem formed by water flow plays a crucial role in greening urban images.

2.1.2 *River bank*

River bank, one of the components of river space, is the transition of water and land areas.

Because of the juncture position between water and land areas, river bank functions as the reinforcing dam to prevent from the attack of flood. Therefore, in water conservancy projects, it is also called the protection bank. In traditional definitions, protection bank features hardness, namely stones or bricks shall be piled up on natural banks to form the hard surface, which can ensure the function of water exchange and adjustment between river bank and river body as well as a certain flood-fighting effect. Each kind of protection bank has its own features and range of application. However, in modern landscape, traditional methods have severely damaged the ecological environment of river landscape. Therefore, it is the key for the construction of ecological image of a modern urban area to transform the artificial protection banks into ecology-oriented protection banks.

2.1.3 *Traffic factor*

The traffic or river area is relatively concentrated in the urban traffic system, with some rivers even functioning as the transference between water and land transportation, linking various factors within the area. Generally, the traffic factors of urban rivers include vehicle lanes, bicycle lanes and pavements. Vehicle lands are linked with the overall urban traffic system and separate the river space from the urban space; bicycle lands are used by urban sightseeing vehicles or bicycles for sightseeing tourism and leisure and relaxation, and usually closer to the river landscape than vehicle lands; pavements are used by residents to walk on the greenbelts along the rivers, in order to ensure the possibility of residents to take waterfront activities. Traffic design shall take urban traffic and internal traffic of waterfront into consideration, to ensure the coordination and transition of spatial environment and architectures, provide the opportunities of water loving for urban residents, and integrate various public activity centers, water-loving platforms and plank roads into an organic whole that is run through people's daily life and leisure routes, aiming at creating a comfortable leisure space.

2.1.4 *Landscape factor*

River landscape factors include greening, waterfront landscape architecture and featured landscape. Green system is the important support of waterscape that can set off the beauty of river landscape. The greening design of urban river space can adjust measures to local conditions, advocate the principle of focusing on local plants, and enrich the environmental landscape of waterfront space along with the change of the seasonal appearances of plants. Waterfront landscape architecture and featured landscape are the public and artistic facilities of river space. In the urban river landscape design, featured landscape can improve the spatial visual quality and enhance the space quality. The appropriate application of featured landscape can remarkably enhance people's overall understanding of waterfront environment, acting as an inevitable factor in the modern urban river landscape.

2.2 *Features of urban river landscape*

Urban river spatial landscape is composed of multiple factors, while river landscape is characterized by multifaceted features, which are mainly shown by the ecological feature, open feature and regional feature.

2.2.1 *Ecological feature*

River channel is the relatively independent and integrated section in the urban ecosystem, whose ecosystem is also more natural, compared with other sections in urban areas. The terminal purpose of urban river landscape design is the high-degree harmony between human and water and to create the vigorous urban river landscape environment featuring harmony, coexistence and soundness. Ecology is the important instruction concept of modern environmental landscape design; therefore, the ecological feature of landscape design is also becoming the key point of landscape design construction.

2.2.2 *Open feature*

Urban rive space is the major component of urban open public space, undertaking the important responsibility of the water-loving entertainment and leisure of surrounded residents. Urban river landscape consists of multiple kinds of spaces, including various open spaces of leisure and entertainment. It is a non-profit space area that can enhance residents' living quality, acting as the ideal area providing rich and varied entertainment and leisure activities for residents and tourists to enjoy the gifts of nature.

2.2.3 *Regional feature*

The regional features of urban river landscape are often ignored by designers and users; however, in the wake of people's deepening demands, the regional features of urban river landscape have been stressed gradually in the recent years.

Urban rivers are usually the cradle of urban civilization. People are living along the rivers; naturally, rivers become the space for men's ideological and cultural communication. The gathering of human beings and materials has formed an all-bracing, open and free culture-waterfront culture. It is easier for men to retrospect the historic footprints and experience the change of time along rivers. China features the extreme difference of climates and cultures in the northern and southern parts. The landscapes of various regions can clearly reflect the geographic features of specific regions. Therefore, the urban river landscape with strong local features and regional features will definitely become the leader of urban river landscape in the future.

3 PRINCIPLES OF URBAN RIVER LANDSCAPE DESIGN

3.1 *Stress on creation of natural environment: principle of ecological balance*

Ecology is one of the important instruction concepts of modern environmental landscape design. The focus of ecological balance design is based on the protection and utilization of the ecosystem balance of existing regions. In any case, it is the primary principle for urban river landscape design. The concrete designing shall be based on the following aspects for research.

First, designers should pay attention to the protection and development of existing natural environment; fully respect the natural forms of rivers; control and protect the original terrain and geologic structure. On this basis, designers should link various landscape nodes together in a harmonious way, to create the landscape visual environment focusing on the original forms of rivers.

Second, designers should pay attention to creating the artificial ecological environment of the greening system. In the concrete designing of green system, as an important factor in the ecosystem of urban river landscape, designers should strengthen the utilization of the original local plants, seek for the multi-level and all-round ecological greening system, to form a harmoniously stable ecological structure, in order to give a full display of greening system in improving the microclimate of urban rivers and attracting the returning of various creatures to urban areas, so as to nurture and maintain the balanced soil ecosystem.

3.2 *Stress on creation of man-water relationship: principle of humanization*

Humanization is the basis of modern urban river landscape design. And it shall follow the principle of human-orientation and focus on the creation of man-water relationship. Water is the most vigorous natural factor that can make a man relaxed, soothed and delightful. The change, flowing, shadow, and sound and color of water can be the most expressive natural landscape of a city. Therefore, urban river landscape design should take the factor of men into concrete consideration; fully utilize the factor of water; stress the participation possibility of the water-loving space; allow residents to participate in activities in various forms; enhance the appealing of the space; and construct the harmonious relationship between man and water.

3.3 *Stress on extension of history and culture: principle of inheritance of regional culture*

The ecological design and urban river should not only follow the natural environment features of the areas, but also inherit and reflect local history and culture. At the current stage, most urban river landscapes pay too much attention to forms, and indiscriminately imitate the design methods and models of western countries, leading to the excessive "modernization" of landscapes without any connotation or regional features and sameness among cities gradually. Consequently, when restructuring urban river landscape, designers should follow the principle of inheriting regional culture and proceed from urban regional culture to protect and effectively utilize those sites, e.g. old architectures with living imprints, symbols or life styles. Mastery of this line can better promote the inheritance of regional culture, demonstrate the characters of cities, and increase the city charms.

4 BASIC METHODS OF URBAN RIVER LANDSCAPE DESIGN

Design of urban river landscape, as a comprehensive special design subject, refers to multiple specialized knowledge and departments, including environment protection agencies, urban planning bureaus, architectural design institutes and landscape design institutes, which should cooperate and coordinate mutually. The river landscapes of different cities feature regional features and differentiations; design departments should conduct concrete analyses for concrete issues. However, the basic methods of urban river landscape design are universal, which can be summarized as follows.

4.1 *Preliminary field survey*

Field survey is at the initial stage of the entire design. First, design departments should make a survey of the general situation of the rivers, such as river width, river section and flow velocity; second, design departments should survey on the existing architectures along the rivers, and make key marks and completed recordings of existing architectures with historic and humanistic or landscape values for protection; third, design departments should investigate on the ecological situation, humanistic and regional characters. After the investigation, responsible design departments should integrate opinions of different departments on the survey results to prepare a feasibility report and set up the project based on the report, and ultimately formulate the policies and principles for the overall planning.

4.2 *Theme and orientation of river landscape*

Design departments should first confirm the theme and orientation of a landscape design. Based on the analysis of site current status, designers should take the basic functions of urban rivers into consideration to set up the themes according to different urban demands and river types under the premise of satisfying the flood and draining water logging. Because of the wide urban river basins, different sections have different orientation demands; therefore, design departments can define the orientations of different river sections under the general design concept.

4.3 *Graphic layout of river landscape*

On the basis of detailed investigation and mastery of various design materials, design departments should focus on the design themes, integrate the actual topography, and proceed from functionality to make an initial graphic planning rationally according to design conceptions. Rivers in different cities are disparate in terms of surrounding environment and vernacular and humanistic characteristics.

Generally, urban river landscapes are in the line style, leading to the importance of the continuity of graphic layout. The landscape of a river can be divided into different sections according to section functions or natural river sections or the different cultural significances of natural resources in different sections. The landscape of a river in different sections should be corresponded and coherent with key designs for nodes. The landscape design should make a distinction between important ones and lesser ones, and build one highlighted theme according to river features and historic and cultural features in each certain length.

5 CONCLUSIONS

Urban rive landscape design is the landscape design and analysis of urban rivers and surrounding environment, including planning layout, landscape transformation and environment, featuring the important values in upgrading urban environment quality, enriching urban landscape contents, and promoting urban social and economic development.

REFERENCES

[1] Liu Binyi. *Modern Landscape Planning and Design* [M]. Nanjing: Southeast University Publishing House, 2000.

[2] Xu Shiguo, Gao Yongmin. *Planning Design and Governance of Urban River* [M]. Beijing: China Water Power Press, 2005.

[3] Tang Zhenyu, Zhang De. *Urban River Landscape Design* [M]. Beijing: China Architecture & Building Press, 2006.

Material Science and Environmental Engineering – Chen (Ed.)

Research on the speed limit for highway traffic safety

Q. Cheng
College of Civil Engineering and Transportation, Hohai University, Nanjing, China
Jiangsu Key Laboratory of Urban ITS, Southeast University, Nanjing, China

C.J. Zheng
College of Civil Engineering and Transportation, Hohai University, Nanjing, China

ABSTRACT: This paper introduces the software architecture and functional model. The system uses the Vissim rendering of expressway network, and builds the network server based on the COM interface, using VB. NET visual programming language for software development. The software realizes the network information maintenance, the condition of the network monitoring and control, entrance ramp control, variable speed control, road traffic event entry and inquiry, and traffic accident emergency management module function. Through the study of different grade highway speed management and control technologies, this paper integrates development speed control equipment, reduces the illegal behavior of highway speeding and speed dispersion, and improves driving safety.

Keywords: variable speed limit control; software development; safety

1 INTRODUCTION

Speed management and control is an important means to ensure the safety of vehicles. The core of the variable speed limit control technology is based on a risk judgment of highway traffic accidents. It can actively intervene by adjusting the speed limit value for dangerous driving traffic condition microcosmic driving behavior, so as to improve the traffic operational conditions and enhance the purpose of safe driving. According to the characteristics of urban expressway reasonable speed limit, it not only meets the traffic demand at the same time, but also ensures the efficiency and safety of the rapid road system, and also can reduce the standard deviation of speed and improve the stability of road traffic, which has become one of the key issues, and variable speed limit control can also provide an effective technical support.

From the perspective of domestic and foreign research situation, the present-stage study on the variable speed limit control strategy is still relatively simple, and the control model is estimated based only on one or two kinds of control strategy, and not on establishing a comprehensive and effective variable speed system of decision-making [1]. The domestic and foreign research on the control strategy is mainly based on traffic flow characteristics, and the study on the weather, traffic accidents, road construction and other factors is limited. The study focused on the highway, but very little research on the city expressway, and traffic flow characteristic of the city expressway has its particularity. This paper through the study of different grade highway speed management and control technologies integrates development speed control equipment, reduces the illegal behavior of highway speeding and speed dispersion, and improves driving safety.

2 SYSTEM FRAMEWORK

2.1 *Overall designs*

2.1.1 *Hardware configuration*

Table 1. System hardware configuration.

Device name	Number
Microwave traffic flow monitoring equipment	9
Environmental testing equipment	1
The variable information board	11
Information prompt board	9

The table lists the hardware and the 1000M Internet connection to fulfill the system.

2.1.2 *Software platform type*

System functions need to ensure that the system hardware is based on the installation of the operating system, as listed in Table 2. The software of

Table 2. System software platform type.

Server type	Corresponding operating system
Database Service	Linux operating system /Windows operating system
Data exchange server	Linux operating system /Windows operating system
Application server	Windows operating system

this system is two times development of microscopic traffic flow simulation software Vissim of PTV Company Germany [2]. The system adopts Vissim as the model of expressway network server, Vissim highway road network under the environment of drawing to provide the GoogleEarth 500 feet under the background of network data layer; using the COM interface of Vissim, based on the Microsoft. NET Framework3.5 framework, using Access database to store the relevant data, the application of VB. NET language development can be achieved in the Visual Studio 2008 environment [3].

2.2 *System structure*

2.2.1 *System level structure*

The system level structure is shown in Figure 1. System is divided into four layers: data collection layer, data layer, variable speed-limit control layer, and information release. Data collection layer is composed of environmental testing equipment and traffic monitoring equipment, to complete the information collection. Environmental data and traffic data, environmental data acquisition unit and traffic parameters are transmitted to the host matrix, and then summarize the facts database. Information is transmitted to the variable speed-limit control layer, and information is released by the message board in the end. System network topology structure is shown in Figure 2.

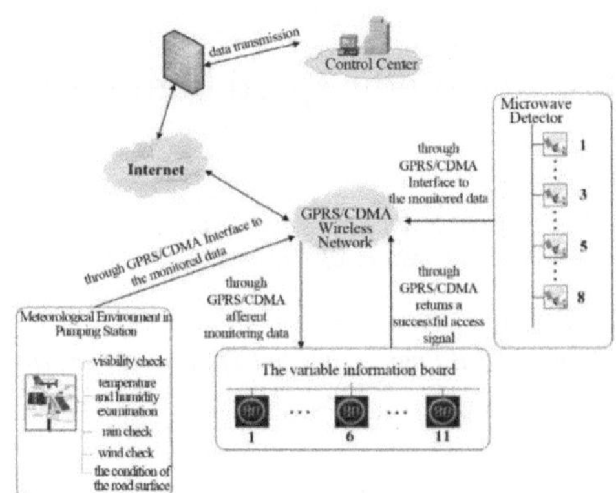

Figure 1. The system level structure.

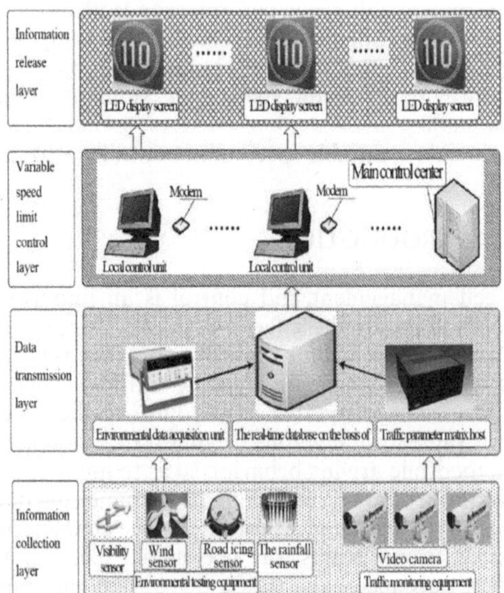

Figure 2. System network topology structure.

2.2.2 *The system function structure*

The system mainly includes five modules, namely network information maintenance, the condition of the network monitoring, network traffic incident management, traffic control system, and traffic accident emergency management [4]. Network information maintenance is mainly management and maintenance for the basic elements of road network, such as query, modify, add and provide functions. The condition of the network monitoring is mainly real-time dynamic within the running state of each section, in the main system of the on-ramp traffic flow status, with access to the network traffic monitoring function. Traffic incident management is the main function of the system modules, including traffic event entry and inquiry function, but also including the network control strategy optimization function under the traffic incident. Network control system mainly includes the ramp control and variable speed limit road information control, and other functions. Traffic accidents management is realized on different kinds of traffic accident under the emergency management capabilities.

2.2.3 *The system database design*

Building a database is the basis of variable speed-limit control system software. Building a database is the basis of variable speed-limit control

Figure 3. Vissim environment highway network diagram.

system software. Software is the main research object for the network environment of highway traffic control. In order to facilitate maintenance, the system software establishes two databases: one is the network information database; another is traffic data information. Network information database is used to store the basic information highway system in the road. Traffic data information is used to store the historical traffic flow data, the control sections of real-time traffic flow data collected information, traffic event information and traffic accident control plan [5].

3 SYSTEM FUNCTION AND IMPLEMENTATION

The implementation of this system is mainly based on the mapping of highway network in Vissim, using the COM interface development of software system based on Windows platform. Road network map is the most important in this system and is also one of the most complex modules. This system operates by using the Vissim rendering of high-speed network, using Google Earth satellite software on 500-foot high-speed highway network images as a background image, thus ensuring the drawing of road network and real network agreement. Vissim environment highway network diagram is shown in Figure 4. Also, in the process of simulation to show the image of more realistic effect, this project uses Vissim to provide a 3 d simulation model of a vivid demonstration of the emergency management system running effect of freeway traffic [6].

Section variable speed limit control implementation module in this system consists of four parts. The first step is through the main line traffic flow condition monitoring identifying road traffic

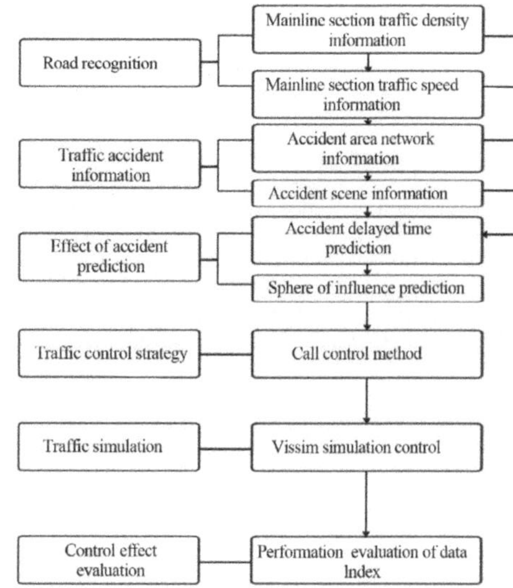

Figure 4. The module software design.

accident. The second step involves the input by the user traffic accident details. The third step is to impact on the traffic accident prediction, including delay time prediction and distance prediction effects on the upstream traffic. The fourth step is through dynamic optimal when the variable speed-limit control method for the variable speed-limit control, and gives the values of speed limit section. Finally, control information is put by the user. The module software design flow is shown in Figure 4.

4 CONCLUSION

Software system is the PTV Company in Germany, on the basis of microscopic traffic flow simulation software Vissim for secondary development. For highway system control problem, researchers provide a set of simulation experiment platform, but also for traffic management decision-making departments to set up a validation control decision-making environment. The research of highway has a certain theoretical and practical significance. The implementation of this system for highway researchers provides a set of experiment platform according to the need, based on the software development of a new function module.

ACKNOWLEDGMENTS

This work was supported by "the Fundamental Research Funds for the Central Universities" (Project No. 2014B27214).

REFERENCES

[1] Wu Zheng. Hydro Mechanical Model of Mixed And Low-speed Metropolitan Transportation [J]. Acta Mechanics Sinica, 1994, 26(2): 149–157.

[2] Boumedience Kamel, Amar Benasser and Daniel Jolly. Flatness Based Control of TrafficFlow for Coordination of Ramp Metering and Variable Speed Limits[C]//Proceedings of the 11th International IEEE Conference on Intelligent Transportation Systems, Beijing, China, 2008: 838–843.

[3] Boris S. Kerner. Introduction to Moder Traffic Flow Theory and Control [M]. Berlin: Spring Press, 2009.

[4] Boris S. Kerner. On-ramp metering based on three-phase traffic theory downstream off-ramp and upstream on-ramp Bottlenecks [J]. Journal of the Transportation Research Board, 2008, 80–89.

[5] Pei-Wei Lin, kyeong-Pyo Kang, Gang-Len Chang. Exploring the Effectiveness of Variable Speed Limit Controls on Highway Work-Zone Operation [J]. Journal of Intelligent Transportation Systems: Technology, Planning, and Operations, 2004, 8(3): 155–168.

[6] David Navon. The paradox of driving speed: two adverse effects on highway accident rate [J]. Accident Analysis & Prevention, 2003, 35(3): 361–367.

Material Science and Environmental Engineering – Chen (Ed.)

Recycling of industrial solid waste in landscape design

W.J. Zhao, H. Deng & L.Y. Zheng
Hunan University of Technology, Zhuzhou, China

ABSTRACT: China's industrialization has to be responsible for the accumulation of industrial solid waste. Some of those solid wastes are valuable resources and can be recycled in landscape design. From the perspective of environmental design, this paper discusses the recycling process of industrial solid waste, and, by case study, presents the reutilization in landscape design in terms of pavement, furniture design, and sculptures, which may be enlightening for recycling of industrial solid waste.

Keywords: recycling; industrial solid waste; landscape design

1 INTRODUCTION

With the rapid progress in China's industrialization, industrial wastes have been increasing with each day, thus placing a heavy health and environmental burden to a developing country like China. However, manufacturers have made insufficient efforts to deal with those solid wastes.

Huge amounts of industrial solid waste, whether hazardous or not, are directly dumped in the fields, rivers, lakes, seas, irrigation canals or along waterways. Those wastes contain many metals, plastics and other substances, some of which may be harmful to the soil and plant/animal survival, and even pollute the underground water, seriously affecting the health of residents located closer to dumpsites and exposing them to poisonous emissions and effluents. On the other hand, some of those solid wastes are valuable resources (e.g. gold, silver, palladium, copper, aluminum, and plastics) that should not be wasted but can be recovered for use.

Landfills are often the most cost-efficient way to dispose of waste, especially in countries with large open spaces. But landfills have the potential of causing pollutions, such as contamination of groundwater or soil contamination. Incineration, once touted as a solution to deal with solid waste, is not so good from an environmental point of view. Incineration of metals or materials with high metal content can lead to the spread of toxic metals in the environment. Composting technology is widely used for soil amendment, but the process may cause potential environmental harm.

Although the above three methods have improved a lot with the technological development, they need great investments and mass processing. Meanwhile, the reusing and recycling of industrial solid waste has been paid much attention to in recent years. This paper aims to discuss the recycling of the industrial solid waste in landscape design. According to attributes of solid waste, many industrial solid wastes can be reused in landscape design and can become a part of urban landscape with zero carbon emission. From the perspective of environmental design, this paper discusses the recycling process of industrial waste, and presents the reutilization in landscape design in terms of pavement, furniture design, and sculptures.

2 RECYCLING PROCESS OF INDUSTRIAL SOLID WASTE

The "3 Rs", namely Reduce, Reuse and Recycle, are crucial. This means reducing, reusing, and recycling most if not all materials that remain after production is a major step in the waste management process. Responsibly managing industrial solid waste is a vital component of a sustainable environment and economic development.

There are many successful cases of recycling industrial solid wastes in urban landscape design, such as Fresh Kills Parkland in New York, Byxbee Park in Pafo Alto, Granville Island in Vancouver, IBA Park in Emscher, Nancuiping Park in Tianjin, Dashahe Park in Shenzhen, the landscape design in the core area of ecological Park of Baoshan in Shanghai, Beijing 798 Art Zone, and Ziyun Park in Yianjin. However, the study of recycling of solid waste in landscape design is insufficient in China, and such successful cases are very few.

As the recycling industry grows, it is developing better processes for recycling different materials. Industrial solid waste, such as smelting slag, mining debris, fuel waste, chemical debris, mechanical

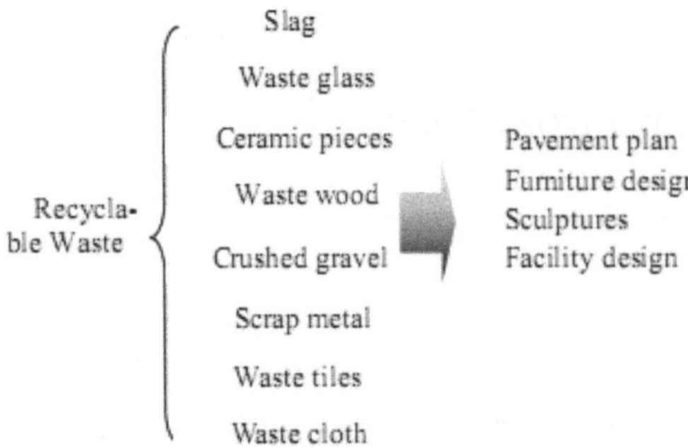

Figure 1. Recycling of industrial solid waste in landscape design.

waste, inorganic nonmetallic materials and polymeric materials, can be recycled. Some residues can be dealt with by incineration and composting, and then changed into energy and soil conditioners. Many recyclable wastes are employed in interior design, landscape design, building appearance design and public art.

Landscape design includes design of pavement in gardens, feature wall, ponds, sculptures, and other public facilities. First, waste hierarchy should be made to extract the maximum practical benefits and to make full use of waste. The classification of industrial solid waste in landscape design is shown in Figure 1. Recyclable waste can be an important material used in pavement plan, furniture design, sculptures and facility design.

3 THE RECYCLING OF INDUSTRIAL SOLID WASTE IN LANDSCAPE DESIGN

3.1 *The pavement plan in landscape design*

The paving materials can be divided into hard ones, such as concrete, stones, gravels, wood and bricks, and soft ones, such as sand, broken fabrics and lawn. The industrial solid wastes used in paving should be safe at first, and then should be strong and tough enough, skid-proof, and comfortable.

Some solid wastes, if processed properly, can be substitution for normal paving materials. For example, slag, a mixture of metal oxides and silicon dioxide, mixed with lime and fly ash, can be changed into burn-free bricks and building bricks; and can also take the place of limestone and clay in the production of cements, and increase the density of concrete if mixed with it. In landscape design, industrial solid waste can be widely applied to paving design, furniture wall building, and artificial hill due to their attributes of skid-proof, permeability, intensity, heat resistance, erosion resistance and stability.

As shown in Figure 2, the paving materials used are waste mining ores, waste wood, broken tiles from buildings, waste ceramic pieces and broken glasses from factories and waste fabrics.

As the hand-painted sketch shown in Figure 3, the vegetation carpet is made from wastes from apparel and textile industries, which can be processed into soft paving materials. Such paving materials measure up to the standards of softness, permeability, intensity and safety, and recycle the discarded resources into paving materials.

In addition, there are many other ways for the recycling of industrial solid waste to pavement plan, such as the reprocessing of waste ceramic pieces into filling materials and glazes, the usage of fly ash and glazes as base materials, and the recycling of iron rods to produce decorative pillars. These processed wastes can be made as paving materials.

3.2 *Recycling in furniture design*

Landscape furniture design, a very important part in landscape design, is characterized as its

Figure 2. The pavement plan employing industrial solid waste in landscape design (I) hand-painted sketch.

Figure 3. The pavement plan using industrial solid waste in landscape design (II) hand-painted sketch.

functionality and role of decoration and partition. It contains the design of different facilities including curtain wall, waterscape, waterside pavilion, boat house, corridor, stepping stone on water surface, flower trellis and terrace.

For example, as the art wall design shown in Figure 4, the wall is reprocessed by and the combination of broken glasses, gravels, waste wood, ceramic pieces and waste iron wire, which not only play a role of partition and guidance, but also can be used as a background and a beautiful decoration from the aesthetical view point. The design presents the original form of the gravels, and uses iron wires to tie them together.

As we can see in Figure 4, the design of the front side of the art wall, suitable for small squares and communities, is space-saving and matches wonderfully with the leisure bench made of waste wood. On the back side of the art wall, as shown in Figure 5, small hanged bonsais act as decorations with containers made of bamboo tubes, which adds some greenness to the dull color of the wall.

In addition, waterscape, which contains revetment in garden, stepping stone, artificial hill, stone layout and pond bottom pavement, is indispensible to landscape design. Some industrial solid waste can be employed in water feature design and may make an unexpected aesthetical effect. For example, broken tiles in the construction industry can be employed in the pond bottom pavement, or be made to act as channels; crushed ceramics and waste wood can be made into stepping stone on water surface; and iron wire and waste fabric can be changed into the revetment. In one word, industrial solid waste can be recycled and reutilized to realize unique decorative effects.

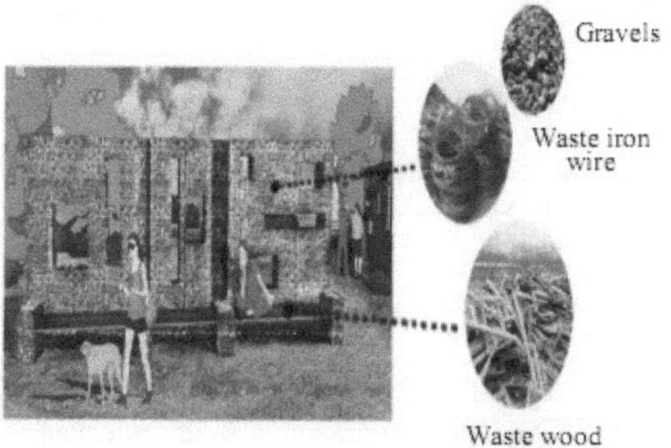

Figure 4. The front side of the art wall (hand-painted sketch).

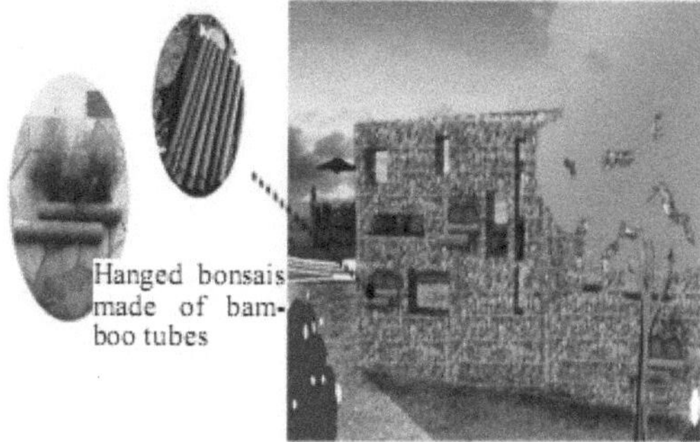

Figure 5. The back side of the art wall (hand-painted sketch).

3.3 *The employment in public facility design and sculptures*

In landscape design, public facility design includes benches, signs, garden lights and dust bins. Metal, plastics, rubber, waste wood and discarded furniture can be used as materials in public facility building.

Sculpture design can be expressed according to solid waste attributes and their different combinations. Waste steel, waste tires or wood pieces can be recycled, transformed, mixed and assembled to manifest themselves in terms of ecological co-existence, culture and harmony.

As shown in Figure 6, waste tires, as the well-known hazardous waste, are made of indecomposable high polymer materials. Meanwhile, they can be reproduced into stools with wood carving, or trimmed and made up to installation artworks, thus adding new, meaningful and cultural features to the landscape.

Figure 6. Recycling of waste tires in sculptures.

Figure 7. Recycling of waste plastic in exterior applications.

Another example is recycling of waste plastic. Some industrial plastic waste is recycled, particularly thermoplastics. Thermoplastics include PET soft-drink containers, most plastic packaging, and a substantial portion of plastic used to make automobiles and other durable goods. Once separated, plastics are heated to be reformed. Thermoplastics can be reheated and reformed several times, making them good candidates for recycling.

Recycled plastics are becoming widely used in landscape design. As shown in Figure 7, plastic chairs or benches made from milk and detergent containers and industrial scraps are now used for decks, fences and other exterior applications. Waste plastic containers can be reformed to a sculpture with the same beauty of ice carving. Furthermore, soft-drink containers, which can be reprocessed into carpeting, can be employed in pavement plan.

4 CONCLUSIONS

For more than three decade's unprecedented growth, China's industrialization has led to economic competitiveness. Nevertheless, this industry also has far-reaching impact on the environment by way of tons of industrial solid waste generated. Responsibly and properly managing industrial solid waste is a vital component of a sustainable environment and economic development. Although there are many ways in dealing with waste, recycling of some industrial solid waste should be employed to minimize its effect as a hazard. The application of recycled industrial solid waste to landscape design will be helpful. This paper points out the recycling process of industrial solid waste, and further discusses the recycling process of industrial waste and presents the reutilization in landscape design in terms of pavement plan, furniture design, sculptures and facility design. In the future, more studies should be made in terms of the reutilization and recycling of industrial solid waste in landscape design.

ACKNOWLEDGMENT

This paper was sponsored by the Project of Philosophy and Social Science of Hunan Province (Grant No. 14YB137).

REFERENCES

[1] Shaw, R. 2002, The International Building exhibition (IBA) Emscher Park, Germany: A model for sustainable restructuring?. *European Planning Studies*, 10(1):77–97. Shen, S. 2011. Analysis of garbage disposal technology in China—Incineration is a wrong way to garbage disposal. *Inheritage* (20): 78–79.

[2] Wang, S.W. Liang, F.Z. & Wang, J.Z. 2003, Recycling technology and reutilization of solid waste, Beijing: Metallurgical Industry Press.

[3] Zhao, Z.C. 2011. Study of landscape regeneration design-from wasteland to City Park, Harbin: Northeast Forestry University.

[4] Li, C. 2013. Application of low-carbon landscape design to regeneration of abandoned industrial zone, Chengdu: Sichuan Agricultural University.

Material Science and Environmental Engineering – Chen (Ed.)
© 2016 Taylor & Francis Group, London, ISBN 978-1-138-02938-5

Ecological town construction strategy based on the analysis of SWOT and AHP: A case study of Panjin-Dawa

S.T. Ma, H.S. Cui & L.X. Yang
College of Forestry, Shenyang Agricultural University, Shenyang, Liaoning Province, China

ABSTRACT: For better control of ecological town construction strategy, we use the SWOT strategic approach combined with the analytic hierarchy process, and specifically discuss the process of building Panjin-Dawa ecological urban advantages and disadvantages, and opportunities and challenges. Through the surveys, we obtain key elements of construction and development and ultimately build strategic quadrilateral, exploring reasonable new ways to ecological urban construction. Furthermore, we put forward an optimization strategy and construction proposals. Sustainable urban planning can be manipulated to provide a theoretical guide and technical support.

Keywords: ecological town; SWOT analysis; strategic approach

1 INTRODUCTION

1.1 *Ecological town*

Ecological town is the inherent requirement to build a resource-saving and environmental-friendly society and a new road to industrialization, and to promote the economic structural change. It is an essential aspect in the process of improving the ecology.

Meanwhile, ecological cities and towns are also facing concepts update, technical progress and policy promotion's pressure and challenge. If the propulsion method is improper and regulations are unreasonable, it will cause a series of more serious problems and consequences, and then ecological towns will likely to be the largest ecological "black hole."

In order to make better ecological urban construction and development, it requires holding its significance meaning and exploring a new reasonable ecological urban construction's way. Lastly, we propose an optimization strategy and construction proposal.

1.2 *Method*

In this paper, we use the SWOT strategic approach, combining analytic hierarchy with Panjin-Dawa for instance, and specifically discuss the process of building Dawa ecological town's advantages and disadvantages, and opportunities and challenges.

Through surveys, we obtain key elements of construction and development, build strategic quadrilateral, and eventually elect locally appropriate strategies to arrive at a reasonable and a sustainable plan of management.

2 DAWA TOWN SWOT QUALITATIVE ANALYSIS

2.1 *Panjin-Dawa town's advantage analysis*

2.1.1 *Rich in natural resources*

Dawa has rich coastal resources. Dawa region has a coastline of 68-km, sandy coastline of nearly 100-km, corresponding to 205 million acres of waters. The beach area of 45 acres can be used.

Dawa have rich biological resources. It is the habitat of bird species diversity, red-crowned cranes, gulls and other 260 kinds of rare birds in protected areas. Red-crowned crane population accounts for 80% of the country, accounting for 75% of the world population of gulls.

Dawa also has rich reed resources. Covering more than 20 million acres of vast natural reed field, it is one of the world's largest preserved wetlands.

Rich hot spring in resources has been further developed throughout the county, which is adapted to successfully establish infirmary, swimming and spa bath.

2.1.2 *Good ecological environment*

Dawa has the Shuangtaizi estuarine National Nature Reserve, a habitat and reproduction space for a variety of wild animals. One hundred million acres of reeds and living creatures developed a unique estuarine wetland ecosystem. We can see the wonders of the world along the shoreline: "Red Beach", high tide the sea, ebb beach for "Bohai Gold Beach".

2.1.3 *Better economic foundation*

Dawa rural per capita income is in the forefront of counties in the province.

Agriculture is the state's key commodity grain production base and export base of high-quality rice. Rice, aquaculture, livestock, vegetables and reed are the commodities that have contributed to the formation of five leading industries.

Dawa has oil and natural gas resources, promoting the rapid development of urban and rural industries. Dawa's equipment manufacturing, petrochemical, food processing, new building materials has become the supportive industry.

2.1.4 *Good ecological county building foundation*

In 1975, the State Council approved Dawa as the county.

In 1988, the State Council regarded Dawa as a coastal county, the Bohai's economic strength county.

In 2000, it named the National Agricultural ecological advanced counties, one of the 12 counties' economic development demonstrated County in Liaoning Province.

2.1.5 *Cultural advantages*

Dawa has fishing and Yan culture, farming culture, is the northern land of plenty. With youth culture as the theme of "youth center", fully tapping the youth's unique lifestyle and characteristics of the times.

Dawa's farmhouse tourism is also a major feature in developing eco-tourism, and has a number of national agricultural tourism demonstration sites.

2.1.6 *Location advantage*

Dawa is located in the northeast of the southern tip of Songliao Plain, the Liaohe River downstream, near the Bohai Sea, is the heart of the Liaohe River Delta and in the Bohai economic circle.

In addition, it consists of 305 state roads, Beijing-Shenyang Expressway, the Sea Camp highway, Qinhuangdao-Shenyang high-speed railway and ditch sea railway passing through it.

In the sea route, the sea also has a harbor, where waterway transportation is convenient with good geographical advantages.

2.2 *Panjin-dawa town weakness analysis*

2.2.1 *Ecosystem vulnerability*

Dawa territory has a fragile ecological environment, wetlands water system is vulnerable to environmental pollution, and the food chain is easily prone to damage or threat. The excessive use of pesticides will have an impact on its environment. It has a large number of rare and valuable species, but the food chain tends to be simple.

2.2.2 *Non-renewable resources reducing year by year*

Dawa underground area is rich in oil, gas and geothermal resources, but due to yearly exploitation, reserves are reducing. As a result, the entire regional economic growth stimulating effect becomes weaker.

2.2.3 *Offshore pollution*

Dawa is located in the Liaohe River downstream, has the lower the average altitude and low-lying. The pollution of Liaohe River upstream water is more serious, leading to a serious pollution to the segment of Panjin Liaohe and Panjin coastal waters in which Dawa is located. Panjin Liaohe segment remains Grade V water now, with Bohai Sea coastal functional areas of low compliance.

2.2.4 *Underdeveloped infrastructure*

Mixing of the industrial and residential lands results in urban industrial layout fragmentation. Infrastructure and public facilities are undeveloped, which cannot meet the needs of residents living in it. Greenery and landscape construction is unsystematic, and land utilization is low.

2.3 *Panjin-dawa town opportunity analysis*

2.3.1 *National policy*

Country has launched a number of policy measures aimed at protecting the environment.

For example, the State Council issued the "Air Pollution Prevention Action Plan implementation assessment methods (Trial)" that was established to implement the most strict liability and appraisal system.

The State Department comments on the proposed Northeast several major policy initiatives in the development of modern agriculture innovation system, to strengthen environmental protection in advance of key ecological function areas, promoting the views of the industrial wasteland and old mining environmental management and other supports.

Eighteen meetings was held in 2012, from four in one became five in one, which add an "ecological civilization" for the construction of eco-towns that has laid a good foundation.

2.3.2 *Sustainable development requirements*

Under the background of new urbanization, sustainable development promoted the necessity to accelerate the construction of eco-towns. Approaches that are taken for the town are rationalizing, minimization, recycling and recovery of the resource use.

2.3.3 *Environmental building requirements*

Dawa is based on the actual characteristics of the local ecological environment that actively explore the new ways of ecological county, and has made remarkable achievements in improving the system of agricultural ecological construction, urban ecological construction system, eco-industrial system, eco-cultural system and other aspects.

2.3.4 *Market demand*

Improving the urban ecological construction system creates demonstration bases. In this way, it promotes ecological civilization, ecological construction combined with economic development, under the market-drive, and rational conduct tourism industry. Through advocacy, people can come close to nature, understand nature and protect nature.

2.3.5 *Government and public support*

In recent years, the city government has strongly supported the Dawa Environmental Protection Agency to put spiritual civilization construction work in an important position, in which everyone was involved to create a good atmosphere. This paved way to vigorously carry forward the concept of ecological civilization, to promote harmony between man and nature.

It received a good response and broad support of the people, through surveys reflecting the degree of satisfaction of the people on the environment up to 98%.

2.4 *Panjin-dawa town threat analysis*

2.4.1 *Impact of tourism development on the environment*

Dawa's existing original ecological state of the natural environment is remarkable, emphasizing that nature reserves and wildlife reserves are more apparent. How to introduce visitors to the Dawa's beautiful natural scenery without destroying the original environmental protection is a major issue put before the people of Dawa.

2.4.2 *Institutional rules*

Imperfect Rules system made a lot of interference, and the impact of external factors hindered or interrupted the promotion of eco-cities and towns. At the moment, the rules need to have a more comprehensive system to ensure its smooth development, up to a viable mode of operation. Not just being a simple theory, it needs to make missing or complex provisions perfect for its smooth operation.

2.4.3 *The environmental impact of economic development*

External investment and economic development give rise to industrial pollution and ecological damage issues, which have a large impact on the original ecological environment. How to reconcile environmental protection and economic and social development is the key issue.

2.4.4 *Ecosystem balanced development*

The complexity and diversity of eco-systems need to take into account the overall ecological factors for the overall control.

Ecological limits should emphasize the issues about the limit height from the supply of ecosystem functions to make plans. Factors between the ecosystem should adapt to each other with balanced development, but one lacking factor should not let the system imperfect, which would ultimately result in ecosystem damage.

3 AHP ANALYSIS OF PANJIN-DAWA ECOLOGICAL TOWN

3.1 *Determining factor model to build chromatography*

In the literature, on the basis of field surveys to determine the main factors to be evaluated, relevant experts and scholars conducted surveys on strategic factors through screening, and ultimately determined their key factors.

Comparisons with increasing frequency will rapidly increase in the number of elements. Because SWOT elements within the group are best not more than 10, so we choose the highest score of three factors within each group.

3.2 *Construction of the judgment matrix*

By the above screened factors, we conduct the surveys to make pairwise comparisons between groups and within the group. Using the questionnaire with the "1–9" scale method for data recovery processing, through questionnaire recovery and data statistics, we obtain the following results.

3.3 *Consistency checking*

We use the square root method to calculate the largest characteristic root and characteristic vector of the judgment matrix. Ultimately by CR = CI/RI < 0.1, we determine whether the matrix can be obtained through consistency checking.

Consistency index calculating

$$C.I = (\lambda_{max} - n)/(n-1)$$

$$\lambda_{max} = \frac{1}{n}\sum_{i=1}^{n}\frac{(BW)i}{W_i}$$

$$C.R = CI/RI$$

Table 1. Key factors.

Strength	Weaknesses	Opportunity	Threat
Rich in natural resources (**S1**) Good ecological environment (**S2**) Better economic foundation (**S3**)	Ecosystem vulnerability (**W1**) Non-renewable resources reducing year by year (**W2**) Offshore pollution (**W3**)	National policy (**O1**) Sustainable development requirements (**O2**) Environmental building requirements (**O3**)	Impact of tourism development on environment (**T1**) Imperfect institutional rules (**T2**) The environmental impact of economic development (**T3**)

Table 2. SWOT matrix comparison between groups.

A	S	W	O	T
S	1	5	3	1
W	1/5	1	1/4	1/5
O	1/3	4	1	1/2
T	1	5	2	1

Table 3. Strength group elements comparison matrix.

S	S_1	S_2	S_3
S_1	1	1/4	1
S_2	4	1	3
S_3	1	1/3	1

Table 4. Weakness group elements comparison matrix.

W	W_1	W_2	W_3
W_1	1	3	1/3
W_2	1/3	1	1/5
W_3	3	5	1

Table 5. Opportunity group elements comparison matrix.

O	O_1	O_2	O_3
O_1	1	1/2	1
O_2	2	1	3
O_3	1	1/3	1

Table 6. Threat group elements comparison matrix.

T	T_1	T_2	T_3
T_1	1	2	5
T_2	1/2	1	2
T_3	1/5	1/2	1

Table 7. Random index references.

Order n	1	2	3	4	5	6	7	8	9	10
RI	0	0.58	0.90	1.12	1.24	1.32	1.41	1.45	1.49	1.51

Table 8. Weight and total ordering.

SWOT group	Group priority	SWOT element	C.R	Group elements weight	Total elements weight
Strength	0.3963	Rich in natural resources	0.88%	0.1744	0.0691
		Good ecological environment		0.6337	0.2511
		Better economic foundation		0.1919	0.0760
Weakness	0.0637	Ecosystem vulnerability	3.7%	0.2583	0.0165
		Non-renewable resources reducing		0.1047	0.0067
		Offshore pollution		0.6370	0.0405
Opportunity	0.1820	National policy	1.76%	0.2402	0.0731
		Sustainable development requirements		0.5499	0.1001
		Environmental building requirements		0.2098	0.0382
Threat	0.3581	Impact of tourism development on environment	0.53%	0.5954	0.1269
		Imperfect institutional rules		0.2764	0.0990
		The environmental impact of economic development		0.1283	0.0459

R.I. (Random Index)

When calculating the proportion of consistency C.R < 0.1, we consider the judgment matrix consistency or through the consistency checking. When using the consistency test, the data is meaningful to be used; otherwise, it needs to return to be revised.

Weight calculation formula is given by

$$W = n\sqrt{\prod_{j=1}^{n} b_{ij}}\,(i = 1, 2, \ldots\ldots, n)$$

$$W_i = \frac{\overline{W_i}}{\sum_{j=1}^{n} \overline{W_j}}(i = 1, 2, \ldots\ldots, n)$$

4 CONSTRUCTION OF SWOT STRATEGIC QUADRILATERAL

4.1 *Group intensity*

Each group strength intensity can be calculated as follows.

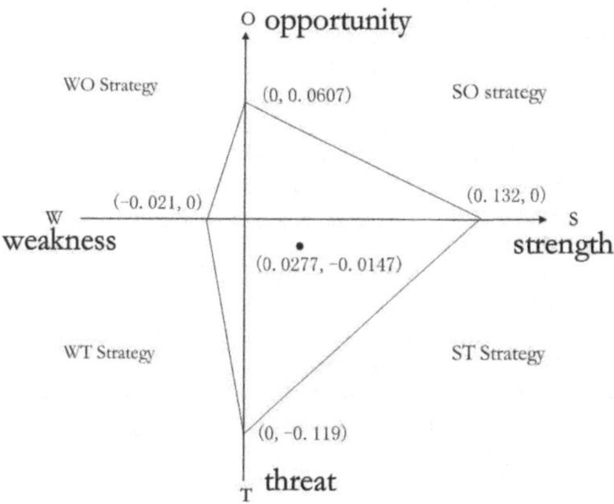

Figure 1. Strategic quadrilateral diagram.

The total strength intensity is given by

$$S = \sum_{i=1}^{n} \frac{s_i}{ns}$$

The total disadvantage intensity is given by

$$W = \sum_{i=1}^{n} \frac{w_i}{nw}$$

The total opportunity intensity is given by

$$O = \sum_{i=1}^{n} \frac{o_i}{no}$$

The total threat intensity is given by

$$T = \sum_{i=1}^{n} \frac{t_i}{nt}$$

The calculated results obtained are as follows: $S = 0.132$, $W = 0.0212$, $O = 0.0607$, $T = 0.119$.

From the above result calculation, we find that the advantages and challenges have a greater impact on urban construction.

4.2 *Structure strategic quadrilateral*

Using strengths, weaknesses, opportunities and threats to the strength of calculations, we construct the town strategic quadrilateral. Then, we determine the strategic position, to ultimately determine the strategic type.

By establishing the SWOT coordinate system, we label the computed velocity value on the coordinate system's S-axis, W-axis, O-axis and T-axis. We then connect the axis of the intensity values to build strategic quadrilateral. In the quadrant, the gravity center's coordinate of the quadrilateral $P(X, Y)$ determines the type of strategy.

Table 9. Quadrilateral and type of strategy.

Quadrant	Strategic area	Position domain	Type
The first quadrant (SO)	The pioneering strategic area	[0, π/4)	Strength type
		[π/4, π/2)	Opportunity type
The second quadrant (WO)	Striving type strategic area	[π/2, 3π/4)	Enterprising type
		[3π/4, π)	Adjustment type
The third quadrant (WT)	Conservative strategic area	[π, 5π/4)	Retreat type
		[5π/4, 3π/2)	Avoidance type
The fourth quadrant (ST)	Fighting type strategic area	[3π/2, 7π/4)	Adjustment type
		[7π/4, 2π)	Enterprising type

The quadrilateral $P(X, Y)$ can be calculated as follows:

$$P(X,Y) = \left(\sum \frac{X_i}{4}, \sum \frac{Y_i}{4}\right)$$

5 CONCLUSIONS AND SUGGESTIONS

Through the Panjin-Dawa eco-towns of SWOT-ahp analysis, we can draw the following conclusions:

Panjin-Dawa town should adopt the ST strategy for ecological construction and for more obvious advantages of enterprising construction. Because the natural resources strategy has obvious advantages in the construction process, we make full use of this advantage, making it to be an important strategic initiative. At the same time, we take advantage of own strengths, to reduce external threats to the ecological cities and towns.

Taking ST measures to minimize the adverse effects brought by disadvantages, we ensure stability in the development process, and determine whether reasonable protection and utilization become its principal contradiction for the development of ecological resources. We thus establish a long-term and stable ecosystem and the planning strategic, to seek complementary to its surrounding towns, establish a good system, and seek broad partnership to strengthen the security system to ensure sustainable development.

REFERENCES

[1] Peng Kang, Linyu Xu. The urban ecological regulation based on ecological carrying capacity [J]. Procedia Environmental Sciences, 2010, 2.

[2] Feng Yuan, Yanping Peng. The Construction of Ecological Management System of Urban Lakes: A Case Study of Wuhan [J]. Management Science and Engineering, 2014, 82.

[3] Guo Yun, Korea, Feng Na. Ecological industrial park in our country enterprise strategy based on SWOT and AHP analysis [J]. Science and technology progress and countermeasures, 2011, 01:63–67.

[4] Qun Hu wen-yun liu. The SWOT method based on analytic hierarchy process (ahp) to improve the instance analysis [J]. Journal of theory and practice of intelligence, 2009 01:68–71.

[5] Daniel T. Heggem et. al. A Landscape Ecology Assessment of the Tensax River.

[6] Basin [J]. Environmental Monitoring and Assessment. 2000, 64:41–54.

[7] Forman RTT. Some General principles of landscape and regional ecology [J]. Landscape Ecology, 1995, 10 (3):133.

[8] Robert Fabac, Ivan Zver. Applying the modified swot–ahp method to the tourism of gornje međimurje [J]. Tourism and Hospitality Management, 2011, 172.

[9] Mikko Kurttila, Mauno Pesonen, Jyrki Kangas, Miika Kajanus. Utilizing the Analytic Hierarchy Process (AHP) in SWOT analysis—a hybrid method and its application to a forest-certification case [J]. Forest Policy and Economics, 2000, 11.

[10] Yonghong Ye. Tourism planning SWOT strategy analysis method to investigate [J]. Journal of hunan college of finance and economics, 2005, 05:53–55.

[11] Behnaz Aminzadeh, Mahdi Khansefid. A case study of urban ecological networks and a sustainable city: Tehran's metropolitan area [J]. Urban Ecosystems, 2010, 131.

[12] Brian Deal. Ecological urban dynamics: the convergence of spatial modelling and sustainability [J]. Building Research & Information, 2001, 295.

[13] Ali Görener, Kerem Toker, Korkmaz Uluçay. Application of Combined SWOT and AHP: A Case Study for a Manufacturing Firm [J]. Procedia—Social and Behavioral Sciences, 2012, 58.

[14] Mohammad Mehdi Tahernejad, Reza Khalokakaie, Mohammad Ataei. Determining proper strategies for Iran's dimensional stone mines: a SWOT–AHP analysis [J]. Arabian Journal of Geosciences, 2013, 61.

[15] LuMao Hou Juan, Liao Bin, I change me. A SWOT analysis and countermeasures of construction of chengdu garden city—based on ecological civilization perspective [16][A]. Logistics enterprise operational research branch, China. Chinese enterprise operational research [2013] (1) [C], 2013:8.

Material Science and Environmental Engineering – Chen (Ed.)
© 2016 Taylor & Francis Group, London, ISBN 978-1-138-02938-5

Urban road traffic early warning classification in adverse weather conditions

P.F. Gong
Jiangsu Police Institute, Nanjing, China

ABSTRACT: In order to overcome the negative effect on the urban road traffic system effectively, it is necessary to determine the urban road traffic early warning level and take some appropriate emergency management measures when adverse weather occurs in urban area. Based on the traffic impact analysis of urban road traffic system in adverse weather conditions, this paper presents a four-level urban road traffic early warning classification method in accordance with the early warning level of the adverse weather and its temporal-spatial distribution characteristics. The classification method refers to the risk matrix, which is a kind of popular evaluation method in the emergency management field because of its simplicity and adaptability. The present work may be very useful for decision-makers and practitioners trying to determine the urban road traffic early warning levels in adverse weather conditions.

Keywords: urban road traffic; warning classification; adverse weather conditions; temporal-spatial distribution characteristics; risk matrix

1 INTRODUCTION

Adverse weather in urban area, such as rain, sleet, snow, fog, high wind, and icy pavement, affects drivers' capability, vehicle performances, pavement friction, roadway infrastructure, crash risk, traffic flow, and agencies' productivity. In a word, it has a negative impact on the Urban Road Traffic (URT) system. Timely and effective URT early warnings not only allow people to take some actions that adjust their travel time and travel model, and reduce unnecessary travel, but also provide aids for the department of urban traffic management to take appropriate emergency transportation operations earlier. It is advantageous to keep the balance of the URT system.

Some studies on early warning have been performed during the last decades. Burnett (1998) outlined a crisis classification matrix that used a sixteen-cell matrix based on threat level (high versus low), response options (many versus few), time pressure (intense versus minimal), and degree of control (high versus low). The Homeland Security Advisory System, a color-coded terrorism threat advisory scale that consists of five color-coded threat levels, is intended to reflect the probability of a terrorist attack and its potential gravity. In the system, red color means severe risk, orange color means high risk, yellow color means significant risk, blue color means general risk, and green color means low risk. Many early warnings have adopted such color codes. For example, it was proposed to evaluate the forewarning grades of emergency in urban mass transit (Gao et al. 2010; Xu et al. 2010). Early warning is a topic widely studied in a variety of emergency management fields including earthquake (Kubo et al. 2011), meteorological disasters (Paulo et al. 2005; Kim Oanh and Leelasakultum 2011), and transportation field. Nowadays, many practical methods have been adopted and applied for early warning classification including Analytic Hierarchy Process (AHP) (Lee et al. 2008), fuzzy evaluation (Xu et al. 2010), decision tree model (Zhu et al. 2011), dynamic event tree model (Mercurio et al. 2009), risk matrix (Zhang 2012), and other intelligent methods (Tchamova and Dezert 2012).

Existing studies have made significant contributions to the early warning classification of URT, providing various analysis methods for classification. These classification studies mainly considered the attributes of emergency, such as casualties, property losses, affected area, time urgency, and degree of control. However, it may be different early warning levels of URT when the same scale rainstorm takes place at different times and areas. Therefore, in order to reduce the negative impact on the URT system in adverse weather conditions, it is necessary to further study the early warning classification of URT, considering both the temporal and spatial distribution characteristics and the attribute of adverse weather.

In the following section of the paper, based on the analysis of adverse weather's impact on the URT system, this paper presents a risk classification matrix method to determine the early warning level, considering the temporal and spatial distribution characteristics and the early warning levels of various forms of adverse weather.

2 ANALYSIS OF ADVERSE WEATHER'S IMPACT ON THE URT SYSTEM

Usually, adverse weather can lead to short-term variations in traffic demand and traffic supply. There are 3 kinds of traffic demand affected by adverse weather: evacuation traffic demand, rescue traffic demand and normal traffic demand (demand of work, travel, and shopping). Adverse weather can make some parts of the urban road network impassable, such as collapsing urban road and bridges, closures due to maintenance activities or handle emergencies, which can reduce the traffic supply capacity of the URT system. The resulting of adverse weather leads to travel delay as well as other extra costs for the overall URT system.

In adverse weather conditions, disturbances have a primary effect on the traffic demand in the URT system. The normal traffic demand varies from the temporal and spatial distribution. Therefore, it is necessary to consider the temporal and spatial distribution characteristics of adverse weather for the early warning classification of the URT system in adverse weather conditions.

2.1 *Temporal distribution of adverse weather in urban area*

Volume varies considerably over 24 hours in the urban road system. According to the difference in volume, it can be divided into rush hours period, normal traffic volume period and the valley period in a day. Adverse weather, which occurs in different periods, has different impacts on the urban road system.

1. Adverse weather occurring in rush hours period
Rush hours refer to the periods of maximum flow occurring during the morning and evening commuters. Usually, it is the periods from 7:00 to 9:00 and from 17:00 to 19:00. In these periods, the proportion of hour volumes and daily volumes is often more than 8%. The flow is close to saturation during the rush hours periods in many urban road networks. Therefore, the urban road system is vulnerable. If the adverse weather occurs in these periods, it is easy to lead to overcrowding and congestion, even to affect the normal operation of the URT system. Traffic administrative departments should focus on these forms of adverse weather.
2. Adverse weather occurring in normal traffic volume period
In the normal traffic volume period, the proportion of hour volumes and daily volumes is between 3% and 8%. Usually, traffic supply capacity is surplus relative to the traffic demand during the normal traffic volume period in many urban road networks. Some adverse weather may impact the localized area of the URT system; however, the URT system can reach the balance between the traffic supply and traffic demand through its self-organizing capacity or by taking some conventional traffic management measures. While some other serious adverse weather occurs, it is necessary to monitor, warn early and take timely traffic emergency management measures.
3. Adverse weather occurring in the valley period
The valley period mainly refers to the period from 0:00 to 6:00. The proportion of hour volumes and daily volumes is often less than 3%. When the adverse weather occurs between midnight and early morning, there are fewer social activities for people. Therefore, the traffic demand is not high, which is much less than the traffic supply capacity. Generally, URT early warning and traffic emergency measures are not required in this period.

2.2 *Spatial distribution of adverse weather in urban area*

According to the area of influence, the adverse weather can be simply divided into 3 categories:

1. The citywide or almost citywide adverse weather
Generally speaking, adverse weather, such as rain, snow, fog, wind, and freeze, which has a significant impact on the city road traffic system, usually occurs and influences the whole city or most of the city. Basically, such adverse weather can be forecasted timely and accurately, and early warning classification is performed. However, the adverse weather in citywide or almost citywide still has an overall influence on the city road traffic system.
2. The adverse weather in the city center
The city center area is usually intensively economic, political, scientific and technological, cultural, and informational area, which is heavily populated and has densely distributed road networks. As a result, the city center affects the surrounding areas, with the traffic demand pressure. Therefore, the adverse weather occurring in the city center area has a relatively large impact on the city road traffic system, and if it

Table 1. Early warning signal classifications of some adverse weather.

Forms of adverse weather	Early warning signal classifications and their definitions			
	Red	Orange	Yellow	Blue
Rainstorm	Rainfall in 3 hours will reach or has reached more than 100 mm, and the rainfall may continue	Rainfall in 3 hours will reach or has reached more than 50 mm, and the rainfall may continue	Rainfall in 6 hours will reach or has reached more than 50 mm, and the rainfall may continue	Rainfall in12 hours will reach or has reached more than 50 mm, and the rainfall may continue
Snow	Snowfall in 6 hours will reach or has reached more than 15 mm, and the snow continues, likely to have or having had a great influence on traffic or agriculture and animal husbandry	Snowfall in 6 hours will reach or has reached more than 10 mm, and the snow continues, likely to have or having had a great influence on traffic or agriculture and animal husbandry	Snowfall in 12 hours will reach or has reached more than 6 mm, and the snow continues, likely to have or having had an influence on traffic or agriculture and animal husbandry	Snowfall in 12 hours will reach or has reached more than 4 mm, and the snow continues, likely to have or having had an influence on traffic or agriculture and animal husbandry
Fog	The fog in 2 hours may appear or has appeared with visibility being less than 50 m, and it will continue	The fog in 6 hours may appear with visibility being less than 200 m or has appeared with visibility between 50 m and 200 m, and it will continue	The fog in12 hours may appear with visibility being under 500 m or has appeared with visibility between 200 m and 500 m, and it will continue	/
Typhoon	Having been affected or may be affected by tropical cyclones in 6 hours, the average coastal or land wind is up to more than 12, or the gust is up to14, which is likely to continue	Having been affected or may be affected by tropical cyclones in 12 hours, the average coastal or land wind is up to more than 10, or the gust is up to12, which is likely to continue	Having been affected or may be affected by tropical cyclones in 24 hours, the average coastal or land wind is up to more than 8, or the gust is up to10, which is likely to continue	Having been affected or may be affected by tropical cyclones in 24 hours, the average coastal or land wind is up to more than 6, or the gust is up to 8, which is likely to continue
Typhoon	It appears precipitation when the temperature of pavement is less than 0°C. Furthermore, icy pavements have appeared or may appear in 2 hours, which have a great influence on transportation	It appears precipitation when the temperature of pavement is less than 0°C. Furthermore, icy pavements have appeared or may appear in 6 hours, which have a great influence on transportation	It appears precipitation when the temperature of pavement is less than 0°C. Furthermore, icy pavements have appeared or may appear in 12 hours, which have an influence on transportation	/

Source: China Meteorological Administration.http://www.cma.gov.cn/2011xwzx/2011xfzjz/. [2014-09-21].

does not get timely early warning and handling, it tends to impact the stability of the overall city road traffic system.

3. The adverse weather in the city non-center area
Compared with the city center area, the city non-center area generally has relatively small traffic demand, so adverse weather occurring in such areas only has a local impact on the city road traffic system. If the adverse weather is handled improperly, it may also transfer to influence the entire city road traffic system, though.

3 EARLY WARNING CLASSIFICATION OF ADVERSE WEATHER

For adverse weather, meteorological departments have a series of sound early warning signal classification systems. Usually, rainstorm, snow, fog, typhoon, and icy pavement have important effects on the URT system in Chinese urban areas. The early warning signal classifications and their definitions of these forms of adverse weather proposed by the China Meteorological Administration are listed in Table 1.

Traffic administrative departments should strengthen the communication with agencies of meteorology and flood control to grasp early warning information of the adverse weather. In addition, in order to detect and warn earlier the adverse weather, the traffic administrative departments should also actively acquire the adverse weather information of local area through traffic technology monitoring equipment, the road patrol, 122 alarm system and network alarm platform.

4 URT EARLY WARNING IN ADVERSE WEATHER CONDITIONS

In essence, URT early warning in adverse weather conditions is a kind of risk evaluation. Risk evaluation methods are mainly of 3 types: qualitative, quantitative and semi-quantitative. Risk matrix method is a typical semi-quantitative evaluation method. Although the risk matrix method lacks the rigorous mathematical foundation, the principle is simple and easy to understand. Furthermore, it is easy to use and has few restrictions. Therefore, it is widely used, especially popular in practice. This paper presents a risk classification matrix method to determine the early warning level, considering comprehensively the temporal and spatial distribution characteristics and the early warning levels of various forms of adverse weather (Table 2). In Table 2, this paper suggests that it is not necessary to release URT early warning and that no traffic emergency management measures should be taken at level 0. It should release urban road traffic early warning with blue, yellow, orange, and red signal from level 1 to level 4. The higher the level is, the higher the early warning classification is.

As seen from Table 2, the early warning about city road traffic in adverse weather is not entirely consistent with the early warning about the adverse weather itself. Of the 36 cells, 8 cells do not need to release traffic warning, accounting for 22.2% of the total. As to the adverse weather occurring in the rush hours period and having a large-scale impact, the URT early warning level even surpasses the early warning level about the adverse weather itself, which requires the special attention of traffic administrative departments. For example,

Table 2. Early warning classification matrix of the URT system in adverse weather conditions.

Early warning signal classifications of adverse weather	Time of occurrence	City areas of influence of the adverse weather		
		City non-center area	City center area	Citywide and almost citywide
Blue	Valley period	(1) Level 0	(2) Level 0	(3) Level 1
	Normal traffic volume period	(4) Level 0	(5) Level 0	(6) Level 1
	Rush hours period	(7) Level 1	(8) Level 1	(9) Level 2
Yellow	Valley period	(10) Level 0	(11) Level 0	(12) Level 1
	Normal traffic volume period	(13) Level 0	(14) Level 1	(15) Level 2
	Rush hours period	(16) Level 1	(17) Level 2	(18) Level 3
Orange	Valley period	(19) Level 0	(20) Level 1	(21) Level 2
	Normal traffic volume period	(22) Level 1	(23) Level 2	(24) Level 3
	Rush hours period	(25) Level 2	(26) Level 3	(27) Level 4
Red	Valley period	(28) Level 1	(29) Level 2	(30) Level 2
	Normal traffic volume period	(31) Level 2	(32) Level 3	(33) Level 4
	Rush hours period	(34) Level 2	(35) Level 4	(36) Level 4

on July 18, 2011, Nanjing was hit by a rainstorm, which almost lead the Nanjing URT system into paralysis. Early at 12:28, the Nanjing City Meteorological Administration released an orange rainstorm warning (upgraded to the red warning at 16:55), because of the almost citywide influence. According to Table 2, if the rainstorm was forecasted to last until the evening rush hours period, traffic administrative departments in Nanjing City should release the URT red early warning and take corresponding emergency management measures, but if the rainstorm was forecasted not to last long or affect the evening rush hours period, orange early traffic warning should be released and corresponding measures of traffic emergency management should be taken. Obviously, however, traffic administrative departments in Nanjing City were not well prepared for the rainstorm and failed to release the URT early warning timely. Although some emergency measures were taken afterwards, the road traffic congestion was not resolved fundamentally.

5 CONCLUSION

Early warning of city road traffic in adverse weather mainly takes account of the adverse weather's impact on the normal operation of the URT system. Therefore, the URT early warning classification should consider not only the warning level of the adverse weather itself, but also the characteristics of the temporal-spatial distribution of the adverse weather. Synthetically considering the early warning level of the adverse weather as well as the temporal and spatial distribution characteristics of the adverse weather, this paper proposes a four-level (4 colors) early warning system for the URT early warning classification in adverse weather conditions, by referring to the analysis method of the risk matrix. Consequently, by using this method, the traffic administrative departments can pick out the early warning level quickly and conveniently, launch the emergency plans rapidly, and take the appropriate emergency traffic management measures, to minimize the influence of all forms of adverse weather on the URT system.

ACKNOWLEDGMENT

This work was funded by the Priority Academic Program Development (PAPD) of Jiangsu Higher Education Institutions, Science and Technology of Public Security of Jiangsu Police Institute the Key Construction Disciplines at the Provincial Level of Jiangsu Province during 12th Five-Year Plan, and Science Research and Innovation Team of Jiangsu Police Institute.

REFERENCES

[1] Burnett, J.J. (1998). "A strategic approach to managing crises." *Public relations review,* 24:4, 475–488.

[2] Gao, J., Xu, R. and Jiang, Z. (2010). "Combinatorial evaluation for forewarning grade of emergency in urban mass transit." *2010 WASE International Conference on Information Engineering,* vol. 3, 443–447.

[3] Han, Y., Su, G., Yuan, H., and Wang. W. (2010). "A collaborative early warning method of transportation during snow disaster based on fuzzy comprehensive evaluation." *2010 Seventh International Conference on Fuzzy Systems and Knowledge Discovery (FSKD),* Vol. 2, 945–951.

[4] Kim Oanh, N.T., and Leelasakultum, K. (2011). "Analysis of meteorology and emission in haze episode prevalence over mountain-bounded region for early warning." *Science of the Total Environment,* 409:11, 2261–2271.

[5] Kubo, T., Hisada, Y., Murakami, M., Kosuge, F., and Hamano, K. (2011). "Application of an earthquake early warning system and a real-time strong motion monitoring system in emergency response in a high-rise building." Soil Dynamics and Earthquake Engineering, 31:2, 231–239.

[6] Lee, S., Lee, M., Kim, C., Park, C., Park, S., Liu, Y., and Kim, B. (2008). "On determination of early warning grade based on AHP analysis in warranty database." In Advanced Intelligent Computing Theories and Applications. With Aspects of Artificial Intelligence, Springer Berlin Heidelberg, 84–89.

[7] Mercurio, D., Podofillini, L., Zio, E., and Dang, V.N. (2009). "Identification and classification of dynamic event tree scenarios via possibilistic clustering: application to a steam generator tube rupture event." Accident Analysis & Prevention, 41:6, 1180–1191.

[8] Paulo, A.A., Ferreira, E., Coelho, C., and Pereira, L.S. (2005). "Drought class transition analysis through Markov and Loglinear models, an approach to early warning." Agricultural water management, 77:1, 59–81.

[9] Tchamova, A., and Dezert, J. (2012). "Intelligent alarm classification based on DSmT." 2012 6th IEEE International Conference on Intelligent Systems (IS), 120–125.

[10] Xu, R., Gao, J. and Jiang, Z. (2010). "Fuzzy evaluation for forewarning grade of emergency in urban mass transit." 2010 International Conference on Intelligent Computation Technology and Automation (ICICTA), vol. 2, 371–374.

[11] Zhang, H. (2012). "Study on grades of freeway meteorological disasters by risk matrix." Applied Mechanics and Materials, 178, 2788–2792.

[12] Zhu, X., Cao, J., and Dai, Y. (2011). "A decision tree model for meteorological disasters grade evaluation of flood." 2011 Fourth International Joint Conference on Computational Sciences and Optimization (CSO), 916–919.

Material Science and Environmental Engineering – Chen (Ed.)
 ISBN 978-1-138-02938-5

Analysis of the urban heat island effect in Shenyang City

J.L. Wang & Y.T. Ma
School of Transportation Engineering, Shenyang Jianzhu University, Shenyang, Liaoning Province, China

ABSTRACT: To study the effect of the Normalized Difference Vegetation Index (NDVI) and Normalized Difference Building Index (NDBI) on the land surface temperature, as well as the characteristics of the spatial distribution of the heat island in Shenyang City, China, the NDVI and NDBI were extracted based on TM remote sensing data of Shenyang in 2010 and 2014, and then, surface temperature analysis was conducted inversely by the image algorithm (IB). The superposition method of temperature difference and temperature range was proposed to determine the heat island intensity of Shenyang in 2010 and 2014, and the characteristics of the spatial distribution of heat island was derived finally by means of the "normalized difference index". The results show that the temperature in the heat island decreases from its center to perimeter zone; the heat island intensity in Shenyang increases from 0.54°C in 2010 to 1°C in 2014. Thus, there is a close negative linear correlation of NDVI with land surface temperature; otherwise, NDBI has a positive correlation with land surface temperature.

Keywords: urban heat island effect; image (IB) algorithm; vegetation index; building index

1 INTRODUCTION

Urban heat island effect is a special city climate formed by urbanization, which belongs to the typical climates of modern city [1, 2]. Urban heat island effect is a kind of public hazard that causes elevation of temperature and hot air rises. At the same time, cold air in the suburb flows into the urban areas, so that city air pollution becomes more and more serious. Finally, it leads to a vicious cycle. With the acceleration of urbanization, the urban heat island effect becomes more and more intense; the impact on people's life and production is larger than before [3]. At present, studies of urban heat island at home have achieved a great progress by using thermal infrared remote sensing data. Liu Enqin obtained the ground heat field distribution map of Chengdu City. From the map, he obtained favorable results through image processing and inversion of thermal infrared radiation by using the ETM+ remote sensing data [4]. Xu Huixi counted the largest heat island intensity of 12 cities in Chengdu Plain, and analyzed the dynamic change rule of maximum intensity of Chengdu City both in daylight and at night, and then he drew a conclusion that fast dynamic monitoring and macro analysis of the urban heat island effect by using NOAA thermal infrared remote sensing data is feasible through the AVHRR fourth channel [5]. Zhou Hongmei analyzed the distribution and variation of the heat island effect in Shanghai by using NOAA/AVHRR thermal infrared data [6]. Runfeng inverted the surface temperature of Shanghai City through split window algorithm quantitatively by using EOS/MODIS data in 2003 and 2005. He concluded that the surface distribution has obvious seasonal variation characteristics relative to the high temperature distribution, and the heat island effect in summer was more significant [7].

In this paper, the land surface temperature of Shenyang City was inverted by using the TM data, and the situation of the heat island effect in Shenyang City was researched according to the surface temperature. Moreover, the relationship between the land surface temperature of Shenyang City and the Normalized Difference Vegetation Index (NDBI) as well as the Normalized Difference Build-up Index (NDBI) was analyzed for proposing relevant measures to improve the situation of the heat island effect in Shenyang and for providing the basis for future scientific and rational planning of Shenyang City.

2 OVERVIEW OF THE STUDY AREA

Shenyang City is located in the northeast China, which lies in the middle part of Liaoning Province. It is a flat plain located between 122° 25′ 9″ E and 123° 48′ 24″ E, 41° 11′ 51″ N and 43° 2′ 13″ N. Mountainous regions are concentrated in the southeast, which belong to the extension parts of Liao dong mountainous. Liao River and Hun River plain are on the west, the terrain tilts slowly from northeast to southwest. Shenyang City has a warm sub-humid continental climate, and its annual average temperature is about 8.1°C, annual rainfall is about 721.9 mm, hours of sunshine is 2289.2 hours, and frost-free days is about 188 days.

3 RESEARCH METHOD AND PROCESS

3.1 *Extraction of the Normalized Difference Vegetation Index (NDVI)*

NDVI (Normalized Differential Vegetation Index) is a remote sensing index that reflects the situation of vegetation cover, which is used to detect the state of vegetation growth as well as vegetation coverage. It is defined as the difference between the near-infrared channel and visible light reflectivity divided by their sum:

$$NDVI = \frac{NIR - R}{NIR + R} \tag{1}$$

where NIR is the reflectance of the near-infrared band and R is the reflectance of the red band.

3.2 *Extraction of the Normalized Difference Build-up Index (NDBI)*

NDBI is put forward by Zhayong, which reflects the construction information more accurately. The higher value of the building land ratio means the higher building density:

$$NDBI = \frac{MIR - NIR}{MIR + NIR} \tag{2}$$

where NIR is the reflectance of the near-infrared band and MIR is the reflectance of the mid-infrared band.

The DN value of urban land between the TM4 band and the TM5 band was big; the other DN values were small. This paper adopted the TM4 band and the TM5 band to extract the Normalized Difference Build-up Index according to the above formula (2).

3.3 *Inversion process of land surface temperature*

3.3.1 *Inversion method of land surface temperature*

Inversion of land surface temperature was based on the image (IB) algorithm. The principle was to convert the DN values of the thermal infrared band into radiation values, and then inverted the radiation values into ground bright temperatures. Land surface temperatures were obtained after the correction of land surface emissivity. Image (IB) algorithm had the following advantages: 1) the inversion process is simple to operate; 2) the dependence of external parameters is less; and 3) land surface emissivity effects are under consideration. The formula is given as follows:

$$T_s = \frac{T_{sensor}}{1 + (\lambda \cdot T_{sensor}/\rho) \ln \varepsilon} \tag{3}$$

$$\rho = h \cdot c/\sigma \tag{4}$$

where T_s is the surface temperature; ε is the land surface emissivity; T_{sensor} is the bright temperature value; λ is the center wavelength of the thermal infrared band; h is Planck's constant; C is the beam; σ is Boltzmann's constant, with $\sigma = 1.38 \times 10^{-23} (J/K)$.

3.3.2 *Estimation of required parameters*

1. Estimation of brightness temperature
   Brightness temperature is given as follows:

$$T_{sensor} = \frac{K_2}{\ln(K_1/L_\lambda + 1)} \tag{5}$$

$$L_\lambda = \frac{L_{max} - L_{min}}{255} \times DN + L_{min} \tag{6}$$

   where L_{max} is the highest radiation value and L_{min} is the lowest radiation value that can be detected. The values can be found in the header files. K1 and K2 are scaling constants.
2. Estimation of land surface emissivity
   Sobrino [8], Vande Gfiend & Owe [9] put forward the empirical formula according to the NDVI value for the calculation of land surface emissivity. Tan Zhihao [10] divided surface cover into water bodies, buildings and natural surfaces (except water and outside buildings). Then, he put forward the following formula of surface radiation:

$$\varepsilon_{water} = 0.995 \tag{7}$$

$$\varepsilon_{surface} = 0.9625 + 0.0614 F_V - 0.0461 F_V^2 \tag{8}$$

$$\varepsilon_{building} = 0.9589 + 0.086 F_V - 0.0671 F_V^2 \tag{9}$$

   where $\varepsilon_{surface}$ is the emissivity of natural surface pixels; $\varepsilon_{building}$ is the emissivity of town pixels; and F_V is the vegetation coverage. The vegetation coverage can be calculated by the following formula:

$$F_V = \frac{NDVI - NDVI_s}{NDVI_V - NDVI_s} \tag{10}$$

where $NDVI$ is the Normalized Difference Vegetation Index, and $NDVI_V$ and $NDVI_s$ are set equal to 0.05. When the $NDVI$ is more than 0.7, then $F_V = 1$. When the $NDVI$ is less than 0.05, then $F_V = 0$.

4 ANALYSIS OF THE HEAT ISLAND EFFECT IN SHENYANG CITY

4.1 *Determination of heat island intensity in Shenyang City*

4.1.1 *Measurement method of heat island intensity*

Urban heat island intensity has been expressed by the temperature difference between the high temperature in urban area and the low temperature in suburb in the former studies. However, the influence area of different temperature level is an important factor. In this paper, the temperature difference and the influence area were taken into consideration to calculate the heat island intensity in Shenyang City by using the following formula:

$$I_{UHI} = \sum_{i=1}^{n}(T_i - T_0) * P_i \quad (11)$$

where I_{UHI} is the heat island intensity (unit: °C); T_i is the average temperature in different temperature zones (unit °C); T_0 is the average temperature in suburb (unit °C); P_i is the area ratio of different levels in the temperature zone; i is the i-th temperature area; and n is the number of temperature zones.

Table 1. Results of land surface temperature grade classification.

Temperature grade	2010-9-29	2014-9-8
High temperature zone	$T_s > 22$	$T_s > 24$
Secondary high temperature	$20 < T_s \leq 22$	$22 < T_s \leq 24$
Medium temperature	$18 < T_s \leq 20$	$20 < T_s \leq 22$
Secondary medium temperature	$16 < T_s \leq 18$	$18 < T_s \leq 20$
Low temperature	$T_s \leq 16$	$T_s \leq 18$

4.1.2 *Division of temperature class*

Division of temperature zone has no standard method. The current method is inversion by using Landsat TM images. The two commonly used methods of surface temperature grade classification of the urban heat island effect are as follows: 1) equal interval classification and 2) mean-standard deviation. In this paper, according to the results of land surface temperature inversion, the average surface temperatures in the suburb of Shenyang City in 2010 and 2014 were found to be 18°C and 20°C. Land surface temperature was divided into equidistance based on the average temperature in suburb in 2010 and 2014, the results of which are summarized in Table 1.

4.1.3 *Measurement results of heat island intensity*

According to the range of temperature rating, the results of the classification of temperature ratings were counted, which were used to calculate the average temperature T_i of different grades. The area ratio (P_i) of different grades in the temperature zone was the quotient between the different grade temperature zone numbers and the total number of pixels. Finally, according to the formula of heat island intensity, the heat island intensity numerical values in September 2010 and September 2014 in Shenyang City were found to be 0.54°C and 1°C (Tables 2 and 3). Therefore, the heat island effect in Shenyang City was enhanced, as reflected from the calculated results.

4.2 *Characteristics of the heat island effect in Shenyang City*

4.2.1 *Characteristics of spatial distribution*

Distribution maps of the heat island effect were generated based on the above tables. According to the maps, a conclusion that there was an obvious urban heat island effect in Shenyang City was obtained. Spatial characteristics of the enhanced urban heat island effect in Shenyang City were expanded outwards from the center. The main areas include the following: Huanggu District,

Table 2. Results of urban heat island intensity measurement in September 2010.

Temperature grade	Temperature range	T_i/°C	P_i/%	ΔT/°C	ΔT P_i/°C	T_0/C	IUHI/°C
High temperature	$T_s > 22$	23.5	9.36	5.76	0.51	18	0.54
Secondary high temperature	$20 < T_s \leq 22$	20.9	12.05	3.23	0.42		
Medium temperature	$18 < T_s \leq 20$	19.2	19.54	1.62	0.38		
Secondary medium temperature	$16 < T_s \leq 18$	17.5	45.36	–0.54	–0.40		
Low temperature	$T_s \leq 16$	15.4	13.69	–2.7	–0.38		

Table 3. Results of urban heat island intensity measurement in September 2014.

Temperature grade	Temperature range	T_i/°C	P_i/%	ΔT/°C	ΔT P_i/°C	T_0/°C	IUHI/°C
High temperature	$T_s > 24$	25.4	7.02	5.52	0.42	20	1
Secondary high temperature	$22 < T_s \le 24$	23.2	15.74	3.16	0.47		
Medium temperature	$20 < T_s \le 22$	21.3	33.03	1.43	0.50		
Secondary medium temperature	$18 < T_s \le 20$	19.5	36.04	–0.39	–0.20		
Low temperature	$T_s \le 18$	17.6	8.17	–2.19	–0.21		

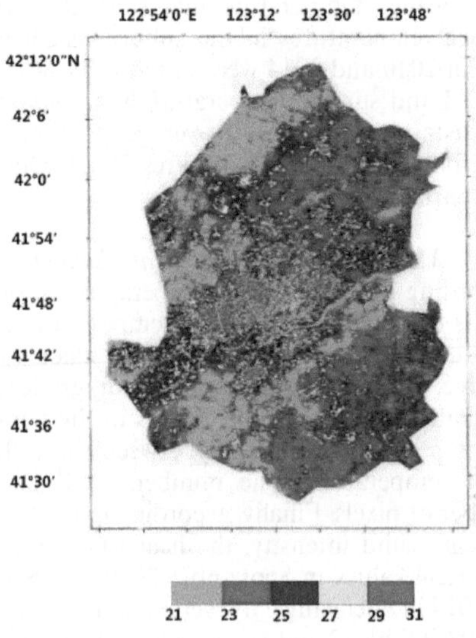

Figure 1. Distribution map of the land surface temperature grade classification on September 29, 2010.

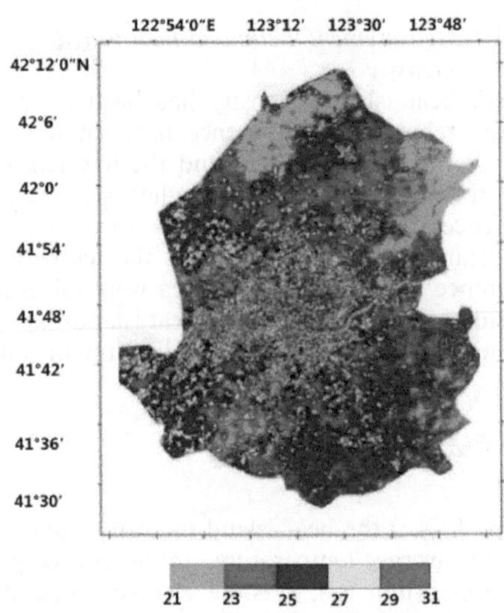

Figure 2. Distribution map of the land surface temperature grade classification on September 8, 2014.

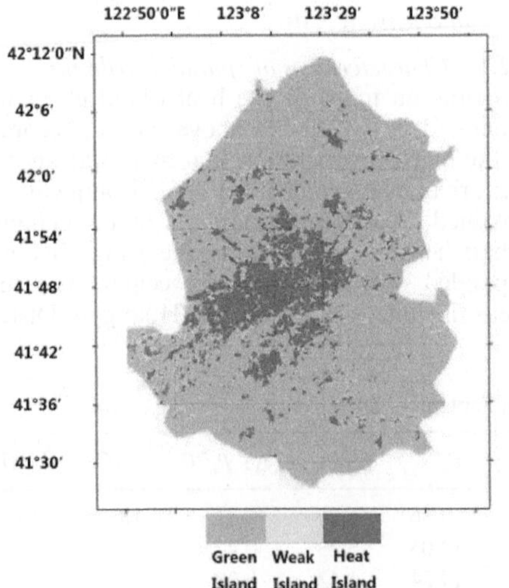

Figure 3. Distribution map of the urban heat island effect in 2010.

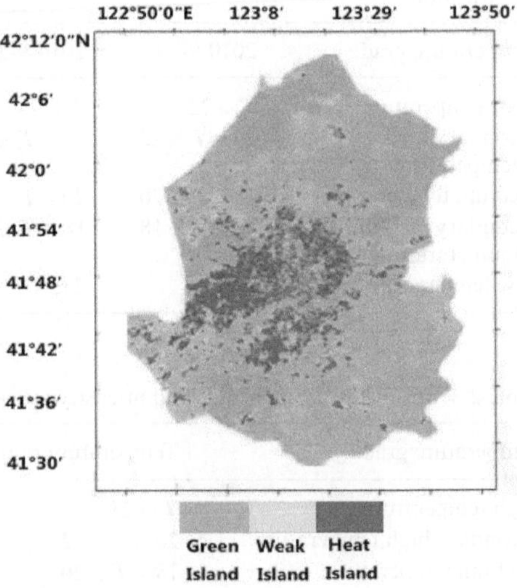

Figure 4. Distribution map of the urban heat island effect in 2014.

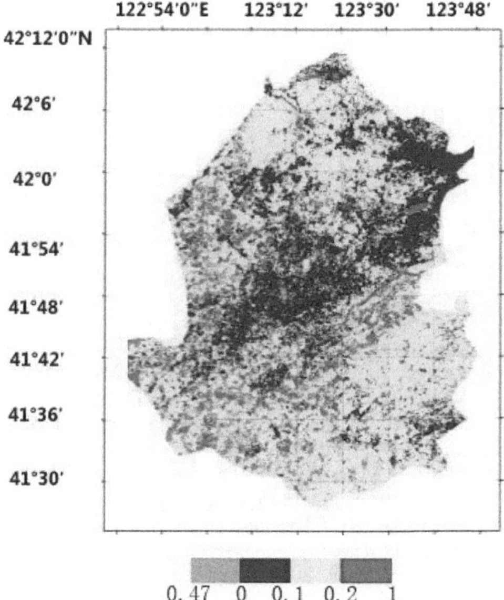

Figure 5. Difference image map of the land surface temperature in 2010 and 2014.

Tiexi District, Heping District, Shenhe District, Dadong District, Dongling District, Sujiatun District, Yuhong District, and Xinchengzi District.

4.2.2 *Characteristics of time variation*

In order to reflect the changing features of the heat island effect more reasonably, the proportion index method was used to invert the land surface temperature according to the following formula, in which values should be in the range between 0°C and 1°C:

$$H_i = \frac{T_{si} - T_{s\min}}{T_{s\max} - T_{s\min}} \tag{12}$$

where H_i is the ratio index of the urban heat island intensity; T_{si} is the surface temperature value of the i-th pixel; $T_{s\max}$ is the maximum surface temperature and $T_{s\min}$ is the minimum surface temperature.

The difference image was generated by the processed land surface temperature in 2014 minus the temperature in 2010. Green (negative value) in the map meant decrease, blue as well as yellow and red (positive value) meant increase. According to the difference image, the regions where the land surface temperature enhanced were significantly larger than the regions where the land surface temperature weakened. From the perspective of the spatial distribution changes, in the downtown areas within the Second Ring Road, increasing land surface temperatures were not very significant, and in some areas, the land surface temperatures weakened. Outside the Second Ring Road, such as east Qipanshan and areas around the Expo garden, increasing land surface temperatures were not very obvious, because the regions were full of mountains and forests. So, the degree of vegetation cover was high. Land surface temperatures of the south and the northern regions as well as part of the western regions were obviously enhanced, and those of the south regions such as Wangjia Town, Guchengzi Town, Baitabao Town, Shenjingzi Town, Taoxian Town, Zhangyi Town, Dapan Town, Yongle Town and other regions enhanced significantly. Land surface temperature enhancement in the Shenbei New Area was more obvious. In the western towns such as Majia Town, Zaohua Town and Daxing Korean rural area, the land surface temperatures enhanced prominently. The above changes in the land surface temperature were closely related to the rapid expansion area of Shenyang construction. From 2010 to 2014, the rapid development of the Third Ring Road areas of Shenyang led to the development of surrounding towns, the increase in construction area and the changes underlying surface, which caused the surface temperature to be enhanced obviously.

4.3 *Relationship between the land surface temperature and the Normalized Difference Vegetation Index (NDVI)*

Random sample points were taken separately from land surface temperature map in 2010 and 2014 as well as the Normalized Difference Vegetation Index (NDVI) map. Sample values of the surface temperature and NDVI values were read from coordinates. Then, analysis of linear regression was conducted.

According to the analysis of linear regression, the formula of linear regression between the land surface temperature and the NDVI in 2010 is as follows: $y = -17.228x + 31.655$, where the square value of (R^2) is 1 and the significant level value (SIG) is 0. The formula of linear regression between the land surface temperature and the NDVI in 2014 is as follows: $y = -11.465x + 21.238$, where the square value of (R^2) is 0.997 and the significant level value (SIG) is 0. Summarizing the two years of the analysis results of the land surface temperature and the Normalized Difference Vegetation Index (NDVI), we can see that the land surface temperature had an obvious negative correlation relationship with the NDVI. As the Normalized Difference Vegetation Index (NDVI) increased, the land surface temperature decreased; while the Normalized Difference Vegetation Index (NDVI) decreased, the land surface temperature increased.

4.4 *Relationship between the land surface temperature and the Normalized Difference Build-up Index (NDBI)*

In order to analyze the relationship between the land surface temperature and the Normalized Difference Build-up Index (NDBI), sample points were took from the NDBI diagram and the land surface temperature in 2010 and 2014, and then a linear analysis of NDBI values and land surface temperature was carried out. According to the analysis results of linear regression, the formula of linear regression between the land surface temperature and the NDBI in 2010 is as follows: $y = 11.127x + 22.984$, where the square value of (R^2) is 1 and the significant level value (SIG) is 0. The formula of linear regression between the land surface temperature and the NDBI in 2014 is as follows: $y = 13.058x + 17.149$, where the square value of (R^2) is 1 and the significant level value (SIG) is 0. Summarizing the two years of the analysis results of the land surface temperature and the Normalized Difference Build-up Index (NDBI), we can see that the land surface temperature had an obvious positive correlation relationship with the NDBI. As the Normalized Difference Build-up Index (NDBI) increased, the surface temperature also increased.

5 CONCLUSIONS

Distribution characteristics of the urban heat island effect in Shenyang and the relationship between the land surface temperature and the NDVI as well as NDBI were analyzed based on TM remote-sensing images in this paper. On the basis of the analysis, the following conclusions were drawn. (1) the intensities of the urban heat island effect of Shenyang City in 2010 and 2014 were 0.54°C and 1°C. (2) Urban heat island effect in Shenyang was enhanced and the greatly enhanced areas mainly concentrated in the southwestern, northwestern and southern regions. (3) Normalized Difference Vegetation Index (NDVI) had a close negative linear relationship with land surface temperature. Normalized Difference Build-up Index (NDBI) had a positive relationship with land surface temperature.

REFERENCES

[1] Morris, C.J.G. et al. 2001 Quantification of the Influences of Wind and Cloud on the Nocturnal Urban Heat Island of a Large City[J]. *Journal of Applied Meteorology articles* 40(2):169–182.

[2] Peng, S.L. & Ye, Y.H. 2007. The Influence of Urban Heat Island on Urban Planning[N] *ACTA Scientiarum Naturalium Universitatis Sunyatseni* (in Chinese) 5(46):59–63.

[3] Rajagopalan, P. & Wong, N.H. 2008. Microclimatic Modeling of the Urban Thermal Environment of Singapore to Mitigate Urban Heat Island[J]. *Solar Energy* 8(28):727–745.

[4] Zhou, H.M. et al. 2001 The Surveying on Thermal Distribution in Urban Based on GIS and Remote Sensing[J]. *ACTA Geographical Sinica* (in Chinese) 56(2):189–197.

[5] Xu, H.X. et al. 2007 Remote Sensing Analysis of Urban Heat Island Effect in Chengdu Plain[J]. *Environmental Science and Technology* (in Chinese) 30(8):21–23.

[6] Liu, E.Q. & Chen, N. 2009. Study of Heat Island Effect Based on Remote Sensing—By the Example of the Chengdu Area[N]. *ACTA Geologica Sichuan* (in Chinese) 29(4):484–487.

[7] Men, F. et al. 2007 on Urban Heat Island of Shanghai City from Modis Data[J]. *Geomatics and Information Science of Wunan University* (in Chinese) 32(7):576–580.

[8] Bai, J. et al. 2008 Inversion and Verification of Land Surface Temperature with Remote Sensing TM/ETM+ Data. *Transactions Of The Chinese Society Of Agricultural Engineering* (in Chinese) 24(9):148–154.

[9] Van de Griend, A.A. & Owe, M. 1993 On the Relationship between Thermal Emissivity and the Normalized Difference Vegetation Index for Natural Surfaces[J]. *International Journal of Remote Sensing* 14(06):1119–1131.

[10] Tan, Z.C. et al. 2004 The Estimation of Land Surface Emissivity For Landsat TM6[J]. *Remote Sensing For Land & Resources* (in Chinese) 16(3):28–32, 36, 41.

[11] Sobrino, J. et al. 2001 A Comparative Study of Land Surface Emissivity Retrieval from NOAA Data [J]. *Remote Sensing of Environment* 75(2):256–266.

Material Science and Environmental Engineering – Chen (Ed.)
 ISBN 978-1-138-02938-5

The sandstorms and environmental governance of Kaifeng City in the Northern Song Dynasty

H.Q. Li & L.X. Si
College of Civil Engineering and Architecture, Henan University, Henan, China

ABSTRACT: Kaifeng City, surrounded by sand, was greatly endangered by sandstorms especially in the spring because the northwest wind blew throughout the year in the Northern Song dynasty. For this purpose, some measures were taken to improve the environment, such as planting trees, sprinkling water on the streets, greening streets, cleaning ditches, and excavating lakes. These measures can serve as a reference for present-day urban environmental regulation.

Keywords: sandstorms; environmental governance; street greening

1 INTRODUCTION

Today's Kaifeng City, as the capital of the Northern Song Dynasty, lasted for as long as 167 years. There are plains around the city and no mountains. Kaifeng City, surrounded by sand, was greatly endangered by sandstorms especially in the spring because the northwest wind blew throughout the year. In order to deal with the environmental problems, the government took various measures, such as planting trees, sprinkling water on the streets, greening streets, cleaning ditches, and excavating lakes.

2 THE SANDSTORMS BLOWING IN THE KAIFENG CITY OF THE NORTHERN SONG DYNASTY

Many people of the Song Dynasty recorded the sandstorms in Kaifeng City. For example, Zhouhui, as a scholar, stated that officials wore thin coats on the horses to prevent sandstorms in Kaifeng City. While walking on the streets, women always covered their upper bodies with purple transparent silk called head cloth. And when eating a steamed cake or bun, one must get rid of the sand-covered surface. It is said that when a vehicle was passing, some people carried jars to sprinkle water onto the street in order to prevent the dust floating in the air [1]. A poem written in the Northern Song dynasty described this situation: the Kaifeng City was full of dust day and night where all kinds of carts ran. What a pity, all the trees on both sides of the streets lost their bright green colors covered by dust. Another poem by Wang'anshi also states that when sunlight was shining, the whole city became yellow because of the blown sand. What's more, the Bianhe River flowed through the Kaifeng City. Every few years, the mud at the river bottom must be dug out and be piled on both banks. Finally, the sand banks rose to the height of eaves. It can be imagined that the near air would be full of sand and dust during heavy wind. In the Riverside Scene at Qingming Festival, in a picture about the Kaifeng City of the Northern Song Dynasty, a rider wore a hat from which a thin silk cloth was hung down for protecting the face from the sand (Fig. 1).

Figure 1. A rider in the Riverside Scene at Qingming Festival.

And at the end of the Northern Song Dynasty, while attacking Kaifeng City, the Jin soldiers failed just because of sandstorm blowing against them. A few days later, the Jin soldiers avoided the wind direction, but still were defeated just because the wind blew against them again. Was it God's will? [2]. Similar incidents have been recorded in historical documents.

3 STREET GREENING AND MANAGING OF THE KAIFENG CITY OF THE NORTHERN SONG DYNASTY

To resist the sandstorms and improve the environment, the government widely planted trees along the streets of Kaifeng City. In all streets and lanes, poplars and willows were planted, casting much shade on the streets, especially during the spring and summer. According to a historical document, the north-south royal street, the only axial line of Kaifeng City, had both ditches in which many lotuses were planted. And near the banks of the ditches, many kinds of trees were grown, such as peaches, pears, apricots and plums, with their flowers blooming luxuriantly just like colorful silk during the spring and summer [3]. According to the statistics, there were over 170 trees in the Riverside Scene at Qingming Festival, locating at both sides of the streets, near houses, and on the river banks. These trees were just in their budding stage in the spring (Fig. 2). In addition to greening, the streets were required to be kept clean. Particle pollution was forbidden to be dumped into the streets through the wall holes; otherwise, the violator will be hit 60 times with a stick. If an official on the duty did not stop the action, he would undergo the same punishment [4].

Figure 2. The trees in the Riverside Scene at Qingming Festival.

What's more, it was also the important content of environmental governance for the government to clean up the rivers in Kaifeng City of the Northern Song Dynasty. There were four main rivers flowing through the city. Besides, many ditches in the city played an important role in the excretion of domestic sewage and rainwater. According to a piece of document, the ditches were so deep and wide that many desperadoes hid there and regarded these places as "safe caves", and even hid women there [5]. These ditches were often blocked due to littering, resulting in poor drainage, giving off an unbearable stink. Even flooding disaster resulted during heavy rainfall. Therefore, in the spring, the government often organized manpower to clean the ditch mud and rubbish. And in order to pile these things, some pits were often dug called the "mud basin", and covered after checking [3]. At the same time, the government actively dug wells to solve the drinking-water problem of the dwellers. At the early time of the Northern Song Dynasty, the wells were very few because of which some people died from thirst in dry seasons. In 1046, another 390 wells were dug far away from official wells, basically satisfying the need of drinking water. There was a well that appeared in the Riverside Scene at Qingming Festival (Fig. 3).

Figure 3. A well in the Riverside Scene at Qingming Festival.

Preventing occupying streets was also one of the environmental management programs of Kaifeng City. At the early time of the Northern Song Dynasty, with more population and prosperous commerce, the dwellers vied with one another in building their houses on the streets. These caused the streets narrower and narrower so that sometimes it was difficult for the carts to pass through. For this reason, the government repeatedly demolished these illegal buildings. Later, on both sides of the streets, each wooden pillar was erected at regular distances as a symbol, forming both boundary lines of the roads. And all the houses inside the pillars were demolished.

4 ROYAL AND PRIVATE GARDENS IN KAIFENG CITY OF THE NORTHERN SONG DYNASTY

The landscaping of Kaifeng City was commendable in the Northern Song Dynasty. Besides extensively planting grasses outside the city, numerous royal and private gardens were located inside and outside the city. It was recorded that the gardens filled the urban outside, with no empty land within a hundred Li (the unit of length in ancient China) [3]. These played a significant role in preventing sandstorms and beautifying the environment. As far as the royal gardens were concerned, there were mainly four famous ones located outside the Kaifeng City, namely Jinmingchi and Chonglinyuan, Yichunyuan, Yujinyuan, Ruishengyuan, as well as the Genyue mountain inside the city.

Jinmingchi, located in the west of the city, was really a pool with a perimeter of nine Li thirteen Bu. Its northern bank was connected to the Bianhe River, providing water. On the eastern and western banks, poplars and willows grew widely, which were visited by tourists during the spring. On the pool lay a central island on which five halls were situated. During each spring, water performances were held in the pool, with the emperor sitting in the hall watching them. During this time, it was open to ordinary dwellers too. This intense scene was displayed in one picture named "championship contests on the Jinming Pool" painted by Zhangzeduan of the Northern Song Dynasty (Fig. 4). On the opposite side of Jinmingchi was Chonglinyuan, mainly including trees, flowers and halls.

Yujinyuan was located to the south of Kaifeng City, which was flowed through by the Huiminhe River. In the garden, there were several squares or round pools in which some lotuses were planted. And willows were around these pool banks. Yichunyuan was located to the east of Kaifeng City, with the characteristics of widely planting flowers and plants.

Figure 4. The championship contest on the Jinming Pool.

Genyue, located to the east of the imperial city, had a perimeter of over nine Li. It was built by piling stones, including all kinds of artificial hills, with the highest one up to ninety Bu (the unit of length of ancient China, where 1 Bu is about 1.5 meters). There were many trees, grass, flowers and waterfalls in the garden. As one largest imperial garden, it was built in the Zhenghe fifth (1115 AD), designed personally by Huizong emperor, who was also the author of "Genyue Ji", introducing its layout and panorama. The water of Genyue came from the Jinglongjiang, entering the northwest corner of the park, and then became a pool where one small island was located. After passing through the pool, the stream flowed southwest. In front of the Wansui mountain and the Wansong ridge, the stream was divided into two parts: one pouring into the Fengchi pool after surrounding the Wansong ridge; the other into one larger pool where two islands, named Luzhu and Meizhu, were located at south and north, respectively. Both the streams flowed into the Yanchi pool: the largest one in the Genyue Mountain and then went out from the southeast corner. This was its complete water system (Fig. 5).

Finally, there was also a pool named Yingxiangchi located inside the outer city wall of Kaifeng City. To the south of it was one temple called Wuyueguan. The pool was full of lotuses with many *Salix babylonica* around its banks, and fish swam in the pool. People visited it throughout the day during the Qingming Festival period.

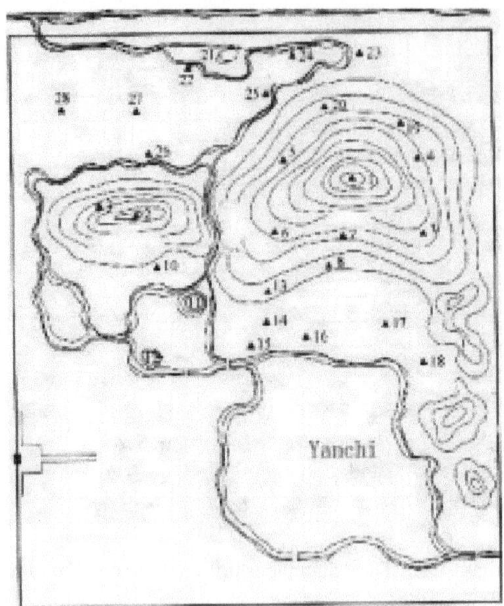

Figure 5. Genyue layout.

5 SUMMARY

The windstorms brought great harm and inconvenience to people's daily life in Kaifeng City of the Northern Song dynasty. For this purpose, the government carried out much environmental management programs, such as planting trees, cleaning ditches and rivers, building gardens, digging pools and piling hills. These measures significantly improved the surrounding residents at that time, and could provide some reference for today's urban construction. Because of numerous rivers and pools, Kaifeng City is known as the northern water city.

ACKNOWLEDGMENT

This article is part of the Preliminary result of humanities and social science project of the Ministry of Education in China named "The research on the spatial distribution and its evolution of the Dongjing Kaifeng city of the Northern Song dynasty" (14YJA780001).

REFERENCES

[1] Zhouhui. Qingbozazhi, Vol, 2, Zhonghua publishing house, (1985, P41).

[2] Shiyanzhi. Beichuangzhiguo, Vol. 2, the commercial press, 1934, P20.

[3] Mengyuanlao. A book about the Kaifeng city of the Northern Song Dynasty, Zhonghua publishing house, (1992, P52); p123; 184.

[4] Douyi. Songxingtong, Vol.26, Zhonghua publishing house, (1984, P417).

[5] Luyou. Laoxueanbiji, Vol.6 p123, Qingdao publishing house, (2002, P123).

Material Science and Environmental Engineering – Chen (Ed.)
ISBN 978-1-138-02938-5

Study on the factors of the coastal city impact on economic developing of development zone

Z.X. Sun
Transport and Management College, Dalian Maritime University, Dalian, Liaoning, China

T.P. Fu
History and Culture College, Tianjin Normal University, Tianjin, China

ABSTRACT: It is known that cities have made great support for the growth of its development zone, while there is still a question that which factors of the city provide these contributions to the economic developing of development zones. This paper used the factor analysis method and multiple regression analysis method to explore the factors of the city impact on development zones, by studying 17 cities and development zones as samples. The analysis revealed that the city's investment and population is the greatest influential factor, and the second factor is the urbanization and urban supportability factor. The urban transport factor is affecting the level of outreach convenience of development zones. The industrial structure factor is a relatively small influential factor, since the development zones have been basically matured and formed a relatively complete industrial architecture. According to the results, it is recommended that when planning the growth of development zones, it should fully consider all the factors' condition of the city to avoid the waste of land resources.

Keywords: city; development zone; economic developing; impact

1 INTRODUCTION

Development zone is one of the indispensable forces and important engine for China's economic development, and it has also made important contributions to the regional economy, in the year 2013, where the GDP of national development zones accounted for 12% of the country's total economic output, tax accounted for about 10% of revenue, the value of industrial output accounted for 22.8% of the country, and the proportion of tertiary industry accounted for 6.1%. To promote local economic development, many cities joined the trend of setting up its own development zones in 1984 to 2014; however, within 30 years, the number of national development zones turned over 30 times, increasing from 14 to 424. In addition, the differences in the speed and quality of development showed that there are big gaps between the development zones, i.e. some of them are booming and become regional growth pole. Taking Tianjin and Dalian as an example, TEDA and Dalian Economic & Technological Development Zone maintain their average annual growth rate at 20% and 10%, with their proportion of total economic output accounting for the city being more than 17%. Still, some of them lack of driving force for development and difficult to break through obstacles, such as Zhangzhou China Merchants Economic and Technological Development Zone (hereinafter referred to as CMZD). CMZD was founded in 1992; after 20 years of development, its economic aggregate reached only up to 3.5 billion yuan, exhibiting slow growth, since the growth of development zone was deeply affected by the city's comprehensive strength. So far, there have been fifty tree coastal cities that contain more than 400 development zones, the average scale of which has expanded to 49 square kilometers, and this kind of development tare converted into problems for the local governments. Since if we neglect the city's comprehensive strength and influence, blind-planning, while reclamation areas and new urban areas and other projects continue to be launched, it will inevitably lead to a waste of resources. This paper focuses on the factors that coastal cities have impacted on the growth of development zones, providing a scientific basis for understanding and rational planning for sustainable growth of development zones.

There are many factors, both external and internal, that will impact on the growth of development zone. Zhang Xiaoping (2002) argued that China's development zones were authorized by policies of government, affected by the production and operation of Multi-National Corporation, relyingied on market forces for promoting the economy, and had the ability of learning and innovation enhanced for competitiveness, so does the ability of social-cultural for strengthening the regional integration and development. Explored from the inside, Bai Jingfeng (2002) pointed out that both high-tech workers and research funds are the two key factors affecting the economic performance of development zones. Moreover, Zhang Yan (2008) saw the developing as the result of a composite effect, including external force, natural endowment and local government policy. Li Guiqing (2012) argued that the development zones were under the combined effect of various factors, both external and internal, such as location, demand, policy or innovation, and industrial cluster, on the basis of the comprehensive study of the Nanjing high-tech development zone. Combining the force study of development zones with the study of its stages, Wang Jiating (2013) found that the factors of force for the growth of development zones were on the dynamic evolution process, which begin with the external force of policy, through the external force of attraction, which then turn into the internal type, and finally attain the force of innovation. Generally speaking, previous research on internal and external factors has been comprehensive, but has only used policies in expressing the city's influences on the development zone, which was not sufficient, nor did it show the importance of the city. Comprehensive strength of the city is the guarantee and supporter for development zones, which includes economic power, industrial structure, social culture, social supportability and education level.

Because of setting up earlier, costal development zones are relatively mature at present, and there is an obvious trend that they are keen on new reclamation construction. With its rich data for studying costal city and its development zones and exploring the relationship between factors related to the city and the economy of development zones, the research is of more practical significance. Therefore, this paper discussed the city's impact on development zones by using the method of factor analysis for providing a reference to the government for reasonable and scientific planning and predicting the growth of development zones.

2 ANALYSIS OF INFLUENCING FACTORS

2.1 *Indicators*

The city's comprehensive strength is an important indicator of urban development at all levels.

Table 1. Indicators of city's comprehensive strength system.

Primary indicator	Secondary indicators
Economic development indicators	GDP (x_1); Investment in fixed assets (x_2); Proportion of secondary industry (x_3); Proportion of third industry (x_4); Quantity of shipments per capita (x_5);
Social development indicators	Population (x_6); Proportion of non-agricultural population (x_7); Employment (x_8); Number of buses in every ten thousand people (x_9); Road area per capita (x_{10}); Pension insurance coverage (x_{11});
Science, education and health indicators	Educational expenditures (x_{12}); Technology expenditures (x_{13}); Students in colleges and universities (x_{14}); Number of doctors in every ten thousand people (x_{15}).

2.2 *Research data and sources*

China has 53 coastal cities, with 42 national economic and technological development zones; however, in fact, there are different types in each city, which lead to statistical difference in the data. This is why we chose National Economic and Technological Development Zones and the cities (a total of 31 cities) as samples, for ensuring the consistency of the data as possible as we can.

The economic growth of development zones can be reflected by regional GDP, Real GDP per Capita, Patient-Centered Data Input and other indicators. However, due to the lack of relevant statistical data and the incomplete of demographic data or other restrictions and limitations, this paper consequently selects Gross regional Domestic Product (GDP) as the variable for the study, for which the data is relatively intact and capable of expressing the level of economic growth of development zones. There are many cities that have more than one development zone, but generally these development zones have over 20 years of development and well developed, so some method is needed to process the GDP. City-related indicators data comes from the 'China City Statistical Yearbook' 2013, and the provinces and municipalities statistical yearbooks.

2.3 *Factor analysis*

To study and construct so many urban indicators that have a certain intrinsic link with each other, this paper aims at fully reflecting the differences in characteristics between cities. Meanwhile, it also increases the difficulty and complexity of the analysis, and inevitably there will be concern over multicollinearity between indicators simultaneously. To find out the overall characteristics of the city data from a number of indicators in the data and to apply indicators independent from each other, this paper uses the method of factor analysis which is suitable for multivariate analysis.

There is a 31×15 data matrix constructed in this paper, by taking China's 31 coastal cities and their National Economic and Technological Development Zones as the research object and statistical data of the year 2012 as the base. With the purpose of discovering the influence factors of mother city impact on the economic growth of development zones, this paper extracted representative factors from the original indicator system by mathematical operation.

The data is analyzed by the KMO test and the result is 0.669, which means there is no big difference in the correlation between each indicator, so the factor analysis can be performed. After calculation, in line with the principle that the eigenvalue is greater than 1.0, we extract four common factors (see Table 2), and the accumulated variance contribution rate is 87.418%, which may fully reflect the original indicators that represent characteristic information of the city.

The correlative degree between common factors and original indicators is revealed by the factor loading matrix: the higher the factor loading matrix figure is, the more information the common factor contained. After varimax rotation, we get the factor loading matrix (see Table 2). The five factors can completely reflect the information of all the indicators, and are independent of each other and represent a class of factors. The first common factor (F1) includes educational expenditures, GDP, technology expenditures, employment, investment in fixed assets and students in colleges and universities. Given the indicator content of the urban investment ability and the quality of population and population size, the first common factor can be summarized as investment and population factor. The second common factor (F2) includes number of doctors in every ten thousand people, number of buses in every ten thousand people, pension insurance coverage, and proportion of non-agricultural population, which can be summarized as urbanization and urban supportability factors. There are two indicators in the third common factor (F3), proportion of third industry and proportion of secondary industry, named the industrial structure factor. And the last one (F4) can be summarized as transportation factor, including quantity of shipments per capita and road area per capita.

Table 2. Eigenvalue and cumulative contribution rate of common factors.

Common factor	Eigenvalue	Contribution rate (%)	Cumulative contribution rate (%)
F1	6.125	40.834	40.834
F2	3.984	26.561	67.395
F3	1.589	10.590	77.986
F4	1.415	9.432	87.418

3 INFLUENCING MECHANISM

3.1 *Model construction*

The above analysis shows that the city's investment and population, urbanization and urban supportability, industry structure, and transportation will impact on the economic growth of development zones. To quantify the quantitative relationship between various factors and economic growth of development zones, we set the GDP of development zones as the dependent variable and the factor score (F1, F2, F3, F4) of 31 coastal cities as the independent variable, then conduct a multivariate regression analysis, and get the multiple regression model of urban factors and economic growth of development zones, which are as follows:

$$GDP = 0.823F1 + 0.311F2 + 0.083F3 + 0.148F4 + 0.086 \quad (1)$$

R^2 equals 0.803 and the fit is good, meaning that the regression equation can express the relationship of the city's impact on the economic growth of development zones, and showing that four common factors play an important role. From the perspective of the correlation coefficient of each factor, four factors are positive, indicating that the impact on the economic growth of development zones is a positive influence; however, the degree of effect is different.

3.2 *Result analysis*

According to the multiple regression models of urban factors and economic growth and the factor score function, the correlation coefficient between urban factors and economic growth of the development zone can be calculated.

Table 3 lists most of the impacts of each urban factor on the economic growth of development zones that have positive effects. Among which, some correlation coefficients are greater than 0.1, such as population, investment in fixed assets, educational expenditures, GDP, technology expenditures and employment of the city. And some correlation coefficients are less than 0.1, such as quantity of shipments per capita, students in colleges and universities, number of doctors in every ten thousand people, number of buses in every ten thousand people, pension insurance coverage, proportion of non-agricultural population, and proportion of third industry. Only the impacts of proportion of secondary industry and road area per capita on development zones are negative.

3.2.1 *Investment and population factors*

The factor of investment and population has the greatest influence, each of which increases by a unit, for example, zone economy grew by 0.823 units. Therefore, the city's financial and human resources are the most effective and direct support for the economic growth of development zones, and they are 'strong push' to the zone growth and provide a solid guarantee to infrastructure construction, environment improvement, and foster and promote the development of high-tech industry of the zone.

The investment of fixed assets and the total economy is the important embodiment of the city's investment ability that directly impacts on the capacity of the development zone to build infrastructures that are the precondition for the economic growth of development zones. Total population and employment reflect the size of the population and the scale of labor, and they provide not only human resources and support, but also the possibility of exploring residential function and transforming to the new urban district. The education expenditures, technology expenditures and the number of students in Colleges and universities reflect the level of urban education, which provide intellectual support for innovations and energy or potential for the economic growth of development zones, which are the characteristics of high-tech industry.

Table 3. Indicators of the city's comprehensive strength and its correlation coefficient.

Indicators	Correlation coefficient
Educational expenditures	0.154
GDP	0.144
Population	0.155
Technology expenditures	0.136
Employment	0.114
Investment in fixed assets	0.162
Students in colleges and universities	0.081
Number of doctors in every ten thousand people	0.081
Pension insurance coverage	0.074
Number of buses in every ten thousand people	0.075
Proportion of non-agricultural population	0.061
Proportion of secondary industry	–0.062
Proportion of third industry	0.031
Road area per capita	–0.086
Quantity of shipments per capita	0.088

3.2.2 *Urbanization and urban supportability factors*

The factor of urbanization and urban supportability has less influence than investment and population, each of which increases by a unit, for example, zone economy grew by 0.311 units. Some western academics consider that the key factor for the growth of development zone is an attractive living environment and a flexible administrative management. A study urban supportability system will create a good humanistic environment for economic growth, which is an important way of development and talent attraction, as well as improve the perfection of the zone's social service functions.

3.2.3 *Transportation factor*

Although transportation is the fourth factor, it has less influence coefficient that is higher than the industrial structure factor, i.e. zone economy grew by 0.311. Convenient traffic not only has city and its development zone closely linked but also shortens the moment of the city's radiation force, so that the zone will get great support and supply from the city. It has also narrowed the distance between the development zone and the consumer market, as a result getting information in time helps zone develop rapidly in the market economy. As research shows, most urban development zones are near the port, so the strong transportation capacity provides the reliable guarantee for the economic growth of the development zone.

Quantity of shipments per capita reflects the city's transportation capacity and the degree of the city's degree of extroversion. Stronger transport capacity and high degree of extroversion are the advantages of development zones for attracting investment and expanding the market for its enterprises. Road area per capita embodies the level of urban road facilities; however, a low rate of road utilization will lead to the waste of land resources. Only a high quantity of shipments per capita and a low road area per capita situation indicate a higher utilization of roads and transportation capacity.

If road area per capita is high but quantity of shipments per capita is low, then the economic efficiency undertaken by the road is not high.

3.2.4 *Industrial structure factor*

As the data show, the factor of industrial structure has the lowest influence on the economic growth, i.e. increasing by a unit will result in the growth of 0.823 units. Although only a little influence is made by the industrial structure factor, the economic growth of development zone still cannot be completely out of the industry; if so, it is difficult to develop. Moreover, in the initial stages of development zones, besides attracting business, its industrial development largely depends on the industrial transfer from the old districts to the development zone as a launcher of its economy. In addition to this, industrial transformation and upgrading of the city's need to increase the proportion of high-tech industries. The development zones that play the role of the scientific and technological incubator lead to the industrial development of high-tech products and promote the sustainable development of high-tech industries.

Industrial restructuring and upgrading is a process that the economic center transfers from the primary industry to the secondary industry, and then to the third industry. In this process, development zone is an important base for the city's industrial transfer, which takes the opportunity to grow the economic aggregate as well as the space for the development of the city's third industry, helping the city to increase its proportion gradually. Accordingly, the proportion of the third industry accounted for an appositive impact on the economic growth of development zones; conversely, the proportion of the secondary industry accounted for a negative effect.

4 CONCLUSIONS

Using the method of factor analysis, this paper presents the various aspects of the city grouped into four factors, namely the factor of investment and population, urbanization and urban supportability, industry structure, and factor of transportation. Based on this, the paper adopts the multiple regression model to investigate the relationship between the above factors and the economic growth of development zones. As the result of the study shows the, influences of all four factors are notable, with the order from highest to relative lowest being as follows: investment and population, urbanization and urban supportability, factor of transportation and industry structure. That is to say, as passing though the early period as the city gradually completes industrial restructuring and upgrading and the development zones' relative maturity, industry structure's influence decreases and is no longer considered as the most important factor. However, in the initial stage of construction, the industrial structure of the city is still an important factor for the growth of development zone.

REFERENCES

[1] Xiaoping Zhang. 2002. Characteristics and Development Mechanism of the Economic and Technological Development Areas in China. *Geographical Research* (21)5:656–666.

[2] Jingpeng Bai, Weixin OU. 2002. Appraisal and Influencing Factors Analyzing on New and Hi-tech Development Zones of Shanxi Province. *Territory & Natural Resources Study* 4:11–13.

[3] Yan Zhang. 2008. *The Practice & Transition of National Development Zones in China From the Public Policy Perspective*. Shang Hai: Tongji University.

[4] Guiqin Li. 2010. *Research on the Dynamic Factors of the Development of Nanjing High-tech Development Zone*. Nan Jing: Nanjing University.

[5] Jiating Wang, Lu Yang. 2013. Evolution Stages and Implementation Mechanism of Chinese Development Zones. *Study and Practice* (2):27–33.

Ecological environment and sustainable development

Material Science and Environmental Engineering – Chen (Ed.)
© 2016 Taylor & Francis Group, London, ISBN 978-1-138-02938-5

A discussion on design method and development trend of eco-architecture

W. Chen
Institute of Art and Design, Nanchang Hangkong University, Nanchang, China

ABSTRACT: Eco-architecture, generated under the background of environmental protection, is the product of the natural integration between men and environmental space, acting as an important form to realize the sustainable development of architectures. This paper introduces the basic concept and design method of eco-architecture nowadays and conducts certain explorations on the prospective development direction of eco-architecture.

Keywords: eco-architecture; sustainable development; architectural design; design method

1 INTRODUCTION

Eco-architecture, as a sustainable architectural form referring to extensive disciplines, is comprehensive system engineering with cross-over of various subjects and diversified extension. Eco-architecture aims at ensuring the harmonious coexistence and a dynamic balance point between men and nature. Its major task is to create a sound living environment for human beings under the premise of less influences and natural damage, in order to reach the coordination between architectural space and natural space ultimately. At present, the field of emerging architectures has become the stage for the development of eco-architecture; concerning eco-architecture design and improvement and development of functions and roles of eco-architecture has become the sustainable development direction of eco-architecture.

2 CONCEPT AND FEATURES OF ECO-ARCHITECTURE

Eco-architecture indicates the modern architectures that aim at environmental protection and stresses the influence reduction in architecture design on natural ecology to the minimum, in order to realize the harmonious uniformity between architectures and nature and the virtuous circle of materials and energy. This architectural form is the perfect integration of environmental science, ecological science, architectures, philosophy, etc. Eco-architecture takes full consideration of the influence of architectures on natural environment and the balanced coordination between natural environment and human beings without excessive dependence on natural environment, in order to make architectures back to nature. This has determined the features of eco-architecture as low energy consumption, environmental protection and small natural influence. Concretely speaking, eco-architecture has the following features:

Firstly, the design concept of eco-architecture is to fully utilize the natural renewable resources and reduce the influence on and damage to nature, this has also reduced the operation cost of eco-architecture. During the design phase, eco-architecture applies a lot of new technologies and materials, especially the targeted structures designed based on local conditions, to realize the effects of lower energy consumption, heat preservation, moisture preservation, ventilation and improvement of energy utilization efficiency.

Secondly, the implementation of eco-architecture takes the minimization of influence on environment into consideration. Measure, such as harmless treatment of building materials, waste control during architecture construction, application of renewable materials during construction, reduction of high energy consumption during construction and adoption of appropriate natural construction materials based on local conditions, can make various environmental-influential factors during the construction process under effective control, realize the orderly circle of matters and ultimately form the sound ecological balance with less waste and pollution between inner and outer space of architectures and environment. Secondly, eco-architecture can from the vernacular ecological expressions to accommodate certain geographic environment and climate environment, according to different geographies and culture and natural environments. Those expressions are actually under the influence of regional cultures on ecological architectures, without dependence on high and new technologies and new materials. Instead, they are formed naturally, with higher comfort level by passive measure. Besides, they are less dependent on energy and resources and more relevant to local climate, culture and geographic features, able to demonstrate the significances of eco-architecture [1].

3 DESIGN METHOD OF ECO-ARCHITECTURE

The design of eco-architecture shall be centered on realizing the balance of natural ecology, the harmonious coexistence of architectures and environment, returning to natural status. In addition, it shall be more rational and scientific as an organic living entity, in order to coordinate the relationship between eco-architecture and urban development at macro level, and focus on the natural properties of eco-architecture at micro level. The design of eco-architecture shall stress the following three aspects.

3.1 *Introduction of concept of natural ecosystem for integration with architectures*

The construction process of eco-architecture is the process of the coordination between architectures and natural ecology. The introduction of the concept and rules of natural ecosystem for eco-architecture design is the primary method to solve the conflict between architectures and environment. Under the instruction of this concept, we shall incorporate architectures into ecosystem into an organic ecological form; make architectures have organic features and a series of biological activities. This cognition can better solve the problems existing in the operation process of architectures in terms of utilization of natural resources and pollutant discharge. In concrete design, designers shall take the following factors into consideration. The first is the ecological property of architectures; the second is the influential factors for those ecological properties; the third is the analysis on expression degrees of the ecological properties to confirm the optimal design project.

3.2 *Optimal architecture design based on natural influential factors*

Nature is the start of eco-architecture design. An architecture shall comply with and adept nature. The eco-architecture design is just to build up an architectural form based on natural data. Therefore, at the initial stage, designers shall have the aid of modern sciences and technologies to analyze the natural environment of architectures. The obtained data act as the foundation for selection of building materials, architectural forms and building constructions, exerting immediate influence on the comfort degree, cost and efficiency of materials of the architectures. This design method centered on nature can make architectures incorporated in the natural ecosphere, and consider the circular relation of matters and energies between architectures and natural environment on the whole.

3.3 *Sustainability-oriented flexible design*

The concept of sustainability is a dynamic idea. In eco-architecture, it indicates the sufficient flexibility of architectures to adept the prospective development. It is shown from the flexible and optional requirements of architectural structures and construction equipments. For instance, the developable capability of buildings: including basic developable allowance, pre-consideration of floor bearing, and backup land for surrounding environment development; reserved pipeline space: including the development space of water, electricity and telecommunication [2]. This sustainable model can provide with plasticity for architectural development in the future. Flexible design can make architectures express their dynamic changes and sustained growth, and extend the development of living space and natural space of architectures. It has shown that the sustainable development design concept demonstrated by eco-architecture can ensure the long-term benefit, realize the cost-saving of eco-architecture design, reduce the consumption of construction materials, lower down the negative influence on environment and achieve the maximum economic and social benefit with minimum input.

4 DEVELOPMENT TREND OF ECO-ARCHITECTURE

Human beings are confronting with a series of problems in their survival and development and forming strong awareness of environmental protection ideologically. Consequently, the development of ecological construction shall respond to our survival problems and environmental problems. Designers shall regard eco-architecture design as a demand and a social responsibility to promote the sound development of eco-architecture in the future.

4.1 *Stress on micro energy consumption design of architectures*

Designers shall lower down the energy consumption for architecture operation to the minimum level. The realization of micro energy consumption operation complies with the demands of architectural economics and environmental protection in the future, to achieve the constant coexistence of human beings and environment and sustainable development. During the process of social development in the future, energy saving and environmental protection will become the guiding ideology for human beings' prospective development. In the field of architecture, all countries have further proposed their energy-saving targets and formulated the assessment systems and demands for eco-architecture development in the future. Measures including setting up quantitative indicators that describe principles of energy saving, water saving, low carbon consumption and renewable resources economic utilization in the eco-architecture development in the future, can ensure the low energy consumption in eco-architecture operation and form the architectural space beneficial to ecological environment.

4.2 *Advocacy of health-centered architectural environment design*

The healthier inner and outer architectural environment is the target pursued by eco-architectures in the future. Eco-architecture indicates the integration of architectures into nature to form a uniform and mutually circulating system for better utilization of natural resources. The healthy outer environment of eco-architectures is the premise to guarantee the stability and health of eco-architectures. Designers shall make selection, process and initiative improvement of the outer environment to form a stable natural ecological environment. The inner environment of eco-architecture is the basis of human beings' healthy lives. The improvement of inner environment is in need of the introduction of nature, stress the natural ecological improvement of inner space to promote the extension of nature in inner spaces and form the integration of inner and outer spaces of eco-architectures; to transform the inner and outer spaces of architecture into an organic entity and realize the consistence of light energy, thermal energy, wind energy and various natural resources in the operation of inner and outer environment.

4.3 *More application of originally ecological and vernacular architecture design factors*

In the prospective development of eco-architecture, the traditionally ecological and vernacular architecture factors will be applied more. The vernacular architectures of original ecology have become a language in the form of ecology; besides, most vernacular architectures enjoy the close symbiotic relation with local environment, storing the experience and consciousnesses accumulated during men's long-term living period. The new eco-architectures have inherited and developed the traditional architectural forms, extracted the ecological design factors from tradition to reappear the cultural and personal characters of vernacular architectures. The search of the harmonious coexistence of natural traditional and temporary civilization has been another interpretation of ecology.

4.4 *Stress on habitability of temperate climates*

The human-oriented nature of eco-architecture design has determined that it has to have the habitability of temperate climates. In the development of eco-architecture, designers have found that making architectures adapt to climate initiatively can not only reduce the operation cost of architectures and better utilize natural resources, but also make them more comfortable for habitation. From the perspectives of ecology and climatology, we can realize that they are mutually independent and influential. Creatures which adapt to natural climate initiatively can make their evolutions more completed and utilize more natural resources more completely; and make their own lives more comfortable. Therefore, the development of eco-architecture shall be based on the study on the internal relationship between climate and architecture. Designers shall make initiative changes of architectures according to local environment and climates, set up appropriate design schemes and strategies and build up rational and scientific operation methods. For instance, the change of architectural structures can lead to efficiently circular utilization of energies; besides, rational structural transformation can also make the architectural space better adapt the slight changes in local climates. This adaptability of architectures to climate is directly correlated with architectural habitability. In addition, architectural materials can be transformed to adapt local climate; materials of eco-architecture pay more attention to the ecological nature, such as the initiative adjustment and control of temperature and humidity, which are the components to construct the comfort of architectures. This has shown that study on the relationship between climate and architecture, utmost utilization of natural climate for adjustment and control of inner environment of architectures, such as ventilation, temperature and humidity, can form more comfortable integral living space and enhance the communication and connection between men and nature.

4.5 *Formation of new mode of artistic appreciation*

The design thinking of eco-architecture, which is extreme different from the design models in the past, pursues the systematic design to integrate architectures and nature together. Therefore, eco-architectures pay more attention to the pursuit of integrity. In such an integrity-oriented, integrated and systematic architectural form, the expression idea of architectures has been changed, leading to the alternated beauty appreciation mode. Concretely speaking, the new beauty appreciation mode is consisted of three aspects. Firstly, it is the multi-center expression of beauty appreciation. In the previous architectural forms, beauty appreciation focused on single center, namely the architecture itself, to set up the diverging expression model centered on architecture, namely the homogeneous model of beauty appreciation. Eco-architecture is an architecture system; the architecture itself is just the basic component instead of the main body of vision and beauty appreciation. The expression of the entire eco-architecture system is the main body of beauty appreciation, which is consisted of various forms of aesthetic tastes. Consequently, it has led to various centers for beauty appreciation with different natures; and beauty appreciation transformed from vision into sense. The beauty appreciation expressed by an eco-architecture doesn't pursue the graphic form of art or the subjective expressions of designers and breaks traditional standard for physical beauty appreciation, but emphasizing on the experience of eco-architecture and feeling the eco-intelligence of designers. This experience concerns the spiritual expressions of users for their spiritual enhancement. Thirdly, it pursues the beauty appreciation of higher level. Eco-architectures have formed the beauty of state. Its integration and interdependence with natural environment have created the more personalized natural survival space for people's returning back to nature, experiencing the true meaning of nature and ultimately realizing their survival values. In addition, this state can change along with the change of nature, so that traditional concepts and values can be more vital in this new state for self improvement and transcendence. This has expressed that the beauty appreciation mode of eco-architecture, as the expression of ecological awareness and intelligence, is determined by the meaning and the natural concept of eco-architecture.

5 CONCLUSION

Eco-architecture design is a complicatedly systematic engineering, which is in need of more considerations to enrich is significance, develop new design thoughts, improve the design scheme of architecture structures to ultimately formulate the more completed and effective method which complies with the natural ecological rules and be able to upgrade the quality of eco-architectures, in order to promote the sustainable development of eco-architectures. At the same time, eco-architecture design shall pay more attention to human orientation, strengthen humanistic caring and respect to nature, to establish the more harmoniously ecological relationship between men and architectures, architectures and nature and nature and men, to provide with a living environment for human beings more close to nature ultimately.

REFERENCES

[1] Wang Hongtao, Wang Xiaofan. *Basic Concept and Vernacular Expression of Eco-architecture* [J]. Chinese and Overseas Architecture, 2007 (5): 57–59.

[2] Fang Jun, Ye Jiong, Zhu Dasha. *Eco-architecture Design Method based on Sustainable Development* [J]. Journal of Wuhan University of Technology, 2009, 31 (12): 91–93, 98.

Material Science and Environmental Engineering – Chen (Ed.)
 ISBN 978-1-138-02938-5

The effect of vegetation restoration on denitrification in Shuangtaizi estuary reed wetland

Y. Zhang
Key Laboratory of Marine Environment and Ecology, College of Environmental Science and Engineering, Ocean University of China, Qingdao, China

K.R. Li
College of Marine Life Sciences, Ocean University of China, Qingdao, China

H.M. Yao, Y.G. Zhao, X. Huang & J. Bai
Key Laboratory of Marine Environment and Ecology, College of Environmental Science and Engineering, Ocean University of China, Qingdao, China

ABSTRACT: We made the study in Shuangtaizi estuary reed wetland from July to October, 2012. The result showed that the denitrifying bacteria value in restored wetland was closed to that in natural wetland. Both the values of the two regions were higher than that of the degraded wetland. The maximum value was reached in October. There values were 19.8×10^4, 24.5×10^4 and 2.1×10^4 cells $\cdot g^{-1}$ for the restored wetland, natural wetland and degraded wetland, respectively. Acetylene inhibition method was used to measure denitrification capacity in Shuangtaizi estuary reed wetland. From the analysis of the results, we found that the Potential Denitrification Rate (PDR) changed similarly to that of the natural wetland. They both reached a maximum in October, with the values being 31.11 ± 13.71 and 38.62 ± 1.05 mg N_2O-N $\cdot kg^{-1}$ soil $\cdot d^{-1}$, respectively. The largest PDR value for the degraded wetland emerged in August was 15.60 ± 10.77 mg N_2O-N $\cdot kg^{-1}$ soil $\cdot d^{-1}$, being significantly lower than the other two types of wetland. All PDR values improved with the depth increasing, indicating that there was a greater capacity for nitrogen removal in Shuangtaizi estuary reed wetland.

Keywords: Shuangtaizi wetland; denitrifying bacteria; restoration; potential denitrification rate

1 INTRODUCTION

Nitrate removal ability is regarded as an important function of wetland. Estuary wetland is more valuable because of its special geographical location. Moreover, wetland is a major place for denitrification owing to its anaerobic environment caused by the overlying water. Denitrification is more than a significant part of nitrogen removal in wetland. In addition, as one of its products, N_2O contributes a lot to the greenhouse effect. The greenhouse effect of N_2O is 310 times that of CO_2. In recent years, because of intensifying human activities and climate change, wetland degradation has become a global environment problem. Today, all countries in the world are implementing the policies of corresponding wetland restoration. However, because of the difference in soil properties and environmental conditions, there are many differences in different parts of the wetland restoration effect, which made coastal wetland restoration difficult to implement.

Denitrification is a key step in the wetland ecosystem nitrogen cycle. Currently, the research on denitrification focuses on four areas in the wetland system, including the denitrification efficiency of nitrogen removal (D.L., et al. 2001; Ullah, et al. 2006), denitrification mechanism in wetland sediment (K.N., et al. 2009), the complexity of the wetland ecosystem (Kumar, et al. 2010) and structural characteristics analysis of microbial communities in denitrification (Song, et al. 2011). During the recovery process of Shuangtaizi estuary reed wetland, degraded areas, restored areas and natural areas are selected in this study. We discussed the relationship between the recovery of estuary sediment and denitrifying bacteria and the denitrification rate changes, in order to make sure the mechanism of ecological restoration for denitrification in wetland sediment. The contact between the preliminary study of vegetation degradation and the ecological restoration of coastal wetland was explored, in order to provide a reference for further research on the coastal wetland nitrogen cycle and the environmental impact mechanism.

2 MATERIALS AND METHODS

2.1 *Experimental design*

In this study, soil samples from degraded severely wetland (S1), restored wetland (S2) and natural wetland were collected (S3). Sampling was carried out in the growing season of reeds, including July, August, September and October. Tube samples were used for each wetland random sample within a radius of 2 m with nine samples in four soil layers (0–10 cm, 10–20 cm, 20–30 cm and 30–40 cm). Each sample was mixed thoroughly to remove reed root aftershave, loaded immediately in a sterilized closure pocket and put in an ice box back to the laboratory for analysis (Yu, et al. 2012).

2.2 *Laboratory simulation cultivation*

Preparation of denitrifying bacterial cultures was as follows: 1 g of fresh soil samples was placed in a 98 mL conical flask containing a medium and 1 mL Tween 80, and mixed well to a concentration of 10^{-2} suspension. After adding to the prepackaged 10 mL liquid medium in a 15 × 100 mm test tube, we diluted them into five gradients with concentrations of 10^{-3}, 10^{-4}, 10^{-5}, 10^{-6} and 10^{-7}. Each of three parallel dilutions was put into a small tube, adding denitrifying bacteria suspension in the culture medium sealed with paraffin. Then, they were placed in a constant temperature incubator at 28°C for 14 days. According to the presence or absence of small tube bubbles, we checked the MPN count table to make sure the value of denitrifying bacteria. All liquid reagents, culture medium and test tubes were sterilized.

Preparation of denitrifying bacteria medium was as follows: sodium citrate 5.0 g, KNO_3 2.0 g, K_2HPO_4 1.0 g, $MgSO_4 \cdot 7H_2O$ 0.2 g and KH_2PO_4 1.0 g. We set the volume to 1000 mL water, adjusting the pH to 7.2–7.5 (Fan, et al. 2011).

2.3 *Determination of potential denitrification rates*

In this study, Potential Denitrification Rate (PDR) was used to measure the denitrification intensity in wetland. For this purpose, we selected a modified acetylene inhibition method. A study has shown that this method measured disturbed mixing soil samples PDR with static soil column measured unanimously (P.G., et al. 2003).

We took 20 g of fresh soil samples and placed it in a 250 mL conical flask with stopper after adding 20 mL of deionized water. Three parallel samples were set for each sample, and a group without the addition of acetylene was added as the control group. All the upper parts of the soil sample space were filled with C_2H_2:N_2 at a 1:5 ratio in the mixed gas. Then, the samples were put in the dark for 5 days at room temperature. Subsequently, after 7 days, acetylene was likely to become a C source, thus affecting the microorganisms' activity (A.R., et al, 2004). Then, 2 mL gas was collected each time from the upper part of the space at 0, 6, 12, 24, 48, 72, 96 and 120 h during the cultivation stage to measure the content of N_2O gas by a gas chromatograph.

Table 1. Distribution of sampling sites in Shuangtaizi estuary reed wetland.

Site	Longitude	Latitude	Community's characteristics
S1	121°35′15.8″	40°51′11.7″	Degraded wetland
S2	121°36′32.9″	40°52′06.9″	Restored wetland
S3	121°47′29.5″	41°09′34.3″	Natural wetland

2.4 *Sample determination*

Total Organic Carbon (TOC) content was determined by taking fresh soil samples using the potassium direct extraction method with a TOC-Vario analyzer (Elementar, Germany). Total Phosphorus (TP) content was determined by taking sediment samples over a 100-mesh sieve, and using perchloric acid to digest. It was measured using a spectrophotometer by Mo-Sb colorimetry. Total Nitrogen (TN) was determined by taking each sediment sample over a 100-mesh sieve, by using the Kjeldahl method. NO_3^--N, NO_2^--N and NH_4^+-N contents were determined by taking sediment samples over a 60-mesh sieve. Ammonium, nitrite and nitrate were extracted from the soil samples with potassium chloride solution, and determined by using spectrophotometric methods with an AA3 continuous Flow Analyzer (BRAN-LUBBE, Germany).

2.5 *Data analysis*

Regression analysis was performed to analyze the data using SPSS 19.0.

3 RESULTS

3.1 *Changes of denitrifying bacteria in the process of wetland restoration*

In natural wetland, the value of denitrifying bacteria was significantly higher than that of the other two soil layers (Fig. 1) in July, in with the layer of 25–40 cm. Denitrifying bacteria in the three soil layers was much less in August. The lowest number was 0.05 × 10^4 cells per gram of dry soil in the layer of 0–10 cm. In October, denitrifying bacteria of

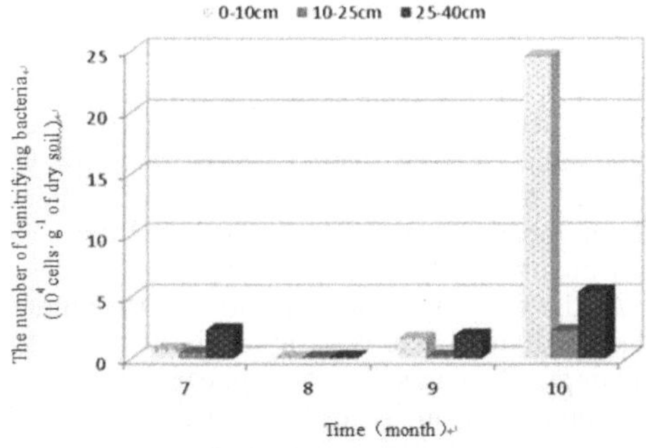

Figure 1. The spatial and temporal distribution of nitrifying bacteria in natural wetland.

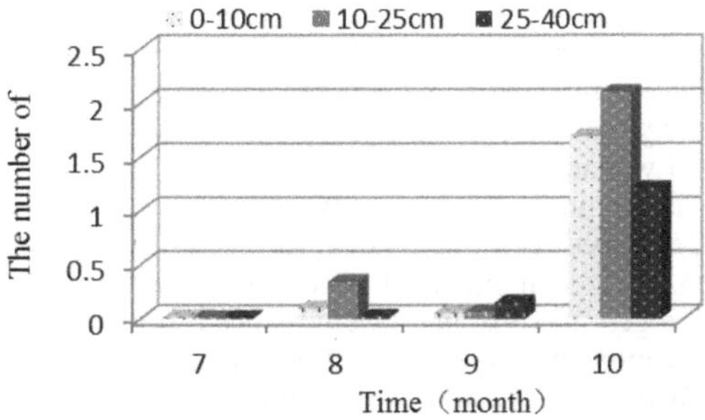

Figure 2. The spatial and temporal distribution of nitrifying bacteria in degraded wetland.

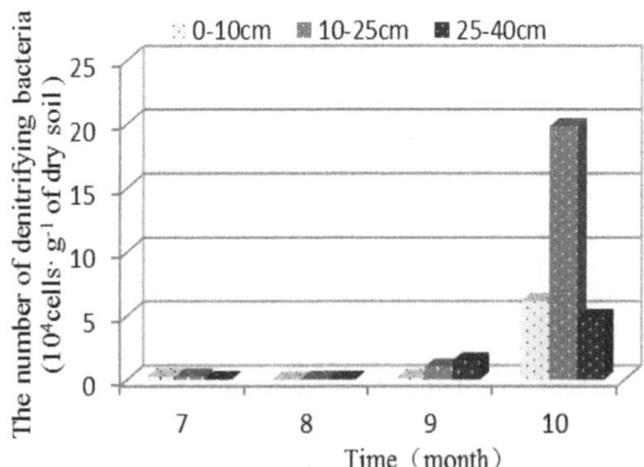

Figure 3. The spatial and temporal distribution of nitrifying bacteria in restored wetland.

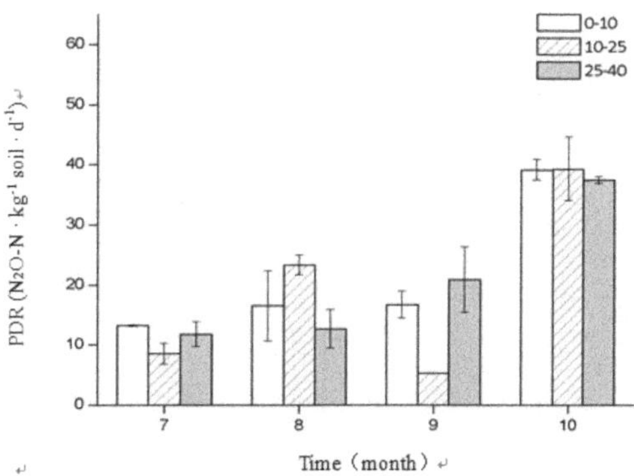

Figure 4. The spatial and temporal distribution of denitrification in natural wetland.

the layer of 0–10 cm was 24.51 × 10^4 cells per gram of dry soil. And it was significantly higher than that of the other soil layers.

The minimum number of denitrifying bacteria in degraded wetland appeared in July (Fig. 2). The number of denitrifying bacteria increased to 10–25 cm in August, and then decreased from September. The maximum value of the 25–40 cm layer was 0.15 × 10^4 cells·g^{-1} of dry soil. The number of denitrifying bacteria was the highest in October compared with the four months.

The layer of 10–25 cm in July showed the maximum value of denitrifying bacteria in restored wetland (Fig. 3). The number of denitrifying bacteria in August was the least in all the months. Then, it increased in September, and a huge increase was observed in October. The largest number of denitrifying bacteria was 19.81 × 10^4 cells·g^{-1} dry soil.

3.2 *The effect of ecological restoration on denitrification*

3.2.1 *Denitrification in natural wetland*

The maximum value of the average PDR was in October (Fig. 4), reaching 38.62 ± 1.05 mg N_2O-N·kg^{-1} soil·d^{-1}. Then, this was followed by 17.51 ± 5.37 mg N_2O-N·kg^{-1} soil·d^{-1} in August, 14.27 ± 8.08 mg N_2O-N·kg^{-1} soil·d^{-1} in September and 11.21 ± 2.43 mg N_2O-N·kg^{-1} soil·d^{-1} in July. The maximum of the PDR appeared in October in the layer of 10–25 cm with 39.31 ± 5.28 mg N_2O-N·kg^{-1} soil·d^{-1}. The minimum value occurred in September with 5.25 ± 0.05 mg N_2O-N·kg^{-1} soil·d^{-1} in the layer of 15–25 cm.

3.2.2 *Denitrification in degraded wetland*

The average PDR value in August was 15.60 ± 10.77 mg N_2O-N·kg^{-1} soil·d^{-1} (Fig. 5). This value was close to the value in September. According to the order from highest to lowest, the next was the PDR in October, with 10.94 ± 3.28 mg N_2O-N·kg^{-1} soil·d^{-1}. The lowest was in July with a value of 6.72 ± 1.81 mg N_2O-N·kg^{-1} soil·d^{-1}. The minimum PDR of the three layers appeared in the layer of 25–40 cm with 4.64 ± 1.06 mg N_2O-N·kg^{-1} soil·d^{-1}.

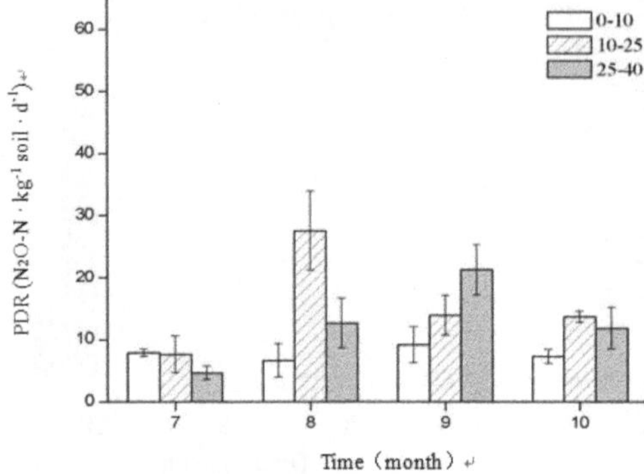

Figure 5. The spatial and temporal distribution of denitrification in degraded wetland.

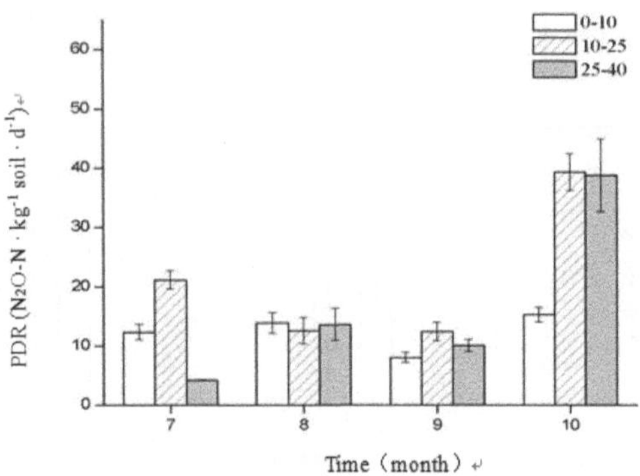

Figure 6. The spatial and temporal distribution of denitrification in restored wetland.

3.2.3 *Denitrification in restored wetland*

The highest average of PDR was in October (Fig. 6), with a value of 31.11 ± 13.71 mg. The lowest average of PDR appeared in September, with 10.12 ± 2.22 mg N_2O-N · kg^{-1} soil · d^{-1}. However, the minimum value of PDR was in July in the layer of 25–40 cm, with a value of 4.18 ± 0.03 mg N_2O-N · kg^{-1} soil · d^{-1}.

4 DISCUSSION

4.1 *The impact of ecological restoration on denitrifying bacteria in spatial and temporal distribution*

There was a fairly significant increase in September and October compared with the number of denitrifying bacteria in July and August ($P < 0.01$). This explained that the value of denitrifying bacteria in Shuangtaizi reed wetland in fall was higher than that in summer. We speculate that due to withered reeds in fall caused a significant increase in soil nutrients, especially TOC, so as to give enough priority to denitrifying bacteria. Significant changes in the number of denitrifying bacteria were observed between restored wetland and degraded wetland. The number was significantly increased especially in September and October ($P < 0.05$).

At the same time, we can find that the number of denitrifying bacteria varied with the increasing soil depth. The most obvious appearance was that the value of denitrifying bacteria in the layer of 10–25 cm in October, which was much higher than that of the other soil layers measured (Figs. 1–3). This could be explained by the fact that most denitrifying bacteria were anaerobic microorganisms, with the increasing soil depth, and the DO content was reduced. It provided favorable conditions for the growth of bacteria. We also observed a similar phenomenon in degraded wetland. However, the situation was exactly opposite to that of natural wetland. The correspondent layer was almost the minimum (Fig. 1). We concluded that the developed reed root system in the region resulted in a large number of soil nutrient absorption. Exogenous substances were reduced so that heterotrophic microorganisms would be hard to survive (Wang, et al. 2007). This could also be observed from the soil nutrient distribution structure in degraded wetland.

4.2 *Contribution of ecological restoration to denitrification*

Overall, the PDR of natural wetland was slightly higher than that of restored wetland, and also higher than that of degraded wetland, especially in July and October. The value of the average PDR in degraded wetland was almost one-half or one-third of natural wetland and restored wetland. This is due to the fact that the restoration of wetland has made significant improvement of function to remove nitrogen. Its strength was similar to that of the natural wetland, and greater than that of the degraded wetland. After ecological restoration, the structure of wetland soil was improved to benefit denitrification.

The largest PDR value in restored wetland occurred in the layer of 10–25 cm, indicating that denitrification intensity was the most violent in the region (Fig. 6). This phenomenon was similar to the degradation of wetland. However, for the natural wetland, the PDR value declined in July and September obviously. This decrease can be explained by the fact that due to the C absorption of plant roots, the organic C required by microbial denitrification reduced. The content of TOC in this layer dropped suddenly in July and September.

The fact was a good explanation. It was worth noting that in all the wetlands that we observed, the PDR increased with the increasing soil depth. According to the study, the PDR profile in the deeper layer was mainly decided by the diffusion of NO_3^- concentration from surface waters to the deep (Venterink, et al. 2003). It showed that N removal function would strengthen with increasing depth in Shuangtaizi reed wetland. Due to the relatively small soil volume-weight and high salinity, low osmotic pressure promotes the diffusion of water in the soil, resulting in a higher concentration of NO_3^- in deeper soil.

5 CONCLUSION

The number of denitrifying bacteria in Shuangtaizi reed wetland was significantly improved after ecological restoration, which was close to the earlier state of natural wetland. However, the gap between the restored wetland and the natural wetland still existed in the soil structure and function. In Shuangtaizi reed wetland, the value of denitrifying bacteria could not directly reflect the intensity of denitrification. However, there was a relationship between the two indicators.

Different types of region in Shuangtaizi reed wetland had different denitrification capacities. During the reeds' growing season, from July to October, the variation trend of denitrification capacity in restored wetland was similar to that in natural wetland. But its denitrification pathway was changed. This study showed that the maximum denitrification in restored wetland appeared in September and October, which was later than that observed in natural wetland. Overall, the PDR value in the region was high. Compared with the contribution of greenhouse gas N_2O, the order of sequence from highest to lowest was as follows: degraded wetland > restored wetland > natural wetland.

Denitrification in the three types of wetland was significantly affected by the content of TOC. The PDR value increased apparently with the increase in the content of TOC.

ACKNOWLEDGMENTS

This work was supported by the National Water Pollution Control and Management Technology Major Project of China (No. 2013ZX07202-007).

REFERENCES

[1] Saunders. D.L., Kalff. J. Nitrogen retention in wetlands, lakes and rivers. Hydrobiologia. 2001, 443: 205~212.

[2] Sami Ullah, Stephen P. Faulkner. Use of cotton gin trash to enhance denitrification in restored forested wetlands. Forest Ecology and Management. 2006, 237:557~563.

[3] Hopfensperger. K.N., Kaushal. S.S., Findlay. S.E.G., et al. Influence of plant communities on denitrification in a tidal freshwater marsh of the Potomac River, United States [J]. Journal of environmental quality, 2009, 38(2):618–626.

[4] Mathava Kumar, Jih-Gaw Lin. Co-existence of anammox and denitrification for simultaneous nitrogen and carbon removal—Strategies and issues. Journal of Hazardous Materials. 2010, 178:1~9.

[5] Keunyea Song, Seung-Hoon Lee, Hojeong Kang, Denitrification rates and community structure of denitrifying bacteria in newly constructed wetland. European Journal of Soil Biology. 2011, 47:24~29.

[6] Huibin Yu, Yonghui Song, Beidou Xi, Erdeng Du, Xiaosong He, Xiang Tu. Denitrification potential and its correlation to physico-chemical and biological characteristics of saline wetland soils in semi-arid regions. Chemosphere. 2012, 89(11):1339–1346.

[7] Fan Jing-feng, Chan Jia-ying, Chen Li-guang, Guan Dao-ming. Research on denitrifying bacteria quantification and diversity in Liaohe Estuary sediments. Acta Oceanologica Sinica. 2011, 3:011.

[8] Hunt. P.G., Matheny. T.A., Szögi. A.A. Denitrification in constructed wetlands used for treatment of swine wastewater. Journal of environmental quality. 2003, 32:727~735.

[9] Hill. A.R., Cardaci. M. Denitrification and organic carbon availability in riparian wetland soils and subsurface sediments. Soil Science Society of America Journal. 2004. 68:320~325.

[10] Wang Bei, Zhang Xu, Li Guanghe, Zhong Yi. Impact of reed roots on the vertical migration and transformation of petroleum in oil-contaminated soil. Acta Scientiae Circumstantiae. 2007, 27(8):1281~1287.

[11] Venterink. H.O., Hummelink. E., Van den Hoorn. M.W. Denitrification potential of a river floodplain during flooding with nitrate-rich water: grasslands versus reed beds. Biogeochemistry. 2003, 65:233~244.

Material Science and Environmental Engineering – Chen (Ed.)
© 2016 Taylor & Francis Group, London, ISBN 978-1-138-02938-5

The ecological factors of arbor forests in saline-alkali soil

W.Z. Huang, Y.J. Qin, J.H. Yang, J.K. Li & W.Q. Gong
The College of Horticulture and Landscape, Tianjin Agricultural University, Tianjin, China

B.B. Wang
The Green Landscape Engineering Corporation, Tianjin, China

ABSTRACT: In order to understand environmental factors of woodland in saline-alkali land, the temperature, humidity, wind speed and light intensity of eight kinds of woodlands were compared with the cleaning places near each woodland. The results show that the temperature in woodlands was higher than the cleaning places (except *Sophora japonica* forest), the temperature of *Robinia pseudoacacia* '*Aurea*' and *Robinia pseudoacacia L* forest was highest (0.27 °C and 0.25 °C, respectively) higher than the cleaning places among them. Humidity of *Sophora japonica*, *Fraxinus chinensis* and *Sophora japonica* forest increased considerably, followed by *Koelreuteria paniculata* forest, and the humidity of *Sophora japonica L.* was the highest. Wind speed of woodland was less than or equal to the control (except *Fraxinus chinensis* forest), *Robinia pseudoacacia* '*Aurea*' reduced the most, between the strains reduced to 0.65 m/s and 0.43 m/s in the lines. The light intensity of all woodlands was decreased. The light intensities of *Koelreuteria paniculata* woodland and *Salix babylonica* forest were the highest, with 71.83–85.1% of full sunlight. This was followed by *Fraxinus chinensis*, *Salix matsudana* and *Robinia pseudoacacia L.*, with 42.03–65.57% of full sunlight. In addition, among the strains of different species, the trends of temperature, humidity, wind speed and photosynthetic intensity were consistent between the lines.

Keywords: ecological factors; different woodlands; strains; lines

1 INTRODUCTION

In recent years, more and more arable land has been salinized. Second soil survey data in China showed that the area of saline soil was approximately 36 million hectares, accounting for 4.88% of available land area of China [1]. Tianjin is a coastal city and its land salinization is more serious [2]. Therefore, it is important to develop saline land and understand the ecological reaction of woodland in saline soil. Ecological factors between different forests on saline land have been researched. Huang Liangmei studied the correlation between the climatic factors and greening in small areas. he found a positive correlation between height and cooling, humidification, shading, windscreen, and between plant diameter and cooling and humidification, and a negative correlation with light shielding, but no correlation with wind [3]. Guo Congjian and Zhang Xinsheng found that utilization of light was better in the middle planting density of Catalpa bungei with different wide rows [4]. Wang Yurong, Liao Kang, Jia Yang, Liu Manman reported the order of the Photosynthetically Active Radiation (PAR) in strain rows, and the canopy center of apricot was found to be rows > strains > the center of canopy [5]. Xiao Sen, Wang Youke reported a better moisture condition between plants than between the rows after the water spatial distribution was analyzed in the soil of high-density red jujube forest [6].

In order to understand the changes in ecological factors further between different forests, the different plant growth conditions in saline soil and woodland eco-efficiency provide the basis for the selection of tree species. On this basis, eight kinds of woodlands were chosen, and the difference between temperature, humidity, light intensity and wind speed was compared.

2 MATERIALS AND METHODS

2.1 *Materials*

The experimental was conducted in the nursery of Tianjin LVYIN Landscape and Ecology Construction Co., Ltd located in Tianjin Jinghai, including artificial forests of 8 woodlands, namely *Koelreuteria paniculata* (Kp), *Sophora japonica* (Sj), *Fraxinus chinensis* (Fc), *Salix matsudan* (Sm), *Platanus* (Pl), *Robinia pseudoacacia L.* (RL), *Robinia pseudoacacia* '*Aurea*' (RA) and *Salix babylonica* (SA).

A 3 × 4 m spacing in row and spacing between rows were kept in each woodland. The plants of *Robinia pseudoacacia* '*Aurea*' were 5 years old and the rest were 10 years old, which were cultivated in loam with 0.21% salinity and 8.2 pH, and managed by the unified conventional method.

2.2 *Methods*

Randomly selected plants were evenly distributed in the sample plots of 0.16 acre each woodland. Each treatment was replicated 3 times. Four sites of each plot were selected for investigation by using Z-shaped sampling. Temperature, humidity, wind speed and light intensity were measured in the sides and in the afternoon on October 8, 2014 on a sunny day. The empty plot next to the woodland was used as the control. Temperature, humidity and wind speed was the difference value between the treatments and the control. Light intensity was calculated by using the following formula: (Light intensity of Control-Light intensity in rows) ÷ Light intensity of Ck × 100%. Temperature and humidity were measured by using a hygrometer (Elitech RC-4HA), wind speed was measured by using an anemoscope impeller (HCJYET HT-8392-type), light intensity was measured with a portable light meter (JD-3 type) (Jiading Xuelian Instrument Factory).

3 RESULTS AND ANALYSIS

3.1 *Temperature changes in different woodlands*

There was a significant difference in temperatures among the different types of woodland, as shown in Figure 1. Temperatures in each woodland were higher than that in the empty plot (CK), except the woodland of *Sophora japonica*. The highest temperature was observed in the forests of *Robinia pseudoacacia* '*Aurea*' and *Robinia pseudoacacia L*, in which it was 0.27 °C and 0.25 °C higher in rows than in CK. In addition, the temperature change tendency of different species in rows was the same as the space between the plants. A less change in the temperature in the forests was reported. For example, Jiang Cuihua found that the diurnal range of temperature outside the forests on early to mid-October was 0.1 °C less than the temperature inside [7], which means a higher temperature within the forest. Dai Renzhi found that the temperature in the forests was lower than that in the open ground during summer. The change and amplitude fluctuations of air temperature were reduced in the artificial Osmanthus forest during summer. This decrease is due to the fact that screen of trees and leaf transpiration increased humidity, thereby increasing heat capacity and other factors [8]. Plants of *Robinia pseudoacacia* '*Aurea*' were shorter, which were cultivated by higher planting density and formed more branches. Due to the high density of branches and leaves in the woodland, the temperature was higher than that in the other species.

Sophora japonica and *Robinia pseudoacacia* had the same plant height, plant shape and leaf canopy in the experiment. However, within the survey, the number of locusts in the woodland grew more vigorously with more number of branches. This may be the reason why there was a higher temperature in the woodland than that in *Sophora japonica*. Therefore, there is a greater relationship between the temperature changes and their growth conditions among different types of woodlands.

3.2 *Humidity changes in different types of woodlands*

There were significant differences in humidity between woodlands and empty plots, as shown in Figure 2. Overall, humidity difference was smaller in the rows and strains of *Salix matsudan,* and in the strains of *Robinia pseudoacacia* '*Aurea*' and *Robinia pseudoacacia L.* than that of the open areas (ck). However, the humidity was higher in other species of woodlands than in CK. The maximum difference in humidity was shown between

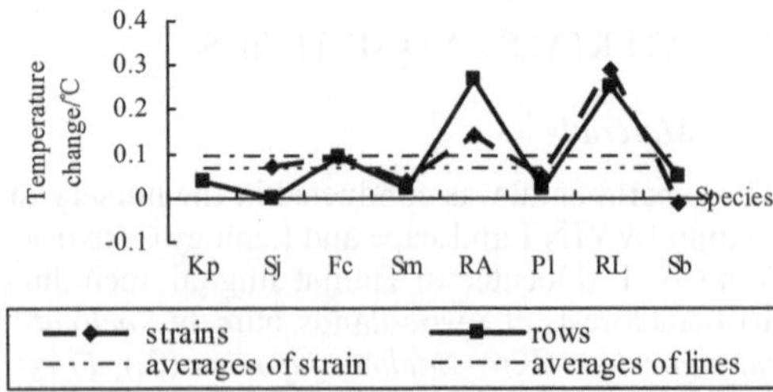

Figure 1. The temperature between woodland and empty lands.

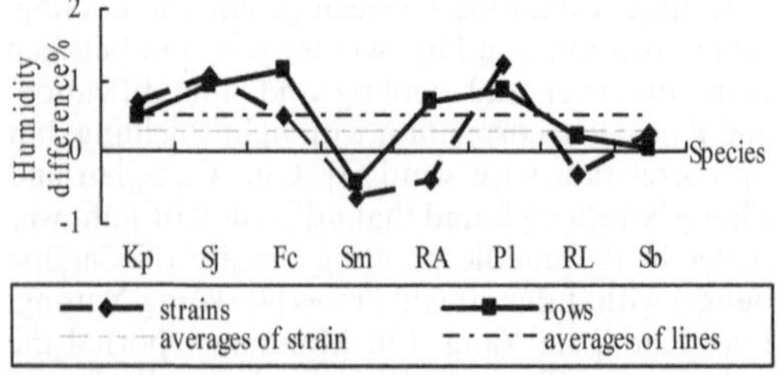

Figure 2. Humidity difference between woodlands and empty lands.

the strains of *Platanus*, which was 7.05% higher than that in CK.

The humidity difference between the rows of *Fraxinus chinensis* was highest among all the species, which was 0.93 and 5.29% higher than that in CK. There was higher humidity in the strains and rows of *Sophora japonica* forest compared with CK. The humidity changed with the same trend in the strain and rows of each species of woodlands, compared with the humidity curve between the strains and rows, except for *Robinia pseudoacacia 'Aurea'*.

Plant transpiration increased the humidity in the forest, as reported in many papers. For example, Li Ke, Pan Xinjun mentioned that transpiration of woody plants increased humidity in the forest, whose effect was 10 times more or several dozen times more than that of herbaceous plants [9]. In this study, seven kinds of plant lands made humidity increase, except *Salix matsudan*, which is consistent with Li' result.

3.3 *Wind speed changes in different woodlands*

There was a significant difference in the wind speed of different woodlands compared with CK, as shown in Figure 3. The wind speed in all forests changed less than or equal to that in CK, except *Fraxinus chinensis*. It reduced most in *Robinia pseudoacacia 'Aurea'* and reduced to 0.65 m/s in strains, and to 0.43 m/s in rows, followed by *Platanus*, *Robinia pseudoacacia L.* and *Salix matsudan*. Overall, change in wind speed trended similarly between lands of different species in rows and strains.

Wind speed was independent of the plant height and was related to plant morphology, branch density compared with different woodlands. The woody forests reduced the wind speed, as reported in many reports. For example, Fan Zhiping, Zeng Dehui found that the wind speed in different sites of vertical direction of forest decreased significantly and varied greatly [10]. On the contrary, plant height and wind speed were independent in

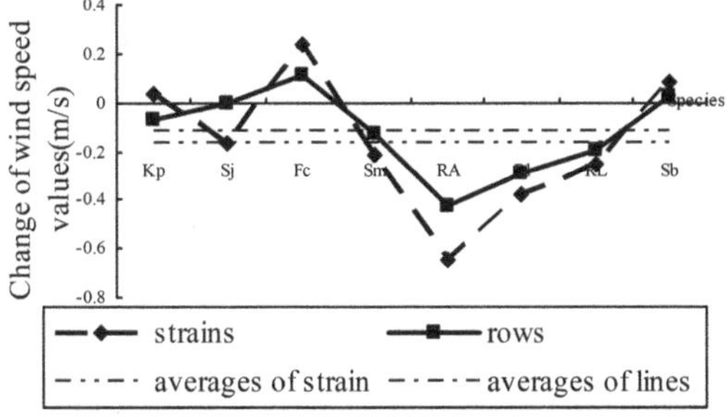

Figure 3. Wind speed difference between open lands and woodlands.

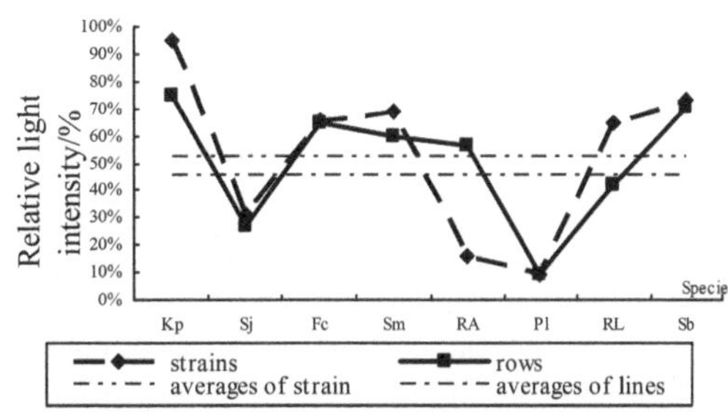

Figure 4. The ratio of the light intensity of different forests and open lands.

the other species [3], which was consistent with the results.

3.4 *Light intensity changes in different woodlands*

The relative light intensity of different forests was significantly different from that shown in Figure 4. Overall, the changes in light intensity had the same trend in the lines and the strains of different forests. Woodland photosynthesis of all the eight species was declined with different degrees. The plots with strong light were lands of *Koelreuteria paniculata* and *Salix babylonica*, with 71.83–85.1% light intensity of full light. This was followed by *Fraxinus chinensis*, *Salix matsudan* and *Robinia pseudoacacia L.* The average value was 42.03–65.57% light intensity of full light. Light intensity was lower in *Robinia pseudoacacia 'Aurea'*, *Sophora japonica* and *Platanus* than that in the other species. The average light intensity was only 8.89–30.88% of full light.

The forests of *Robinia pseudoacacia 'Aurea'* and *Robinia pseudoacacia L.* had increased temperature and reduced wind speed and *Platanus* had reduced wind speed and increased humidity more than the other species according to each indicator.

4 DISCUSSIONS

When the wind speed reduced, the temperature was increased in species such as *Robinia pseudoacacia 'Aurea'* and *Robinia pseudoacacia L.* Contrarily, at a higher wind speed, the temperature decreased in species such as *Koelreuteria paniculata*, *Sophora japonica*, *Fraxinus chinensis* and *Salix babylonica*. At strong light, photosynthesis got stronger, stomata opened more and transpiration was also stronger, so that the humidity correspondingly increases in species such as *Koelreuteria paniculata* and *Fraxinus chinensis*. All ecological indicators of *Robinia pseudoacacia 'Aurea'* had a greater difference than those of the other species in the same forest. However, the woodland of *Robinia pseudoacacia* '*Aurea*' decreased most in wind speed and more in light and increased more in temperature according to the average value of each index between the rows and strains, which could be the closed leaf stomata, leading to the decrease in transpiration and humidity also low under low light intensity.

However, the relationship between each ecological indicator was much complex since more change in ecological factors did not follow any rules among the species of woodlands.

ACKNOWLEDGMENTS

The authors thank the National Agricultural Science and Technology Achievements Transformation Fund (Grant No. 2012GB2A100015) and Major Scientific and Technological Projects in Tianjin Science and Technology Commission (Grant No. 12ZCDZNC04800 and 14JCTPJC00530) in People's Republic of China.

REFERENCES

[1] Wang Jiali, Huang Xianjin, Zhong Taiyang, Chen Zhigang. Review on Sustainable Utilization of Salt-affected Land [J]. *Acta GeographicA Sinica*, 66(5), pp. 673–684, 2011.

[2] Zhang Daguang, Yang Nan. On saline—alkali soil forestation project in Tianjin Development Zone [J]. *Engineering Technology and Management*, 29, pp. 208, 2013.

[3] Huang Liangmei. *Study on the Correlativity between the Nanning City's Greening and the Environmental Quality* [D]. Guangxi University: Guangxi University, 2003.

[4] Guo Congjian, Zhang Xinsheng. Research on Catalpa bungei forest planting density and its effects [J]. *Journal of Henan Forestry Science and Technology*, 4, pp. 11–15, 1994.

[5] Wang Yurong, Liao Kang, Jia Yang, Liu Manman, Liao Xiaolong, Zhao Shirong, Peng Xiao. Study on light change rule in Apricot Orchards with different planting densities [J]. *Northern Horticulture*, 14, pp. 6–9, 2014.

[6] Xiao Sen, Wang Youke, LI Penghong, Lin Jun, Xin Xiaogui, LU Junhuan, Zhao Xia. Soil moisture among plants and in rows of the close planting Jujube forest in the loess hilly and gully region [J]. *Agricultural Research in the Arid Areas*, 30 (6), pp. 83–87, 2012.

[7] Jiang Cuihua, Wu Xinsheng, Wang Wenqing. Temperature changes in the forest mushroom spawn of double spore Growth and fruiting body period [J]. *Jiangsu Agricultural Sciences*, 39 (5), pp. 345–347, 2011.

[8] Dai Renzhi. *Study on Osmanthus Forest of Microclimate Characteristics in Changsha City of Green Artificial* [D]. Central South University of Forestry and Technology: Central South University of Forestry and Technology, 2013.

[9] Li Ke, Pan Xinjun. plant selection of City forest construction [J]. *Hunan Forestry Science and Technology*, 32 (1), pp. 73–76, 2005.

[10] Fan Zhiping; Zeng Dehui; Liu Dayong; Yu Xinxiao; Niu Jianzhi. Characteristics of windspeed distribution.

Material Science and Environmental Engineering – Chen (Ed.)
 ISBN 978-1-138-02938-5

Distribution features of nitrogen and phosphorus in different grain-size surface sediment in the Yellow River Estuary

J.J. Chang & Y. Liu
College of Environment Sciences and Engineering, Ocean University of China, Qingdao, China

X.X. Luo
College of Environment Sciences and Engineering, Ocean University of China, Qingdao, China
Key Laboratory of Marine Environmental Science and Ecology, Ministry of Education, Ocean University of China, Qingdao, China

ABSTRACT: Nitrogen (N) and Phosphorus (P) in marine sediments are important in dynamic cycle of ocean biogenic elements. Concentrations and distribution characteristics of N, and P in different grain-size sediment were discussed in this study. Results showed: Concentrations of N and P in the <63 μm sediment samples were much higher than those in the >63 μm sediment samples, this indicated sediment particles with smaller grain size had a larger absorption capacity of N and P. TN and TP concentrations showed a negative correlation with sediment grain size and were at safety levels in the sediment assessment guidelines. From fresh water to sea, N and P concentrations increased gradually, showing significant region characteristics of river-sea.

Keywords: the Yellow River Estuary; surface sediment; TN; TP; distribution characteristics

1 INTRODUCTION

Nitrogen (N) and Phosphorus (P) in marine sediments play an important role in dynamic cycle of ocean biogenic elements. Numerous studies show potential bio-available N and P of suspended matter and surface sediments take a relatively high proportion of total bioavailable N and P in estuarine and nearshore waters, and are significant sources and sinks for biological required N and P (Williams et al. 1980; He et al. 1992). N and P concentrations in the sediment of different grain size vary greatly (Ma et al. 2003). Figuring out geochemical characteristics of N and P and estimating their roles and contribution in different grain-size sediments, are the premise of thoroughly discussing N and P biogeochemical cycling. The study area was the Yellow River Estuary, analysis and determination of concentrations and distribution characteristics of N, and P in the Yellow River Estuary surface sediment are of great significance for evaluating dynamic cycles of biogenic elements.

2 MATERIALS AND METHODS

2.1 *Sample collection*

Along the direction of stream channel to Yellow River Estuary, 7 stations' surface sediment samples were collected, and specific sampling sites were shown in Figure 1. Collection and pre-treatment of samples were carried out in accordance with the specification for Marine Monitoring (GB17378-2007). Water salinity was determined by using the CTD and salinity of all sites were shown in Figure 2. Salinity of P1 was 0, and for P7, the salinity was 28.8. From P1 to P7, water salinity showed an increasing trend, and from P4 site to P5 site, salinity changed drastically, showing typical river-sea mixing feature.

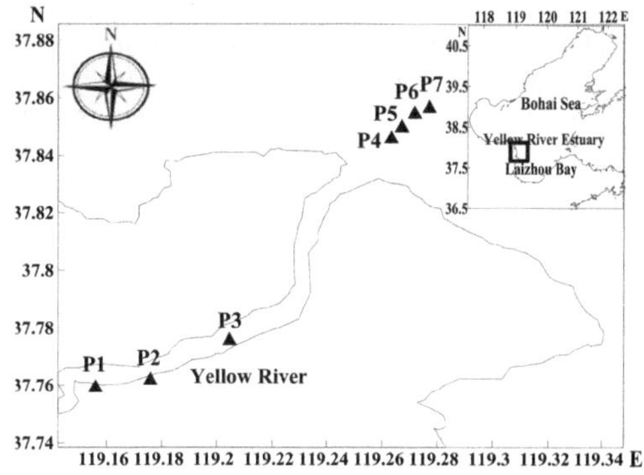

Figure 1. Sampling sites of the Yellow River Estuary.

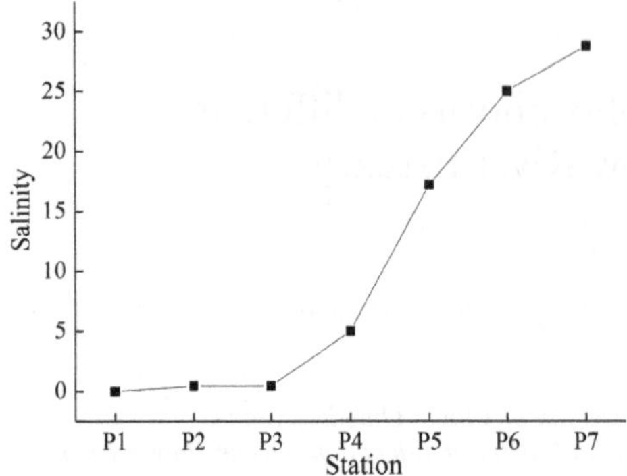

Figure 2. Water salinity of sample sites.

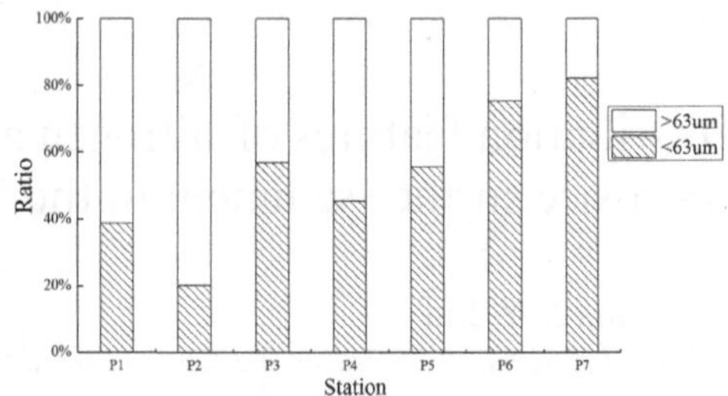

Figure 3. The proportions of sediments grain sizes.

2.2 *Sample analysis and data processing*

Sediment samples were ground to pass a 250-mesh sieve after ultrasonic oscillation, then the sediment samples were divided into three groups: grain sizes > 63 μm, grain sizes < 63 μm and unsized samples. Sediment samples were dried for future use. TN was determined by Kjeldahl method, TP was determined by phosphorus molybdenum blue Spectrophotometric method (specification for Marine Monitoring) (GB17378-2007), recoveries were controlled within the scope of 95%~100% (Luo et al. 2011).

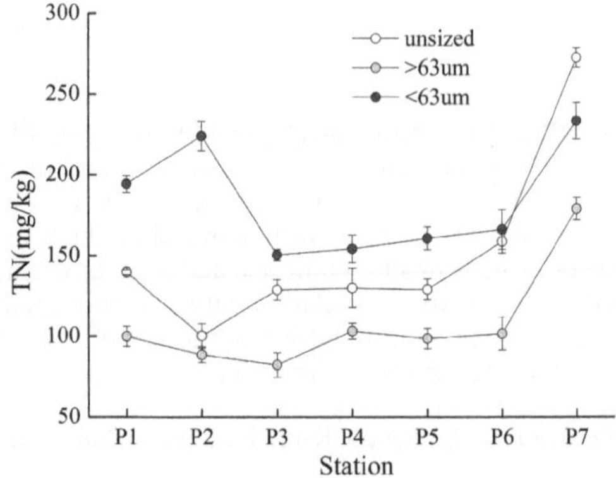

Figure 4. TN concentration of different surface sediments grain sizes. Error bars represent standard error (n = 3).

3 RESULTS AND DISCUSSION

3.1 *Surface sediment particle size distribution*

From Figure3, along the river to the sea, the proportion of the small grain-size content in total sample gradually increased. Far from the sea, sediments with grain size <63 μm of P2 site accounted for only 20.17%, that meant grain size of sediments here was coarse and mainly were sandy sediments; however, for P7 sites at depths of the sea, the content of particles which < 63 μm was in high level, accounting for 82.2%, and this meant sediment particles in this region were fine and sediments here were mainly argillaceous sediments. From the river to the sea, content of fine components in sediments increased gradually.

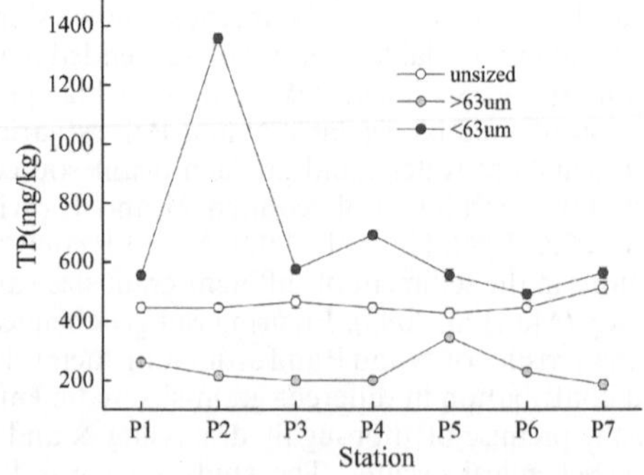

Figure 5. TP concentration of different surface sediments grain sizes. Error bars represent standard error (n = 3).

3.2 *Distribution and concentration of TN in surface sediments*

From Figure 4, TN concentration in the >63 μm sediment samples was in the range of 82. 01~178.75 mg/kg and average concentration was 107.37 mg/kg; TN concentration in the <63 μm sediment samples was in the range of 150.36~233.20 mg/kg and average concentration was 183.19 mg/kg; TN concentration of unsized samples was in the range of 99.99~272.43 mg/kg and average concentration was 151.18 mg/kg.

TN mass in the <63 μm sediment samples dominated TN mass in the total samples with

Table 1. Assessment results between sediment quality guideline and the Yellow River Estuary sediment quality.

Parameter	Security level	Lowest level	Severity level	Experimental data	Evaluation
TN (μg/g)	<550	550	4800	99.99–272.43	Security
TP (μg/g)	<600	600	2000	427.02–512.73	Security

a 39.02%~85.76% ratio and average ratio was 65.21%, that showed TN mass in small particles remained a greater ratio of TN in all the samples, and the proportion showed a significant increase from the river to the sea. A reasonable explanation may be that particles of small size had a large specific surface area and its adsorption capacity of N was strong, which finally resulted in a higher level of N concentration (Christensen et al. 1985; Chichester et al. 1969). Therefore, small grain-size sediment samples in Yellow River Estuary contributed more to the N biogeochemical cycling.

3.3 *Distribution and concentration of TP in surface sediments*

From Figure 5, TP concentration in the >63 μm sediment samples was in the range of 184.97~346.22 mg/kg and average concentration was 728.25 mg/kg; TP concentration in the <63 μm sediment samples was in the range of 491.55~1357.68 mg/kg and average concentration was 728.25 mg/kg; TP concentration of unsized samples was in the range of 427.02~512.73 mg/kg and average concentration was 455.60 mg/kg.

TP mass in the <63 μm sediment samples dominated TP in the total samples with a 61.43%~93.58% ratio and average ratio was 75.33%, that showed TP mass in small particles remained a greater ratio of TP in the total samples, and the proportion showed a significant increase from the river to the sea. A possible explanation may be: particles of small size had a large specific surface area; clay mineral has a large number of cation and can combine with solid organic particles (Christensen et al. 1985; Chichester et al. 1969). Therefore, P concentration in small grain-size sediment samples was higher and contributed more to the P biogeochemical cycling.

3.4 *Quality evaluation of surface sediments in Yellow River estuary*

This paper was based on sediment quality guidelines (Mudroch et al. 1995) made by the Canada Ontario environment and energy department, according to the eco-toxicity of pollutants on benthic organisms in the sediments, the guide was divided into three levels, including security level, the lowest level and severity level. TN concentration in this paper was significantly lower than the lowest level of sediment quality guidelines. TP concentration was close to the lowest level of sediment quality guidelines. Therefore, pollution of N, P in the Yellow River Estuary sediments was relatively lighter (Table 1).

4 CONCLUSIONS

From fresh water to sea, N and P concentrations in sediments with different fractions increased gradually, but change was not obvious. Relationships between TN, TP concentrations and grain-size sediments were: TN and TP concentrations in the <63 μm sediment samples were much higher than those in unsized samples, while TN and TP concentrations in the >63 μm sediment samples were the lowest. TN concentrations showed a significantly negative correlation with sediment grain size, and TP concentrations showed a weak correlation with sediment grain size. TN and TP concentrations were at safety levels in the sediment assessment guidelines.

ACKNOWLEDGMENTS

This research was funded by the NSFC-Shandong Joint Fund for Marine Science Research Centers (Grant No. U1406403) and the National Marine Public Welfare Research Project of China (Grant No. 201405007).

REFERENCES

[1] Chichester, F.W. 1969. Nitrogen in soil organo-mineral sedimentation fractions. *Soil Science* 107(5): 356~363.

[2] Christensen, B.T. 1985. Carbon and nitrogen in particle size fractions isolated from Danish arable soils by ultrasonic dispersion and gravity-sedimentation. *Acta Agriculturae Scandinavica* 35(2): 175–187.

[3] General Administration of Quality Supervision, Inspection and Quarantine of China, Standardization Administration of China, 2007, GB17378. The specification for marine monitoring. Beijing: Standards Press of China.
[4] He, Q.X. 1992. Distribution characteristics of the environmental geochemical forms for nitrogen and phosphorus in the sediments of Daya Bay. *Journal of Tropic Oceanology* 2: 005.
[5] Ma, H.B. 2003. Nitrogen forms and their functions in recycling of the Bohai Sea sediments. *Geochimica* 32(1): 48–54.
[6] Mudroch, A. & Azcue, J.M. 1995. *Manual of aquatic sediment sampling*. Boca Rat on: Lewis Publications, CRC Press.
[7] Williams, J.D.H. 1980. Availability to Scenedesmus quadricauda of different forms of phosphorus in sedimentary materials from the Great Lakes. *Limnology and Oceanography* 25(1): 1–11.
[8] Luo, X.X. 2011. Distribution characteristics of heavy metals and nutrient elements in inter-tidal surface sediments of the Yellow River estuary. *Ecology and Environmental Sciences* 20(5): 892–897.

Material Science and Environmental Engineering – Chen (Ed.)

Activated carbon impregnated with zirconium hydroxide for toxic chemical removal

L. Li, K. Li, L. Ma & Z.Q. Luan
Research Institute of Chemical Defense, China
State Key Laboratory of NBC Protection for Civilian, China

ABSTRACT: Zirconium hydroxide and activated carbon impregnated with zirconium hydroxide were evaluated for their ability to remove toxic chemical gases, namely SO_2, ClCN and C_6H_6 from steams of air in respirator applications. Zirconium hydroxide impregnated with TEDA displayed a high capacity for the removal of SO_2 and ClCN; however, the ability of zirconium hydroxide to remove C_6H_6 was relatively low. Utilizing the porosity of activated carbon, a new method was put forward for the in-situ synthesis of Zirconium hydroxide, which displays a comprehensive balanced capacity for the removal of acidic and organic toxic chemicals.

Keywords: zirconium hydroxide; activated carbon; protective material

1 INTRODUCTION

Investigators and researchers are leading efforts to increase filter and mask efficiency, broaden filter capabilities to meet emerging threats and reduce the burden to the warfighter that include both organic and inorganic gases.

Zirconium hydroxide is an amorphous material that can be used as catalyst supports. Previous studies have shown that zirconium hydroxide is an effective material for the removal of a wide range of toxic chemicals, especially acidic/acid-forming gases, including cyanogen chloride [1], chlorine, phosgene, hydrogen chloride [2], and sulfur dioxide [3]. Furthermore, zirconium hydroxide is known to provide reactive removal of chemical warfare agents such as VX and GD [4].

However Zirconium hydroxide, due to its basic nature, is not expected to provide adequate removal capabilities against organic compounds, such as benzene and sarin; however, activated carbon can provide enough porosity for adsorption and retention. Activated carbon was selected because it was a common substrate that is used in the preparation of general-purpose filtration media; furthermore activated carbon is less expensive than zirconium hydroxide.

Of particular concern is the removal of highly toxic acid gases such as cyanogen chloride, and sulfur dioxide, all of which are typical of challenges faced in a chemical release scenario [5]. Unimpregnated carbon had a rather low removal capacity for acidic toxic chemical because pure physical adsorption is ineffective for the removal of this type of chemicals. Although the impregnation of activated carbon with 10%CuO can prolong the SO_2 breakthrough time and removal capacity, the performance of this material is less than that of Zirconium hydroxide [3]. An effort is currently underway aimed at impregnating other catalysis components that are capable of removing a wide range of acidic and organic toxic chemical.

Utilizing the porosity of activated carbon and the reactive substrate of zirconium hydroxide, we put forward a new method for the in-situ synthesis of zirconium hydroxide in activated carbon. The objective of this effort is to evaluate the removal capability of porous activated carbon impregnated with zirconium hydroxide for the chemical warfare agent and acid gas, at the same time maintain a certain adsorptive capacity.

2 EXPERIMENTAL SECTIONS

2.1 *Materials*

Activated carbon was obtained in the form of 12 × 30 mesh granules; the density of the formed activated carbon was 0.41 g/cm^3. The surface area and pore volume, determined by N_2 adsorption, following off-gassing at 200°C for ~16h, were 1650 m^2/g and 0.893 cm^3/g, respectively.

A ZrC activated carbon was prepared by in-situ synthesis method dissolving ammonium hydroxide-ammonium carbonate and zirconium sulfate solution. Following impregnation, the granules were spread across a stainless steel sieve tray, dried in a forced convection oven for 2h at 120°C.

As compared sample, nanosized zirconium hydroxide has been prepared combining sol-gel and hydrothermal method, which the surface area was 364 m^2/g; C-Cu was prepared by dissolving copper carbonated in a 15% ammonium hydroxide-5% ammonium carbonate solution.

The composite samples were also loaded with triethylenediamine (TEDA) to 6 wt% (nominal) to produce ZrC-T and $Zr(OH)_4$-T. TEDA was sublimed onto the samples in an oven at 60°C for 4 h. The weight percentage was based on the mass of the individual components.

2.2 *Nitrogen adsorption*

Samples were evaluated for nitrogen adsorption using a Quantachrome Autosob-1 instrument at 77 K. Samples were off-gassed at 110°C for ~16h, achieving a final vacuum of less than 10 μm Hg. The surface area was calculated using the BET method.

2.3 *Breakthrough testing*

Sulfur dioxide and cyanogen chloride breakthrough time was examined by storm filling 12 × 30 mesh granules of filtration media into a 2.0 cm inside diameter jacketed stainless steel tube. All tests were performed employing a 2.0 cm deep bed of filtration material, with testing performed at 25°C employing a feed concentration of 2000 mg/m^3 of SO_2 and 9000 mg/m^3 of CNCl. Details of the testing methodology have been showed in Table 1.

2.4 *Benzene adsorption*

Samples were evaluated for dynamic benzene adsorption using benzene adsorption instrument at 20°C under 0.175 and 0.95 relative pressures.

3 RESULTS AND DISCUSSION

3.1 *Structure and surface analysis*

From Figure 1, zirconium hydroxides are all amorphous by X-ray diffraction, with very broad peaks

Table 1. Breakthrough test parameters.

Parameter	Value
Particle size	12 × 30 mesh
Bed depth	2.0 cm
Air flow velocity	ClCN: 4.17 cm/s; SO_2: 9.6 cm/s
Challenge concentration	ClCN: 9000 mg/m^3; SO_2: 2000 mg/m^3 (765ppm)
Breakthrough concentration	ClCN: MJ-7–82 (titration); SO_2:13 mg/m^3 (5ppm)

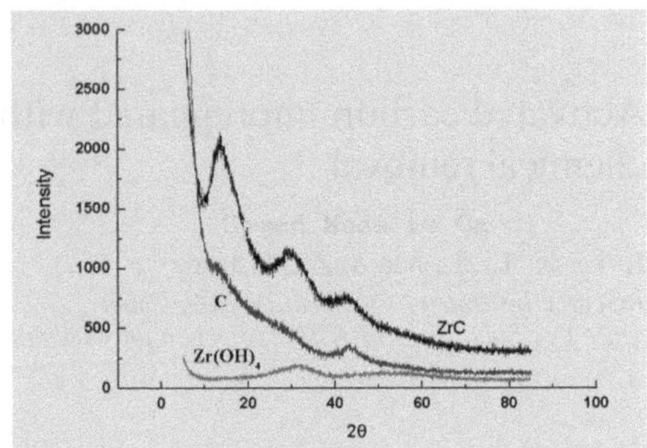

Figure 1. XRD patterns for zirconium hydroxide sample.

at approximately 30° and 50–60° 2θ, but will crystallize as zirconium oxide at calcination temperatures in excess of 350°C (static air). The activated carbon also exhibits broad peaks and shows amorphous character. After impregnating $Zr(OH)_4$, ZrC specie exhibits the corresponding peak at 30° at the same site of $Zr(OH)_4$ and shows to generate a mixture of amorphous $Zr(OH)_4$ and some impurity, according the following equation.

$$Zr^{4+} + 4\,NH_3 \cdot H_2O \rightarrow Zr(OH)_4 + 4\,NH_4^+$$

Nitrogen isotherm data were collected to determine the changes in porosity due to calcination. Isotherm data is shown in Figure 2, and BET surface area and porosity measurements are summarized in Table 2. Compared to $Zr(OH)_4$, the activated carbon samples show significantly higher nitrogen uptake, which assume they have better adsorptive capacity. After impregnating $Zr(OH)_4$,Cu and TEDA, the activated carbon samples have a slightly lower nitrogen uptake but still exhibits extensive total porosity. Yet the microporosity decreases and the mesopore increases in the ZrC material compared to activated carbon, which the result shows some $Zr(OH)_4$ maybe generate in microporosity as nano size, and some $Zr(OH)_4$ maybe generate in macroporosity that make the mesopore increasing. The porosity structure of ZrC is helpful to disperse the active components as a catalyst substrate and can furthermore impregnate other active components such as TEDA.

3.2 *Gaseous contaminants removal*

Removal capacities for SO_2, ClCN and C_6H_6 using samples studied are presented in Table 3. From the results presented in Table 3, zirconium hydroxide displays the lowest C_6H_6 capacity as a result of the smallest BET surface. And zirconium hydroxide impregnated TEDA also shows lower

CNCl capacity comparing the activated carbon samples.

SO_2 was selected as a representative acid gas for testing, and it is a weak acid that will interact with basic metal oxides and hydroxides to form the corresponding sulfites (SO_3^{-2}) and sulfates (SO_4^{-2}). [3] As results presented in Table 3 demonstrate, pure zirconium hydroxide displays the greatest SO_2 removal capacity. Peterson and co-workers investigated the SO_2 reactive adsorption on porous zirconium hydroxide [3]. It was found that SO_2 was strongly adsorbed via formation of sulfites, as shown the following equation.

$$Zr(OH)_4 + SO_2 \rightarrow Zr(OH)_2(SO_3) + H_2O$$

$$Zr(OH)_2(SO_3) + 2SO_2 \rightarrow Zr\ (SO_3)_2 + H_2O$$

But the pure zirconium hydroxide has poor removal capacity for ClCN and C_6H_6.

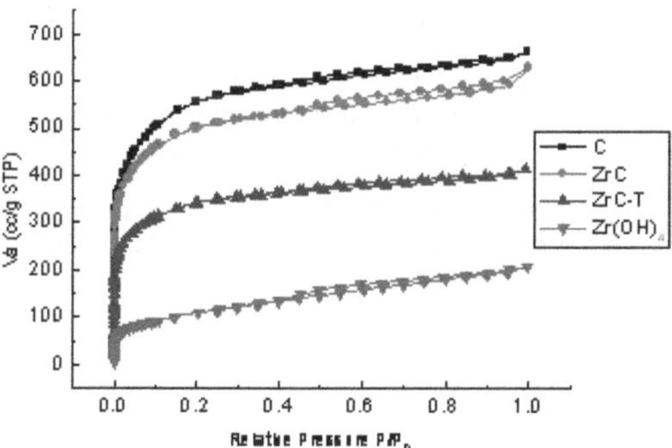

Figure 2. Nitrogen isotherms for samples studied.

Table 2. Surface area and pore volumes for samples studied.

Material	S_{BET} (m^2/g)	V_{Total} (ml/g)	$V_{micro,MP}$ (ml/g)	$V_{meso,BJH}$ (ml/g)
C	1650	0.893	0.766	0.127
ZrC	1427	0.975	0.753	0.222
ZrC-T	1241	0.644	0.513	0.131
$Zr(OH)_4$	364	0.437	0.235	0.202

Triethylenediamine (TEDA), a widely used carbon impregnate, was a strongly alkaline compound which can react with acidic compound theoretically [1]. Data were recorded for the purpose of further assessing the role of TEDA. The TEDA-impregnated $Zr(OH)_4$ displayed a significantly greater capacity for the removal of ClCN, increasing the breakthrough time from 2 to 25 min, which Peterson reported the breakthrough time could reach 87 min at the lower feed concentration of 4000 mg/m^3 [1] (see Table 3). The enhanced capacity results that cyanogen chloride is first hydrolyzed, facilitated by TEDA, yielding a mixture of inorganic and organic byproducts. But with the addition of TEDA, $Zr(OH)_4$ shows a decreasing capacity for SO_2 from 69 to 30 min (see Table 3). And $Zr(OH)_4$ also shows a limited capacity for physical absorption, such as C_6H_6, however TEDA exists or not. Therefore, Zirconium hydroxide was in-situ synthesized in the porosity of activated carbon, expecting to obtain comprehensive balanced protection.

In Table 3, the abilities of ZrC to remove SO_2 and ClCN were both twice higher than C-Cu. ClCN cannot be effectively removed by physical adsorption but required reactive sites within the sorbent for permanent removal. For ZrC series samples, the SO_2 and ClCN were removed via reactions involving surface hydroxyl groups of zirconium hydroxide [1,3].

Figure 3 compared the SO_2 breakthrough curves recorded using activated carbon impregnated with different components, which breakthrough concentration was 5 ppm. In Figure 3, the ability of ZrC to remove SO_2 was twice higher than C-Cu. Consistent with results reported in the literature [6–8], impregnation of carbon with CuO increased the SO_2 breakthrough time via an oxidation mechanism [8].

Table 3 presented the addition of TEDA to the ZrC improves the performance versus cyanogen chloride, increasing the breakthrough time about 7 minutes. But the addition of TEDA to the ZrC hardly improved the performance versus SO_2 and C_6H_6, as illustrated by the data of ZrC-T. This behavior is alike in $Zr(OH)_4$-T. One possible explanation is that TEDA is physically blocking pores and/or reactive sites, thereby neutralizing its contributions to chemical reactivity.

Table 3. SO_2,CNCl and C_6H_6 removal capacities for samples studied.

Gas	ZrC	ZrC-T	C-Cu	$Zr(OH)_4$	$Zr(OH)_4$-T
SO_2 (min)	56	35	25	69	30
ClCN (min)	34	41	19	2	25
C_6H_6 (mg/g; P/Ps = 0.175)	385.9	338.6	345.6	41.2	36
C_6H_6 (mg/g; P/Ps = 0.95)	391.6	364.8	355.8	110.1	58

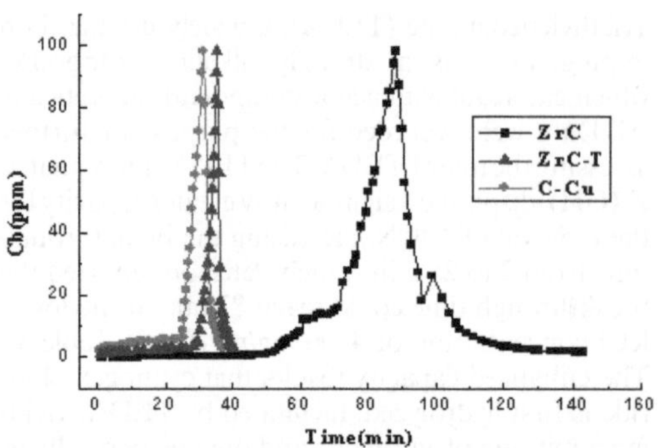

Figure 3. SO_2 breakthrough curves recorded using activated carbon impregnated with different components.

4 SUMMARY

Impregnated with TEDA, the zirconium hydroxide displays a significantly greater capacity for the removal of ClCN but shows a limited capacity for physical absorption, such as C_6H_6. Activated carbon impregnated with $Zr(OH)_4$ displays a comprehensive balanced capacity for the removal of acidic and physical adsorption toxic chemical. The addition of TEDA in ZrC series samples can enhance the capability for the removal of ClCN but is no effects for SO_2.

ACKNOWLEDGEMENTS

This work was financially supported by the State Key Laboratory of NBC Protection for Civilian.

REFERENCES

[1] Peterson G.W., Wagner G.W. & Jennifer H. Keller. 2010. Enhanced Cyanogen Chloride Removal by the Reactive Zirconium Hydroxide Substrate. Industrial & Engineering Chemistry Research. 49(22): 11182–11187.

[2] Peterson G.W. & Joseph A.R. 2012. Removal of Chlorine Gases from Streams of Air Using Reactive Zirconium Hydroxide Based Filtration Media. Industrial & Engineering Chemistry Research. 51(10): 2675–2681.

[3] Peterson G.W., Christopher J.K. 2009. Zirconium Hydroxide as a Reactive Substrate for the Removal of Sulfur Dioxide. Industrial & Engineering Chemistry Research. 48(4): 1694–1698.

[4] Bandose T.J, Laskoski M. & Mahle J. et al. 2012. J. Phys. Chem. C, 116: 11606.

[5] Smith J.W., Westreich P. & Croll L.M. et al. 2009. Understanding the Role of Each Ingredient in a Basic Copper Carbonate Based Impregnation Recipe for Respirator Carbons. J. Colloid Interface Sci. 337:313.

[6] Waqif M., Saad A.M., Bensitel M. et al. 1992. Comparative Study of SO2 Adsorption of Metal Oxides. J. Chem. Soc. Faraday. Trans. 88: 293.

[7] Carabineiro S.A.C., Ramos A.M, Vital J. et al. 2003. Adsorption of SO2 using Vanadium and Vanadium Copper Supported on Activated Carbon. Catal. Today. 78: 203.

[8] Tseng H.H., Wey M.Y., Fu C.H., 2003. Carbon Materials as Catalyst Supports for SO2 Oxidation: Catalytic Activity of CuO-AC. Carbon 41: 139.

Material Science and Environmental Engineering – Chen (Ed.)

Research on the resource-constrained government investment multi-project scheduling problems

Y.N. Wu & Y. Ding
North China Electric Power University, Beijing, China

ABSTRACT: The government investment projects are often carried out at the same time, but there are lots of resource-constrained problems among the agent construction projects. Therefore, the project scheduling management of multiple agent projects has a vital significance for government construction project management. In this paper, based on the government investment agent, we studied the government investment multiple project scheduling. We first proposed to establish the mechanism to solve the communication problem between the agent and the government. Based on this, we built the resource-constrained project scheduling model, so that we can make full use of the limited resources among multiple projects from the agent. The application of the mechanism and model in real projects confirmed that we can scientifically achieve the optimization scheduling goal.

Keywords: government investment projects; agent construction system; scheduling mechanism; multi-project scheduling model

1 INTRODUCTION

In recent years, the government investment projects have gained growing attention from the public. Thus, as the investor, government also needs to plan and schedule the projects more reasonably. The current government investment projects mainly use the agent construction system, but the resources of the agent are always limited, and multi-agent resource allocation is easy to create the waste of reusable resources. Therefore, scheduling the resources reasonably in a number of government investment projects is particularly important.

Chen Ning [1] proposed an efficiency model of resource allocation and dynamic alliance among enterprises, which solved single enterprise resources constrained problems and partner communication problem between the enterprises, and constructed an efficiency reconstructed model across the enterprise resource allocation. Chen Zhonglin [2] proposed the owner multi-project management organizational maturity evaluation index system and resource allocation optimization model on the basis of performance evaluation and dynamic programming. This solved the rigid organizational structure, irrational allocation of resources and supply chain management confusion, irrational allocation of resources and confused supply chain management problems in multi-project management. Deng Jing [3] proposed the project priority assessment method to solve the problem of shortage of resources. With financial indicators, customer indicators, internal business process, and learning and growth indicators index built system, with resource allocation in the evaluation results, this model is only suitable for single enterprise resources scheduling. Based on the research of the project organization structure setting, Mats Engwall [5] emphasized that the reasonable allocation of resources is the core problems of project management, and also discussed the mechanisms behind the multi-project scheduling. Scott E. Fricke [6] analyzed the public factors that affected the success of project management, and used the empirical analysis method to overview the many problems in project management. By comparing the factors that affected the success of multi-project and single project management, he pointed out the differences. S.M.T Fatemi Ghomi [7] proposed that the scheduling project as soon as possible was approached to conflict and contradiction problem of limited resource allocation in the process of multi-project management; he simulated a multiple channel queuing system, and set up a mathematical model and carried out the simulation calculation.

In summary, the allocation of resources has become the core of the multi-project scheduling problem. Currently, research on multi-project scheduling mainly focuses on single enterprise resource scheduling, that is the model is built based on multi-project schedule arrangements and cost requirements, and genetic and simulation calculation methods are used to solve the multi-project scheduling problem of single enterprise. However, unlike these studies, the government investment projects are based on agent-dependent government schedule multiple construction parties. This article studies the government investment construction project scheduling problem, and proposes to establish a resource scheduling mechanism between the agent and the government. It clears the scheduling responsibilities of the government and the construction agent in the form of a contract, and builds a mathematical model of resource scheduling to schedule the resource in multiple agents.

2 THE ESTABLISHMENT OF THE GOVERNMENT INVESTMENT PROJECT SCHEDULING MECHANISM

The agent model government investment multi-project scheduling is a multi-project scheduling application in government projects, in which the body of scheduling is the government; the objects of scheduling are projects built by the construction agent. The construction agents are from different organizations, have their respective interests, and undertake an independent project task in a single project. Government party will inevitably face all kinds of contradictions in multi-project scheduling. Therefore, scheduling mechanism is very important.

Scheduling mechanism needs to establish a corresponding scheduling process to guarantee the mechanism of effective operation. In the resource-constrained conditions, the construction side, according to their own strength, develops the resources scheduling and timing plan of the construction project and upload to the government, which include resource requirements plan and resource scheduling plan. The government approves the single project scheduling scheme. When the scheme has achieved the optimal and has a little impact on the multi-project scheduling, the project is not involved in scheduling. When the scheme has not achieved the optimal or has a great impact on multiple projects, the government should build a multi-project scheduling configuration to solve the problems. Model optimization results in combining with scheduling mechanisms to feed back the construction side, construction side in the process of execution continuously upload effective information, the government party to continuously adjust and optimize the multi-project scheduling plan until the end of the project. The specific processes are as shown in Figure 1.

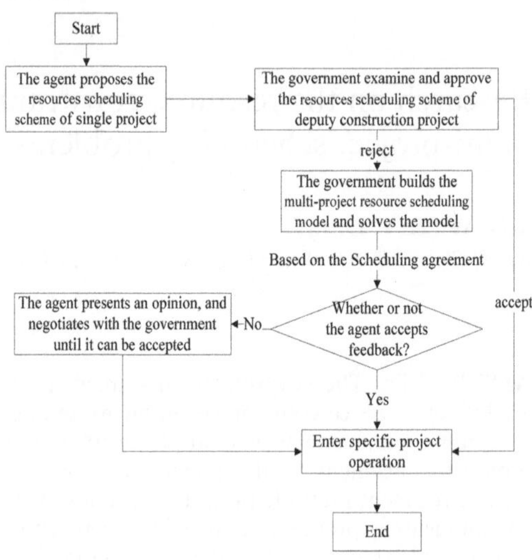

Figure 1. Government investment project resource scheduling process.

Considering the feasibility of the scheduling command, the government investment project scheduling objects should be at the same time similar types of projects. Similar types of projects are required for similar types resources, which is convenient to schedule resource in the construction agent. The scheduled resource should be reusable, which the individual construction agent cannot have, such as senior talents and large equipment. The government investment project scheduling mechanism principle is shown in Figure 2.

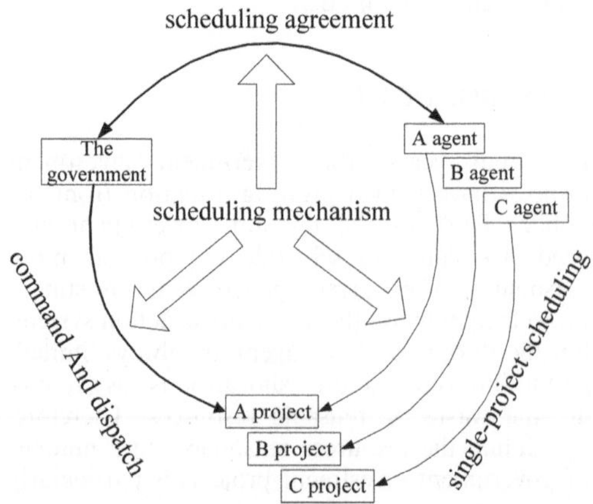

Figure 2. Government investment project scheduling mechanism principle.

In the government scheduling mechanism, the establishment for scheduling agreement is the premise. That is, the government and the construction agent shall join the scheduling agreement in the construction contracts. First, the government evaluates the priority of the scheduling project, in the form of weights visually representing the level of importance of multiple projects. Second, the government gives each agent to build different scheduling agreements according to the priority of each project. Finally, the government and the construction agent sign scheduling agreement through consistent consultation. The scheduling agreement is mainly for incentive and constraint, to make the construction agent behavior consistent with the expectations of the government.

3 THE GOVERNMENT INVESTMENT PROJECTS RCMPSP PROBLEM MODEL

3.1 *Symbol description*

The letter i represents the construction project number, i = 1, 2, ……, N, There are N independent subprojects, where j is the task number, j = 1, 2, ……, Ji; A_{ij} is the j task in the i subproject; d_{ij} is the duration of task A_{ij}; S_{ij} is the start time of task A_{ij}; F_{ij} is the end time of task A_{ij}; A_t is a collection of tasks in working conditions at time t; P_{ij} is a collection of preceding activity of task A_{ij}; w_i is the priority of project i, which is different according to the importance of the project; k is the type of renewable resources, k = 1, 2, ……, K; k_i is the i-th agent that possesses k resource quantity at a time; R_k is the amount of the k kinds of resources at a time; and r_{ijk} is the k-th resource demand for A_{ij} task at each stage.

3.2 *Model creation*

Before the building model, this paper makes the following assumptions: ① schedulable resources are reusable resources in the single construction possess; ② resource scheduling time-transfer cost between different projects is neglected. According to this hypothesis, the RCMPSP model can be established as follows:

$$\min \sum_{i=1}^{N} w_i F_{iJ_i} \tag{1}$$

$$S_{ij} \geq \max A_{ih} \in P_{ij}\ F_{ih} \quad \forall i, j \tag{2}$$

$$\sum A_{ij} \in A_t r_{ijk} \leq R_k \quad \forall k, t \tag{3}$$

$$A(\mathrm{t}) = \left\{ A_{ij} \in U_{i=1}^{N} A_i \middle| S_{ij} \leq t \leq S_{ij} + d_{ij} \right\} \tag{4}$$

$$R_{\mathrm{k}} = \sum_{i=1}^{N} k_i \tag{5}$$

$$\mathrm{r}_{\mathrm{ijk}} \geq 0 \tag{6}$$

$$\sum_{i=1}^{N} w_i = 1 \tag{7}$$

Formula (1) shows the weighted sum of the independent investment project period as the objective function, where F_{iJ_i} shows the end time of the last task for the project i. Formula (2) is logic relationship constraints, where $A_{ih} \in P_{ij}$ shows A_{ih} included in a collection of preceding activity of task A_{ij}, for any task A_{ih}, if A_{ih} is not complete, then A_{ij} would not be able to start, and the start time of A_{ij} must not be less than the maximum of F_{ih}. Formula (3) embodies the resource constraints in the process of scheduling. It shows that the resource supply R_k at a moment must not be less than the sum of the r_{ijk} demand for the resources for the ongoing task at the same time. Formula (4) expresses a collection of the ongoing task at hand at some time in the mathematical equation. Formula (5) constraints on the total resources that are from the sum of each agent resource. Formula (6) constraints on the task of resource demand that it is unlikely to be negative. Formula (7) shows that the sum of the weighted coefficient of each component is 1.

4 THE GOVERNMENT INVESTMENT PROJECTS RCMPSP CASE STUDY

4.1 *Case introduction*

This article uses the multi-project instance from the literature [4] to modify it, which contains four single project instances that have the same preceding activity structure and logical relations, and set the four project priority as 0.4, 0.25, 0.2 and 0.15. Then, they set resources gross, time duration of task, and the needed resources type and demand by each task resource to complete the project instance. According to the resource constrained project scheduling model in the above section, they solve this instance with the simulated annealing algorithm and get the project scheduling optimization solution of the problem.

Figure 3 shows a double code network diagram of a single component. Each component contains 13 tasks, and shares two renewable resources. The limit amount of resources 1 and 2 is 15 and 17, respectively at a time, and these two kinds of resource gross are the sum of the agent resources. We construct

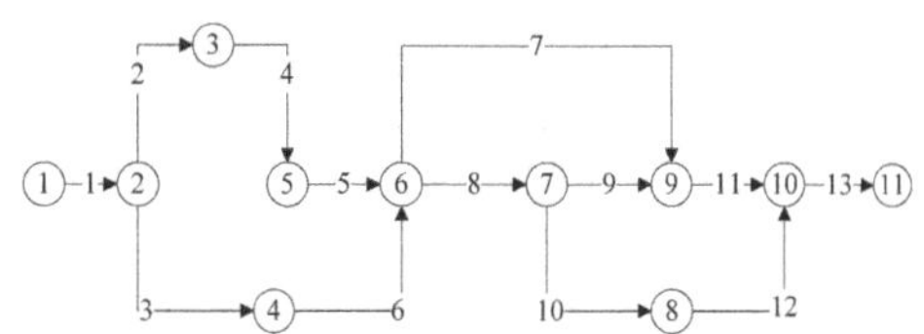

Figure 3. Double code network diagram of a single component.

an example through using random access in the 0–9 series, which takes an integer value randomly for the duration and needed resource amount of different project tasks. Finally, the duration setting for every project task is presented in Table 1, and the resource demand is given in Table 2.

4.2 *Analysis of the test results*

Because the established mathematical model is a nonlinear programming model, the traditional methods are difficult to solve it, thus this article uses the simulated annealing algorithm to solve it. The initial parameters of the algorithm are set, such as initial temperature, coefficient of temperature decreasing, termination of the temperature, and length of Markov chain. A feasible scheduling plan of the instance is obtained by the parameter, as given in Table 3. ST indicates the scheduled start time of the task and FT indicates the scheduled end time of the task.

According to the feasible scheduling plan in Table 3, we can draw the distribution of task renewable resources in the project schedule, as shown in Figure 4 and Figure 5. From the figures, we can examine whether the project timing relationship conforms with the basic requirements of the project logic relationship, and we can see that the unit time consumption of renewable resources has

Table 1. The duration setting for every project task.

Task number	Task duration (day)			
	Project 1	Project 2	Project 3	Project 4
1	3	5	4	4
2	6	1	3	5
3	4	3	7	4
4	7	4	3	3
5	1	5	2	2
6	3	4	2	3
7	4	6	4	6
8	2	3	1	3
9	4	2	5	2
10	5	4	3	4
11	5	3	4	3
12	6	3	4	2
13	3	5	4	4

Table 2. The resource demand for every project task.

Task number	Resource type	Resource demand (unit)			
		Project 1	Project 2	Project 3	Project 4
1	1	8	7	9	4
2	2	3	2	6	7
3	2	5	6	4	4
4	1	9	6	5	6
5	2	8	6	7	5
6	1	5	9	4	9
7	1	4	5	8	4
8	2	5	4	7	6
9	1	6	4	5	2
10	1	5	2	5	5
11	2	8	6	9	7
12	1	7	5	7	6
13	2	8	8	3	5

Table 3. A scheduling plan with resource constraint.

Task number	Project 1		Project 2		Project 3		Project 4	
	ST	*FT*	*ST*	*FT*	*ST*	*FT*	*ST*	*FT*
1	0	3	0	5	5	9	3	7
2	3	9	5	6	9	12	7	12
3	3	7	5	8	9	16	12	16
4	9	16	7	11	15	18	12	15
5	16	17	12	17	18	20	18	20
6	20	23	16	20	18	20	20	23
7	23	27	29	29	33	28	23	29
8	23	25	20	23	20	21	23	26
9	29	33	27	29	33	38	27	29
10	33	38	23	27	33	36	38	42
11	33	38	29	32	38	42	29	32
12	38	44	44	44	44	48	42	44
13	44	47	47	52	48	52	44	48

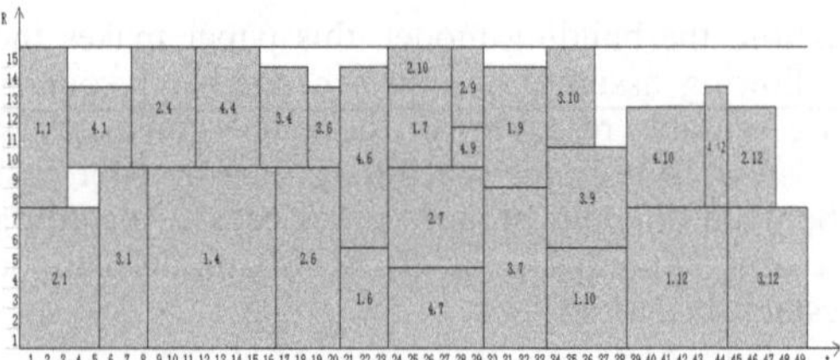

Figure 4. The unit time distribution of resource 1.

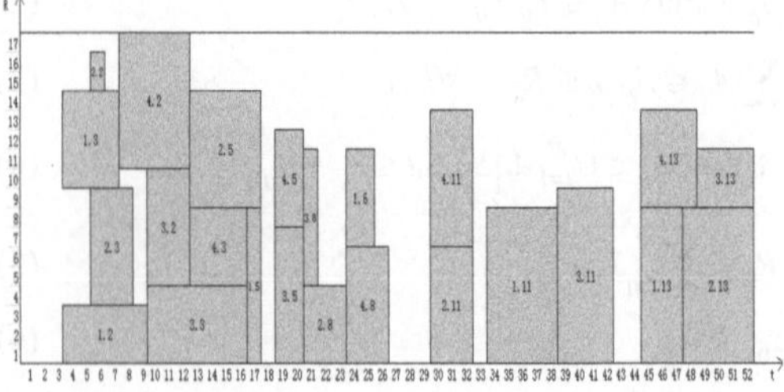

Figure 5. The unit time distribution of resource 2.

not exceeded the restraint of renewable resource 1 and renewable resource 2. It can be seen that the plan completely meets all the constraints requirements of multi-project scheduling.

5 CONCLUSION

Based on the feature of the government investment project that the agent and the investor are separate, this paper first establishes a mechanism between the construction agent and the government to solve the communication problem between them. Second, it builds the minimum weighted time limit for the objective function of the multi-project scheduling model. Finally, it uses the simulated annealing algorithm to calculate the resource scheduling between the construction projects.

The innovation of this paper lies in the following: the mechanism is proposed to coordinate the communication problem between the agent and the government, which can increase the scheduling work efficiency. However, there are still shortcomings in this paper; the resource scheduling time-transfer cost is neglected when we establish the model. Therefore, combining resource transfer with scheduling can be used as a further research direction.

ACKNOWLEDGMENTS

This work was financially supported by the National Nature Science Foundation of China (No. 71271085) and the Beijing Twelfth Five Year Plan Project of Philosophy and Social Sciences (No. 12JGB044).

REFERENCES

[1] Ning Chen & Xueyan Zhang. 2007. Research On the allocation of resources for dynamic enterprise alliance multiple project operation [J]. *Science and Technology Management Research, 01:121–123+126.*

[2] Zhonglin Chen. 2013. Construction project owner multi-project management study [D]. *Beijing Jiaotong University.*

[3] Jing Deng & Jibing Xue. 2010. The research of allocation resources within the group of enterprise projects [J]. *Project Management Technology, 10:65–69.*

[4] Ming Li. 2008. Based on genetic algorithm resource constrained multi-project scheduling problems research [D]. *Zhe Jiang University.*

[5] Mats Engwall & Anna Jerbrant. 2003. The resource allocation syndrome: the prime challenge of multi-project management [J]. *International Journal of Project Management, 10:403–409.*

[6] Scott E. Fricke & Aron J. Shenhar. 2000. Managing Multiple Engineering Projects in a Manufacturing Support Environment [J]. *IEEE Transactions on Engineering Management, 47(2):258–268.*

[7] S.M.T. Fatemi Ghomi & B. Ashjari. 2002. A simulation model for multi-project resource allocation [J]. *International Journal of Project Management, 20:127–130.*

Material Science and Environmental Engineering – Chen (Ed.)
© 2016 Taylor & Francis Group, London, ISBN 978-1-138-02938-5

Characterization of the sediment pore in Chaohu Lake with a new *in situ* sampling method

S.F. Wen
China Institute for Rural Studies, School of Public Policy and Management, Tsinghua University, Beijing, China

ABSTRACT: An integrated approach with the *in situ* sampling method and micro-CT was used to characterize the sediment pore on the μm scale in Chaohu Lake. 3D pore images of the sediments with a resolution of 4.8 μm $pixel^{-1}$ were obtained and used to calculate the sediment microscale pore parameters, such as porosity, connectivity, tortuosity and fractal dimensions, by a novel approach to model ion diffusion in a real sediment micropore. An exponential relationship between tortuosity and porosity was elucidated ($\theta^2 = 0.94 \times \varphi^{-1.19}$, R2 = 0.9828) in Chaohu Lake.

Keywords: unlithified sediments; micro-CT; 3D pore; μm scale; tortuosity

1 INTRODUCTION

Knowledge regarding the chemistry, physics and processes of surface sediments is essential for characterizing pollutant fluxes in sediments (Fernandez, Lao et al. 2014). Research regarding biogeochemical processes in surface sediments has progressed greatly in terms of the heterogeneity of the innovative methods used for determining pore water concentrations and characteristics, including the use of microelectrodes, planar optodes, and diffusive gradients in thin films (Stockdale et al. 2009, Amato et al. 2014). Compared with pore water determinations, the evaluation of microscale sediment pore parameters has progressed slowly because no appropriate method is available (Bowles et al. 1969, Uramoto et al. 2014).

Pore structure contains macroscopic and microscopic parameters. Macroscopic pore structure parameters represent the average behavior of a large sample, which are mostly measured by indirect methods. Macroscopic parameters, such as porosity, permeability, specific surface area, and formation resistivity factor, are completely determined by the pore structure of the medium, and do not depend on any other property (Dullien 1991). Microscopic pore structure is an extremely difficult subject because of the great irregularity of pore geometry, but important in explaining and correlating various phenomena in porous media. With the development of tomography and computation technology, a great deal of work has been done to improve the understanding of the microscopic pore structure, e.g. fractal geometry and tortuosity model.

Fractal theory is an effective and available method for quantifying the complex and irregular pore structure (Sarkar & Chaudhuri 1992). Fractal dimension is a method-dependent concept for different destinations to describe the self-similarity (Mandelbrot 1983, Falconer 2013). Fractal dimension calculated by pore images can be used to characterize roughness (surface fractal dimension/Ds) and pore size distribution (volume fractal dimention/Dv) (Sarkar & Chaudhuri 1992). With the same porosity, fractal dimension can indicate the difference in pore complexity. Tortuosity is a term used to describe the sinuosity and interconnectedness of the pore space as it affects transport processes through porous media (Clennell 1997). There are experimental, empirical and theoretical methods to calculate tortuosity. One method is to measure the diffusion coefficient of a chosen non-reactive species both in the free solution and in a sediment of known porosity (Sweertz et al. 1991), which is a time-consuming process. Another method is to relate tortuosity to a measured quantity called the formation factor, f, that is obtained from electrical resistivity measurements (McDUFF & Ellis 1979). Random walk in pore to simulate the diffusion is another method based on the 3D micropore model. The pore modeling offered a way to the verification of the empirical formula between tortuosity and porosity (Boudreau 1996, Shen & Chen 2007, Matyka et al. 2008, Stockdale et al. 2009).

In the sediment pore structure research, the breakthrough of the sampling and processing method makes it possible for the visualization and numerical technologies conducted to calculate complex microscope 3D pore parameters, such as fractal dimension, connectivity, pore size distribution, and tortuosity. Microtomography yields three-dimensional images of sediment, with a digital resolution of 4.8 μm. Although several sediment macropores were studied by computed tomography (Orsi et al. 1994, Perez et al. 1999), this is the first application of microtomography for examining the sediment pore in such a high resolution. The high-resolution images make it possible for the general pore parameters to change from average in centimeters to spatial heterogeneity in microns, for pore geometry change from a single parameter to an integrated parameter, and pore-size distribution to fractal dimension, which can indicate the complexity and irregularity of the pore structure.

The present paper used the computed numerical method to describe the macro- and microscale sediment pore structures based on the high resolution 3D images, leading to a new understanding of the sediment pore structure. Samples with different particle sources in Chaohu Lake were analyzed to check the empirical formula between tortuosity and porosity given by previous research.

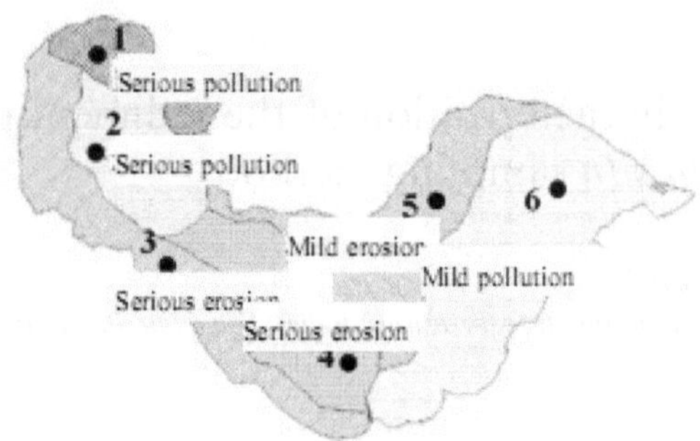

Figure 1. Sampling sites in Chaohu Lake.

Table 1. General description of the sediment sampling sites.

No	Particle size/μm	Organic matter mg/Kg	Water content %
1	30–45	>20	78.8%
2	30–45	>20	82.5%
3	45–100	<10	69.5%
4	30–45	15–20	70.2%
5	<20	<10	70.9%
6	<20	15–20	72.6%

2 METHODS

2.1 *Sampling site*

The Chaohu Lake watershed (30°58′–32°06′N, 116°24′–118°00′E) is located along the left bank of the Yangtze River and covers an area of 13,486 km^2. The watershed is subject to a transitional subtropical–warm temperate monsoon climate with 1,000–1,158 mm annual average precipitation. Chaohu Lake is a typical shallow lake with a depth of 3 m and surface area of 760 km^2. There are eight important rivers in the lake, with Nanfei-Dianbu pollution being heaviest, and Fengle-Hangbu and Baishishan having serious erosion. The lake is divided into an eastern lake and a western lake, with the western part suffers from more serious eutrophication than the eastern part. Taking the water input and flow distribution into account, six sediment cores with different clay sources were selected to analyze the sediment pore structure (Fig.1 and Table 1).

2.2 *Sampling method*

To obtain 3D sediment pore images, the following four approaches (Fig. 2) should be considered to maintain an intact pore structure. (1) Shock-freezing the sediment *in situ*: during sampling, it is important that the surface sediment, especially the floc-bed sediment, does not move. A novel piece of sampling equipment was developed to satisfy this requirement. (2) Embedding the sediment: before scanning with micro-CT, the sediment should be sufficiently small and stable to endure 2 hours of scanning. A series of processing steps, including precutting, pre-embedding, freeze-drying, and impregnation, were conducted to transform the shock-frozen sediment cores to a stable sample with a suitable size. (3) CT scanning and reconstruction: to obtain a 3D gray image of the sediment on the μm scale, micro-CT was used to scan and reconstruct the sediment. (4) Image interpretation: a threshold method was used to separate the pore space from the sediment particles in the gray images. The reference samples were composed of resin and squeezed sediment particles. After completing the above-described procedures, the sediment pore structural parameters, including porosity, fractal dimensions, connectivity, and tortuosity, were calculated based on the 3D sediment pore images.

2.3 *Calculation method of pore structure parameters*

Density, porosity, fractal dimension, connectivity, and tortuosity were chosen to analyze the sediment pore structure (Table 2) on the basis of 3D sediment gray and binary images.

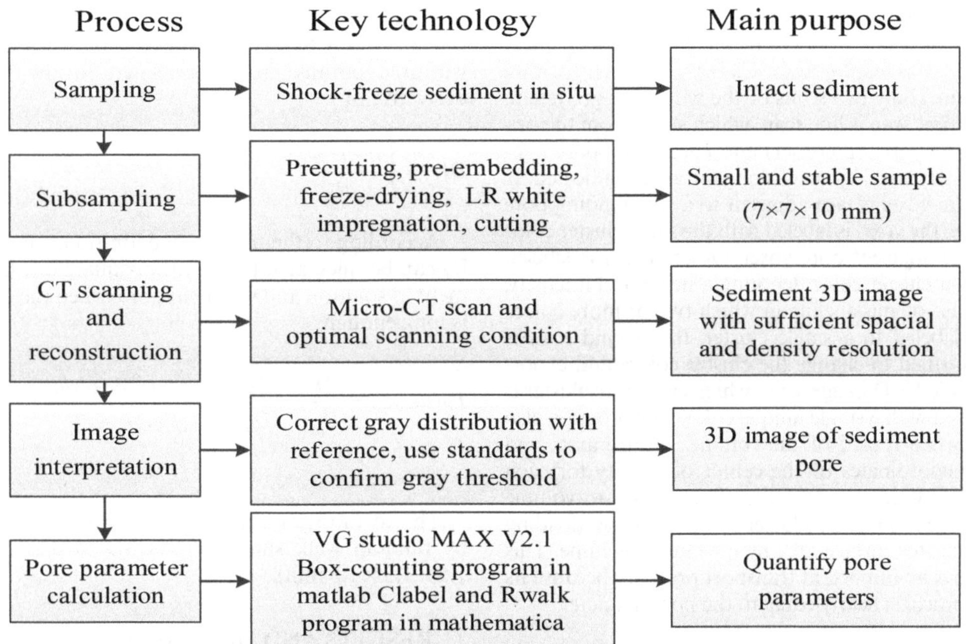

Figure 2. Integrated approaches for characterizing sediment 3D pore structural parameters on the μm scale.

Table 2. Calculation method of pore structure parameters.

Parameter	Physical meaning	Calculation method	Software	Program
Density	Sediment component	Statistic voxel with different gray values in 3D images	VG studio MAX	
Porosity	Pore volume proportion	Statistics pore voxel in 3D binary images	VG studio MAX	
Volume fractal dimension/Dv	Pore-size distribution	Box-counting method in the binary images of the pore	Matlab	FractalDv.m
Surface fractal dimension/Ds	Pore surface roughness	Box-counting method in the binary images of the pore edge	Matlab	FractalDs.m
Connectivity	Transport properties	Cluster labeling of the pore	Mathematica	Clabel.nb
Tortuosity	Diffuse properties	Simulate diffusion of non-sorbing species in a pore cluster by discrete lattice walk	Mathematica	Rwalk.nb

2.3.1 *Fractal dimension*

Fractal dimensions of the parameters were measured to describe the complexity of the pore structure. Box-counting fractal dimension definitions were employed in this paper (Russell et al. 1998). They involve three major steps: (1) meshing the image using various step sizes; (2) counting the number of boxes that contain the object of interest; (3) performing least-squares regression of log (Nr) versus log (1/r) to obtain the FD by the slope of the fitted line. A program based on 2D and 3D binary images was written and tested prior to examination using Euclidean structures, deterministic fractals Koch curves and Sierpinski carpet. Selected box sizes, image resolution, and threshold value were considered to reduce the influence on the fractal dimensions (Wang et al. 2012). The method was compared with other existing methods to show that our method is both efficient and accurate.

2.3.2 *Connectivity*

Connectivity and tortuosity calculation programs adopted here, such as Clabel.nb and Rwalk.nb, was written by Nakashima and Kamiya, which is available on the website (Nakashima et al. 2007). The fast algorithm proposed by Hoshen and Kopelman (1976) was employed for Clabel. nb. This algorithm requires only two scans of the whole image system. The first scan is line scan, which starts from the origin (the left top corner) and checks the pore connectivity voxel by voxel along the arrow indicated. If a pore voxel is face-adjacent to a surrounding pore voxel, the voxel is labeled with the same cluster color as the adjacent pore voxel. Otherwise, it is labeled with a cluster color denoting a new voxel intensity. In case of mislabeling in which two or more colors are labeled to a single cluster, the second scan is performed to change the cluster colors and export a labeled 3-D image set in which each pore cluster is labeled with a single unique color. The program also exports a record of the volume, surface area, and 3-D coordinates of the center of gravity for each pore cluster as a text file. The surface-to-volume ratio of each pore cluster is obtained by dividing the cluster surface area by the cluster volume. This ratio is an important transport property because its reciprocal is nearly equal to the pore diameter.

2.3.3 *Tortuosity*

Random walk simulation of pore molecules diffusion using Rwalk.nb is processed in this labeled pore cluster. The random walk performed by Rwalk. nb is a discrete lattice walk in a simple cubic lattice (Stauffer & Aharony 1994). A pore voxel is chosen randomly as the start position of the lattice walk at _$\tau = 0$, where τ is the dimensionless integer time. The walker executes a random jump to one of the nearest pore voxels (the maximum number of the nearest pore voxels is six for a 3-D simple cubic lattice); τ is incremented by a unit time after the jump so that the time becomes $\tau + 1$. If the randomly selected voxel is a solid voxel, the jump is not performed, but the time still becomes $\tau + 1$. The main output of Rwalk. nb is the mean-square displacement, $<r^2>$ of the walkers as a function of τ.

The exact solution of the mean-square displacement for a lattice walk in a free space (i.e., porosity = 100 vol. %), $<r^2>$ free, is given by (Stauffer & Aharony 1994)

$$<r^2>_{free} = 6D_0 t = a^2\tau \tag{1}$$

where t is the time; D_0 is the diffusion coefficient of the walker in the free space without solids; a is the lattice constant of the simple cubic lattice (i.e., the dimension of a cubic CT voxel); and τ is the dimensionless integer time.

For diffusion in sediment pores, $<r^2>$ is reduced compared with $<r^2>$ *free* owing to the obstruction effects of solids. The degree of the reduction is measured quantitatively by tortuosity as follows. The mean-square displacement is important because the (scalar) diffusion coefficient, D, of the non-sorbing species in the three-dimensionally isotropic porous media is related to the time derivative of

$$D(t) = \frac{1}{6}\frac{d<r^2>}{dt} \tag{2}$$

According to the tortuosity definition, tortuosity can be calculated by the mean-square displacement in solution and sediment pore, when the time is long enough:

$$Tortuosity = \frac{D_0}{D_{(s)}} = \frac{D_0}{D_{(t)}} = \frac{d<r(t)^2> free/dt}{d<r(t)^2>/dt}\ t \to \infty \tag{3}$$

The tortuosity of the sediment was calculated by random walk simulation in 3D images (size $1.5 \times 1.5 \times 1.5$ mm).

3 RESULTS AND DISCUSSION

3.1 *Pore characteristics of surface sediments from Chaohu Lake*

The pore cluster-labeled images of the sediment samples from different sites captured their varying complexity (Fig. 3 and Table 3). The sediments from sites 2 and 5 had greater numbers of large pores dispersed throughout the sediment, which were reflected by their higher Dv and lower Ds values. In contrast, the sediments from sites 1, 3, 4, and 6 had relatively homogeneous pore distributions. The differences in the pore distributions may be related to the sediment source. Sites 2 and 5 are located in the center of the lake and are subject to more detritus deposition from lake biota (C/N ratio 6.4–7.0, unpublished data). The other sediments receive greater mineral inputs from watershed soil erosion30 (estuary: sites 1, 3, and 6) or from lakeshore collapse31 (middle lake: site 4).

The connected pore volume in the surface sediments accounted for more than 99.8% of the total pore volume. In addition, the connected pores at

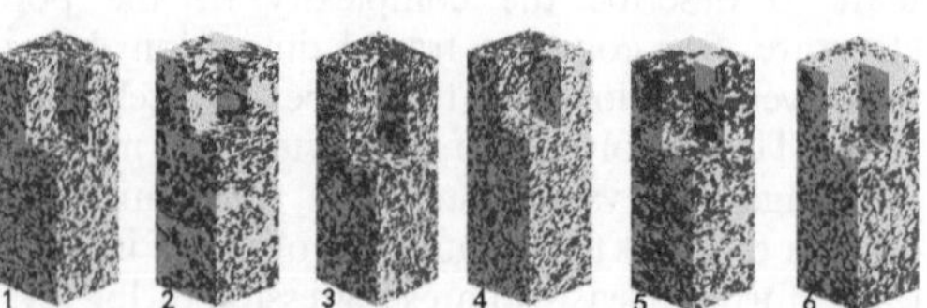

Figure 3. 3D images of connective micropores (diameter > 4.8 μm) in surface sediments from different sites in Chaohu Lake (blue represents particle; yellow represents connective pore; sediment size 1.5 × 1.5 × 4.5 mm).

Table 3. Pores in the surface sediment in Chaohu Lake (300 × 300 × 300 voxels, 3.0 mm^3).

		Conn_pore		Isol_pore					
Site	Porosity	V mm^{-3}	S/V mm^{-1}	Num	$\bar{V}$ μm^3	S/V mm^{-1}	Dv	Ds	Tortuosity
1	0.62	1.86	158	203	1060	703	1.96	1.85	1.74
2	0.58	1.69	160	796	1529	668	1.97	1.78	1.79
3	0.50	1.48	233	1534	1248	681	1.95	1.86	2.16
4	0.60	1.80	155	858	1995	604	1.95	1.80	1.61
5	0.56	1.68	157	1037	1466	658	1.98	1.76	1.74
6	0.57	1.69	73	662	1101	672	1.94	1.77	1.75

* Conn_pore means connective pore, while Isol_pore means isolated pore, with a volume of more than 4 voxels.

site 3 had the largest surface area-to-volume ratio and the largest Ds value. Isolated pores were most abundant in the sediments from site 3. In addition, the sediments from site 3 had a smaller average pore volume and the largest surface area-to-volume ratio. The tortuosity calculated by simulating the non-sorbing species and performing random walks on the largest pore cluster indicated that the pore water species at site 3 traveled the longest path through the sediment during diffusion. Unlithified/unconsolidated sediments have greater porosities than rock and more homogenous pore size distributions than soil. However, the microscale pore parameters of materials were not compared with the macroscale pore characteristics because such a comparison is invalid.

3.2 *Pore profiles of sediments from varying sources on the cm scale*

According to the visualization of the surface sediment pore system, the sediment samples from sites 3 and 5 were chosen for analyzing the porosity changes with depth under sediment diagenesis (Fig. 4). The sediment from site 3 contained materials that originated from soil erosion, while the sediment from site 5 mainly contained materials from biological residues.

Sediment density is represented by the gray value of the sediment CT image. The distributions of gray values covered similar ranges across depth for the sediment from site 3. The domain gray value of the sediment from site 5 (700) was lower than that of site 3 (1200) and increased to 1400 with increasing depth. This result indicated that the sediment at site 5 underwent an obvious diagenetic process, and the sediment at site 3 underwent less diagenesis. In addition, the clear variations in the porosity, Dv, and Ds in the sediments from site 3 indicated that the sediment was heterogeneous with less diagenesis. The decreasing porosity, Dv, and Ds with depth at site 5 indicated that the pore volume distribution became less homogeneous with depth.

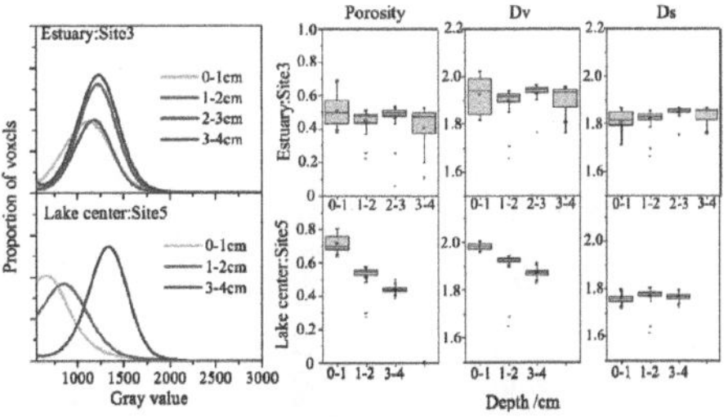

Figure 4. Sediment density and pore parameters on the cm scale.

3.3 *Microscale pore parameter profiles of the sediments*

The changes in the pore characteristics of the sediment near the water-sediment interface should be observed on the microscale. The microscale pore characteristics of a 3 mm-deep portion of sediment were obtained by analyzing a 2D horizontal cross-section image at an interval of 4.8 μm. All of the sediments from the six sites were characterized by decreases in porosity and Dv values with increasing depth. In addition, the Ds value increased rapidly with depth below the interface before stabilizing. The most significant change occurred at the water-sediment interface within the upper range of 0–1 mm.

3.4 *Relationship between tortuosity and porosity in Chaohu Lake*

The tortuosity of the sediment samples from Chaohu Lake ranged from 1.7 to 3.5 across sites and depths, and the porosity ranged from 0.34 to 0.63. The tortuosity and porosity exhibited the following relationship at Chaohu Lake: $\theta^2 = 0.94 \times \varphi^{-1.19}$ ($R^2 = 0.9828$). Two relationships were extensively considered in marine and lacustrine sediment research, Archie's curve ($\theta^2 = \varphi^{1-m}$, $\phi \le 0.7$, m = 2; $\phi >$ 0.7, m = 2.5~3) and Boudreau's cure ($\theta^2 = 1 - \ln\varphi^2$) (Boudreau, 1996; Shen & Chen, 2007). However, the tortuosity and porosity data for Chaohu Lake did

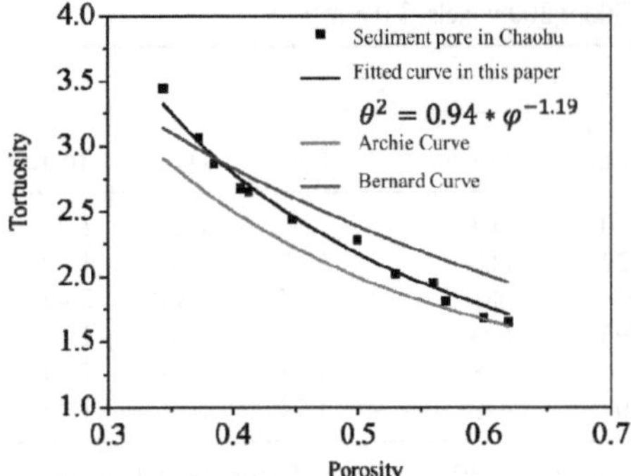

Figure 5. Relation between tortuosity and porosity in Chaohu Lake.

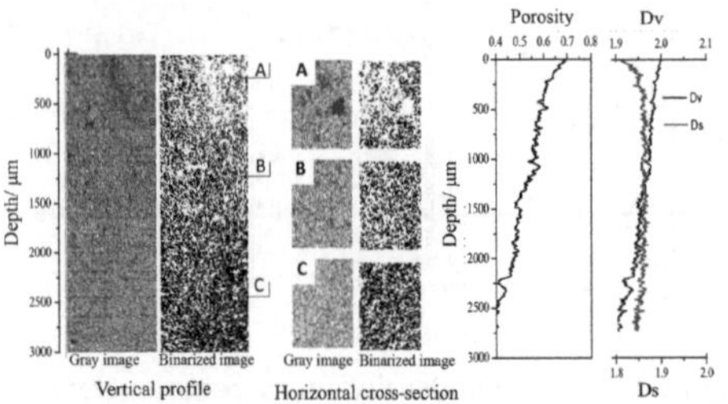

Figure 6. Sediment pore parameter profile on the microscale (black in the binary image represents particle; white represents pore).

not fit these curves well, especially at low porosity values. The random walk simulation of the 3D images produced the following relationship: $\theta^2 = A \times \varphi^B$. It is unknown whether this relationship is applicable to other sediments; thus, this relationship should be evaluated at other locations.

REFERENCES

[1] Amato, E.D., Simpson, S.L., Jarolimek C.V. and Jolley D.F. 2014. Diffusive Gradients in Thin Films Technique Provide Robust Prediction of Metal Bioavailability and Toxicity in Estuarine Sediments. *Environmental Science & Technology* 48(8): 4485–4494.

[2] Boudreau, B.P. 1996. The diffusive tortuosity of fine-grained unlithified sediments. *Geochimica Et Cosmochimica Acta* 60(16): 3139–3142.

[3] Bowles, F.A., Bryant W.R. and Wallin C. 1969. Microstructure of unconsolidated and consolidated marine sediments. *Journal of Sedimentary Petrology* 39(4): 1546-&.

[4] Clennell, M.B. 1997. Tortuosity: a guide through the maze. *Geological Society, London, Special Publications* 122(1): 299–344.

[5] Dullien, F.A. 1991. *Porous media: fluid transport and pore structure*, Academic press.

[6] Falconer, K. 2013. *Fractal geometry: mathematical foundations and applications*, John Wiley & Sons.

[7] Fernandez, L.A., Lao W.J., Maruya K.A. and Burgess R.M. 2014. Calculating the Diffusive Flux of Persistent Organic Pollutants between Sediments and the Water Column on the Palos Verdes Shelf Superfund Site Using Polymeric Passive Samplers. *Environmental Science and Technology* 48(7): 3925–3934.

[8] Hoshen, J. and Kopelman, R. 1996. Percolation and cluster distribution. I. Cluster multiple labeling technique and critical concentration algorithm. *Physical Review B* 14(8): 3438.

[9] Mandelbrot, B.B. 1983. *The fractal geometry of nature,* Macmillan.

[10] Matyka, M., Khalili A. and Koza Z. 2008. Tortuosity-porosity relation in porous media flow. *Physical Review E* 78(2).

[11] Mcduff, R.E. and Ellis R.A. 1979. Determining diffusion coefficients in marine sediments; a laboratory study of the validity of resistivity techniques. *American Journal of Science* 279(6): 666–675.

[12] Nakashima, Y. and Kamiya, S. 2007. Mathematica programs for the analysis of three-dimensional pore connectivity and anisotropic tortuosity of porous rocks using X-ray computed tomography image data. *Journal of Nuclear Science and Technology* 44 (9):1233–1247.

[13] Orsi, T.H., Edwards C.M. and Anderson A.L. 1994. X-Ray Computed Tomography: A Nondestructive Method for Quantitative Analysis of Sediment Cores. *Journal of Sedimentary Research* 64(3).

[14] Perez, K.T., Davey, E.W., Moore, R.H., Bunn, P.R., Rosol, M.S., Cardin, J.A., Johnson R.L. and Kopans D.N. 1999. Application of Computer-aided Tomography (CT) to the study of estuarine benthic communities. *Ecological Applications* 9(3): 1050–1058.

[15] Russell, D.A. Hanson, J.D. and Ott, E. 1980. Dimension of strange attractors. *Physical Review Letters* 45: 1175–1178.

[16] Sarkar, N. and Chaudhuri B.B. 1992. An efficient approach to estimate fractal dimension of textural images. *Pattern recognition* 25(9): 1035–1041.

[17] Shen, L. and Chen Z. 2007. Critical review of the impact of tortuosity on diffusion. *Chemical Engineering Science* 62(14): 3748–3755.

[18] Stockdale, A., Davison W. and Zhang H. 2009. Micro-scale biogeochemical heterogeneity in sediments: A review of available technology and observed evidence. *Earth-Science Reviews* 92(1–2): 81–97.

[19] Stauffer, D. and Aharony, A. 1994. *Introduction to percolation theory*. Taylor and Francis.

[20] Sweertz, J.P., Kelly, C.A., Rudd, J.W., Hesslein R. and Cappenberg, T.E. 1991. Similarity of whole-sediment molecular diffusion coefficients in freshwater sediments of low and high porosity. *Limnology and oceanography* 36(2): 335–342.

[21] Uramoto, G.I., Morono Y., Uematsu K. and Inagaki F. 2011. An improved sample preparation method for imaging microstructures of fine-grained marine sediment using microfocus X-ray computed tomography and scanning electron microscopy. *Limnology and Oceanography-Methods* 12: 469–483.

[22] Wang, H. Liu, Y. Song, Y. Zhao, Y. Zhao, J. and Wang, D. 2012. Fractal analysis and its impact factors on pore structure of artificial cores based on the images obtained using magnetic resonance imaging. *Journal of Applied Geophysics* 860: 70–81.

Material Science and Environmental Engineering – Chen (Ed.)
© 2016 Taylor & Francis Group, London, ISBN 978-1-138-02938-5

China home kitchen waste disposal and recycling way in O2O mode

L.W.J. Zhang, X.Q. Feng & J.J. Wang
Graduate Student of Donghua University, Shanghai, China

ABSTRACT: This paper describes a new way to solve the garbage siege problem. There are two main parts involved in this process. The first part is the preparatory work that has three steps. It details the way to design the garbage size and code number, and how to clean and distinguish the home kitchen waste. In the second part, it introduces the process of how to mark and enter the information, and pay or earn the money online in detail. This method is a new way in O2O mode to help citizens and government to manage the home kitchen waste and reduce the impact on the environment.

Keywords: kitchen waste; waste disposal; waste recycling; O2O mode

1 PRESENT SITUATION AND THE PROBLEMS

1.1 *Kitchen waste disposal and recycling problem*

In China, a large number of cities face the garbage siege problem. But other countries, such as Germany or other Europe countries, do not have these problems, because they have garbage disposal policy and the scientific recycling method (Weiming, 2010).

The kitchen waste is a big part of the rubbish in China; for example, in Shanghai, the kitchen waste accounts for about 59% of the living rubbish (Song, 2013). In kitchen waste, most of them are food waste, compared with the plastic waste that was used for packing the food. Also, the waste includes paper, glass or metal.

Kitchen waste is different from other living garbage. First, kitchen waste is wet because they are used for packing the food. They produce a smell and attract a lot of worms. Second, most of the living waste is recyclable, but the kitchen waste also includes the recyclable part and the non-recyclable part. Third, in China, kitchens have a lot of oil stain.

There is a difference in disposing the rubbish between China and other counties. In Europe, the recyclable waste will be washed when they are used. Then, they take the food waste into the wet garbage bin. In China, most of the citizens will take all kinds of the waste into one plastic bag when they cook in the kitchen.

So in China, the big problem is how to dispose the kitchen wastes. In the past days, some local governments introduced policies for this, which is complex and needs a lot of human resources to explain how to use it.

However, there is another popular way to clean the kitchen waste. In the USA and Japan, some families use the family kitchen garbage grinder. They, put the waste into the machine and use it to crush the waste, and then washed them away. But this method has some problems. First, it will increase water consumption and pollution. Second, it will easily plug the sewer. Third, it is hard to deal with the waste in the terminal of the system.

1.2 *About the O2O mode*

O2O means online to offline, which indicates combining the business opportunities offline and the Internet. China's home kitchen waste disposal and recycling way in O2O mode means to use the Internet to encourage citizens to throw the home kitchen waste in a right way.

2 PREPARATORY WORK

2.1 *Overview*

There are 3 steps that need to be followed before throwing the waste. The first step is the need for a suitable garbage to pack the rubbish, and also each garbage will have a code number. The second step is to clean the waste. The third step is to distinguish the waste. In China, the big problem is how to distinguish the waste, because at most time, citizens just put all the waste together. This will pollute the environment.

2.2 *Garbage size and code number design*

Garbage size is important for cleaning. Different kinds of waste have different sizes. For example, when putting the waste into a very small garbage, the waste cannot hold all the waste, and the cap of the garbage cannot be closed. In this condition, the smell and the germs will spread around. It will also affect children's health. But if the garbage is too big, the garbage will be very hard to clean for a family. So, it is important to design a suitable size.

Design the garbage size according to the kitchen waste disposal and recycling way in O2O mode. In O2O mode, the modular size is easier to run. First, it is easier to calculate the volume of the waste. Then, when the waste is collected together, it is easy to transport and store up. And the modular size can improve the recovery efficiency. Normally, the garbage size can be 150 mm modulus. It can fit the furniture size and run in O2O mode. For instance, the garbage size can be 150 mm × 150 mm × 300 mm or 300 mm × 300 mm × 600 mm. Different kinds of garbage have different sizes according to this rule.

Garbage code number is a necessary part. Today, everybody just look the waste like an unprofitable thing, but effectively they are big wealth for everyone. The garbage can look like a saving box. And the family or the government should understand how much money is in their box now. May be the garbage is in the home or on the way of transport, but each garbage should have a code number. So that, it can have an effective management. On the other hand, if the garbage is broken or it needs to be cleaned, the code number is only used, and it can be found and replaced. The code number should have these important elements. First, it is the location information, different cities or streets will have different numbers, just like the telephone number, i.e. when you call it, you can understand where it is. Second, it is the kind of the waste. Perhaps some garbage has the same color, but we can distinguish them. For example, the food waste can use the number 1; the plastic can use the number 2. Third, the code number should include the special number. The special number is the garbage self-number. It is just like our name. To sum up, the code number can be like "AAABCCC". "AAA" refers to the location information, "B" refers to the kind of the waste, and "CCC" refers to the special number. This will be like our ID number, each one just having one number in the world. Also, the code number can be made into the QR code so that it is easier to distinguish.

2.3 *Cleaning the waste*

The waste should be cleaned before throwing them except the food waste. If we get a broken money, because it cannot be used, we will throw it. The broken money is like the food waste. But if we pick up a dirty money, most of us will clean it before use. This is a question of ideas; if we think the waste is money, we will take it seriously. But in fact, the waste needs to be cleaned before recovering them. And cleaning the waste is not a very difficult thing for anybody. After cleaning, the waste becomes a kind of resources.

The waste maybe can be used in other function. Most waste from the kitchen is plastic and glass, and they can be easily reused. For example, the glass can be used to grow plants. The plastic waste can be used to make some artwork to decorate the home. The waste can be sent to the recycling factory, which cannot be reused in home.

It can save money, manpower and material resources. The waste is washed and can be used directly: they are not only easy to be distinguish as materials, but also can be processed. If the waste is not clean, it will cost more money to hire workers to waste more water to wash them. Sometimes, the waste is very hard to clean because the stain is mixed. Also, it will involve the professional cleaning agent to help them. So. before throwing the waste, everybody should use one or two minutes to clean them, because they know why the waste is dirty and which kind of the stain it is.

2.4 *Distinguishing the waste*

Before throwing the waste into the different kinds of garbage, they should be distinguished. First, some waste can be recovered in home. Second, in China, the plastic bottle and the papers can be sold after use, which have different recycling methods. Third, the biological waste, such as food waste, has the special way to recycle.

In the kitchen waste, there are two big recyclable parts. First, is the food waste. They also can be the pet's food. The other way is to put the food waste into a bottle; after a long time, it will become a good fertilizer for the plants. This is because in most Chinese home, everyone will grow plants, not only for the beautiful feeling, but also the plants will purify air. Second part is the plastic. The plastic is used to pack the food, such as plastic bag and preservative film. It is a good way to use plastic bags to throw the rubbish.

The biological waste has the special way to recycle them. They can be sent to the factory to generate electricity. But before using the plastic bags to pick them, the best way is to make them dry. This is because the dirty water also leaks out from the plastic bags and it will pollute the other wastes or the garbage.

3 KITCHEN WASTE DISPOSAL AND RECYCLING IN O2O MODE

3.1 *Overview*

In home, we cook or eat the food, so that we make the kitchen waste. And every day, the kitchen waste is the big part of the waste. So. using the O2O mode to encourage the citizens and manage the waste is one of the best ways. After the preparatory work, there are three main steps to be followed. First, it should make the waste information, and then enter the information online, and pay or earn the fee online.

3.2 *Mark the waste information*

In China, it is a hard way to require the citizens to mark the garbage bags, because they do not want to do it and do not know how to take papers and write down some words.

But every day, the newspapers are sent from door to door. The date information (Fig. 1) printed on the newspapers can be easily taken out and pasted on to the plastic bag, which will be more effective.

3.3 *Enter the information online*

Website can be found easily, which is the most important thing, and the most useful function can be run on it. After building it, it should be easier to find it. There are some ways for this purpose. First, on the search engine, the main page can be clicked; second, there is much advertisement in the community or somebody can tell the method from door to door; third, the website address can be printed on each package.

The real information can be used to register the users. Everyone should use the real information to register the users because, first, it can manage the waste more efficiently and, second, if somebody fills the false information, he or she will be found quickly. If someone has the problem, the recycling company assistant can go to his or her place.

Entry the waste information when throwing the waste. When throwing the waste, everybody can use the website and enter the details of the information, such garbage code number, waste kind and information. If we want to leave the message, we also can enter it on the website. This process is the most important one, which it will help to add up the volume of the waste.

Checking the waste handling state. The waste recycling process is just like the process of delivery. Everyone can find where it is or which step it is working in real time. Perhaps nobody will check the waste state because they will think it is rubbish and do not need to care them. But if this process can earn the money for them, maybe they will change their ideas.

Jan.1 2016

Garbage Code Number AAABCCC

Garbage O2O Number XXXXXXXXXXXXXXXX

Figure 1. The waste information mark on the newspaper.

3.4 *Pay or earn the fee online*

The website can be used to pay the waste tax. In some countries or cities, citizens should pay the waste tax each month. But how much they should pay for it is a big problem. For rich people, the tax is little money, but for the poor, it is a large fee. Paying the waste tax does not mean that we have a good chance to make a lot of money, as some think that this is their right. So, using the new method will change this situation. For example, if somebody enters the information online each time, the website can calculate the tax automatically.

Earn the waste recycling reward. It is an encourage mechanism to encourage everyone to have a serious attitude towards disposal of waste. According to the analysis of the authoritative data, if China manages the spontaneously recycling process, it will create a very considerable benefit. We take Beijing as an example, in which the public in the waste recycling industry will create a direct economic benefits of more than 11.2 billion yuan (Fuxing, 2002). In some cities in China, if the refuse classification is done, everyone will get the integral, and then use the integral to buy something. But the integral is not the real money, and cannot be used everywhere. If citizens use the China home kitchen waste disposal and recycling way in O2O mode, they will earn the money immediately. They can use it to pay the tax or buy other gifts. In this way, each user should have a cyber-bank. This process will tell everyone that the waste is money.

4 CONCLUSIONS

In China, each city faces the garbage siege problem. Home kitchen waste disposal and recycling way in O2O mode can process the waste more efficiently and benefit the citizens and the government. Everybody can manage the waste, and the environment will be more tidy and beautiful.

REFERENCES

[1] Fuxin. Yang. 2002. Recycling of packaging materials and urban environment. Beijing: Chemical Industry Press.

[2] Song. Lin. 2013. The key technology and equipment of the kitchen garbage processing. Beijing: China Machine Press.

[3] Weiming. Li. 2010. Recycling of waste. Beijing: Chemical Industry Press.

Material Science and Environmental Engineering – Chen (Ed.)
© 2016 Taylor & Francis Group, London, ISBN 978-1-138-02938-5

Comparison of non-linear models for determining the growth rhythm of stem height in *Populus deltoides* seedlings

Q.Y. Ma, J. Hou, C.M. Qian & S.X. Li
The Southern Modern Forestry Collaborative Innovation Center, Nanjing Forestry University, Nanjing, China

ABSTRACT: Seven different non-linear models were constructed for stem heights of 200 *Populus deltoides* progeny in a full-sibs family in Nanjing, China. Based on the determination coefficient (R^2), the best R^2 estimate was obtained by the logistic model, followed by the Gompertz and Von Bertalanffy (>0.990). Residual Mean Square (RMS) and Mean Absolute Deviation (MAD) analyses revealed the logistic model produced the lowest RMS (29.95) and MAD (13.17). Meanwhile the parameter A of the logistic model was notably close to the ending observation value. Considering all statistical analyses, the logistic model presented the best fit in describing the growth pattern in *P. deltoides* seedlings, and then the Gompertz and Von Bertalanffy models. Additionally, we established an ideal statistical model to track the changes in growth rates of a *Populus* pedigree. This study provides essential information and biological interpretation for understanding the dynamic growth rates for *Populus* spp.

Keywords: age; growth curve; non-linear model; *Populous deltoids*; stem height

1 INTRODUCTION

In many temperate regions, Short Rotation Woody Crops (SRWC) such as poplar are one of the most appealing sources because of their high yield and high ecological value; they also provide a renewable biomass feedstock in terms of low input requirements and are prized for their abilities to help land managers maintain ecosystem biodiversity [3]. The increasing attention given to SRWC plantations for the production of biomass has resulted in extensive interest in the genetic analysis of their growth curves. Therefore, biologists are currently very interested in the description of biological growth of these species and are trying to understand the underlying biological processes.

At present, estimates for growth curves tend to be based on non-linear models. Although non-linear models are more difficult to fit than linear models, parameter estimates can be obtained by iteration [18]. Growth studies of complex non-linear functions have been demonstrated to be justified in many branches of science and may be required if the range of independent variable includes juvenile, adolescent, and mature stages of growth [7]. An abundance of analysis of growth curves has been applied to lactation curves of dairy cattle [24, 19], body weight-age curves in poultry, growth curves in sheep, body length-age curves in fish [1, 14, 22]. In addition, some research studies provide estimates of growth curves for woody perennial species, such as *Pinus taeda* L. and *Salix* spp. [9, 5]. As for studies related to poplar, Rizvi [21] constructed non-linear models for aboveground biomass of *P. deltoides* planted in agroforestry areas in Haryana, India.

Of various indicators of biomass, stem height is a notable index that can represent growth performance, especially in seedlings. Furthermore, stem height is easily and accurately measured. The objective of this study was to describe the growth pattern of one-year-old *P. deltoides* using seven non-linear models, the logistic, Gompertz, Von Bertalanffy, Mitscherlich, Korf, Schumacher and Hossfeld models, to provide a biological interpretation for the growth of this species. The results may provide a theoretical basis for exploring the best methods of raising seedlings.

2 MATERIALS AND METHODS

2.1 *Plant materials and origin of data*

A male and a female plant of *P. deltoides* were collected from Nanjing Forestry University, China. A hybrid was created in the spring of 2012, and 200 progeny were planted with 0.5 m × 0.5 m spacing in Baima Nursery in Nanjing (118°46′E, 32°03′N), China. The moderately fertile lateritic soil had a pH of 5.2–6.0. During the experimental period, the mean temperature was 16.4°C, and the maximum and minimum temperatures were 39.7°C and

−6.1°C, respectively. The stem heights of 200 progeny were measured with a steel tape at eight consecutive times from 23 July 2012 to the end of the growing season.

2.2 *Statistical models*

Table 1 presents all seven sigmoidal models that were considered in this study. Initially, several versions of the models were tested to fit the data and the versions gained good fits for all the seven models. Biologically, growth curve parameters were interpreted using the following variables: the height at t (Y), the asymptotic height (A) and integration constant related to initial height (B). The value of B is defined by the initial values for Y and t; the growth rate (k) indicated how fast the plant approached the tallest height as time changes. Individual estimates for growth parameters were obtained by the Gauss–Newton model modified using the NLIN procedure and were calculated with SPSS 17.0 [8]. Convergence was assumed when the difference in the sum of residual squares between the i–1 and ith iteration was $<10^{-8}$. The statistic used to evaluate fit quality included parameter a values for each model, the Residual Mean Square (RMS), the adjusted coefficient of determination (R^2) and the Mean Absolute Deviation (MAD). In theory, when the models are compared, when R^2 is closer to 1, a better fit is indicated [25]. R^2 was calculated as the square correlation between observed and estimated heights, which was equivalent to 1–(RSS/TSSc) [26] where RSS is the residual sum of squares and TSSc is the total sum of squares based on the mean. MAD is calculated using Equation (1):

$$MAD = \sum ni = 1\left|Y_i - \hat{Y}_i\right|77n \quad (1)$$

78

where Y_i is the observed value, $\hat{Y}_i$ is the estimated value, and n is the sample size. A lower value obtained for MAD indicates a better fit.

The Absolute Growth Rate (AGR) was calculated based on the first derivative from the adjusted function in relation to time ($\partial Y/\partial t$) after selecting the function. In fact, the AGR represented the height gained per time unit; therefore, it corresponds to the mean plant growth rate within a population.

Table 1. Non-linear models used in this study to describe the growth of *Populus deltoides*.

Models	Formula	References
Logistic	$Y = A/(1 + Be^{-kt}) + \varepsilon$	Hutchinson [11]
Gompertz	$Y = Ae^{-Be(-kt)} + \varepsilon$	Causton & Venus [4]; Zullinger et al. [30]
Von Bertalanffy	$Y = A(1 - Be^{-kt})^3 + \varepsilon$	Bertalanffy [27]
Mitscherlich	$Y = A(1 + Be^{-kt}) + \varepsilon$	Mitscherlich [15]
Korf	$Y = Ae^{-Bt^{(-k)}} + \varepsilon$	Korf [12]
Schumacher	$Y = Ae^{-B/t} + \varepsilon$	Schumacher [23]
Hossfeld	$Y = A/(1 + Be^{-k}) + \varepsilon$	Hossfeld [10]

3 RESULTS

3.1 *Changes of stem heights during different stages of growth*

The Absolute Growth Rate (AGR) provided some information on plant performance during plant growth. In addition, it indicated the maximum period at which a plant presented satisfactory gains during the one-year growth. The stem heights of 200 progeny at different stages of growth were measured and the means and AGR were calculated and listed in Table 2. The AGR first showed a great increase, especially from day 20 to day 37 and day 52 to day 67, with a gain of approximately 1.6 cm d^{-1} detected. However, it decreased rapidly after age 90 days, especially

Table 2. Average growth of height increment of one-year-old *Populus deltoides* seedlings during different periods.

	Date (month/day)							
	7/23	8/5	8/22	9/7	9/22	10/15	10/30	11/29
Growth periods (d)	7	20	37	52	67	90	105	135
Total growth (cm)	15.6	28.4	56.2	74.3	98.5	121.5	127.7	130.0
Net growth (cm)	–	12.8	40.6	58.7	82.9	105.9	112.1	114.4
Absolute Growth Rate (AGR)*	–	1.0	1.6	1.2	1.6	1.0	0.4	0.1

*Absolute Growth Rate (AGR) = $(H_{i+1}-H_i)/(t_{i+1}-t_i)$, H_{i+1} is the total growth height of $i + 1$, and H_i is the total growth height of i ($i \geq 1$).

after 105 days, which showed an increase of only 0.1 cm d^{-1}. Plotting the growth of 200 individuals against the growth period followed similar S-shaped growth curves (Fig. 1). The growth curves consisted of two phases, one exponential and the other asymptotic which was after 90 days.

3.2 *Comparison and analyses of different models*

Table 3 provides estimates of the parameters for each function and the criteria adopted to evaluate the best adjusted models to fit the growth curves. Based on the coefficient of determination (R^2), we observed that almost all models obtained good results, especially the logistic model which obtained the largest R^2 estimate (0.998). The Gompertz and Von Bertalanffy functions also presented good adjustments with R^2 higher than 0.990, following by the models of Mitscherlich, Korf, and Schumacher that provided an R^2 of 0.982, 0.977, and 0.971, respectively. Among the seven models, the Hossfeld model produced the smallest R^2, which was only 0.942. The analysis of the RMS revealed the smallest value for the logistic model (29.95) when compared with the Gompertz (64.25), Von Bertalanffy (103.92), Mitscherlich (262.43), Korf (333.53), Schumacher (411.81) and Hossfeld (825.59) models. Regarding the MAD values, the logistic model also had the smallest magnitude followed by the Gompertz and Von Bertalanffy functions, while the Hossfeld model produced the largest magnitude.

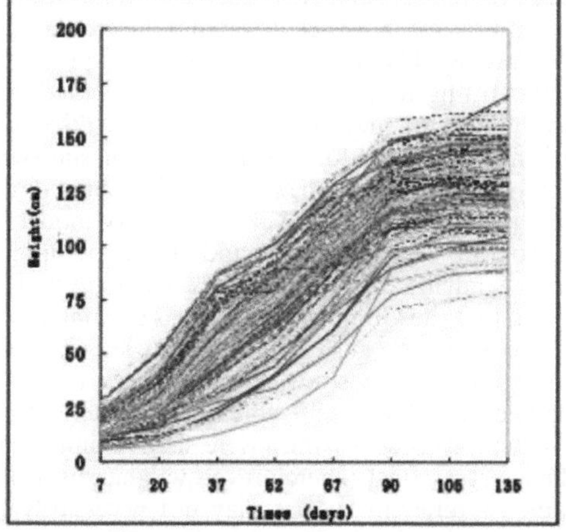

Figure 1. Plots of stem heights against time for two-hundred one-year-old *Populus deltoids* seedlings.

Table 4 shows the predicted versus observed mean stem height values for each measured stage of growth. We found that the logistic model accurately predicted stem heights at each specific measured day; therefore, we concluded that the logistic model provided a valid measure that could be used to evaluate stem height. For the Gompertz and Von Bertalanffy models, the differences between the predicted and observed mean of stem heights were larger than for the logistic model, especially at days 7, 52 and 90.

Parameter A is an estimate of asymptotic height. For the dataset studied here we obtained high asymptotic stem height estimates of 132.9–213292.9 cm (Table 3), Supporting the results above, the parameter A of the logistic model (132.9 cm) was notably close to the ending observed value (131.0 cm) (Table 4); the differences of A of the Gompertz and Von Bertalanffy functions were 8.6 and 13.4, respectively. However, A of the Hossfeld model was too large, which was 213292.9 cm.

Moreover, by the better adjustments of the logistic, Gompertz and Von Bertalanffy models in this study (Fig. 2), we can see that the pattern of the height–time relationship was sigmoidal pattern and all the predicted values were very close to the observed mean stem heights and the largest growth period, which was between days 20 and 67.

Table 3. Parameter estimates, Residual Mean Square (RMS), determination coefficient (R^2) and Mean Absolute Deviation (MAD) for non-linear models describing growth in height for *Populus deltoides*.

	Parameters					
Models	A	B	K	RMS	R^2	MAD
Logistic	132.9	9.81	0.05	29.95	0.998	13.17
Gompertz	139.6	2.96	0.03	64.25	0.995	13.21
Von Bertalanffy	144.4	0.68	0.03	103.92	0.993	14.02
Mitscherlich	168.4	1.03	0.01	262.43	0.982	14.32
Korf	329.9	11.15	0.52	333.53	0.977	14.40
Schumacher	184.0	42.09	–	411.81	0.971	14.83
Hossfeld	213292.9	26333.04	0.58	825.59	0.942	15.02

Table 4. Observed and predicted mean heights of one-year-old *Populus deltoides* seedling, computed using specified non-linear models.

		Time (Days)							
	Models	7	20	37	52	67	90	105	135
Predicted stem height (cm)	Logistic	16.8	28.9	52.5	77.2	99.2	120.1	126.5	131.5
	Gompertz	13.0	28.9	55.6	78.7	97.8	117.5	125.4	133.9
	Von Bertalanffy	11.5	29.7	57.1	79.1	97.0	116.4	124.8	132.0
	Mitscherlich	9.6	34.0	60.3	79.2	94.8	113.6	123.2	129.6
	Richards	7.3	29.3	58.3	79.8	96.9	115.7	124.3	130.4
	Korf	6.4	32.2	60.2	79.2	94.4	112.9	122.7	129.9
	Schumacher	0.5	22.4	59.0	81.9	98.2	115.3	123.3	134.8
	Hossfeld	24.99	45.89	65.53	79.79	92.40	109.61	119.84	130.60
Observed mean height (cm)	–	15.60	28.40	56.20	74.30	98.50	121.50	127.70	131.00

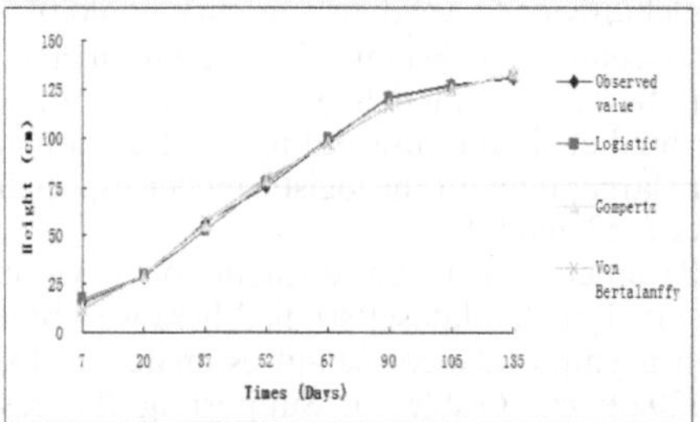

Figure 2. Growth curves of stem heights based on three preferred models and experimental observation of 1-year-old *Populus deltoides* seedling.

4 DISCUSSION

Genetic analysis of plant growth curves has received some attention. Growth curves are considered infinite-dimensional traits or function-valued traits because the traits can be described by an infinite set of measurements [9]. In our study, the growth increments for one-year-old seedlings were periodically measured in a full-sibs family of 200 progeny in *P. deltoides*. We found that each of the individuals followed a slow–rapid–slow growth tendency which was regarded as an S-shaped growth curve (Fig. 1), so non-linear models were chosen to describe the growth pattern in *P. deltoides* in our study. In general, the R^2, RMS and MAD values are the key factors in assessing goodness of fit [16]. In this study logistic, Gompertz and Von Bertalanffy functions presented good adjustments, based on their higher R^2 measurements (higher than 0.990) as well as their lower RMS and MAD values. In addition, among the three models the logistic model produced the largest R^2 (0.998) as well as the lowest RMS (29.95) and MAD (13.17). Considering all analyses of these three statistics and the parameter A (notably close to the ending observation value), the logistic model obviously presented the best fit in describing the growth pattern in *P. deltoides* seedlings. This result was similar to the findings of Rizvi et al. [21] on diameter at breast height for *P. deltoides*. The logistic curve is regarded as the most important one to capture the age-specific change in growth [17, 29, and 28]. The logistic curve not only produced better fitting results in plants, it also worked best for animals [6], and the study of viruses [2]. Thus, it is a good model for a wide variety of statistical analyses. However, Chen et al. [5] achieved the best fitting results with Von Bertalanffy model by comparing with the logistic and Gompertz models when studying *Salix suzhouensis*. Ismail et al. [13] found that the Gompertz model was a suitable tool in modeling the growth of tobacco leaves, stems and roots. In our study, the Gompertz and Von Bertalanffy models also presented better fitting results; nevertheless, they were less suitable than the logistic model. Therefore, growth curves of the same or different species may not necessarily be best described by the same equation [20].

For one-year-old seedlings, the growth of stem heights was more significant than the diameter growth; therefore, in our study we only studied the stem heights of *P. deltoides* with different non-linear models. The goodness-of-fit obtained by our data probably does not provide a sufficiently precise description of the stem heights in the subsequent years. However, it is sufficient to represent the observed field variability and to determine the dynamic growth rates for *Populus* spp. Furthermore, the growth in diameter as determined with different models still needs additional study.

5 CONCLUSIONS

The genus *Populus* (and specifically *P. deltoides)* is a widely used genus (and species) for biomass production in short rotation woody crops. To analyze its growth regularity, the stem heights of two-hundred one-year-old *P. deltoides* seedlings were measured, and we found that each followed an S-shaped growth curve. Seven different growth models were evaluated with regard to their ability to describe the relationship between stem height and time. The studies revealed that the logistic, Gompertz and Von Bertalanffy models were adequate in describing the growth pattern in one-year-old *P. deltoides* progeny. However, the logistic model was considered the most suitable model. This paper may provide a theoretical basis for intensive management of *Populus* plantations at the juvenile stage and help in the exploration of the best seedling-raising technology.

ACKNOWLEDGMENTS

The Key Forestry Public Welfare Project (201304102), the Natural Science Foundation of China (31270711) and Jiangsu Province (BK20130968), the Program for Innovative Research Team in Universities of Jiangsu Province, the Educational Department of China, and the Priority Academic Program Development (PAPD) of Jiangsu Higher Education Institutions provided funding for this work.

REFERENCES

[1] Barbato, G.F. 1991. Genetic architecture of growth curve parameters in chickens. *Theoretical and Applied Genetics*, 83(1): 24–32.

[2] Bergua, M., Luis-Arteaga, M. & Escriu, F. 2008. Comparison of logistic regression and growth function models for the analysis of the incidence of virus infection. *Spanish Journal of Agricultural Research*, 6(S1): 170–176.

[3] Börjesson, P. 1999. Environmental effects of energy crop cultivation in Sweden: identification and quantification. *Biomass Bioenerg*, 16(2): 137–154.

[4] Causton, D.R. & Venus, J.C. 1981. The Biometry of Plant Growth. Edward Arnold. London, UK.

[5] Chen, Y.N., Ma, Q.Y., Qu, J.T. & Dai, X.G. 2013. Growth trajectory modeling for a full-sib pedigree of *Salix suchouenesis*. *Advanced Materials Research*, 733: 784–788.

[6] da Silva, L.S.A., Fraga, A.B., da Silva, F.D.L., Guimarães Beelen, P.M., de Oliveira Silva, R.M., Tonhati, H. & Barros, C.D.C. 2012. Growth curve in Santa Inês sheep. *Small Ruminant Research*, 105(1): 182–185.

[7] Fekedulegn, D., Mac Siurtain, M.P. & Colbert, J.J. 1999. Parameter estimation of non-linear growth models in forestry. *Silva Fennica*, 33(4): 327–336.

[8] George, D. Mallery, P. 2009. SPSS for Windows Step by Step: A Simple Study Guide and Reference, 17.0 Update.

[9] Gwaze, D.P., Bridgwater, F.E. & Williams, C.G. 2002. Genetic analysis of growth curves for a woody perennial species, *Pinus taeda* L. *Theoretical and Applied Genetics*, 105: 526–531.

[10] Hossfeld, J.W. 1822. Mathematik für Forstmänner. Ökonomen und Cameralisten, Gotha, p.4.

[11] Hutchinson, G.E. 1978. An Introduction to Population Ecology. Yates University Press, New Haven, CT., USA.

[12] Korf, V. 1939. Příspěvek kmatematické definici vzrůstového zákona hmot lesních porostů. Lesnická práce. XVIII, s. 339–379.

[13] Ismail, Z., Khamis, A. & Jaafar, M.Y. 2003. Fitting nonlinear Gompertz curve to tobacco growth data. *Journal of Agronomy*, 2: 223–236.

[14] Malhado, C.H.M., Carneiro, P.L.S., Affonso, P.R.AM., Souza Jr, A.A.O. & Sarmento, J.L.R. 2009. Growth curves in dorper sheep crossed with the local Brazilian breeds, Morada Nova, Rabo Largo, and Santa Inês. *Small Ruminant Research*, 84(1–3): 16–21.

[15] Mitscherlich, E.A. 1919. Das Gesetz des Pflanzenwachstums. *Landwirtsch Jahrb*. 53: 167–182.

[16] Motulsky, H.J. & Ransnas, L.A. 1987. Fitting curves to data using nonlinear regression: a practical and nonmathematical review. *The FASEB Journal*, 1(5): 365–374.

[17] Niklas, K.L. 1994. Plant Allometry: The scaling of form and process. University of Chicago, Chicago.

[18] Ratkowsky, D.A. 1990. Handbook of Nonlinear Regression Modeling. Marcel Dekker, New York.

[19] Rekaya, R., Carabano, M.J. & Toro, M.A. 2000. Bayesian analysis of lactation curves of Holstein-Friesian cattle using a nonlinear model. *Journal of Dairy Science*, 83(11): 2691–2701.

[20] Ricklefs, R.E. 1967. A graphical method of fitting equations to growth curves. *Ecology*, 48(6): 978–983.

[21] Rizvi, R.H., Khare, D. & Dhillon, R.S. 2008. Statistical models for aboveground biomass of *Populus deltoides* planted in agroforestry in Haryana. *Tropical Ecology*, 49(1): 35–42.

[22] Rocchetta, G., Vanelli, M.L., Pancaldi, C. 2000. Analysis of inheritance of growth curves in laboratory populations of guppy-fish. *Growth Development and aging*, 64(3): 83–90.

[23] Schumacher, F.X. 1939. A new growth curve and its application to timber yield studies. *Journal of Forestry*, 37: 819–820.

[24] Shanks, R.D., Berger, P.J., Freeman, A.E. & Dickinson, F.N. 1981. Genetic Aspects of Lactation Curves. *Journal of Dairy Science*, 64(9): 1852–1860.

[25] Sit, V. & Melanie, P.C. 1994. Catalog of Curves for Curve Fitting. *Crown Publications, England*, 108 p.

[26] Souza, G.S. 1998. Introduction to linear and nonlinear regression models. Department of Information Production, Brasília.

[27] Von Bertalanffy, L. 1957. Quantitative laws in metabolism and growth. *Quarterly Review of Biology*, 32(3): 217–231.

[28] West, G.B., Brown, J.H. & Enquist, B.J. 2001. A general model for ontogenetic growth. *Nature*, 413: 628–631.

[29] Wu, R.L., Ma, C.X., Yang, M.C.K., Chang, M., Littell, R.C., Santra, U., Wu, S.S., Yin, T.M., Huang, M.R., Wang, M.X. & Casells, G. 2003. Quantitative trait loci for growth trajectories in *Populus*. *Genetics Research*, 81(1): 51–64.

[30] Zullinger, E.M., Ricklefs, R.E., Redford, K.H. & Mace, G.M. 1984. Fitting sigmoid equations to mammalian growth curves. *Journal of Mammalogy*, 65(4): 607–636.

Power engineering and green energy

Material Science and Environmental Engineering – Chen (Ed.)
© 2016 Taylor & Francis Group, London, ISBN 978-1-138-02938-5

Considering the economic benefits of wind power planning in power system

Y.D. Ling & W.X. Liu
North China Electric Power University, Beijing, China

ABSTRACT: This paper analyzes the cost/benefit model of wind farms and the grid company in order to provide a general framework for wind capacity planning. Under the coordination of wind farms and the grid company, our goal is to obtain the optimal wind capacity while achieving the total maximum net benefit of wind farms and the power system. Thus, it will promote the healthy development of renewable energy.

Keywords: wind power planning; economic benefit

1 INTRODUCTION

By the end of 2011, China's energy consumption amount has already surpassed the United States and ranks the highest in the world. Energy demand will continue to grow in the next ten years (Kang et al. 2013). In order to alleviate the dependency on fossil fuel, adjust the unreasonable energy structure, reduce environmental pollution, and realize sustainable development (Zhang et al. 2010), developing clean and renewable energy is the necessary choice of the Chinese government. Among these renewable energy, wind power is one of the most mature in power generation technology, the most large-scale in development conditions and owns the commercial prospects for development (Logan et al. 2008). However, it is obvious that with the rapid increase of wind power installed capacity and the expansion of the scale of wind farm, the phenomenon of wind curtailment becomes more and more serious. In order to solve this problem, it has more realistic significance to study how much wind can be best accepted (Optimal acceptance) in regional power grid. Researchers have explored the acceptance problem in great depth from the perspective of integration factors, integration technology, integration policy, integration capacity and other issues. Considering the stability of the steady-state voltage constraints and thermal stability constraints of lines, it determines the biggest scale of wind farm at a given point by means of successive increases in capacity of wind farm in the simulation calculation (Wiik et al. 2002). Then, it determines the acceptance of wind power capacity of power system based low load peaking capacity of conventional units (Sun et al. 2011).

When considering the ability to accept wind power in the regional power grid, it relates both the wind farms and grid issues. Only when the two companies are in a win-win situation can it really promote the coordinated development of wind power industry. In order to explore the best wind power integration capability this paper proposed an evaluation model with the net income of the combined system (wind farms and the grid) as the objective function from the view of economic benefit, considering transmission security and peak regulation capacity constraints of the power system.

2 OPTIMAL CAPABILITY OF WIND POWER INTEGRATION IN REGIONAL POWER GRID

2.1 *Economic benefit analysis of wind farm*

Wind power investments consist of two parts, which are the initial construction cost and the operation and maintenance cost. A single wind farm's annual net benefit is:

$$I_w = L_{w,su} + L_{w,s} - \frac{C_{w,con} i}{1-(1+i)^{-T_w}} \tag{1}$$

where I_w = annual net benefit of wind farm; $L_{w,su}$ = the annual subsidies for the wind farm, which is determined by renewable energy regulation policy; $L_{w,s}$ = wind farm annual sale-profit; $C_{w,con}$ = the initial construction cost of the wind farm; i = bank interest rate, T_w = the lifetime of the wind farm. $i/(1-(1+i)^{-T_w}$ = the equivalent annual value coefficient.

2.2 *Economic benefit analysis of power grid with wind power integration*

The integration of wind power will inevitably cause certain economic loss to the grid, including the cost of ancillary service, etc. Thus, the economic benefit must be taken into account before wind power integration. It acts as a fundamental factor to find the acceptability of wind power integration into grid, determine whether it is feasible to accept.

The grid's annual net benefit caused by wind power integration is shown in Equation (2), as expressed as:

$$I_g = L_{g,su} + L_{g,selling} - C_{g,op} - C_{g,an} - L_{w,s} - \frac{C_{g,tr} i}{1-(1+i)^{-T_l}} \tag{2}$$

where $L_{g,su}$ = annual subsidies of grid, $L_{g,selling}$ = the annual benefits of electricity selling; $C_{g,an}$ = annual cost of ancillary services for the grid caused by the wind power integration; $C_{g,op}$ = annual operating cost of the grid, here it refers the cost of network loss; $C_{g,tr}$ = the cost of new transmissions lines for wind power integration; T_l = operating life of the new lines.

Ancillary services caused by integration of wind power include peak regulation, spinning reserve and so on. Therefore, the ancillary service cost is expressed as:

$$C_{g,on} = C_{re} \times P_{re} + C_{peak} \times P_{peak} \tag{3}$$

where C_{re} = the unit reserve cost; P_{re} = the reserve capacity needed; C_{peak} = the unit peaking cost; P_{peak} = the deep peak regulation capacity needed.

The peak capacity needed only for wind power is expressed as:

$$P_{peak} = \int_{K_B \times P_{Gn} \geq P_{G,actual}} (K_B \times P_{GN} - P_{G,actual}) \tag{4}$$

where K_B = the factor of peak regulation, generally ranging from 50% to 60%; P_{GN} = rated capacity of the peaking units; $P_{G,actual}$ = actual power output of the thermal units.

The additional reserve capacity is related to the actual power generation of wind power, which can be expressed as:

$$P_{re} = (v_1 + v_2) \times P_{wind,actual} \tag{5}$$

where v_1 and v_2 = the up and down coefficient of reserve capacity, which range from 0.1 to 0.2. $P_{wind,actual}$ = the actual wind power output.

2.3 *Objective function*

The planning of wind power capacity is determined by the combined system (wind farms and the grid) in order to reach the maximum profit. Thus the optimization objective function is:

$$P^*_{CO} = \arg\max_{P_{co}}(I_w + I_g) \tag{6}$$

where P^*_{CO} = the optimized wind power capacity, which will make the combined system get maximum profit; I_w = the net profit of wind farms; I_g = the net profit of the power system.

2.4 *Constraints*

1. Active power balance constraint:

$$\sum_{j=1}^{N} P_{wind,actual,j} + \sum_{i=1}^{M} P_{G,actual,i} = P_{Load} + \Delta P_{Loss} \tag{7}$$

where M = the number of thermal power plants; $P_{G,actual,i}$ = the actual output of thermal power plant i; N = the number of wind farms; $P_{wind,actual,j}$ = the active power output of wind farm j; P_{load} = active load of power system; ΔP_{loss} = active power loss.

2. Output constraint of thermal units:

$$P_{G,i,\min} \leq P_{G,actual,i} \leq P_{G,i,\max} \tag{8}$$

where $P_{G,i,\min}$ = the lower output of thermal power units in technology and $P_{G,i,\max}$ = the upper one.

3. Peak regulation capacity constraint:

$$P_{load,p} - P_{load,v} - (P_{wind,j,p} - P_{wind,j,v}) \leq P^{pr} \tag{9}$$

where $P_{load,p}$ = the peak load; $P_{load,v}$ = the valley load; $P_{wind,j,p}$ = wind power output at system peak time; $P_{wind,j,v}$ = the one of valley time; P^{pr} = the peak regulation capacity of the whole system.

For a generator, it can be operated within the scope from minimum technical output to rated output, and the capacity difference between rated and minimum output is called peak regulation capacity. Thus, for a power system consist of many generators; the whole peak regulation capacity can be described below.

$$P^{pr} = \sum_{i \in M} (P_{i,rated} - P_{Gi,\min}) \tag{10}$$

where P^{pr} = the whole peak regulation capacity of the system; $P_{i,rated}$ = the rated output of the generator and i.

3 IMPROVED DIFFERENTIAL EVOLUTION (DE) ALGORITHM

3.1 *DE algorithm*

The model of optimal capability of wind power integration determines the scale of wind farm construction by optimizing calculation. The computation is a typical optimization problem which is high-dimensional, nonlinear and multi-constrained. As there is an implicit differential equation for calculating the reserve cost, traditional methods based on gradient is unsuitable to solve this model, such as Inter Point. *DE* has fast convergence, fewer adjustable parameters and good robustness, and it is easy to operate. Therefore, *DE* is used to solve complex optimization problems in this paper.

3.2 *Control parameters*

The *DE* optimization process is mainly influenced by weighting factor *F* and crossover constant *CR*, which directly determine the wind farm planning scheme in the evolutionary process. *F* represents amplification factor of the difference vector, and *CR* represents crossover control parameter. In this paper, the parameter control technique is based on the self-adaptation of two parameters (*F* and *CR*), associated with the evolutionary process. The better values of these (encoded) control parameters lead to better individuals, which, in turn, are more likely to survive and produce offspring and, hence, propagate these better parameter values. Though it is difficult to select control parameters reasonably in the absence of prior knowledge, *F* and *CR* can be estimated by the adaptive updating and selection, which can effectively increase the robustness of parameter control. *DE* is used to solve unconstrained problems, therefore, but it is not adapted to the model in this paper. Usually the strategy for constraint problems is Penalty Function, including static and dynamic Penalty Function. The adaptable restricted algorithm of *DE* is proposed to apply *DE* to the solution of the optimal model. New control parameters $F_{i,G+1}$ and $CR_{i,G+1}$ are calculated as:

$$F_{i,G+1} = \begin{cases} r_1 F_{\max} + F_{\min}, & r_2 < \beta_1 \\ F_i, & else \end{cases} \tag{11}$$

$$C_{Ri,G+1} = \begin{cases} r_3, & r_4 < \beta_2 \\ C_{Ri,G}, & else \end{cases} \tag{12}$$

where r_1, r_2, r_3, r_4 = random numbers uniformly distributed between 0 and 1; β_1 and β_2 = the probabilities of *F* and *CR* of corresponding adjustment probability at individuals in a population. Equation (11) and Equation (12) produce factors in a new parent vector. The new *F* takes a value from [0.1, 1] in a random manner. The new *CR* takes a value from [0, 1]. $F_{i,\ G+1}$ and $CR_{iF,\ G+1}$ are obtained before the mutation is performed. So, they influence the mutation, crossover, and selection operations. We have made a decision about the range for *F*, which is determined by values F_{min} and F_{max}. It seems that our self-adaptive DE has even more parameters, but please note that we have used fixed values for F_{min}, F_{max}, β_1 and β_2 and for all benchmark functions in our self-adaptive *DE* algorithm. In this paper, the F_{min}, F_{max}, β_1 and β_2 are set as 0.1, 0.9, 0.1 and 0.1 respectively.

3.3 *The optimization flow*

The optimization flow of the model is as follows:

Step 1 Initialize installed capacity of each wind farm. Obtain the local wind speed data, thermal power units installed capacity and load forecasting data. Then, get wind power output of the whole year, according to the output module.

Step 2 Calculate the generated energy of wind farms in a whole year, and obtain the sub-components of the objective function which are directly related to it. The sub-components include the annual benefits of electricity selling of the grid, annual subsidies of the grid, annual subsidies of the wind farm, the initial construction cost of the wind farm, annual operation and maintenance cost, and connection cost for wind power accommodation.

Step 3 Turn into real-time simulation, that is, the practice flow calculation for each hour and verify whether it satisfy the constraints. If yes, get all the sub-components of the objective function in this hour and accumulate hour by hour. If no, it is required to process generation economic dispatching considering economic constraints and to search new operating point for the system.

Step 4 Considering 24 hours a day as scheduling period, repeat steps (1)–(3). Do the simulation of wind power output for 24 hours continuously, and calculate the cumulative value of every sub-component.

Step 5 Turn to the unit combination of the next dispatch period. Repeat the process above until the end of the year and obtain the system objective function, which is the net benefit of the combined system referred from the previous subsection.

Step 6 Update the wind farm installed capacities and repeat steps (1)–(5) above. Get the installed capacities of two wind farms when the net income of the combined system is maximized based on DE algorithm. This is the optimal capacities.

4 CASE STUDY

In this research, the construction cost of the wind farm is 1400 \$/kW, and average price of electricity selling is 8 cents/kWh in the region under investigation. For the integration two new wind farms, two new lines with a specification of LGJ-185 and a total length of 15 km should be set up, for which the unit price is 20800 \$/km. The subsidies for Wind farms and grid are 0.3 cents/kWh and 0.2 cents/kWh respectively. The Peaking factor is 55%. Unit cost of reserve and peak regulation is taken 0.3 cents/kWh and 0.15/kWh respectively. The equivalent annual value coefficient is 0.08. This paper uses RTS79 system as the example. Wind farm parameters are shown in Table 1.

4.1 *Optimal installed capacity of single-point integration*

Figure 1 shows net benefit curves of wind farm 1, the power grid and the combined system when only wind farm 1 is allowed to integrate into the RTS79 power grid on bus 7. The abscissa in the figure represents the installed capacity of wind farm 1 (MW), while the ordinate represents each subject's annual net benefit. In the programming, the initial value of installed capacity is 50 MW, while the step length is 1.5 MW.

According to Figure 1, the net benefit of wind farm 1 increased first and then declined with the increase of the installed capacity of wind farm 1, it is noticed that the net benefit curve of wind farm 1 trends to reduce at last. And this reduction is due to the wind curtailment in the computing program of this paper, which was established according to the shortage of transmission capacity and peak regulation capacity. The net benefit curve of the grid declines with the increase of installed capacity of wind farm 1. In order to accept wind power and ensure the stable operation of power system, the power grid needs to bear a lot of extra costs, including construction cost, operation and maintenance cost, extra reserve cost and peak regulation cost, etc. Sum of these costs are far larger than the electricity selling benefit caused by wind power integration, even the subsidies offered by the government are unable to make a balance, which is the reason why the power grid has little enthusiasm in implementing wind power.

Table 1. Wind farm parameters.

	Wind farm 1	Wind farm 2
Scale parameter (m/s)	6.13	7.11
Shape parameter	1.96	2.83
Cut-in wind speed (m/s)	2.33	3.05
Rated wind speed (m/s)	11.35	12.14
Cut-out wind speed (m/s)	21.47	22.07

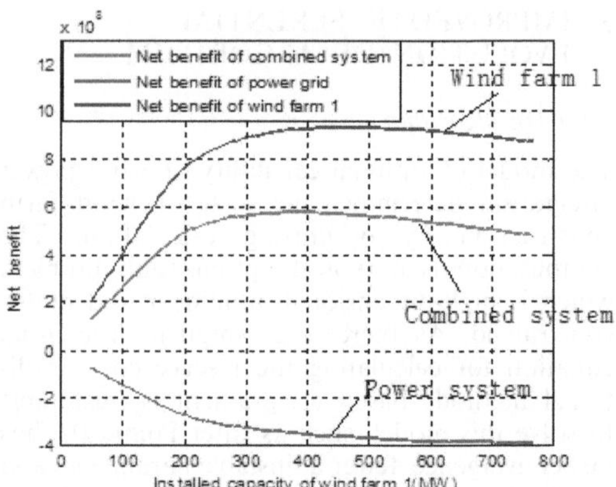

Figure 1. The result of single-point integration.

Figure 1 also shows that the net benefit of the combined system (including wind farm and power grid) increases first, and then declines along with the increasing installed capacity of wind farm 1. Therefore, there should be a certain installed capacity value maximizing the net benefit of the combined system and this capacity value can be the optimal capacity of wind power integration on bus 7.

Similarly, we can get the optimal capacity of wind farm 2 when it is only allowed to integrate into the grid on bus 14. The result shows that the optimal capacity of wind farm 2 is 201 MW, which is smaller than that of wind farm 1. To the different access point, the optimal installed capacity is different due to power supply, power load and the transmission capacity of lines around. In RTS79 system, there is a thermal power on bus 7 and the power load there is larger than that of bus 14, which is probably the reason why the optimal capacity of wind farm 1 is larger than wind farm 2.

4.2 *Optimal installed capacity of two-point integration*

Figure 2 shows optimal acceptance capacity by maximizing the net benefit of the combined system when two different wind farms are allowed to integrate in RTS79 system on bus 7 and bus 14. The result is obtained based on Differential Evolution algorithm programming. Optimal capacities of two wind farms for 100 iterations are shown in Figure 2. In the program, the initial values of two wind farms' installed capacities are set to be 1500 MW considering the installed capacity of thermal power is 3405 MW for RTS79 system.

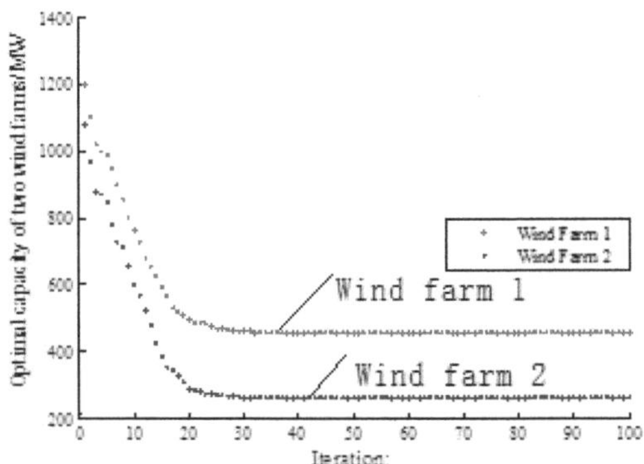

Figure 2. The result of two-point integration.

According to the Figure 2, the algorithm gradually begins to converge after 20 iterations and the optimal capacity of wind farm 1 is 457.5 MW and 264 MW to wind farm 2 after 100 iterations. The total installed capacity here by this method is 721.5 MW, which is much larger than the optimal capacity of single-point integration in section 4.1. Therefore, the decentralized multi-point integration of wind farms can improve the optimal installed capacity, which is also an effective way to increase the capacity of wind power integration.

5 CONCLUSIONS

This paper analyzes the costs and benefits of wind farms and power grid in detail with the purpose of designing the plan for large-scale wind power integration. Moreover, the models used to calculate the costs and benefits of wind farms and power grid are given, respectively. And the evaluation model of optimal acceptance capacity of wind power is proposed and it performed on RTS79 system. The simulation results show that the proposed model can maximize the comprehensive benefits of combined system and ensures the profitability and security of the system.

According to the results obtained in this paper, we are trying to put forward some related suggestions on China's wind power development in china as follows.

1. Policies and measures for wind power development should be improved. The reasonable compensation mechanism for power grid should be established, including subsidies for construction cost, operation cost and so on. Moreover, a reasonable benefit allocation method between wind farms and power grid should also be explored. All of these measures will raise the enthusiasm of the power grid to accept wind power. Therefore, it will promote the renewable energy develop.
2. Multi-point integration of wind power should be developed and encouraged, which will promote field application and the probability reduction of transmission congestion of power flow as well as, improvement of the acceptance capacity of wind power.
3. It is necessary to make reasonable planning for the power structure and increase the peak regulation capacity of power system, which will reduce wind curtailment and improve the acceptance capacity of wind power in the future.

REFERENCES

[1] Chongqing Kang, Xinyu Chen, et al. 2013. Balance of Power Toward a More Environmentally, Friendly, Efficient, and Effective Integration of Energy Systems in China, Power and Energy Magazine, IEEE 9:57–64.

[2] Zhang X., Ruoshui W., Molin H., Martinot E. 2010. A study of the role played by renewable energies in China's sustainable energy supply. Energy 35 (11): 4392–9.

[3] Logan J., Kaplan S.M. 2008. Wind power in the United States: technology, economic, and policy issues, Tech. rep. Congressional Resources, Science, and Industry Division 8.

[4] J. Wiik, G.O. Gjerde, T. Gjengedal, and M. Gustafsson. 2002. Steady state power system issues when planning large wind farms," Power Engineer Society Winter Meeting, IEEE, pp. 199–204, Vol. 1.

[5] Sun R., Zhang T., Liang J. 2011. Evaluation and application of wind power integration capacity in power grid. Automation of Electric Power Systems 35 (4): 70–76.

Material Science and Environmental Engineering – Chen (Ed.)
© 2016 Taylor & Francis Group, London, ISBN 978-1-138-02938-5

Innovation of the Chinese photovoltaic enterprises business model

L. Yang
North China Electric Power University, Beijing, China

ABSTRACT: This paper first analyzes the difficult position of Chinese PV enterprises and the existing problems in the process of development. It then puts forward the main ways for the business model innovation of photovoltaic enterprise by combining with the related theory of business model innovation that includes value innovation, value network innovation and profit model innovation. Finally, it brings forward the main countermeasures of photovoltaic enterprise business model innovation.

Keywords: photovoltaic enterprise; innovation of business model; countermeasure

1 INTRODUCTION

In recent years, with the increasing energy crisis and environmental pollution, the development of new energy is receiving more and more attention. Few years ago, photovoltaic enterprise, as the representative of new energy in the international policy of positive cases at home and abroad, achieved swift and violent development. Nevertheless, our PV enterprises have been hit severely since 2009 because of economic crisis, debt crisis and the influence of European and American policies. By the second half of 2012, under the influence of Euramerican "double turn over" and its debt crisis, the development of Chinese photovoltaic industry was in trouble and quite a few PV enterprises went bankrupt. The excess capacity led by blind investment and the potential problems under photovoltaic enterprise business model are gradually exposed. Therefore, for PV enterprises in China, exploring the current situation of the development of business model suitable to China's development of photovoltaic enterprises is of great significance.

2 BACKGROUND: CHINESE PHOTOVOLTAIC ENTERPRISES

2.1 *Market demand atrophy and serious overcapacity*

In the "five-year plan", the government put the PV enterprise as the primary industry in strategic emerging industries to develop, so the local governments responded to the call. By the driving of the GDP, local governments had to jump on the bandwagon to take the photovoltaic industry as the key industry in succession. Because PV modules' production is mostly a simple labor-intensive industry and the threshold is low, a large number of enterprises began to invest in the photovoltaic industry, leading to the mass production of middle photovoltaic products. What's more, the government published various preferential policies. As a result, China would become the first country of photovoltaic manufacturing industry soon. But 90% of China's photovoltaic products export to overseas countries, which mainly include Euramerican countries; thus, the survival and development of companies relies heavily on exports. With the economic crisis and the outbreak of the crisis, many European countries cut subsidies on photovoltaic enterprises. From 2011 until now, the Occident has brought about the "double reverse" survey to Chinese PV enterprises. Photovoltaic product export has been hindered fearfully. The overseas market has been hit seriously. And the domestic market has not yet been fully started, so the photovoltaic industry is in a state that supply is much greater than the demand and there exists excess capacity.

2.2 *Low level of technology and imbalances of industrial chain's development*

China is a manufacturer of photovoltaic enterprises, but as a result of features such as upstream polysilicon material preparation, funds and technology have become intensive. The main production preparation technology is mainly composed of the United States, Japan and Germany, forming serious technical barriers and market monopoly. In recent years, there have been domestic enterprises producing crystalline silicon. But owing to the high cost of production, the market is lacking competitiveness significantly. In addition, foreign crystalline silicon prices have dropped and the crystalline silicon production costs have become significantly higher than those of China, and competing among the international countries is becoming more difficult. Downstream of photovoltaic system development is slow because of the technology and market. PV enterprises are mainly concentrated on the production of crystalline silicon cells and components but the techniques and financial thresholds are low. The industrial chain presents the phenomena of double ends in the outside, and the phenomena of middle and unbalanced development. Moreover, the abilities of Chinese photovoltaic enterprises to create and apply are rather inferior; furthermore, they are at the bottom of the "smiling curve". It appears "three out" trends: crystalline silicon materials and production equipments are imported from abroad; the product also relies on export. Blocked by overseas market, the development of Chinese PV enterprises has now become more difficult.

2.3 *Serious debt crisis and the enterprise financing difficulties*

The chain of photovoltaic industry spans wide. And the companies belong to capital-intensive businesses, which need a lot of money. Early photovoltaic enterprises financed through overseas listings, trust, bank loans, and convertible bond issues. Among those businesses, bank loans were the main financing way. But with the "double reverse" survey of the Occident, photovoltaic enterprise overseas market shrank and the banks 'attitudes toward PV companies with excess capacity label got cold feet, directly leading to the consequence of rejecting and withdrawing loans that became an ordinary activity. Implementing IPO in the capital market and financing by bond in the interbank market also became quite difficult. As photovoltaic giant Sunte chin Wuxi fell due to the debt crisis, a large number of photovoltaic enterprises were on the verge of bankruptcy because of the difficult financial problem.

3 PHOTOVOLTAIC ENTERPRISE INNOVATION PATH

Value Innovation: the value innovation concept was first proposed by Professor W. Chan Kim and Professor Rennie Mauborgne in the European School of International Business Administration. Value innovation means a change of strategic thinking about business growth. In essence, it puts the starting point of enterprise's strategic thinking transformation from the competitors to create a new market or re-interpret the existing market. Also, enterprise value innovation takes the customer demand as the fundamental starting point. The management guru Peter Drucker (1954) was the first to put the customer value into the company strategy. In describing the definition, he said that the purpose of business exists outside of the enterprise itself, and the purpose of business has only one definition, which is "to create a customer" [1]. It means enterprises must provide products and services with the greatest value and innovative from the level of value for customers, and innovative value while creating markets.

Value Network Innovation: the network value is the relationship and structure of value generation, distribution, transfer and use, which have been formed under the mutual effect of stakeholders. The value network improves the value recognition system, and expands the value influence of resources. In addition, it breaks through the point to maintain the relationship between the value chain of the enterprise and the stockholder, putting the interests of suppliers, other partners, clients and stakeholders together as a whole. What's more, they share resources and complementary resources, and maximize the overall interests of the value network there.

Profit model innovation: the scholar Paul Timmers (1998) thought that the profit model is a system framework for product (service) and information flow, and described as a potential profit that the enterprise can obtain, but also a description of their potential to generate revenue and income. He proposed that the way of system identification of the profit model structure is deconstruction and reconstruction based on value chain. This approach is also a possible method to identify the value chain elements and the information of integration chain.

4 THE PV BUSINESS MODEL INNOVATION STRATEGIES

PV companies need to combine business model innovation and technology innovation in order to achieve effective results.

4.1 *Improving technology, and innovative customer business model*

Customer business model innovation is the center of the business model. Companies must be based on the customer-centric, and create value for customers. (4) The core of creating value for customers is technological innovation. Companies may develop new products, improve product performance and reduce product prices to better meet customer needs, which will surely improve customer satisfaction. Taking distributed photovoltaic power plants as an example, whether PV companies are in the development and application of the distributed PV power plant, or at the time of signing a letter of intent with customers, it must combine the different needs of consumers in different regions to develop different strategies. Thus, photovoltaic enterprises may achieve the goals of the producing utility maximization, price rationalization, and customer satisfaction by strengthening staff training to improve the pre-service level, innovating technology to reduce product costs and improve product's quality.

4.2 *Integrating innovative internal industries, optimizing and upgrading the industrial chain*

In the condition of international market shrinking, and the domestic market not fully developed, photovoltaic enterprises' severe overcapacity, combined with the industry's enterprises' vicious competition lead to serious internal friction. So, the consolidation of the internal photovoltaic industry is imminent. PV enterprise may be survival of the fittest, horizontal and vertical integration of the industry under the market condition. However, when experiencing the industry restructuring, from the long-term interests, photovoltaic companies must either develop their own vertical integration, or hold together, in the form of industrial army development. Vertical integration can be carried at the cost supply of raw materials in each product line to maximize profits, while a complete production line, which costs slightly less, can be affected by the market price, which is more competitive in the market. However, it can reduce or enhance production rapidly and get stronger activity through the form of link together to heat itself. Each enterprise should combine their own reality, under the guidance of policy, and realize the optimization and upgrading of the industrial chain.

4.3 *Innovating public relations and optimizing the network value*

Public relation innovation includes the innovation of partner network public relations, the innovation of public relations within the enterprise, the innovation of governmental public relations and international public relations [5]. Partner network public relation innovation refers to the photovoltaic enterprises that must be friendly to each other when dealing with the relationships with various suppliers on the industry chain, customers and investors. Both sides should benefit from the cooperation realize long-term cooperation and common development. Public relation innovation requirements of photovoltaic enterprises deal with the relations between shareholders, managements and employees, especially to study how to introduce strategic investors and how to leave the talents in key positions properly. With photovoltaic enterprises as a strategic emerging industry, the development in the early stage depends on the government's support to a large extent, so it is especially important to innovate the public relations with governments. Photovoltaic enterprises must strengthen the contact with the local government and make full use of the government's relevant supporting policies of photovoltaic enterprises. At the same time, they should fulfill their duties of enterprise actively. The government also ought to coordinate financial institutions for financial innovation and solve the problem of photovoltaic enterprises' financing difficulties as soon as possible. International relation innovation requires that photovoltaic enterprises deal with the relationships with foreign suppliers and competitors well in the process of going out, and to understand the relevant international laws and regulations. Enterprises are supposed to not only obey the law, but also use legal means to safeguard their legitimate rights and interests.

4.4 *Developing new markets and innovating market relations*

Under the condition of the implementation of the "double reverse" investigation that Europe and the United States contra pose to Chinese PV enterprises, it costs prohibitively for Chinese PV enterprises to enter the European market. So, photovoltaic enterprises should actively explore a new international market such as in Africa and Southeast Asia, with rich solar energy resources. At the same time, photovoltaic enterprises can use the domestic market as a key development object because its development potential is great and the national policy supports the industry actively.

REFERENCES

[1] Drucker Peter F. The Practice of Management. New York: Harper & Row, 1954.

[2] Lihuan, Liuyi: "Based on the value of the value creation network management: the features and formation", <Journal of management engineering>, in 2001, 15.

[3] GaoTingting: "the explore of the Internet profit model", in 2012, Capital economic and trade university master's degree thesis.

[4] Hongchang: "the explore of Strategic emerging industry business model innovation", <Science and technology management>, in 2011, 3.

[5] Shiyanxin: the explore of Chinese PV enterprises business model innovation, Research on Economics and Management, in 2013, 10.

[6] Fujing: international competitiveness present situation and the path of ascension of Chinese photovoltaic industry, Journal of Hebei University (Philosophy and Social Science), in 2013, 2.

[7] DuFanghua: The research on electric car industry business model innovation, in 2012, master's thesis.

Material Science and Environmental Engineering – Chen (Ed.)

Chaotic differential evolution algorithm for low-carbon power dispatch with wind power integration

B. Cao, R.Z. Wang & J.H. Zhang
North China Electric Power University, Beijing, China

ABSTRACT: Integrating the power generated by wind farm into power grid could bring new changes to the power system operations. Low-carbon power dispatch model with wind power integration is formulated; the fuel cost of thermal generators is treated as objective function. Also, the emission of CO_2 is inserted as a constraint. In view of the intermittent nature of wind power, the power balance constraint was described in the form of probability. The stochastic constraint is transformed to a deterministic one according to the cumulative distribution function of wind power. Based on Tent map, a novel chaotic differential evolution algorithm for power dispatch model is proposed. In this algorithm, chaotic sequence is applied to initialize the population and obtain the dynamic parameter settings, so the global searching ability is enhanced. The effectiveness of the proposed power dispatch strategy is verified on a system consisting of 10 thermal generators and one wind farm.

Keywords: wind power; low-carbon; power dispatch; tent map; chaotic differential evolution

1 INTRODUCTION

As the problems of the global energy supplying security and the climate change become increasingly serious in recent years, the wind power develops quickly and has been an imperative part of world energy because of its lower cost, more mature technologies and higher reliability compared with other renewable energies. The wind power installed capacity of 44.73GW has been the first in the world until the end of 2010, and a large number of wind power generation plants of megawatt level have been planned. The wind power characteristics of randomness and intermittent result in the fluctuation of the wind generation power, so the large-scale grid-connected wind power generation will take new challenges into the traditional power system considering the economical and secure operation. As well as, the grid dispatching strategy adapting of the wind power integration is an imperative problem to be solved.

The power system generation dispatching issues considering the grid-connected wind power have been explored by scholars in the world. According to the reference, the wind power generation is added directly to the system balance constraint conditions as negative load by the dispatching model. This cannot reflect the fluctuation of wind power generation and cannot adapt of the grid power generation dispatch in the condition of the high penetration rate of wind power; according to the reference, the influence of the wind power fluctuation to generation plan is handled by introducing the positive and negative rotation spare capacity constraints; a generation dispatching model is built in the reference based on fuzzy theory, and the penalty cost when wind power is not used fully is added to the objective function considering both operation risk and economic benefit; in the reference, the saved coal consumption due to wind power is seen a measurement of wind power value, and the dispatching strategy is built on the base of this in order to coordinate the coal consumption with the capacity of the grid acceptance to wind power.

However, any above research considers no influence of the CO_2 emission in the process of thermal generation to climate warming when making the grid generation plan. The government of China promises that the CO_2 emission of per unit of GDP should decrease 40%~45% from 2005 to 2020. The power industry should be positive to the tasks of the CO_2 emission reduction as a large consumption industry considering that the energy structure of this country is mainly coal. So introducing low carbon power dispatch is an effective solution in short time to control the grid power CO_2 emission.

This paper constructs a low-carbon generation dispatching model containing wind power plant, of which the objective function is the minimum fuel cost of coal-fired thermal power units and considers the CO_2 emission in the constraint conditions.

Due to the inherent uncertainty of wind power, power balance constraints are described as a form of probability. Aiming at the disadvantage of the differential evolution algorithm which is easy to fall into local convergence and prematurity, a Chaos Differential Evolution algorithm (CDE) is provided to solve this model. Finally, an example containing 10 conventional thermal power units and a grid-connected wind power plant is used to verify the effectiveness of the described power generation dispatching strategy.

2 THE LOW-CARBON GENERATION DISPATCHING MODEL CONSIDERING WIND POWER CONNECTION

2.1 *Objective function*

The objective function is the minimum fuel cost of the coal-fired units, which can be described as follows

$$\min F = \sum_{i=1}^{N_G} F_i(P_i) \tag{1}$$

Type in; F is the whole fuel cost. N_G is the number of coal-fired units in the system. P_i is the positive output of the coal-fired units. $F_i(P_i)$ is the consumption characteristic of the coal-fired unit i. Considering the valve point effect, the generator consumption characteristic is depicted as follow.

$$F_i(P_i) = a_i + b_i P_i + c_i P_i^2 + |e_i \sin(f_i(P_{\min,i} - P_i))| \tag{2}$$

Type in, a_i, b_i, c_i, e_i, f_i are the fuel cost coefficient. $P_{min,i}$ is the lower limit of active power output of generation unit i.

2.2 *Constraint condition*

Power balance constraint. The system power balance constraint is as follows.

$$\sum_{i=1}^{N_G} P_i + \sum_{i=1}^{N_W} W_{Ti} = P_D + P_L \tag{3}$$

Type in, N_w is the number of wind power generators. W_{Ti} is the positive output of the wind power generators. P_D is the demand of system load. P_L is the loss of the system grid, which can be solved by B coefficient method, namely,

$$P_L = \sum_{i=1}^{N_G} \sum_{j=1}^{N_G} P_i B_{ij} P_j \tag{4}$$

Type in, B_{ij} is losing coefficient. Because the wind power output is random variable, the type (3) can be described as follows in the form of probability.

$$P_r \left\{ \sum_{i=1}^{N_G} P_i + W_F \geq P_D + P_L \right\} \geq p \tag{5}$$

Type in, W_F is the positive wind power output. $P_r\{\cdot\}$ is the probability of the event $\{\cdot\}$; p is the reliable degree to meet the system load demand, of which value should meet the following,

$$0 < p < 1 - P_r\{W_F = 0\} \tag{6}$$

According to the cumulative distribution function of the wind power output, a deterministic inequality constraint can be shown,

$$P_D + P_L - \sum_{i=1}^{N_G} P_i \leq \frac{2 v_m W_{Fr}}{\sqrt{\pi}(v_r - v_{ci})} \sqrt{\ln\left\{ p + \exp\left[-\frac{\pi}{4}\left(\frac{v_{co}}{v_m}\right)^2 \right] \right\}} - \frac{v_{ci} W_{Fr}}{v_r - v_{ci}} \tag{7}$$

Type in, W_{Fr} is the rated output of the wind power.

The CO_2 emission constraint. The coal-fired units can emit lots of CO_2, SOx as well as NOx in the process of generation. The model in this paper will only consider the reduction of the CO_2 emission, which can be described as the following quadratic function

$$E = \sum_{i=1}^{N_G} \alpha_i + \beta_i P_i + \gamma_i P_i^2 \tag{8}$$

Type in, E is the total amount of the CO_2 emission; $\alpha_i, \beta_i, \gamma_i$ are the characteristic factors which can be got through the monitoring data of the harmful gas emissions in the power plant using fitting method. The CO_2 emissions constraints can be expressed as

$$E \leq E_{\text{limit}} \tag{9}$$

Type in, E_{limit} is the maximum permitted CO_2 emission.

The constraint of the generators output. The output constraint of the coal-fired units can be expressed as

$$P_{\min,i} \leq P_i \leq P_{\max,i} \tag{10}$$

$P_{max,i}$ is the active output upper limit of the coal-fired unit i. The output constraint of the wind power units can be expressed as

$$0 \leq W_{Ti} \leq W_{ri} \tag{11}$$

Type in, W_{ri} is the rated output of the wind power unit i.

3 CHAOTIC DIFFERENCE EVOLUTION ALGORITHM

3.1 *The standard differential evolution algorithm (DE)*

The basic thought of the DE algorithm is: firstly, a test population can be got by regrouping based on the difference of individuals; secondly, a new population can form through the competition survival strategy between the test populations with the original population. The overall structure of the DE algorithm is similar to the genetic algorithm, and here only the mutation and crossover operation process is given.

In the generation G, the crossover operation is finished to every object individual $\boldsymbol{u}_{i,G}$, in order to get the corresponding mutation individual $\boldsymbol{v}_{i,G+1}$:

$$\boldsymbol{v}_{i,G+1} = \boldsymbol{u}_{r1,G} + F(\boldsymbol{u}_{r2,G} - \boldsymbol{u}_{r3,G}) \tag{12}$$

In this type $\boldsymbol{u}_{r1,G}$ is the parent individual, $(\boldsymbol{u}_{r2,G} - \boldsymbol{u}_{r3,G})$ is the differential vector, and $r1 \neq r2 \neq r3 \neq i$. F is the zoom factor. After the mutation operation, if the decision variables cross border, the boundary value can replace it. Crossover operation can be expressed as,

$$z_{ij,G+1} = \begin{cases} v_{ij,G+1} & \text{if } r_j \leq C_R \text{ or } j = k \\ u_{ij,G+1} & \text{other conditions} \end{cases} \tag{13}$$

In this type, $z_{i,G+1}$ is the test individual; C_R is the crossover probability. r_j is the random number between 0 and 1. k is the a random integer from 1 to D. D is the variable dimension.

3.2 *Chaotic difference elevation algorithm*

Tent mapping. As DE exists defects that the algorithm will fall into premature convergence and its control parameter cannot be selected easily like other algorithms, this paper proposes Tent chaotic mapping into DE algorithm. Chaos is a kind of nonlinear phenomenon existing widely in nature, which has randomness, egocentricity and sensitivity to initial conditions and other characteristics. Most of studies used the chaos search mechanism based on Logistic mapping. But Logistic mapping has the defect that it cannot traverse evenly and thus affects the algorithm speed. Literature points out that Tent mapping have better traversal uniformity than Logistic mapping.

The mathematical expression for Tent Mapping as,

$$x_{k+1} = \begin{cases} 2x_k & 0 \leq x_k \leq 0.5 \\ 2(1 - x_k) & 0.5 < x_k \leq 1 \end{cases} \tag{14}$$

Tent mapping iterative sequence exist small cycle, unstable periodic point and other defects. For instance, the value 0.25, 0.5, 0.75 will all iterate to a fixed point 0. So this paper uses the improved measures mentioned in the literature. When the Tent mapping reaches the point of small cycles or fixed point, the mapping will apply some perturbation to re-enter a state of chaos. Figure 1 shows the chaotic trajectory with the initial value 0.2819 and the 100 iterations.

Population initialization based on Tent mapping. Using the Tent mapping to create initial population, which will not change random nature of initialization, and also can maintain the diversity of initial population. The specific steps are as follows:

a. Randomly generating D numbers that exist between 0 and 1, then forming the initial sequence number $\boldsymbol{x}_1 = (x_{1,1}, x_{1,2}, \ldots, x_{1,D})$.
b. According to the chaotic sequence generated by equation (15), the chaotic matrix is

$$\begin{bmatrix} x_{1,1} & x_{1,2} & \cdots & x_{1,D} \\ x_{2,1} & x_{2,2} & \cdots & x_{2,D} \\ \vdots & \vdots & \ddots & \vdots \\ x_{N_P,1} & x_{N_P,2} & \cdots & x_{N_P,D} \end{bmatrix} \tag{15}$$

In the matrix, N_P is population size.

c. According to the equation (16), Map the matrix elements to the range of control variables

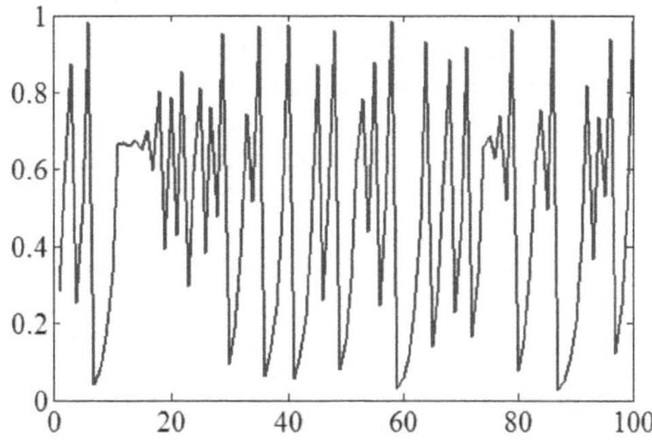

Figure 1. Tent map beginning with 0.2819.

$[u_{j\min}, u_{j\max}]$, the initiative population is obtained as U_0.

$$u_{ij,0} = u_{j\min} + x_{ij}(u_{j\max} - u_{j\min}) \tag{16}$$

The dynamic adjustment of control parameters based on Tent mapping. In the standard DE algorithm, the control parameters of F and CR keep static in the whole process of search, which makes the algorithm easy to fall into local optimum in the later evolution. By using chaotic sequence based on Tent map in the evolutionary process to dynamically adjust control parameters, the optimization traverse can guarantee complete, so as to improve the global convergence of the algorithm. The strategy of dynamically adjusting control parameters as following:

$$F^{G+1} = \begin{cases} 2F^G & 0 \le F^G \le 0.5 \\ 2(1-F^G) & 0.5 < F^G \le 1 \end{cases} \tag{17}$$

$$C_R^{G+1} = \begin{cases} 2C_R^G & 0 \le C_R^G \le 0.5 \\ 2(1-C_R^G) & 0.5 < C_R^G \le 1 \end{cases} \tag{18}$$

The initial value of F and C_R both random numbers that exist between 0 and 1.

4 EXAMPLE

4.1 *Example description*

A 10-machine system that contains 10 coal-fired units and 1 wind farm is calculated and analyzed in this paper. The system load is 1800 MW and the fuel cost coefficient of coal-fired generation units and system B-coefficient are shown in literature. CO_2 emission characteristic coefficient and active power output constraints are shown in Table 1. The wind farm consists of 60 same type turbines and their wind parameters are shown in Table 2.

4.2 *Method validation*

The model parameters are set as: the confidence level of meeting system load demand is 0.85, the average wind speed is 20 m/s, the CO_2 maximum allowable emission is 16 t/h. CDE algorithm parameters are set as: the population size takes 50, the maximum number of iterations takes 2000. The calculated optimal dispatch scheme is shown in Table 3. For comparison, Table 3 also displays the generation units output scheme under no CO_2 emission constraints.

According to the analysis of dispatch schemes, the wind farm output are the same in both schemes (72.82MW). The grid loss in both schemes are nearly the same. The load distribution among each generation units that is beard by coal-fired units cause a large difference between the two schemes in system fuel cost and CO_2 emission. Under the CO_2 emission constraints, the fuel cost and CO_2 emission value in the optimal dispatch scheme is respectively about 94525 $/h and 20.65 t/h. If ignoring the CO_2 emission constraints, the fuel cost in the optimal dispatch scheme increases to 100278 $/h (increase by 6.09%), while the CO_2 emission value is reduced to 16.00 t/h (reduce 22.52%). Thereby,

Table 1. Parameters of the thermal generators.

Units number	α/ (kg/h)	β/(kg/ (MW·h))	γ/(kg/ (MW²·h))	P_{max}/ MW	P_{min}/ MW
1	130.00	−2.86	0.022	470	150
2	132.00	−2.72	0.020	470	135
3	137.70	−2.94	0.044	340	73
4	130.00	−2.35	0.058	300	60
5	125.00	−2.36	0.065	300	73
6	110.00	−2.28	0.080	260	57
7	135.00	−2.36	0.075	240	20
8	157.00	−1.29	0.082	180	47
9	160.00	−1.14	0.090	120	20
10	137.70	−2.14	0.084	120	20

Table 2. Parameters of the wind turbine.

v_{ci}/(m/s)	v_r/(m/s)	v_{co}/(m/s)	W_r/MW
5	15	45	3

Table 3. Optimal dispatch schemes.

	Optimal dispatch schemes	
	Considering CO_2 emission constraints	Neglecting CO_2 emission constraints
P1/MW	165.60	150.00
P2/MW	222.27	135.00
P3/MW	276.91	203.77
P4/MW	180.83	241.25
P5/MW	222.60	272.47
P6/MW	182.81	253.35
P7/MW	166.12	239.18
P8/MW	161.94	180.00
P9/MW	116.17	52.06
P10/MW	86.84	53.42
WF/MW	72.82	72.82
F/($/h)	100278	94525
E/(t/h)	16.00	20.65
PL/MW	54.90	53.31

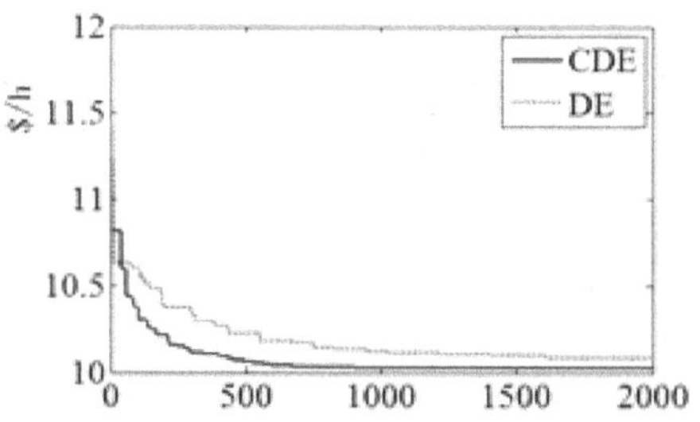

Figure 2. Convergence characteristics of the CDE and DE.

to take consideration of CO_2 emission constraints in generation dispatch model can effectively control CO_2 emission level in electrical industry.

To validate the performance of CDE algorithm, this paper also adopts standard DE algorithm to solve the model and its control parameter F and CR are respectively 0.5 and 0.3. The convergence curves of each algorithm are shown in Figure 2. It can be seen that the standard DE algorithm in the later evolution falls into premature and the optimal fuel cost searched is only 100858 $/h that is obviously more inferior than CDE. The results shows the superiority of the CDE algorithm.

5 SUMMARY

Large scale of power generation integration as well as the development of low-carbon power technique is an effective way to alleviate the energy crisis and environment problems, but also an important component of strong & smart grid our country is constructing.

a. This paper has established a low-carbon generation dispatch model with wind power integration. The model is to minimize the coal-fired unit fuel cost as the objective function, and takes consideration of CO_2 emission constraint. According to the inherent fluctuation of wind power, the power balance constraints are described by a probability form in this paper and determinedly transformed by wind power output cumulative distribution function.
b. This paper has purposed a chaotic differential evolution algorithm to solve the model. The algorithm utilizes the egocentricity of Tent chaotic mapping to initialize population and dynamical adjust control parameters, which can enhance population diversity and improve the global searching ability.
c. This paper take a system with 10 coal-fired units and 1 wind farm as an example. The result shows that, the dispatch proposed is feasible and effective.
d. As a preliminary study, this paper takes hour-dispatch as research objective, and the next job is the dynamic generation scheduling taking account into the coupling of different time sections.

ACKNOWLEDGEMENT

This paper is sponsored by the National High Technology Research and Development of China 863 Program (2012AA050201).

REFERENCES

[1] Basu M. 2008. Dynamic economic emission dispatch using non-dominated sorting genetic algorithm-II, International Journal of Electrical Power & Energy Systems, 30(2): 140–149.
[2] Bai Y.X. & Fang D.Z. & Hou Y.H. 2010. Real-time online monitoring and control system in dispatch center for large-scale wind-farms, Electric Power Automation Equipment, 30(11): 6–9.
[3] Chen Q.X. & Zhou T.R. & Kang C.Q. 2009. An assessment model of low-carbon effect and its application to energy saving based generation dispatching, Automation of Electric Power Systems, 33(16): 24–29.
[4] Farhat I.A. & El-Hawary M.E. 2010. Dynamic adaptive bacterial foraging algorithm for optimum economic dispatch with valve-point effects and wind power, IET Generation, Transmission & Distribution, 4(9): 989–999.
[5] Jiang W. & Yan Z. & Hu Z. 2011. A novel improved particle swarm optimization approach for dynamic economic dispatch incorporating wind power, Electric Power Components and Systems, 39(5): 461–477.
[6] Li J.F. 2011. Unlimited wind power and solar: the development report of Chinese wind power in 2011, Chinese Environment Science Press: 2–4.
[7] Li Z. & Han X.S. & Yang M. 2010. Power system dispatch considering wind power grid integration, Automation of Electric Power Systems, 34(19): 15–19.
[8] Liu B. & Wang L. & Jin Y.H. 2007. Advances in differential evolution, Control and Decision, 22(7): 721–729.
[9] Liu X & Xu W & Huang C. 2010. Economic load dispatch with stochastic wind power: model and solutions[C]//Proceedings of 2010 IEEE PES Transmission and Distribution Conference and Exposition, New Orleans, LA, USA: IEEE, 2010: 1–7.
[10] Miranda V. & Hang P.S. 2005. Economic dispatch model with fuzzy wind constraints and attitudes of dispatchers, IEEE Transactions on Power Systems, 20(4): 2143–2145.
[11] Miranda V. & Hang P.S. 2005. Economic dispatch model with fuzzy wind constraints and attitudes of dispatchers, IEEE Transactions on Power Systems, 20(4): 2143–2145.
[12] Zhang H. & Zhang T.N. 2008. Research on decision-makings of structure optimization based on improved Tent PSO, Control and Decision, 23(8): 857–862.

Figure 2. Convergence characteristics of the CDE and DE.

to take consideration of CO_2 emission constraints in generation dispatch model can effectively control CO_2 emission levels in electrical industry.

To validate the performance of CDE algorithm, this paper also adopts standard DE algorithm to solve the model and its control parameter F and CR are respectively 0.5 and 0.3. The convergence curves of each algorithm are shown in Figure 2. It can be seen that the standard DE algorithm in the later evolution falls into premature and the optimal fuel cost searched is only 100855.2$, that is obviously more inferior than CDE. The results shows the superiority of the CDE algorithm.

5 SUMMARY

Large-scale wind power generation integration as well as the development of low-carbon power technology is an effective way to alleviate the energy crisis and environment problems, but also an important component of strong & smart grid, or security is emphasized.

a. This paper has established a low-carbon generation dispatch model with wind power integration. The model is to minimize the conventional unit fuel cost as the objective function, and takes consideration of CO_2 emission constraint. According to the inherent fluctuation of wind power, the power balance constraints are described by a probability form in this paper and deterministically transformed by wind power output cumulative distribution function.

b. This paper has proposed a chaotic differential evolution algorithm to solve the model. The algorithm utilizes the ergodicity of chaotic mapping to initialize population, and dynamical adjust control parameters, which can enhance population diversity and improve the global searching ability.

c. This paper take a system with 10 coal-fired units and 1 wind farm as an example. The result shows that the dispatch proposed is feasible and effective.

d. As a preliminary study, this paper takes hour-dispatch as research objective, and the next step is the dynamic generation scheduling taking account into the coupling of different time sections.

ACKNOWLEDGEMENT

This paper is sponsored by the National High Technology Research and Development of China (863 Program) (2012AA050201).

REFERENCES

[1] Peng M. 2008. Dynamic economic emission dispatch using an adjustment setting adaptive algorithm[J]. International Journal of Electrical Power & Energy Systems, 30(2):140–149.

[2] Bai Y.X. & Fang D.Z. & Hou Y.H. 2010. Real-time online monitoring and control system in dispatch center for large-scale wind farms. Electric Power Automation Equipment 30(11): 6–9.

[3] Chen H. & Zhou T.R. & Kang C.Q. 2009. An assessment model of low-carbon effect and its application in energy saving based generation dispatching. Automation of Electric Power Systems 33(16): 24–29.

[4] Bahan T.A. & El-Hawary M.E. 2010. Dynamic adaptive bacterial foraging algorithm for optimum economic dispatch with valve-point effects and wind power. IET Generation, Transmission & Distribution 4(9): 989–999.

[5] Liang W. & Yu Z. & He Z. 2011. A novel improved particle swarm optimization approach for dynamic economic dispatch incorporating wind power. Electric Power Components and Systems 39(5): 461–477.

[6] Li J.F. 2014. Unlimited wind power and solar thermal development report of Chinese wind power in 2014. Chinese environment Science Press: 45–54.

[7] Li Z. & Han X.S. & Yang M. 2010. Power system dispatch considering wind power grid integration. Automation of Electric Power Systems 34(19): 15–19.

[8] Liu B. & Wang L. & Jin Y.H. 2007. Advances in differential evolution. Control and Decision 22(7): 721–729.

[9] Liu X. & Xu W. & Huang C. 2010. Economic load dispatch with stochastic wind power model and solutions[C]//Proceedings of 2010 IEEE PES Transmission and Distribution Conference and Exposition, New Orleans, LA, USA: IEEE, 2010: 1–7.

[10] Miranda V. & Hang P.S. 2005. Economic dispatch model with fuzzy wind constraints and attitudes of dispatchers. IEEE Transactions on Power Systems 20(4): 2143–2145.

[11] Miranda V. & Hang P.S. 2005. Economic dispatch model with fuzzy wind constraints and attitudes of dispatchers. IEEE Transactions on Power Systems 20(4): 2143–2145.

[12] Zhang H. & Zhang P.J. 2008. Research on decision-makings of structure optimization based on improved Tent PSO. Control and Decision 23(8): 857–862.

Material Science and Environmental Engineering – Chen (Ed.)
ISBN 978-1-138-02938-5

Study on the application of green energy-saving technology in the civil construction environment

W. Wu
The Department of Architecture and Engineering Qinhuangdao College, Northeast Petroleum University, Qinhuangdao, China

ABSTRACT: With the promotion of sustainable development of the times, green energy-saving technology plays a more and more important role in civil engineering. This paper first limits the concept of green technology, and then analyzes the present situations of both energy saving and environmental protection in construction engineering. Four kinds of green energy technology, namely water-saving technology, light, doors and windows, and ground energy-saving technology, are introduced. The target is to provide a reference for energy-saving design in civil construction.

Keywords: civil construction; green energy; construction engineering

1 INTRODUCTION

As the best choice to improve the quality of the project, energy-saving and environmental protection technology is one of the most advanced development technology, which complies with the time development of advanced technology. In civil construction, if the construction period is long, building energy consumption will increase exponentially. Using energy-saving and environmental protection technology can reduce the construction period and save the energy consumption of buildings.

With the development of modernization, people urgently need to raise the standard of comfortable living. Clean building environment has already become the needs of people's lives to create a good environment of architecture. Innovation and development of building materials in civil construction and material application of new environmental protection has become an urgent need. These factors cannot only reduce environmental consumption, but also improve people's living standard (Yang 2010). The rational use of environmental-friendly building materials requires that the new construction change their energy consumption in the original way. This can be achieved through the energy-saving emission reduction and environmental protection technology, which play the role of new energy materials, so as to improve people's living environment of architecture.

2 ANALYSIS OF ENERGY-SAVING CONSTRUCTION AND ENVIRONMENTAL PROTECTION SITUATION

In civil building energy consumption in the process, there are two kinds of energy: building materials and building energy consumption of construction energy consumption. The former can reduce the energy consumption of buildings through new energy environmental protection materials discovery and use (Fu 2010). The latter increase the need for long time efforts, which according to statistics, consume a lot of energy construction and increase the construction waste emissions more drastically with each passing day, and the construction personnel's awareness of environmental protection cannot win the support of people.

2.1 *Energy saving and environmental protection consciousness*

In the present civil construction, there is still some construction enterprise construction of energy-saving and environmental protection consciousness of the deviation, to energy conservation and environmental protection laws and regulations on the development of the enterprise. The managers of the enterprise are too lazy to implement the provisions of the state on the construction, and ineffectively perform energy-saving and environmental protection technology use of measures, environmental protection and energy-saving consciousness, and carrying out the staff and some construction is weak, not effectively implementing the provisions of energy-saving and environmental protection technology program (Yang 2013).

2.2 *Management mechanism of energy saving and environmental protection*

In the mechanism of management of civil construction units of energy saving and environmental protection, there is a big loophole. The management responsibility is not clear as well as the evaluation mechanism of energy saving and environmental protection is not perfect. The supervision mechanism is not implemented and fails to form a scientific and rational management mechanism.

Although energy saving and environmental protection has been put on the agenda for the development of government, there are still products, construction methods and equipment of some construction enterprises with high energy consumption. The government in the supervision and punishment is not clear, which seriously affect the promotion and implementation of energy-saving and environmental protection technology in construction.

2.3 *Construction equipment and construction technology*

In the civil construction, part of the backward construction units still uses aged equipment, causing a great waste of energy. Management of some enterprises, in order to reduce the construction cost, ignores the construction of infrastructure renewal fee, and some electricity water equipment is still used in construction, to reduce the construction cost. But in long-term consideration, a serious waste of energy, such as electricity and resources of water resources, is made and some of the old equipment, power, and the effect are small. The efficiency is not high, which not only affects the construction process, but also leads to a waste of resources.

Due to personal reasons for construction habits, construction technology is backward, seriously restricting the implementation of the construction of energy saving and environmental protection. Some small construction habits may lead to a huge waste of resources, for example, water-saving habits, in which closing the switch can save precious water resources.

3 GREEN ENERGY-SAVING TECHNOLOGY

3.1 *Water-saving construction technology*

In the construction project construction program where a lot of water is used, most of the water after use enters the sewage treatment equipment, and then gets discharged to the outside environment, which does not meet the economic requirement. So, the introduction of water-saving measures in the construction process promotes water reciprocating use and water use, in order to improve water utilization efficiency.

Not as in groundwater dewatering of foundation pit when the extraction is directly discharged to the outside, but when using the collected and concrete mixing, because the groundwater temperature is not high, and can reduce the concrete work of the temperature to a large extent. For example, it can also use the rainwater collection measures to collect rainwater, and the use of building site to reduce dust spray or flushing construction equipment, in addition to the flow into the waste water processing device such as the formation of drilling program in wastewater by precipitation. Filtration and other procedures can be directly reused in the procedure of production or other drilling mud on the water quality demand is not high in the program, as shown in Figure 1.

3.2 *Top surface of building lighting energy saving*

The top surface of the building is affected by natural conditions in building structures, which has the greatest impact site. The entire top almost all the day is exposed to light conditions, and easy snow accumulation in the winter. So, the top surface of the insulation, moisture, and heat insulation performance must meet the requirements; otherwise, there will be a top-level housing in hot summer and cold winter conditions. In the actual construction, the top surface can be considered by poor heat conductivity of building materials, or in the top surface between the insulation layer and moisture-proof layer multi applied a layer of moisture-proof

Figure 1. Construction of building a water-saving cistern.

Figure 2. The top surface of the building lighting energy-saving equipment.

insulation composite materials, improving the top surface of the insulation and moisture proof performance, as shown in Figure 2.

But the buildings will be designed into the top surface by energy-saving building solar panels, and the large area and long time illumination into the building energy can be used as the building interior air-conditioning refrigeration or heating. Water heater is a key measure for the development of modern buildings. In addition, it also can use the roof planting technology, and the top building planting vegetables such as green plants. On the one hand, it can make full use of the excellent natural conditions of the top surface, which is conducive to the growth of vegetable; on the other hand, it makes the top surface of the building. More than a layer of insulation or heat insulation layer is an important manifestation of the concept of green energy-saving construction. But in the construction process of the top surface, the moisture-proof performance should be fully ensured in order to avoid leakage due to irrigation.

3.3 *Construction technology of doors and windows*

Doors and windows in the building construction project are the key components for the use of energy-saving technology. Because of its poor performance of thermal insulation door and window section, the light projection is very good characteristics. So, it will make efficient use of the energy-saving effect on the doors and windows of the building, and it can also improve the building itself to use the energy-saving effect to a great extent.

First, we can choose a better economy, with strong practicability of building materials. For example, energy saving and environmental protection is the most commonly used glass, Low-E glass. The glass is ordinary glass formed on the surface of coating a layer of semiconductor oxide film, which can effectively reduce the reflectivity of the doors and windows. At present, the energy conservation environmental protection glass has been widely applied in the construction industry. But because the related supervision mechanism is not perfect, it makes a lot of building materials with poor quality. This does not cause the energy-saving and environmental protection effect in the practical application of delta; thus, the technology greatly reduces green energy-saving construction. So, enhanced detection is related to the intensity. Second, we must grasp the doors and windows of the ratio between the doors and windows. If the whole occupied proportion is too large, the adverse effects of insulation conditions would definitely be on buildings, so according to the relevant provisions to the north and east of the building cannot be more than 20%. The west of the building cannot be more than 30%, and the south of the building cannot be more than 35%, as shown in Figure 3.

3.4 *Ground construction technology*

The key to the building on the ground is to ensure that the ground insulation and moisture proof building materials would not be destroyed. It is necessary to strengthen the cover on the insulation and ground materials building materials business to have a good moisture resistance impact and pressure function.

The insulation structure and moisture-proof material damage and then change thus consume more construction labor and material resources, which do not meet the green energy-saving construction requirements. So, in the ground

Figure 3. Energy-saving design of doors and windows.

Figure 4. Ground energy-saving and environmental protection design.

construction of insulation structure and selection of moisture-proof material in accord with the fundamental energy demand as much as possible, the use of materials mechanics performance is good, which can also enhance the durability of ground, as shown in Figure 4.

4 CONCLUSION

With the domestic city township construction speed becoming faster and faster, the number of housing construction is increasing, and the scale is also expanding. The construction industry is facing a rare good opportunity of development, to complete the unit's sustainable and healthy development, to have a place must follow social the progress of the times the pace in the future challenges. The use of green energy-saving technology is necessary in the construction process, so that it can not only minimize the pollution degree of the environment, but also save the resources.

Therefore, it can reduce the construction cost of the unit. But all sorts of reasons should also be clearly understood, and thus now the construction technology of green energy use is not much in practice. So, in the process of practice, the relevant agencies must advocate the use of construction technology of green energy. This will not make the construction technology of green energy stagnate at the ideological level as before, but it can really be applied to the construction program, to provide services for the building, and then make a contribution to our living environment, the nature and improvement of social environment.

REFERENCES

[1] Yang, G.J. 2010. The research and application of green energy technology in the residential district. Shanxi Building, 19(7): 233–234.

[2] Fu, C.Q. 2010. Analysis of green energy saving technology in intelligent building. Intelligent Building, 6(8): 17–20.

[3] Hu, L. 2013. Green energy technology in building doors and windows. Value Engineering, 10(9): 142–143.

Material Science and Environmental Engineering – Chen (Ed.)
© 2016 Taylor & Francis Group, London, ISBN 978-1-138-02938-5

Research on the credit rating model of energy contract management mode in power grid enterprises

Y.N. Wu, M. Yang & K.F. Chen
North China Electric Power University, Beijing, China

ABSTRACT: With the current development of energy contract management mode, credit evaluation for the power grid enterprises has become more and more important, so power grid enterprises put forward some new requirements. There is no effective credit evaluation index system for the energy contract management mode aimed at the Chinese national condition currently, while the credit evaluation index system for other industries is not suitable for the energy contract management mode. First, it ignores the fuzzification of the credit evaluation index, and the scoring method is relatively rough; second, it pays too much attention to the quantitative indicators, but pays little attention to the qualitative indicators; finally, there is no credit evaluation system for the energy contract management pattern in power grid enterprises. In this paper, we hope to solve the credit problems of the energy contract management mode, and reduce the risk of energy-saving renovation project that is carried out by power grid enterprises.

Keywords: power grid enterprises; energy contract management pattern; fuzzy comprehensive evaluation method

1 INTRODUCTION

The characteristic that "investment and then profit, initial investment is large, and payoff period is long" of energy contract management pattern leads to the credit of energy-using enterprises is vital for power grid enterprises, deciding the final success or failure of project directly [1]. The reasons why the credit evaluation index system of other industries is not suitable for the energy contract management pattern of China are as follows: (1) the evaluation methods are relatively rough because of ignoring the characteristic of fuzziness about the credit evaluation index; (2) indicator selection overemphasizes on the quantitative indicators, while not paying enough attention to the qualitative indicators; (3) grid enterprises have not yet obtained the credit evaluation system about energy contract management pattern at present [2]. This paper hopes to solve the credit barrier in the energy contract management pattern in order to reduce the risk of power grid enterprises that implement energy conservation and transformation project.

In order to solve the above problems, the credit evaluation index system of the energy contract management pattern in power grid enterprises is built in this paper by using the fuzzy comprehensive evaluation method based on previous studies. Finally, practicability and feasibility is proved by a case study.

2 INDICATORS

2.1 *Indicators*

There is no effective credit evaluation indicator system for the energy contract management pattern at present. This paper refers to the credit evaluation indicator system in electricity charge for power clients because the characteristic that is "pay after use" of power grid enterprises is similar to the characteristic which is "share after transform" of energy contract management pattern, and combines with the credit evaluation index system of the enterprise in the bank and "The Enterprise Performance Evaluation Standard (2013)" in order to ensure scientificity and operability.

2.2 *Weight value*

The weight that reflects the importance of each indicator plays an important role in the whole calculation process. Combining the literature [3] with the actual need of project, the indicator system weight value is determined by using the arithmetic mean method and frequency statistical method, as shown in Figure 1.

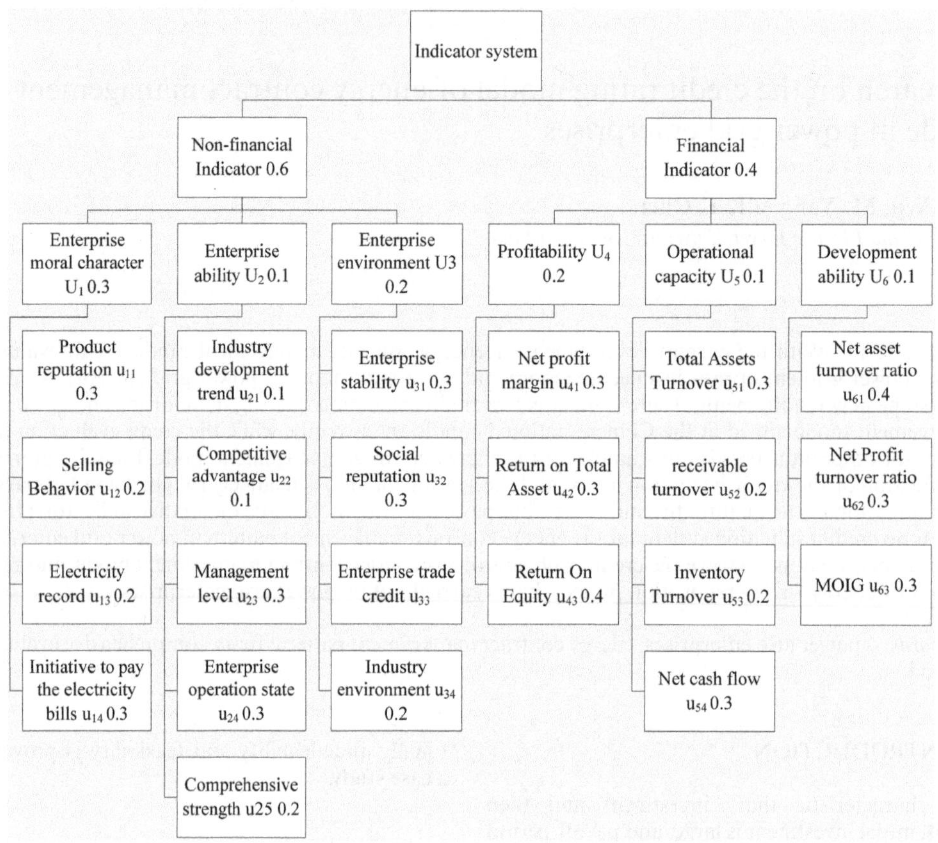

Figure 1. Indicator system and the weight value.

3 METHODOLOGY

3.1 *Remark set*

Remark set is determined to be nine classes as {AAA, AA, A, BBB, BB, B, CCC, CC, C} according to the credibility of power users. Here P means the number of grade that is equal to 9 in this paper.

AAA means the management will be relatively stable for quite a long time, and the enterprise operators with high quality and strong speciality are rigorous and standard for the management and the financial operation. The enterprise that has paid electricity bills actively for nearly 3 years, attaches great importance to talents training, talents attraction and talents utilization, and makes great contributions to the development of the industry that developed relatively stable.

On the contrary, C signifies that the enterprise cannot share the benefit that is agreed in the contract in the future because it has been well proven that the financial capacity is poor, the management faces difficulties and bankruptcy is a real possibility.

3.2 *Weight set*

The weight of each indicator means that the importance degree to the previous level is given to make up the weight set.

The first level: $A = (0.3, 0.1, 0.2, 0.2, 0.1, 0.1)$

The second level:

$$A_1 = (0.3, 0.2, 0.2, 0.3)\ A_2 = (0.1, 0.1, 0.3, 0.3, 0.2)$$
$$A_3 = (0.3, 0.3, 0.2, 0.2)\ A_4 = (0.3, 0.3, 0.4)$$
$$A_5 = (0.3, 0.2, 0.2, 0.3)\ A_6 = (0.4, 0.3, 0.3)$$

3.3 *The process*

Fuzzy evaluation vector is given as follows:

$$R_{ij} = \left(\mu_{ij}^{(1)}, \mu_{ij}^{(2)}, \ldots\ldots, \mu_{ij}^{(p)}\right) \tag{1}$$

Fuzzy evaluation matrix R_i is given as follows:

$$R_i = \begin{bmatrix} \mu_{i1}^{(1)} & \mu_{i1}^{(2)} & \cdots & \mu_{i1}^{(p)} \\ \mu_{i2}^{(1)} & \mu_{i2}^{(2)} & \cdots & \mu_{i2}^{(p)} \\ \vdots & \vdots & \ddots & \vdots \\ \mu_{in}^{(1)} & \mu_{in}^{(2)} & \cdots & \mu_{in}^{(p)} \end{bmatrix} \quad (2)$$

where $\mu_{ij}^{(m)}$ is the membership of indicator u_{ij} about the grade v_m, and $\sum_{m=1}^{p} \mu_{ij}^{(m)} = 1$.

3.4 *Computing method of* $\mu_{ij}^{(m)}$

For non-financial indicators,

$$\mu_{ij}^{(m)} = \frac{T_m}{t} \quad (3)$$

Here, T_m is the frequency, and t is the total number of experts, as listed in Table 2.

For financial indicators, the ceiling limit value is a, the lower limit value is b, and the actual value is c according to the "standard value of performance appraisal". The financial Indicator numerical value in recent 3 years is X_{t-2}, X_{t-1}, X_t, so the counting equation of c is given as follows:

$$c = 0.2X_{t-2} + 0.3X_{t-1} + 0.5X_t \quad (4)$$

Suppose that $d = \left[\frac{c-b}{a-b} \times 9\right]$, where [ ] is the integer notation.

Membership function $\mu_{ij}^{(m)}$ is given in Equation (5) if there is a positive correlation between the financial indicator and the evaluation level.

Membership function $\mu_{ij}^{(m)}$ is given in Equation (6) if there is a negative correlation between the financial indicator and the evaluation level:

$$\begin{cases} \text{if } d \geq 9, \text{ then } \mu_{ij}^{(m)} = \begin{cases} 1 & m = 1 \\ 0 & m \neq 1 \end{cases} & m = (1, 2, \ldots, 9) \\ \text{if } 0 \leq d < 9, \text{ then } \mu_{ij}^{(m)} = \begin{cases} 1 & m = 9 - d \\ 0 & m \neq 9 - d \end{cases} & m = (1, 2, \ldots, 9) \\ \text{if } d < 0, \text{ then } \mu_{ij}^{(m)} = \begin{cases} 1 & m = 9 \\ 0 & m \neq 9 \end{cases} & m = (1, 2, \ldots, 9) \end{cases} \quad (5)$$

$$\begin{cases} \text{if } d \leq 0, \text{ then } \mu_{ij}^{(m)} = \begin{cases} 1 & m = 1 \\ 0 & m \neq 1 \end{cases} & m = (1, 2, \ldots, 9) \\ \text{if } 0 < d \leq 9, \text{ then } \mu_{ij}^{(m)} = \begin{cases} 1 & m = d + 1 \\ 0 & m \neq d + 1 \end{cases} & m = (1, 2, \ldots, 9) \\ \text{if } d \geq 9, \text{ then } \mu_{ij}^{(m)} = \begin{cases} 1 & m = 9 \\ 0 & m \neq 9 \end{cases} & m = (1, 2, \ldots, 9) \end{cases} \quad (6)$$

The calculation formula for the first level of the fuzzy comprehensive evaluation set B_i is given as follows:

$$B_i = A_i \cdot R_i = \{a_{i1}, a_{i2}, \ldots a_{in}\} \cdot \begin{bmatrix} \mu_{i1}^{(1)} & \mu_{i1}^{(2)} & \cdots & \mu_{i1}^{(p)} \\ \mu_{i2}^{(1)} & \mu_{i2}^{(2)} & \cdots & \mu_{i2}^{(p)} \\ \vdots & \vdots & \ddots & \vdots \\ \mu_{in}^{(1)} & \mu_{in}^{(2)} & \cdots & \mu_{in}^{(p)} \end{bmatrix} = \{b_{i1}, b_{i2}, \ldots, b_{ip}\} \quad (7)$$

The calculation formula for the second level of the fuzzy comprehensive single factor assessment matrix R is given as follows:

$$R = \begin{bmatrix} B_1 \\ B_2 \\ \vdots \\ B_m \end{bmatrix} = \begin{bmatrix} A_1 \times R_1 \\ A_2 \times R_2 \\ \vdots \\ A_m \times R_m \end{bmatrix} \quad (8)$$

The calculation formula for the second level of the fuzzy comprehensive evaluation set B is given as follows:

$$B = A \cdot R = A \cdot \begin{bmatrix} A_1 \times R_1 \\ A_2 \times R_2 \\ \vdots \\ A_m \times R_m \end{bmatrix} = (b_1, b_2, \ldots b_p) \quad (9)$$

3.5 *Credit rating*

The score set can be expressed as follows:

$$F = (f_1, f_2, \ldots, f_9)^T = (95, 85, 75, 65, 55, 45, 35, 25, 10)^T$$

The calculation formula for the comprehensive evaluation score of credit rating is given as follows:

$$Z = BF = \sum_{j=1}^{p} (b_i f_i) \quad (10)$$

Table 1. Credit rating of the grid user.

Grade	Interval
AAA	[90, 100)
AA	[80, 90)
A	[70, 80)
BBB	[60, 70)
BB	[50, 60)
B	[40, 50)
CCC	[30, 40)
CC	[20, 30)
C	[0, 20)

Enterprise credit grade is determined by the corresponding grade evaluation remark, as listed in Table 1.

4 CASE STUDY

A large-scale construction enterprise is discussed in this paper to make the credit evaluation model more specific.

4.1 *Indicators*

Non-financial indicators and financial indicators of the enterprise are listed, respectively, in Tables 2 and 3.

4.2 *The first level of fuzzy comprehensive evaluation*

$$R_1 = \begin{bmatrix} 0 & 0.2 & 0.3 & 0.4 & 0.1 & 0 & 0 & 0 & 0 \\ 0.3 & 0.4 & 0.3 & 0 & 0 & 0 & 0 & 0 & 0 \\ 1 & 0 & 0 & 0 & 0 & 0 & 0 & 0 & 0 \\ 0.4 & 0.3 & 0.3 & 0 & 0 & 0 & 0 & 0 & 0 \end{bmatrix}$$

$$R_2 = \begin{bmatrix} 0.4 & 0.4 & 0.2 & 0 & 0 & 0 & 0 & 0 & 0 \\ 0.2 & 0.4 & 0.4 & 0 & 0 & 0 & 0 & 0 & 0 \\ 0.3 & 0.4 & 0.3 & 0 & 0 & 0 & 0 & 0 & 0 \\ 0 & 0.4 & 0.3 & 0.3 & 0 & 0 & 0 & 0 & 0 \\ 0 & 0.3 & 0.4 & 0.3 & 0 & 0 & 0 & 0 & 0 \end{bmatrix}$$

$$R_3 = \begin{bmatrix} 0.2 & 0.4 & 0.3 & 0.1 & 0 & 0 & 0 & 0 & 0 \\ 0.2 & 0.4 & 0.4 & 0 & 0 & 0 & 0 & 0 & 0 \\ 0 & 0.3 & 0.4 & 0.3 & 0 & 0 & 0 & 0 & 0 \\ 0.2 & 0.4 & 0.4 & 0 & 0 & 0 & 0 & 0 & 0 \end{bmatrix}$$

$$R_4 = \begin{bmatrix} 0 & 0 & 0 & 1 & 0 & 0 & 0 & 0 & 0 \\ 1 & 0 & 0 & 0 & 0 & 0 & 0 & 0 & 0 \\ 0 & 1 & 0 & 0 & 0 & 0 & 0 & 0 & 0 \end{bmatrix}$$

$$R_5 = \begin{bmatrix} 0 & 0 & 1 & 0 & 0 & 0 & 0 & 0 & 0 \\ 1 & 0 & 0 & 0 & 0 & 0 & 0 & 0 & 0 \\ 1 & 0 & 0 & 0 & 0 & 0 & 0 & 0 & 0 \\ 0 & 1 & 0 & 0 & 0 & 0 & 0 & 0 & 0 \end{bmatrix}$$

$$R_6 = \begin{bmatrix} 1 & 0 & 0 & 0 & 0 & 0 & 0 & 0 & 0 \\ 1 & 0 & 0 & 0 & 0 & 0 & 0 & 0 & 0 \\ 1 & 0 & 0 & 0 & 0 & 0 & 0 & 0 & 0 \end{bmatrix}$$

The first level of the fuzzy comprehensive evaluation set B_i can be deduced as follows:

$$B_1 = A_1 \cdot R_1$$

$$= \{0.3, 0.2, 0.2, 0.3\} \cdot \begin{bmatrix} 0 & 0.2 & 0.3 & 0.4 & 0.1 & 0 & 0 & 0 & 0 \\ 0.3 & 0.4 & 0.3 & 0 & 0 & 0 & 0 & 0 & 0 \\ 1 & 0 & 0 & 0 & 0 & 0 & 0 & 0 & 0 \\ 0.4 & 0.3 & 0.3 & 0 & 0 & 0 & 0 & 0 & 0 \end{bmatrix}$$

$$= (0.38, 0.23, 0.24, 0.12, 0.03, 0, 0, 0, 0)$$

The same procedure may be easily adapted to any other evaluation set as follows:

$$B_2 = A_2 \cdot R_2 = (0.15, 0.38, 0.32, 0.15, 0, 0, 0, 0, 0)$$
$$B_3 = A_3 \cdot R_3 = (0.16, 0.38, 0.37, 0.09, 0, 0, 0, 0, 0)$$
$$B_4 = A_4 \cdot R_4 = (0.3, 0.4, 0, 0.3, 0, 0, 0, 0, 0)$$
$$B_5 = A_5 \cdot R_5 = (0.5, 0.3, 0.3, 0, 0, 0, 0, 0, 0)$$
$$B_6 = A_6 \cdot R_6 = (1, 0, 0, 0, 0, 0, 0, 0, 0)$$

Table 2. Expert evaluation of non-financial indicators.

The first level	The second level	Grade								
		AAA (v_1)	AA (v_2)	A (v_3)	BBB (v_4)	BB (v_5)	B (v_6)	CCC (v_7)	CC (v_8)	C (v_9)
U_1	u_{11}		2	3	4	1				
	u_{12}	3	4	3						
	u_{13}	10								
	u_{14}	4	3	3						
U_2	u_{21}	4	4	2						
	u_{22}	2	4	4						
	u_{23}	3	4	3						
	u_{24}		4	3	3					
	u_{25}		3	4	3					
U_3	u_{31}	2	4	3	1					
	u_{32}	2	4	4						
	u_{33}		3	4	3					
	u_{34}	2	4	4						

Table 3. Analysis of financial indicators.

No.	$c = 0.2X_{t-2} + 0.3X_{t-1} + 0.5X_t$	a	b	$d = \left[\frac{c-b}{a-b} \times 9\right]$
u_{41}	7.49%	12%	0	5
u_{42}	7.7%	8%	0	8
u_{43}	9.95%	12%	0	7
u_{51}	65.02%	80%	20	6
u_{52}	861.61%	300%	100	34
u_{53}	367.26%	300%	100	12
u_{54}				7
u_{61}	41.07%	10%	0	36
u_{62}	32.61%	8%	0	36
u_{63}	14.01%	11%	0	11

4.3 *The second level of fuzzy comprehensive evaluation*

$$R = \begin{bmatrix} B_1 \\ B_2 \\ B_3 \\ B_4 \\ B_5 \\ B_6 \end{bmatrix} = \begin{bmatrix} 0.38 & 0.23 & 0.24 & 0.12 & 0.03 & 0 & 0 & 0 & 0 \\ 0.15 & 0.38 & 0.32 & 0.15 & 0 & 0 & 0 & 0 & 0 \\ 0.16 & 0.38 & 0.37 & 0.09 & 0 & 0 & 0 & 0 & 0 \\ 0.3 & 0.4 & 0 & 0.3 & 0 & 0 & 0 & 0 & 0 \\ 0.5 & 0.3 & 0.3 & 0 & 0 & 0 & 0 & 0 & 0 \\ 1 & 0 & 0 & 0 & 0 & 0 & 0 & 0 & 0 \end{bmatrix}$$

$$B = A \cdot R = (0.3, 0.1, 0.2, 0.2, 0.1, 0.1) \cdot R$$
$$= (0.371, 0.293, 0.208, 0.129, 0.009, 0, 0, 0, 0)$$

$$Z = BF = (0.371, 0.293, 0.208, 0.129, 0.009, 0, 0, 0, 0)$$
$$\cdot (95, 85, 75, 65, 55, 45, 35, 25, 10) = 84.63$$

The grade of this enterprise is AA, as given in Table 1.

5 CONCLUSIONS

In consideration of a variety of factors that are suitable for credit rating in the power grid enterprises, and with reference to the indicator system of credit rating in the bank, credit rating model of the energy contract management pattern in power grid enterprises is established by using the fuzzy comprehensive evaluation method, combined with the characteristic of a long investment payout period, to eliminate credit disorders and reduce the risk.

The indicator system that not only pays attention to the financial indicators, but also strengthens the effect of non-financial indicators meets the requirement of credit rating and the thinking habit, thus the reference for the results given by the indicator system is strong.

The fuzzy comprehensive evaluation that is easy and convenient to understand could synthesize the indicators and quantify the factors so that the results are obvious and rational, which makes the energy contract management pattern easy-to-market the application.

ACKNOWLEDGMENTS

This work was financially supported by the National Nature Science Foundation of China (No. 71271085) and the Beijing Twelfth Five Year Plan Project of Philosophy and Social Sciences (No. 12JGB044).

REFERENCES

[1] Fabozzi F.J, Vink D. The information content of three credit ratings: the case of European residential mortgage-backed securities [J]. The European Journal of Finance, 2015, 21(3): 172–194.

[2] Tewari M., Byrd A., Ramanlal P. Callable bonds, reinvestment risk, and credit rating improvements: Role of the call premium [J]. Journal of Financial Economics, 2015, 115(2): 349–360.

[3] Baorui R.E.N. Credit Evaluation of Electricity Customers Based on Fuzzy Synthetic Evaluation Method [J]. Modern Electric Power, 2011, 4: 020.

Material Science and Environmental Engineering – Chen (Ed.)

A real-time failure model for transmission lines considering healthy index

P.J. Shi, Y.B. He & C.X. Guo
College of Electrical Engineering, Zhejiang University, Hangzhou, China

H.J. Fu & J.G. Wang
Power Dispatch and Control Center, Henan Electric Company, Zhengzhou, China

ABSTRACT: Traditional failure model for transmission lines only concerns the degradation of equipment and the environment they are in. In order to consider the healthy index of transmission lines, which can be acquired in today's power PMS, this paper proposed a new real-time failure model for transmission lines based on PHM (Proportional Hazard Model). Its hazard function contains two parts: the baseline failure rate which represents the healthy index, and the state description function which reflects how the weather factors influence real-time failure rate. In this paper, we use the idea of pooling to classify discrete variables and least squares method to estimate the model's parameters. A practical case is tested to prove the ability of the proposed model to characterize the real-time failure rate.

Keywords: failure rate; transmission line; healthy index; weather conditions; PHM; pooling

1 INTRODUCTION

The stable operation of the transmission lines is the foundation of power grid to supply power reliably. With in-time knowledge of the failure rate for transmission lines, we can make a quantitative assessment to the system's operation risk, which helps a lot in dispatching, maintenance decision and grid planning. As information like load data, on-line monitoring, inspection records and equipment testing data is regularly collected and integrated, the diversity and precision of input for transmission line failure rate has improved drastically in these years. The failure model can be sorted into two categories, repairable failure model caused by permanent aging and repairable failure model based on healthy index.

Healthy index model fully quantifies the online monitoring information, inspection records and testing results of equipment, focusing on the relationship between failure rate and the healthy state transmission lines are in. With adequate data source, we can get the value of model's parameters by back-calculation. However, the error grows as the data source reduces [4]. Nowadays, healthy index and outage data can be acquired in power Production Management System (PMS), which makes it possible to build an accurate model.

In practice, real-time weather conditions have significant effects on transmission lines' failure rate, which makes the lines in poor or extreme weather conditions more likely to break down [6]. In [1–2], a multi-state weather model is proposed. Paper [3], proposed a failure model concerning weather conditions and online monitoring by fuzzy set. In order to consider both the healthy index and weather condition, this paper proposed a new real-time failure model for transmission lines based on PHM (Proportional Hazard Model). Its hazard function contains two parts: the baseline failure rate which represents the healthy index, and the state description function which reflects how the weather factors influence real-time failure rate. Inspired by [5], the idea of pooling has been used to classify discrete variables and least squares method is used to estimate the model's parameters.

The remainder of this paper is structured as follows: Section II shows the mathematic models. Section III introduces the method of parameters estimation. In section IV, an example is discussed to verify the effectiveness of the proposed model. The conclusion is drawn in Section V.

2 MATHEMATIC MODELS

2.1 *PHM (Proportional hazard model)*

The main idea of PHM model is to find a connection between hazard function and all kind of factors, like healthy index of equipment, weather, and degradation and so on.

$$h(t) = h_0(t) \cdot \exp(\gamma Z(t)) \tag{1}$$

where t represents the performance period of equipment; $h_0(t)$ is the baseline failure rate; $\exp(\gamma Z(t))$ is the state description function, which takes kinds of factors into consideration and reflects their effect on real-time failure rate. $\gamma Z(t)$ is a linear combination of concomitant variable, like $(\gamma_1 Z_1(t) + \gamma_2 Z_2(t) + \ldots + \gamma_n Z_n(t))$. Each concomitant variable represents a particular measured or state value. With the ability to reflect the time-varying interior factors like healthy index, and external factors like temperature and humidity, it can give a reasonable and comprehensive real-time failure rate.

2.2 *Failure model based on healthy index*

In the process of power system operation, the operating condition of the system is changing time by time. When the power system is suffering from a disturbance or some equipment is broken down, either the network structure or the operating conditions will change. Traditional reliability assessment methods for power system are mostly based on fixed operation mode and constant equipment parameters and models, ignoring the effect of bus voltage, power flow, system frequency and other real-time operating conditions on the failure rate, neglecting impact of operation mode of the unit, the real-time load change, the network structure change on the failure consequences. So the traditional reliability evaluation can't represent the real-time level of safety and reliability of power system.

With the application and the development of state detection technology, the effectiveness of the device data accuracy has improved greatly. Based on the results of state detection, operators can obtain the quantized real-time assessment of equipment according to the "Power transmission and transformation equipment state overhaul test procedures" guideline, which is published by State Grid Corporation of China, and then we get the corresponding healthy index of equipment. Health index is a numerical parameter to describe the state of degradation. Studies show that the relationship between equipment failure rate and its healthy index is an exponential function as follows:

$$\lambda = Ke^{-C*HI} \tag{2}$$

where λ is the failure rate of equipment; K is the proportionality coefficient; C is the curving coefficient; HI is the healthy index. The guideline distinguishes the healthy states of lines into normal, attention, abnormal and critical, whose healthy indexes are 95, 85, 75 and 50 accordingly. In principle we can get the proportionality and curving coefficient by back calculation as long as we have two-year data. The failure model based on healthy index is a comprehensive assessment taking real-time equipment state into consideration.

Based on the idea of PHM model, the comprehensive real-time model for transmission line considering healthy index and weather condition is in the shape of

$$\lambda(t) = \lambda_0(t) \cdot \exp(b_0 + b_1 x_1(t) + b_2 x_2(t) + \cdots b_n x_n(t)) \tag{3}$$

where $\lambda(t)$ is the real-time failure rate; $\lambda_0(t)$ is the failure rate based on healthy index; $b_0 + b_1 x_1(t) + b_2 x_2(t) + \ldots b_n x_n(t)$ represents all kinds of parameters related to weather condition.

3 SOLUTION METHODOLOGY

As the mathematic model has been given above, what we should do next is to define the value of all parameters according to healthy index data, transmission line-outage data and weather data. Here, we need to do estimation for b_0, b_1, b_2 related to weather effect and K, C related to equipment's healthy index.

3.1 *Pooling*

In fact, the values of weather condition like temperature, humidity and wind speed are discrete, which means that for each particular situation, the sample set is too small to give a reasonable estimation. Aiming to solving this problem, we improve the estimator by grouping the transmission line by pooling their outage history data so that data from different components having similar characteristics are combined. This increase the size of sample set on which we operate to compute the parameter. What's more, it helps to reduce the variance of estimation. The idea of pooling outage history data is shown in two aspects:

1. Group the transmission lines by healthy state. We assume that outage rates for the lines in the same healthy state are similar.
2. Group the transmission lines by weather condition. We assume that outage rates for the lines in the same healthy state are similar. By analyzing the weather condition when lines break down, we obtain the relationship between weather condition and outage rate.

Grouping by healthy state is easy and we only illustrate how to group the data by weather condition here.

For simplification and amplification of sample set, we only concern temperature and humidity in this paper. Assuming that the highest and lowest temperatures in database are T_{max} and T_{min}, we divide the whole interval $[T_{max}, T_{min}]$ into N_v small intervals with the same size, S_v, where N_v*S_v equals to $T_{max}-T_{min}$. Therefore, each small temperature interval turns out to be $[T_i, T_{i+1}]$, $1 \leq i \leq N_v$, $V_1 = V_{min}$, $V_2 = V_{max}$. We deal with the humidity interval similarly and get N_w small intervals. Finally we have N_v*N_w intervals which are the combination of temperature and humidity.

We add up all the outage events happened in interval k, $[T_i, T_{i+1}; H_j, H_{j+1}]$, and get the total number n_k. It can be proved by maximum likelihood method that Eq. (10) is an unbiased estimation for the failure rate of interval k.

$$\lambda_k = n_k / (D_k \cdot L_k) \ \ (times/(100km \cdot year)) \tag{4}$$

where D_k is the duration of weather condition corresponding to interval k in history database; L_k is the total length of transmission line.

When S_v, S_w is small enough, we can use $\{X_{1k}, X_{2k}\} = \{(T_i+T_{i+1})/2, (H_j+H_{j+1})/2\}$ to represent the environmental characteristic of interval k. So far, we have already extracted $\{X_{1k}, X_{2k}, \lambda(t)\}_{k=1\ldots Nv*Nw}$.

3.2 *Modeling procedure*

The procedure of getting the real-time failure model for transmission line considering healthy index can be summarized as the following steps in Figure 1.

Step 1: Group the transmission line into normal, attention, abnormal and critical set according to healthy index,

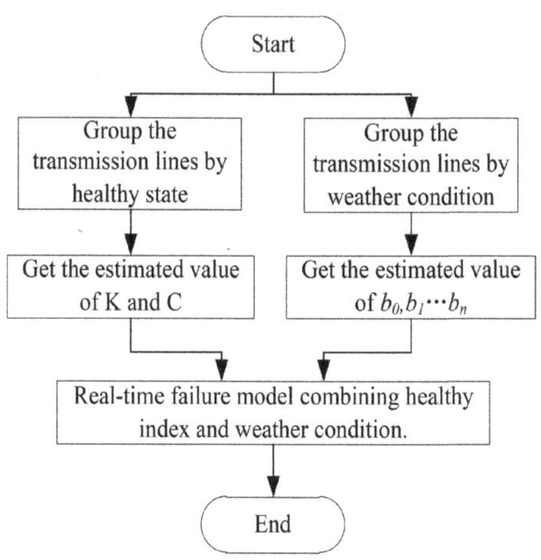

Figure 1. Flow chart of the modeling process.

Step 2: Get the estimated value of proportionality coefficient K and curving coefficient C by Using Least squares method.

Step 3: Group the transmission lines by weather condition. It should be mention that the basic failure is different for lines in different healthy state, so we only group the normal state lines in this step.

Step 4: Get the estimated value of $b_0, b_1 \ldots b_n$ which reflect the effect of weather condition on failure rate by Using Least squares method.

Step 5: Combine the result of Step 2 and Step 4 and get a real-time failure model for transmission lines considering healthy index and weather condition.

4 ILLUSTRATED CASE

In order to demonstrate the accuracy of the proposed real-time failure model for transmission line considering healthy index, a practical case is given here. The outage data is collected from province C from year 2009 to 2014. For privacy protection, we can only give rough information here. Table 1 shows the healthy index data and outage data transmission lines in province C in these 5 years.

By using least squares minimum method, we solve out the estimated value of K and C are 3.437 and –0.026, so the failure rate for transmission line considering healthy index turns out to be:

$$\lambda_0(t) = 3.437e^{-0.026*HI(t)} \tag{5}$$

By applying Eq. (5), we calculate the failure rate of lines corresponding to different healthy index in Table 2.

The differences between Table 1 and 2 is relatively slightly, which means that the equation we get in Eq. (5) is reasonable. Failure model concerning healthy index reflects the fact that the worse

Table 1. Transmission line healthy index and outage data of province C (2009–2014).

Healthy state	Normal	Attention	Abnormal	Critical
Healthy index	95	85	75	50
Length [100 km]	99.48	31.20	20.88	7.20
Duration [year]	5	5	5	5
Outage number	124	56	62	32
Failure rate [time/100 km*y]	0.249	0.359	0.594	0.889

Table 2. Failure rate of transmission lines corresponding to different healthy index.

Healthy state	Normal	Attention	Abnormal	Critical
Healthy index	95	85	75	50
Failure rate [time/ 100 km*y]	0.291	0.377	0.489	0.937

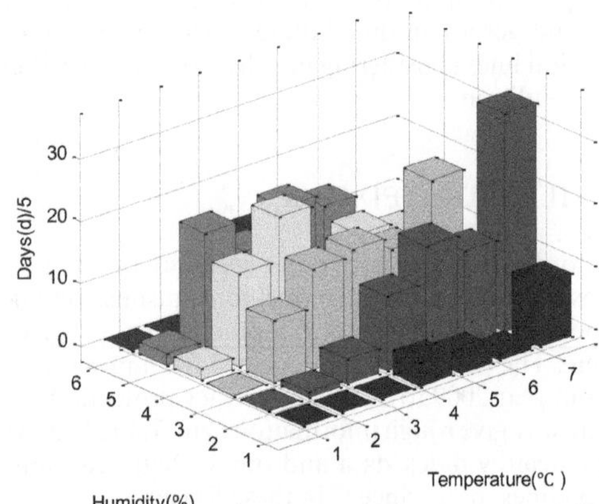

Figure 2. Duration of each weather block (2009–2014).

healthy condition transmission lines are in, the more easily for lines to break down. Results also show that the failure rate of lines in critical state is close to 1 time per 100 km per year, which is relatively high and operators should pay much attention to them.

As the sample set of normal transmission lines is much larger than others, here we only use normal lines, whose total length is 9948 km to find out the relationship between weather conditions and real-time failure rate. We divide the temperature scale into 7 intervals, which are [0, 5), [5, 10), [10, 15), [15, 20), [20, 25), [25, 30), [30, 35] and the units are °C. Similarly, the humidity scale is divided into 6 intervals, which are [40, 50), [50, 60), [60, 70), [70, 80), [80, 90), [90, 100) and the units are %. It should be mentioned that because the cities in one province are geographically close, some weather conditions like temperature and humidity have something in common. In this way, we can use the weather of provincial capital to represent that of the whole province.

By gathering the duration data in Figure 2 and outage number in Figure 3, the failure rate of each weather block is obtained, shown as Figure 4. Figure 4 also shows that when the real-time temperature is in the fourth interval and humidity in

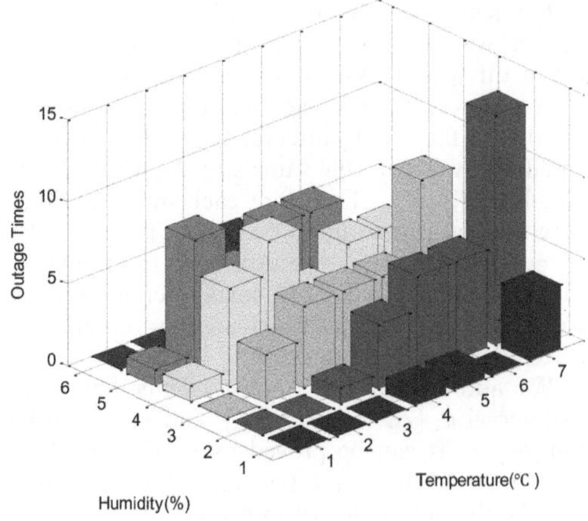

Figure 3. Outage number of each weather block (2009–2014).

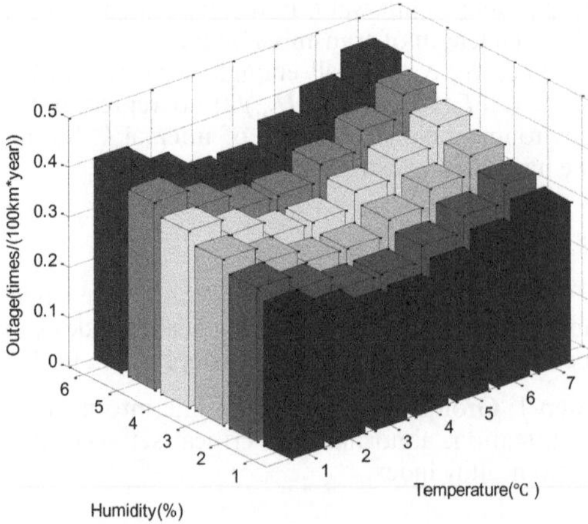

Figure 4. Failure rate of each weather block (2009–2014).

the first, the rate is lowest; and when temperature is in the seventh and humidity in the sixth, the rate is highest.

Then we use least squares minimum method to match the information given by Figure 4. Noticing the fact that the rate is relatively low in 15°C, so we replace x_1 by $|x_1-15|$ in Eq. (3) the real-time failure model for transmission line in normal state is given as:

$$\lambda(t)=\lambda_0(t)\cdot\exp(-0.568+0.0211|x_1-15|+0.0066x_2) \tag{6}$$

where x_1 is the temperature and x_2 is the humidity.

Figure 5 gives us the real-time failure rate of transmission lines in common state in 2014,

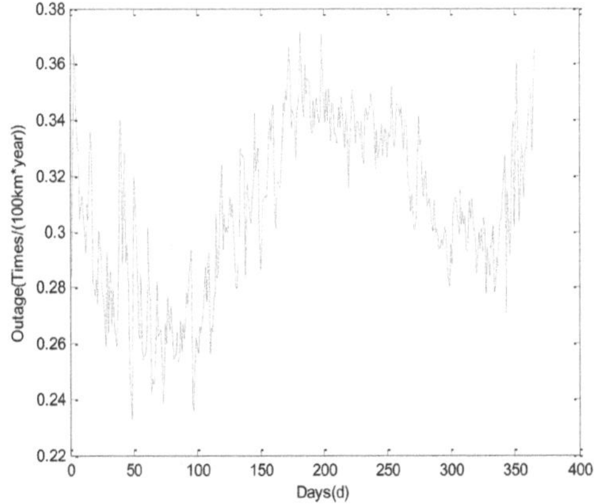

Figure 5. Failure rate of each day in 2014.

according to Eq. (6). It reveals that the transmission line is more likely to break down in summer and winter, rather than spring and fall, which is in line with the general operating experience. In this way, the accuracy of the suggested model has been proved in some extend.

We suspect that weather conditions like temperature and humidity affect all the transmission lines in the same way, no matter which healthy state they are in. Therefore, if we want to calculate the real-time failure rate of lines in critical healthy state, what we should do is get a new $\lambda_0(t)$ by Eq. (5) and replace that in Eq. (6). However, as we do not have enough outage data for lines in other healthy state, this idea has not been proven yet.

5 CONCLUSION

This paper proposed a real-time failure model for transmission lines based on PHM model. Comparing to traditional ones, the proposed model can take both healthy index of transmission lines and the weather conditions like temperature and humidity into consideration, so the failure rate obtained is comprehensive as it reflects both the external and interior factors. Case study has proved the efficiency of this model, which also provides a new methodology to calculate the real-time failure rate of transmission lines, which can further be used in operation risk assessment and reliability evaluation for power system.

REFERENCES

[1] Billinton R., Acharya J. 2005. Consideration of multi-state weather models in reliability evaluation of transmission and distribution systems [J], *Canadian Journal of Electrical and Computer Engineering, 916–922.*

[2] Billinton R., Singh G. 2006. Application of adverse and extreme adverse weather: modeling in transmission and distribution system reliability evaluation [J], *IEE Proceedings of Generation, Transmission and Distribution, 153(1):115–120.*

[3] Duan Tao, Luo Yi, Shi Lin, et al. 2013. Application of adverse and extreme adverse weather:modeling in transmission and distribution system reliability evaluation [J], *Power System Protection and Control, 41(15):59–67(in Chinese).*

[4] Wang Yi, He Bengteng, Wang Huifang. 2011. Research on equipment reliability based on life cycle states [J], *Power System Technology, 35(8):207–211 (in Chinese).*

[5] Xiao F., McCalley J.D., Ou Y., et al. 2006. Contingency probability estimation using weather and geographical data for on-line security assessment [C], *Probabilistic Methods Applied to Power Systems.*

[6] Zhang Yongjun, Xu Liang, Wu Chengwen. 2012. Research on Ice Disaster Risk Evaluation Model of Power System Considering Multi-Factors [J], *Power System Protection and Control, 40(15):12–17 (in Chinese).*

Material Science and Environmental Engineering – Chen (Ed.)
 ISBN 978-1-138-02938-5

Transmission line maintenance based on the Proportional Hazards Model

H.J. Fu & J.G. Wang
Power Dispatch and Control Center, Henan Electric Company, Zhengzhou, China

C.X. Guo
College of Electrical Engineering, Zhejiang University, Hangzhou, China

ABSTRACT: A new maintenance strategy based on proportional hazard model is proposed for transmission line, considering the aging process, weather factors and equipment status of transmission line fault rate. Model parameters are estimated by using the maximum likelihood estimation method. Using the minimum maintenance cost method to determine the optimal maintenance strategy threshold, and then working out the optimal maintenance strategy. The example analysis shows that the proposed method can provide decision support for the transmission line maintenance, and have important practical value.

Keywords: Proportional Hazard Model; two-state weather model; equipment state; maximum-likelihood estimation; maintenance strategy

1 INTRODUCTION

Transmission line is an important part of power system. The safe and stable operation of the transmission line have an important relationship with the security and stability of power system. Timely and accurate understanding of the fault rate of the transmission line can be used to evaluate the reliability and risk system, and provide sufficient basis for the operation, maintenance planning and the power system planning.

Many literatures study the effects of weather conditions on the reliability of power transmission and transformation equipment, often using two state or polymorphic model to describe the equipment failure rate under different weather conditions. With the development of condition based maintenance technology, the study found that the equipment state has a direct impact on the equipment failure rate. Transmission line fault is affected by the combination of internal factors (equipment aging equipment, etc.) and external factors (weather conditions). Only considering the effects of unilateral factor will make the transmission line failure rate calculation results to be too optimistic. How to integrate the influence of various factors on the line failure rate is still a difficult problem, on which this paper will discuss and study.

In terms of the equipment maintenance, traditional maintenance modes tend to be blind and waste, which have been unable to meet the requirement of the new situation. Condition Based Maintenance (CBM) is a strategy that can monitor the state parameters of a system, timely detect the fault and take preventive measures to achieve "maintenance according to need", which can avoid the serious consequences caused by excessive or insufficient maintenance. In order to reflect the comprehensive effect of various factors on the failure rate of equipment, a lot of research have made on this issue both at home and abroad. Calman-filtering, Markov process, information fusion and artificial intelligence are much used methods to build model of the system according to the test information, but these models are lack of practicality in the optimization maintenance decision.

This paper establishes a failure rate model based on the PHM (Proportional Hazards Model) for transmission line, which has a comprehensive consideration of equipment aging, weather factors and the state of the equipment. The model is more suitable for the decision-making of maintenance of the transmission line. The benchmark failure rate function of the model is temperature aging model. The state covariate function chooses weather conditions which can reflect the running environment and equipment state which reflect the internal operating situation of the selected transmission line. Parameters are estimated by maximum likelihood estimate method. Monitoring data are analyzed for a certain type of transmission line fault data. The simulation results validate that the model and method proposed in this paper is effective and reasonable.

2 PROPORTIONAL HAZARDS MODEL

The PHM was proposed by D.R. Cox in 1972, the earliest application of which was biomedical research and economics. Now it is getting more and more application in the reliability research.

The failure rate function for the proportional hazards model is

$$h(t) = h_0(t) \cdot \exp(\gamma Z(t)) \tag{1}$$

Among them, t is the working time of the equipment. $h_0(t)$ is the basic failure rate function (baseline hazard function), which is used to represent the aging process of the equipment, often using Weibull distribution. g is the covariate coefficients to be estimated. $\exp(gZ(t))$ is a state description function, which reflects in the influence of different state on the equipment failure rate.

The benchmark failure rate function in the model is used to describe the process of aging equipment, and many failure model can describe the situation, such as the Poisson distribution, exponential distribution, Weibull distribution and lognormal distribution. The benchmark failure rate function commonly used Weibull distribution, as follows:

$$h_0(t) = \frac{\beta}{\eta}\left(\frac{t}{\eta}\right)^{\beta-1} \tag{2}$$

Among them, β is the shape parameter, η is characteristic parameters of life.

Proportional hazards model assume that state variables have multiplicative effect on the failure function, in which $gZ(t)$ are a linear combination of the covariates, just like $g_1Z_1(t) + \ldots + g_iZ_i(\mathrm{t}) + \ldots + g_nZ_n(t)$ and each covariate can characterize a specific value or state value. The advantages of proportional hazards model is that covariate is time dependent, in practical application covariate can react with equipment state information such as internal variables, or detection information of apparatus body, and can also be the influence of external variables of equipment operation, such as environmental conditions. It is a comprehensive consideration of all kinds of state information when the equipment is running, and can effectively use the state information for the assessment of failure.

3 PHM OF TRANSMISSION LINE

The frame is shown in Figure 1. The input information used in the model includes two parts: one part is the service time of the transmission line; another part is the weather conditions and equipment conditions at time t.

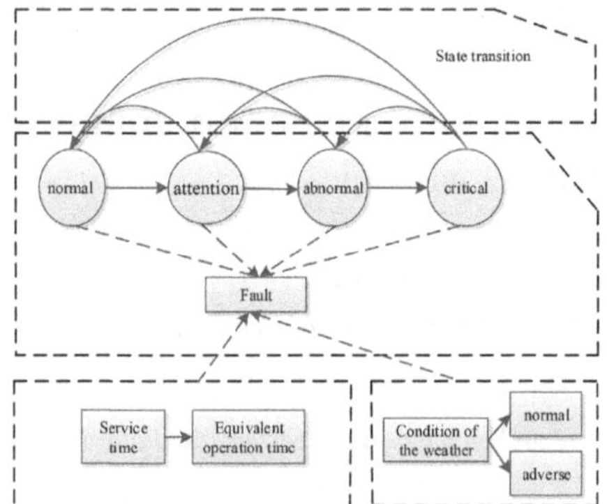

Figure 1. The framework of proposed failure rate model.

Table 1. The covariate values of weather related factors.

Condition of the weather	Normal	Adverse
Value of Z_1	1	2

Table 2. The covariate values of equipment state related factors.

State of the line	Normal	Attention	Abnormal	Critical
Value of Z_2	1	2	3	4

On the basis of the two state weather models, weather condition can be divided into two states: normal and adverse, of which severe degree increased gradually. When the weather is respectively in the good condition and the adverse condition, the information of the state description function covariate is shown in Table 2.

According to "the guidelines for the state assessment of overhead transmission line" by the China State Grid Corp, transmission line state can be evaluated for the four types: normal, attention, abnormal, critical state. Using equipment condition monitoring information to determine the running state of transmission line, the corresponding value of state variables in describing function model is shown in Table 2.

Proportional hazards model of transmission line is shown below:

$$\mathrm{h}(t;Z) = \frac{\beta}{\eta}\left(\frac{t}{\eta}\right)^{\beta-1} \exp(\gamma_1 Z_1 + \gamma_2 Z_2) \tag{3}$$

Among them, β is the shape parameter, η is characteristic parameters of life. γ_1, γ_2 are co factor related covariates, Z_1, Z_2 are covariates.

4 LIKELIHOOD PARAMETER ESTIMATION

In order to evaluate the parameters in the proportional hazards model, characteristics data of transmission line needs to be collected when it is in the given operation condition. These data includes a certain amount of life time data, censored time data, the number of failure data. Assuming $(t_1, t_2 \ldots t_i)$ as the data column, $\theta\,(\beta,\gamma)$ as the parameters to be evaluated, n_f as the failure characteristics in the n data samples. Then we can get the general form of the likelihood function:

$$L(\theta)=\prod_{i\in F} f(t_i,\theta)\prod_{i\in C} R(t_i,\theta) \tag{4}$$

Among them, $f(t_i,\theta)$—Failure probability density function

$R(t_i,\theta)$—The reliability function

F,C—Failure set and censoring set.

The probability density function of maximum likelihood estimation is:

$$f(t)=h(t)R(t)=h(t)\exp\left\{-\int_0^t h(t)dt\right\} \tag{5}$$

And then get the maximum likelihood function

$$\begin{aligned} L(\theta)&=\prod_{i\in F}h(t_i,\theta)R(t_i,\theta)\prod_{i\in C}R(t_i,\theta)\\ &=\prod_{i\in F}h(t_i,\theta)\prod_{i\in N}R(t_i,\theta)\\ &=\prod_{i\in F}h(t_i,\theta)\prod_{i\in N}\exp\left(-\int_0^{t_i}h(t_i,\theta)dt\right)\\ &=\left\{\prod_{i\in F}\frac{\beta}{\eta}\left(\frac{t_i}{\eta}\right)^{\beta-1}\exp(\gamma_1 Z_1+\gamma_2 Z_2)\right\}\\ &\cdot\left\{\prod_{i\in N}\exp\left[-\int_0^{t_i}\frac{\beta}{\eta}\left(\frac{s}{\eta}\right)^{\beta-1}\exp(\gamma_1 Z_1+\gamma_2 Z_2)ds\right]\right\} \end{aligned} \tag{6}$$

The logarithmic form of Maximum likelihood function is:

$$\begin{aligned} \ln L(\theta)&=n_f\ln\left(\frac{\beta}{\eta}\right)+\sum_{i=1}^{n_f}\ln\left(\frac{\beta}{\eta}\right)^{\beta-1}\\ &+\sum_{i=1}^{n_f}(\gamma_1 Z_1(t_i)+\gamma_2 Z_2(t_i))\\ &-\sum_{i=1}^{n}\int_0^{t_i}\frac{\beta}{\eta}\left(\frac{s}{\eta}\right)\exp(\gamma_1 Z_1(t_i)+\gamma_2 Z_2(t_i))ds \end{aligned} \tag{7}$$

The value z of state function generally is not continuous monitoring, therefore it need subsection integral. Comparing with the monitoring interval selected, rate of metastasis status of transmission line in operation process is very slow, and it will not change in the vast majority of monitoring interval. Therefore state in the monitoring interval can do the following approximation: when $t_i < t < t_{i+1}$, $Z(t) = Z_{ti}$, namely, before arriving at next monitoring point, the state of the transmission line is assumed to be unchanged.

According to the maximum likelihood parameter assessment method, $\ln L$ will do partial derivative separately with parameter β and γ, and make each equation be zero to get a set of nonlinear equations.

By solving this equation can obtain the parameter estimation θ

$$\frac{\partial}{\partial\theta}\ln[L(\theta)]=0 \tag{8}$$

According to the Nelder-Mead's algorithm, the algorithm is an iterative optimization algorithm without derivation. And then use the optimization function of fminsearch optimization toolbox in Matlab software, for the solution of parameters β, and the regression vector γ.

5 THE OPTIMAL MAINTENANCE STRATEGY

The optimal maintenance strategy of transmission lines is different with the decision goal. The paper chooses the classical method in the maintenance strategy—the lowest cost method, namely within unit time using repair cost minimum as target to develop the optimal repair strategy. The total cost per unit time expectations can be expressed as:

$$E\left(\frac{C}{t}\right)=\frac{C_{PM}+C_{CM}\cdot N(t)}{t} \tag{9}$$

Among then, C_{PM} is the prevention maintenance cost; C_{CM} is the repair cost; C is the total cost; $N(t)$ is the average number of failure in time t. According to the minimum of the total cost per unit time, we can develop an optimal prevention interval T_{PM}.

$$N(t)=\int_0^t\exp[\gamma\cdot Z(u)]d\left(\frac{u}{\eta}\right)^{\beta} \tag{10}$$

By putting (10) into (9), we get:

$$E\left(\frac{C}{t}\right)=\frac{C_{PM}+C_{CM}\cdot\int_0^t\exp[\gamma\cdot Z(u)]d\left(\frac{u}{\eta}\right)^{\beta}}{t} \tag{11}$$

It is difficult to directly calculate the integral, usually using numerical simulation method to get the minimum cost per unit time, and get the optimal preventive maintenance time interval, as is shown in Figure 2.

Using the optimal preventive maintenance time interval, finding a group of the most close to failure time data and the adjoint variable in the fault data, then we can seek the threshold of the failure rate h^*, which meet below in the normal operation condition:

$$\frac{\beta}{\eta}\left(\frac{t}{\eta}\right)^{\beta-1}\exp[\gamma\cdot Z(u)]\le h^* \tag{12}$$

The logarithmic form of the above formula is:

$$\gamma\cdot Z(t)\le \ln(\eta^{\beta}h^*/\beta)-(\beta-1)\ln t \tag{13}$$

Then we get the optimal maintenance threshold curve. If the $\beta > 1$, the curve will show as Figure 3.

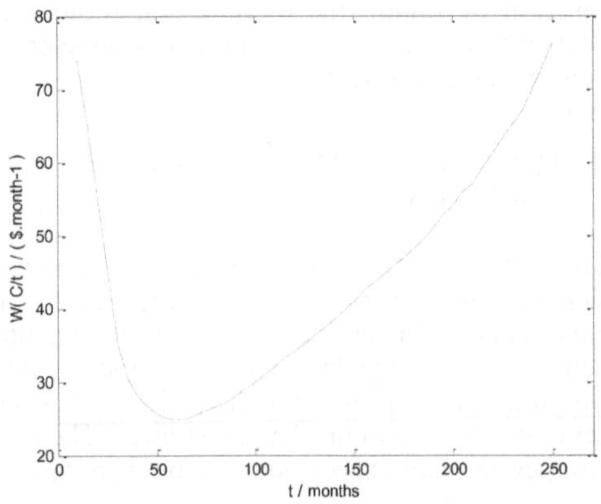

Figure 2. The optimal repair interval.

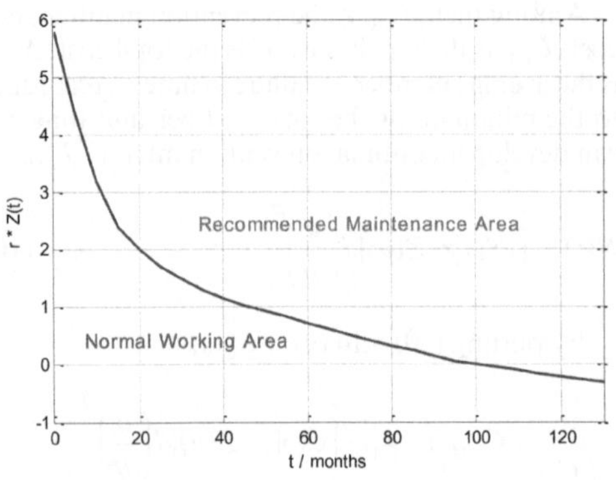

Figure 3. The optimal repair threshold curve.

And then monitoring data gotten in the operation process is brought into $\gamma Z(t)$ to draw point on the figure. If the point is below the curve, the transmission line is in the normal area, can continue to run. If the point is above the curve, the transmission line needs to take immediate maintenance measures. If the point is near the curve, it should strengthen the monitoring and take maintenance measures recently.

6 INSTRUCTIONS

6.1 *Data analysis and parameter estimation*

In this paper, a district grid of Guangdong province is used as an example. By getting the weather data of the recent ten years in this area, and selecting the corresponding transmission line state evaluation results, we get a certain number of sample data. Part of the sample data is shown in Table 3.

Using the sample data and the maximum likelihood parameters estimation method to estimate the parameters of the proposed model, parameter estimation results are shown in Table 4.

After getting the parameters of the model, the expression of the proportional hazards model of the transmission line can be shown below:

$$\mathrm{h}(t;Z)=\frac{7.996}{386.67}\left(\frac{t}{386.67}\right)^{6.996}\exp(2.365Z_1+1.763Z_2) \tag{14}$$

Table 3. Part of the sample data.

	Transmission line 1			Transmission line 2		
Number	T	Weather	Status	T	Weather	Status
1	12	1	1	15	1	1
2	19	2	1	29	2	1
3	26	1	1	36	1	2
4	39	1	1	56	2	2
5	55	1	1	75	1	2
6	68	1	2	90	2	3
7	81	2	2	102	2	4
8	99	1	1	–	–	–
9	105	1	1	–	–	–

Table 4. The results of parameter estimation.

Parameter	η	β	γ_1	γ_2
Value	386.67	7.996	2.365	1.763

6.2 *Maintenance decision*

Using the minimum cost method described above (CPM/CCM = 0.025 set according to the actual project experience), the optimal preventive repair interval can be calculated to be 68 months. Using the formula (13) we can be get maintenance decision threshold curve as shown in Figure 3.

For the actual monitoring data of transmission lines, bringing it into $\gamma_1 Z_1(t) + \gamma_2 Z_2(t)$, and drawing point in this diagram. If the point is below the curve, the transmission line is in the normal area, can continue to run. If the point is above the curve, the transmission line needs to take immediate maintenance measures.

For monitoring data of two transmission lines in Table 3, using the method described above, we can draw maintenance decision diagram in Figure 4 and Figure 5. As we can be seen from Figure 4, until the last moment of monitoring, the points of monitoring data in the decision diagram are below the repair decision threshold curve, which are in working area, so the line can continue to run. As can be seen from Figure 5, parts of the points of the latest monitoring data have come into the recommended maintenance area, so the line should take immediate maintenance measures.

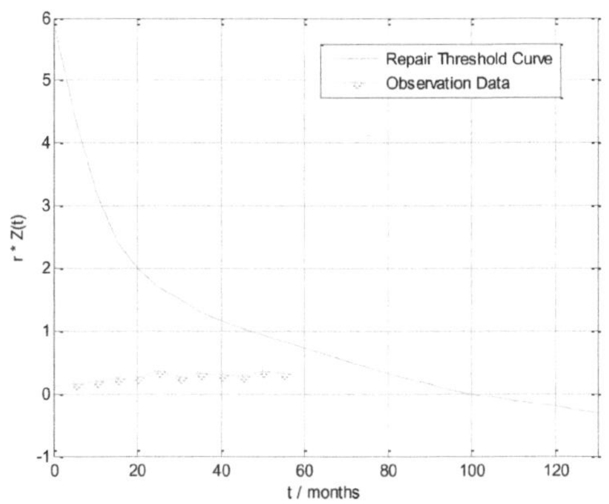

Figure 4. Maintenance decision diagram of transmission line 1.

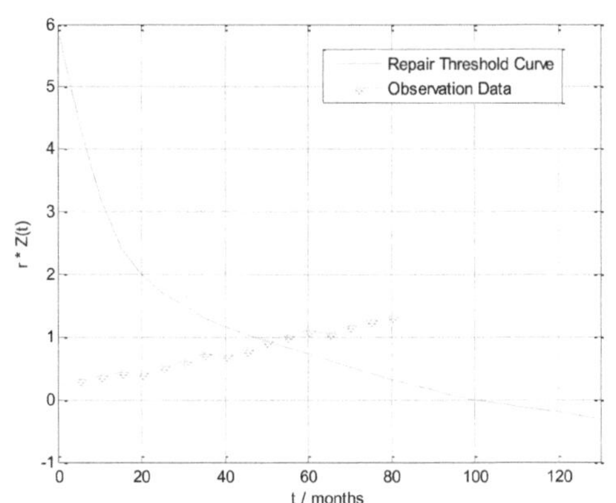

Figure 5. Maintenance decision diagram of transmission line 2.

7 CONCLUSION

This paper establishes a failure rate model based on the proportional hazards model for transmission line, which has a comprehensive consideration of equipment aging, weather factors and the state of the equipment. Comparing with the traditional model only considering the weather factor or equipment fault rate, this model can reflect the influence of the combination factors on transmission line, reflecting the influence of internal and external causes. Aiming at the failure of transmission line is very rare, and there are a large number of censored data conditions, the method of maximum likelihood is used for parameter estimation.

At the same time, the paper uses the minimum cost method which target is minimizing the time unit repair cost to determine the repair decision threshold. The method used as a quantitative calculation method, can determine the maintenance decision thrshold reasonably, which can develop the optimal maintenance strategy. Finally through the example simulation, it proves the validity of this method. At the same time, this method can also be extended to be used in other equipment's state based maintenance, so it has more practical value.

REFERENCES

[1] Chen Yongjin, Ren Zhen, Huang Wenying. 2004. Model and analysis of power system reliability evaluation considering weather change [J], *Automation of Electric Power Systems, 28(21):17–21(in Chinese).*

[2] Cox D.R. 1972. Regression models and life-tables [J], *Journal of the Royal Statistical Society, 187–220.*

[3] Gaver D.P, Montmeat F.E, Patton A.D. 1964. Power system reliability: Part I measures of reliability and methods of calculation[J], *IEEE Trans on Power Apparatus and Systems, 83(7):727–737.*

[4] He Jian, Cheng Lin, Sun Yuanzhang, et al. 2009. Condition dependent short-term reliability models of transmission equipment [J], *Proceedings of the CSEE, 29(7):39–46(in Chinese).*

[5] Qu Jing, Guo Jianbo. 2004. Statistics and analysis of faults in main domestic power systems from 1996 to 2000[J], *Power System Technology, 28(21):60–63(in Chinese).*

[6] Wang P, Billinton R. 2002. Reliability cost/worth assessment of distribution systems incorporating time-varying weather conditions and restoration resources [J], *IEEE Trans. on Power Systems, 17(1):260–265.*

[7] Wang Yi, He Bengteng, Wang Huifang. 2011. Improvement of state failure rate model for power transmission and transforming equipment [J], *Automation of Electric Power Systems, 35(16):27–31(in Chinese).*

[8] Xu Jing, Wang Jing, Gao Feng, et al. 2000. A survey of condition based maintenance technology for electric power equipments [J], *Power System Technology, 24(8):48–52.*

[9] Zhang Xiang, Yang Zhihui, Song Zitong, et al. 2013. A Failure Model for Oil-Immersed Transformer Based on Load Factor and Equipment Inspection[J], *Information power system technology, 37 (4):39–45(in Chinese).*

[10] Zuo Hongfu, Cai Jing, Wang Huawei, et al. 2008. The theory and method of maintenance decision [M], *Beijing: Aviation Industry Press, 2008:8(in Chinese).*

Material Science and Environmental Engineering – Chen (Ed.)
 ISBN 978-1-138-02938-5

Simulation on erosion surface of transmission line by sand particles based on LS-DYNA software

G.R. Hua, Y.Y. Guo & W.H. Li
Department of Mechanical Engineering, North China Electric Power University, Baoding, China

ABSTRACT: Surface condition of transmission line is an important factor to affect the corona characteristics. To research on its changing rules of LGJ-400/50 ACSR under sand and dust storms environments, the impact process of single sand particle on cable surface was simulated based on LS-DYNA software under the condition of different angles and particle sizes combined with the physical experimental model using LGJ-400/50 ACSR cable and indoor closed wind tunnel experimental system. The results show that: maximum impact force of 0.25 mm sand particle size is 0.372 MPa; maximum impact force of 0.5 mm is 2.138 MPa; the surface stress under 30° case is smaller than that of under 90° case, but its direction is tangential and the impact stress area is larger.

Keywords: sand particles; erosion angle; finite element simulation

1 INTRODUCTION

Sand and dust storms are natural events that occur in arid regions where the land is not protected by a covering of vegetation. They are reported frequently in Northwest China, which belong to the areas of the Central Asia storms. Wind erosion process mainly manifested in the sand erosion on material surface. In these areas, the main reason of material failure is erosion wear [1–4]. By far, study on the field is mainly focusing on effects of water conservancy projects by solid-liquid two phase flow (for example the influence of sediment laden flow on the concrete). The scholars in the field have made some progress on the aspect of research methods, estimation of the erosion, anti-erosion of materials and etc [5]. However, for the wind and sand environment, study on erosion mechanism and characteristics of engineering materials is not deep enough. So, it has important academic value to research on surface erosion rule of transmission line in wind sand environment and to find out the feasible measures to reduce the erosion effect for prolonging the working life of the transmission line and reducing the corona losses.

2 PHYSICAL EXPERIMENTAL MODEL

This experiment uses LGJ-400/50 Aluminum Conductor Steel Reinforced (in short, ACSR) cable as the test line. Take the above line 0.6 m long to make the sample. Totally four samples were made, for this experiment involves 90°,60°,45°,30° four different attack angles. Wire suspension mode inside the wind tunnel is shown in Figure 1. Wind speed is set to 21.23 m/s. Particle sizes involve 0.25 mm and 0.5 mm two different kinds. The concentration value is set to175 mg/m^3.

Figure 1. LGJ-400/50 ACSR cable suspension diagram inside the wind tunnel.

3 FINITE ELEMENT SIMULATION OF THE IMPACT PROCESS BY LS-DYNA SOFTWARE

So far, most of scholars use the analysis method of SEM surface morphology to study the erosion effect by particles. But the analysis results is based on the images after the erosion, however the erosion process can not be involved. So this paper uses ANSYS/LS-DYNA finite element software to simulate the erosion process.

3.1 *Simulation process of two particle diameters*

This paper uses the Lagrange method to simulate the impact in LS-DYNA software. Contact between the particle and the surface uses *CONTACT_ERODING_SURFACE_TO_SURFACE algorithm. Unit type uses 3D solid 164 units. The grid meshing uses three dimensional mapping methods by which the model can be divided into hexahedron element. The solution time is set to 100 μs, every 2.0 μs output a computation results.

Sand particles use the rigid material model, its density is 2.6 g/m^3, elasticity modulus is 50 GPa, Poisson's ratio is 0.3. Wire materials uses mat-Johnson Cook model, its density is 2.7 g/m^3, elasticity modulus is 70 GPa, Poisson's ratio is 0.3.

This paper uses 0.25 mm and 0.5 mm two different kinds of sand particles to simulate, both the attack angles are set to 90°. Velocity of sand particles is set to 25.7 m/s. Establish the 3D model of the particles and the target material. As particles are very small, in terms of the wire surface, it can approximately used plane instead. So, the target can be established into a cuboids model, sand can be established into a ball, as shown in Figure 2.

Post processor LS-PREPOST can observe the dynamic impact process of stress nephogram. Screenshot images at the typical time to analyze the stress nephogram, the results are as follows:

3.1.1 *Particle size of 0.25 mm erosion simulation results*

As shown in Figure 3, when T = 60 μs, the particles just leaving the contact surface and it formed stress concentration region in the impact center position, the maximum stress value is up to 0.372 MPa, stress at the periphery gradually reduced in the form of gradient grade. The stress concentration appears in the circular domain surround the impact region.

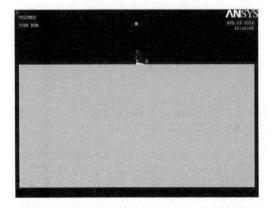

Figure 2. Three dimensional model of particles and target.

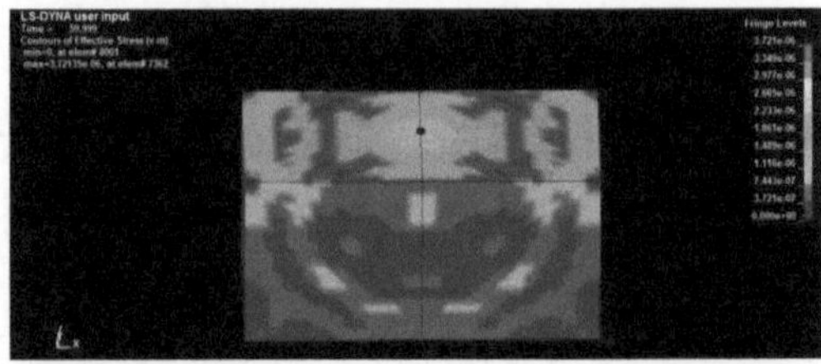

(1)T=60μs

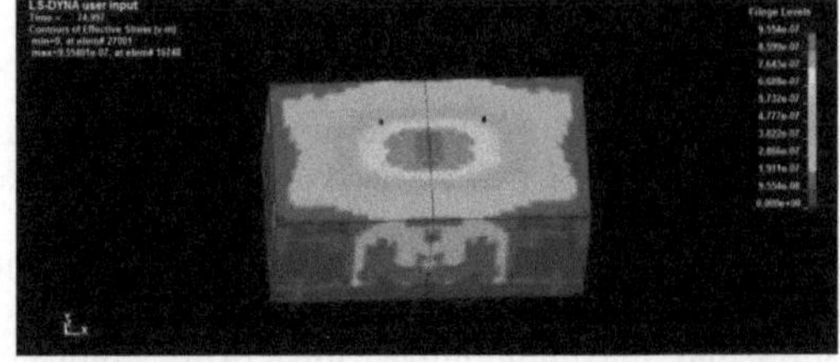

(2)T=75μs

(3)T=95μs

Figure 3. Simulation results of 0.25 mm particle.

When T = 75 μs, particles continued to leave the target, the stress concentration area expand than before. The moment it becomes to form circular ring distribution. In the middle of the circular ring has maximum stress and its value is 0.0955 MPa. The press value reduced than the moment of T = 60 μs. At this moment the stress has already spread to the entire surface of target material. When T = 95 μs, the center region of stress concentration has completely disappeared, instead it evolved into a long strip of dissipating stress zone. Beside the zone stress is small; some places even have zero press.

3.1.2 *Particle size of 0.5 mm erosion simulation results*

When the particle size is 0.5 mm, Screenshot images at the typical time as before. The results of the simulation are as follows:

As shown in Figure 4, when T = 55 μs the moment particles impact the target, wire surface appeared stress concentration zone and began to spread around. Maximum stress value is located in the regional center and its value is about 2.138 MPa, this value is 5.74 times the results of

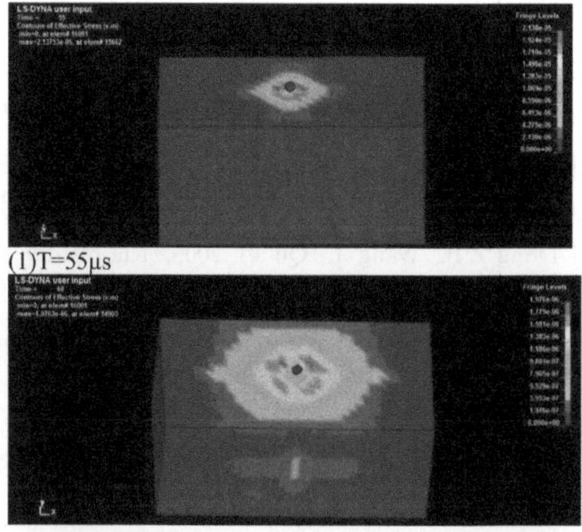

(1)T=55μs

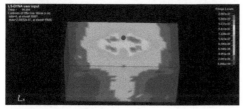

(2)T=60μs

(3)T=100μs

Figure 4. Simulation results of 0.5 mm particle.

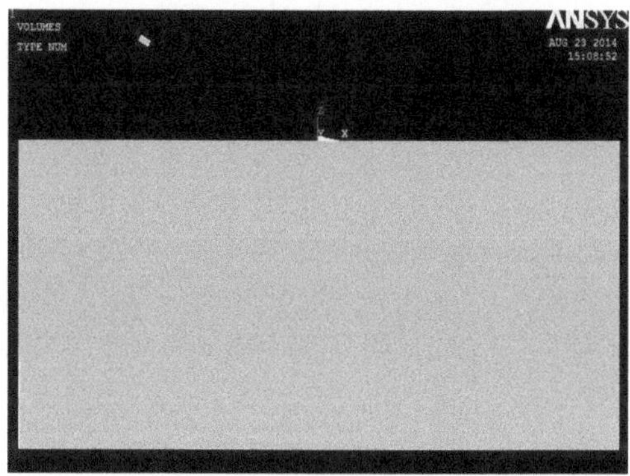

Figure 5. Model of particle and target.

0.25 mm diameter. When T = 60 μs, the stress concentration region evolved into four small stress concentration zone which becomes symmetrical distribution. When T = 100 μs, the surface stress zone has spread to the entire surface.

3.2 *Finite element simulation of two different angles*

The comparative angle is 90° and 30°. Taking into account of the attack angle of 30° case, there exist the cutting effect of polygonal particles on the target, so, for the model, particles use the cuboids model and ensure that the cuboids bottom surface forms a 30 degree angle with the surface of target material. The built model is shown in Figure 5. Velocity of sand particles is set to 25.7 m/s. Here uses 0.25 mm diameter sand particles to simulate. The solution time is set to 200 μs. When the attack angle is 90° the erosion situation results is already described in the above section. So, the following only gives the simulation results when the attack angle is 30°.

When the attack angle is 30°, typical time erosion results are shown as follows:

As shown in Figure 6, when T = 120 μs, the moment particles impact the target, stress area throughout a wide range of impact zone, this kind of stress distribution is different from that when the attack angle is 90°, the stress value is very small and ranges in 4.818e-3 to 5.782e-3 MPa. Considering the stress direction is tangential direction, the erosion effect can not be ignored. When T = 130 μs, the surface stress region decreases from the outside to the inside, and continue to shrink to a circular domain, this is the difference with the attack angle of 90° case.

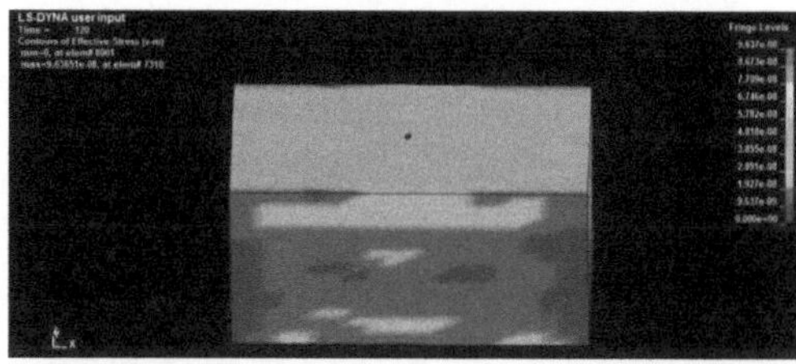

(1)T=120μs

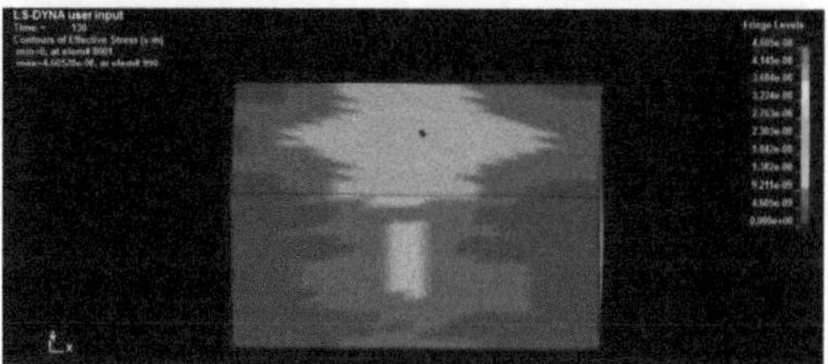

(2)T=130μs

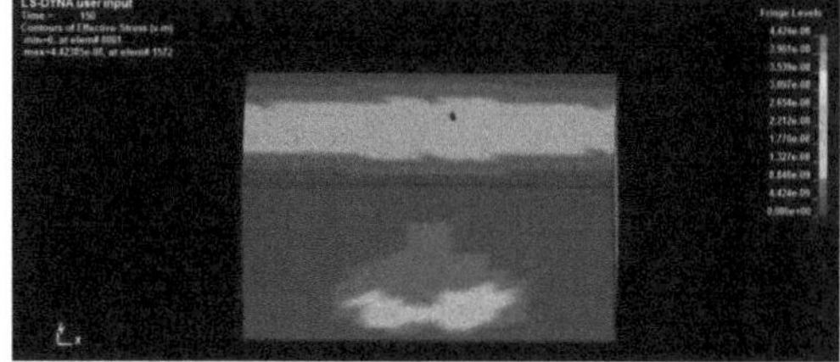

(3)T=150μs

Figure 6. Attack angle of 30° simulation results.

4 CONCLUSIONS

Through the establishment of three-dimensional model of particle and the wire surface, LS-DYNA software was used to simulate the impact process on wire surface by single particle at different attack angles and particle sizes. The results are as follows:

1. The maximum impact force of 0.25 mm sand particle is 0.372 MPa, and the maximum impact force of 0.5 mm is 2.138 MPa.
2. Although the surface stress under 30° case is smaller than that of under 90° case, its direction is tangential and the impact stress area is larger. Under the two kinds of attack angles, stress dissipates in different ways.

ACKNOWLEDGEMENT

This research was financially supported by Beijing Natural Science Foundation (2132038).

REFERENCES

[1] Zheng X.J., Zhou Y.H. 2003. Some key problems in the research of wind-blown and sands, *Mechanics in Engineering*, 25(2): 1–6.

[2] Zheng X.J., Huang N. 2004. Advances in investigation on electrification of wind-blown sands and its effects, *Advances in Mechanics* 34(1): 77–86.

[3] Dong Z.B., Wang T., Qu J.J. 2003. The history of desert science over the last 100 years, *Journal of desert research* 23(1): 1–5.

[4] Zheng X.J. 2007. On mechanism of wind-blown sand movement, *Science & technology review* 25(14): 22–27.

[5] Huang N., Zheng X.J. 2000. Experiment test of electrification of wind-blown sands, Chinese science bulletin 45(20): 2232–2235.

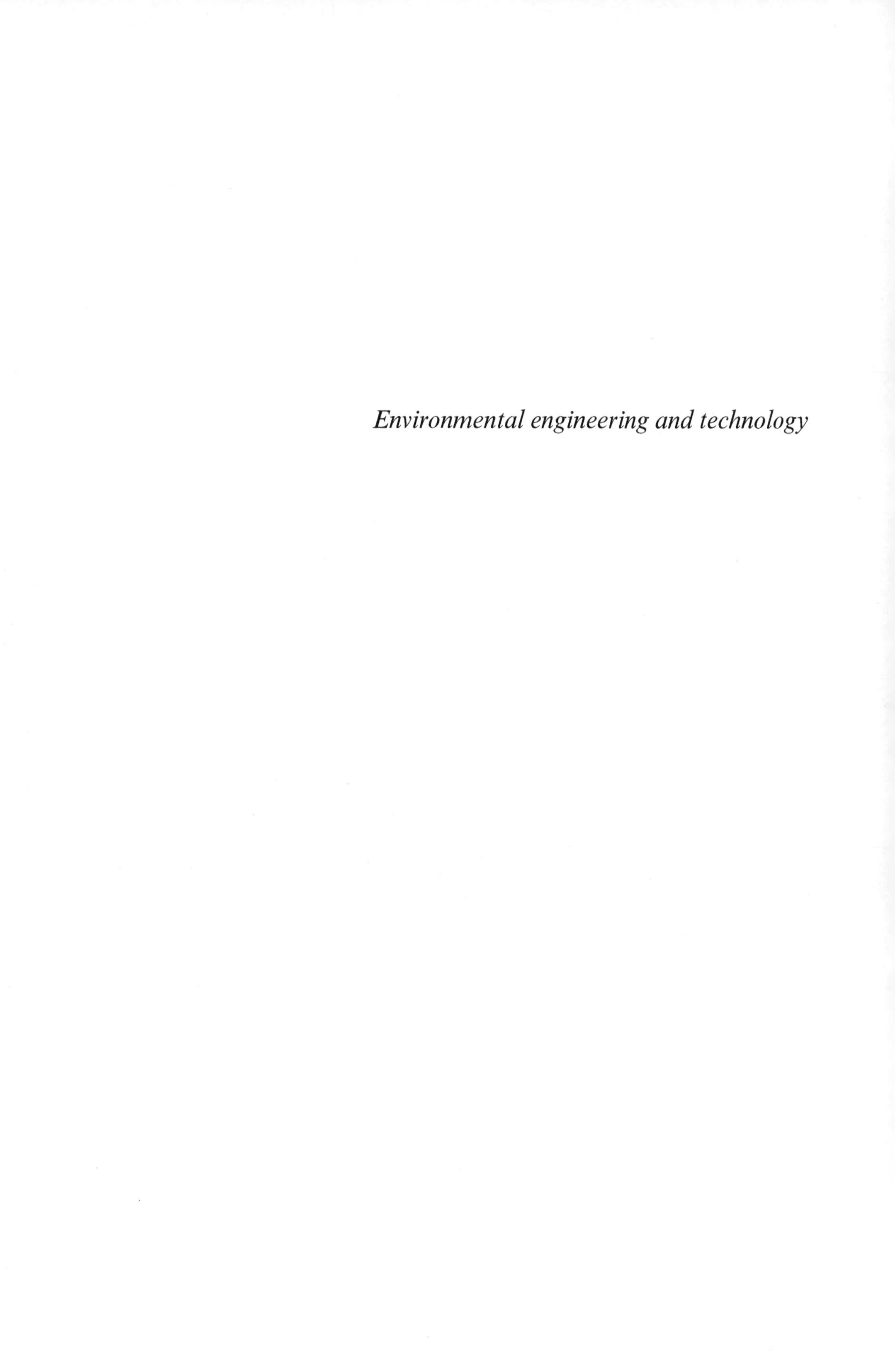

Environmental engineering and technology

Material Science and Environmental Engineering – Chen (Ed.)
 ISBN 978-1-138-02938-5

Preparation of nanometer Fe-doped SnO_2 photocatalyst and photocatalytic oxidation application in aquaculture wastewater

X.C. Yu, X.J. Jin, X.L. Shang, Y.Q. Liu, D.N. Yin & G.H. Xue
College of Ocean Technique and Environment Department, Dalian Ocean University, Dalian, China

ABSTRACT: Fe-doped SnO_2 photocatalysts were synthesized by co-precipitation method. The morphology and crystal pattern of some prepared catalysts were characterized by XRD and SEM techniques which demonstrated that the prepared catalysts were quartet rutile structure. An orthogonal experiment was carried out to investigate the optimal combination of factors and the influence order of factors. Under UV-light, the removal rate of NH_4^+-N can up to 81.6% when the experiment was undertaken under the optimization of conditions: the dosage of catalyst 0.2 g/L, calcination time 3.0 h, the molar ratio of Fe_2O_3 and SnO_2 1:3, initial concentration of ammonia—N 60 mg/L, photocatalytic degradation reaction time 3.0 h. Under visible light with the same conditions, The optimal combination is as follows: the dosage of catalyst 0.2 g/L, calcination time 2.0 h, the molar ratio of Fe_2O_3 and SnO_2 1:2, initial concentration of ammonia-N 50 mg/L, photocatalytic degradation reaction time 2.0 h, the removal rate of ammonia-N in seawater is expected to reach 77.74%.

Keywords: nano Fe-doped Sno_2; photocatalytic degradation; co-precipitation method; preparation of catalysts

1 INTRODUCTION

Aquaculture has been a fast-going industry because of significant increases in demand for fish and seafood through-out the world. Along with the development of intensive aquaculture in China, concerns are evoked about the possible effects of ever-increasing aquaculture waste both on productivity inside the aquaculture system and on the ambient aquatic ecosystem (Cao et al. 2007).

Thus, research on new methods for aquaculture wastewater treatment is under way. Over the past few decades, the photocatalytic technique has been shown to be one of the most promising processes for wastewater treatment because of its advantages over traditional techniques (Yu et al. 2014). Ammonia nitrogen is the main pollutant of the wastewater.

SnO_2 is an n-type semiconductor with a large band gap, and is well known for its applications in gas sensors and dye-based solar cells (Wu et al. 2009). It has aroused great interest to synthesize SnO_2 nanostructures in recent years.

It is relatively easy to prepare SnO_2 ultrafine powder, which is a favorable factor for its application in catalysis and gas sensors. However, due to the poor thermal stability of the pure ultrafine powder, its applications confined to a great extent. Therefore, it is very meaningful to improve the thermal stability of SnO_2 ultrafine powder. Since it was reported that a small amount of additives can improve the gas-sensing nature greatly, a lot of metal or non-metal oxides such as NiO, CuO, ZnO, Bi_2O_3, MoO_3, Cr_2O_3 and Sb_2O_3 have been used in manufacturing gas sensors with different characteristics. According to the literature (Gao et al. 2000), the functions of additives can be classified into three ways: Controlling the size of crystallite SnO_2 grains, improving the photocatalytic performance and changing the electronic structure of SnO_2. SnO_2 is a potential semiconductor as photocatalyst. Nevertheless, its main absorbance wavelength is in the range of UV light, which consist only a small proportion of solar. In order to utilizing the large proportion of solar light, the dispersion behavior of Fe_2O_3 on the surface of SnO_2 and the effects on the Catalytic oxidation performance of SnO_2 have been studied in this article.

2 EXPERIMENTAL PROCEDURE

2.1 *Chemicals and apparatus*

$SnCl_4 \cdot 5H_2O$ and $FeCl_3 \cdot 6H_2O$ are Stannic source and Ferric source, respectively. A certain concentration of $NH_3 \cdot H_2O$ is used as the precipitation agent. Deionized water is employed throughout the preparation and is introduced as a washing reagent, and NaOH and HCl solutions of certain concentration are utilized to adjust pH value of seawater, which is get from Dalian Black Reef coast. A certain concentration of $(NH_4)_2 \cdot SO_4$ was used to preparation of simulate aquaculture wastewater. SEM (Tecnai G2F20, PHILIPS, Netherlands) and XRD (D/MAX2500, Rigaku, Japan) equipments are utilized to characterize the morphology and crystal structure of as-prepared SnO_2 and Fe doped SnO_2.

2.2 *Nano-Fe_2O_3-SnO_2 catalyst synthesis*

Nano-Fe-SnO_2 particles were synthesized by co-precipitation method (Zhang et al. 2007; Chen et al. 2007).According to different proportion dissolved ferric chloride and stannic chloride were dissolved in a small amount of deionized water completely. $NH_3 \cdot H_2O$ solution was added dropwise to the solution, and reddish brown precipitation generated immediately, until precipitated completely. Then, the mixture was washed several times through the deionized water and washing liquid was tested with $AgNO_3$ until no white precipitation generated. The mixture was dried at 110°C in blast oven and introduced into a muffle furnace and calcination at different temperature for different hours. Finally, grounded into powder. Thus, nano-Fe_2O_3-SnO_2 were prepared.

2.3 *Photocatalytic study*

An expected proportion of aquaculture wastewater was produced as a sort of simulated polluted seawater. Different kinds of photocatalysts were added into the mixture respectively for different experiments. Each experiment carried out in an enclosed box with UV lamps and visible light and the contains mixture were stirred by magnetic stirrers under the same conditions. The pollutant load used in the experiments was about 50 mg/L NH_4^+-N. The method of measurement was spectrophotometer for ammonia nitrogen (Jiang et al.1997).

3 RESULTS AND DISCUSSION

3.1 *X-Ray Diffraction (XRD)*

Characterization of photocatalysts. The XRD patterns of nanometer Fe_2O_3-SnO_2 prepared by co-precipitation method are depicted in Figure 1. Five patterns (a), (b), (c), (d) and (e) stand for five different Fe loadings, calcinations temperature or calcinations time prepared Fe_2O_3-SnO_2. They show the crystal structures of the doped oxides as primitive rutile structure with crystal constants a and b as 3.811 Å and c as 3.483 Å. The diffraction

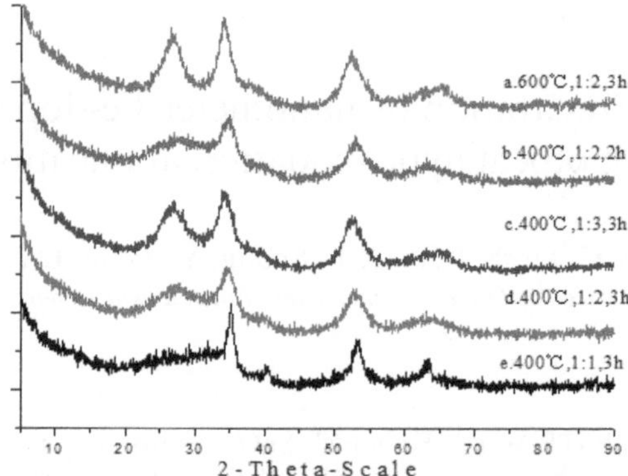

Figure 1. XRD pattern of as-prepared Fe_2O_3-SnO_2 particles.

patterns match with the standard JCPDS pattern of SnO_2(41–1445). However, due to join the iron, the dispersion of catalyst will be enhanced, the pure SnO_2, as shown in figure, the types of five patterns as a whole are not sharp, the peak value decreases, there are significant peaks appeared only at 24.93°, 34.87°, 52.60°, 64.76°. With the same temperature, the strength of the pattern (b), is the weakest, it's may be the calcination time is too short. There are the obvious characteristic peaks of SnO_2 in the pattern(c), but the type of the SnO_2 is narrow and small in the patterns (e). (a) Compared with them, the strength of peaks increase. It shows that the crystallinity of catalyst gets better, the grain size of each component is increased, and the corresponding specific surface area is decreased. The type of (d) becomes moderate, the dispersion is the best, and this is consistent with the results of SEM. The crystallite size of the Fe_2O_3-SnO_2 has been deduced from the Half-Width of the Full Maximum (HWFM) of the most intense peaks of the crystal utilizing the Scherrer formula and the crystal size in the basis of the peak (101) is 17.91 nm. Consequently, loading iron into SnO_2 by coprecipitation method will decrease crystallite size (Zhang et al. 2007; Li et al. 2005).

3.2 *Scanning Electron Microscope (SEM)*

The SEMs of as-prepared Fe_2O_3-SnO_2 are displayed in Figure 2. They reveal that the morphologies of the photocatalysts are rutile structure. Comparing the five patterns, it is evident that the catalyst type of (d), when the molar ratio of Fe_2O_3 and SnO_2 is 1:2, which has the most uniform shape and the best dispersion. Contrasting the pattern (a), (b), (c) and (e), the Shorter Calcination time, the higher calcination temperature and the more or less Fe loads, the more aggregation

Figure 2. SEM image of Fe doped SnO_2 particle.

of SnO_2 occurs. The aggregation might attribute to a higher crystallite size which match with the XRDs. Combined with the XRD patterns, we can see Nano-Fe_2O_3 powder is evenly dispersed on the surface of SnO_2 (Liu et al. 2012).

4 RESULTS OF SINGLE-FACTOR EXPERIMENTS

4.1 *Effect of dosage of photocatalyst on degradation of NH_4^+-N*

The effect of dosage of different photocatalyst on photocatalytic degradation of NH_4^+-N is shown in Figure 3(a). In each experiment, pH value is 8.0, concentration of H_2O_2 is 2.0 g/L, reaction system is carried out under visible light and the photocatalysts are fired under 400 °C, the photocatalyst dosage is used as a single factor variable, which is range from 0.2 g/L to 1.2 g/L. With the dosage increasing, the dosages of photocatalysts rise before the dosages reach around 0.4 g/L. It is acceptable that the more powders add into seawater, the less active photocatalysts exist, and this phenomenon might contribute to the fact that light diffusion and hindering take place after too much powders add, and inhibit the photocatalytic degradation.

4.2 *Effect of Fe loadings of Fe_2O_3-SnO_2 on degradation of NH_4^+-N in aquaculture wastewater*

The effects of pure SnO_2 and the molar ratio of Fe_2O_3 and SnO_2 (1:1, 1:2, 1:3, 2:1 and 3:1) (Zhang et al. 2007) on degradation of NH_4^+-N under UV.

Light for 2 h are demonstrated in Figure 3(b). The results show that the Fe loading has a significant effect on the degradation process. When the molar ratio of Fe_2O_3 and SnO_2 is 1:2, the NH_4^+-N removal rate reach the top. Compared the removal

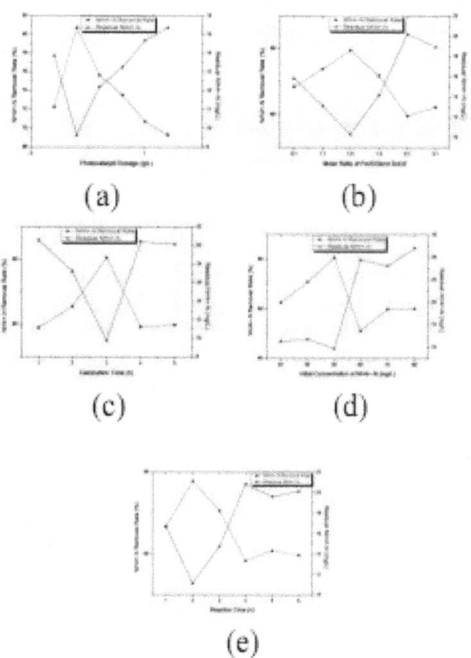

Figure 3. Effect of (a) photocatalyst dosage; (b) molar ratio of Fe_2O_3 and SnO_2; (c) calcination time; (d) initial concentration of NH_4^+-N pH value; (e) reaction time on removal rate.

rate of undoped SnO_2. It is obvious that Fe doped SnO_2 manifests a better removal rate, and too much Fe loading exhibit a barrier to photocatalytic degradation. It is said that Fe particles can hinder the recombination of photoelectron-hole pairs, thus more hydroxide radicals (·OH) can generate to oxide ammonia—N.

4.3 *Effect of calcination time on catalytic oxidation degradation*

Figure 3(c) depicts how the calcination time of photocatalyst affects the photocatalytic degradation of NH_4^+-N It is evident that the removal rate increase with the calcination time rise. Until the spots of calcination time reaches 3 h, there is the best photocatalytic degradation ability. When the times reach between 4 h and 5 h, the removal rates become steady or even slightly declining. This phenomenon may result from the fact that shorter or longer calcination time to the worse crystallization, the crystallization of samples with the shorter time may be not formed completely, and the structure of the longer time may be barricaded or destroyed, and then hinders the light absorbance of photocatalyst. Then the crystal of right time with a better crystallization can absorb light efficiently which can benefit the generation of electron-hole pairs, the pairs possess the oxidation ability.

4.4 *Effect of initial concentration of NH_4^+-N on photocatalytical degradation*

Figure 3(d) illustrates the relationship between initial concentration of NH_4^+-N and degradation under ultraviolet light. The line displays that NH_4^+-N removal rate increases when the initial concentration of NH_4^+-N is in the range from 30 to 50 mg/L, then the lines turn into stable or declining. In all, the best removal rate can achieve 80.74%. Appropriate concentration is disadvantage to NH_4^+-N removal. Inadequate NH_4^+-N might lead to insufficient driving power, excessive NH_4^+-N might surround photocatalyst which will block the light illuminating into the surface of photocatalyst.

4.5 *Effect of reaction time on degradation of NH_4^+-N*

Figure 3(e) reveals the influence of reaction time during the degradation process. From the Figure 3(e), it is obvious that NH_4^+-N removal rate increases firstly and turns into steady after reaction time is more than 2.0 h. There may be the following reasons that to begin with, UV-light provide enough energy to exciting the electrons and thus electron-hole pairs generate quickly, so the removal rate increases continuingly. However, along with the declination of residual NH_4^+-N, the photocatalytic power decreases which is a barrier for degradation reaction, thus the removal rate decrease slightly or even becomes steady (Yu et al. 2014).

4.6 *Optimisation of reaction condition*

An orthogonal experiment is implemented to optimize the experimental condition on the basis of

Table 1. The results of orthogonal test under UV-light and visible light.

实验	Photocatalyst dosage (g/L)		Calcination time (h)		Molar ratio of Fe_2O_3 and SnO_2		Initial concentration of NH_4^+-N (mg/L)		Reaction time (h)		Removal Rate (%)	
	UV-light	visible light	UV-light	visible light	UV-light	visible light	UV-light	visible light	UV-light	visible light	UV-light	visible light
1	0.2		1		1:1		40		1		68.21	63.8
2	0.2		2		1:2		50		2		70.15	77.74
3	0.2		3		1:3		60		3		81.59	71.67
4	0.2		4		2:1		70		4		59.81	58.86
5	0.4		1		1:2		60		4		67.07	68.18
6	0.4		2		1:1		70		3		69.10	65.48
7	0.4		3		2:1		40		2		62.69	62.91
8	0.4		4		1:3		50		1		67.86	70.86
9	0.6		1		1:3		70		2		75.40	73.67
10	0.6		2		2:1		60		1		46.31	62.11
11	0.6		3		1:1		50		4		61.51	67.33
12	0.6		4		1:2		40		3		70.63	71.29
13	0.8		1		2:1		50		3		32.05	46.17
14	0.8		2		1:3		40		4		60.27	64.24
15	0.8		3		1:2		70		1		70.83	72.41
16	0.8		4		1:1		60		2		68.54	65.05
K1	279.76	272.07	242.73	251.82	267.36	261.66	261.8	262.24	253.21	269.18		
K2	266.72	267.38	245.83	269.57	278.68	289.62	231.57	262.1	276.78	279.37		
K3	253.85	274.4	276.67	274.32	285.12	280.44	263.51	267.01	253.37	254.61		
K4	231.69	247.87	266.84	266.06	200.86	230.05	275.14	270.42	248.66	258.61		
R	48.07	26.53	33.89	22.5	84.26	59.57	43.57	8.32	28.12	24.76		

single-factor experiments above. The detail factors and levels selected and results are listed in table, which are reacted under ultraviolet and visible light respectively. With the certain calcinations temperature 400°C and pH value 8, concentration of H_2O_2 2.0 g/L, the removal rate of NH_4^+-N can up to 81.6% when the experiment was undertaken under the optimization of conditions: The dosage of catalyst 0.2 g/L, calcination time 3.0 h, the molar ratio of Fe_2O_3 and SnO_2 1:3, initial concentration of ammonia-N 60 mg/L, reaction time 3.0 h. It reveals that the influence of five factors on degradation of NH_4^+-N is ranked in decreasing order is: molar ratio of Fe_2O_3 and SnO_2 > photocatalyst dosage > initial concentration of NH_4^+-N > calcination time > reaction time. However, under visible light with the same conditions, through the orthogonal experiment and the result was as follows: molar ratio of Fe_2O_3 and SnO_2 > photocatalyst dosage > reaction time > calcination time > initial concentration of ammonia-N. The optimal combination of the five factors is as follows, the dosage of catalyst 0.2 g/L, calcination time 2.0 h, the molar ratio of Fe_2O_3 and SnO_2 1:2, initial concentration of ammonia-N 50 mg/L, reaction time 2.0 h, the removal rate of ammonia-N in seawater is expected to reach 77.74%. Compared with ultraviolet light, the ammonia nitrogen removal rates were Similar or increased slightly. There may be red shift appeared in Photocatalytic reaction (Wang et al. 2014).

5 CONCLUSIONS

Photocatalysts Fe-doped SnO_2 particles are successfully fabricated through co-precipitation method, and they are rutile structure. doping Fe appropriately can improve the photocatalystic ability of SnO_2 under ultraviolet light, even under visible light, which can have a great effect on photocatalytic. The degradation removal rate can up to 81.6% and 77.74%, respectively. In the abovementioned experiments, molar ratio of Fe_2O_3 and SnO_2 is also an evident factor to affect the photocatalytic ability of photocatalyst.

ACKNOWLEDGMENTS

This work has been supported by a grant from Science Foundation of Department of Ocean and Fisheries of Liaoning Province (2011024), Science and Technology Department of Liaoning Province (201301) and State Oceanic Administration People's Republic of China (201305002).

REFERENCES

[1] Cao, L. Wang, W.M. and Yang, Y. 2007. Microbial studies and technologies supporting waste disposal, management, and remediation of municipal and industrial hazardous wastes. *Environment Science Pollution Research* 14(7): 452–462.

[2] Yu, X.C. Hu, D.D. Chen, J.F. 2014. Photocatalytic degradation of diesel pollutants in seawater by using nano Li-doped Zinc Oxide under visible light. *Key Engineering Materials* 609–610: 317–323.

[3] Wu, S.S. Cao, H.Q. Yin, S.F. Liu, X.W. Zhang, X.R. 2009. Amino Acid-Assisted Hydrothermal Synthesis and Photocatalysis of SnO_2 Nanocrystals. *The Journal of Physical Chemistry C* 113: 17893–17898.

[4] Gao, Y. Zhao, H.B. and Zhao, B.Y. 2000. Monolayer dispersion of oxide additives on SnO_2 and their promoting effects on thermal stability of SnO_2 ultrafine particles. *Journal of Materials Science* 35(4): 917–923.

[5] Zhang, T. Xia, H.L. Xiao, D.C. and Zhuang, H.S. 2007. Photocatalytic degradation of acid blue over Fe_2O_3-SnO_2 nanocomposite oxide photocatalyst under irridiation of simulated sunlight, *Chemical Industry and Engineering Progress* 26(1): 47–50.

[6] Chen, A.D. Lin, X.L. Wang, C. and Zhong, A.G. 2007. The Studies of Photocatalytic Degradation of Methylene Blue over Fe_2O_3-SnO_2. *Zhejiang Chemical Industry* 38(12): 11–14.

[7] Jiang, Y. Chen, W.S.M. and Ma, Y. 1997. Selection of favorable condition for the measurement of ammoniain seawater with spectrophotometer. *Marine Environmental Science* 16(4): 44–74.

[8] Li, J.J. Zhang, X.T. Chen, Y.H. Li, Y.C. Huang, Y.B. Du, Z.L. and Li, T.J. 2005. Synthesis of highly ordered SnO_2 /Fe_2O_3 composite nanowire arrays by electrophoretic deposition method. *Chinese Science Bulletin* 50(10): 1044–1047.

[9] Liu, S.L. Li, M.M. Li, S. Li, H.L. and Yan, L. 2012. Synthesis and characterization of SnO_2 and SnO_2/ Fe_2O_3 mico-maerials. *Journal of Shaanxi University of Science & Technology* 30(6): 53–56.

[10] Wang, Z. Lu, Y. and Wang, P.J. 2014. The Electronic Structure and Optical Properties of SnO_2 Superlattices Doped with Fe. *Journal of Jinan University (Sci. &Tech.)* 28(2): 127–132.

Material Science and Environmental Engineering – Chen (Ed.)

Adsorption of harmful gases by the activated carbon catalyst impregnated with Pd

L. Ma, P.W. Ye, L. Li, Z.N. Chen & L.Y. Wang
Research Institute of Chemical Defense, China
State Key Laboratory of NBC Protection for Civilian, China

ABSTRACT: A novel activated carbon catalyst was introduced that impregnated multiple active components, such as Pd, in the coconut. It can purify multiple toxic gases at the same time. The activated carbon impregnated with Pd (namely PdC) was evaluated for their ability to remove CO, NH_3, NO_2, H_2S and HCN from streams of air. The results revealed that PdC had a good removal capacity towards the aforementioned toxic gases, and the effective sorption capacity matched at high and low concentrations, Furthermore, it can effectively catalyze CO for a long period of time.

Keywords: activated carbon impregnated with Pd; CO; air-purifying

1 INTRODUCTION

A variety of harmful hazards in the closed space require the use of filtration to ensure that breathing air meets appropriate health specifications. In general, the activated carbon can work as an effective air filtration material for organic compounds except for CO_2. However, the traditional activated carbon shows a limited capacity for some adsorbates, such as CO, H_2S, SO_2, NH_3 and NO_X. Especially for CO, it is difficult to purify at room temperature, and CO exceeds the standard in the re-entry of Shenzhou No. 4. Therefore, the purpose of this work is to investigate the ability of the activated carbon impregnated with Pd (PdC) to remove some toxic chemicals from air, including CO, H_2S, HCN, NH_3 and NO_2.

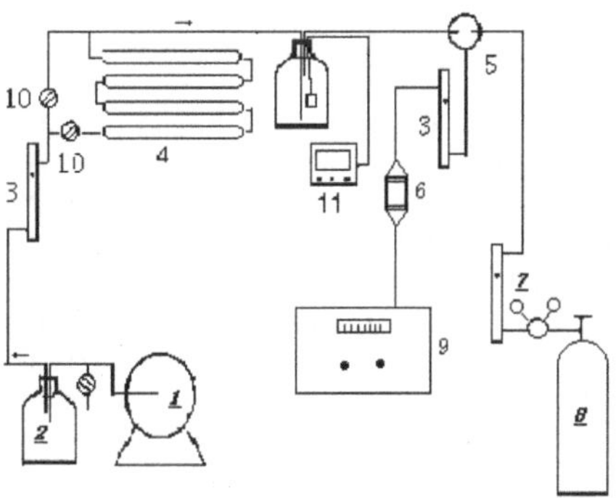

Figure 1. Breakthrough test apparatus.
1. Air compressor 2. Surge flask 3. Flowmeter 4. Humidity controller 5. Mixing ball 6. Sample 7. Microflowmeter 8. Gas source 9. Detector 10. Control valve 11. Humiture meter.

2 EXPERIMENTAL SECTIONS

2.1 *Materials*

PdC was prepared by dissolving palladium bichloride in an acidic solution containing activated carbon (namely C). Following impregnation, the granules were dried in a forced convection oven for 2 h at 210°C.

2.2 *Breakthrough test system*

Breakthrough curves for the adsorption of NH_3, H_2S, NO_2, HCN and CO in PdC were recorded, which the traditional activated carbon cannot adsorb effectively. The termination was confirmed by the respective analyzer for NH_3, H_2S, NO_2 and CO. A gas chromatograph spectrometer was used to analyze the feed and effluent streams for the concentration of HCN. Details of the breakthrough test system are reported in the following section.

3 RESULTS AND DISCUSSION

3.1 *Carbon monoxide filtration*

The breakthrough curves were recorded by storm filling 12 × 30 mesh granules of filtration media into a jacketed stainless-steel tube of inner diameter 0.8 cm. All tests were performed employing a

3.0 cm deep bed of filtration material, with testing performed at 14°C and 80% Relative Humidity (RH) and employing a feed concentration of 2500 ppm.

Figure 2 illustrates the dynamic conversions for CO. In the initial period of 5~7 minutes, the percent conversions of CO slightly decreased to 92%, and then reached 100% and maintained until thousands of minutes. The results show that PdC can effectively and purify CO for a long period of time under the high humidity environment.

3.2 *HCN filtration*

As given in Table 1, the breakthrough time of C was only 20 min for HCN at 1000 ppm concentration. But PdC showed a significantly greater capacity for the removal of HCN, increasing the breakthrough time from 20 to 330 min. Furthermore, the saturation capacity of PdC at 300 ppm concentration was lower than that at 1000 ppm concentration, because except the reactive mechanism, there was a physical adsorption that would be less at low concentration on porous PdC.

3.3 *H_2S filtration*

As given in Table 2, the breakthrough time of C sample was only 12 min for H_2S at 1000 ppm concentration. The ability of ZrC to remove H_2S was 18 times higher than that of C. Unlike HCN adsorption, the saturation capacity of PdC at 200 ppm concentration of H_2S was higher than that at 1000 ppm concentration. The result shows the different adsorption mechanism in

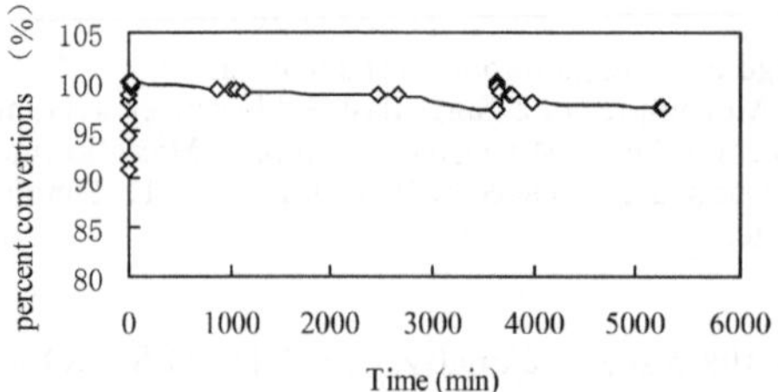

Figure 2. CO dynamic adsorption curves for the PdC studied.

Table 1. HCN breakthrough time and saturation capacity of PdC and C.

Challenge concentration (ppm)		1000	300
PdC	Breakthrough time (min)	330	1040
C		20	/
PdC	Saturation capacity (mg/g)	146.1	138.1

Table 2. H_2S breakthrough time and saturation capacity of PdC and C.

Challenge concentration (ppm)		1000	200
PdC	Breakthrough time (min)	210	1080
C		12	/
PdC	Saturation capacity (mg/g)	117.1	120.4

Table 3. NO_2 breakthrough time and saturation capacity of PdC and C.

Challenge concentration (ppm)		1000	200
PdC	Breakthrough time (min)	100	540
C		11	
PdC	Saturation capacity (mg/g)	75.4	81.4

Table 4. NH_3 breakthrough time and saturation capacity of PdC and C.

Challenge concentration (ppm)		1000	140
PdC	Breakthrough time (min)	94	720
C		10	/
PdC	Saturation capacity (mg/g)	26.2	28.1

Table 5. Formaldehyde breakthrough time of PdC and C.

Challenge concentration (mg/m^3)		130
PdC	Breakthrough time (min)	60
C		180

which only chemical reactivity occurs. Saturation capacity at low concentration was higher because the transfer thickness was thinner.

3.4 *NO_2 filtration*

As given in Table 3, the breakthrough time of the C sample was only 11 min for NO_2 at 1000 ppm concentration. The ability of ZrC to remove NO_2 was 10 times higher than that of C, reaching 100 minutes.

3.5 *NH_3 filtration*

As given in Table 4, the PdC showed a significantly greater capacity for the removal of NH_3 than C, increasing the breakthrough time from 10 to 94 min. Saturation capacity was only 2.8% because it was calculated by weight, and the molecular weight of NH_3 was small.

3.6 *Formaldehyde filtration*

As given in Table 5, the breakthrough time of C was 60 min for formaldehyde at 130 mg/m^3 concentration. But PdC showed obviously greater capacity for the removal of formaldehyde, increasing the breakthrough time to 180 min, which was three times greater than that of C.

4 SUMMARY

PdC has a good removal capacity towards the aforementioned toxic gases, and the effective sorption capacity matched at high and low concentrations. Furthermore, it can effectively catalyze CO for a long period of time.

REFERENCES

[1] Zhao Zhuo: Ship Science & Technology Vol. 49 (1995), p. 17.

[2] Tang Lanxiang, Gao Feng, Deng Yibing et al: Space Medicine & Medical Engineering Vol. 21 (2008), p. 167.

Material Science and Environmental Engineering – Chen (Ed.)
 ISBN 978-1-138-02938-5

Anaerobic ammonia oxidation (anammox) technology: A review of application issues

W.B. Hou & W.J. Zhang
College of Environmental Science and Engineering, Guilin University of Technology, Guilin, P.R. China

Y. Jin
College of Civil Engineering and Architecture, Guilin University of Technology, Guilin, P.R. China

Y. Ye
Ministry of Environmental Protection Water Project Management Office, Beijing, P.R. China

ABSTRACT: Anaerobic ammonia oxidation (anammox) is a novel autotrophic denitrification process, which has outstanding advantages in treating low C/N ratios and high concentrations of ammonia wastewater. This paper reviewed the research issues of anammox application, including the principles, microbial classification and properties, seed sludge, reactor types and processes. The optimum running parameters should be investigated in detail due to the complicated application cases. Especially in industrial wastewater treatment, anammox is deemed as the substitute process for the conventional nitrification-denitrification process, and the integrated process is expected to realize the industrialized expansion application. On the other hand, bacterial diversity and functional analysis are extremely important; however, the co-existing bacteria are still largely unknown in the anammox reactor.

Keywords: anammox; application; wastewater treatment; nitrogen

1 INTRODUCTION

Recently, new biological denitrification technique, namely anaerobic ammonia oxidation (anammox), has attracted the attention of researchers. Compared with the traditional biological denitrification technology, the anammox process has the advantages of being economic, highly efficient, and having no secondary pollution.

2 ANAMMOX PROCESS AND ANAMMOX BACTERIA

Anammox is a biological process, which uses NO_2-N as an electron donor and NH_4^+-N as an electron acceptor to generate nitrogen under anaerobic conditions (Gaaaf A.A. et al. 1996). According to the chemical measurement and material balance, Strous et al. summarized the anammox reaction equation as follows (Strous M. et al. 1998):

$$1NH_4^+ + 1.32NO_2^- + 0.066HCO_3^- + 0.13H^+ \rightarrow 1.02N_2 + 0.26NO_3^- + 0.066CH_2O_{0.5}N_{0.15} + 2.03H_2O \quad (1)$$

In the equation, HCO_3^- or CO_2 (CO_2 dissolves in water to generate HCO_3^-) can be used as the carbon source for anammox; therefore, anammox bacteria is classified as autotrophic microorganisms.

With the development of genetics and enzymology, researchers have gained a deeper understanding of the main function enzymes of genes and cellular aspects of anammox bacteria. Hooper et al. put forward the biochemical reaction model as follows (Jetten M.S.M. et al. 2001, Strous M., et al. 2006, Shimamura M. et al. 2008, Van Niftrik L. et al. 2008).

The description of the notations is as follows: NiR—Nitrite reductase; HH—Hydrazine

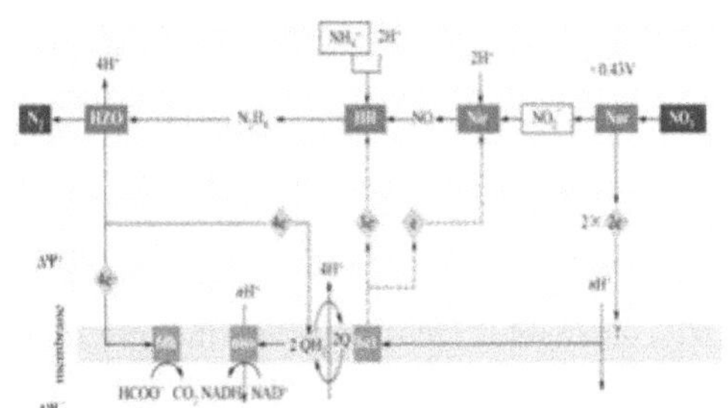

Figure 1. Anammox model based on *Kuenen stuttgartiensis*.

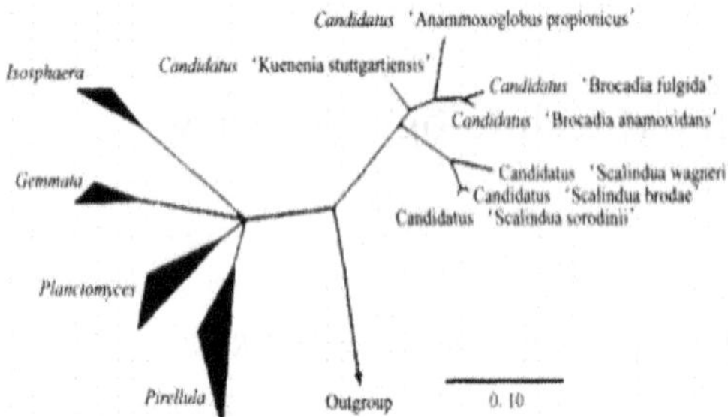

Figure 2. Anaerobic ammonia oxidation bacteria's system tree (Jetten M.S.M., et al. 2005).

hydrolase; HZO—Hydrazine redox enzyme; Nar—Nitrate reductase; Q—Ubiquinone; fdh—Formate dehydrogenase; nuo—Ubiquinone oxidoreductase; QH2—Reducing ubiquinone; bc1—Cytochrome bc1 complex.

This model can explain not only the need of ATP and $NADPH_2$ sources in the growth of anammox bacteria, but also the nitrate generation phenomenon in the process of anaerobic ammonia oxidation.

Mulder et al. discovered the anammox phenomenon in a fluidized bed; thereafter, researchers have found that anammox bacteria widely exist in the sewage treatment system, as well as in oceans and lakes (Mulder A et al. 1995).

The morphology of anammox bacteria is spherical and oval generally, which belongs to Gram-negative bacteria.

According to the GenBank database, anammox microorganisms are composed of 2 families, including *Planctomyces, Pirellula, Gemmata, Isosphaera, Candidatus Brocadia, Candidatus Kuenenia, Candidatus Scalindua, Candidatus Jettenia, and Candidatus* (Fig. 2).

3 APPLICATION

3.1 *Seed sludge*

Sludge cultivation is the biggest challenge in the application of the anammox process. It has been proved that nitrification sludge, denitrification sludge, aerobic sludge, anaerobic sludge, river sediment, sludge, mixed sludge and anammox bacteria-rich sludge can be used as the seed sludge in the anammox reactor (Tang et al. 2008, Ruan et al. 2010, Yang et al. 2004, Xie et al. 2009, Liu et al. 2006). Zhang et al. (2004) cultivated the autotrophic denitrification biofilm of anammox bacteria. After 110 days of cultivation, the anammox reactor was successfully achieved with a nitrogen loading rate of 0.145 $kg/(m^3 \cdot d)$. Zhao et al. (2007) used anaerobic granular sludge as the seed sludge to start a SBR reactor. After 58 days, the granular sludge changed from black to brown, and then the sludge changed from brown to reddish brown on day 110, with the total nitrogen loading rate reaching to 34.3 $g/(L^{-1} \cdot d^{-1})$. It has been proved that anammox bacteria widely exist in nature and in the sewage treatment system. Considering the start-up time, the inoculation sludge should be chosen, where the growth conditions should be similar to those of anammox bacteria.

3.2 *Reactor*

The double time of anammox is extremely long. Therefore, sludge retention time is very crucial in running the anammox reactor. In order to apply the anammox technology, it is important to develop and choose a suitable reactor. Numerous studies have reported different types of anammox reactors (Ahn Y.H. et al. 2004, Ahn Y.H. & Kim H.C. 2004, Zhang & Xiao 2005, Zuo et al. 2003, Tang et al. 2009), including fixed bed reactor, anaerobic sequencing batch reactors, fluidized bed reactors, up-flow anaerobic sludge blanket reactor and gas-lift reactor.

The gas-lift reactor can improve the mass transfer process by introducing gas to mix sludge and substrate. Thereafter, the nitrogen loading rate of the gas-lift reactor is very high, which can reach up to 8.9 $kg/(m^{-3} \cdot d^{-1})$ (Ahn Y.H. et al. 2004). Compared with the other anammox reactor, the gas-lift reactor can reach the highest nitrogen loading rate by means of internal adjustment of gas stripping. However, the structure of the gas-lift reactor is complex, and the operation and control requirements are very high.

Anammox phenomenon was first discovered in a fluidized bed reactor. Since then, many researchers have cultivated anammox bacteria in the fluidized bed reactor, and conducted an in-depth study. Studies have shown that the fluidized bed reactor could obtain a higher total nitrogen volume loading, which can be up to 1.5–4.8 $kg/(m^{-3} \cdot d^{-1})$. This is lower than the total nitrogen volume load, which is obtained from the gas-lift reactor. The reason for this is the presence of sludge loss, and it is difficult to maintain a long-term stable operation in the sludge bed reactor.

Anaerobic sequencing batch reactor, also called the ASBR reactor, has the characteristics of simple structure and good solid-liquid separation. Therefore, the operation and control of the ASBR anammox reactor is simple. The anammox sludge could retain in the reactor with a good type of anammox granules. Nitrogen volume load is at the middle level of the anammox reactor. It is relatively simple to apply in practical application.

Table 1. Anammox reactor.

Reactor	Advantages	Disadvantages
Anaerobic sequencing batch reactors	Operation and control is simple, good type of anammox granules, relatively simple to apply in practical application	Nitrogen volume load is at the middle level of the anammox reactor
Fluidized bed reactors	Nitrogen volume load is relatively high, lower than the gas-lift reactor	Difficult to maintain a long-term stable operation
Up-flow anaerobic sludge blanket reactor	Good mass transfer effect and simple operation, widely used at present	Nitrogen volume load of UASB reactor is quite different, need changing the structure to improve the efficiency
Gas-lift reactor	High nitrogen volume load	High operation and control requirements
Fixed bed reactor	Carrier blockage or sludge loss accidents are difficult to occur in the up-flow biofilm reactor and push streaming biofilm reactor	Hydraulic retention time is long, volume total nitrogen load is relatively low in the up-flow biofilm reactor and the push streaming biofilm reactor

Up-flow Anaerobic Sludge Bed Reactor (UASB reactor) is widely used at present. UASB has the advantages of high biomass, good mass transfer effect and simple operation, but the nitrogen volume load of the UASB reactor is quite different, ranging from 0.38 to 6.39 kg/($m^{-3} \cdot d^{-1}$). Changing the structure (e.g. adding filler, setting back) can improve the efficiency of the anammox reactor (Ahn Y.H. & Kim H.C. 2004, Zhang & Xiao 2005, Zuo et al. 2003).

4 PROCESS

The anammox technology mainly includes SHARON-anammox process and OLAND process. The anammox bacteria need a substrate in the form of ammonia nitrogen and nitrite nitrogen. To provide a good living environment for nitrite-oxidizing bacteria and anammox bacteria, the SHARON-anammox process generally has two reactors to achieve the shortcut nitrification and anammox process. By controlling the temperature and hydraulic retention time, the nitrite-oxidizing bacteria cannot prevail in the SHARON reactor (about 50% influent ammonia is converted into nitrite nitrogen), and the effluent from the SHARON reactor can be the influent of the anammox reactor. Then, ammonia and nitrite nitrogen are converted into N_2 by anammox bacteria. Compared with the traditional nitrification-denitrification process, the SHARON-anammox process has the advantages of high loading rate, less oxygen consumption, no external carbon source and less sludge generation. The SHARON-anammox process has already been applied in full-scale treatment plants.

OLAND process is also known as the limited oxygen nitrification-denitrification process. The key point of the OLAND process is to control dissolved oxygen in wastewater. When the dissolved oxygen is low, anammox bacteria are more likely to form affinity with dissolved oxygen than nitrite-oxidizing bacteria, thus the accumulation of nitrite can be achieved. The application of the OLAND process is almost the same as that of the SHARON-anammox process at present. Despite complicated controlling requirements, the OLAND process is preferred due to the simple infrastructure; consequently, it is easy to upgrade the existing wastewater treatment plant.

5 FULL-SCALE APPLICATION

Although numerous anammox processes are reported in full-scale treatment plants, in China, the application of the anammox process is still in its embryonic stage. The full-scale anammox treatment plants are listed in Table 2.

In China, the research on anammox application have mostly been at the pilot-scale or in the laboratory stage. Paques introduced the anammox process in the food factory in Shandong Province in China. However, the running conditions are not reported in detail. Tang et al. inoculated a mixed sludge in the up-flow anaerobic ammonia oxidation reactor to treat simulation wastewater. After 255 days of operation, the start-up of the pilot anaerobic ammonia oxidation reactor was accomplished successfully, and the nitrogen removal rate reached up to 1.30 kg/($m^3 \cdot d$) (Tang et al. 2009). The double time of anaerobic ammonia oxidation bacteria is extremely long. Therefore, the start-up period generally takes 3–5 months, even up to 200 days (W.R.L. et al. 2007, Abma W. et al. 2007). In 2002, the world's first plant of the anammox process, the Netherlands Dokhaven Sewage Treatment Plant, was designed to deal with ammonia in digest liquor. The volume of the reactor was 70 m^3, and the original design load was 490 kg/($m^3 \cdot d$).

Table 2. Anammox reactor start-up time, design and processing capacity.

No.	Plants	Volume (m^3)	Wastewater	The design load ($kg \cdot m^{-3} \cdot d^{-1}$)	Actual processing capacity/ ($kg \cdot m^{-3} \cdot d^{-1}$)	Running time/ months	References
1	Rotterdam, the Netherlands Dokhaven municipal sewage treatment plant	70	Sludge digestion liquid	490	750	42	[21]
2	Netherlands Lichtenvoorde industrial wastewater project	70	Tannery wastewater	325	150	12	[21]
3	The Netherlands Oburgen industrial wastewater project	600	Potato processing wastewater	1200	700	6	[22]
4	Japan, Mie-ken semiconductor plant	58	Semiconductor wastewater	220	220	2	[22]

After 3.5 years of operation, the actual load reached up to 750 kg/($m^3 \cdot d$). Following the experience of the world's first anammox plant, the start-up time of the subsequent anammox plant is getting shorter (Table 1). For example, Japan's Mie-ken semiconductor plant took only 2 months to start one 58 m^3 anammox reactor, and the actual load reached up to 220 kg/($m^{-3} \cdot d^{-1}$). However, given its long start-up time and complex control parameters, the large-scale applications of the anammox process are not boomed as expected.

6 CONCLUSIONS AND PROPOSALS

As reviewed in the above sections, the low growth rate of anammox bacteria is the biggest ban, which restricts the application of the anammox process in full scale. On the other hand, the anammox process must be combined with other processes to remove ammonia nitrogen. Therefore, special requirement must be met to create appropriate conditions for the anammox process. Finally, the running parameters of the anammox process must be strictly controlled to reduce the sludge loss. The mechanism of co-existing anammox bacteria is not well understood, and further research is expected to carry out to illuminate the mechanism of their metabolism.

ACKNOWLEDGEMENTS

This research was supported by the National Natural Science Foundation of China (No. 51108108), Guangxi Natural Science Foundation (2013GXNSFCA019018;2014GXNSFBA118265), Research Projects of the Education Department of Guangxi Government (2013ZD031; 2013ZL076; ZL2014051; KY2015ZL118), Guangxi Key Laboratory of New Energy and Building Energy Saving (12-J-21-2).

REFERENCES

[1] Abma W., Schultz C., Mulder J.W., et al. The advance of Anammox. Water 21, 2007(2): 36–37.

[2] Ahn Y.H., Hwang I.S., Min K.S. Anammox and partial denitritation in anaerobic nitrogen removal from piggery waste. Waster Sci & Technol, 2004, 49(5~6): 145~153.

[3] Ahn Y.H., Kim H.C., Nutrient removal and microbial granulation in an anaerobic process treating inorganic and organic nitrogenous wastewater. Waster Sci & Technol, 2004, 50(6): 207~215.

[4] Jetten M.S.M., Wagner M., Fuerst J., et al. Microbiology and application of the anaerobic ammonium oxidation ('anammox') process. Current Opinion in Biotechnology, 2001, 12(3): 283–288.

[5] Jetten M.S.M., Cirpus I., Kart al B., van Niftrik L., et al, 1994–2004: 10 years of research on the anaerobic oxidation of ammonium. Biochem. Soc. Tran, 2005, 33: 119–123.

[6] Liu Yin, Du Bing, Si Ya-an, et al. Cultivation of ANAMMOX Bacteria and the Ammonium Anaerobic Oxidation Technology in the Plug Flow Bio-Reactor. Environmental Science. 2005, 26(2), 137–141.

[7] Mulder A., Vande Graff A.A., Robertson L.A., et al. 1995. Anaerobic ammonium oxidation discovered in a denitrifying fluidized bed reactor. FEMS Microbiology Ecology, 16(3): 177–183.

[8] Ruan Xiaohong, Zhang Ying, Zhang Yaping, et al. Preliminary cultivation and characteristics of the ammonium oxidation of anammox bacteria in freshwater sediments. Acta Scientiae Circumstantiae, 2010, 30(10): 1999–2003.

[9] Strous M., Heijnen J.J., Kuenen J.G., et al. The sequencing batch reactor as a powerful tool for the study of slowly growing anaerobic ammonium oxidizing microorganism. Appl Microbiol Biotechnol, 1998, 50(5): 589–596.

[10] Strous M., Pelletier E., Mangenot S., et al. Deciphering the evolution and metabolism of anammox bacterium from a community genome. Nature, 2006, 440(7085): 790–794.

[11] Shimamura M., Nishiyama T., Shinya K., et al. Another multiheme protein, hydroxylamine oxidoreductase, abundantly produced in an anammox bacterium besides the hydrazine-oxidizing enzyme. Journal of Bioscience and Bioengineering, 2008, 105(4): 432–432.

[12] Tang Chong-jian, Zheng Ping, Chen Jian-wei, et al. Performance of Anammox bioreactors started up with different seeding sludges. China Environmental Science, 2008, 28(28): 683–688.

[13] Tang Chongjian, Zheng Ping, Chen Jianwei, et al. Start-up and process control of a pilot-scale Anammox bioreactor at ambient temperature. Chinese Journal of Biotechnology. 2009, 25(3): 406–412.

[14] Vande Gaaaf A.A., De Bruijin P, Robertson L.A., et al. Autotrophic growth of aerobic ammonium.

[15] Oxidizing microorganisms in a fluidized bad reactor. Microbiology, 1996, 142(8): 2187–2196.

[16] Vander Star W.R.L., AbmaW.R., Blommers D., et al. Startup of reactors for anoxic ammonium oxidation: experiences from the first fullscale anammox reactor in Rotterdam [DB/OL]. http://www:elsevler.com/locate/waters, 2007-04.

[17] Van Niftrik L., Geerts W.J.C., Van Donselaar E.G., et al. Linking ultrastructure and function in four genera of anaerobic ammonium-oxidizing bacteria: Cellplan, glycogen storage and localization of cytochromec proteins. Journal of Bacteriology, 2008, 190(2): 708–717.

[18] Xie Qinglin, Li Xiaoxia, Li Yanhong, et al. Start up of an anaerobic ammonium oxidation process in an anaerobic sequencing batch reactor. Acta Scientiae Circumstantiae. 2009, 29(4), 759–763.

[19] Yang Yang, Zuo Jian-e, Shen Ping, et al. Start-up of Anaerobic Ammonia Oxidation Reactors Inoculated with Different Sludge. Environmental Science. 2004, 25(vol), 39–42.

[20] Zhang Long, Xiao Wen-de. Cultivation of Anaerobic Ammonium Oxidation Bacteria and Their Performance in an Up-Flow Anaerobic Sludge Bed Reactor. Journal of East China University of Science and Technology. 2005, 31(1): 99~102.

[21] Zhang Shao-hui, Zheng Ping. Studies on start-up technique of anammox bioreactor. China Environmental Science. 2004, 24(4): 496–500.

[22] Zhao Zhi-hong, Liao De-xiang, Li Xiao-ming, et al. Cultivation and Nitrogen Removal Characteristics of ANAMMOX Granules, Environmental Science, 2007, 28(4): 800–804.

[23] Zuo Jiane, Yang Yang, Meng Aihong, et al. Study on Start-up of Anaerobic Ammonia Oxidation Process in UASB Reactor. Shanghai Environmental Sciences. 2003, 22(10): 66~669.

Material Science and Environmental Engineering – Chen (Ed.)
© 2016 Taylor & Francis Group, London, ISBN 978-1-138-02938-5

Cyclic measurement of the carbon blank value with the carbon/sulfur determinator

J.H. Wei
AVIC Beijing Institute of Aeronautical Materials, Beijing Key Laboratory of Aeronautical Materials Testing and Evaluation, Beijing, China

ABSTRACT: After reducing and stabilizing the blank value, and having two to three times of cyclic measurements by using the cyclic measurement method, the carbon/sulfur determinator is able to measure the carbon blank value with good precision and accuracy.

Keywords: carbon/sulfur determinator; carbon, blank value; analysis

1 INTRODUCTION

Carbon/sulfur determinator is a high-tech product integrated with mechanical, optical, electrical and computer, heating, and analytical technologies, and it can quickly and accurately detect the mass fraction of the carbon element in steel, non-ferrous metals, cement, ore, glass, pottery and porcelain, as well as a in lot of other materials [1]. The determinator consists of three parts, namely combustion system, mechanical and electrical control system, and infrared detection system, which takes computation as its core. Since the blank value is affected by many factors, the determinator is not able to eliminate the impact of blank values. If measuring low carbon contents without deducting the blank value, it may result in large error [2]. Therefore, measurement of the carbon blank value is of great significance.

Since many factors may affect the measurement of carbon, a simple and easy-to-operate carbon blank value measurement method is not currently available in practice.

Based on the experiment, this paper proposes a blank value measurement method that could be applied to the regular measurement operation. With 2–3 cyclic measurements, this method is able to measure the blank value, which is called the Carbon Blank Value Cyclic Measurement Method (CBVCMM).

2 SOURCES OF BLANK VALUES

Blank value is a measurement result when the sample is absent, under the same conditions of the experimental environment and experimental reagent when the sample is present.

In the real analysis process, fluxing agent, gas circuit, crucible, and combustion improver are the major blank value sources [4]. The magnitude and stability of blank values are affected by the drying agent and the dust in the combustion tube, which vary along with the channels and measurement data.

3 REDUCING AND STABILIZING THE BLANK VALUE

During the actual analysis process, it is not possible to control the sources of the blank value with one-to-one correspondence. However, it is possible to uniformly reduce and stabilize the blank value.

We choose fluxing agent, crucible, and pure oxygen with small blank values. An accurate amount of fluxing agents should be added each time.

Desiccant should be kept dry, so as to effectively purify carrier gas and eliminate the impact of water on measurements. Since CO_2 is soluble in water to generate H_2CO_3, it may disturb the measurement of carbon contents.

Burner must be kept clean. It is found that in practice, long operation of instrument, and analysis of humid or high carbon content samples may result in a larger blank value of the system. The reason for this is that, during the process of melting, powdery basic oxide and vapor are generated, and they a have strong adsorption capacity to gas, which leads to low analytical results; when adsorption is saturated, they are gradually released, which may lead to higher analytical results. In order to maintain a stable blank value, it is necessary to regularly clean the combustion furnace, pipes and filters.

When blank values are reduced and stabilized, the blank value cyclic measurement could be conducted.

4 CARBON BLANK VALUE CYCLIC MEASUREMENT METHOD (CBVCMM)

4.1 *First time measurement*

Instrument conditions are as follows: a channel is chosen, and the instrument blank value is set to 0.00000; other conditions are in accordance with the operation requirements, keeping consistent.

A standard material calibration instrument with low C contents is chosen. By taking iron-based standard substance AR 870 (C: 0.032 ± 0.002%) as an example, this paper illustrates the specific operation processes of the cyclic measurement method.

First time instrument calibration. Approximately 0.500 g of the standard substance of AR 870 (C: 0.032 ± 0.002%) are taken and added to the determinator. Fluxing agent with consistent mass is added, for instance, adding 2.0 g of tungsten granular and 0.5 g of scrap iron. In practice, a spoon of tungsten granular and a relatively equal amount of scrap iron can be directly added.

Through analysis, the first time measurement value of standard substance is obtained. It is the value calculated by the instrument based on the coefficient retained from the previous analysis (which is certainly different from the C standard value of the standard substance). According to the C standard value of the standard substance, the standard substance measurement value is calibrated. After calibration, the first time display value of the standard substance is obtained (which should be within the permissible deviation range of the standard value). First time displayed value of the standard substance is the integrated display of C and Blank in the standard substance. The values are given in Table 1.

Measurement control sample. Taking approximately 0.500 g of the standard substance of LECO 502-402 (C: 0.00352 ± 0.00010), the analysis conditions are the same as those described above, as given in Table 2.

It can be seen from Table 2 that the average of the first time measurement value of the control sample is 0.00400%, and the difference between this measurement value and the C standard value of 0.00352 ± 0.00010% exceeds the permissible range. It indicates that the measurement of the instrument, after first time calibration, is not accurate.

First time measurement of the blank. The same amount of fluxing agents is added. Nominal mass entered into the instrument is 0.5000 g. The first time blank measurement value is given in Table 3.

The measurement of the instrument, after first time calibration, is still not accurate. Consequently, the first time measurement values of the blank listed in Table 3 are also not accurate.

Table 1. First time calibration data of the instrument (%).

Standard substance	C standard value	Standard substance First time measurement value	Standard substance First time calibration display value
AR 870	C: 0.032 ± 0.002%	0.0275, 0.0278, 0.0286, 0.0285 0.0276, 0.0279, 0.0288, 0.0281	0.0314, 0.0317, 0.0325, 0.0325 0.0315, 0.0318, 0.0327, 0.0319
Average value		0.0281	0.0320
Deviation		0.000484	0.000499
Standard deviation		1.72	1.56

Table 2. First time measurement value of the control sample (%).

	C standard value	First time measurement values of the control sample
LECO 502-402	0.00352 ± 0.00010	0.00400, 0.00395, 0.00389 0.00386, 0.00402, 0.00409 0.00413, 0.00406
Average value		0.00400
Deviation		0.0000950
Standard deviation		2.39

Table 3. First time measurement value of the blank value (%).

	First time measurement value of the blank value
	0.00091, 0.00100, 0.00087, 0.00071 0.00117, 0.00095, 0.00081, 0.00086
Average value	0.00091
Deviation	0.000137
Standard deviation	15.1

Table 4. Second time instrument calibration data.

Standard substance	C standard value	Second time measurement value	Second time calibration display value
AR 870	0.032 ± 0.002%	0.0330, 0.0334, 0.0339, 0.0328 0.0329, 0.0329, 0.0339, 0.0344	0.0316, 0.0320, 0.0325, 0.0314 0.0315, 0.0315, 0.0325, 0.0330
Average value		0.0334	0.0320
Deviation		0.000600	0.000600
Standard deviation		1.80	1.87

Table 5. Second time measurement value of the control sample (%).

	C standard value	Second time measurement value
LECO 502-402	0.00352 ± 0.00010	0.00350, 0.00369, 0.00370 0.00348, 0.00355, 0.00366 0.00361, 0.00354
Average value		0.00359
Deviation		0.0000859
Standard deviation		2.39

Table 6. Impact analysis values of the blank (%).

	Impact analysis value
	0.00002, −0.00007, −0.00004, −0.00008 −0.00007, −0.00005, −0.00003, 0.00000
Average value	−0.00004
Deviation	0.0000354
Standard deviation	–

Table 7. Second measurement value of the blank (%).

	Second measurement value
	0.00078, 0.00099, 0.00081, 0.00086 0.00085, 0.00079, 0.00090, 0.00087
Average value	0.00086
Deviation	0.0000680
Standard deviation	7.91

4.2 *Second time measurement*

Average value of the first time measurement value of the blank of C: 0.00091% is entered into the instrument by using the manual blank operation. The instrument will automatically deduct 0.00091% later in the analysis.

The same channel is used, and other operation conditions are the same as those of the first measurement.

Second time instrument calibration. Data are summarized in Table 4.

Measurement control sample. Data are summarized in Table 5.

It can be seen from Table 5 that the second measurement values of the control sample are within the permissible range of standard deviation value.

Analysis of the impact of carbon blank. The same amount of fluxing agents is added. Nominal mass entered into the instrument is 0.5000 g. The analytical result is summarized in Table 6.

The impact analysis value of the blank is −0.00004%, which indicates that the impact of carbon blank on the measurement value of the samples is very small.

Second measurement of the blank.

Under the condition of without deducting the blank, the same weight of fluxing agents is added into the instrument. Nominal mass entered into the instrument is 0.5000 g. The result of the second measurement value is summarized in Table 7.

After the second calibration of the instrument, the second measurement values of the control sample are within the permissible range of standard value error. Consequently, the second measurement values of the blank listed in Table 7 are accurate. It can be seen from Table 7 that the precision of the measurement values is high.

If the cyclic measurement is conducted again, the measurement value of the blank will be closer to the real value. After entering this blank value into the instrument, it will automatically deduct the blank value, so that the impact of the blank value on the measurement of the samples can be totally eliminated.

5 CONCLUSION

Carbon/sulfur determinator, after reducing and stabilizing the blank value, and having two to three times of cyclic measurements by using the cyclic measurement method, is able to measure the carbon blank value with good precision and accuracy.

Entering this blank value into the instrument, it will automatically deduct the blank value, so that the impact of the blank value on carbon measurement can be eliminated.

REFERENCES

[1] Tian Yingyan, Ye Fanxiu, Shen Yongxiang, "Monograph on Carbon and Sulfur Analysis", Metallurgical Industry Press, 2010, pp7.

[2] Wan Xiuzhi, Yu Rurong, "Measuring Ultra-low Carbon by Using Infra-red Analyzer" (J), Metallurgical Analysis, 1995, 3(15):54.

[3] Cheng Ying, "Influence factors in High Frequency Infrared Analysis of Low Carbon and Low Sulfur", Physical Testing and Chemical Analysis Part B Chemical Analysis, May 2004, Volume 40(5). PTCA (Part B: Chemical Analysis) Vol. 40 No. 5 May 2004, P279.

[4] Li Quan, Influence Factors in High Frequency Infrared Analysis of Low Carbon and Low Sulfur, Inner Mongolia Science Technology & Economy, No. 19, the 173th issue.

Material Science and Environmental Engineering – Chen (Ed.)
 ISBN 978-1-138-02938-5

Effects of O_3 and drought on the leaf gas exchange parameters of three seedling types in China

X.L. Hou & L. Xue
College of Forestry, South China Agricultural University, Guangzhou, China

ABSTRACT: Photosynthesis physiology of *Michelia macclurei*, *Cinnamomum camphora* and *Rhodoleia championii* seedlings was studied under ozone stress and drought stress at Open-Top Chambers (OTC). Leaf gas exchange parameters of the three seedling types were determined 0 d (control), 15 d, and 30 d after treatment. Compared with the control under ozone, drought, and ozone-drought intercross stresses, P_n of the three seedling types declined at 30 d after the treatment. In addition to *C. camphora* treated with ozone-drought intercross stress, G_s and T_r of other treatments declined; C_i and L_s increased or declined. The decrease in P_n was mainly resulted from stomatal limitation for *C. camphora*, and from non-stomatal limitation for *M. macclurei* and *R. championii* under ozone stress. The decrease in P_n was mainly resulted from stomatal limitation under drought stress for *M. macclurei*, and from non-stomatal limitation for *C. camphora* and *R. championii.*

Keywords: O_3; drought; seedling; gas exchange parameters; principal component analysis

1 INTRODUCTION

Ozone has a strong oxidizing effect, and high concentration ozone affects the growth of plants (Wang et al., 1995). If average ozone reaches 40 ppb, some visible injurious symptoms and physiological changes will occur in sensitive flora species (Wittig et al., 2007), which induces stomatal closure of plant leaves, limits carbon dioxide (CO_2) entering the plant leaves, inhibits carbon assimilation, decreases the net photosynthesis of plants, and inhibits their growth (Karonen et al., 2006). Long-time exposure to ozone pollution can also increase stomatal conductance or lose control of the stoma of plants (Reiner et al., 1996).

Ozone pollution is becoming more and more serious in the Pearl River Delta. In this study, *Michelia macclurei, Cinnamomum camphora* and *Rhodoleia championii* seedlings were placed in a simulated natural ozone and drought environment, and then their gas exchange parameters were measured to provide the basis for screening tree species with strong ozone and drought resistance.

2 MATERIALS AND METHODS

2.1 *Experimental material*

One-year-old seedlings of *M. macclurei*, *C. camphora* and *R. championii* were individually planted in nutrition bags with the diameter of 12 cm and height of 10 cm. Fumigation test equipment with Open-Top Chamber (OTC), hexagonal cross-section, glass wall and a diameter of 3 m and a height of 2.4 m was located in Yuejinbei Nursery in South China Agricultural University, Guangzhou City, Guangdong Province, China. Ozone was produced by CFY5 oxygen manufacture equipment and COM-AD-01 ozone generator (range 0~1 ppm, resolution 0.001 ppm). And ozone concentration was detected by a GT901-03 Portable ozone detector continuously. The seedling growth status is given in Table 1.

2.2 *Test method*

The OTC design consisted of two ozone concentrations: E40 (ozone concentration is the same as the natural ozone concentrations of atmosphere in the OTC through only ventilation, about 40 ppb) and E150 (elevated ozone concentration is about 150 ppb). Each chamber was designed with

Table 1. Growth status of the three seedling types.

Tree species	DBH (mm)	Height (cm)	Crown width (cm)
M. macclurei	4.50 ± 0.18	32.97 ± 0.89	14.77 ± 0.64
C. camphora	4.16 ± 0.18	38.96 ± 1.12	13.55 ± 0.97
R. championii	3.15 ± 0.18	32.97 ± 0.95	12.67 ± 0.48

Notice: The data is "average + standard error" in the table.

two moisture gradients: control (the relative water content of soil was 75–80%) and drought treatment (the relative water content of soil was 40–45%). So, ozone and drought stress were combined into three treatments: E40+D (drought stress), E150 (ozone stress) and E150+D (ozone and drought stress). Ventilation and air system could meet the gas in the chamber exchange more than one time per min, kept almost consistent between temperature, humidity and CO_2 concentration in the chamber and the outside environment. Seedlings were watered regularly for a week in the chambers to accommodate the chambers, and ventilation had begun after the soil moisture became stable. The fumigation lasted for 30 days. Aeration time was 9:00–17:00 every day. At the beginning of the experiment, the 15th day and the end of the experiment, photosynthesis physiology was measured separately.

2.3 *Determination method for index*

The determination methods of gas exchange parameters were as follows: five seedlings of coordinated growth status were selected, with each seedling between the third and eighth functional leaves being marked, to determine the net photosynthetic rate (P_n), stomatal conductance (G_s), intercellular CO_2 concentration (C_i) and transpiration rate (T_r) of marked seedling leaves in the open air way. Leaf chamber temperature was 20~25°C, light intensity 1000 μmol·m^{-2}·s^{-1}, CO_2 concentration 400 μmol·mol^{-1} and low rate 500 μmol·s^{-1} under the determined condition, and each leaf was measured three times. The formula of stomatal limitation (L_S) is as follows:

$$L_S = (1 - C_i/C_a),$$

where C_a is the atmospheric CO_2 concentration.

2.4 *Data analysis*

All statistical analyses were performed using Excel 2010 and the Statistical Analysis System (SAS 9.0). After each index was standardized using the range method, principal component analysis was used to analyze the data, and the score of resistance ozone and drought of each tree species in the treatments E150, E40+D and E150+D were calculated.

3 RESULTS

3.1 *The effects of ozone and drought stress on the net photosynthetic rate (P_n) of seedlings*

The P_n of *M. macclurei* continuously decreased in the treatments E150, E40+D and E150+D with increasing ozone and drought stress time (Fig. 1).

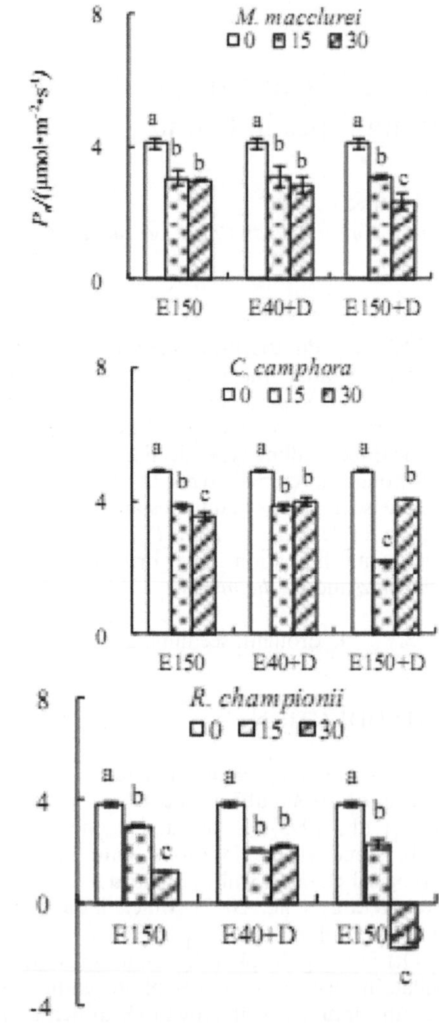

Figure 1. Change in the net photosynthetic rate under O_3 stress and drought stress.
Significant differences between treatments were indicated by different letters above bars, $P < 0.05$.

The P_n of *C. camphora* had a continuous decrease in the treatment E150, a significant decrease ($P < 0.05$) and then undulation in the treatment E40+D, while it had a significant decrease followed by a significant increase in the treatment E150+D ($P < 0.05$). The P_n of *R. championii* had a significant decrease in the treatments E150, E40+D and E150+D ($P < 0.05$).

3.2 *The effects of ozone and drought stress on stomatal conductance (G_s) of seedlings*

With increasing ozone and drought stress time, the G_s of *M. macclurei* decreased in the treatment

E150, increased and then decreased in the treatment E40+D, and significantly dropped and then increased in the treatment E150+D ($P < 0.05$) (Fig. 2). The G_s of *C. camphora* decreased and then increased in the treatments E150 and E150+D ($P < 0.05$), whereas it increased and then decreased in the treatment E40+D ($P < 0.05$). The G_s of *R. championii* had a continuous decrease in the treatments E150 and E150+D, while it had a significant decrease and then significantly increased in the treatment E40+D ($P < 0.05$).

3.3 *The effects of ozone and drought stress on intercellular CO_2 concentration (C_i) of seedlings*

With increasing ozone and drought stress time, the C_i of *M. macclurei* had a significant increase and then significantly dropped in the treatments E150 and E40+D ($P < 0.05$), while it significantly dropped and then significantly rose in the treatment E150+D ($P < 0.05$) (Fig. 3). The C_i of *C. camphora* had a significant decrease ($P < 0.05$) and then had a slight increase in the treatment E150, while it

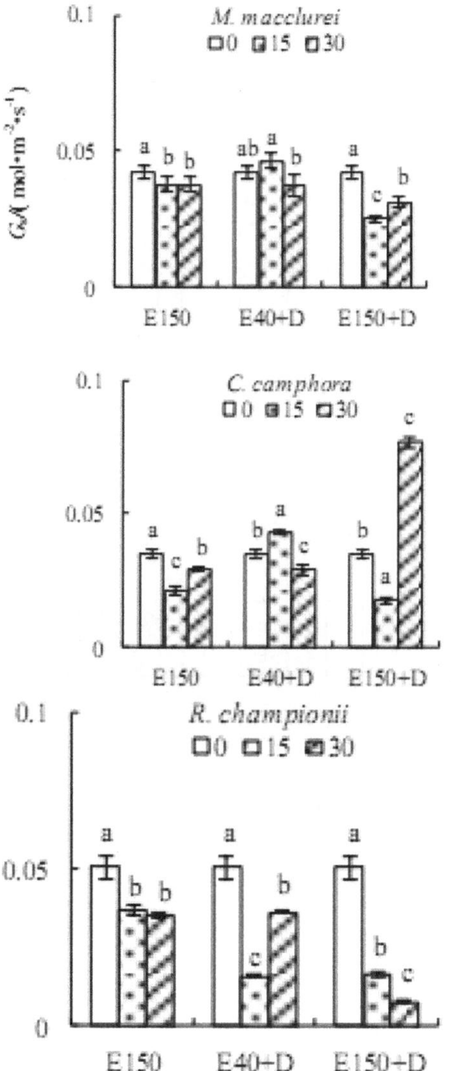

Figure 2. Change in stomatal conductance under O_3 stress and drought stress.
Significant differences between treatments were indicated by different letters above bars, $P < 0.05$.

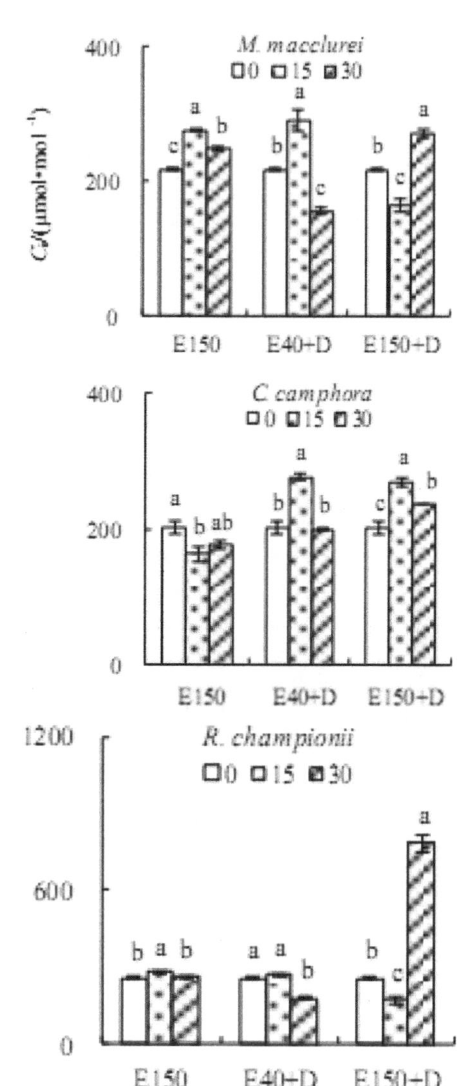

Figure 3. Change in intercellular carbon dioxide concentration under O_3 and drought.
Significant differences between treatments were indicated by different letters above bars, $P < 0.05$.

significantly rose and then significantly dropped in the treatments E40+D and E150+D ($P < 0.05$). The C_i of *R. championii* had an increase and thereafter a decrease in the treatment E150, a slight undulation first and then a significant decrease ($P < 0.05$) in the treatment E40+D, and a significant decrease and then a significant increase in the treatment E150+D ($P < 0.05$).

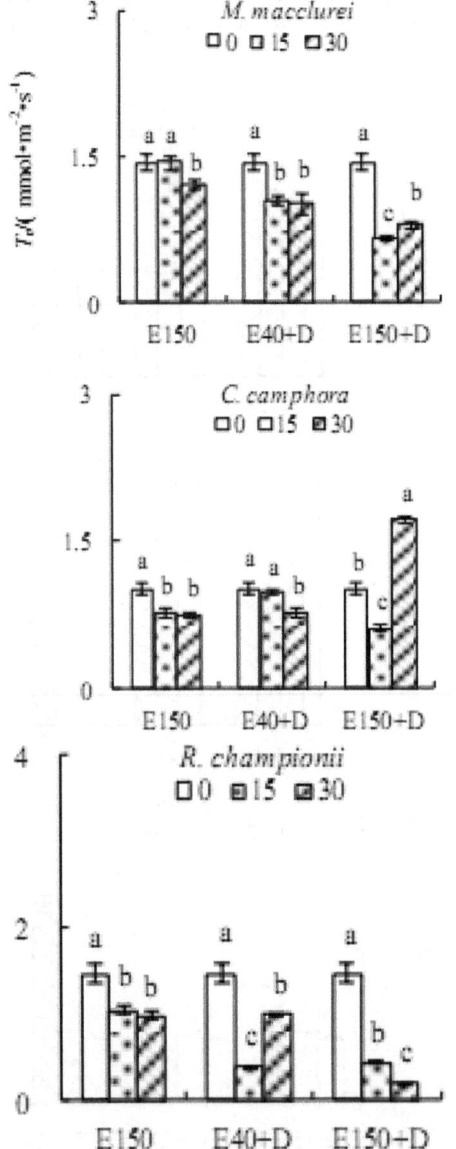

Figure 4. Change in transpiration rate under O_3 and drought stress.
Significant differences between treatments were indicated by different letters above bars, $P < 0.05$.

3.4 *The effects of ozone and drought stress on transpiration rate (T_r) of seedlings*

With increasing ozone and drought stress time, the T_r of *M. macclurei* tended to decrease in the treatments E150 and E40+D, whereas it significantly decreased followed by a significant increase in the treatment E150+D. The T_r of *C. camphora* continuously decreased in the treatments E150 and E40+D, while it had a significant decrease followed by a significant increase ($P < 0.05$). The T_r of *R. championii* had a significant decrease ($P < 0.05$) and then kept stable in the treatment E150, a significant

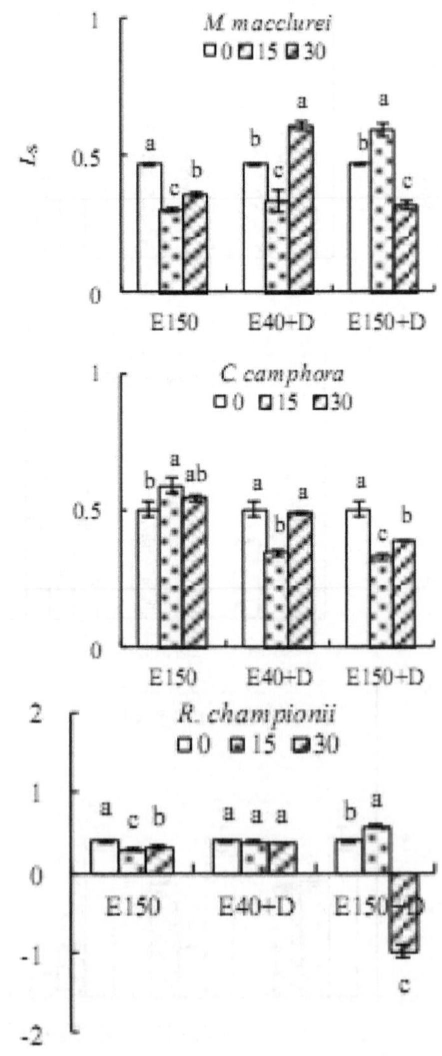

Figure 5. Change in stomatal limitation under O_3 and drought stress.
Significant differences between treatments were indicated by different letters above bars, $P < 0.05$.

Table 2. Comprehensive assessment on O_3 + drought resistance of the three seedling types.

Testing indices	E150		E40+D		E150+D	
	Score	Rank	Score	Rank	Score	Rank
C. camphora	1.32	1	1.22	1	1.32	1
M. macclurei	1.00	2	0.95	2	1.09	2
R champions	0.97	3	0.90	3	0.86	3

decrease and then a significant increase ($P < 0.05$) in the treatment E40+D, and a continuous decrease in the treatment E150+D ($P < 0.05$).

3.5 *The effects of ozone and drought stress on stomatal limitation (L_S) of seedlings*

With increasing ozone and drought stress time, the L_S of *M. macclurei* had a significant decrease and then a significant increase in the treatments E150 and E40+D ($P < 0.05$), while it had a significant increase and then a significant decrease in the treatment E150+D ($P < 0.05$). The L_S of *C. camphora* had a significant increase ($P < 0.05$) and then a decrease in the treatment E150, while it had a significant decrease and then a significant increase in the treatments E40+D and E150+D. The L_S of *R. championii* had a significant decrease and then significantly rose in the treatment E150 ($P < 0.05$), an undulation in the treatment E40+D, while it had an increase and then a significant decrease in the treatment E150+D ($P < 0.05$).

3.6 *Principal component analysis of the ability to resist ozone and drought of three seedling types*

Principal component analysis was used to evaluate the resistance ability to ozone and drought of the three seedling types. The cumulative contribution rate of the first two principal components was greater than 85%, so they were used for calculating. The calculating score decreased in the order of *C. camphora* seedlings > *M. macclurei* seedlings > *R. championii* seedlings in the treatments E150, E40+D and E150+D, indicating that the resistance abilities of the three seedling types to O_3, drought or intercross stress decreased in the order of *C. camphora* seedlings > *M. macclurei* seedlings > *R. championii* seedlings (Table 2).

4 DISCUSSION AND CONCLUSIONS

Photosynthesis is the most basic physiological process for plants, and is sensitive to environmental factors. A large number of experiments showed that ozone caused the net photosynthetic rate of plants to decrease significantly. For example, Wittig (2009) found that the photosynthetic rate of the trees decreased by 11% when ozone concentration increased from 44 ppb to 81 ppb. This study shows that the P_n of *M. macclurei*, *C. camphora* and *R. championii* decreased significantly in the treatment of ozone stress (E150), drought stress (E40+D) and ozone and drought intercross stress (E150+D), indicating that the three seedling types are sensitive to ozone stress or drought stress.

Stomatal conductance is usually considered the most important factor to decide the stress tolerance of plants (Elagoz et al., 2006), because it limits ozone uptake of plants, and directly affects C_i, photosynthesis, transpiration and other important physiological processes (Zheng et al., 2010). In this study, the G_s and T_r of *M. macclurei* and *R. championii* in each treatment and *C. camphora* in the treatments E150 and E40+D decreased. Usually, plants showed lower stomatal conductance in the high concentration of O_3 environment (Pääkkönen et al., 1998), and the T_r usually decreases with the deepening of the ozone or drought stress (Li et al., 2012).

The reason of P_n decrease can be classified as stomatal limitation and non-stomatal limitation, which can be judged based on the change in C_i and L_s (Li et al., 2012). In the case of P_n and C_i decrease and L_s increase, P_n decrease is mainly due to the stomatal limitation. In the case of P_n decrease, C_i increase or without change, and L_s decrease, P_n decrease is mainly due to the non-stomatal limitation (Farquhar et al., 2002). With increasing experimental time, P_n and C_i of *M. macclurei* in the treatment E40+D and those of *C. camphora* in the treatment E150 decreased, and L_s increased, indicating that the P_n decrease in *M. macclurei* under drought stress and in *C. camphora* under ozone stress was mainly due to stomatal limitation. The P_n of *C. camphora* decreased in the treatment E40+D, and C_i and L_s kept stable, indicating that P_n decrease in *C. camphora* in the treatment E40+D was due to stomatal and non-stomatal limitation. The P_n of *M. macclurei* decreased in the treatments E150 and E150+D. In *C. camphora* in the treatment E150+D and in *R. championii* in each treatment, C_i and L_s increased, indicating that the P_n decrease in *M. macclurei* under ozone stress and intercross stress, in *C. camphora* under intercross stress and in *R. championii* under each stress was due to non-stomatal limitation.

C. camphora may avoid the disadvantage through stomatal limitation to decrease P_n under ozone stress. The decrease in stomatal conductance can reduce the ozone amount entering the leaves (Wieser et al., 2007). Paoletti (2006) found that high ozone concentration decreased G_s, promoted the emissions of volatile organic compounds of leaves, and improved the content and activity of antioxidants. The P_n decrease in *M. macclurei* and *R. championii* under ozone stress was mainly due to non-stomatal limitation. The P_n decrease in *M. macclurei* under drought stress was mainly due to stomatal limitation. Stomatal closure caused by drought can prevent further water evaporation (Ahmeda et al., 2009). The P_n decrease in *R. championii* under drought stress was mainly due to non-stomatal limitation, which showed that its photosynthesis may be more controlled by the capacity of chloroplast-fixing CO_2.

High ozone concentration may slow down the stomatal reaction and weaken the ability to control transpiration, and therefore drought stress increases (Onandia et al., 2011). In this study, the P_n decrease in the three seedlings under ozone and drought intercross stresses was mainly due to the non-stomatal limitation, which showed that ozone and drought intercross stress resulted in the excessive production of reactive oxygen species (Tausz et al., 2007) and change in stomatal function (Li et al., 2014), so that the seedlings suffered from serious drought stress (Onandia et al., 2011).

In this study, principal component analysis indicated that the ability to ozone and drought resistance of the three seedling types decreased in the order of *C. camphora* seedlings > *M. macclurei* seedlings > *R. championii* seedlings, so that among the three tree species, *C. camphora* was the best species for ozone and drought resistance in the forestation and landscaping.

ACKNOWLEDGMENTS

This work was financially supported by the Foundation of Guangdong Forestry Department (No. 4400-F11031, 4400-F11055).

REFERENCES

[1] Ahmeda C.B., Rouina B.B., Sensoy S., *et al.* 2009. Changes in gas exchange, proline accumulation and antioxidative enzyme activities in three olive cultivars under contrasting water availability regimes. *Environmental and Experimental Botany*, 67(2): 345–352.

[2] Elagoz V., Han S.S., & Manning W.J., 2006. Acquired changes in stomatal characteristics in response to ozone during plant growth and leaf development of bush beans (*Phaseolus vulgaris* L.) indicate phenotypic plasticity. *Environmental Pollution*, 140: 395.

[3] Farquhar G.D., & Sharkey T.D., 2002. Stomatal conductance and photosynthesis. *Annual Review of Plant Physiology and Plant Molecular Biology*, 33: 317–345.

[4] Karonen M., Ossipov V., Ossipova S., *et al.* 2006. Effects of elevated carbon dioxide and ozone on foliar proanthocyanidins in *Betula platyphylla, Betula ermanii,* and *Fagus crenata* seedlings. *Journal of Chemical Ecology*, 32(7): 1445–1458.

[5] Li Q.J., Lu G.C., & Xue L. 2014. Effects of ozone and drought on photosynthetic physiology of three seedling types in South China. *Guangdong Forestry Science and Technology*, 30(2): 45–52.

[6] Li Y., Chen S.L., Li Y.C. *et al.* 2012. Photosynthetic physio-responsse of *Oligostachyum lubricum* to atmospheric ozone stress. *Ecological Science*, 31(4): 390–395.

[7] Li Y.X., Shen S.H., Li L. *et al.* 2012. Effect of soil moisture on leaf gas exchange and chlorophyll fluorescence parameters of winter wheat during its late growth stage. *Chinese Journal of Ecology*, 31(1): 74–80.

[8] Onandia G., Olsson A., Barth S., *et al.* 2011. Exposure to moderate concentrations of tropospheric ozone impairs tree stomatal response to carbon dioxide. *Environmental Pollution*, 159: 2350–2354.

[9] Pääkkönen E., Vahala J., Pohjola M. *et al.* 1998. Physiological, stomatal and ultrastructural ozone response in birth (*Betula pendula* Roth.) are modified by water stress. *Plant, Cell & Environment*, 21(7): 671–684.

[10] Paoletti E. 2006. Impact of ozone on Mediterranean forests: a review. *Environmental Pollution*, 144(2): 463–474.

[11] Reiner S., Wiltshire J.J.J., Wright C.J., *et al.* 1996. The impact of ozone and drought on the water relations of ash trees (Fraxinus excelsior L.). *Journal of Plant Physiology*, 148: 166–171.

[12] Tausz M., Grulke N.E., & Wieser G. 2007. Defense and avoidance of ozone under global change. *Environmental Pollution*, 147(3): 525–531.

[13] Wang C.Y., & Guan F.L., 1995. O_3 concentration change has an effect on main crops in China. *Journal of Applied Meteorological Science*. 6(1): 69–74.

[14] Wieser G. & Matyssek R. 2007. Linking ozone uptake and defense towards a mechanistic risk assessment for forest trees. *New Phytologist*, 174(1): 7–9.

[15] Wittig V.E., Ainsworth E.A. & Long S.P. 2007. To what extent do current and projected increases in surface ozone affect photosynthesis and stomatal conductance of trees A meta-analytic review of the last 3 decades of experiments. *Plant, Cell & Environment*, 30: 1150–1162.

[16] Wittig V.E., Ainsworth E.A., Naidu S.L, *et al.* 2009. Quantifying the impact of current and future tropospheric ozone on tree biomass, growth, physiology and biochemistry: a quantitative meta-analysis. *Global Change Biology*, 15(2): 396–424.

[17] Zheng Y.F, Hu C.D., Wu R.J., *et al.* 2010. Experiment with effects of increased surface ozone concentration upon winter wheat photosynthesis. *Acta Ecologica Sinica*, 30(4): 0847–0855.

Material Science and Environmental Engineering – Chen (Ed.)
© 2016 Taylor & Francis Group, London, ISBN 978-1-138-02938-5

Photodynamic effect on Human Esophageal Carcinoma Cell Eca-109 and tissue distribution *in vivo* of Pyropheophorbide a

Y. Ye, Y.J. Yan, Z.L. Chen
Department of Pharmaceutical Science and Technology, College of Chemistry and Biology, Donghua University, Shanghai, P.R. China

J.W. Li
Yiwu City Central Hospital, Zhejiang, P.R. China

D. O'Shea
Center for Synthesis and Chemical Biology, University College Dublin, Belfield, Dublin, Ireland

ABSTRACT: This study was designed to investigate the efficacy of Photodynamic Therapy (PDT) in treating esophageal cancer in a preclinical study. Pyropheophorbide a (PPa), a photosensitizer, was tested on Human Esophageal Carcinoma Cell Eca-109 in culture. The cell photo damage occurred since the time the concentrations of PPa reached 0.01 μM (20 J/cm^2). The in vivo distribution of PPa (30 mg/kg, intraperitoneal) was investigated on Eca-109 tumor-bearing nude mice. The tumor vs skin concentration ratio was 2.7:1 at 24 h. Our results showed that PPa was an effective photosensitizer with good photo toxicity and tumor targeting distribution.

Keywords: Eca-109; PPa

1 INTRODUCTION

Esophageal cancer remains a leading cause of cancer-related deaths. The challenge of improving survival in patients suffering from esophageal cancer remains a major concern. In spite of a reduction in postoperative mortality, the surgical management of these patients is still associated with a high morbidity, especially in older patients. Moreover, chemotherapy and external radiotherapy are suitable for only a small proportion of patients. Therefore, Photodynamic Therapy (PDT), a new modality of treating tumor by the combined use of locally or systemically administered photosensitizers and local application of light, may have a role in the management of these patients.

A number of PDT protocols have been developed for the treatment of superficial lesions of the skin [1], age-related macular degeneration [2] and solid tumors located in oral cavity, head/neck, breast, lung, gastrointestinal and genitor-urinaty regions. The efficacy of the photodynamic treatment depends on various factors including the chemical nature of the photosensitizer, its intracellular localization, type of target cells and PDT dose, i.e. light flounce and photosensitizer concentration [3]. A number of porphyrin photosensitizers have been approved for clinical treatments in oncology. One promising photosensitizer for PDT is a derivative of the photosynthetic pigment chlorophyll a named Pyropheophorbide a (PPa).

The interest in this compound has recently increased, since new studies have revealed that PPa adheres most of the criteria that a good photosensitizer should satisfy. In this review we study the photophysical, phototoxicity and tissue distribution of PPa in vitro and in vivo.

2 MATERIALS AND METHODS

2.1 *Photosensitizer*

The photosensitizer PPa has been described previously. Its molecular weight is 534.66. For this paper, it was synthesized in our laboratory. All other chemicals and reagents were of analytical grade and used without any purification.

2.2 *Cell lines and cell culture*

The human esophageal carcinoma cell line Eca-109 was purchased from Chinese Academy of Sciences. The cells were cultured in RPMI 1640 medium, containing 10% (v/v) Fetal Bovine Serum (FBS; Lonza, Germany), in a humidified atmosphere at 37 °C and 75% CO_2. Cell viability was detected via 3-[4, 5-dimethylthiazol-2-yl]-2, 5-diphenyltetrazolium bromide (MTT) assay and expressed as the percentage of control cells. All cells were incubated at 37°C in 5% CO_2 in a humidified incubator. All experimental steps after seeding the cells, including illumination and post-illumination incubation were performed in the same medium. Photosensitizer incubation was done in 1640 but without serum or antibiotics. For measurement of cytotoxicity, cells were seeded in 35 mm plates (Greiner Bio-One, Solingen, Germany) at a density of 100 cells/mm^2 for Eca-109 cells in 2 ml medium.

2.3 *Animals*

Nude BALB/C mice weighing 20–22 g were housed in an authorized facility and maintained under standards complying with the National Research Council's guidelines.

2.4 *In vitro photosensitizing efficacy*

2.4.1 *Absorption spectra*

UV–vis absorption spectrum was recorded on an ultraviolet visible spectrophotometer (Model V-530, Japan). All the measurements were carried out at room temperature in quartz cuvettes with path length of 1 cm. PPa was dissolved in N,N-dimethyl formamide (DMF) as 5 μM.

2.4.2 *Fluorescence excitation–emission spectra*

Fluorescence spectra were measured on a Fluorescence Spectrometer (FluoroMax-4, France). PPa was dissolved in DMF as 5 μM. All the measurements were carried out at room temperature in quartz cuvettes with path length of 1 cm. Use right angle to detect and in emission the slits were kept narrow to 1 nm in excitation and 1 or 2 nm.

2.4.3 *Photodynamic effect*

The MTT assay was used to monitor the cytotoxicity mediated by PPa–PDT. Briefly, Eca-109 cells, which were incubated in media containing 10, 1, 0.1, 0.01, 0.001 μM PPa for 24 h, were irradiated at 650 nm by a Nd:YAG laser and the MTT assay was then used to analyze cellular sensitivity. The dose of illumination was 0, 5, 10, 20, 40 J/cm^2 respectively. All cells were seeded in 96-well plates, each group in triplicate at least, and then incubated in a cell culture incubator with 5% CO_2 at 37°C. Five microliter of MTT (5 mg/ml) reagent (Sigma, MO, USA) was added to each well and incubated for 4 h at 37°C. At the end of the incubation period, the medium was removed and the formazan complex was solubilized with 100 μl DMSO. Absorbance of the complex was measured with a micro-plate reader (Bio-Rad, California, USA) at a wavelength of 570 nm.

2.5 *In vivo tissue distribution*

2.5.1 *Tumor system*

Mice were injected subcutaneously with 1.7×10^7 Eca-109 tumor cells. Investigations were carried out 10 days following inoculation, when the tumors reached 8–10 mm in diameter (i.e.0.25–0.50 cm^3 in volume).

2.5.2 *Tissue concentrations of PPa*

Mice bearing Eca-109 tumor were injected intraperitoneally with PPa (30 mg/kg). Animals were sacrificed in sequence (three animals at each time) from 0 (no injection) to 48 h. The tumor, the non-adjacent skin and all visceral tissues were excised, weighed and assayed for PPa concentrations by spectrofluorophotometer determinations, as previously described [4,5].

3 RESULTS

3.1 *In vitro experiments*

3.1.1 *UV-vis absorption spectrum*

The UV–vis absorption spectrum of Pyropheophorbide a (PPa) was determined in DMF at 5 μM. As can be seen, the PPa had an elevated absorption peak at 668 nm (Fig. 1), which suggesting that PPa might be an effective photosensitizer.

3.1.2 *Fluorescence excitation–emission spectra*

PPa can be excited at 415 nm, and its emission was monitored at wavelength 664 nm (Fig. 2).

3.1.3 *Photodynamic effect*

Eca-109 cells were incubated with PPa for 24 h and exposed to light as described above. As shown in Figure 3. Subjection of Eca-109 cells to different concentrations of PPa without subsequent

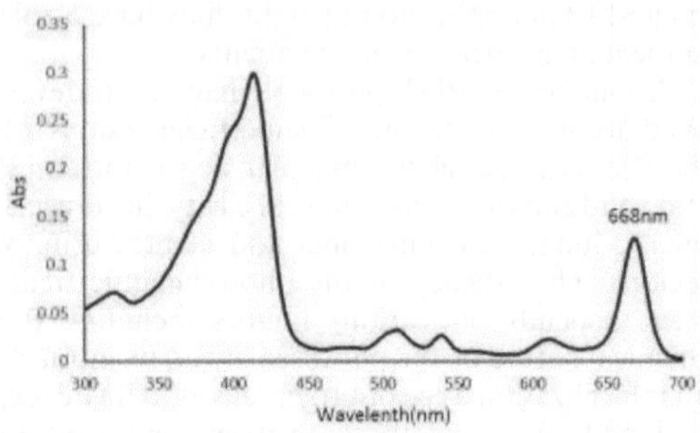

Figure 1. UV-vis absorption spectrum of PPa in DMF.

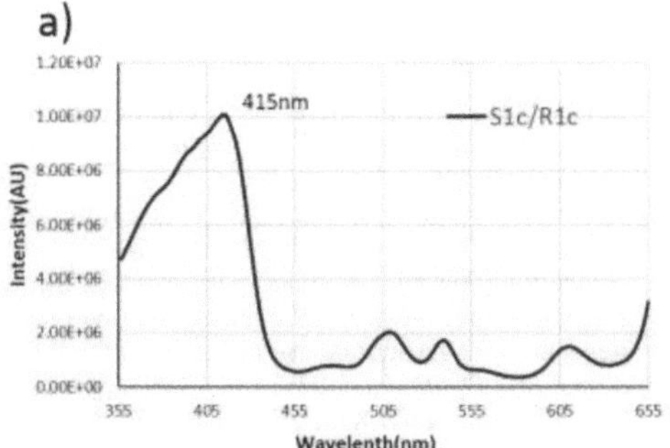

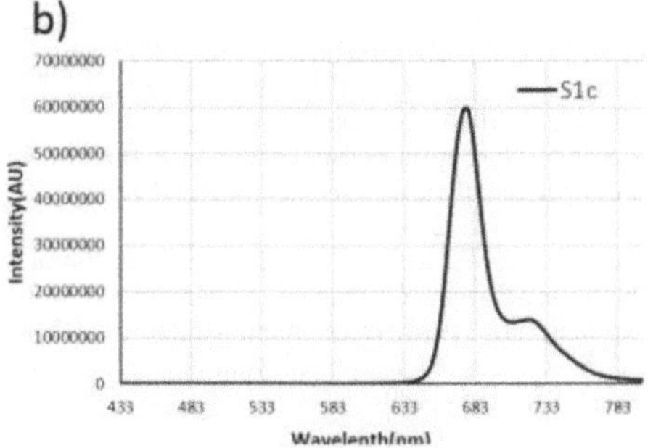

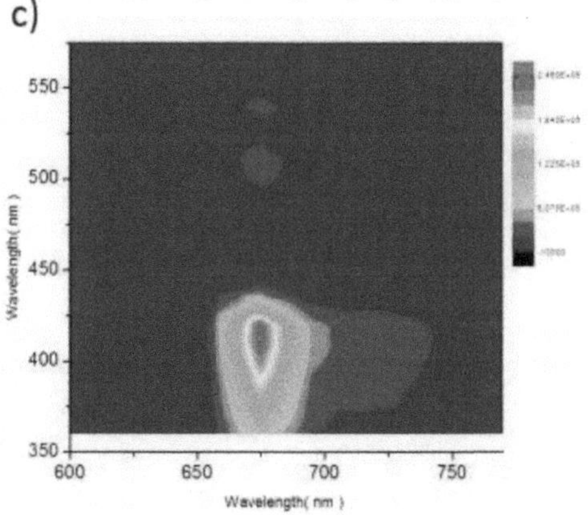

Figure 2. Fluorescence spectrum of PPa. a) absorption spectra (λ_{ex} = 415 nm); b) emission spectra (λ_{em} = 664 nm); c) 3D fluorescence spectra of PPa.

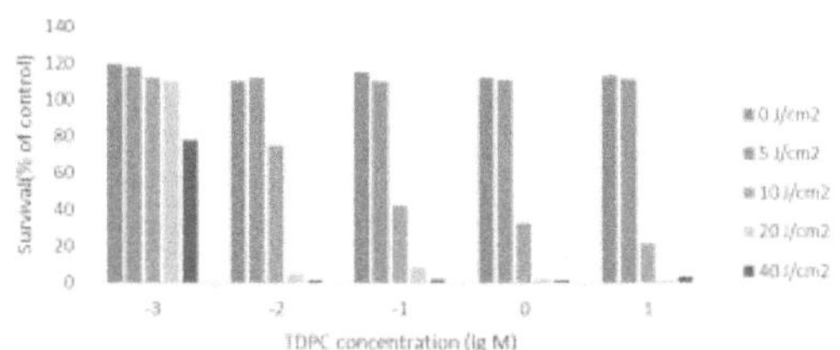

Figure 3. Cytotoxicity to Eca-109 cells with different concentrations of PPa and different laser dose.

Table 1. Eca-109 tumor and tissue concentrations of photosensitizers in nude mice*.

	PPa (μg/g of tissue)		
	6 h	24 h	48 h
Tumor	4.8±0.5	3.6±0.3	1.7±1.2
Liver	11.8±0.7	0.6±0.5	0.2±0.1
Kidney	0.0±0.0	0.0±0.0	0.0±0.0
Lung	0.0±0.0	0.0±0.0	0.0±0.0
Skin	3.3±0.9	1.3±0.1	0.2±0.1

*Values are means 6 s.e.m. of six animals bearing one subcutaneously grafted HT29 tumor. Photosensitizers (30 mg/kg) were administered via an i.p. route.

irradiation (termed dark toxicity experiment) demonstrated no change in viability compared to normal untreated cells. The survival fraction reached 96.20 ± 0.043% when concentration of drug was 10 μM. Survival of cells dropped only by a combination of the photosensitizer and light. Following PDT, when the concentration of PPa was increased the viability of cells was declined. However the PDT effect was not strong at 0.01 μM and 0.001 μM concentration but opposite at the concentration 0.1 μM, 1 μM, 10 μM. $P < 0.0001$, $R_2 = 0.9763$. It was indicated that PPa may be an excellent photosensitizer in biological applications.

3.2 *In vivo experiments*

The concentrations of photosensitizer in engrafted human esophageal tumor and in host nude mouse tissues were assayed 6, 24 and 48 h after i.p. injection (Table 1). PPa was absent from lung and kidney. The main PPa targets was the liver but only for one day. After one day, PPa concentration dropped in all tissues, but the drop was slightly less in tumor, giving a satisfactory tumor vs skin concentration ratio which is 2.7:1.

4 DISCUSSION

Pyropheophorbide a (PPa) is a chlorophyll derivative characterized by adequate photo physical parameters to be used as a photosensitizer in the photodynamic therapy. Compared to widely used Photofrin, PPa is activated at a higher wavelength, 668 versus 630 nm. Considering that PPa promotes a strong PDT effect on Eca-109 cells and that it is activated at wavelengths that are highly tissue penetrating, this photosensitizer is suitable to target deep intraperitoneal tumors. PDT using PPa inhibits its tumor cell growth. Its in vitro effectiveness was confirmed by evidence of a strong phototoxicity involving tumor cell apoptosis. The tissue distribution investigations of this study gave a satisfactory result that PPa can target tumor and the skin phototoxicity is weak. So PPa is worth further study as an effective photosensitizer.

5 CONCLUSIONS

Our results showed that PPa was an effective photosensitizer with good photo toxicity and tumor targeting distribution.

ACKNOWLEDGEMENTS

This work was supported by Chinese National Natural Science Foundation (No. 21372042, 21402236, 81101298, 81301878), Foundation of Shanghai government (No. 14431906200, 14140903500, 13431900700, 13430722300, 13ZR1441000, 13ZR1440900, 14ZR1439800, 14ZR1439900, 15ZR1439900, 15XD1523400, 14SJGGYY08, 201370), International Cooperation Foundation of China and Croatia (6-11) and Foundation of Yiwu Science and Technology Bureau (No. 2012-G3-02, 2013-G3-03).

REFERENCES

[1] Babilas P, Schreml S, Landthaler M, Szeimies R.M. 2010, Photodynamic therapy in dermatology: state-of-the-art. *Photodermatol Photoimmunol Photomed*, 26, 118–132.

[2] Iwama D, Otani A, Sasahara M, Yodoi Y, Gotoh N, Tamura H, Tsujikawa A, Yoshimura N. 2008, Photodynamic therapy combined with lowdose intravitreal triamcinolone acetonide for age-related macular degeneration refractory to photodynamic therapy alone. *Br J Ophthalmol*, 92, 1352–1356.

[3] Yoo J.O., Lim Y.C., Kim Y.M., Ha K.S. Differential cytotoxic responses to low- and high-dose photodynamic therapy in human gastric and bladder cancer cells. 2011, J *Cell Biochem*, 112, 3061–3071.

[4] Lantz, J.M., C. Meyer, C. Saussine, C. Leberquier, F. Heysel, J. Miehe, J. Marescaux, R. Sultan and M. Kedinger. Experimental photodynamic therapy with a copper metal vaporlaser in colorectal cancer. 1992, *Int. J. Cancer*, 52, 491–498.

[5] Evrard, S., P. Keller, A. Hajri, G. Balboni, L. Mendoza-Burgos, C. Damge′, J. Marescaux and M. Aprahamian. 1994, Experimental pancreatic cancer in the rat treated by photodynamic therapy. *Br. J. Surg.*, 81, 1185–1189.

Material Science and Environmental Engineering – Chen (Ed.)
© 2016 Taylor & Francis Group, London, ISBN 978-1-138-02938-5

Active control of transformer noise with adaptive filtering algorithm

J.B. Liang & T. Zhao
School of Electrical Engineering, Shandong University, Jinan, China

X.H. Zhao, J.Q. Liu, Z.H. Qi & X.H. Li
State Grid Laiwu Power Supply Company, Laiwu, China

ABSTRACT: This paper describes the adaptive filtering algorithm based on LMS aiming at the low-frequency noise to realize the transformer active noise control. Additionally, the effects of the step side and the filter order of the LMS adaptive filtering algorithm on noise reduction are analyzed by simulation. Finally, the experiment of the transformer noise reduction is conducted with hardware devices and the software platform. The noise sound level decreases nearly 10 dB, which verifies the practicability of the LMS adaptive filtering algorithm on the active control of transformer noise.

Keywords: active noise control; transformer noise; adaptive filtering algorithm; LMS

1 INTRODUCTION

Noise generated by large transformers in urban areas has a serious impact on people's work and rest, where the low frequency is produced by the vibration of stator core mainly consisting of harmonics of 100 Hz, 200 Hz, 300 Hz, 400 Hz, 500 Hz and 600 Hz. Low-frequency noise may cause human chronic injury and result in neurological diseases. Especially, it will affect fetal development if pregnant women are exposed to low-frequency noise for a long time. Therefore, it is necessary to perform transformer noise reduction.

Improving manufacturing processes to decrease the transformer vibration and traditional passive noise control by absorbing and obstructing sound is difficult to quantitatively analyze in practical engineering. In addition, these measures are so easily constrained by the cost, technology and other conditions that cannot achieve significant noise reduction. Instead, the active noise control generates the sound with an opposite phase through the electronic circuit and the sound reinforcement equipment to offset the original noise, which can better reduce low-frequency noise.

Early, active noise reduction depends on manually adjusting the amplitude and phase. In order to obtain better noise control and improve the stability of the system, it is essential to use the adaptive filtering algorithm. This paper introduces an adaptive filtering algorithm based on LMS, and studies the effects of the step size and the filter order on transformer noise reduction by LabVIEW simulation. In addition, the experiment platform of active noise control for the transformer combining with hardware devices is built up to verify the feasibility of the algorithm.

2 THE PERFORMANCE OF THE LMS ALGORITHM

2.1 *Theory*

The adaptive filtering algorithm uses the Least Mean Square (LMS) as the criterion, which means using the steepest descent method to get the weight coefficient. The block diagram of the LMS algorithm is shown in Figure 1, where $x(n)$ is the reference signal; $d(n)$ is the desired signal; $e(n)$ is the adaptive filter output signal (the secondary signal); and $e(n)$ is the error signal. The secondary signal can be expressed as follows:

$$y(n)=\sum_{l=1}^{L} w_l(n)x(n-l+1) \tag{1}$$

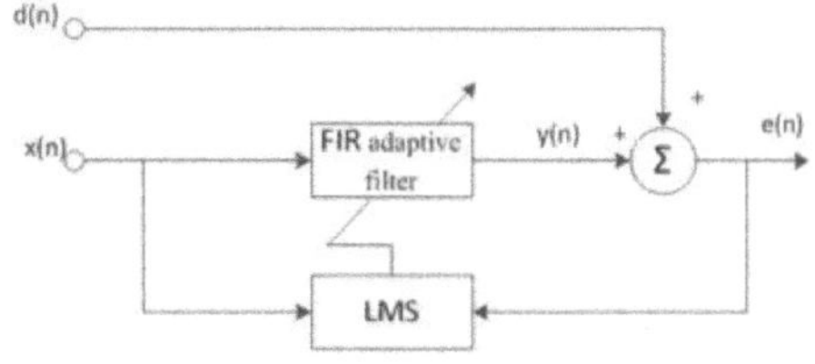

Figure 1. Block diagram of the LMS algorithm.

where L is the filter order and $W_l(n)$ ($l = 1, 2, \ldots, L$) is the weighted coefficient.

The weight vector and the reference vector of the adaptive filter are as follows:

$$W = [w_1, w_2, \ldots, w_L]^T \quad (2)$$

$$X(n) = [x(n), x(n-1), \ldots, x(n-L+1)]^T \quad (3)$$

Then, $y(n)$ can be expressed as follows:

$$y(n) = X^T(n)W = W^T X(n) \quad (4)$$

The update equation of the weighted coefficient $W(n)$ can be expressed as follows:

$$W(n+1) = W(n) + 2\mu e(n)X(n) \quad (5)$$

where μ is the step side.

2.2 *The performance of the step size μ on the noise reduction*

LabVIEW, as the virtual instrument software development platform, can exploit a friendly interactive interface, and analyze and process complex data with high speed and high accuracy. The programming of LabVIEW is simple because its graphical programming combining with a variety of encapsulated modules can achieve to compile the algorithm quickly. Therefore, it is suitable to emulate different kinds of algorithms.

The primary noise is assumed to be the superposition of sinusoids with a frequency of 100 Hz, 200 Hz and 300 Hz, and a amplitude of 0.3, 0.1 and 0.05, and 0.002 Gaussian.

L is assumed to be 250, and μ is assumed to be 0.0003, 0.003, 0.01 and 0.09. The noise reduction and the learning curve are shown in Figures 2–5.

According to Figures 2–5, when $\mu = 0.0003$, the system converges slowly. With a rising of μ, the error signal accelerates convergence and the noise reduction increases from 20 dB to 40 dB. When μ increases to 0.09, the system diverges.

Thus, when the step size μ increases with other conditions keeping constant, the secondary signal gets closer to the reference signal, which leads to the acceleration of convergence of the system and obtains larger reduction of noise. However, when μ continues to increase, the system stability will become poor and the speed of convergence will decrease. Especially, when μ increases to a certain extent, the system will diverge.

2.3 *The performance of the filter order L on noise reduction*

The step size μ is assumed to be 0.09, and L is set to be 20, 100, 210 and 250. The noise reduction and the learning curve are shown in Figures 6–8.

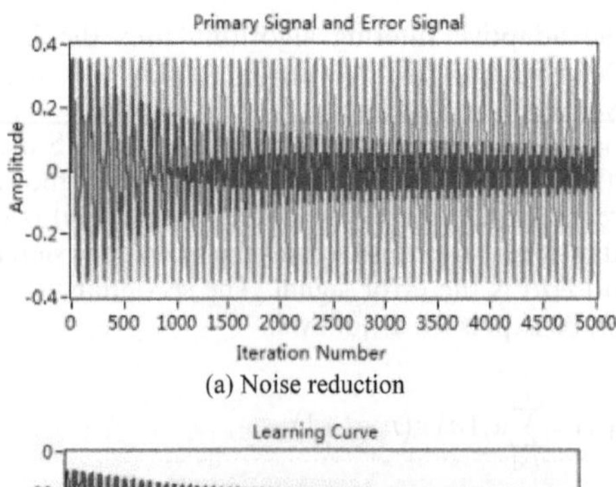

(a) Noise reduction

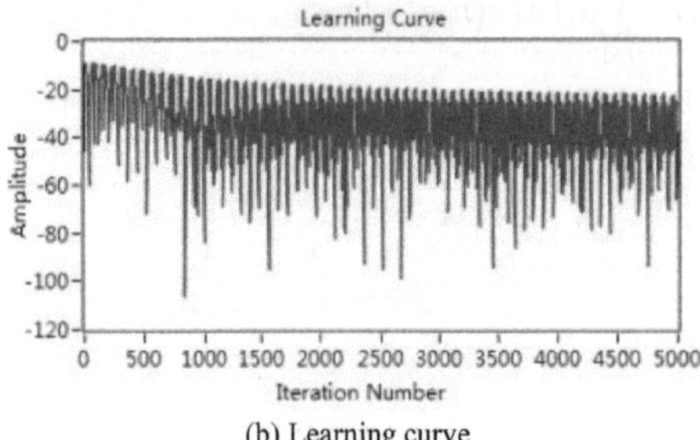

(b) Learning curve

Figure 2. L = 250 and μ = 0.0003, the noise reduction and the learning curve.

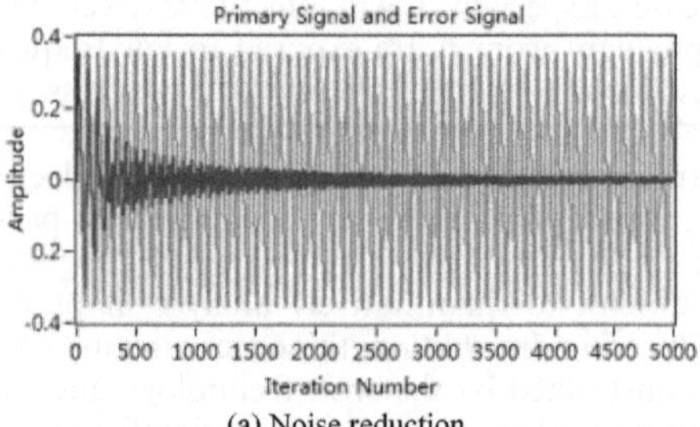

(a) Noise reduction

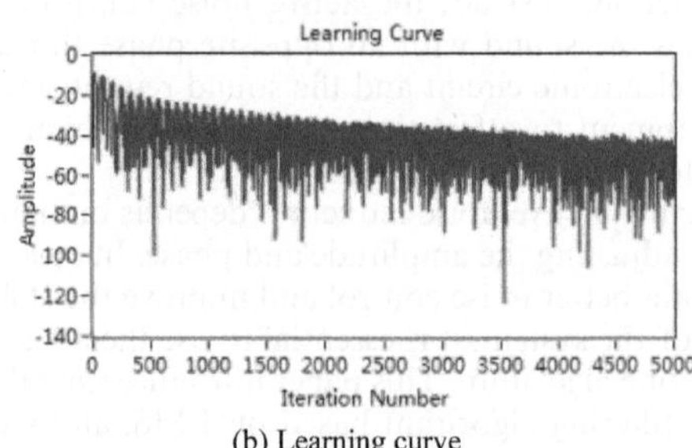

(b) Learning curve

Figure 3. L = 250 and μ = 0.003, the noise reduction and the learning curve.

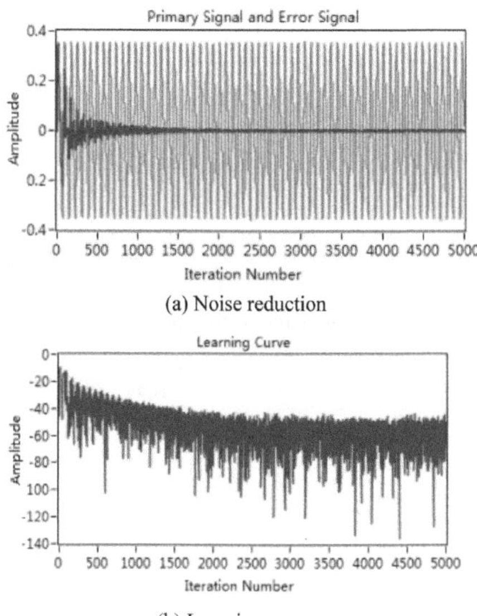

(a) Noise reduction

(b) Learning curve

Figure 4. L = 250 and μ = 0.01, the noise reduction and the learning curve.

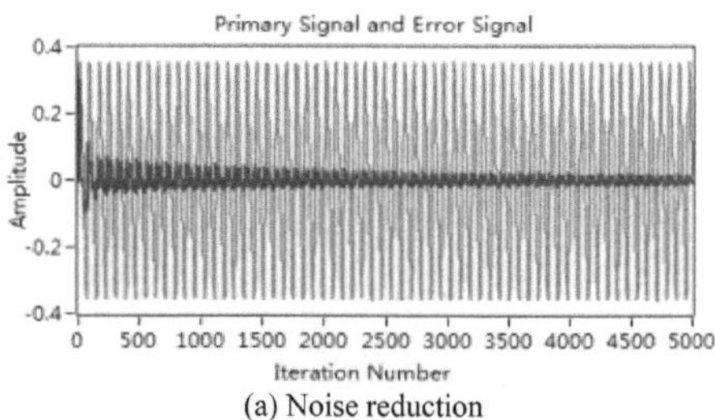

(a) Noise reduction

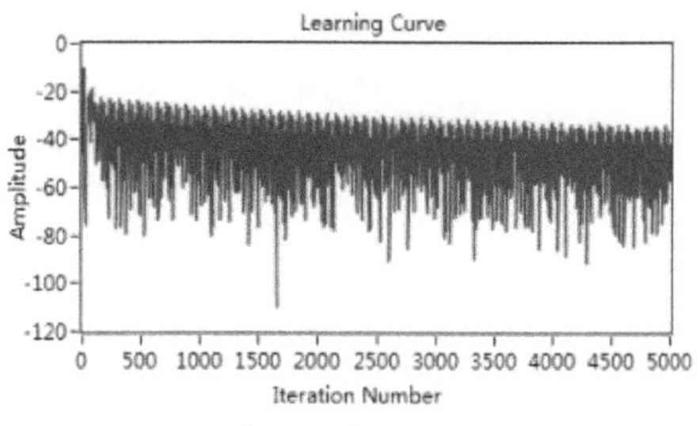

(b) Learning curve

Figure 6. μ = 0.09 and L = 20, the noise reduction and the learning curve.

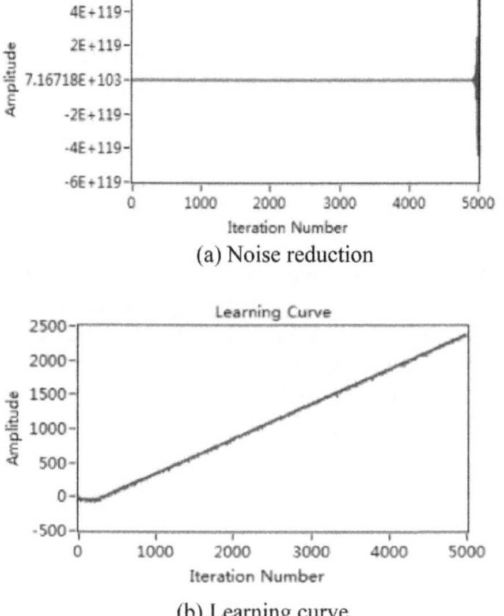

(a) Noise reduction

(b) Learning curve

Figure 5. L = 250 and μ = 0.09, the noise reduction and the learning curve.

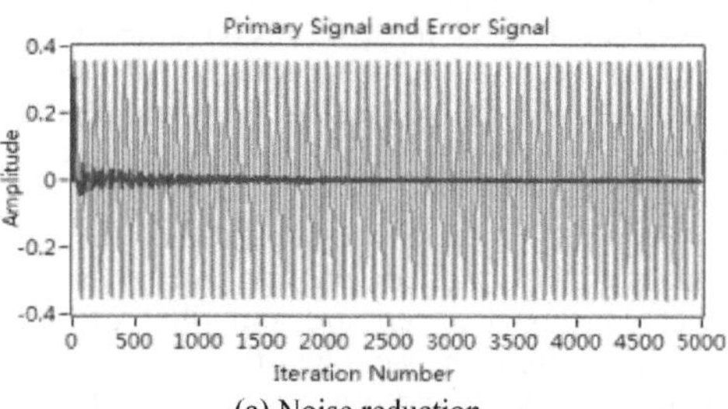

(a) Noise reduction

(b) Learning curve

Figure 7. μ = 0.09 and L = 100, the noise reduction and the learning curve.

According to Figures 5–8, when $L = 20$, the system converges slowly. With a rising of L, the error signal accelerates convergence and the noise reduction increases from 30 dB to 40 dB. While $L = 210$, the convergence rate decreases, the noise reduction returns to 30 dB and the system appears to have a divergent trend. When L increases to 250, the system diverges.

Thus, when the filter order L increases with other conditions keeping constant, the secondary

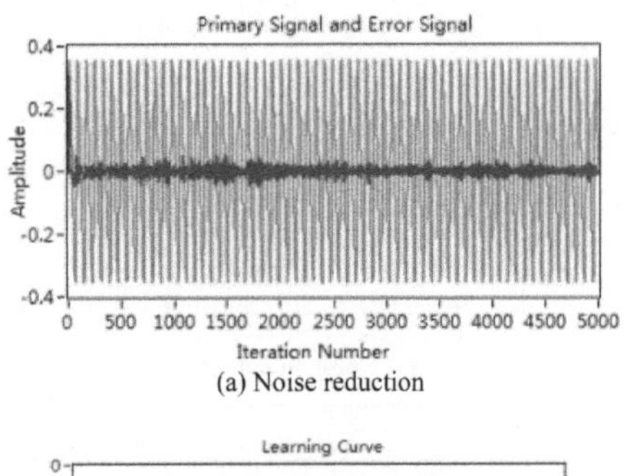

(a) Noise reduction

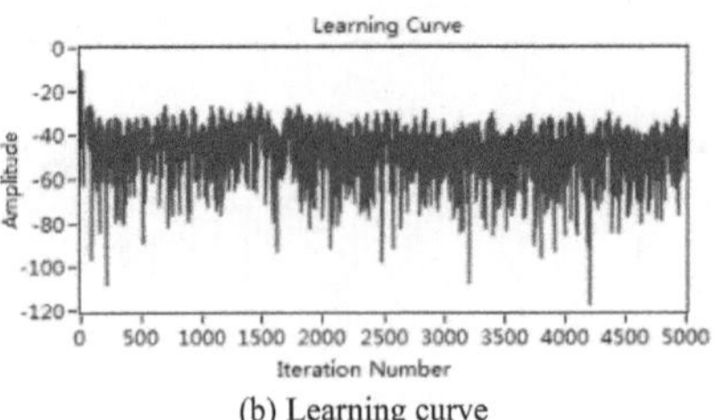

(b) Learning curve

Figure 8. $\mu = 0.09$ and $L = 210$, the noise reduction and the learning curve.

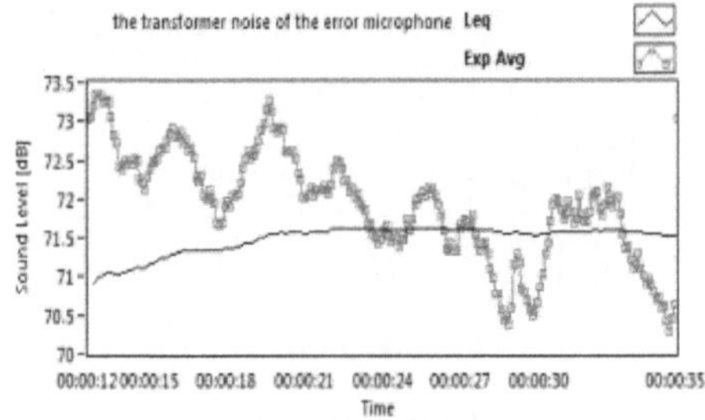

(a) Noise before active noise control

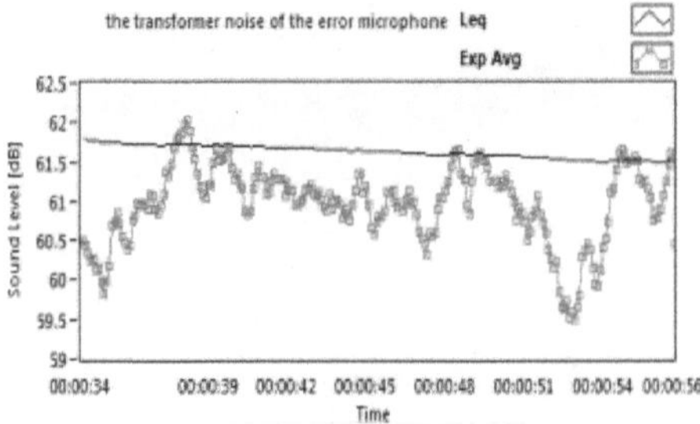

(b) Noise after active noise control

Figure 9. Reduction of the transformer noise.

signal gets closer to the reference signal, which accelerates convergence of the system and obtains larger reduction of noise. However, when L continues to increase, the computation will increase, which results in the system in a poor real time and stability. Especially, when L attains to a certain extent, the system will diverge.

According to the above simulations, the noise reduction performance will improve by using a larger convergence step size and filter order. In addition, there is the sampling frequency, the secondary path delay and other factors affecting the noise reduction. Therefore, in practical applications, when to select parameters, it is necessary to consider convergence speed, stability, noise reduction and other various factors in order to obtain better stability and higher noise reduction.

3 THE EXPERIMENT OF ACTIVE NOISE CONTROL FOR THE TRANSFORMER

3.1 *The reduction of the transformer noise*

The reduction of the transformer noise is showed in the Figure 9, where figure a shows the noise before active noise control and figure b shows the noise after the control. As shown in Figure 9, before the running of the adaptive filtering algorithm, the sound level of the error microphone is about 72 dB. After the running of the algorithm, the sound level decreases to 62 dB. There is about 10 dB noise reduction, which shows that the LMS adaptive filtering algorithm can do well in noise reduction.

4 CONCLUSION

This paper describes the LMS adaptive filtering algorithm in the active transformer noise control, and discusses the influence of step size and filter order on noise reduction by simulation and experiment. Within a certain range, when the step size and the filter order increase, the system will converge quickly and obtain larger noise reduction. However, when one of these two parameters exceeds, the system convergence speed will slow down, and noise reduction will decrease. Finally, the system will diverge. In the experiment, based on the LabVIEW platform and hardware devices, the algorithm can achieve to make the noise sound level decrease 10 dB, so it is feasible to apply the algorithm to transformer noise control.

REFERENCES

[1] Ross, C.F. 1978. Experiments on the active control of transformer noise. *Journal of Sound and Vibration* 61(4): 473–480.

[2] Qiu, X. & Li, X. 2002 A waveform synthesis algorithm for active control of transformer noise: implementation. Applied Acoustics 63(5): 467–479.

[3] Elliott, S.J. & Nelson, P.A. 1993. Active noise control, *Signal Processing Magazine, IEEE* 10(4): 12–35.

[4] Glentis, G.O. & Berberidis, K. 1999. Efficient least squares adaptive algorithms for FIR transversal filtering, Signal Processing Magazine, IEEE 16(4): 13–14.

Material Science and Environmental Engineering – Chen (Ed.)
 ISBN 978-1-138-02938-5

Apoptosis effect of photosensitizer Hematoporphyrin Monomethyl Ether on human hepatoma cell SMMC-7721

Y. Ye, Y.J. Yan & Z.L. Chen
Department of Pharmaceutical Science and Technology, College of Chemistry and Biology, Donghua University, Shanghai, P.R. China

J.W. Li
Yiwu City Central Hospital, Zhejiang, P.R. China

D. Margetić
Laboratory for Physical-Organic Chemistry, Division of Organic Chemistry and Biochemistry, Ruđer Bošković Institute, Zagreb, Croatia

ABSTRACT: The effectiveness of PDT depends strongly on the type of photosensitizers. Hematoporphyrin Monomethyl Ether (HMME) is a promising second-generation porphyrin-related photosensitizer for PDT. HMME-PDT-induced cell death and its mechanisms were investigated in human hepatoma cell SMMC-7721. The survival rate of SMMC-7721 cells determined by MTT assay decreased with the increasing HMME concentration and laser dose. The expression level of TNFSF13, a member of tumor necrosis factor superfamily, increased significantly. These results imply that photodynamic therapy with HMME may therefore be a useful clinical treatment for liver cancer.

Keywords: SMMC-7721; PDT

1 INTRODUCTION

Primary liver cancer is globally the sixth most frequent cancer, and the second leading cause of cancer death[1]. In 2012 it occurred in 782,000 people and resulted in 746,000 deaths. Higher rates of liver cancer occur where hepatitis B and C are common, including East-Asia and sub-Saharan Africa. Five year survival rates are 17% in the United States. During the past decade, advances in surgical technique and chemotherapy have resulted in response rates that exceed 70%; however, most patients with advanced-stage liver cancer recur and ultimately died of their disease. These ominous statistics justify the search for effective new therapies, such as photodynamic therapy, for patients afflicted with liver cancer.

Photodynamic Therapy (PDT) is a promising new cancer treatment strategy that involves the combination of visible light and a photosensitizer[2, 3]. Each factor is harmless by itself, but when combined with oxygen, they can produce lethal cytotoxic agents that can inactivate tumor cells. This enables greater selectivity towards diseased tissue as only those cells that are simultaneously exposed to the photosensitizer, light, and oxygen are exposed to the cytotoxic effect. The dual selectivity of PDT is produced by both a preferential uptake of the photosensitizer by the diseased tissue and the ability to confine activation of the photosensitizer to this diseased tissue by restricting the illumination to that specific region. Therefore, PDT allows for the selective destruction of tumors while leaving normal tissue intact.

Hematoporphyrin Monomethyl Ether (HMME) is a second-generation, porphyrin-related photosensitizer that has recently been developed. HMME consists of two monomeric porphyrins, i.e., 3-(1-methyloxyethyl)-8-(1-hydroxyethyl) deuteroporphyrin IX and 8-(1-methyloxyethyl)-3-(1-hydroxyethyl) deuteroporphyrin IX (Fig. 1). Experimental studies and clinical trials have demonstrated that HMME which can be selectively taken by tumor tissues has a stronger photodynamic effect, lower toxicity and shorter-term skin photosensitizations. Moreover, HMME is less costly compared with other drugs.

It is therefore of interest to know whether HMME is capable of inducing photodynamic cytotoxic effects on human hepatoma cell SMMC-7721. In this work, SMMC-7721 was used to examine the possible effects and mechanisms of HMME-mediated PDT in vitro.

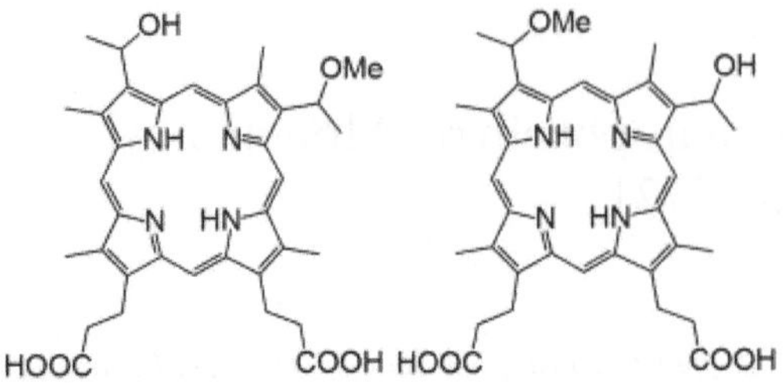

Figure 1. Chemical structure of hematoporphyrin monomethyl ether.

2 MATERIALS AND METHODS

2.1 *Preparation of HMME solutions*

Drug grade powder form HMME ($C_{35}H_{36}N_4O_6$, synthesized in our laboratory and its structure was determined by spectral analysis) was reconstituted initially at 100μM in N, N-dimethyl formamide (DMF) and kept in dark at 4°C. HMME solutions were then diluted to an appropriate working concentration based on the sensitivity of the instrument in use. Freshly prepared HMME solutions were kept from light at all times except during actual measurement.

2.2 *Visible absorption spectra*

Visible absorption spectra of hematoporphyrin monomethyl ether in DMF as 5 μM were investigated. Absorption spectra were scanned in 300–700 nm with the ultraviolet visible spectrophotometer (Model V-530, Japan).

2.3 *Fluorescence excitation–emission spectra*

Fluorescence spectra were measured on a Fluorescence Spectrometer (FluoroMax-4, France). Hematoporphyrin monomethyl ether was dissolved in DMF as 5 μM. All the measurements were carried out at room temperature in quartz cuvettes with path length of 1 cm. Use right angle to detect and in emission the slits were kept narrow to 1 nm in excitation and 1 or 2 nm.

2.4 *Cell lines and cell culture*

Human hepatoma cell SMMC-7721 cell line was purchased from Chinese Academy of Sciences. SMMC-7721 cells were maintained in a humidified 5% CO_2 incubator at 37°C in RPMI 1640 (Gibco) medium containing 10% fetal bovine serum (FBS, Sijiqin Co. Hangzhou, China), 20 μg/ml penicillin and 20 μg/ml streptomycin.

2.5 *Phototoxicity assay*

The survival rate of the cells after PDT was measured by MTT assay. During the experiments, 1×10^5/ml cells were incubated with different concentrations of HMME (0–10 μM) in a 96-well culture plate for 3h. Then the drug-containing medium was aspirated and the cells were rinsed with PBS. The medium was replaced with 200 μl RPMI 1640 before illumination. The laser source was Nd:YAG laser with a maximal output of 500 mW. The irradiation was carried out for 0.5–3 min by 635 nm light with an output power of 0–40 mW/cm^2. Following PDT, the medium was replaced with 10% FBS RPMI 1640 and the cells were able to proliferate. At 12 h post irradiation, 20 ml of MTT (final concentration: 0.5 mg/ml) was added to each well and the cells were incubated for 4h at 37 °C. Afterwards, the culture medium was replaced with 200 ml DMSO. The Optical Density (OD) of the 96-well culture plate was examined immediately at 490 nm with a micrometer reader (BioTek ELx800). The cell survival rate was calculated as the ratio of the absorbance of the treated cells over untreated cells.

2.6 *Real-time PCR*

Total RNA was extracted from cells using Trizol reagent (Invitrogen, Carlsbad, CA) according to the manufacturer's protocol. cDNAs were synthesized using M-MLV reverse transcriptase (Promega) for real-time PCR analysis. Real-time PCR was performed using SYBR Master Mixture (TAKARA, Japan) to measure mRNA levels of selected genes. Data were analyzed using the standard curve analysis method with Light-Cycler 480 Software (Roche Applied Science).

3 RESULTS

3.1 *Absorption spectrum of hematoporphyrin monomethyl ether*

The DMF was irradiated with UV-A, no signals was observed. However, when hematoporphyrin monomethyl ether was dissolved in DMF, signals were observed in the solution containing HMME with UV-A irradiation. The results indicated that the signals were derived from HMME. HMME exhibited four absorption bands in the wavelength range from 300 to 700 nm, and the peak on 621 nm could be chosen as light source for PDT. (Fig. 2).

3.2 *Fluorescence excitation–emission spectra*

Hematoporphyrin monomethyl ether can be excited at 400 nm, and its emission was monitored at wavelength 623 nm (Fig. 3).

3.3 *Phototoxicity study*

The survival rate of the SMMC-7721 cells after PDT was determined by MTT assay. If the cells

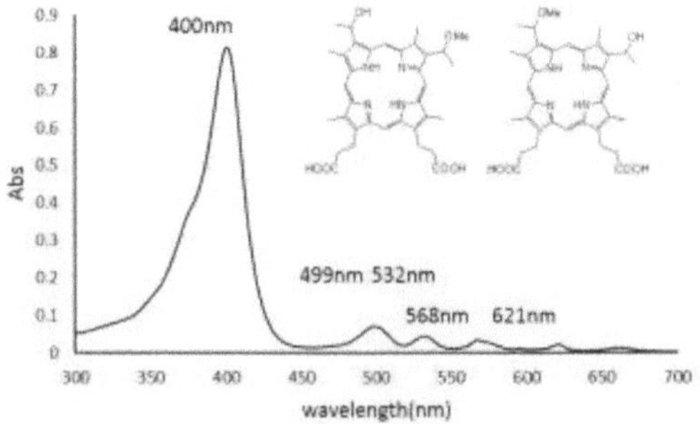

Figure 2. UV-vis absorption spectrum of hematoporphyrin monomethyl ether in DMF.

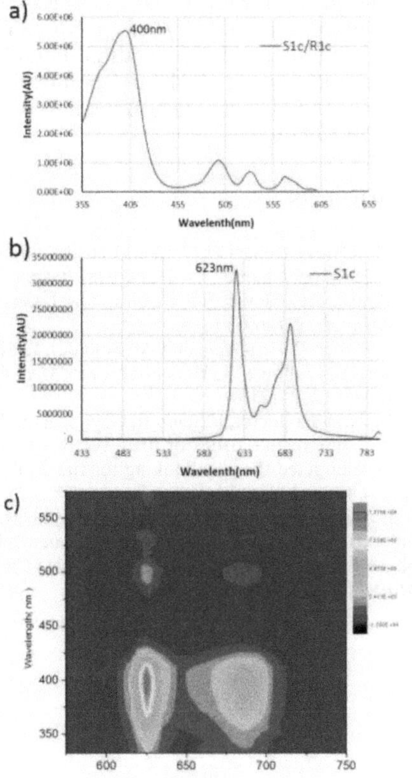

Figure 3. Fluorescence spectrum of hematoporphyrin monomethyl ether in DMF. a) absorption spectra ($\lambda_{ex} = 400$ nm); b) emission spectra ($\lambda_{em} = 623$ nm); c) 3D fluorescence spectra.

were not exposed to light, lower concentrations (≤10 μM) of HMME did not influence cell survival rate.

The survival rate of the PDT group was significantly lower than that of the untreated group, although the use of low power light irradiation alone did not inhibit cell growth. The survival rate of the human hepatoma cell SMMC-7721 decreased as the increase of HMME concentration and the laser dose. Our results indicated that HMME-mediated PDT was capable of inducing cell death significantly. (Fig. 4).

3.4 *The changes of gene expression in SMMC-7721 cells treated by HMME–PDT*

To explore the changes of gene expression after HMME–PDT, total cellular RNA was harvested from SMMC-7721 cells after 24h incubation in medium containing 2 or 4 μM HMME followed by irradiation with a dose of 0.18 J/cm^2 light ($\lambda = 635$ nm). The expression levels of genes involved in apoptosis and cell growth were then detected by real-time PCR assay. During HMME–PDT, some genes involved in apoptosis such as X-linked inhibitor of apoptosis (XIAP) was significantly up-regulated, and the tumor necrosis factor superfamily member 13 (TNFSF13) also greatly increased after

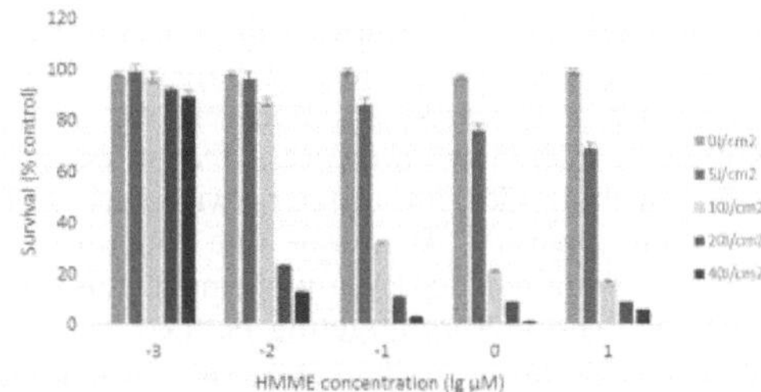

Figure 4. The photocytotoxicity of HMME towards SMMC-7721. The cell survival rates are shown at 12h after treatment with PDT using different concentrations of HMME and laser dosess. The data are representative of independent experiments. *$P < 0.05$; **$P < 0.005$ vs. the untreated controls.

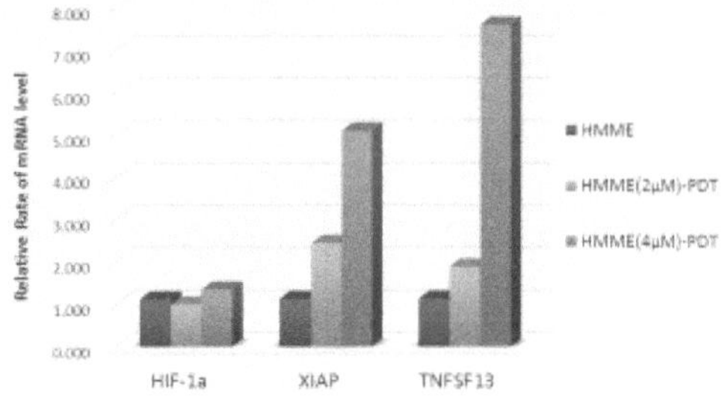

Figure 5. The Changes of gene expression in SMMC-7721 cells after HMME-PDT. RT PCR results showed the expression changes of 3 genes in SMMC-7721 cells treated by PDT with 2 or 4 μM HMME, or with 4 μM HMME with no photo-excitation after 24h.

PDT with the higher dose of HMME (4 μM). HMME–PDT at the dose of 2 μM caused hypoxia-inducible factor-1a (HIF-1a) to be down-regulated; however increasing the dose of HMME further more could promote HIF-1a expression. (Fig. 5).

4 DISCUSSION

The poor prognosis of advanced liver cancer and recent developments in photomedicine have generated a considerable interest in PDT for this disease. Hematoporphyrin monomethyl ether is a novel second generation of photosensitizers, we provided the evidences that HMME is a novel photosensitizer for liver cancer treatment. To identify possible absorption peaks, we used hematoporphyrin monomethyl ether for visible absorption spectra assay, and we demonstrated that HMME has four absorption peaks. An absorption peak at 621 nm is an ideally matched laser wavelength for PDT, especially for intraperitoneal tumors. The MTT assay for cell toxicity confirmed that cell death was dependent on both drug concentration and light dose. At similar drug concentration, higher light dose was required for the same level of cell death. Disease-associated cells expressing molecules, including proteases, receptors, and adhesion molecules, are different with their normal counterparts[4]. Therefore one intention in targeting therapy is to develop chemically derived drugs or drug vectors targeting defined cells via specific recognition mechanisms and overcoming biological barriers[5]. It was found that the expression of some apoptosis factors including XIAP and TNFSF13 were up-regulated significantly during HMME–PDT, and it potentially contributed to cell apoptosis.

5 CONCLUSIONS

This study is preliminary and the treatment effectiveness needs to be verified in vivo in the future, but it strongly suggests that HMME-mediated PDT might be a promising method for liver cancer and merits further evaluation in experimental studies and clinical practices.

ACKNOWLEDGEMENTS

This work was supported by Chinese National Natural Science Foundation (No. 21372042, 21402236, 81101298, 81301878), Foundation of Shanghai government (No. 14431906200, 14140903500, 13431900700, 13430722300, 13ZR1441000, 13ZR1440900, 14ZR1439800, 14ZR1439900, 15ZR1439900, 15XD1523400, 14SJGGYY08, 201370), International Cooperation Foundation of China and Croatia (6-11) and Foundation of Yiwu Science and Technology Bureau (No. 2012-G3-02, 2013-G3-03).

REFERENCES

[1] World Cancer Report 2014. World Health Organization. 2014. pp. Chapter 1.1. ISBN 9283204298.

[2] L. Ayaru, S.G. Bown, S.P. Pereira, Int. J. Gastrointest. Cancer 35 (2005) 1–13.

[3] M. Schaffer, B. Ertl-Wagner, P.M. Schaffer, U. Kulka, G. Jori, E. Duhmke, A. Hofstetter, *Curr. Med. Chem.* 12 (2005) 1209–1215.

[4] L. Juillerat-Jeanneret, F. Schmitt, Chemical modification of therapeutic drugs or drug vector systems to achieve targeted therapy: looking for the grail, *Med. Res. Rev*. 27 (2007) 574–590.

[5] F. Bordi, S. Sennato, D. Truzzolillo, Polyelectrolyte-induced aggregation of liposomes: a new cluster phase with interesting applications, *J. Phys.: Condens. Matter.* 21 (2009) 203102 (26pp).

Author index

T - #0320 - 101024 - C0 - 246/174/35 [37] - CB - 9781138029385 - Gloss Lamination